ENCYCLOPEDIA OF

MEDICAL DEVICES AND INSTRUMENTATION

Second Edition
VOLUME 2

Capacitive Microsensors for Biomedical Applications – Drug Infusion Systems

ENCYCLOPEDIA OF MEDICAL DEVICES AND INSTRUMENTATION, SECOND EDITION

ENCYCLOPEDIA OF

MEDICAL DEVICES AND INSTRUMENTATION

Second Edition
Volume 2

Capacitive Microsensors for Biomedical Applications – Drug Infusion Systems

Edited by

John G. Webster

University of Wisconsin–Madison

The *Encyclopedia of Medical Devices and Instrumentation* is available online at
http://www.mrw.interscience.wiley.com/emdi

⊛WILEY-INTERSCIENCE

A John Wiley & Sons, Inc., Publication

Library of Congress Cataloging-in-Publication Data:

Encylopedia of medical devices & instrumentation/by John G. Webster,

 editor in chief. – 2nd ed.
  p. ; cm.
 Rev. ed. of: Encyclopedia of medical devices and instrumentation. 1988.
 Includes bibliographical references and index.
 ISBN-13 978-0-471-26358-6 (set : cloth)
 ISBN-10 0-471-26358-3 (set : cloth)
 ISBN-13 978-0-470-04067-6 (v. 2 : cloth)
 ISBN-10 0-470-04067-x (v. 2 : cloth)

 1. Medical instruments and apparatus–Encyclopedias. 2. Biomedical engineering–Encyclopedias. 3. Medical physics–Encyclopedias. 4. Medicine–Data processing–Encyclopedias. I. Webster, John G., 1932- . II. Title: Encyclopedia of medical devices and instrumentation.

 [DNLM: 1. Equipment and Supplies–Encyclopedias–English. W 13

E555 2006]
R856.A3E53 2006
610.2803–dc22
             2005028946

Printed in the United States of America

10 9 8 7 6 5 4 3 2 1

ENCYCLOPEDIA OF

MEDICAL DEVICES AND INSTRUMENTATION

Second Edition
VOLUME 2

Capacitive Microsensors for Biomedical Applications – Drug Infusion Systems

C

CAPACITIVE MICROSENSORS FOR BIOMEDICAL APPLICATIONS

D. Tsoukalas
S. Chatzandroulis
D. Goustouridis
NTUA
Athens, Attiki
Greece

INTRODUCTION

The use of reliable, high performance miniature sensors in the medical field is of growing importance for patient health monitoring. Batch sensor fabrication, as this has been introduced by Integrated Circuit (IC) manufacturing, is an efficient way to produce silicon sensors with desirable characteristics. Since microsensors combine small size with electrical and mechanical principles of operation they constitute together with microactuators what is usually called Microelectromechanical Systems (MEMS). These components are mainly made from silicon and other related materials to explore existing know-how and infrastructure of silicon technology. All of the above factors have allowed for fast growth of the microsensor field during the last years. This article focuses on physical microsensors used in the medical field that are based on the capacitive approach.

Historically, silicon piezoresistive devices were first introduced in the early 1960s to monitor pressure related variations (1). Piezoresistive silicon sensors take advantage of the piezoresistance effect observed in Si and Ge crystals (2). During this phenomenon, the value of a resistor realized from these semiconductors changes when mechanical strain is applied.

Piezoresistive devices have since then been investigated and used with catheters or as implantable units to monitor blood pressure variations (3,4). Such devices are already present in the market (5,6). Problems related to the nature of the piezoresistance effect include the relatively important temperature sensitivity of piezoresistive sensors that have to be compensated for with rather sophisticated electronics as well as the increased power consumption when compared to capacitive devices (7). These drawbacks of piezoresistive devices have accelerated the investigation of other device options for pressure measurements. It was in the 1980s that research on capacitive sensors began.

THE CAPACITIVE APPROACH

The capacitive approach is based on the parallel plate capacitor principle. In that structure the capacitance between two parallel electrodes is given by

$$C = \varepsilon_r \varepsilon_0 \frac{A}{d} \qquad (1)$$

where ε_r, ε_0 is the relative and vacuum permittivity constant, respectively, A is the plate surface area, and d is the plate distance.

When an applied external force acts in a way to change the distance between the two electrodes, the displacement is detected as a capacitance change. In many applications when a micromechanical structure is realized to detect a capacitance change, one of the two electrodes remains fixed and the other is flexible (Fig. 1a). The flexible electrode that is rigidly supported on its edges deflects so it does not remain parallel to the fixed electrode (Fig. 1b). In that case, the previous simple expression is modified and the capacitance is given by

$$C = \varepsilon_r \varepsilon_0 \int_A \int \left(\frac{1}{d - w(x,y)} \right) \qquad (2)$$

where $w(x,y)$ is the displacement of the flexible electrode.

Since the measured capacitance depends on the distance between the two electrodes, it can be used to calculate

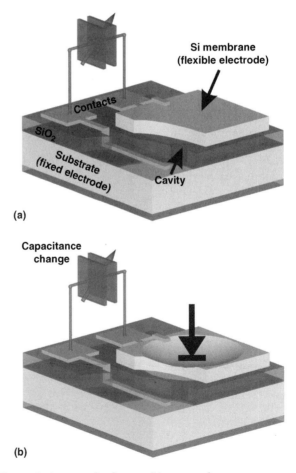

Figure 1. An example of a capacitive sensor is a pressure sensor. In parts a, the thin sensor diaphragm remains parallel to the fixed electrode and in part b, the diaphragm deflects under applied pressure resulting in capacitance change.

the value of the stimulus modifying the interelectrode spacing. Different approaches for capacitive micro devices have appeared that make use of the above principle.

One major application in the medical field for capacitive transducers is pressure measurement. First explorative research of capacitive microdevices made using silicon technology was initiated in the early 1980s (8–11). These initial studies have demonstrated that capacitive devices exhibit superior properties from their piezoresistive competitors in terms of pressure sensitivity, scalability, manufacturing simplicity, as well as process variation tolerance.

APPLICATIONS IN THE MEDICAL FIELD

The manufacturing advantages of micromachined silicon sensors have made them very attractive in the field of minimally invasive therapy. Catheters used in this field need to penetrate into small blood vessels putting stringent requirements on sensor size, which should be one-half of the size of the vessel to be accessed. In small vessels, catheters usually have a diameter of 2 mm, while the smallest catheter at the moment has a diameter of 0.36 mm (12). Miniaturization, therefore, of both sensor and packaging size drives this kind of application. Capacitive devices within the above size requirements have been demonstrated in silicon (13) and are mainly competing with fiber optic pressure sensors (14), or a free hanging strain gauge (15).

Catheters, however, are unsuitable for long-term monitoring of blood pressure. With AAA (abdominal aortas aneurism) and CHF (congestive heart failure) being the major cardiovascular diseases in our days, an implantable pressure monitoring system that would tailor treatment medication by measuring blood pressure appears very attractive. These implantable applications will require a miniature batteryless telemetric sensor able to communicate with an external handheld unit. In such applications, power consumption becomes an additional issue and capacitive devices offer a clear advantage over piezoresistive ones. These types of systems have been under investigation (16,17) and have already demonstrated good performance.

Capacitive pressure sensors have also been successfully applied in the following areas. (1) Intraocular pressure monitoring (10,18). Glaucoma is, for example, a serious disease characterized by an increased pressure in the eye that may result in blindness. In that case, a sensitive capacitive device has been developed for remote sensing of eye pressure (normally 10–20 mmHg 1.33–2.66 kPa above atmospheric pressure, but much increased in glaucoma). (2) Intracranial pressure monitoring (19), as well as for clinical assessment of prosthetic socket fit (20) and pressure distribution in artificial joints (21). Intracranial pressure monitoring is an implant used in the treatment of patients with trauma of the head as well as in neurological patients.

Although pressure is the field where capacitive sensors have been applied more extensively, there are also other applications under development in the medical field. Accelerometers, for example, are used for measuring inclination of body segments and activity of daily living, with application in patient rehabilitation (22), but also register the kind of movements that occur in healthy persons during normal standing (23,24). Physical activity as well as energy expenditure as these can be followed by accelerometers proves to be useful information for personal status monitoring. An accelerometer design includes the fabrication of a proof mass that is displaced in proportion with acceleration. The use of capacitance to measure that displacement significantly improves sensitivity.

Recently, ultrasound imaging technology has also exploited the advantages of capacitive sensors for both transmission and detection purposes (25). Such capacitive devices can be batch fabricated to form a transducer array with array elements that can be as small as 50 μm diameter. Ultrasound devices are made of a thin flexible electrode facing a rigid electrode. For transmission purposes, the membranes are driven into vibration by the electrostatic force exerted between the two electrodes. For reception, the membrane vibration is excited by an impinging acoustic wave that is converted by the capacitive device to electrical signal. This as well as other efforts (26,27) are driven from the need to obtain in the future high resolution images within the body using three-dimensional 3D echographic probes.

Miniature capacitive transducers, known in low frequency applications as condenser microphones, are used in hearing aids and have been reported from different research groups (28,29). In their work, Rombach et al. developed a low noise capacitive microphone with higher sensitivity and broader bandwidth than those used in traditional hearing aids. This device consists of two backplates with an intermediate membrane made of a low stress silicon-rich nitride and B^+ polysilicon multilayer. Pedersen et al., on the other hand, presented an integrated capacitive microphone based on polyimide technology and realized by postprocessing on a CMOS wafer at low processing temperatures.

Capacitive sensors have also been used as humidity sensors for the diagnosis of pulmonary diseases. In this device, a chemically absorbing layer (usually a polymer) is placed between the parallel electrodes of a capacitor. Then, humidity is detected as a change in capacitance because of the dielectric constant change as water molecules are absorbed in the polymer. In a similar configuration, hydrogel has been used between the electrodes of a capacitive sensor to measure body analytes from the capacitance variation occurring due to the swelling of the polymer (30).

FABRICATION TECHNOLOGIES FOR CAPACITIVE SENSORS

Silicon is widely accepted as the material of choice for microsensor fabrication. Known for its good mechanical properties, high mechanical strength, and light weight it's an ideal material for physical sensors (31). More importantly, the existing know-how from IC manufacturing made the use of silicon technology for transducers quite straightforward during the last decades.

This section describes the main silicon technologies that are used for the fabrication of capacitive devices. Most of the capacitive devices used in medical technology, namely, pressure sensors, accelerometers, and ultrasound sensors, have been developed using two major technology platforms.

Bulk Micromachining Technologies

Bulk micromachining has historically been developed first. It consists of engineering a silicon wafer by a series of lithographic processes followed by wet or dry etching, in order to form thin membranes or other free standing structures that can move upon an external stimulation. In bulk micromachining, the micromechanical silicon structures are fabricated by selectively removing whole sections, or in some cases, all but a small part of the silicon wafer. Thus the structures fabricated in this way are made of single- crystal silicon and have excellent mechanical properties. In bulk micromachining, processing may take place in either the front or the back side of the silicon wafer.

Wet etching of silicon is based on a chemical reaction of silicon with a base solution (KOH) in order to remove silicon material and form the intended 3D structure. Dry etching is a physicochemical process that was initially developed for the etching of thin films. During the last decade, however, this technique has also been used for the removal of thick Si material (32).

The above techniques combined with others usually used in IC manufacturing, like ion implantation, thermal processing, and thin-film (metal or silicon insulator) deposition, constitute the backbones of capacitive sensor fabrication. A detailed description of these processes is beyond the scope of this article and can be found elsewhere (33).

Apart from these processing technologies it is necessary in many applications to use other specific processes. For example, in pressure sensor fabrication it is always desirable to have a reference pressure enclosed in the sensor body. This requires a reliable technology for sealing a cavity with known pressure. For that reason, bulk micromachining techniques are combined with technologies usually referred to as bonding technologies. Two are the most established technologies in this area: anodic bonding and fusion bonding, in chronological order of their discovery and application. Anodic bonding is achieved between silicon and a glass substrate at medium temperature ($\sim 400\,°C$) under the application of a direct current (dc) voltage across the two substrates (34). Prior to contact of the two substrates, their surfaces are adequately cleaned to remove any particle that can inhibit their good contact. Bulk micromachining combined with anodic bonding has been successfully used for the realization of medical capacitive pressure sensors by a couple of research groups (35,36). In the process developed at the University of Michigan, boron etch-stop techniques and silicon-glass anodic bonding is used to fabricate a capacitive pressure sensor. In this process, KOH is used to initially form a recess in the surface of a silicon wafer, followed by a deep boron diffusion to define the rim of the transducer, and a shallow diffusion defining the eventual thickness of the diaphragm. The completed silicon wafer is finally electrostatically bonded to a glass wafer. The silicon wafer is then dissolved in EDP letchant, leaving only the silicon

transducer islands bonded to the glass. In this way, a thin silicon membrane structure over a sealed cavity is fabricated that exhibits high pressure sensitivity adequate for use in blood pressure monitoring.

A more recently discovered technique that has been applied for sealing of a pressure reference cavity is fusion bonding. This technique does not require the application of a voltage across the bonded substrate. Instead, it includes a high temperature heat treatment. So after a thorough wafer cleaning and drying process of two silicon wafers that renders their surface hydrophilic, they are brought into contact at room temperature. The two wafers are initially drawn together at room temperature by van de Waals forces developed between the hydrogen atoms of water molecules covering their surfaces. This initial attraction is commonly known as prebonding. Prebonding follows a high temperature heat treatment that during which a hydrogen atom is removed transforming the hydrogen bonds to covalent bonds between oxygen atoms thus increasing the bonding strength (37). The temperature necessary to achieve high strength bonding is $> 800\,°C$. A successful example of this technology together with bulk micromachining is applied for the realization of a capacitive-type pressure sensor for blood pressure monitoring developed by Goustouridis et al. (13). This simple process results in robust capacitive sensors with low parasistics (Fig. 2). It involves two silicon wafers that are silicon fusion bonded to form the final 3D structure, and a thick oxide in between in which a sealed cavity is formed. The sensor diaphragm is formed by creating a heavily boron-doped region in the cavity bottom. After bonding, the wafer stack is first mechanically ground and then etched in EDP etchant to leave the sensor diaphragm on top of the cavity thus creating the pressure sensor. The sealed cavities are then metalized and packaged. Figure 3 depicts the complete pressure sensing element as it appears after the metallization step.

Hydrophobic bonding is also applied when we need to bond two bare silicon surfaces without a SiO_2 layer on the surface. This technique requires an HF final step to remove any oxide layer from the surface. In hydrophobic bonding, the water molecules necessary to complete the prebonding step are substituted from the HF molecules, while the rest of the bonding process is similar to the hydrophilic one.

More recently, plasma activated bonding (38) or other low temperature bonding ($< 400\,°C$) using spin on glass (SOG) (39) have been applied. These techniques are more appropriate for wafer level packaging and have to be more developed in the future for use in sensor fabrication.

Recently, wafer bonding has also been used for demonstration of capacitive ultrasound devices (25), as well as of capacitive accelerometers (40).

Surface Micromachining and SOI Technologies

Surface micromachining technologies have been developed with the primary goal of cointegrating the electronic and mechanical parts on the same silicon chip. Since the active and passive layers of a surface micromachined structure are realized using the same conductive and insulating thin-film layers as in IC manufacturing, it is possible by appropriately designing the mask sequence to realize both

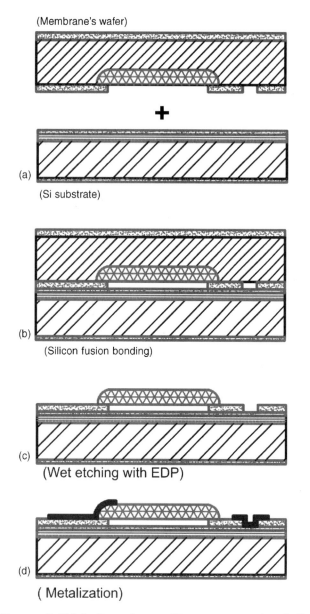

Figure 2. Fabrication of capacitive sensors using bulk micromachining and silicon fusion bonding. Two wafers are used for this purpose. (a) Each wafer is processed independently before the bonding; (b) after the bonding process the two wafers are permanently stacked together; (c) selective wet etching releases the boron doped silicon diaphragms; (d) metallization is performed for electrical connects of the membrane and the substrate.

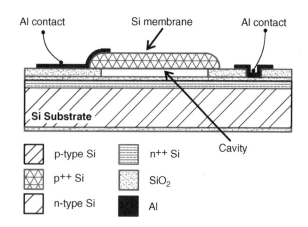

Figure 3. This figure illustrates the final capacitive pressure sensor structure (not to scale).

components in parallel by adding a few mask levels after the completion of the electronic circuit. This particular feature is behind the drive for the development of this technology.

In the case of surface micromachining, all of the processing takes place only on the surface of the front side of the silicon wafer. The micromechanical structures fabricated in this manner are made out of polycrystalline silicon deposited by low pressure chemical vapor deposition (LPCVD) techniques over a sacrificial layer. The sacrificial layer is a deposited oxide, usually phosphorsilicate glass (PSG). This layer is subsequently removed by an HF solution through narrow access channels to release the poly-

silicon structure (Fig. 4). Sealing of the channels is necessary and it is realized by a deposition step. With surface micromachining, it is possible to fabricate far smaller microelectromechanical devices than with bulk micromachining. Depositing processes allow for very good control over the dimensions of the deposited materials.

Surface micromachining is a very important technology with a demonstrated potential. Problems related to stiction of membrane structures on the substrate during aqueous removal of sacrificial layers have been resolved by the application of other etching and drying techniques like gas-phase etching or freeze-drying techniques using cycloexane (41). A typical process sequence developed (42) for capacitive ultrasound transducers is shown in Fig. 5.

There is continuing discussion on the advantages–disadvantages of the cointegration of sensors with electronics. Although it is considered that surface micromachining can result in a higher packing density, and consequently smaller and cheaper components, it appears that yield issues still need to be overcome until this technology can be definitively adopted in preference to hybrid fabrication technologies.

A recent variation of the two technologies employs silicon-on-insulator (SOI) technology with some unique features. The SOI technology offers the possibility of using crystalline silicon as the active part of a capacitive-type structure with more predictable mechanical behavior than the polysilicon film used in surface micromachining. In fact, the internal stresses (either compressive or tensile) developed during the growth of the polysilicon film, which evolved during subsequent thermal treatment because of change of the grain size, is a source of potential uncertainties for the behavior of these films.

Silicon-on-insulator is a mature technology and nowadays can offer crystalline silicon structures of varying thicknesses over a variety of SiO_2 buried layer thicknesses. This has become possible after the discovery of wafer bonding technology, which enables the development of back-etch-silicon-on-insulator (BESOI) structures.

Thin as well as thick silicon structures allow for capacitive pressure sensors development, ultrasound capacitive sensors using rather thin membranes of some micron thickness as well as capacitive-type accelerometers, where

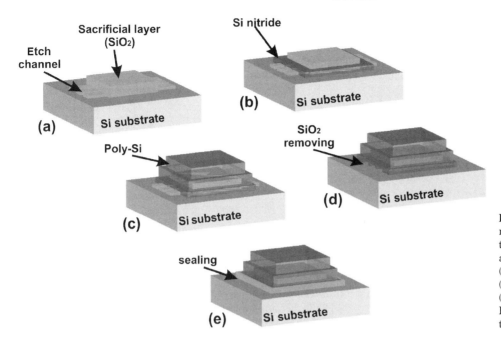

Figure 4. Typical surface micromachining process. (a) Deposition of the sacrificial layer (silicon oxide) and etch channels formation, (b) deposition of silicon nitride ring, (c) deposition of polysilicon film, (d) removal of the sacrificial layer by HF in wet or vapor form, (e) sealing of the device with deposited silicon oxide.

Si structures exceed some hundreds of micrometers. The introduction of deep reactive ion etching (DRIE) in silicon processing especially during the last few years, has enabled the vertical anisotropic etching of Si at the rate of several microns per minute, and the fabrication of thick proof masses on BESOI or glass structures for capacitive accelerometers (43). This accelerometer uses a silicon proof mass of 0.5 mg with 120 μm thickness formed by DRIE and measures in plane (x or y axis) acceleration. It uses a sense gap of only 2 μm between sense fingers and the electrodes (Fig. 6). As the proof mask moves under acceleration, the distance between sense fingers and fixed electrodes change, which consequently modifies the capacitance.

Finally, a different but similar approach results in capacitive-like pressure sensors based on a field effect transistor (44). In this case, the usual dielectric of a MOSFET is replaced by a vacuum cavity. The external pressure variations deflect the gate electrode, and consequently influence the capacitive coupling of the flexible membrane (gate) with the channel thus modifying the current flow in the device. Although these devices present an attractive design, there have not seen any new developments.

OPERATION ISSUES OF CAPACITIVE SENSORS

As introduced in the first paragraph a capacitive sensor is an equivalent parallel plate capacitor with clamped edges where the diaphragm deforms during the application of a deferential pressure across the two sides in the case of capacitive pressure devices.

In the normal operation mode of a capacitive pressure sensor, the diaphragm does not contact the substrate electrode. The capacitance response of a typical pressure sensor is shown in Fig. 7. The output capacitance is nonlinear because of its inverse relationship with the electrode gap d_0-w (eq. 2), which is a function of pressure, P. This nonlinearity becomes more significant for large membrane

deflections. At the point when the sensor diaphragm touches the cavity bottom (Fig. 7), the behavior of the sensor changes and enters "touch mode" operation (45). In this operating region, linearity increases and the sensor capacitance is dominated by the area touching the cavity bottom, since the gap there is replaced from the very thin and high dielectric constant SiO_2 layer of the bottom electrode.

Sensitivity of the Capacitive Sensor

Because of the nonlinearity of the response, the sensitivity of a capacitive sensor is not constant. For example, the sensitivity of the capacitive pressure sensor for blood pressure monitoring (defined as $1/C_0(\Delta C/\Delta P)$ with the response shown in Fig. 7 is 1.5 fF/mmHg for the low pressure range and increases to > 18 fF/mmHg for the upper part of the measurement range (> 200 mmHg, 26.66 kPa). In the touch mode operation region, (> 300 mmHg, 39.99 kPa) the sensor has a nearly linear response with a sensitivity of 12 fF/mmHg.

Temperature variation, although not a critical issue for medical application (because of the small variation of the body temperature), can affect the accuracy of the measurements. The temperature influence on the capacitive sensor response is due to either the mismatch of thermal expansion of dissimilar materials used in the fabrication process (usually Si and SiO_2), or to gas expansion in case gas is trapped in a sealed cavity of the sensor. It can of course be due to both of the above reasons. In the case of a sealed cavity, the temperature influence, because of the expansion of the gas trapped inside the cavity, can be eliminated if the cavity is sealed in vacuum. On the other hand, the effect of the different thermal expansions of the materials used is always a problem that requires appropriate design in order to reduce or even eliminate the effect. In the cases of capacitive devices fabricated with surface micromachining,

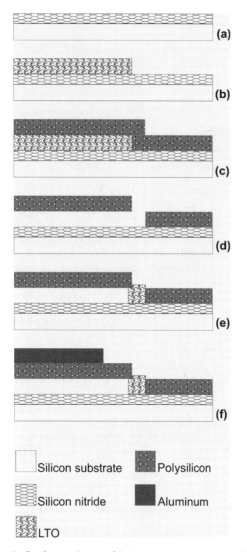

Silicon substrate   Polysilicon

Silicon nitride   Aluminum

LTO

Figure 5. Surface micromachining process sequence for the fabrication of capacitive ultrasonic transducers as taken from X. Jin et al. (42).

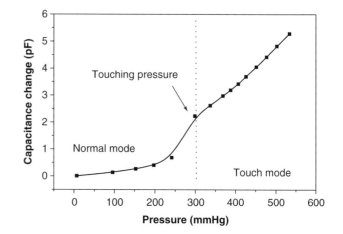

Figure 7. Typical response of a capacitive pressure sensor with 325 μm OD circular diaphragm. The diaphragm touches the cavity bottom at 300 mmHg ($\sim$ 40 kPa). A pressure sensor must be designed to operate either for lower values than 300 mmHg or for higher pressure values. Otherwise an hysteresis phenomenon can be observed.

or by using anodic bonding of silicon with a glass substrate, the influence of temperature becomes more complicated.

Figure 8 shows the influence of temperature on the response of a capacitive pressure sensor, not sealed in a vacuum. The variation of the distance between the curves with temperature represents the temperature coefficient of the pressure sensitivity (TCS) while the slope of the lower curve is the temperature coefficient of zero pressure offset (TCO). The TCS is defined as $\Delta S/S\Delta T$, where S is the pressure sensitivity defined as $\Delta C/C_0\Delta P$ and TCO is defined as $\Delta C/C_0\Delta T$.

This simple configuration of a parallel plate capacitor is used mainly in pressure sensors and ultrasound imaging devices. Accelerometers are usually designed with comb-shaped electrodes that result in increased linearity and sensitivity (40,43).

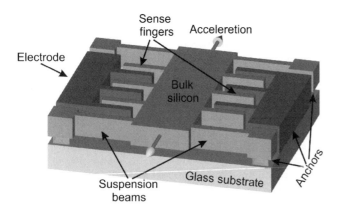

Figure 6. Comb-shaped accelerometer structure fabricated using a combination of processes like bonding silicon with glass substrate and Deep Reactive Ion Etching (43).

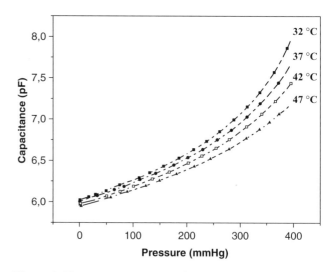

Figure 8. Temperature variation of a capacitive pressure sensor in the range 0–400 mmHg ($\sim$53.3 kPa). The effect is due mainly to trapped gas expansion inside the cavity.

CAPACITIVE SENSOR ELECTRONIC INTERFACES

Although capacitive sensors offer advantages with respect to high sensitivity and low power operation, they also present difficulties in the design of electronic interfaces to convert capacitance changes into electrical signals. Parasitic capacitances often dominate system performance by reducing sensitivity and increasing nonlinearity (46). Therefore, it becomes absolutely essential for sensor systems incorporating capacitive sensors to either integrate the MEMS component with the electronic interface or place the two chips at close proximity to each other to reduce parasitics. At the same time, it is important that the electronic interface is designed to suppress the remaining parasitic capacitances and provide for a large zero capacitance range. A good starting point for the study of capacitance measurement circuits may be found in Ref. 47, where a number of basic circuits are discussed.

Depending on the technological steps used to fabricate the sensor, several parasitic capacitances $C_{p1,2}$ and conductances G_p may be present in the device and increase in importance as devices get smaller and the sensing capacitance C_s gets in the pF range. These parasitic effects may originate from various sources depending on the fabrication process used (i.e., stray capacitances of metal lines and connecting pads in parallel with the sensing capacitance or leakage resistors). Further variations may also be observed, due to technological reasons, between batches of the same sensor. In Fig. 9, a simple electrical model of a typical capacitive sensor, including parasitic effects, is shown. The spread in sensor characteristics (e.g., zero capacitance, sensitivity) resulting from these effects puts a great strain in the design of electronic interfaces for capacitive sensors. For this reason, most of the capacitive interfaces to date have been designed by taking into consideration the particular sensor and application that they are going to be used for. The design is a trade-off between power consumption, interface accuracy, and resolution, or even die size in some applications in the medical field.

A number of architectures have been proposed to date to convert capacitive changes into electrical signal. Some are built around a relaxation oscillator (46,48), others use switch capacitor techniques to convert capacitance changes into voltage, and others interface the sensor directly into a sigma–delta modulator. A few attempts to develop a generic interface have also been reported (49,50). However, the power consumed is too high for remote sensing applications.

Van Der Goes and Meijer (49) presented a universal transducer interface for the read out of capacitors, platinum resistors, thermistors, resistive bridges, and potentiometers. The circuit uses the three signal technique in which the sensor signal E_x, a reference signal E_{ref}, and the offset E_{off} of the whole interface are measured in a identical way to achieve continuous autocalibration of offset and gain. The two port measurement technique is used to eliminate sensor parasitic capacitance. In this technique, a testing voltage, V, is forced on one capacitor electrode while current I is sensed on the other (51). The current I then depends only on the applied voltage and sensing capacitance. However, the technique is not energy efficient, since it requires four measurement cycles and three external voltage sources to determine the sensor capacitance.

Yazdi et al. (50) also developed a standardized switch capacitor interface for capacitive sensors, as shown in Fig. 10. This interface is capable of interfacing through a standard bus with a microcontroller that collects sensor data through the interface, calibrates data, and either stores or transmits it wirelessly or through a serial port. The readout circuit utilizes a low noise front-end charge integrator to read out the difference between the sensor capacitance and a reference capacitor. An input multiplexer allows for interfacing with up to six capacitive sensors. Finally, the chip can be digitally programmed to operate with one of three external or internal reference laser trimmable capacitors.

Bracke et al. (52) reports on a low power generic switched capacitor interface for capacitive sensors. The circuit uses a special clocking scheme, in which the analog sensor circuit block is clocked at a low 8 kHz frequency, while the sigma–delta modulator is clocked at 128 kHz. The technique ultimately reduces power consumption to 90 µW on the ON-state.

CAPACITIVE ELECTRONIC INTERFACES FOR IMPLANTABLE APPLICATIONS

In medical applications, where the diagnostic device is to be implanted inside the human body, only limited power is available for the electronic circuits and sensor. In such cases, power can only be found via a battery implanted together with the sensing system or through passive telemetry. In passive telemetry, energy may be harvested from a remote electromagnetic field transmitted outside the body. The same field may also be used to receive control data and transmit sensor data to a data logger. In both cases, minimizing power consumption is essential.

A simple low power interface for biomedical applications could be realized by using a simple relaxation oscillator (46,48). A capacitance-to-frequency modulated output was first proposed by Hanneborg et al. (46). This circuit delivers a digital pulse trail with frequency dependent on the sensor

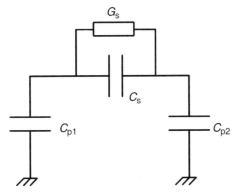

Figure 9. Simple electrical model of a capacitive sensor. Parasitic capacitances is designated by $C_{p1,2}$, conductances by G_p and the sensing capacitance by C_s.

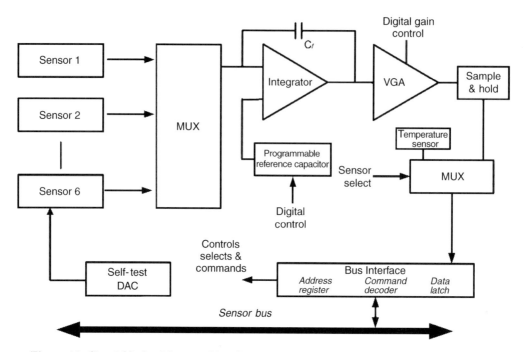

Figure 10. Circuit blocks of the capacitive electronic interface as described by W. Bracke et al. (50).

capacitance. The circuit consists of two current sources: a switch and a Schmitt trigger. The unknown capacitance is periodically charged and discharged by flipping the switch between current $I+$ and $I-$ (Fig. 11).

An implementation of this type of capacitance-to-frequency converter is presented in Fig. 12. The circuit converts capacitance at its input into a frequency signal that can be readily fed to a digital microcontroller for processing. Either a capacitive sensor or a reference may be

switched in, and the frequency delivered at the circuit output is dependant on the input capacitance according to

$$f = \frac{I_0}{2C_x V_h} \tag{3}$$

where I_0 is the current by which the sensor capacitance C_x is charged or discharged, and V_h is the hysteresis of the Schmitt trigger. The parameter V_c is a control signal that allows for switching between the unknown sensor capacitance C_x and a reference capacitor C_{ref}. The output frequency when the reference is selected is independent of pressure, thus compensation of temperature and long-term drifts are possible by taking the ratio of the reference frequency and sensor frequency, $F_{\text{ref}}/F_{\text{sens}}$.

The response of a pressure measuring system based on this circuit realized in $1.2\,\mu\text{m}$ of CMOS technology and a capacitive pressure sensor (53) is shown in Fig. 13. It operates at 4 V and draws $20\,\mu\text{A}$ of average current. The system exhibits a sensitivity of 36 Hz/mmHg and is able to resolve pressure changes of $5\,\text{mmHg}$ ($0.66\,\text{kPa}$). A photograph of this system is shown in Fig. 14.

In medical science, however, there is often the need for long-term monitoring of vital life parameters. A good example is abdominal aortic aneurysm (AAA), which is a ballooning of the abdominal aorta. Patients who suffer from this condition need to undergo a procedure during which a stent graft is inserted. After the operation, however, it is possible that the aneurysmal sac is not completely isolated, leading to recurrent pressurization of the sac, a complication that, if left undetected, may lead to rupture of the sac and patient death. Long-term, postoperation monitoring of the patients is therefore necessary.

Early efforts for systems for the monitoring of blood pressure (54) used miniature active transmitters to transfer measured data and were battery powered, which limited

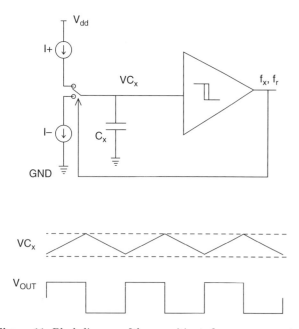

Figure 11. Block diagram of the capacitive to frequency converter proposed by Hanneborg et al. (46). The circuit consists of two current sources, a switch and a Schmitt trigger.

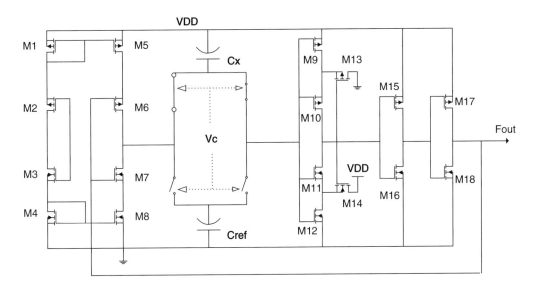

Figure 12. Schematic of Schmitt-trigger based oscillator used as a capacitive interface.

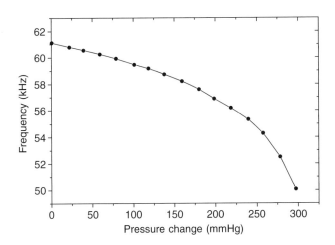

Figure 13. Frequency response of a pressure measuring system consisting of the simple circuit of Fig. 12 and a capacitive pressure sensor in the range 0–300 mmHg (~40 kPa).

their lifetime. An alternative approach to power these modules is through induction coupling, while the same radio frequency (RF) field can be used to transfer data out of the implanted module (55). The implanted circuit should be virtually immune to supply fluctuations arising from random misalignment of the implanted and the external coil. A new circuit was developed based on the previous architecture, but in which each circuit block was redesigned.

The block diagram of the passive telemetry system is depicted in Fig. 15. It consists of an external control unit (the base unit) and an implantable transponder. Wireless communication can then be established between the two units, based on an absorption modulation mechanism. The transponder receives power and external control data

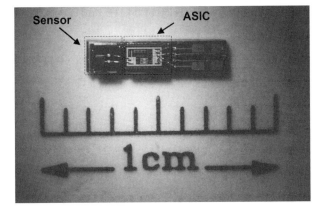

Figure 14. A photograph of a hybrid pressure measuring system consisting of a capacitive sensor and associated signal conditioning electronic circuit.

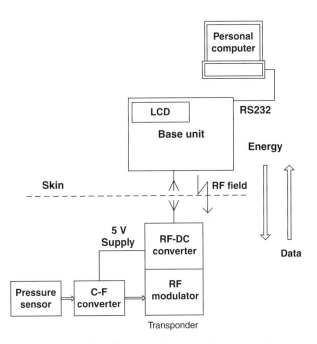

Figure 15. Block diagram of passive telemetry system.

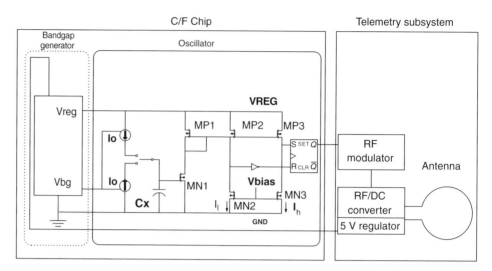

Figure 16. Transponder electronics.

through an RF field, while it can transmit data by modulating the absorption rate. The base unit, on the other hand, demodulates the transmitted data and processes it though a microcontroller to convert the signal into an eight bit unsigned byte array. The resulting byte can then be sent to a PC through a serial output.

The block diagram of the transponder electronics is shown in Fig. 16. The improved capacitance-to-frequency circuit is used to interface a capacitive pressure sensor. The circuit consists of two basic blocks: a bandgap reference voltage generator and an oscillator. In order to achieve independence of the output frequency from the received power, this time the oscillator operates on an internally stabilized voltage generated from a bandgap reference. Voltage regulation is a critical part of telemetric systems as the induced power, and thus the voltage output, of the RF/dc converter in the transponder can greatly fluctuate in an actual implanted system becaused the relative position changes of the two antennas. In addition, to further immune the system from supply fluctuations a current mode comparator is used in the oscillator.

The bandgap reference voltage circuit is capable of operating at a low power supply as it operates on an internally regulated voltage VREG (56). This same node is also used for the supply of the oscillator circuit after the contributions of the extra branches of the oscillator are accounted for.

The oscillator itself is designed around a current mode comparator that results in an output frequency independent of power supply fluctuations and with small temperature drift. Triggering levels of this oscillator are defined by two currents: I_h and I_l. The period of the output pulse can then be shown to be equal to

$$T = \frac{2C_x}{I_0}(V_{bias} - V_{TN})(\sqrt{n} - 1) \qquad (4)$$

where n stands for the ratio of I_h to I_l, and V_t is the threshold voltage. By taking the inverse of eq. 4 and substituting for I_0, we obtain

$$f = \frac{k\frac{W_0}{L_0}(V_{bias} - V_{TN})}{2C_x(\sqrt{n} - 1)} \qquad (5)$$

Equation 5 implies that the output frequency is independent of the supply voltage and is dependent on temperature through the mobility term in k and the threshold voltage V_t. Note also that the bias voltage V_{bias} is chosen to be equal to the bandgap reference voltage produced from the previous stage, and is thus considered independent of voltage and temperature variations.

The C/F converter was designed and fabricated in 0.8 μm CMOS technology. A hybrid pressure measuring system composed of a capacitive pressure sensor (16) and the C/F chip. The system remains operational for a supply voltage down to 2.7 V and exhibiting high immunity to voltage variations from 3.7 to 5.5 V (Fig. 17). Simulated pressure pulses as those present in the aorta were measured using passive telemetry (Fig. 18).

CONCLUSIONS

Capacitive microsensors are in progressively increasing use in medical applications because of their advantages, such as small size, high sensitivity, and low power consumption. Silicon is the material of choice for these devices, which are finding applications for measuring pressure and

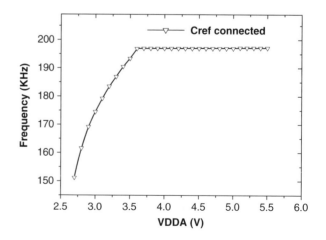

Figure 17. Pressure measuring system frequency output.

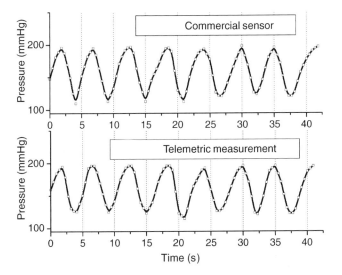

Figure 18. Simulated pressure pulses as those present in the aorta measured using passive telemetry.

acceleration, but also in ultrasound imaging and microphones and for monitoring body analytes. Actually, although most of the existing devices are not integrated together with signal conditioning electronic circuits on the same chip, in the future it is expected that there will be an increasing integration scheme driven by the need of higher level miniaturization. Because of their low power consumption advantage, capacitive microsensors are of interest as implantable monitoring devices. An attractive application appears to be a miniaturized telemetry system that combines techniques for wireless power and data transfer to a capacitive sensor integrated with signal conditioning electronics.

BIBLIOGRAPHY

1. Tufte ON, Chapman PW, Long D. Silicon Diffused-Element Piezoresistive Diaphragm. J Appl Phys 1962;33:3322–3327.
2. Smith CS. Piezoresistance Effect in Germanium and Silicon. Phys Rev 1954;94:42.
3. Samaun X, Wise KD, Angel JB. An IC Piezoresistive Pressure Sensor for Biomedical Instrumentation. IEEE Trans Biomed Eng 1973;BME-20(2):101–109.
4. Ko WH, Hynelek J, Boettcher SF. Development of a miniature Pressure Transducer for Biomedical Application. IEEE Trans Electron Dev 1979;ED-26:1896.
5. Data Science Int., Roseville Minnesota.
6. Konigberg Instruments Inc. Pasadena, California.
7. Clark SK, Wise KD. Pressure sensitivity in anisotropically etched thindiaphragm pressure sensors. IEEE Trans Electron Dev 1979;ED-26(12):1887–1896.
8. Sander GS, Knutti JW, Meindl JD. A monolithic Capacitance Pressure Sensor with Pulse Periodic Output. IEEE Trans Electron Dev 1980;ED-27(5):927–930.
9. Lee YS, Wise KD. A batch-fabricated Silicon Capacitive Pressure Transducer with low temperature Sensitivity. IEEE Trans Electron Dev 1982;ED-29(1):42–47.
10. Backlund Y, Rosengren L, Hök B. Passive Silicon Transensor Intended for biomedical, Remote Pressure monitoring. Sens Actuators 1990;A21(1–3):58–61.
11. Puers R, Peeters E, Van den Bossche A, Sansen W. A capacitive pressure sensor with low impedance output and active supression of parasitic effects. Sens Actuators 1990;A21(1-3):108–114.
12. Goosen JFL. Design considerations for silicon sensors for use in catheters and guide wires. Smart Mater Struct 2002; 11:804–812.
13. Goustouridis D, Chatzandroulis S, Normand P, Tsoukalas D. A miniature self-aligned pressure sensing element. J Micromech Microeng 1996;6:33–35.
14. Zhu YZ, Wang AB. Miniature fiber optic pressure sensor. IEEE Photonic Technol 2005;17(2):447–449.
15. Melvås P, Kälvesten E, Enoksson P, Stemme G. A free-hanging strain-gauge for ultraminiaturized pressure sensors. Sens Actuators A 2002;97(8):75–82.
16. Chatzandroulis S, Tsoukalas D, Neukomm PA. A Miniature Pressure System with a Capacitive Sensor and a Passive Telemetry Link for Use in Implantable Applications. J Micro-ElectroMech S 2000;9(1):18–231.
17. Najafi N, Ludominky A. Initial Animal Studies of a Wireless, Batteryless, MEMS Implant for Cardiovascular Applications. Biomed Microdevices 2004;6(1):61–65.
18. Coosemans J, Catrysse M, Puers R. A readout circuit for an intra-ocular pressure sensor, Sens. Actuators A 2004;110(1-3):432–438.
19. Hierold C, et al. Low power integrated pressure sensor for medical application. Sens Actuators A 1999;73(1–2):58–67.
20. Polliack AA, et al. Laboratory and clinical tests of a prototype pressure sensor for clinical assessment of prosthetic socket fit. Prosth Ortho Inter 2002;26(1):23–24.
21. Müller O, Parak WJ, Wiedemann MG, Martini F. Three-dimensional measurements of the pressure distribution in artificail joits with a capacitive sensor array. J Biomech 2004;37:1623–1625.
22. Luinge HJ, Veltink PH. Inclination measurement of human movement using a 30D accelerometer with autocalibration. IEEE T Neur Sys Reh 2004;12:112–121.
23. Puers R, Reyntjen S. Design and processing experiments of a new miniaturized capacitive triaxial accelerometer. Sens Actuators A 1998;68(1-3):324–328.
24. Lötters JC, Olthuis W, Veltink PH, Bergveld P. Design, realization and characterization of a symmetrical triaxial capacitive accelerometer for medical application. Sens Actuators A 1997;61(1-3):303–308.
25. Huang Y, et al. Fabricating capacitive micromachined ultrasonic transducers with wafer-bonding technology. J Microelectromech S 2003;12(2):128–137.
26. Cianci E, et al. One-dimensional capacitative micromachined ultrasonic transducer arrays for echographic probes. Microelect Eng 2004;73–74:502–507.
27. Knight J, McLean J, Degertekin FL. Low Temperature Fabrication of Immersion Capacitive Micromachined Ultrasonic Transducers on Silicon and Dielectric Substrates. IEEE Trans UFFC 2004;51:1324–1333.
28. Rombach P, Müllenborn M, Klein U, Rasmussen K. The first low voltage, low noise differential silicon microphone, technology development and measurement results. Sens Actuators A 2002;95:196–201.
29. Pederson M, Olthuis W, Bergveld P. High-performance condenser microphone with fully integrated CMOS amplifier and DC-DC voltage converter. J MicroElectroMech S 1998;7(4):387–394.
30. Strong ZA, Wang AW, McConaghy CF. Hydrogel-actuated capacitive transducer for wireless biosensors. Biomed Microdevices 2002;4(2):97–103.
31. Petersen KE. Silicon as a mechanical material. Proc IEEE 1982;70(5):420–457.
32. de Boer MJ, et al. Guidelines for etching silicon MEMS structures using fluorine high-density plasmas at cryogenic temperatures. J MicroElectroMech S 2002;11(4):385–401.

33. Madou M. Fundamentals of Microfabrication: The Science of Miniaturization. 2nd ed. Boca Raton (FL): CRC Press; 2002.

34. Wallis G, Pomerantz DI. Field assisted glass-metal sealing. J Appl Phys 1969;40:3946–3949.

35. Ji J, Cho ST, Najafi K, Wise KD. An ultaminiature CMOS pressure sensor for a multiplexed Cardiovascular Catheter. IEEE Trans Electron Dev 1992;39:2260–2267.

36. Puers R, Van den Bossche A, Peeters E, Sansen W. An implantable pressure sensor for use in cardiology. Sens Actuators A 1990;23:944–947.

37. Tong QY, Gosele U. Semiconductor Wafer Bonding. New York: Wiley-Interscience; 1999.

38. Henttinen K, Suni I, Lau SS. Mechanically induced Si layer tranfer in hydrogen-implanted Si wafers. Appl Phys Lett 2000;76:2370–2372.

39. Goustouridis D, et al. Low temperature wafer bonding for thin silicon film transfer. Sens Actuators A 2004;110: 401–406.

40. Tsuchiya T, Funabashi H. A z-axis differential capacitive SOI accelerometer with vertical comb electrodes. Sens Actuators A 2004;116:378–383.

41. Kim C, Kim JY, Shridharan B. Comparative evaluation of drying techniques for surface micromachining. Sens Actuators A 1998;64:17–26.

42. Jin X, et al. Fabrication and characterization of surface micromachined capacitive ultrasonic immersion transducers. J Microelectromech S 1999;8:100–114.

43. Chae J, Kulah H, Najafi K. A CMOS-compatible high aspect ratio silicon-on-glass in-plane micro-accelerometer. J Micromech Microeng 2005;15:336–345.

44. Lysko JM, Jachowisz RS, Krzycki MA. Semiconductor pressure sensor based on FET structure. IEEE T Instrum Meas 1995;44:787–790.

45. Ko WH, Wang Q. Touch mode capacitive pressure sensors. Sens Actuators A 1999;75:242–251.

46. Hanneborg A, et al. An integrated capacitive pressure sensor with frequency-modulated output. Sens Actuators 1986; 9(4):345–351.

47. Senturia S. Microsystem Design. Boston: Kluwer Academic Publishers; 2000.

48. Matsumoto Y, Esashi M. Integrated silicon capacitive accelerometer with PLL servo technique. Sens Actuators A 1993; 39:209–217.

49. Van Der Goes FML, Meijer GCM. A universal transducer interface for capacitive and resistive sensor elements. Analog Integr Circuits Signal Process 1997;14:249–260.

50. Yazdi N, Mason A, Najafi K, Wise KD. A generic interface chip for capacitive sensors in low-power multi-parameter Microsystems. Sens Actuators A 2000;84:351–361.

51. Van Der Goes FML, Meijer GCM. A novel low-cost capacitive-sensor interface. Trans Instr Meas 1996;45(2):536–540.

52. Bracke W, Merken P, Puers R, Van Hoof C. On the optimization of ultra low power front-end interfaces for capacitive sensors. Sens Actuators A 2005;117(2):273–285.

53. Chatzandroulis S, Goustouridis D, Normand P, Tsoukalas D. A solid-state pressure-sensing microsystem for biomedical applications. Sens Actuators A 1997;62:551–555.

54. Casadei FW, Gerold M, Baldinger E. Implantable Blood Pressure Telemetry System. IEEE T Biomed Eng 1972;BME-19(5): 334–338.

55. Neukomm PA, Kuendig H. Passive wireless actuator control and sensor signal transmission. Sens Actuators 1990;A21-A23:258–262.

56. Tham KM, Nagaraj K. A low Supply Voltage High PSRR Voltage Reference in CMOS Process. IEEE J Solid-St Circ 1995;30(5):586–590.

See also BIOELECTRODES; INTEGRATED CIRCUIT TEMPERATURE SENSOR.

CARBON.   See BIOMATERIALS: CARBON.

CARDIAC CATHETERIZATION.   See CORONARY ANGIOPLASTY AND GUIDEWIRE DIAGNOSTICS.

CARDIAC LIFE SUPPORT.   See CARDIOPULMONARY RESUSCITATION.

CARDIAC OUTPUT, FICK TECHNIQUE FOR

STEVEN C. FADDY
University of Sydney
Darlinghurst, Australia

INTRODUCTION

Cardiac output (CO) is an important measurement in many medical investigations. It is the amount of blood pumped by the ventricles of the heart and can be defined as the product of stroke volume (SV) and heart rate (HR), where stroke volume is the amount of blood expelled by the ventricle with each contraction and the HR is the number of contractions per minute:

$$CO = SV \times HR$$

Cardiac output gives an indication of ventricular function and is also used in the calculation of a number of flow-dependent parameters, such as cardiac index, systemic vascular resistance, pulmonary vascular resistance, valve areas, and intracardiac shunt ratios.

The Fick technique is the gold standard in CO measurement. It relies on direct measurement of oxygen consumption and expenditure to derive the rate of blood flow throughout the individual.

HISTORY

In 1870, the German physiologist, Adolf Fick (1829–1901), described a novel method of determining cardiac output based on diffusion of respiratory gases in the lungs. This came after almost 30 years of work by Fick and numerous others, who reasoned that diffusion was one of the most essential events within the living organism. In 1855, Fick had published his findings relating to diffusion of gas across a fluid membrane. These became known as Fick's law of diffusion and stated that the rate of diffusion of a gas is proportional to the partial pressures of the gas on either side of the membrane, the area across which diffusion is taking place and the distance over which diffusion must take place. As an aside, Fick also invented contact lenses in 1887.

The 1870 publication by Adolf Fick stated: "It is astonishing that no one has arrived at the following obvious method by which [the amount of blood ejected by the ventricle of the heart with each systole] may be determined directly, at least in animals. One measures how much oxygen an animal absorbs from the air in a given time, and how much carbon dioxide it gives off. During the experiment one obtains a sample of arterial and venous blood; in both the oxygen and carbon dioxide content are measured. The difference in oxygen content tells how much oxygen each cubic centimeter of blood takes up in its passage through the lungs. As one knows the total quantity of oxygen absorbed in a given time one can calculate how many cubic centimeters of blood passed through the lungs in this time. Or if one divides by the number of heart beats during this time one can calculate how many cubic centimeters of blood are ejected with each beat of heart. The corresponding calculation with the quantities of carbon dioxide gives a determination of the same value, which controls the first (1)."

In simplest terms, cardiac output can be calculated as a ratio of the amount of oxygen consumed through breathing and the rate in which oxygen is taken up by the tissues.

Cardiac Output

= Oxygen consumption/Arteriovenous oxygen dierence

It was not until the 1930s that quantitative measurement of the components allowed confirmation of the Fick equation as a means of calculating cardiac output.

PHYSIOLOGY OF THE FICK TECHNIQUE

Oxygen Consumption (VO_2)

The first step in calculating CO by the Fick technique is to determine the amount of oxygen consumed by the individual over a period of time. This is best done in the resting state so that there is constant oxygen consumption over the collection period. The traditional method is collection of expired gases in a Douglas bag over a period of ~ 3 min. Then, from the volume of expired gas, the oxygen content of the expired gas and the oxygen content of the inspired room air, it is possible to calculate the amount of oxygen taken up by the individual.

Subtracting the oxygen content of expired gas from that of the inspired room air (%v/v or mL of $O_2 \cdot 100\,\mathrm{mL}^{-1}$ of the gas) gives the oxygen difference between the inspired and expired gases, expressed in mL of $O_2 \cdot 100\,\mathrm{mL}^{-1}$ of expired gas. Applying a factor of 10 gives this figure as milliliters of O_2 per liter of expired gas, which are the units used later in the calculation. Dividing the total volume of expired gas by the collection time gives the minute ventilatory rate, expressed in liters per minute $L \cdot \mathrm{min}^{-1}$.

The product of the O_2 difference ($\mathrm{mL} \cdot \mathrm{L}^{-1}$) and the minute volume ($\mathrm{L} \cdot \mathrm{min}^{-1}$) is the oxygen consumption (VO_2) expressed in milliliters of oxygen absorbed per minute.

$$VO_2 = (O_{2\,\text{Room air}} - O_{2\,\text{expired}}) \times (\text{volume/time})$$

Example: Inspired $O_2 = 21.0\,\mathrm{mL} \cdot 100\,\mathrm{mL}^{-1}$ room air

Expired $O_2 = 16.7\,\mathrm{mL} \cdot 100\,\mathrm{mL}^{-1}$ expired gas

O_2 difference $= 21.0 - 16.7 = 4.3\,\mathrm{mL} \cdot 100\,\mathrm{mL}^{-1}$

Total volume expired $= 26.1\,\mathrm{L}$

Collection time $= 3$ min

Minute volume $= 26.1\,\mathrm{L} \cdot 3\,\mathrm{min}^{-1} = 8.7\,\mathrm{L} \cdot \mathrm{min}^{-1}$

Therefore,

O_2 consumption $= (4.3 \times 10)\mathrm{mL} \cdot \mathrm{L}^{-1} \times$

$8.7\,\mathrm{L} \cdot \mathrm{min}^{-1} = 374\,\mathrm{mL} \cdot \mathrm{min}^{-1}$

An alternative to the Douglas bag method is the use of a metabolic rate meter with a hood or facemask, a variable-speed blower and a servocontrol loop with an oxygen sensor. This method employs essentially the same principle as the Douglas bag method, but gives a real-time measurement of VO_2. The variable-speed blower maintains a flow of room air through the hood or facemask past the patient into a polarographic oxygen sensor (gold and silver–silver chloride electrode), varying the flow in order to keep the oxygen concentration at the measuring electrode constant. By keeping the oxygen concentration at the measuring electrode constant, the only variable is the flow rate through the system. Under steady-state conditions, this is the only variable determining the oxygen consumption (VO_2).

Although this method provides a real-time measurement of VO_2, thus excluding the need for collection of a Douglas bag, it is still rather time and labor intensive. In addition, it has been suggested that it is difficult to obtain reproducible results and the method gives consistently lower results then the Douglas bag technique.

Arteriovenous Difference

As with oxygen consumption, measurement of oxygen uptake by the body involves measuring blood oxygen content before and after entering the lungs. The arteriovenous oxygen difference (AV_{diff}) is the difference between the content of oxygen (ctO_2) in the oxygenated arterial blood leaving the lungs and the deoxygenated venous blood returning to the lungs (mL O_2 per 100 mL of blood). The AV_{diff} represents the volume of oxygen delivered to meet the body's metabolic demands. Again, this figure is multiplied by 10 to give the AV_{diff} in units of mL O_2 per liter of blood.

$$AV_{\text{diff}} = ctO_{2(\text{Arterial})} - ctO_{2(\text{Venous})}$$

Example: Arterial O_2 content $= 19.5\,\mathrm{mL} \cdot \mathrm{dL}^{-1}$ blood

Venous O_2 content $= 13.2\,\mathrm{mL} \cdot \mathrm{dL}^{-1}$ blood

$AV_{\text{diff}} = 19.5 - 13.2 = 6.3\,\mathrm{mL} \cdot \mathrm{dL}^{-1} = 63\,\mathrm{mL} \cdot \mathrm{L}^{-1}$

Typically, a sample from the main pulmonary artery is used for venous blood and a sample from the left ventricle or aorta is used for arterial blood oxygen content measurements.

Cardiac Output

The rate at which oxygen is taken up by the lungs and the rate at which it is taken up by the body is now known from the above calculations. The ratio of these two figures gives the cardiac output. The examples above show that the lungs take up 374 mL of oxygen each minute and that the blood takes up 63 mL of oxygen for each liter that passes through the lungs. How many lots of 63 mL (1 L aliquots of blood) must pass through the lungs to take up 374 mL of oxygen each minute? The answer is 5.9 L of blood must pass through the lungs each minute in order to absorb this amount of inspired oxygen.

Example: $VO_2 = 374\,\mathrm{mL} \cdot \mathrm{min}^{-1}$

$AV_{\text{diff}} = 63\,\mathrm{mL} \cdot \mathrm{L}^{-1}$

$CO = 374/63 = 5.9\,\mathrm{L} \cdot \mathrm{min}$

PRACTICAL CONSIDERATIONS FOR USING THE FICK TECHNIQUE

Oxygen Consumption

The measured volume of a gas is affected in part by the ambient temperature and atmospheric pressure in which it is collected. Obviously, these will vary from day to day, leading to a potential source of variation in the calculation of oxygen consumption, and ultimately, cardiac output. The combined gas law (a combination of Boyle's and Charles' law) describes the relationship of pressure, temperature, and volume in a gas. This law can be used to correct measured gas volumes to standard temperature and pressure (STP). This means that the measured volume of gas is standardised to 273 K and 760 mmHg (101.32 kPa).

As well as correcting for variations in atmospheric pressure, it is also necessary to correct for water vapor pressure. Dalton's law tells us that the pressure of a gas mixture is equal to the partial pressures of all of the components of the mixture. Water vapor is present in the atmosphere and in exhaled gas and its partial pressure contributes to the total atmospheric pressure. Water vapor exerts a constant pressure at a given temperature, regardless of the atmospheric pressure. Water vapor pressure is 47 mmHg (6.26 kPa) at normal body temperature and 17.5 mmHg at (2.33 kPa) 20 °C. Before correcting for STP it is necessary to subtract the water vapor pressure from the total atmospheric pressure to obtain the dry gas pressure at the ambient temperature. This is known as 'standard temperature and pressure, dry' (STPD), which is used for the correction.

Example :   Atmospheric pressure = 762 mmHg (6.26 kPa)
Ambient temperature = 23 °C
Water vapor pressure at
23 °C = 21 mmHg (2.79 kPa)
Dry gas pressure = 762 − 21 mmHg
= 741 mmHg (98.79 kPa)
STPD correction factor for 741 mmHg and
23 °C = 0.8991{from standard tables}
Volume of expired gas = 9.68 L · min^{-1}
STPD corrected volume = 9.68 × 0.8991
= 8.7 L · min^{-1}

By standardizing to STPD, we have removed the effect of water vapor pressure, ambient pressure, and temperature on the volume measurement and, hence, potential sources of day to day variation in measurement of the cardiac output.

Arteriovenous Oxygen Difference

Although many current generation analyzers can calculate oxygen content (ctO_2), earlier models did not. It may be necessary to manually calculate oxygen content from the hemoglobin level (Hb) and oxygen saturation of a sample. Hemoglobin is able to carry 1.36 mL of oxygen per gram of hemoglobin. Therefore, by multiplying the hemoglobin

level by 1.36 it is possible to calculate the oxygen carrying capacity of the individual. Simply stated, this is the maximum amount of oxygen that can be carried by 100 mL of the individual's blood and is dependent on the hemoglobin level. Some textbooks have quoted the constant as 1.34 and others add a value of 0.03 to account for oxygen dissolved in plasma, but 1.36 is the generally accepted constant for calculation of oxygen carrying capacity.

$$\text{Oxygen carrying capacity} = \text{Hgb} \times 1.36 \, \text{mL} \cdot \text{dL}^{-1}$$

If the total amount of oxygen that the blood is capable of carrying and the saturation of the sample is known, it is possible to calculate the oxygen content of that sample.

$$ctO_2 = \text{oxygen carrying capacity} \times \% \, \text{saturation}$$

The arteriovenous oxygen difference is the difference in oxygen content between arterial and venous blood.

Example :   Hb = 14.5 g · dL^{-1}
Oxygen carrying capacity = 14.5 × 1.36
= 19.72 mL · dL^{-1}
Arterial saturation = 98.9%
Arterial oxygen content = 19.72 × 98.9%
= 19.5 mL · dL^{-1}
Venous saturation = 66.9%
Venous oxygen content = 19.72 × 66.9%
= 13.2 mL · dL^{-1}

Therefore,

$$\text{AV}_{\text{diff}} = 19.5 - 13.2 = 6.3 \, \text{mL} \cdot \text{dL}^{-1} = 63 \, \text{mL} \cdot \text{L}^{-1}$$

In this example, each liter of blood leaving the lungs delivers 63 mL of oxygen to the tissues.

Figure 1 shows a complete example of CO measurement using the Fick technique.

ASSUMPTIONS WHEN USING THE FICK TECHNIQUE FOR CARDIAC OUTPUT

Absence of Intracardiac Shunt

The method of calculating cardiac output described above uses the amount of oxygen absorbed by the blood as it travels through the lungs. We then assume that the amount of blood pumped by the right ventricle through the lungs is equal to the amount pumped by the left ventricle through the systemic vessels since the cardiovascular system is a closed system. This assumption does not always hold true and it is sometimes necessary to alter the calculation.

The term shunt describes the condition where a communication exists between the left- and right-sided chambers of the heart. If this condition results in shunting of blood between the venous and arterial circulation, the assumption becomes invalid because some blood is being recirculated through part of the circuit and the two ventricles are pumping unequal volumes. If an intracardiac shunt is known or suspected, it is necessary to collect blood samples at different points than the standard arterial and pulmonary artery sites. Calculation of cardiac output and

A. Standard Temperature and Pressure

Atmospheric pressure = 762 mmHg
Ambient temperature = 23 °C
    Water vapor pressure at 23 °C = 21 mmHg
Dry gas pressure = 762 − 21 mmHg = 741 mmHg
STPD correction factor for
    741 mmHg and 23 °C = **0.8991**

B. Volume measurement

Total volume expired = 28.14 L
Collection time = 3 min
Minute volume = 28.14 L·3 min^{-1} = **9.68** L·min^{-1}

STP corrected volume
0.8991 × 9.68 L·min^{-1} = **8.7** L·min^{-1}

C. Oxygen Difference

Inspired O_2 = 21.0 mL 100 mL^{-1}
Expired O_2 = 16.7 mL /100 mL^{-1}
O_2 difference = 21.0 − 16.7 = **4.3** mL/100 mL^{-1}

D. Oxygen Consumption

O_2 consumption = (4.3 × 10) mL·L^{-1} × 8.7 L·min^{-1} = **374** mL·min^{-1}

E. Arteriovenous O_2 Difference

Arterial O_2 content = 19.5 mL·dL^{-1} blood
Venous O_2 content = 13.2 mL·dL^{-1} blood
AV_{diff} = 19.5 − 13.2 = 6.3 mL·dL^{-1} = **63** mL·L^{-1}

F. Cardiac Output

VO_2 = 374 mL·min^{-1}
AV_{diff} = 63 mL·L^{-1}
Cardiac output = 374 / 63 = **5.9** L·min^{-1}

Figure 1. Example of cardiac output measurement using the Fick technique. The arrows indicate how the various parameters discussed in the text are interrelated in the various calculations.

shunt ratios in the presence of an intracardiac shunt is discussed later in this article.

Collection of True Arterial Sample

In a normal heart, it is not easy to gain access to the pulmonary veins to collect an arterial sample as the blood leaves the lungs. As a result, left ventricular or aortic blood is used to measure arterial oxygen content. This method ignores the small amount of venous admixture from bronchial and thebesian venous drainage into the left atrium.

Direct Measurement of Oxygen Consumption

Owing to the time- and labor-intensive methods of measurement of oxygen consumption (VO_2), there is often a temptation to use an estimate of oxygen consumption rather than direct measurement. Standard formulas and nomograms are used to estimate VO_2 from height, weight, age, and sex. The body surface area (BSA) is calculated from height and weight and expressed in units of square meters (m^2).

$$BSA = 0.007184 \times weight^{0.425} \times height^{0.725}$$

Age, sex, and basal metabolic rate are used to determine heat production from standard nomograms. Finally, heat production and BSA are used to estimate the oxygen consumption.

$$VO_2 = [BSA \times Heat\ Production]/291.72$$

This method estimates the basal oxygen consumption at rest. It does not make allowances for any pathological conditions, including those being investigated, that may affect the resting oxygen consumption. Studies comparing measured and estimated VO_2 have shown that estimating VO_2 from the various available formulas can lead to large and unpredictable errors in both VO_2 and cardiac output values (2,3). The practice of estimating VO_2 is strongly discouraged.

DETECTION AND ASSESSMENT OF INTRACARDIAC SHUNTS

Earlier in this article it was seen how the oxygen content of blood entering and leaving the lungs was used to calculate the cardiac output. It was assumed that blood flowing through the lungs is equal to blood flowing through the systemic circulation (since the cardiovascular system is a closed circuit). Several conditions may result in blood being recirculated between the left and right sides of the heart, leading to unequal flow in the pulmonary and systemic circulation. These conditions include atrial septal defects, patent foramen ovale, ventricular septal defects, and patent ductus arteriosus. Patent foramen ovale has been estimated to be present in 27.3% of the population (4), but the presence of a defect does not necessarily result in intracardiac shunting.

A communication between the left- and right-sided chambers of the heart can result in blood being shunted from right to left (venous blood being mixed into the arterial circulation), left to right (arterial blood being mixed into the pulmonary circulation), or as a bidirectional shunt (blood moves back and forth across the communication at different stages of the cardiac cycle). Although the method of calculating cardiac output remains essentially the same, the sites of blood collection are different in cases where intracardiac shunting exists. The following passages describe the methods for calculating systemic and pulmonary flow in the presence of different intracardiac shunts.

Left-to-Right Shunt

In a left-to-right shunt, arterial blood is pushed across the defect into the pulmonary circulation. This will artificially elevate the oxygen saturation and oxygen content in the pulmonary artery. To avoid error in the calculation of systemic cardiac output in the presence of a left-to-right shunt, it is necessary to collect the venous blood sample in the chamber immediately proximal to the shunt. In the case of atrial defects, blood is collected from both the inferior and superior vena cavae. Oxygen content (ctO_2) from these sites is used in the calculation of mixed venous oxygen content (MVO_2). The individual values are weighted and averaged according to the relatively higher flow from the superior vena cava and the absence of coronary sinus blood in the measurements. The generally accepted formula used for estimation of mixed venous oxygen content is

$$MVO_2 = [3 \times ctO_{2(SVC)} + ctO_{2(IVC)}]/4$$

MVO_2 becomes the venous component of the arteriovenous difference calculation and cardiac output (systemic flow, Q_s) is calculated as described earlier.

It is also possible to calculate pulmonary flow (Q_p) by using the pulmonary artery oxygen content as the venous component (blood entering the lungs) and left ventricular or aortic oxygen content as the arterial component (blood leaving the lungs). The pulmonary flow is equal to the systemic flow returning to the heart plus the volume being recirculated from the left-sided chambers via the shunt. This is reflected in the CO calculation. Because of the recirculated arterial blood, the venous oxygen content in the pulmonary artery will be elevated, leading to a decrease in the arteriovenous difference and, hence, a higher pulmonary flow.

Right-to-Left Shunt

The opposite occurs in a right-to-left shunt. Venous blood is mixed into the arterial circulation leading to a decrease in systemic arterial oxygen saturation. The calculation of systemic flow (Q_s) uses the arterial and venous (pulmonary artery) samples as usual. The calculation of pulmonary flow (Q_p) requires a sample to be taken after the blood leaves the lungs, but proximal to the shunt. In a right-to-left shunt it is necessary to sample blood from the pulmonary veins. In practical terms, this requires the catheter to pass from the right atrium across the defect to the left atrium and then into a pulmonary vein. Pulmonary vein oxygen content becomes the arterial component of the arteriovenous difference and pulmonary flow is calculated as usual.

In the presence of a right-to-left shunt, systemic flow is equal to the pulmonary flow leaving the lungs plus the amount that passes across the shunt directly from the right-sided chambers. Sampling in the left ventricle or aorta distal to the shunt will therefore give a lower arterial oxygen content than would be measured in the pulmonary veins, leading to an decrease in the arteriovenous difference and, hence, a higher systemic flow compared to the pulmonary flow.

A right-to-left shunt should be suspected in any patient who has an arterial oxygen saturation less than 95%. Investigation of these patients should include assessment for the presence of a bidirectional shunt.

Bidirectional Shunt

The presence of a bidirectional shunt complicates the calculation of pulmonary and systemic flow. Neither of the methods described above is suitable since both assume shunting in only one direction. The systemic blood flow (SBF) is calculated using oxygen contents sampled at the sites where blood enters and leaves the systemic circulation (arterial and mixed venous sites, respectively). Pulmonary blood flow (PBF) is calculated using sampling sites where blood enters and leaves the lungs (pulmonary artery and pulmonary vein, respectively). Finally, effective blood flow (EBF) is calculated using samples taken where the blood enters the heart and leaves the lungs (mixed venous and pulmonary vein oxygen contents). The mixed venous oxygen values should be the same as the pulmonary artery sample if no shunts are present. Similarly, the pulmonary vein should be the same as the left ventricular or aortic

$VO_2 = 201$ mL·min^{-1}

Hb = 13.9 g·dL^{-1}

Oxygen carrying capacity = $13.9 \times 1.36 = 18.9$ mL·dL^{-1}

Site	Saturation %	ctO$_2$ mL·dL^{-1}
Arterial		
Pulmonary vein (PV)	94.1	17.8
Radial artery (RArt)	83.0	15.7
Venous		
Superior vena cava (SVC)	58.6	11.1
Inferior vena cava (IVC)	61.0	11.6
Right atrium (RA)	68.8	13.0
Pulmonary artery (PA)	63.4	12.0
Mixed venous (MV) = (3 × SVC + IVC)/4	59.2	11.2

Pressure measurements
Mean Pulmonary Artery pressure = 43 mmHg (5.73 kPa)
Mean Left Atrial pressure = 4 mmHg (0.53 kPa)

$$\text{PBF} = \frac{VO_2}{ctO_{2\,(PV)} - ctO_{2\,(PA)}} = \frac{201}{17.8 - 12.0} = 3.5 \text{ L·min}^{-1}$$

$$\text{SBF} = \frac{VO_2}{ctO_{2\,(R\,Art)} - ctO_{2\,(MV)}} = \frac{201}{15.7 - 11.2} = 4.5 \text{ L·min}^{-1}$$

$$\text{EBF} = \frac{VO_2}{ctO_{2\,(PV)} - ctO_{2\,(MV)}} = \frac{201}{17.8 - 11.2} = 3.0 \text{ L·min}^{-1}$$

Left-to-right shunt = PBF – EBF = 3.5 – 3.0 = 0.5 L·min^{-1}

Right-to-left shunt = SBF – EBF = 4.5 – 3.0 = 1.5 L·min^{-1}

$$\text{PVR} = \frac{80 \times (PA_m - LA_m)}{Q_p} = \frac{80 \times (43 - 4)}{3.5} = 891 \text{ dyn·s·cm}^{-5}$$

Figure 2. Bidirectional shunt calculation in a 55 year old woman with Eisenmenger's syndrome secondary to an atrial septal defect (ASD). Note the predominantly right-to-left shunt due to increased pulmonary vascular resistance. These results show little deterioration compared to measurements taken six months earlier [$Q_p = 3.7$ L·min^{-1}, mean PA pressure = 39 mmHg (5.19 kPa) and PVR = 800 dyn·s·cm^{-5}].

sample in the absence of any shunts. Therefore, by sampling at these sites, the pulmonary (and hence, systemic) flow that would normally occur if no shunts were present is being calculated

$$\text{SBF} = VO_2/[ctO_{2(Art)} - ctO_{2(MV)}]$$
$$\text{PBF} = VO_2/[ctO_{2(P\,vein)} - ctO_{2(P\,art)}]$$
$$\text{EBF} = VO_2/[ctO_{2(P\,vein)} - ctO_{2(MV)}]$$

Since the systemic flow (SBF), pulmonary flow (PBF), and the flow that would occur in the absence of any shunts (EBF) is known, the size of the shunts can also be calculated

Right to left = SBF – EBF
Left to right = PBF – EBF

Figure 2 shows calculations for a bidirectional shunt based on the principles discussed above. Note the changes in oxygen saturation and content as blood passes the atrial defect. In the left heart, oxygen saturation decreases as blood passes from pulmonary veins to the left ventricle due to mixing of deoxygenated blood being shunted across the defect. Conversely, oxygen saturation in the right heart increases as blood passes from the vena cavae (mixed venous) to pulmonary artery due to oxygenated blood being shunted across the defect from the left atrium.

Pulmonary–Systemic Flow Ratio

An alternative to calculating flow across a defect is to calculate the pulmonary/systemic flow ratio (P/S ratio). This value is a ratio of the pulmonary flow relative to the systemic flow. Calculation of the P/S ratio does not involve calculation of actual flows, so it is not necessary to collect expired gases to calculate the VO_2. The P/S ratio is calculated using only the oxygen content (or saturation) from arterial, pulmonary artery, mixed venous, and pulmonary vein samples.

The AV_{diff} for the pulmonary component is calculated by subtracting mixed venous oxygen content from systemic arterial oxygen content. The systemic AV_{diff} is calculated by subtracting the pulmonary artery oxygen content from the pulmonary vein oxygen content. If a pulmonary vein sample is not possible, use an estimate of 98% unless the arterial saturation is higher. Using arterial oxygen content to estimate the pulmonary vein content will assume that there is no right-to-left shunt. The P/S ratio (or P:S ratio) is the calculated by dividing the pulmonary component by the systemic component.

The P/S ratio is the proportion of flow through the pulmonary circulation relative to the systemic circulation. Therefore, a value > 1.0 indicates left-to-right shunting. An arbitrary value of between 1.5 and 2.0 is often used to determine the need for definitive treatment to correct the defect, in order to avoid late sequelas from prolonged pulmonary vascular overload. A P/S flow ratio < 1 indicates right-to-left shunting and may be a sign of irreversible pulmonary vascular disease.

Example: $Hb = 14.5 \, g \cdot dL^{-1}$

$$\text{Oxygen carrying capacity} = 19.72 \, mL \cdot dL^{-1}$$

Arterial oxygen content = 98.9%
$$ctO_{2(Art)} = 19.5 \, mL \cdot dL^{-1}$$
Pulmonary artery oxygen content = 66.9%
$$ctO_{2(PA)} = 13.2 \, mL \cdot dL^{-1}$$
Mixed venous oxygen content = 63.1%
$$ctO_{2(MV)} = 12.4 \, mL \cdot dL^{-1}$$

Pulmonary: $ctO_{2(Art)} - ctO_{2(MV)}$ = 19.5 − 12.4
$$= 7.1 \, mL \cdot dL^{-1}$$
Systemic: $ctO_{2(PV)} - ctO_{2(PA)}$ = 19.5 − 13.2
$$= 6.3 \, mL \cdot dL^{-1}$$

$$\text{P/S ratio}: \ 7.1/6.3 = 1.13$$

FLOW-DEPENDENT PARAMETERS

A number of frequently used parameters in cardiovascular medicine are dependent on knowing systemic or pulmonary flow. Calculation of these parameters, and the effect that the cardiac output has on each, is discussed. Table 1 lists expected normal ranges for a number of common flow-dependent parameters.

Cardiac Index

Cardiac output is often corrected for patient's size, based on body surface area (BSA). Cardiac index (CI) is calculated by dividing the cardiac output by the body surface area.

$$CI = CO/BSA \ L \cdot min^{-1} \cdot m^{-2}$$

Table 1. Expected Ranges for Common Flow-Dependent Parameters

$$\text{Cardiac Output} = \frac{\text{Oxygen consumption}}{\text{Arteriovenous oxygen difference}}$$

$$VO_2 = \frac{\text{BSA} \times \text{Heat Production}}{291.72}$$

$$MVO_2 = \frac{3 \times ctO_{2(SVC)} + ctO_{2(IVC)}}{4}$$

$$SBF = \frac{VO_2}{ctO_{2(Art)} - ctO_{2(MV)}}$$

$$PBF = \frac{VO_2}{ctO_{2(Pvein)} - ctO_{2(Part)}}$$

$$EBF = \frac{VO_2}{ctO_{2(Pvein)} - ctO_{2(MV)}}$$

$$CI = \frac{CO}{BSA} \, L \cdot min^{-1} \cdot m^{-2}$$

$$\text{Area} = \frac{\{\frac{CO}{(SEP \times HR)}\}}{(44.3 \times \sqrt{\text{gradient}})}$$

$$\text{Area} = \frac{\text{Cardiac output}}{\sqrt{\text{Gradient}}}$$

$$\text{Area} = \frac{(CO \times DFP \times HR)}{37.7 \times \sqrt{\text{gradient}}}$$

$$SVR = \frac{80 \times (Ao_m - RA_m)}{Q_s}$$

$$PVR = \frac{80 \times (PA_m - LA_m)}{Q_p}$$

$$CO = \frac{VCO_2}{ctCO_{2(Ven)} - ctCO_{2(Art)}}$$

$$CO = \frac{\Delta VCO_2}{\Delta ctCO_{2(Art)}}$$

$$CO = \frac{\Delta VCO_2}{S \times \Delta ETCO_2}$$

Some believe cardiac index is a more useful parameter than cardiac index because it accounts for the patient's size. A large person (as approximated by BSA) would be expected to have a higher cardiac output while a low cardiac output in a smaller person may not necessarily be indicative of a poorly functioning ventricle. Many authors only express cardiac output in terms of cardiac index for this reason.

Valve Areas

Basic fluid dynamic principles state that a fluid exerts pressure equally in all directions. Therefore, when the valves of the heart are open they should allow equalisation of pressure in the two chambers that they separate. Sometimes the valves of the heart become stiff, thickened or do not open properly, inhibiting flow through the valve. This is referred to as valve stenosis. A result of this process is a pressure gradient, a difference in pressure on either side of the valve. Take an example of aortic valve stenosis. When the left ventricle contracts, it is pushing against an obstruction. The systolic pressure will be higher in the ventricle that in the aorta. The difference in pressure is referred to as a pressure gradient (expressed in mmHg) and can be measured during cardiac catheterisation. The pressure gradient across a valve is often used to determine the severity of a valve stenosis. However, the main parameter that should be considered is the cross-sectional area of the valve. A pressure gradient of 20 mmHg (2.66 kPa) is often considered an indication of mild aortic stenosis. However, in the presence of low cardiac output, it is necessary to have quite a narrow valve orifice to achieve this gradient. Conversely, a less severe stenosis could achieve a gradient of 50 mmHg (6.66 kPa) in a patient with a high cardiac output.

Both the cardiac output and mean pressure gradient are used in the calculation of valve area. The Gorlin formula is used for calculating valve area of the aortic or pulmonary valve:

$$\text{Area} = [\{\text{CO}/(\text{SEP} \times \text{HR})\}/(44.3 \times \sqrt{\text{gradient}})]$$

where HR is the heart rate and SEP is the systolic ejection period (since gradients in these valves are measured during systole). However, a shorter formula is often used as an approximation:

$$\text{Area} = \text{Cardiac output}/\sqrt{\text{Gradient}}$$

Taking the example of aortic stenosis, a mean gradient of 20 mmHg (2.66 kPa) in a patient with a normal cardiac output of 4.5 L·min^{-1} would give a valve area of $4.5/\sqrt{20} = 1.00$ cm^2. In a patient with a low cardiac output of 3.1 L·min^{-1}, the valve area ($3.1/\sqrt{20} = 0.69$ cm^2) would be much smaller to achieve this gradient. Similarly, our patient with a mean gradient of 50 mmHg (6.66 kPa) would not have such a severe narrowing if a high cardiac output (e.g., 7.1 L·min^{-1}) is present ($7.1/\sqrt{50} = 1.00$ cm^2).

The Gorlin formula for calculating mitral or tricuspid valve area is slightly different:

$$\text{Area} = (\text{CO} \times \text{DFP} \times \text{HR})/(37.7 \times \sqrt{\text{gradient}})$$

where DFP is the diastolic filling period (since gradients in these valves are measured during diastole).

Vascular Resistance

Measurements of vascular resistance are based on principles of fluid dynamics where resistance is defined as the decrease in pressure between two points in a vascular segment divided by the flow through that segment. While this simplification does not account for pulsatile flow, calculation of vascular resistance in this way is useful in a number of clinical settings.

In the past, Wood units (mmHg·L·min^{-1}) were used to express vascular resistance. Today, vascular resistance is more commonly expressed in absolute resistance units (dyn.s.cm^{-5}), which are derived from the mean pressure gradient (dyn·cm^{-2}) divided by the mean flow (cm^3·s^{-1}). A constant of 80 is used to convert values from traditional units (mmHg and L·min^{-1}) to absolute resistance units.

Systemic vascular resistance (SVR) is therefore defined as the difference in pressure between blood entering the systemic circulation (mean aortic pressure) and blood leaving the systemic circulation (mean right atrial pressure) divided by the systemic blood flow:

$$\text{SVR} = [80 \times (AO_m - RA_m)/Q_s]$$

Similarly, pulmonary vascular resistance (PVR) is defined as the difference between mean pulmonary artery pressure and mean left atrial pressure divided by the pulmonary flow:

$$\text{PVR} = [80 \times (PA_m - LA_m)/Q_p]$$

In the absence of intracardiac shunting both SVR and PVR can be calculated using the standard cardiac output measurement. If intrapulmonary shunting is present, systemic and pulmonary flow must be individually calculated for use in the SVR and PVR calculations, respectively.

There are a number of causes of increased systemic or pulmonary vascular resistance, some reversible and some permanent. The use of serial cardiac output and pressure measurements during drug challenges can assist in identifying management strategies that may be helpful in reducing vascular resistance.

VARIATIONS OF THE FICK METHOD

The Fick principle can be applied to any gas involved in diffusion, including carbon dioxide. Such variations to the classic Fick formula are often referred to as the indirect Fick principle.

By measuring the difference between inspired and expired CO_2 and the minute ventilation volume we can calculate CO_2 production (VCO_2). Arteriovenous CO_2 difference is calculated from the measured values of arterial and venous carbon dioxide content ($ctCO_2$). The ratio of VCO_2 and the arteriovenous CO_2 difference gives the cardiac output.

Earlier in this article it was seen how the oxygen content (ctO_2) of blood is calculated from the amount of hemoglobin and oxygen saturation of the sample, since nearly all of the oxygen is bound to hemoglobin. A relatively smaller proportion of CO_2 is bound to hemoglobin. About 70% is transported in the blood as bicarbonate. Only 23% is bound to hemoglobin and 7% is transported as dissolved CO_2. Therefore, the calculation of CO_2 content is not dependent on hemoglobin level.

Carbon dioxide content ($ctCO_2$) is calculated from the formula:

$$ctCO_2 = 11.02 \times PCO_2^{0.396}$$

Thus, if we have partial CO_2 pressure of arterial ($PaCO_2$) and venous ($PvCO_2$) samples, we can calculate the arteriovenous carbon dioxide difference as

$$ctCO_{2(Ven)} - ctCO_{2(Art)} = 11.02(PvCO_2^{0.396} - PaCO_2^{0.396})$$

Then cardiac output is calculated with the formula:

$$CO = VCO_2/(ctCO_{2(Ven)} - ctCO_{2(Art)})$$

There are some advantages to using the carbon dioxide method. When a patient is receiving high concentrations of supplemental oxygen, analysis of inspired and expired oxygen will give a small difference between two relatively large values. Even a small error in the estimation of either value will yield an inaccurate VO_2. Additionally, some oxygen analysers (e.g., paramagnetic analyzers) have poor accuracy at high oxygen concentrations. Measurement of cardiac output in patients receiving high concentrations of supplemental oxygen may be erroneous and the Fick principle using carbon dioxide may prove more accurate.

Applying the Fick principle to carbon dioxide involves the same steps as using oxygen for the calculation. There is still the requirement for analysis of expired gases, as well as collection and analysis of arterial and mixed venous blood samples. However, there are a number of ways of estimating, rather than directly measuring, the various parameters necessary to calculate cardiac output using the Fick CO_2 technique.

Infrared (IR) light absorption sensors in the breathing circuit can measure inspired and expired CO_2 content. Alternatively, an assumption can be made about the content of CO_2 in the inhaled gas (especially if the patient is being ventilated with 100% oxygen) and only expired CO_2 needs to be measured. Along with an airflow sensor (e.g., as a differential pressure pneumotachometer), these measurements can provide real-time estimation of VCO_2.

There is a logarithmic relationship between cardiac output and end-tidal CO_2 (ETCO$_2$). At normal or high cardiac output the respiratory rate determines the amount of CO_2 that is eliminated by the lungs with each breath. If it is assumed that CO_2 exchange at the alveolar–arterial membrane reaches equilibrium, then ETCO$_2$ can be used to estimate $PaCO_2$. In this way, it is possible to estimate cardiac output without subjecting the patient to unnecessarily invasive procedures.

The critically ill patient presents a number of challenges. These patients are usually intubated and manually ventilated with high concentrations of inspired oxygen. While many will have arterial lines for blood pressure monitoring, those that do not are exposed to added risk of morbidity if arterial access is necessary to determine cardiac output. In addition, placement of a pulmonary artery catheter for the measurement of mixed venous gas tension exposes the patient to significant risk of sepsis, pneumothorax, thrombosis, or pulmonary artery rupture. However, cardiac output is often vital in determining end-organ perfusion.

Recently, a system for noninvasive measurement of cardiac output using the Fick principle and carbon dioxide was developed for use with ventilated patients in the intensive care unit, based on the estimations described above. A number of assumptions allow this system to be used without the need for arterial or mixed venous blood samples. The technique involves measuring changes in carbon dioxide production and arterial CO_2 content between normal breathing conditions and under rebreathing conditions with 10–15% CO_2 and a reservoir with a volume 1.5 times the tidal volume. Carbon dioxide production (VCO_2) is calculated from the minute ventilation and expired CO_2 content under normal breathing conditions. Arterial CO_2 content [$ctCO_{2(Art)}$] is estimated from the end-tidal CO_2 (ETCO$_2$) with adjustments for the slope of the CO_2 dissociation curve and degree of dead space ventilation.

During partial rebreathing, carbon dioxide elimination from the blood is reduced, but ETCO$_2$ increases and reaches a plateau within a few breaths. Studies conducted in anaesthetised dogs have showed that during a brief period of CO_2 rebreathing there is a change in $PaCO_2$ and in calculated VCO_2, but little or no change in venous carbon dioxide content ($ctCO_{2(Ven)}$) (5). It is believed that this finding is due to the quantity of CO_2 stores in the body being large and new equilibrium levels not being attained for 20–30 min. This finding becomes highly important in the noninvasive estimation of cardiac output. Any change in the arteriovenous CO_2 difference during the brief rebreathing period can be attributed to changes in the arterial CO_2 component alone.

If it is assumed that the cardiac output and $ctCO_{2(Ven)}$ remain constant during normal breathing (N) rebreathing (R):

$$CO = VCO_{2(N)}/(ctCO_{2(Ven)(N)} - ctCO_{2(Art)(N)})$$
$$= VCO_{2(R)}/(ctCO_{2(Ven)(R)} - ctCO_{2(Art)(R)})$$

From basic algebra it is known that

$$X = A/B = C/D = (A - C)/(B - D)$$

Then,

$$CO = (VCO_{2(N)} - VCO_{2(R)})/[(ctCO_{2(Ven)(N)} - ctCO_{2(Art)(N)}) - (ctCO_{2(Ven)(R)} - ctCO_{2(Art)(R)})]$$

Rearranging this equation:

$$CO = (VCO_{2(N)} - VCO_{2(R)})/[(ctCO_{2(Ven)(N)} - ctCO_{2(Ven)(R)}) - (ctCO_{2(Art)(N)} - ctCO_{2(Art)(R)})]$$

Since it has been assumed that $ctCO_{2(Ven)}$ does not change during rebreathing ($ctCO_{2(Ven)(N)}$ is equal to $ctCO_{2(Ven)(R)}$), these values cancel each other out and the equation becomes:

$$CO = (VCO_{2(N)} - VCO_{2(R)})/(ctCO_{2(Art)(R)} - ctCO_{2(Art)(N)})$$

In other words, cardiac output is equal to the change in VO_2 divided by the change in arterial CO_2 content between the normal and rebreathing states:

$$CO = \Delta VCO_2/\Delta ctCO_{2(Art)}$$

As $ctCO_{2(Art)}$ is estimated from ETCO$_2$ and the slope (S) of the CO_2 dissociation curve:

$$CO = \Delta VCO_2/S \times \Delta ETCO_2$$

This method of cardiac output estimation gives a measure of the pulmonary capillary blood flow (Q_{PCBF}). Changes in VCO_2 and ETCO$_2$ only reflect the blood flow that participates in gas exchange. An intrapulmonary shunt occurs when venous blood passes through unventilated areas of the lungs and moves into the arterial circulation without taking up oxygen or releasing carbon dioxide. A large intrapulmonary shunt will not be reflected in the changes seen in VCO_2 and ETCO$_2$. Therefore, it is necessary to estimate the degree of shunting and correct the cardiac output estimation accordingly. For example, if only 80% of pulmonary blood flow is participating in gas exchange, the Q_{PCBF} estimated by this method will be 80% of the total cardiac output. The shunting fraction is calculated using the arterial oxygen saturation (SaO_2, from a pulse oximeter), the fraction of inspired oxygen (FiO_2), arterial oxygen tension (PaO_2) and standard isoshunt tables. The requirement of an arterial blood gas sample for PaO_2 means that this method is not truly noninvasive.

As mentioned previously, this noninvasive method involves a number of assumptions. In summary, these assumptions are the CO_2 exchange at the alveolar–arterial membrane reaches equilibrium. Therefore, $ETCO_2$ is equal to $PaCO_2$; cardiac output remains constant during rebreathing; venous CO_2 content does not change during a brief period of rebreathing; and there is little or no intrapulmonary shunting.

SUMMARY

The Fick method remains the gold standard of cardiac output measurement. While technically challenging, it relies on direct measurement of oxygen consumption and uptake to determine the rate of blood flow through the lungs and around the body. Accurate measurement of cardiac output is necessary for the estimation of several important parameters and assessment of complex congenital cardiac conditions. Variations of the classical Fick principle allow estimation of cardiac output in patients who might otherwise be unsuitable.

BIBLIOGRAPHY

1. Vandam LD, Fox JA, Fick A. (1829–1901), Physiologist: A heritage for anaesthesiology and critical care medicine. Anaesthesiology 1998;88(2):514–518.
2. Kendrick AH, West J, Papouchado M, Rozkovec A. Direct Fick cardiac output: Are assumed values for oxygen consumption acceptable? Eur Heart J 1988;9(3):337–342.
3. Wolf A, et al. Use of assumed versus measured oxygen consumption for the determination of cardiac output using the Fick principle. Catheterization Cardiovascular Diagnosis 1998;43(4): 372–380.
4. Hagen PT, Scholz DG, Edwards WD. Incidence and size of patent foramen ovale during the first 10 decades of life: An autopsy study of 965 normal hearts. Mayo Clinic Proc 1984; 59(1):17–20.
5. Tachibana K, et al. Effect of ventilatory settings on accuracy of cardiac output measurement using partial CO(2) rebreathing. Anesthesiology 2002;96(1):96–102.

Reading List

Grossman W. Blood flow measurement: The cardiac output. In: Baim DS, Grossman W. Cardiac Catheterization, Angiography and Intervention. 5th ed. Baltimore: Williams and Wilkins; 1996: pp 109–120.
Davidson CJ, Fishman RF, Bonow RO. Cardiac Catheterization (Ch 6). In: Braunwald E, editor. Heart Disease: A textbook of cardiovascular medicine. 5th ed. Philadelphia: WB Saunders; 1997: pp 177–203.
Grossman W. Shunt detection and measurement. In: Baim DS, Grossman W. Cardiac Catheterization, Angiography and Intervention. 5th ed. Baltimore: Williams and Wilkins; 1996: pp 167–180.
Feneley MP. Measurement of cardiac output and shunts. In: Boland J, Muller DWM, editors. Cardiology and cardiac catheterisation: The essential guide. Amsterdam: Harwood Academic Publishers; 2001: pp 197–205.

See also BLOOD GAS MEASUREMENTS; CARDIAC OUTPUT, INDICATOR DILUTION MEASUREMENT OF; CARDIAC OUTPUT, THERMODILUTION MEASUREMENT OF; PERIPHERAL VASCULAR NONINVASIVE MEASUREMENTS; RESPIRATORY MECHANICS AND GAS EXCHANGE.

CARDIAC OUTPUT, INDICATOR DILUTION MEASUREMENT OF

F. M. DONOVAN
University of South Alabama

B. C. TAYLOR
The University of Akron
Akron, Ohio

INTRODUCTION

Cardiac output is defined as the volume of blood pumped by the left or right ventricle per unit of time and is normally expressed in $L \cdot min^{-1}$ (1). For the average 70 kg adult male, the cardiac output is $\sim 5\ L \cdot min^{-1}$, however, exercise can cause this figure to increase as much as six times the resting value in well-trained athletes (2). Knowledge of cardiac function is an important tool for determining the hemodynamic status of an individual whether he/she is a trained athlete or a patient in a critical care setting. Accurate direct measurement of cardiac output is a rather difficult task since to obtain a direct measurement would require collecting and measuring all of the blood pumped from the heart into the aortic outflow tract. It is necessary, therefore, to develop indirect methods for the measurement of cardiac output that would provide equivalent accuracies. The Fick (2) and other indicator dilution methods (3) are two of the invasive procedures that provide good reasonable results. More recently, echo cardiography and other noninvasive techniques have been gaining in popularity, yet the Fick and Indicator Dilution methods remain the "Gold Standards" against which all other methods are compared because of their accuracy, safety, reproducibility, and relative simplicity (1).

The Fick principle (4) is based on the fact that the amount of an indicator taken up (or released) by an organ is the product of its blood flow and the difference in concentration of the substance between the organ's arterial and venous blood. Cardiac output can be determined by dividing the amount of oxygen consumed by the arterial-venous oxygen difference (AVO$_2$ difference). The theory behind this procedure is explained more fully below.

The indicator dilution method became widely accepted after Hamilton, in 1948, demonstrated that this technique agreed with the Fick method. In the indicator dilution method (5) an indicator(dye, thermal, saline) is injected into the venous blood and its concentration is measured continuously in the arterial blood as it passes through the circulatory system The cardiac output is determined by analyzing the resulting time-dependent concentration curve.

INDICATOR DILUTION METHOD FUNDAMENTAL EQUATIONS

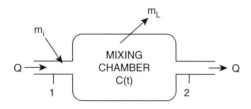

Figure 1. Simple mixing chamber.

Consider the simple mixing chamber shown in Fig. 1, where Q is the constant volumetric flow rate into and out of the chamber, m_i is the mass of indicator injected into the inflow stream, $C(t)$ is the concentration of indictor in the chamber at any instant, and m_L is the mass of indicator that leaves the chamber due to diffusion to the wall and/or metabolism. The differential mass of indicator that leaves the chamber at point 2 per differential time is given by

$$dm_2 = C(t)Qdt$$

where $C(t)$ at point 2 is the same as the concentration of indictor in the mixing chamber assuming complete mixing in the chamber. The total mass of indicator that leaves the chamber is determined by integrating the above equation with the result shown below.

$$m_2 = Q \int_0^\infty C(t)dt$$

The differential mass of indicator that leaves the chamber due to diffusion to the chamber wall and/or metabolism is proportional to the concentration of indicator, $C(t)$, and the surface area of the chamber, A.

$$dm_L = C(t)ADdt$$

where D is the proportionality constant for diffusion and/or metabolism.

The total mass of indicator leaving the chamber due to diffusion is

$$m_L = AD \int_0^\infty C(t)dt$$

The total mass of indicator leaving the chamber is equal to the mass of indicator entering the chamber minus the mass loss of indicator due to diffusion

$$m_2 = m_i - m_L$$

which leads to

$$Q \int_0^\infty C(t)dt = m_i - AD \int_0^\infty C(t)dt$$

and subsequently to the equation for volumetric flow rate

$$Q = \frac{m_i}{\displaystyle\int_0^\infty C(t)dt} - AD$$

The integral of $C(t)dt$ is determined from the area under the indicator dilution curve. Note that if the area of the

chamber wall is large and/or the diffusion coefficient is large, then the flow rate will be overestimated unless the diffusion is taken into account. In practice, the effect of diffusion is taken into account by a multiplying calibration factor (K) as shown in equation 1.

$$Q = \frac{m_i}{\displaystyle\int_0^\infty C(t)dt} K \qquad (1)$$

This is the familiar Stewart–Hamilton equation for calculating cardiac output from the indicator dilution curve (6).

INDICTOR DILUTION CURVE FUNDAMENTAL EQUATIONS

The equations for the indicator dilution curve are determined by the indicator mass flow rate conservation, which states that the mass flow rate of indicator into the mixing chamber must equal the mass flow rate of indicator out of the mixing chamber plus the mass rate of removal by diffusion plus the rate of change of indictor stored in the chamber.

$$C_iQ = C(t)Q + C(t)AD + V\frac{dC(t)}{dt}$$

The parameter V is the volume of the chamber and complete mixing is assumed so that the concentration of indicator leaving the container at point 2 is equal to the concentration of indicator in the chamber at any instant.

Rearranging this equation results in the first-order differential equation

$$\left(\frac{V}{Q+AD}\right)\frac{dC(t)}{dt} + C(t) = \frac{Q}{Q+AD}C_i$$

which has a time constant of

$$\tau = \frac{V}{Q+AD}$$

After the injection of the indicator is complete, the concentration of indicator flowing into the chamber becomes zero resulting in the following equation during the washout phase.

$$\tau\frac{dC(t)}{dt} + C(t) = 0$$

The solution to this equation is

$$C(t) = C(t_1)e^{-[(t-t_1)/\tau]}$$

where $C(t_1)$ is the indicator concentration at time t_1 on the washout part of the indicator dilution curve.

Taking the natural log of this equation results in

$$\ln C(t) = \ln C(t_1) - [(t-t_1)/\tau]$$

This shows that if the indicator dilution curve is plotted as natural log of C versus t, then the curve will become a straight line during the washout phase and the slope of the curve is the negative reciprocal of the system time constant.

Rearranging this equation yields

$$\tau = \frac{t_1 - t_2}{\ln C(t_2) - \ln C(t_1)}$$

where $C(t_1)$ and $C(t_2)$ are indicator concentrations at time t_1 and t_2, respectively, all located on the washout part of the indicator dilution curve.

This equation can be rearranged to the following:

$$\tau = -C(t_1) \Big/ \left(\frac{dC}{dt}\right)_{t_1}$$

The total area under the indicator dilution curve from t_1 to infinity is given by

$$\int_{t_1}^{\infty} C(t)dt = \tau\, C(t_1) \int_{t_1}^{\infty} e^{-[(t-t_1)/\tau]} d\left(\frac{t-t_1}{\tau}\right)$$

which results in the equation for the remaining area under the curve.

$$\int_{t_1}^{\infty} C(t)dt = C(t_1)\tau = \frac{(t_1-t_2)}{\ln C(t_2) - \ln C(t_1)} C(t_1) \quad (2)$$

This equation can be rearranged to the following:

$$\int_{t_1}^{\infty} C(t)dt = \frac{-[C(t_1)]^2}{[C(t)/dt]_{t_1}} \quad (3)$$

APPLICATION OF THE INDICATOR DILUTION EQUATIONS AND RECIRCULATION

The following results are from a computer simulation in which 6 mg of indicator are injected into the right atrium of an average male with indicator concentrations being read from the radial artery. The volumes used by the simulation for the chambers involved are shown in Figure 2.

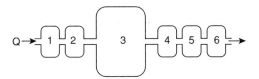

Figure 2. Schematic of simulation system.

Q is cardiac output (6 L·min^{-1})

1 is the right atrium ($V1 = 100$ **mL**)

2 is the right ventricle ($V2 = 100$ mL)

3 is the pulmonary circulatory system ($V3 = 600$ mL)

4 is the left atrium ($V4 = 100$ mL)

5 is the left ventricle ($V5 = 100$ mL)

6 is the systemic artery volume from the left ventricle to the radial artery ($V6 = 100$ mL)

The total circulatory system volume is taken to be 6 L.

The diffusion coefficient D is taken to be zero in the simulation. The resulting indicator dilution curve as measured in the radial artery is shown in Fig. 3.

The dashed line beginning at ~ 22 s shows what the curve would do if there were no recirculation of the indicator through the circulatory system back to the right atrium. The solid line shows the actual curve with recirculation.

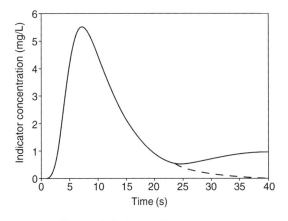

Figure 3. Indicator dilution curve.

The area under the indicator dilution curve can not be determined directly from the dilution curve. The dilution curve is plotted on a semilog graph in Fig. 4.

During the washout phase, the concentration curve approaches a straight line (dashed line) before the recirculation distorts the plot (solid line). This indicates that the system is behaving as a first-order decay, and we can use equation 2 to determine the area under the curve from any chosen time on the straight-line portion of the curve to infinity.

In this simulation, the area under the indicator dilution curve that would occur if there were no recirculation from 0 to 40 s is found to be 59.8 mg·s·L^{-1}, which results in a calculated cardiac output of 6.02 L·min^{-1}.

With recirculation we determine the area under the curve from 0 to 15 s to be 47.387 mg·s·L^{-1} and use equation 2 to calculate the area from 15 s to infinity.

For example, use the concentrations at 15 and 20 s that are in the straight-line portion of the semilog plot.

$$C_{t=15} = 2.096\,\text{mg}\cdot\text{L}^{-1} \quad C_{t=20} = 0.908\,\text{mg}\cdot\text{L}^{-1}$$

Equation 2 gives the area from 15 s to infinity as 12.528 mg·s·L^{-1} so the total area from 0 to infinity is calculated to be 59.915 mg·s·L^{-1}. Now using equation 1, the cardiac output is found to be 6.01 L·min^{-1}.

Using equation 3 at 15 s yields the same result.

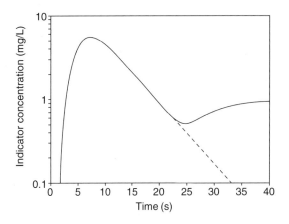

Figure 4. Semilog plot of indicator dilution curve.

FICK PRINCIPLE

If the indicator is supplied to the mixing chamber shown in Fig. 1 at a steady rate until steady state is reached, then the concentration of indicator in the stream leaving the chamber will be constant and is given by the equation

$$m_{f_2} = C_2 Q$$

where m_{f_2} is the mass flow rate of indicator flowing out of the chamber.

Under steady flow conditions the mass flow rate of indicator into the chamber in the inflow stream plus the mass flow rate of indicator entering the chamber by injection will be equal to the mass flow rate of indicator flowing out of the chamber in the outflow stream.

$$m_{f_1} + m_{f_i} = m_{f_2}$$

In terms of inflow and outflow concentration of indicator, this equation is

$$C_1 Q + m_{f_1} = C_2 Q$$

Solving for Q yields the equation on which the Fick method is based.

$$Q = \frac{m_{f_i}}{C_2 - C_1}$$

In practice, the indicator used in the measurement of cardiac output by the Fick method is oxygen so that m_{f_i} is the consumption rate of oxygen in the lungs, C_1 is the concentration of oxygen in the venous blood, and C_2 is the concentration of oxygen in the arterial blood.

THERMAL DILUTION METHOD FUNDAMENTAL EQUATIONS

For thermal dilution measurement of cardiac output, a warm or cold injectate is injected into the right atrium and the temperature of blood in the pulmonary artery is measured by a thermistor as shown in Fig. 5. A warm injectate would need to be considerably warmer than the blood that might be hot enough to denature proteins ($60°C$). If it were not very warm, the poor signal/noise ratio would render the method unusable. Therefore cold injectate is the only practical thermal indicator (7,8).

A bolus of cold fluid is injected into the right atrium and the resulting temperature is recorded from the pulmonary artery. Conservation of energy requires that the total thermal energy entering the system during the procedure must be equal to the total thermal energy that leaves the system as the system returns to normal temperatures.

The thermal energy carried across a boundary by a differential volume of fluid is given by

$$dE = \rho C_P T(t) Q \, dt$$

where ρ is the density of the fluid, C_P is the specific heat of the fluid, $T(t)$ is the temperature of the fluid at any instant, Q is the volumetric flow rate of fluid crossing the boundary, and dt is the differential time.

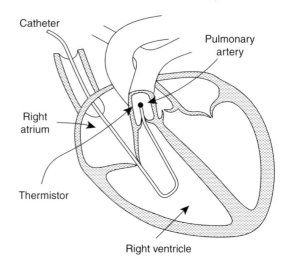

Figure 5. Catheter in place for thermal dilution measurement.

The total thermal energy that crosses a boundary for a constant volumetric flow rate with constant thermal properties is

$$E = \rho C_P Q \int_0^\infty T(t) dt$$

The total thermal energy that enters the system is a combination of energy carried into the system by blood flow and the thermal energy carried into the system by the injectate, where the subscript b refers to blood and subscript i refers to injectate.

$$E_i = \rho_b Q C_{P_b} \int_0^\infty T_b \, dt + \rho_i C_{P_i} V_i T_i$$

The total thermal energy that leaves the system is a combination of the energy carried out of the system by blood flow and thermal energy loss to the walls of the atrium and ventricle.

$$E_o = \rho_b Q C_{P_b} \int_0^\infty T(t) dt + \rho_i C_{P_i} V_i T_b + hA \int_0^\infty (T - T_b) dt$$

where $T(t)$ is the temperature recorded by the thermistor at any instant, h is the thermal convection coefficient, and A is the internal surface area of the right atrium and right ventricle.

Using these equations in the thermal energy conservation equation

$$E_i = E_o$$

and rearranging gives

$$\rho_b C_{P_b} Q \int_0^\infty (T(t) - T_b) dt = \rho_i C_{V_i} V_i (T_i - T_b)$$

$$- hA \int_0^\infty (T(t) - T_b) dt$$

Solving for Q yields the thermal dilution equation for volumetric flow rate.

$$Q = \frac{\rho_i C_{V_i}}{\rho_b C_{V_b}} \frac{V_i (T_i - T_b)}{\int_0^\infty (T(t) - T_b) dt} - \frac{hA}{\rho_b C_{P_b}}$$

The term on the right represents the heat loss to the walls of the atrium and ventricle. In practice, there would be an additional heat loss in the catheter. The heat loses are normally accounted for by a correction term (K) that is a function of the catheter type being used as shown below.

$$Q = \frac{\rho_i C_{V_i}}{\rho_b C_{V_b}} \frac{V_i(T_i - T_b)}{\int_0^\infty (T(t) - T_b)dt} K$$

The thermal dilution method has the advantage that recirculation is not a problem due to the large surface available in the circulation to bring the injectate temperature to body temperature. A disadvantage is that the injection site and the sensing site must be close together to avoid large heat losses and the absence of total mixing in the ventricle can cause inaccuracies.

BIBLIOGRAPHY

1. Yang SS, Bentivoglio LG, Maranhao V, Goldberg H. From Cardiac Catheterization Data To Hemodynamic Parameters. 2nd ed. Philadelphia: F.A. Davis Company; 1980. p 55.
2. Geddes LA. Cardiovascular Devices and Their Applications. New York: John Wiley & Sons, Inc.; 1984. p 102–106.
3. Stewart GN. The output of the heart in dogs. Am J Physiol 1921;57:27–50.
4. Fick A. Uber die Messung des Blutstroms in den Herzventrikeln. Verhandl Phys Med Ges Zu Wurzburg 1870;2:XVI.
5. Hamilton W, et al. Comparison of the Fick and dye injection methods of measuring the cardiac output in man. Am J Physiol 153:309–321.
6. Valentinuzzi ME, Posey JA. Fast estimation of the dilution curve area by a procedure based on a compartmental hypothesis. Med Ins Sept-Oct1972;(6)5.
7. Taylor BC, Sheffer DB. Understanding Techniques for Measuring Cardiac Output. Biomedi Inst Technol May/June, 1990; 188–197.
8. Swinney RS, Davenport MW, Wagers P, Sebat F, Johnston W. Iced versus room temperature injectate for thermal dilution cardiac output. Ninth Annual Scientific and Educational Symposium. Soc Crit Care Med, May12–16,1980;137.

See also CARDIAC OUTPUT, FICK TECHNIQUE FOR; CARDIAC OUTPUT, THERMODILUTION MEASUREMENT OF; ECHOCARDIOGRAPHY AND DOPPLER ECHOCARDIOGRAPHY; FLOWMETERS, ELECTROMAGNETIC; TRACER KINETICS.

CARDIAC PACEMAKER.   See PACEMAKERS.

CARDIAC OUTPUT, THERMODILUTION MEASUREMENT OF

EDWIN D. TRAUTMAN
RMF Strategies
Cambridge, Massachusetts

MICHAEL N. D'AMBRA
Harvard Medical School
Cambridge, Massachusetts

INTRODUCTION

The amount of blood pumped by the heart each minute, the cardiac output, provides a measure of the body's potential for supplying oxygen and nutrients and is relevant to assessing the condition of the heart. Taken together with various pressures, it is a key clinical indication of the heart's ability to meet the body's needs and an indirect indication of the status of those needs. In the clinical setting, measurement of cardiac output is required to guide drug therapy aimed at manipulating the function of the cardiac muscle (inotropic drugs) and the state of the systemic and pulmonary vascular resistance (vasoconstrictor and vasodilator drugs). Combined with ultrasound velocity data, cardiac output allows precise assessment of the status of mitral and aortic valve stenosis and regurgitation.

G. N. Stewart articulated the basic principle of indicator- dilution measurement of cardiac output in a landmark paper in 1897 (1). Stewart stated that if a substance was introduced at a constant rate into the flowing bloodstream and allowed to mingle with the blood, then the measured steady-state concentration of that substance downstream of the site of introduction would be inversely proportional to the flow rate (cardiac output). Of greater practical importance was his additional observation that if a small amount of the substance was introduced rapidly, then the cardiac output could still be computed. To do that one would divide the average rate at which the substance is introduced (total amount divided by measurement time) by the average concentration. Stewart called this technique the "sudden injection" method. In the late 1920s, W. F. Hamilton and his colleagues further investigated Stewart's sudden injection method (2,3). They found that the concentration curve from timed samples did not simply return to baseline, but exhibited a secondary rise. This was attributed to fast physiologic recirculation of an unknown amount of the indicator. To eliminate the influence of any recirculating indicator on measurement calculation, they proposed extrapolating the original down slope of concentration to zero using an exponential function. This method proved successful in validation studies both in mechanical models and animal experiments, and the sudden injection method with exponential extrapolation is commonly referred to as the Stewart–Hamilton method.

Various indicators have been used to measure cardiac output with the Stewart–Hamilton method, notably saline (detected by its effect on electrical conductivity) and optical dye (detected by its effect on optical absorption), but G. Fegler's proposal in the mid-1950s that heat could be used has proved the most convenient, although initially controversial (4,5). His earliest report "was received with polite incredulity" (6). Fegler rapidly injected a small amount of cold Ringer's solution into the vena cava and recorded the transient decrease in temperature in the aortic arch and in the right ventricle. He computed both left and right heart outputs from these data.

Concerns were voiced regarding the ability to quantify this "negative" indicator, the stability of the baseline temperature, and the background noise (6). These are all valid concerns, but concerns that have been successfully addressed with clinical technology. Current practice is to introduce a small bolus of cold solution into the right atrium (via a venous catheter) and to measure the consequent transient temperature decrease in the pulmonary artery. With the advent of balloon-flotation pulmonary-artery pressure measurement catheterization techniques in the late 1960s (7), small thermistor sensors could be readily placed in the pulmonary artery, and the thermal dilution measurement became clinically accepted even as validation experiments progressed. The pulmonary-artery catheters provide important hemodynamic pressure information and, for that reason alone, are placed in many patients.

Swan and Ganz are credited with developing and popularizing the pulmonary-artery pressure catheter containing a thermistor and injection port for thermal dilution (8), and these catheters are commonly called Swan–Ganz catheters, although Swan–Ganz, strictly speaking, is a registered trademark of Edwards Life-sciences Corporation. The instrumentation required to process the temperature signal and determine cardiac output is modest, fitting into either a small, battery-operated instrument easily used at the patient's bedside, or into a module component of a bedside workstation in the ICU or operating room, making the method easy and convenient. The additional "invasion" of a thermistor is negligible, and the additional value of cardiac output measurements is great. And since the indicator is a physiologically innocuous solution, thermal dilution measurement of cardiac output has become an important part of clinical care. Today, bolus thermal dilution cardiac output is considered the gold standard against which other methods are compared.

THEORY

Principle of Indicator Dilution

The basic principle of indicator dilution is quite simple: If the concentration of a uniformly dispersed indicator in an unknown volume is measured, then the unknown volume can be simply determined by dividing that concentration into the total amount of indicator. If the volume is flowing past a sensor, then the volume in any given period of time will equal the amount of indicator in that period of time divided by the concentration. If the rate at which the indicator is flowing past the sensor is controlled and known, then the amount of indicator in a period of time is also known and the volume flow rate can be determined. Alternatively, if the total amount of indicator over a larger period of time is known, such as when a bolus is introduced all at once, then average flow rates can be determined. Each approach has strengths and weaknesses in technique, necessary assumptions, and equipment. We focus on the popular bolus technique but, particularly with thermal techniques, the theory can be extended.

In the case of thermal dilution, the indicator is introduced into the right atrium and its concentration is measured in the pulmonary artery. We assume that all of the indicator introduced, an amount I, eventually passes into the pulmonary artery at some rate $i(t)$. If we assume no indicator recirculates, we may write this

as

$$I = \int_0^\infty i(t)\, dt \tag{1}$$

$$= \int_0^\infty F(t)c(t)dt \tag{2}$$

where F is volumetric flow and c is concentration. When the flow is constant, F can be moved out of the integral and we can solve for F as

$$F = \frac{I}{\int_0^\infty c(t)dt} \tag{3}$$

Several assumptions have been made in arriving at this equation. Equation 1 is a statement of conservation and requires that all indicators pass the sensor exactly once. Equation 2 requires that the concentration in the pulmonary artery be uniform across the area where the concentration is being measured, and equation 3 requires that the flow rate be constant. All but the first requirement can be satisfied for the pulmonary artery catheter-based measurements if we consider the right heart to be a perfect mixing chamber, with a competent valve at the outflow, and a pumping rate that is constant over the integration time. The mixing chamber guarantees that the blood will be equivalently labeled at its outflow so that each flow stream is representative of the total, and the valve guarantees that the concentration changes in a stepwise fashion in all flow streams, which allows legitimate averaging of pulsatile variations in flow. A rigorous proof can be found in Perl et al. (9) and the assumptions and necessary conditions are discussed in Trautman and Newbower (10). Each of these articles contains numerous relevant references.

Heat as an Indicator: "Thermal" Dilution

In thermal dilution, the indicator is caloric, introduced as a known volume of a cold physiologic solution whose concentration is measured via the induced temperature change. The relationship between temperature and the concentration of heat in a solution, the amount of heat in a unit volume, involves the specific heat and density of the solution. When two solutions at different temperatures, such as the indicator and blood, are mixed, the temperature of the mixture may be predicted by

$$T = \frac{T_1 C_{p1} m_1 + T_2 C_{p2} m_2}{C_{p1} m_1 + C_{p2} m_2} \tag{4}$$

where C_p and m are, respectively, the specific heat and mass of the solutions. If we take the first solution to be the indicator solution and the second to be the blood, then the difference in temperature due to the indicator will be predicted by

$$T - T_2 = \frac{(T_1 - T_2)C_{p1} m_1}{C_{p1} m_1 + C_{p2} m_2} \tag{5}$$

A very good assumption, at least prior to significant heat exchange with tissue, is that the indicator solution and the temperature transient travel together. In this case, equation 5 holds for all instances of time, and the mass concentration of the indicator solution may be predicted from the temperature transient by

$$c(t) = \frac{[T(t) - T_2]\rho_1}{T(t) - T_2 - (C_{p1}\rho_1/C_{p2}\rho_2)[T(t) - T_1]} \quad (6)$$

where ρ is the density of the solution. The amount of indicator is equal to the volume of the physiologic solution V times its density ρ_1 (giving its mass) and equation 3 becomes

$$F = \frac{V\rho_1}{\int_0^\infty c(t)\,dt} \quad (7)$$

$$F = \frac{V\rho_1}{\int_0^\infty \dfrac{[T(t) - T_2]\rho_1}{T(t) - T_2 - (C_{p1}\rho_1/C_{p2}\rho_2)[T(t) - T_1]}\,dt} \quad (8)$$

Equation 8 forms the basis for the thermal dilution method for measuring cardiac output. This equation may be easily programmed, but it generally has been approximated to simplify implementation, The most common approximation is based on the assumption that the indicator solution has no effect on the thermal properties of the blood. In this case, the increment in the amount of heat leaving the heart due to the indicator is equal to

$$H = \rho_2 C_{p2} F \int_0^\infty [T(t) - T_2]dt \quad (9)$$

This is equivalent to equation 2 and requires the same assumptions. The amount of heat added in a certain volume of an indicator solution is equal to

$$H = \rho_1 C_{p1} V (T_1 - T_2) \quad (10)$$

Equating equations 9 and 10 leads to the simpler formula for flow:

$$F = \left(\frac{C_{p1}\rho_1}{C_{p2}\rho_2}\right) \frac{V[T_1 - T_2]}{\int_0^\infty [T(t) - T_2]dt} \quad (11)$$

Equations 8 and 11 are equivalent only when cool blood is used as the indicator. Equation 11 is simpler than equation 8 and can be implemented in an analog circuit. However, it is based on the implausible condition that the indicator solution carries heat (or cold) into the blood and then is either transported completely apart from the thermal transient or has no thermal effect on the blood. Fortunately, the practical difference between flow estimates based on these two equations is small. The ratio of thermal properties for a dextrose-in-water (D5W) indicator solution is approximately equal to 1.08, and for a normal (0.9%) saline solution it is approximately equal to 1.10. Since the expected temperature transient is 0.5–1.0 °C, the expected difference between equations 8 and 11 is only 1–2% for these indicators.

If the indicator is not introduced as a finite bolus, then the conservation statement of equations 1 and 2 needs to be generalized and other assumptions made. The product of flow rate and concentration at the outflow of the mixing chamber will still be equal to the amount of indicator passing by, but its relationship to the input indicator can be more complex. A simple example is where the indicator is infused at a constant rate in which case, absent recirculation, the flow rate will be inversely proportional to measured concentration. If the infusion rate is not constant then the transient response of the heart system needs to be considered.

Heat can be introduced by direct energy transfer, such as from an electrical heater. In this case, the volume factor in the numerator of equation 11 is not relevant and must be replaced by a measure of the amount of heat introduced. It is, however, impractical to introduce a large bolus (impulse) of heat comparable to the 750 W of 10 mL of iced saline: The surface temperature would be dangerously high. Instead, the heater is pulsed at low power, the resulting temperature changes measured with a fast-response thermistor, and sophisticated signal processing used to extract the dilution signal from the baseline. Such techniques have the potential to measure cardiac output continuously and were introduced in the early 1980s (11). Catheters with heating elements (10 cm long filaments in the right ventricle) have been produced since the early 1990s (12). The surface temperature and thus the amount of heat that can be introduced are limited by physiological concerns (4–7 °C), and therefore the technique is sensitive to background thermal noise. The heater is typically pulsed, and the accuracy is dependent on processing the correlation between the heating waveform and the measured temperature response (11–13). These techniques are entering clinical use as a companion to bolus thermal dilution, but are not considered here.

Necessary Conditions

For the preceding development, we assumed that the flow is constant; the volume and temperature of the solution are known, the indicator does not recirculate, and perfect mixing occurs somewhere between injection and sampling. Little can be done to control the variation of flow; it is flow that is being measured. (In those situations where flow is *not* constant, it can be shown that the computed result will be a concentration-weighted average of the true cardiac output over the period of measurement.) The other assumptions are usually reasonable, although in practice it is difficult to have an accurate measure of the volume or temperature of the injected solution since heat will exchange with all material contacting the solution. There is also a lost volume in the dead space of the catheter used to introduce the solution, and it is impossible to eliminate the physiologic recirculation of indicator. In addition, the integrals must be truncated to permit a practical measurement. These and other practical issues are covered next.

Notably absent from these formulas is the time response of the thermal sensor. Although not intuitively appealing, it can be shown that this response is of little importance as long as the curve does not become distorted by the effects of noise and recirculating indicator. The area under the thermal curve is preserved even with slow-responding thermal sensors. The operator should, however, be aware that the thermal curve obtained with a slow thermistor is not necessarily a high fidelity representation of the temperature transient. The observed or recorded temperature curve will be smoothed and filtered over time.

PRACTICAL APPLICATION OF THE THEORY

The practical application of thermal dilution theory is simple; all that is typically necessary to measure cardiac output is to reset the "cardiac output computer," inject 2–10 mL of an ice-cold or room temperature solution into the catheter port, and wait for the answer to appear on the computer. The thermal sensor is contained on a special pulmonary artery catheter that also provides the injection lumen into the right atrium. The temperature curve may be recorded to reassure the operator that a reasonable signal was obtained. Typical curves for various flow rates are shown in Fig. 1.

When the computer is reset, it samples the baseline temperature, integrates the processed temperature curve until recirculation is detected or assumed, and calculates the cardiac output assuming predefined conditions. It applies a correction for the portion of the curve that is lost by the truncation of the integral (to avoid influence of recirculation). The predefined conditions include the volume and temperature of the injectate, the thermal characteristics of the fluids, and a correction factor that corrects for the physical properties of the particular injection catheter employed. Most computers make approximations and apply corrections. The most common are listed below.

Thermal Properties of Blood Are Approximately Constant

The thermal properties of blood vary with hematocrit. However, the convenience of assuming a normal hematocrit—and thus not requiring knowledge of the actual hematocrit and entering it—far outweighs the importance of the potential error. The specific heat-density product for erythrocytes is ~ 3.52 $J\cdot K^{-1}\cdot mL^{-1}$, and for plasma it is $\sim 4.03 J\cdot K^{-1}\cdot mL^{-1}$ (14). Therefore, for blood with a hematocrit of 40% this product is 3.83 $J\cdot K^{-1}\cdot mL^{-1}$, with a hematocrit of 30% it would be 3.88 $J\cdot K^{-1}\cdot mL^{-1}$, and with 50% it would be 3.78 $J\cdot K^{-1}\cdot mL^{-1}$. Thus, the nominal value assumed for blood could be in error by 1–2% causing an error in the cardiac output measurement of the same value.

Indicator Does not Affect Thermal Properties of Blood

The specific heat-density product for normal (0.9%) saline is $\sim 4.19 J\cdot K^{-1}\cdot mL^{-1}$, and that for 5% dextrose-in-water is ~ 4.11 $J\cdot K^{-1}\cdot mL^{-1}$ (15). These are significantly different from the nominal 3.81 $J\cdot K^{-1}\cdot mL^{-1}$ for blood. The thermal properties of the indicator-blood mixture will thus vary with the level of dilution. However, most computers assume that the indicator does not affect the thermal properties of blood. This assumption leads to the simpler equation [Eq. 11] derived above. Cardiac output computers using this approximation can be expected to overestimate the cardiac output by 1–2%.

Heat Loss Is Predictable

When the syringe containing the cold solution is taken from the ice bath it immediately begins to warm. The solution warms further as it is injected through caloric exchange with the walls of the catheter. Only a negligible amount of heat is gained during manipulation of the syringe before injection. However, the exchange with the walls of the catheter can account for several percentages with a 0 °C solution (16). In addition to those conductive losses of indicator, a significant amount of solution is left in the catheter after the injection has been terminated. The typical dead space volume is 0.9 mL so that only 91% of the solution is injected into the bloodstream. However, the solution that filled the dead space prior to injection is pushed into the blood stream and, if not at blood temperature, can add to the effective indicator volume. In addition, some of the "cold" left in the dead space after injection can leak through the catheter wall and add to the injectate.

Empirical studies have shown that the combination of these losses and gains can be grouped into a single correction factor, multiplying the total indicator volume. This correction factor varies only a few percentages with catheter insertion length and other mechanical factors (17). The correction factor does depend on the temperature and volume of the injected solution and on the design of the catheter. Catheter manufacturers generally provide, with their package inserts, a table of values for the correction factor or "computation constant" under various typical conditions. This factor is determined by measuring the average temperature of the injectate, as it emerges from the injectate, lumen of the catheter, while the appropriate length of catheter is immersed in a 37 °C bath. The amount of injectate that emerges is a reasonably constant fraction of the amount introduced.

Devices can be used to measure the temperature of the injected solution as it enters the injection catheter. This reduces the need for precisely controlling the initial temperature of the solution and reduces errors due to warming of the solution during handling. These devices do not improve knowledge about the unknown heat loss during injection. Catheters have also been fabricated with thermistors in the distal port of the injection lumen, to measure true injectate temperature. These catheters demonstrate better reproducibility particularly with room temperature injectates, but have yet to win clinical acceptance due to cost and complexity. Note also that the rate at which the indicator is introduced must be controlled and consistent to allow inferring amount of heat from temperature.

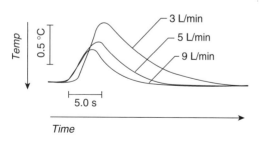

Figure 1. Typical thermal dilution curves, taken at different flows and superimposed to illustrate variations in shape and area.

Decay of Dilution Signal Is Exponential

Recirculation of the thermal indicator is relatively small in humans since there is ample opportunity for exchange with the tissue beds. In smaller animals, recirculation is more apparent. In either case, the decay of the temperature signal measured in the pulmonary artery approximates an exponential (the result of the mixing chamber) and, once truncated, the true but obscured curve can be mathematically extrapolated with reasonable accuracy. The relatively small amount of curve area being estimated limits the significance of errors in extrapolation. Some cardiac output computers actually fit an exponential to the uncorrupted curve and use the parameters of fit to extrapolate the curve, while most assume that curves generally have the same shape and integrate to a fraction of the peak temperature and multiply this area by a constant. Either method appears to result in a reliable measure of cardiac output. Certain pathologies can alter the shape of the thermal dilution curve and reduce the effectiveness of these extrapolation procedures.

Baseline Temperature Is Constant

The baseline temperature is not constant, varying with the respiratory cycle and subject to the fluid infusions from other sources (i.e., intravenous fluid administrations). In addition, the heart itself generates heat that can be observed as very small pulsatile variations in temperature in the pulmonary artery. Fortunately, these variations and the baseline shifts are usually small compared to the ~ 0.5–$1.0\,^{\circ}C$ dilution signal. Cardiac output computers thus assume that the baseline acquired prior to the arrival of the dilution signal remains constant during the course of the measurement. (Note that shifts in baseline can adversely affect the extrapolation procedure used to reduce the effects of recirculation.)

Flow Rate Is Constant Throughout the Integral

Cardiac output can vary by as much as 10–20% over the respiratory cycle. Since thermal dilution measurement integrals typically average only 5–10 s of the cardiac output, the measured cardiac output could vary by as much as 10–15% depending on where in the cycle the injection is made. There is really nothing the computer can do about this without information about the phases of the respiratory cycle. Cardiac arrhythmias, which can result from the cold injection, can cause dramatic errors in the measured output. The clinical practice of averaging several separate cardiac output determinations helps to average out some of the potential variation from both of these sources. See section on *Measurement Performance* for more discussion on accuracy and reproducibility.

EQUIPMENT

The thermal dilution method for cardiac output measurement is popular because it is easily performed with a minimum of equipment and little additional invasion of the patient. The basic equipment consists of a pulmonary artery catheter to position a thermistor or other temperature-sensitive element in the pulmonary artery, a means for making thermal indicator (usually saline) injections into the right atrium (usually a syringe), typically through a separate lumen in the catheter, a source of measured volumes of a cold solution, and an electronic instrument to determine the blood temperature from the thermistor signal, to determine and integrate the dilution signal, and to compute a final result. Each of these elements is described separately.

Pulmonary Artery Catheters

The pulmonary artery (PA) catheter generally contains several lumens (channels) that terminate at measured distances from the tip. A balloon, located at or near the tip, is inflated during catheter insertion to carry the tip through the heart and into the pulmonary artery (flow directed). One lumen terminates at the tip and is used to measure the pressure during catheter insertion to follow its position relative to the heart; later it measures pulmonary artery pressure and, intermittently, pulmonary capillary wedge pressure (with the balloon inflated). A second lumen typically terminates in the right atrium and is used to monitor right atrial pressure (central venous pressure). Indicator solutions are injected either through the right atrial port or through a second atrial lumen intended for drug infusion. Catheters can have several additional lumens (e.g., atrial and RV pacing wires) and sensors (e.g., mixed venous oxygen saturation). The pulmonary artery catheter provides important hemodynamic information and may be inserted in patients for that purpose alone.

For use with thermal dilution, the pulmonary artery catheter is augmented by adding a thermistor proximal to (before) the balloon (typically, 4 cm from the tip). A thermal dilution catheter is illustrated in place in Fig. 2. The

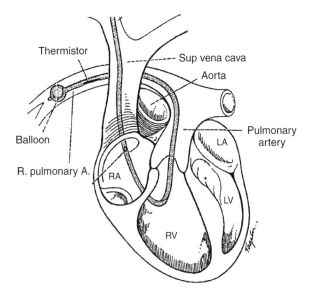

Figure 2. A pulmonary artery catheter in place in the right heart. The balloon, shown inflated here for wedge pressure measurement, normally remains deflated during pressure monitoring and cardiac output measurements. The cold injectate enters the bloodstream through the injection port, and the temperature transient is sensed by the thermistor.

thermistor typically is encapsulated in glass and coated with epoxy to fully insulate it electrically from the blood. The relatively slow time response of this encapsulated sensor does not affect the accuracy of the measurement since the area under a temperature curve is preserved. Wires connecting the thermistor are contained in a separate lumen. The thermistor wires terminate in an external connector that typically contains an electrical resistance used to standardize the response of the thermistor. Thus the catheter contains one-half of a Wheatstone bridge. The overall length is ∼ 100 cm, with distance marks every 10 cm to guide insertion. The typical size is 7–8.5 Fr although pediatric catheters may be 5–6 Fr. (One French is equal to one millimeter in circumference.)

Edwards Lifesciences was the original commercial manufacturer of the thermal dilution catheter, basing it on a concept acquired from Swan and Ganz, researchers involved with validation experiments. Their Swan–Ganz catheter was introduced in 1971. Since that time several manufacturers (e.g., Instrumentation Laboratories, Cobe, Abbott, Arrow) produce thermal dilution catheters, disposable items selling for ∼ $50–80 each, although catheters with heating filaments and multiple sensors can cost $200 and up. A thermal dilution catheter is shown in Fig. 3.

Pulmonary artery catheters are not without clinical complications. The threat of infection is always present. Clots can form on the poly vinyl chloride (PVC) catheter surface, but this complication has been mostly eliminated with anticoagulant coatings. Unfortunately, there are patients who have severe reaction to heparin (i.e., heparin induced thrombosis) and in these patients heparin-coated catheters need to be avoided. The catheter can become knotted in the right heart, a complication that requires a trip to the cardiac catheterization laboratory to resolve.

Most dangerous is the rare complication of the catheter tip puncturing the pulmonary artery. In normal use, the balloon is temporarily inflated, the balloon wedges into a branch of the pulmonary artery, flow through that branch of the pulmonary circulation is stopped and the pressure measured from the distal lumen will approximate the left atrial pressure. It is critical that nurses and physicians understand the waveforms associated with "permanent wedge" position to avoid pulmonary rupture. This complication is almost always associated with erosion of the

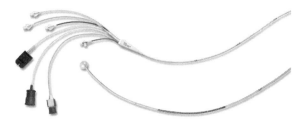

Figure 3. A Swan–Ganz pulmonary artery catheter produced by Edwards Lifesciences. The various lumens are accessed through individual Luer-Lok connectors fanning out from an external divider on the main catheter. The electrical connector for the thermistor wires is also connected to the main catheter at the same point. (Photograph courtesy of Edwards Lifesciences Corporation, Irvine, CA.)

pulmonary artery from a catheter permanently in the wedge position or with inflating a catheter that is in the distal pulmonary arterial position.

The balloon is an important feature of pulmonary artery catheters since it plays a key role in the acquisition of pressure information in addition to facilitating placement of the catheter. Balloons are generally made from latex and designed to inflate beyond the tip of the catheter while not occluding the distal pressure lumen. This shields the tip reducing the tip trauma to the pulmonary artery. Manufacturers attach the balloon to the base catheter in a way that minimizes rough surfaces and overall size of the catheter while being durable. Latex-free balloon catheters are available for use in patients with latex allergies, but are expensive and have limited functionality, usually having only a single right atrial port. The non-latex balloon is also not as durable and measurements of wedge pressure must be kept to a minimum.

The size and material of the catheters can vary among manufacturers and models, both of which can affect their stiffness and thus the ease of insertion, and the size of the dead space in the injection lumen and, thus, the heat loss correction factor. The injection port may be larger in some catheters, reducing injection effort. The frequency response of the pressure measurement lumens may be different due to attention to details of fluid mechanics. Personal preference, reliability, and economic concerns are also clearly important in purchase decisions. Some catheters offer other capabilities such as continuous cardiac output measurements based on advanced signal processing algorithms, mixed venous oximetry, and the ability to electrically pace right atrium and ventricle.

Cold Solution

The indicator solution is typically an isotonic saline or dextrose solution cooled to 0 °C by placing the bottle or prefilled capped syringes in an ice bath. This makes the injected indicator ∼ 37 °C cooler than body temperature. This 10 mL of 37 °C difference injected in 2 s represents extraction of thermal energy from the bloodstream at a rate of ∼ 750 W. Cold indicator is most useful in the operating room where patient temperatures can vary rapidly and dramatically. In the non-OR setting, or in operative patients where normothermia is expected, room-temperature solutions are frequently used because they are more convenient, but the variable room temperature must be monitored by the computer. The injected energy is reduced by a factor of ∼ 3, reducing the signal-to-noise ratio and, thus, expected measurement performance. Most cardiac output computers provide a temperature probe to measure the actual temperature of the bath or of the room, which is presumably the temperature of the injectate. Manufacturers also produce an optional temperature-measuring probe that attaches to the injection port on the thermal dilution catheter and measures the temperature of the injectate as it enters the catheter. This further reduces concern about the actual room or bath temperature. Some catheters also have thermistors at the injection port itself.

The ice bath is the subject of some unproved concerns regarding infection. Undesirable organisms could remain

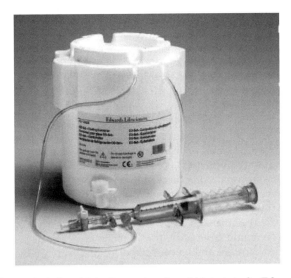

Figure 4. A closed injection system for cold injectate, by Edwards Lifesciences. A cooling coil rests in a Styrofoam ice bucket, the syringe serves as a piston pump to draw up a known volume of precooled injectate from the coil and to force it into the injection lumen of the catheter. The closed system reduces the risk of nosocomial contamination associated with traditional injectate delivery methods. CO-Set+ System improves reproducibility and accuracy through its in-line temperature probe and volume-limited syringe. (Photograph courtesy of Edwards Lifesciences, Irvine, CA.)

or grow in the capped syringes when left in the bath for long periods of time. A cleaner alternative to the ice bath is offered by a closed injectate system with a cooling coil, offered as an optional accessory by some manufacturers. One such device is shown in Fig. 4. The coil sits in an ice bath keeping the solution cold. The syringe is used to draw up solution and then immediately introduce it into the catheter port. Fig. 5 shows the complete system.

Although these solutions are generally benign, in some circumstances, such as in small pediatric patients, there is a risk of volume overload from frequent measurements. In these situations, smaller volumes are used (e.g., 3 mL) and fewer measurements are made.

Cardiac Output Computers

When thermal dilution measurements were first introduced in the early 1970s, manufacturers produced catheters with nonstandardized thermistors. Each manufacturer then produced a computer to mate with its catheter. In addition to generic differences in thermistor types, each individual thermistor of a given type can have a different temperature response requiring the operator to enter a calibration constant, idiosyncratic to the specific catheter, its thermistor response, and even the patient's blood temperature. In current products, the thermistor connector contains an electrical resistance selected to match the particular thermistor and complete a half Wheatstone bridge with a standard response. The value of the resistance in this circuit is chosen such that the voltage response of the half bridge will be the same for all thermistors of a given family and also will be linear near 37 °C. The nearly linear range is ~20 °C. With these catheters, the catheter is

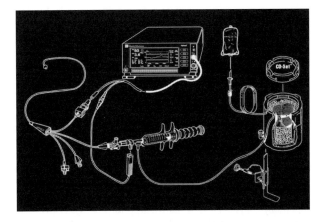

Figure 5. Basic thermal dilution system, with computer, syringe, catheter, and source of cold injectate, in this case from a closed system. (Courtesy of Edwards Lifesciences, Irvine, CA.)

merely connected to the computer's electronics, and the pulmonary artery blood temperature and the cardiac output can be measured. The electronic circuitry used to measure the thermal response is electrically isolated, since the thermistor is in the conductive bloodstream quite close to the heart and insulation failures can conceivably occur. Since the catheter functions as half of a Wheatstone bridge, the electronics merely mimic the other half of the bridge circuit, excite the bridge with a low level of current, and amplify the voltage difference proportional to temperature change.

When the operator signals that a measurement is to be made, the baseline temperature is acquired and the indicator concentration is computed (or approximated) and integrated. When the "end of curve" criterion is reached, the integration is stopped, the integral is adjusted for area lost by truncation, and the final area is inverted and multiplied by the appropriate constants for the measurement conditions. This constant is entered in the computer and only changed when conditions change. Cardiac output computers differ both in the method they use to truncate the integration and in the options they offer, such as, syringe size, injectate temperature, and integration into bedside systems.

Calibration of Equipment

Each catheter is individually calibrated by the manufacturer to give a standard response (as described previously), and the heat loss correction factor is determined also by the manufacturer for the particular catheter model for a variety of measurement conditions. No operator calibrations are necessary or practical.

In certain research settings (e.g., custom-made catheters), it is desirable to add the calibration resistor to the catheter. The value of the resistance is given by

$$R = R_0(\beta - 2T_0)/(\beta + 2T_0) \qquad (12)$$

where R_0 is the thermistor resistance at 37 °C, T_0 is 310 K, and β is the characteristic temperature (a gain constant) for the thermistor, equal to 3500 K for those used in thermal dilution catheters compatible with the Edwards Lifesciences standard.

Example Equipment

Most cardiac output computers are fully integrated into hemodynamic monitoring systems. The cardiac output component is usually part of the temperature-sensing module. Data from the measurements are acquired into the system's data recording and analysis packages and automated calculations of systemic and pulmonary vascular resistances are obtained.

In the typical computer, the preamplifier is fully isolated and uses a conservative $7 \mu A$ to sense the thermistor resistance. When a new cardiac output is desired, the operator presses a button and rapidly injects the cold solution. The dilution curve is typically shown as it is measured and the result is displayed once the curve has finished. Analog and digital outputs may be provided for integrating into a larger measurement or workstation system. In a typical computer the temperature difference is integrated from baseline up to its peak, then down to 30% of the peak value, and multiplied by 1.22. This integral is then inverted and multiplied by the computation constant provided by the catheter manufacturer. The computation constant is the product of all constants (e.g., the ratio of thermal constants, the injectate volume, $60 \text{ s}\cdot\text{min}^{-1}$, and $0.001 \text{ L}\cdot\text{mL}^{-1}$) and the catheter heat-loss correction factor (e.g., 0.825). Although the method for extrapolating the integral appears overly simple, it is quite effective. As pointed out earlier, the dilution curve has a consistent shape, and amount of area obscured by recirculation is small and comes relatively late in time (in humans). Some computers use more elaborate methods.

A typical computer and display is shown in Fig. 6. This example from Philips Medical Systems integrates into the monitoring system and computes several derived values as well as displaying hemodynamic information.

MEASUREMENT PERFORMANCE

Direct measurement of cardiac output is quite difficult given the location of the measurement site and the necessity

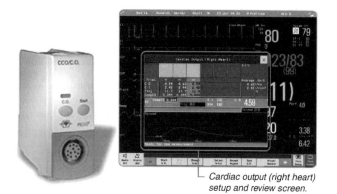

Cardiac output (right heart) setup and review screen.

Figure 6. A modern cardiac output computer, integrated into a monitoring system, by Philips Medical Systems. The thermal dilution curve is displayed for inspection, and the cardiac indices can be automatically computed from the cardiac output and body parameters entered by the clinician. (Photograph courtesy of Philips Medical Systems, Andover, MA.)

to divert flow in some manner. The performance of thermal dilution measurements of cardiac output has been assessed in a number of less direct ways. Validation studies have been performed in mechanical flow models to assure that the measurement theory is sound in practice (and to determine the heat-loss corrections appropriate to specific catheters). Simultaneous thermal dilution and dye- dilution measurements, and also thermal dilution and direct Fick measurements, have been performed in animals and humans. Comparisons have also been made with electromagnetic flowmeters in animal preparations. All of these methods have shown thermal dilution to be effective for measuring cardiac output and as accurate as these other methods. In addition, an important consideration in the clinic is the reproducibility of the measurement over time and from operator to operator. Clinical studies of this sort have shown thermal dilution to be reliable and it is now considered the gold standard against which other measurements are compared.

It is interesting to note that the dye-dilution method was the incumbent standard, using indicators, such as indocyanine green dye measured by withdrawing blood from an artery through an optical sensor. This technique measures somewhat different flows—left-heart output rather than right-heart output, for example—and is subject to other issues of physiology and technique, such as greater recirculation and accumulation of indicator. Nevertheless, dye-dilution was clinically useful and thermal dilution was shown to be better and more convenient. By the early 1980s, thermal dilution was the technique of choice.

Accuracy

The accuracy of thermal dilution is degraded by the various assumptions and approximations discussed previously. Thus, even in the absence of physiologic noise, this measurement can only be expected to be within 2–7% of the true value without other measurements and specific corrections relevant only to a research setting. This accuracy is, however, well within a clinically acceptable range and is no worse than that of other methods. In his first trials with this technique, Fegler compared thermal dilution with standard direct Fick measurements in animals, finding a discrepancy of $< 7\%$. Early validation studies with thermal dilution catheters were performed by Ganz and colleagues in the early 1970s (8,17,18). In addition to supporting the overall accuracy of thermal dilution, they determined that the sensitivity of the result to mechanical and technique-dependent factors, such as catheter insertion length and speed of injection, was within 3%. This they considered to be biologically insignificant (17).

Others have since obtained good correlation with simultaneous dye dilution and other methods for measuring flow, if not slopes of identity. It is interesting to note that since the dye concentration is usually measured in a systemic artery, dye dilution will provide a measurement of left heart output that is $\sim 4\%$ higher than the right heart output measured by thermal dilution, due to the bronchiolar circulation bypassing the right heart. In addition, all of these reference methods have some of their own uncertainty in calibration, and conclusions are thus necessarily limited.

Exploration of the heat loss during injection has yielded interesting information on variability (16,19,20) but has not quantified the systematic loss to the point of accurate prediction of total injected heat (cold). The *in vitro* studies of effective losses and determination of correction constants provide adequate foundation for an accurate measurement.

Reproducibility

Cardiac output need not be known to great accuracy (within 10% is quite adequate) as long as the measurements are reproducible and can be used to track therapies. The reproducibility (variance) of the measurement with 10 mL of iced solution is generally accepted to be in the range of 10–15%. This is higher with room temperature solutions, and with smaller volumes (21). The reproducibility can be improved by 20–40% with a thermistor sensing the injectate temperature at the injection port in the right atrium. (20,22)

Some factors that can affect the reproducibility of this measurement derive from physiology and some from technique.

Physiology. As noted previously, the cardiac output can be expected to vary over the respiratory cycle, particularly with positive-pressure-assisted ventilation. This is shown quite succinctly in a careful study in animals by Jansen et al. (23) where the injection was made at random, but at known phases of the ventilation cycle. When plotted sequentially in time, the results span a range of ±15% and appear randomly distributed. When ordered according to the phase of the ventilator, the cardiac output result varies cyclically over the course of ventilation, Therefore, a determination of cardiac output using an arbitrary injection time could differ from another determination at another arbitrary time by as much as 30% due, presumably, to real physiologic variations, with flow modulated by intrathoracic and intra-abdominal pressure. Some contribution of baseline drift and thermal noise cannot be discounted by this study.

Clinically, the baseline temperature is usually assumed to be constant during the period of the measurement. Yet blood temperature varies by as much as 0.1 °C over the ventilatory cycle due to differential blood return from the upper and lower extremities. This fluctuation in baseline temperature is typically small compared with the ~ 1 °C thermal dilution signal obtained with 0 °C injectate, but extends over significant time. It is more significant with room temperature injectates and with heated-filament (continuous cardiac output) signals. The magnitude of the baseline drift can be much greater, particularly with patient movement. Of note, intravenous fluid infusions will affect the blood temperature enormously, particularly during flushes of the lines. The heat output from the heart itself returns into the right atrium from the coronary veins in synchrony with the heartbeat. These pulsations in temperature are less pronounced, being smoothed by the mixing volume in the right ventricle, and cause little practical difficulty.

Since the indicator is introduced immediately upstream of the heart, the solution can, conceivably, transit the heart in a single beat (or very few beats) if the ejection fraction is high. (Thermal dilution curves obtained with fast-response thermistors can be used to determine the ejection fraction by quantifying this washout time when the injection is made directly into the ventricle.) Therefore, these few beats must be representative of the average output for the measurement to be useful. An arrhythmia at the time of injection, occasionally caused by the injection, can lead to a single very large ejection with a very good ejection fraction. In this situation, the measured output will be much larger than the true average cardiac output. The method is not in error, but the measured output is not the steady-state output. Therefore, if arrhythmias are suspected, the measurement should be discarded or a very large variation in results anticipated.

Certain pathologies can affect thermal dilution cardiac output measurements. Tricuspid regurgitation will increase the effective mixing volume for the indicator, thus increasing the extent and decreasing the magnitude of the thermal transient. However, it is important to note that TR does not invalidate the fundamental physical principles upon which the measurement is based. If the computer can wait long enough, the CO measurement in the face of TR should be accurate. In order to be sure this is the case, the practitioner must watch the thermal dilution curve as it evolves on the monitor screen.

Very low ejection fractions can have a similar effect. Although the basic assumptions underlying the measurement remain intact, the curve can be distorted to an extent that makes the practical measurement unreliable. More serious problems are caused by an incompetent pulmonic valve. This valve is necessary to minimize the nonlinear averaging effects of the pulsatile flow, and any flow reversal at the thermistor can lead to multiple re-measurement of the thermal transient. Either of these effects degrades the cardiac output determination.

Technique. Thermal dilution measurements are reasonably insensitive to variations in operator technique. As noted above, the injectate will not warm significantly as the syringe is handled briefly prior to injection. And if this is a concern, probes can be used which measure the temperature of the injectate as it is injected. The content of the injection catheter dead space must be considered to achieve a high level of reproducibility. If multiple measurements are made over a short period of time, that is, to average several serial determinations, sufficient time must be allowed for the residual injectate to return to blood temperature. A couple of minutes appears sufficient to warm the dead space as well as to allow the blood temperature to return to a stable baseline. One strategy is to discard the first measurement, using it merely to fill the dead space with a cool solution. Another strategy for assuring a consistent effect from the dead space is to withdraw blood immediately following the injection. The potential for blood clotting, however, limits the applicability of this procedure. As noted earlier, some catheters can measure the temperature of the injectate at the point of injection (20,22) thus minimizing these effects.

In summary, judicious choice of the time of injection can improve reproducibility. Attention should be paid to the phase of ventilation, to changes in any concomitant intravenous fluid infusions, and to any concurrent cardiac arrhythmias. The temperature curve can be recorded from most cardiac output computers. This curve, and the prior baseline, can give the knowledgable operator evidence on which to judge the validity of a particular result. Recall, however, that the time course of the thermal curve is not necessarily the same as the time course of the thermal transient in the flow stream. Most clinicians use a single measurement to guide therapy although in many settings, such as in studies, it is still common practice to use the average of three serial cardiac output determinations or to discard the outlier and average the remaining two.

Ease of Use

The ease and robustness of thermal dilution measurements of cardiac output are probably responsible for its clinical popularity. The equipment is straightforward to operate, and specialized technicians are not needed to acquire reliable data. The right heart catheters may be placed for other clinical reasons without fluoroscopy. When a measurement of cardiac output is indicated, all that is necessary is to attach a computer and inject cold solution.

FUTURE DEVELOPMENTS

This measurement is simple, fundamentally inexpensive, and has remained popular for several decades. It is, however, moderately invasive. If the need for the pressure information from the pulmonary artery catheters was reduced or supplanted, the ease of making a thermal dilution measurement would diminish. There are liabilities and contraindications associated with pulmonary artery catheters and the injection of cold solutions, and this measurement is not always prescribed in critical care. Use of PA catheters is falling somewhat as other methods of assessing cardiac filling and function become more widely available, such as ultrasound-based measurements and central venous lines. However, several million pulmonary artery catheters are used each year in North America and their widespread use is likely to continue. It is interesting that many surgeons who must manage their patients via phone consultations rely heavily on PA catheter measurements, especially when the ICU team does not include experienced physicians.

Indicator-dilution measurements of the sort described in this article are fundamentally intermittent. In many cases, a continuous measurement would be favored. Continuous cardiac output measurement with the heated filament paired with advanced signal processing is becoming popular, and other techniques such as analyzing pulse contours are also becoming more accepted. Thermal dilution with 10 mL of iced solution is the standard against which these techniques are compared, and periodically calibrated (13). Catheters will continue to improve, with better clot resistance, materials, additional lumens, heating elements and sensing elements as measurements demand. Cardiac output is integrated into measurement systems forming part of derived parameters and important correlations, a trend that will continue to follow medical instrumentation and healthcare information technology in general. And as the reliability of the measurements increases with experience and technology, the long-promised closed-loop therapies may become a reality.

BIBLIOGRAPHY

1. Stewart GN. Researches on the circulation time and on the influences which affect it. IV. The output of the heart. J Physiol (London) 1897;22:159–183.
2. Hamilton WF, Moore JW, Kinsman JM, Spurling RG. Simultaneous determination of the pulmonary and systemic circulation times in man and of a figure related to the cardiac output. Am J Physiol 1928;84:338–344.
3. Kinsman JM, Moore JW, Hamilton WF. Studies on the circulation. I. Injection method; physical and mathematical considerations. Am J Physiol 1929;89:322–330.
4. Fegler G. Measurement of cardiac output in anaesthetized animals by a thermo-dilution method. Q J Exp Physiol Cogn Med Sci 1954;39:153–164.
5. Fegler G. The reliability of the thermodilution method for determination of the cardiac output and the blood flow in central veins. Q J Exp Physiol Cogn Med Sci 1957;42:254–266.
6. Dow P. Estimations of cardiac output and central blood volume by dye dilution. Physiol Rev 1956;36:77–102.
7. Swan HJC et al., Catheterization of the heart with use of a flow-directed balloon-tipped catheter. N Engl J Med 1970;283:447–451.
8. Ganz W, Swan HJ. Measurement of blood flow by thermodilution. Am J Cardiol 1972;29:241–246.
9. Perl W, Lassen NA, Effros RM. Matrix proof of flow, volume and mean transit time theorems for regional and compartmental systems. Bull Math Biol 1975;37:573–588.
10. Trautman ED, Newbower RS. The development of indicator-dilution techniques. IEEE Trans Biomed Eng 1984;BME-31:800–807.
11. Philip J et al., Continuous thermal measurement of cardiac output. IEEE Trans Biomed Eng 1984;BME-31:393–400.
12. Yelderman ML et al., Continuous thermodilution cardiac output measurement in intensive care unit patients. J Cardiothorac Vasc Anesth 1992;6:270–274.
13. Schmid ER, Schmidlin D, Tornic M, Seifert B. Continuous thermodilution cardiac output: clinical validation against a reference technique of known accuracy. Intensive Care Med 1999;25:166–172.
14. Spector WS, editor. Handbook of Biological Data. Philadelphia: Saunders; 1956; Mendlowitz M. The specific heat of human blood. Science 1948;107:97–98.
15. Diem K, editor. Documenta Geigy, Scientific Tables. Ardsley (NY): Geigy Pharmaceuticals; 1962.
16. Meisner H et al., Indicator loss during injection in the thermodilution system. Res Exp Med 1973;159:183–196.
17. Forrester JS et al., Thermodilution cardiac output determination with a single flow-directed catheter. Am Heart J 1972;83:306–311.
18. Ganz W et al., A new technique for measurement of cardiac output by thermodilution in man. Am J Cardiol 1971;27:392–396.
19. Vliers ACAP, Visser KR, Zijlstra WG. Analysis of indicator distribution in the determination of cardiac output by thermal dilution. Cardiovasc Res 1973;7:125–132.
20. Lehmann KG, Platt MS. Improved accuracy and precision of thermodilution cardiac output measurement using a dual thermistor catheter system. J Am Coll Cardiol 1999;33:883–891.

21. Bourdillon PD, Fineberg N. Comparison of iced and room temperature injectate for thermodilution cardiac output. Cathet Cardiovasc Diagn 1989;17:116–120.

22. Williams JE Jr., Pfau SE, Deckelbaum LI. Effect of injectate temperature and thermistor position on reproducibility of thermodilution cardiac output determinations. Chest 1994; 106:895–898.

23. Jansen JRC et al., Monitoring of the cyclic modulation of cardiac output during artificial ventilation. In: Nair S, editor. Critical Care and Pulmonary Medicine. New York: Plenum; 1980. p 59–68.

See also CARDIAC OUTPUT, FICK TECHNIQUE FOR; CARDIAC OUTPUT, INDICATOR DILUTION MEASUREMENT OF; CORONARY ANGIOPLASTY AND GUIDEWIRE DIAGNOSTICS; MICROPOWER FOR MEDICAL APPLICATIONS; THERMISTORS.

CARDIOPULMONARY BYPASS.   See HEART-LUNG MACHINES.

CARDIOPULMONARY RESUSCITATION

EDWARD GRAYDEN
Mayo Health Center
Albertlea, Minnesota

INTRODUCTION

Cardiopulmonary resuscitation (CPR) may be defined as the emergency restoration of vital functions in a person who has undergone a life-threatening event. The term "cardiopulmonary resuscitation" is actually misleading since the goal of all CPR is to return the victim to appropriate cerebral function; cardiopulmonary resuscitation is the vehicle by which the rescuer attempts to reach this goal. The process of resuscitation may be viewed as a continuum where at one end of the spectrum psychomotor skills of CPR may be initiated by a lay bystander who might be the first rescuer on the scene of an accident, witness to someone choking on food at a restaurant, or perhaps is present when a family member succumbs to a heart attack. Cardiopulmonary resuscitation may also be viewed in a more general and organizational sense to encompass the entire process of the emergency response to victims. The education and training of the public and first responders in basic life support, such as policeman and firefighters, is the cornerstone in an attempt to reduce sudden death through lifesaving skills. Training in basic life support focuses on providing the rescuer with the ability to recognize emergencies, activate the Emergency Medical System (EMS, 911), maintain an airway, provide effective rescue breathing and cardiac circulation. American Heart Association sponsored programs also focus on prevention of risk through education of the public regarding the etiologies of coronary artery disease, myocardial infarction (heart attack), and cerebrovascular disease (stroke). Information presented through these programs attempts to modify lifestyle patterns and behaviors, such as smoking, known to cause or exacerbate these events. The new focus in community emergency response is in the training of laypersons in the use of the Automatic External Defibrillator

(AED). Documentation of successful resuscitation in communities with high proportions of the public trained in CPR and use of an AED reach 49% in out-of-hospital victims known to have suffered ventricular fibrillation (a terminal cardiac dysrhythmia that is a common endpoint in the progression toward death) (1,2). Currently, there has been significant progress made in making these automatic defibrillators present in communities and in public places, such as shopping centers, sporting event facilities and mass transportation. The American Heart Association "ABCs" of CPR (airway, breathing, circulation) have now been supplanted with the "ABCDs" (airway, breathing, circulation, defibrillation). The progression of CPR continues into Advanced Cardiac Life Support (ACLS) supervised by a physician and consists of BLS as well as sophisticated adjuncts to provide oxygenation and ventilation, intravenous access with administration of drugs that support circulation, monitoring of cardiac rhythms with rapid interpretation of dysrhythmias and subsequent maneuvers to terminate or suppress these harmful cardiac electrical abnormalities, and postresuscitation care.

This article will first review the history of CPR followed by a detailed analysis of the pulmonary and cardiac physiology relevant to the application of these resuscitative functions. An overview of Emergency Cardiac Care (ECC) will be undertaken to enlighten the reader about the organizational process guiding CPR. The actual mechanism of BLS and ACLS will be then addressed with a brief overview of defibrillators. Finally, the salient points of this article will be summarized and future directions of resuscitation will be explored.

HISTORICAL PERSPECTIVE

Restoration of life to the dying has been a common action from antiquity to the present time. Ancient attempts at artificial respiration have been described by the prophet Elisha in the Bible (3). Galen was able to observe the inflation of a dead animal's lungs in the second century, but there has been no recording of this significant finding applied to early attempts at resuscitation (4). Resuscitation methods during this time were futile—such as applying hot materials to the abdomen or whipping the victim; animal bladders were expanded with smoke and then the outlets of these bladders placed into the dying person's rectum (5). Centuries later Paracelsus, a Swiss physician (1493–1591), first reported the use of a fireplace bellows to ventilate a dying patient. In 1740, the Paris Academy of Sciences recommended the instillation of air into a victim through a mouth-to-mouth technique and within 4 years Tossach used this method successfully to revive a person (4). Ironically, this technique was lost, only to be rediscovered some 200 years later. During the eighteenth century, multiple new attempts at artificial respiration occurred. The "Inversion Method" practiced in Europe and America was used for drowning whereby the victim was hung upside-down in an effort to drain water from the lungs and many successful attempts have been recorded for this maneuver. The "Barrel Method" as well as the "Trotting

Horse Method" were also used at this time consisting of rotating the prone drowning victim over a barrel that alternated chest compression (expiration) and chest relaxation (inspiration) or placing the drowned individual prone on a horse, with the bouncing incurred during the trot inducing the same rhythmic compression and relaxation (5). The realization that alternating compression and relaxation of the chest could induce expiration and inhalation, respectively, led to direct manual efforts by the rescuer. DeHaen in 1783 first described a chest compression, arm-lift combination (6). Leroy reported the first use of the supine victim ventilation position ∼ 1830 and later in this century (∼ 1860–1870s) Silvester's, Howard's, and Schafer's prone methods of manual compression became popular and persisted into the twentieth century. The familiar Schafer–Emerson–Ivy ventilation method of scapular compression combined with pelvic-lift emerged in the United States at the beginning of this century.

The efficacy of these various methods of manual artificial respiration was resolved in the 1950s by Gordon, who performed experiments upon fresh corpses prior to rigor mortis and then on volunteers who underwent general anesthesia and paralysis by curare. Ventilatory volumes were measured and the "push–pull" maneuvers that caused active inspiration and expiration were at least twice as effective as the Schafer method or other procedures that only produced either active inspiration or expiration (7–9). The Holger–Nielsen method (prone back-pressure, arm-lift) for resuscitation became the standard of care.

At the time of these scientific studies attempting to clarify manual methods of artificial respirations, Elam elected to evaluate the physiology of mouth-to-mouth ventilation. As an anesthesiologist, Elam had serendipitously performed mouth-to-mouth ventilation to paralyzed polio patients, for as long as several hours. Though mouth-to-mouth or mouth-to-nose ventilation had been know to have been practiced by midwives for the newborn, the question posed by this physician was, "What was the mechanism involved in the success of exhaled-air ventilation?" (10). The answers to this question came from a series of experiments where volunteers allowed themselves to be paralyzed while awake, and then ventilated by mouth-to-mouth, mouth-to-mask, or mouth-to-endotracheal tube by Elam and his colleagues until the paralyzing agent was allowed to wear off. Blood gas values were analyzed and the conclusion was that normal physiological parameters could be maintained by exhaled-air ventilation (11). This landmark study brought forth the subsequent challenge to the current back-pressure, arm-lift mode of artificial ventilation. In an effort to answer the question of which form of artificial oxygenation and ventilation would prove superior, a series of controlled experiments was then conducted by Elam and Safar. The various lung volumes with blood gas analysis for the back-pressure, arm-lift was compared with mouth-to-mouth ventilations. These two methods were used on awake, paralyzed volunteers and patients without any mask, endotracheal tubes or adjunctive airway support! These experiments also investigated the mechanisms of soft tissue airway obstruction and the effectiveness of head-tilt and jaw-thrust in maintaining the airway in rescue breathing (the jaw-thrust was first

described in Germany by Esmarch and Heiberg in the nineteenth century). The data and conclusions of these studies were published and within one year a dramatic change was made within the American and International Red Cross, global medical associations and the Armed Forces. Modern resuscitation through mouth-to-mouth oxygenation and ventilation was born through these landmark investigations (12–20). "Airway, Breathing" of the "ABCs" for current CPR principles had been founded.

The advent of electrical energy production in the eighteenth century made possible the first recorded successful defibrillation by Squires in 1775; a landmark publication came later in 1809 when Burns hypothesized that effective resuscitation would occur with the combination of artificial ventilation and electric shock (6). Even though a primitive "shock instrument" was fabricated by Aldini (6) in the 1830s, there did not appear to be any significant research into electrical cardiac excitation until much later in the century. The miraculous discovery of anesthesia in the 1840s unfortunately led to catastrophic complications. Documentation of the first case of cardiac arrest was reported in 1848 when a child died under chloroform anesthesia while having a superficial procedure completed (21). As this type of complication became more commonplace, research began to focus upon cardiac physiology and mechanisms to restore the normal heart rhythm and function. Open-chest cardiac compression was first reported by Schiff in 1847 during unsuccessful attempts to circulate blood in dogs and 2 years later Niehans reported an emergency attempt at open cardiac compression in a patient who arrested during an induction of general anesthesia using chloroform. Cardiac contractions reoccurred for a brief time prior to the patient's death (21). Interestingly, in 1847 Boehm reported the first study of closed-chest cardiac compressions in cats (22). The chest was compressed with a rhythmic motion and a cardiac pressure was sustained. In the next 10 years, Koenig and Maass reported eight successful closed-chest cardiac compressions in humans (23) secondary to anesthetic-initiated cardiac standstill; one of these resuscitations lasted for more than 1 h (24). Unfortunately, the open-thorax mode of direct cardiac massage was to be the predominant form of attempted circulatory support for the next 60 years despite these reports.

Alternating current, brought forth by the investigations of Tesla, was first reported by Prevost and Batelli to stop dog heart fibrillation in 1899 (25). Intense research into terminal cardiac dysrhythmias and electrical termination of these lethal rhythms was started in the United States by Kouwenhoven, a professor of electrical engineering, in 1928. The funding for this project was undertaken by the Consolidated Edison Company because of the numerous fatalities induced by electrocution of its employees. Termination of ventricular fibrillation through electrical countershock was confirmed and the effects of both alternating and direct current were investigated in the dog open heart model. By 1933, this group had described the principles necessary for successful open heart alternating current (ac) defibrillation (26). In 1939, the Russians Gurvich and Yuniev were the first to describe successful external defibrillation and reported that direct current (dc) countershock was superior to ac generated currents. They reported that a capacitor discharge applied to the exterior of the dog's chest would stimulate a cardiac rhythm if only applied no later than 1.5 min after the induction of ventricular fibrillation; however, they noted that the time to successful defibrillation could be extended to as long as 8 min by the application of external chest compressions. There was no description as to how these chest compressions were done (27). Unfortunately, their report was not available to western researchers until 1947 and substantiation of the benefits of dc versus ac would not be made for a number of years.

The research of Kouwenhoven at the Johns Hopkins Hospital continued in defibrillation experiments and in 1958 Knickerbocker, a research fellow, made an astute observation; during a defibrillation experiment he noted a pressure wave form being generated by the application of external electrodes on the dog's thorax (28). During a later, but similar study, Knickerbocker had a dog unexpectedly start to fibrillate and since defibrillation electrodes were not immediately available, he employed the same type of pressure upon the dog's sternum that he had found to generate a systolic pressure. After ~5 min of chest compression, the animal was successfully defibrillated into a normal sinus rhythm. A surgeon, Dr. James Isaacs, who was also conducting experiments in the same laboratory, became aware of this incident and had the foresight to encourage new research by this group into the generation of circulatory blood pressures by external cardiac massage (29). During these subsequent studies, arterial-venous pressure gradients were found to be generated and carotid artery flow was documented. Data that was reproducible indicated that if chest compressions were initiated within 1 min of ventricular fibrillation and continued for as long as 20 min, dogs could be resuscitated by defibrillation and appeared to have no deficits in central nervous system function. Further experimentation on dogs led to the conclusion that the optimum location for chest compressions was on the distal one-third of the sternum with a force of between 35 and 45 newtons (30). Even though postmortem studies revealed numerous injuries, such as rib fractures to these animals, the life-saving benefits were very apparent. Soon the practicality of closed-chest compressions became evident when Kouwenhoven and Isaacs made these laboratory observations available to the surgical staff and, in the same year, a 2-year old child was successfully resuscitated in the operating room at Johns Hopkins Hospital. An organized approach directed at patient resuscitation followed, resulting in 118 cases of successful restoration of life by chest compression following documented ventricular dysrhythmia (31).

Further collaboration at this time by Safar, Elam, and Kouwenhoven resulted in the basic tenets of modern CPR. Since external cardiac chest compressions were found not to produce adequate tidal volumes from airway obstruction (32), control of the airway confirmed by head-tilt data became the "A" in the "ABCs" of CPR. Exhaled air ventilation would become the "B" for rescue breathing. The addition of cardiac compressions, the "C" in the rudiments of basic cardiopulmonary resuscitation was then combined to produce what is now the standard protocol of care in basic life support. The final studies determined what ratio for breathing and chest compressions would be used; one rescuer CPR utilized 2 breaths for every 15 compressions while the addition of a second rescuer could increase the ratio to 1 ventilation per 5 chest compressions (33).

While the first open chest defibrillation in an operating room was reported by Beck at Case Western University in 1947, Zoll reported the first successful closed-chest or external defibrillation in humans (34). This early defibrillator utilized 60 Hz ac current of 1.5 A at a range of 120–150 V. A 6:1 isolation step-up transformer converted the 120-V line current to a range of 0–720 V with the duration of current set at 0.15 s by a condenser-relay circuit. The machine was capable of producing 12,000 W during this time interval. The copper electrodes were 7.5 cm in diameter. This paper described the successful countershock for terminating ventricular fibrillation in four patients. The advent of external cardiac defibrillation would now usher in modern cardiopulmonary resuscitation when conjoined with airway manipulation, rescue breathing, and closed cardiac chest compressions.

The historical evolution for understanding the mechanisms of cardiopulmonary resuscitation has been paradoxical; the physiology of rescue breathing appears to have been well understood versus the mechanisms of cardiac flow due to chest compressions. Positive pressure ventilation, of which mouth-to-mouth resuscitation is an example, utilizes different mechanical principles to expand the lungs versus normal breathing, but the gas exchange once in the alveoli is very similar. The action of chest compressions, however, has remained controversial. After the serendipitous finding of increased blood pressure upon application of defibrillator paddles, Kouwenhoven hypothesized that sternum compression of the heart against the spine forced blood out of the ventricles (28), but no hemodynamic studies supported this claim. Further research demonstrated an increased venous pressure equal to arterial pressure during chest compression that brought into question whether the heart ejected blood in the normal manner (35). A study 1 year later actually measured cardiac output in patients being resuscitated utilizing external cardiac compressions. The ejected blood was found to have flows approximately one-quarter of normal even though systolic blood pressures appeared to be adequate (36). An investigation of actual intravascular pressures during external cardiac compressions determined that left atrial (venous) pressure was very close to arterial pressure, which argued against a projectile expulsion of blood by the heart. The hypothesis of this study was that the requisite flow needed for organ perfusion was driven by the action of the cardiac valves. This action was thought to account for the arterial-venous pressure gradient to sustain oxygen delivery (37). The cardiac compression–cardiac flow hypothesis was further contested with a series of studies generated by the observation that coughing by patients sustained blood pressure. Reports of successful resuscitation in documented ventricular fibrillation by coughing led to research that compared arterial pressures produced by chest compressions to that produced by cough. The conclusion was that improved hemodynamic parameters occurred with coughing CPR (38). Further interest into these mechanisms was induced by a number of reports whereby trauma patients with a flail chest were not able to be resuscitated through closed-chest compressions; a flail chest results when the thoracic cage is compromised during rib fracture. Direct cardiac compression should be easier to produce since the ribs offer no resistance. Evidence appeared to support increased intrathoracic pressure rather than direct cardiac compression as the mechanism producing blood flow (39,40). Echocardiography was also utilized in several studies where CPR was initiated in humans; the cardiac valves were visualized and noted to be in the open position. Additionally, the left ventricle did not appear to be compressed, again lending credence to the "thoracic pump" theory of blood flow (41,42). Unfortunately, this theory could not account for coronary circulation blood flow or as to the mechanism of blood flow during disruption of intrathoracic pressure, such as when a pneumothorax (collapsed lung) occurs. Subsequent research utilizing very sophisticated instrumentation determined that, indeed, pressure gradients were generated with chest compressions in animals relative to aortic and thoracic venous vessels, data not supported by the thoracic pump theory. Contrast dye echocardiography demonstrated typical opening and closure of the mitral valve with projection of the contrast being propelled throughout the heart and then into the aorta (43,44). The momentum changed with these studies in elucidating the exact mechanism for blood flow, resupporting the cardiac compression hypothesis. What is currently hypothesized today is that both mechanisms seem to operate relative to resuscitation–generated cardiac ejection of blood. The key to understanding this paradox is that chest compressions involve two forces: compression and release of pressure upon the sternum. Compression of the heart forces blood through the atria and ventricles with flow generated, as evidenced by arterial and venous pressure gradients. Release of sternum pressure appears to augment venous return, supporting the thoracic pump theory. Therefore, it appears at this time that the current literature supports both mechanisms in CPR generated blood flow (45).

PULMONARY PHYSIOLOGY

Pulmonary function provides for the oxygenation of tissues and the removal of carbon dioxide from cell metabolism; human's survival is dependent on this function. It is by no coincidence that the first two actions of cardiopulmonary resuscitation, airway establishment and then rescue breathing, must be accomplished prior to chest compressions. Resuscitation is hopeless unless oxygenation and ventilation can be established. It is easiest to appreciate pulmonary function as a progression of air transport from the airway into the lungs, with an overview of lung mechanics and the molecular basis for oxygen and carbon dioxide transport.

After a volume of air is breathed through the oral or nasal passages, this inspired gas passes to the lungs by way of the trachea, bronchi, and bronchioles. Muscular tone in the soft palate and pharynx maintain this anatomical area of the airway. The trachea is supported by numerous cartilaginous rings. At the bronchiole and alveolar level, transpulmonary pressures are responsible for patency. Cardiopulmonary resuscitation of the unconscious victim demands that the first action taken by the rescuer is to make sure that the airway is open. The usual cause is obstruction of the airway by the tongue or soft tissues. Maneuvers to open the airway are the first line treatment in CPR when a person is found to be unresponsive.

The lungs function by expanding through a negative pressure pump mechanism causing inspiration of air by two mechanisms. The diaphragm, a large muscle located at the lung bases, contracts increasing the subatmospheric pressure and thus producing a pressure gradient relative to ambient air. Movement of the rib cage acts in conjunction with the diaphragm, as lung expansion occurs during elevation of the ribs. Normally, the ribs are positioned in a superior–inferior dimension; as the thoracic cage is raised, the ribs move in an anterior–posterior direction, increasing the intrathoracic lung compartment by $\sim 20\%$. The lung expansion through this mechanism also acts to produce a subatmospheric gradient, drawing air into the lungs. This occurs because the lung volumes increase at a more rapid rate than gas flow through the airway. As energy is utilized to cause this expansion, expiration during normal breathing is simply the result of the elastic recoil of the lungs and air is expelled, as now the pressure gradient reverses. During episodes of rapid oxygen metabolism, the work of breathing increases and thus the rapidity of chest wall movement requires a forceful expiration. The abdominal musculature functions in this manner to compress the diaphragm. It should be apparent that pressure–volume relationships establish the adequacy of lung mechanics. Transmural pressures, that is, the difference between the interior of the lung minus the lung exterior (or the pleural space, which separates the lung from the chest wall), define the various lung volumes as well as being a measure of elastic forces on the lung (the force tending to cause lung collapse). The slope of the P–V curve at any point represents the lung compliance; in the normal adult lung this averages 200 mL of air/cm of water, that is, when transpulmonary pressure increases by 1 cm of water, the lungs expand by 200 mL. Lung compliance is not only affected by the elastic force of the lung tissue, but also by the forces generated by surface tension in lung and pleural fluids. This surface tension elastic force is reduced in the lung by surfactant, a complex molecule primarily composed of phospholipids, which has hydrophilic and hydrophobic moieties.

When a rescuer determines that a person is unconscious and begins CPR, the airway is first opened and then rescue breathing is attempted. As mouth-to-mouth ventilations are instituted, now the lungs are expanded by positive pressure, quite different than the previously described normal mechanism. The intraalveolar as well as intrapleural pressure will rise above atmospheric pressure. The diaphragm is progressively pushed toward the abdomen in contradistinction to this muscle's upward or cephalad movement with contraction. Upon expiration, the intrapleural pressure, which is positive, decreases to subatmospheric pressure upon end-expiration and the diaphragm also moves away from the abdomen. When

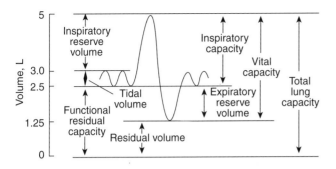

Figure 1. The dynamic lung volumes that can be measured by simple spirometry are the tidal volume, inspiratory reserve volume, expiratory reserve volume, inspiratory capacity, and vital capacity. The static lung volumes are the residual volume, functional residual capacity, and total lung capacity. Reprinted from Anesthesiology, 4th ed., Benumof: Respiratory Physiology and Respiratory Function During Anesthesia, p. 590, 1981, with permission from Elsevier Science.

positive pressure ventilation is employed in other clinical settings, such as with ventilator therapy, a constant concern is that with any damage to the lungs, gases will be propelled into the pleural space. If there is no egress of these gases, a ball-valve mechanism ensues, and the increasing positive pressure in this pleural space will compress the lung, causing hypoxemia and death (pneumothorax).

The lungs are subdivided into four static volumes and four capacities (Fig. 1). A capacity is the combination of two or more lung volumes; capacities are helpful in describing the pulmonary function and disease processes. A device called a spirometer, invented in 1846 by Huntchinson for amusement purposes, is used for these measurements (46). The original machine was a watertight bell emersed in a water tank and connected by tubing to the patient's airway. As this bell moves with inhalation or exhalation, an attached writing instrument marks these volumes on a chart. The current spirometers utilize a bellows or piston with electronic circuitry. All measurements are representative of the average adult man. These volumes and capacities are 20–25% less in women.

1. Tidal Volume: The volume of air either inspired or expired with a normal breath; This is ~500 mL; these are minimum volumes that are typically attempted in rescue breathing.
2. Inspiratory Reserve Volume: This is the maximum volume of air that can be inspired after a normal tidal volume; it is ~3000 mL.
3. Expiratory Reserve Volume: The maximum volume of air that can be ejected after expelling the tidal volume; it is ~1100 mL.
4. Residual Volume: The volume of air remaining in the lungs after a maximal expiration; this volume is ~1200 mL.

The four lung capacities consist of the following:

1. Inspiratory Capacity–Tidal Volume plus Inspiratory Reserve Volume: This volume represents the maximum amount of air that can be inspired after a normal expiration and represents ~3500 mL.
2. Functional Residual Capacity–Expiratory Reserve Volume plus Residual Volume: The volume of air in the lungs after a normal expiration; ~2300 mL.
3. Vital Capacity–Inspiratory Reserve Volume plus Tidal Volume plus Expiratory Reserve Volume: This volume is the maximum amount of air that can be expelled after a maximum inspiration and is ~4600 mL.
4. Total Lung Capacity–Vital Capacity plus Residual Volume: The maximum volume of air that can be expired after greatest possible inspiration.

The minute respiratory volume is equal to the tidal volume as a product of the respiratory rate. Since the normal tidal volume is ~500 mL and the normal respiratory rate is ~12–15 breaths/min, the minute respiratory volume is ~6–7.5 L/min. The inspired and expired volumes are not quite equal since the volume of oxygen absorbed through the alveoli is slightly greater than the volume of carbon dioxide that is expired. Only the inspired air that reaches the alveoli can participate in oxygenating the blood. There is a portion of a normal inspiration that does not reach the alveoli and this volume of gas is referred to as dead space ventilation. Anatomic dead space refers to the volume of gas from the nose, mouth, and trachea to the respiratory bronchioles. This volume averages ~2.2 mL/kg. Thus in a normal tidal volume of 500 mL, only 350 mL of air and thus 72 mL of oxygen, is available for gas exchange.

The tidal volume and the respiratory rate have a profound effect upon the total alveolar ventilation. This fact has been reflected in the revisions of CPR literature over the years. Suppose patients all have the same total minute ventilation of 5000 mL. The first patient has only a small tidal volume of 150 mL and is breathing 33 times/min, producing a minute ventilation of ~5000 mL. Recall that not all of the air in a breath reaches the alveoli; dead space is ~150 mL. The total dead space ventilation would be equivalent to the total minute ventilation. The actual alveolar ventilation would be zero. This patient will become hypoxic very quickly. The second patient has a tidal volume of 250 mL and is breathing at a rate of 20 times/min. The total minute ventilation will be again 5000 mL. The alveolar ventilation will be 2000 mL. The third patient has a tidal volume of 500 mL and a breathing rate of 10 times/min; again the total minute ventilation is 5000 mL, but in this case the actual alveolar ventilation is 3500 mL. The conclusion that should be drawn from these examples is that the efficiency of ventilation is greater when the tidal volume is increased versus the equivalent change in respiratory rate relative to total alveolar ventilation.

The composition of air that one breathes changes significantly from the atmosphere to the alveolus. At sea level, nitrogen produces a partial pressure of ~597 mmHg and composes ~78% of room air. Oxygen has a partial pressure of 159 mmHg and represents almost 21% of the total for atmospheric gas. Carbon dioxide and water make up the remaining partial pressures and percentages. Once the air is humidified by the nasal and oral airways, water vapor comprises 47 mmHg and increases to ~6% of the mixture with a corresponding reduction for nitrogen and oxygen. The alveolar air has a reduction in both nitrogen (569 mmHg, 75%) and oxygen (104 mmHg and 13%). In the clinical setting, the alveolar oxygen tension is an extremely useful measurement to evaluate the variables in pulmonary mechanics and gas exchange. The ideal alveolar gas equation is useful approximation and is expressed as follows:

$$PA_{O_2} = [(P_B - P_{H_2O})(F_IO_2)] - \frac{PA_{CO_2}}{R} + F$$

Where PA_{O_2} is the partial pressure of oxygen in the alveoli; P_B is the barometric pressure; P_{H_2O} is the partial pressure of the water vapor in the alveoli at 37 °C; F_IO_2 is the partial pressure of oxygen; $PACO_2$ is the partial pressure of alveolar carbon dioxide; R is the ratio between the volume of carbon dioxide diffusing from the pulmonary blood to the alveoli and the oxygen diffusing from alveoli into pulmonary blood. Approximately 200 mL/min of carbon dioxide versus 250 mL of oxygen exchange, so the ratio 0.8. F is a small correction factor that can be ignored clinically. Therefore, for example, suppose that a patient has been medicated with opioids after a painful operation and the alveolar partial pressure rises to 65 mmHg since these drugs reduce the respiratory sensitivity to carbon dioxide. The barometric pressure is 760 mmHg.

Therefore,

$$PA_{O_2} = [(760 - 47)](0.21) - \frac{65}{0.8}$$

$$PA_{O_2} = 68 \text{ mmHg}$$

These figures have a profound influence upon oxygenation in resuscitation. A simplified example will enlighten the reader; from the previous review of lung volumes, the total lung capacity is ~5000 mL. If roughly 20% of the atmosphere is oxygen, then 20% of the total lung volume, 1000 cm^3, will contain oxygen. As mentioned earlier, the basal metabolic rate for oxygen consumption is ~250 mL/min. Therefore, the quotient of the 1000 cm^3 relative to the oxygen consumption of 250 mL/min yields 4 min until hypoxia ensues from lack of oxygen. This is reason why time is so critical for the rescuer; unfortunately, the brain is the most oxygen-sensitive organ in the body and cerebral function diminishes rapidly after this critical four minutes. In ACLS, supplemental oxygen is immediately made available to the victim. Given the previous example, if 100% oxygen is administered without entrainment of room air (and nitrogen), now the total lung volume of oxygen would be 5000 cm^3. At the same basal metabolic rate for oxygen utilization, 250 mL/min, theoretically the patient could remain apneic for 20 min before hypoxia would ensue! Practically, this does not occur because of the metabolic byproduct of carbon dioxide diffusing into the alveoli as well as the tremendously increased energy requirements caused by the ventricular dysrhythmias; however, the point to be made here is how the atmospheric composition of gases can easily be altered by the addition of supplemental oxygen to improve the mortality and morbidity of cardiopulmonary resuscitation.

Alveolar ventilation is the ultimate endpoint with respect to lung mechanics. Air must be transmitted throughout the respiratory passages until oxygen can be absorbed by the blood. As a person inspires a normal tidal volume, the contained oxygen reaches the terminal bronchioles. Interestingly there is no organized flow of gas from this point to the alveoli; the oxygen traverses the respiratory bronchiole and alveolar duct into the alveolus for gas exchange by simple diffusion. Once the oxygen reaches the alveolar membrane, the diffusing capacity, which averages 21 mL/min per mmHg, causes the 250 mL of oxygen to traverse the respiratory membrane since the driving oxygen pressure difference is ~12 mmHg. The basic metabolic rate for oxygen utilization is equal to 250 mL/min. Therefore, during quiet respiration, with normal tidal volumes, oxygen intake is appropriate for oxygen utilization. When physical work or exercise increases the metabolic requirements for oxygen, the diffusing capacity can increase threefold in a young healthy adult male. The egress of carbon dioxide through the alveolar membrane is also crucial for survival. The diffusing capacity has never been measured accurately for carbon dioxide due to the rapidity with which this gas passes from red blood cell to alveolus; however, since the diffusion coefficient of carbon dioxide is ~20 times that of oxygen, a range of between 400 and 1200 mL/min per mmHg would be expected for this gas.

OXYGEN AND CARBON DIOXIDE TRANSPORT

Once oxygen diffuses through the alveolar membrane and enters the venous pulmonary blood, it is primarily carried in combination with hemoglobin encased in the red blood cells and secondarily in solution. Hemoglobin is a tetramer molecule consisting of four amino acid polypeptide chains and four heme groups. The globin, or protein portion, consists of two pairs of identical alpha chains and, in the adult hemoglobin, two beta chains. The locus for the alpha chains is located on chromosome 16. The alpha chain is always present; however, there may be some variety in the non-alpha chain. Fetal hemoglobin, for example, has two gamma chains, which increases the hemoglobin binding of oxygen, increasing the efficiency of maternal oxygen transport across the placenta. The four heme moieties are located in the center of each globin molecule. Heme is synthesized from glycine and succinyl coenzyme A to form a tetrapyrrol ring. Subsequent enzymatic reactions produce a protoporphrin and, finally, ferrous iron is inserted into the center of this ring as a function of mitochondrial synthesis. Since there are four heme-combining sites in each hemoglobin molecule, a maximum of four oxygen molecules can attach to the receptors. When all four receptor sites are combined with oxygen, the hemoglobin has a 100% saturation. If only three molecules of oxygen are bound, the hemoglobin is 75%, and so forth. Oxyhemoglobin is hemoglobin that has oxygen bound to the heme sites (HbO$_2$); unbound hemoglobin is termed "reduced hemoglogin" or "deoxyhemoglobin" (Hb). The key principle to

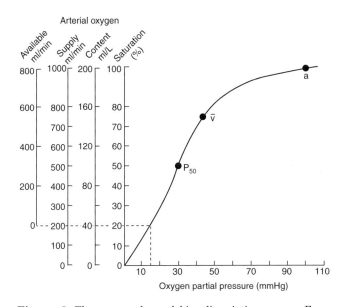

Figure 2. The oxygen–hemoglobin dissociation curve. Four different ordinates are shown as a function of oxygen partial pressure (the abscissa). In order from right to left, they are: saturation (%), O_2 content (mL of O_2/0.1 L) of blood; deoxygen (O_2) supply to the peripheral tissues (mL/min); and O_2 available to the peripheral tissues (mL/min), which is the O_2 supply minus ~200 mL/min that cannot be extracted below a partial pressure of 20 mmHg. Three points are shown on the curve: a, normal arterial: v, normal mixed venous; and P_{50}, the partial pressure (27 mmHg) at which hemoglobin is 50% saturated. Reprinted from Anesthesiology, 4th ed., Benumof: Respiratory Physiology and Respiratory Function During Anesthesia, p. 596, 1981, with permission from Elsevier Science.

understand is that oxygen binding to hemoglobin is directly related to the partial pressure of oxygen. As the inhaled air reaches the alveoli and participates in gas exchange, hemoglobin becomes fully saturated with oxygen relative to the partial pressure at the alveolar membrane. Oxygen delivery and unbinding occurs at the tissue partial pressure. The initial binding of the first oxygen molecule to hemoglobin facilitates the further binding of the second molecule, and in turn, these first two molecules facilitate further binding of the third oxygen molecule. This interaction occurs until the fourth oxygen molecule is bound, and this characteristic of changing oxygen affinity of hemoglobin is reflected in a sigmoid curve when the percent saturation of hemoglobin is plotted against the partial pressure of oxygen (Fig. 2).

The curve has a steep and flat portion. The steep slope of the curve reflects the rapid combination of oxygen with hemoglobin as the partial pressure increases. Beyond ~60 mmHg, the curve flattens, reflecting very low increases in saturation relative to increases in oxygen partial pressures. The clinical significance of this flat portion of the curve can be observed by noting that a fall from 100 to 60 mmHg only decreases the oxygen saturation from near ~100–90%. This zone of the curve provides for a safe range of minimal saturation and decreases relative to great decreases in partial pressure during oxygen loading. Furthermore, increasing the partial pressure beyond 100 mmHg of O_2 does not really oxygenate the blood to any

significant degree; since the hemoglobin is fully saturated, only the dissolved plasma oxygen will increase.

Another significant property of hemoglobin is the fact that the oxygen affinity of this molecule changes with intracellular pH (Bohr effect). As the end product of metabolism, carbon dioxide is present at the tissue level and is converted to a weak acid by the red blood cell catalyst, carbonic anhydrase. This weak acid ionizes to hydrogen ion and lowers the intracellular pH, which decreases the oxygen affinity of hemoglobin, and thus facilitates the unloading of oxygen at the tissue level where it is precisely needed. Since reduced hemoglobin is a weaker acid than hemoglobin, the hydrogen ions are bound and thus deoxyhemoglobin returns to the lungs, where the reverse situation occurs. Carbon dioxide is reconverted in the red blood cell, and with the diffusion of this CO_2 into the alveoli, the pH rises and the affinity of hemoglobin increases for oxygen.

PULMONARY CIRCULATION

Pulmonary blood flow begins with ejection of venous blood from the right ventricle into the pulmonary arteries. Successive arterial branching occurs so that at the level of the alveolar circulation the capillaries lie in intimate contact with the alveoli allowing for a very efficient and exceedingly large surface area for gas exchange. Since the pulmonary arterial pressure is only 20% or so of the systemic circulation, with a mean pressure of ~18 mmHg, these arterioles do not require significant amounts of smooth muscle. Thus the walls of these vessels are extremely thin, allowing for the diffusion of oxygen and carbon dioxide. This characteristic makes these capillaries very susceptible to distortion relative to alveolar pressure. Since the arterial pressure is so low, alveolar pressure may at times exceed pulmonary capillary pressure and this transmural pressure will cause these tiny vessels to collapse. In the upright lung, this situation occurs where pulmonary blood flow pressure is minimal, that is, at the superior aspect of the lungs. This pressure gradient scenario may be observed in Fig. 3.

In *zone 1*, where pulmonary pressure can fall below alveolar pressure, the potential exists for no flow to occur in the capillary. Any situation that decreases systemic blood pressure and thus pulmonary blood flow such as hemorrhage, or increases alveolar transmural pressure, such as might positive pressure ventilation encountered in rescue breathing, might cause this change. The alveolar pressure exceeds pulmonary arterial pressure and, in turn, pulmonary venous pressure.

In *zone 2*, the pulmonary arterial pressure increases due to the elevated hydrostatic pressure as a function of position relative to the column of blood. The alveolar pressure exceeds pulmonary arterial pressure in this zone; however, the pulmonary venous pressure relative to alveolar pressure is low and thus the gradient in this zone is the difference between arterial and alveolar pressure. The analogy to this unique lung region has been described as the vascular waterfall effect (47). The elevation of the river above the dam is described as pulmonary arterial pressure

The Four Zones of the Lung

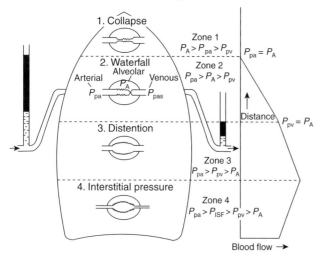

Figure 3. The Four Zones of the Lung. Schematic diagram showing distribution of blood flow in the upright lung. In zone 1, alveolar pressure (P_A) exceeds pulmonary artery pressure (P_{pa}), and no flow occurs because the intraalveolar vessels are collapsed by the compressing alveolar pressure. In zone 2, arterial pressure exceeds alveolar pressure, but alveolar pressure exceeds venous pressure (P_{pv}). Flow in zone 2 is determined by the arterial–alveolar pressure difference (P_{pa}–P_A) and has been likened to an upstream river waterfall over a dam. Since Ppa increases down zone 2 and P_A remains constant, the perfusion pressure increases, and flow steadily increases down the zone. In zone 3, pulmonary venous pressure exceeds alveolar pressure, and flow is determined by the arterial–venous pressure difference (P_{pa}–P_{pv}), which is constant down this portion of the lung. However, the transmural pressure across the wall of the vessel increases down this zone so that the caliber of the vessels increases (resistance decreases), and therefore flow increases. Finally, in zone 4 pulmonary interstitial pressure becomes positive and exceeds both pulmonary venous pressure and alveolar pressure. Consequently, flow in zone 4 is determined by the arterial interstitial pressure difference (P_{pa}–P_{ISF}). Reprinted from Anesthesiology 4th ed., Benumof: Respiratory Physiology and Respiratory Function During Anesthesia, p. 578, 1981, with permission of Elsevier Science. Diagram modified and reprinted with permission from West JB: Ventilation/Blood Flow and Gas Exchange, 4th ed., Blackwell Scientific Publishers, Oxford, 1970.

and the dam height analogous to alveolar pressure. The downstream river is equivalent to pulmonary venous pressure. Pulmonary blood flow is relative only to the difference between the height of the river upstream and the elevation of the dam. The distance that the water falls over the dam is immaterial to flow rate. Since the alveolar pressure tends to remain constant throughout this zone, but the pulmonary alveolar pressure increases secondary to the gravity, flow increases linearly. *Zone 2* circulation is unique in that ventilation and cardiac changes may alter flow dynamics, shifting these relationships into a momentary *zone 1* or *3* picture.

The dynamics in *zone 3* are straightforward. Here pulmonary venous pressure exceeds alveolar pressure and blood flow is governed by the arterial–venous gradient, which occurs in the systemic circulation. Blood flow never ceases and all capillaries remain patent, with the

additional feature of decreasing alveolar pressure maximizing vessel diameters and decreasing pulmonary vascular resistance. The rate of pleural pressure rises as a function of the transmural pressure gradient between lung apex and base; this pressure does not increase as rapidly as the pulmonary artery–venous difference that optimizes blood flow.

Zone 4 is ordinarily not present in normal lung physiology. Some pathological process is required to increase fluid pressure between cells where pulmonary venous and alveolar pressure is exceeded. Conditions such as iatrogenic fluid overload, pulmonary embolism, high levels of negative pleural pressure encountered with airway obstruction in a spontaneously breathing patient, or thoracentesis maneuvers causing profound negative pleural pressures (48,49) may cause this situation. Pulmonary arterial pressures exceed interstitial pressures, which, in turn exceeds venous and alveolar pressures. Since interstitial pressures are greater than venous pressures, regional blood flow is decreased relative to *zone 3*, and flow is governed by the pulmonary arterial-to- interstitial gradient.

In conclusion, it should be evident that both alveolar ventilation and pulmonary blood flow have a variable distribution throughout the lung. The lung base not only receives more blood flow than the apex but, because the compliance of the basal alveoli is greater than the apical alveoli, the lung base receives a greater amount of the tidal volume. Since the blood flow gradient is steeper than the ventilation gradient, the base is relatively overperfused and thus hypoventilated; the reverse situation occurs in the apex where the lung is overventilated and hypoperfused. These conditions have a profound effect upon end-organ oxygen transport. The first scenario refers to physiologic shunt blood flow; should absolutely no ventilation occur, a true shunt occurs. Decreased ventilation relative to perfusion increases alveolar carbon dioxide and thus, as seen in the alveolar gas equation, alveolar oxygen concentration will decrease. The oxygen content of the systemic arterial blood is decreased and thus oxygen transport to the tissue results in hypoxemia. A ventilated alveoli that is not perfused, as in *zone 1*, does not participate in gas exchange. Alveolar carbon dioxide decreases and alveolar oxygen increases due to the absence of blood flow. This situation is termed "alveolar dead space ventilation". The composition of alveolar gas is essentially equal to atmospheric gas. The extremes of alveolar dead space ventilation and shunt are ends of a continuum in lung ventilation and perfusion dynamics. Ventilation and perfusion ratios will vary throughout the lung both on an anatomical and physiological basis. The total effective gas exchange can thus be seen as the complex interplay between lung mechanics, ventilation, perfusion, and molecular interactions.

CARDIAC PHYSIOLOGY

The heart is an extremely efficient pump, which results in the progressive pulsatile ejection of blood to the organs. The heart is composed of four chambers: two atria and two ventricles. As blood enters the right atrium from the large veins, passive flow continues into the right ventricle. The right atrium then contracts, forcefully ejecting the remaining 25% of this blood into the right ventricle. After a delay, this right ventricular blood flow is directed into the pulmonary arteries. A progressive reduction in vessel size results in a capillary meshwork intimately in contact with the alveoli whereby the gas exchange mechanisms function. Pulmonary venous blood, now oxygenated and devoid of carbon dioxide, enters the left atrium. This blood is pumped to the left ventricle, and into the systemic circulation where the cycle is continuously repeated.

Cardiac muscle has some similarities to skeletal muscles, but also some very significant differences as well. Cardiac muscle is arranged in a striated latticework with actin and myosin filaments, which lie adjacent to one another and contract in the same manner as skeletal muscle. However, cell membranes separate these fibers yet allow ionic diffusion between these membranes or *intercalated disks*. Thus during a chemical depolarization resulting in an action potential, unimpeded progression of this electrical current flows with minimal resistance throughout the heart. The *intercalated disks* allow for the heart to actually act as two separate systems. The two atria are electrically excited as a unit, as are the ventricles. The anatomical division of atria and ventricles by nonconducting fibrous tissue does not allow conduction to occur between the atrial and muscle in an unorganized manner. A very specialized conduction system ensures that atria and ventricles are depolarized in a progressive manner.

The atrioventricular valves close during ventricular contraction (systole) preventing the backflow of blood into the atria. The tricuspid valve lies between the right atrium and right ventricle; the mitral valve is located between the left atrium and left ventricle. As blood is ejected out of the right and left ventricles, the semilunar valves open; the pulmonary and aortic valves, respectively, then close during cardiac relaxation (diastole) to prevent blood from returning from the pulmonary and systemic circulation. Note that the first arterial branches off the aorta are the coronary arteries.

The specialized conducting system of the heart that produces a progressive, rhythmical contraction of atria and ventricles has several components. The sinus node provides the genesis for cardiac depolarization. This specialized cardiac muscle is located in the right atrium just below and lateral to the superior vena caval ostium. This strip of tissue, ~ 15 mm long, connects directly to the atrial musculature. Generation of action potentials in the sinus node progresses directly to the entire atria causing a unified contraction of all muscle fibers at once. The resting membrane potential of the sinus node fibers is ca. -60 mV compared with the ca. -90 mV for cardiac muscle. This difference in the sinus node electronegativity is due to the fact that sodium ions with their positive charge progressively "leak" intracellularly. A progressive rise in threshold voltage occurs until ca. -40 mV; opening of the rapid sodium channels at this point then produces the initial cardiac depolarization. The sustained contraction of the cardiac muscles is due to the secondary influx of calcium ions, followed by the influx of potassium ions, which exchange with the outward diffusion of the sodium ions.

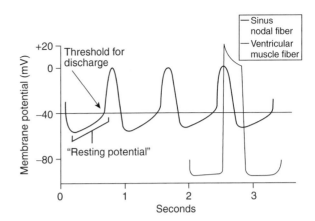

Figure 4. Rhythmical sinus node action potential compared with ventricular muscle fiber. Reprinted from Textbook of Medical Physiology, 10th ed., Guyton and Hall, p. 108, 2000, with permission from Elsevier Science.

This last ion counterexchange of potassium for sodium limits the induced hyperpolarization of the cell allowing repolarization. This phenomenon of "leaky" sodium channels produces the rhythmic excitation, which initiates the cardiac cycle. The rate of this sinus node depolarization is controlled by the autonomic nervous system through the interaction of the para-sympathetic (acetylcholine) and sympathetic (norepinephrine) fibers. Generally, the length of time for this activation is on the order of 10 ms. Drugs utilized in ACLS, such as atropine and epinephrine, affect the firing interval of the sinus node.

Once the atrial muscle fibers are activated, the action potentials cause a generalized contraction of all of these fibers at once, again due to the unique anatomy of the cardiac musculature. Activation of the left atrium occurs through the specialized fibers termed the "anterior interatrial band". The anterior, middle, and posterior internodal pathways transmits the pacemaker impulses to the atrioventricular node in ~ 0.03 s. The AV node is essentially a junction box that has two unique features; a delay in the pacemaker action potential occurs here, which affords a delay in ventricular contraction so that the blood is allowed to empty from the atria to both ventricles and normally action potentials can only travel in one direction. This atrioventricular node is positioned in the right atrial posterior wall just behind the tricuspid valve. The delay in the ventricular depolarizing impulse is ~ 0.13 s.

The final pathway for the activation of the ventricles occurs through the Purkinje fibers, which terminate in the left and right bundle branches. These branches run in the ventricular septum separating the right and left ventricle and then terminate into progressively smaller branches throughout the ventricular muscle. The Purkinje fibers act in contradistinction to the AV node; action potentials are transmitted at a velocity 100-fold allowing rapid excitation and contraction of both ventricles. Transit through the Purkinje fibers is only ~ 0.03 s with the same approximate time necessary for complete ventricular muscle activation.

The electrical activity described in Fig. 5 can be measured at the skin such that electrical potentials are recorded as the ECG. A normal ECG consists of several

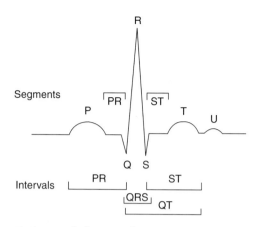

Figure 5. A normal electrocardiogram (ECG) cycle with wave segments and intervals. Reprinted from Anesthesiology, 5th Edition, Hillel and Thys: Electrophysiology, p. 1232, 2000, by permission from Elsevier Science.

waves of depolarization and repolarization. The P wave is produced from the summed action potentials generated during atrial depolarization. It is upright and, after returning to baseline, a pause is observed reflecting the progressive depolarization through the AV node. This P–R interval (actually the P–Q interval, but often the Q-wave is not visualized) begins at the initiation of the P wave and ends at the beginning of the QRS complex. The normal duration of the P–R interval is ~0.12–0.21 s or three to five of the small squares on the ECG graph paper. During the P–R interval atrial depolarization occurs as well as the electrical activity generated in the AV node. The QRS complex, which consists of a Q wave, R wave, and S wave, represents the electrical activity causing ventricular depolarization. The Q wave is seen as a negative deflection from the baseline, which is followed by the large positively deflected R wave. The S wave follows the R wave and, like the Q wave, has a negative deflection. Often the Q wave and S wave may not be observed in the complex. The normal QRS duration is usually no >0.10 s or 2.5 of the small ECG squares. The S–T segment begins at the end of the S wave (commonly termed the "J point") and ends at the onset of the T wave. This interval is usually isoelectric, but can have a normal variance of ca. −0.5 to +2.0 in the precordial leads (see below). The normal duration is <0.12 s and ~2.5 ECG squares. The T wave is a repolarization wave and reflects potentials generated with ventricular recovery. This wave usually has a positive deflection of ~0.5 mV, but often is not observed because of decreased amplitude. The T wave represents a continuum of absolute to a relative refractory period of ventricular depolarization. It is important to note that the electrical activity observed in the electrocardiogram represents electrical activation of the atrial and ventricular muscles and not the actual contractions themselves. A standardized method for recording ECGs consists of graph paper upon that positivity is reflected with upward deflections and negativity, downward deflections. Ten small divisions represent 1 mV. The large vertical lines represent 0.20 s with the smallest intervals representing 0.04 s. These voltages recorded at the skin are very small relative to the actual potentials of ~110 mV at the heart. Proximity of the recording electrodes

as well as angular direction from the heart then will affect the size and shape of the ECG, respectively.

The standard electrocardiogram utilizes 12 leads (electrodes) to view the electrical activity of the heart. Six standard limb leads are combined with six chest (precordial) leads. Initial resuscitation events that require dysrhythmia interpretation usually view the bipolar leads I, II, III. These leads utilize two electrodes, one positive and one negative, to monitor the heart and record potential differences. Electrodes are applied through an adherent conductive gel to the left shoulder, right shoulder and left leg essentially forming a triangle (Einthoven's triangle, Fig. 6). The ground lead is usually placed on the right leg. Lines that bisect the sides of the triangle have their origin at the heart, the center of the triangle, which is the zero axis of each side. In lead I, the left shoulder is connected to the positive electrode and the right shoulder electrode is negative; the recording is the potential difference between the left shoulder and right shoulder. In lead II, which is the most common limb lead to be monitored, the negative electrode is placed on the right shoulder and the positive terminal to the left leg. Depolarization in the heart follows this same electrical vector as lead II and thus optimizes P wave height and shape as well as QRS morphology. Lead III places the positive terminal on the left leg and the negative terminal on the left arm. These three leads are very similar to one another in that the P, Q, and T waves are positive. They are excellent for dysrhythmia interpretation; since electrical activity from atria to ventricles is displayed, waveform and time related changes are very apparent, both to the diagnostician and computer (such as in an automatic external defibrillator).

CARDIAC ARREST DYSRHYTHMIAS

The previous section presented information relevant to understanding and interpreting the normal ECG. This

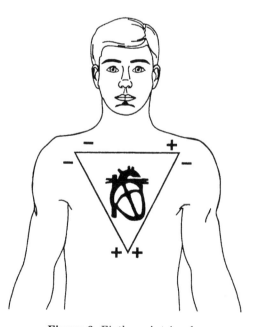

Figure 6. Einthoven's triangle.

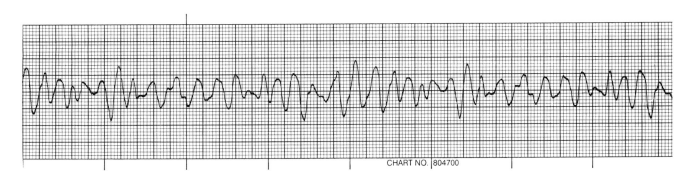

CHART NO. 804700

Figure 7. The fatal rhythm of ventricular fibrillation.

primer of basic electrocardiography will enable the reader to now have some ability to differentiate the lethal rhythms that cardiopulmonary resuscitation demands for optimizing treatment.

The fatal dysrhythmia of ventricular fibrillation (Fig. 7) is a common endpoint in cardiac arrest. There are no QRS complexes, P waves or T waves that are identifiable. Scientific evidence over the years has conclusively established that only early defibrillation has any hope of restoring a normal sinus rhythm to the heart and subsequent survival. The fibrillating heart has no ability to eject blood since there is no coordination of heart muscle and no progressive flow of blood from atria to ventricles to systemic circulation. Cardiac depolarization and repolarization within the ventricular muscle occurs in a chaotic manner; ventricular muscle is activated in an unorganized fashion. This electrical activity sustains a vicious cycle of reexcitation, never allowing the return of normal cardiac function. The ventricles neither relax nor contract and in this ventricular fibrillating state consume massive amount of energy. Since there is no ejection of blood, unconsciousness from lack of cerebral blood flow occurs within seconds and death ensues from hypoxia.

After a normal cardiac sequence there is a refractory period (as mentioned previously with the beginning of the T wave, which is at the end of the cardiac cycle), whereby this cardiac impulse cannot reexcite the heart until a new electrical stimulus is generated from the sinoatrial node. However, the underlying etiology of ventricular fibrillation appears to be due to electrical reentry or "circus movements", where the normal termination of depolarization does not occur. These abnormal electrical pathways may be generated in several ways: a shortened refractory period, decreased depolarization velocities, or increased distance for the normal electrical impulse to travel. Recall that the progression of depolarization takes place only in one direction and essentially travels almost in a circle with excitation of the ventricles. If this normal impulse reaches cardiac muscle that has already been depolarized, the refractory time will not allow another depolarization. Stimulation of cardiac muscle will not occur until the entire myocardium is ready to be energized as one unit. Suppose that one of the three abnormal conditions were present; any ventricular muscle that was not refractory could be stimulated to contract in an unorganized manner. In a clinical setting, many individuals have hypertension and develop enlarged hearts. A large ventricular muscle mass

would create an increased distance for the normal electrical impulse to follow, thus creating the potential for a "circus movement" to initiate reexcitation of muscle fiber. Rates of depolarization from the sinoatrial node through the AV node may result from blockade of this specialized system from a variety of causes. Electrolyte imbalance, as well as coronary artery disease, are common factors in inducing conduction block. Alterations in the sympathetic nervous system as well as drugs may act in sensitizing the heart, allowing more rapid conduction of impulses and increased susceptibility to fibrillation. Once the ventricular muscle begins this chaotic activity, a chain reaction phenomena begins: conduction velocities throughout the heart decrease, allowing even more time for reentrant depolarizations to occur and the actual muscle refractory time is decreased, increasing the opportunity for these impulses to propagate this dysrhythmia.

The cornerstone of cardiopulmonary resuscitation is early defibrillation. The previous American Heart Association mnemonic of CPR, the "ABCs", which consisted of Airway, Breathing, Circulation, has been changed to "ABCDs" to include defibrillation. After one shock, 60% of all victims succumbing to ventricular fibrillation will survive; after two shocks, 80% survive; after three shocks, 90% will be successfully resuscitated (48). Electrical countershock utilizing high-voltage current can inhibit defibrillation by instantaneously depolarizing all cardiac muscle tissue. The myocardium then is totally refractory to any reentry currents. The electrocardiogram will typically record asystole, or no evidence of electrical activity, from the heart for several seconds. Resumption of the normal cardiac pacemaker will resume, and organized contraction should reoccur. Time is of the essence, since as a heart continues in fibrillation, the rapid utilization of high-energy phosphates depletes this "fuel" for resumption of normal cardiac activity. It is obvious that delay in defibrillation induces a state whereby even successful technique in countershock will be not be able to sustain a normal cardiac rhythm due to the lack of substrate for myocardial energy consumption. The underlying philosophy of cardiopulmonary resuscitation now is early access to defibrillation for the victim.

Pulseless ventricular tachycardia (Fig. 8) is the other malignant dysrhythmia that requires immediate external countershock. Unlike ventricular fibrillation, the electrocardiogram displays a rapid regular rhythm with a widened and abnormal appearing QRS complex. Usually,

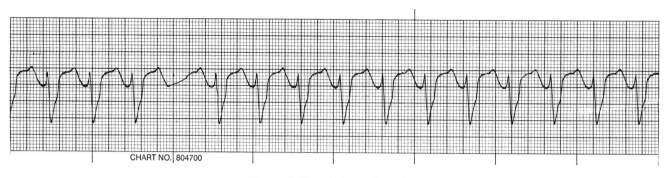

Figure 8. Ventricular tachycardia.

the rate is very rapid, with a range of ∼ 100–250 beats/min, contrasted to the normal sinus rhythm rate of 60–100 beats/min. Three or more of these bizarre appearing complexes define this dysrhythmia. Usually, the P wave and T wave are obscured and thus the P–R interval cannot be measured. This rapid rate does not allow adequate filling of the ventricles, and causes a ventricular lack synchrony with the atria; the result is dramatic loss of cardiac output and blood pressure. Since there is no effective cardiac output and these dysrhythmias degenerate into ventricular fibrillation, external countershock is mandated. The sinus node functions normally in VT, whereby the atria are properly depolarized. There can be a retrograde depolarization of the atria from the ventricles in some instances of ventricular tachycardia and there will be a definite P wave associated with the abnormal QRS complex. Usually, these retrograde P waves have a negative (downsloping) peak. Another unusual feature of this dysrhythmia is that at certain time intervals the atria may be able to initiate an impulse completely through the AV node and Purkinje system at the instant where the ventricular-initiated depolarizations leave the conduction system vulnerable. The result is termed a "capture beat". If this normally conducted impulse occurs at the same time that a ventricular depolarization is generated, a fusion beat is generated, which appears as a cross between the normal- and ventricular-originated complex. This dysrhythmia may either be paroxysmal or sustained and the shape of these QRS waves either monomorphic or polymorphic. Degeneration into ventricular fibrillation is a common course; for this reason countershock is required. The mechanism for pulseless ventricular tachycardia is hypothesized to be a reenty depolarizing current due to delayed conduction. The site of occurrence for this aberrant mechanism would localize to the Purkinje system and ventricles.

An unusual form of ventricular tachycardia (Fig. 9) is termed "Torsade de Pointes" or twisting of points (49). Notice the QRS morphology viewing the rhythm strip from left to right. On first appearance, this dysrhythmia appears to be ventricular tachycardia. The QRS complexes are wide (versus the normal narrow QRS shape), but constantly changing in shape and amplitude yet there appears to be a rhythmic oscillation about the baseline. The depolarization wave appears to twist around the central axis or helix. This dysrhythmia is triggered by electrical potentials that either occur before or after the normal spontaneous depolarization of the heart or is associated with a prolonged QT interval. Should this dysrhythmia be associated with no evidence of effective blood flow, immediate countershock would be the treatment of choice after the "ABCs" of CPR have been accomplished. However, if Torsade de Pointes is misdiagnosed as ventricular tachycardia and a pulse is present, the potential exists for the wrong treatment and lethal results. This dysrhythmia usually occurs when there is an underlying prolongation of the QT interval. Since many of the antiarrhythmic agents will prolong this interval, it should be apparent that selection of one of these drugs to terminate a misdiagnosed ventricular fibrillation when Torsade is present has the potential to produce a nontreatable dysrhythmia. Antidepressants, antipsychotics, and electrolyte abnormalities (particularly hypokalemia and hypomagnesemia) will produce the underlying QT prolongation that serves as the catalyst for Torsade

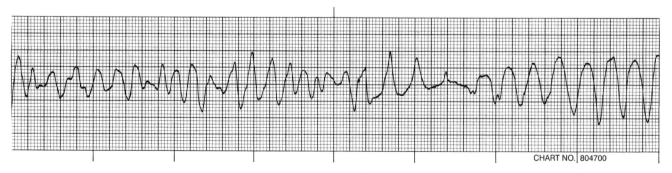

Figure 9. Torsade de Pointes.

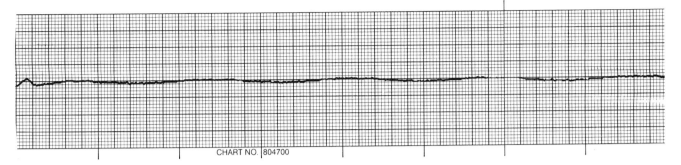

Figure 10. Asystole.

to appear (50). Strategies for treating this rhythm are to correct the factors providing the substrate for dysrhythmia; magnesium supplementation, and correction of all electrolyte abnormalities as well as what is termed "overdrive pacing" to externally pace the heart to rates of between 100 and 120 beats/min (51).

Asystole (Fig. 10) is the terminal arrhythmia that represents a dying heart. Observation of the above example shows the absence of electrical activity (flat line), but occasionally P waves, or what is termed "ventricular escape beats", may be present. Usually, the presence of asystole means death for the victim, so rescuers are taught to view this rhythm in more than one lead since fine ventricular fibrillation may actually be the abnormal rhythm rather than asystole. This distinction is extremely important since VF needs to be immediately treated with defibrillation, whereas in asystole the possible causes need to be identified, and defibrillation is contraindicated for resuscitation. Why not shock the heart that is in asystole? The answer lies in the fact that rarely a victim may experience high levels of parasympathetic nerve input to the heart. This parasympathetic stimulation can produce complete cessation of the atrial and ventricular pacemaker. Defibrillation also produces an intense parasympathetic nerve discharge, which, in this specific situation could terminate any chances for the heart to recover normal pacing function (52). Evidence substantiates that defibrillation for asystole does not improve the survival rates in out-of-hospital arrest scenarios (53). There have been rare occasions reported in the literature where victims have positively responded to transcutaneous cardiac pacing for which many defibrillators now have this

capability. The heart is stimulated externally to produce an effective cardiac output. The caveat is that this maneuver can only be effective if there are only several minutes of asystole. A specific situation that lends itself to transcutaneous pacing is the asystole that occurs after defibrillation. This witnessed arrhythmia occurs immediately after some countershock attempts and thus there can be a window of < 1 min for pacing to be initiated. There may be certain types of patients who also might respond to asystolic pacing, such as those with Stokes–Adams–Morgagni syndrome, where intermittent atrioventricular block produces asystole or the hypoxia initiated P wave asystole (54).

The last terminal rhythm group that occurs frequently in resuscitation efforts is pulseless electrical activity (PEA). Observe the rhythm strip in Fig. 11; this is an example of a normal sinus rhythm. At first, one would question why a normal ECG would be included in emergency cardiac care and treatment. The key to understanding pulseless electrical activity is that the rhythm may appear to be entirely normal; however, upon further physical assessment of the victim, no pulse is detectable. Historically, cardiologists have referred to this state as electromechanical dissociation, where the normal atrial and ventricular depolarization progresses without any myocardial muscle contraction. Recent scientific evidence utilizing echocardiography and invasive pressure monitoring catheters have found that minute myofibril contractions do occur with depolarization; however, there is not enough effective projectile pressure generated to produce any external signs of perfusion, thus the term pseudoelectromechanical dissociation has been proposed (55,56).

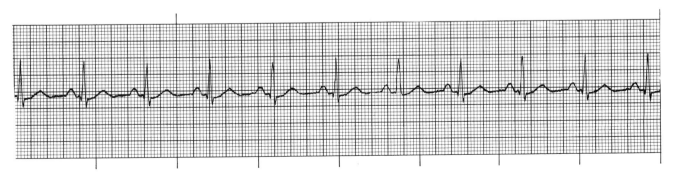

Figure 11. Pulseless electrical activity.

Other rhythms are also included in this group; ventricular escape rhythms, postdefibrillation rhythms, and bradyasystolic rhythms have different QRS morphology than the normal sinus rhythm; however, the same underlying problem of no pulse and no organ perfusion is the defining issue. The key to resuscitation of the victim in PEA is to quickly recognize and treat the underlying problem that has created this scenario. Since there is no perfusing rhythm it should be obvious that the "ABC" basics must be immediately made available to the victim while the differential diagnosis prioritized to the specific situation is reviewed.

There have been a number of mnemonic memory aids suggested to recall the most common conditions provoking pulseless electrical activity. Two common lists are the '5 Hs' and '5Ts' and the "MATCH-HHH-ED" (57,58).

Hypovolemia; Hypoxia; Hydrogen ion excess (acidosis).

Hyper or Hypokalemia (excess or loss of potassium ions).

Hypothermia; "Tablets" (referring to medications or drug overdose).

Tamponade (blood trapped around the heart chamber).

Tension pneumothorax (trapped air causing lung over-inflation and resultant loss of cardiac perfusion).

Thrombosis (secondary to myocardial infarction) and Thrombosis (secondary to pulmonary embolism).

M—Myocardial injury or infarction

A—Acidosis

T—Tension pneumothorax

C—Cardiac tamponade

H—Hypothermia

H—Hypo or Hyperkalemia

H—Hypoxia

H—Hypovolemia

E—Embolism

D—Drug overdose or chemical toxins.

The most common cause of PEA is hypovolemia: inadequate intravascular fluid volume. These fluid volume losses may be real, such as in acute hemorrhage or relative, wherein the capacitance vessels (veins) vasodilate, leaving inadequate vascular blood and fluids to compensate. A classic evolution to pulseless electrical activity would ensue from hemorrhage, producing normal electrical complexes, with progression to a tachycardia where the heart rate increases to compensate for pressure loss. Peak pressure for each cardiac ejection of blood (systolic pressure) would decrease, but increased constriction of the circulatory vessels would increase (diastolic pressure). As the blood pressure continues to drop to extremely low levels secondary to the hemorrhage, the sinus tachycardia continues. At this point, the astute clinician would consider this cause for PEA and immediately give the victim intravascular fluid followed by blood products to stabilize the victim. It is obvious that the definitive treatment in this case would be surgery; the goal of the resuscitative response is to maintain perfusion and oxygen delivery to the tissues. Time again is critical since without effective cardiac flow, perfusion of the myocardium is nonexistent and hypoxia develops quickly. Progression toward ventricular fibrillation, asystole, and death are the end result. An example of relative hypovolemia is septic shock. In this situation, a bacterial infection produces endotoxins, which causes the resistance vessels to dilate. There is normal circulating blood volume, but increased vessel capacitance resulting in a progressive blood pressure drop. Due to the underlying pathology, the vessels are unable to adequately constrict and the evolution to cardiac dysfunction secondary to poor or absent coronary perfusion occurs, eventually producing the terminal rhythms of ventricular fibrillation and asystole. Treatment in this scenario consists of increasing the fluid volume as well as to utilize drugs that act to constrict the blood vessels.

Each one of the other conditions mentioned above that can cause pulseless electrical activity also has an immediate treatment response, after the appropriate steps in basic life support have been met. Thus while the victim is being oxygenated and ventilated, and perfusion accomplished through chest compressions, the clinician is rapidly focusing upon the conditions that would be most causative and then initiating treatment. In hypoxia, treatment consists of supplemental oxygen and adequate mechanical ventilation of the lungs. In cardiac tamponade or tension pneumothorax, needle decompression of the pericardial sac or lung is the required emergency management. Electrolyte imbalances require immediate infusion or neutralization of the ion. A preexisting acidosis might require exogenous sodium bicarbonate. An acute myocardial infarction (heart attack) causing PEA would need supplemental oxygen, pain management, fluid resuscitation and immediate infusion of coronary artery thrombolytics. Pulseless electrical activity is always associated with an underlying clinical condition that, when identified and treated early, has the potential to be reversed and thus effective perfusion restored to the victim.

MECHANISMS FOR COMMUNITY CPR

The American Heart Association is a volunteer organization that has as its goal the reduction of mortality and morbidity secondary to cardiovascular and cerebrovascular disease. The AHA established a scientific mission in 1963 to evaluate and promulgate standards of CPR. In 1971, the Emergency Cardiac Care Committee was founded. Since the first conference on CPR in 1966 sponsored by the National Academy of Sciences and the National Research Council, guidelines for CPR have been based upon the available scientific evidence. Since this time there have been six conferences, the last in 2000 consisting of resuscitation experts worldwide. All aspects of CPR have been evaluated, including the actual access for emergency care in the community, education of the layperson, and delivery of basic CPR and ACLS. "Emergency Cardiovascular Care" is the term used to describe the organized response to life-threatening emergencies for the adult, pediatric, and neonatal victim. Other ECC provisions for care include educating the layperson to recognize the signs and symptoms of myocardial ischemia and infarction, stroke, activation of the EMS system (911) as well as early implementation of basic life support, defibrillation, and advanced cardiac life support with immediate victim transfer to the hospital. The goal of ECC is to save lives and this also includes educating the public regarding cardiovascular risk and how to maintain healthy lifestyles. The cornerstone of all emergency care is the layperson. Effective emergency care can only occur with public awareness of these events as well as prompt administration of CPR and use of the automatic external defibrillator. Once the EMS system has become involved with the victim, progression of care is dictated by a physician and ACLS protocol. The optimization of oxygenation and ventilation through supplemental oxygen and adjunctive respiratory devices, electrocardiographic monitoring with rhythm assessment and treatment, establishment of intravenous access for appropriate medications and post-resuscitation management are all elements of ACLS.

The American Heart Association "Chain of Survival" is a crucial concept to be understood in the context of CPR and is taught to the layperson in the basic lifesaving courses. There are four critical links in this chain that require specific actions on the part of the public. In the event of an emergency the first link involves activating the emergency medical services system. Obviously, the layperson must be able to recognize that a true emergency exists, and this underscores the efforts by the community organizations to educate the public regarding signs and symptoms of heart attack, stroke, and loss of the airway. The key is unresponsiveness; anyone found unconscious should initiate the first link by way of immediately involving EMS through calling 911. Once an emergency medical dispatcher is contacted, immediate aid in the form of paramedicals is sent to the scene. The dispatcher is also taught how to aid the layperson in how to provide basic CPR, which is the second link in the chain. Only until the EMS personnel arrive and begin managing care for the victim does the dispatcher terminate the call. In some communities, "enhanced 911" is available in which a computer will provide the dispatcher with the address from which the call is made in case communication is difficult or there is a premature telephone disconnection. This second link of CPR is the most critical phase in the Chain of Survival since the rescuer provides oxygen to the victim and, if needed, circulates this oxygen by chest compressions to the brain, heart, and vital organs. If bystander CPR is initiated within 4 min after a victim collapses, the odds of survival, after discharge from the hospital, are doubled (59).

The third link in the chain is early defibrillation, either by the public use of the automated external defibrillator or by the paramedic on the scene. Early access to defibrillation for the victim of cardiac arrest will significantly improve the chance for survival: each minute of delay for defibrillation for the victim in ventricular fibrillation decreases the chance of survival by 7–10%, and if defibrillation is provided within the first 5 min, chance of survival is 50% (60). Unfortunately, if >12 min of delay from collapse to initial resuscitation is encountered, the survival rate only ranges between 2 and 5%, and intact neurological function is compromised (61).

Recently, gambling casinos have implemented access for defibrillators and for victims who received a shock within 3 min had a 74% survival rate since a low response time of between 2 and 3 min was documented (62). Since time to defibrillation is so critical, automatic external defibrillators have been made much more accessible to the public. These devices can now be found in large gathering places such as stadiums, golf courses, airports and airplanes, shopping malls, large grocery stores, and other facilities where people in great numbers tend to congregate. So important is the early access to defibrillation that a great majority of states have enacted legislation to encourage use of these devices. The Cardiac Arrest Survival Act provides legal immunity for the layperson and the public business or corporate entity that uses or provides an automatic external defibrillator for resuscitation, which is essentially an expansion of the "Good Samaritan" type legislation. This immunity should encourage active participation by the public for involvement in victim resuscitation. Public access defibrillation has been described as the second most significant advance, compared to CPR, in the pre-hospital rescue scenario.

The final fourth link in the Chain of Survival is early ACLS by highly trained paramedical personnel. Emergency medical technicians expand (EMTs) incorporate basic CPR with interpreting cardiac dysrhythymias and, if required, defibrillation. Emergency medical technicians expand the immediate life-saving care by providing supplemental oxygen, intubation and control of the airway, gaining intravenous access and administering pharmacologic medications while in contact with a physican. This process occurs at the scene, and once the victim is stabilized, advanced cardiac life support continues through transport to the hospital emergency room. The most significant impact of early ACLS is to prevent the catastrophic progression of lack of oxygenation and cardiac arrest rather than to treat the terminal conditions inherent in this process.

The rescuer, whether a lay person of EMT, who begins CPR in the "field" must continue BLS (63) until one of the following events occurs:

1. The victim begins to show signs of spontaneous ventilation and perfusion.
2. Care is transferred to another qualified BLS responder, EMT, or ALS medical providers; or to a physician who makes the determination that resuscitation should be terminated.
3. The rescuer cannot continue resuscitation due to exhaustion or to hazards that may jeopardize the rescuer's life or the lives of others in the team.
4. An authentic no-CPR order is presented to the responders.

The determination to discontinue resuscitation depends on a stepwise evaluation of the efforts made during BLS and ACLS. A review of the process should ensure that each step in the resuscitation has been carried out in a flawless manner. Successful ventilation and intubation, intravenous access and the administration of appropriate medications as well as countershock should be achieved according to ACLS protocol. Electrocardiography evaluation should render a conclusion of no reversibility for the underlying agonal rhythm. Recently, the determination of end-tidal carbon dioxide has been advocated as a potential predictor of death (64). During resuscitation end-tidal carbon dioxide reflects the adequacy of cardiac output generated during chest compressions. This study suggested that after standard ACLS protocols had been followed for 20 min, a persistent end-tidal carbon dioxide level of 10 mmHg or less predicted nonsurvival in the victim with electrical activity, but without a pulse.

CARDIOVASCULAR DISEASE

Every year ~500,000 people are hospitalized for treatment of chest pain secondary to cardiac origin and 1.5 million victims will experienced a heart attack (65,66).

Some 500,000 people a year who have a myocardial infarction (heart attack) will die from this insult and ~225,000 of these deaths will occur within the first hour after symptoms and prior to reaching a hospital (67,68). In 17% of the victims, chest pain is the first and only symptom (67). It is again significant that time to intervention for the patient experiencing a myocardial infarction is crucial to survival; treatment must be undertaken within the first several hours after the symptoms occur (70,71). Early treatment underscores the necessity for rapid recognition of a cardiac event, followed by rapid CPR, defibrillation, ACLS and transport to the hospital.

The most common cause of a heart attack is ischemic atherosclerotic disease. The essential pathophysiology is the narrowing of the coronary artery lumens by deposits of fat-substrate such as cholesterol and lipids, which eventually retain calcium. The actual process of the accumulation of these plaques occurs very slowly, but has been demonstrated to begin at an early age. This same process affects the cerebral arteries as well and is the etiology for an ischemic stroke. As the coronary artery lumen narrows, a dynamic situation develops where blood flow and thus oxygen supply will not meet with increased demand for oxygen by the cardiac muscle fibers. Typically the coronary artery will have a circumference reduction of 70% for symptoms to occur. This condition of ischemia will produce a characteristic constellation of transient symptoms, referred to as angina pectoris. Chest pain is the most common sign of an acute cardiac ischemic event, occurring in 70–80% of the population (72). This pain appears to have several different components: transmission of dull, poorly localized pain occurs through the sympathetic visceral nerve fibers; a somatic pain generator produces the sharp and dermatomal aspects; and psychological input gives rise to the sense of impending doom (73,74). This cerebral input to the event may significantly exacerbate the ischemia since activation of sympathetic nervous system will increase heart rate and contractile force, further tipping the scale toward more energy consumption and thus oxygen demand.

Paradoxically, the majority of episodes of an acute coronary event (angina and or infarction) occur during periods of rest or mild to moderate exercise; profound physical exercise is associated with the minority of events (75). The victim will experience an intense, dull, crushing pressure sensation in the chest, most commonly behind the breastbone (retrosternally) and/or pain in the back, arms, shoulders, or mandible. Often nausea, vomiting, sweating, and shortness of breath (dyspnea) may accompany the pain. There appears to be a circadian rhythm regarding the occurrence of angina and the progression to infarction. Two daily peaks in incidence have been noted with the first pattern beginning from awakening to about noon, and the second peak occurring in the early evening (76,77). There are atypical presentations to unstable angina or myocardial infarction that will delay access for the victim. This subset of the population may have only vague, mild discomfort, which can be confused with a myriad of medical complaints. Diabetics, women, and the elderly all have a higher incidence of nonclassical presentations for cardiovascular ischemia (78,79). Diabetics are prone to neurological dysfunction and thus may have no sensation of the pain associated with angina. A retrospective review found that 30% of first heart attacks in men and 50% of first infarctions in women did not present with classical signs and symptoms and were clinically not recognized (80). When oxygen demand decreases, such as when the physical activity is discontinued, the decreased oxygen supply secondary to the narrowed lumen will be adequate and the symptoms will usually resolve. Progression of the disease, however, results in a much more severe mechanism for ischemia. The plaque is predisposed to rupture and when this occurs, activation of the coagulation system releases mediators that form a clot or thrombus over the plaque, which can further limit blood flow, or catastrophically, stop all blood flow completely. If the partial occlusion from the thrombus is severe enough, what is termed "unstable angina" develops. Though there have been many definitions of unstable angina, the main characteristic is that this type of angina occurs at rest and is progressive or prolonged in nature. Nocturnal angina, again with the victim at rest, would be classified as unstable angina. The heart is consuming the least amount of oxygen yet there is a lack of supply due to lumen reduction from the plaque. Clot enlargement provides the mechanism for dislodgement of particles or emboli, which then travel downstream, lodging in the microvasculature and these individuals are at a very high risk for progression to irreversible cardiac damage.

Complete occlusion of the coronary artery results in a myocardial infarction. Deprivation of oxygen results in the death of cardiac myofibrils and induces irritability in the cardiac conduction system, setting the stage for the initiation of lethal dysrhythmias such as ventricular fibrillation. Where and how severe the damage is to the heart depends on what coronary artery has the occlusion. If the left main coronary artery has an acute total obstruction, mortality is very high since no blood flow will occur through the two distal branches, the left anterior descending and circumflex arteries. Blood flow will be blocked to the entire left ventricle, the septum between the left and right ventricle, and the bundle branches. Even if the patient receives timely CPR, and clot lysis, a significant amount of heart may be destroyed resulting in scar tissue and a drastic reduction in blood flow, resulting in what is termed "congestive heart failure". If the thrombus were to lodge and occlude the right coronary artery, hypoxia would occur in the AV node, right ventricle, and in the majority of individuals, the posterior and inferior aspect of the left ventricle.

The continuum of acute cardiac injury, from unstable angina to myocardial infarction typically presents with characteristic electrocardiographic signs. Recall what the normal elements are to the ECG. During episodes of unstable angina where cardiac muscle demand for oxygen exceeds supply, ischemic changes to the ECG are seen as S–T segment depression, defined as $\geq$ to a 1 mm change from baseline on the standard graphpaper or changes in the T waves. When this S–T segment depression is downsloping, this is a sensitive sign of ischemia. The T waves may appear inverted or enlarged and symmetrical. When actual occlusion of the artery occurs, myocardial injury to tissues ensues and both muscle contraction and conduction are decreased from normal. The ECG change characteristic of injury is S–T segment elevation in contradistinction to changes of ischemia. When there is ≥ 1 mm above baseline for the S–T segment elevation, significant cardiac injury has occurred. This group of patients who exhibit S–T segment elevation on the ECG in two contiguous leads may be salvaged through reperfusion therapy. Restoration of blood flow and the course of injury is very dependent on early administration of thrombolytics or percutaneous transluminal coronary angioplasty (PTCA). The greatest improvement in mortality and morbidity occurs when reperfusion therapy is administered within the first 3 h after onset of symptoms (81,82). Conjoined with early CPR and defibrillaton, reperfusion therapy stands as one of the greatest advances in acute coronary syndromes. Current regimens include the fibrinolytics streptokinase and alteplase, as well as numerous other similar agents that act by inducing fibrinolysis through interactions with tissue plasminogen activator. The PTCA is a mechanical procedure whereby a catheter is guided through the coronary arteries into the area of stenosis, and then a balloon is inflated to expand the vessel. This procedure is restrictive in that only specialized centers have the capability to utilize this regimen, although it may be superior to thrombolytics.

Myocardial infarction defines the actual death of cardiac tissue. This is the end result of the process of ischemia with myocardial cell injury. The infarcted tissue area, again representative for the specific coronary artery occluded, will exhibit characteristics associated with loss of cellular life. Intracellular contents are released after loss of cell wall integrity. Some of these enzymes, such as creatine phosphokinase and the troponins, can be measured in the bloodstream to confirm infarction. The classic ECG changes for myocardial infarction are the presence of abnormal Q is waves. When a Q wave is or ≥ 1 mm in width and the height is >25% of the R wave height, the diagnosis can be made. An abnormal Q wave reveals the existence of dead cardiac tissue, but does not reveal anything about when the infarction happened. The assumption can be made for a recent infarction if the Q wave is associated with S–T segment changes and/or T wave changes. A non-Q wave infarction can also occur: There is myocardial cell wall dissolution with the release of cardiac enzymes, but only accompanied by S–T segment changes or T wave abnormalities. There is a lower mortality rate for non-Q wave heart attacks, but, unfortunately, an increased incidence of future reinfarction or

death (83,84). Fibrinolytics are contraindicated in patients with non-Q wave infarction since the clot occlusion may be paradoxically aggravated by release of thrombin, which further activates platelets (85).

Prehospital intervention for the victim suffering an acute cardiac event is based upon the "Chain of Survival". Once the signs and symptoms of a heart attack are recognized, early access to the emergency medical system is imperative. A common problem encountered is denial either from the victim or the rescuer, which impedes response time. Once the EMS personnel arrive at the scene a pertinent medical history is obtained and physical examination completed. A complete 12 lead ECG is obtained and then transmitted to the physician who is dictating care. Oxygen is the first line treatment for anyone complaining of chest pain. It should be recalled that supplemental oxygen substantially increases the oxygen tension in the blood and significantly improves tissue oxygenation. A critical blood flow restriction may be palliated by improving oxygen supply in this manner. The administration of the drug nitroglycerin is quickly administered for the victim symptomatic for angina in conjunction with oxygen. Nitroglycerin is delivered sublingually for rapid absorption into the bloodstream. Nitroglycerin is effective in relieving the symptoms of angina in several ways: relaxation of venous smooth muscle occurs due to binding of specific vascular receptors. As relaxation of the venous capacitance vessels occurs, venous return to the heart is decreased, thereby relieving ventricular wall tension, which ultimately decreases ventricular work and oxygen consumption. The nitrates also dilate the large coronary arteries as well as increasing blood flow through collateral vessels, which improves ischemic blood flow (86,87). Aspirin is the third drug that should be administered immediately by either the BLS provider or EMT when symptoms suggest a cardiac event (and the victim is not allergic to aspirin). A regular tablet of aspirin (325 mg), when ingested, will cause an immediate anticlotting mechanism by way of platelet inhibition. There is evidence that suggests aspirin decreases coronary artery reocclusion and future coronary symptoms, with reduction of death and furthermore, the effects of aspirin appear to be additive to fibrinolyis (88). The fourth drug that is administered during episodes of chest pain secondary to unstable angina or myocardial infarction is morphine. Although morphine is a narcotic analgesic, it produces beneficial hemodynamic effects in addition to profound pain relief. Morphine causes decreased vascular tone in the venous capacitance vessels, thus reducing myocardial wall tension, much like nitroglycerin. The mechanism, however, is different in that the action appears to be mediated through central nervous system reductions in sympathetic tone (89). A convenient mnemonic has been utilized, "MONA", for recall of these four immediate effective therapies for pre-hospital treatment of the acute coronary syndrome.

CEREBROVASCULAR DISEASE

There has been a concerted effort in the Emergency Cardiovascular Care system to improve pre-hospital recognition of the warning signs of stroke and provide rapid access to the Emergency Medical System. Public awareness of the issues regarding a "brain attack" has lagged relative to the exposure and education afforded cardiovascular disease. Stroke ranks third behind heart disease and cancer for morbity in the United States; 500,000 Americans a year suffer from a cerebrovascular accident and 125,000 of these victims will die (90). Until recently, stroke victims were only offered supportive and rehabilitative therapy for the complications experienced if they survived the initial insult. However, advances in fibrinolytic therapy, as in treatment of cardiovascular disease, dramatically improves outcome for the patient who has experienced an ischemic stroke (91). Fibrinolytic treatment reduces stroke disability and significantly improves quality of life after hospital discharge (92,93). The caveat is that the cerebrovascular accident must be recognized and treatment initiated in a timely manner; fibrinolytics need to be provided within 3 h after the onset of an ischemic stroke (94). Thus there is a narrow window of opportunity to limit cerebral damage, which underscores how important the role is for the public in providing immediate access to the EMS system for the victim.

The underlying cause for an ischemic stroke is comparable to the etiology for myocardial infarction. There is a disruption to cerebral blood flow due to the presence of an occlusive clot. The oxygen supply to the particular area of the brain supplied by the blocked artery does not meet the tissue demand and the same process of ischemia, injury, and cell death will occur. The thrombus, which occludes the vessel, is the end result of atherosclerotic changes to the artery. However, due to the unique anatomical positions of the cerebral arterial system, a blood clot formed elsewhere in the body can embolize to disrupt blood flow to the brain. Approximately 75–85% of all strokes are of this type, and defined as ischemic, and furthermore can be classified as to the arterial system that is affected. The two major arterial conduits to the brain are the carotid arteries and vertebrobasilar arteries, which affect the cerebral hemispheres or brain stem–cerebellum, respectively. Typically, a person who is at risk will develop what is termed "a transient ischemic attack" (TIA) prior to a full-blown stroke. Essentially, a TIA is a reversible mini-stroke that may affect specific brain function or eyesight and will last anywhere from minutes to hours (95). The TIA is a harbinger of a future "brain attack" much like unstable angina will forecast a heart attack. About 5% of those persons presenting with a TIA will end up with a stroke in 1 month; the risk will increase to ~12% after 1 year and an extra 5%/year thereafter (96). Fortunately, the symptoms from a TIA will bring the patient into the medical system for evaluation whereby treatment regimens clearly reduce risk for ischemic stroke. The surgical procedure of carotid endarterectomy in which the carotid artery plaque is removed has been proven very beneficial for patients that have had a recent TIA and a >70% stenosis of the carotid artery (97). In those individuals who are not operable candidates, aspirin and the specific platelet inhibitor types of drugs have been shown to be successful in preventing subsequent stroke in patients presenting with TIA (98).

The minority of acute strokes are due to hemorrhage of cerebral artery. The bleeding may occur in the subarach-noid space, which is in the superficial exterior aspect of the brain, or in the brain tissue itself, defined as an intracerebral hemorrhage. The common etiology to a subarachnoid hemorrhage is an aneurysm where the arterial wall weakens, and eventually a disruption occurs (99). In the case of a hemorrhage into the brain tissue itself, high blood pressure appears to be the major causative factor (100). While there are similar signs and symptoms in both types of stroke, there are also distinct differences in findings, which aids in the diagnosis. In general, the presentation for a subarachnoid hemorrhagic stroke is more severe with a very common complaint of an extremely painful headache, which tends to be global, and may have radiation of pain into the face or neck. This headache is often accompanied by mental status changes, nausea, vomiting, photophobia, or cardiac dysrhythmias. In a minority of patients, a prodromal episode of these symptoms may be caused by leakage of the aneurysm offering a warning sign (101). While the victim suffering from an intracerebral hemorrhage may also present with a severe headache, these patients tend to have a greater neurological insult with significantly depressed mental status function. The signs and symptoms of an ischemic versus a hemorrhagic stroke overlap and diagnosis may be difficult based upon the medical history and physical findings. Since radiological imaging offers the greatest aid in differentiating these two types of cerebrovascular accidents, time is very critical to clinch the diagnosis to offer the appropriate treatment. Fibrinolytics would obviously be a catastrophic therapy in the mistaken treatment of what appears to be an ischemic stroke when the etiology is a ruptured blood vessel that requires surgery.

The American Heart Association "Chain of Survival" that has been implemented and associated with cerebrovascular disease has been applied to the pre-hospital care of the stroke victim. Early recognition of a stroke and activation of the EMS system are paramount in initial therapy, which is often problematic, since, unlike a heart attack, stroke may be difficult to detect. While early defibrillation is not ordinarily indicated for the stroke victim, the possibility always exists that coincidental lethal dysrhythmias may be present during the initial presentation (102). The last link in the chain is early hospital care. The common theme regarding out-of-hospital management for cardiac or stroke victims is rapid entry for the victim into advanced life saving. The '7-D' mnemonic has been recommended as an aid for care in the stroke patient: Detection; Dispatch; Delivery: Door; Data; Decision; Drug (103). Early detection, with an accurate recall of the initial signs of a stroke are critical to care and must be accomplished by the immediate family member or layperson, with immediate access to "Dispatch", the EMS personnel. An important point to emphasize is that the majority of strokes occur at home (104). Paramedics who arrive at the scene must confirm a rapid, tentative diagnosis through focused medical history and physical examination, and then "Deliver" the patient rapidly to the hospital. Once the patient is through the Emergency Department "Door" the medical history and physical examination are further refined along with radiography (computerized tomography). A "Decision" is made regarding whether fibrinolytic therapy is indicated for an ischemic stroke and the "Drug" treatment is initiated. The drug therapy must be initiated within 3 h after the onset of an ischemic stroke.

The changes in mental status and/or sensorimotor function in a cerebrovascular accident may range from minor, almost unrecognizable changes, to loss of consciousness and seizures. A person may exhibit grades of confusion, with a progression to stupor or coma where the airway is obtunded and basic life support is required. Comprehension of language often occurs with inappropriate responses to simple questions. Physical manifestations are often present unilaterally. Paralysis in either the face, upper, or lower extremity may range from slight weakness to frank inability to exercise any muscular control. Since the face is always exposed, muscle weakness is exhibited by loss of tone and sagging of the muscles in facial expression. Difficulty writing (aphasia) or speaking (dysarthria) occurs due to loss of appropriate motor input from the brain. Loss of sensation is another common sign of a stroke (or a TIA). Visual disturbances, including blindness are much more obvious and usually involve only one eye. If the location of the ischemia is in the vertebrobasilar arterial system, centers of the brain controlling coordination are involved and signs such as gait disturbances (gait ataxia) are common. The dilemma of pre-hospital rapid neurological assessment to evaluate the possible stroke victim when the presentation is varied has been improved by several instruments. The Cincinnati Pre-hospital Stroke Scale (104) is very effective in identifying the stroke victim. Three physical findings are assessed: facial droop; arm drift; and speech. Abnormal features in any one category is very predictive for cerebrovascular accident. The Los Angeles Pre-hospital Stroke Screen also is extremely useful for assessment. Six criteria are first evaluated in the medical history: (1) age > 45 years, (2) absent history of seizures or epilepsy, (3) no history of motor loss, (5) serum glucose not < 60 g/dL nor > 400 g/dL, and (6) asymmetry in any one of the three categories of facial musculature, grip strength, and arm strength. If all criteria are positive, there is a 97% chance of an acute stroke (105). These tests have streamlined the response time and have allowed the hospital emergency room to prepare for rapid definitive diagnosis.

THE "ABCs" OF ADULT CPR

The evolution of the current basics of life-support has resulted in a streamlined set of actions that has standardized the initial care for the victim, whether accomplished by a public bystander or emergency room physician. Assessment of the victim always precedes a physical maneuver on the part of the rescuer; constant appraisal of the effectiveness of CPR and the response of the victim is a core principle in American Heart Association Basic Life Support tactics. The initial steps of resuscitation are never bypassed; for example, if the airway is not established, the single rescuer would never start chest compressions, or begin intravenous access. The stepwise process in the algorithm ensures that an orderly process occurs in a situation where chaos and a high degree of emotional turmoil exist for the rescuer. Since there are some basic differences in how resuscitation is administered to the adult versus a child, anyone 8 years or older is considered an adult.

When one encounters a potential victim, the first assessment is to determine unresponsiveness. "Shake and shout" has been a common first action to determine that the victim is really unconscious (there no doubt has been a number of resuscitations initiated upon someone who was sleeping, assuming unconsciousness)! Once there is no doubt that a true emergency exists, the rescuer sends another member of the group to activate the EMS system by phoning 911 and to obtain an AED. If the rescuer is alone, he or she must leave the victim momentarily to call 911 and get the AED; these automated defibrillators have a standardized placement near a telephone. After accessing the EMS and obtaining a defibrillator, the rescuer places the victim in the supine position, and kneels at the head (the left side is suggested when utilizing an AED). The "ABCs" of CPR now are initiated.

A = Airway

The unconscious person has a generalized relaxation of all muscles and in the throat this causes the tongue to move in a posterior direction, occluding the airway. Since the tongue is attached to the mandible, manipulating the jaw and head will retract anteriorly. Two methods are utilized to open the airway; the "head-tilt and chin-lift" or the "jaw-thrust" maneuvers. Tilting the head backward by placing one hand on the forehead and raising the chin with the two fingers of the other hand is the most commonly used technique. An important feature of this technique is to make sure that the fingers are placed on the inferior surface of the mandibular bone and not the soft tissue under the tongue, as the later placement will worsen airway compromise. In a situation where a neck injury is suspected with possible spinal cord compromise, extension of the head is contraindicated; the jaw-thrust is utilized to open the airway. The head is held in the neutral position while applying forward pressure with both hands at the angle of the jaw, just below the ears. In this way, there is no change in head position. At this point inspection of the mouth is important to remove any secretions, vomitus or foreign bodies that may be an impediment to air exchange. Once the airway is opened, the rescuer "looks, listens, and feels" for breathing by placing his or her cheek and ear close to the victim's mouth. The chest is examined for movement while feeling and listening for air passage. In the case of a partial obstruction of the airway, the victim will tend to make high-pitched "crowing" noises that may be accompanied by cyanosis of the skin (due to unoxygenated hemoglobin). Instead of the chest expanding with an inspiration, retraction of thorax or lung compartment will occur. It is imperative that the airway be opened and maintained in this situation since ineffective ventilations will invariably lead to hypoxia.

B = Breathing

Once it is ascertained that the victim is not breathing (this should occur within ~10 s), the rescuer places his mouth around the victim's mouth and pinches the nose shut with one hand while maintaining chin-lift with the other hand. Two long, extended breaths are given, each ~ 2 s, with the goal of providing ventilation to the lungs while minimizing the egress of air into the stomach. Since during unconsciousness there is a relaxation of all muscles, the lower esophageal sphincter will relax and thus any air that enters the stomach may force gastric contents into the esophagus and then into the trachea. Aspiration of these highly acidic stomach contents into the lungs may occur. The complex interplay between rescuer, positive pressure ventilation, peak airway pressure, tidal volume, and inspiratory flow rate has had a considerable degree of scientific evaluation (106–109). The consensus supports a tidal volume of between 800 and 1000 mL to maintain adequate oxygenation when only room air is provided in the rescue breathing. This volume is slightly less than the 1992 ECC Guidelines of a rescue tidal volume of 800–1200 mL. A slow prolonged breath over 2 s decreases peak positive pressure and thus entry of air into the stomach while providing the optimum tidal volume. When supplemental oxygen is available, evidence has confirmed that a tidal volume of 500 mL provides effective oxygenation and ventilation in the unintubated patient as long as the inspired oxygen fraction is >40% (110,111). Once rescue breathing is commenced, one should assess effectiveness by noting whether the chest rises with each breath. If there is no change or the rescuer observes that significant effort is required to minimally expand the chest, the airway step has not been optimized and the rescuer has to reopen the airway with additional head extension and chin-lift (or jaw-thrust if a head or neck injury is suspected). If readjustment of the airway does not provide the ability to ventilate, a foreign body lodged in the airway should be suspected and the rescuer should proceed through the algorithm specific for dealing with this issue. Victims with dentures may prove difficult to ventilate; generally dentures should be left in place since it is easier for the rescuer to form a seal around the mouth. However, loose dentures may be aspirated and should be removed if their retention is inadequate. In the case where the rescuer cannot maintain an adequate seal, or if the mouth is unavailable for airway exchange secondary to trauma, mouth-to-nose breathing should be attempted (112). A deep breath should be inhaled by the rescuer prior to respiratory exchange since this maneuver optimizes the maximum amount of oxygen made available for each tidal volume (113). Should the victim only require oxygenation and ventilation, rescue breathing provides one breath every 5 s or 12 breaths/min (114).

There is always the concern regarding exposure to an infectious organism when performing rescue breathing. At this point in time, there has not been any evidence documenting the transmission of human immunedeficiency virus (HIV), hepatitis, or tuberculosis when mouth-to-mouth resuscitation has been instituted in an emergency (115). However, reluctance upon the part of any lay rescue person to perform this action is understandable and there is no moral or legal duty to do so. Barrier devices have been developed that prevent intimate contact with the victim and there are two basic types: face shields and masks. This adjunctive equipment has been made available in the healthcare environments due to the requirements of the Occupational Health and Safety Administration. The face shield has a flexible plastic covering with a one way circular valve that, when placed over the victim, separates the rescuer from contact and from exhaled gases. Mouth-to-mask rescue ventilation provides a better seal and further distance from the victim's mouth than the face shield, which is advantageous should vomiting occur. Some of these masks have a port where supplemental oxygen can be provided and entrained with the rescue breathing. A flow rate of 10 L/min through one of these masks will increase the inspired concentration of oxygen to at least 40% (116). When supplemental oxygen is supplied in this manner, smaller tidal volumes, on the order of 400–600 mL, will maintain oxygenation while decreasing the risk of gastric insufflation (117).

C = Circulation

After delivery of two rescue breaths, the next step in basic CPR is to assess for signs of circulation. For many years the layperson was taught to feel for the presence or absence of a carotid pulse. Research in the 1990s found numerous pitfalls with the pulse check that appeared to have a negative impact on survival and, since 2000, this task is not taught to the lay responder anymore. Significant time delays in trying to determine if a pulse was present delayed time to defibrillation and thus survival (118). The accuracy of the pulse test revealed a sensitivity of only 55% and a specificity of 90% and overall the accuracy was 65% (119). At this time, the lay rescuer is instructed to look for signs of perfusion, such as movement, breathing, or coughing and if unsure, to begin chest compressions.

Correct positioning of the hands and compression skills are easily learned by the layperson. A simplified method for hand placement has been taught for several years and consists of placing the heel of one hand over the center of the breastbone (sternum) between the nipples and then interlocking the fingers of the remaining hand over the first, so that pressure will be transmitted through the heels of both hands. Effective compressions are generated by positioning the rescuer's shoulders over the hands and sternum and depressing the sternum from 1.5 to 2 in. Release after compression must be complete without taking the hands off the chest to prevent "bouncing". Chest compressions should be similar in action to that of a piston in a reciprocating engine with half of the cycle spent in compression and the other half spent in relaxation. The effectiveness of this ratio has been documented with regard to both cerebral and coronary perfusion pressures (120). The recommended rate for chest compressions is 100/min, which has been substantiated by numerous studies (121,122).

The single rescuer initiates chest compressions after providing the victim with two rescue breaths and assessing for signs of effective cardiac blood flow. A ratio of 15 compressions followed by 2 rescue breaths continues for four cycles and then the victim is reassessed for spontaneous circulation. Chest compressions should be resumed within 10 s after noting no signs of perfusion and the 15:2 cycle continued with an interruption for assessment of vital signs in several minutes, followed by the same ratio.

When additional responders are present during resuscitation of a cardiac arrest victim, immediate activation of the EMS system must be accomplished and a defibrillator brought to the scene, if these actions have not already been completed by the lone rescuer. The second rescuer should then assess the adequacy of ventilations and chest compressions and reassess for signs of a pulse and breathing within 10 s while CPR is halted. Though it is not expected that the layperson be able to engage in two-person resuscitation, the process is included here for completeness. Medical professionals as well as the paramedical caregivers should all be able to demonstrate this skill. The compressor is positioned in the normal manner, at the side of the victim. The second rescuer is stationed at the victim's head, maintaining the airway, monitoring for effective compression by carotid artery pulse check, and providing rescue breaths. Previous scientific guidelines utilized a compression:ventilation ratio of 5:2 (123), which has now been changed in light of recent scientific evidence. Currently, a ratio of 15 compressions to 2 ventilations is recommended for both one and two rescuer CPR (124–126) since it appears that improved survival occurs as a result of the higher rate in spite of a decreased number of ventilations. The effectiveness of chest compressions relative to coronary perfusion pressure (the difference between aortic diastolic pressure and the left ventricular end-diastolic pressure) suggests that, as the number of compressions increases, so does the perfusion pressure; therefore, 15 chest compressions improves and sustains blood pressure more effectively than the previous recommendation of 5 compressions to 2 ventilations. The pauses with the previously recommended 5:2 compression:ventilation scheme had more drops in cerebral and coronary perfusion and therefore decreased oxygen delivery compared to the new scheme. Therefore, whether a one- or a two-rescuer resuscitation occurs, the preferred compression/ventilation ratio is 15:2. When advanced life support is initiated and the patient is intubated (a breathing tube place through the mouth and into the trachea) there is no pausing for ventilations; chest compressions continue at 100/min and ventilations are provided at a rate of 12 times a minute (127).

Despite the fact that there has never been evidence to suggest that transmission of disease occurs through mouth-to-mouth exchange of air or secretions, studies have demonstrated a lack of enthusiasm upon the part of both the layperson and professional rescuers to perform this maneuver on strangers (128,129). Current guidelines, as of 2001, now indicate that if the rescuer is unable to perform mouth-to-mouth ventilations, chest compressions should be started for the victim (130). The Cerebral Resuscitation Group of Belgium concluded that there was no difference in outcome for the victim if chest compressions were or were not accompanied by mouth-to-mouth rescue breathing (124). Since any resuscitation attempt utilizing chest compressions without ventilation may provide a better outcome for the victim than no action at all, education regarding this tactic in resuscitation has been made available to the lay responder. While it appears contrary to basic physiological principles that resuscitation could be successful without providing oxygen to the blood, evidence suggests that agonal breathing mechanisms are able to maintain adequate PaO_2 and $PaCO_2$ during CPR without rescue breathing (131). The etiology for this paradox appears to be due to the decreased perfusion from chest compressions; since the cardiac output is only one-fourth that of normal, ventilation perfusion mismatch does not occur due to low rates of blood flow through the lungs. In essence there is a decreased requirement for oxygen; any excess ventilation is wasted due to this decreased perfusion and the lack of oxygen transport by the available red blood cells (132,133). This form of CPR is only recommended for the public rescuer since paramedical personnel should always have adjunctive airway devices available for resuscitation.

AIRWAY OBSTRUCTION

The tongue is the most common cause of airway obstruction and basic life support addresses this issue with various maneuvers to open the airway thus allowing either spontaneous respirations to resume or mouth-to-mouth ventilations to be initiated for the victim. Foreign body airway obstruction is the cause of ~3000 deaths a year (134).

In perspective, there are 198 deaths per 100,000 persons for coronary artery disease, 16.5 deaths per 100,000 individuals for motor vehicle accidents, and 1.2 deaths per 100,000 due to foreign body obstruction (135). The "cafe coronary" (where choking was mistaken for an acute coronary event) appears to be the most common cause of choking in adults since this emergency usually happens during eating and meat seems to be the culprit for most occurrences (136). A foreign body lodged in the airway can either completely occlude or partially occlude any segment of the respiratory passages. The key to distinguishing these two scenarios is that the victim is able to continue to breath, albeit with difficulty, during a partial obstruction, and therefore the rescuer should not attempt any rescue attempt that potentially could convert a partial to a complete obstruction. As in all basic life support, it is crucial to activate the emergency medical system to get assistance. When a victim begins to make high pitched "crowing" sounds, cannot speak, or becomes cyanotic, hypoxia quickly ensues and this person needs immediate aid. The public is taught the universal chocking sign where the neck is clutched with both hands. The first question to ask the choking victim if, in fact, he or she is unable to breathe and if they can speak. The next immediate step is the Heimlich maneuver (137), which should be attempted in anyone between the ages of one and adulthood. This action is not indicated in infants <1 year old (138). Forceful external elevation of the diaphragm utilizes the remaining volume of air in the lungs to expel a foreign body. Placement of the hands is very important to minimize injury to the internal organs; when the victim is standing or sitting the rescuer wraps both arms around the victim's body and clenches the hand to make a fist. This fist, with the thumb compressed against the abdominal skin is placed above the umbilicus and below the xiphoid process (the distal end of the breastbone). A rapid thrust is made in a superior–posterior direction and continued until either the foreign object is displaced or the person becomes unconscious.

Once unconscious the EMS system must be activated by calling 911 and CPR is initiated for the victim. In this circumstance, after the airway is opened, mouth-to-mouth ventilation may be possible due to the muscle relaxation that occurs with unconsciousness, converting a complete obstruction to a partial obstruction, thus allowing rescue breaths to provide oxygenation to the blood. When opening the airway and during subsequent ventilations, the only change in basic CPR is to open the mouth and look for the presence of the offending obstruction, and if visible, grasp the object with the fingers and remove it (139). Blind finger sweeps are prohibited since injury can easily occur to the soft tissues of the mouth and throat. The initiation of chest compressions may also create enough intra-thoracic pressure to expel the foreign object (140,141). The recommended maneuver for the obese or pregnant patient is to utilize chest thrust since the increased abdominal girth in either type of victim makes the Heimlich procedure difficult in terms of finding land-marks and avoiding injury, especially to the fetus. The victim is grasped from behind and the arms encircle the chest, just under the armpits, with the fist placed upon the breastbone. Again the thumb of the first hand is placed next to the skin overlying the breastbone and the second hand is then placed over the first, and with the hands interlocked, rapid compressions are performed in a pos-terior manner.

Once the foreign body is expelled from the unconscious victim, the basic "ABCs" are initiated as in any life-support situation. "Look, listen, and feel" for signs of breathing and if no excursion of air is present, the two rescue breaths are immediately provided for the victim. The next action in the sequence is to observe signs of adequate circulation, and if none are present, chest compressions are initiated and an AED is attached to the patient (142).

THE AUTOMATIC EXTERNAL DEFIBRILLATOR

The key issue in cardiopulmonary resuscitation is rapid initiation of the "Chain of Survival" and the length of time between victim collapse and defibrillation. Public access defibrillation (PAD) has the capability of significantly improving survival rates from cardiac arrest, in some cases to almost 50% (143). Since ventricular fibrillation has the highest frequency of occurrence in cardiac arrest and can only be terminated by countershock, it is clear that early defibrillation is will dramatically improve survival rates. The ECC guidelines of 2000 have a goal defibrillation within five minutes of the EMS activation (911 call).

The first automated cardiac resuscitator was described by Diack et al. in 1979 (144). This 19 lb battery-powered and most importantly, portable device sensed respiratory rate and the ECG. Ventricular fibrillation and asystole could be diagnosed. An oropharyngeal airway utilized a transducer to detect respiratory pressure as well as an electrode, which was applied to the base of the tongue. An electrode was placed over the chest wall (xiphoid area), which completed the electrical circuit for defibrillation. After initial trials in animals, the device was used at St. Vincent Hospital, Portland, Oregon on a 49-year-old male

who arrived to the ED in ventricular fibrillation. The patient had failed to respond to CPR, chest defibrillation and medications. The patient was actually seizing from massive doses of lidocaine injected to decrease the auto-maticity of cardiac condition. The automatic resuscitator converted this lethal rhythm with one 335 J shock via the tongue-epigastric pathway.

There are actually no fully automatic external defibril-lators available to the public; this is a misnomer since some actions are required upon the rescuer for the device to work. Once the AED is attached to the patient by adhesive pads and turned on, electronic evaluation occurs of the victim's underlying rhythm, and if a lethal dysrhythmia is evident, the AED will advise the rescuer to activate the "shock" button. These devices first record and then inter-pret the ECG. Narrow range bandwidth amplifiers first filter out various artifacts such as powerline transmission, high or low frequency radio transmissions, and extraneous "noise" that occurs from poor connections. Successive seg-ments of the filtered ECG are than analyzed through a mathematical algorithm with each segment further tested in a sequential matter. The specifics of these algorithms are not made public due to the competitive nature of the various companies manufacturing these devices, but, essentially the rate, amplitude, waveform, frequency, and baseline variability of the ECG are analyzed and mathe-matical integration results to identify the rhythm that would or would not be treatable by defibrillation. The accuracy of these devices is extremely high after extensive testing both *in vitro* and *in vivo* to an early skeptical audience of medical personnel. The rare errors encoun-tered with AED function appears to be related to move-ment of the patient, such as when the patient is being ventilated or repositioned during the analysis mode. Agonal respirations by the victim also create movement artifact that will interfere with rhythm analysis. The errors have generally been of omission in that dysrhy-thmias that would benefit from defibrillation have not been recognized, such as very coarse or fine ventricular fibrillation (145).

Since the first biphasic waveform was used in an AED in 1996 there has been a progression toward utilizing this new technology in future defibrillators. The current mono-phasic waveform devices deliver current that is unipolar and either damped sinusoidal or truncated relative to the rate at which the pulsed current decreases to zero. The biphasic defibrillators utilize a biphasic wave form of which each are exactly opposite in polarities. Optimum defibrilla-tion is a balance between producing enough current to terminate VF without extensive damage to the heart. When a monophasic defibrillator is used during CPR, 200 J of energy is recommended for the first shock, with two succeeding shocks of between 200 and 300 J suggested in the AHA VF algorithm. These increases in energy with each shock have been demonstrated to optimize defibril-lation success while minimizing tissue damage (146,147). Numerous studies have confirmed that biphasic wave-form energy utilizing shocks as low as 115–130 J were as effective in terminating VF as the monophasic 200 J shock (148,149). The current scientific evidence supports conclusions that low-energy biphasic waveforms have at least the equivalent effectiveness as the monophasic waveforms in defibrillating VF; it appears that biphasic waveform defibrillation will be the standard of care in the future.

Operation of the various AEDs is very straightforward and, although there are different models, the devices have very similar controls and functions. Once the AED is brought to the scene of a victim that is not breathing and has no effective pulse, it should be positioned at the victim's left side for easier electrode placement and to allow the "ABCs" of CPR to continue without interruption on the victim's right side. The machine is turned on, and then a series of prompts by an electronic voice guide the rescuer through the remaining steps. The electrodes are next placed on the skin, with the right pad positioned just under the right collar bone (clavicle) and the left pad placed lateral to the left nipple. The third step requires that everyone involved in the resuscitation desist from any contact with the victim while the AED analyzes the rhythm. This is extremely important in that any movement generated by the rescuers will induce an artifact error. Depending on the particular device, some AEDs will automatically analyze while some machines require manual selection by the operator. If VF or VT is present, an electronic voice will indicate that a shock should be delivered; everyone should clear the victim and there should be no one in contact with the body. Once the operator is sure that everyone is clear of the victim, the "shock" button is depressed and the victim's body will exhibit the generalized musculature contracture observed with a defibrillating current. Cardiopulmonary resuscitation is not resumed since the AED must again be ready to deliver another shock after analyzing the postshock rhythm. In this way, three shocks are successively delivered before resumption of "airway, breathing, and chest compressions". If the "no shock indicated" message is transmitted to the rescuers, either there is now a pulsatile, perfusing rhythm, or there is a lethal dysrhythmia that would not improve survival with a defibrillating shock (such as asystole). If signs of circulation become present, then rescue breathing should continue, unless of course the victim also has a recurrence of adequate ventilations (150).

There are some specific circumstances that must be considered with the use of an AED. Children that are < 8 years of age or victims weighing < 25 kg may not benefit from the current AED since the energy delivered in the monophasic devices are in excess of the recommended 2:4 J/kg. Though it appears that VF can be detected accurately in children, there is insufficient evidence that the algorithms devised for the adult cardiac arrest patient will always determine an arrest rhythm that will benefit from defibrillation, particularly with regard to pediatric tachy-arrhythmias (151,152). However, in the pediatric arrest patient the potential benefit greatly outweighs the risk of AED–individual injury, and therefore this device should be readily available for early application. Another special situation occurs when the victim has been in the water. Even when the victim has been removed from water, there is the risk of induced shock to the rescuer from the moisture present on the victim's body. The other concern, which has the most

potential for occurrence, is that the shock would not be conducted along the appropriate pathway from the electrodes, instead bypassing the heart by the water present upon the skin. The third issue regarding AEDs is that many older victims may have a pulse generator present for cardiac rate control or for internal defibrillation. These devices are usually implanted in the superficial tissues of the chest and appear as an elevated, hard lump, which may or may not be easily observed, depending on the obesity of the patient. What can be observed is the scar from implantation; avoidance of pad placement over the implanted device would mitigate any interference for the AED to detect and provide a defibrillating shock. The last situation that may be encountered is the medication patches, which many patients utilize for a variety of conditions. Pad placement over these patches could significantly decrease the energy delivered to the heart; the recommended procedure is to remove the medication patch and wipe the medication away from the skin prior to pad placement (153).

The AED has the potential to be a strategic intervention to increase out-of-hospital survival in the cardiac arrest patient. Public access defibrillation will continue to be more available due to the efficacy already demonstrated in the previously mentioned studies as well as progressive legislation. The continued technology to produce smaller devices at less cost will undoubtedly increase access both in public places and into the homes of those individuals at risk for sudden cardiac death.

PEDIATRIC AND INFANT CARDIOPULMONARY RESUSCITATION

There are significant differences in the etiology for cardiac arrest in the adult compared to children. In adults, a terminal dysrhythmia is usually caused by progression of coronary artery disease where an acute hypoxic event produces hypoxia, injury, and then disruption of organized depolarization of myofibrils. Children are entirely different in that a hypoxic pulmonary event is the usual cause of the progression of a normal sinus rhythm to an extremely slow heart rate (bradycardia) and then to asystole. The majority of these arrests occur from foreign body obstruction, drowning, trauma, or Sudden Infant Death Syndrome (SIDS), poisoning, asthma, or pneumonia (154). This is where the American Heart Association CPR algorithm for the child is different than that of the adult. Since respiratory failure is such a significant etiology for cardiac arrest, the rescuer is instructed to provide CPR first then "phone fast" rather than the adult rescue scenario where activation of the EMS system (phone first) occurs prior to CPR. The remaining steps in the "ABCs" of life support are similar to the adult. Once unresponsiveness is determined for the victim and the rescuer shouts for help, the child should be placed supine, being careful to move the entire body as one if a head or neck injury is suspected. The airway is opened with the "head-tilt, chin-lift" as in the adult, or the "jaw-thrust maneuver" if there is suspicion for head or neck trauma. After opening the airway and inspecting the mouth for a foreign body, "look, listen, and feel" for the passage of air by the victim within the time frame of no more than ten seconds. If the assessment reveals lack of adequate ventilation, two slow rescue breaths with an adequate pause between them to allow for exhalation are delivered to the victim; the endpoint for tidal volume is a visible rise in the thoracic chest wall. As in the adult, if no air enters the lungs, the most common cause is an inadequate airway lumen secondary to either tongue obstruction or a foreign body. Repositioning of the head should be immediately attempted, and if the rescuer can still not ventilate, entry into the foreign body obstruction algorithm should be the next action undertaken. Should the rescuer provide effective ventilation for the two initial breaths and signs of circulation exist, a rate of 20 breaths/ min is recommended for an infant ($<$ 1 year of age) and for the child (1–8 years of age.) If no movement, spontaneous breathing, or other signs of circulation exist, chest compressions should be started within 10 s. The depth for compressions in an infant or child should be about one-third to one-half of the total distance from the child's anterior chest wall to the back, and at a rate of 100 times/min. In a child (1–8 years of age), the use of only one hand is recommended, with placement between the nipple line and the bottom of the breastbone (sternum.) As in the adult, the heel of the hand is used. A ratio of five compressions to one ventilation is recommended, with reassessment of the child after 20 cycles of compressions and ventilations have occurred. Chest compressions in an infant ($<$ 1 year of age) are similar except that chest compressions are accomplished with two fingers to a depth of one-third to one-half the distance from the anterior chest to the back. The location for compression is one finger-width below an imagined line between the nipples. The rate is 100 times/min and the compression to ventilation ratio is 5:1, as in the child. Reassessment for signs of circulation and spontaneous ventilations should occur after $\sim$ 1 min or 20 cycles of compressions and ventilations (155).

Foreign body obstruction is treated in a similar manner to the adult in a child between 1–8 years of age when assessment by the responder indicates a severe or complete airway occlusion. After immediate activation of the EMS system, the Heimlich maneuver is instituted with the same adult landmarks where the rescuer's fists are place above the umbilicus, being careful to stay away from the inferior aspect of the breastbone, where the xyphoid process is located. Should the child become unresponsive from respiratory arrest, CPR should be initiated, with the initial airway assessment focused upon looking for the foreign object, and, if visible, removing it. Two rescue breaths should be initiated and then chest compressions begun. As in the adult, chest compressions may cause the foreign object to be dislodged where it can be extricated. The infant (under the age of 1 year) is treated quite differently when foreign body aspiration is suspected, since concerns have been raised regarding intraabdominal injury and the Heimlich maneuver (156,157). The infant is cradled with one hand and arm, turned prone, and then five back blows are delivered between the shoulder blades (scapulae) using the heel of the hand. Immediately the victim is turned supine and cradled with the opposite hand and arm. Up to

five chest-thrusts are delivered in the same location for chest compressions. This scheme is alternated until either the obstruction is dislodged or the infant becomes unconscious. Once unresponsive, the next action, as in the adult and child, is to look in the airway and determine if the obstruction can be removed. Cardiopulmonary resuscitation is then initiated and activation of the EMS system undertaken (158).

CONCLUSION

Cardiopulmonary resuscitation has made gigantic strides since the scientific community integrated the rudiments of airway maintenance, rescue breathing and external chest compressions in the 1960s. Once these procedures were developed, the foresight by healthcare professionals to implement CPR in the community and educate the public in the early 1970s has contributed greatly to the success of this program. Currently, the strategy is to simplify the training for Basic Life Support as well as to increase the number of lay responders. Since 1973 $>$ 40 million people have learned the basic life-saving skills taught in CPR classes (156). The recognition that emergency cardiac care should not only provide an organized structure for resuscitation, but also to incorporate education in the prevention of risk factors for coronary artery disease, stroke, and pediatric mortality has the capability of dramatically reducing future morbidity and death. For example, $\sim$ 30% of the deaths from atherosclerotic vascular disease are attributable to smoking, and in those individuals who quit smoking, the death rate declines to almost near normal (160,161). Risk factors, such as smoking, that can be changed by providing both education and interventional guidelines, are a continual focus for organizations like the American Heart Association to address at the present and in the future. As previously mentioned, public access for the automated defibrillator has the potential to greatly reduce deaths in the prehospital arrest scenario. As the time element for CPR and defibrillation has been proven to be so critical for reducing morbidity and mortality from cardiac arrest, those communities that have incorporated aggressive public and paramedical training for these two modalities have reported an almost 50% resuscitation rate for victims documented to have had ventricular fibrillation (1,2). The proliferation of AEDs both in public gathering places and into the home, certainly has the capability to improve these statistics. The continued focus upon rapid recognition of stroke victims in lay responder courses will make a dramatic improvement in rapid treatment and neurological salvage for these individuals. It has been stated that the community will be "the ultimate coronary care unit"(162) since the majority of cardiac arrests occur in the out-of-hospital setting and, therefore, public involvement is crucial to survival. The expectation for the future is that the public "coronary care unit" will be expanded to include a "neurological care unit" as well. The challenge for the future will be to continue to expand the public awareness and involvement in these programs as new scientific evidence continues to guide the evolution of cardiopulmonary resuscitation.

HISTORICAL EVENTS IN CPR

Antiquity	Prophet Elisha describes attempts to revive the dead.
Second century	Galen observed the inflation of a dead animal's lungs.
Middle Ages	Hot materials to the abdomen, whipping, rectal smoke.
Paracelsus, sixteenth century	Fireplace bellows ventilated a patient.
Tossach 1744	First recorded mouth-to-mouth resuscitation.
Squires, 1775	First successful defibrillation.
DeHaen, 1783	Chest compression, arm lift technique.
Leroy, 1830	First description of supine ventilation.
Schiff, 1847	Open chest cardiac compression.
Silvester; Howard, twenteeth century	Alternating arm position ventilation; back, abdominal and chest pressure, respectively.
Koenig; Maass, 1850s	Reports of eight successful closed-chest cardiac compressions in humans.
Holger-Nielson, twenty-first century	Prone back pressure, arm-lift.
Gurvich; Yuniev, 1939	First successful external defibrillation by a device.
Kouwenhoven, Knickerbocker, Isaacs, 1950s	Defibrillation experiments coupled with chest compressions.
Elam, 1960s	Physiology of rescue breathing.
Elam, Safar, Kouwenhoven, 1960s	Principles of modern CPR.
Zoll, 1956	First successful external defibrillation in humans.

BIBLIOGRAPHY

1. White RD, Asplin BR, Bugliosi TF, Hankins DG. High discharge survival rate after out of-hospital ventricular fibrillation by police and paramedics. Ann Emerg Med 1996;28: 480–485.
2. White RD, Hankins DG, Bugliosi TF. Seven years experience with early defibrillators by police and paramedics in an emergency medical services system. Resuscitation 1998;39: 145–151.
3. Holy Bible, Kings II 4:34–35 (King James Version).
4. Baker AB. Artificial respiration: the history of an idea. Med Hist 1971;15:336–346.
5. [Anonymaus]. Cardiopulmonary Resuscitation Conference Proceedings. The Ad Hoc Committee on Cardiopulmonary Resuscitation; Division of Medical Sciences of the National Research Council; 1967. p 7.
6. Julian DG. Cardiac resuscitation in the eighteenth century. Heart Lung 1975;4:46–48.
7. Gordon AS. The principles and practice of heart-lung resuscitation. Acta Anaesth Scand Suppl 1961;9:134–147.
8. Gordon AS, Affeldt JE, Sadove M, Raymon F, Whittenberger JL, Ivy AL. Air-flow patterns and pulmonary ventilation during manual artificial respiration on apneic normal adults. J Appl Physiol 1955;4:408–420.
9. Gordon AS, Fainer DC, Ivy AL. Artificial respiration: a new method and a comparative study of different methods in adults. JAMA 1950;144:1455–1464.
10. Elam JO. Rediscovery of expired air methods for emergency ventilation. In: Safar P, Elam JO, editors. Advances in Cardiopulmonary Resuscitation. Chapt. 39, New York: Springer-Verlag; 1977. p 263–265.
11. Elam JO, Brown ES, Elder JD Jr. Artificial respiration by mouth-to-mask method. A study of the respiratory gas exchange of paralyzed patients ventilated by operator's exhaled air. New Engl J Med 1954;250:749–754.
12. Safar P, Elam J. Manual versus mouth-to-mouth methods of artificial respiration. Anesthesiology 1958;19:111–112.
13. Safar P, Escarraga LA, Elam JO. A comparison of the mouth-to-mouth and mouth-to-airway methods of artificial respiration with the chest-pressure arm-lift methods. New Engl J Med 1958;258:671–677.
14. Safar P. Failure of manual respiration. J Appl Physiol 1959;14:84–88.
15. Safar P. Ventilatory efficacy of mouth-to-mouth artificial respiration. Airway obstruction during manual and mouth-to-mouth artificial respiration. JAMA 1958;167:335–341.
16. Safar P, Aguto-Escarraga L, Chang F. Upper airway obstruction in the unconscious patient. J Appl Physiol 1959;14: 760–764.
17. Morikawa S, Safar P, DeCarlo J. Influence of head-jaw position upon upper airway patency. Anesthesiology 1961; 22: 265–270.
18. Safar P, Redding J. The "tight jaw" in resuscitation. Anesthesiology 1959;20:701–702.
19. Elam JO, Greene Dg, Brown ES, Clements JA. Oxygen and carbon dioxide exchange and energy costs of expired air resuscitation. JAMA 1958;167:328–324.
20. Gordon AS, Frye CW, Gittelson L, Sadove MS, Beattie EJ. Mouth-to-mouth versus manual artificial respiration for children and adults. JAMA 1951;147:1444–1453.
21. Jude JR, Kouwenhoven WB, Knickerbocker GG. Cardiac arrest. JAMA 1961;128:1063.
22. Boehm RV. Arbeiten aus dem pharmakologischen Institue der Universitat Dorpat:13. Ueber Wiederbelebung nach Vergiftungen un Asphysie. Arch Exper Path U Pharmakol 1878;8:68–101.
23. Safar P. History of cardiopulmonary resuscitation. In: Kaye W, Bircher N, editors. Cardiopulmonary Resuscitation. New York: Churchill Livingstone; 1989.
24. Maass Die Methode der. Wiederbelebung bei Herztod nach chloroformeinathmung, Berlin Klin. Wochschr 1892;29: 265–268.
25. Crile WG. Surgical Anemia and Resuscitation. New York: D Appleton and Co; 1904. p 220–244.
26. Hooker DR, Kouwenhoven WB, Langworthy OR. The effect of alternating currents on the heart. Am J Physiol 1933;103: 444–454.
27. Gurvich HL, Yuniev GS. Restoration of heart rhythm during fibrillation by condenser discharge. Am Rev Soviet Med 1947;4:252–256.
28. Kouwenhoven WB, Jude JR, Knickebocker GG. Closed-chest cardiac massage. JAMA 1960;173:1064–1067.
29. Safar P. Initiation of closed-chest cardiopulmonary resuscitation basic life support. A personal history. Resuscitation 1989;18:7–20.
30. Kouwenhoven WB, Langworth OR. Cardiopulmonary resuscitation. An account of forty-five years of research. JAMA 1973;226:877–886.

31. Jude JR, Kouwenhoven WB, Knickerbocker GG. Cardiac arrest: report of application of external cardiac massage on 118 patients. JAMA 1961;178:1063–1070.

32. Safar P, Brown TC, Holtey WH. Ventilation and circulation with closed chest cardiac massage in man. JAMA 1961;176:574–576.

33. Harris LC, Kirimli B, Safar P. Ventilation-cardiac compression rates and ratios in cardiopulmonary resuscitation. Anesthesiology 1967;28:806–813.

34. Zoll PM, et al. Termination of ventricular fibrillation in man by externally placed electric countershock. New Engl J Med 1956;254:727.

35. Weale FE, Rothwell-Jackson RL. The efficiency of cardiac massage. Lancet 1962;1:990–992.

36. Del Guercio LR, Coomraswamy RP, State D. Cardiac output and other hemodynamic variables during external cardiac massage in man. New Engl J Med 1963;269:1398–1404.

37. Thomsen JE, Stenlund RR, Row GG. Intracardiac pressures during closed chest cardiac massage. JAMA 1968;205:116–118.

38. Criley JM, Blaufuss AH, Kissel GL. Cough induced cardiac compression. JAMA 1976;236:1246–1250.

39. Taylor GJ, Tucker WM, Greene HL, Rudikoff MT, Weisfeldt ML. Importance of prolonged compression during cardiopulmonary resuscitation in man. New Engl J Med 1977;296:1515–1517.

40. Chandra N, Rudikoff MT, Weisfeldt ML. Simultaneous chest compression and ventilation at high airway pressure during cardiopulmonary resuscitation. Lancet 1980;1:175–178.

41. Werner JA, Greene HL, Janko CL, Cobb LA. Visualization of cardiac valve motion in man during external compression using two-dimensional echocardiography: implications regarding the mechanism of blood flow. Circulation 1981;63:1417–1421.

42. Rich S, Wix HL, Shapiro EP. Clinical assessment of heart chamber size and valve motion in man during external compression using two-dimensional echocardiography: implications regarding the mechanism of blood flow. Am Heart J 1981;102:368–373.

43. Maier GW, et al. The physiology of external cardiac massage: high impulse cardiopulmonary resuscitation. Circulation 1984;70:86–101.

44. Feneley MP, et al. Sequence of mitral valve motion and transmitral flow during manual cardiopulmonary resuscitation in dogs. Circulation 1987;76:363–375.

45. Porter TR, et al. Transesophageal echocardiography to assess mitral valve function and flow during cardiopulmonary resuscitation. Am J Cardiol 1992;70:1056–1060.

46. West JB. Physiological Basis of Medical Practice. 12th ed. Baltimore: Williams & Wilkins; 1990. p 522.

47. Permut S, Bromberger-Barnea B, Bane HN. Alveolar pressure, pulmonary venous pressure and the vascular waterfall. Med Thorac 1962;19:239.

48. Cummins RO, editor. ACLS Provider Manual. American Heart Association; 2002. p 78.

49. Vukmir RB. Torsades de pointes: a review. Am J Emerg Med 1991;9:250–255.

50. Cummins RO, editor. Advanced Cardiac Life Support. American Heart Association; 1997-1999. p 39–40.

51. American Heart Association in collaboration with International Liaison Committee on Resuscitation and Emergency Cardiovascular Care: International Consesus on Science Part 3: Adult Basic Life Support. Circulation 2000; 102(Suppl I): I–226.

52. Brown DC, Lewis AJ, Criley JM. Asystole and its treatment: the possible role of the parasympathetic nervous system in cardiac arrest. J Am Coll Emerg Phys 1979;8:448–542.

53. Thompson BM, Brooks RC, Pionkowski RS, Aprahamian C, Mateer JR. Immediate countershock treatment of asystole. Ann Emerg Med 1984;13:827–829.

54. Bocka JJ. External transcutaneous pacemakers. Ann Emerg Med 1989;18:1280–1286.

55. Paradis NA, Martin GB, Goetting MG, Rivers EP, Feingold M, Nowak RM. Aortic pressure during human cardiac arrest: identification of pseudo-electromechanical dissociation. Acta Anaesthesiol Scand 1991;35:253–256.

56. Bocka JJ, Overton DT, Hauser A. Electromechanical dissociation in human beings: an echocardiographic evaluation. Ann Emerg Med 1988;17:450–452.

57. Cummins RO editor. ACLS Provider Manual, American Heart Association. 2001. p 103–104.

58. Rosenberg D, Levin E, Myerburg RJ. A mnemonic for the recall of causes of electro-mechanical dissociation (EMD). Resuscitation 1999;40:57.

59. Cummins RO, Eisenberg MS. Prehospital cardiopulmonary resuscitation: is it effective?. JAMA 1985;253:2408–2412.

60. Cummins RO. From concept to standard-of-care? Review of the clinical experience with automated external defibrillators. Ann Emerg Med 1989;18:1269–1275.

61. Cummins RO, Eisenberg MS. Prehospital cardiopulmonary resuscitation: is it effective?. JAMA 1985;253:2408.

62. Valenzuela TD, Roe DJ, Nichol G, Clark LL, Spaite DW, Hardman RG. Outcomes of rapid defibrillation be security officers after cardiac arrest in casinos. N Engl J Med 2000;343:1206–1209.

63. Cummin RO.. Advanced Cardiac Life Support, American Heart Association. 1997–1999. p 16–4.

64. Levine RL, Wayne MA, Miller CC. End-tidal carbon dioxide and outcome of out-of hospital cardiac arrest. New Eng J Med 1997;337:301–306.

65. Graves EJ. National hospital discharge survey: annual summary 1991. Vital Health Statistics 1993;13:1–62.

66. 2000 Heart and Stroke Statistical Supplement. Dallas: American Heart Association; 1999.

67. Eisenberg MD, Mengert TJ. Cardiac Resuscitation. New Engl J Med 2001;344:1304–1313.

68. Gillum RF. Trends in acute myocardial infarction and coronary heart disease death in the United States. J Am Coll Cardiol 1994;23:1273–1277.

69. Kannel WB, Schatzkin A. Sudden death: lessons from subsets in population studies. J Am Coll Cardiol 1985;5(Suppl 6); 141B–149B.

70. Mathey DG, Sheehan FH, Schofer J, Dodge HT. Time from onset of symptoms to thrombolytic therapy: a major determinant of myocardial salvage in patients with acute transmural infarction. J Am Coll Cardiol 1985;6:518–525.

71. Anderson JL, Karagounis LA, Califf RM. Metaanalysis of five reported studies on the relation of early coronary patency grades with mortality and outcomes after acute myocardial infarction. Am J Cardiol 1996;78:1–8.

72. Kannel WB. Prevalence and clinical aspects of unrecognized myocardial infarction and sudden unexpected death. Circulation 1987;75(Suppl 2, pt 2): II–4–II–5.

73. Cummins RO. Advanced Cardiac Life Support, American Heart Association. 1997–1999. p 9–4.

74. Herrick J. Clinical features of sudden obstruction of the coronary arteries. JAMA 1912;59:2015–2020.

75. Smith M, Little WC. Potential precipitating factors of the onset of myocardial infarction. Am J Med Sci 1992;303:141–144.

76. Muller JE, Tofler GH, Stone PH. Circadian variation and triggers of onset of acute cardiovascular disease. Circulation 1989;79:733–744.

77. Peters RW, et al. Identification of a secondary peak in myocardial infarction onset 11 to 12 hours after awakening: the Cardiac Arrhythmia suppression Trial (CAST) experience. J Am Coll Cardiol 1993;22:998–1003.

78. Solomon CG, et al. Comparison of clinical presentation of acute myocardial infarction in patients older than 65 years of age to younger patients: the Multicenter Chest Pain Study experience. Am J Cardiol 1989;63:772–776.

79. Douglas PS, Ginsburg GS. The evaluation of chest pain in women. New Engl J Med 1996;334:1311–1315.

80. Brand FN, Larson M, Friedman LM, Kannel WB, Castelli WP. Epidemiologic assessment of angina before and after myocardial infarction: the Framingham Study. Am Heart J 1996;132(pt 1): 174–178.

81. Mathey DG, Sheehan FH, Schofer J, Dodge HT. Time from onset of symptoms to thrombolytic therapy: a major determinant of myocardial salvage in patients with acute transmural infarction. J Am Coll Cardiol 1985;6:518–525.

82. Newby LK, et al. Time from symptom onset to treatment and outcomes after thrombolytic therapy: GUSTO-1 Investigators. J Am Coll Cardiol 1996;27:1646–1655.

83. Berger CJ, et al. Prognosis after first myocardial infarction: comparison of Q-wave and non-Q-wave myocardial infarction in the Framingham Heart Study. JAMA 1992;268:1545–1551.

84. Gibson RS. Non-Q-wave myocardial infarcton: pathophysiology, prognosis, and therapeutic strategy. Annu Rev Med 1989;40:395–410.

85. TIMI investigators. Effects of tissue plasminogen activator and a comparison of early invasive and conservative strategies in unstable angina and non-Q-wave myocardial infarction: results of the TIMI IIIB Trial. Thrombolysis in Myocardial Ischemia. Circulation 1994; 89:1545–1556.

86. Cohen MV, et al. The effects of nitroglycerin on coronary collaterals and myocardial contractility. J Clin Invest 1973;52:2836–2847.

87. Malindzak GS. Jr., Green HD, Stagg PL. Effects of nitroglycerin on flow after partial constriction of the coronary artery. J Appl Physiol 1970;29:17–22.

88. ISIS-2 (Second International Study of Infarct Survival) Collaborative Group. Randomized trial of intravenous streptokinase, oral aspirin, both or neither among 17,187 cases of suspected acute myocardial infarction: ISIS-2. Lancet 1988;2:349–360.

89. Lee G, et al. Comparative effects of morphine, meperidine and pentazocine on the cardiocirculatory dynamics in patients with acute myocardial infarction. Am J Med 1976; 60:949–955.

90. 2000 Heart and Stroke Statistical Update. Dallas, TX: American Heart Association; 1999.

91. The National Institute of Neurological Disorders and Stroke rt-PA Stroke Study Group. Tissue plasminogen activator for acute ischemic stroke. New Engl J Med 1995;333:1581–1587.

92. Bendzus M, Urbach H, Ries F, Solymosi L. Outcome after local intra-arterial fibrinolysis compared with the natural course of patients with a dense middle cerebral artery on early CT. Neuroradiology 1998;40:54–48.

93. Kwiatkowski TG, et al., for the NINDS r-tPA Stroke Study Group. Effects of tissue plasminogen activator for acute ischemic stroke at one ear. New Engl J Med 1993;340:1781–1787.

94. Albers GW, et al. Intravenous tissue-type plasminogen activatior for treatment of acute stroke: the Standard Treatment with Alteplase to Reverse Stroke (STARS) Study. JAMA 2000;283:1145–1150.

95. Barnett HJ. The pathophysiology of transient cerebral ischemic attacks. Med Clin North Am 1979;63:649–679.

96. Viitanen M, Eriksson S, Asplund K. Risk of recurrent stroke, myocardial infarction and epilepsy during long- term follow-up after stroke. Eur Neurol 1988;28:227–231.

97. North American Symptomatic Carotid Endarterectomy Trial Collaborators. Beneficial effect of carotid endarterectomy in symptomatic patients with high-grade stenosis. New Engl J Med 1991;325:445–453.

98. Antiplatelet Trialists' Collaboration. Secondary prevention of vascular disease by prolonged antiplatelet treatment. BMJ 1988;296:320–331.

99. Weir B, editor. Aneurysms Affecting the Nervous System. Baltimore: Williams & Wilkins; 1987.

100. Brott T, Thalinger K, Hertzberg V. Hypertension as a risk factor for spontaneous intracerebral hemorrhage. Stroke 1986;17:1078–1083.

101. Waga S, Otsubo K, Handa H. Warning signs in intracranial aneurysms. Surg Neurol 1975;3:15–20.

102. Korpelainen JT, Sotaniemi KA, Makikallio A. Dynamic behavior of heart rate in ischemic stroke. Stroke 1999;30: 1008–1013.

103. Hazinski MF. Demystifying recognition and management of stroke. Curr Emerg Cardiac Care, Winter 1996;7:8.

104. Lyden PD, Rapp K, Babcock T, Rothcock J. Ultra-rapid identification, triage, and enrollment of stroke patients into clinical trials. J Stroke Cerebrovas Dis 1994;4:106–107.

105. Kidwell CS, et al. Design and retrospective analysis of the Los Angeles prehospital stroke screen (LAPSS). Prehosp Emerg Care 1998;2:267–273.

106. Wenzel V, Idris Ah, Lindner KH. Ventilation with an unprotected airway during cardiac arrest. In: Vincent JL, editor. Yearbook of Intensive Care and Emergency Medicine. Berlin: Springer-Verlag; 1997: 483–492.

107. Wenzel V, Idris AH. The current status of ventilation strategies during cardiopulmonary resuscitation. Curr Opin Crit Care 1997;3:206–213.

108. Stalinger A, et al. Effects of different mouth-to-mouth ventilation tidal volumes on gas exchange during simulated rescue breathing (abstract). Crit Care Med 2001. Forth-coming.

109. Dorges V, et al. Smaller tidal volumes with room air are not sufficient to ensure adequate oxygenation during bag-valve-mask ventilation. Resuscitation 2000;44:37–41.

110. Baskett P, Nolan J, Parr M. Tidal volumes which are perceived to be adequate for resuscitation. Resuscitation. 1996;31:231–234.

111. Wenzel V, et al. Effects of smaller tidal volumes during basic life support ventilation in patients with respiratory arrest: good ventilation, less risk?. Resuscitation 1999;43: 25–29.

112. Heartsaver CPR: A comprehensive course for the lay responder. Dallas: American Heart Association; 2000. p 23–29.

113. Htin KJ, et al. Rescuer breathing pattern significantly affects O_2 and CO_2 received by patient during mouth-to-mouth ventilation. Crit Care Med 1998;26: (Suppl 1) A56.

114. Stapleton ER, Auferheide TP, Hazinski MF. BLS for Healthcare Providers. Dallas: American Heart Association; 2001. p 68–70.

115. Cummins RO, editor. ACLS Provider Manual. Dallas: American Heart Association; 2001. p 208.

116. Johannigman JA, Branson RD. Oxygen enrichment of expired gas for mouth-to mask resuscitation. Respir Care 1991;36:99–103.

117. Idris AH, Wenzel V, Banner MJ, Melker RJ. Smaller tidal volumes minimize gastric inflation during CPR with an unprotected airway. Circulation 1995;92(Suppl I): I–759.

118. Cummins RO, Hazinski MF. Cardiopulmonary resuscitation techniques and instruction: when does evidence justify revision?. Ann Emerg Med 1999;34:780–784.

119. Eberle B, et al. Checking the carotid pulse check: diagnostic accuracy of first responders in patients with and without a pulse. Resuscitation 1996;33:107–116.

120. Handley AJ, Handley JA. The relationship between rate of chest compression and compression:relaxation ratio. Resuscitation 1995;30:237–241.

121. Swenson RD, et al. Hemodynamics in humans during conventional and experimental methods of cardiopulmonary resuscitation. Circulation 1998;78:630–639.

122. Kern KB, et al. A study of chest compression rates during cardiopulmonary resuscitation in humans. Arch Intern Med 1992;152:145–149.

123. Guidelines for cardiopulmonary resuscitation and emergency cardiac care. Emergency Cardiac Care Committee and Subcommittees, American Heart Association. JAMA 1992;268: 2171–2295.

124. Van Hoeyweghen RJ, et al., Quality and efficiency of bystander CPR: Belgian cerebral resuscitation. Resuscitation 1993;26: 47–52.

125. Kern KB, et al. Efficacy of chest compression—only BLS/CPR in the presence of an occluded airway. Resuscitation 1998;39: 179–188.

126. Wik L, Steen PA. The ventilation-compression ratio influences the effectiveness of two rescuer advanced cardiac life support on a manikin. Resuscitation 1996;31:113–119.

127. Cummins RO, editor. ACLS Provider Manual. Dallas: American Heart Association; 2001. p 30.

128. Hew P, Brenner B, Kaufman J. Reluctance of paramedics and emergency medical technicians to perform mouth-to-mouth resuscitation. J Emerg Med 1997;15:279–284.

129. Locke CJ, et al. Bystander cardiopulmonary resuscitation: concerns about mouth-to-mouth contact. Arch Intern Med 1995;155:938–943.

130. Stapleton ER, Aufderheide TP, Hazinski MF, editors. BLS for Healthcare Providers. Dallas: American Heart Association; 2001. p 80.

131. Tang W, et al. Cardiopulmonary resuscitation by precordial compression but without mechanical ventilation. Am J Resp Crit Care Med 1994;150:1709–1713.

132. Weil MH, et al. Differences in acid- base status between venous and arterial blood during cardiopulmonary resuscitation. New Engl J Med 1986;315:153–156.

133. Sanders AB, et al. Acid-base balance in a canine model of cardiac arrest. Ann Emerg Med 1988;17:667–671.

134. National Safety Council. Injury Fact. 1999. Itasca: National Safety Council; 1999.

135. National Safety Council. Accident Facts. 1997. Chicago: National Safety Council; 1997.

136. Ekberg O, Feinberg M. Clinical and demographic data in 75 patients with near-fatal choking episodes. Dysphagia 1992;7:205–208.

137. Heimlich HJ, Hoffmann KA, Canestri FR. Food-choking and drowning deaths prevented by external subdiaphragmatic compression: physiological basis. Ann Thoracic Surg 1975;20:188–195.

138. American Heart Association, International Liaison Committee on Resuscitation (ILCOR). Guidelines 2000 for Cardiopulmonary Resuscitation and Emergency Cardiovascular Care. International Consensus on Science. Circulation 2000;102 (Suppl I): I46–I48.

139. Heartsaver CPR: A comprehensive course for the lay responder. Dallas: American Heart Association; 2000. p 31–33.

140. Langhell A, Sunde K, Wik L, Steen PA. Airway pressure with chest compression versus Heimlich manoeuvre in recently dead adults with complete airway obstruction. Resuscitation 2000;44:105–108.

141. Skullberg A. Chest compressions: an alternative to the Heimlich manoeuvre?. Resuscitation 1992;24:91.

142. Stapleton ER, Auferheide TP, Hazinski MF. BLS for Healthcare Providers. Dallas: American Heart Association; 2001. p 128.

143. White RD, Vukov LF, Bugliosi TF. Early defibrillation by police; initial experience with measurement of critical time intervals and patient outcome. Ann Emerg Med 1996;28:480–485.

144. Diack AW, Welborn WS, Rullman RG, Walter CW, Wayne MA. An automatic cardiac resuscitator for emergency treatment of cardiac arrest. Med Instrum 1979; Mar–Apr; 13(2): 78–83.

145. Cummins RO, Eisenber M, Bergner L, Murray JA. Sensitivity, accuracy, and safety of an automatic external defibrillator. Lancet 1984;2:318–320.

146. Kerber RE, et al. Energy, current, and success in defibrillation and cardioversion: clinical studies using an automated impedence-based method of energy adjustment. Circulation 1988;77:1038–1046.

147. Dahl C, et al. Myocardial necrosis from direct current countershock. Circulation 1974;50:956.

148. Bardy GH, et al. Truncated biphasic pulses for transthoracic defibrillation. Circulation 1995;91:1768–1774.

149. Bardy GH, et al. Transthoracic Investigators. Multicenter comparison of truncated biphasic shocks and standard damped sine wave monophasic shocks for transthoracic ventricular defibrillation. Circulation 1996;94:2507–2514.

150. Auferheide TP, Stapleton ER, Hazinski MF. Heartsaver AED for the Lay Rescuer and First Responder: Adult Cardiopulmonary Resuscitation and Automated External Defibrillation. Dallas: American Heart Association; 2002. p 4–11.

151. Cecchin F, et al. Accuracy of automatic external defibrillator analysis algorithm in young children. Circulation 1999; 100:I–663.

152. Hazinski MF, Walker C, Smith J, Deshpande J. Specificity of automatic external defibrillator (AED) rhythm analysis in pediatric tachyarrhythmias. Circulation 1997;96(Suppl I); I561.

153. Stapleton ER, Auferheide TP, Hazinski MF. BLS for Healthcare Providers. Dallas: American Heart Association; 2001. p 95–98.

154. Sirbaugh PE, et al. A prospective, population-based study of the demographics, epidemiology, management, and outcome of out-of-hospital pediatric cardiopulmonary arrest. Ann Emerg Med 1999;33:174–184.

155. Heartsaver CPR: A comprehensive course for the lay responder. Dallas: American Heart Association; 2000. p 101–107.

156. Hazinski MF, editor. PALS Provider Manual. Dallas: American Heart Association; 2002. p 64.

157. Fink JA, Klein RL. Complications of the Heimlich maneuver. J Pedatr Surg 1989;24:486–487.

158. Heartsaver CPR: A comprehensive course for the lay responder. Dallas: American Heart Association; 2000. p 101–107.

159. Heartsaver CPR: A comprehensive course for the lay responder. Dallas: American Heart Association; 2000. p 39.

160. Gordon T, Kannel WB, McGee D, Dawber TR. Death and coronary attacks in men after giving up cigarette smoking: a report from the Framingham Study. Lancet 1974;2:1345–1348.

161. US Dept of Health and Human Services. Reducing the Health Consequences of Smoking: 25 Years of Progress: A Report of the Surgeon General. US Dept of Health and Human Services, Public Health Service, Centers for Disease Control, Center for Chronic Disease Prevention and Health Promotion, Office on Smoking and Health. 1989. DHHS Publication (CDC); 89–8411.

162. McIntyre KM. Cardiopulmonary resuscitation and the ultimate coronary care unit. JAMA 1980;244:510–511.

See also CARDIAC OUTPUT, THERMODILUTION MEASUREMENT OF; RESPIRATORY MECHANICS AND GAS EXCHANGE; SHOCK, TREATMENT OF; VENTILATORS, ACUTE MEDICAL CARE; VENTILATORY MONITORING.

CARTILAGE AND MENISCUS, PROPERTIES OF

JEREMY J. MAO
ALEXANDER J. TROKEN
NICHOLAS W. MARION
University of Illinois
Chicago, Illinois

LEO Q. WAN
VAN C. MOW
Columbia University
New York, New York

INTRODUCTION

Synovial or diarthrodial joints are created to enable the movement between bones. Articular cartilage on the end of articulating bone, therefore, must accomplish two functions: (1) absorb, distribute, and transmit mechanical loading, and (2) create a low friction and wear surface for movement over decades of mammalian life. Three cartilage phenotypes exist: hyaline cartilage, fibrocartilage, and elastic cartilage. In the older literature, articular cartilage is often referred to as hyaline cartilage due to its glassy appearance; this appearance is derived from its high proteoglycan content. Indeed, articular cartilage has the highest proteoglycan content of all biological tissues, while, at the same time, it has the lowest cellular content. The chondrocytes not only secret and control collagen and proteoglycan contents in the extracellular matrix, but also are responsible for regulating the elaborate molecular architecture of these macromolecules and their ultrastructural organization (1–3). Throughout life, the healthy chondrocytes under normal conditions secrete and elaborate sufficient amounts of the extracellular matrix macromolecules and completely encase themselves in an environment that possesses truly remarkable biomechanical mechanisms that protect them against the mechanical insults associated with joint loading, and thus survive for long periods of time under normal health conditions (4–6).

Cartilage in a small number of joints in humans, such as the knee meniscus, temporo-mandibular joint, and intervertebral discs, is fibrocartilage. The intervertebral disc, besides its complex macromolecular architectural and ultrastructural organization, also has a complex macrostructural organization; the latter is manifested in the macro-layering of the outer rings of collagen-rich annulus fibrosis and an inner core of a proteoglycan-rich "kidney-shaped" nucleus pulposus. The cells of these fibrocartilaginous tissues are fibroblasts and chondrocytes, some of which are called fibrochondrocytes. Whereas the genotype and phenotype of cartilage cells determine the biochemical and molecular properties of cartilage, the mechanical properties of articular cartilage are largely dependent on the constituents of extracellular matrix (3). This divergence in the determination of biological and mechanical properties is attributed to the scarce cellularity in adult cartilage, with chondrocytes that account for only less than 10% of the adult cartilage volume (4,7). Comprehensive reviews of hyaline and fibrocartilage can be found elsewhere (1,3,8).

An average human takes approximately 2 million steps per year. The joints in the lower limbs, therefore, can undergo 1–4 million cyclic loads from physical activities (9,10). These loads can peak 4–5 times body weight (11,12), and can cause both macro- and micro-structural changes in articular cartilage that may ultimately lead to degenerative diseases such as osteoarthritis (4,7,13–15). Arthritis, which encompasses more than 100 diseases and conditions, is recognized as among the leading causes of physical disability worldwide (16). Thus, investigation of the properties of normal and arthritic cartilage is essential not only for the understanding of the etiology of arthritis (6), but also devising possible approaches toward the tissue engineering of cartilage and meniscus for clinical treatment modalities (17).

ARTICULAR CARTILAGE AND MENISCUS: COMPOSITION AND STRUCTURE

Chondrocytes and fibrochondrocytes are responsible for the morphogenesis, matrix synthesis, and maintenance of articular cartilage and meniscus as functional tissues. However, these cells only account for approximately 10% of the total cartilage volume in adults (3,7,18,19). Chondrocytes receive nutrients and shed metabolic waste products largely from convective transport and diffusion, either from/to the synovial fluid or from subchondral bone. In the adult, articular cartilage is generally aneural and avascular. Vasculature is present only in the periphery of mature meniscus (20).

Hyaline Cartilage and Articular Cartilage

Hyaline cartilage is present on the articulating surfaces of the bones in most, but not all, synovial joints. Articular cartilage serves to bear and distribute load and contribute to joint lubrication. It serves these different purposes through varying the amount of water relative to the amounts of Type I collagen and proteoglycans and the molecular and ultrastructural organizations of these structural molecules. Healthy articular cartilage appears smooth, bluish white, glistening, and intact. Osteoarthritic articular cartilage appears dull and coarse and may have tears and frays. Hyaline cartilage is also present in the growth plate at the metaphyseal region of long bones and in the cranial base and serves to enable longitudinal bone growth by endochondral ossification (21–23). A review of growth plate cartilage is beyond the scope of this chapter, but can be found elsewhere (24–26).

Articular cartilage has two immiscible phases—a solid phase and a fluid phase. Small electrolytes such as Na^+ and Cl^- are dissolved in the fluid phase and are freely mobile by diffusion and convection through the porous-permeable solid phase. The fluid and solid phases have been modeled in the now classic biphasic theory developed by Mow et al. (27). Normal fluid component ranges from 75 to 80% by wet weight, and the remaining 20 to 25% of the organic matrix forms solid material with complex material properties (3,18,19). Up to 65% of the solid ECM by dry weight is made of collagen, whereas proteoglycans constitute up to 25%; other glycoproteins, chondrocytes, and lipids can generally make up 10% (3,28,29). Collagen fibers are classified on the basis of their amino acid composition and molecular structure. Although an assortment of collagens exist in both hyaline cartilage and fibrocartilage, Type II collagen is most prevalent in articular cartilage, whereas Type I collagen is most common in the meniscus (3,30). The collagen fibers are assembled as tight triple-helical structures made from three polypeptide alpha-chains. The triple helices are then arranged as tropocollagen molecules, which are wound in a helical manner to form larger collagen fibers that are, in turn, are organized into a strong cohesive collagen network (3,31,32). This arrangement allows for considerable tensile stiffness and strength (33–38). The collagen also serves to restrain the swelling pressure created from the surrounding embedded proteoglycans (2,18,19,39). Proteoglycans (PGs) are hydrophilic macromolecules with numerous glycosaminoglycans (chondroitin and keratin sulfates) attached to a protein core; the protein core of this bottle-brush-shaped molecular is, in turn, attached to a hyaluronan (mw: molecular weight $\sim 0.5 \times 10^6$) resulting in a supra-macromolecule with an approximate molecular weight ranging from (200 to 300) $\times 10^6$ (3,29,40–42). These enormous, negatively charged molecules are trapped in the fine porous meshwork of collagen by frictional and electrostatic forces and by steric exclusion; thus, in the ECM, PGs function largely to generate osmotic pressure (2,39) and to resist the compressive stresses of articulation acting on the cartilaginous surface. Although various PGs exist in cartilage, the one that constitutes up to 80–90% of the total PGs in cartilage is aggrecan (3). As the name implies, aggrecan facilitates the formation of large aggregates. Like collagen,

aggrecan, as with all PGs, maintains a structure that is directly correlated with its function. The general structure of PGs occurs through noncovalent bonding of aggrecans to the hyaluronan via link proteins, thus securing firm linkages. Attached to the protein core are the glycosaminoglycan side chains (GAGs) that are vital for biological and biomechanical functions of the tissues; indeed, they are hallmarks of chondrogenic activity in tissue engineering. These GAGs bear the necessary physical properties that ultimately confer onto these tissues their hydrophyllic tendencies and compressive load-carriage abilities (3,18,19). The presence of large numbers of sulfate and carboxyl groups on the GAGs gives rise to a high negative-charge density in the ECM (2). This anionic nature attracts positively charged ions, creating an osmotic pressure, known as Donnan osmotic pressure, that favors tissue hydration (3,39,43). The fixed-negative charges also create intense repulsive forces of the GAGs against each other. This expansion force of the PG molecules causes tensile stresses to be developed within the surrounding collagen network surrounding the PGs. This swelling pressure thus resists the compressive forces against the cartilage without volume loss.

Articular cartilage is organized into three layers or zones (3,44) as shown in Fig. 1. The superficial zone forms the articular surface, whereas the deep zone is anchored to calcified cartilage and subchondral bone; both zones have well-defined collagen architectures. An intermediate zone exists in between with a random collagen fiber ultrastructural organization. These layered structural arrangements have long been hypothesized to be important in cartilage function (45–47). The overall thickness of articular cartilage, including the three zones, varies between joints, age, individuals, and species from less than a millimeter to a few millimeters, with the thickest being measured at the retro-surface of the human patella and femoral trochlea (3,13,48).

The superficial zone has the highest collagen, water content, and chondrocyte density, but the lowest proteoglycan content among all three zones (2,49–51). The abundant collagen fibrils are aligned parallel to the articular surface and provide the superficial zone with substantial tensile strength in an orthotropic manner (37,38,52,53). The chondrocytes in this zone are flattened and are apparently polarized to be parallel to the surface (Fig. 2a) (7,54). Recently, intricate 3D images have been taken to

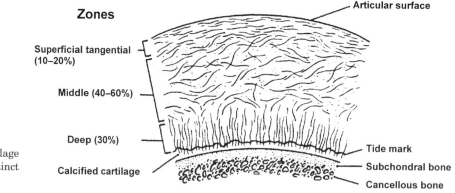

Figure 1. Layered structure of cartilage collagen network showing three distinct regions (3).

Zones

Superficial tangential (10–20%)

Middle (40–60%)

Deep (30%)

Calcified cartilage

Articular surface

Tide mark

Subchondral bone

Cancellous bone

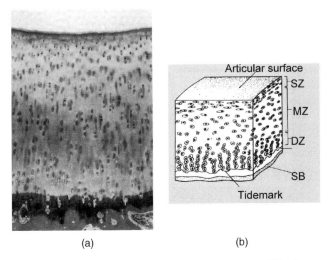

(a)   (b)

Figure 2. (a) Articular cartilage from a mature rabbit femur showing typical zonal arrangement of chondrocytes (polished saw-cut of resin-embedded tissues, surface-stained with basic fuchsine and toluidine blue) (54). (b) Schema of chondrocyte organization in the superficial zone (SZ), middle zone (MZ), and deep zone (DZ) (7).

view the discoid-shaped cells as they are maintained in this layer. Methods have ranged from digital volumetric imaging (55) to atomic force microscopy (56).

The intermediate, or middle, zone is generally the thickest amongst the three uncalcified zones of articular cartilage. Collagen fibrils, although less dense, have a greater diameter than the superficial zone, but appear to be more randomly oriented (47,57). The intermediate zone also has the highest proteoglycan content (3). Chondrocytes are more rounded, although cell density is not as high as in the superficial zone (3,58) (Fig. 2b).

The deep zone is relatively thin and the collagen are intertwined to form larger fiber bundles, and, from this zone, they insert perpendicularly into the calcified zone, and thus anchor the uncalcified tissue to the bony ends as required by joint articulation. This organization allows the bundles to firmly anchor the articular cartilage to the underlying subchondral bone. In general, chondrocyte density decreases from the middle zone to the deep zone, where they are similarly aligned as the collagen bundles, arranging into columns perpendicular to the uncalcified-calcified cartilage intersurface (3,55).

Several recent studies have investigated the pericellular matrix (PCM) and the interterritorial matrix (ITM) of chondrocytes. Using algorithms to account for fluid flow and differences in the relative stiffness between the PCM, the ITM, and the chondrocyte, different elastic moduli between PCM and ITM have been found to have a significant effect on chondrocyte's mechanical environment (59). Gradient distributions of charges and material densities relative to chondrocyte surface are important in cartilage fluid flow dynamics and deformation behavior (60). Using micropipette isolation of chondrocytes and nuclei, chondrocyte nuclei have been found to be stiffer than intact chondrocytes (59,61,62). Cultured chondrocytes are able to elaborate a PCM rich in Type VI collagen; however, intact chondron pellets accumulate significantly more

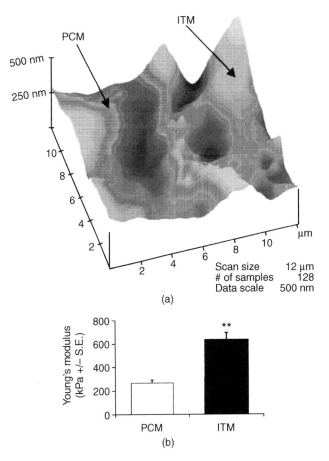

(a)

(b)

Figure 3. A representative height map of the PCM and ITM chondrocytes obtained through force mode of atomic force microscopy. Qualitatively, the ITM showed greater peak and valley contours than the topographic contour of the PCM (a). (b) presents the average Young's moduli of the PCM and ITM attained via nanoindentation. The average Young's modulus of the ITM (636.1 ± 124.91 kPa) was significantly greater than the PCM (265.1 ± 52.76 kPa) ($p < 0.01$) ($N = 19$) (64).

proteoglycans and Type II collagen than chondrocytes without a native PCM (63). Following a few weeks of accumulation of the ITM and PCM by isolated chondrocytes, a rapid increase in compressive stiffness occurs in both the chondron and the chondrocyte pellets (63). Using atomic force microscopy (AFM), the ITM is found to be stiffer to nanoindentation than the PCM (Figs. 3a and 3b) (64).

Meniscus and Fibrocartilage

The meniscus in the knee joint is a fibrocartilage. The two menisci (lateral and medial) in each knee joint are crescent or semi-lunar shaped and are attached to the joint capsule. The triangular cross section of the meniscus tapers radially inward from the periphery, and the center of the meniscus is thin and unattached. Thus, the cross section of the meniscus is wedge-shaped. The central region is a vascular and has more proteoglycans, hence more hyaline in appearance. The anterior and posterior horns of the meniscus form the tips of the crescents. The anterior horn of the lateral meniscus is attached to the tibia in front of the intercondylar eminence, partially blending with the anterior cruciate ligament. The posterior horn is attached to the tibia near the intercondylar eminence as well as to the femur via the meniscofemoral ligament. The anterior and posterior horns of the medial meniscus are attached to the tibia near their respective intercondylar fossae. The anterior horns of the lateral and medial menisci are connected by the transverse ligament. The thick peripheral borders and associated horns of the meniscus are vascularized by blood supply predominantly from the genicular arteries surrounding the joint. The thinner central portions of the meniscus are aneural and avascular, a region very much like hyaline cartilage (20). The meniscus is lubricated with synovial fluid (65), probably by the same lubrication mechanisms known to exist in articular cartilage (66). Fibrocartilage is found in a small number of other joints. The disk of the temporomandibular joint (TMJ) and the intervertebral disks are both composed of fibrocartilage, although they are drastically different structures with different distributions of cartilage and PGs and ultrastructual organization.

The fibrocartilagenous structure of the meniscus differs from that of hyaline cartilage in many ways. The cells of the meniscus are sometimes called fibrochondrocytes, although it is probable that some cells are more like fibroblasts, whereas others are more like chondrocytes (65,67). The peripheral two-thirds of the meniscus are primarily composed of a randomly oriented mesh-like, coarse, collagen fibrillar matrix (68–71). In deeper portions, large rope-like collagen fiber bundles are arranged circumferentially, retaining the overall semi-lunar shape of the meniscus and providing tensile strength. Smaller fibers are also found radially and connect to the larger circumferential collagen fiber bundles (72). As mentioned above, the inner portion of the meniscus resembles that of hyaline cartilage, containing a higher percentage of proteoglycans enmeshed within a randomly arranged collagen fibrillar matrix (71,73,74).

The function of the meniscus is to enhance higher congruity of the articulating surfaces of the distal femur and proximal tibia, to accommodate the range of motion, in addition to the same functions of load bearing and load distribution to that of articular cartilage (20). The previous assumption that the menisci are functionless, evolutionary remains of leg muscles is erroneous and that menisectomy (a common clinical procedure) is indeed a common procedure in animal models to study the etiology of osteoarthritis (75–78).

CARTILAGE AND MENISCUS: MECHANICAL PROPERTIES

Articular cartilage and meniscus are both important load-bearing tissues and vital to the maintenance of normal joint functions (3,18,19). Articular cartilage can absorb mechanical shock of joint motion and spread the applied load onto the subchondral bone. It also contributes to the lubrication mechanism and provides a surface with low friction, enabling repetitive gliding motion between articulating surfaces (7,66). The meniscus of the knee has important biomechanical functions such as load transmission at the otherwise highly incongruent tibiofemoral

articulation, shock absorption, joint congruity, and stability (18,19). The salient biomechanical functions of articular cartilage and meniscus are dependent on their biological structure, composition, and the intrinsic material properties of the ECM. The knowledge of their material properties such as tensile, compressive, and shear moduli is essential to understand not only their biomechanical functions, but also in the tissue engineering of articular cartilage and meniscus to produce in a biomimetic manner artificial-biological replacements (17,79).

Tensile Properties

When cartilage is tensed, the tensile stress-strain behavior is nonlinear. A typical nonlinear stress-strain (σ-ϵ) curve for cartilage, meniscus, and other soft tissues is depicted in Fig. 4 (3). For small deformations, a 'toe-region' is seen in the stress-strain curve, in which the collagen fibrils will primarily realign in the direction of the externally applied force instead of being stretched (elongation per unit length). For larger deformations, the collagen fibrils are stretched, and a larger tensile stress is generated within the collagen fibers (e.g., 35–38,52,53,80–83). In this linear region, the stress is proportional to the applied strain, and their ratio is known as Young's modulus in tension, or tensile modulus. This tensile modulus is a measure of the stiffness of the collagen-PG solid matrix and is primarily dependent on the density of collagen fibrils, fibril diameter, and type or amount of collagen cross-linking (3,18,19,52). Beyond the linear region, the cartilage strip will rupture abruptly, and the tensile failure stress is a measure of the strength of the collagen fibrillar network.

In general, the tensile modulus of articular cartilage will be in a range of 1–30 MPa, which is much larger than the compressive modulus of cartilage (~ 0.5 MPa), which is known as tension-compression nonlinear property of the

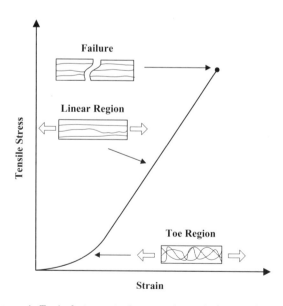

Figure 4. Typical stress-strain curve for articular cartilage and meniscus in a uniaxial and uniform strain rate experiment. The toe region is marked by an increasing slope, whereas the linear region appears to be a straight line (3).

cartilage (84–86). The tensile properties are also known to vary with location, depth, and orientation of test specimens of cartilage and meniscus. Hultkrantz (45) demonstrated the anisotropic organization of the collagen network by puncturing holes in the surface of articular cartilage with a round pin. He found that round puncture holes will form elongated splits, analogous to splits formed in lumber when a large round awl pierces it. The split-line patterns were, to him, evidence of collagen fiber orientation, which is still an enigma today because electron microscopy has not found such surface collagen anisotropy. (Nevertheless, the pattern of split lines is similar to Langer lines formed in the skin in a similar manner.) Much later, Woo et al. (38), Kempson et al. (52), and Roth and Mow (37) showed that the tensile strength and stiffness of the samples cut parallel to the split-line direction were higher than those cut perpendicular to it. The cartilage strips from high weight-bearing areas of human knee joints exhibit larger tensile modulus than those from low weight-bearing areas (53,82) because high weight-bearing areas generally have a relatively higher proteoglycan content. The adult human femoral articular cartilage exhibits a gradual decrease in tensile strength and stiffness as the distance from the articular surface increases (36,81), while this functional dependence was not observed for young bovine humeral joints (38). A dependency of cartilage tensile properties with skeletal maturation was found by Roth and Mow (37). These investigators found that, with the closing of the growth plate (indicative of skeletal maturity), the strength and stiffness of cartilage are much less that those properties of immature cartilage (open-physis). The effects of age on the tensile properties of adult cartilage were extensively studied by Kempson (81), and the results showed that the tensile modulus decreases with age, and that the modulus of the hip cartilage decreases more markedly than that of ankle cartilage. This finding may explain the relatively high occurrence of osteoarthritis in the hip compared with the ankle. Like articular cartilage, the tensile properties of meniscus vary with respect to the location (anterior, central, and posterior) and specimen orientation relative to the predominant collagen fiber direction (circumferential and radial) (18,19,74). Specimens from the posterior half of the medial meniscus have been shown to be significantly less stiff and less strong in tension than specimens from all other regions (87). This experimental result agrees with the ultrastructural findings using polarized light; In the posterior half of the medial meniscus, collagen fiber bundles have significantly reduced circumferential organization (87).

Numerous experiments have shown that the tensile modulus is correlated with the collagen content or the ratio of collagen content to proteoglycan content in articular cartilage (e.g., 82). The tensile modulus of articular cartilage decreases to only 1% after disruption of collagen cross-linking by elastase (88). In contrast, no significant correlations have been found between the tensile property of the cartilage and proteoglycan content (36,89). These findings indicate that collagen content, organization, and cross-linking play significant roles in generating high tensile modulus of articular cartilage. For meniscus, although the tensile properties show significant regional

and directional variations, little difference appeared in the biochemical composition with site, and no significant correlation exists between tensile property and chemical contents (74). The variation of tensile properties seems to reflect local differences in collagen ultrastructure and fiber bundle direction as described above.

Compressive Properties

The compressive behavior of cartilage and meniscus has been extensively studied under various configurations, such as confined compression, unconfined compression, and indentation (see Fig. 5). Most of the earliest studies (e.g., 90–94) used the indentation technique to determine the mechanical property of articular cartilage and modeled the cartilage to be a single-phase, elastic body with the assumption of the Poisson's ratio ranging between 0.4 and 0.5 (e.g., 91–96). However, this single-phase elastic model cannot describe the time-dependent viscoelastic behavior of the tissue nor the role played by cartilage's major component (i.e., water). Cartilage and meniscus exhibit a viscoelastic creep in response to a constant load (i.e., its deformation will increase with time). Conversely, if a constant displacement is applied, the force response will decrease gradually with time to a constant value (i.e., a stress-relaxation will be observed).

These viscoelastic behaviors derive from the friction of water flowing through solid matrix (27,97), as well as the flow-independent intrinsic energy dissipation inside the macromolecular solid matrix during mechanical loading (98–101). As mentioned, articular cartilage and meniscus can be regarded as biphasic materials: a fluid phase composed of water and electrolytes, and a solid phase mainly composed of collagen and proteoglycans (27). The solid matrix is considered as being porous and permeable. Water resides in the microscopic pores and flows through the matrix during joint loading. Under a slow ramp loading, the observed viscoelastic behaviors are usually dominated by the large drag forces generated by the flow of interstitial fluid through the porous-permeable solid matrix, and therefore, the flow-independent intrinsic energy dissipation is negligible. However, osteoarthritic cartilage has higher permeability and lower ECM stiffness; in such tissues, the intrinsic viscoelastic behavior becomes the dominating component governing their mechanical behaviors.

The transient behavior of the tissue under compression is primarily determined by the mechanism of fluid pressurization because of high friction between solid and fluid phases, which is also known as flow-dependent viscoelastic behavior (27). Figure 6 (3) shows the stress-relaxation behavior of a tissue specimen under confined compression. In this experiment, before time t_0, the tissue is compressed with a constant rate, and the interstitial fluid inside the tissue will be pushed out through upper porous platen. As a result of the distributive fluid drag force, a larger deformation can be seen at the downstream side (Fig. 6a and 6b). During the relaxation phase (after t_0), no fluid exudation occurs, but the fluid needs to redistribute inside the tissue before the equilibrium is reached (Figs. 6c, 6d, and 6e). Although the velocity of fluid flow is very low, the friction force, or the drag force, could be very large because the pore

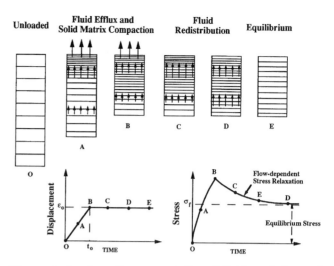

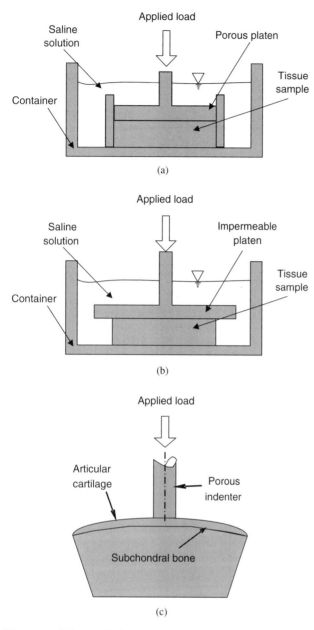

Figure 6. A schematic representation of fluid exudation and redistribution within cartilage during a rate-controlled confined compression stress-relaxation test (lower left). The horizontal bars in the upper figures. indicate the distribution of strain in the tissue. The lower graph (right) shows the stress response during the compression phase (O, A, B) and relaxation phase (B, C, D, E) (3).

Figure 5. Schema of three configurations frequently used to study the compressive properties of articular cartilage. (a) In the confined compression configuration, a load is applied to the cartilage sample via a rigid porous permeable platen. The side walls are assumed to smooth, impermeable, and rigid, thereby preventing lateral expansion and fluid flow. (b) In the unconfined compression configuration, the cartilage sample is compressed between two rigid, smooth, and impermeable platens. The lateral side allows fluid flow. (c) In the indentation configuration, the cartilage is compressed via a rigid porous permeable indenter. The porous indenter allows the fluid exudation to occur freely into the indenter tip and, therefore, creep of the cartilage layer.

size inside the tissue is very small (~ 50–65 nm for articular cartilage), and the permeability of the tissue is as low as 10^{-15} N·s/m^4. Therefore, the generated fluid pressure can be remarkably high inside the tissue during the transient state, which also means that chondrocytes encased within the ECM will normally be bathed in a highly

pressurized fluid. It has been estimated that this fluid pressure could be 30 times more than the elastic stress generated in the solid matrix of articular cartilage (3). Considering that the equilibration process usually takes several hours, no real equilibrium state occurs in joints under physiological conditions because the joints are moving virtually at all times, even during sleep. Thus, the mechanism for fluid pressurization is likely to be the major physiological load-supporting mechanism in diarthroidal joints, and it plays an important role in shielding the solid matrix from large compressive stresses during the joint function (7).

At equilibrium, the fluid flow stops, no fluid pressure gradient exists inside the tissue, and the applied load is entirely supported by the solid matrix of the tissue. Thus, the compressive property can be obtained from the relations between stress and strain. It has been found that the equilibrium strain is proportional to the applied load. Typically, the equilibrium aggregate modulus (27) for normal articular cartilage ranges from 0.4 to 1.5 MPa, whereas the average equilibrium aggregate modulus for the meniscus is about 0.4 MPa. Table 1 shows the equilibrium aggregate moduli of lateral condyle and patellar groove cartilage and meniscus, showing considerable variation among the species and tissue location (3).

Tissue mechanical properties are highly dependent on their composition and structure. It has been shown that the equilibrium aggregate modulus for human articular cartilage correlates in an inverse manner with water content and in a direct manner with PG content (27,102,103). The highly loaded regions of articular cartilage generally have larger compressive modulus and greater PG content (53,104,105). In contrast, no correlation is found between the compressive stiffness and collagen content. Removal of PGs from articular cartilage samples dramatically decreases the compressive modulus, whereas trypsin

Table 1. Equilibrium Aggregate Modulus of Lateral Condyle, Patellar Groove Cartilage and Meniscus (MPa) (3)

	Human[a]	Bovine[b]	Canine[c]	Monkey[d]	Rabbit[e]
Lateral condyle	0.70	0.89	0.60	0.78	0.54
Patellar groove	0.53	0.47	0.55	0.52	0.51
Meniscus	NA	0.41	NA	NA	NA

[a]Young normal.
[b]18 months to 2 years old.
[c]Mature beagles and greyhounds.
[d]Mature cynomologus monkeys.
[e]Mature New Zealand white rabbits.
[f]Not available.

digestion of collagen fibrils has little effect on compressive modulus (80,106).

The biphasic theory has been the most successful model for the compressive viscoelastic behaviors of cartilage and meniscus under various conditions (27). This theory assumes that (1) the solid matrix and interstitial fluid are immiscible and incompressible; (2) viscous dissipation is due to the fluid flow between water and the porous-permeable solid matrix; and (3) the frictional drag is proportional to the relative velocity and can be affected by ECM compression. This biphasic theory further assumes that the solid matrix experiences infinitesimal strain and that the stress-strain relations can be described by the generalized Hooke's law. Despite its simplification, as biological models typically are, the isotropic form of the linear biphasic theory has been shown to provide an accurate description of the compressive creep and stress relaxation behavior of these tissues. In particular, a numerical algorithm based on this biphasic theory was developed and accurately predicted the aggregate modulus, Poisson's ratio, and permeability of articular cartilage from the indentation creep experiment (13,100,107). The biphasic theory has also been extended by employing higher levels of tissue complexities, including material inhomogeneities (108–110), material symmetries (33,34,85,86,111), and matrix viscoelasticities (33,34,84,98–100).

Shear Properties

The intrinsic viscoelastic properties of the solid matrix of cartilage and meniscus can only be determined in a *pure* shear experiment and under small strain conditions. In pure shear, the kinematics of deformation does not permit volumetric change, and hence, no interstitial fluid flow is possible when no pressure gradients are applied. Under these three conditions, the tissue deforms without change in volume, and therefore, the interstitial fluid pressure and fluid flow are minimal. As a result, the flow-dependent viscoelastic properties are excluded, and the measured physical parameters will be independent on the friction or drag force between fluid phase and solid phase, which often occurs in compressive configurations, thus directly reflecting the intrinsic viscoelastic property of solid matrix. This flow-independent viscoelastic behavior of the collagen-PG matrix derives from the internal friction between collagen and PG molecules (3,101).

The first shear properties measurement was reported by Hayes and Mockros (112), and later, nonlinear viscoelastic and fatigue properties of bovine articular cartilage were investigated (113,114). However, all these tests were performed in a simple shear configuration, and dynamic shear properties of these studies were reported at frequencies (e.g., 20–1000 Hz) much higher than the physiological range (e.g., 1 Hz). Pure shear tests of articular cartilage and meniscus have been performed under transient, equilibrium, and dynamic conditions to characterize the intrinsic or flow-independent viscoelastic behavior (e.g., (101,115–117)). When a circular cartilage specimen is subject to a sudden change of angular displacement, the shear stress will increase instantaneously, followed by a rapid decay before equilibrium is reached. The quasilinear viscoelastic theory (118) has been shown to provide an excellent description of this intrinsic stress-relaxation behavior of normal human patellar cartilage (116). The equilibrium shear modulus for normal human, bovine, and canine articular cartilage has been found to vary in a range of 0.05–0.25 MPa. Values for the magnitude of the dynamic shear modulus $|G^*|$ of normal cartilage are in the range of 0.2–20 MPa and vary with both the frequency and magnitude of the normal stress. The phase shift angle (δ) for cartilage lies between $9°$ and $20°$ over a frequency range of 0.01 Hz to 20 Hz (101). Please note that δ is a measure of matrix dissipation, with a loss angle of $0°$ corresponding to a perfectly elastic material, and $90°$ to a perfectly dissipative material. The viscoelasticity of meniscus in response to shear is qualitatively similar to that exhibited by articular cartilage, although the magnitudes of the material coefficients of these tissues are significantly different. Meniscus shear properties exhibit an orthotropic symmetry (i.e., the three planes of symmetry defined by its fibrous architecture dominate the shear properties of the meniscus). The equilibrium shear moduli are 36.8 kPa, 29.8 kPa, and 21.4 kPa in the circumferential, axial, and radial directions, respectively (18,19,119). These shear modulus values are ten times less than those observed for articular cartilage. For dynamic tests, the magnitude of the complex shear modulus $|G^*|$ and phase shift angle δ for circumferential, axial, and radial specimens reflect orthotropic collagen fiber organizational symmetry as well (Fig. 7) (3).

The collagen network plays an active mechanical role in contributing to the shear stiffness and energy storage in cartilage (101). Conceptually, the role played by collagen when the specimen is in shear may be visualized as shown in Fig. 8 (101). The tension in the diagonally oriented collagen acts to increase the shear stiffness of the solid matrix. This effect is confirmed by the experimental result that $|G^*|$ is directly and significantly related to the collagen content of articular cartilage and also by the fact that

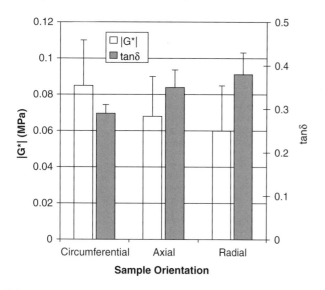

Figure 7. The magnitude of dynamic shear modulus $|G^*|$ and $\tan\delta$ for meniscal specimens with normal vectors oriented circumferentially, axially, and radially at 1 rad/s (3).

cartilage has a relatively small loss angle and large shear modulus (101) compared with that of PG solutions at physiological concentrations (107,120,121). The depletion of the PG content has been shown to decrease the dynamic shear modulus up to 55% (101,122), which is considered as a result of the decrease of tensile stress inside the collagen fibrils due to the decrease of PG swelling pressure.

Swelling Properties

Swelling in articular cartilage derives from the presence of negatively charged groups (SO_3^- and COO^-) along the GAG chains of PG molecules. For normal and degenerative femoral head cartilage, the fixed-charge density (FCD) ranges from 0.04 to 0.18 mEq/g wet tissue at physiological pH (2,123). These fixed charges will require a high concentration of counter-ions (Na^+) to maintain electroneutrality, and the concentration, along with that of co-ions (Cl^-), is governed by the Donnan equilibrium ion distribu-

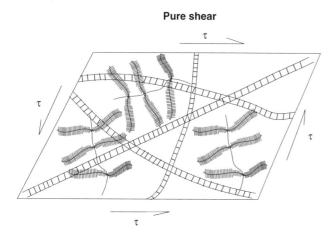

Pure shear

Figure 8. A scheme representation of cartilage in pure shear. The tensile stress inside collagen fibrils provides shear stiffness (101).

tion law (124). This excess of freely mobile ions will introduce an imbalance of ion concentration between the fluid compartment inside the tissue and the bathing fluid outside the tissue, giving rise to a higher pressure in the interstitial fluid than the ambient pressure in the external bath, known as the Donnan osmotic pressure. This osmotic pressure decreases with FCD and has a range of 0.1 to 0.25 MPa for normal articular cartilage. With the increase of the external saline concentration, the osmotic pressure will decrease. For the very large external saline concentration (e.g., 2 M), the osmotic pressure is considered to be extremely small, or zero.

This osmotic pressure causes the tissue to swell, as measured by both weight change (2,123) and dimensional change (125,126). These latter experimental results show that the swelling of articular cartilage is inhomogeneous and anisotropic. The swelling ability increases with depth, with the largest dimensional change for the deep zone and almost no change for the superficial layer. The magnitude of swelling is the largest in the thickness direction and the smallest in the split-line direction. In articular cartilage, the osmotic pressure is restrained by the surrounding collagen network. Therefore, residual stress or pre-stress exists inside the solid matrix even before external load is applied. Articular cartilage will warp or curl toward its articular surface upon its removal from the subchondral bone, and the curvature or the extent of curling will decrease with the increase of external saline concentration (126). It has been hypothesized that this curling is caused by the combination effects of swelling pressure and inhomogeneity inside the tissue (126–129). Recently, a three-layer orthotropic model based on triphasic theory (39) has been developed to describe the curling behavior of cartilage strip by considering its layered structure that includes depth-dependent collagen fibril orientation and chemical content distributions (128). The predicted curvature change with external saline concentration agrees well with previously published experimental results. This model has also suggested that the large stiffness of the superficial layer and high swelling pressures play key functional roles in the development of pre-stress in cartilage and in its curling behavior.

Quantification of morphological changes has been extensively used to study changes in cartilage swelling with osteoarthritis (OA) (44). With OA, compositional and microstructural changes will occur, which includes the fibrillation of the superficial zone of articular cartilage, the decrease of the PG concentration, and the imbibitions of water; these items are the earliest indicators of the OA degeneration of cartilage (18,19,29,102). The elevated water content or swelling has been shown to be very sensitive to collagen fibrillation. Experimental results also suggest that the water content of the tissue increases after digestion with collagenase (82,83,101). Physically, collagen fibrillation decreases the stiffness of the solid matrix, specifically the elastic bulk modulus, which allows the tissue to imbibe more water (52,104).

Triphasic Mixture Theory

To account for the swelling behavior, Donnan osmotic effects, ion transport, and electrical potentials inside the tissue, Lai et al. (39) developed the triphasic theory to incorporate the effects of negatively charged groups on the PGs of solid matrix. In this theory, the electrolytes (mainly Na^+ and Cl^-) within the interstitial fluid are considered as a separate phase, and the solid phase is charged. This triphasic theory was further extended to account for multiple species of ions in the tissue (130). Note that Huyghe and Janssen (131) developed an equivalent theory, named the quadriphasic theory, in which ion species Na^+ and Cl^- were treated as two separate phases.

Using the triphasic theory, the electrokinetic coefficients such as electrical conductivity has also been derived in terms of the physical parameters of charged tissues (39,130,132–134). Furthermore, a theoretical analysis (134,135) showed that the electrical potential inside the tissue comes from two competing sources: a diffusion potential deriving from the FCD inhomogeneity and a streaming potential resulting from the fluid flow within a charged material. These two sources of electrical potential have different polarity and compete against each other. Within the physiological range of material properties of articular cartilage, the polarity of electrical potential inside the tissue depends on the stiffness of the tissue. For softer tissues (such as OA tissue), the diffusion potential tends to dominate, whereas the streaming potential tends to dominate for stiffer tissues (such as normal tissue).

Numerical methods, such as finite difference and finite element formulations, have been developed to demonstrate the contributions of the FCD in mechano-electrochemical (MEC) behaviors of charged, hydrated soft tissues (43,134,136,137). These studies showed that that higher FCD decreases the characteristic time (gel time) and causes the tissue to reach equilibrium in a shorter amount of time, and showed that the osmotic effects can contribute up to 50% of the equilibrium confined compression stiffness (136) and about 30% in unconfined compression (43). With the finite element formation, Lu et al. (138) successfully correlated the predicted FCD with the biochemical measurements, while simultaneously measuring the apparent mechanical properties from the indentation creep experiment.

Recently, the triphasic formulations have been linearized, and the analytic solutions for the MEC response of the tissue under unconfined compression have been obtained both for transient state and at equilibrium (139). With a regular perturbation, simple relations have been derived to describe how the apparent properties, such as Young's modulus and Poisson's ratio, change with the fixed negative-charge density (FCD) at equilibrium (139,140). These relations actually are applicable to various testing configurations, even for steady permeation, and they indicate the correspondence of mechanical properties between an elastic body and a charged triphasic material such as articular cartilage and meniscus (139–142).

CARTILAGE WEAR AND DEGENERATION

Cartilage wear and degeneration have been extensively studied due to their significant roles in physically debilitating diseases such as osteoarthritis and rheumatoid arthritis (1). Two types of wear occur in synovial joints: fatigue wear and interfacial wear (66). Fatigue wear is independent of the lubrication within the joint and is caused by functional activities such as cyclic, repetitive loading. A balance is presumably maintained under the normal physiological condition whereby tissue turnover is maintained by cells in various components of the synovial joint. A number of factors may contribute to cartilage wear and degeneration. Collagen fibers can be severed by excessive functional activities, leading to a compromise in tensile strength. The normally tight collagen fiber bundles can be unwound and loosened (47,143). When inflammatory cytokines are released, proteoglycans are lost rapidly, leading to a breakdown of the ECM (144–146). Fibrillated cartilage from osteoarthritic patients shows an increase of apoptotic chondrocytes deeper than in the normal articular cartilage, which generally has apoptotic cells only near the surface (145,147). Collective loss of chondrocyte and ECM may lead to microcracks and fissures, which may further grow with functional loading. Thus, fatigue wear of cartilage is a mechanism dependent on biological synthesis and mechanical loading (5).

Interfacial wear can result from physical contact loading of articulating surfaces. Interfacial wear has been categorized into two classes: adhesive and abrasive wear (66). Adhesive wear is more common and occurs when a junction is created when the two solids are in contact. As the opposing surfaces continue to move past the junction, fragments from the weaker surface may be torn off and adhere to the stronger material. The concept is analogous to rubber skid marks left on a road from a braking car. The car is able to move past the formed junction only by having elements of the weaker material, the rubber tire, come off and adhere to the stronger material, which is, in this case, the road. Abrasive wear occurs when a harder material comes into contact with a softer material. No junction is formed. While in contact and rubbing against each other, the harder material cuts or plows into the softer material. The harder material can be either one of the opposing surfaces or loose particles caught between two softer opposing surfaces, cutting into both of them (66).

Pain, stiffness, swelling, and reduced range of motion are the common phenotypic characteristics of osteoarthrosis. Typically, clinical diagnosis is only made after significant cartilage deterioration. A number of methods have been formed to monitor the development of osteoarthrosis. Radiography has been the conventional method for both diagnosis and monitoring. The topographical variation in degenerative cartilage has been used in designing an arthroscopic indentation instrument in which osteoarthrosis could be diagnosed *in vivo* and possibly treated before the diminishing qualities of the disease (148). Other manners that have been proposed for detection are chondrocalcin measurement (149) and knee wear particle analysis, which is derived from the concept of abrasive wear (150).

CURRENT CARTILAGE REPAIR STRATEGIES

Cartilage's poor capacity for self-regeneration is well known. The poor regenerative capacity of cartilage has been contrasted to bone, because bone readily regenerates (unless it is a critical size bone defect) and has a relatively rich blood supply (151). A lack of angiogenesis has been cited as the primary cause for cartilage's poor capacity for regeneration. However, normal cartilage is avascular. Vascular supply to cartilage likely will turn it into bone. Thus, a lack of vascularization is not the direct cause for cartilage's poor regenerative capacity (152).

The regenerative ability of cartilage in response to injury also depends on factors such as joint loading, the degree of injury, the location of injury, and whether it is cartilage lesion alone or osteochondral lesion (153,154). Cartilage responds differently to slowly or rapidly applied loads. For example, loading causes fluid movement in the matrix and may serve to counteract the deformation and to distribute the loads throughout the tissue. However, rapid compressive loading may not allow the fluid to infuse matrix, thus transferring excessive loading to the cells and the ECM macromolecules. Should excessive force be sustained, the chondrocytes and ECM molecules may undergo rupture or degradation. Another factor in determining the cartilage's regenerative capability is the chondrocytes' intrinsic ability to replenish the supply of matrix molecules, as well as the approaches to remove degraded materials (17,153,155).

Articular cartilage injuries can be classified as follows: (1) cartilage matrix and cell injuries without substantial tissue defects; (2) defects, fissures, or ruptures in articular cartilage only; and (3) osteochondral lesions. The first category of cartilage injuries without substantial tissue defects, nonetheless, is associated with a decreased concentration of matrix macromolecules such as PGs and collagen. Albeit without tissue-level defects, loss of PGs and collagen results in a decrease in mechanical strength. Unless repaired, even the first category of cartilage injuries can lead to more substantial defects as described in the second and third categories (153,155).

The second category of articular cartilage injuries is localized within cartilage. They include focused mechanical disruption of the matrix including fissures, tears, incisions, or interruption of the integrity of articular surface. Chondrocytes not only attempt to replace the loss of matrix macromolecules but also proliferate to fill the voids created during injury (153,156). However, the rates of chondrocyte proliferation and matrix synthesis may not be sufficiently high to match the rate of cartilage degradation. As articular cartilage has no nerve supply, except at the very periphery, even substantial cartilage lesions may not elicit pain. A few types of synovial joint injuries likely exist that elicit pain, such as osteochondral injuries, synovial membrane injuries, and injuries to the periphery of articular cartilage. These injuries are usually not repaired by the articular chondrocytes (4).

The third category is osteochondral injuries that involve both articular cartilage and subchondral bone and elicit inflammatory responses such as an influx of blood-borne cells, platelets, and cytokines. Hemorrhaging or fibrin clot formation may occur and later develop into a fibrous mass. The influx of cytokines may induce migration of progenitor cells, although no guarantee exists that these progenitor cells, likely mesenchymal stem cells that are capable of

differentiating into all connective tissue lineage cells, will differentiate into chondrocytes. In fact, osteochondral lesions are likely repaired, if reparable, by fibrocartilage or fibrous tissue instead of hyaline cartilage (154), and rarely possess the complex zonal structures of native articular cartilage (153,157). The mechanical strength of fibrocartilage is approximately one-third of the strength of native hyaline articular cartilage, and thus may not be able to fulfill the weight-bearing and load-bearing functions of normal articular cartilage. Over time, osteochondral lesions may undergo further degradation, leading to the exposure of subchondral bone, which results in osteoarthrosis and can lead to joint immobility.

Current clinical treatments for articular cartilage injuries have several deficiencies. Depending on the degree of injury and whether the defect is partial- or full-thickness, the treatments generally involve surgical irrigation, debridement, and tissue augmentation. Partial thickness injuries of the articular cartilage involving clefts and fissures, often in the early stages of osteoarthrosis, are most commonly treated with arthroscopic surgery such as lavage or debridement. Arthroscopic lavage involves the irrigation of the joint, whereas debridement is the arthroscopic removal of damaged tissue. By performing these treatments either alone or in combination, a decrease in joint pain usually results. However, lavage or debridement treatments rarely induce the repair process of cartilage (155,157).

Full-thickness injuries refer to lesions in both articular cartilage and subchondral bone. Although a large number of treatments are empirical, several procedures have taken the advantage to simulate the native repair process. Arthroscopic treatments such as abrasion arthroplasty, Pridie drilling, and microfracture are commonly used, and all include the further perforation of the subchondral bone to induce bleeding and further fibrous tissue formation. Abrasion arthroplasty and microfracture are used in conjunction with debridement to reduce the amount of damaged tissue within the joint. The outcome of these treatments is variable, largely because the healing and repair process within the articular surface are somewhat unpredictable. Furthermore, factors such as the patient's age, postoperative activity level, and overall heath also affect the outcome (158).

Recently, soft tissue grafts such as the transplantation of the periosteum or perichondrium have been used clinically to repair the articular surface for cylindrical, full-thickness defects. The rationale for using periosteum is its observed chondrogenic potential during development and fracture repair (159). The periosteum consists of a fibrous and cambial layer. The cambial layer contains precursor cells that are capable of differentiating into osteoblasts and perhaps chondrocytes. The process of periosteum transplantation involves the creation of a defect spanning the full thickness of articular cartilage and penetrating the subchondral bone, and then placement of the periosteum graft within the defect. However, much debate has occured as to which layer of the periosteum should lay adjacent to the bone and which layer should face the articular surface, as the cambial layer can form cartilage, whereas the fibrous layer forms fibrous tissue. Larger full-thickness defects are generally repaired using allogenic or autogenic osteochondral tissue plugs, called mosaicplasty, excised from nonload-bearing regions of the joint and inserted into the full-thickness defect. Reports exist of fibrous tissue formation and chondrocyte death at the interface between the plug and surrounding tissue, which may lead to further degeneration of the joint. Furthermore, donor site morbidity remains a drawback of the periosteum graft or mosaicplasty (160).

For substantial osteochondral lesions, total joint replacement using metallic condyle and plastic socket is most commonly in practice. Current modalities of total joint replacements suffer from drawbacks such as donor site morbidity, pathogen transmission, wear and tear, and a limited life span (161). Secondary surgeries are necessary in 10–15% of the cases and suffer from substantial difficulties such as scar tissue formation and loss of host tissue (162). More importantly, current total joint replacement therapies fail to yield biological regeneration.

CURRENT MENISCAL REPAIR STRATEGIES

Prior to the recognition of the importance of the meniscus in the biomechanics of the knee joint in the 1970s, the preferred treatment for meniscal injury such as tear was total excision of the meniscus or open meniscectomy. Despite some reports of temporary relief of symptoms, the long-term outcome of excision or meniscectomy was poor. In the 1980s, understanding of the material properties and biomechanical roles of the meniscus led to more conservative treatments for meniscal tears. Furthermore, the development of arthroscopy enabled more accurate diagnosis of meniscal tears and, subsequently, more precise surgical treatment that has substantially reduced the amount of damage to the surrounding tissue in comparison with open-joint surgery (163).

Partial-thickness split tears and small (< 5 mm) full-thickness split tears, vertically or obliquely, are usually left alone and without surgical intervention. The inner wall of the meniscus must be stable during probing, which is commonly performed arthroscopically during diagnosis. Follow-up arthroscopic examinations are often necessary to monitor tissue healing. These injuries are usually associated with ligament tears such as the anterior cruciate ligament (ACL). Ligament tears in conjunction with meniscal tears drastically reduce the stability of the knee and can lead to further meniscal damage.

Meniscal injuries that require surgical repair or excision are large defects with compromised vascular supply, large meniscal deformations, or damage to the peripheral or circumferential collagen fibers. In need of excision, it is preferable to leave a partial meniscus intact as opposed to full meniscectomy. The surgical approaches are either open-joint surgeries or arthroscopic surgery to suture tears within the meniscus (163).

TISSUE ENGINEERING OF ARTICULAR CARTILAGE AND MENISCUS

The rapidly evolving field of tissue engineering has promised to deliver biological replacements of damaged articular cartilage and meniscus. In comparison with current treatment modalities, tissue engineering represents a shift in the paradigm. Whereas current treatment modalities improve articular cartilage and meniscal injuries by increments, the end goal of tissue engineering is to generate or regenerate articular cartilage and meniscus.

Previous investigations of the structural, biochemical, and mechanical properties of articular cartilage and meniscus serve as the necessary foundation and have set the stage for the tissue engineering of these structures. For example, a commonly stated long-term goal of a biochemical study to investigate the PG distribution in various zones of articular cartilage was to improve the treatment of cartilage defects in arthritis, which was all too familiar for a published study or a grant proposal decades ago.

To tissue-engineer articular cartilage or meniscus, one has the conceptual liberty of selecting cell sources, scaffolds, or growth factors. Another essential choice is whether mechanical stress is to be applied to the tissue-engineered articular cartilage or meniscus prior to or after *in vivo* implantation. Thus, the initial stage of tissue engineering of articular cartilage and meniscus is an optimization process of cells, scaffolds, growth factors, or mechanical stimulus.

Cells Capable of Generating Articular Cartilage and Meniscus

Articular chondrocytes are the obvious choices for articular cartilage regeneration (164–166). From the standpoint of scientific discovery, articular cartilage or meniscal regeneration from articular chondrocytes or meniscal fibrochondrocytes has revealed a wealth of information (e.g., 167–169). From the standpoint of alignment with eventual therapeutic regeneration of synovial joint condyle in arthritis patients, the selection of articular chondrocytes is problematic. The essential problem at this time is that articular chondrocytes are not very expandable *ex vivo*. Thus, relatively large donor site defects are necessary to harvest a large amount of tissue(s) from the patient in order to obtain sufficient numbers of articular chondrocytes or meniscal fibrochondrocytes for healing substantial articular cartilage or meniscal defects. In contrast, mesenchymal stem cells, whose natural progeny includes both chondrocytes and fibrochondrocytes, can be obtained in small quantities (e.g., a few cc of bone marrow content) (170) or from other connective tissue sources such as adipose tissue (170,171), readily expanded *ex vivo* in cell culture and reliably differentiated into chondrocytes cells (170,172). Embryonic stem cells may turn out to be a viable cell source for synovial joint regeneration, especially in consideration of the recent demonstration of the differentiation of embryonic stem cells into osteogenic cells (173,174). Embryonic stem cells are likely to be of greater significance in synovial joint condyle regeneration if the isolation and expansion of adult MSCs encounter substantial difficulties. However, thus far, it does not appear to be the case for synovial joint condyle regeneration (172). Chondrogenic and osteogenic cells derived from MSCs appear to be the logical choices at this time for exploring clinically applicable approaches toward regenerating the synovial joint condyle (152,172,175–180).

Biomaterial Scaffolds Are often Necessary for the Engineering of Structural Tissues such as Articular Cartilage and Meniscus

The optimal scaffolds for articular cartilage and meniscal regeneration are yet to be determined. An increasing number of meritorious studies have reported a wide range of natural and synthetic polymers for articular cartilage regeneration. Many of the tested synthetic polymers are biocompatible and biodegradable, two desirable features for cartilage regeneration (181). A model scaffold should allow effective diffusion of essential nutrients and metabolic wastes, given that chondrocytes and fibrochondrocytes both rely on diffusion for survival.

For cartilage regeneration, natural and synthetic polymers may need to simulate the extracellular matrix environment of chondrocytes that are created by Type II collagen and PGs in a highly aqueous matrix. Several hydrogels simulate cartilage matrix to various degrees, such as alginate, hyaluronate, chitosan, and polyethylene glycol-based polymers (175,176,181–183). The organization of chondrocyte phenotypes in various zones of articular cartilage may also need to be simulated, as demonstrated in recent reports by encapsulating articular chondrocytes from various zones of bovine articular cartilage into different hydrogel layers (184–186). The water content and diffusion properties of hydrogels mimic an ECM to allow tissue-forming cells to obtain systemic nutrients (187). The initial viscous liquid form of several hydrogel materials provides a unique capability to form complicated shapes while maintaining uniform cell distributions. For examples, an aqueous-derived silk scaffold also encompasses hydrogel-type properties and has been shown to support chondrogenesis (188). Pellet culture is a practice devoid of scaffolds that takes advantage of the dense, avascular, and aneural condition of cartilage. This system is done simply by centrifuging chondrocytes or MSCs into a pellet and incubating them in desired conditions. Although pellets do not provide efficient shape retention, they do bestow valuable information *in vitro* as models for chondrogenic MSC differentiation. Despite various levels of reported success with *in vitro* models, cell-hydrogel interactions need to be better understood, along with optimization of hydrogel composition, cross-linking, and degradation behavior as a function of the *in vivo* regenerative outcome.

In engineering an osteochondral construct, it is essential to construct a mold that is specific to the joint. Computer-aided approaches have been developed to construct molds that will both accurately replicate the anatomy of the joint as well as preserve the intricate architectural integrity of the interior of the scaffold. These intricate details can range from pore size to channel orientation to surface texture and can effectively contribute to the synthesis, or lack thereof, of the tissue one is trying to engineer. Common methods for 3D mold fabrication are based on software programs that read and digitize computerized tomography or magnetic resonance imaging. Solid free-form fabrication (SFF) technology can be used to produce an actual 3D scaffold through combining the interior architecture image and the external scaffold image, which is done via a layering process from computer-aided design files. SFF shows excellent promise due to the possibility of controlling the aforementioned necessary parameters needed in scaffold fabrication.

Growth Factors Are Necessary for Modulating Cell Behavior

Growth factors are proteins and polypeptides capable of modulating all aspects of cell behavior such as proliferation, differentiation, and apoptosis. Cells can be regulated by self-released growth factors (autocrine effect), or by growth factors released by other cells (paracrine effect). For chondrogenic differentiation, TGF-β superfamily is frequently used (e.g., 176,189). The de-differentiation and re-differentiation of chondrocytes are regulated by combinations of TGF-β1, fibroblast growth factor-2, and sequential exposure to IGF-I (190). TGF-βs also stimulate GAG synthesis in isolated cultured meniscus cells (191). Platelet-derived growth factor (PDGF) promotes chondrogenesis in chick limbs both *in vitro* and *in vivo* (191). Resting zone chondrocytes treated with PDGF shows hypertrophic activity in forming new cartilage (192). Although many growth factors have provided positive results, the optimal scheme of their application is not yet fully understood. The reader is referred to several recent in-depth reviews of growth factor delivery in cartilage regeneration (154,183,193).

Functional Tissue Engineering of Articular Cartilage and Meniscus

The field of functional tissue engineering has been proposed in response to the need to engineer tissues that have not only the appropriate cellular and matrix structures, but also the necessary physical properties (194,196). The aforementioned mechanical properties of both articular cartilage and meniscus are quite complex and intricate to replicate. The rather severe loading environment of cartilage in a synovial joint attributes to poor repair capabilities as well as degeneration. It has been proposed that a potential solution to engineer tissues with the appropriate physical properties lies in the physical factors, among which mechanical stress is most well studied (194–196). Prior to the conception of functional tissue engineering, it is widely acknowledged that mechanical factors can effectively influence cell behavior such as proliferation, differentiation, and matrix synthesis. Significant evidence exists that physical stress can accelerate or improve tissue regeneration and repair *in vitro*. The preference of chondrocytes to synthesize proper ECM components at an accelerated pace can help engineer cartilage and meniscus in an efficient manner. This concept can be used to enhance engineered grafts to more accurately represent native tissue. In order to better provide these stresses, a movement has begun to construct mechanical bioreactors that increase matrix synthesis through different approaches. Examples in which mechanical stimulation has proven advantageous in synthesizing cartilage has been through fluid flow (197,198), simulated hypogravity (199), simulated microgravity (200), static and cyclic compression (201–203), and hydrostatic pressure (204). Recent studies have suggested that external mechanical loading of cells or cell-polymer constructs may enhance their mechanical strength (e.g., 186,205,206). Although practical in theory, the type, frequency, area, and amount of loading still need to be determined to furnish the optimal results of engineering each individual tissue.

The mechanical properties of engineered articular cartilage and meniscus can be readily tested, in many ways similar to the mechanical testing of native articular cartilage and meniscus. The results of compressive moduli of native and engineered articular cartilage or meniscus tissue using excised tissue plants and *in vitro* testing can be directly compared. However, loading measurements of native or engineered articular cartilage and meniscus have not been accomplished *in vivo*. Therefore, the ideal mechanical properties of tissue engineered articular cartilage can only be estimated, not accurately determined. Various biomaterials have been investigated for articular cartilage repair and have various mechanical properties that must be considered when designing a suitable scaffold. Engineering a scaffold that is too stiff may detrimentally affect cell viability and matrix synthesis. Stress shielding may occur and surrounding tissue may degrade as physiological loads are transferred to the more mechanically stiff implant. In contrast, a scaffold too soft will not exhibit the needed mechanical stiffness for the applied loads, causing physical breakdown and degradation of the material, leaving a void in the host tissue. Various zones of articular and fibrocartilage have different mechanical properties (56,207). Whether regional differences in the mechanical properties of articular cartilage and meniscus need to be simulated in tissue engineering remains to be explored (79).

SUMMARY AND CONCLUSIONS

Articular cartilage and meniscus are load-bearing and hydrated tissues that are key structures of diarthrodial joints. The similarities between the two structures include their remarkable capacity for resistance to and transmission of mechanical stress and their ability to enable joint lubrication. However, articular cartilage and meniscus have many important differences. Articular cartilage, in the overwhelming majority of human synovial joints, consists of hyaline cartilage, whereas the meniscus is composed of fibrocartilage. The mechanical properties of native articular cartilage and meniscus have been studied extensively and shown to vary with species, location, and even the orientation of test specimens. Motivated by the concept of functional tissue engineering, the mechanical properties of engineered articular cartilage from cells and biomaterials have also been investigated in recent years. As a result of drastically different structural and mechanical properties between articular cartilage and meniscus, the engineering challenges of the two tissues are different. Mesenchymal stem cells, or other stem cells, need to be differentiated into chondrocytes for the engineering of articular cartilage, whereas these stem cells may need to be differentiated into fibroblasts and chondrocytes, or perhaps fibrochondrocytes, for the engineering of meniscus. The optimal biomaterials for the engineering of articular cartilage and meniscus remain to be identified or fabricated. At least, the engineered articular cartilage and meniscus must adapt to possess similar mechanical properties of their native target tissues. The existing knowledge on the biological and mechanical properties of articular cartilage and meniscus provides the necessary foundation for the eventual goal to regenerate or replace diseased or lost articular cartilage and knee meniscus with engineered tissue analogs.

BIBLIOGRAPHY

1. Buckwalter JA, Mankin HJ. Instructional course lectures, the American Academy of Orthopaedic Surgeons—articular cartilage. Part II: Degeneration and osteoarthrosis, repair, regeneration, and transplantation. J Bone Joint Surg 1997;79:612–632.
2. Maroudas A. Physicochemical properties of articular cartilage. In: Freeman MAR, ed. Adult Articular Cartilage. Kent, UK: Pitman Medical Publishing; 1979. pp 215–290.
3. Mow VC, Gu WY, Chen FH. Structure and function of articular cartilage and meniscus. In: Mow VC, Huiskes R, eds. Basic Orthopaedic Biomechanics and Mechano-Biology; Third ed. New York: Lippincott Williams & Wilkins; 2005. pp 181–258.
4. Buckwalter JA, Mow VC. Cartilage repair in osteoarthritis. In: Moskowitz RW, Howell DS, Goldberg VM, Mankin HJ, eds. Osteoarthritis: Diagnosis and Management, 2nd ed. Philadelphia, PA: Saudners; 1992.
5. Howell DS, Treadwell BV, Trippel SB. Etiopathogenesis of osteoarthritis. In: Moskowitz RW, Goldberg VM, Howell DS, Altman RD, Buckwalter JA, eds. Osteoarthritis: Diagnosis and Medical/Surgical Management. 2th ed. Philadelphia, PA: WB Saunders; 1992. pp 233–252.
6. Poole AR, Howell DS. Etiopathogenesis of osteoarthritis. In: Moskowitz RW, Howell DS, Altman RD, Buckwalter JA, Goldberg VC, editors. Osteoarthritis, Diagnosis and Management. 3rd ed. Philadelphia, PA: WB Saunders Publishers; 2001. pp 29–47.
7. Mankin HJ, Mow VC, Buckwalter JA. Articular cartilage structure, composition and function. In: Buckwalter JA, Einhorn TA, Simon SR, eds. Orthopaedic Basic Science: Biology and Biomechanics of the Musculoskeletal System. Rosemont, IL: American Academy of Orthopaedic Surgeons; 2000. pp 443–470.
8. Benjamin M, Evans EJ. Fibrocartilage. J Anat 1990;171: 1–15.
9. Weightman B. Tensile fatigue of human articular cartilage. J Biomech 1976;9:193–200.
10. Seedhom BB, Wallbridge NC. Walking activities and wear of prosthesis. Ann Rheum Dis 1985;44:838–843.
11. Morrison JB. Bioengineering analysis of force actions transmitted by the knee joint. J Biomech Eng 1968;3:164–170.
12. Morrison JB. The mechanics of the knee joint in relation to normal walking. J Biomech 1970;3:51–61.
13. Athanasiou KA, Rosenwasser MP, Buckwalter JA, Malinene TI, Mow VC. Interspecies comparisons of in situ intrinsic mechanical properties of distal femoral cartilage. J Orthop Res 1991;9:330–340.
14. Ewers BJ, Dvoracek-Driksna D, Orth MW, Haut RC. The extent of matrix damage and chondrocyte death in mechanically traumatized articular cartilage explants depends on rate of loading. J Orthop Res 2001;19:779–784.
15. Torzilli PA, Grigiene R, Borrelli J Jr, Helfet DL. Effect of impact load on articular cartilage: Cell metabolism and viability, and matrix water content. J Biomech Eng 1999;121: 433–441.
16. Centers for Disease Control and Prevention (CDC). Prevalence of self-reported arthritis or chronic joint symptoms among adults- United States, 2001. MMWR 2002;51:948–950.
17. Vunjak-Novakovic G, Goldstein SA. Biomechanical principles of cartilage and tissue engineering. In: Mow VC, Huiskes R, eds. Basic Orthopaedic Biomechanics and Mechano-Biology,

3rd ed. New York: Lippincott Williams & Wilkins; 2005. pp 343–407.

18. Mow VC, Ratcliffe A, Chern KY, Kelly MA. Structure and function relationships of the meniscus of the knee. In: Mow VC, Arnoczky SP, Jackson DW, eds. Knee Meniscus: Basic and Clinical Foundations. New York: Raven Press; 1992. pp 37–57.

19. Mow VC, Ratcliffe A, Poole AR. Cartilage and diarthroidal joints as paradigms for hierarchical materials and structures. Biomaterials 1992;13:67–97.

20. Arnoczky SP. Gross and vascular anatomy of the meniscus and its role in meniscal healing, regeneration, and remodeling. In: Mow VC, Arnoczky SP, Jackson DW, eds. Knee Meniscus, Basic and Clinical Foundations. New York: Raven Press; 1992. pp 1–14.

21. Uhthoff HK, Wiley JJ. Behavior of the Growth Plate. New York: Raven Press; 1988.

22. Cohen B, Chorney GS, Phillips DP, Dick HM, Mow VC. Compressive stress-relaxation behavior of bovine growth plate may be described by the nonlinear biphasic theory. J Orthop Res 1994;12:804–813.

23. Williams PL. Gray's Anatomy. New York: Churchill Livingston; 1995.

24. Shimazu A, Nah HD, Kirsch T, Koyama E, Leatherman JL, Golden EB, Kosher RA, Pacifici M. Syndecan-3 and the control of chondrocyte proliferation during endochondral ossification. Exp Cell Res 1996;229:126–136.

25. Mao JJ, Nah HD. More research needed to understand how orthodontists communicate with cells. Am J Orthod Dentofacial Orthop 2004;125:676–689.

26. Tamamura Y, Otani T, Kanatani N, Koyama E, Kitagaki J, Komori T, Yamada Y, Costantini F, Wakisaka S, Pacifici M, Iwamoto M, Enomoto-Iwamoto M. Developmental regulation of Wnt/beta-catenin signals is required for growth plate assembly, cartilage integrity, and endochondral ossification. J Biol Chem 2005;280:19185–19195.

27. Mow VC, Kuei SC, Lai WM, Armstrong CG. Biphasic creep and stress relaxation of articular cartilage in compression: Theory and experiments. J Biomech Eng 1980;102:73–84.

28. Eyre DR. Structure and function of the cartilage collagen: Role of type IX in articular cartilage. In: Brandt KD, ed. Cartilage Changes in Osteoarthritis. Indianapolis, IN: Ciba-Geigy; 1990. pp 12–16.

29. Heinegard D, Bayliss M, Lorenzo P. Biochemistry and metabolism of normal and osteoarthritic cartilage. In: Brandt KD, Doherty M, Lohmander LS, eds. Osteoarthritis. New York: Oxford University Press; 2003. pp 73–82.

30. Eyre DR, Wu JJ. Collagen of fibrocartilage: A distinctive molecular phenotype in bovine meniscus. FEBS Lett 1983;158:265–270.

31. Eyre DR, Oguchi H. The hydroxypyridinium crosslinks of skeletal collagens: Their measurement, properties and a proposed pathway of formation. Biochem Biophys Res Commun 1980;92:402–410.

32. Eyre DR, Dickson IR, Van Ness K. Collagen cross-linking in human bone and articular cartilage. Age-related changes in the content of mature hydroxypyridinium residues. Biochemistry 1988;252:495–500.

33. Huang CY, Mow VC, Ateshian GA. The role of flow-independent viscoelasticity in the biphasic tensile and compressive responses of articular cartilage. J Biomech Eng 2001;123:410–417.

34. Huang CY, Soltz MA, Kopacz M, Mow VC, Ateshian GA. Experimental verification of the roles of intrinsic matrix viscoelasticity and tension-compression nonlinearity in the biphasic response of cartilage. J Biomech Eng 2003;125:84–93.

35. Kempson GE, Freeman MA, Swanson SA. Tensile properties of articular cartilage. Nature 1968;220:1127–1128.

36. Kempson GE, Muir H, Pollard C, Tuke M. The tensile properties of the cartilage of human femoral condyles related to the content of collagen and glycosaminoglycans. Biochim Biophys Acta 1973;297:456–472.

37. Roth V, Mow VC. The intrinsic tensile behavior of the matrix of bovine articular cartilage and its variation with age. J Bone Joint Surg 1980;62:1102–1117.

38. Woo SL-Y, Akeson WH, Jemmott GF. Measurements of nonhomogeneous directional mechanical properties of articular cartilage in tension. J Biomechan 1976;9:785–791.

39. Lai WM, Hou JS, Mow VC. A triphasic theory for the swelling and deformation behaviors of articular cartilage. J Biomech Eng 1991;113:245–258.

40. Buckwalter JA, Rosenberg LC. Electron microscopic studies of cartilage proteoglycans: Direct evidence for the variable length of the chondroitin sulfate-rich region of proteoglycan subunit core protein. J Biol Chem 1982;257:8930–8939.

41. Muir H. Proteoglycans as organizers of the intercellular matrix. Biochem Soc Trans 1983;9:613–622.

42. Muir H. The chondrocyte, architect of cartilage: Biomechanics, structure, function and molecular biology of cartilage matrix macromolecules. Bioessays 1995;17:1039–1048.

43. Sun DN, Guo XE, Likhitpanichkul M, Lai WM, Mow VC. The influence of the fixed negative charges on mechanical and electrical behaviors in articular cartilage under unconfined compression. J Biomech Eng 2004;126:1–11.

44. Hunziker EB, Michel M, Studer D. Ultrastructure of adult human articular cartilage matrix after cryotechnical processing. Microsc Res Tech 1997;37:271–284.

45. Hultkrantz W. Ueber die spaltrichtungen der gelenkknorpel. Anat Anzeig verhandl anat Gesellsch 1898;14:248–256.

46. Benninghoff A. Form und bau der gelenkknorpel in ihren beziehungen zu funktion. II. Der aufbau des gelenkknorpel in seinen beziehungen zu funktion. Z Zellforsch 1925;2:783–862.

47. Broom ND. The collagen framework of articular cartilage: Its profound influence on normal and abnormal load-bearing function. In: Nimni ME, ed. Collagen: Chemistry, Biology and Biotechnology. Boca Raton, FL: CRC Press; 1988. pp 243–265.

48. Ceohn ZA, Henry JH, McCarthy DM, Mow VC, Ateshian GA. Computer simulations of patellofemoral joint surgery. Am J. Sports Med 2003;31:87–98.

49. Lipshitz H, Etheredge R 3rd, Glimcher MJ. Changes in the hexosamine content and swelling ratio of articular cartilage as functions of depth from the surface. J Bone Joint Surg 1976;58:1149–1153.

50. Mitrovic D, Quintero M, Stankovic A, Ryckewaert A. Cell density of adult human femoral condylar articular cartilage. Joints with normal and fibrillated surfaces. Lab Invest 1983;49:309–316.

51. Aydelotte MB, Schumacher BL, Kuettner KE. Heterogeneity of articular chondrocytes. In: Kuettner KE, ed. Articular Cartilage and Osteoarthritis. New York: Raven Press; 1992. pp 237–249.

52. Kempson GE. Mechanical properties of adult articular cartilage. In: Freeman MAR, ed. Adult Articular Cartilage. Kent, UK: Pitman Medical Publishing; 1979. pp 215–290.

53. Akizuki S, Mow VC, Muller F, Pita JC, Howell DS. Tensile properties of human knee joint cartilage. II. Correlations between weight bearing and tissue pathology and the kinetics of swelling. J Orthop Res 1987;5:173–186.

54. Hunziker EB, Tyler JA. Articular cartilage repair. In: Brandt KD, Doherty M, Lohmander LS, eds. Osteoarthritis. Oxford, UK: Oxford University Press; 2003. pp 93–101.

55. Jadin KD, Wong BL, Bae WC, Li KW, Williamson AK, Schumacher BL, Price JH, Sah RL. Depth-varying density and organization of chondrocytes in immature and mature bovine articular cartilage assessed by 3-D imaging and analysis. J Histochem Cytochem DOI: 10.1369/jhc.4A6511. 2005 (In Press).

56. Tomkoria S, Patel RV, Mao JJ. Heterogeneous nanomechanical properties of superficial and zonal regions of articular cartilage of the rabbit proximal radius condyle by atomic force microscopy. Med Eng Phys 2004;26:815–822.

57. Broom ND, Silyn-Roberts H. Collagen-collagen versus collagen-proteoglycan interactions in the determination of cartilage strength. Arthritis Rheum 1990;33:1512–1517.

58. Quinn TM, Hunziker EB, Hauselmann HJ. Variation of cell and matrix morphologies in articular cartilage among locations in the adult human knee. Osteoarthritis Cartilage 2005;13:672–678.

59. Guilak F, Mow VC. The mechanical environment of the chondrocyte: A biphasic finite element model of cell-matrix interactions in articular cartilage. J Biomech 2000;33:1663–1673.

60. Guilak F. The deformation behavior and viscoelastic properties of chondrocytes in articular cartilage. Biorheology 2000;37: 27–44.

61. Guilak F. Compression-induced changes in the shape and volume of the chondrocyte nucleus. J Biomech 1995;28:1529–1542.

62. Knight MM, Ross JM, Sherwin AF, Lee DA, Bader DL, Poole CA. Chondrocyte deformation within mechanically and enzymatically extracted chondrons compressed in agarose. Biochem Biophys Acta 2001;1526:141–146.

63. Graff RD, Kelley SS, Lee GM. Role of pericellular matrix in development of a mechanically functional neocartilage. Biotechnol Bioengin 2003;20:457–464.

64. Allen DM, Mao JJ. Heterogeneous nanostructural and nanoelastic properties of pericellular and interterritorial matrices of chondrocytes by atomic force microscopy. J Struct Biol 2004;145:196–204.

65. Stockwell RA. Biology of Cartilage Cells. Cambridge, UK: Cambridge University Press; 1979.

66. Ateshian GA, Mow VC. Friction, lubrication, and wear of articular cartilage and diarthroidal joints. In: Mow VC, Huiskes R, eds. Basic Orthopaedic Biomechanics and Mechano-Biology, Third ed. New York: Lippincott Williams & Wilkins; 2005. pp 447–494.

67. McDevitt CA, Miller A, Spindler KP. The cells and cell matrix interaction of the meniscus. In: Mow VC, Arnoczky SP, Jackson DW, eds. Knee Meniscus, Basic and Clinical Foundations. New York: Raven Press; 1992. pp 29–36.

68. Bullough PG, Munuera L, Murphy J, Weinstein AM. The strength of the menisci of the knee as it relates to their fine structure. J Bone Joint Surg Br 1970;52:564–567.

69. Aspden RM, Yarker YE, Hukins DW. Collagen orientations in the meniscus of the knee joint. J Anat 1985;140:371–380.

70. Kelly MA, Fithian DC, Chern KY, Mow VC. Structure and function of the meniscus: Basic and clinical implications. In: Mow VC, Ratcliffe A, Woo SLY, eds. Biomechanics of Diarthroidal Joints. New York: Springer-Verlag; 1990. pp 191–214.

71. Yasui K. Three-dimensional architecture of human normal menisci. J Jpn Orthop Assoc 1978;52:391–399.

72. Skaggs DL, Warden WH, Mow VC. Radial tie fibers influence the tensile properties of the bovine medial meniscus. J Orthop Res 1994;12:176–185.

73. Nakano T, Thompson JR, Aherne FX. Distribution of glycosaminoglycans and the nonreducible collagen crosslink, pyridinoline in porcine menisci. Can J Vet Res 1986;50: 532–536.

74. Fithian DC, Kelly MA, Mow VC. Material properties and structure-function relationships in the menisci. Clin Orthop Relat Res 1990;252:19–31.

75. Walker PS, Erkman MJ. The role of the menisci in force transmission across the knee. Clin Orthop Relat Res 1975;109:184–192.

76. Smith MM, Ghosh P. Experimental models of osteoarthritis. In: Moskowitz RW, Howell DS, Altman RD, Buckwalter JA, Goldberg VC, eds. Osteoarthritis, Diagnosis and Management. 3rd ed. Philadelphia, PA: WB Saunders Publishers; 2001. pp 171–199.

77. Moskowitz RW, Davis W, Sammarco J, et al., Experimentally induced degenerative joint lesions following partial meniscectomy in the rabbit. Arthritis Rheum 1973;16:397–404.

78. Dehaven KE. The role of the meniscus. In: Ewing JW, ed. Articular Cartilage and the Knee Joint Function: Basic Science and Arthroscopy. New York: Raven Press; 1990. pp 103–115.

79. Guilak F, Butler DL, Goldstein SA. Functional tissue engineering: The role of biomechanic in articular cartilage repair. Clin Orthop Relat Res 2001; (391 Suppl):S295–S305.

80. Kempson GE. Mechanical properties of articular cartilage and their relationship to matrix degeneration and age. Ann Rheum Dis 1975;34:111–113.

81. Kempson GE. Age-related changes in the tensile properties of human articular cartilage: A comparative study between the femoral head of the hip joint and the talus of the ankle joint. Biochim Biophys Acta 1991;1075:223–230.

82. Akizuki S, Mow VC, Muller F, Pita JC, Howell DS, Manicourt DH. Tensile properties of human knee joint cartilage: I. Influence of ionic conditions, weight bearing, and fibrillation on the tensile modulus. J Orthop Res 1986;4:379–392.

83. Setton LA, Mow VC, Muller FJ, Pita JC, Howell DS. Mechanical properties of canine articular cartilage are significantly altered following transection of the anterior cruciate ligament. J Orthop Res 1994;12:451–463.

84. Setton LA, Zhu W, Mow VC. The biphasic poroviscoelastic behavior of articular cartilage: role of the surface zone in governing the compressive behavior. J Biomech 1993;26: 581–592.

85. Soltz MA, Ateshian GA. A conewise linear elasticity mixture model for the analysis of tension-compression nonlinearity in articular cartilage. J Biomech Eng 2000;122:576–586.

86. Huang CY, Stankiewicz A, Ateshian GA, Mow VC. Anisotropy, inhomogeneity, and tension-compression nonlinearity of human glenohumeral cartilage in finite deformation. J Biomech 2005;38:799–809.

87. Fithian DC, Zhu WB, Ratcliffe A, Kelly MA, Mow VC. Exponential law representation of tensile properties of human meniscus. Proc Inst Mech Eng 1989;c384/058:85–90.

88. Bader DL, Kempson GE, Barrett AJ, Webb W. The effects of leucocyte elastase on the mechanical properties of adult human articular cartilage in tension. Biochemica et Biophysica Acta 1981;677:103–108.

89. Schmidt MB, Mow VC, Chun LE, Eyre DR. Effects of proteoglycan extraction on the tensile behavior of articular cartilage. J Orthop Res 1990;8:353–363.

90. Hirsh 1944.

91. Sokoloff L. Elasticity of aging cartilage. Fed Proc 1966;25: 1089–1095.

92. Kempson GE, Freeman MA, Swanson SA. The determination of a creep modulus for articular cartilage from indentation tests of the human femoral head. J Biomech 1971;4: 239–250.

93. Kempson GE, Spivey CJ, Swanson SA, Freeman MA. Patterns of cartilage stiffness on normal and degenerate human femoral heads. J Biomech 1971;4:597–609.

94. Hoch DH, Grodzinsky AJ, Koob TJ, Albert ML, Eyre DR. Early changes in material properties of rabbit articular cartilage after meniscectomy. J Orthop Res 1983;1:4–12.

95. Parsons JR, Black J. The viscoelastic shear behavior of normal rabbit articular cartilage. J Biomech 1977;10:21–29.

96. Altman RD, Tenenbaum J, Latta L, Riskin W, Blanco LN, Howell DS. Biomechanical and biochemical properties of dog cartilage in experimentally induced osteoarthritis. Ann Rheum Dis 1984;43:83–90.

97. Armstrong CG, Lai WM, Mow VC. An analysis of the unconfined compression of articular cartilage. J Biomech Eng 1984;106:165–173.

98. Mak AF. The apparent viscoelastic behavior of articular cartilage— The contributions from the intrinsic viscoelasticity and interstitial fluid flow. J Biomech Eng 1986;108:123–130.

99. Mak AF. Unconfined compression of hydrated viscoelastic tissues: A biphasic poroviscoelastic analysis. Biorheology 1986;23:371–383.

100. Mak AF, Lai WM, Mow VC. Biphasic indentation of articular cartilage—I. Theoretical analysis. J Biomech 1987;20:703–714.

101. Zhu W, Mow VC, Koob TJ, Eyre DR. Viscoelastic shear properties of articular cartilage and the effects of glycosidase treatments. J Orthop Res 1993;11:771–781.

102. Armstrong CG, Mow VC. Variations in the intrinsic mechanical properties of human articular cartilage with age, degeneration, and water content. J. Bone Joint Surg 1982;64:88–94.

103. Roth V, Mow VC, Lai WM, Eyre DR. Correlation of intrinsic compressive properties of bovine articular cartilage with its uronic acid and water content. Trans Orthop Res Soc 1981;6:21.

104. Froimson MI, Ratcliffe A, Gardner TR, Mow VC. Differences in patellofemoral joint cartilage material properties and their significance to the etiology of cartilage surface fibrillation. Osteoarthritis Cartilage 1997;5:377–386.

105. Jurvelin J, Kiviranta I, Arokoski J, Tammi M, Helminen HJ. Indentation study of the biochemical properties of articular cartilage in the canine knee. Eng Med 1987;16:15–22.

106. Stahursky TM, Armstrong CG, Mow VC. Variation of the intrinsic aggregate modulus and permeability of articular cartilage with trypsin digestion. Proc Biomech Symp Trans ASME 1981;AMD43:137–140.

107. Mow VC, Zhu W, Lai WM, Hardingham TE, Hughes C, Muir H. The influence of link protein stabilization on the viscometric properties of proteoglycan aggregate solutions. Biochim Biophys Acta 1989;992:201–208.

108. Schinagl RM, Ting MK, Price JH, Sah RL. Video microscopy to quantitate the inhomogeneous equilibrium strain within articular cartilage during confined compression. Ann Biomed Eng 1996;24:500–512.

109. Chen AC, Bae WC, Schnagl RM, Sah RL. Depth- and strain-dependent mechanical and electromechanical properties of full-thickness articular cartilage in confined compression. J Biomechan 2001;34:1–12.

110. Wang CC, Hung CT, Mow VC. An analysis of the effects of depth-dependent aggregate modulus on articular cartilage stress-relaxation behavior in compression. J Biomech 2001;34:75–84.

111. Cohen B, Lai WM, Mow VC. A transversely isotropic biphasic model for unconfined compression of growth plate and chondroepiphysis. J Biomech Eng 1998;120:491–496.

112. Hayes WC, Mockros LF. Viscoelastic properties of human articular cartilage. J Appl Physiol 1971;31:562–568.

113. Spirt AA, Mak AF, Wassell RP. Nonlinear viscoelastic properties of articular cartilage in shear. J Orthop Res 1989;7:43–49.

114. Simon WH, Mak A, Spirt A. The effect of shear fatigue on bovine articular cartilage. J Orthop Res 1990;8:86–93.

115. Roth V, Schoonbeck JM, Mow VC. Low frequency dynamic behavior of articular cartilage under torsional shear. Trans Orthop Res Soc 1982;7:150.

116. Zhu W, Lai WM, Mow VC. Intrinsic quasilinear viscoelastic behavior of the extracellular matrix of cartilage. Trans Orthop Res Soc 1986;11:407.

117. Setton LA, Mow VC, Howell DS. Mechanical behavior of articular cartilage in shear is altered by transection of the anterior cruciate ligament. J Orthop Res 1995;13:473–482.

118. Fung YC. Mechanical Properties of Living Tissues. New York: Springer-Verlag; 1981.

119. Zhu W, Chern KY, Mow VC. Anisotropic viscoelastic shear properties of bovine meniscus. Clin Orthop Relat Res 1994;306:34–45.

120. Mow VC, Mak AF, Lai WM, Rosenberg LC, Tang LH. Viscoelastic properties of proteoglycan subunits and aggregates in varying solution concentrations. J Biomech 1984;17:325–338.

121. Zhu W, Mow VC, Rosenberg LC, Tang LH. Determinations of kinetic changes of aggrecan-hyaluronan interactions in solution from its rheological properties. J Biomech 1994;27: 571–579.

122. Hayes WC, Bodine AJ. Flow-independent viscoelastic properties of articular cartilage matrix. J Biomechan 1978;11: 407–419.

123. Maroudas A, Muir H, Wingham J. The correlation of fixed negative charge with glycosaminoglycan content of human articular cartilage. Biochim Biophys Acta 1969;177:492–500.

124. Donnan FG. The theory of membrane equilibria. Chem Rev 1924;1:73–90.

125. Myers ER, Lai WM, Mow VC. A continuum theory and an experiment for the ion-induced swelling behavior of articular cartilage. J Biomech Eng 1984;106:151–158.

126. Setton LA, Tohyama H, Mow VC. Swelling and curling behaviors of articular cartilage. J Biomech Eng 1998;120: 355–361.

127. Setton LA, Lai WM, Mow VC. Swelling-induced residual stress and the mechanism of curling in articular cartilage in vitro. In: Tarbell JM, ed. Advances in Bioengineering. 1993. pp 59–62.

128. Wan LQ, Miller C, Guo XE, Mow VC. A three-layer orthotropic model for swelling and curling of articular cartilage. ASME-BED. Vail, Colorado, 2005.

129. Olsen S, Oloyede A. A finite element analysis methodology for representing the articular cartilage functional structure. Comput Meth Biomech Biomed En 2002;5(6):377–386.

130. Gu WY, Lai WM, Mow VC. A mixture theory for charged-hydrated soft tissues containing multi-electrolytes: Passive transport and swelling behaviors. J Biomech Eng 1998;120: 169–180.

131. Huyghe JM, Janssen JD. Quadriphasic mechanics of swelling incompressible prorous media. Int J Eng Sci 1997;35: 793–802.

132. Gu WY, Lai WM, Mow VC. Transport of fluid and ions through a porous-permeable charged-hydrated tissue, and streaming potential data on normal bovine articular cartilage. J Biomech 1993;26:709–723.

133. Gu WY, Yao H. Effects of hydration and fixed charge density on fluid transport in charged hydrated soft tissues. Ann Biomed Eng 2003;31:1162–1170.

134. Lai WM, Mow VC, Sun DD, Ateshian GA. On the electric potentials inside a charged soft hydrated biological tissue: streaming potential versus diffusion potential. J Biomech Eng, 2000;122:336–346.

135. Lai WM, Sun DD, Ateshian GA, Guo XE, Mow VC. Electrical signals for chondrocytes in cartilage. Biorheology 2002;39: 39–45.

136. Mow VC, Ateshian GA, Lai WM, Gu WY. Effects of fixed charges on the stress-relaxation behavior of hydrated soft tissues in a confined compression problem. Int J Solids Structures 1998;35:4945–4962.

137. Sun DN, Gu WY, Guo XE, Lai WM, Mow VC. A mixed finite element formulation of triphasic mechano-electrochemical theory for charged, hydrated biological soft tissues. Int J Num Methods Eng 1999;45:1375–1402.

138. Lu X, Sun DD, Guo XE, Chen FC, Lai WM, Mow VC. Indentation determined mechano-electrochemical properties and fixed charge density of articular cartilage. Ann Biomed Eng 2004;32:370–379.

139. Wan LQ, Miller C, Guo XE, Mow VC. Fixed electrical charges and mobile ions affect the measurable mechano-electrochemical properties of charged-hydrated biological tissues: The articular cartilage paradigm. Mechan Chem Biosyst 2004;1: 81–99.

140. Ateshian GA, Chahine NO, Basalo IM, Hung CT. The correspondence between equilibrium biphasic and triphasic material properties in mixture models of articular cartilage. J Biomechan 2004;37:391–400.

141. Likhitpanichkul M, Miller C, Guo XE, Mow VC. A triphasic model of cell under micropipette aspiration: The osmotic effect on cell mechanical properties. ASME-BED. Vail, Colorado, 2005.

142. Lu X, Miller C, Guo XE, Mow VC. A new correspondence principle for triphasic materials: Determination of fixed charge density and porosity of articular cartilage by indentation. ASME-BED. Vail, Colorado, 2005.

143. Broom ND. Structural consequences of traumatizing articular cartilage. Ann Rheum Dis 1986;45:225–234.

144. Cawston T. Matrix metalloproteinases and TIMPs: Properties and implications for the rheumatic diseases. Mol Med Today 1998;4:130–137.

145. Lotz M, Hashimoto S, Kuhn K. Mechanisms of chondrocyte apoptosis. Osteoarthritis Cartilage 1999;7:389–391.

146. Meredith Jr JE , Fazeli B, Schwartz MA. The extracellular matrix as a cell survival factor. Mol Biol Cell 1993;4:953–961.

147. Hashimoto S, Ochs RL, Komiya S, Lotz M. Linkage of chondrocyte apoptosis and cartilage degradation in human osteoarthritis. Arthritis Rheum 1998;41:1632–1638.

148. Lyyra T, Kiviranta I, Vaatainen U, Helminen HJ, Jervelin JS. In vivo characterization of indentation stiffness of articular cartilage in the normal human knee. J Biomed Mater Res 1999;48:482–487.

149. Kobayashi T, Yoshihara Y, Samura A, Tanaka O, Shimmei M. Chondrocalcin as a marker of articular cartilage degeneration in anterior cruciate ligament-deficient knees. Orthopedics 1998;21:773–776.

150. Kuster MS, Posdiadlo P, Stachowiak GW. Shape of wear particles found in human knee joints and their relationship to osteoarthritis. Br J Rheumatol 1998;37:978–984.

151. Hollinger JO, Winn S, Bonadio J. Options for tissue engineering to address challenges of the aging skeleton. Tissue Eng 2000;6:341–350.

152. Mao JJ. Stem-cell driven regeneration of synovial joints. Biol Cell 2005;97:289–301.

153. Buckwalter JA. Articular cartilage injuries. Review Clin Orthop Relat Res 2002;3:257–264.

154. Hunziker EB. Articular cartilage repair: basic science and clinical progress. A review of the current status and prospects. Osteoarthritis Cartilage 2002;10:432–463.

155. Martin JA, Buckwalter JA. The role of chondrocyte-matrix interactions in maintaining and repairing articular cartilage. Biorheology 2000;37:129–140.

156. Redman SN, Oldfield SF, Archer CW. Current strategies for articular cartilage repair. Eur Cell Mater 2005;9:23–32.

157. Johnson LL. Arthroscopic Abrasion Arthroscopy. In: McGinty JB, ed. Operative Arthroplasty. Philadelphia, PA: Lippincott-Raven; 1996.

158. Shapiro F, Koide S, Glimcher MJ. Cell origin and differentiation in the repair of full-thickness defects of articular cartilage. J Bone Joint Surg Am 1993;75:532–553.

159. O'Driscoll SW. Articular cartilage regeneration using periosteum. Clin Orthop 1999;367:S186–S203.

160. Ahmad CS, Guiney WB, Drinkwater CJ. Evaluation of donor site intrinsic healing response in autologous osteochondral grafting of the knee. Arthoscopy 2002;18:95–98.

161. NIH Consensus Panel. NIH Consensus Statement on total knee replacement December 8–10, 2003. J Bone Joint Surg Am 2004;86-A: 1328–1335.

162. Haydon CM, Mehin R, Burnett S, Rorabeck CH, Bourne RB, McCalden RW, MacDonald SJ. Revision total hip arthroplasty with use of a cemented femoral component. Results at a mean of ten years. J Bone Joint Surg Am 2004;86-A: 1179–1185.

163. DeHaven KE. Meniscectomy versus repair: Clinical experience. In: Mow VC, Arnoczky SP, Jackson DW, eds. Knee Meniscus: Basic and Clinical Foundations. New York: Raven Press; 1992. pp 131–139.

164. Jadlowiec JA, Celil AB, Hollinger JO. Bone tissue engineering: Recent advances and promising therapeutic agents. Expert Opin Biol Ther 2003;3:409–423.

165. Vacanti JP, Langer R. Tissue engineering: The design and fabrication of living replacement devices for surgical reconstruction and transplantation. Lancet 1999;354(Suppl)1: S132–S134.

166. Zhang JY, Doll BA, Beckman EJ, Hollinger JO. Three-dimensional biocompatible ascorbic-acid containing scaffold for bone tissue engineering. Tissue Eng 2003;9:1143–1157.

167. Weng Y, Cao Y, Silva CA, Vacant MP, Vacanti CA. Tissue-engineered composites of bone and cartilage for mandible condylar reconstruction. J Oral Maxillofac Surg 2001;59: 185–190.

168. Niederauer GG, Slivka MA, Leatherbury NC, Korvick DL, Harroff HH, Ehler WC, Dunn CJ, Kieswatter K. Evaluation of multiphase implants for repair of focal osteochondral defects in goats. Biomaterials 2000;21:2561–2574.

169. Freed LE, Grande DA, Lingbin Z, Emmanuel J, Marquis JC, Langer R. Joint resurfacing using allograft chondrocytes and synthetic biodegradable polymer scaffolds. J Biomed Mater Res 1994;28:891–899.

170. Caplan AI. Mesenchymal stem cells. J Orthop Res 1991;9: 641.

171. Gimble J, Guilak F. Adipose-derived adult stem cells: Isolation, characterization, and differentiation potential. Cytotherapy 2003;5:362–369.

172. Alhadlaq A, Mao JJ. Mesenchymal stem cell: Isolation and therapeutics. Stem Cells Develop 2004;13:436–448.

173. Buttery LD, Bourne S, Xynos JD, Wood H, Hughes FJ, Hughes SP, Episkopou V, Polak JM. Differentiation of osteoblasts and in vitro bone formation from murine embryonic stem cells. Tissue Eng 2001;7:89–99.

174. Sottile Thomson VA, McWhir J. In vitro osteogenic differentiation of human ES cells. Cloning Stem Cells 2003;5:149–155.

175. Alhadlaq A, Mao JJ. Tissue-engineered neogenesis of human-shaped mandibular condyle from rat mesenchymal stem cells. J Dent Res 2003;82:950–955.

176. Alhadlaq A, Elisseeff JH, Hong L, Williams CG, Caplan AI, Sharma B, Kopher RA, Tomkoria S, Lennon DP, Lopez A, Mao JJ. Adult stem cell driven genesis of human-shaped articular condyle. Ann Biomed Eng 2004;32:911–923.

177. Gao J, Dennis JE, Solchaga LA, Awadallah AS, Goldberg VM, Caplan AI. Tissue-engineered fabrication of an osteochondral composite graft using rat bone marrow-derived mesenchymal stem cells. Tissue Eng 2001;7:363–371.

178. Gao J, Dennis JE, Solchaga LA, Goldberg VM, Caplan AI. Repair of osteochondral defect with tissue-engineered two-phase composite material of injectable calcium phosphate and hyaluronan sponge. Tissue Eng 2002;8:827–837.

179. Gao J, Caplan AI. Mesenchymal stem cells and tissue engineering for orthopaedic surgery. Chir Organi Mov 2003;88:305–316.

180. Rahaman MN, Mao JJ. Stem cell based composite tissue constructs for regenerative medicine. Biotechnol Bioeng 2005;91:261–284.

181. Lee KY, Mooney DJ. Hydrogels for tissue engineering. Chem Rev 2001;101:1869–1879.

182. Anseth KS, Metters AT, Bryant SJ, Martens PJ, Elisseeff JH, Bowman CN. In situ forming degradable networks and their application in tissue engineering and drug delivery. J Control Release 2002;78:199–209.

183. Randolph MA, Anseth K, Yaremchuk MJ. Tissue engineering of cartilage. Clin Plast Surg 2003;30:519–537.

184. Klein TJ, Schumacher BL, Schmidt TA, Li KW, Voegtline MS, Masuda K, Thonar EJ, Sah RL. Tissue engineering of stratified articular cartilage from chondrocyte subpopulations. Osteoarthritis Cartilage 2003;11:595–602.

185. Kim TK, Sharma B, Williams CG, Ruffner MA, Malik A, McFarland EG, Elisseeff JH. Experimental model for cartilage tissue engineering to regenerate the zonal organization or articular cartilage. Osteoarthritis Cartilage 2003;11:653–664.

186. Williams CG, Kim TK, Taboas A, Malik A, Manson P, Elisseeff J. In vitro chondrogenesis of bone marrow-derived mesenchymal stem cells in a photopolymerizing hydrogel. Tissue Eng 2003;9:679–688.

187. Peppas NA, Huang Y, Torres-Lugo M, Ward JH, Zhang J. Physicochemical foundations and structural design of hydrogels in medicine and biology. Annu Rev Biomed Eng 2000;2:9–29.

188. Wang Y, Kim U, Blasioli DJ, Kim H, Kaplan DL. In vitro cartilage tissue engineering with 3D porous aqueous-derived silk scaffolds and mesenchymal stem cells. Biomaterials 2005;26:7082–7094.

189. Pittenger MF, Mackay AM, Beck SC, Jaiswal RK, Douglas R, Mosca JD, Moorman MA, Simonetti DW, Craig S, Marshak DR. Multilineage potential of adult human mesenchymal stem cells. Science 1999;284:143–147.

190. Pei M, Seidel J, Vunjak-Novakovic G, Freed LE. Growth factors for sequential cellular de- and re-differentiation in tissue engineering. Biochem Biophys Res Commun 2002;294:149–154.

191. Collier S, Ghosh P. Effects of transforming growth factor beta on proteoglycan synthesis by cell and explant cultures derived from the knee joint meniscus. Osteoarthritis Cartilage 1995;3:127–138.

192. Lohmann CH, Schwartz Z, Niederauer GG, Boyan BD. Degree of differentiation of chondrocytes and their pretreatment with platelet-derived growth factor. Regulating induction of cartilage formation in resorbable tissue carriers in vivo. Orthopade 2000;29(2):120–128.

193. Almarza AJ, Athanasiou KA. Design characteristics for the tissue engineering of cartilaginous tissues. Ann Biomed Eng 2004;32:2–17.

194. Butler DL, Shearn JT, Juncosa N, Dressler MR, Hunter SA. Functional tissue engineering parameters toward designing repair and replacement strategies. Clin Orthop Relat Res 2004; Suppl: S190–S199.

195. Guilak F, Fermor B, Keefe FJ, Kraus VB, Olson SA, Pisetsky DS, Setton LA, Weinberg JB. The role of biomechanics and inflammation in cartilage injury and repair. Clin Orthop Relat Res 2004;423:17–23.

196. Wang CC, Guo XE, Sun D, Mow VC, Ateshian GA, Hung CT. The functional environment of chondrocytes with cartilage subjected to compressive loading: A theoretical and experimental approach. Biorheology 2002;39:11–25.

197. Freed LE, Martin I, Vunjak-Novakovic G. Frontiers in tissue engineering. In vitro modulation of chondrogenesis. Clin Orthop Relat Res 1999; (367 Suppl):S46–S58.

198. Pazzano D, Mercier KA, Moran JM, Fong SS, DiBiasio DD, Rulfs JX, Kohles SS, Bonassar LJ. Comparison of chondrogenesis in static and perfused bioreactor culture. Biotechnol Prog 2000;16:893–896.

199. Freed LE, Langer R, Martin I, Pellis NR, Vunjak-Novakovic G. Tissue engineering of cartilage in space. PNAS USA 1997;94:13885–13890.

200. Freed LE, Vunjak-Novakovic G. Microgravity tissue engineering. In Vitro Cell Dev Biol Anim 1997;33:381–385.

201. Buschmann MD, Gluzband YA, Grodzinsky AJ, Hunziker EB. Mechanical compression modulates matrix biosynthesis in chondrocyte/agarose culture. J Cell Sci 1995;108(Part 4):1497–1508.

202. Mauck RL, Soltz MA, Wang CC, Wong DD, Chao PH, Valhmu WB, Hung CT, Ateshian GA. Functional tissue engineering of articular cartilage through dynamic loading of chondrocyte-seeded agarose gels. J Biomech Eng 2000;122:252–260.

203. Seidel JO, Pei M, Gray ML, Langer R, Freed LE, Vunjak-Novakovic G. Long-term culture of tissue engineered cartilage in a perfused chamber with mechanical stimulation. Biorheology 2004;41:445–458.

204. Saris DB, Sanyal A, An KN, Fitzsimmons JS, O'Driscoll SW. Periosteum responds to dynamic fluid pressure by proliferating in vitro. J Orthop Res 1999;17:668–677.

205. Simmons CA, Matlis S, Thornton AJ, Chen S, Wang CY, Mooney DJ. Cyclic strain enchances matrix mineralization by adult human mesenchymal stem cells via the extracellular signal-regulated kinase (ERK1/2) signaling pathway. J Biomech 2003;36:1087–1096.

206. Grodzinsky AJ, Levenston ME, Jin M, Frank EH. Cartilage tissue remodeling in response to mechanical forces. Annu Rev Biomed Eng 2000;2:691–713.

207. Hu K, Radhakrishnan P, Patel RV, Mao JJ. Regional structural and viscoelastic properties of fibrocartilage upon dynamic nanoindentation of the articular condyle. J Struct Biol 2001;136:46–52.

Further Reading

Archard JF. Wear theory and mechanisms. In: Peterson MB, Winder WO, eds. *Wear Control Handbook*. New York: ASME Publications; 1980. pp 35–80.

Dowson D. Basic tribology. In: Dowson D, Wright V, eds. *Introduction to the Biomechanics of Joints and Joint Replacement*. London: Mechanical Engineering Publications, Ltd.; 1981. pp 120–145.

Mow VC, Setton LA, Howell DS, Buckwalter JA. Structure-function relationships of articular cartilage and the effects of joint instability and trauma on cartilage function. In: Brandt KD, ed. *Cartilage Changes in Osteoarthritis*. Indianapolis, IN: Ciba-Geigy; 1990. pp 22–42.

Mow VC, Sun DD, Guo XE, Likhitpanichkul M, Lai WM. Fixed negative charges modulate mechanical behavior and electrical signals in articular cartilage under unconfined compression: The triphasic paradigm. In: *Porous Media, Proc Tribute to Professor Reint de Boer*. Berlin: Springer Verlag; 2002. pp 227–247.

See also Biomechanics of exercise fitness; joints, biomechanics of; ligament and tendon, properties of.

CATARACT EXTRACTION. See LENSES, INTRAOCULAR.

CELL COUNTER, BLOOD

YI ZHANG
SRIRAM NEELAMEGHAM
University of Buffalo
Buffalo, New York

INTRODUCTION: NATURE OF BLOOD CELLS (1,2)

Cells compose $\sim 50\%$ of the volume of normal human blood, while plasma constitutes the remaining volume. Generally, cells in blood are divided into three categories: platelets, erythrocytes (or red blood cells, RBCs) and leukocytes (or white blood cells, WBCs) (Table 1) (3). Among these, the platelets or thrombocytes are small, irregular, disk-shaped cells that lack a nucleus. They are of size 2–3 μm in diameter. These cells primarily function to stop bleeding or hemorrhage, and they also participate in coronary artery disease. They do so by being part of the blood coagulation cascade and by aggregating with each other. Platelets are found in blood at a concentration of 0.15–0.5×10^6 cells·mm^{-3}. The second type of cells in blood, erythrocytes, contains a red respiratory protein called hemoglobin. These are disk-shaped, biconcave cells without nuclei. Their diameter ranges from 6 to 8 μm and their thickness is 1.5–2.5 μm. The primary function of erythrocytes is to transport oxygen and carbon dioxide between the lung and body tissues. Erythrocytes are the most numerous blood cells at concentrations of $4–6 \times 10^6$ cells·mm^{-3}. Mature erythrocytes emerge from precursors that are called reticulocytes. Erythrocyte counts are on average $\sim 10\%$ higher in the human adult male population than those in the female population. Lack of iron and hemoglobin in erythrocytes can lead to anemia, a pathological deficiency in the oxygen-carrying component of blood. The third type of blood cells is the leukocytes, whose primary function is to provide the body with immunity and to protect it from infection. Leukocytes are fewer in number than the erythrocytes with a concentration of $\sim 5–10 \times 10^3$ cells·mm^{-3}. These cells are roughly spherical in shape and they contain nuclei, and considerable internal and cell-surface structures.

Leukocytes are categorized in various ways depending on their function and differentiation pathway. One common method subdivides these cells into myeloid and lymphoid cells. Myeloid cells differentiate into phagocytes, while lymphoid cells primarily produce lymphocytes. The phagocytes include polymorphonuclear granulocytes and monocytes–macrophages. Of these, the former have lobed, irregular shaped (polymorphic) nucleus. They are further subdivided into neutrophils (55–70% of all leukocytes), eosinophils (2–4%), and basophils (0.5–1%), on the basis of how the cellular cytoplasmic granules are stained with acidic and basic dyes. Leukocytes are also commonly characterized based on particular cell-surface receptors that are expressed by them, since these are specific to a particular subpopulation. Many of these receptors are recognized by monoclonal antibodies. A systematic nomenclature has now evolved in which the term CD (Cluster Designation) refers to a group of antibodies that recognize a particular cell-surface antigen. The CD classification is thus often used to classify and identify particular leukocyte subpopulations. A method of blood cell counting called flow cytometry (described below), often uses the fluorescence of labeled antibodies to distinguish between various cell types.

Among the granulocytes, neutrophils are the most abundant cell type. These represent the body's first line of defense during immune response. They are characterized by a number of segmented nucleus lobes connected by fine nuclear strands or filaments. Immature–young neutrophils have a band- or horseshoe-shaped nucleus. Thus, while the younger neutrophils are known as band neutrophils, the mature cells are the segmented neutrophils. Segmented neutrophils are the predominant species in human blood, while band neutrophil levels are elevated following bacterial infection or acute inflammation. The term left shift is used to indicate an increase in the number of circulating immature neutrophils. Condition under which the number of circulating neutrophils is increased is called neutrophilia, while a decrease in this cell type results in neutropenia. Another important morphological characteristic of neutrophils is the virtual lack of endoplasmic reticulum and mitochondria. Mature neutrophils have short lifetimes in circulation and they migrate into tissues to defend against invading microbes during inflammation. Eosinophils, like other granulocytes, possess a polymorphous nucleus, generally with two lobes. They can response to allergy and parasitic infection. They attack large parasites such as helminthes via their C3b receptors. Eosinophils release various substances from

Table 1. Characteristic of Normal Blood Cells[a]

Cell Type	Concentration	Size, μm	Density, g·mL	Shape	Nucleus	Cytoplasm
Platelet(thrombocyte)	$0.15–0.5 \times 10^6 \cdot \mu L^{-1}$	2–3	1.03–1.06	Small disk shape	None	Granular
Erythrocyte (RBC)	$4–6 \times 10^6 \cdot \mu L^{-1}$	6–8	1.09–1.11	Biconcave disk	None	Hemoglobin
Leukocyte (WBC)	$5–10 \times 10^3 \cdot \mu L^{-1}$	8–20	1.05–1.10			
Neutrophil	55–70% of WBC	9–15	1.08–1.10	Various	Lobed	Granular
Eosinophil	2–4% of WBC	9–15	1.08–1.10	Various	Lobed	Granular
Basophil	0.5–1% of WBC	10–16	1.08–1.10	Various	Lobed	Granular
Monocyte	3–8% of WBC	14–20	1.05–1.08	Various	Round	Fine
Lymphocyte	20–40% of WBC	8–16	1.05–1.08	Round	Round	Clear

[a](Adapted from Ref. 3. pp 3–6.)

their eosinophilic granules. These include major basic proteins, plus cationic proteins, peroxidase, phospholipase D and histaminase. The number of eosinophils can augment in blood during allergy (eosinophilia), dermatological disorder and parasitic infection. Basophils have a two-lobe nucleus. They release inflammatory mediators, such as histamine and bradykinin, and prostaglandins and leukotrienes. Basophils play an important role in inflammatory and allergic response. Their number is increased in patients with hypoactive thyroid conditions and during certain malignancies like chronic myeloid leukemia.

Monocytes and macrophages (tissue monocytes) represent the second type of phagocytes and these are relatively large, long-lived cells compared to polymorphonuclear granulocytes. Their cytoplasm is transparent with typically a horseshoe-shape nucleus. Monocytes are involved in both acute and chronic inflammation. The transformation from monocytes to macrophages is controlled by different cytokines. When responding to chemical signals at the inflammation site, monocytes quickly migrate from the blood vessels and start to perform phagocytotic activity. These cells also have an intense secretory activity that results in the production and secretion of chemical mediators such as lysozymes and interferons. The number of monocytes in circulation is increased whenever there is increased amount of cell damage, such as during recovery from infection.

Lymphocytes are mononuclear cells that constitute ~ 20–40% of all leukocytes. These cells have a round or oval shaped nucleus that is typically large in comparison to the overall cell size. Besides circulating in blood vessels, lymphocytes also populate the lymphoid organs, as well as the lymphatic circulation. The specificity of immune response is due to lymphocytes, since these cells can distinguish between different antigenic determinants. Lymphocytes are subdivided into three main categories: (1) T-cells, (2) B-cells, and (3) natural killer (NK) cells. T-Cells are responsible for cellular immune response and are involved in the regulation of antibody reactions by either helping or suppressing the activation of B lymphocytes. B cells are the primary source of cells responsible for humoral–antibody responses. These are responsible for the production of immune antibodies. The NK cells destroy target cells via nonphagocytic reaction mechanisms that are termed cytotoxic reaction. Lymphocyte number may increase in blood in patients with skin rashes from certain viral diseases such as measles and mumps, in patients with thyrotoxicosis, and in patients recuperating from certain acute infections.

RATIONALE FOR CELL COUNT

Blood cell count is achieved by determining the concentrations and other parameters of different cell types in a unit volume of circulating blood. It can be a complete blood count (CBC, defined later in this article), which examines every blood component, or it may measure only one element. Table 1 demonstrates the characteristic of normal blood cells from a healthy adult. These values vary with age, sex, race, living habit, and health status. Further, under pathological conditions, the distribution of blood cells may be perturbed and

thus blood cell counting can aid diagnostics. For example, a typical symptom of common anemia is an inadequate amount of RBCs. The increase or decrease in the numbers of the different types of WBCs may indicate infection and inflammation as discussed above. In addition to the cell numbers, other information regarding blood cells, such as cell size, is also important. For example, in patients with anemia caused by vitamin B_{12} deficiency, the average size of the RBCs is larger than normal and this disease state is called macrocytic anemia. On the contrary, if red blood cells are smaller than normal, as in the case of microcytic anemia, the condition may be indicative of iron deficiency. Therefore, a routine blood test, which includes not only blood cell count but also measurement of other parameters, can aid disease diagnosis and treatment by health professionals.

HISTORY AND BASIC PRINCIPLES FOR BLOOD CELL COUNTING (4–10)

Cell counting has evolved over the centuries from a manual method that heavily relied on microscopic examination, to one where electrical and optical measurement strategies, along with computer automation, are playing an increasingly important role. Indeed, manual methods are still important in research laboratories that study a wide variety of animal systems and cell types, in addition to human blood. On the other hand, automated systems are typically used in clinical studies. In this context, modern technology has automated the process of blood cell counting and the assessment of various blood cell parameters. A CBC, thus, not only provides a panel of tests to quantify the composition of whole blood, it may also include more detailed information regarding the cell profile. Based on technical feasibility and cost, either a simple manually operated cell counter or a more advanced automated blood count platform can be applied to serve the specific medical diagnose need. A brief history and rationale that has lead to current strategies for blood cell counting is outlined below.

Microscopy Coupled with Manual Visualization

Notable, among the early attempts to count blood cells, was the work by Anton van Leeuwenhoek in the seventeenth century who counted the number of chicken erythrocytes in a glass capillary of known dimensions using his microscope. Later, in the nineteenth and early twentieth century, Burker employed a shallow rectangular chamber with a thin coverglass as a counting chamber. Advances in this basic design have now resulted in the laboratory hemocytometer, which is commonly employed for manual cell counting. Ehrlich's classical work on the staining of white blood granules laid the foundations of hematology and differential cell counting. In these studies, he demonstrated that it is possible to distinguish between the various blood cell subgroups using acidic and basic dyes that differentially stained the cellular granules and nucleus.

Hemocytometer

One device that utilizes the light microscope for cell counting is the hemocytometer. This is a commercially available counting chamber that is used for manual blood counting. It consists of two parts: a microscopic slide with improved Neubauer ruling, and a special thick flat cover slip (Fig. 1a). Both the hemocytometer slide and cover slip must meet specifications of the National Bureau of Standards. The slides have two raised surfaces for duplicate cell counting, each of them bearing square-shaped grids of dimensions 3×3 mm. The two raised surfaces are separated by an H-shaped moat. Each of the 3×3 mm squares has a central area of size 1×1 mm that is further subdivided by 25 groups of 16 smaller squares (Fig. 1b and c). During cell counting, the coverslip is placed on top of the counting surfaces such that the distance between the counting surface and the coverslip is 0.1 mm. Thus, the total volume in the space between the central 1×1 mm area and the coverslip is fixed at 0.1 mm^3. Samples to be studied can be loaded into the chamber using a standard laboratory pipette placed at the point labeled V-slash.

A phase contrast microscope is used to view blood on a hemocytometer slide. In such runs, a sample of diluted blood mixture is placed in a hemocytometer. For a proper count, cells should be evenly distributed. In a white cell count, blood is typically diluted 1:20 in a solution that lyses red cells and stains white cells. Because red cells are so much more numerous than white cells, blood is normally diluted 1:200 for red cell counts. The total number of cells in the central area with fixed volume of 0.1 mm^3 is counted and this measurement is used to estimate the concentration of cells per cubic millimeter (mm^3) according to, cell concentration = number of cells counted × dilution factor / volume under central grid. For simplicity, instead of counting all the cells in the central 1×1 mm area, counting cells present in a sufficient number of representative squares is also reasonable as long as the acceptable level of accuracy can be ensured. A suitable convention should be applied to avoid counting cells twice, for example, by counting only those cells that touch the top and right-hand margins of a square and omitting cells that touch the bottom and left margins. The World Health Organization (WHO) has recommended methods for the visual determination of WBC count and platelet count using hemocytometer (Recommended methods for the visual determination of WBC count and platelet count. Geneva: World Health Organization, 2000. WHO/DIL/00.3). It describes the detailed sample preparation procedure and counting techniques, and this could be used as a basic protocol for cell counting using the hemocytometer.

Electrooptic Measurements

Advances in electronics and electrooptics in the twentieth century have dramatically simplified blood cell counting and made automation of these processes possible. Some examples of early advances are illustrated in Fig. 2. Panel a describes a method developed in the 1940s where cells

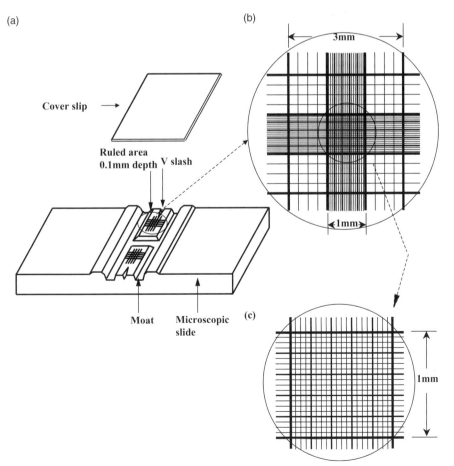

(a)

(b)

(c)

Cover slip

Ruled area 0.1mm depth   V slash

Moat   Microscopic slide

3mm

1mm

1mm

Figure 1. (a) Diagram of hemocytometer with cover slip. (b and c) Expanded view of ruled area as seen under a microscope.

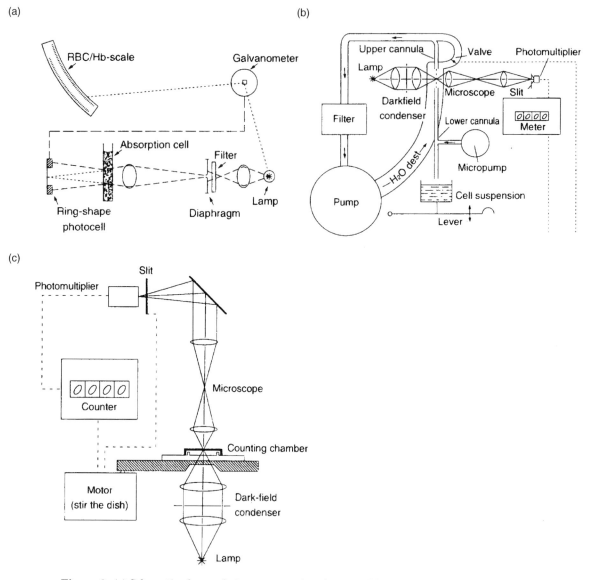

Figure 2. (a) Schematic of an early instrument using the ensemble method to obtain a blood cell count. The intensity of light that is scattered onto a ring-shaped photodetector is measured. The intensity is proportional to cell concentration. (b) Schematic of a photoelectric device that optically counts cells under flow. Here, a fluid stream containing cells is passed through a microscope viewing station. The photomultiplier detects the passage of cells. (c) Schematic diagram of a device using a photoelectric spot scanning method. Mechanical motion is provided to scan cells contained in the counting chamber (3).

were electrooptically measured based on turbidimetry. Here, light lost per millimeter of path length based on scattering or absorbance by blood cells was related to cell concentration. Using cell reference or artificial standards, thus, RBC concentrations could be determined. In this approach, instead of detecting cells individually, the cell concentrations were measured using principles analogous to Beer's law. Panel b illustrates a photoelectric device from 1953, where a thin fluid stream was created such that single cells passed via a microscope viewing station. Images of these cells were magnified and detected using a photomultiplier tube. Panel c illustrates another early instrument where erythrocytes could be counted automatically by means of photoelectric spot-scanning of a thin

layer of diluted blood. Here the manual visual counting chamber technique discussed above was improved by introducing a photomultiplier and an electronic counting unit. A motor drives the counting chamber. An instrument based on this principle is the Casella Counter shown in Fig. 2c.

Electronic Cell Counter

In 1950s, Wallace Coulter (Founder of Coulter Company, now Beckman-Coulter Co.) developed a method for cell counting based on electric impedance. This method now forms the basis of most particle size analysis methods in the world. This method, also called low voltage direct

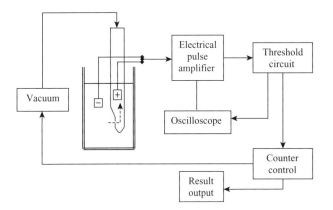

Figure 3. Schematic diagram of electronic cell counter using electric resistance method.

current (dc) method, is based on the measurement of changes in electrical resistance as cells pass through a small orifice that separates two electrodes. In this type of device (Fig. 3), cells are suspended in an electrically conductive diluent, such as saline. Low frequency electrical current is applied between two electrodes; one of them being placed in the cell medium and the second within the aperture tube. The aperture tube has a small orifice or sensing aperture that is typically of size 50–200 µm in diameter. During the measurement, cells are drawn through the aperture using a pressure gradient that is either generated by a mercury manometer or oil displacement pump. Cells are assumed to be non-conducting. Electrical resistance between the two electrodes or impedance in the current occurs as the cells pass through the sensing aperture, causing voltage pulses that are measurable. The number of pulses is proportional to the number of cells counted. The size of the voltage pulse is directly proportional to the size (volume) of the cell. This principle allows discrimination between cells of different sizes. Counting of specific-sized cells is also possible using threshold circuits that cut-off voltage pulses above and below predetermined values. The quantity of suspension drawn through the aperture is precisely controlled to allow the system to count and size particles precisely. Finally, several thousand particles are individually counted within seconds in this device. Measurements are independent of particle shape, color, and density.

Analogous to the above method is the radiofrequency (RF) resistance method where high voltage electromagnetic current is flown between the two electrodes instead of dc. This current circuits the cell membrane lipid layer and penetrates into the cell. While the dc method defines the volume of the cell, changes in conductivity measured using the RF method correlate with the cell's interior structure including the nucleus volume and density, and cytoplasm granule composition. Both dc and RF may be applied simultaneously and this can yield different information about cell size and cellular structure. Such a dual measurement strategy is employed by Sysmex cell counters to quantify the differential leukocyte counts (DLC) as discussed later.

Several factors affect the precision and accuracy of measurements made using the electric impedance meth-

ods. First, the aperture size is critical. The instrument is set to count only particles within the proper size range. The upper and lower levels of the size range are called size exclusion limits. Any cell or material larger or smaller than the size exclusion limits will not be counted. Sample must also not contain other material that might erroneously be counted as cells. In practice, erythrocyte and platelet aperture should be smaller than leukocyte aperture in order to increase platelet count sensitivity. Besides the size exclusion limits and aperture size, cell shape and physical properties are also important in determining the shape factor or the ratio of electrically measured volume to the geometric volume. Erythrocytes may result in different signals depending on their orientation with respect to the aperture in the sensing zone. Simultaneous passage of more than one cell at a time through aperture may also cause artificially large pulses, and thus circuits to correct for this coincidence error are required. The magnitude of the coincidence error increases with cell concentration. Correction should be completed by the countercomputer based on the relationship of cell count with cell concentration and aperture size. Finally, an internal cleaning system to prevent or slow down protein buildup in aperture is beneficial in minimizing aperture blockage.

Hydrodynamic focusing as discussed below helps to solve many of the problems above and it provides improved cell counting and characterization. This has been developed and assembled in many cell counters today and this feature dramatically improves the cell volume distribution resolution.

Laser Light Scattering and Fluorescence Detection

Optical scattering can be used alone or in combination with other electrical measurement strategies discussed above for cell counting and characterization. A key feature of such instruments is hydrodynamic focusing where an external sheath flow allows alignment of blood cells one-at-a-time in the path of a light beam, usually within a quartz flow cell (Fig. 4a). Incident light on cells within this flow stream are scattered or redirected in a manner that is dictated by the size of the cell and the intracellular distribution of refractive index. Lasers are generally preferred as the light source since it produces monochromatic light that has a small spot size. Photomultiplier tubes (PMTs) are used to collect the weak signal of scattered light. The light scattered at angles from 5–10° (forward scatter) in general correlates with cell size. Light scattered at 90° is called side-scatter and this is related to the cell shape, orientation, and cellular content. A cell with many complex intracellular organelles will also give a larger side-scatter signal than a cell with fewer intracellular organelles. The design of precise angles where scattered light signals are measured is specific to instrument manufacturers and the cell-enumeration strategy employed. In general, these scatter data together allow reliable identification of distinct populations, such as platelets, RBCs, monocytes, and neutrophils in a mixture. They can also allow enumeration of lymphocyte subsets and reticulocytes. While, optical scattering methods reveal information about cells that is

(a)

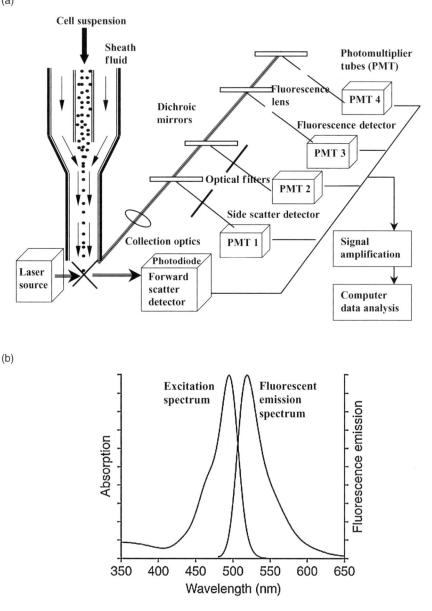

(b)

Figure 4. (a) Schematic of flow cytometry showing hydrodynamic focusing of cells by sheath fluid that brings the cells in the path of a laser beam. Light scattered by cell at various angles is collected. These are passed through an arrangement of optical filters to yield measures of forward scatter, side scatter and particle fluorescence. (b) Stokes shift is depicted for the fluorescent probe fluorescein where the wavelength of the absorbed and emitted quanta are shifted, with the emitted wavelength being longer than the absorbed light.

distinct from that obtained from the above electrical methods, their estimates of cell volume are not as accurate as the electrical methods.

A major advantage of optical methods using lasers is that such methods can be readily coupled with fluorescence detection (Fig. 4b). Fluorescent conjugated antibodies to specific cell-surface CD markers or specific ligands can be used not only to identify particular blood cells, but also to label cellular components that may be indicative of disease states. In such work, when the laser light reaches these cells, a fraction of the photons are absorbed by the fluorescent probe, which then reemit the photon at a longer wavelengths. The quantity of this emitted light (both scattered and fluorescent) is measured using photomultiplier tubes that are arranged in conjunction with a series of optical filters and dichroic mirrors as shown in Fig. 4a. The detection and conversion of scattered or fluorescent light into electrical signals is accomplished by photodetectors

that capture photons on a light sensitive surface that elicits an electron cascade. The signal output from such detectors is amplified (either linearly or logarithmically) and then converted from analog to digital form for computer analysis. Multidimensional plots of various scattering properties with fluorescent signals can thus be generated to individually characterize each cell in a complex mixture.

COMPLETE BLOOD CELL COUNT (5,10)

Computer Blood Cell Count is a series of tests that result in the quantitation of the number of erythrocytes, leukocytes, and platelets in a volume of blood. The measurements also estimate the hemoglobin content and packed cell volume (or hematocrit) of erythrocytes. This can be done manually using a microscope along with cytochemical dyes, such as Wright–Giemsa stain. The combination of acidic and basic dyes here can differentially stain the granules, cytoplasm, and nuclei of various blood cell types. Alternatively, in clinical laboratories an automated cell counter can be used to count cells in a given volume. Low end instruments offer RBC and platelet analysis with three-part differential leukocyte count (DLC) while higher end instruments may include a five-part differential count along with reticulocyte analysis. The speed of the instrument and level of automation varies with the class of instrument. The analysis thus obtained is compared with the normal range and assessed for clinical or research purposes. A complete blood cell count mainly includes the following parameters:

Hemoglobin

Hemoglobin concentration (HGB) is reported in grams per deciliter ($g \cdot dL^{-1}$) of blood. This parameter typically varies in proportion to erythrocyte concentration in blood. The normal range for hemoglobin is age and sex dependent. Traditionally, hemoglobin is measured using the cyanmethemoglobin method, as recommended by the International Council for Standardization in Hematology (ICSH). Here, a lysing agent is added to disrupt RBCs and to release cellular hemoglobin. This hemoglobin is converted into a stable form called cyanmethemoglobin (see reaction below), the quantity of which can be measured using a spectrophotometer for absorbance measurement at $\sim 540\,nm$.

$$Hb(Fe^{2+}) \xrightarrow{K_3Fe(CN)_6} methemoglobin(Fe^{3+})$$
$$\xrightarrow{KCN} cyanmethemoglobin$$

Since cyanmethemoglobin measurements contain poisonous cyanide reagent, other more environmentally friendly methods for automated HGB measurement have been developed. Among them, sodium lauryl sulfate-hemoglobin (SLS-Hb) method is used by Sysmex automated cell counters. Here, the lauryl group of the ionic surfactant, which is hydrophobic, binds strongly with hemoglobin. This binding leads to rapid globin molecular conformation change and conversion of hemoglobin from the ferrous (Fe^{2+}) to the ferric (Fe^{3+}) state. The hydrophilic group of SLS now binds with Fe^{3+} to form a stable SLS-Hb. The absorption maximum of SLS-Hb occurs at $535\,nm$ with a shoulder at $560\,nm$, and this feature is used to determine hemoglobin content. This reaction mechanism is useful since conversion to SLS-Hb occurs rapidly within $10\,s$.

Platelet Count

Platelet count (PLT) is normally expressed as thousands per microliter (μL) and can be measured manually using the hemocytometer. Care must be taken during such measurements to avoid platelet clumps that can occur in the absence of appropriate anticoagulant. Electronic counting of platelets can also be performed using electric impedance or light scattering methods. Such measurements are typically performed in channels that are designed to discriminate between erythrocytes and platelets. Size distributions resulting from platelet counts can be used to estimate the mean platelet volume (MPV), which is a measure of the platelet volume variation. In general, increased MPV

may be expected in regenerative thrombocytopenia, which is accompanied by an increased production of platelets by bone marrow.

Red Blood Cell Count

The RBC or erythrocyte count is expressed in millions per microliter of whole blood. Such counts can be measured manually using the hemocytometer. In hematology analyzers, RBC content is typically measured using either the dc impendence method, light scattering analysis, or a combination of the two. Attention is placed during these measurements to discriminate between small RBCs and platelets. Results of such analysis typically result in a RBC size distribution plot from which other indices can be estimated. These indices include: (1) Hematocrit (HCT), which is also called packed cell volume (PCV). This is a measure of the volume fraction of RBCs in whole blood expressed in %vol/vol. Normal adult hematocrit ranges from 35 to 50%, and this is both sex and age dependent. Traditionally, hematocrit is determined by monitoring the height of packed RBCs after centrifugation in a standard microhematocrit tube, relative to the column length. Electronic cell analyzers can also estimate hematocrit by measuring the individual volumes of RBCs (also called MCV as described below) and determining the product of RBC count and MCV. (2) Mean corpuscular volume (MCV) is the mean volume of RBCs expressed in femtoliters (fl). The normal range is ~ 80–$100\,fL$. The MCV can be experimentally determined from the RBC size distribution height. Alternatively, if HCT value is known, MCV is calculated based on the ratio of hematocrit and RBC count. This parameter is analogous to MPV, which can be derived from platelet data. When the MCV is low with normal HCT, the blood is said to be microcytic. (3) Mean corpuscular hemoglobin concentration (MCHC) is the mean concentration of hemoglobin in the RBCs in grams per deciliter ($g \cdot dL^{-1}$). This is calculated based on the ratio of HGB by HCT. Red cell populations with normal, high, or low values of MCHC are referred to as normochromic, hyperchromic, or hypochromic, respectively. The last case can occur during strongly regenerative anemia, where an increased population of reticulocytes with low HGB content pulls the average value down (an increased MCV would be expected under this scenario). (4) Mean corpuscular hemoglobin (MCH) is a measure of the mean mass of hemoglobin (HGB) in RBC, and is expressed in picograms (pg). (5) Red cell distribution width (RDW) is an index of the variation in cell volume within the RBC population. It is mathematically determined by (Standard deviation of RBC volume/ MCV) × 100. The normal range for RDW is 11–15%. While, red cell populations with normal RDW are called homogeneous, those with higher than normal are termed heterogenous. For example, increased number of reticulocytes, which is associated with erythropoiesis, will cause increased RDW values. The RDW index may be an early indicator of changes in red cell population sizes, for example, during anemia caused by iron deficiency. In this case, the presence of few microcytic RBCs may increase the standard deviation of the cell distribution even before marked changes in MCV are observed.

White Blood Cell Count

White blood cell or leukocyte count is measured in thousands per microliter. During manual WBC count, RBCs in blood are lysed and diluted sample is charged into the hemocytometer. Nucleated cells are counted and WBC concentration is determined. Alternatively, impedance-based electronic cell counters can be used to measure WBC count. Besides these basic methods, in automated cell counters, one of many technologies can be applied for WBC differential count. Beckman–Coulter instruments employ the VCS (volume, conductivity, and scattering) technology. In this method, the dc impedance principle is used to physically measure the volume of the cell that displaces the isotonic diluent. Alternating current in the RF range short circuits the bipolar lipid layer of the cell membrane allowing energy penetration into cell. This probe provides information on cell size and internal structure. This data is adjusted by the cell volume measurement to obtain an index called opacity. Finally, coherent light scattering from an incident laser beam is collected to obtain information on cellular granularity and cell surface structure. In Sysmex instruments both dc and RF methods are employed along with differential lysis of cells using lysis solution and temperature treatment. In CELL-DYN instruments from Abbott laboratories, the Multi-Angle Polarization Scattering Separation (M.A.P.S.S.) technology is used to obtain the differential count. Here light scattered by cells localized in a hydrodynamically focused flow stream is measured at three angles (0, 10, and 90°). Polarized light at 90° is also measured. Together these four parameters are used to perform the five-part differential count. Two methods are employed in the Bayer cell counters for differential leukocyte count. In the first method called the peroxidase method, RBCs are lysed and white cells are stained with peroxidase. These cells are counted based on size by forward scatter analysis, and absorbance using dark field optics. The second method, called the basophil method, involves stripping the cells using a non-ionic surfactant in acidic solution. Basophils are resis-

tant to lysis while RBCs and platelets are lysed and other leukocytes are stripped of their cytoplasm. Light scattering analysis distinguishes basophils from other polymorphonuclear and mononuclear cells. The above peroxidase and basophil methods thus provide automated differential cell count by separating the cells into clusters.

Reticulocyte Count (RTC, RET, or RETIC)

Reticulocytes are formed in the last stages of erythropoiesis. These cells spend ~2 days in the bone marrow and 1–2 days in peripheral blood prior to maturing into RBCs. These are nonnucleated RBC, which by definition upon staining with supravital dyes contain two or more particles of blue-stained material that correspond to ribosomal RNA (ribonucleic acid). With new methylene blue, reticulocytes stain bluish-purple. Reticulocyte count as a percentage of RBCs is a measure of the erythropoietic activity in the bone marrow. This is a useful marker of bone marrow suppression following chemotherapy, recovery from anemia, and so on. Reticulocyte counts may be high when the body is replenishing the RBCs in circulation. Reiculocyte counts can be performed using microscopy and supravital stains, such as new methylene blue or brilliant cresyl blue. Reticulocyte counts can also be done using automated instruments. Here light scattering is typically applied to detect cell size and cell fluorescence–absorbance measurements in conjunction with dyes like Auramine O and new methylene blue for quantitation of reticulocytes. Such methods provide good discrimination between reticulocytes and mature RBCs, with greater accuracy than microscopy examination.

AUTOMATED CELL COUNTERS (4,5)

Manufacturers of automated cell counters typically present a vast product line with varying levels of sophistication to meet the market needs. Although the analysis principles may differ, all cell counters have some common basic components, specifically hydraulics, pneumatics and electrical systems (Fig. 5). Among these, the hydraulic

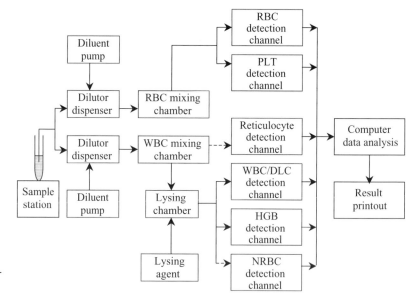

Figure 5. Flow diagram of an automated multi-channel cell counter. (Adapted from Ref. 3).

system is designed to dispense, dilute and mix samples prior to analysis. The pneumatic system operates various valves and drives the sample through the hydraulic system. The electrical system controls the operation sequences including optical–electrical detection of signals and computer-assisted data analysis. Instrument electronic analyzers typically have at least two channels. In one channel a diluent is added and RBCs are counted and sized. In the second, lysing agent is added to remove red blood cells and leave WBC intact for counting. These also produce a solution in which hemoglobin can be measured. Platelet count may be performed in either of these two channels or in a different channel. Normally, a separate channel will be required for reticulocyte count measurement. Analysis of a single blood specimen can be performed rapidly within 1 min, and results are presented in the form of numerical tables, histograms, or cytograms. The degree of analysis is both software and user dependent. Upon comparison with standard values, the software may also place flags on the output data that indicate either potential problem with analysis or deviation from cell count characterization of normal controls.

Numerous companies manufacture automated cell counters. Table 2 presents the characteristics of four high end instruments manufactured by some of them.

The Beckman–Coulter LH750 (Beckman Coulter Inc., Fullerton, CA) is a new instrument that provides CBC and five-part DLC. Additionally, it provides automated detection of subpopulations of pathological cells, such as immature granulocytes and atypical lymphocytes. It uses the three-dimensional (3D) Volume, Conductivity, Scatter (VCS) technology to probe hydrodynamically focused cells. A helium–neon laser and multiangle light scattering analysis provide information about cellular internal structure, granularity and surface morphology.

Abbott Cell-DYN 4000 (Abbott Laboratories, Abbott Park, IL) is capable of providing 41 parameters, including fully automated reticulocyte and immature granulocyte count. It uses four-angle argon-laser light scattering (M.A.P.S.S. technology) and two-color fluorescence flow cytometry (two fluorescence emission laser optics) to perform automated leukocyte counts, reticulocyte count, and DLC analysis. Both hydrodynamically focused impedance count and optical method are used for optimal erythrocyte and platelet size distribution analysis. Hemoglobin concentration is measured in a separate sample aliquot based on spectrophotometry. Immature granulocyte and variant lymphocytes are detected by a multiparameter, multiweighted discriminant function: This function generates a flag and reports a confidence fraction (i.e., the probability that these cells are classified correctly).

Sysmex XE-2100 (Sysmex Corporation, Japan) provides analysis of 32 parameters including simultaneous WBC, five-part DLC, human progenitor cell and reticulocyte analysis. Using flow cytometry with a semiconductor laser, RF, and dc measurements, this instrument analyzes the size and the structural complexity of cells. Selective dyes and reagent assist in differentiating the WBC, nucleated RBCs and reticulocyte. The RBC and platelet counts are measured using sheath flow dc detection method. Hemoglobin concentration is measured using a non-cyanide hemoglobin method.

The Bayer ADVIA 120 hematology system (Bayer Diagnostics, Tarrytown, NY) is an automated analyzer with four independent measurement channels. The peroxidase [PEROX and basophil-lobularity (BASO)] channels determine WBC and DLC count. Hemoglobin channel is used to measure HGB. The last channel is the RBC/PLT channel that provides information on platelet activation in addition to measuring PLT and RBC indices. This instrument measures the intensity of light scattered by platelets at low angles (2–3°) to obtain cell volume/size data and high angles (5–15°) for information on internal complexity. From these paired intensities the instrument computes platelet volume (MPV) and platelet component concentration on a cell-by-cell basis. The mean platelet component concentration (MPC) is indicative of platelet activation state. Mean platelet mass can also be computed from the MPV and MPC.

CONCLUDING REMARKS

This article discussed the basic principles of hematology with emphasis on humans. Enumeration of cell population

Table 2. Characteristics of Hematology Analyzers[a]

Instrument	Beckman–Coulter LH 750	Abbott Cell-Dyn 4000	Sysmex XE-2100	Bayer ADVIA 120
Number of parameters	28	41	32	30
HGB	Modified cyanmethemoglobin method	Spectrophotometry	Non-cyanide hemoglobin method	Modified cyanmethemoglobin method
Platelet	VCS	Optical method and impedance count	Hydrodynamic focusing with dc detection	Light scattering
RBC	VCS	Impedance count and optical method	Hydrodynamic focusing with dc detection	Light scattering
WBC and DLC	Five-part DLC VCS technology	Five-part DLC Light scatter and fluorescence flow cytometry	Five-part DLC Flow cytometry, RF and dc detection	Five-part DLC Peroxidase staining optics system, light scattering
Reticulocyte Count	New Methylene blue staining and VCS	Fluorescent dye CD4K530 staining and flow cytometry	Auramine O staining, light scattering, flow cytometry	Oxazin 750 staining and optical scatter

[a]Adapted from Ref. 3.

distribution in peripheral blood is examined using optical, electrooptical and light scattering techniques. As seen, such experimental modalities can be automated and the resulting hematology analyzers can be used for clinical application. Even though the exact strategy of cell counting varies between various manufacturers of automated cell counters, performance standards for such instrumentation have been established by the National Committee for Clinical Laboratory Standard (NCCLS) and the International Council for Standardization in Hematology (ICSH). The parameters evaluated here include (1) accuracy in measurement within a single batch and between batches of blood samples; (2) carryover of parameters between consecutive samples; (3) linearity or the ability to get similar measurements when the sample is diluted to different levels before being read; and (4) clinical sensitivity or the specificity and efficiency with which flags are generated during analysis to detect abnormal readouts. In order to evaluate the above and to tune the instrument for higher accuracy and sensitivity, blood count calibrators are also available from instrument manufacturers. Suitable preparations of preserved blood can also be made by individual laboratories as described by WHO document LAB/97.2 (Calibration and control of basic blood cell counters. Geneva: World Health Organization, 1997. WHO/DIL/97.2). Besides automated counting, manual and semiautomated methods are also applied by research laboratories. Establishment of such methods requires optimization of blood anticoagulant [ethylenedramenatetraacetic acid (EDTA), heparin, or sodium citrate typically], definition of appropriate electrolyte for sample dilution and design and optimization of lysis reagents required for specific experimental systems.

While the last 50 years have seen the automation of blood counting using hematology analyzers, a plethora of cell-specific antibodies have also been developed more recently. While some of these reagents are already being applied in the modern blood analyzer, their application may increase in the future. Such development can not only increase the range of parameters measured by the analyzer, they can also improve the accuracy and sensitivity of today's instrumentation.

BIBLIOGRAPHY

1. Armitage JO, editor. Atlas of Clinical Hematology. 2004; Philadelphia: Lippincott Williams & Wilkins; p 266.
2. Stiene-Martin EA, Lotspeich-Steininger CA, Koepke JA, editors. Clinical Hematology: Principles, Procedures, Correlations. 2nd ed. 1998; Philadelphia: Lippincott Williams & Wilkins Publishers; p 817.
3. Webster JG, editor. Encyclopedia of Medical Devices and Instrumentation, 4 Volume Set. 1988; New York: John Wiley & Sons; p 3022.
4. Rodak BF. Hematology: Clinical Principles and Applications. 2nd ed. 2002; Philadelphia: WB Saunders.
5. Bain BJ. Blood Cells A Practical Guide. 3rd ed. 2002; Oxford: Blackwell Science Ltd.
6. Fujimoto K. Principles of Measurement in Hematology Analyzers Manufactured by Sysmex Corporation. Sysmex J Iner 1999;9(1):31–44.
7. Groner W, Kanter R. Optical Technology in Blood Cell Technology. Sysmex J Iner 1999;9(1):21–30.
8. Shapiro HM. Practical Flow Cytometry. 4th ed. 2003; New York: John Wiley & Sons, Inc.; p 736.
9. Tatsumi N et al. Principle of Blood Cell Counter-Development of Electric Impedance Method. Sysmex J Iner 1999;9(1):8–20.
10. Hamaguchi Y. Overview of the Principles of Sysmex's Hemoglobinometry. Sysmex J Iner 1999;9(1):45–51.

Further Reading

Lotspeich-Steininger CA, Stiene-Martin EA, Koepke JA. Clinical Hematology: Principles, Procedures, Correlations. 1992; Philadelphia: Lippincott. xix; p 757.
Carr JH, Rodak BF. Clinical Hematology Atlas. 2nd ed. 2004; St. Louis (MO): Elsevier Saunders.
Brown BA. Hematology: Principles and Procedure. 5th ed. 1988; Philadelphia: Lea & Febiger.

See also ANALYTICAL METHODS, AUTOMATED; BLOOD COLLECTION AND PROCESSING; CYTOLOGY, AUTOMATED; DIFFERENTIAL COUNTS, AUTOMATED.

CELLULAR IMAGING

AMMASI PERIASAMY
University of Virginia
Charlottesville, Virginia

INTRODUCTION

For decades, autoradiography has been used widely to follow the synthesis of macromolecules by using radioactive isotopes (1). Interpretation of autoradiograms depends on knowledge of biochemical pathways and precursors and are carefully chosen so that they are used by the cell to build only one kind of molecule. On the other hand, light microscopy techniques have become a powerful tool for cell biologists to study cells live or fixed noninvasively (2–4). Fixed cells can also be studied using electron microscopy, which provides higher resolution than the light microscopy system (5,6). However, the light microscopy system allows studying live cells in physiological conditions.

The microscope has been an essential tool found in virtually every biological laboratory after the observation and description of protozoa, bacteria, spermatozoa, and red blood cells by Antoni van Leeuwenhoek, in the 1670s (7,8). The ability to study the development, organization, and function of unicellular and higher organisms and to investigate structures and mechanisms at the microscopic level has allowed scientists to better grasp the often misunderstood relationship between microscopic and macroscopic behavior. Further, the microscope preserves temporal and spatial relationships that are frequently lost in traditional biochemical techniques and gives two- (2D) or three-dimensional (3D) resolution that other laboratory methods cannot. The benefits of fluorescence microscopy techniques are also numerous (3,9). The inherent specificity and sensitivity of fluorescence, the high temporal, spatial, and 3D resolution that is possible, and the enhancement of contrast resulting from detection of an absolute rather than relative signal (i.e., unlabeled features do not emit) are several advantages of fluorescence techniques. Additionally, the plethora of well-described spectroscopic techniques providing different types of information, and the commercial availability of fluorescent probes, many of which exhibit an environment- or analytic-sensitive response, broaden the range of possible applications. Recent advancements in light sources, detection systems, data acquisition methods, and image enhancement, analysis, and display methods have further broadened the applications in which fluorescence microscopy can successfully be applied (2,3). Particularly, the fluorescent probes can be used to target many cellular components to follow the cell signaling in space (nm to m) and time (ns to days).

There are a number of microscopic techniques that have been established for cellular imaging including transmitted light–differential interference contrast microscopy (DIC)–phase contrast, reflection contrast microscopy, polarization microscopy, luminescence microscopy, and fluorescence microscopy (2,3,10). Fluorescence microscopy has been categorized into wide-field fluorescence microscopy, laser scanning confocal microscopy, multiphoton excitation microscopy, Förster (or fluorescence) resonance energy-transfer (FRET) microscopy, fluorescence lifetime imaging (FLIM) microscopy, fluorescence correlation spectroscopy (FCS), total internal reflection fluorescence (TIRF) microscopy, and fluorescence recovery after photobleaching (FRAP) microscopy (2–4,10–15). Some of the other advanced microscopy techniques include near-field microscopy (16,17), atomic force microscopy (18), scanning force–probe microscopy (19), X-ray microscopy (20), and Raman microscopy (21,22). In this article, selected fluorescence microscopy techniques (see Fig. 1) used for cellular imaging such as wide-field, confocal, multiphoton, FRET, FLIM, and CARS microscopy with biological examples are described.

BASICS OF FLUORESCENCE

Fluorescence is one of the many different luminescence processes in which molecules emit light. Fluorescence is the emission of light from the excited singlet state. Since this type of transition is usually allowed within the molecular orbitals, the emission rates of fluorescence are in the order of 10^8 s^{-1}, and fluorescence lifetimes are in nanoseconds. In contrast, phosphorescence is the emission of light from the triplet excited state and this transition is typically forbidden. The emission rates are much in the order of 10^0–10^3 s^{-1}, and phosphorescence lifetimes are typically milliseconds to seconds.

The excitation of molecules by light occurs via the interaction of molecular dipole transition moments with the electric field of the light and, to a much lesser extent, interaction with the magnetic field. The fluorescence processes following light absorption and emission are usually illustrated by a Jabłoński diagram shown in Fig. 2. Examination of the Jabłoński diagram in Fig. 2 reveals that the energy of the emitted photon is typically less then that of the absorbed photon. Hence, the fluorescence occurs at lower energy (longer wavelength) and this process is called Stokes' shift. The reasons for the Stokes' shift are rapid transition to the lowest vibrational energy level of the excited state S_1, and decay of the fluorophore to a higher vibrational level of S_0. The excess of the excitation energy is typically converted to the thermal energy.

Very intense radiation fields, such as those produced by ultrafast femtosecond lasers, can cause simultaneous absorption of two or more photons (two-photon, three-photon absorption, etc.). This phenomenon was originally predicted by Maria Göppert-Mayer in 1931 (30). It is important to realize that fluorescence intensity resulting from one- and two-photon excitation has a different dependence on excitation light intensity (power). Consequently, the fluorescence intensity depends on the squared laser power for two-photon excitation, the cubed (3rd) power for three-photon excitation, and the fourth power for four-photons excitation. This very strong dependence of fluorescence signal on the excitation power is frequently used to control the mode of excitation.

FLUOROPHORES AND FLUORESCENCE MICROSCOPY

Fluorescence microscopy (FM) plays a vital role in the biological and biomedical sciences, where fluorescence

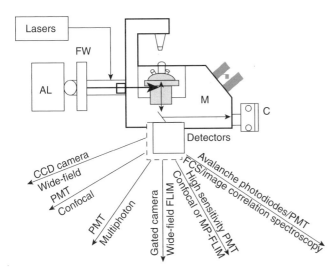

Figure 1. Illustration of various fluorescence microscopy techniques that could be coupled to any upright or inverted epifluorescent microscope. The respective instrumentations are described in the literature. Wide-field (23); confocal (2); multiphoton (3); wide-field FLIM (24); confocal FLIM (25); MP-FLIM (26); FCS and image correlation spectroscopy (27,28).

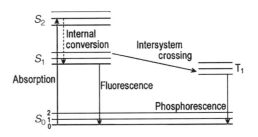

Figure 2. Jabłoński energy level diagram. S_0, S_1, and S_2 are singlet ground, first, and second electronic states, respectively; T1 = triplet state (4,29).

probe specificity and sensitivity can provide important information regarding the biochemical, biophysical, and structural status of cells. The continuing development of fluorescent probes, such as various mutant forms of green fluorescent proteins (GFPs) or fluorophores in conjunction with the strong emergence over the past two decades of confocal and multiphoton microscopy (and specialized applications, such as FRAP, FRET, and FLIM), has been a major contributor to our understanding of dynamic processes in cells and tissue (3,4,12,30–34).

Florescence microscopy can be applied noninvasively to the study of living cells in tissue down to detection levels corresponding to single molecules. Fluorescent probes bound to cellular components with monoclonal antibodies, specific ligand affinities, or covalent bonds allow us to measure chemical properties, such as ion concentrations, membrane potential, and enzymatic activity (35). They allow the experimenter to observe the distribution and function of macromolecules (proteins, lipids, nucleic acids) in living cells and tissues. Techniques have been developed that allow the investigator to place, in cells and tissues, chemically blocked (caged) molecules that can be released or activated (uncaged) by a pulse of light (photolysis) (36). Therefore, a variety of ions, metabolites, drugs, and peptides can be released at carefully controlled times and locations within the specimen. Fluorescent probes and reagents are available from Amersham Pharmacia Biotech, Calbiochem, Fluka, Jackson ImmunoResearch Laboratories, Molecular Probes, Polysciences, Serotec, Sigma-Aldrich, and others. Fluorescent proteins (GFP) and GFP vectors are available from Clontech Laboratories, Quantum Biotechnologies and Life Technologies. Details regarding the selection of fluorophores, labeling, and loading conditions for live cell imaging have been described in the literature (37).

Wide-Field Fluorescence Microscopy

Wide-field fluorescence microscopy is a conventional fluorescence microscope equipped with a movable *xyz*-axis stage that permits imaging of the specimen at different focus and lateral positions, a higher quantum efficiency CCD camera for quantification of the light emitted by the specimen at different spectra, excitation and emission filter wheels, and an appropriate software package that is capable of synchronizing hardware, acquiring images, and correcting them for distortions and information loss inherent in the imaging process (22). To allow simultaneous monitoring of spectral emissions at two or three wavelengths, a dichroic, double, or triple pass filter is used that reflects the respective excitation wavelength to excite the double- or triple-labeled cells and transmit the respective emission bands (www.chromatech.com; www.omegaoptical.com).

Wide-field microscopy is the simplest and most widely used technique. It is used for quantitative comparisons of cellular compartments and time-lapse studies for cell motility, intracellular mechanics, and molecular movement (www.api.com). For example, new fluorescent indicators have allowed the measurement of Ca^{2+} signals in the cytosol and organelles that are often localized (38,39) and nondestructive imaging of dynamic protein tyrosine kinase activities in

single living cells (40). This microscope has also been used for localizing protein molecules in living cells (22,41–43). Moreover, it is essential to implement digital deconvolution approaches to remove the out-of-focus information from the images collected in wide-field microscopy (22) (www.api.com).

Laser Scanning Confocal Microscopy (LSCM)

Wide-field microscopy, however, suffers from a major drawback due to the generation of out-of-focus fluorescent signals. Laser scanning confocal microscopy (LSCM) provides the advantage of rejecting out-of-focus information, and also allows associations occurring inside the cell to be localized in three dimensions. A confocal image with improved lateral resolution yields a wealth of spectral information with several advantages over a wide-field image including controllable depth of field and the ability to collect serial optical sections from thick specimens. Owing to its nanometer depth resolution and nonintrusiveness, confocal provides a new approach to measure viscoelasticity and biochemical responses of living cells and real-time monitoring of cell membrane motion in natural environments (2). The LSCM has been widely used in many biological applications, such as calcium, pH, and membrane potential imaging (2,35).

Confocal microscopy was introduced in 1957. Since then, the technique has gained momentum, particularly after the invention of lasers in the 1960s. Commercially available LSCM generate a clear, thin image (512×512) within 1–3 s or less, free from out-of-focus information. A single diffraction-limited spot of laser or arc lamp light is projected on the specimen using a high numerical aperture objective lens. The light reflected or fluorescence emitted by the specimen is then collected by the objective and focused upon a pinhole aperture where the signal is detected by a photomultiplier tube (PMT). Light originating from above or below the image plane strikes the walls of the pinhole and is not transmitted to the detector (see Fig. 3). To generate a 2D image, the laser beam is scanned across the specimen pixel-by-pixel. To produce an image using LSCM, the laser beam must be moved in a regular 2D raster scan across the specimen. Also, the instantaneous response of the photomultiplier must be displayed with equivalent spatial resolution and relative brightness at all points on the synchronously scanned phosphor screen of a CRT monitor. For a 3D projection of a specimen, one needs to collect a series of images at different *z*-axis planes. The vertical spatial resolution is ~ 0.5 μm for a 40×1.3 NA objective; for lenses with higher magnification, the vertical spatial resolution is even smaller. Three-dimensional image reconstruction can be accomplished with many commercially available software systems. Another alternative is a commercially available spinning disk based confocal microscope that can be used for cellular imaging (44) (www.perkinelmer.com).

The LSCM has been widely used in many biological applications and as an example here we describe protein localization using Förster resonance energy transfer (FRET) (4,43,45–48). FRET is a distance-dependent physical process by which energy is transferred nonradiatively from an excited molecular fluorophore (the donor) to

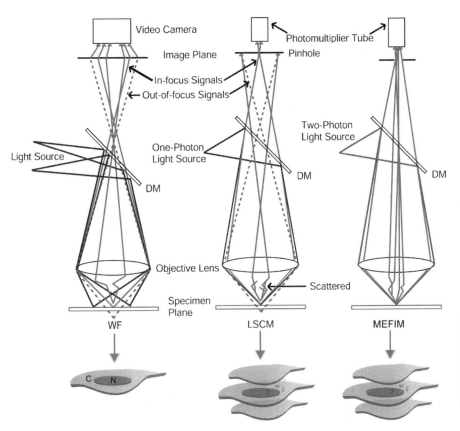

Figure 3. Illumination and detector configuration for wide-field, confocal, and multiphoton microscopy systems. DM = dichroic mirror, WF = wide-field, LSCM = laser scanning confocal microscope, MEFIM = multiphoton excitation fluorescence imaging microscopy, N = nucleus, C = Cytoplasm (22).

another fluorophore (the acceptor) by means of intermolecular long-range dipole–dipole coupling. It can be an accurate measurement of molecular proximity at nanometer distances (1–10 nm) and highly efficient if the donor and acceptor are positioned within the Förster radius (the distance at which half the excitation energy of the donor is transferred to the acceptor, typically 3–6 nm). The efficiency of FRET is dependent on the inverse sixth power of intermolecular separation (29,49,50) making it a sensitive technique for investigating a variety of biological phenomena that produce changes in molecular proximity (51). As an example Fig. 4 shows acquisition and data analysis for localization of CFP- and YFP-C/EBPα proteins expressed in live mouse pituitary GHFT1-5 cell nucleus.

Multiphoton Excitation Microscopy

The instrumentation configuration of multiphoton excitation microscopy (MEM) is generally the same as the LSCM with the exceptions of the excitation light source and the optics. In the LSCM, a visible or ultraviolet (UV) light source is used and an infrared (IR) light source is used for MEM system [see Fig. 3; (52)]. In one-photon (wide-field or confocal) fluorescence microscopy, the absorption of laser energy excites the fluorescent molecules to a higher energy level and results in the emission of one-photon fluorescence. The fluorescence intensity increases at a linear rate with the excitation intensity. Typically, some of the absorbed light energy is dissipated as heat, so the emission wavelength is longer than the absorption wavelength. For example, a fluorophore might

absorb one photon at 365 nm and fluoresce at a blue wavelength ∼ 420 nm.

The fluorophores exhibit two-photon absorption at approximately twice (730 nm) their one-photon absorption wavelengths, while two-photon (2p) emission is the same as that of one photon (420 nm), allowing the specimen to be imaged in the visible spectrum. When an IR laser beam is focused on a specimen, it illuminates at a single point and the fluorescence emission is localized to the vicinity of the focal point. The fluorescence intensity then falls off rapidly in the lateral and axial direction. In one-photon (1p) microscopy, illumination occurs throughout the excitation beam path, in an hourglass-shaped path (22). This results in absorption along the excitation beam path, giving rise to substantial fluorescence emission both below and above the focal plane. Excitation from other focal planes contributes to photobleaching and photodamage in the specimen planes that are not involved in imaging. The IR illumination in 2p excitation also penetrates deeper into the specimen than visible light excitation due to its higher energy, making it ideal for cellular imaging involving depth penetration through thick sections of tissue.

Two-photon absorption was theoretically predicted by Göeppert-Mayer in 1931 and was experimentally observed for the first time in 1961 using a ruby laser as the light source (30,53). Denk and others have experimentally demonstrated 2p imaging in a laser scanning confocal microscopy (54). Two-photon excitation occurs when two photons of $h\omega$ and $h'\omega'$ are absorbed simultaneously and a molecule is excited to the state of energy $E = h\omega + h'\omega'$ (h = Planck's constant, ω = frequency). The probability that

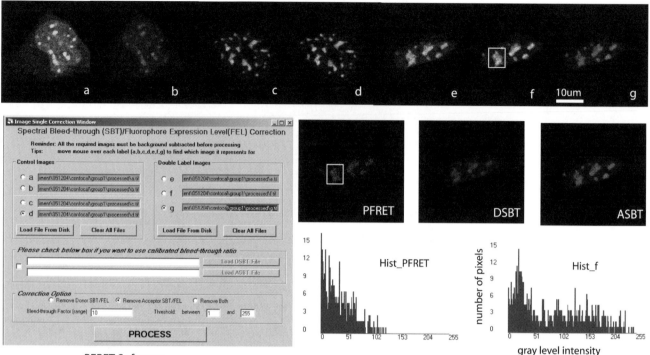

Figure 4. Localization of CFP- and YFP-C/EBPα proteins expressed in live mouse pituitary GHFT1-5 cells studied using confocal-FRET microscopy. Seven images (a–g) are required to remove the contamination in the FRET image (f). The PFRET (processed FRET) image was obtained after removing the donor (DSBT) and acceptor (ASBT) spectral bleedthrough using the PFRET software (shown on the left panel, www.circusoft.com). The spectral bleedthrough varies depending on the excitation power for the donor and acceptor molecules. The respective histogram for the processed (Hist_PFRET) and the contaminated FRET (Hist_f) demonstrates the importance of removing the spectral bleedthrough signals. The energy-transfer efficiency ($E = 20\%$) was estimated after implementing the detector spectral sensitivity correction for the donor and acceptor channel (43).

2p absorption will occur depends on the colocalization of two photons within the absorption cross-section of the fluorophore. The rate of excitation is proportional to the square of the instantaneous intensity. This extremely high local instantaneous intensity is produced by the combination of diffraction-limited focusing of a single laser beam in the specimen plane and the temporal concentration of a femtosecond (fs) mode-locked laser (typically of the order of 10^{-50}–10^{-49} cm$^4 \cdot$ s^{-1}/photon^{-1}/molecule) (55). Three- or four photon (or multiphoton) is the extension of two-photon excitation (56).

Two-photon excitation microscopy has been widely used in the area of biomedical sciences including tissue engineering, protein–protein interactions, cell, neuron, molecular, and developmental biology (3,13,22,57–60). Here, we demonstrate as an example the importance of MEM in drug molecule cellular uptake, where MEM is the ideal system for monitoring cellular drug uptake. The separation between excitation and emission wavelengths is considerably more than the 1p (wide-field and confocal) excitation and emission. For example, the excitation for the YK-II-140 drug molecule is 416 nm and emission is at 528 nm and a Stokes shift is ∼ 112 nm. In the case of MEM, the excitation for the same drug molecule is 770 nm and the Stokes shift

separation is wider than 112 nm. Moreover, in MEM we were able to detect 100 μM drug cellular uptakes compared to 1.0 mM in the wide-field microscopy. The sensitivity of drug detection is improved largely due to the advantage of the MEM (see Fig. 5 for details).

Spectral Imaging Microscopy

Human color vision is a form of imaging spectroscopy, by which we determine the intensity and proportion of wavelengths present in our environment. Spectral imaging improves on the eye in that it can break up the light content of an image not just into red, green, and blue, but into an arbitrarily large number of wavelength classes. Furthermore, it can extend the range to include the invisible UV and IR regions of the spectrum denied to the unaided eye; this type of imaging is usually known as hyperspectral (61). The result of (hyper) spectral imaging is a data set, known as a data cube, in which spectral information is present at every picture-element (pixel) of a digitally acquired image. Integration of spectral and spatial data in scene analysis remains a challenge.

These multispectral imaging approaches have been used to analyze multiple dyes within a sample. Recently,

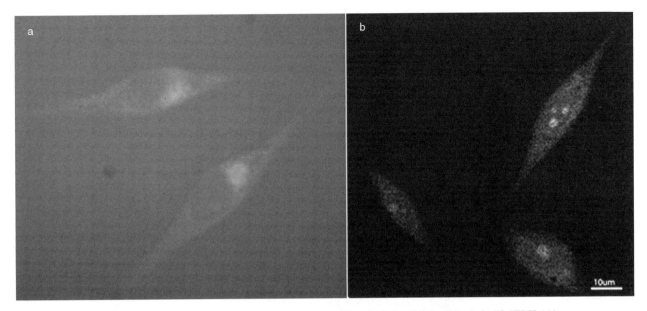

Figure 5. Comparison of one- and two-photon excitation of a living PC-3 cell loaded with YK-II-140 anticancer drug (1.0 mM concentration). Wide-field microscopy provides more autofluorescence from the cell and media (a) compared to the two-photon microscopy (b) The less autofluorescence improves the detection sensitivity of the drug cellular uptake. Wide-field (Ex 416 nm and Em 528 nm). MEFIM or two-photon (Ex 770 nm and Em 528 nm). Biorad Radiance 2100 confocal–multiphoton microscopy was used for the data acquisition.

Carl Zeiss (www.zeiss.de) introduced the Laser Scanning Microscope (LSM) 510 META system with the revolutionary emission fingerprinting technique permitting the clean separation of several even spectrally overlapping fluorescence signals of a specimen (62). The number of dyes that can be used and detected in the experiment is almost unlimited. The new system overcomes the limits of existing detection methods and permits both qualitative and quantitative analyses quickly and precisely *in vitro* and *in vivo*. Furthermore, it is beneficial in many cases for the elimination of unwanted signals, such as background noise or autofluorescence.

The Zeiss 510 Meta system scan head contains two conventional photomultiplier tube detectors (PMT), where the wavelength of the emission light is selected by means of either bandwidth or long passes filters. In the third detector, emission light is passed through a prism and the resulting spectrum is projected onto a detector consisting of a linear array of 32 PMTs, thus enabling the spectral detector to detect a full emission spectrum from a given fluorophore (www.zeiss.com). The advantage of detecting a broad spectrum of emissions is fully realized by the process of linear unmixing (63). This is an image analysis technique that is intrinsic to the LSCM controlling software that compares the experimentally derived emission data to a previously recorded reference spectrum for that fluorophore. In a situation involving samples with multiple overlapping spectra, linear unmixing allows the resolution of fluorophores with closely related emissions, the accurate distinction between GFP and FITC or GFP and YFP being the most often cited example of this feature.

There are other commercial spectral imaging units are available including Leica AOBS (www.leicamicrosystems.com) and Olympus FV1000 (www.olympus.com). The main differences between these three commercial systems are FV1000 based on Grating/slit/PMT two-channel bidirectional scanning mode; Leica system based on Prism/slit/PMT; and the Zeiss system based on Grating/multi-anode PMT.

Here, as an example we provided the data acquired using the Zeiss multiphoton Meta system to measure FRET signals resulting from protein–protein interactions involving C/EBPα. The GHFT1-5 mouse cells that expressed either the CFP- or YFP-C/EBPα fusion protein were used to collect the reference spectra. These reference spectra were used for the linear unmixing of spectra from cells expressing both proteins. Images were then acquired of cells expressing both the CFP- and YFP-C/EBPα bound as dimers to DNA elements in regions of heterochromatin that form clearly defined focal bodies in the nuclei of the mouse cells used here (Fig. 6). Images collected (ex 820 nm; em 545 nm) from cells expressing both the CFP- and YFP-C/EBPα lambda (λ) stacks were spectrally unmixed to reveal the donor bleed-through into the FRET channel (Fig. 6a), allowing the FRET signal to be corrected for the bleed-through signal (Fig. 6b). The emission spectrum for the signal in the FRET channel (Fig. 6c) was determined (b-FRET in panel d) and was then reacquired after selective photobleaching of YFP (a-FRET) using 514 nm, showing the unquenched donor signal. These results demonstrate the power of spectral FRET imaging using the Meta system to detect protein–protein interactions in a single living cell.

Fluorescence Lifetime Imaging Microscopy

Each of the fluorescence microscopy techniques described above uses intensity measurements to reveal fluorophore concentration and distribution in the cell. Recent advances in camera sensitivities and resolutions have improved the

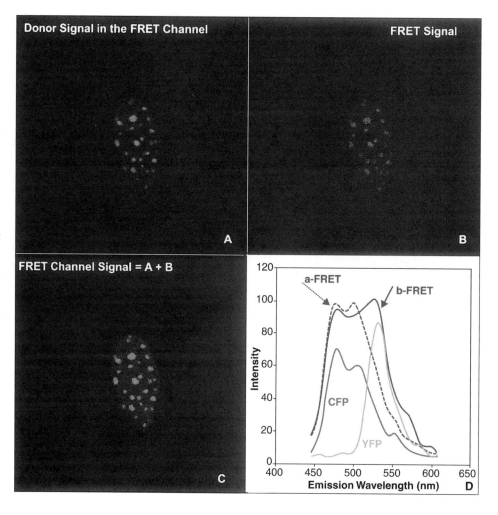

Figure 6. Spectral FRET imaging microscopy. Reference spectra (CFP, YFP) were established using cells expressing either CFP- or YFP-C/EBPα; alone, and the emission spectra are shown in panel D. Images were then collected from cells expression both the CFP- and YFP-C/EBPα, and the signals were spectrally unmixed to reveal the donor bleed-through (a) into the FRET channel (c), the corrected FRET signal is shown in b. The emission spectrum for the signal in the FRET channel (b-FRET) is shown in panel D, and was reacquired after selective photobleaching of YFP (a-FRET). CFPex 820 nm; YFPex 920 nm. Zeiss510 META system was used for the data acquisition.

capability of these techniques to detect dynamic cellular events (3). Unfortunately, even with the improvements in technology, these fluorescence microscopic techniques do not have high speed time resolution to fully characterize the organization and dynamics of complex cellular structures. In contrast, the time-resolved fluorescence (lifetime) microscopic technique allows the measurement of dynamic events at very high temporal resolution (nanoseconds). Fluorescence lifetime imaging microscopy (FLIM) merges the information of the spatial distribution of the probe with probe lifetime information to enhance the reliability of the concentration measurements. This technique monitors the localized changes in probe fluorescence lifetime (14,24,25,29,43,64–67) and provides an enormous advantage for imaging dynamic events within the living cells.

The fluorescence lifetime (τ) is defined as the average time that a molecule remains in an excited state prior to returning to the ground state. In practice, the fluorescence lifetime is defined as the time in which the fluorescence intensity decays to $1/e$ of the initial intensity (I_0) immediately following excitation (i.e., 37% of I_0). If a laser pulse excites a large number of similar molecules with a similar local environment and as long as no interaction with another protein or cell organelles occurs, the lifetime is the "natural fluorescence lifetime", τ_0. If energy is transferred, however, the actual fluorescence lifetime, τ, is less than the natural lifetime, τ_0, because an additional path for deexcitation is present (28).

Conventional fluorescence microscopy provides images that reveal primarily the distribution and amount of stain in the cell based on measurements of intensity. In contrast, the time-resolved fluorescence microscopic (or FLIM) technique allows the measurement of dynamic events at very high temporal resolution and can monitor interactions between cellular components with very high spatial resolution, as well. A fluorophore in a microscopic sample may exist, for example, in two environmentally distinct regions and have a similar fluorescence intensity distribution in both regions, but different fluorescence lifetimes. Measurements of fluorescence intensity alone would not reveal any difference between two or more regions, but imaging of the fluorescence lifetime would reveal such regional differences (52).

Instrumental methods for measuring fluorescence lifetimes are divided into two major categories: frequency-domain (29,65) and time-domain (52). With the time-domain method (or pulse method), the specimen is excited with a short pulse and the emitted fluorescence is integrated in two or more time windows (24). The relative intensity captured in the time windows is used to calculate the decay characteristics. The determination of prompt fluorescence with lifetime in the range of 0.1–100 ns requires elaborate fast excitation pulses and fast-gated detection circuits. As an alternative to the time-domain method, the frequency-domain method uses a homodyne detection scheme and requires a modulated light source and a modulated detector. The excitation light is modulated in a sinusoidal fashion. The fluorescence intensity shows a delay or phase shift with respect to the excitation and a smaller modulation depth (29).

The FLIM system can be coupled to any wide-field microscope (24,29,65) (www.tautec.com; www.lambert-instruments.nl). The lifetime method can also be applied to a laser-scanning confocal microscope (www.coord.nl; www.picoquant.com) and multiphoton microscopy (26) (www.becker-hickl.com). The FLIM techniques measure environmental changes within the living cells and can be used in multilabeling experiments. An important advantage of FLIM measurements is that they are independent of change in probe concentration, excitation intensity, and other factors that limit intensity based steady-state measurements. Additionally, FLIM enables the discrimination of fluorescence coming from different dyes, including autofluorescent materials that exhibit similar absorption and emission properties but show a difference in fluorescence lifetime. The FLIM system is not only used for protein–protein interactions, but also for various biological applications from single cell to single molecule as well as deep tissue cellular imaging (3,26,66,68,69).

The data provided here were collected using the two-photon FLIM system to demonstrate the feasibility of implementing the lifetime imaging technique for drug uptake in live cells. The intensity and the lifetime image are shown in Fig. 7 of a prostate cancer (PC-3) cell after adding the drug (1.0-mM concentration) for ∼2 min. The data clearly demonstrate that there is a considerable difference in lifetime distribution in the cytoplasmic area versus the nucleus, thus allowing the quantitation of the dynamic process of drug molecule uptake in different cellular organelles. The lifetime distribution in the cytoplasm (2 ns) and nucleus (4 ns) clearly reflects differences in molecule uptake between them and both are considerably reduced compared to the natural lifetime (20 ns) of the drug molecule. The FLIM system would reduce background interference and thus enhance measurement precision to yield more accurate understanding of drug molecule associations involved in living cells. These technologies will significantly improve and expand existing capabilities for understanding the drug molecules interactions and for characterizing their binding properties as an ensemble and at the single molecule level.

Fluorescence Correlation Spectroscopy (FCS)

Fluorescence correlation spectroscopy (FCS) is a technique in which spontaneous fluorescence intensity fluctuations are measured in a microscopic detection volume of ∼10^{-15} L as defined by a tightly focused laser beam. This spectroscopy is a special case of fluctuation correlation techniques where the laser induced fluorescence from a very small probe volume is autocorrelated in time. Fluorescence intensity fluctuations measured by FCS represent changes in either the number or the fluorescence quantum yield of molecules resident in the detection volume (27). Small,

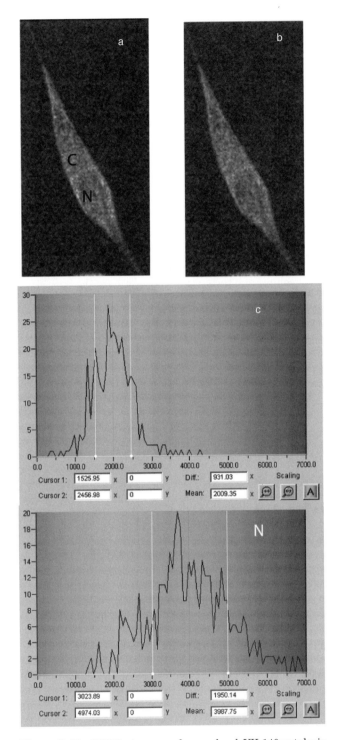

Figure 7. The FLIM microscopy-drug moleculeYK-140 uptake in a single living cell. Multiphoton excitation time-resolved intensity (a) and lifetime (b) images of PC-3 cell after adding the YK-II-140 drug (1.0-mM concentration). There is a clear difference in distribution of lifetime in the nucleus ('N') (mean $\tau_N = 3.967$ ns) versus cytoplasm ('C') (mean $\tau_C = 2.009$ ns). Moreover, considerable amount of quenching of the drug molecule in the cellular environment demonstrates that the drug molecules were interacting with various cell organelles. Consequently, the lifetime was considerably reduced from its natural lifetime 20 ns. Ex–770 nm; Em–528/30 nm. Becker and Hickl board was used in the Biorad Radiance system to acquire the data.

rapidly diffusing molecules produce rapidly fluctuating intensity patterns, whereas larger molecules produce more sustained bursts of fluorescence. If no further effects on fluorescence characteristics are present, fluctuations in the emission light simply arise from occupation number changes in the illuminated region by random particle motion. Excellent article on FCS basics was written by Petra Schwille can be seen in the URL http://www user.gwdg.de/~pschwil/BTOL_FCS.pdf.

Image Correlation Spectroscopy (ICS) was developed as the imaging analog of FCS for measuring protein aggregation in biological membranes. The ICS method entails collecting fluorescence intensity fluctuations as a function of position by using a laser scanning microscope imaging system and analyzing the imaged intensity fluctuations by spatial autocorrelation analysis (28,70). The amplitude of the normalized spatial autocorrelation function is directly related to the absolute concentration of fluorophore in the focal volume and the state of aggregation of the fluorescent entities. Extension of ICS to temporal autocorrelation analysis of image time series also permits measurement of molecular transport occurring on slower time scales characteristic of macromolecules within the plasma membrane. The other related technique, Image Cross-Correlation Spectroscopy (ICCS) allows direct measurement of the interactions of two colocalized proteins labeled with fluorophores having different emission wavelengths. Both ICS and ICCS involve the use of laser scanning confocal microscopy to obtain fluorescence images of fluorescently labeled cell membranes.

RAMAN AND CARS MICROSCOPY

Confocal, multiphoton, and fluorescence lifetime imaging microscopy have become powerful techniques for revealing 3D imaging of molecular distribution and dynamics in living specimens. This followed the development of various natural and artificial fluorophores. For chemical species or cellular components that cannot be fluorescently labeled, Raman microscopy, which measures vibrational properties and does not require molecules to have a fluorescent label, can be used to identify specific signatures of cellular or chemical components (71,72). Raman spectroscopy is an extremely powerful tool for characterizing the physical and chemical properties of the biological molecules. Raman spectroscopy is based upon the Raman effect, which may be described as the scattering of light from a molecule with a shift in wavelength from that of the usually monochromatic excitation wavelength from ultraviolet to infrared light (21). The Raman shifts are thus measures of the amounts of energy involved in the transition between initial and final states of the scattering molecule. Resonance Raman can provide more specific molecular information by working on resonance with particular electronic transitions in the protein (73). Resonant Raman spectroscopy of neutrophilic and eosinophilic granulocytes provided very clear fingerprints of the presence of oxidizing enzymes that these cells require for their functionality (74). With the use of advanced detector technology, single-cell vibrational Raman spectroscopy proved to be sufficiently

sensitive to show the typical spectra of the cell nucleus and cytoplasm in human white blood cells (75). The low scattering cross-section of naturally occurring compounds, such as DNA, RNA, and proteins can be overcome by high peak powers in the laser beams used to generate the Raman signal (76,77). Using the principle of Raman microscopy for cellular imaging, several systems have already been realized: Resonant Raman spectroscopy (75), surface-enhanced Raman spectroscopy (SERS) (78), coherent anti-Stokes Raman spectroscopy (CARS) (71,72), and Fourier transform infrared absorption (FTIR) (79).

CARS microscopy relies on the Raman Effect (80). In the spontaneous Raman process, molecules scatter photons, modifying the photon energy with energy quanta that corresponds to the molecule's vibrational modes. Vibrational contrast in CARS microscopy is inherent to the cellular species, thus requiring no endogenous or exogenous fluorophores that may also be prone to photobleaching. For CARS, two optical beams of frequencies ω_p (1064 nm) and ω_s (tunable 770–900 nm) interact in the sample to generate an anti-Stokes optical output at $\omega_{as} = 2\omega_p - \omega_s$ in the phase matched or a specific direction and is resonantly enhanced if $\omega_p - \omega_s$ coincides with the frequency of a Raman active molecular vibration across the entire focal volume. This nonlinear process uses pulsed laser sources. The signal intensity has quadratic and linear dependence on pump and stokes powers, respectively. As a result, it generates signal only within the focus, where the laser intensity is the highest, enabling 3D resolution. The molecular vibrational information obtained by CARS provides a detailed fingerprint of different bonds, functional groups, and conformations of molecules, biopolymers and even microorganisms (71,72). For example, the Raman shift at 2845 cm^{-1} was used to collect the lipid signal (bright red dots, as shown by arrow in Fig. 8a). When the frequency

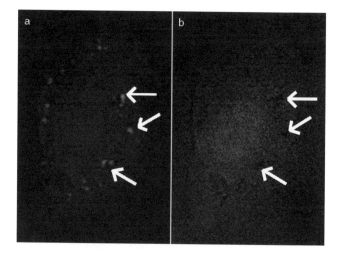

Figure 8. Demonstration of CARS and non-CARS image. (a) CARS image of lipids in 3T3 fibroblast cells excited at the vibrational frequency of 2845 cm^{-1}. (b) The frequency was tuned away from the lipids vibrational modes, 2947 cm^{-1}. No lipid is observed as pointed by the arrows. This CARS image was collected at Prof. Sunney Xie's laboratory (Harvard University) and the CARS microscopy is based on Olympus Fluoview single beam scanning system using synchronizely pumped High Q laser system.

was tuned to 2947 cm^{-1}, the lipid signal disappeared as shown by arrow in Fig. 8b. As described in the Research Activity section above we propose to study the Raman C–H stretching modes and C–C stretching modes in lipid-phase transitions for which we need to tune to different vibrational frequency to calibrate the system. If a particular molecule vibrational frequency is not known, it can be determined by conventional Raman spectrometry. Therefore, vibrational spectroscopy has found wide application in structural characterization of biological materials and in probing interaction dynamics.

CONCLUSION

Multifaceted microscopy technology moved to the center stage in cellular imaging. There is no question that the described microscopy approaches in this paper will continue to increase in all directions, driven by advances in technological development and the growing number of cell biologist researchers who will routinely use this technology for cellular imaging. Even though some of the microscopy techniques are somewhat more complex, they provide an unprecedented level of information about the micromolecular interactions in cells under physiological conditions at a very high temporal and spatial resolution. New fluorophores such as green fluorescent proteins (GFPs) and in particular Quantum Dots will expand the usefulness of cellular imaging qualitatively and quantitatively and that will lead to more detailed insights in studying the cellular dynamics. Raman and CARS microscopy techniques would allow characterizing the physical and chemical properties of the biological without the fluorophore labeling.

ACKNOWLEDGMENTS

The author would like to thank Ms. Ye Chen, Jalan Washington, and Erica Caruso for their help provided in preparation of the manuscript. The author also would like to thank Dr. Milton Brown for providing the drug compounds and Ms. Elise Shumsky, Carl Zeiss for her help in spectral FRET imaging. This work is supported by funds from National Center for Research Resources (NCRR-NIH) and Funds for Excellence in Science and Technology (FEST) at the University of Virginia.

BIBLIOGRAPHY

1. Prescott D. Methods in Cell Physiology. New York: Academic Press; 1968.
2. Pawley JB, editor. Handbook of Biological Confocal Microscopy. 2nd ed. New York: Plenum Press; 1995.
3. Periasamy A, editor. Methods in Cellular Imaging. New York: Oxford University Press; 2001.
4. Periasamy A, Day RN, editors. Molecular Imaging: FRET Microscopy and Spectroscopy. New York: Oxford University Press; 2005.
5. Grimstone A. The Electron Microscope in Biology. New York: St. Martins; 1968.
6. Frank J. Three-dimensional imaging techniques in electron microscopy. Biotechniques 1989;7(2):164–173.
7. Hogg J. The Microscope: History, Construction, and Application. London: George Routledge and Sons; 1871.
8. Jones T. 1997. History of the Light Microscope. Available at http://www.utmem.edu/personal/thjones/hist/c1.htm.
9. Periasamy A, Herman B. Computerized fluorescence microscopic vision in the biomedical sciences. J Computer-Assisted Microsc 1994;6:1–26.
10. Inoue S, Spring KR. Video Microscopy: The Fundamentals. New York: Plenum Press; 1986.
11. Wang XF, Herman B, editors. Fluorescence Imaging Spectroscopy and Microscopy. New York: John Wiley & Sons, Inc.; 1996.
12. Lippincott-Schwartz J, Snapp E, Kenworthy A. Studying protein dynamics in living cells. Nat Rev Mol Cell Biol 2001;2(6):444–456.
13. Diaspro A, editor. Confocal and Two-photon Microscopy: Foundations, Applications, and Advances. New York: John Wiley & Sons, Inc.; 2002.
14. Marriott G, Parker I. Biophotonics, Part A and B. Methods in Enzymology. San Diego: Academic Press; 2003.
15. Prasad PN. Introduction to Biophotonics. New York: Wiley-Interscience; 2003.
16. Betzig E, Trautman JK. Near-field optics: microscopy, spectroscopy, and surface modification beyond the diffraction limit. Science 1992;257:189–195.
17. Lewis A, Radko A, Ben Ami N, Palanker D, Lieberman K. Near-field scanning optical microscopy in cell biology. Trends Cell Biol 1999;9(2):70–73.
18. Lal R, John SA. Biological applications of atomic force microscopy. Am J Physiol 1994;266(1 Pt. 1):C1–C21.
19. Driscoll RJ, Youngquist MG, Baldeschwieler JD. Atomic-scale imaging of DNA using scanning tunnelling microscopy. Nature (London) 1990;346(6281):294–296.
20. Jacobsen C, Lindaas S, Williams S, Zhang X. Scanning luminescence X-ray microscopy:imaging fluorescence dyes at sub-optical resolution. J Microsc 1993;172(2):121–129.
21. Hanlon EB, Manoharan R, Koo TW, Shafer KE, Motz JT, Fitzmaurice M, Kramer JR, Itzkan I, Dasari RR, Feld MS. Prospects for *In vivo* Raman spectroscopy. Phys Med Biol 2000;45(2):R1–59.
22. Periasamy A, Skoglund P, Noakes C, Keller R. An evaluation of two-photon excitation versus confocal and digital deconvolution fluorescence microscopy imaging in Xenopus morphogenesis. Microsc Res Tech 1999;47(3):172–181.
23. Periasamy A, Day RN. Visualizing protein interactions in living cells using digitized GFP imaging and FRET microscopy. Methods Cell Biol 1999;58:293–314.
24. Elangovan M, Day RN, Periasamy A. Nanosecond fluorescence resonance energy transfer-fluorescence lifetime imaging microscopy to localize the protein interactions in a single living cell. J Microsc 2002;205(Pt. 1):3–14.
25. Gerritsen HC, Asselbergs MA, Agronskaia AV, Van Sark WG. Fluorescence lifetime imaging in scanning microscopes: acquisition speed, photon economy and lifetime resolution. J Microsc 2002;206(Pt. 3):218–224.
26. Chen Y, Periasamy A. Characterization of two-photon excitation fluorescence lifetime imaging microscopy for protein localization. Microsc Res Tech 2004;63(1):72–80.
27. Berland KM, So PT, Gratton E. Two-photon fluorescence correlation spectroscopy: method and application to the intracellular environment. Biophys J 1995;68(2):694–701.
28. Petersen N. FCS and spatial correlations on biological surfaces. In: Rigler R, Elson EL, editors. Fluorescence Correlation Spectroscopy. Berlin: Springer Verlag; 2001. p 2–35.
29. Lakowicz JR. Principles of Fluorescence Spectroscopy. 2nd ed. New York: Plenum Press; 1999.
30. Göppert-Mayer M. Ueber Elementarakte mit Quantenspreungen. Ann Phys 1931;9:273–295.

31. Haugland RP. Molecular Probes: Handbook of Fluorescent Probes and Research. Eugene, Oregon: Molecular Probes Inc.; 1989.

32. Tsien RY, Waggoner A. Fluorophores for confocal microscopy: photphysics and photochemistry. In: Pawley JB, editor. Handbook of Biological Confocal Microscopy. New York: Plenum Press; 1995. p 267–279.

33. Tsien RY. Imagining imaging's future. Nat Rev Mol Cell Biol 2003;(Suppl):SS16–SS21.

34. Wallrabe H, Periasamy A. Imaging protein molecules using FRET and FLIM microscopy. Curr Opin Biotechnol 2005;16: 19–27.

35. Lemasters JJ, Qian T, Trollinger DR, Muller-Borer BJ, Elmore SP, Cascio WE. Laser scanning confocal microscopy applied to living cells and tissues. In: Periasamy A, editor. Methods in Cellular Imaging. New York: Oxford University Press; 2001. p 66–87.

36. Corrie JET, Katayama Y, Reid GP, Anson M. The development and application of photosensitive caged compounds to aid time-resolved structure determination of macromolecules. Philos Trans R Soc London A Ser 1992;340:233–236.

37. Harper IS. Flurophores and their Labeling Procedures for Monitoring Various Biological Signals. In: Periasamy A, editor. Methods in Cellular Imaging. New York: Oxford University Press; 2001. p 20–39.

38. Miyawaki A, Llopis J, Heim R, McCaffery JM, Adams JA, Ikura M, Tsien RY. Fluorescent indicators for Ca^{2+} based on green fluorescent proteins and calmodulin. Nature (London) 1997;388(6645):882–887.

39. Miyawaki A, Griesbeck O, Heim R, Tsien RY. Dynamic and quantitative Ca^{2+} measurements using improved cameleons. Proc Natl Acad Sci USA 1999;96(5):2135–2140.

40. Ting AY, Kain KH, Klemke RL, Tsien RY. Genetically encoded fluorescent reporters of protein tyrosine kinase activities in living cells. Proc Natl Acad Sci USA 2001;98 (26):15003–15008.

41. Day RN. Visualization of Pit-1 transcription factor interactions in the living cell nucleus by fluorescence resonance energy transfer microscopy. Mol Endocrinol 1998;12(9): 1410–1419.

42. Chen Y, Periasamy A. Time-correlated single photon counting (TCSPC) FLIM-FRET microscopy for protein localization. In: Periasamy A, Day RN, editors. Molecular Imaging: FRET Microscopy and Spectroscopy. New York: Oxford University Press; 2005. Chapt.13.

43. Chen Y, Elangovan M, Periasamy A. FRET data analysis: The algorithm. In: Periasamy A, Day RN, editors. Molecular Imaging: FRET Microscopy and Spectroscopy. New York: Oxford University Press; 2005. Chap. 7.

44. Maddox P, Desai A, Salmon ED, Mitchison TJ, Oogema K, Kapoor T, Matsumoto B, Inoue S. Dynamic confocal imaging of mitochondria in swimming Tetrahymena and of microtubule poleward flux in Xenopus extract spindles. Biol Bull 1999;197(2):263–265.

45. Elangovan M, Wallrabe H, Chen Y, Day RN, Barroso M, Periasamy A. Characterization of one- and two-photon excitation fluorescence resonance energy transfer microscopy. Methods 2003;29(1):58–73.

46. Mills JD, Stone JR, Rubin DG, Melon DE, Okonkwo DO, Periasamy A, Helm GA. Illuminating protein interactions in tissue using confocal and two-photon excitation fluorescent resonance energy transfer microscopy. J Biomed Opt 2003;8(3):347–356.

47. Sekar RB, Periasamy A. Fluorescence resonance energy transfer (FRET) microscopy imaging of live cell protein localizations. J Cell Biol 2003;160(5):629–633.

48. Wallrabe H, Elangovan M, Burchard A, Periasamy A, Barroso M. Confocal FRET microscopy to measure clustering of ligand-receptor complexes in endocytic membranes. Biophys J 2003; 85(1):559–571.

49. Forster T. Delocalized excitation and excitation transfer. In: Sinanoglu O, editor. Modern Quantum Chemistry Part III: Action of Light and Organic Crystals. New York: Academic Press; p 93–137.

50. Clegg RM. Fluorescence resonance energy transfer. In: Wang XF, Herman B, editors. Fluorescence Imaging Spectroscopy and Microscopy. Volume 137, New York: John Wiley & Sons, Inc.; 1996. p 179–251.

51. dos Remedios CG, Miki M, Barden JA. Fluorescence resonance energy transfer measurements of distances in actin and myosin: A critical evaluation. J Muscle Res Cell Motil 1987;8: 97–117.

52. Periasamy A, Wodnicki P, Wang XF, Kwon S, Gordon GW, Herman B. Time-resolved fluorescence lifetime imaging microscopy using a picosecond pulsed tunable dye laser system. Rev Sci Instrum 1996;67(10):3722–3731.

53. Kaiser W, Garrett CGB. Two-photon excitation in CaF2: Eu^{2+}. Phys Rev Lett 1961;7:229–231.

54. Denk W, Strickler JH, Webb WW. Two-photon laser scanning fluorescence microscopy. Science 1990;248(4951):73–76.

55. Denk W, Piston DW, Webb WW. Two-photon molecular excitation in laser-scanning microscopy. In: Pawley JB, editor. Handbook of Biological Confocal Microscopy. New York: Plenum Press; 1995. p 445–458.

56. Szmacinski H, Gryczynski I, Lakowicz JR. Three-photon induced fluorescence of the calcium probe Indo-1. Biophys J 1996;70(1):547–555.

57. Svoboda K, Helmchen F, Denk W, Tank DW. Spread of dendritic excitation in layer 2/3 pyramidal neurons in rat barrel cortex In vivo. Nat Neurosci 1999;2(1):65–73.

58. Bacskai BJ, Hickey GA, Skoch J, Kajdasz ST, Wang Y, Huang GF, Mathis CA, Klunk WE, Hyman BT. Four-dimensional multiphoton imaging of brain entry, amyloid binding, and clearance of an amyloid-beta ligand in transgenic mice. Proc Natl Acad Sci USA 2003;100(21):12462–12467.

59. Konig K, Riemann I. High-resolution multiphoton tomography of human skin with subcellular spatial resolution and picosecond time resolution. J Biomed Opt 2003;8(3): 432–439.

60. Soeller C, Jacobs MD, Jones KT, Ellis-Davies GC, Donaldson PJ, Cannell MB. Application of two-photon flash photolysis to reveal intercellular communication and intracellular Ca^{2+} movements. J Biomed Opt 2003;8(3):418–427.

61. Farkas D. Spectral microscopy for quantitative cell and tissue imaging. In: Periasamy A, editor. Methods in Cellular Imaging. New York: Oxford University Press; 2001. Chap. 20.

62. Dickinson ME, Simbuerger E, Zimmermann B, Waters CW, Fraser SE. Multiphoton excitation spectra in biological samples. J Biomed Opt 2003;8(3):329–338.

63. Nashmi R, Dickinson ME, McKinney S, Jareb M, Labarca C, Fraser SE, Lester HA. Assembly of alpha4beta2 nicotinic acetylcholine receptors assessed with functional fluorescently labeled subunits: effects of localization, trafficking, and nicotine-induced upregulation in clonal mammalian cells and in cultured midbrain neurons. J Neurosci 2003;23(37): 11554–11567.

64. Gadella TWJ, Jovin TM, Clegg RM. Fluorescence Lifetime Imaging Microscopy (Flim) - Spatial-Resolution of Microstructures on the Nanosecond Time-Scale. Biophys Chem 1993;48(2):221–239.

65. Gratton E, Breusegem S, Sutin J, Ruan Q, Barry N. Fluorescence lifetime imaging for the two-photon microscope: time-domain and frequency-domain methods. J Biomed Opt 2003;8(3):381–390.

66. Krishnan RV, Masuda A, Centonze VE, Herman B. Quantitative imaging of protein-protein interactions by multiphoton fluorescence lifetime imaging microscopy using a streak camera. J Biomed Opt 2003;8(3):362–367.

67. Redford G, Clegg RB. Real-Time Fluorescence Lifetime Imaging and FRET using Fast Gated Image Intensifiers. In: Periasamy A, Day RN, editors. Molecular Imaging: FRET Microscopy and Spectroscopy. New York: Oxford University Press; 2005. Chapt. 11, in press.

68. Bastiaens PI, Squire A. Fluorescence lifetime imaging microscopy: spatial resolution of biochemical processes in the cell. Trends Cell Biol 1999;9:48–52.

69. Hohng S, Joo C, Ha T. Single-Molecule Three-Color FRET. Biophys J 2004;87(2):1328–1337.

70. Petersen NO, Hoddelius PL, Wiseman PW, Seger O, Magnusson KE. Quantitation of membrane receptor distributions by image correlation spectroscopy: concept and application. Biophys J 1993;65(3):1135–1146.

71. Cheng JX, Volkmer A, Book LD, Xie XS. Multiplex coherent anti-Stokes Raman scattering microspectroscopy and study of lipid vesicles. J Phys Chem B 2002a 106:8493–8498.

72. Cheng JX, Jia YK, Zheng G, Xie XS. Laser-scanning coherent anti-Stokes Raman scattering microscopy and applications to cell biology. Biophys J 2002b;83(1):502–509.

73. Carey PR. Raman spectroscopy, the sleeping giant in structural biology, awakes. J Biol Chem 1999;274(38):26625–26628.

74. Salmaso BL, Puppels GJ, Caspers PJ, Floris R, Wever R, Greve J. Resonance Raman microspectroscopic characterization of eosinophil peroxidase in human eosinophilic granulocytes. Biophys J 1994;67(1):436–446.

75. Puppels GJ, de Mul FF, Otto C, Greve J, Robert-Nicoud M, Arndt-Jovin DJ, Jovin TM. Studying single living cells and chromosomes by confocal Raman microspectroscopy. Nature (London) 1990;347(6290):301–303.

76. Volkmer A, Cheng JX, Xie XS. Vibrational imaging with a high sensitivity via epidetected coherent anti-Stokes Raman scattering microscopy. Phys Rev Lett 2001;87:023901.

77. Uzunbajakava N, Lenferink A, Kraan Y, Volokhina E, Vrensen G, Greve J, Otto C. Nonresonant confocal Raman imaging of DNA and protein distribution in apoptotic cells. Biophys J 2003;84(6):3968–3981.

78. Hawi SR, Rochanakij S, Adar F, Campbell WB, Nithipatikom K. Detection of membrane-bound enzymes in cells using immunoassay and Raman microspectroscopy. Anal Biochem 1998;259(2):212–217.

79. Diem M, Chiriboga L, Lasch P, Pacifico A. IR spectra and IR spectral maps of individual normal and cancerous cells. Biopolymers 2002;67(4–5):349–353.

80. Nan X, Yang WY, Xie XS. CARS microscopy: lights up lipids in living cells. Biophoton Int 2004;11(8):44–47.

See also CYTOLOGY, AUTOMATED; MICROSCOPY, CONFOCAL; MICROSCOPY, ELECTRON.

CEREBROSPINAL FLUID. See HYDROCEPHALUS, TOOLS FOR DIAGNOSIS AND TREATMENT OF.

CHEMICAL ANALYZERS. See ANALYTICAL METHODS, AUTOMATED.

CHEMICAL SHIFT IMAGING. See NUCLEAR MAGNETIC RESONANCE SPECTROSCOPY.

CHROMATOGRAPHY

THAYNE L. EDWARDS
University of Washington
Seattle, Washington

INTRODUCTION

Chromatography is the process of separating a mobile phase mixture into its individual components using the relative interactions of the components with a stationary phase. Chromatography is often used as a method of purification, even when the components are closely related. When used in conjunction with a concentration or mass-based sensor, it can also be a power analytical method. This chapter explores the basic processes of the various types of chromatography and the closely related techniques of field-flow fractionation (FFF) and electrophoresis. Some of the basic theory of chromatography will be given in terms of retention mechanism and separation performance. In addition, a description of the basic types of chromatography, electrophoresis, and FFF will be given in relation to their theory and application.

Chromatography literally means "color writing" because of an observation in the early 1900s by Mikhail Tswett in separating pigments of plants into various color bands using $CaCO_3$ (1). Although Tswett is considered the father of chromatography, he was not the first to observe the chromatographic process in an experimental setting. Pliny the Elder (ca. 79 AD) recorded a crude paper chromatography experiment. Several others between this time and the realization of the usefulness of chromatography by Tswett also observed the process in the laboratory. It was not until the 1930s that it was recognized generally as an analytical technique (2). Since then, many advances, discoveries, and inventions have made chromatography an indispensable laboratory and industrial technique.

The purpose of chromatographic methods falls under either purification or analysis. Gas chromatography is almost invariably an analytical method whereas liquid chromatography is used for either. As a result of the extensive research into various methods of sample retention, even samples with relatively little difference in their physical or chemical structure can be separated. Chromatographic systems also range in size and complexity from a simple, gravity-fed packed column to a complex high pressure industrial-sized system with integrated components for sample introduction and fraction detection and collection. Often, chromatography is used in conjunction with another analytical method for determination of fraction purity and composition. For medical device instrumentation, chromatography is an invaluable technique that may find its application in a wide variety of ways such as protein purification and sample contamination detection.

Numerous journal articles and books have been written on the subject, as well as reviews (3) and a host of information published on the Internet (2,4–6). Several journals dedicated to chromatography and related fields also exist (Advances in Chromatography, Chromatographia, Biomedical Chromatography, Journal of Chromatography, Journal of Liquid Chromatography and Related Technologies, Journal of Planar Chromatography, Journal of Separation Science). This information is well accessible to anyone who has interest in pursuing the methods and techniques described in this chapter. No attempt has been made here to review all this material. For this reason, this chapter will not focus on the details and extensive literature, but on the

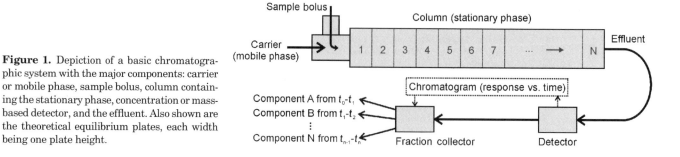

Figure 1. Depiction of a basic chromatographic system with the major components: carrier or mobile phase, sample bolus, column containing the stationary phase, concentration or mass-based detector, and the effluent. Also shown are the theoretical equilibrium plates, each width being one plate height.

basics and presenting what is generally possible with chromatography.

Chromatographic systems have a mobile phase and a stationary phase. The mobile phase is used to transport the samples through the stationary phase. The mobile phase is inert and does not interact with the sample or stationary phase. The stationary phase is unique for each separation because its purpose is to provide selective interaction with the samples. A simple chromatographic system is depicted in Fig. 1. Apart from the two fundamental required components just mentioned, a system must typically contain a sample injection port and detector in order to be useful as an analytical system. If it is to be used as a purification method, then it must also contain some method of recovering the fractions in the effluent.

The basic operating principle is that samples in the mixture, which interact to a higher degree with the stationary phase, travel slower and are retained longer in the system. This interaction, speed, and retention is relative to samples that interact to a lesser degree and so travel faster and are retained for a shorter duration. The spatial separation occurs in the mobile phase flow direction.

Chromatographic methods are grouped either into *liquid* or *gas* chromatography, based on the carrier phase and further identified by its particular method of sample retention. For example, in size-exclusion chromatography, the stationary phase is a bed of porous beads. The pores allow only samples smaller than a particular size to enter and then exit again. As a result of the pores tortuosity, the travel distance for these smaller samples is longer and thus increases the residence time relative to the larger samples. Some other methods include gas-adsorption, gas-liquid, capillary gas, liquid-adsorption, liquid-liquid, supercritical fluid, ion exchange, and affinity. In addition, the closely related fields of electrophoresis and FFF need to be mentioned as they typically complement chromatographic methods. In order to more fully understand the methods mentioned here, the basic theory will be mentioned first and then a more detailed description of the methods will follow.

GENERAL THEORY

The basic measurement in chromatography is the retention factor. This measurement is calculated from the measured retention times (t_0, t_A, t_B, ..., t_n) in the chromatogram output, as shown in Fig. 2. The void time t_0 is the elution time of an unretained sample. It is the time required to sweep one-column volume. All retention values

are calculated relative to this time. The retention factor for sample A, R_A, is then

$$R_A = \frac{t_A - t_0}{t_0}$$

where t_A is the sample residence time. The ability of the separation column to distinguish between two components is quantified in the selectivity term S

$$S_{AB} = \frac{R_S}{R_A}$$

For mixtures, the degree of separation between two components is termed the resolution R_S. In most cases, the resolution is the most important factor because that is the purpose of performing chromatographic separations. It is calculated from the retention times t and the peak widths w for the two samples as

$$R_S = \frac{2(t_B - t_A)}{W_B + W_A}$$

The relative distance in the system at which a degree of equilibrium between the two phases occurs is referred to as a theoretical plate. Chromatographic systems usually have many theoretical plates. By dividing the length of the column L with the number of plates N gives the plate height, or height of an equivalent theoretical plate H. These numbers characterize the quality of the retention and can be determined from a chromatogram by assuming

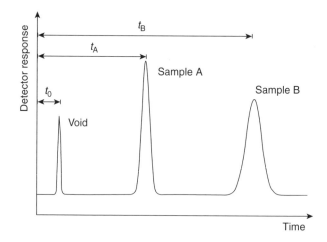

Figure 2. Chromatogram of separation of a two-component mixture. The retention times are measured from the injection time.

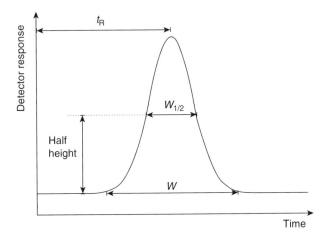

Figure 3. Chromatogram of single peak from response of mass- or concentration-based detector showing measurements for determining the number of theoretical plates.

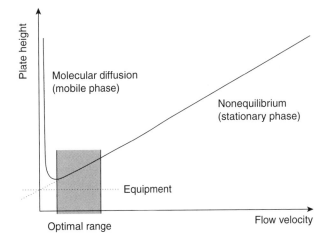

Figure 4. The van Deemter plot, based on the van Deemter equation for determining the optimal flow rate for a given chromatographic separation.

that the peaks are Gaussian shaped, as shown in Fig. 3. It is noted that this method is for an ideal situation and that other methods are available for peaks of other shapes. The sample peak width is measured either at baseline w or at half-height, $w_{1/2}$. In practice, it is more convenient and accurate to measure at half-height

$$N = 5.54 \left(\frac{t_R}{W_{1/2}} \right)^2$$

and

$$H = \frac{L}{N}$$

Plate height is a summation of three effects, two of which are dependent on flow rate. This theory was proposed by van Deemter et al. (7). The first effect is due to equipment and users, such as column-packing and injection variabilities, and is not a function of the flow rate. This term can be minimized through careful design and manufacturing of the column. The use of automated sample handling also helps to reduce this effect. At low flow velocities u, molecular diffusion of the sample in the carrier dominates and peak broadening rises quickly. At higher flow velocities the sample plug broadens due to nonequilibrium effects such as eddy diffusion and multiple paths in the stationary phase. The van Deemter equation is

$$H = H_0 + H_1(u^{-1}) + H_2(u)$$

This equation is easily graphed (Fig. 4) for a visual indication of the optimal flow rate. In practice, it is best to have the flow rate slightly higher than at the optimum to avoid the region of rapid plate height increase. Each of the terms is dependent on the type of chromatography used.

Now, the resolution of the separation can be put in terms of plate height or the equivalent number of theoretical plates,

$$Rs_{AB} = \frac{\sqrt{N}}{4} \left[\frac{S_{AB} - 1}{S_{AB}} \right] \left[\frac{1 + R_B}{R_B} \right]$$

This form is useful for optimizing a separation based on the three groups in the equation. Resolution can be increased by increasing the number of theoretical plates, which can be accomplished by both increasing the column length and minimizing the plate height using the van Deemter plot. However, increasing the column length will also cause proportional band broadening, and so it is not always ideal. The second term involves adjusting the selectivity through column modifications such as changing either or both of the phases and the temperature. The temperature also plays an important role in the last term as well. By adjusting and tuning these parameters, nearly every difficult separation can be successfully resolved. Theoretically, choosing the proper stationary phase for the column and an appropriate mobile phase, any two materials can be separated. These phases are at the heart of chromatography.

TYPES OF CHROMATOGRAPHY AND RELATED TECHNOLOGIES

The mobile phase is either a gas, a liquid, or a supercritical fluid, whereas the stationary phases can be either a solid or a liquid. The types of chromatography are typically named by their phases or interaction process and are categorically divided by their stationary phase into either gas chromatography (GC) or liquid chromatography (LC). Another variation on LC is high performance liquid chromatography (HPLC), in which the carrier is driven by pressure. Some common subclasses of LC and GC are based on ion-exchange, phase change, adsorption, size exclusion, partitioning, and absorption. Specific types and hybrid systems also exist that fall under each of these categories. Each of these complements the others to build a broad spectrum of types of chromatographic separation technique available (8–11).

Gas Chromatography (GC)

Gas chromatography makes use of a pressurized gas cylinder and a carrier gas, such as helium, to carry the solute through the column. GC can be used for both purification and analysis, when a detector is used in tandem. Common detectors used in GC are thermal conductivity and flame ionization detectors. Many more types of detectors exist, each with its advantages and disadvantages. Three types of GC that are among the more common methods are gas adsorption, gas-liquid, and capillary gas chromatography.

Gas Adsorption.   Gas adsorption chromatography has a solid stationary phase packed bed. The samples selectively adsorb and desorb to the stationary phase, effectively increasing each sample's retention time based on its isotherm. Some of the more common adsorbents used are silica, zeolite, and activated alumina. This method is the primary method for separating mixtures of gases.

Gas-Liquid.   Separation in gas-liquid chromatography is based on the gaseous samples partitioning with a viscous liquid stationary phase. This liquid is supported in the column by coating a solid, most commonly diatomaceous earth. The solid support to the stationary phase liquid is inert to the samples. The sample retention time is governed by the rate at which it dissolves into and vaporizes out of the liquid. Thus, the relative partitioning of each of the samples in the liquid stationary phase is the basis of the separation.

Capillary Gas.   In this method, the stationary phase is a capillary coated with a liquid (wall-coated open tubular) or a solid-coated capillary onto which the liquid is adsorbed (support-coated open tubular), as has been described in the previous two methods. Liquid or gum temperature stable polymers are used as the stationary phase. Most common polymers used are poly-ethylene glycol or poly-siloxanes. Also used are molecular sieves and alumina particles. Unlike the previous methods, the stationary phase has a small volume due to the capillary geometry and is thus limited to the amount of sample that can interact. However, because of the small column geometry, the partitioning or adsorption of the sample is relatively fast. The capillary is typically glass or fused silica coated with polyimide for support. The tubing (column) can be long and also be wound into tight areas for compactness and good temperature control. It is the most common gas chromatography analytical method.

Liquid Chromatography (LC) and High Performance Liquid Chromatography (HPLC)

As the carrier phase in LC is a liquid, it is naturally more amenable for biological separations and analysis, such as the purification of proteins. However, it is also amenable to any sample dissolved in a liquid. In LC, the carrier is driven by gravity through the column. These columns, made of glass or plastic and sometimes disposable, are typically used for lab-scale preparative work. For analysis of samples, the carrier is pressurized for increased speed and sample resolution. This variation is termed high performance liquid chromatography or HPLC. These systems are much more complex and costly. The columns are made of steel to withstand high pressures and are reused a

number of times. Detectors are also placed inline with these columns for analysis, although HPLC is also used in preparative work as well (3,6,11–14).

Liquid Adsorption.   Liquid adsorption, also termed liquid-solid chromatography, uses a solid stationary phase made of particles such as alumina or silica. In particular, this method is used in separating isomers. The retention is based on the adsorption/desorption kinetics of each sample onto the particles. Liquid adsorption is often found in large-scale applications because the adsorbent beds are relatively inexpensive.

Liquid-Liquid.   In liquid-liquid chromatography (LLC), also called partition chromatography, the stationary phase is a liquid-coated solid surface. This liquid is immiscible with the liquid solvent mobile phase. Retention is based on partitioning of the sample between the two phases. LLC can be accomplished in either *normal* phase or *reverse* phase. Normal phase has a nonpolar mobile phase and polar stationary phase. Reverse phase is the opposite of having a polar mobile phase and nonpolar stationary phase. It is used primarily in separating nonvolatile components of mixtures and is similar to a chemical extraction process.

Size Exclusion.   This method was described briefly in the introduction. It is somewhat unique because the stationary phase is inert to the sample. The increased path length due to tortuous pores that exclude large samples causes an increased retention time for samples smaller than the cutoff size. It is also referred to as filtration, gel permeation, or molecular-sieve chromatography. This method is useful for protein separation and purification such as in antibody production and buffer exchange applications.

Supercritical Fluid.   Unlike the other methods, supercritical fluid chromatography is characterized by its unique carrier fluid. Supercritical fluids used to carry the sample have very high viscosities and molecular diffusivities compared with liquids but with densities on the same order. One type of supercritical fluid used is a mixture of carbon dioxide and modifiers. Implementation of this technique is difficult because of the high temperature and pressures to reach the supercritical fluid state.

Ion-Exchange.   Ion-exchange chromatography is commonly used in the purification of biological materials, such as amino acids and proteins, and also ions in solution. This method is capable of quantifying samples in the ppb to ppm concentration range. The stationary phase is an ion-exchange resin that is either cationic or anionic. Charged atoms or molecules in the liquid phase sample bind to the stationary phase as they are passed through the column. The sample is released by adjusting the carrier pH or ionic strength. Separation by this method is highly selective and especially useful for anions in which separations are typically slow. The resins are typically high capacity and inexpensive.

Affinity.   Affinity chromatography has a stationary phase that is highly selective to one particular sample.

Unlike other chromatographic methods, the sample is highly bound to the stationary phase until the carrier solution is changed and the sample released. To accomplish this selective release, the stationary phase is engineered using an inert affinity matrix, such as agarose or cellulose derivative, and infused with ligand molecules that are design to bind only the sample of choice. Immunologic interactions of specific antibody-antigen pairs are particularly useful because of the high specificity that can be obtained and the reversibility of the binding event. The addition of a high salt concentration or low pH to the stationary phase reverses the selectivity, similar to ion-exchange chromatography, and allows the release of the sample after the other components of the mixture have been washed away. Care must be taken to ensure that impurities do not foul the matrix. Some preprocessing is typically accomplished prior to the separation to remove the potential fouling components. This method is used often with biological samples.

Electrophoresis

Electrophoresis is a separation method using the transport of electrically charged compounds in a conductive liquid environment under the influence of an electric field (15,16). Positively charged molecules migrate toward a negative electrode, and negatively charged molecules migrate toward a positive electrode. It is regularly applied in analytical chemistry to determine the constituent molecules of a compound. It is also widely used in medical diagnostics and other biological areas to determine molecules within biological samples, such as protein and DNA. From the various modes of electrophoresis, capillary electrophoresis (CE) is the most widely used separation method used in a modern analytical laboratory (17). High separation speed, excellent resolution power, and low consumption of buffer and sample are some of the advantages. Typically, samples are injected into a capillary tube with diameters ranging between 25 and 100 μm, and an electrical field is applied along the capillary tube to separate compounds based on the differences in charge to mass ratio. Negatively ionized surface silanol group of the capillary creates an electrical double layer at the solid/liquid interface to preserve electroneutrality, and this mobile layer is pulled toward the negatively charged electrode when an electric field is applied. These ion layers drag the bulk buffer solution, causing an electro-osmotic flow. Compared with HPLC, which has a parabolic flow profile due to the laminar flow inside the channel, the flow profile is flat in the electro-osmotic flow, which helps the detection peak to be very sharp, increasing its sensitivity. Laser-induced fluorescence detection is the most widely used method for detecting the separated molecules (18). Over 100 review articles exist covering capillary electrophoresis, and Beale wrote an excellent review categorizing these articles (19).

Development in microfabrication technologies and the lab-on-a-chip concept in the early 1990s further expanded the role of this powerful analysis technique (20–22). Smaller sample injection into a microchannel and higher electric field result in short analysis time with excellent resolution. Applying higher electric field is possible due to the high surface-to-volume ratio of a microchannel that can dissipate the heat produced during electrophoresis faster. Automated sample injection and capability to perform the separation on arrays of microchannel in conjunction with the short analysis time enables high throughput analyses. Low manufacturing cost due to the batch fabrication capability of the microchips is another advantage over conventional separation technologies. Fig. 5 shows a typical channel configuration for a capillary electrophoresis microchip. High voltages are applied between reservoir 1 and 2 so that the sample in reservoir 1 fills the injection channel. Once the injection channel is filled, the high voltages are switched to reservoir 3 and 4, and the sample plug in the cross section gets injected into the separation channel. As the sample plug moves down the separation channel, separation occurs depending on the charge to mass ratio of the compounds being analyzed. In fluorescence detection, detection typically occurs at the end of the separation channel by illuminating the fluorescence-tagged sample with laser or UV light, followed by light detection using a photomultiplier tube (PMT).

Typical channels are several tens to hundreds of microns wide, a couple tens of microns deep, and the separation channel lengths are in the centimeter scale. Electric fields applied to these channels range in kV/cm scale. This high electric field and small sample size enables separation within several seconds compared with tens of minutes required in conventional chromatography, which can be of great benefit when monitoring time-dependent reactions, such as conducting an enzyme kinetic study (23). This system also enables automatic sample injection because voltages can be simply switched between reservoirs to inject samples into the separation channels without the need for manual sample loading, which further reduces the time associated with sample analyses. Several other sample injection schemes have been also studied. Instead of using a simple cross-channel injector, twin-T injectors can be used to further control the sample plug size (24). Electrical biasing of the different reservoirs such as

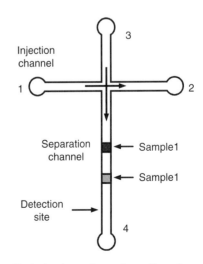

Figure 5. Typical channel configuration for a capillary electrophoresis microchip including loading or injection channel and separation channel.

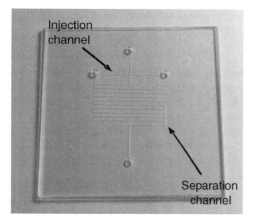

Figure 6. CE microchip fabricated in borosilicate glass. Serpentine separation channel can be observed.

pinch injection can reduce band broadening of the sample plugs caused by leakage at the intersection of the injection channel (25). Fig. 6 shows a CE microchip made in borosilicate glass. Serpentine separation channel was created to have a longer separation channel within a compact geometry.

The most common material used for the microchip is glass. Some of the earliest capillary electrophoresis microchips were fabricated in glass due to the good optical properties, a surface that enables electro-osmotic flow, and well-developed microfabrication techniques. To further reduce the microchip fabrication cost, polymer materials have been used due to the low material cost and easy mass-fabrication processes. Injection molding or hot embossing of plastics and polymer casting of polydimethylsiloxane (PDMS) are some of the methods used (26–28).

The application of this technology has been expanded from simple chemical analysis to biological applications, such as DNA sequencing, immunoassay, and biological particle separation (viruses, bacteria, and eukaryotic cells) (22,29–31). DNA sequencing on microchips was first reported by Mathies in 1995 (32). Single base resolution reached 150–200 bases in 10–15 min. Some of the advantages on top of the excellent resolution and short analysis time are the capability for high throughput. The compact size of the separation channels enable a large number of channels to be placed close together, which also facilitates fast detection using optical imaging or scanning lenses. To pack even more channels into a small area, a 6 inch circular glass plate carrying 96 radical channels conversing at the center of the chip was also developed (33). Detection occurred using a spinning confocal system. Separation of antigen and antibody from the corresponding antibody-antigen complex can be separated using microchip CE for immunoassays (34). Automatic sample injection capability can possibly eliminate the need of conventional robotic sample injection, which is commonly used in life science laboratories. More complex operations, by using electro-osmotic flow in conjunction with other operations such as lysing and concentration, have been demonstrated to show transport and analysis of biological particles (35–37). Dual injection has been also used where sample and reagents can be mixed on-column and analyzed to provide information about reaction kinetics, to

perform on-column derivatization for improved separation and termination, and to develop methods for simultaneous analysis of anionic and cationic compounds (38). Electro-osmotic flow in microchannels has been used in numerous applications to transport and mix fluid and particles but is beyond the subject of this chapter (39,40).

Electrophoresis with Other Separation/Detection Methods

One of the advantages of capillary electrophoresis is that this technique is easy to combine with different separation and detection methods to provide even more versatile, powerful, and efficient analysis tools. Isoelectric focusing (IEF) is a separation technique to resolve amphoteric molecules based on their isoelectric points (pI) (41). Isoelectric point is the pH at which a molecule carries no net electric charge. In capillary isoelectric focusing (CIEF), the capillary is first filled with a mixture of ampholytes and samples (42). When an electric field is applied to the capillary, a pH gradient is formed inside the capillary and the sample molecules migrate and stop at a position where the pH equals the pI of the sample molecules due to the loss of their net charges. In a one-step process, the entire capillary is illuminated to obtain images of the separation. In a two-step process, the separated samples are mobilized to the detection point using chemical, hydrodynamic, or electro-osmotic flow mobilization to simplify the detection equipments. When analyzing mixture of peptides in a microchip-based CIEF, focusing time of less than 30 seconds and total analysis time as short as 5 minutes is possible. As a result of the high resolving power, this method is most commonly used for studying peptides, proteins, recombinant products, cell lysates, and other complex mixtures (43,44).

Although capillary electrophoresis can provide excellent resolution, it is challenging to identify unknown substances. Mass spectrometry is a technique used for separating ions by their mass to charge ratios that enables identification of compounds by the mass of one or more elements in the compounds and enables determination of isotopic composition of one or more elements in the compound. By coupling capillary electrophoresis with mass spectrometry, direct identification of analytes by molecular mass, selectivity enhancement, and insight into the molecular structures are possible (43,44). The most prominent application of this combination is in proteomics (45). Interfacing these two techniques is of great importance because mass spectrometry requires ionized gas as samples whereas the output from a CE system is fluid (46). Electrospray ionization (ESI) and matrix-assisted laser desorption/ionization (MALDI) are some of the most widely used ionization methods. ESI, first developed in the 1980s, is the softest ionization technique currently available. It transforms ions in solution into ions in gas phase based on the electrostatic effects in solutions. An electric potential applied to an electrospray tip breaks the solution containing mixture of samples and solvents into small charged droplets. The shrinkage of the charged droplets by solvent evaporation further disintegrates the drops and forms gas-phase ions. MALDI uses laser beams to ionize samples located inside a crystallized bimolecular matrix that is used to protect the sample from being destroyed by direct laser beam. For microchip-based capillary electrophoresis systems, efforts have been focused on developing on-chip electrospray ionization techniques so that coupling to MS systems are efficient (47). The ultimate goal of such a coupled system is to use the microchip for fast and convenient sample preparation followed by online sample introduction for MS analysis. Beyond proteomics, the CE-ESI-MS combination has been widely used for drug analysis, food analysis, achiral and chiral solutes analysis, glycoscreening, and metabolic disorder screening (48–52).

Other detection methods used include electrochemical detection (53), electrochemiluminescence, nuclear magnetic resonance (NMR), ultraviolet resonance Raman spectroscopy (54,55).

Field-Flow Fractionation

Another technique similar in many ways and complementary to chromatography is field-flow fractionation (FFF). It is relatively young compared with chromatography, proposed by Giddings in 1968. This technique is always performed in an open channel (no packing or coatings) that is usually, but not restricted to, a wide, flat geometry with the breadth to height ratio being greater than 100 (Fig. 7). The purpose of this geometry is to take advantage of the laminar velocity parabolic flow profile of the carrier while minimizing secondary dispersion effects from the sidewalls. Circular channels can also be used in this way but are difficult to implement a uniform field in. Just as in chromatography, the samples spatially separated along the length of the channel are eluted discretely at the outlet, if enough column length and separating power are provided. For this reason, chromatographic principles and basic theory also apply to FFF. For example, FFF system retentions are often characterized by the number of theoretical plates and the van Deemter theory and the separations are characterized by resolution. However, some distinct differences exist that demonstrate the unique and complementary characteristics of FFF.

The first difference between FFF and chromatography is that a force field is applied to affect the samples instead of a stationary phase. The second difference is that the field is applied normal to the flow and separation direction. In chromatography, this field always acts opposite to the direction of the separation. The third difference is that, in most simple FFF systems, a direct mathematical relationship exists between the field and sample elution time.

The mechanism of retention and separation in FFF systems is based on compartmentalizing the various samples in the mixture to velocity zones in the parabolic flow profile of the carrier. The samples are selectively perturbed using a field applied normal to the carrier flow that are then concentrated at the accumulation wall. Normal diffusion opposes this movement until an equilibrium condition is established. Each sample will form a layer thickness based on its degree of perturbation. The sample specific velocity is then obtained from the first moment of the concentration and velocity profiles. The exact mathematical relationship among the variables in each of these steps allows for the determination of specific sample properties based solely on the elution time. A simple concentration or mass-based detector is sufficient without the need for calibration as in chromatography. In practice, this type of analysis is complicated by other factors leading to the need for calibration or more complicated detectors.

Any type of field or combination of fields can, in theory, be used to drive the separation. Some of the common fields used are sedimentary, flow, thermal, and electric. Some less common fields include magnetic, acoustic, dielectric, and others.

Recent advancements in FFF have included miniaturization of FFF channels. Miniaturization not only reduces sample and carrier volume requirements but also enhances

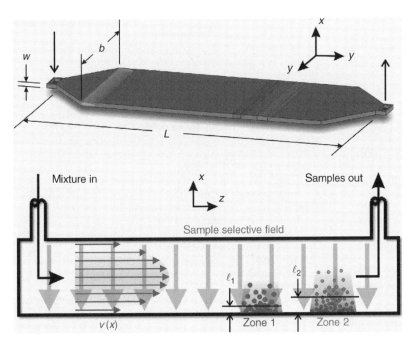

Figure 7. *Top*: Representation of a typical FFF channel. *Bottom*: Cross-sectional slice of channel in the center along the length and showing the operational principles of a separation in FFF. The required components for retention are a parabolic velocity profile and sample selective field.

retention and separation in some cases. Some of the systems that benefit through field enhancement from reducing the scale are thermal, electric, magnetic, and dielectrophoretic. In addition, microscale and nanoscale fabrication technologies have also made possible the implementation of systems that were previously impossible, difficult, or unreasonable to try, such as acoustic, magnetic, and dielectrophoretic FFF, as well as combinations of these systems.

Several microscale systems have been successfully implemented: micro-electric FFF (56–60); micro-thermal FFF (61–66); and AcFFF by Edwards and Frazier (67). Gale also integrated an electrical impedance detector within the channel to minimize the plate height due to extracolumn volumes in the detector and between the column and detector (56). Both Gale and Edwards also developed sample injection methods to minimize plate height further (56,61).

Systems manufactured using microelectricalmechanical system (MEMS) technologies are not only suitable for creating an inexpensive, disposable analysis device, but also for integrating with other methods such as chromatography, sensors, fluid handling devices, and actuators to create a total analysis system, or lab-on-a-chip. FFF can be used as a sample preparation tool, analytical device, or both in these systems.

CONCLUSION

Chromatography and the closely related fields of FFF and electrophoresis have proven to be valuable methods over the last century for separation, purification, and analysis. As a result of the wide variety and combinations of phases used in chromatography and fields applied in FFF, the number of types of samples that have been or can be retained and separated in these systems appears to be endless. Further advances in column, carrier, and detector technology, such as miniaturization, will continue to push the limits outward and make available faster and higher quality separations. In turn, researchers and industries in fields such as chemical engineering, bioengineering, chemistry, and pharmaceutics that rely on these techniques will also advance.

BIBLIOGRAPHY

1. Tswett M. Physical chemical studies on chlorophyll adsoptions. Berichte der Deutschen botanischen Gesellschaft 1906;24: 316–323.
2. Lesney MS. A brief history of "color writing." Today's Chemist at Work 1998;7(8):67–72.
3. Dorsey JG, et al. Liquid chromatography: Theory and methodology. Analyt Chem 1998;70(12):591R–644R.
4. Carrier R, Yip K. Intro to Chromatography. Available http:// www.rpi.edu/dept/chem-eng/Biotech-Environ/CHROMO/ chromintro.html.
5. Hardy JK. Analyt Chem. Available http://ull.chemistry.vakron.edu/analytical/.
6. Kazakevich Y, McNair H. Basic liquid chromatography. Available http://hplc.chem.shu.edu/NEW/HPLC_Book/.
7. van Deemter J, et al. Longitudinal diffusion and resistance to mass transfer as causes of nonideality in chromatography. Chem Eng Sci 1956;5:271–289.
8. Brede C, Pedersen-Bjergaard S. State-of-the art of selective detection and identification of I⁻, Br⁻, Cl⁻, and F-containing compounds in gas chromatography and liquid chromatography. J Chromatogr A 2004;1050(1):45.
9. Gonzalez FR. Application of capillary gas chromatography to studies on solvation thermodynamics. J Chromatog A 2004; 1037(1-2):233.
10. Roubani-Kalantzopoulou F. Determination of isotherms by gas-solid chromatography: Applications. J Chromatogr A 2004;1037(1-2):191.
11. He L, Beesley TE. Applications of enantiomeric gas chromatography: A review. J Liquid Chromatogr Related Technol 2005; 28(7-8):1075.
12. Abraham MH, et al. Hydrogen bonding. 42. Characterization of reversed-phase high-performance liquid chromatographic C-18 stationary phases. J Phys Organ Chem 1997;10(5):358–368.
13. Sherma J. High-performance liquid chromatography/mass spectrometry analysis of botanical medicines and dietary supplements: A review. J AOAC Int 2003;86(5):873.
14. Petrovic M, et al. Liquid chromatography-tandem mass spectrometry for the analysis of pharmaceutical residues in environmental samples: A review. J Chromatogr A 2005;1067(1-2):1.
15. Kuhn R. Hoffstetter-Kuhn S. Capillary Electrophoresis: Principles and Practice. Berlin, Germany: Springer-Verlag; 1993.
16. Baker DR. Capillary Electrophoresis. New York: Wiley-Interscience; 1995.
17. Horvath C, Nikelly JG. Capillary Electrophoresis and Chromatography. Washington, DC: American Chemical Society; 1990.
18. Nouadje G, et al. Capillary electrophoresis with laser-induced fluorescence detecion: Optical design and applications. Progress in HPLC-HPCE 1997;5:49–72.
19. Beale SC. Capillary electrophoresis. Analyt Chem 1998;70:279R–300R.
20. Effenhauser CS. Integrated chip-based microcolumn separation systems. Topics Curr Chem 1998;194:51–81.
21. Regnier FE, et al. Chromatography and electrophoresis on chips: critical elements of future integrated, microfluidic analytical systems for life science. Trends Biotechnol 1999;17(3): 101–106.
22. Dolnik V, et al. Capillary electrophoresis on microchip. Electrophoresis 2000;21:41–54.
23. Starkey DE, et al. A fluorogenic assay for b-glucuronidase using microchip-based capillary electrophoresis. J Chromatogr B 2001;762:33–41.
24. Effenhauser CS, et al. Glass chips for high-speed capillary electrophoresis separations with submicrometer plate heights. Analyt Chem 1993;65:2637–2642.
25. Jacobson SC, et al. Effects of injection schemes and column geometry on the performance of microchip electrophoresis devices. Analyt Chem 1994;66:1107–1113.
26. Effenhauser CS, et al. Integrated chip-based capillary electrophoresis. Electrophoresis 1997;18:2203–2213.
27. Martynova L, et al. Fabrication of plastic microfluidic channels by imprinting methods. Analyt Chem 1997;69:4783–4789.
28. McCormick RM, et al. Microchannel electrophoretic separations of DNA in injection-molded plastic substrates. Analyt Chem 1997;69:2626–2630.
29. Colyer CL, et al. Clinical potential of microchip capillary electrophoresis systems. Electrophoresis 1997;18:1733–1741.
30. Effenhauser CS, et al. Integrated capillary electrophoresis on flexible silicone microdevices: Analysis of DNA restriction fragments and detection of single DNA molecules on microchips. Analyt Chem 1997;69:3451–3457.
31. Carrilho E. DNA sequencing by capillary array electrophoresis and microfabricated array systems. Electrophoresis 2000;21: 55–65.
32. Woolley AT, Mathies RA. Ultra-high-speed DNA sequencing using capillary electrophoresis chips. Analyt Chem 1995;67: 3676–4086.

33. Simpson PC, et al. High-throughput genetic analysis using microfabricated 96-sample capillary array electrophoresis microplates. Proc Nat Acad Sci USA 1998;95:2256–2261.

34. Chiem N, arrison HDJ. Microchip-based capillary electrophoresis for immunoassays: analysis of monoclonal antibodies and theophylline. Analyt Chem 1997;69:373–378.

35. Li PCH, Harrison DJ. Transport, manipulation, and reaction of biological cells on-chip using electrokinetic effects. Analyt Chem 1997;69:1564–1568.

36. Cabrera CR, Yager P. Continuous concentration of bacteria in a microfluidic flow cell using electrokinetic techniques. Electrophoresis 2001;22:355–362.

37. Kremser L, et al. Capillary electrophoresis of biological particles: Viruses, bacteria, and eukaryotic cells. Electrophoresis 2004;25:2282–2291.

38. Priego-Capote F, Castro MDLd. Dual injection capillary electrophoresis: Foundations and applications. Electrophoresis 2004;25:4074–4085.

39. Debesset S, et al. An AC electro-osmotic micropump for circular chromatographic applications. Lab Chip 2004;4:396–400.

40. Glasgow I, et al. Electro-osmotic mixing in microchannels. Lab Chip 2004;4:558–562.

41. Hofmann O, Che D, Cruickshank KA, Muller UR. Adaptation of capillary isoelectric focusing to microchannels on a glass chip. Analyt Chem 1999;71:678–686.

42. Tan W, Fan ZH, Qiu CX, Ricco AJ, Gibbons I. Miniaturized capillary isoelectric focusing in plastic microfluidic devices. Electrophoresis 2002;23:3638–3645.

43. Shimura K. Recent advances in capillary isoelectric focusing: 1997–2001. Electrophoresis 2002;23:3847–3857.

44. Kilar F. Recent applications of capillary isoelectric focusing. Electrophoresis 2003;24:3908–3916.

45. Simpson DC, Smith RD. Combining capillary electrophoresis with mass spectrometry for applications in proteomics. Electrophoresis 2005;26:1291–1305.

46. Schmitt-Kopplin P, Frommberger M. Capillary electrophoresis-mass spectrometry: 15 years of development and applications. Electrophoresis 2003;24:3837–3867.

47. Sung W-C, Makamba H, Chen S-H. Chip-Based microfluidic devices coupled with electrospray ionization-mass spectrometry. Electrophoresis 2005;26:1783–1791.

48. Smyth WF. Recent applications of capillary electrophoresis-electrospray ionization-mass spectrometry in drug analysis. Electrophoresis 2005;26:1334–1357.

49. Simo C, Barbas C, Cifuentes A. Capillary electrophoresis-mass spectrometry in food analysis. Electrophoresis 2005;26:1306–1318.

50. Shamsi SA, Miller BE. Capillary electrophoresis-mass spectrometry: Recent advances to the analysis of small achiral and chiral solutes. Electrophoresis 2004;25:3927–3961.

51. Zamfir A, Peter-Katalinic J. Capillary electrophoresis-mass spectrometry for glycoscreening in biomedical. Electrophoresis 2004;25:1949–1963.

52. Senk P, Kozak L, Foret F. Capillary electrophoresis and mass spectrometry for screening of metabolic disorders in newborns. Electrophoresis 2004;25:1447–1456.

53. IV WRV, Pasas-Farmer SA, Fischer DJ, Frankenfeld CN, Lunte SM. Recent developments in electrochemical detection for microchip capillary electrophoresis. Electrophoresis 2004;25:3528–3549.

54. Qiu H, Yan J, Sun X, Liu J, Cao W, Yang X, Wang E. Microchip capillary electrophoresis with an integrated indium tin oxide electrode-based electrochemiluminescence detector. Analyt Chem 2003;75:5435–5440.

55. Wolters AM, Jayawickrama DA, Webb AG, Sweedler JV. NMR detection with multiple solenoidal microcoils for continuous-flow capillary electrophoresis. Analyt Chem 2002;74:5550–5555.

56. Gale BK, et al. A micromachined electrical field-flow fractionation (mu-EFFF) system. IEEE Trans Biomed Eng 1998;45(12):1459–1469.

57. Gale BK, et al. Geometric scaling effects in electrical field flow fractionation. 1. Theoretical analysis. Analyt Chem 2001;73(10):2345–2352.

58. Gale BK, et al. Geometric scaling effects in electrical field flow fractionation. 2. Experimental results. Analyt Chem 2002;74(5):1024–1030.

59. Gale BK. Novel techniques and instruments for field flow fractionation of biological materials. Abstr Papers Am Chem Soc 2003;225:U138–U138.

60. Gale BK. Miniaturized field flow fractionation systems. Abstr Papers Am Chem Soc 2004;227:U116–U116.

61. Edwards TL, et al. A microfabricated thermal field-flow fractionation system. Analyt Chem 2002;74(6):1211–1216.

62. Schimpf ME, Polymer analysis by thermal field-flow fractionation. J Liquid Chromatogr Related Technol 2002;25(13-15):2101–2134.

63. Janca J. Micro-channel thermal field-flow fractionation: High-speed analysis of colloidal particles. J Liquid Chromatogr Related Technol 2003;26(6):849–869.

64. Janca J, Ananieva IA. Micro-thermal field-flow fractionation in the characterization of macromolecules and particles: Effect of the steric exclusion mechanism. E-Polymers 2003.

65. Janca J, et al. Effect of channel width on the retention of colloidal particles in polarization, steric, and focusing micro-thermal field-flow fractionation. J Chromatogr A 2004;1046(1-2):167–173.

66. Bargiel S, et al. A micromachined system for the separation of molecules using thermal field-flow fractionation method. Sens Actuators A-Phys 2004;110(1-3):328–335.

67. Edwards TL. Microfrabricated acoustic and thermal field-flow fractionation systems. Electrical and computer engineering. Atlanta, GA: Georgia Institute of Technology; Ph.D. thesis, 2005. p 300.

See also ANALYTICAL METHODS, AUTOMATED; PHARMACOKINETICS AND PHARMACODYNAMICS; TRACER KINETICS.

CO$_2$ ELECTRODES

JOHN W. SEVERINGHAUS
University of California in San Francisco
San Francisco, California

METHODS OF MEASURING BLOOD pCO$_2$ BEFORE DISCOVERY OF THE pCO$_2$ ELECTRODE

Bubble Equilibration Methods

Carbon dioxide in blood is largely in the form of the bicarbonate ion, which could be converted to CO$_2$ gas by adding acid and extracting the gas in a vacuum. The concept of partial pressures gradually stimulated interest in measuring pCO$_2$ in the late nineteenth century. Gas analysis had been developed earlier, so the first method was to equilibrate a small gas bubble with a large volume of blood sample at body temperature, and then remove the bubble for gas analysis. Pflüger developed a tonometer for this purpose in the 1870s, and August Krogh used this method in fish in the early twentieth century. It was developed into a clinical and laboratory method by Richard Riley, using a specially adapted syringe with a capillary attached, which was invented by F. J. W. Roughton and P. Scholander during World War II. Riley's bubble method worked well for pCO_2, but poorly for pO_2 especially when blood was saturated with oxygen.

The Henderson–Hasselbalch Method

The most accurate early method was made possible by L. J. Henderson's discovery of buffering and his equation in 1908, its logarithmic modification by K. A. Hasselbalch in 1916, and P. T. Courage's design of a glass pH electrode (1925) in which blood could be measured with little loss of CO_2 to air. Blood pH was determined, usually at room temperature, there being no thermostated electrodes, and the Rosenthal temperature correction ($-0.0147\,pH$ units/ $°C$) was used to compute pH at $37\,°C$. Plasma CO_2 content was determined in the Van Slyke manometric apparatus that used 1 mL of plasma (after carefully centrifuging blood under a gas tight seal, a floating cork). This method reached a precision of $0.3\,mmHg\ pCO_2$ in studies of the arterial to alveolar pCO_2 difference during surgical hypothermia at The National Institute of Health (NIH) (5).

Astrup and The Equilibration Method

Hundreds of patients with polio needed artificial ventilation in the communicable disease hospital in Copenhagen during epidemics in 1950–1952. Poul Astrup, M.D. (Professor of Clinical Chemistry, University of Copenhagen, Copenhagen, Denmark, and Director of the Clinical Laboratory, Rigshospitalet, Copenhagen, Denmark) and his associates, particularly Ole Siggaard Andersen, Ph.D., M.D. (Professor of Clinical Chemistry, University of Copenhagen, and Director, Clinical Chemistry Laboratory, Herlev Hospital, Copenhagen, Denmark), devised a way of determining blood pCO_2 using only a pH electrode to measure pH before and after equilibration of a blood sample with two known concentrations of pCO_2(6). Astrup made use of the little known fact that, as pCO_2 is changed, the relationship of pH to pCO_2 in a given blood sample is semilogarithmic (Fig. 1). By plotting the two measured values of pH at the known equilibrated pCO_2, he could graphically interpolate the pCO_2 from the original sample pH.

From 1954 until the mid-1960s, Astrup's method was made widely available by the Radiometer Co. of Copenhagen. The device had a thermostated capillary pH electrode, reference electrode, and tiny shaking equilibrator through which humidified gas flowed. Astrup's apparatus and method became obsolete with the introduction of the CO_2 electrode.

Ole Siggaard Andersen, Astrup, and others used the values obtained for pH and pCO_2 to calculate bicarbonate, total CO_2, and base excess, a term they introduced as a quantitative measure of the nonrespiratory or metabolic abnormality in a whole blood sample. Base excess proved to be the first accurate index of the nonrespiratory component of acid–base balance (7). Its first application was only for blood, but by 1966, it was shown to apply to the extracellular fluid of the entire body if one assumed an average extracellular fluid hemoglobin concentration of 5 g/dL.

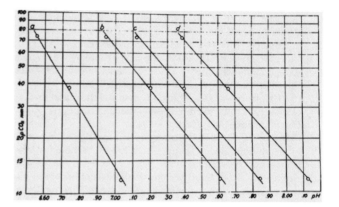

Figure 1. Equilibration method for measuring arterial pCO_2 introduced by Astrup during the Copenhagen polio epidemic, 1952–1954. Log pCO_2 plotted versus pH results in straight lines with varying pCO_2, and shifts of pH and slope when blood is acidified or alkalinized. The shift gave rise to the concept of base excess.

THE CO₂ ELECTRODE

History

A carbon dioxide (CO_2) electrode was first described by physiologists Gesell and McGinty at the University of Michigan in 1926, for use in expired air, but not in blood (8). It used the effect of CO_2 on the pH of a film of peritoneal membrane wet with a salt solution. Their paper was rediscovered 40 years later by M. Laver at Massachusetts General Hospital who informed Trubohovich of this effect (9).

In August 1954, Richard W. Stow, Ph.D. (Associate Professor of Physical Medicine, Ohio State University, Columbus, Ohio) (Fig. 2), a physical chemist, reported the design of a CO_2 electrode at the fall meeting of the American Physiologic Society in Madison, Wisconsin (10).

Figure 2. Richard Stow, invented the CO_2 electrode in 1954 to assist in managing polio patients on ventilators (10).

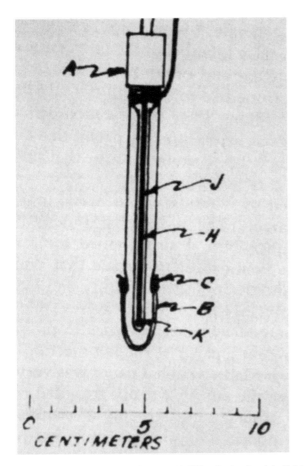

Figure 3. Stow's sketch of his 1954 CO_2 electrode. (a) Cable connection enclosure. (b) Rubber membrane. (c) Retaining O ring. (h) Chamber for internal pH electrolyte. (j) Reference electrode of silver chloride, not in contact with internal electrolyte, but opening to exterior through port K.

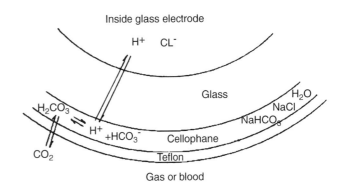

Figure 4. The concept of a pCO_2 electrode. The pH sensitive surface of a pH glass electrode is covered by a layer of electrolyte (here in cellophane), and then by a thin layer of a membrane permeable to CO_2, but not to hydrogen ions (here Teflon). The pH in that film is controlled by the partial pressure of CO_2 on the outside of the outer membrane as CO_2 dissolves and reacts with water to form carbonic acid. The carbonic acid dissociates into hydrogen and bicarbonate ions. Because the electrolyte has 5–20 mM HCO_3^- ions, changes of pCO_2 have no measurable effect on HCO_3^-. The mass law then requires H^+ to change in direct proportion to change in pCO_2. A doubling of pCO_2 doubles H^+ concentration, which is seen as a 0.3 pH unit fall.

The polio epidemic was raging at the time, and as part of the physical therapy faculty he had sought some way to measure pCO_2 in the victims. He read in the library about specific ion electrodes, and conceived the electrode idea. He had wrapped a thin rubber membrane wet with distilled water over a homemade combined pH and reference electrode (Fig. 3). When he changed gas pCO_2 outside the device, the pH inside changed as a log function of gas pCO_2. However, he was unable to get stable readings and said he doubted it could be made useful.

After his talk, Severinghaus asked him why he did not try adding sodium bicarbonate ($NaHCO_3$) to the water film in the electrode. He replied that he believed this would abolish the signal because bicarbonate would buffer the effect of pCO_2 on pH. Severinghaus replied that he was confident that bicarbonate would not block the sensitivity. Stow agreed that Severinghaus would further investigate this idea. In September, 1954, after returning from Madison to the National Institutes of Health, Severinghaus confirmed the advantage of adding bicarbonate ions. A schematic diagram of his modification of Stow's electrode is shown in Fig. 4. He used a Beckman bulb-type pH electrode, a chloride-coated silver wire reference, and a Beckman pH meter. He tied a film of cellophane over the

pH electrode soaked in 25 mM $NaHCO_3$ and then covered the entire tip with a thin rubber dam, later from a surgical glove. The bicarbonate not only made the device stable, but doubled the pCO_2 sensitivity compared with an electrolyte of distilled water (or 1% NaCl). Salt was added to help stabilize the silver chloride reference electrode.

In 1957, Stow. Baer and Randall (11) published their discovery of the CO_2 electrode without mentioning the need to add bicarbonate ion, and took no further interest in this idea. Stow had no interest in a patent, thinking it would distract him from his job, and also because his university only allowed inventors 10% of royalties. As a U.S. government employee, Severinghaus was not permitted to patent it, certainly not with a reluctant coinventor.

Severinghaus and co-worker A. Freeman Bradley proceeded to investigate and optimize the electrode design and to test its performance, linearity, drift, and response time. They constructed electrodes for laboratory use by several colleagues, but unfortunately made no attempt at commercial development for 4 years.

Between 1958 and 1960 several other investigators constructed and published similar CO_2 electrodes, in several instances without being aware of the Stow–Severinghaus electrode (12–14).

CO₂ ELECTRODE DESIGN DETAILS

A CO_2 electrode consists of a slightly spherical surfaced glass pH electrode and a silver chloride reference electrode. Both are mounted in a glass or plastic sleeve holding a Teflon or silicone rubber membrane, typically 12 μm thick, over the glass surface, in some cases with a spacer of very thin lens-cleaning paper between membrane and glass to

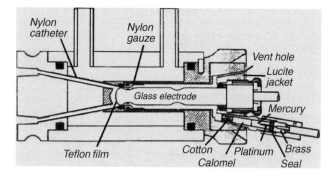

Figure 5. Cuvette with blood inlet and outlet connections in a thermostated water jacket made for the Stow–Severinghaus pCO_2 electrode (National Welding Co, San Francisco, 1959).

insure a uniform distribution of the electrolyte that is NaCl or KCl with ~5–20 mequiv/L of $NaHCO_3$. For use in blood, the electrode is mounted in a 37 °C cuvette into which a small sample of blood can be injected (typically 50 μL) (Fig. 5).

The electrode output voltage is a logarithmic function of pCO_2, ~60 mV for a 10-fold change of pCO_2, which induces a pH change of ~1 pH unit. Sensitivity is defined as Δ pH/ $\Delta \log pCO_2$, where S reaches nearly the ideal maximum value of 1.0 with HCO_3^- concentrations of 5–25 mM (Fig. 6). At higher bicarbonate levels, carbonate acts as a buffer, and reduces both sensitivity and speed of response. Response is faster at lower bicarbonate concentration, but carbonic acid pK' is 6.1, resulting in some change of bicarbonate as pCO_2 changes, reducing sensitivity. As bicarbonate concentration is lowered, sensitivity falls to 30 mV/decade pCO_2 change, or $S = 0.5$, at zero bicarbonate.

The log response is almost linear from 5 to 700 mmHg pCO_2. The response time to a step change of pCO_2 is exponential with a 95% response time of ~30 s, depending on the membrane thickness and material, bicarbonate ion concentration and the thickness of the electrolyte layer over the glass electrode surface. It can be made to respond

in <1 s by using thin silastic (silicone rubber) membrane, low bicarbonate concentration (i.e., 1 mequiv/L), and adding carbonic anhydrase to the electrolyte, but the downside is loss of stability and signal amplitude.

The CO_2 electrode is usually calibrated to read in millimeters of mercury. It reads the same value for gas and liquid equilibrated with that gas at the electrode temperature, usually 37 °C.

A useful test of a leaking membrane is to equilibrate a dilute solution (e.g., 1 mequiv/L) of HCl, or lactic acid with a known calibration gas, and test its reading. Any leak will permit acid entry and an erroneously high pCO_2.

Maintenance requires replacement of the membrane and electrolyte when errors are detected or when drift has driven the electrode beyond the ability of the apparatus to compensate its potential. The pH glass may become so impermeable to hydrogen ions that it shows low sensitivity or slow response after years of use.

The amplifier circuit must be electrically isolated from the ground because any ground path leakage will draw current through the silver chloride reference and changes its potential causing drift. The input impedance of all modern pH and pCO_2 meter amplifiers is >10^{11} Ω.

The Combined Blood Gas Analysis Apparatus

In 1956, Leland Clark disclosed his invention of the oxygen electrode at a meeting in Atlantic City to which Severinghaus had invited physiologists interested in measuring pO_2. That invention made a huge difference in blood gas analysis.

While Severinghaus completed his anesthesia residency at the University of Iowa, with help from the physiology workshop, he constructed a thermostat into which he mounted both the Stow–Severinghaus CO_2 electrode and the Clark O_2 electrode in a stirred cuvette with a small blood tonometer. That apparatus was exhibited at the meeting of the American Society of Anesthesiologists in October 1957 and at the meeting of the Federation of American Societies of Experimental Biology in Atlantic City in the spring of 1958 and published in 1958 (15) (Fig. 7).

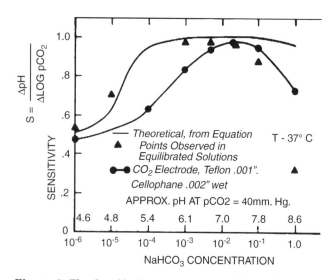

Figure 6. The first blood gas apparatus, with the Clark pO_2 electrode (below) in a stirred cuvette, and the Stow–Severinghaus pCO_2 electrode above, tilted to keep the internal air bubble of the pH electrode away from the tip (1957).

Figure 7. The first three-function blood gas analyzer, using a McInnes Belcher pH electrode (1930) with the pCO_2 and pO_2 electrodes in a 37 °C bath.

The Three-Function Blood Gas Analyzer

In 1958, after moving from the National Institutes of Health to the University of California, San Francisco, Severinghaus and Bradley added a pH electrode to the blood gas electrode waterbath, making the first three-function blood gas apparatus (Fig. 8). Forrest Bird, Ph.D., M.D. (President, Bird Corporation, Palm Springs, California) had designed popular positive-pressure ventilators, manufacturing them at the National Welding Co. in San Francisco. He proposed to manufacture the CO_2 electrode and to make it commercially available. From 1959–1961 the National Welding Co. sold the only available pCO_2 electrode. The design concept was soon copied and marketed by Beckman, Radiometer, Instrumentation Labs and later by several other firms.

Impact of Blood Gas Analysis

During the 1960s, blood gas analysis became widely available in anesthesia, intensive and critical care facilities, and cardiorespiratory research laboratories. For several years, the Severinghaus paper (15) was among the most quoted articles in biologic literature, and blood gases were called the most important laboratory test for critically ill patients. Blood gas apparatus now uses automatic self-calibration and automatic transport of sample and washing of cuvettes, printing of results, and often sending the values to remote terminals. In the United States, regulations have been used by pathologists to require that these automated instruments can only be used by licensed technicians, usually meaning that the income flows to pathologists. Gone are the days when students, nurses, residents, and faculty all took part in doing blood gas analysis.

A more complete history of the CO_2 electrode and related blood gas technology is available in References (9,16).

Figure 8. Relationship of pCO_2 electrode sensitivity to its internal electrolyte bicarbonate ion concentration. Maximum sensitivity occurs at $\sim 20\,mM$ HCO_3^-, but for faster response, most electrodes operate at 5–10 mM.

HISTORY AND THEORY OF TRANSCUTANEOUS BLOOD OXYGEN MONITORING

From 1951 to 1952, the discovery of oxygen related blindness in premature infants created an urgent need for continuous noninvasive monitoring of blood oxygen. A new solution to the problem came from physiologists studying skin respiration. Human skin breathes, taking up oxygen and giving off CO_2 to the air. If skin is covered (as by a flat unheated pCO_2 electrode) the surface $tcpO_2$ falls to zero in a few minutes. However, in 1951 Baumberger and Goodfriend showed that if skin blood flow is greatly increased by the highest tolerable heat (45 °C), the surface pO_2 rises to about paO_2 (arterial blood) (17).

Within a year after Clark's invention of the membrane covered platinum polarographic electrode (18,19), Rooth used polarography to confirm the Baumberger report (20). Researchers tried unsuccessfully to use chemical vasodilators to make skin pO_2 a monitor of paO_2. Kwan and Fatt (21) noted that pO_2 of the palpebral conjunctiva measured with an unheated tiny Clark electrode mounted facing outward on a contact lens over the cornea simulated paO_2. This device was briefly marketed a decade later, but discontinued due to the danger of infection.

In Marburg, Germany, Professor of Physiology Dietrich Lübbers and students, especially Renate Huch, pursued the concept of heating the skin under an oxygen electrode by heating the electrode itself to as high as 45 °C. They were joined by Patrick Eberhard, and the group soon found ways of making electrically heated, thermostated oxygen surface electrodes. By 1972, they had shown a good relationship between heated skin and arterial blood pO_2 in infants (22). Several firms began to design electrodes for this purpose.

DEVELOPMENT OF METHODS AND UNDERSTANDING OF THEORY

By 1977, the Marburg group had published at least 11 papers documenting the validity of transcutaneous oxygen measurement. At least three commercial $tcpO_2$ electrode systems were available (Helige, Roche, Radiometer). In November 1977, some 18 research teams joined for a workshop on transcutaneous blood gas methods in San Francisco, assessing the theory, problems, possibilities, and progress (23–30). The following summer (1978) many of these workers joined the Marburg team and others for the first international congress on transcutaneous blood gas monitoring, establishing the technology as an essential tool in neonatology and as useful in many other fields (31,32).

The agreement of $tcpO_2$ with paO_2 proved to be a cancellation of two opposing effects illustrated in Fig. 9 (27,33).

1. Heating of desaturated blood raises its pO_2 by 7%/°C, or 50% at 43 °C, but in saturated blood, as in water, pO_2 rises only 1.3%/°C (35);
2. Skin metabolism at the high temperature consumes O_2 as it diffuses outward from capillaries through living cells, reducing the value to about paO_2.

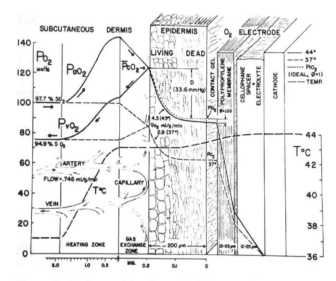

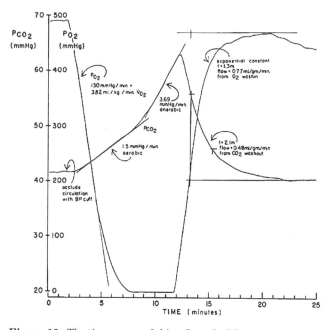

Figure 9. A schema of the effect of both heating of skin surface by a transcutaneous electrode, and of local metabolism, on the tissue internal oxygen tension from the arteries out past the capillaries and the living and dead epidermis to the surface, and through the electrode membrane into the cathode that keeps its surface $pO_2 = 0$ by its electrical negative potential (39).

Figure 10. The time course of skin pO_2 and pCO_2 on an arm after sudden circulatory occlusion with a blood pressure cuff. The rate of fall of pO_2 from a high level is a measure of the skin metabolic rate. The pCO_2 rises at first from metabolic CO_2 production, but later at a steeper rate as skin generates lactic acid when skin pO_2 reaches zero. With release of occlusion, the electrode recovery time is delayed by both the skin washin and washout, and by electrode equilibration (34).

The outward oxygen diffusion is facilitated by heat that proved to "melt" some skin diffusion barriers (33,36). Skin O_2 conductivity C (adult volar forearm) was determined by two groups by comparison of flux with two membranes (teflon and mylar) of very high and low conductivity. With a large gold cathode Clark electrode, $C = 15\,nL \cdot cm^{-2} \cdot s^{-1} \cdot atm^{-1}$(37) and with a mass spectrometer $C = 10\,nL \cdot cm^{-2} \cdot s^{-1} \cdot atm^{-1}$(38).

Skin O_2 consumption (VO_2) was determined after thermal vasodilation by the rate of fall of $tcpO_2$ with circulatory occlusion (arm cuff) (Fig. 10) (27). Relative skin blood flow under the heated electrode was estimated by measuring the required heating power (39). Analysis of data collected at two levels of pO_2 and two temperatures permitted calculation of blood flow, capillary temperature under a heated electrode, and diffusion gradient from capillary to surface (40). Mean adult volar forearm skin VO_2 was $4.2 \pm 0.4\,\mu L \cdot g^{-1} \cdot min^{-1}$ at 44 °C and $2.8 \pm 0.3\,\mu L \cdot g^{-1} \cdot min^{-1}$ at 37 °C. At 44 °C, skin blood flow averaged $0.64 \pm 0.17\,mL \cdot g^{-1} \cdot min^{-1}$, capillary temperature was 43 °C and the diffusion gradient was $32 \pm 7\,mmHg$.

TRANSCUTANEOUS CO₂

In 1959, Severinghaus constructed a 37 °C thermostated open tipped CO_2 electrode to determine pCO_2 of various tissue surfaces in animals (Fig. 11) (41). Without heating the skin well above body temperature, skin pCO_2 at 37 °C climbed steadily over one-half of an hour to $> 80\,mmHg$. Dog intestinal mucosa and liver surfaces were very high.

Fifteen years later, the success in transcutaneous measurement of oxygen led to design and testing of electrodes to measure $tcpCO_2$ by Beran et al. (42,43), Huch et al. (44) and Severinghaus et al. (45,46). Combined $tcpO_2$–$tcpCO_2$ electrodes were initially described by Parker et al. (47) and

Severinghaus (48). Figure 12 schematically shows the internal design of an early Radiometer combined electrode. Figure 13 shows the electrode with a membrane mounted.

When a heated combined pO_2–pCO_2 electrode is first attached to skin, the time needed to equilibrate is $\sim 5\,min$ for both electrodes, although the pO_2 electrode may show later small changes as thermal vasodilation slowly develops (Fig. 14). The response to step changes in alveolar and arterial pCO_2 is slower as seen in Fig. 15. Here the response is delayed both by the washout or washin of CO_2 into the tissue by blood flow, and the electrode's own delay. The response is pseudoexponential, a combination of the two delays, resulting in a 95% response times of $\sim 10\,min$.

Without correction, $tcpCO_2$ is not similar to $paCO_2$. Heating of blood (and water) raises pCO_2 $\sim 4.6\%/°C$

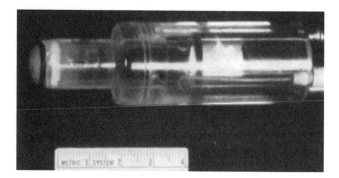

Figure 11. The first tissue surface pCO_2 electrode (41) with a circulating temperature controlled water jacket.

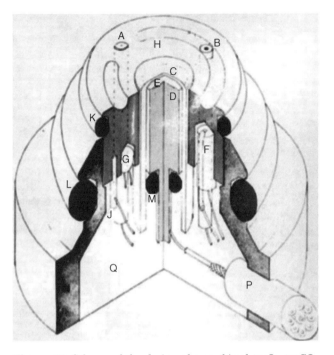

Figure 12. Schema of the design of a combined tc$p$O₂–tc$p$CO₂ electrode. (a) pO₂ cathode, the end of a 12 μm platinum wire fused in glass. (b) A silver wire reference electrode. (c) pH glass electrode surface. (d) solid silver internal pH electrode (used to improve heat transfer to skin). (e) Internal pH electrolyte. (f) Heater Zener diode. (g) Thermistor. (h) Silver body, and reference electrode. J, K, L, M: O rings. N: Lexan jacket. Q: epoxy. P: Cable (48).

Figure 13. Photograph of combined pO₂–pCO₂ electrode with teflon membrane (48).

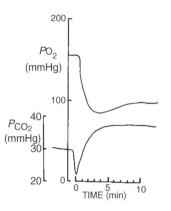

Figure 14. Initial responses of a combined pO₂–pCO₂ electrode when first mounted on skin. Both electrodes need ~5 min to equilibrate, the pO₂ showing a small late rise as skin hyperemia develops from the heating (49).

(41), metabolism adds ~3 mmHg pCO₂, and the cooling by skin and blood of the electrode surface further raises the electrode reading. The effect of heating on blood pCO₂ may be computed as $\Delta p\mathrm{CO}_2 = \exp(0.046[T - 37])$ (51). The net effect at 43 °C was found to be tc$p$CO₂ = 1.33paCO₂ + 4 mmHg (48,52) or tc$p$CO₂ = 1.4paCO₂ (53). This form of temperature-dependent correction factor was later incorporated in most commercial transcutaneous blood gas monitoring apparatus.

With this correction factor, the relationship of tc$p$CO₂ to paCO₂ is excellent, as shown in Fig. 16. The previous correction factors appear to have become incorrect for a second generation of the Radiometer tc$p$CO₂ electrodes, due to a design change in the internal temperature coefficient of the glass pH electrode. The additive factor of 4 mmHg changed to ~8 mmHg in the newer instruments (Kagawa S, personal communication).

Although tc$p$CO₂ appears to work at 42–43 °C, Tremper et al. showed that 44 °C was a better temperature when

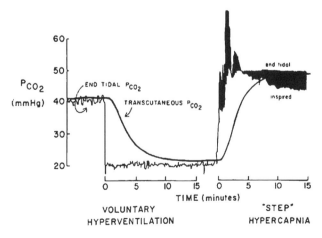

Figure 15. Transcutaneous pCO₂ electrode response to a step increase of ventilation adjusted by the subject to reduce end tidal pCO₂ suddenly from 40 to 20 mmHg and hold it at a constant level for 18 min, followed by addition of enough CO₂ to inspired gas to raise end tidal pCO₂ as quickly as possible to ~50 mmHg. The response time constants (63%) are ~5 min (45).

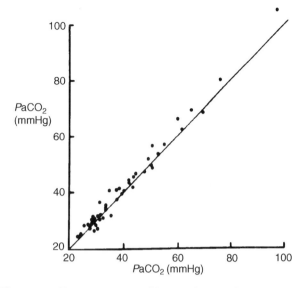

Figure 16. Transcutaneous pCO_2 correlates well with arterial pCO_2 in patients during anesthesia or intensive care (48).

blood pressure was or had been low (54). The $tcpCO_2$ value was better than the $P_{ET}CO_2$ value (end-tidal or end-expired air) in predicting $paCO_2$ (bias and s.d. -1.6 ± 4.3 mmHg) in anesthetized adults ($n = 24$) (55).

A special advantage of $tcpCO_2$ is that it averages out breath-by-breath variations, and has almost no inherent "noise" or variability, such that it often is found to be the best trend monitor for detecting small changes in $paCO_2$ such as those induced by experimental variations (anesthesia, ventilatory settings, posture, F_IO_2, F_ICO_2, blood pressure, pharmacologic agents, etc).

APPLICATIONS

Transcutaneous technology is used in many ways, some of which are discussed in accompanying papers:

1. Neonatology: Guidance of O_2 therapy remains the most common use of transcutaneous monitoring (56–58). The suspected etiologic role of hyperoxia ($tcpO_2 > 80$ mmHg) in retinitis of premature infants has been confirmed in a cohort study (59). The $tcpO_2$ value can be measured above and below the ductus to demonstrate closure (60). In low birth weight infants, $tcpCO_2$ (at 40 °C!) is the best available monitor of ventilation (61).

2. Fetal Monitoring: Using specially designed electrodes attached to the fetal scalp, intrapartum monitoring revealed some important new pathophysiologic understanding (62–65). As hoped, changes in $tcpO_2$ rapidly reflected changing maternal and fetal conditions (66). The $tcpO_2$ value fell and $tcpCO_2$ rose with contractions during the second stages of labor (67). The $tcpCO_2$ value closely followed fetal $paCO_2$ (68). When there were signs of fetal distress, fetal scalp $tcpO_2$ was < 15 mmHg (69). Surprisingly, O_2 administration to mothers with fetal distress did not alter

fetal pCO_2 or raise pO_2 (70). During maternal hypocapnia, fetal $tcpO_2$ fell due to the Bohr effect, whereas it rose during hypercapnia (71). Fetal $tcpO_2$ was considered influenced by local scalp blood flow (72). Repeated episodes of asphyxia were reported to express catecholamines, which reduced blood flow to the fetal skin, artifactually reducing $tcpO_2$ (73,74). Fetal $tcpCO_2$ may have failed to disclose severe acidosis or circulatory impairment (75).

3. Sleep Studies: Combined pO_2–pCO_2 electrodes are used in sleep studies in combination with pulse oximetry, because nostril sampling of end-tidal pCO_2 is somewhat annoying and more apt to become plugged or dislodged (76–83). The combined $tcpO_2$–$tcpCO_2$ electrode made it possible to show that the ventilatory response to induced mild hypoxia in sleeping infants changes with age from acute depression at 1–5 days, to stimulation at 4–8 weeks, and mild or no stimulation at 10–14 weeks (84). A method was designed for estimating the ventilatory response to CO_2 during sleep using capnography and $tcpCO_2$ (79).

4. Peripheral Circulation: The $tcpO_2$ electrodes are extensively used in evaluating arterial disease in the peripheral circulation (85–88). A test of adequacy of peripheral circulation, "initial slope index" (ISI) was suggested by Lemke and Lübbers (89). Blood flow is stopped by an arm cuff above the electrode and restarted when $tcpO_2 = 0$. The initial rate of rise should be a slope per min of at least 75% of the preocclusion $tcpO_2$.

5. Skin Circulation: Monitoring the viability of skin after injury or transplant or flap movement (90,91).

6. Ventilatory Control: In intensive care, transcutaneous electrodes greatly increased the safety and simplicity of PEEP optimization and respiratory management of adults with respiratory distress syndrome (92). They are widely used simply to reduce arterial blood sampling.

7. Hyperbaric Oxygen: Monitoring and guiding hyperbaric oxygen therapy, primarily for infections and wound healing (93,94). The $tcpO_2$ tracked paO_2 up to 4-atm hyperbaric pressure in normal subjects (95). Surprisingly, no one has reported using $tcpO_2$ in hyperbaric treatment of CO poisoning despite the demonstration by Barker and Tremper in experimental CO administration that transcutaneous pO_2 falls linearly as COHb increases, and reaches about one-fifth of its initial value at the highest COHb levels despite the maintenance of constant arterial pO_2 (96). It is thus unknown whether HBO can normalize tissue pO_2 in the presence of high levels of COHb.

8. Clinical Physiology: Transcutaneous monitoring has found use in exercise tolerance studies (97,98). End-tidal CO_2 is not exactly equal to $paCO_2$ and the difference between them varies with posture and inspired oxygen concentration. When testing hypoxic ventilatory responses by monitoring $P_{ET}CO_2$, $tcpCO_2$ helps to correct these small errors (99).

9. Pharmacologic Research: Transcutaneous monitoring may be the simplest monitor of the depressant effects of opiates, sedatives, and anesthetics especially in awake children (100).

10. Animal Studies: Intestinal or other tissue animal experimental ischemia has been found to be better detected by the rise of the organ or tissue surface pCO_2 using $tcpCO_2$ electrodes at body temperature than by gastric tonometry (101). Both $tcpO_2$ and $tcpCO_2$ have been widely used in small and large animal studies (102) and to assess the effect of cardiopulmonary resusitation (CPR) (103).

ACCURACY

With the widespread use of $tcpO_2$ and $tcpCO_2$ came concern about its accuracy and the possible sources and effects of errors, especially with severe hypotension (28,104). Peabody et al.(25) identified two groups of infants in whom $tcpO_2$ was lower than paO_2. These were infants receiving an intravascular infusion of tolazoline and infants with mean arterial blood pressures > 2.5 s.d. below the predicted average value. Vasoconstrictors also lower $tcpO_2$(105). Both of these situations represent extreme alterations in peripheral blood flow. Mild hypotension, hypothermia, anemia, radiant warmers, and bilirubin lights did not adversely affect transcutaneous accuracy (106). In a large multiinstitutional study of 327 patients older than 1 month, when paO_2 was between 80 and 220 mmHg, Palmisano found the mean bias $\pm$ s.d. of $tcpO_2$ was -43 ± 40 mmHg, and the slope of the regression was 0.65 (107). It was determined that $tcpCO_2$ correlated far better with $paCO_2$: $R = 0.929$, slope 1.052, bias and s.d. $= 1.3 \pm 4.0$ mmHg ($n = 756$).

Defining a $tcpO_2$ index as $tcpO_2/paO_2$, Tremper and Shoemaker (108) studied the effect of shock. For 934 data sets taken on 92 patients not in shock, there was a correlation coefficient (r) of 0.89 and a $tcpO_2$ index 0.79 ± 0.12 (SD). In five patients with moderate shock, the r was 0.78 and the $tcpO_2$ index was 0.48 ± 0.07. In nine patients with severe shock, there was no correlation between $tcpO_2$ and paO_2 and the $tcpO_2$ index was 0.12 ± 0.12.

LIMITATIONS

Skin burns may occur after an electrode has been in one place over several hours at 44–45 °C, and sometimes even at 43 °C. Long-term monitoring requires site changes, or a dual electrode alternating system (109). There may be problems with drift of calibration, membrane failure, partial loss of skin contact giving errors in both O_2 and CO_2 readings. Maintenance of these electrodes requires training and some technical proficiency.

IMPACT OF PULSE OXIMETRY

Pulse oximetry came into widespread use in 1985–1987, and quickly replaced transcutaneous blood gas analysis in many situations. However, after an initial switch to oximetry, neonatologists found that oximetry failed to detect hyperoxia adequately (110) and now mostly use both technologies (111–115). In neonatology, a significant problem is that the inherent errors of pulse oximetry are ∼ 3%, which could fail to warn of $paO_2 > 80$ unless a set point of ∼ 90% S_pO_2 is chosen (116). Some have arbitrarily dismissed transcutaneous monitoring as "...plagued by technical problems, ...Its use in efforts to prevent retinopathy of prematurity, an eye disease of preterm newborns often leading to blindness, proved disappointing" (117). To them, the transcutaneous field served as a model of problems in medical innovation, new technology, and personnel training. Not everyone agrees with this pessimism. Most technical problems have been solved, and the occurrence of blindness in very premature infants is now believed to be multifactorial, not just due to hyperoxia. Therefore when it occurs, it is not appropriate to attribute it to failed transcutaneous methodology.

CONCLUSIONS

The enthusiasm for transcutaneous blood gas analysis of the period 1976–1986 was followed by a decrease due to the advent of pulse oximetry. The number of papers per year listing medline keywords "transcutaneous blood gas" reached an early peak of 75 in 1979, when the first international symposium was devoted to this field, in Marburg and 200 in 1987. However, after 1986 many papers used the keywords "transcutaneous blood gas" when writers meant to refer to pulse oximetry.

Transcutaneous technology is inherently somewhat complicated. Users must change membranes and calibrate, change skin sites periodically to avoid burns, beware of drift or error due to poor circulation or poor skin attachment, and take account of the slower response than given by oximetry. Nonetheless, transcutaneous blood gas measurement continues to be used because of its unique ability to meet many special situations needing its characteristics of noninvasively and continuously determining partial pressures of O_2 and CO_2. Several professional organizations have published guidelines for use of these monitors (118,119).

BIBLIOGRAPHY

1. Severinghaus JW. The current status of transcutaneous blood gas analysis and monitoring. Blood Gas news (Radiometer house organ) 1998;7:4–9.
2. Severinghaus JW. The Invention and Development of Blood Gas Apparatus. Anesthesiology 2002;97:253–256.
3. Severinghaus JW. Severinghaus electrode. In: JR Maltby, editor. Notable Names in Anaesthesia. London: Royal Society of Medicine Press Ltd.; 2002.
4. Severinghaus JW, Astrup P, Murray J. Blood gas analysis and critical care medicine. Am J Respir Crit Care Med 1998;157:S114–S122.
5. Severinghaus JW, Stupfel MA, Bradley AFJ. Accuracy of blood pH and pCO_2 determinations. J Appl Physiol 1956;19:189–196.
6. Astrup P. A simple electrometric technique for the determination of carbon dioxide tension in blood and plasma, total

content of carbon dioxide in plasma and bicarbonate content in "separated" plasma at a fixed carbon dioxide tension. Scand J Clin Lab Invest 1956;8:33–43.

7. Siggaard-Andersen O, Engel K, Jorgensen K, Astrup P. A micro method for determination of pH, carbon dioxide tension, base excess and standard bicarbonate in capillary blood. Scand J Clin Lab Invest 1960;12:172–176.

8. Gesell R, McGinty DA. Regulation of respiration: VI. Continuous electrometric methods of recording changes in expired carbon dioxide and oxygen. Am J Physiol 1926;79:72–90.

9. Trubuhovich RV. History of pCO$_2$ electrodes. Br J Anaesth 1970;42:360–362.

10. Stow RW, Randall BF. Electrical measurement of the pCO$_2$ of blood (abstract). Am J Physiol 1954;179:678.

11. Stow RW, Baer RF, Randall B. Rapid measurement of the tension of carbon dioxide in blood. Arch Phys Med Rehabil 1957;38:646–650.

12. Gertz KH, Loeschcke HH. Elektrode zur bestimmung des CO$_2$ drucks. Naturwissenschaften 1958;45:160–161.

13. Hertz CH, Siesjo B. A rapid and sensitive electrode for continuous measurement of pCO$_2$ in liquids and tissue. Acta Physiol Scand 1959;47:115–123.

14. Snell FM. Electrometric measurement of carbon dioxide and bicarbonate ion. J Appl Physiol 1960;15:729–732.

15. Severinghaus JW, Bradley AF. Electrodes for blood pO$_2$ and pCO$_2$ determination. J Appl Physiol 1958;13:515–520.

16. Severinghaus JW, Astrup P. History of blood gas analysis. Int Anesthesiol Clin 1987;25:69–95.

17. Baumberger JP, Goodfriend RB. Determination of arterial oxygen tension in man by equilibration through intact skin. Fed Proc 1951;10:10.

18. Clark LC. Monitor and control of tissue O$_2$ tensions. Trans Am Soc Artif Intern Organs 1956;2:41–48.

19. Clark LC, Clark EW. Personalized history of the Clark oxygen electrode. Inter Anesthesiol Clin 1987;25:1–30.

20. Rooth G, Sjostedt S, Caligara F. Bloodless determination of arterial oxygen tension by polarography. Sci Tools LKW Instr J 1957;4:37.

21. Kwan M, Fatt I. A noninvasive method of continuous arterial oxygen tension estimation from measured palperal conjunctival oxygen tension. Anesthesiology 1971;35:309–314.

22. Huch R, Lübbers DW, Huch A. Quantitative continuous measurement of partial oxygen pressure on the skin of adults and new-born babies. Pflügers Arch 1972;337:185–198.

23. Vesterager P. Transcutaneous pCO$_2$ electrode. Scand J Clin Lab Invest 1977;37:27–30.

24. Friis Hansen B. Transcutaneous measurement of arterial blood oxygen tension with a new electrode. Scand J Clin Lab Invest 1977;37:31–36.

25. Peabody JL, Willis MM, Gregory GA, Tooley WH, Lucey JF. Clinical limitations and advantages of transcutaneous oxygen electrodes. Acta Anaesthesiol Scand Suppl 1978;68:76–82.

26. Tremper KK, Huxtable RF. Dermal heat transport analysis for transcutaneous O$_2$ measurement. Acta Anaeshesiol Scand Suppl 1978;68:4–8.

27. Severinghaus JW, Stafford MJ, Thunstrom AM. Estimation of skin metabolism and blood flow with tcPo$_2$ and tcPco$_2$ electrodes by cuff occlusion of the circulation. Acta Anaesth Scand Suppl 1978;68S:9–15.

28. Versmold HT, Linderkamp O, Holzmann M, Strohhacker I, Riegel KP. Limits of tcpO$_2$ monitoring in sick neonates: Relation to blood pressure, blood volume, peripheraal blood flow and acid base status. Acta Anaesthesiol Scand Suppl 1978;S68:88–90.

29. Kimmich HP, Kreutzer F. Model of oxygen transport through skin as basis for absolute transcutaneous measurement of PaO$_2$. Acta Anaesthesiol Scand Suppl 1968;S68:16–19.

30. Fatt I. Transmucosal measurement of blood pH at the palpebral conjunctiva. Acta Anaesthesiol Scand Suppl 1978;S68:142–144.

31. Huch A, Huch R. The development of the transcutaneous pCO$_2$ technique into a clinical tool. In: Huch R, Huch A, Lucey JR, editors. Continuous Transcutaneous Blood Gas Monitoring, Birth Defects: Original Article Series. Volume XV-No. 4. New York: A.R.Liss; 1979.

32. Lübbers DW, Cutaneous and Transcutaneous pO$_2$ and pCO$_2$ and their measuring conditions. In: Huch R, Huch A, Lucey JF, editors. Continuous Transcutaneous Blood Gas Monitoring, Birth Defects: Original Article Series. Volume XV-No. 4. New York: A. R. Liss; 1979.

33. Lübbers DW. Theoretical basis of the transcutaneous blood gas measurements. Crit Care Med 1981;9:721–733.

34. Severinghaus JW. Transcutaneous Blood Gas Analysis. Respir Care 1982;27:152–159.

35. Severinghaus JW. Simple, accurate equations for human blood O$_2$ dissociation computations. J Appl Physiol 1979;46: 599–602.

36. Lübbers DW. Theory and development of transcutaneous oxygen pressure measurement. Int Anesthesiol Clin 1987; 25:31–65.

37. Eberhard P, Severinghaus JW. Measurement of heated skin O$_2$ diffusion conductance and pCO$_2$ sensor induced O$_2$ gradient. Acta Anaesthesiol Scand Suppl 1978;68:1–3.

38. Hansen TN, Sonoda Y, McIlroy MB. Transfer of oxygen, nitrogen and carbon dioxide through normal adult human skin. J Appl Physiol 1980;49:438–443.

39. Parker D, Delpy D, Reynolds EOR, St. Andrew D. A transcutaneous pO$_2$ electrode incorporating a thermal clearance local blood flow sensor. Acta Anaesthesiol Scand Suppl 1978; S68:33–39.

40. Thunstrom AM, Stafford MJ, Severinghaus JW. A two temperature, two pO$_2$ method of estimating the determinants of tcpO$_2$. In: Huch R, Huch A, Lucey JR, editors. Continuous Transcutaneous Blood Gas Monitoring, Birth Defects: Original Article Series. Volume XV-No. 4, New York: A. R. Liss; 1979.

41. Severinghaus JW. CO$_2$ Spannung und Perfusion in Gewebe. Anaesthetist 1960;9:50–55.

42. Beran AV, Huxtable RF, Sperling DR. Electrochemical sensor for continuous transcutaneous pCO$_2$ measurement. J Appl Physiol 1976;41:442–447.

43. Beran AV, Shigezawa GY, Yeung HN, Huxtable RF. An improved sensor and a method for transcutaneous CO$_2$ monitoring. Acta Anaesthesiol Scand Suppl 1978;S68:111–117.

44. Huch A, Seiler D, Meinzer K, Huch R, Galster H, Lübbers DW. Transcutaneous pCO$_2$ measurement with a miniaturised electrode. Lancet 1977;1:982–983.

45. Severinghaus JW, Stafford M, Bradley AF. tcpCO$_2$ electrode design, calibration and temperature gradient problems. Acta Anaesthesiol Scand Suppl 1978;68:118–122.

46. Severinghaus JW, Bradley AF, Stafford MJ. Transcutaneous pCO$_2$ electrode design with internal silver heat path. In: Huch A, Huch R, Lucey JF, editors. Continuous Transcutaneous Blood Gas Monitoring, Birth Defects: Original Article Series. Volume XV-No. 4, New York: A.R. Liss, Inc.; 1979.

47. Parker D, Delpy D, Reynolds EOR. Single electrochemical sensor for transcutaneous measurement of pO$_2$ and pCO$_2$. In: Huch R, Huch A, Lucey JF, editors. Continuous Transcutaneous Bloo Gas Monitoring, in Birth Defects: Original Article Series. Volume XV-No. 4, New York: A. R. Liss; 1979.

48. Severinghaus JW. A combined transcutaneous pO$_2$—pCO$_2$ electrode with electrochemical HCO$_3$- stabilization. J Appl Physiol 1981;51:1027–1032.

49. Severinghaus JW. Transcutaneous monitoring of arterial pCO$_2$. Resp Monit Int Care 1982; 85–91.

50. Gothgen I. Heat-indued changes in pO$_2$ and pCO$_2$ of blood. Acta Anaesthesiol Scand 1984;28:447–451.

51. Jacobsen E, Gothgen I. Relationship between arterial and heated skin surface carbon dioxide tension in adults. Acta Anaesthesiol Scand 1985;29:198–202.

52. Hazinski TA, Severinghaus JW. Transcutaneous analysis of arterial pCO$_2$. Med Instrum 1982;16:150–153.

53. Wimberley PD, Pedersen KG, Thode J Fogh-Andersen, Sorensen AM, Siggaard-Andersen O. Transcutaneou and capillary pCO$_2$ and and pO$_2$ measurements in healthy adults. Clin Chem 1983;29:1471–1473.

54. Tremper KK, Mentelos RA, Shoemaker WC. Effect of hypercarbia and shock on transcutaneous carbon dioxide at different electrode temperatures. Crit Care Med 1980;8:608–612.

55. Phan CQ, Tremper KK, Lee SE, Barker SJ. Noninvasive monitoring of carbon dioxide: A comparison of the partial pressure of transcutaneous and end-tidal carbon dioxide with the partial pressure of arterial carbon dioxide. J Clin Monit 1987;3:149–154.

56. Hoppenbrouwers T, Hodgman JE, Arakawa K, Durand M, Cabal LA. Transcutaneous oxygen and carbon dioxide during the first half year of life in premature and normal term infants. Pediatr Res 1992;31:73–79.

57. Huch R. Review: Perinatal monitoring. Acta Anaesthesiol Scand Suppl 1995;S107:91–94.

58. Huch A. Transcutaneous blood gas monitoring. Acta Anesthesiol Scand Suppl 1995;107:87–90.

59. Flynn JT, et al., A cohort study of transcutaneous oxygen tension and the incidence and severity of retinopathy of prematurity [see comments]. New Engl J Med 1992;326: 1050–1054.

60. Schmidt S, Kakatschikaschwili T, Langner K, Dudenhausen JW, Saling E. [Circulatory adaptation of the newborn infant immediately post partum by biolocal measurement of transcutaneous pCO$_2$]. Z Geburtshife Perinatol 1984;188: 21–23.

61. Binder N, Atherton H, Thorkelsson T, Hoath SB. Measurement of transcutaneous carbon dioxide in low brithweight infants during the first two weeks of life. Am J Perinatol 1994;11:237–241.

62. Huch A, Huch R, Schneider H. Fetal transcutaneous pO$_2$—current knowledge. In: Huch R, Huch A, Lucey JF, editors. Continuous Transcutaneous Blood Gas Monitoring, Birth Defects: Original Article Series. Volume XV-No. 4, New York: A.R.Liss; 1979.

63. Huch R, Huch A. Fetal and maternal PtcO$_2$ monitoring. Crit Care Med 1981;9:694–697.

64. Lofgren O. Continuous transcutaneous carbon dioxide monitoring in the fetus during labor. Crit Care Med 1981;9:750–751.

65. Okane M, Shigemitsu S, Inaba J, Koresawa M, Kubo T, Iwasaki H. Non-invasive continuous fetal transcutaneous pO$_2$ and pCO$_2$ monitoring during labor. J Perinat Med 1989;17:399–410.

66. Antoine C, Young BK, Silverman F. Simltaneous measurement of fetal tissue pH and transcutaneous pO$_2$ during labor. Eur J Obstet Gynecol Reprod Biol 1984;17:69–76.

67. Schmidt S, Langner K, Dudenhausen JW, Saling E. Reliability of transcutaneous measurement of oxygen and carbon dioxide partial pressure with a combined pO$_2$–pCO$_2$ electrochemical sensor in the fetus during labor. J Perinat Med 1985;13:127–133.

68. Bergmans MG, van Geijn HP, Weber T, Nickelsen C, Schmidt S, van den Berg PP. Fetal transcutaneous pCO$_2$ measurements during labour. Eur J Obstet Gynecol Reprod Biol 1993;51:1–7.

69. Kaneoka T, Kobayashi H, Uchida K, Shirakawa K. [Continuous fetal biochemical monitoring and cardiotocography]. Nippon Sanka Fujinka Gakkai Zasshi 1988;40:721–728.

70. Bartnicki J, Langner K, Harnack H, Meyenburg M. The influence of oxygen administration to the mother during labor on the fetal trasncutaneously measured carbon-dioxide partial pressure. J Perinat Med 1990;18:397–402.

71. Aarnoudse JG, Oeseburg B, Kwant G, Zwart A, Zijlstra WG, Huisjes HJ. Influence of variations in pH and pCO$_2$ on scalp tissue oxygen tension and carotid arterial oxygen tension in the fetal lamb. Biol Neonate 1981;40:252–263.

72. Smits TM, Aarnoudse JG, Zijlstra WG. Fetal scalp blood flow as recorded by laser Doppler flowmetry and transcutaneous pO$_2$ during labour. Early Hum Dev 1989;20:109–124.

73. Jensen A, Kunzel W, Kastendieck E. Fetal sympathetic activity, transcutaneous pO$_2$, and skin blood flow during repeated asphyxia in sheep. J Dev Physiol 1987;9:337–346.

74. Paulick R, Kastendieck E, Wernze H. Catecholamines in arterial and venous umbilical blood: placental extraction, correlation with fetal hypoxia, and transcutaneous partial oxygen tension. J Perinat Med 1985;13:31–42.

75. Braems G, Kunzel W, Lang U. Transcutaneous pCO$_2$ during labor—a comparison with fetal blood gas analysis and transcutaneous pO$_2$. Eur J Obstet Gynecol Reprod Biol 1993;52: 81–88.

76. Fukui M, Ohi M, Chin K, Kuno K. The effects of nasal CPAP on transcutaneous pCO$_2$ during non-REM sleep and REM sleep in patients with obstructive sleep apnea syndrome. Sleep 1993;16:S144–5.

77. Manning DJ, Stothers JK. Sleep state, hypoxia and periodic breathing in the neonate. Acta Paediatr Scand 1991;80:763–769.

78. Morielli A, Desjardins D, Brouillette RT. Transcutaneous and end-tidal carbon dioxide pressures should be measured during pediatric polysomnography. Am Rev Respir Dis 1993;148: 1599–1604.

79. Naifeh KH, Severinghaus JW. Validation of a maskless CO$_2$-response test for sleep and infant studies. J Appl Physiol 1988;64:391–396.

80. Naughton M, Benard D, Tam A, Rutherford R, Bradley TD. Role of hyperventilation in the pathogenesis of central sleep apneas in patients with congestive heart failure [see comments]. Am Rev Respir Dis 1993;148:330–338.

81. Naughton MT, Benard DC, Rutherford R, Bradley TD. Effect of continuous positive airway pressure on central sleep apnea and nocturnal pCO$_2$ in heart failure. Am J Respir Crit Care Med 1994;150:1598–1604.

82. Schafer T, Schafer D, Schläfke ME. Breathing, transcutaneous blood gases, and CO$_2$ response in SIDS siblings and control infants during sleep. J Appl Physiol 1993;74:88–102.

83. Schlaefke ME, Schaefer T, Kronberg H, Ullrich GJ, Hopmeier J. Transcutaneous monitoring as trigger for therapy of hypoxemia during sleep. Adv Exp Med Biol 1987;220:95–100.

84. Milerad J, Hertzberg T, Lagercrantz H. Ventilatory and metabolic responses to acute hypoxia in infants assessed by transcutaneous gas monitoring. J Dev Physiol 1987;9: 57–67.

85. White RA, Nolan L, Harley D, Long J, Klein S, Tremper K, Nelson R, Tabriski J, Shoemaker W. Noninvasive evaluation of peripheral vascular disease using transcutaneous oxygen tension. Am J Surg 1982;144:68–75.

86. Kram HB, Shoemaker WC. Diagnosis of major peripheral arterial trauma by transcutaneous oxygen monitoring. Am J Surg 1984;147:776–780.

87. Padberg FT, Back TL, Thompson PN, Hobson RW. Transcutaneous oxygen (TcpO$_2$) estimates probability of healing in the ischemic extremity. J Surg Res 1996;60:365–369.

88. Wutschert R, Bounameaux H. Determination of amputation level in ischemic limbs. Reappraisal of the measurement of TcpO$_2$. Diabetes Care 1997;20:1315–1318.

89. Lemke R, Klaus D, Lübbers DW, Oevermann G. Noninvasive $ptCO_2$ initial slope index and invasive $ptCO_2$ arterial index as diagnostic criterion of the state of peripheral circulation. Crit Care Med 1988;16:353–357.

90. Keller HP, Klaue P, Hockerts T, Lübbers DW. Transcutaneous pO_2 measurement on skin transplants. In: Huch R, Huch A, Lucey JF, editors. Continuous Transcutaneous Blood Gas Monitoring, Birth Defects: Original Article Series. Volume XV-No. 4, New York: A.R.Liss; 1979.

91. Lübbers DW. Transcutaneous measurements of skin O_2 supply and blood gases. Adv Exp Med Biol 1992;316:49–60.

92. Tremper KK, Waxman K, Shoemaker WC. Use of transcutaneous oxygen sensors to titrate PEEP. Ann Surg 1981;193:206–209.

93. Dooley J, Schirmer J, Slade B, Folden B. Use of transcutaneous pressure of oxygen in the evaluation of edematous wounds. Undersea Hyperb Med 1996;23:167–174.

94. Wattel F, Pellerin P, Mathieu D, Patenotre P, Coget JM, Schoofs M, Leps P. [Hyperbaric oxygen therapy in the treatment of wounds, in plastic and reconstructive surgery]. Ann Chir Plast Esthet 1990;35:141–146.

95. Huch A, Huch R, Hollmann G, Hockerts T, Keller HP, Seiler D, Sadzek J, Lübbers DW. Transcutaneous pO_2 of volunteers during hyperbaric oxygenation. Biotelemetry 1977;4: 88–100.

96. Barker SJ, Tremper KK. The effect of carbon monoxide inhalation on pulse oximetry and transcutaneous pO_2 [see comments]. Anesthesiology 1987;66:677–679.

97. Sridhar MK, Carter R, Moran F, Banham SW. Use of a combined oxygen and carbon dioxide transcutaneous electrode in the estimation of gas exchange during exercise. Thorax 1993;48:643–647.

98. Breuer HW, Skyschally A, Alf DF, Schulz R, Heusch G. Transcutaneous pCO_2-monitoring for the evaluation of the anaerobic threshold. Comparison to lactate and ventilatory threshold [see comments]. Int J Sports Med 1993;14:417–421.

99. Sato M, Severinghaus JW, Powell FL, Xu FD, Spellman MJJ. Augmented hypoxic ventilatory response in men at altitude. J Appl Physiol 1992;73:101–107.

100. Alswang M, Friesen RH, Bangert P. Effect of preanesthetic medication on carbon dioxide tension in children with congenital heart disease. J Cardiothorac Vasc Anesthesiol 1994;8: 415–419.

101. Rozenfeld RA, Dishart MK, Tønnessen TI, Schlichtig R. Methods for detecting intestinal ischemic anaerobic metabolic acidosis by local pCO_2. J Appl Physiol 1996;81:1834–1842.

102. Keller HP, Klaue P, Lübbers DW. Transcutaneous pO_2 measurements on rats and rabbits. In: Huch R, Huch A, Lucey JR, editors. Continuous Transcutaneous Blood Gas Monitoring, Birth Defects: Original Article Series. Volume XV-No. 4, New York: A.R.Liss; 1979.

103. Tremper KK, Shoemaker WC. Continuous CPR monitoring with transcutaneous oxygen and carbon dioxide sensors. Crit Care Med 1981;9:417–418.

104. Versmold HT, Linderkamp O, Holzmann M, Strohhacker I, Riegel K. Transcutaneous monitoring of pO_2 in newborn infants: where are the limits? Influence of blood pressure, blood volume, blood flow, viscosity, and acid base state. In: Huch R, Huch A, Lucey JF, editors. Continuous Transcutaneous Blood Gas Monitoring, in Original Article Series. Volume XV-No. 4, New York: A.R. Liss; 1979.

105. Wendling P, Fussinger R, Schmidt HD, Stosseck K. [Validity of the transcutaneous pO_2-measurement during pharmacologically induced changes of skin perfusion (author's transl)]. Anaesthesist 1982;31:135–138.

106. Ewald U, Huch A, Huch R, Rooth G. Skin reactive hyperemia recorded by a combined $TcpO_2$ and laser Doppler sensor. Adv Exp Med Biol 1987;220:231–234.

107. Palmisano BW, Severinghaus JW. Transcutaneous pCO_2 and pO_2: a multicenter study of accuracy. J Clin Monit 1990;6: 189–195.

108. Tremper KK, Shoemaker WC. Transcutaneous oxygen monitoring of critically ill adults, with and without low flow shock. Crit Care Med 1981;9:706–709.

109. Fallenstein F, Ringer P, Huch R, Huch A. A new system for $tcpO_2$ long-term monitoring using a two-electrode sensor with alternating heating. Adv Exp Med Biol 1987;220: 285–289.

110. Paky F, Koeck CM. Pulse oximetry in ventilated preterm newborns: reliability of detection of hyperoxaemia and hypoxaemia, and feasibility of alarm settings. Acta Paediatr 1995;84:613–616.

111. Baeckert P, Bucher HU, Fallenstein F, Fanconi S, Huch R, Duc G. Is pulse oximetry reliable in detecting hyperoxemia in the neonate?, Adv Exp Med Biol 1987;220:165–169.

112. Bragiroli A, Sacco C, Carone M, Donner CF. Pulse oximeter and transcutaneous O_2 monitoring: criteria for a choice. Eur Respir J Suppl 1990;11:515s–517s.

113. Fallenstein F, Baeckert P, Huch R. Comparison of in-vivo response times between pulse oximetry and transcutaneous pO_2 monitoring. Adv Exp Med Biol 1987;220:191–194.

114. Wimberley PD, Helledie NR, Friis-Hansen B, Fogh-Andersen N, Olesen H. Pulse oximetry versus transcutaneous pO_2 in sick newborn infants. Scand J Clin Lab Invest Suppl 1987;188:19–25.

115. Wimberley PD. Oxygen monitoring in the newborn. Scand J Clin Lab Invest Suppl 1993;214:127–130.

116. Poets CF, Southall DP. Noninvasive monitoring of oxygenation in infants and children: practical considerations and areas of concern [see comments]. Pediatrics 1994;93:737–746.

117. Mike V, Krauss AN, Ross GS. Doctors and the health industry: a case study of transcutaneous oxygen monitoring in neonatal intensive care. Soc Sci Med 1996;42:1247–1258.

118. American Academy of Pediatrics Committee on Drugs: Guidelines for monitoring and management of pediatric patients during and after sedation for diagnostic and therapeutic procedures. Pediatrics 1992;89:1110–1115.

119. Wimberley PD, Burnett RW, Covington AK, Maas AHJ, Mueller-Plathe O, Siggaard-Andersen O, Weisberg HF, Zijlstra WG. Guidelines for transcutaneous pO_2 and pCO_2 measurement. IFCC document. Ann Biol Clin 1990;48:39–43.

See also BLOOD GAS MEASUREMENTS; CARDIOPULMONARY RESUSCITATION; RESPIRATORY MECHANICS AND GAS EXCHANGE.

COBALT-60 UNITS FOR RADIOTHERAPY

JOHN R. CUNNINGHAM
Camrose, Alberta, Canada

INTRODUCTION

Cobalt is a metal, between iron and nickel, in the periodic table. It resembles them and occurs fairly commonly in iron and nickel ores, such as those found near Sudbury, Ontario, Canada. Cobalt as a substance has been known since about the mid-1700s. It was discovered in 1735 by a Swedish chemist named Brandt and was named after Kobald, a goblin from Germanic legends, known for stealing silver. Its salts were used in ancient days for making pigments, which produced brilliant blue colors in pottery.

The ancient Egyptians used it in painting murals in tombs and temples. It is necessary, in trace amounts, for proper nutritional balance.

The nucleus of ^{59}Co, which is the only isotope of cobalt found in Nature, has 27 protons and 32 neutrons. It happens to have an unusually large neutron capture cross-section, which means that bombardment with neutrons turns many of its atoms into ^{60}Co, which is very highly radioactive. ^{60}Co has a relatively long half-life (5.26 years) and it decays to ^{60}Ni by the emission of a beta particle (an electron). ^{60}Ni is also radioactive and emits two energetic gamma rays with energies 1.17 and 1.33 MeV. Million electron volts = MeV. An electron volt is the amount of energy an electron has when it is accelerated through a voltage of 1 MV. It is very small: 1 MeV = 1.602×10^{-13} J. These gammma-rays are produced in almost equal number and the pair of them can be approximated by their average 1.25 MeV, to form radiation that has high penetration in matter.

Cobalt has an atomic weight of 58.933 atomic mass units (amu), a mass density of 8900 kg/m^3, and melts at $\sim 1500\ °$C. All of these properties combine to make it unique as a practical source of radiation for cancer treatment, industrial radiography, sterilization of food, and other purposes requiring intense but physically small sources of radiation.

It was not isolated as a metal until early in the eighteenth century and was not used for its metallic properties until the twentieth century. Its most important modern use is in the production of alloys of steel that are very hard and very resistant to high temperatures. These alloys find their uses in cutting tools and such diverse products as jet engines and kitchen cutlery.

HISTORY

Sampson et al. (1) noted the interesting radioactive properties of ^{60}Co at least as early as 1936. Livingood and Seaborg (2) described its properties in 1941. W.V. Mayneord, of the Royal Cancer Hospital in London (later the Royal Marsden Hospital), and A. J. Cipriani, then Head of the Biology Division at Chalk River, Ontario, Canada, described its production by neutron bombardment of ^{59}Co in a nuclear reactor in 1947 (3).

In June of 1949, H.E. Johns, then professor of physics at the University of Saskatchewan, Canada, and physicist to the Saskatchewan Cancer Commission, visited the NRX nuclear reactor at Chalk River, Ontario, to discuss, with Cipriani and others, the possibilities of irradiating a sample of cobalt in order to produce a ^{60}Co source. The theoretical advantages of using the energetic gamma rays of ^{60}Co to destroy cancer cells had been known for some time, but practical problems of source production centered on the availability of a reactor with a sufficiently high neutron flux combined with a facility to handle and prepare the resulting highly radioactive source. Earlier, in 1945, J.S. Mitchell of Cambridge and J.V. Dunworth of Chalk River had discussed the possibilities of producing ^{60}Co using the high neutron flux expected to be available from NRX, a nuclear reactor being built at the Chalk River site. At that time NRX was not yet operating, but in 1949, when

Johns visited it, it was. The NRX is a heavy water reactor and at that time had the highest available neutron flux in the world ($\sim 3 \times 10^{13}$ neutrons/cm^2/s). The reactor was heavily involved in a program of radioisotope production and the irradiation of cobalt was taken to be part of this program.

Arrangements were made to irradiate three samples of cobalt and they were placed in the reactor in the fall of 1949. They were removed ~ 1.5 years later. The first source was destined for a cobalt unit being designed and built by Dr. Johns and his students in Saskatoon (4). It was delivered there in July of 1951 and on the 18th of August it was installed in the cobalt unit that had been prepared for it. The second source was sent to the Victoria Hospital in London, Ontario, where it was installed on the 23rd of October 1951 in a unit that had been designed and built by Eldorado Mining and Refining Company (later Atomic Energy of Canada Ltd.). Dr. Ivan Smith treated the first patient in London on 27th of October 1951, just 4 days after the installation of the source. The first patient treated on the Saskatoon unit, by Dr. T.A. Watson, was on the 8th of November 1951. Some mystery surrounds the details of the third source. There is some evidence that it was originally intended to go to Mayneord in England, but that in 1951 it was considered that postwar reconstruction was not yet sufficiently advanced there so it was diverted to the M.D. Anderson Hospital in Houston, Texas. It was to be installed in a unit designed by L.G. Grimmett, who had recently been hired by Dr. Gilbert Fletcher largely for this task (5). Part of the mystery concerns the fact that it was delayed in its irradiation and was actually removed from the reactor for a time and later replaced. Some have suggested that this may have been related to the outbreak of the Korean War and the general sensitivity concerning nuclear matters. Whatever the reason, it was not actually shipped until July of 1952, almost a full year later than the other two sources. The M.D. Anderson unit was then at Oak Ridge Tennessee for experimental purposes and was transferred, with its source, to the M.D. Anderson Hospital in Houston in September of 1953. The first patient was treated, in Houston, on the 22nd of February 1954. Pictures of these three cobalt units are given in Fig. 1. Roger F. Robinson has told an informative and interesting history, which includes many details about the original sources, as well as stories about a number of the people involved(6).

Each of these three sources was used in cobalt units for the treatment of cancer for many years. The two Canadian units became prototypes for units that were subsequently sold commercially. The unit in London, built by Atomic Energy of Canada Ltd., was the first of a long series of machines manufactured by them. The first series was known as the "Eldorado" series. A later series of units went under the name "Theratron". The descendant of that company: MDS Nordion, is still building and selling cobalt units. The Saskatoon unit, designed by H. E. Johns and several of his students at the University of Saskatchewan, was made by John MacKay of the Acme Machine and Electric Co. Ltd. in Saskatoon (7) and later commercially by Picker X-ray of Cleveland, Ohio. Each of these units is pictured in Fig. 1 near the times of their source installations.

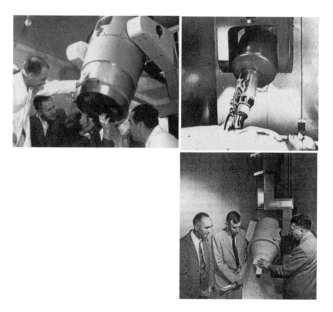

Figure 1. The worlds first three cobalt units. Clockwise from above London, Ont., Canada, Saskatoon, Sask., Canada and Houston, Texas.

Before 1951, radiation therapy had been carried out almost exclusively by X-ray machines operating at tube voltages of 400,000 V or less. Such machines produce X-ray beams having a broad spectrum of X-ray energies with an average of one-third or less of the maximum. Thus, a 400 kV machine would correspond to a single energy of ~ 133 keV. Cobalt-60, with its average photon energy of 1.25 MeV, is the equivalent of an X-ray machine operating about six times the old value. As will be seen later, cobalt units are also mechanically and electrically simple devices and, following their introduction, rapidly became the standard machine for treating nearly all cancers other than that of the skin. Cobalt units have now been almost completely replaced by linear accelerators, which produce X rays having still greater penetration and higher outputs allowing shorter treatment times.

THE PHYSICS OF ACTIVATION: EXPOSURE AND DOSE

Only the physics directly related to the description of ^{60}Co sources and units will be discussed here. More detailed information can be found in standard textbooks such as those of Attix (8), Greening (9), and Johns and Cunningham (10).

Almost any material placed within the neutron radiation field of a nuclear reactor will become radioactive. The probability of this happening is determined by the cross-section of the material for capturing a neutron. The cross-section is the equivalent of a probability, although it is usually expressed as an area. Many atoms have neutron capture cross-sections, of the order of 10^{-24} cm^2 around 1935, Enrico Fermi, then in Rome, was measuring these cross-sections. When he found one of about this size he exclaimed, "That's as big as a barn!" 1 barn = 10^{-24} cm^2 is the common measure of nuclear cross-section and its use is permitted by the International System (SI) of units, and if a neutron passes through this area it is "captured" by the

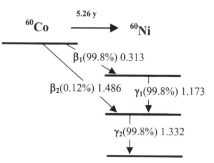

Figure 2. The decay schemes of ^{60}Co and ^{60}Ni, showing the beta particle energies of ^{60}Co and the gamma-ray energies from ^{60}Ni.

nucleus to form a new nuclear species, which usually is radioactive.

The interaction of the neutron with a nucleus is quite complex, and a number of different products may be formed. The nucleus may capture the neutron to produce a new species that is stable, or the neutron may be re-emitted at the same or a different energy. In the latter case, we refer to the process as neutron scattering. The production of ^{60}Co is an example of neutron capture. A nucleus of ^{59}Co absorbs a neutron and forms ^{60}Co, which is radioactive and decays with a half-life of 5.26 years by the emission of an electron that turns it into an isotope of nickel, ^{60}Ni. The decay scheme of ^{60}Co and ^{60}Ni is shown in Fig. 2. The two gamma rays mentioned earlier are actually emitted by the Nickel nucleus ^{60}Ni. Some properties of cobalt and its radiation are given in Table 1.

The ^{60}Co activity produced is determined by the neutron flux density in the reactor, the neutron capture cross-section, the amount of ^{59}Co inserted into the reactor, and the length of time it is left there. The rate of production of radioactive atoms can be expressed as

$$\frac{N}{t} = N \sigma \phi \qquad (1)$$

where N is the number of ^{59}Co atoms placed in the reactor, σ is the neutron capture cross-section per atom, ϕ is the flux density of neutrons, and Δt is a time interval. The

Table 1. Properties of Cobalt and Its Radiation

Property	Value
Cobalt-59	
Atomic number	$Z = 27$
Atomic weight	$A = 58.933$ amu
Mass density	$\rho = 8900$ kg/m^3
Melting point	1500 K
Neutron capture cross-section	$\sigma = 37 \times 10^{-24}$ cm^2
Cobalt-60	
Half-life	$T_{1/2} = 5.26$ years
Bata energies	0.313 MeV (99.8%)
	1.486 MeV (0.12%)
Nickel-60	
Photon energies	$\gamma_1 = 1.733$ MeV
	$\gamma_2 = 1.332$ MeV
Interaction coefficient in water	$(\mu/\rho) = 0.0698$ cm^2/g
Average Energy Absorbed in water	$E_{ab} = 0.456$ MeV
Half-value layer in Pb	$X_{1/2} = 11$ mm

parameter ΔN will be the number of activations that take place in this time interval.

As an illustrative numerical example, consider a sample of 15 g of ^{59}Co to be located in a nuclear reactor at a point where the neutron flux density is 10^{14} cm^{-2}/s. This represents a source that is 1.5 cm in diameter and ~ 1 cm high and is fairly representative of sources and neutron fluxes that have been used. The original two Canadian sources were 2.54 cm in diameter and composed of ~ 26 disks each 0.5 mm thick. The American source was square in cross-section. From Eq. 1, and with the use of some of the information given in Table 1, we calculate the number of atoms of ^{59}Co that are converted to ^{60}Co during a period of time Δt. We also require a value for Avogadro's Number N_A, so that we can calculate the number of atoms (at) of ^{59}Co in 1 g of the substance.

$$N_A = 6.023 \times 10^{23} \text{ atoms/mol}$$

The number of ^{59}Co atoms in our 15 g sample is

$$N_{59_{Co}} = 15 \text{ g} \times \frac{6.023 \times 10^{23} \text{at}}{\text{mol}} \times \frac{1 \text{ mol}}{58.933 \text{ g}} = 1.533 \times 10^{23} \text{ at}$$

From Table 1, we see that the cross-section for neutron capture in ^{59}Co is 37×10^{-24} cm^2/atom. If the 15 g of cobalt were left in the reactor at this location for a period of 1 h, the number of atoms (at) converted to ^{60}Co, following Eq. 1, would be

$$N = 1.533 \times 10^{23} \times \frac{37 \times 10^{-24} \text{cm}^2}{\text{at}} \times \frac{10^{14} \text{ cm}^{-2}}{\text{s}} \times \frac{3600 \text{ s}}{\text{h}}$$
$$= 2.042 \times 10^{18} \text{ at}$$

Although this appears to be a very large number of atoms it represents only ~ 0.2 mg of ^{60}Co. It does, however, represent a considerable amount of radioactivity and would be easy to measure.

The most fundamental parameter for the specification of the strength of a radioactive source is activity. Activity is defined as the number of decay processes that occur per second and its special unit is the bequerel (Bq), which is defined to be an average of one nuclear disintegration each second. Activity is easy to describe theoretically, but is very difficult to determine experimentally. It can be inferred from the number of atoms of the substance and the value of its half-life, which for ^{60}Co is given in Table 1 as 5.26 years.

Activity can be calculated from the simple relation

$$A = N\lambda \tag{2}$$

where λ is a constant of proportionality known as the transformation constant. It is related to the half-life $T_{1/2}$, of the radioactivity by

$$\lambda = \frac{0.693}{T_{1/2}} \tag{3}$$

where the number 0.693 is the natural logarithm of 2.

For example, the activity of ^{60}Co that would result from the above irradiation of 15 g of ^{59}Co would be

$$A = 2.04 \times 10^{18} \times \frac{0.693}{5.26 \text{ year} \times 3.1557 \times 10^6 \text{ s/year}}$$
$$= 0.0852 \times 10^{12} \text{ s}^{-1} = 85.2 \times 10^9 \text{/s} = 85.2 \text{ GBq} \tag{4}$$

where the half-life $T_{1/2}$ has been expressed in seconds. The activity that is actually produced in a reactor irradiation is considerably less than this theoretical amount. This is largely due to attenuation of the neutron flux by the considerable mass of the cobalt.

The more traditional unit of activity has been the curie (Ci), which corresponds to 3.7×10^{10} nuclear decays/s. The activity of the above source, stated in curies would be

$$A = \frac{85.2 \times 10^9}{\text{s}} \times \frac{1 \text{ Ci}}{3.7 \times 10^{10}\text{/s}} = 2.30 \text{ Ci}$$

The specification of a commercial source of radiation in terms of activity is not very practical because activity does not uniquely relate to the radiation output when an individual source is loaded into a treatment unit. The output will depend on the physical size and configuration of the source and the design of the collimator of the treatment unit.

This problem was solved by the use of a quantity called exposure. Exposure is defined in terms of the amount of ionization that is produced in air by the radiation. The special unit is the roentgen (R). One roentgen corresponds to the release of 2.58×10^{-4} C/kg of air.

For gamma-ray emitters, such as this one, a quantity known as the exposure rate constant (Γ), has been defined that relates the activity in curies to the exposure rate in roentgen/hour at a point in air 1 m from the source. It is calculated from the gamma-ray spectrum using the interaction coefficients of air (the required data are given in Table 1). For a ^{60}Co source, Γ, is

$$\Gamma = 1.29 \text{ R} \cdot \text{m}^2/\text{h} \cdot \text{Ci}^{-1}$$

This allows calculation of the parameter that is frequently used to specify source strength: the "roentgens per hour at a meter" (Rmm). For our 2.30 Ci source it is

$$\text{Rmm} = 2.30 \text{ Ci} \times \frac{1.29 \text{ R} \cdot \text{m}^2}{\text{h} \cdot \text{Ci}} \frac{1}{1 \text{ m}^2} = 2.97 \text{ R} \cdot \text{h}$$

A much more practical quantity, from the point of view of radiotherapy, is the absorbed dose rate produced at some agreed distance. To explain this, it will be useful to first define absorbed dose and to go through some approximate calculations connecting activity and absorbed dose rate.

Absorbed dose is the physical quantity that most closely correlates with the biological effect of the radiation and it is defined (11) as the amount of energy absorbed per unit mass of an irradiated material. The special unit of absorbed dose is the gray (Gy), which is defined as 1 joule (J) of energy imparted to 1 kg of matter.

A ^{60}Co activity of 85.2×10^9 Bq, as derived above, would give rise to the following photon fluence rate at a distance of 1 m.

$$\psi = 2\frac{A}{4}\frac{1}{100^2} = \frac{85.2 \times 10^9 \text{ Bq}}{2 \; 10^4 \text{ cm}^2} = 1.356 \times 10^6 \text{ cm}^{-2}/\text{s} \tag{5}$$

The rate of photon interactions with a mass M of the water is given by

$$N' = \sum_i \psi_i \left(\frac{\mu}{\rho}\right)_i M \tag{6}$$

where ψ_i is the fluence (number crossing an area equal to 1 cm^2) of each of the photon energies, $(\mu/\rho)_i$ is the mass interaction coefficient for each of them. The parameter (μ/ρ) expresses the cross-section, or probability of interaction of photons with 1 g of material and M is the mass of the material in grams. The summation in Eq. 6 is over the two components of the photon spectrum as depicted in Fig. 2.

Since the photon energies are so close together, we can use the average value of the interaction coefficients, which is given in Table 1 as 0.0698 cm^2/g. The rate of photon interactions, calculated from Eq. 6, would then be

$$N' = \frac{1.356 \times 10^6}{\text{cm}^2\text{s}} \times 0.0698 \, \frac{\text{cm}^2}{\text{g}} \times 1 \, \text{g} = 94.6 \times 10^3/\text{s} \qquad (7)$$

Each photon that interacts imparts an average of 0.456 MeV (Table 1) of energy so the rate of energy absorbed E', from this irradiation would be

$$E' = \frac{94.6 \times 10^3}{\text{s}} \times 0.456 \, \text{MeV} = 43.1 \times 10^3 \, \frac{\text{MeV}}{\text{s}} \qquad (8)$$

This is a very tiny amount of energy. It was deposited in 1 g of water. Its value can be converted to a more familiar energy unit by using the relation 1 MeV $= 1.6022 \times 10^{-13}$ J. The absorbed dose rate from these photons would then be

$$D' = 43.1 \times 10^3 \, \frac{\text{MeV}}{\text{g}\,\text{s}} \times \frac{1.6022 \times 10^{-13}\text{J}}{1 \, \text{MeV}} \times \frac{10^3 \, \text{g}}{\text{kg}} \qquad (9)$$

$$= 69.1 \times 10^{-7} \, \frac{\text{J}}{\text{kg}\,\text{s}} = 6.91 \times 10^{-6} \, \text{Gy/s}$$

$$D = 6.91 \times 10^{-6} \, \frac{\text{Gy}}{\text{s}} \times 3600 \, \frac{\text{s}}{\text{h}} = 0.025 \, \text{Gy} \qquad (10)$$

A simple radiation treatment for cancer typically involves an absorbed dose at the tumor of 2.0 Gy (in the old units; 200 rad), and because of attenuation in the tissues, and various other factors, this implies, for say a 2 min treatment, an activity almost 5000 times stronger than in our example source. The distance from the source to the tumor has typically been 80 cm. This would call for a source activity of $\sim 25 \times 10^{13}$ Bq or 250 TBq or $\sim$ 7500 Ci.

To attain this, the cobalt must be left in the reactor for a much longer time than in our example above. With a longer activation, one must note that while ^{60}Co is being formed it is also decaying. The resulting activity would be the sum of that which is being produced, as described by Eq. 1, and the amount that decays. This can be written as

$$\frac{dN}{dt} = N_0 \, \sigma \, \phi - \lambda N \qquad (11)$$

where N_0 is the initial number of ^{59}Co atoms present and λ is the transformation constant (see Eq. 2) for the ^{60}Co decay. The other symbols have the same meaning as for Eq. 1. The solution to this equation, expressed in terms of activity, is

$$A(t) = A_{max}(1 - e^{-\lambda t}) \qquad (12)$$

where $A_{max} = N_0 \, \sigma \, \phi$ is the maximum activity attainable for an infinitely long irradiation. For the neutron irradia-

tion conditions of our example, the maximum activity attainable is

$$A_{max} = 1.533 \times 10^{23} \, \text{at} \times 37 \times 10^{-24} \, \frac{\text{cm}^2}{\text{at}} \times \frac{10^{14}}{\text{cm}^2 \cdot \text{s}} \qquad (13)$$

$$= 56.72 \times 10^{13}/\text{s} = 567.2 \, \text{TBq} = 15,000 \, \text{Ci}$$

It would require 5 years in the reactor to produce a source half this strong, that is, 7500 Ci.

This is not strong enough for modern treatment requirements and a higher neutron flux is required. As time has passed, reactor fluxes have increased considerably, and this has allowed both the irradiation times to be shortened and the sources to be made smaller.

There are a number of advantages to making cobalt sources as small as possible. One of these has to do with the sharpness of the edges of the radiation beam. This is known as penumbra and will be discussed later under that topic. It will be seen that a small diameter source is desirable. Another reason for a small source has to do with the amount of self-absorption and photon scattering that will take place within it. The source that we have been considering was a cylinder 1.0 cm in height, and for the gamma rays of cobalt this is almost a half value layer even in lead (Table 1), let alone in cobalt. It must be expected that the radiation emitted by such a source would be accompanied by considerable attenuation and would include an appreciable component of scattered photons. Because of the attenuation and scatter that takes place in the source, the dose rate is greatly overestimated in the calculations made above. The larger the source physically, the greater the activity required to give a desired dose rate.

SPECIFICATION OF SOURCE STRENGTH

In actual practice, the strength of the source is stated in terms of exposure rate at 1 m (Rmm). This is a measured quantity and is determined by the vendor of the source. Sources delivering up to 250 R/min at a meter are now available.

One way of judging the "efficiency" of the neutron irradiation is by stating the specific activity of the source produced. This is the activity, expressed in becquerel (or curie) per gram of cobalt. The specific activity of a 7500 Ci source that weighed 15 g would be 7500 Ci/15 g = 500 Ci/g. In modern reactors, the neutron flux density can be greater than the $10^{14}/\text{cm}^{-2} \cdot \text{s}^{-1}$ that we assumed, sources can be irradiated for longer times than in the example. Specific activities of up to 500 Ci/g have been produced. Cost goes up linearly with irradiation time, but, as can be seen, activity does not, and source strengths actually produced are decided by economic considerations. In actual practice, sources are not irradiated as solid cylinders, as has been assumed for this example, but rather they are made up into a capsule on demand from stocks or pellets that were preirradiated to a selection of specific activities. Pellets are shown in Fig. 3 along with a pair of stainless steel containers into which they will be placed. The pellets will be loaded into the cylinder shown in the center of the picture, then spacers, such as those shown on the right, are inserted to hold the pellets in position, and finally this cylinder, when capped, is inserted into the cylinder shown on the left and cold-welded shut. All

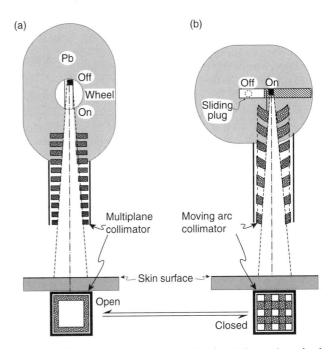

Figure 3. Cobalt pellets and source capsule with components.

of these operations are carried out remotely in a hot cell. Finally, the source is shipped in a well-protected and shielded container to be loaded into a cobalt unit.

COBALT UNIT DESIGN

The first cobalt units went into operation in 1951. Very soon after that they became available commercially, and the production of cobalt sources and cobalt units expanded to such an extent that, for 30 years, more radiotherapy was carried out with ^{60}Co than with all other types of radiation combined. Cobalt machines have the tremendous advantages of producing a completely predictable, steady, reliable beam of relatively high energy radiation, being mechanically simple, rarely needing repair, and being easy to repair when required.

HEAD DESIGN

In all cobalt units, the source is placed near the center of a large, lead-filled steel container. A device is provided for moving the source from a position where it is "Off", because it is shielded in all directions, to a position opposite an opening through which the useful beam may emerge. A number of mechanisms have been devised for moving the source, and two of them are shown in Fig. 4. In Fig. 4a, the source is mounted in a heavy metal (mostly tungsten) wheel that may be rotated through 180° to carry it from the Off position to the On position. In Fig. 4b, the source is mounted in a sliding plug or drawer that carries it from the Off to the On position. In one of the first cobalt units (the Eldorado A), the source did not move at all. The beam opening was filled with a tank of mercury that was pumped out of the way by air pressure to turn the machine On and then the mercury returned by gravity to turn the beam Off.

Figure 4. Two designs for cobalt unit heads. (a) A rotating wheel carries the source to the "on" position. A multiplane collimator controls the size of the rectangular beam. (b) A sliding drawer moves the source and a multileaf collimator moves on an arc to control the beam.

The sliding drawer mechanism shown in Fig. 4b has tended to be the more commonly used.

All machines must be arranged so that they fail "safe". That is, the source must be held in the On position by the continuous application of a force so that if the power fails, it must return quickly to the Off position. For both a and b in Fig. 4, this is provided by a strong spring. The lead-filled container, or "head" of the unit, must be of the order of 25 cm thick in all directions from the source. The design criteria will depend on the regulations in force where it is to be used, but basically it must be such that the leakage radiation emerging from the shield would not cause an overexposure to anyone staying at its surface for prolonged periods of time. This would imply, for example, a yearly equivalent dose of not >5 mSv (or - 500 mrem) at a distance of 1 m from the source. This exposure level is greater than the average in low natural background areas, but is less than the exposure in many other regions of the world where people live. The sievert (Sv) is the special unit of equivalent dose. One sievert will result in the same biological effect as 1 (Gy) gray of conventional X rays. If we assume a maximum source strength of 10,000 Ci, and again use the exposure rate constant of 1.29 R m^2/h·Ci^{-1}, and assume that 1 R corresponds to an equivalent dose of 0.01 Sv, this would imply a thickness of 20–30 half-value layers. The half-value layer in lead for cobalt radiation is ∼ 1.1 cm (Table 1), and this calculation would imply a thickness of ∼ 30 cm. In actual practice a much more detailed calculation would be done, augmented by measurement.

This simple calculation can serve as a guide only. The half-value layer for a broad beam of radiation, such as in this case, would be > 1.1 cm. On the other hand, it is unlikely that anyone would remain for a whole year just beside the head of the cobalt unit. In fact, 20–25 cm is about the thickness of the heads of most cobalt units.

Figure 4 also shows two types of collimators. Both consist of sets of bars that can be adjusted to produce a radiation beam with a rectangular cross-section. The diagrams at the bottom of Fig. 4 show an end-on view of the appearance of both collimator bars in the open and the closed positions.

MOUNTS

There are only two basic ways of mounting and "porting" radiation treatment units. One of the two oldest designs is illustrated in Fig. 5 and is an example of the so-called SSD mount. The head of the unit was held in a yoke, which was suspended by a column from a set of rails attached to the ceiling. It could be moved up and down or back and forth and the head could be rotated about the horizontal axis seen. The unit was also equipped with a treatment applicator, which in this case was mounted on the end of the collimator. The motions of the mount allowed the unit to "point" over a wide range of directions and enabled the operator to place the end of the treatment applicator against the skin of the patient at a prescribed location. The floor was left clear to allow easy and full movement of the treatment couch. The distance from the source to the skin of the patient (SSD) was thus a fixed quantity, usually 80 cm, and the focus of the "setup" was the surface of the patient. The size of the beam was defined there, and the reference point for dosimetry was just under the skin.

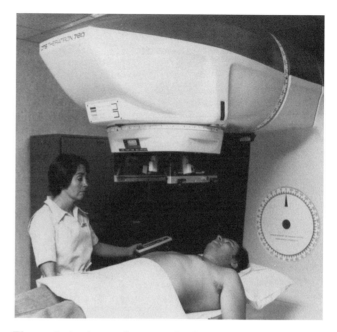

Figure 6. An isocentric mounted cobalt unit of the Theratron series produced by Atomic Energy of Canada Ltd., installed at the Ontario Cancer Institute in Toronto in the 1970s and 1980s.

The alternative mount is the so-called isocentric or fixed source-axis-distance (SAD) mount. An example, dating from the 1970s and 1980s is shown in Fig. 6. The head, encased in a streamlined plastic cover, is mounted on a gantry that can rotate about a horizontal axis. The patient lies on a couch as shown and is raised, lowered, moved sideways, or lengthways so that the tumor is positioned on the intersection of the gantry axis and the collimator axis. This means that for any angle of the gantry, the beam will pass through the tumor. This point is called the isocenter and was a fixed distance from the source, usually 80 cm, in later units 100 cm. The beam is specified by its size at the isocenter. The focus of attention is now at the tumor rather than the surface. In addition, the couch can usually be rotated about a vertical axis, also passing through the isocenter. Virtually all modern treatment units are mounted in the isocentric manner.

The procedures for treatment planning and dosimetry are somewhat different for each of these two types of mount. Treatment planning is discussed in several standard textbooks such as those by Bentel (12), Johns and Cunningham (10), and Kahn (13).

In 1956, an early and innovative symposium was held at Oak Ridge Institute of Nuclear Studies, just before the Eighth International Congress of Radiology, which was held in Mexico City. Problems of source production, machine design and installation, dosimetry, and source specification were discussed. The title of the publication that resulted from this symposium "Roentgens, rads and Riddles", largely reflected the uncertainties of the day in dosimetry. It also includes some history to that time and descriptions of a variety of cobalt units that had been made experimentally and by commercial suppliers.

Cobalt units are inherently simple machines and can be designed and constructed by relatively unsophisticated

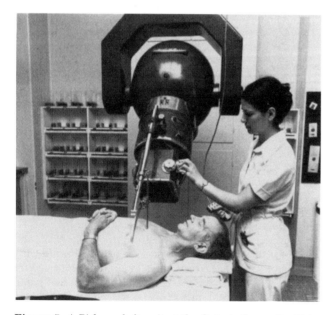

Figure 5. A Picker cobalt unit at the Ontario Cancer Institute, Toronto in the 1960s–1980s. The unit was mounted on a column suspended from rails on the ceiling leaving the floor clear. A protractor allows the angle to be set carefully using the "SSD" technique. The rack on the wall holds "wedge filters" that shape the beam intensity.

(a)

(b)

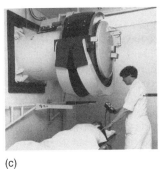

(c)

Figure 7. Three experimental cobalt unit designs: (a) a unit with a number of special features, (b) a double-headed unit, and (c) a unit for half-body irradiation.

engineering facilities. This is illustrated by Fig. 7, which shows three quite different units that were designed and built at the Ontario Cancer Institute in Toronto. The unit in (a) was built in 1959 and had a number of special experimental features (14). These included a diagnostic X-ray tube installed in the head of the unit so that good quality placement films could be taken of patients undergoing treatment. This facility is now standard equipment in all modern radiation treatment machines. The films are called "port films". It was isocentrically mounted and was capable of full 360° rotation about the patient. This allowed continuous rotation during treatment or easy set up for the use of several fixed fields from different angles. The latter feature too, is standard on modern machines. The unit also had a large (95 cm) source-to-axis distance, which improved the depth dose characteristics (see the following section). This unit also had an ionization chamber in the counterweight so that the effective thickness of the patient could be determined. This did not prove to be as useful as expected and was not adopted by unit manufacturers.

The unit in Fig. 7b contained two sources and was called the Double-Header (15). The sources were arranged to be very nearly equal in strength and the beams were directed opposite to each other. This provided an automatic "parallel pair" of beams, which forms a component of many multiple field treatments. The real reason for the two sources, however, was to extend their useful life. The Ontario Cancer Institute had, at different times, as many as eight other cobalt units and two of the sources, after each had been used for ~ 5 years (approximately one half-life) in

one or another of them were transferred to the Double-Header for another 5 years of use.

The third cobalt unit depicted in Fig. 7c, was especially designed for "half-body" treatments. It was equipped with a special collimator to provide radiation fields up to 150 cm long and 50 cm wide. It was fitted with a compensating filter so that a uniform dose distribution could be achieved (16).

CHARACTERISTICS OF THE RADIATION BEAM

The decay scheme for ^{60}Co is shown in Fig. 2. There are two γ rays of photon energies 1.17 and 1.33 MeV, respectively. These energies are very close to each other, so ^{60}Co is almost a monoenergetic emitter with energy 1.25 MeV. The actual beam from a cobalt source also contains lower energy photons, which come from the scattering processes that take place within the source. It is also inevitably contaminated with photons scattered from the mechanism that holds the source in position as well as from the various collimator components that are "in view" of the source. That the beam is not purely that from ^{60}Co is attested to by the fact that the linear attenuation coefficient for 1.25 MeV photons in water is 0.0698 cm^{-1} (Table 1), while the experimentally determined coefficient for a cobalt unit beam in water is closer to 0.063 cm^{-1}. A rather more "realistic" spectrum of the radiation for a cobalt unit has been determined by Rogers et al. (17) by Monte Carlo calculations. The low energy components contribute up to ~ 15% to the dose received by the patient.

In a patient, the intensity of a radiation beam falls off approximately exponentially. This can be seen from the data plotted in Fig. 8, where percentage depth doses for cobalt-60 radiation, and a few other radiations used in radiotherapy, are shown plotted against depth. Percentage depth dose is the single most important quantity in choosing a radiation for radiotherapy. The radiations shown vary from that produced by 100 kV X rays to 25 MV. The depth at which the percentage depth dose falls to

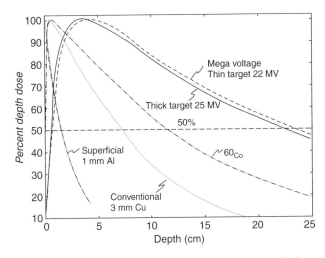

Figure 8. Percentage depth doses plotted against depth for a series of beam energies from superficial (low energy X rays) to megavoltage radiation. All curves are for a 10 × 10 cm field and the depth to 50% dose can be easily determined.

50% can be seen for each radiation by reference to the horizontal dashed line. It varies from < 2 cm for the superficial radiation through ~7 cm for "conventional" or 250 kV radiation, ~12 cm for ^{60}Co radiation, to > 22 cm for the 26-MV radiation. Cobalt-60 is right in the middle of this range. The graphs in Fig. 8 also show that for the higher energies, the dose at the surface is low and rises as penetration increases. For ^{60}Co radiation, it reaches its maximum at a depth of 0.5 cm and falls off relatively slowly from there. This low dose on the surface, the so-called skin sparing effect, was one of the important properties cobalt radiation had for radiotherapy.

When the cross-sectional area of a radiation beam is small, the dose received at a point below the surface is due almost entirely to primary radiation. As the area of the field is increased, the doses will increase due to an increase in scattered radiation. The greater the depth, the greater the increase, with the result that percentage depth dose increases with field size.

CALIBRATION

Calibration of the output of a cobalt unit is normally done with the use of an ionization chamber that has been calibrated against a standard exposure reference at a standardization laboratory. A calibration factor N_X, is determined by the laboratory and its meaning is that $N_X = X/M$, where X is a known exposure and M is the reading of the electrometer monitoring the ionization produced in the chamber by the radiation.

The traditional and simplest method for calibrating the output of a cobalt unit has been to measure exposure rate in air at a chosen distance and field size, and to derive from this the absorbed dose rate that would occur at the center of a small mass of tissue-like material located at this point. An alternative, but equivalent, method is to determine the dose at a chosen position at a specified depth in a water phantom, again for a specified beam size.

Procedures for calibration, and the mathematical formalism required, to determine absorbed dose from exposure measurements are given in textbooks (9,10), as well as in various dosimetry protocols, both national and international. Examples are those of the American Association of Physicists in Medicine (18) and the International Atomic Energy Agency (19). Since the calibration procedures will only be outlined here, these sources should be consulted for more detailed procedures.

Calibration in Air

A number of physical arrangements for making measurements in a radiation beam are illustrated in Fig. 9. The diagram on the left can be used to refer to calibration in air. An ionization chamber, which has been calibrated in terms of exposure, is placed at point P′, free in air, and a reading, M, is taken for a specified "source-on" time T. This exposure time must be the actual exposure time; that is, it must be exclusive of a time, if any, taken for the source mechanism to move the source from the off to the on position. The reading, M, must also include any adjustment required for atmospheric conditions if the temperature and pressure

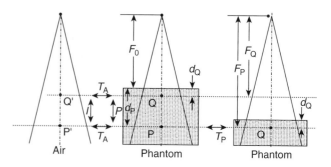

Figure 9. Diagrams showing the meaning of a number of functions used for calibration and dose calculation for treatment planning.

differ from those that pertain to the exposure calibration factor. This would normally be 22°C and 101.3 kPa (equivalent to 1 atm, or 760 mmHg). The parameter M must also be corrected for any small loss of charge that might occur due to charge recombination in the ion chamber during the exposure. Methods for making all of these corrections are discussed in Ref. 9,10,14, and 15. The ion chamber must also have been fitted with a buildup cap, if this is required to make its walls sufficiently thick to provide electronic equilibrium in them. The buildup cap must be made of water-like material. With these precautions, the exposure rate at the point designated in Fig. 9 as P′ would be

$$X = N_X \frac{M}{T} \quad (12)$$

If the cobalt unit is "isocentric" in mount, point Q′ would be on the axis of rotation of the gantry, a distance F_P from the source and the field size would be specified at this point. If the unit were operated in an SSD mode, the calibration point would be the one shown as Q′ in Fig. 9 and would be at a distance F_Q from the source. The absorbed dose rate, free in air, may be calculated from the exposure by the following relationship:

$$\dot{D}_{P'} = N_X \left(\frac{M}{T} \right) \left[0.00876 \frac{J}{kg\,R} \right] \left(\frac{\bar{\mu}_{en}}{\rho} \right)^{wat}_{air} k(d_Q) \quad (13)$$

The term in square brackets is derived from the definition of the roentgen, which is the release of a certain electrical charge per kilogram of air, and the average energy required to release 1 C of this charge. (One roentgen is defined as the release of 2.58×10 C/kg of air, and each coulomb released requires an average 33.85 J. Thus, 1 R corresponds to 0.00876 J/kg of air.) The next term is the ratio of mass energy absorption coefficients averaged over the radiation spectrum for water to air, and the final term is a correction factor to account for the fact that in order to characterize a dose rate at a point in air, it must be surrounded by at least enough phantom (water-like) material to produce electronic equilibrium. This material will attenuate and scatter radiations, and $k(d_Q)$, the allowance for this, is estimated to be 0.985.

Although the size of the beam at point P′ is larger than it is at point Q′, the collimator opening is the same for both,

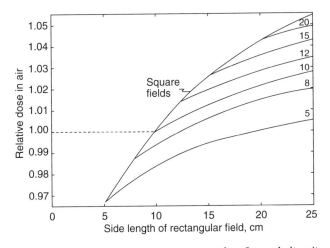

Figure 10. Graphs showing relative output data for a cobalt unit. The output is measured in air and is expressed relative to that of a 10 × 10-cm field.

and so the source self-absorption and scatter, and collimator scatter, would be expected to be essentially the same. Consequently, the dose rate at P′ should be related to that at Q′ by the inverse square law. For any given cobalt unit this must be tested experimentally, but would be expected to be valid except for distances F, of Fig. 9, that are < 50 cm or so

This is indicated by I in Fig. 9 and by the relation

$$\frac{\dot{D}_{Q'}}{\dot{D}_{P'}} = \frac{F_P^2}{F_Q^2} \qquad (14)$$

On the other hand, if the collimator opening is changed, the dose rate at points such as P′ or Q′ will change, due principally to a change in the amount of collimator scatter reaching them. The way this output changes for an example cobalt unit is shown in Fig. 10, where relative dose rates measured on the axis (point P′ of Fig. 9) of an isocentric cobalt unit are plotted against the side length of a rectangular field. The data are normalized to 1.00 for a 10 × 10 cm field. From this diagram, it can be seen that

the dose rates differ by > 8% from a small, 5 × 5 cm field to a large 25 × 25 cm field. The family of curves shown represents rectangular fields, and it can be seen that a rectangular field gives approximately the same relative dose rate, as does a square field of the same area. For example, a 5 × 20 cm field shows a relative dose rate of almost exactly 1.00, as does the square field, 10 × 10 cm, of the same area. Curves such as these are specific to a particular collimator design and must be determined as part of the procedure of commissioning a new treatment unit.

Calibration in a Phantom

The right-hand diagram in Fig. 9 shows the arrangement for calibration in a phantom. The procedure is essentially the same as that for calibration in air; Q in this diagram has the same location and field size as does P′. The same precautions must be taken with the ion chamber reading and the same calibration factor, N_X, is used. The dose rate at depth d_Q in a water phantom is given by an expression that is very similar to that in Eq. 13:

$$\dot{D}_{P'} = N_X \left(\frac{M}{T}\right) \left[0.00876 \frac{J}{kg \cdot R} \right] \left(\frac{\bar{\mu}_{en}}{\rho}\right)_{air}^{wat} k(c) \qquad (15)$$

$(\bar{\mu}_{en}/\rho)_{wat}^{air}$ is, as before, the ratio of averaged mass-energy absorption coefficients, but in this case they should be averaged over the photon spectrum that is present in the phantom. Values for this ratio are given in Table 2. It is generally assumed to be the same in the phantom as in air, although this cannot be quite correct, as shown by Cunningham et al. (20), Eq. 12. The factor $k(c)$ is very similar to $k(d_Q)$ of Eq. 13, except that c is the radius of the ion chamber as it was configured when the calibration factor was obtained. This factor will be the same whether or not a buildup cap is actually in place in the phantom. The dose rate in a phantom, like that in air, varies with the field size, and a set of data like that shown in Fig. 10 can be compiled. The variation is greater, however, because the beam intensity incident on the phantom changes with collimator opening, as discussed previously, but in

Table 2. Dosimetry Factors for ⁶⁰Co Radiation[a]

	$(\bar{\mu}_{en}/\rho)_{med}^{wat}$				$(\bar{\mu}_{en}/\rho)_{air}^{med}$			
Spectrum	Graphite	Bakelite	Lucite	Polystyrene	Water	Muscle	Fat	Bone
	Ratios of averaged mass energy absorption coefficient for a few materials							
Primary[b]	1.111	1.051	1.029	1.032	1.112	1.103	1.113	1.061
Primary plus scatter[c]	1.116	1.055	1.032	1.037	1.111	1.102	1.107	1.105
	Ratios of averaged mass stopping powers							
Primary[b]	1.009	1.071	1.099	1.105	1.129			
Primary plus scatter[c]	1.011	1.073	1.101	1.109	1.131			
Average energy required to cause ionization in air, $W = 33.85$ (dry air)								
$= 33.7$ (ambient air)								

[a]From Ref. 10, page 230.
[b]Assuming monoenergetic 1.25-MeV photons.
[c]Spectrum derived by Monte Carlo calculation for depth 10 cm in a 20 × 20 cm beam.

addition, the scatter generated within the phantom changes with a change in irradiated volume.

General Calibrations

Radiation beams of energy lower than that of ^{60}Co are most frequently calibrated in air. Radiation beams higher in energy should always be calibrated in a phantom. Cobalt units, because of their energy and constancy of output, form a natural reference for all radiotherapy calibration procedures.

RELATIVE DOSE FUNCTIONS THAT ARE USED IN TREATMENT PLANNING

Over the years, a set of functions has been defined that make possible accurate point dose calculations as part of treatment planning. These are "tissue air ratio", "percentage depth dose", "backscatter factor", and "tissue phantom ratio". They are also used with radiations other than that from Co-60, but several of them were derived or refined for use with cobalt therapy. They will be discussed briefly. They can all be clarified by reference to Fig. 9.

Tissue Air Ratio (TAR)

Tissue air ratio, first called "tumor air ratio", was introduced by Johns to facilitate the calculation of tumor dose for rotation therapy. This type of treatment uses the isocentric mode of operation in that the tumor is placed on the axis of rotation of the treatment unit and the beam may be pointed toward the tumor from a selection of angles. The tissue air ratio, which may be defined by referring to Fig. 9, is the quotient formed by the dose, as determined for point P, on the central ray of the beam in a water phantom to the dose determined at the same point P′, with the water phantom removed. The dose at point P would be determined from Eq. 13 and the dose at P′ by Eq. 15, both exposures being made for the same time interval. In practice, it is assumed that all factors except the ion chamber readings will cancel, and tissue air ratios are actually taken to be

$$T_a(d, W_d) = M_P/M_{P'} \qquad (16)$$

In this expression d, is the depth below the surface of the phantom and W_d is the field size at that depth. Tissue air ratio is an expression of the way the radiation beam is attenuated and scattered by the material of the phantom. It is the most fundamental of the relations discussed, and all of the others can be derived from it. Numerical data for this quantity for Co-60 are readily available.

Backscatter Factor

The ratio of doses determined from points Q and Q′ of Fig. 9 is a special value of the tissue air ratio. The depth, d_Q, is the special depth just needed to produce electronic equilibrium at the point of dose measurement. At this point primary attenuation is the same in the phantom at Q and in the small mass of phantom-like material placed at Q′ in order to make the measurement. Most of the scattered radiation

reaching point Q is scattered backward from within the phantom. For the range of X rays that were in use before the advent of ^{60}CO, the depth d_Q, was very small and the point Q, was considered to be on the surface, hence the name backscatter factor. This quantity is also called "peak scatter factor" because the depth at which electronic equilibrium is attained also tends to be the depth of peak dose in the phantom. For ^{60}Co radiation, the depth of electronic equilibrium is taken to be 0.5 cm.

Percent Depth Dose

Whereas tissue air ratios relate doses in the phantom to doses free in air, percent depth doses interrelate doses at points within the phantom. Again referring to Fig. 9, the dose at point P is related to that at point Q by the percentage depth dose.

$$P(d, d_Q, W, F_0) = 100 M_P/M_Q \qquad (17)$$

For this quantity, the field size is defined at the surface, and the distance F_0 from the source to the surface must be stated. The doses at points P and Q should be determined from ion chamber measurements by the factors indicated in Eq. 15, and, as for tissue air ratios, it is generally assumed that all factors, except for instrument readings, cancel between the numerator and denominator.

Since point P is farther from the source than is Q, part of the falloff in dose with depth is due to the inverse square attenuation. Because of this, percentage depth doses increase with SSD. For example, the most common source–surface distance in use for Co-60 has been 80 cm. This was chosen as a compromise between increasing percentage depth dose and decreasing output. If the surface distance is increased from 80 cm to 1 m, the percentage depth dose at 10 cm in a 10×10 cm beam will increase from 55.6 to 57.8. This change is just slightly less than would be entirely accounted for by the inverse square law.

Tissue Phantom Ratios

For radiation of energy higher than that of cobalt, the dosimeter must be equipped with thick walls, and its size makes it inconvenient for use in air—particularly for small field sizes. It becomes convenient, therefore, to make the reference measurement in a phantom rather than in air. This is indicated in the right side of Fig. 9 by the point indicated by Q, which is the same distance from the source as is P (and P′), but is in a phantom at some chosen reference depth. The tissue phantom ratio is then the ratio D_Q/D_P and is entirely analogous to tissue air ratio and has many of the same properties. This quantity is, for example, also independent of distance from the source.

Tissue phantom ratios were introduced by Karzmark et al. (21) for use with high energy radiation, but can be applied equally well to Co-60 radiation.

Relationships between the Dose Calculation Functions

From Fig. 9, one can easily see the relationships between the various doses. For example, D_P can be related to D_Q directly by a percentage depth dose. It could also be expressed by means of two tissue air ratios and the inverse square law:

$$D_P = D_Q \frac{T(d_P, W_{d_P})}{T(d_Q, W_{d_Q})} \left(\frac{F_Q}{F_P}\right)^2 = \frac{D_Q}{100} P(d_P, d_Q, W_{d_Q}, F_0) \quad (18)$$

The tissue phantom ratio is a combination of two tissue air ratios:

$$Tp = \frac{T(d_Q, W_Q)}{T(d_P, W_P)} \quad (19)$$

PENUMBRA

All of the previous considerations of dosimetry have been for points on the axis of the beam. Treatment planning is a 3D process, and regions not on the axis must also be considered. The behavior of the dose at points off the beam axis can be discussed by referring to Fig. 11.

In Fig. 11a, the radiation beam is incident on a point X′, in air. The conditions are the same as for the left side of Fig. 9. Consider a small dosimeter to be moved laterally across the beam from A to F. At A it is shielded by the collimator, while at X′ it is in the middle of the beam, in full "view" of the source. The dose will be at its greatest value at X′. At C it would still be in full view of the source, but it is slightly further away from the source than it is at X′ and the dose will be slightly lower. The expected doses at A, X′ and C, as well as other points on the line are shown by the dashed lines in Fig. 11b. At point D, the collimator blocks off half of the source and the dose would be expected to be one-half of its value at C. The point at E is just out of view of the source, and ideally the dose here should sink to zero. The portion of the line A–F between C and E is called the geometric penumbra. It is dependent on the diameter of the source, the distance f_c, from the source to the end of the collimator, and the distance $(f - f_c)$, from the end of the collimator to the line A–F. The geometrical penumbra is given by the very simple relation:

$$p = s \frac{(f - f_c)}{f_c} \quad (20)$$

The actual measured penumbra differs somewhat from this and is always a little larger. The source does not behave like a sharp, well-defined disk because of its volume, and the radiation therefore scattered within it and the radiation scattered from the structures that hold it in place, and from the beam collimating apparatus. There is also, inevitably, some transmission through the collimator and some scattering from its lower end. The result is that the dose outside of the beam at points A and F is not zero, and the real dose profile is rounded off as depicted by the solid curve in Fig. 11b.

The shape of the dose profile in a phantom for ^{60}Co radiation is only slightly different from that observed in air. The penumbral region is broadened somewhat by the transport of energy along the tracks of the electrons that are set into motion by the photons near the edge of the beam.

The meaning of field size can also be derived from Fig. 11. It is, by convention, taken to be the distance between points B and D. It is indicated as W_d in that diagram. This is the distance between the points that are at 50% of the dose on the axis at the same depth. It is also the full width at half maximum (fwhm) of the dose profile. Normally, the measurement of field size would be made in a phantom.

ISODOSE CHARTS

A more complete description of the dosage pattern of the beam is by means of an isodose chart. An isodose chart is a map of the distribution of the dose in a plane. Such charts are found in many books and papers in the literature and only one example will be given here. In Fig. 12, a small

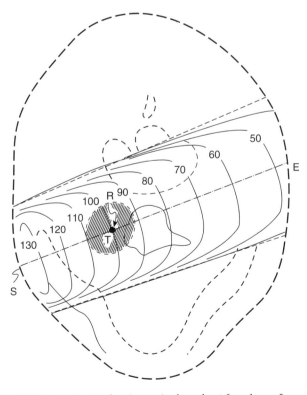

Figure 12. Diagram showing an isodose chart for a beam from a cobalt unit treating a tumor in the neck of a patient. The target and some structures are shown.

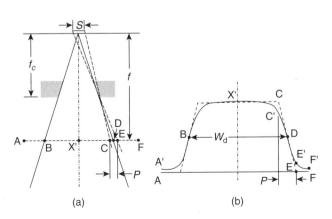

Figure 11. Diagrams showing the geometrical considerations involved in describing the shape of a cross-beam profile for a cobalt unit. (a) Shows the source and the collimator and (b) side shows a dose profile line A–F.

beam from an isocentric cobalt unit is treating a tumor in the neck region of a patient. This is an application for which cobalt radiation is still ideal. The target area, which is shown by the cross-hatched region, has been chosen by a radiation oncologist. A safety region has been allowed for and the beam is planned to be directed as shown. The dose at the target will be calculated from the calibration information as described above and (in this case) a tissue air ratio. On the diagram, it has been given the value 100. The solid lines in Fig. 12 show the distribution of percentages of the dose planned for the center of the target. In this case, the dose distribution has not been corrected for air cavities that might be in the path of the beam but modern treatment planning methods, carried out by computers, would include such considerations.

The dose at the center of the tumor is $\sim 77\%$ of the maximum dose, and a single beam like this would not be deemed suitable. The planning process would be carried further by the addition of at least one more beam from another direction, so that the two would cross at the tumor and produce the maximum dose there. Such a treatment plan might even call for four or even more beams, all of which would be arranged to cross at the target.

Complete isodose distributions drawn for individually designed treatments for individual patients are part of the normal procedures of treatment planning. Dose calculation functions that have been discussed in this article have been incorporated into computer programs and enhanced into procedures that allow calculation of dose distributions for complicated treatment conditions. These calculation procedures have been refined to the extent that the 3D shape of the patient and tissue inhomogeneities can be accounted for. The development of these calculation methods has a lengthy history. Suffice it to say that current methods of calculation use Monte Carlo procedures and are quite precise.

NOTABLE FIRST CLINICAL APPLICATIONS

The use of large irregularly shaped radiation beams was introduced for use with cobalt units in the late 1950s. The task was irradiation of chains of lymph nodes for treatment of Hodgkin's disease. This was so successful that a diagnosis of this disease went from a virtual "death sentence" to one that was highly curable. These "mantle" fields, frequently shaped from low melting point lead alloys for individual patients, are still in use. One of the earliest developments of a computer program for the calculation of the dose distribution was introduced with these treatments in mind (22). This program too, is still in use.

A precursor to today's intensity modulation radiation therapy was instituted in the 1960s in Japan by Takahashi who described the use of multileaf collimators and dynamic treatment delivery with a cobalt unit in 1963 (23).

A group under A. Green at the Royal Northern Hospital in London, England pioneered conformal radiation therapy by developing cobalt machines in which the patient was automatically positioned during rotational therapy by moving the treatment couch and machine gantry by electromechanical systems. It was given the name "The

Tracking Cobalt Project" because it attempted to make the dose distribution conform to the spread of the disease. With a similar intent Proimos in Patras, Greece and later Rawlinson and Cunningham in Toronto (24), described the use of synchronous shielding in a Co-60 beam to make the radiation beam conform to the target while avoiding critical normal tissues.

SUMMARY AND CONCLUSIONS

It is still likely, even now, that more cancer patients have been treated by radiation from cobalt units than by any other kind of radiation. The number is estimated to be > 30 million (25). The cobalt unit was the backbone of radiation therapy for over four decades. The cobalt unit is mechanically simple and its output is totally predictable and reliable. Sources with sufficient strength to enable practical, short treatment times can easily be produced. Because of the source decay, sources must be renewed at intervals of 5 years or so, but this procedure is quite straightforward and its expense is more than offset by the low maintenance cost of the machine.

The beam characteristics are well known and relatively easy to measure. It is also easy to make special filters and beam modifiers for individual treatment needs. The energy is high enough to provide skin sparing. The most important single parameter in choosing a radiation energy for therapy is depth dose and the depth dose of cobalt radiation is quite satisfactory for treating tumors that are within 10 cm or so of the surface. This includes head and neck tumors and all but deep-seated lesions in very large patients. With respect to this quantity ^{60}Co is in the middle ground. It remains the unit of choice as a first unit in a developing department and is a must as part of the equipment for any large radiotherapy department. Cobalt units are still being manufactured and sold at about half the rate that obtained at the peak of their use. Modern cobalt units include many of the technological innovations, such as computer control, that are part of the more modern treatment machines. An excellent chapter on Co-60 and its role in modern times has been written by Glenn Glasgow (26). This is recommended to the interested reader.

BIBLIOGRAPHY

1. Sampson M, Ridenouri LN, Bleakney W. A long lived radio-cobalt produced by irradiating cobalt with neutrons. Phys Rev 1936;50:382.
2. Livingood JJ, Seaborg GT. Radio-active isotopes of Cobalt. Phys Rev 1941;60:913.
3. Mayneord WV, Cipriani AJ. The absorption of gamma-rays from ^{60}Co. Can J Res Sec A: Phys Sci 1947;25:303.
4. Johns HE, Bates LM, Watson TE. 1000 curie cobalt units for radiation therapy. The Saskatchewan cobalt-60 unit. Br J Radiol, 1952;25:296.
5. Grimmett LG, Kerman HD, Brucer M, Fletcher GH, Richardson JE. Design and construction of a multicurie cobalt teletherapy unit. A preliminary report. Radiology (Easton Pa) 1952;59:19.
6. Robinson RF. The race for Megavoltage. Acta Oncol 1995; 34:1055.

7. Johns HE, MacKay JA. A collimating device for ^{60}CO tele-therapy units. J Fac Radiol, London 1953-1954;5:239.

8. Attix FH. Introduction to Radiological Physics and Radiation Dosimetry. New York: John Wiley and Sons Inc.; 1986.

9. Greening JR. Fundamentals of Radiation Dosimetry. Medical Physics Handbook 6. Bristol, England: Adam Hilger; 1981.

10. Johns HE, Cunningham JR. The Physics of Radiology. 4th ed. Springfield, (IL): Charles C. Thomas; 1983.

11. ICRU Report 33. Radiation Quantities and Units. Bethesda, (MD): International Commission on Radiation Units and Measurements; 1980.

12. Bentel GC. Radiation Therapy Planning. 2nd ed. New York: McGraw-Hill; 1996.

13. Khan FH. The Physics of Radiation Therapy. 3rd ed. Philadelphia: Lippincott Williams and Wilkins; 2003.

14. Johns HE, Cunningham JR. A precision cobalt 60 unit for fixed field and rotation therapy. Am J Roentgenol 1959;81:4.

15. Cunningham JR, Ash CL, Johns HE. A double headed cobalt 60 teletherapy unit. Am J Roentgenol 1964;92:202.

16. Leung PM, Rider WD, Webb HP, Aget H, Johns HE. Cobalt-60 therapy unit for large field irradiation. Int J Radiat Oncol Biol Phys 1981;7:705.

17. Rogers DWO, Bielajew AF, Ewart GM. Co beam contamination from the source capsule (Abstr.). Med Phys 1984;11:401.

18. American Association of Physicists in Medicine (AAPM), Task Group 51, A protocol for the determination of absorbed dose from high energy photon and electron beams. Med Phys 1983;120:741.

19. A Code of Practice for Absorbed Dose Determination in Photon and Electron Beams. Vienna: International Atomic Energy Agency (IAEA); 1987.

20. Cunningham JR, Woo M, Rogers DWO. The dependence of mass energy absorption coefficient ratios on beam size and depth in a phantom. Med Phys 1986;13:496.

21. Karzmark CJ, Deubert A, Loevinger R. Tissue-phantom ratios-an aid to treatment planning. Br J Radiol 1965;38:158.

22. Cunningham JR, Shrivastava PN, Wilkinson JM. Program IRREG–Calculation of dose from irregularly shaped radiation beams. Comp Prog Biomed 1972;2:192.

23. Takahashi S. Conformation radiotherapy-rotation techniques as applied to radiography and radiotherapy of cancer. Acta Radiol 1965;242 (Suppl): 1.

24. Rawlinson JA, Cunningham JR. An Examination of Synchronous Shielding in 60-Co Rotation Dose Distributions. Radiology. 1972;102:667.

25. Battista JJ. Cobalt-60 Radiation Therapy: Fifty Years Review and More. London: Ontario; October 27th 2001.

26. Glasgow GP. Cobalt-60 Teletherapy. Chapt. 10. In: Van Dyk J, editor. The Modern Technology of Radiation Oncology. Madison (WI): Medical Physics Publishing; 1999.

See also PHANTOM MATERIALS IN RADIOLOGY; RADIATION DOSIMETRY FOR ONCOLOGY; RADIOTHERAPY TREATMENT PLANNING, OPTIMIZATION OF; X-RAY THERAPY EQUIPMENT, LOW AND MEDIUM ENERGY.

COCHLEAR PROSTHESES

FRANCIS A. SPELMAN
University of Washington
Seattle, Washington

INTRODUCTION

Cochlear prostheses (also called *cochlear implants*) bypass acoustic processing of sound by the cochlea and convert acoustic signals into electrical currents. These currents are delivered via intracochlear electrodes, which directly stimulate the auditory nerve fibers that connect the cochlea to the central nervous system. Cochlear prostheses convert auditory signals into minute electrical currents that stimulate auditory nerve cells via electrodes placed near viable nerve cells. Cochlear implants differ profoundly from acoustic hearing aids. They stimulate the cells of the auditory nerve directly, bypassing the hair cells of the organ of Corti. Acoustic aids increase the mechanical signals that are delivered to the hair cells, aiding their depolarization and the delivery of signals to the auditory nerve. Since the introduction of commercial implants nearly 30 years ago, cochlear prostheses have become one of bioengineering's prominent success stories: > 60,000 people use cochlear implants worldwide. The devices provide patients with a means to overcome deafness. Their success is such that, since the time that the article was written about cochlear implants in the first edition of this Encyclopedia, the cochlear implant has been recommended for people who are severely deaf, rather than reserving the implant for the profoundly deaf (1). Cochlear prostheses provide the standard treatment for people who are profoundly deaf.

In addition to cochlear prostheses, some prostheses are implanted surgically in the central nervous system as auditory brainstem implants, in the cochlear nucleus, or as mid-brain implants in the inferior colliculus.

This article is an update of the article *Cochlear Prosthesis* in the 1st ed. of this Encyclopedia (2).

CANDIDATES FOR IMPLANTS

Hearing loss can occur in either one or both ears. The common classifications of hearing impairment are mild (21–40 dB), moderate–severe (61–70 dB), severe (71–81 dB), and profound (90+ dB) (3). Here dB ($20 \log_{10}[P_2/P_1]$) is the sound pressure, P_2, referenced to normal hearing thresholds, P_1, usually measured at 500, 1000, and 2000 Hz. It refers to the increase of sound pressure that must be used for a subject to reach hearing threshold. Blanchfield et al. number the severely to profoundly deaf between 464,000 and 738,000, all of whom are candidates for cochlear implants (4).

Some prostheses are implanted surgically in the cochlear nucleus. The numbers of patients receiving those devices are much smaller than those who receive implants in the cochlea, $\sim$ 300 people (5). The candidates come primarily from subjects with neurofibromatosis (6–8). The morbidity and mortality with central implants is small, and the success is reasonable. The subjects do not do as well as those with the cochlear prostheses described below, but are able to decode speech (6). The emerging field of central auditory implants will not be covered further in this article because the numbers of users are relatively small at this time.

THE AUDITORY SYSTEM

A complete description of the functioning of the peripheral auditory system is beyond the scope of this article. However, to understand the operation of the prosthesis, one

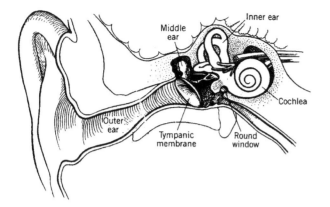

Figure 1. A sketch of the peripheral auditory system, external ear, ear canal, eardrum, middle ear, and inner ear (2).

must know a little about the anatomy and physiology of the peripheral auditory system, which consists of the external ear, the middle ear, and the inner ear (9). Figure 1 shows the auditory system in a simplified form. Sound impinges on the external ear and is guided by way of the ear canal to the tympanic membrane (eardrum). The tympanic membrane vibrates with a relatively large displacement and low pressure. The ossicles (bones) of the middle ear act as an acoustic impedance transformer to change the vibration to relatively small displacement and high pressure at the oval window. The cochlea, the spiral-shaped organ of the inner ear, contains the cells that convert mechanical motion into the electrochemical signals that are recognized by the nervous system (9,10). Several sites on the World Wide Web provide animations of the operations of the components of the auditory system. One such site may be found at, http://www.neurophys.wisc.edu/animations. Other

animations and data are maintained in a "virtual library" that has been assembled by the Association for Research in Otolaryngology at its web site http://www.aro.org. Geisler refers to both sites in his work, From Sound to Synapse (9).

THE AUDITORY PERIPHERY

The peripheral auditory system (Fig. 1) consists of the external ear, the middle ear and the inner ear (9). The external ear guides acoustic waves through the external auditory meatus to the tympanic membrane, which vibrates in response to air moving in the ear canal. The middle ear acts as a mechanical transformer, a system of levers and pistons, to match the air-driven tympanic membrane to the fluid-filled inner ear, the cochlea (9).

Figure 2 shows a cutaway view of the inner ear and its three chambers or scalae, that is, the scala vestibuli and the scala tympani, which communicate via the helicotrema, an opening at the apical end of the cochlea, and the scala media, which is isolated from the other two scalae by membranes (9,10). The stapes (stirrup) of the middle ear drives the fluids of the scala vestibuli and in doing so deflects the membranes of the scala media (10). One of these membranes, the basilar membrane, bears the hair cells, the motion-sensitive cells that excite the VIII cranial nerve (9,10).

The inner ear acts as a transduction and signal processing mechanism. Auditory information is decomposed into its fundamental frequencies by the frequency-sensitive basilar membrane. Amplitude, phase, and frequency information is carried by the cells of the auditory (VIII cranial) nerve. Simplistically, sounds are decomposed into their spectral peaks (11).

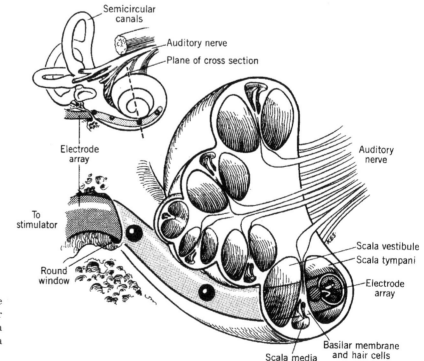

Figure 2. Cutaway view of the cochlea of the inner ear, showing the three chambers or scalae of the ear, and an artist's conception of a cochlear electrode array inserted into the scala tympani of the cochlea (2).

Each of the 30,000-odd fibers of the auditory nerve has an auditory threshold function that is sensitive to a small range of frequencies. All threshold minima lie within 10–15 dB; fibers have dynamic ranges that can be as much as 30–40 dB at their characteristic frequencies (9,12). The rate at which a single peripheral fiber fires is a monotonically increasing function of the acoustic stimulus at its characteristic frequency. The dynamic range of a fiber depends on a number of factors, including its threshold and its spontaneous firing rate, the latter of which can be as large as 100 spikes/s (9).

The responses of auditory nerve fibers are nonlinear. At low intensities, the responses of single nerve fibers mimic the frequency spectra of the complex sounds that stimulate the ears of experimental animals (13). At higher intensities, the spectra produced by the responding fibers are dominated by the low frequency component of the speech sound (its first formant) and the distortion products of that frequency (13). Recent evidence provides strong support for nonlinear system to preserve speech sounds at low and high intensity, in quiet and in noise (9).

In summary, the auditory system has a number of features that enable it to decode sound: (1) specific cells are excited at threshold by specific acoustic frequencies; (2) increasing intensity of an acoustic signal causes an increasing spread of influence from cells for which it is the best frequency, to cells that respond at threshold to other frequencies; (3) the intensity of a particular signal appears to be coded both by the rate at which cells fire and by the numbers of cells excited by a particular stimulus; (4) nonlinear properties of the auditory system cause the suppression of one cell's response to one frequency by stimulation with another frequency, by saturation of rate and by the production of distortion products in the system's response to high intensity excitation; and (5) frequency information contained in complex stimuli is preserved in the temporal responses of auditory neurons.

HISTORY OF COCHLEAR PROSTHESES

The first report of electrical stimulation of the ear is attributed to Volta, in a paper read to the British Royal Society in July of 1800 (14,15). He reported that his approach, using perhaps 50 V excitation, was uncomfortable, sounding like the boiling of fluid. He did not repeat the study. More recently, Djuorno and Eyries (16) reported the first attempt to excite the auditory nerve directly with electricity. Later, Doyle et al. reported results with electrical stimulation of the auditory nerve (17). Simmons performed an experiment a year later in which he went further, stimulating the VIII auditory nerve and the inferior colliculus of a human patient, showing that it was possible for the subject to distinguish frequencies well below 900 Hz, but not > 1000 Hz (18). Simmons demonstrated that both peripheral and central stimulation of the auditory system was possible. In 1964, the House Ear Institute began an extensive series of surgeries to implant cochlear prostheses, reporting on their long-term effects in 1973 (19). The first experiments on multichannel cochlear prostheses were initiated by Simmons et al. in 1979 (20). Their results were promising, and now multichannel implants are the standard of the industry.

Since the first experiments, cochlear prostheses have been built and applied worldwide, receiving approval from governmental agencies and remarkable success in > 60,000 patients. Indeed, cochlear prostheses are considered the standard treatment for profoundly and severely deaf adults. Three commercial firms, Cochlear Corp. (Sydney, Australia), Advanced Bionics Corporation (Valencia, CA; recently purchased by Boston Scientific Corporation), and Med-El Corporation (Innsbruck, Austria) produce cochlear implants successfully. The early cochlear implants were single-channel devices (21), but all of the cochlear prostheses that are implanted today are multichannel devices (22).

THEORY OF OPERATION

The cochlear implant operates on the premise that, if the hair cells of the auditory system are damaged, they can be bypassed and that neurons can be driven directly with very small electrical signals. Figure 3 shows a greatly simplified block diagram of a cochlear implant. Acoustic signals are transduced by a microphone, whose small electric signal is amplified. An external processor decomposes the electrical analogue of the acoustic signal. In the processors that are produced today, processing is digital, with the processor analyzing the instantaneous frequency content of the acoustic signal in the frequency domain. The signals are sent across the skin via a radio frequency link (in the VHF band) that transmits both information and power from the outside of the subject to the inside. These transcutaneous signals are shown with bidirectional paths. Data can be transferred in both directions, providing information to therapists about the condition of the electrodes and the state of the auditory system of the patient. The data flowing to and from the external signal processor are serial bit streams.

The transcutaneous bit streams can have rapid rates: consider that the sampling rate of the audio signal can

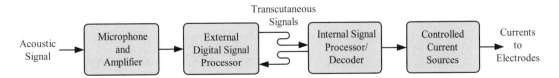

Figure 3. Simplified block diagram of a cochlear implant. Four blocks are shown, a microphone and amplifier, external digital signal processor; internal signal processor/decoder; and, controlled current sources (see text; after Ref. 23).

exceed 20,000 samples/s, and that updates of information delivered to the internal processor may present data at 5000 or more data points per second. The data must include the electrode(s) that are being driven, the amplitude and the pulse width of the current pulse that is applied. Rates can exceed 80,000 pulses/s (25). The internal signal processor–decoder decodes the incoming bit stream. It distributes drive signals to the current sources, selecting specific sources to drive, and setting the amplitude and duration of the control signals.

The current sources drive the electrodes of the cochlear implant's electrode array. Those electrodes are placed in the scala tympani of the cochlea, and direct the current drive signals to the neurons of the auditory nerve, the VIII cranial nerve. The electrodes of the array are placed in proximity to the neurons, in order to reduce the threshold currents necessary to excite the cells, and to reduce current spread within the inner ear (26).

Figure 4 shows one of the cochlear electrode arrays that is produced commercially by Cochlear Corporation (Sydney, Australia). There are 24 contacts in all, two of which are placed outside of the cochlea, leaving 22 contacts that may be driven to excite neurons of the auditory nerve. The contacts may be driven as monopoles (single internal current sources, referenced to an return contact external to the inner ear), as dipoles (pairs of internal current sources) or as combinations of three or more contacts. The contacts are placed along the inside of the spiraling, silicone carrier. The carrier is shaped to fit snugly against the modiolus of the scala tympani. The contacts can be driven singly or in combinations, for example, as dipoles or as multiple sources.

The three manufacturers of cochlear implants use scala tympani arrays that are similar to the array shown in Fig. 4. However, other approaches to stimulate the auditory nerve cells are possible. Normann and his colleagues have tested monolithic electrode arrays that are designed to penetrate the auditory nerve directly (27). Like others before him, Normann realized that bringing electrode contacts near the neurons will reduce thresholds and

Figure 4. Picture of the Nucleus 24 Contour electrode array, showing 24 contacts and a shape that is designed to appose the modiolar wall of the scala tympani (24).

limit the spread of excitation (28–30). The concept has not been tested chronically in human subjects.

The concept of an *information channel* is critical to the understanding of the cochlear prosthesis. A channel may drive current to one electrode, but it often distributes drive to two or more electrodes. Field shaping and steering techniques suggest the use of multiple electrodes for each channel (31–33). Indeed, demonstrations by Bierer and Middlebrooks (33,34) showed that the quadrupolar configurations (called tripolar in the Bierer paper) produce more focused stimuli than either monopolar or bipolar excitations. Recent experiments in cats have upheld the finding, showing that multipolar stimulation allows two triads of electrodes to be driven simultaneously without significant crosstalk (35). It is clear that a channel may involve several electrodes driven simultaneously, and cannot be defined as the information conveyed by a single contact on an electrode array. While one contact may be driven at a time, bipolar stimulus configurations are common and multipolar configurations may emerge soon.

Signal processing techniques have changed dramatically since the time that the first version of this article was written. The number of available electrodes has more than doubled. In common to most processors is a bank of filters, analogue or, commonly, digital. The filtered signals are decomposed into time-varying envelope signals that are compressed and delivered as either amplitude modulated pulses or width modulated pulses. The pulses are delivered with a variety of strategies.

Continuous interleaved stimulation (CIS) is a technique by which a single electrode is stimulated at a time in order to eliminate field interactions between and among channels when electrodes are driven as monopoles (31). The electrodes receive signals from specific filters. The signals are converted to symmetrical, rectangular, biphasic current pulses whose amplitudes may be proportional to the envelope of the filter signal and whose width is invariant. Conversely, amplitude can be held constant and width can be varied. More recently, Advanced Bionics Corporation has used a processor whose repetition rate can be 5800 pulses/s per channel, to develop rapid updates of channels in the CIS paradigm. In a recent processor, HiRes, stimulation rates can be as much as 5800 pps when two widely spaced channels are driven simultaneously, and drops by one-half when the two channels are driven sequentially (36).

The *n*-of-*m* strategy employs a larger number of filters, *m*, than there are electrodes, *n* (37). Depending on which filters contain the maximum acoustic energy, pulses are delivered to appropriate electrodes. The cochlea is organized tonotopically along the basilar membrane. Hence, each electrode's field excites a specific group of characteristic frequencies in perceptual space. Those filters that exhibit the maximum energy determine the electrodes that will be driven by a given temporal sample of the acoustic signal. Biphasic, symmetrical, rectangular pulses are delivered to specific electrodes, *n*, at particular sample times. Because of the field interactions between electrodes no more than two channels are driven during a given sample.

Other techniques include simultaneous analog stimulation (SAS), in which widely separated electrodes are driven simultaneously to increase the rate at which information is transferred to the auditory nerve. The field interactions are reduced by driving electrodes that are separated by several millimeters in the inner ear (38). Simultaneous analog stimulation is a special case of the "filters with compression" technique described by Eddington > 20 years ago (39). Today, fewer electrodes are driven simultaneously, but they are updated more rapidly (40). Thus, SAS is a variation of both Eddington's filters and CIS. Eddington described a means by which electrodes were assigned the compressed analog outputs of filters. Those analogue signals were delivered continuously to the electrodes.

A potentially exciting new technique of stimulation takes advantage of the stochastic behavior of auditory neurons. If a stimulator provides high rate conditioning pulses to its electrode array, it is possible to simulate the stochastic firing frequencies of the cells of the auditory nerve (41). This approach has been tested in small numbers of European patients with what appears to be dramatic success, particularly with auditory signals in noise (Rubinstein, personal communication; see below).

Despite the richness of the processing techniques that have been employed, there are still hurdles to be overcome. The number of true, simultaneous channels is too small. It should be at least 16; there is often a mismatch between the frequency assigned to an electrode and its position in the cochlea; the signals that are delivered to the neurons do not contain fine temporal information; the phase information between channels is not preserved; and, there may be neurons missing, causing some electrodes and critical frequencies to be missing as well (42). Future implants may be able to address some of the concerns that are raised here.

EVALUATION OF COCHLEAR PROSTHESES

When human subjects first used cochlear implants, the numbers of subjects were small and tests were not standardized. As the devices improved, standard tests were developed and used across the centers at which implantation was being done (15). The tests include materials that are open and closed set. The test subjects do not review open set materials prior to the test, whereas closed set materials are reviewed before testing takes place. Subjects participate in word tests and sentence tests. In the former, single words are presented while in the latter sentences are presented and the subjects can deduce parts of the sentence logically.

In addition to providing word and sentence tests, consonant (C) and vowel (V) discrimination tests are included in the test batteries. In these tests, nonsense utterances, CVCs or VCVs, are presented and the subject must identify the appropriate vowel or consonant.

Open word tests are difficult while sentence tests are relatively easy. For example, implant users have steadily increased their comprehension of sentences from much < 10% with early single channel devices to 80% or above with today's multichannel devices (22). Many implant users are able to converse on the telephone, a significant result, since they cannot rely on the cues presented by lip reading in that situation. Still, word comprehension from open-word sets remains relatively low, between 40 and 50%, and most users dislike listening to music (43). Clearly, the context that comes from sentence structure and content is important to comprehension, and the complex spectral content of music makes it difficult.

Cochlear implants are a great bioengineering success. Wilson used an aviation metaphor recently, likening the cochlear implant to a DC-3, a reliable workhorse of an aircraft without the sophistication of a twenty-first century transport airplane (43). The implant has advanced from the single-channel stage of the Wright flyer, but has yet to reach its pinnacle.

THE BENEFITS AND RISKS OF IMPLANTATION

Cochlear implants provide clear benefits to their users. For example, hearing-impaired children learn language more rapidly with cochlear implants than they do with hearing aids (44). Adults do well and benefit from their implants, particularly when they are dealing with speech in quiet. However, for patients to achieve the greatest benefits from the device, their prostheses should be adjusted individually for the minimum and maximum stimulation levels for each electrodes in the array, the stimulation rate, and the speech processing strategy (45). Skinner suggests that for best results the parameters should be adjusted for the maximum dynamic range: from quiet sounds to maximum sounds that are "...not too loud..." (45).

A recent survey of patients from Toronto, Ontario, Canada, was taken of 42 early deafened adult users. Of the 30 who responded, > 96% said that they were satisfied with the implant, > 93% would undergo the procedure again, and 90% said that they would recommend the implant to another person in the same situation (46). The subjects were encouraged by family and peer support and bolstered by having a positive attitude before, during and after the process of implantation and therapy.

There are risks associated with the surgery, but they are quite small. Cunningham et al. (47) reviewed the cases of 462 adults and 271 children in a private tertiary care center for the years 1993–2002. They found that the overall incidence of infection postoperatively was 4.1%. Major infectious complications occurred in 3.0% of the cases; those complications required surgical intervention (47). Bacterial meningitis was found in 26 of 4264 children receiving cochlear implants in the United States (48). That was found to be associated with a particular electrode array that used a positioner to place it near the modiolar wall. The array was subsequently withdrawn from the market (http://www.fda.gov/cdrh/safety/cochlear.html), and there have been no other reports of the occurrence of meningitis. Cunningham et al. (47) recommended that children undergo vaccination before implantation to prevent bacterial infections.

THE COST OF IMPLANTATION

A recent article cited the cost of cochlear implant treatment as > $40,000.00, of which $20,000.00 is the approximate cost of the device itself (49). Despite the high cost of the device and the surgery, the cochlear prosthesis is beneficial when compared with the long-term costs of other medical device procedures (50,51). Garber et al. (49) asked why the cochlear implant has limited access despite its success and the likely market, and surveyed 25 of 231 practices and 96 of 213 hospitals to try to learn what caused the limits of availability. They concluded that both the practitioners and hospitals lose money when they provide cochlear implants, limiting access to the devices. The cochlear implant is approved in the United States for Medicare, Medicaid, and insurance reimbursement.

THE FUTURE OF COCHLEAR PROSTHESES

In a recent review, Wilson et al. (52) suggested that the future held combined acoustic and electrical stimulation, bilateral implants, new electrode designs and closer mimicking of processing in the normal cochlea. This article discusses electrode designs, combined acoustic, and bilateral stimulation and the closer mimicking of processing in the normal cochlea.

HIGH DENSITY ELECTRODE ARRAYS

Electrode arrays have remained much the same for more than a decade. They are built manually on substrates of silicone, using Pt–Ir (90–10%) alloyed electrodes. The group of Dr. Kensall Wise at the University of Michigan has proposed the use of high density arrays that are made on silicon substrates using IrO contacts (54,55). If such arrays can be built for human use, they will reduce the cost of building electrode arrays while increasing the specificity of excitation of cells. Another approach to the problem is to build electrode arrays on multilayered polymer substrates (Fig. 5). Sample arrays have been used to demonstrate the use of high density arrays in animal studies, with clear

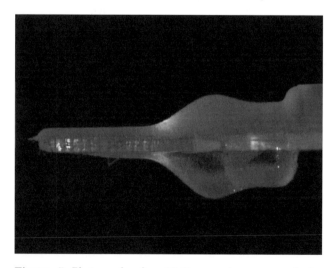

Figure 5. Photograph of a 12-site sample array made by Advanced Cochlear Systems (Snoqualmie, WA) to insert into the scala tympani of a cat (53). The width of each gold electrode contact is 100 μm

independence of channels driven in the first turn of the scala tympani (Snyder, Corbett, Bonham, Rebscher, and Johnson, personal communication).

The goal driving the development of these high density arrays is to increase the specificity of stimulation and to allow several independent groups of cells to be driven simultaneously (32,56). The work of Jolly (32) and Bierer and Middlebrooks (33,57) suggested that this might be the case. More recent work has confirmed the earlier results and extended them (35,34). The benefit of focused multipolar stimulation and of simultaneous excitation of several independent groups of neurons is not without cost. More driven electrodes require greater current consumption. Current consumption is increased with focusing, since focused stimuli require more applied current to reach the same potential fields in conducting media (58). It is likely that high density electrode arrays will be a part of cochlear implants, but there are engineering challenges to be met before it will happen.

COMBINED ACOUSTIC AND ELECTRICAL STIMULATION

Preliminary studies of combined electrical and acoustic stimulation have been done successfully in both Europe and the United States (52,59,60). The subjects come from the substantial population of people who preserve some hearing for frequencies < 1 kHz, but who are severely impaired for frequencies > 1 kHz. Two questions arise immediately. (1) Can low frequency hearing be preserved after an electrode array has been placed in the high frequency regions of the inner ear? (2) Can acoustic and electrical stimuli be applied simultaneously and successfully?

The likelihood of success is great, particularly if patients have short electrode arrays implanted, avoiding damage to the delicate structures of the inner ear. That concern is critical in the case of the hybrid stimulation scheme, since low frequency information will come via the normal, albeit amplified, acoustic pathway. Two manufacturers, Cochlear Corporation (Sydney, Australia) (60) and Med-El (Innsbruck, Austria) (59) have produced electrode arrays for the purpose and have tested them in clinical settings. The Med-El array has an implanted length of 31.5 mm (59), while the Cochlear Corporation array's length is 10 mm in its latest version (60). Both have had extensive laboratory tests and have been used clinically. Clinical tests confirm the initial hypothosis: when patients suffer primarily from high frequency hearing loss, the use of hybrid stimulation is likely to provide great benefit, and may well increase the numbers of people who can have near-normal hearing (52). Electrical and acoustic stimuli can be combined by implanting one ear with a cochlear prosthesis and using a hearing aid in the contralateral ear. This approach has had some reports of success and is currently under study in research laboratories.

NORMAL PROCESSING: CONDITIONING PULSES

In the 1990s, investigators began to consider the issue of the stochastic behavior of neurons (61) and that high rate conditioning stimuli might improve the behavior of cells in the auditory nerve, decreasing thresholds and increasing dynamic ranges (52). They proposed to use electric currents with 5 kHz pulse trains of brief pulses, biphasic rectangular pulses of 40 µs duration for each phase (62). Rubinstein and various colleagues pursued the idea further, suggesting that high rate stimuli might mimic stochastic resonance in neurons and improve signal processing in cochlear implants (60). Computer models validated the concept, as did initial tests in a human subject (63). An extensive neurophysiological study confirmed the idea in experimental animals (62).

Rubinstein and Frijns did preliminary tests for the use of high rate, low amplitude conditioning pulses in the processors of some human subjects, reporting success in the majority of their subjects (Rubinstein, personal communication). The concept is certainly a logical and promising idea; whether it will provide a dramatic improvement to cochlear implants is something that will be learned from further experiments in human subjects.

NORMAL PROCESSING: FINE STRUCTURE

Present cochlear implants impose low pass filter functions on the acoustic signals that they decode. Signals are filtered, and their envelopes detected, with a concomitant loss of fine structure. Fine structure is defined as information spanning frequencies from 500 to 10 kHz (22). Speech can be well understood in quiet environments. The users of present-day cochlear implants rarely enjoy music. Some of that may be improved by increasing the fine structure of the signals delivered to the ear via the cochlear implant. The Hilbert transform provides a potential approach to providing both amplitude information and fine structure (22,64,65). Smith et al. (63) determined that the envelope of the transform was important for speech perception, while the fine structure determines localization and pitch.

Processors that employ Hilbert transforms have yet to be produced in quantity. Although prototypes exist, they have not made their way into cochlear implants (64). The development of implants that can reproduce the fine structure of signals is likely to improve cochlear prostheses.

BILATERAL IMPLANTS

Binaural hearing is critical to sound localization and the extraction of auditory signals in noise. In addition, binaural implants may allow listeners to employ the "head shadow" benefit to hear a specific voice in the face of sounds produced by a competing crowd of people (52). Wilson notes promising results from several centers at which patients have received bilateral implants (52). He reports improvements in speech comprehension, as well as the results of several careful psychophysical studies that were focused on the balance between the prostheses that were implanted. Wilson and his colleagues concluded that bilateral implants are likely to provide clear benefits. While users are tolerant of some timing and amplitude mismatches, the careful matching of stimulus sites, that is, electrode locations, may be necessary for success (52). Another issue to consider is the cost of bilateral implantation. Bilateral implantation incurs the cost of two cochlear prostheses and two surgeries. Does the benefit accrued by the patient double? That remains to be seen at the time of this writing.

CONCLUSION

Cochlear prostheses are a clear bioengineering success story. More than 60,000 patients have benefited worldwide. Many users can talk on the telephone and communicate effectively without visual aids, like lipreading. The design of the cochlear prosthesis is likely to improve, even as the number of implantees grows rapidly, indeed, at double-digit rates. With that rich background and rapid growth, there are opportunities for bioengineers to produce even better cochlear prostheses.

ACKNOWLEDGMENTS

This work was sponsored by grants R43DC000531 and R43DC04614 of the National Institutes of Health.

BIBLIOGRAPHY

1. NIH NIH Consensus Statement: cochlear Implants in Adults and Children. Bethesda, MD, National Institutes of Health; 1995.
2. Webster JG. Encyclopedia of medical devices and instrumentation. New York: John Wiley & Sons; 1988.
3. Blanchfield BB, Feldman JJ, et al. The severely to profoundly hearing impaired population in the United States: Prevalence and demographics. Policy Anal Brief H Ser 1999; 1 (October): 1–4.
4. Blanchfield BB, Feldman JJ, et al. The severely to profoundly hearing-impaired population in the United States: prevalence estimates and demographics. J Am Acad Audiol 2001;12(4): 183–189.
5. Kuchta J. Neuroprosthetic hearing with auditory brainstem implants. Biomed Tech (Berlin) 2004;49(4):83–87.
6. Otto SR, Brackmann DE, et al. Multichannel auditory brainstem implant: update on performance in 61 patients. J Neurosurg 2002;96(6):1063–1071.
7. Schwartz MS, Otto SR, et al. Use of a multichannel auditory brainstem implant for neurofibromatosis type 2. Stereotact Funct Neurosurg 2003;81(1–4):110–114.
8. Kanowitz SJ, Shapiro WH, et al. Auditory brainstem implantation in patients with neurofibromatosis type 2. Laryngoscope 2004;114(12):2135–2146.
9. Geisler CD. From Sound to Synapse: Physiology of the Mammalian Ear. New York: Oxford University Press; 1998.
10. Dallos P, Popper AN, Fay RR, editors. The Cochlea. Springer Handbook of Auditory Research. New York: Springer; 1996.
11. Sachs MB, Young ED. Encoding of steady-state vowels in the auditory nerve: representation in terms of discharge rate. J Acoust Soc Am 1979;667(2):470–479.
12. Liberman MC. Auditory-nerve response from cats raised in a low-noise chamber. J Acoust Soc Am 1978;63(2):442–455.
13. Sachs MB, Young ED. Effects of nonlinearities on speech encoding in the auditory nerve. J Acoust Soc Am 1980; 68:858.
14. Volta A. On the electricity excited by mere contact of conducting substances of different kinds. R Soc Philos Trans 1800;90: 403–431.

15. Clark GM. Cochlear Implants: Fundamentals and Applications. New York: Springer-Verlag; 2003.
16. Djourno A, Eyries C. Prothese Autitive par excitation electrique a distance du nerf sensoriel a l'aide d'un bobinage inclus a demcure. Presse Med 1957;35:14–17.
17. Doyle JB, Doyle HD, et al. Electrical stimulation in eighth nerve deafness. Bull LosAngeles Neurol Soc 1963;18:148.
18. Simmons FB, Mongson CJ, et al. Electrical stimulation of the acoustic nerve and inferior colliculus in man. Arch Otolaryngol Head Neck Surg 1964;79:559.
19. House WF, Urban J. Long term results of electrode implantation and electronic stimulation of the cochlea in man. Ann Otolaryngol Rhinol Laryngol 1973;82:504.
20. Simmons FB, Mathews RG, et al. A functioning multichannel auditory nerve stimulator. A preliminary report on two human volunteers. Acta Otolaryngol 1979;87(3–4):170–175.
21. House WF, Berliner KI. Cochlear Implants: from idea to clinical practice. In: Cooper H, editors. Volume 1, Cochlear Implants: A Practical Guide. San Diego, CA: Singular Publishing Group, Inc.; 1991. p 9–33.
22. Zeng F-G. Trends in cochlear implants. Trends Amplif 2004;8(1):1–34.
23. Spelman F. Cochlear Prostheses. In: Ratner BD, Hoffman AS, Schoen FJ, Lemons JE, editors. Volume 1, Biomaterial Science: An Introduction to Materials in Medicine. Amsterdam: The Netherlands Elsevier Academic Press; 2004. p 658–669.
24. Anonymous. Nucleus 24 Contour, Cochlear Americas. 2004.
25. Kessler DK. The CLARION Multi-Strategy Cochlear Implant. Ann Otol Rhinol Laryngol Suppl 1999; 177(Apr.):8-16.
26. Jolly CN, Gstöttner W, et al. Principles and outcome in perimodiolar positioning. Ann Otolaryngol Rhinol Laryngol Suppl 2000;185(12):20–23.
27. Hillman T, Badi AN, et al. Cochlear nerve stimulation with a 3-dimensional penetrating electrode array. Otolaryngol Neurotol 2003;24(5):764–768.
28. Simmons FB. Electrical Stimulation of the Auditory Nerve in Man. Arch Otolaryngol 1966;84 (July, 1966):24–76.
29. White MW, Merzenich MM, et al. Multichannel cochlear implants: Channel interactions and Processor design. Arch Otolaryngol 1984;110:493–501.
30. Arts HA, Jones DA, et al. Prosthetic stimulation of the auditory system with intraneural electrodes. Ann Otolaryngol Rhinol Laryngol Suppl 2003;191:20–25.
31. Wilson BS, Finley CC, et al. Better speech recognition with cochlear implants. Nature (London) (July 18, 1991); 352:236–238.
32. Jolly CN, Spelman FA, et al. Quadrupolar stimulation for cochlear prostheses: Modeling and experimental data. IEEE Trans Biomed Eng 1996;43(8):857–865.
33. Bierer JA, Middlebrooks JC. Cortical responses to cochlear implant stimulation: Channel interactions. J Assoc Res Otolaryngol. 2004;5(1):32–48.
34. Snyder RL, Bierer JA, et al. Topographic spread of inferior colliculus activation in response to acoustic and intracochlear electric stimulation. J Assoc Res Otolaryngol 2004;5(3):305–322.
35. Bonham B, Snyder RL, et al. The neurophysiological effects of simulated auditory prosthesis stimulation: channel interaction, current steering and channel morphing. San Francisco, CA: University of California at San Francisco; 2004.
36. Anonymous. New methodology for fitting cochlear implants. Valencia, CA: Advanced Bionics Corporation. 1–5; 2003.
37. McDermott HJ, McKay CM, et al. A new portable sound processor for the University of Melbourne/Nucleus Limited multielectrode cochlear implant. J Acoust Soc Am 1992;91: 3367–3371.
38. Anonymous. Clarion S-Series. Sylmar, CA, Advanced Bionics, Inc.; 1997.
39. Eddington DK. Speech discrimination in deaf subjects with cochlear implants. J Acoust Soc Am 1980;68:885–891.
40. Anonymous. PULSARci Cochlear Implant, Med-El; 2004.
41. Rubinstein JT, Hong R. Signal coding in cochlear implants: Exploiting stochastic effects of electrical stimulation. Ann Otol Rhinol Laryngol 2003;112(9, Part 2):14–19.
42. Moore BCJ. Coding of sounds in the auditory system and its relevance to signal processing and coding in cochlear implants. Otolaryngol Neurotol 2003;24(2):243–254.
43. Wilson BS. The History of Cochlear Implants. Neural Interfaces Workshop, Hyatt Regency Bethesda Hotel, Bethesda, MD: NIDCD, National Institutes of Health; 2004.
44. Skinner MW. Cochlear implants in children: What direction should future research take? 2001 Conference on Implantable Auditory Prostheses, Pacific Grove CA; 2001.
45. Skinner MW. Optimizing cochlear implant speech performance. Ann Otolaryngol Rhinol Laryngol Suppl 2003;191: 4–13.
46. Chee GH, Goldring JE, et al. Benefits of cochlear implantation in early-deafened adults: the Toronto experience. J Otol 2004;33(1):26–31.
47. Cunningham CD, 3rd, Slattery WH, 3rd, et al. Postoperative infection in cochlear implant patients. Otolaryngol Head Neck Surg 2004;131(1):109–114.
48. Reefhuis J, Honein MA, et al. Risk of bacterial meningitis in children with cochlear implants. N Engl J Med 2003;349(5): 435–445.
49. Garber S, Ridgely MS, et al. Payment under public and private insurance and access to cochlear implants. Arch Otolaryngol Head Neck Surg 2002;128(10):1145–1152.
50. Cheng AK, Niparko JK. Cost-utility of the cochlear implant in adults. Arch Otolaryngol Head Neck Surg 1999;125(11): 1214–1218.
51. Niparko JK, Kirk KI, et al. Cochlear Implants: Principles and Practices. Baltimore MA: Lippincott Williams & Wilkins; 2000.
52. Wilson BS, Lawson DT. Ann Rev Biomed Eng 2003;5:207–249.
53. Corbett SS, III, Johnson T, Rebscher S, Carson M, Ketterl J, Snyder R. unpublished results.
54. Weiland JD, Anderson DJ. Chronic neural stimulation with thin-film, iridium oxide electrodes. IEEE Trans Biomed Eng 2000;47(7):911–918.
55. Weiland JD, Anderson DJ, et al. *In vitro* electrical properties for iridium oxide versus titanium nitride stimulating electrodes. IEEE Trans Biomed Eng 2003;49(12): 1574-1579.
56. Clopton BM, Spelman FA. Technology and the future of cochlear implants. Ann Otolaryngol Rhinol Laryngol Suppl 2003;191:26–32.
57. Bierer JA, Litvak L, et al. Effects of electrode configuration on psychophysical measures of channel interaction in cochlear implant subjects. Soc Neurosci 2003.
58. Spelman FA, Pfingst BE, et al. The effects of electrode configuration on potential fields in the electrically-stimulated cochlea: models and measurements. Ann Otol Rhinol Laryngol 1995;104(Suppl. 166):131–136.
59. Adunka O, Kiefer J, et al. Development and evaluation of an improved cochlear implant electrode design for electric acoustic stimulation. Laryngoscope 2004;114(7):1237–1241.
60. Gantz BJ, Turner C. Combining acoustic and electrical speech processing: Iowa/Nucleus hybrid implant. Acta Otolaryngol 2004;124(4):344–347.
61. Rubinstein JT, Abbas PJ, et al. Stochastic Resonance: Can it be exploited by speech processors? Conference on Implantable Auditory Prostheses, Pacific Grove, CA; 1997.

62. Runge-Samuelson CL, Abbas PJ, et al. Response of the auditory nerve to sinusoidal electrical stimulation: effects of high-rate pulse trains. Hear Res 2004;194(1–2):1–13.

63. Rubinstein JT, Wilson BS, et al. Pseudospontaneous activity: stochastic independence of auditory nerve fibers with electrical stimulation. Hear Res 1999;127(1–2):108–118.

64. Clopton BM, Lineaweaver SKR, et al. Method of processing auditory data. United States Patent and Trademark Office. Advanced Cochlear Systems.

65. Smith ZM, Delgutte B, et al. Chimaeric sounds reveal dichotomies in auditory perception. Nature (London) 2002; 416: 87–90.

See also AUDIOMETRY; COMMUNICATIVE DISORDERS, COMPUTER APPLICATIONS FOR.

CODES AND REGULATIONS: MEDICAL DEVICES

MORRIS WAXLER
PATRICIA J. KAEDING
Godfrey & Kahn S.C.
Madison Wisconsin

INTRODUCTION

The U.S. Food and Drug Administration (FDA or agency) regulates medical devices according to specific definitions, classifications, requirements, codes, and standards. The FDAs authority and framework for medical device regulation are specified in the Federal Food, Drug, and Cosmetic Act of 1938, as amended (FDCA). The FDCA is codified at Title 21, Chapter 9, United States Code (21 USC) (1). For purposes of medical device regulation, several acts of Congress amending the FDCA are especially significant: the Medical Device Amendments of 1976, the Safe Medical Devices Act of 1990, the Food and Drug Administration Modernization Act of 1997, and the Medical Device and User Fee and Modernization Act of 2002. The FDA has promulgated regulations for the efficient enforcement of the FDCA. These regulations, which generally have the force of law, are codified in Title 21 of the Code of Federal Regulations (21 CFR or the regulations) (2). The agency also has issued guidances and guidelines to assist in the regulation of medical devices (3).

Pursuant to the FDCA, the FDA determines the entities subject to regulation (e.g., manufacturers, specifications developers), evaluates whether products and regulated entities are in compliance, and initiates appropriate regulatory and enforcement actions to impose penalties for violations. The FDA's requirements affect each stage of a medical device's lifecycle. Some FDA requirements apply to particular periods of a medical device's lifecycle. Others apply more broadly. Design, technical development, preclinical testing, clinical study, market authorization, market approval, postmarket assessment, modification, obsolescence, redesign, and labeling requirements are part of this regulatory framework for medical devices. The FDA's Center for Devices and Radiological Health (CDRH) is the FDA component with primary responsibility for medical device regulation.

WHAT IS A MEDICAL DEVICE?

The FDCA contains definitions for the various product areas the FDA regulates, including medical devices. Under the FDCA, a "device" must be

- "an instrument, apparatus, implement, machine, contrivance, implant, *in vitro* reagent, or other similar or related article, including any component, part, or accessory"
- which is either "intended for use in the diagnosis of disease or other conditions, or in the cure, mitigation, treatment, or prevention of disease, in man or other animals," or "intended to affect the structure or any function of the body of man or other animals," and
- "which does not achieve its primary intended purposes through chemical action within or on the body of man or other animals and which is not dependent upon being metabolized for the achievement of its primary intended purposes" [21 USC § 321(h)].

To be a medical device, a product must achieve its "primary intended purpose" without chemical or metabolic action within or on the body. This characteristic distinguishes "devices" from "drugs". For example, perfluorocarbon gas is injected into the human eye to hold a detached retina in place. The gas has no metabolic reaction with the body and thus is regulated as a medical device. But determining whether the FDA would consider a product, a "device", or a "drug" can be difficult. Products can be medical devices even if there is some chemical or metabolic reactions within or on the body. For example, the body often reacts metabolically to hip and other implants. Because these reactions are side effects rather than the primary intended purpose of these implants, the products are medical devices.

The FDCA's definition of medical device includes a concept that is a key part of the FDA's regulatory framework: A medical device is both the physical product and its intended use or uses. "Intended use" is sometimes described as the express and implied claims made for a product. This concept means, for example, that a manufacturer (and his representatives) cannot, without penalty, label, or promote a laser for refractive correction eye surgery if it is legally marketed only for cardiac surgery. The manufacturer must apply to the FDA for authorization or approval to use the laser for a new indication. Changes in indications or uses can create regulatory hurdles for a manufacturer.

MEDICAL DEVICE CLASSIFICATION

Prior to 1976, the FDCA did not contain any specific provisions for medical device regulation. The Medical Device Amendments (MDA) of 1976 greatly expanded the FDA's statutory authority over medical devices and established a comprehensive regulatory scheme for medical devices. The MDA established three classes of medical devices based on the potential risk of the device to patients or users. Devices with greater potential risks are subject to more regulatory controls.

Since 1976, the FDA has established classification regulations for > 1700 different generic types of devices, and grouped them into 16 medical specialties, such as cardiovascular, respiratory, general hospital, infection control, and restorative (4). Each of these generic types of devices is assigned to one of three regulatory classes depending on the level of controls needed to provide a reasonable assurance of the devices' safety and effectiveness. Unclassified devices and new devices are automatically Class III medical devices. But not all medical devices that a layperson likely would understand to be new remain "new" for purposes of the FDCA. If a manufacturer can show that its device is "substantially equivalent" to a device that was legally marketed in 1976, often referred to as a "predicate device", then the device becomes subject to the classification and requirements that apply to that predicate device.

Class I devices are those posing the least amount of risk. Examples include elastic bandages, examination gloves, and hand-held surgical instruments. Class I devices do not require FDA review prior to marketing. However, Class I devices are subject to the FDCA's general controls for all medical devices. These general controls are the regulatory common denominator for all medical devices, and include do not distribute adulterated or misbranded devices; register the commercial establishment with the FDA; list the marketed devices with the agency; label the devices in accordance with applicable labeling regulations; manufacture the devices in accordance with the quality system and good manufacturing practices regulations (many Class I devices, however, are exempt from this requirement); permit FDA inspection. The FDA has the authority to ban medical devices under appropriate circumstances; restrict the sale, distribution, or use of some devices; and require the submission of records and reports.

Class II medical devices have an intermediate level of risk. General controls alone are not sufficient to address the risks of Class II devices. Examples include powered wheelchairs, infusion pumps, and surgical drapes. Class II devices are subject to special controls that are developed to control risks specific to particular devices. Examples of the types of special controls used by FDA include performance standards, guidelines, postmarket surveillance, and patient registries. Most Class II devices require 510(k) premarket notification. The "510(k)" refers to FDCA section 510(k), codified at 21 USC § 360(k). A 510(k) submission contains information and data to show that the device is "substantially equivalent" to a legally marketed predicate device. Clinical data is usually not required for the FDA to clear a 510(k) submission for marketing. Some Class II devices are exempt from 510(k) clearance.

Class III medical devices are those presenting the greatest risks. Examples include replacement heart valves, silicone gel-filled breast implants, and implanted brain stimulators. In general, Class III devices are subject to premarket approval prior to marketing. General and special controls alone are insufficient to provide a reasonable assurance of the devices' safety and effectiveness. Class III devices are usually devices that are life sustaining, life supporting, or implantable, or have the potential for serious injury (e.g., sight threatening). New devices that are not substantially equivalent to a legally marketed device also are usually subject to premarket approval. A premarket approval application (PMA) contains extensive scientific and technical evidence that demonstrates that a reasonable assurance of safety and effectiveness exists for the device. Clinical studies are usually required to support FDA approval of a PMA.

Under the 1997 amendments to the FDCA, manufacturers of certain devices that have been found to be not substantially equivalent can request immediate reclassification into Class I or II based on the device's low risk level. This process is called *de novo* classification. If the FDA agrees, then the device becomes subject to the requirements of either Class I or II, and a PMA is not required.

FDA-REGULATED ENTITIES

The FDA regulates manufacturers, specification developers, distributors, contract manufacturers, sterilization facilities, importers, exporters, contract research organizations, and clinical researchers of medical devices. In addition, the FDA regulates, and otherwise influences, the use and nonuse of voluntary standards by these organizations and individuals to support their regulatory activities and submissions to the agency. The manner in which parties are regulated depends on their role in the distribution of the device and on the stage of the device's lifecycle. For example, the FDA requires preapproval of medical device clinical trials that present significant risks to patients. On the other hand, establishments must register with the FDA only after the FDA authorizes marketing of the device.

USE OF STANDARDS

The FDA recognizes that a device's conformance with recognized consensus standards can be used to support a PMA, 510(k), or other submissions to the agency (5). The FDA maintains a list of officially recognized standards (6). Some domestic and international standards focus on specific medical devices (e.g., respirators). Others characterize an important aspect of many medical devices, (e.g., electrical safety). The former is sometimes called a "vertical" standard. The latter is called a "horizontal" standard. The agency also issues guidance documents for specific devices that refer to the FDA-recognized standards or to other standards. Standards should be used consistent with FDAs guidances because there can be a considerable delay between the development of consensus standards and the agency's recognition of them.

ENFORCEMENT AND PENALTIES

The FDCA authorizes civil and criminal penalties for violations (21 U.S.C. §§ 331-337). The statute, for example, prohibits the adulteration or misbranding of medical devices as well as the introduction or delivery for introduction into interstate commerce, or the receipt in interstate commerce, of any adulterated or misbranded device. The

FDCA also prohibits the submission of false or misleading information to the agency, including the withholding of material or relevant information. For example, a failure to report to the FDA all device failures that occurred during the clinical trial of a Class III medical device is a violation. Such actions can lead to not only disapproval or withdrawal of the PMA for the device, but also civil and criminal penalties on manufacturer.

The FDCA authorizes the FDA to pursue some remedies administratively, including clinical investigator disqualifications, temporary detention of medical devices, and certain civil money penalties. Other remedies, including product seizures, injunctions, criminal charges, and some civil money penalties, require judicial proceedings in federal court. The FDA refers judicial enforcement actions to the U.S. Department of Justice, and works closely with the Justice Department to prosecute these actions. The FDCA is a strict liability statute, which means that a company's management may be prosecuted for a failure to detect, prevent, or correct violations. Knowing and following the rules is important.

REQUIREMENTS GENERALLY

Marketing safe and effective medical devices in the United States requires an understanding of FDA requirements that govern the entire life cycle of the device. These include requirements for conducting nonclinical laboratory studies and clinical trials, bringing a product to market, manufacturing practices, labeling, reporting device problems and patient injuries, carrying out recalls and corrective actions, and making modifications to the device.

NONCLINICAL LABORATORY STUDIES

Manufacturers and other entities must comply with the FDA's Good Laboratory Practices (GLP) regulations when conducting nonclinical laboratory studies that are going to be used to support any regulatory submission to the FDA (21 CFR Part 58). Good Laboratory Practices regulate the organization and personnel of the laboratory as well as the facilities, equipment, test operations and study protocols, and records and reporting. Failure to comply with these regulations may invalidate data submitted to the agency. Contract research organizations used to obtain data for regulatory submissions must comply with GLP regulations.

In addition, the study should conform to FDA-recognized standards that are relevant to particular aspects of the studies, for example, laser safety, toxicity, and biocompatibility. Also, the study's documentation should specifically identify and conform to those parts of FDA performance standards and guidance documents relevant to the device rather simply state overall compliance with the standard or guidance. Whenever particular laboratory study practices will not conform to relevant guidance, the manufacturer or study sponsor should, prior to conducting the studies, discuss the discrepancies with knowledgeable FDA staff, obtain a variance from the GLP regulations if necessary, and document the reasons for the discrepancies.

CLINICAL TRIALS

The FDA regulates clinical trials of medical devices under its investigational device provisions [21 USC § 360j(g), 21 CFR Part 812]. Also important are the regulations for institutional review boards [21 CFR Part 56] and the protection of human subjects [21 CFR Part 50], and the consolidated guidance for good clinical practice [ICH E6]. Different Part 812 procedures apply depending on whether the device study presents "significant risk" or "nonsignificant risk" (7). A significant risk device presents a potential for serious risk to the health, safety, or welfare of a subject. Significant risk devices can include implants, devices that support or sustain human life, and devices that are substantially important in diagnosing, curing, mitigating or treating disease, or in preventing impairment to human health. Examples include sutures, cardiac pacemakers, hydrocephalus shunts, and orthopedic implants. Nonsignificant risk devices are devices that do not pose a significant risk to human subjects. Examples include most daily-wear contact lenses and lens solutions, ultrasonic dental scalers, and urological catheters. Although these latter devices generally are nonsignificant risk devices, the FDA could consider a particular clinical trial using these devices to be a significant risk study and regulate the trial accordingly.

An institutional review board (IRB) may approve a nonsignificant risk device study, and the study may proceed without FDA approval. But clinical studies involving significant risks must receive FDA approval prior to IRB approval. Sponsors, usually investigators or manufacturers, apply for this FDA approval through submission of an Investigational Device Exemption (IDE) application. Although IRBs are to evaluate whether a study is a nonsignificant risk, the FDA has final authority and does determine, from time to time, that an FDA-approved IDE is needed even though an IRB approved a clinical trial protocol as being a nonsignificant risk study.

The FDA's IDE regulations set forth the requirements for submitting IDEs and conducting device clinical trials. These regulations are first and foremost designed to protect human subjects from unnecessary risk. In addition, the IDE regulations are designed to guide the development and documentation of evidence needed to evaluate a device's safety and effectiveness in a PMA application, or a device's substantial equivalence in a 510(k) submission. An IDE is a request for an exemption from the restriction that only legally marketed medical devices can be distributed.

The FDA has a pre-IDE meeting program that can be extremely valuable (8). These meetings usually include FDA review of some portions of a planned IDE submission. Pre-IDE meetings can be requested in a variety of circumstances, and are intended to provide the sponsor with preliminary FDA input related to the device. For example, the pre-IDE meeting should help clarify whether any additional preclinical or technical data are needed, what concerns FDA reviewers may have, whether the proposed protocols are adequate from the FDA's perspective, and the appropriate regulatory path to market for the device. Sponsors planning to conduct nonsignificant risk studies sometimes request a pre-IDE meeting to whether deficiencies exist in the protocols that might preclude marketing approval. Other sponsors find it useful to discuss issues related to ongoing preclinical testing.

An IDE sponsor must submit a detailed description of the device, including its intended use and indication for use, that is, what does the device do and in what kind of patients or user. The sponsor must submit an investigational plan and a detailed protocol for the proposed clinical trial, including proposed informed consent documents. An IDE application also requires other documentation, including results from all laboratory and animal studies conducted with the medical device proposed for the clinical study. These laboratory and animal studies must be conducted in conformity with GLPs. The sponsor must report all relevant published studies, both nonclinical and clinical, regarding the device. Information on all medical uses of the device, and on any clinical trials conducted outside the United States may also be required. If consensus standards exist for the device, the sponsor must identify them and explain whether the device conforms with them. If previous clinical trials were conducted under IRB-only approval, then that data must be submitted in the IDE.

The IDE regulations include an IDE application template (21 CFR 812.20). The FDA also has issued a number of guidance documents on IDE processes and specific types of medical devices (3). Prior to submitting an IDE application to FDA, agency guidance documents relevant to the medical device at issue should be reviewed, and relevant aspects of those guidances implemented. These guidances often recommend specific preclinical tests for categories of devices and can include template investigational plans. But these recommendations and templates are not always suitable for particular devices. Also, guidances are not binding on the FDA and may not fully reflect current agency thinking. Consultation with the FDA may be appropriate where a sponsor believes that modifications are needed for its device.

Once a sponsor submits a complete IDE, FDA must make a decision regarding the IDE submission no later than 30 calendar days from the stamped date of arrival of the IDE application at CDRH headquarters. The FDA's initial decision letter usually lists deficiencies in the IDE, even when FDA approves the IDE. A disapproval letter is rare, especially if the sponsor had a pre-IDE meeting with the FDA. Sponsors receiving a disapproval letter may find it useful to seek assistance from an experience regulatory affairs professional to help evaluate and resolve these deficiencies. If the FDA conditionally approves the IDE, but with deficiencies that have major impact on the clinical trial or the device's indications, these deficiencies should be resolved with the FDA before the clinical trial is started. The FDA usually "conditionally" approves an IDE application, meaning that the applicant may start the clinical trial immediately, but that the applicant must answer the deficiencies satisfactorily within a short time period (e.g., 30–45 days). When the FDA perceives a high risk to human subjects, it will initially approve the IDE for a limited number of subjects and study sites, and then approve expansion of the study after the preliminary data demonstrates reasonable safety. The FDA also typically provides a list of deficiencies that do not have to be answered to conduct the clinical trial, but must be responded to in the marketing application [e.g., 510(k) or PMA]. If any aspect of the FDA's response letter is unclear, clarification should be sought from the FDA or an experienced regulatory affairs professional, or both.

Responsibilities of a clinical trial sponsor, include, but are not limited to, obtaining IRB approval, providing adequate informed consent, and ensuring that the investigators are trained and follow the approved protocol. Adequate record keeping, especially of adverse events, and study site monitoring are critical to success. Annual reports of the clinical study must be submitted to the FDA on the anniversary of the FDA's initial approval of the IDE. Also, serious adverse events must be reported to the FDA within five working days of their occurrence. All adverse events must be reported to the FDA even if the sponsor does not believe the event is related to use of the medical device being studied. Sponsors should also consult medical practice specialty standards and international standards that may be relevant to the study.

Although IDE sponsors (and their agents) may conduct limited advertising for subjects, they must not claim or suggest that the device is safe and effective for the uses it is being studied for. When discussing the device with potential investors, issuing reports on the company, and conducting similar activities, sponsors must carefully avoid making any conclusory statements regarding the device's safety and effectiveness. These restrictions continue until the FDA authorizes or approves the device for marketing. Sponsors also may not charge subjects, investigators, hospitals, or other entities a price for the device that is larger than that necessary to recover costs for manufacture, research, development, and handling. These costs should be documented in the event of an FDA inspection or audit.

Although clinical investigations of medical devices generally must comply with IDE requirements, some limited exemptions exist. For example, a diagnostic device that is noninvasive, does not require an invasive sampling procedure that poses significant risk to the subject, does not introduce energy into a subject, is not used as a diagnostic procedure without confirmation by another medically established diagnostic device, and meets certain other requirements, is exempt from IDE requirements. But the study must still comply with IRB and informed consent requirements.

REGULATORY PATHWAYS TO MARKET

Some medical devices require clearance through premarket "510(k)" notification, some medical devices require premarket approval, and others are exempt from premarket notification and premarket review. The majority of devices—more than 75%—have entered the market through 510(k) premarket notification.

Premarket notification is a process under which the FDA decides whether the evidence demonstrates substantial equivalence between a new device and a legally marketed (predicate) device. If the FDA decides that the device is substantially equivalent to the predicate device, then the device is "cleared" for market. If the FDA decides that the device is not substantially equivalent, it is sometimes appropriate for a manufacturer to request *de novo* classification into Class I or II based on the device's low potential risks. But if the FDA denies that request, the only pathway to market is the PMA approval process. Typically, the FDA will determine, in discussions with a manufacturer or sponsor, which of the three pathways to market is required: (1) 510(k) → substantial equivalence; (2) 510(k) → nonequivalence → *de novo*; (3) PMA. But, as noted, some medical devices are exempt from even 510(k) requirements.

The Medical Device User Fee and Modernization Act of 2002 (MDUFMA) authorizes user fees for premarket reviews of PMAs, PDPs, certain supplements, 510(k)s, and certain other submissions (21 USC §§ 379i-379j). The MDUFMA also set agency performance goals for many types of premarket reviews. These goals become more demanding on the FDA over time. User fees must be paid at the time a submission is sent to the agency or the agency will not file or review it. The MDUFMA includes some fee exemption, waiver, and reduction provisions, including a fee waiver for the first premarket application by a small business.

PREMARKET NOTIFICATION EXEMPTIONS

Class I medical devices are exempt from 510(k) notification unless the FDA has by regulation stated that a particular medical device type is not exempt, or has specified conditions under which it is exempt. But the exemption applies only where the device is intended and indicated for the use or uses specified in the applicable regulation. If the device is to be marketed for a different use or medical condition, then the device is not exempt from 510(k) notification. If the new use presents extremely high risks or involves particularly vulnerable patients, a PMA may be required instead of a 510(k).

The same basic exemption rules apply to Class II devices, except that few Class II devices are exempt from premarket review. For devices exempt from premarket review by regulation, some changes in uses or indications do not require premarket review of the device because certain uses or indications are sufficiently similar to legally marketed intended uses. But in other instances, the FDA decides that an otherwise exempt device must receive premarket notification even though the uses or indications seem very similar. Although the FDCA provides a means for manufacturers to obtain a formal opinion from the FDA where uncertainty exists about the regulatory status of a device, an informal opinion may be sufficient, and preferable, in some situations. Manufacturers should consult an experienced regulatory affairs professional to evaluate how best to proceed in these circumstances.

PREMARKET "510(K)" NOTIFICATION

A 510(k) submission → substantial equivalence decision requires a determination by the FDA that

1. The intended use of the sponsor's device is the same as that of the predicate device(s). Predicate devices

may be any Class I or II device with the same intended use. (A limited number of Class III devices marketed before 1976 also can be predicate devices if the FDA has not yet called for a PMA.)

2. The technological characteristics of the sponsor's device must be either

(a) The same as the predicate device.

(b) Have performance characteristics that demonstrate that it is as safe and effective as the predicate device.

A substantially equivalent device is not "approved" for market. Instead, a 510(k) "clearance" decision is based on the FDA's evaluation of whether the device is substantially equivalent to a legally marketed device for which a reasonable assurance of safety and effectiveness exists. "Substantial equivalence" is a term of art, and does not require that a sponsor's device look or even operate the same as a predicate device. Two devices that visually appear dissimilar can be substantially equivalent under the FDCA. For example, the FDA cleared laser-light and water-jet microkeratomes as equivalent to vibrating steel blades to cut the cornea even though the former products use completely different cutting mechanisms than the latter.

The FDA has issued many guidance documents on various medical device types requiring premarket notification (3). The agency also has issued guidance documents for the 510(k) notification process. The FDA will provide prenotification consultation in telephone or in-person conferences to discuss a sponsor's medical device and answer questions regarding written guidance documents and applicable standards. The FDCA requires the FDA to consider, in consultation with a sponsor, the "least burdensome", appropriate means of evaluating a device (9). To maximize this requirement, a sponsor should understand, as much as possible, the requirements, guidances, and standards that apply to its medical device before meeting with FDA staff. As noted, guidances do not "bind" the FDA. But they can provide valuable information on the agency's thinking on particular topics. Also, a sponsor should try to understand how similar devices have been regulated by the FDA.

510(k) Flow Chart

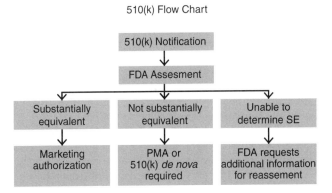

The 510(k) submission → nonequivalence → *de novo* process use the same 510(k) processes to try to establish that substantial equivalence exists and obtain FDA clearance for marketing. But when the sponsor is unable to do so, despite thorough efforts to do so, then the objective becomes convincing the FDA that a PMA is not necessary for regulatory control of the device. This requires a showing that the risks from the device are minimal, that the device is effective for its intended use, and that general controls and, in some cases, special controls will be sufficient to mitigate the product's risks. A request for *de novo* classification must be made within 30 days of receiving a not substantially equivalent determination, describe the device in detail, and provide a detailed recommendation for classification. The FDA then has 60 days to respond to that request with a written order classifying the device and identifying any special controls that may be needed if the device is in Class II. The device is then considered cleared and may be marketed. If the FDA keeps the device in Class III, PMA approval will be required before marketing.

MARKETING APPROVAL

Class III medical devices generally are high risk devices that cannot be regulated adequately by general and special controls alone. In other words, the FDA must review the safety and effectiveness data for these devices to determine if they should be approved for the treatment or diagnosis of diseases or other conditions in humans, and under what conditions. Class III devices may be approved for marketing under the humanitarian use device exemption (HDE), product development protocol (PDP), or premarket approval application (PMA) requirements.

HUMANITARIAN USE DEVICES

The FDCA's humanitarian use device exemption provision is narrow in that the objective is to provide rapid access to new therapeutic or diagnostic devices for patients with rare diseases or conditions, that is, so-called "orphan" devices (21 USC § 360j(m), 21 CFR Part 814, Subpart H). The humanitarian use device (HUD) process is relatively rapid because the applicant does not have to conduct clinical trials to demonstrate reasonable assurance of safety and effective, and the statute allows the FDA significantly less time to act on an HUD application than the agency has for a PMA. Rather than provide data to determine the safety and effectiveness of the device, the applicant has only to satisfactorily explain to the FDA why the probable benefit of the device outweighs the risks to patients in the context of other treatments for the disease. However, this regulatory pathway has many requirements, including the disease or condition affects fewer than 4000 patients/year, the device would not otherwise be available for persons with this disease or condition, the device and will not expose patients to unreasonable or significant risks, and the benefits to health from the device's use must outweigh the risk. Because of the provision's narrow scope and limitations, the humanitarian use device exemption is not used frequently. But it can be very valuable in some instances.

PRODUCT DEVELOPMENT PROTOCOL

The product development protocol (PDP) is an alternative to the PMA process, but is rarely used [21 USC § 360e(f)]. The PDP's distinguishing feature is that it involves a close relationship between the FDA and the sponsor in designing appropriate preclinical and clinical investigations to establish the safety and effectiveness of a device. The PDP requires multiple levels of review and approval of study protocols. The requirements for proof of safety and effectiveness are the same as for a PMA. The PDP process thus offers few advantages for a manufacturer over premarket approval processes, particularly for a device that has undergone significant evaluation and investigation. The PDPs also have required much more FDA staff time than PMA processes.

PREMARKET APPROVAL (PMA)

The FDCA's requirements for PMA approval apply to most Class III medical devices, except for a few devices marketed before the 1976 MDA and those being used consistent with an investigational device exemption (IDE) in order to obtain clinical data to establish the device's safety and effectiveness (21 USC § 360e). The FDA has promulgated regulations on PMA requirements and processes (21 CFR Part 814). These regulations include the FDA's procedures for reviewing and acting on a PMA application. Other important sources for information on PMA issues include general guidances, guidances for specific devices, meetings with the agency and advisory panels, and correspondence from the agency.

The regulations specify and describe the general categories of required information in a PMA [21 CFR 814.20(b)]. These categories include an "indication for use" statement, a device description, and data from nonclinical and clinical studies of the device. The foreign and U.S. marketing history, if any, of the device by the applicant or others must be described in the PMA, including a list of countries in the device has been withdrawn from marketing.

INDICATION FOR USE

A PMA's "indication for use" statement must provide a general description of "the disease or condition the device will diagnose, treat, prevent, cure, or mitigate" and "the patient population for which the device is intended" [21 CFR 814.20(b)(3)]. The "indication for use" statement is key to the device's labeling and, if the device is approved, the uses for which it can be legally marketed. In addition to this statement, the application must include a separate description of existing alternative procedures and practices for the indicated use.

DEVICE DESCRIPTION

The device must be described in summary form and then in detail, including manufacturing and trade secret information where necessary, to allow FDA specialists to evaluate the risks associated with the device. The summary must explain "how the device functions, the basic scientific concepts that form the basis for the device, and the significant physical and performance characteristics of the device" [21 CFR 814.20(b)(3)]. The full device description must include detailed drawings, and details of each functional component or ingredient of the device, all properties of the device relevant to the indication for use, the scientific and technical principles of operation of the device, and the quality control methods (good manufacturing practices) used in the manufacture, processing, packing, storage, and installation of the device [21 CFR 814.20(b)(4)]. In addition, the applicant must reference any standard, mandatory or voluntary, that is relevant to the device for the indicated use. If applicable, the applicant must identify how the device deviates from the standard and demonstrate, to the FDA's satisfaction, how the applicant resolves these deviations.

NONCLINICAL STUDIES

A PMA must include summaries of nonclinical laboratory studies appropriate to the device, including, but not limited to, microbiological, toxicological, immunological, biocompatibility, stress, wear, shelf life studies. The PMA must also include a statement that each study was conducted in accordance with the FDA's good laboratory practices regulations, or explanations as to why not. The study summaries must include descriptions of the objectives, experimental design, data collection and analysis, and results of each study. The results should be described as positive, negative, or inconclusive with regard to the objectives of each study. After each of the studies is summarized, it must be described in sufficient detail to enable the FDA to determine the adequacy of the information for FDA review of the PMA.

CLINICAL STUDIES

Clinical studies involving human subjects with the device must be conducted in accordance with IDE regulations or, if they are conducted outside the United States without an FDA-approved IDE, they must be conducted in accordance with special requirements discussed with the FDA before the PMA is submitted. IRB and human subjects protection requirements and the ICH guidance for good clinical practice also apply. The results of these clinical studies must be summarized first and then discussed in sufficient detail to enable the FDA to determine the adequacy of the information for FDA approval of the PMA. Clinical trial summaries must include the following:

"...a discussion of subject selection and exclusion criteria, study population, study period, safety and effectiveness data, adverse reactions and complications, patient discontinuation, patient complaints, device failures and replacements, results of statistical analyses of the clinical investigations, contraindications and precautions for use of the device, and

other information from the clinical investigations as appropriate...." [21 CFR 814.20(b)(3)].

Discussion of the results of the clinical investigations must include details regarding:

"...the clinical protocols, number of investigators and subjects per investigator, subject selection and exclusion criteria, study population, study period, safety and effectiveness data, adverse reactions and complications, patient discontinuation, patient complaints, device failures and replacements, tabulations of data from all individual subject report forms and copies of such forms for each subject who died during a clinical investigation or who did not complete the investigation, results of statistical analyses of the clinical investigations, device failures and replacements, contraindications and precautions for use of the device, and any other appropriate information from the clinical investigations...." [21 CFR 814.20(b)(6)].

The applicant must identify any investigation conducted under an FDA-approved IDE and provide a written statement with respect to compliance with IRB requirement, or explain the noncompliance. In addition to submitting the data for all the studies conducted by the applicant (or on the applicant's behalf), the applicant is responsible for submitting a bibliography of all studies (nonclinical as well as clinical) relevant to the device and copies of any studies requested by the FDA or the advisory panel. Also, the applicant must identify, discuss, and analyze:

"...any other data, information, or report relevant to an evaluation of the safety and effectiveness of the device known to or that should reasonably be known to the applicant from any source, foreign or domestic, including information derived from investigations other than those proposed in the application and from commercial marketing experience." [21 CFR 814.20(b)(8)].

LABELING

The applicant must submit copies of all proposed labeling for the device, including contraindications, warnings, precautions, and adverse reactions. Labeling typically includes, but is not limited to, physician instructions, an operation manual, a patient brochure, and all applicable information, literature, or advertising materials that constitutes labeling [21 CFR 814.20(b)(10)]. The FDA reviews and revises the proposed labeling prior to PMA approval.

REVIEW STANDARD

The applicant must demonstrate that the nonclinical, clinical, and technical data submitted in the PMA embody valid scientific evidence of reasonable assurance that the device is safe and effective for its intended use. In addition, the applicant must discuss the benefits and risks (including any adverse effects) of the device, and describe any additional studies or surveillance the applicant intends to conduct following approval of the PMA. In evaluating safety and effectiveness, the FDA defines "valid scientific evidence" broadly and retains final authority on what is acceptable [21 CFR 860.7(c)]. The agency considers a variety of factors in deciding whether reasonable assurance of safety and effectiveness has been submitted for a medical device, including intended use (indication), use conditions, benefit–risk considerations, device reliability, and generally requires well-controlled clinical investigations (21 CFR 860.7).

PMA flow chart

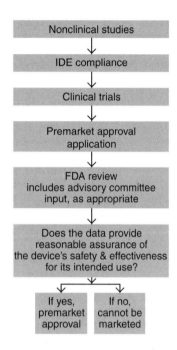

UPDATE REPORT REQUIREMENTS

While the FDA is reviewing a PMA application, the applicant must update it. Such updates are required 3 months after the PMA filing date, following the applicant's receipt of an FDA letter stating the PMA is "approvable", and at any other time as requested by the FDA. An "approvable" letter is a decision by the FDA that the PMA will be approved after the applicant resolves minor deficiencies.

After a device is approved, periodic and other reports are required. The owner of an FDA-approved PMA device is responsible for periodically updating any safety and effectiveness information on the device that may reasonably affect the FDA's evaluation of the device's safety or effectiveness, or that may reasonably affect statements of contraindications, warnings, precautions, and adverse reactions. If a PMA owner becomes aware of off-label (unapproved) uses of its device that may be unsafe or ineffective, then it is responsible for reporting these unauthorized uses to the FDA, especially if adverse events are associated with them.

POSTMARKET RULES

Major postmarket requirements include adequate labeling, medical device reporting, corrections and removals, and device modifications integrated into a system for manufacturing quality medical devices.

LABELING OVERVIEW

Labeling of medical devices is one of a manufacturer's key postmarket responsibilities. Each device must comply with general labeling requirements, and with the specific requirements and limits identified in the FDA's authorization or approval to market the device.

LABELING: GENERAL REQUIREMENTS

Many general labeling requirements exist for medical devices (21 CFR Part 801, Subpart A). The regulations include details on issues such as how a manufacturer's name is to be listed on a device package label. This article focuses on key concepts in the FDA's regulation of device labeling. These concepts include the FDCA's definition of "labeling", the regulation's definition of "intended use", and "adequate directions for use" requirements.

Under the FDCA, "labeling means all labels and other written, printed, or graphic matter (1) upon any article or any of its containers or wrappers, or (2) accompanying such article" [21 USC § 321(m)]. This definition is very broad and includes promotional and advertising materials and oral statements about the device. The FDA's regulation of advertising and promotion presents many challenges for medical device companies. Three basic principles are critical: materials must be truthful and not misleading, must contain a fair balance of benefits and risks, and must provide full disclosure for use.

The "intended use" of a medical device is the objective intent of the product as expressed by the manufacturer or distributor of the device (21 CFR 801.4). It includes all conditions, uses, or purposes stated by the manufacturer or distributor orally or in written form. As discussed earlier, "intended use" is an integral part of the FDCA's "medical device" definition. If a manufacturer or distributor promotes an "intended use" different from the one authorized by the FDA, then the device is adulterated and misbranded until and unless the FDA authorizes the new use. This is often referred to as "off-label" use. The regulation further provides that "if a manufacturer knows, or has knowledge of facts that would give him notice that a device introduced into interstate commerce by him is to be used for conditions, purposes, or uses other than the ones for which he offers it, he is required to provide adequate labeling for such a device which accords with such other uses to the article is to be put." "Intended use" is how the manufacturer intends the device to be used. "Indication for use" is a subset of "intended use" that usually represents a narrowing of the intended use to a specific patient population. In short, why a patient, or a practitioner on a patient's behalf, would use a particular device. Indications for use include a general description of the disease or condition the device will diagnose, treat, prevent, cure, or mitigate, including a description of the patient population for which the device is intended. If differences related to gender, race, ethnicity, age, or other factors exist, they should be reflected as well in the product's labeling.

The FDCA requires device labeling to bear "adequate directions for use," unless the FDA has promulgated regulations exempting a particular device [21 USC § 352(f)]. Under the regulations, "[a]dequate directions for use means directions under which the layman can use a device safely and for the purposes for which it is intended..." (21 CFR 801.5). These directions include specification of all applicable use conditions, dose quantity, use frequency, use duration, time of use, method of use, and use preparation. Adequate directions for use on over-the-counter devices must include a statement of indication for use [21 CFR 801.61(b)].

Prescription devices are exempt from the adequate directions for use requirement because, by definition, such directions cannot be prepared for a prescription device (21 CFR 801.109). However, prescription devices must have adequate instructions for the device's use by practitioners, including, but not limited to information on its use and indications, and any adverse events, contraindications, and side effects that may accompany the use of the device. In addition, to qualify for the adequate directions for use exemption for prescription devices, the device must meet other conditions, such as being in the possession of the practitioner. The regulations authorize other exemptions from the adequate directions for use requirement, including ones for medical devices that have common uses known to ordinary individuals, for medical devices used in certain teaching not involving clinical research, and for medical devices used in manufacturing, processing, and repacking (21 CFR Part 801, Subpart D).

LABELING: SPECIFIC DEVICES

Sources for labeling requirements for particular devices include labeling regulations for a few specific kinds of devices, classification regulations that provide "indications for use" statements for most Class I and Class II devices, guidance documents on specific devices, the FDA marketing authorization and approval letters, and approved labeling for PMA-approved devices.

The FDA has issued specific labeling regulations for dentures, eyeglasses and sunglasses, hearing aids, menstrual tampons, latex condoms, and devices that contain natural rubber (21 CFR Part 801, Subpart H). It also has specific labeling regulations for *in vitro* diagnostic devices (21 CFR Part 809, Subpart B). Each approved PMA includes labeling requirements for the device specified in the approval letter, in the summary of safety and effectiveness, and in written instructions for physicians (and other appropriate professionals) and patients.

REPORTING, CORRECTIONS, AND REMOVALS

Entities that manufacture, prepare, process, package, and/or distribute medical devices are subject to certain requirements regarding device reporting, corrections, and removals. They must track, document, investigate, take action on, and report on events associated with their medical devices. Device user facilities (e.g., hospitals) and importers of medical devices also have responsibilities for reporting certain medical device events (21 CFR Part 803, Subparts A-B and C-D). This article focuses on FDA reporting requirements for device manufacturers (21 CFR Part 803, Subparts A-B and E).

Device manufacturers must report medical device reportable (MDR) events to the FDA with five workdays of becoming aware of a reportable incident if remedial action to prevent an unreasonable risk of substantial harm to the public health, or the event is of a type that the FDA has designated as requiring a report within five work days. Otherwise, MDR events must be reported to the FDA within 30 calendar days. An MDR event is any information that a manufacturer becomes aware of that reasonably suggests that the device marketed by the manufacturer may have "caused or contributed to a death or serious injury" or "malfunctioned...and would be likely to contribute to a death or serious injury, if the malfunction were to recur" [21 CFR 803.3(r), 803.50(a)]. By "any information", the regulations mean all information in the manufacturer's possession or that the manufacturer could obtain from user facilities, distributors, initial reporters of the information, or by analysis, testing, or evaluation of the device. The FDA's regulations specify that manufacturers "become aware" of a reportable event when any employee and any manager or supervisor of employees with responsibility for MDR events acquires information reasonably suggesting that a reportable adverse event has occurred. Moreover, MDR events include any information that necessitates "remedial action to prevent an unreasonable risk of substantial harm to the public health", including, but not limited to, trend analysis [21 CFR 803.3(c)].

Manufacturers should be very inclusive of potential MDR reportable events because the regulations define "caused or contributed" factors very broadly to include events due to user error and labeling misunderstandings in addition to manufacturing and design problems, and device failure and malfunction. In addition, the regulations define malfunction to mean the failure of the device to meet performance specifications of the device for the labeled intended use of the device. "Remedial action" means "any action other than routine maintenance or servicing, of a device where such action is necessary to prevent recurrence of a reportable event" [21 CFR 803.3(z)]. For MDR purposes, the regulations define "serious injury" more broadly than a life-threatening illness or injury. Serious injuries are also those that produce permanent functional impairment, or damage to, body structure or that requires treatment to preclude such impairment [21 CFR 803.3(bb)].

Manufacturers should have written procedures in place to identify, evaluate, and document potential MDR reportable events so that reports can be submitted to the agency accurately and within the required timeframes. A manufacturer must maintain files and records of all events associated with its medical devices whether the manufacturer decided that such events were MDR reportable, and the FDA must be given access to these records upon request. A manufacturer must maintain records of MDR reportable events for ready access by FDA inspectors, and also coordinate these files with the complaint files required by the FDA's Quality System Regulations.

In order to comply with MDR reporting and general record keeping requirements (21 CFR 803.17), a manufacturers must have a system of written procedures to identify, communicate, and evaluate events subject to MDR reporting requirements that is timely and effective; transmit medical device reports to the FDA that are complete and timely; document and record all information that was evaluated in determining if an event was MDR reportable, submitted to the FDA (including MDR reports), used in preparing semiannual reports or certifications to the FDA, and ensure that this documentation and record keeping is readily and promptly accessible to the FDA upon inspection.

Manufacturers also must submit reports to the FDA about medical devices that the manufacturer has corrected in, or removed from, the marketplace to reduce the risk to public health (21 CFR Part 806). A "corrected" medical device is one that the manufacturer has repaired, modified, destroyed, adjusted, relabeled, or inspected at the user location. This includes patient monitoring. A "removed" medical device is one that the manufacturer has physically moved from the user facility to repair, modify, destroy, adjust, relabel, or inspect. Corrections or removals do not have to be reported for devices that have not been distributed to users (stock recovery) or for routine maintenance. However, corrections or removals must be reported for "repairs of an unexpected nature, replacement of parts earlier than their normal life expectancy, or identical repairs or replacements of multiple units" of the device [21 CFR 806.2(k)]. The manufacturer must explain to the agency the reasons for, and estimate the risk to public health of, each correction and removal action within 10 days of initiating the action. The manufacturer must keep records of all corrections and removals, including those not reportable to the FDA, such as those for routine maintenance and stock recovery.

MODIFICATIONS TO MEDICAL DEVICES

Manufacturers must ensure that modifications to their marketed devices are made using design control requirements of the FDA's Quality System Regulations, including, but not limited to, verification and validation processes and updates of the design history file. Manufacturers should also have procedures in place to evaluate whether particular device modifications need to be reported to the FDA. All device modifications must be documented in the company's design and device history files. But some device modifications require prior approval by the agency, some require the opportunity for FDA disapproval prior to implementation, and still others may be reported after the company has implemented the changes. Because a medical device is defined as the physical apparatus and its intended use, significant changes to the product's intended use can require prior authorization from the FDA, even if no physical modification is made to the apparatus; the claim is only implied by the physical modification made to the apparatus; the manufacturer does not make the change but is aware that an entity to which the company sold the device is making additional substantial claims for the device. In other words, the FDA authorized manufacturer of a medical device can be responsible for the device it manufactures for the entire life cycle of the device.

The FDA's guidances for reporting device modifications for 510(k)-cleared devices and PMA-approved devices are summarized in Table 1 (10,11). Manufacturers should establish policies and principles for the company's medical devices based on these guidance documents and agency guidances specific to the company's devices.

QUALITY SYSTEM REGULATIONS

The two main objectives of the FDA's Quality System Regulations (QSR) (21 CFR Part 820) are to ensure (1) that quality in designed into medical devices, and (2) that management is responsible for the device throughout its life cycle and will be held accountable for shortcomings. The QSR sets forth the agency's current good manufacturing practices (cGMP) requirements for medical devices. Each manufacturer must integrate processes for controlling device modifications, labeling, and actions, reports, and record keeping regarding MDR events, corrections and removals into a quality system that is compliant with QSR. The QSR requires manufacturers to integrate all events associated with the manufacture and distribution of the medical devices into a corrective and preventive action (CAPA) subsystem linked to a record keeping subsystem that includes complaint files. The manufacturer must establish standard operating procedures that define, for example, the criteria for MDR reportable events for each kind of medical device that are manufactured, what actions are required, and the processes that must be followed. The manufacturer is responsible not only for maintaining complaint files and device history files, but for actively evaluating this information to maintain the medical device quality. This system involves using diverse information, including device maintenance, modifications, malfunctions, and failures with complaints from users, off-label (unapproved) use, and adverse reactions for continuous evaluation to ensure that the device is performing as designed. Corrective actions are to be taken as appropriate.

The corrective and preventive action subsystem is only one subsystem in a manufacturer's quality system. A quality system should be formed during the establishment of a company's management responsibilities and reviewed and revised during the initial design phase of device development. In addition to management and design control requirements, the QSR requires systems to control documents, purchasing, identification, traceability, production, processing, acceptance, nonconforming products, labeling, packaging, handling, storage, distribution, installation, servicing, and statistics. The regulations give a manufacturer the flexibility to develop a quality system for its medical devices that is tailored to the characteristics of these medical devices. However, the manufacturer's management team must justify and document the quality system, usually in

Table 1. Device Modification Reporting

Types of Modification	Premarket Notice [510(k)]	PMA Supplement
Changes due to recall or corrective action	Recall or corrective actions imply a safety or effectiveness problem with the device. Therefore, the FDA usually requires submission of a 510(k) notice if a device modification is necessary as part of the corrective action	Submit a "180-Day PMA Supplement" for design changes due to recall or corrective action even if the device still meets design specifications. Submit a "Special PMA Supplement-Changes Being Effected" for manufacturing changes that result from the corrective action
Changes that significantly affect safety or effectiveness	Use quality system, especially design controls, to determine if changes that significantly affect safety or effectiveness and if they do then submit a 510(k) notice	Submit a "180-Day PMA Supplement" for changes in, but not limited to, indications for use, labeling, new facilities, sterilization method, packaging, performance or design specifications, and the expiration date that affect safety or effectiveness. Use the quality system, especially design controls, to determine if changes affect safety or effectiveness. The FDA may issue a formal opinion that permits certain changes to be submitted in a "30-Day Supplement" rather than a "180-Day Supplement"
Labeling changes	Most, but not all, changes in intended use/indication for use require submission of a 510(k) notice. For example, if the device will be indicated for use in a subset of patients for which the device is already cleared, then a 510(k) may not have to be submitted. Or no notice may be needed if a risk analysis demonstrates no additional risk by expanding the patient population being treated	Submit a "180-Day PMA Supplement"
Technology or performance specifications	Changes in a device's control mechanism, principles of operation, or energy source usually requires submission of a 510(k) notice. Changes in sterilization method usually do not require 510(k) notification if design verification and validation is adequate	Submit a "180-Day PMA Supplement"
Materials changes	Evaluate the effects of materials changes on the performance characteristics of the device. If the performance characteristics are changed significantly or new labeling must be added then perhaps a 510(k) notice should be submitted to the FDA	Submit a "180-Day PMA Supplement"
Minor incremental changes or changes that do not affect safety or effectiveness	Use design controls to evaluate risks associated with "minor" evolutionary changes in the device. Proactively develop a decision rule about when these incremental changes should be reported to the FDA	Usually does not require FDA approval prior to implementation but describe the modifications in the Annual Report required for the PMA
Minor changes to the manufacturing process	Notice to the FDA not required	File a "30-Day Notice" to the FDA describing the changes in detail. Implement the changes at the end of the 30-day period unless the changes require submission of a "135-Day Supplement" because the 30 day notice to the FDA was inadequate
Changes that improve the safety of the device	Notice to the FDA not required	File a clearly marked "Special PMA Supplement—Changes Being Effected." The changes that enhance safety include, but are not limited to, changes that strengthen a contraindication, an instruction, or quality controls. They must be described in detail

a quality system manual that specifies each of the subsystems as identified in the QSR and any deviations from it.

IMPLEMENTATION

Bioengineers and informed specialists developing innovative medical devices must understand the regulatory implications of their scientific and technical innovations in order to develop a realistic business plan for their product. Sometimes the innovations are considerable yet the agency regulatory pathways remain simple. For example, as discussed earlier, CDRH cleared laser-light and water-jet microkeratomes as equivalent to vibrating steel blades to cut the cornea even though the former products use completely different cutting mechanisms than the latter. Similarly, FDA decided that a manufacturer's microscopic dermal fragments should be regulated as human tissues under the same tissue bank rules used to regulate its macroscopic sheets of dermis. The FDA could have decided to regulate microscopic dermis as a medical device because of the additional processing (a decision that would have required requiring premarket authorization of the dermal fragments), but instead decided both were human tissues from a regulatory point of view. On the other hand, innovative products can be subject to profoundly different regulatory pathways. For example, external kidney dialysis products have almost always been regulated as medical devices by the CDRH using the 510(k) process, an efficient process. However, the FDA decided to use the drug–biologics review process (IND/NDA) to regulate an external kidney dialysis filter using human cells, a more complex and costly review process than for devices.

The following analysis of a hypothetical medical device illustrates some of the regulatory implications of innovative medical devices. The hypothetical device is an implanted artificial kidney that can continuously dialyze the human body. Currently, 90% of patients that require kidney dialysis are treated with an external device in which the patient's blood is dialyzed outside the body, an external kidney dialysis device. Some patients are treated with an external kidney dialysis device that infuses the dialysate (the dialysis solution) into the abdominal cavity (the peritoneum) and then drains the waste products out of the peritoneum ~45 min later or continuously overnight. As mentioned above, the FDA currently is regulating an external kidney dialysis product using more burdensome drug–biologic regulatory requirements rather than the simpler 510(k) process used for other dialysis machines. Therefore, if metabolic interaction and/or cells are used in an implanted artificial kidney devices, it is likely that either CDER or CBER will lead the review of the combination product through the FDA's drug–biologics approval process. However, if the implanted artificial kidney device achieved its intended use of dialysis without primarily biochemical or metabolic interaction with the human body, then the implanted artificial kidney likely would be regulated as a medical device by the CDRH. Filter material and microscopic control elements such as valves and motors are likely critical components. This example illustrates that issues imbedded in the scientific and technical characteristics of an innovative medical product could result in a regulatory pathway that is more complex, and costly, than already marketed alternative products.

Regardless of whether the innovative medical product, an implanted artificial kidney in this example, is reviewed by the FDA as a device, drug, or biologic, or a combination product, agency reviewers may or may not have expertise or knowledge directly relevant to the critical science. In fact, it is unlikely. Therefore, very early in product development the manufacturer should engage FDA reviewers in a dialogue about the cutting edge science or technology used in the device so that a common understanding evolves about key safety and effectiveness issues. This approach should help reduce misunderstandings about necessary nonclinical laboratory studies, animal study protocols, key safety and effectiveness endpoints, and fail-safe mechanisms so that agency reviewers will be comfortable with the risks associated with initial pilot study in humans. Also, the manufacturer should dialogue with agency reviewers about the scientific, clinical, and ethical issues associated with an initial clinical study in humans. In order to maximize control, manufacturers should take the initiative in making study proposals to the FDA rather than simply ask the agency for advice.

REGULATORY CHALLENGES

Medical device developers, academic researchers and engineers, start-up companies, research and development departments of large manufacturers, and other innovators are at the leading edge of scientific, technological, and medical product development, not the FDA. They therefore should be proactive with regard to the issues critical to the development and eventual marketing of the medical device. Developers of medical devices should take advantage of the opportunities to establish conditions for efficient FDA regulation of their devices before making regulatory submissions to the agency by developing a detailed quality system manual tailored to development and manufacture of the company's medical device; implementing good laboratory practices and specific standard operating procedures for nonclinical studies; identifying existing technical standards that are applicable to manufacturing quality devices, developing applicable standards where none exist; identifying the best clinical practices for clinical trials with the device; communicating the science and technology of the device to FDA reviewers; proposing a specific regulatory pathway to the agency based on a risk-benefit analysis of the device; incorporating feedback from discussions with the FDA. These proactive steps are particularly important for devices that are very innovative, and where scientific consensus may not exist on procedures and new standards needed to verify and validate the design of the device.

BIBLIOGRAPHY

1. Online access to the United States Code. Available at http://www.gpoaccess.gov/uscode/index.html. Accessed 2005 Feb 11.
2. Online access to U.S. Food and Drug Administration regulations. Available at http://www.accessdata.fda.gov/scripts/cdrh/cfdocs/cfcfr/CFRSearch.cfm. Accessed 2005 Feb 11.

3. U.S. Food and Drug Administration Guidance Documents. Available online at http://www.fda.gov/cdrh. A searchable database of FDA guidances involving medical devices is available at http://www.accessdata.fda.gov/scripts/cdrh/cfdocs/cfggp/search.cfm. Accessed 2005 Feb 11.

4. A searchable database of U.S. Food and Drug Administration Medical Device Classification Regulations. Available at http://www.accessdata.fda.gov/scripts/cdrh/cfdocs/cfPCD/PCDSimpleSearch.cfm. Accessed 2005 Feb 11.

5. U.S. Food and Drug Administration, Center for Devices and Radiological Health (2001, June 20). Recognition and Use of Consensus Standards; Final Guidance for Industry and FDA Staff. [Online version]. USFDA. http://www.fda.gov/cdrh/ost/guidance/321.html. Accessed 2005 Feb 11.

6. Standards recognized by the U.S. Food and Drug Administration. Available in a searchable online database at http://www.accessdata.fda.gov/scripts/cdrh/cfdocs/cfStandards/search.cfm. Accessed 2005 Feb 11.

7. U.S. Food and Drug Administration, Office of the Commissioner (2001, April 18). Information Sheets: Guidance for Guidance for Institutional Review Boards and Clinical Investigators: Medical Devices. [Online] USFDA. Available at http://www.fda.gov/oc/ohrt/irbs/devices.html. Accessed 2005 Feb 11.

8. U.S. Food and Drug Administration, Center for Devices and Radiological Health. (1999, March 25). IDE Guidance Memorandum-Pre-IDE Program: Issues and Answers. [Online version]. USFDA. Available at http://www.fda.gov/cdrh/ode/d99-1.html. Accessed 2005 Feb 11.

9. U.S. Food and Drug Administration. Center for Devices and Radiological Health. (2002, October 4). The Least Burdensome Provisions of the FDA Modernization Act of 1997: Concept and Principles; Final Guidance for FDA and Industry. [Online version]. USFDA. Available at http://www.fda.gov/cdrh/ode/guidance/1332.html. Accessed 2005 Feb 11.

10. U.S. Food and Drug Administration, Center for Devices and Radiological Health. (1997, Jan 10). Deciding When to Submit a 510(k) for a Change to an Existing Device. [Online version]. USFDA. Available at http://www.fda.gov/cdrh/ode/510kmod.html. Accessed 2005 Feb 11.

11. U.S. Food and Drug Administration, Center for Devices and Radiological Health. (1998, Feb 19). 30-Day Notices and 135-Day PMA Supplements for Manufacturing Method or Process Changes. [Online version]. USFDA. Available at http://www.fda.gov/cdrh/modact/daypmasp.html. Accessed 2005 Feb 11.

See also CODES AND REGULATIONS: RADIATION; HOME HEALTH CARE DEVICES; HUMAN FACTORS IN MEDICAL DEVICES; SAFETY PROGRAM, HOSPITAL.

CODES AND REGULATIONS: RADIATION

BRUCE THOMADSEN
GLENN GLASGOW
BENJAMIN EDWARDS
RALPH LIETO
University of Wisconsin-Madison
Madison, Wisconsin

INTRODUCTION

Every country develops its own regulations governing radiation. Because this text is coming from the United States, the regulations considered here mostly apply to that country. "Radiation" in this article always refers to electromagnetic radiation and to energies higher than that used for communications (radio and microwave): particularly to higher-energy, directly and indirectly ionizing radiation [referred to as ionizing radiation (IR) hereafter] and medium-energy, nonionizing radiation (NIR). The discovery of ionizing radiations (i.e., those forms of radiation with sufficient energy to directly or indirectly *ionize* atoms by stripping away one or more electrons, thereby producing an ion pair consisting of the freed electron and charged atom) at the end of the nineteenth century (c. 1895) was followed quickly by observations of radiation injury.The first *recommendations* on IR dose limitation and personnel protection appeared shortly thereafter. The first general *regulations* for ionizing radiation came with the advent of the program to develop nuclear weapons during World War II. Most of this article deals with IR only because those regulations are more complex and voluminous. In addition, because this Encyclopedia focuses on biomedical applications, this text will concentrate most on those regulations most pertinent to medical settings, particularly those that have changed since the original edition of the Encyclopedia (1).

Although the electrical and magnetic fields associated with NIR were well known long before the discovery of ionizing radiation, a lack of significant observable health effects and the scarcity of powerful NIR sources delayed the development of NIR exposure standards until much later. The development of NIR standards was further complicated by the very wide range of wavelengths and photon energies covered by the NIR designation, and by the consequently wide variety of NIR tissue interaction mechanisms associate with each spectral region. Despite these obstacles, a comprehensive framework of NIR safety guidance now exists, but generally with less regulatory rigor and compulsion than for IR. Efforts to harmonize the exposure limits offered by various standard setting organizations have improved consistency, although disparities remain in some spectral regions.

ORGANIZATIONS INVOLVED IN IONIZING RADIATION PROTECTION RECOMMENDATIONS AND REGULATIONS

Sources of Guidance

The U.S. government relies on guidance from scientific organizations in the development of regulations. None of these organizations have any regulatory authority in the United States, but supply information and recommendations for the regulation-making processes. The most important organizations include the following:

International Commission on Radiation Protection (ICRP): an international organization founded in 1928 under the International Congresses of Radiology (currently called the International Society of Radiology) that occasionally establishes panels to review the published literature on an issue concerning radiation protection and make recommendations.

International Commission on Radiation Units and Measurements (ICRU): An international organization organized in 1925 also under the International Congresses of Radiology that, like the ICRP, occasionally establishes panels to review the published literature on an issue concerning radiation units, measurement or dosimetry, and make recommendations.

National Council on Radiation Protection and Measurement (NCRP): A committee organized in 1929 as an informal gathering of radiation scientists to represent radiation-related organizations in the United States, and then formally chartered by Congress in 1964. As with the two international commissions, the NCRP establishes panels and writes reports on radiation related topic, and serves as the main source for guidance to the US government in the formulation of radiation regulations.

International Atomic Energy Agency (IAEA): An agency of the United Nations, the IAEA provides guidance documents and expert consultation on radiation safety issues, particularly for developing countries.

United Nations Committee on the Effects of Atomic Radiation (UNSCEAR): A committee under the United Nations established in 1955 to study the biological effects of radiation. Periodically this committee publishes report on their findings.

National Academy/Board on Radiation Effects Research (BRER): The BRER was established in 1981 to coordinate activities of the National Research Council involving the biological effects of radiation. Periodically, the BRER establishes panels to review the literature on the Biological Effects of Ionizing Radiation (BEIR) and issue reports bearing that acronym.

Joint Commission on Accreditation of Healthcare Organizations (JCAHO): A commission established by many medical organizations, such as the American Hospital Association and the American Medical Association. The Joint Commission establishes some standards for the use of radioactive materials and radiation in medical settings. Their standards, as of this writing, tend to be broad and vague statements on quality.

Professional Organizations: Organizations of professionals that may make recommendations, guidance documents or standards for various aspects of their profession. Often these documents form the basis for regulations. Some of the major organizations that influence radiation regulations include: The American Association of Physicists in Medicine; The American College of Interventional Cardiologist/The American College of Cardiology; The American College of Medical Physics; The American College of Nuclear Physicians; The American College of Radiology; The American Nuclear Society; The Health Physics Society/American Academy of Health Physics.

International Electrotechnical Commission (IEC)/ American National Standards Institute (ANSI): organizations that establish standards mostly pertaining to industry and manufacturers, their recommendations sometimes find their way into U.S. regulations aimed toward manufacturers of radiation-producing equipment.

Divisions of the U.S. Government Regulating Ionizing Radiation

In the United States, no one governmental agency regulates radiation and radioactive materials. Rather, aspects of radiation regulation fall under several agencies. Some of the major agencies are listed below, although the list is not exhaustive.

Nuclear Regulatory Commission. The Nuclear Regulatory Commission(NRC) is headed by a five-member Commission appointed by the President. The authority for the NRC comes from the Atomic Energy Act of 1954 (as the Atomic Energy Commission), as amended. The NRC was established by the Energy Reorganization Act of 1974. Because of the historical development of radiation regulations, the NRC formerly only exercised control over reactors and reactor byproduct materials. Thus, naturally occurring radioactive material, radioactive materials produced in particle accelerators and machine produced radiation fell outside the purview of the NRC. By these acts, the NRC regulates:

Special nuclear material, which is uranium-233, or uranium-235, enriched uranium, or plutonium.

Source material, which is natural uranium or thorium or depleted uranium that is not suitable for use as reactor fuel.

Byproduct material, which is, generally, nuclear material (other than special nuclear material) that is produced or made radioactive in a nuclear reactor.

Most recently, the Energy Policy Act of 2005 extended NRC authority to include naturally occurring and accelerator-produced radioactive materials (NARM). Before this time, the individual States regulated NARM with a somewhat non-uniform array of regulations.

The relevant NRC rules governing the authorized use of radioactive materials for medical applications are found in Title 10 Code of Federal Regulations. The specific divisions of Title 10 with a significant impact on medical uses are the regulations in Part 19—Notices, instructions and reports to workers: inspection and investigations; Part 20—Standards for protection against radiation; Part 21—Reporting of defects and noncompliance; Part 30—Rules of general applicability to domestic licensing of byproduct material; Part 31— General domestic licenses for byproduct material; Part 32— Specific domestic licenses to manufacture or trade certain items containing byproduct material; Part 33— Specific domestic licenses of broad scope for byproduct material; Part 35— Medical use of byproduct material; Part 71— Packaging and transportation of radioactive material.

The rules in Part 19, Part 20, and, most of all, Part 35 dominate the daily activity of medical licensees (2–5).

Since the late 1990s, NRC regulation changes have been performance-based rather than risk-based only. This was largely in response to the wide criticism by the medical community of regulations and enforcement activity. This resulted in an Institute of Medicine–National Academy of Science report (6) that made several recommendations for improvement to the agency, and the subsequent NRC Strategic Assessment and Rebaselining Initiative. These initiated a major revision of the medical use rules of Part 35 (2). The NRC regulations attempt to protect workers and patients while minimizing its imposition on the practice of medicine. The last major change was completed in March 2005 that addressed training and experience of users, which demonstrates the lengthy federal rulemaking process (5).

Department of Transportation. The Department of Transportation (DOT) regulates (in Title 49 of the Code of Federal Regulations) shipping or carrying radioactive materials, be it by air or surface, including any radioactive materials on public streets or highways.

Environmental Protection Agency. The Environmental Protection Agency (EPA) regulates the allowed levels of radioactive materials in the air, water, and landfills, as well as radiation exposures to the public outside nuclear power reactors. In some cases their regulations also covers occupational exposures to radiation. The rules enforced by the EPA do not all come from a single section of the Code of Federal Regulations.

Department of Energy. The Department of Energy (DOE) is charged with leading the energy development in the United States. A large part of their work involves reactors, and for DOE funded projects and facilities, particular radiation regulations apply.

Department of Defense. The Department of Defense (DOD) establishes radiation regulations for DOD facilities.

Food and Drug Administration. Department of Health and Human Services (DHHS) enters into the radiation regulation field mostly through one of its 10 agencies, the Food and Drug Administration (FDA). The FDA is responsible for protecting the public health by assuring the safety, efficacy, and security of human and veterinary drugs, biological products, medical devices, our nation's food supply, cosmetics, and products that emit radiation, either ionizing or nonionizing. Accordingly, by Title 21 of the Code of Federal Regulations, Food and Drugs, Revised April 1, 2004, the FDA approves and regulates the testing, manufacture, and approved use of a radioactive drug, also called radiopharmaceutical, or a medical device containing a radioactive source. However, the radiation safety regulations of who is authorized to use such drugs or devices and the conditions for use are the responsibility of the NRC or its Agreement States. Regulations in Title 21 can be found on line at http://www.accessdata.fda.gov/scripts/cdrh/cfdocs/cfcfr/cfrsearch.cfm [5 October 2005]

Radiopharmaceuticals. Radiopharmaceuticals are used for diagnostic purposes such uptake, dilution, or imaging, or for therapy applications. The relevant FDA regulations applicable to approving a radioactive drug are found in 21CFR 200–680 (References to the Code of Federal Regulations are usually written with the number of the Title before "CFR" followed by the part number, so this reference is Title 21 of the Code of Federal Regulations, Parts 200 through 680.):

Subchapter C—Drugs: General

Part 201: Labeling

Part 211: Current Good Manufacturing Practice for Finished Pharmaceuticals

Subchapter D—Drugs for Human Use

Part 310: New Drugs

Part 312: Investigational New Drug Application

Part 361: Prescription Drugs for Human Use Generally Recognized as Safe and Effective and Not Misbranded: Drugs Used in Research

Subchapter F—Biologics

Part 600: Biological Products: General

Part 601: Licensing

Part 610: General Biological Products Standards

Part 660: Additional Standards for Diagnostic Substances For Laboratory Tests

A rapidly increasing aspect of nuclear medicine is the use of radioactive drugs employing positron emitters for diagnostic imaging. Positron emitters have a physical characteristic of very short half-lives (less than a few hours). The dominant radiopharmaceutical is F-18 tagged to fluorodeoxyglucose (FDG). They are used to perform positron emission tomography (PET). A problem with the production of PET drugs is meeting the FDA current good manufacturing practices (CGMP) regulation, which ensures that PET drug products meet safety, identity, strength, quality and purity requirements. The cause is their short half-lives prevent completing the current good manufacturing practices (CGMP) in a manner to allow distribution and administration. Current good manufacturing practices (CGMP) covers items such as control of ingredients used to make drugs, production procedures and controls, recordkeeping, quality system and product testing. The FDA and professional societies, such as the Society of Nuclear Medicine, are working to achieve a resolution. (http://www.fda.gov/cder/regulatory/pet/default.htm).

Machines and Devices. It is the responsibility of the FDA to determine if a submission is a device or a drug. With the increasing complexity and miniaturization of technology this is becoming increasingly difficult. Nevertheless, the FDA must approve any device that will be used on–in humans. Examples of such radioactive medical devices are high dose rate (HDR) remote afterloaders, intravascular brachytherapy devices, and radioactive-liquid filled balloons for the treatment of brain tumors. In addition, either the NRC or an Agreement State must perform engineering and radiation safety evaluations of the ability of the device to safely contain radioactivity under the conditions of their possession and use. If deemed satisfactory, the regulatory authority issues a registration certificate. The evaluations are summarized in the registration that the NRC maintains in the National Sealed Source and Device Registry (NSSDR). The registration certificates contain detailed information on the sources and devices, such as how they are permitted to be distributed and possessed (specific license, general license, or exempt), design and function, radiation safety, and limitations on use. Either the NRC or Agreement States can issue a registration certificate for distributors and manufacturers within their jurisdiction, but only the NRC is responsible for devices distributed as exempt products (i.e., smoke detectors) and issues those registration certificates.

Analogous to drugs, the Radiation Control for Health and Safety Act of 1968 established the requirements and responsibility for the FDA to administer an electronic product radiation control program to protect the public health and safety. As part of that program, FDA has authority to issue regulations prescribing radiation safety performance standards for electronic products, most importantly including diagnostic X-ray systems. This gives the FDA the authority to promulgate regulations on the manufacture and assembly of such machines. Again, it is the individual state that regulates who can operate such machines and the radiation safety conditions of use. The exception to this is diagnostic mammography. For mammography, under the Mammography Quality Standards Act (MQSA) of 1992, the FDA approves the accrediting bodies that accredit the facilities to be eligible to perform screening or diagnostic mammography services. It also establishes minimum national quality standards for mammography facilities to ensure safe, reliable, and accurate mammography. These standards address the physician interpreters, the radiologic technologists performing the imaging, the medical physicists performing the testing, and machine performance and testing for mammography *only*. The Center for Devices and Radiological Health (CDRH) is the agency within the FDA that has responsibility for radiation machines and machines (http://www.fda.gov/cdrh/). The relevant FDA regulations applicable to the manufacture and performance of radiation machines are found in 21 CFR 900-1050:

Subchapter I—Mammography Quality Standards Act (MQSA)
 Part 900-Mammography
Subchapter J-Radiological Health
 Parts 1000–1050

For radioactive pharmaceuticals, implantable radioactive sources, radiation producing machines, or computer software that may be used with humans, approval must first be obtained by the vendor from the FDA. Before such approval, any use must be performed under an Investigational Drug Exemption (IDE) from a facility's Institutional Review Board (IRB), and if the drug or device poses significant risk, by the FDA also. After demonstration of the safety of the investigational drug or device, the FDA may approve general use following the manufacturer's instructions as given in the premarket approval (PMA) documentation. Use other than as described is considered "off-label," and, while allowed, imposes increased liability on the institution should something go wrong.

Occupational Safety and Health Administration. The Occupational Safety and Health Administration (OSHA) administers the Occupational Safety Health Act to assure safe and healthful working conditions. The health standard 29CFR 1910.1096 governs employee exposure to ionizing radiation from X-ray equipment, accelerators, accelerator-produced materials, electron microscopes, betatrons, and technology-enhanced naturally occurring radioactive materials not regulated by the NRC (7,8). The OSHA encourages states to develop and operate their own programs, which OSHA approves and monitors.

Their rules have not been revised since 1971 (9). and essentially reflect the NRC regulations at that time. At the time of writing, OSHA is considering revising its regulations.

States. Regulation of radioactive materials and radiation producing machines that are not covered by any federal rules fall to the individual states to regulate. However, the states often enter into agreements with federal agencies to assume the federal regulatory role. This is discussed in greater detail in the section on Regulatory Standards for Radioactive Byproduct Material.

Conference of Radiation Control Program Directors. There is one organization that needs to be noted especially with regard to the establishment of regulations by the individual states. This organization is the Conference of Radiation Control Program Directors (CRCPD). In the early 1960s many states were developing radiation control programs. Such programs included, but were not limited to, regulating the use of diagnostic and therapeutic X ray, environmental monitoring, and regulating the use of certain radioactive materials including NARM. Simultaneous to the development of these early state and local radiation control programs were similar activities at the federal level. Many of these and varied state, local, and federal programs and activities in radiation control were being developed independent of each other.

A need for uniformity was identified to avoid inconsistencies and conflicts of rules and regulations throughout the country regarding radiation users. As a result, the CRCPD was established in 1968 to (*1*) serve as a common forum for the many governmental radiation protection agencies to communicate with each other; and (*2*) promote uniform radiation protection regulations and activities. To achieve these purposes, the CRCPD developed the Suggested State Regulations (SSR) for radiation control, which it regularly updates as federal or industry changes occur at its websitw (10). The SSR address both radioactive materials and radiation machines in medicine and industry. Although the SSR are only recommendations, their importance is that many states have, and continue to, adopt them as their state regulations giving them the force of law. These suggested rules are discussed in detail below. The primary membership of CRCPD is radiation professionals in state and local government who regulate the use of radiation sources. But it works closely with all relevant federal agencies, (NRC, FDA, DOT, EPA, etc).

U.S. Divisions Regulating Nonionizing Radiation

The key U.S. government agencies involved in regulating NIR are listed in Table 1. U.S. regulatory guidance specifically addresses some kinds of NIR sources in some spectral regions while omitting direct mention of other sources and spectral regions. For example, Curtis (11) acknowledges that the exposure standards in the OSHA are dated, noting the following weaknesses: the construction industry standard does not include laser classification and controls; the radio frequency (RF) exposure limit is from the 1966 ANSI standard (it has no frequency dependence and does not address induced current limits); The RF Safety Program Elements are incomplete.

However, the obligation of employers under the General Duty Clause of OSHA [Occupational Safety and Health Act of 1970, 29 USC 654, section 5(a)(1)] to protect workers from recognized hazards compels the control of all potentially harmful NIR hazards, whether specifically regulated or not. Various government agencies also provide a wealth of guidance beyond the requirements specified in the regulations. As noted in Table 1, the FDA regulations apply primarily to manufacturers, so although much FDA guidance clearly pertains to the end users, the FDA typically does not inspect healthcare providers or enforce compliance with FDA guidance by healthcare facilities. However, other organizations that do routinely audit healthcare providers, including in particular the JCAHO, refer to and hold hospitals accountable for compliance with FDA guidance. Table 2 summarizes the requirements of those states having comprehensive laser safety regulations, adapted and updated from Ref. (12). Many of these states have also passed regulations for the control of other NIR hazards as well (see e.g. Article 14 in Chapter 1 of Title 12, Arizona Administrative Code). Several nonregulatory organizations have established exposure limits covering the entire NIR spectrum. The primary industry consensus

Table 1. U.S. Government Agency Nonionizing Radiation Regulations

Agency	Role	NIR Related Regulations/Guidance	Created by
FAA	Responsible for the safety of civil aviation; includes the safe and efficient use of navigable airspace	14cfr91.11: prohibits interference with aircrew FAA Order 7400.2 Part 6 Chapter 29 [Outdoor Laser Operations]; limits laser exposure levels near airports	Federal Aviation Act (1958)
FCC	Responsible for regulating interstate and international communications by radio, television, wire, satellite, and cable	To comply with the National Environmental Policy Act of 1969 (NEPA), established limits for Maximum Permissible Exposure (MPE) to RF radiation based on NCRP and ANSI/IEEE criteria, in 1996 Report and Order, and 1997 Second Memorandum Opinion and Order. In addition, per 47cfr18, industrial, scientific, and medical equipment manufacturers must comply with requirements designed to reduce electromagnetic interference	Communications Act (1934)
FDA and CDRH	Protecting the public health by assuring the safety, efficacy, and security of human and veterinary drugs, biological products, medical devices, our nation's food supply, cosmetics, and products that emit radiation	The following regulations apply primarily to manufacturers: 21cfr1040.10 and 11—laser products 21cfr1040.20—sunlamp products and ultraviolet lamps intended for use in sunlight products 21cfr1040.30—high intensity mercury vapor discharge lamps. 21cfr1030.10—microwave ovens	Food and Drugs Act (1906)
OSHA	Ensure the safety and health of America's workers by setting and enforcing standards; providing training, outreach, and education; establishing partnerships; and encouraging continual improvement in workplace safety and health	29cfr1910.97 nonionizing radiation 29cfr1910.268 telecommunications	Occupational Safety and Health Act (1970)

Table 2. Representative Sample of State Laser Regulations[a]

State	Regulation	Exemptions	Training Required	Warning Signs Required	Controls Required	Registration Required	ANSI or FDA Based	Outdoor or Light Show Requirements
AK	18AAC35, Art. 7, Sec. 670-730	Stored, Inoperable, Enclosed and below MPE (e.g., Class 1)	No	Yes	Yes	No	No	Yes
AZ	AAC Title 12, Chpt. 1, Article 14, Sec. R12-1-1421-1444	None, but focus on control of Class 3b and 4	Yes	Yes	Yes	Yes	ANSI/FDA	Yes
FL	FL Code: Chap. 64-E4	Stored, Class 1, 2, and 3a	Yes	Yes	Yes	Yes	ANSI/FDA	Yes
IL	Chapter 420 ILSC 56	Transported, negligible hazard	No	No	No	Yes	FDA	No
MA	105 CMR 121	Transit and storage	Yes	Yes	Yes	Yes	ANSI	Yes
NY	Title 12 NYCRR Part 50	Non-R&D Class 1, 2, and 3a	Yes	Yes	Yes	Yes	FDA	Yes
TX	Title 25 TAC Part 1 Rule 289.301	Transit, stored, inoperable	Yes	Yes	Yes	Yes	ANSI/FDA/IEC	Yes
WA	WAC 296-62-09005	None, but focus on control of Class 3b and 4	Yes	Yes	Yes	No	ANSI/FDA	No
HI	HAR 12-201-3	None	Yes	Yes	Yes	No	No	Yes

[a]Adapted and updated from Ref. 12.

standard organizations and international standard-setting agencies appear in Table 3. Some of these voluntary standards carry more weight than others, especially internationally. For example, all member countries of the European Union are required to adopt the laser safety standard, IEC/EN 60825-1, of the International Electrotechnical Commission (IEC), which has also been adopted by Japan, Australia, Canada, and nearly every other nation that publishes a laser standard (13). In addition, the FDA now accepts conformance with the IEC/EN 60825-1 in lieu of conformance with most (but not all) of the requirements of the U.S. Federal Laser Product Performance Standard (14). Similarly, the FDA, OSHA, and JCAHO all reference the ANSI Z136 series of standards.

REGULATORY STANDARDS FOR RADIOACTIVE BYPRODUCT MATERIAL

Use of IR in medical, dental, and veterinary facilities is governed by either federal (e.g., NRC, OSHA) or state regulations. The NRC, drawing its authority from the Atomic Energy Act of 1954, regulates byproduct material, source material, and special nuclear material, and their uses. Here OSHA controls IR sources (X-ray equipment, accelerators, accelerator-produced materials, electron microscopes, betatrons, and technology-enhanced naturally occurring radioactive materials) not covered by the Atomic Energy Act of 1954 and not regulated by the NRC. A 1989 "Memorandum of Understanding..." defined responsibilities and authorities of each agency (7). Each agency has arrangements with some states for regulatory enforcement. NRC has an Agreement State Program, by which a State can sign a formal agreement with the NRC to assume

NRC regulatory authority and responsibility over certain byproduct, source, and small quantities of special nuclear material. There are 33 States, listed in Table 4, with two (Pennsylvania and Minnesota) in the process of becoming Agreement States. The Atomic Energy Act of 1954 provides a statutory basis under which NRC relinquishes to the states portions of its regulatory authority to license and regulate byproduct materials (radioisotopes); source materials (uranium and thorium); and certain quantities of special nuclear materials. The mechanism for the transfer of authority to a state is an agreement signed by the Governor of the State and the Chairman of the Commission. The NRC has established compatibility obligations with the Agreement State regarding its current rules and future regulations that it may promulgate. Because two-thirds of the states have assumed Agreement status, the NRC has provided them increasing voice in their activities. This is done through the NRC Office of Tribal and State Programs and the independent Organization of Agreement States (OAS). Both can be accessed via the URL, http://www.nrc.gov/what-we-do/state-tribal/agreement-states.html. The NRC regulations apply in federal facilities directly holding federal licenses and in the nonagreement states. Agreement states have certain periods (3 years or more) within which state regulations must become compliant, at certain levels of compliance, with NRC regulations. During this transition period state regulatory agencies enforce their current state regulations, based on NRC regulations in force prior to the regulatory changes, as they prepare new state regulations compliant with the recent revisions changes in federal codes. Twenty-six states, also in Table 4, have OSHA-approved state plans with their individual state standards and enforcement policies.

Table 3. Selected Organizations Publishing Voluntary NIR Safety Standards

Organization	Role	Significant NIR Standards
American Conference of Governmental Industrial Hygienists (ACGIH)	Professional society devoted to the administrative and technical aspects of occupational and environmental health	TLVs and BEIs (Threshold Limit Values for Chemical Substances and Physical Agents; Biological Exposure Indices)
American College of Radiology (ACR)	Maximize radiology value by advancing science, improving patient care quality, providing continuing education and conducting research	White Paper on MR Safety
American National Standards Institute (ANSI)	Promoting and facilitating voluntary consensus standards and conformity assessment systems	Z136.1—Safe Use of Lasers Z136.2—Safe Use of Optical Fiber Communication Systems Utilizing Laser Diode and LED Sources Z136.3 Safe Use of Lasers in Health Care Facilities Z136.5 Safe Use of Lasers in Educational Institutions Z136.6 Safe Use of Lasers Outdoors B11.21 Machine Tools Using Lasers—Safety Requirements for Design, Construction, Care and Use ANSI/IESNA RP-27.1 Recommended Practice for Photobiological Safety for Lamps and Lamp Systems—General Requirements; RP-27.3 Risk Group Classification and Labeling ANSI/IEEE 95.6 Safety Levels With Respect to Human Exposure to Electromagnetic Fields, 0–3 kHz ANSI/IEEE C95.1 Safety Levels with Respect to Human Exposure to Radio Frequency Electromagnetic Fields, 3 kHz–300 GHz
International Commission on Non-Ionizing Radiation Protection (ICNIRP)	Disseminate information and advice on the potential health hazards of exposure to nonionizing radiation to everyone with an interest in the subject	Guidelines on Limits of Exposure to Ultraviolet Radiation of Wavelengths Between 180 nm and 400 nm (Incoherent Optical Radiation) Guidelines on Limits of Exposure to Laser Radiation of Wavelengths between 180 nm and 1 mm Revision of the Guidelines on Limits of Exposure to Laser radiation of wavelengths between 400 nm and 1.4 μm Guidelines on Limits of Exposure to Broad-Band Incoherent Optical Radiation (0.38–3 μm) Guidelines for Limiting Exposure to Time-Varying Electric, Magnetic, and Electromagnetic Fields (up to 300 GHz) Guidelines on Limits of Exposure to Static Magnetic Fields
International Electrotechnical Commission (IEC); European Committee for Electrotechnical Standardization. (CENELEC)	Prepares and publishes international standards for all electrical, electronic and related technologies; these serve as a basis for national standardization and as references when drafting international tenders and contracts	60601-2-33: Particular Requirements for the Safety of Magnetic Resonance Equipment for Medical Diagnosis 60825-1: Equipment Classification, requirements, and user's guide 60825-2: Safety of Optical Fibre Communication Systems 60825-3: Guidance for laser displays and shows 60825-4: Laser guards 60825-5: Manufacturer's checklist for IEC 60825-1 60825-6: Safety of products with optical sources, exclusively used for visible information transmission to the human eye 60825-7: Safety of products emitting infrared optical radiation, exclusively used for wireless 'free air' data transmission and surveillance 60825-8: Guidelines for the safe use of medical laser equipment 60825-9: Compilation of maximum permissible exposure to incoherent optical radiation 60825-10: Application guidelines and explanatory notes to IEC 60825-1 60825-12: Safety of Free Space Optical Communication Systems used for the Transmission of Information TR60825-14: A User's Guide

In a few instances, the DOE operates research programs at national laboratories under DOE supervision and governs occupational exposure under 10CFR 835 (Occupational Radiation Protection). Because of its limited role, DOE regulations are not further discussed. Table 5 lists the web sites of these federal agencies and other organizations with interests in regulation of radiation in its many forms.

Table 4. NRC Agreement States[a] and States, Commonwealths, and Territories[b] with OSHA-Approved State Plans[c]

Alaska (O)	**Iowa** (A,O)		Rhode Island (A)
Alabama (A)	Kansas (A)	*New Hampshire (A)*	**South Carolina** (A,O)
Arizona(A,O)	**Kentucky**(A,O)	New Jersey (O)	**Tennessee**(A,O)
Arkansas (A)	Louisiana (A)	**New Mexico**(A,O)	Texas (A)
California(A,O)	Maine (A)	*New York*(A,O)[d]	**Utah**(A,O)
Colorado (A)	**Maryland**(A,O)	**North Carolina** (A,O)	Vermont (O)
Connecticut (O)[d]	Massachusetts (A)	North Dakota (A)	*Virgin Islands*(O)[d]
Florida (A)	Michigan (O)	Ohio (A)	Virginia (O)
Georgia (A)	**Minnesota**(A[e],O)	Oklahoma (A)	**Washington** (A,O)
Hawaii (O)	Mississippi (A)	**Oregon** (A,O)	Wisconsin (A)
Illinois (A)	Nebraska (A)	Pennsylvania (A[e])	Wyoming (O)
Indiana (O)	**Nevada** (A,O)	Puerto Rico (O)	

[a]Designated A.

[b]Designated O.

[c]Those that are both (A,O) are in bold print.

[d]These in italics have plans that cover public sector (State and local government) employment only.

[e]These are not yet agreement states, but have filed intent to become agreement states.

NRC Regulations—Summary of Changes to U.S. Regulations

The NRC has actively revised their governing regulations. The most recent revisions (4-24-02) were to four parts of the federal code: *10 CFR 19 (Notices, Instructions, and Reports to Workers; Inspections); 10 CFR20 (Standards for Protection Against Radiation); 10CFR32(Specific Domestic Licenses to Manufacture or Transfer Certain Items Containing Byproduct Material), and 10CFR35 (Medical Use of Byproduct Material)*(2–5). The Occupational Safety and Health Act enforces the terms of the OSHA promulgated in the federal code *29 CFR 1910.1096 (Ionizing Radiation)* which appear to date to 1970 (9). Indeed, current OSHA standards are based on original terms, definitions, and units historically used by the NRC in 1971, many of which were changed in the 2002 NRC revisions. While OSHA websites allude to potential code revisions under active internal review, none are posted for public comment. Hence, we focus on a synopsis of NRC revisions.

10 CFR 19 (Notices, Instructions, and Reports to Workers; Inspections). This long-standing regulation (2), with 14 sections, issued 12/18/1981, unrevised, remains in force. Table 6 lists seven important sections with brief comments

about their content. Insuring that all current members of a constantly changing workforce receive initial and timely recurrent annual instruction is a significant regulatory compliance challenge for radiation safety officers (RSO) charged with their instruction.

10 CFR20 (Standards for Protection Against Radiation). These standards, consisting of 69 sections, are, with one exception, discussed later, mostly unchanged from the 5/21/1991 release (3). Tables 7a,7b lists 10 key headings with brief comments about their content.

Three sections, *10CFR20.1002/Scope; -0.1003/Definitions, and -.1301/Dose Limits for Individual Members of the Public* (4) were revised. 20.1002/Scope now states [conventional radiation units deleted] that "…limits in this part do not apply …to exposures from individuals administered radioactive materials (RAM) and released under §35.75…" 20.1003/ Definitions adds "Occupational dose does not include…dose… dose… from individuals administered RAM and released under §35.75…" "Public dose does not include…dose… from individuals administered RAM and released under §35.75…"

20.1301/ Dose Limits for Individual Members of the Public now adds to the exclusion of dose from RAM in sanitary sewers, the following: "…does not exceed … 1 mSv in a year

Table 5. Useful Web Sites with Information About Radiation, Regulations, and Regulatory Issues

Agency	Internet Address, http://www.	Electronic Mail Address
Conference Radiation Control Program Directors	crcpd.org	Not given on web page
Department of Energy	energy.gov	Not given on web page
Department of Transportation	dot.gov	dot.commentsost.dot.gov
Environmental Protection Agency	epa.gov	Not given on web page
Food and Drug Administration	fda.gov	Not given on web page
Health Physics Society	hps.org	hpsBurkInc.com
Idaho State University	physics.isu.edu/radinf/rso toolbox	Not given on web page
International Atomic Energy Agency	iaea.org	official.mailiaea.org
International Commission on Radiological Protection	crp.org	scient.secretaryicrp.org
International Commission Radiation Units and Measurements	icru.org	icruicru.org
National Council on Radiation Protection and Measurements	ncrp.com	Not given on web page
Nuclear Regulatory Commission	nrc.gov	Not given on web page
Occupational Safety and Health Administration	osha.gov	Numerous information-specific links on web page

Table 6. Partial Contents of 10 CFR 19[a]

Section Major	Major contents of section
0.3/Definitions	Workers, licenses, restricted areas defined
0.11/Postings notices to workers	(a) Post regulations, (i) license and its conditions; (ii) operating procedures; (iii) violations; (b) Documents, forms must be conspicuous
0.12/Instructions to workers	Inform about: (a) storage, use RAM; (b) health protection problems; (c) procedures to reduce exposures; (d) regulations; (e) report conditions, violations; (f) response to warnings; (g) their exposures
0.13/Notification and reports to individuals	(a) Written exposure reports; (b) annual exposure reports per workers request; (c) other provisions not stated here
0.14/Presence of licensee's and workers representatives during inspections	(a) Licensee to allow inspections; (b) inspectors may meet workers; (c) reps may accompany inspectors during inspections; (d) other provisions not stated here
0.15/Consultations with workers during inspections	(a) Inspectors may consult privately with workers; (b) workers may consult privately with inspectors
0.16/Requests by workers for inspections	Workers may request inspections without retribution

[a]Notices, Instructions, & Reports to Workers; Inspections.

exclusive of the dose contributions from background radiation, from any medical administration to the individual, from individuals administered RAM and released under §35.75, from voluntary participation in medical research programs..."

Also, added: "...a licensee may permit visitors to an individual ...to receive a radiation dose greater than ...

1 mSv if 1. the radiation dose ...does not exceed ... 5 mSv and 2. the authorized users has determined *before the visit* that it is appropriate."

Security of RAM is addressed in §20.1801. A new international and national concern is the security of byproduct sources in medical facilities. Most medical licensees have small (tenths of GBq) quantities of long-lived byproduct

Table 7a. Unchanged Components of CFR 20[a]

Section	Major contents of section
0.1101/Radiation Protection Program (RPP)	(a) RPP must be developed, documented, implemented, commensurate with extent and scope of licensed activities; (b) ALARA for occupational and public doses; (c) Annually review RPP content and implementation
0.120/Occupational Dose Limits Dose Equivalent (DE); Deep Dose Equivalent (DDE); Cumulative Dose Equivalent (CDE); Total Effective Dose Equivalent (TEDE)	(a) Annual TEDE 0.05 Sv; sum of DDE and CDE of organs 0.5 Sv; eye DE 0.15 Sv; shallow skin or extremity DE 0.5 Sv; (b) Excess DEs must be planned; (c) Other provisions not stated here
0.1208/Dose to an Embryo/Fetus	(a) 5 mSv dose to embryo/fetus, entire pregnancy, occupational exposure of mother; (b) Avoid variations in uniform monthly doses; (c) Dose is sum of DDE of mother and radionuclides in mother and embryo/fetus; (d) Other provisions not stated here
0.1502/Individual Monitoring of External/Internal Occupational Doses Cumulative Effective Dose Equivalent (CEDE)	(a) Those likely DE 10% of limits; (b) Those in high and very high radiation areas; (c) Those likely to receive CEDE of 10% from radionuclides; (d) Other provisions not stated here
0.1801/Security of radioactive materials	(a) Secure from unauthorized removal or access licensed material stored in controlled or unrestricted areas; (b) Licensed material not in storage shall have control and constant surveillance

[a]Standards for Protection Against Radiation.

Table 7b. Unchanged Components of CFR 20[a]

Section Major	Major contents of section
0.1901/Caution Signs	Radiation symbol (trefoil) color schema (magenta, purple, black) on yellow and design defined;
0.1904/Labeling Containers Radioactive Materials	(a) Containers of RAM must be marked either "CAUTION" or "DANGER", RADIOACTIVE MATERIAL; (b) Label must identify quantity, date, radiation levels, kind of material; (c) Remove/deface labels on empty containers
0.1906/Receiving/Opening Packages	(a) Package receipt and monitoring procedures; (b) Carrier notified if wipe test or radiation levels exceed limits; (c) Package opening procedures; (d) Other provisions not stated here
0.1501/Surveys and Monitoring	(a) Make necessary surveys; (b) Equipment used for surveys calibrated; (c) Excluding direct/indirect pocket dosimeters, NVLAP accreditation for badge processor
0.2001/Waste Disposal	(a) By transfer to authorized recipient; (b) By decay in storage; (c) By effluent release within limits; (d) Others provisions not stated here

[a]Standards...Protection...Radiation.

materials (^{137}Cs, ^{60}Co, etc.), ideal components for dispersal "dirty bomb". The IAEA has developed an action plan to combat nuclear terrorism (15,16). These international efforts likely will lead to new national and state regulations requiring greater security for radioactive sources.

Listed in Table 7b as unchanged is signage. Radiation areas and places or contains that hold radioactive materials must be posted as such. Fig. 1 shows a typical radiation area sign, and gives the criteria for each of the types of signs required.

The 10CFR32 (Specific Domestic Licenses to Manufacture or Transfer Certain Items Containing Byproduct

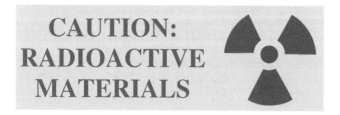

Figure 1. A typical "Caution: Radioactive Materials" sign. The wording on the sign follows the criteria: Rooms containing more than the quantity listed in Part 20 Appendix C–"CAUTION: RADIOACTIVE MATERIALS"; 2. Areas with exposure rates > 0.05 mSv in 1 h, 30 cm from a source (or surface that radiation penetrates) – "CAUTION: RADIATION AREA" or "DANGER: RADIATION AREA"; 3. Areas with exposure rates greater than 1 mSv in 1 h, 30 cm from a source (or surface that radiation penetrates) - "HIGH RADIATION AREA"; 4. Areas with exposure rates greater than 5 GY in 1 h, 1 METER from a source (or surface that radiation penetrates)– "GRAVE DANGER: RADIATION AREA"; 5. Areas where the derived air concentrations exceeds values in appendix B, to 20.1001–20.2401, or where an individual without respiratory protection could exceed, during the hours an individual is present in a week, an intake of 0.6% of the annual intake limit —"DANGER: AIRBORNE RADIOACTIVITY AREA".

Material) revisions (4-24-2002) are only notational bookkeeping, changing the paragraphs numbers and sections in Part 32 to correspond with the corresponding sections of the revised 10CFR35.

Revisions in 10CFR 35 (Medical Use of Byproduct Material). With 126 sections, we focus only on those of direct interest or applicability to medical byproduct material. Tables 8a–e summarizes, using some shorthand notations, the major contents of the important sections. The bulk of regulatory changes relative to byproduct material occur in these sections.

Components of CFR 35 Applicable to All Forms of Brachytherapy (Tables 8a, b). A new term, Authorized Medical Physicist (AMP), and the training thereof, is defined, as well as types (low dose rate, LDR; pulsed dose rate, PDR; and high dose rate, HDR) of remote afterloading units (RAU), including medium dose rate (MDR). Mobile services and medical events are new additions. Roles of management, the RSO, and authorized users (AU) supervision of individuals are explained. Dose prescriptions, or written directives (WD) details and procedures are enumerated.

Table 8b notes source inventories are now at 6 month intervals. §35.75 explains new release criteria for patients (4). Some requirements for mobile medical services are in this section, as well as rules for decay-in-storage of RAM.

Some Components of CFR 35 (F) Applicable to Manual Brachytherapy (Table 8c). One major change is a requirement to decay output or source activities in 1% intervals. Another section adopts AAPM good practices, per various protocols, for quality assurance of therapy planning systems, as a regulation.

Some components of 10CFR 35 (H) for Photon-Emitting Remote Afterloaders (Tables 8d, e). In the nine sections, the most significant change is the requirements for MDR

Table 8a. Components of CFR 35 (A, B) Applicable to All Forms of Brachytherapy

Section	Major Contents of Section
0.2/Definitions	(a) Authorized medical physicist defined; (b) LDR, MDR, HDR, PDR defined; (c) Mobile medical service defined; (d) Medical event (no more misadministration's! explained; (e) Manual prescribed dose (total sources strength and time, or dose per WD) given; (f) Remote prescribed dose (total dose and dose per fraction per WD) given
0.24/Authority Radiation Protection Program	(a) Defines a stronger management role; (b) Defines and strengths RSO role
0.27/Supervision	Explains role of authorized user (AU) and supervised individuals with respect to process and procedures with RAM
0.40/Written Directives (WD)	(a) Written directives required or oral directives with 48 h for written; (b) HDR: radionuclide; site, fx dose, #fxs, total dose; (c) Others; before tmt: radionuclide; site, dose; before finish: # sources, total source strength and time (or total dose); revisions allowed during treatment.
0.41/Procedures...written directives	(a) ID patient; (b) administration per WD; (c) Check manual, computer dose calculations; (d) confirm console data
0.51/Training authorized medical physicist	(a) Board certifications; (b) degrees +1 year training + 1 y experience; (c) preceptor's written statement regarding training

Table 8b. Some Components of CFR 35 (C) Applicable to All Forms of Brachytherapy

Section	Major Components of Section
0.67/Requiremenst for possession	(a) Leak tests (5 nCi) before 1st use, 6 mos.; (b) exempt Ir-192 seeds in ribbons and unused sources; (c) 6 months. inventory
0.75/Release...patients containing...RAM	(a) OK if others TEDE < 5 mSv·year^{-1}; (b) Instruction if others TEDE > 1 mSv/year;
0.80/Mobile medical services	(a) Facility agreement letters; (b) on-site, before use survey meter checks; (c) Post-treatment surveys; (d) possession licenses required for all sites
0.92/Decay in storage	(a) $T_{1/2} < 120$ day; decay to background level; (b) remove labels; keep records

Table 8c. Some Components of CFR 35 (F) Applicable to Manual Brachytherapy

Section	Major Contents of Section
0.404/Surveys after... implant and removal	(a) After implant; source accountability; (b) After source removal; keep records
0.406/Source accountability	(a) ...at all times...in storage and use; record
0.410/Safety instructions	(a) Initially, annually...to caregivers; (b) Size, type, handling, shielding, visitor
0.415/Safety precautions	(a) No room sharing with regular patients; (b) Post-room (RAM) and visitor limits; (c) Emergency equipment for source retrieval from or in patient
0.432/Source calibrations (post-10/24/04)	(a) Determine output or activity; (b) positioning in applicators per "protocols"; (c) decay outputs/activities at 1% intervals; keep records
0.433/Decay Sr-90 sources	Only AMP shall calculate decayed activity and keep records
0.457/Therapy-related computer systems	(a) Acceptance testing per "protocols"; (b) Source input parameters; (c) accuracy of dose/time at points; isodose and graphics plots; (d) localization image accuracy

Table 8d. Some Components of 10CFR 35 (H) for Photon...Remote Afterloaders

Section	Major Components of Section
0.604/Surveys of patients	Before releasing patient...survey patient and RAU to confirm ...returned to safe
0.605/Installation, ..., repair	(a) Certain source work, that is install, adjust, and so on, by licensed person; (b) For LDR RAU, licensed person or AMP can do certain source work; record
0.610/Safety procedures	(a) Secure unattended RAU; (b) only approved individuals present in room; (c) No dual operations; (d) written procedures for abnormal situations; posted copies; initial/annual instructions with drills; records
0.615/Safety precautions	(a) Control access with interlock; (b) area monitors; (c) CCTV/audio for all except LDR RAU; (d) for MPD/PDR an AMP and AU **or** operator-emergency response MD present at initiation and immediately available during treatments; (e) for HDR an AU and AMP physically present at initiation, but, during continuation, AMP and AU or operator-emergency response MD; (f) emergency equipment for unshielded source or source in patient.
0.657/Therapy-related computer system	(a) Acceptance testing per "protocols"; (b) Source input parameters; (c) accuracy of dose/time at points; isodose and graphics plots; (d) localization image accuracy; (e) electronic transfer to RAU accuracy

Table 8e. Some Components of 10CFR 35 (H) for Photon...Remote Afterloaders

Section	Major Contents of Section
0.630/Dosimerty system (DS) equipment	(a) Except for LDR RAUs, NIST/ ADCL calibrated DS; (b) 2 year and after service; or, (c) 4 year, if intercom pared with calibrated DS within 18–30 month.and $< 2\%$ change
0.633/Full calibrations (FC) of RAUs	(a) Before 1st use; at source exchanges and/or repairs to exposure assembly; (b) for $T_{1/2} > 75$ days, excluding LDR RAUs, quarterly; (c) LDR RAUs yearly; (d) FC: 5% output/1 mm positions, source retraction, timer accuracy/linearity; (e) tube lengths and functions; (f) quarterly autoradiographs of LDR RAU sources; (g) decay outputs/activities at 1% intervals; (h) FC and decay by AMP; keep records; for LDR RAU can use manufacturer's data for FC
0.643/Periodic spot-checks (SC) of RAUs	(a) For LDR RAUs, before 1st treatment; for other RAUs 1st use daily; (b) per WP by AMP; (c) AMP review by 15 day; (d) SC includes: interlocks, status lights, audio and CCTV, emergency equipment, source position monitors, timer, clocks, decayed source activity
0.647/Additional requirements...mobile RAUs	(a) Survey meter checks; (b) source inventory; (c) all 0.643 checks; (d) interlocks, status lights, radiation monitors, source positioning, before 1st use, simulated treatment at each address

and PDR units. Physicians other than AUs, trained in MDR and PDR operation, emergency procedures, and source removal, may work under the supervision of an AU. Note: We denote them as substitute authorized users (SAU). For the initial treatment, the AMP and AU *or* SAU must be present; during subsequent (continuation) treatments, the AMP, AU, *or* SAU must be *immediately available*. These requirements are less onerous than the prior requirements of the AU *always* being present during all treatments. Another section adopts AAPM good practices, per various protocols, for quality assurance of RAU therapy planning systems, as a regulation.

Requirements for dosimetry systems (DS), full calibrations (FC), and spot-checks (SC) are described, including those for mobile services.

Some Components of 10CFR35 (J)(Recognition of Specialty Boards). The 2002 revisions in 10CFR 35 did not address personnel training. On March 30, 2005, the NRC published the final rule (5) regarding specialty boards and personnel training. The rule identifies (on the NRC web site, *not* in the published rule) various approved specialty boards and describes pathways for approvals of RSOs, AMPs, authorized nuclear pharmacists, and physicians using many forms of byproduct materials. The rule offers multiple pathways by which individuals may achieve authorizations to perform various tasks or assume authorized titles, (e.g., RSO, AMP, authorized nuclear pharmacist, or physician authorized user). One pathway is the educational degree -> experience -> specialty examination -> certification path. Another pathway is the supervised experience -> preceptor statement path. For example, Table 9 shows five ways an individual, depending on their education, experience, and certification status, can qualify to be an RSO. This flexible approach offers individuals multiple pathways to achieve authorization, which maintaining the integrity of the approval process. For those not physicians, the education requirements are either (a) a bachelor or graduate degree in physical science, or, engineering or biologic science with 20 college credits in physical science, or, (b) a master's degree or PhD in physics, medical physics, or physical science, engineering, or applied mathematics. Experience requirements vary from 1 to 5 years depending on the authorization, and are shorter for those with higher degrees. Generally, experience must be gained under a certified medical physicist or authorized individual, and documented. Preceptors must document the successful completion of any structured training programs and attest to the individual's competency and ability to perform learned tasks independently. In some instances, structured didactic training programs including classroom and laboratory training are allowed. Table 10 show similar requirements for becoming an AMP or ANP.

Training requirements for physicians, Tables 11 and 12, generally offers physicians two options: Completing requirements for medical specialty board certification and passing a certification examination, or, completing a structured educational program with a specific number of classroom and laboratory hours and work experience. In some instances, a preceptor must provide a written statement attesting to the satisfactory completion of the requirements and to the individual's "...competency sufficient to function independently..." (5) In other instances, a certain number of cases must be performed. The classroom and laboratory training requirements are specific to each specialty. Tables 11 and 12 only broadly describe training requirements; details of each specialty training program as described in USNRC 2005.

Table 9. Training Requirements for Radiation Safety Officers

Person	Degree	Experience	Examination	Classroom Laboratory Training	Preceptor Statement	Special Training
(1) Radiation Safety Officer	*and* B or GD in PS; or, E or BS w 20 cc in PS;	*and* 5 or more years in HP including 3 years in AHP	*and* Passes Exam			
or, (2) Radiation Safety Officer	*and* M or PhD in P, MP, or PS, E, AM	*and* 2 years full-time training in MP under supervision by CMP, *or*, in CNM, by physician AU	*and* Passes Exam			
Or, (3) Radiation Safety Officer		1 year full-time RS under supervision by RSO		200 h in topical areas		
Or (4) Radiation Safety Officer	*and* is a CMP	*and* applicable experience			*and* has written attestation by preceptor	*and* training in RS, regulatory issues, and emergency procedures
or (5) Radiation Safety Officer	*and* is AU, AMP, or ANP on license	*and* applicable experience			*and* has written attestation by preceptor	*and* training in RS, regulatory issues, and emergency procedures

ANP = Authorized nuclear pharmacist; PS = Physical science; B = Bachelor's degree; CMP = Certified medical physicist; BS = Biological science; RS = Radiation safety; CC = College credits; AU = Authorized user; E—Engineering; CNM = Clinical nuclear medicine; GD = Graduate degree; MP = Medical Physicist; M = Master's degree; RSO = Radiation safety officer; Ph.D = Doctoral degree; AMP = Authorized medical physicist.

Table 10. Training Requirements for Authorized Medical Physicist and Nuclear Pharmacist

Person	Degree	Experience	Examination	Classroom Laboratory Training	Preceptor Statement	Special Training
(1) Authorized Medical Physicist	*and*; M or Ph.D. in P, MP, or PS, E, AM	*and* 2 years under supervision by CMP, *or*, …	*and* Passes			
or (2) Authorized Medical Physicist	*and*; M or Ph.D. in P, MP, or PS, E, AM	*and* 2 yrs in CRF under supervision by AU eligible physician	*and* Passes			
or (3) Authorized Medical Physicist	*and*; M or Ph.D. in P, MP, or PS, E, AM	*and* 1 year full-time training in MP and 1 year full-time experience by AMP eligible MP			*and* has written attestation of "competency and independency" by MP preceptor	*and* training in device operation, clinical use, and treatment planning systems
(1) Authorized Nuclear Pharmacist	Pharmacy; or, passed FPGEC exam	4000 h in nuclear pharmacy	*and* Passes			Current, active license
or (2) Authorized Nuclear Pharmacist				700 h in structured program with 200 h in topical areas	*and* has written attestation of "competency and independency" by preceptor ANP	

AM = Applied mathematics; ANP = Authorized nuclear pharmacist; PS = Physical science; CMP = Certified medical physicist; AMP = Authorized medical physicist; RS = Radiation safety; FPGEC = Frgn pharm.grad exam comm.; AU = Authorized user; CRF = Clinical radiation facility; E = Engineering; P = Physics; MP = Medical physicist; M = Master's degree.

Licensure. There are two categories of NRC licenses: General License and Specific License. General licenses have been issued for non-medical uses, such as fixed gauges containing sealed radioactive sources. Medical licenses are for specific uses of a licensed material in a medical program, for example, diagnostic nuclear medicine program. Specific licenses control manufacture, production, acquisition, receipt, possession, preparation, use, and transfer of byproduct material for medical uses. A Type A license of broad scope, often held by university medical facilities, exempts the licensee from certain requirements of a specific license, but requires the facility to assume responsibility by certain administrative processes for the radiation protection program. NRC license requirements, applications, renewals, amendments, notifications, exemptions, and issuances are described in *10CFR 35.11–19* (2). Licenses categories exists for the use of unsealed byproduct material for uptake, dilution, excretion studies without a written directive (§35.100), unsealed byproduct material for imaging and localization studies without a written directive (§35.200), unsealed byproduct material requiring a written directive (§35.300), manual brachytherapy sources (§35.400), sealed sources in teletherapy units, and stereo tactic radiosurgery units (§35.600), and for other uses of byproduct materials (§35.1000).

Some components of 10CFR 35 (L) (Record retentions) (Table 13). Table 13 summarize the duration (for license, for program, and for 5 and 3 years) requirements for the retention of records.

Some components of 10CFR35 (M) (Reports …Medical Events … Sources) (Table 14). The term *misadministration* is replaced with the term *medical event* (ME). The ME depends, in some cases, on the *difference* (presumably lower or higher) in delivered dose and prescribed dose (PD), and in other cases, in *exceeding* the PD. Moreover, the definitions are not in medical physics terms of absorbed dose in gray (Gy); rather, they are in health physics terms of effective dose equivalent (EDE), shallow dose equivalent (SDE), in sievert (Sv). Recall that in partial organ irradiation in health physics, organ or tissue weighting factors apply in calculating DE. As a brachytherapy ME will likely involve adjacent organs, some judgment may be required in deciding on the correct DE in an ME.

Table 14 summarizes the reporting of medical events; Reporting requirements are similar to pre-2002 regulations.

Transport

Every day thousands of packages containing radioactive material move via public transportation routes—roads, airplane, and railway. Of all the hazardous material shipments, it is estimated that ~1%, nearly 3 million packages annually, involve radioactive materials (17). These packages are needed for medicine, industry, and research.

Shipments can be made only to persons who are licensed by the Nuclear Regulatory Commission (NRC) or appropriate Agreement State to receive radioactive materials.

Table 11. Some Training Requirements for Physicians Using Sealed Sources and Medical Devices

Person	Certification Examination	Education	Experience	Laboratory Training	Preceptor Statement	Authorized User
(1) Physician (Manual brachytherapy and sources), *or,*	Passes examination by medical specialty board	3 year MR in Rad Onc				
(2) Physician (Manual brachytherapy and sources)		Structured educational program with 200 h topical classroom and laboratory and 500 h work experience	3 years clinical supervision by AU in Rad Onc		*and* has AU preceptor's written attestation of competency sufficient to function independently	
(1) Physician (Ophthalmic use Sr-90)		Active practice and 24 h classroom and laboratory training applicable to medical use of Sr-90,	*and* AU supervised clinical training		*and* has AU preceptor written certification of completed requirements *and* attestation of competency sufficient to function independently	
(1) Physician, dentist, or podiatrist (Sealed sources for diagnosis), *or,*	Passes examination by medical specialty board					
(2) Physician (Sealed sources for diagnosis)		8 h classroom and laboratory training applicable to medical use of Sr-90,		Has completed training in use of device for uses requested		
(1) Physician (RA, T, GSR units, TMD), *or*	Passes examination by medical specialty board	3-year MR in Rad Onc				
(2) Physician (RA, T, GSR units, TMD)		Structured educational program with 200 h topical classroom and laboratory and 500 h work experience	3 year clinical supervision by AU in Rad Onc		*and* has AU preceptor written certification of completed requirements *and* attestation of competency sufficient to function independently	

RA = Remote afterloader; MR = Medical residency; T = Teletherapy unit; AU = Authorized user; GSR = Gamma stereotactic radiosurgery Unit; TM = Therapeutic medical device

The shipment must be made in accordance with procedures established by the recipient. Prior to shipping radioactive materials, a copy of the recipient's radioactive materials license should be on file with the shipper's Radiation Safety Office to document what radionuclides, forms, and quantities the recipient is authorized to receive.

There are five categories of radioactive material packages. Development of the technical criteria for each packaging category is correlated to certain general and performance requirements. The categories include (1) excepted or limited quantity packaging; (2) type A packaging; (3) type B packaging; (4) industrial packaging; (5) fissile material packaging. All medical shipments occur in the first two categories. Figure 2 illustrates the "spectrum" of increasing package hazard with activity.

Both the Department of Transportation (DOT) and the Nuclear Regulatory Commission are responsible for the regulations governing a package containing hazardous materials that are intended for transport on public routes(18). The DOT regulations are found in 49 CFR 107, 172-178. The NRC regulations are found in Title 10 CFR Part 71. In 1979, the DOT and NRC agreed to a Memorandum of Understanding under which the DOT regulates Type A packages and below, carriers, and has authority for international shipments. The NRC regulates Type B and fissile packages, investigates incidents and accidents, and provides technical advise to DOT.

The transportation requirements were revised in October 2004 to bring U. S. standards into consistency with the latest international transportation safety regulations (19). The

Table 12. Some Training Requirements for Physicians Use of Radiopharmaceuticals

Person	Certification Examination	Education	Experience
(1) Physician (Uptake, Dilution, and Excretion Studies), *or,*	Passes examination by medical specialty board	Satisfies board education requirement	
(2) Physician (Uptake, Dilution, and Excretion Studies), *or,*		40 h topical classroom and laboratory	*and,* 20 h clinical supervised by AU
(3) Physician (Uptake, Dilution, and Excretion Studies)		Successfully completed 6 month NM training	
(1) Physician (Imaging and Localization Studies), *or,*	Passes examination by medical specialty board	Satisfies board education requirement	
(2) Physician (Imaging and Localization Studies), *or*		200 h classroom and laboratory training applicable to medical use and 500 h supervised work	*and* 500 h AU supervised clinical training
(3) Physician (Imaging and Localization Studies)		Successfully completed 6 month NM training	
(1) Physician (Therapy use Unsealed Byproduct Material), *or,*	Passes examination by medical specialty board	Satisfies board education requirement	
(2) Physician (Therapy use Unsealed Byproduct Material)		80 h topical classroom and laboratory	*and,* clinical supervision by AU for specific number of cases
Physician (Only I-131 for Hyperthyroidism, Thyroid Ca)		Special experience and 80 h classroom and laboratory training	*and,* clinical supervision by AU for specific number of cases
(1) Physician (Sealed Sources for Diagnosis), *or,*	Passes examination by medical specialty board	Satisfies board education requirement	
(2) Physician (Sealed Sources for Diagnosis)		8 h classroom and laboratory training	

international regulations follow the International Atomic Energy Agency (IAEA) report Safety Series ST-1-R, which most foreign countries have adopted (20). This is important because most radioactive materials for medical use are produced outside U.S. borders, for example sealed sources, ^{99}Mo/^{99m}Tc generators.

There are four essential elements that are the shipper's responsibility to properly providing packages for transport that contain radioactive, or any other hazardous, materials. These are proper containment, labeling/marking, documentation, and training. The major factors affecting these requirements for these elements are the radionuclide, physical form, and quantity (activity). The specific requirements for packaging containment, labeling, and documentation are in the relevant sections of 49CFR 172-177. This information can be found at the website http://hazmat.dot.gov.

Table 13. Some Components of 10CFR 35 (L) (Record Retentions)

Record Retention Requirement	Section
Duration of license	0.2024/RPP (b) RSO authority
Duration of program (device)	0.2610/Safety procedures for device
5 years	0.2041/Procedures for WP; 0.2026/RPP changes
3 years	0.2040/WDs;.2061/Meter calibrations; 0.2067/Leak tests and inventories; 0.2070/Surveys; 0.2075/Patient release; 0.2080/Mobile services; 0.2092/Decay in storage; 0.2310/Safety instructions; 0.2404/Implants and source removals;0.2406/Source accountability; 0.2432/Source calibrations; 0.2433/Sr-90 decays; 0.2605/RAU installation, repairs;0.2632/Full calibrations; 0.2643/Spot checks; 0.2647/Additional mobile records;

Table 14. Some Components of 10 CFR35 (M)[a]

Section	Major Contents of Section
0.3045/Report/notification medical event (excluding patient intervention) (1)	Dose differs from PD > 0.05 Sv EDE, 0.5 Sv organ/tissue and SDE skin, and, TD, *and*, TD delivered differs from PD by $+20\%$ or falls outside PD range; or single fraction delivered dose differs from single fraction PD $+50\%$
0.3045/Report/notification medical event (excluding patient intervention) (2)	Dose *exceeds* 0.05 Sv EDE, 0.5 Sv organ/tissue and SDE skin, and, TD from wrong: (a) byproduct material; (b) administration route; (c) person; (d) treatment mode; (e) leaking source
0.3045/Report/notification medical event (excluding patient intervention) (3)	Excluding migrating permanent implant seeds, dose to skin/organ/tissue *other* than treatment site that exceeds 0.5 Sv organ/tissue and $+50\%$ dose expected from WD
0.3045/Report/notification medical event (excluding patient intervention) (3) (b)	Report any patient interventions producing permanent/physiological damage
0.3045/Report/notification medical event (excluding patient intervention) (3) (c, d)	Notify NRC next calendar day after ME with written report in 15 days; notify referring MD and patient unless referring MD chooses not to for medical reasons; details of reports omitted here
0.3067/Report leaking source	Report > 5 nCi removal contamination within 5 days

[a]Reports ...Medical Events ... Sources.

THE TRANSPORT PACKAGE ACTIVITY SPECTRUM
(With Packaging References)

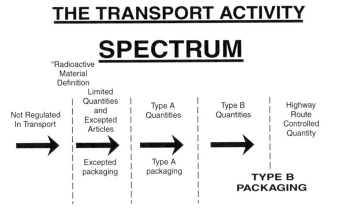

Figure 2. The "spectrum" of increasing radioactive package hazard with activity.

All shipments of radioactive material, with the exception of those containing very small, limited quantities must have labels bearing the word "Radioactive" and affixed to opposite sides of the outer package. There are three different labels: White-I, Yellow-II, or Yellow-III, as shown in Fig. 3. The criteria for the three labels are given in Table 15.

Training must be provided for those that prepare for transport, transport, or receive packages of hazardous materials. The training must be commensurate with the duties involved. For medical facilities, the training involves the proper receipt of radioactive packages and preparing packages for return to the vendor or manufacturer. For shippers, the training must be done at least every three years and be certified by the employer.

The receipt of labeled radioactive packages must be handled according to the procedures in NRC regulations (10 CFR 20.1906). This requires assessing radiation levels and removable contamination within 3 h of taking possession. Examples of returned packages are residual radiopharmaceuticals in syringes or vials from nuclear medicine studies or sealed sources after radiation therapy use. Any returned package must be prepared for transport in accordance with DOT requirements.

It is important to note that *anyone shipping radioactive materials must receive training from an approved program beforehand!*

Figure 3. Labels for radioactive packages based on the activity contents and radiation levels outside as given in Table 15.

Table 15. Shipping Label Criteria[a]

Label	Transportation Index, TI	Maximum Radiation Level on Surface, X
White–I	$TI < 0.05$	$X < 5\,\mu Sv \cdot h^{-1}$
Yellow–II	$0.05 \leq TI < 1$	$5\,\mu Sv \cdot h^{-1} \leq X < 500\,\mu Sv \cdot h^{-1}$
Yellow–III	$1 \leq TI < 10$	$500\,\mu Sv \cdot h^{-1} \leq X < 2\,Sv \cdot h^{-1}$
Yellow–III exclusive use of vehicle	$TI \geq 10$	$2\,mSv \cdot h^{-1} \leq X < 10\,mSv \cdot h^{-1}$

[a]Transportation index $= 100$ the maximum reading in $mSv \cdot h^{-1}$ at 1 m from the surface.

Disposal of Radioactive Material

Radioactive materials used in medicine can be solid, liquid, or gaseous. Some solids are specially encapsulated and called sealed sources. When the material is no longer useful or in a form or presence that is undesirable, the radioactive material is considered waste, and the licensee must dispose of it. All waste generated from medical use is categorized as low level radioactive waste. Depending on various factors, radioactive waste can be disposed by (1) decay-in-storage (DIS); (2). discharge into the environment (3) transfer for land burial; (4) return to the vendor/manufacturer.

Disposal by Decay-in-Storage. This is the dominant disposal method for radioactivity used in nuclear medicine. The container of radioactivity is stored and simply allowed to radioactively decay to background level. Currently, this disposal method is only available for radionuclides with a physical half-life <120 days. Depending on the size of the operations, the limiting factor with this method is adequate space for the waste volumes generated during the time that the waste is segregated to "decay". Shielding of the container(s) may be another consideration depending on the radiation levels involved from the waste during radioactive decay. Procedures must be developed to comply with 10 CFR 35.92 for the decay-in-storage (DIS) of waste. Release into the general medical waste stream requires that the radiation level be at background level when measured at the surface of the unshielded waste container. The survey meter or instrument used to measure the radiation level must be capable of detecting the radiation being emitted from the radionuclide(s) being stored.

Disposal into the Environment. In some circumstances, especially with medical research, the disposal of liquids into the sanitary sewer or by evaporative release to the atmosphere or the discharge of volatile gases may intentionally occur. This is permissible and safe providing that compliance is maintained with other regulations regarding toxic or hazardous properties of these materials.

Disposal in the sanitary sewer or into the atmosphere must comply with 10 CFR 20, Subpart K. The NRC release or discharge limits have a concentration and annual aggregate activity limit. Records of each release is kept, such that a periodic (at least annual) assessment can be performed to confirm compliance with release limits required in Part 20. These limits apply at the facility boundary of the radioactive material licensee. The release limits are radionuclide specific such that exposure or intake would not exceed the applicable occupational or general public dose limits. The current release limits are derived from the scientific basis presented in 1990 by the ICRP (21). Patient excreta containing readily dispersible forms of radioactivity are exempt from these release limits.

Waste from *in vitro* laboratory kits that use radioactive materials under a general licensed pursuant to 10 CFR 31.11 is exempt from waste disposal regulations. Radioactive labels must be defaced or removed, but there is no requirement to keep a record of release or make any measurement. A standard of good practice is to do a radiation survey of any general waste from an area where radioactive materials are used, such as a nuclear medicine radiopharmacy or an inpatient therapy room, to confirm that it is at background levels before release into the general waste.

Transfer for Burial. For some medical facilities, radioactive material may require disposal by transfer to a burial site because the volume of waste generated requires removal or the waste contains long-lived (>120 day half-life) items that cannot be decayed in storage. Medical facilities use a broker licensed by the NRC or Agreement State to receive the material. Packaging will follow instructions received from the broker and the burial site operator. Records of the transfer to the broker must be maintained to comply with 10 CFR 20. Because this is the most expensive means of disposal, most generators of waste also employ volume reduction, (e.g. compaction) to reduce costs.

At the time of writing, there are only three burial sites for low level radioactive waste in the United States Richland, Washington, Barnwell, South Carolina, and Tooele, Utah. All are commercially operated and regulated by the respective state. The facilities are designed, constructed, and operated to meet safety standards. The operator of the facility must also extensively characterize the site on which the facility is located and analyze how the facility will perform for thousands of years into the future. In 1985, the Low-level Radioactive Waste Policy Amendments Act gave the states responsibility for the disposal of their low-level radioactive waste by encouraging the states to enter into compacts that would allow them to dispose of waste at a common disposal facility. While most states have entered into compacts, but no new disposal facilities have been built since the Act was passed, or are any expected to be.

Since the 1985, the volume of medical low level radioactive waste shipped for burial has dropped dramatically because of the cost of disposal, employment of volume reduction methodologies, and the conversion to short half-life or nonradioactive agents.

Return Sources to the Vendor or Manufacturer. For solid or sealed sources especially, a viable means of disposal for a medial facility is return to the vendor or a manufacturer. This is common with items that have exceeded their useful activity or shelf-life, such as $^{99}Mo/^{99m}Tc$ generators or brachytherapy sealed sources (e.g., ^{192}Ir, ^{125}I) or quality control calibration sources (e.g., ^{57}Co, ^{153}Gd). For such package, the packaging, labeling, and surveys must comply with the instructions of the vendor/manufacturer and 10 CFR 71 (NRC) and 49 CFR 173 (DOT) regulations. Currently, there is no distinct time at which a sealed source might be considered waste. The licensee determines when a material is no longer usable and becomes considered part of the radioactive waste stream. For solid sources with >120 day half-life, because land burial is very expensive, many licensees choose to simply "store" sources under their control. Such sources require routine inventory and periodic leak-testing. The current standards for stored, *unused* sealed sources require inventory every 6 months and leak test within 10 years.

COMMUNICATIONS FROM THE NRC

The NRC issues to licensee's bulletins, directives, guidance's, information notices, newsletters, and regulatory summaries as new issues not covered in regulations arise and must be addressed. In some cases, these documents endure for many years, and may actually be incorporated by agreement states into their regulatory statutes.

Bulletins

Bulletins provide information to NRC licensees. Apparently there are no recent bulletins pertaining to medical Applications; the last one was in 1997 (22).

Directives

Directives appear in several forms. FC86-4, Revision 1— *Information Required for Licensing Remote Afterloading Devices*, a long-standing (1986) policy and guidance directive, explained the contents for NRC license applications for RAUs. While it is not current on the NRC web site, some states have adopted it, with some changes, into their licensing process for RAUs. FC83-20, Revision 2-*Facility Interlocks and Safety Devices for High, Medium, and Pulsed Dose-Rate Afterloading Units*, is not on the NRC web site. As the title implies, this release clarified the requirements for interlocks and safety devices. It appears that issues raised are addressed in the 2002 10CFR 35 revisions.

Guidances

Guidance's often discuss evolving technologies. For example, as intravascular brachytherapy developed, the NRC issued several guidance documents (23,24). These were necessary as the new 10CFR35 applies only to photon-emitting RAUs; beta-emitting RAUs fall into the emergent technology category evaluated on a case-by-case basis.

Information Notices

Information Notices advise licenses of recent concerns usually arising from medical events reported to the NRC. A recent notice discussed failures of HDR RAUs (25).

Newsletters

Newsletters, notable, Nuclear Materials Safety and Safeguards (NMSS), announce medical events and enforcement actions against those who violate regulations. A recent one reported on a hospital's failure "...to secure...licensed material..." (26).

Regulatory Summaries

Regulatory Summaries often clarify issues about the interpretation of regulations, such as the calibration measurements for brachytherapy sources (4).

Recent NRC Activities

Recent or current NRC activities are posted on the website www.nrc.gov. For those interested in commenting on proposed NRC regulations, a site, www.ruleform.llnl.gov, is available.

REGULATORY STANDARDS : OSHA

Tables 16–18 offer a limited synopsis of the major components of the OSHA regulations. As noted earlier, current enforceable OSHA regulations, *29 CFR 1910.1096 (Ionizing Radiation)*, dating from the 1970s, are now at variance with the recent NRC regulations.

A recent supporting statement (Fed Register [07/23/2004]) for information-collection requirement offers some insight into OSHA regulation terms, definitions, and their application.

As with the NRC, over the years OSHA has issued Directives, Standard Interpretations, and Compliance Letters regarding regulations. They are available on the website www.osha.gov.

While a few cover general radiation topics, most relate to non-medical (nuclear power plant) radiation issues. There appear to be no releases within the last 5 years that relate to medical uses of radiation under OSHA standards.

NONBYPRODUCT MATERIALS AND MACHINE-PRODUCED RADIATION

As noted above, the NRC was authorized only to oversee the use of fissile and byproduct material. Regulation of naturally occurring or accelerator produced radionuclides, or of radiation from machines fell to the individual states. Since every state develops their own regulations, the depth and coverage of those regulations very widely. Often, states with smaller populations and small non-federal radionuclide programs tended to have less complete or in-depth regulations than states with larger populations and programs. Two developments have been working to change the wide variations in regulations between states.

Table 16. Partial Contents of 29 CFR 1910.1096(a), (b), and (c) of OSHA Regulations[a]

Section	Major Contents of Section
(a) Definitions—Radiation and areas	(1) Radiation; (2) Radioactive materials; (3) Restricted area; (4) unrestricted areas;
(a) Definitions—Quantities and equivalencies	(5) Dose; (6) Rad; (7) Rem; (1 R X or γ- ray; 1 rad X- or γ- ray or beta particle; 0.1 rad high energy proton; (8) Air dose
(a) Definitions—Neutron flux or equivalent	Neutron flux dose equivalency table
(b) (1) Exposure to employed individuals 18 years age or older in restricted areas (Rem/calendar quarter)	Whole body; Head and truck; active blood forming organs; eye lens, gonads: 1.25
	Hands and forearms; feet and ankles: 18.75 Skin of whole body: 7.5
(b)(2) Greater quarterly whole body doses allowed based on individual's age"N"	Whole body dose shall not exceed 3 rem per quarter *and* shall not exceed 5(N-18)
(b) (3) Exposure to employed individuals *under* 18 years age in restricted areas (Rem/calendar quarter)	Quarterly calendar dose limited to 10% of that allowed those 18 years of age
(c) Exposure of employed individuals 18 years age or older to airborne radioactive material in restricted areas shall not exceed	Limits in 10CFR Part 20 Table I, Ax.B (1971); for 40 h workweeks, 7 consecutive days; time proportionately applicable
(c) Exposure of employed individuals *under* 18 years age to airborne radioactive material in restricted areas shall not exceed	Limits in 10CFR Part 20 Table II, Ax.B (1971); for 40 h workweeks, 7 consecutive days; time proportionately applicable

[a]The use of conventional (old) units in this table reflects the fact that these regulations are outdated and lag behind the NRC regulations.

Agreement State Status

The first unifying factor is the growing trend toward agreement state status. An agreement state enters into an agreement with the NRC to take over for the NRC regulation and control of byproduct material. Before doing so, the state must demonstrate that the state regulations for byproduct material are compatible with those of the NRC. "Compatibility" varies based on guidelines from the NRC as to how important the NRC feels that the state regulations agree with the federal, according to the following scale (27):

A Basic radiation protection standard or related definitions, signs, labels or terms necessary for a common understanding of radiation protection principles. The State program element should be essentially identical to that of NRC;

B Program element with significant direct transboundary implications. The State program element should be essentially identical to that of NRC;

C Program element, the essential objectives of which should be adopted by the State to avoid conflicts, duplications or gaps. The manner in which the essential objectives are addressed need not be the same as NRC, provided the essential objectives are met;

D Not required for purposes of compatibility.

For example, occupational exposure limits fall under category A, requiring congruence between the state and federal regulations. On the other extreme, most application and recording regulations are left to the states' discretion. For the most part, the laxer categories are those with less impact. Thus, as states have adopted agreement status, the variation in regulations between states has decreased. Table 4 lists the agreement states as of 2005.

Conference of Radiation Control Program Directors

Established in 1968, the Conference of Radiation Control Program Directors (CRCPD) is an organization of representatives of state radiation control programs. The organization shares information useful to state radiation control agencies, and has educational meetings focused on topics of current interest to state regulators. The CRCPD also distributes to its members model state regulations, so when states revamp their respective radiation safety codes, they need not start from nothing(28). The contents of the model regulations are discussed below. Because many state agencies use these models as a guide for their radiation regulations, increasingly the various states' regulations have been converging. Still, many important aspects of regulations remain, for example, the allowed radiation limit to the general public. While most states follow the federal rules, some use more restrictive levels based (sometimes erroneously) on recommendations of the ICRP.

CRCPD Model Regulations

Since the CRCPD model program serves as the basis for many of the state rules, we will consider the provisions here for regulations dealing with ionizing radiation not from byproduct material. Because of the compatibility

Table 17. Partial Contents of 29 CFR 1910.1096(d) and (e) of OSHA Regulations

Section	Major Contents of Section
(d) (1) Definition of a survey	"An evaluation …radiation hazards…production, use, release, disposal, or presence …radioactive material or …radiation…"
(d) (2) Employer responsibility for monitors	"…shall provide…personnel monitoring equipment…"
(d) (2) (i) 18 year age or older employee use of monitors in restricted areas;	"…employeee …restricted area likely to receive a quarterly dose > 25% that in (b)(1); or, enters a high radiation area
(d) (2) (ii) *Under* 18 year age employee use of monitors in restricted areas	"…employeee …restricted area likely to receive a quarterly dose > 5% that in (b)(1)
(d) (3) Personnel monitoring equipment	"e.g., film badges & rings, pocket chambers and dosimeters
(d) (3) Area definitions	Radiation area…could receive > 5 mrem in any 1 h; or, > 100 mrem in 5 consecutive days; High radiation area… could receive > 100 mrem in any 1 h; Airborne radioactivity area…concentrations in excess 10CFR Part 20 Table I, column 1, Ax.B (1971)
(e) Caution signs, labels, signals	Radiation symbol(trefoil) described;
(e) (2) Radiation area posting	Radiation caution symbol with "Caution–Radiation Area"
(e) (3) (i) High radiation area posting	Radiation caution symbol with "Caution–High Radiation Area"
(e) (3) (ii) High radiation area control	"…equipped with control device …cause radiation levels to be reduced < 100 mrem in 1 h, or, …energize …alarm system… individual entering …supervisor…made aware of entry."
(e) (4) Airborne radioactivity area posting	Radiation caution symbol and "Caution–Airborne Radioactivity Area"
(e) (5) (i) Radioactive materials posting (excluding natural uranium or thorium)	Areas/rooms > 10 times quantities in 10CFR Part 20 Apx.C (1971)
(e) (5) (ii) Radioactive materials posting for natural uranium or thorium	Areas/rooms > 100 times quantities in 10CFR Part 20 (1971)
(e) (6) (i) Container labeling (excluding natural uranium or thorium)	Containers …transported, stored, used…> quantities in 10CFR Part 20 Apx.C (1971)…Radiation symbol and "Caution-Radioactive Materials"
(e) (6) (ii) Container labeling for natural uranium or thorium	Containers …transported, stored, used … > 10 times quantities in 10CFR Part 20 Apx.C (1971)…Radiation symbol and "Caution-Radioactive Materials"

The use of conventional (old) units in this table reflects the fact that these regulations are outdated and lag behind the NRC regulations.

requirement to become an agreement state, those parts of the CRCPD model regulations that deal with material under NRC oversight follow the federal rules as discussed above. Thus, these need not be considered here. The FDA does impose some requirements on the *manufacturers* of radioactive materials and radiation-producing machines intended for human use, but that leaves the *use* of machine-produced radiation and naturally occurring and accelerator-produced radionuclides only under the control of individual states.

The model regulations fall into many sections, with each section covering a particular part of radiation safety. General rules that apply to all applications and follow the NRC notably Parts 19 and 20 come in sections in the beginning. In addition to the general provisions, each of the parts that deal with particular applications all have sections addressing shielding and survey requirements for the modality (such that the radiation levels satisfy Part 20 limits), safety requirements for operation (such as door interlocks to prevent walking in during irradiation), ventilation if airborne radionuclide production is possible, record retention requirements and training and experience.

Machine-Produced Radiation

While much of machine-produced radiation is covered by state regulations, when used on humans applications manufacture of the units falls under the auspices of the FDA. The FDA rules can be found in 21 CFR 1020. For the most part, the state regulations follow the FDA guidances when applicable, but sometimes with a sizable delay.

Mammography forms a notable exception to the general lack of federal control over machine-produced radiation in medicine. Based on the MQSA, as noted above, the FDA sets requirements for practitioners on mammography, and failure to satisfy the requirements prevents providers from obtaining reimbursement from government sources. The requirements for mammography equipment are given below in the section on Diagnostic Units. In addition, there are considerable requirements placed on the training and experience of the persons involved: the radiologist, the radiographer, and the medical physicist [21 CFR 900.12 (a)].

Radiation producing machines fall into three main categories discussed in the following sections.

Table 18. Partial Contents of 29 CFR 1910.1096(f), (g), (h), (i), (j), and (k) of OSHA Regulations[a]

Section	Major Contents of Section
(f) Immediate evacuation warning signal	34 subsections regarding the signal characteristics, design, and testing requirements
(g) (i) Exceptions from posting requirements for sealed sources	Room/area with sealed source...not required...if radiation levels $< 5\,\mathrm{mrem \cdot h^{-1}}$ at 12 in from source container/housing
(g) (ii) Exceptions from posting requirements for rooms housing radioactivity patients	Rooms...not required to be posted...personnel in attendance who shall ...prevent individual exposure above limits
(g) (iii) Exceptions from posting requirements for rooms containing radioactive materials	Cautions signs not required for rooms containing radioactive materials for $< 8\,\mathrm{h}$ and provided materials constantly attended...
(h) Exemptions for radioactive materials packaged for shipment	Radioactive materials packaged and labeled per DOT 49CFR Chp. I are exempt provided containers inside properly labeled
(i) (2) Instruction of personnel, postings	Individuals working in or frequenting any portion of a radiation area shall be informed of radioactive materials and radiation; instructed in safety...; instructed in applicable provisions of regulations...; advised of radiation exposure reports
(i) (3) Posing regulations and operating procedures	Employer...shall post...current copy of regulations and operating procedures
(j) Storage of radioactive materials	...shall be secured against unauthorized removal....
(k) Waste disposal	...by transfer to an authorized recipient...
(l) (i) Notification (immediate) of incidents	...any incident involving radiation which may have caused ...$> 25\,\mathrm{rem}$ whole body, 150 rem skin, or 375 rem to feet, ankles, hands, or forearms, or, release of radioactive materials > 5000 applicable limits averaged over 24 h
(l) (ii) 2Notification (24 h) of incidents	...any individual ...5 rem or more total body; 30 rem skin, 75 rem to feet, ankles, hands, forearms,
(m) Reports of overexposure and excessive levels and concentrations	....written report in 30 days...to OSHA; notification of individual exposed
(n) Records	Advise employees of annual exposures; provide employees exposure records
(p) Definitions of agreement states	List of current agreement states

[a]The use of conventional (old) units in this table reflects the fact that these regulations are outdated and lag behind the NRC regulations.

Radiotherapy Units. Radiotherapy units consist of two major categories: orthovoltage X-ray units (i.e., conventional X-ray machines that treat with bremsstrahlung beams produced with tube potentials from 10 kVp to 300 kVcp) and those from accelerators (from electron beams with energies from 2 to 45 MeV). The regulations use as a diving line between the modalities a photon beam energy of 500 kV, which clearly delineates units since no machines currently in use run close to that specification. Table 19 lists the requirements for an orthovoltage unit, and Table 20 those for an accelerator.

Regardless of the machine type, the regulations require the output of the unit be determined using dosimeters calibrated at either the National Institute of Standards and Technology or at one of the Accredited Radiation Dosimetry Calibration Laboratories. The calibration procedure must follow a protocol established by a recognized national professional society. Also for either type of unit (except for contact therapy units), the facility design requires: the ability to monitor the patient aurally and visually; interlocks on the door to prevent entry during irradiation; beam-on indicators; and emergency power cutoffs by the control panel or door.

Radiography (Imaging) Units. Regulations for diagnostic radiographic units actually exceed those for the therapy units, even though the latter produce much greater quan-

tities of radiation. Tables 21a,b and 22 give *highlights* of the regulations for radiographic and fluoroscopic units. The regulations also contain many points on how the specifications should be measured as well as cover other aspects not included in the tables. Table 21b gives values referred to in Table 21a. As an important factor in patient dose, the regulations also address exposure control for the various types of equipment.

In addition to the regulations for the radiographic and fluoroscopic units, there are also sections on radiotherapy simulators; computed tomography units; mammography units; mobile units; and veterinary units.

As noted above, mammography units have special requirements according to the MQSA. The requirements for these units are given in Table 23, and the special quality assurance requirements in Table 24. The quality assurance summary greatly simplifies the actual requirements, which have undergone some modifications to adapt better to various imaging systems and practice conditions. All persons involved in mammography, including the radiologist, radiographer and the medical physicist performing the quality measurements, must satisfy specific training and experience guidelines, as well as continuing education requirements.

Nonmedical Radiation-Producing Equipment. Nonmedical radiation producing equipment actually finds its way

Table 19. Requirements for Orthovoltage X-Ray Units

Leakage Radiation [air kerma rates]	< 1 mGy·h^{-1} 5 cm from housing
5–50 kV Systems	< 10 mGy·h^{-1} 1 m from target;
> 50 and < 500 kV Systems	< 300 mGy·h^{-1} 5 cm from housing
Permanent Beam Limiting Devices	Same attenuation as housing
Adjustable or Removable Beam Limiting Devices	Transmission $< 5\%$ of useful beam Opening indicated by light beam
Beam Filter System	Cannot be displaced Interlocked to prevent beam use with filter absent Slot provides same shielding as housing Filters clearly identified
Tube Immobilization	Cannot move when locked
Source Marking	Indicated to within 5 mm
Contact units beam blocking	Equivalent to 0.5 mm Pb at 100 kV
Timer	Unit has presetable timer and show elapsed or remaining time Retains reading with interruptions Terminates exposure after set time Precision of at least 1 s or 1% Prevents exposures with zero time Begins with shutter or is compensated for lags
Control Panel Functions	Displays indicate ac power, X-rays possible, X-rays on, shutter condition and tube potential and filter Termination button Locking device
Multiple Tubes	Only one used at a time Indication of which is in use
Target-to-Skin Distance	Accurate to within 1 cm, reproducible to within 2 mm
Shutters	Required if beam takes > 5 s to come on
Low Filtration X-ray Tubes	Permanent warning label

into medical application, for example, as cyclotrons making radioactive materials for imaging or analytic X-ray units to assess kidney stones. Much of the operation of such equipment would be covered by the general radiation safety provision of the regulations. Most of the additional rules deal with preventing the accidental irradiation of a person in a high radiation area.

Particle Accelerators (e.g., Cyclotrons). The main concern for a particle accelerator would be a staff member being in either the accelerator room or one of the rooms served by the beam lines. To prevent such occurrences, the rules require: interlocks on doors to prevent accidental entry with the beam on or inhibit the beam initiation with the door open; buttons to stop the beam from within the room; radiation-detector warning devices in the room; and handheld Geiger counters carried when entering the room. The regulations also include the requirement for periodic testing of the safety devices to assure proper function.

Analytic X-Ray Units. The X-ray units covered under this heading usually are small devices (often fitting on a desktop), used for analysis of small samples, such as for crystallography or pathologic X rays of surgical samples. These devices are usually enclosed within a shielding box. While small, accidents that involve an operator's hand being in the box during beam production frequently lead to loss of fingers or hands. Thus, similarly to the particle accelerator, the rules try to keep hands out with the beam

on, or prevent the beam if the doors are open. Rules include interlocks to prevent beam with doors open; warning lights indicating the status of the beam and shutters; and warning labels.

Nonbyproduct Radionuclides

State regulation of byproduct material must follow closely the NRC regulations. However, before the 2005 agreement, the states have been responsible originating their own regulations for NARM. Much of the suggested regulations (Part C) define quantities of NARM below regulatory concern. Table 31 gives a brief, and not nearly complete listing of some exemptions as examples. Appendix A of Part C of the suggested regulations gives air and water concentrations exemptions.

The regulations go on to exempt devices such as static eliminators containing less than specified amounts of radioactive materials (on the order of 20 MBq for heavy nuclides of 2 MBq for tritium). For clinical laboratories, small quantities of material (generally ~ 0.4 MBq except tritium at 1.85 MBq and ^{59}Fe at 0.7 MBq) used in assay kits and 1.85 MBq check sources are also exempt. The remainder of Part C addresses licensing and labeling. Medical use of radioactive materials is covered under Part G, which mostly mirrors the federal 10CFR35.

What is not clearly addressed in the model regulations is regulation on accelerator-produced radioactive materials. Some states have taken the tack that the same regulations

Table 20. Requirements for a Radiotherapy Accelerator

Leakage Radiation	Maximum $< 0.2\%$, average $< 0.1\%$ useful beam -2 m radius of central ray at isocenter $< 0.5\%$ 1 m from electron path Neutron dose compliant with IEC standard
Collimator leakage	$< 2\%$ of useful beam for photon beams Maximum $< 2\%$, average $< 0.5\%$ useful beam for electron beams, outside 7 cm of beam $< 10\%$ 2 cm outside of field
Filters/Wedges	Identification clearly marked Interlocked requiring selection Panel indicates wedge identification
Stray Radiation	Compliant with IEC standards
Beam Monitors	Redundant independent systems required Both systems show on control panel until reset Retrievable in case of power failures Count up
Beam Symmetry Monitor	Can detect asymmetry $> 10\%$ Terminates beam with asymmetries $> 10\%$
Beam Control	Beam initiation requires monitor setting Preset displayed Reinitiation requires clearing of setting Monitor unit rate is displayed Provides termination for excess dose rate
Termination of beam	By monitor systems at respective preset Manually at panel By timer after preset time with reset necessary
Radiation selection	Type of radiation must be selected (if more than one available) Interlocks prevent simultaneous types Type displayed on panel Interlocks prevent inappropriate beam type and accessories Special mode allows X-rays for imaging with electron applicators
Energy Selection	Energy selection required Energy displayed on panel Interlock prevents beam without appropriate mechanical conditions
Stationary or moving beam	Selection required Mode indicated on panel Interlocks prevent beam in inappropriate condition
Moving beams	Beam controlled for dose per degree Interlocks stop beam if dose per degree off

Table 21a. Highlights of Requirements for All diagnostic X-Ray Units[a–c]

Warning label	Attached to the control panel containing main power switch
Battery charge indicator	Visual on control panel if relevant
Source leakage radiation	< 1 mGy/m^2 at maximum technique for 1 h
Radiation other than tube	< 20 µGy/h 5 cm
Half-value layer[d]	$>$ values in Table 2dx including all material between tube and patient For variable filter units, control prevents incorrect selection
Multiple tubes	Selection of tube clearly indicated
Mechanical Support of the tube head	Hear remains stable during exposure (except dynamic studies)
Technique indication	Technique factors shown before exposure
Locks	Function properly

[a]Table by Tim Burns and Mark Geurts.
[b]For new units. Older units have some allowances made for regulations in effect at manufacture.
[c]Details for such units should be found in 21CFR1020 or in the particular state's regulations.
[d]Based in FDA regulations in 21CFR 1020.

Table 21b. Half–Value Layer Requirements

Operating Range	Measured Potential, kVp	Half-Value Layer in mm Aluminum	
		Diagnostic X-Ray Systems	Dental Intraoral
<51	30	0.3	N/A[a]
	40	0.4	N/A[a]
	50	0.5	1.5
51–70	51	1.2	1.5
	60	1.3	1.5
	70	1.5	1.5
>70	71	2.1	2.1
	80	2.3	2.3
	90	2.5	2.5
	100	2.7	2.7
	110	3.0	3.0
	120	3.2	3.2
	130	3.5	3.5
	140	3.8	3.8
	150	4.1	4.1

[a]Not available = NA.

apply to all radioactive materials regardless of their origin. Others recognize that most accelerator-produced radionuclides tend to have shorter half-lives, and therefore require less control. Thus, when dealing with accelerator-produced material, consultation with the particular state's regulations becomes imperative.

REGULATIONS FOR NONIONIZING RADIATION

Understanding and applying NIR regulatory standards and guidance requires careful attention to the spectral characteristics of the radiation source(s) involved. The situation is probably most clearly described by dividing

Table 22. Highlights of Requirements for Fluoroscopic Units[a-c]

Primary barrier	Primary barrier intercepts entire beam Transmission $\leq 0.2\%$

Beam limitation	Beam not exceed visible area by $>3\%$SID[d]
	    Sum of excess $<4\%$SID
	    Beam $<$ largest spot-film size
	    Units with visible area $>300\,cm^2$ shall have
	    continuously adjustable collimators, down
	    to $5 \times 5\,cm^2$, or, if fixed SID, to $125\,cm^2$

Spot-film beam limitation	Beam automatically limited to film size
	    Beam adjustable to fields smaller than film
	    size down to $5 \times 5\,cm^2$
	    Beam not exceed visible area by $>3\%$SID[d]
	    Sum of excess $<4\%$SID
	    Misalignment of the centers of beam and film $<2\%$ SID

Activation of fluoroscopy	Requires continuous press on switch Serial exposures may be terminated at any time

Entrance exposure rates	≤ 50 mGy/minute, except
	    1. if unit has no high mode for automatic exposure
	    control (AEC) units, then ≤ 100 mGy·min^{-1};
	    2. during image recording

Indications	Panel shows kVp and mA during exposure
Source-to-skin distance	≤ 38 cm for stationary units ≤ 30 cm for mobile units ≤ 20 for mobile, special surgical units
Fluoro timer	Maximum time without resetting ≤ 5 min Signals during fluoroscopy after time until reset

Control of scatter	Unit and table design prevent exposure of persons to
	    scatter, except extremities, without ≥ 0.25 mm
	    Pb equivalent attenuation or 1.2 m from beam

[a]Table by Tim Burns and Mark Geurts.
[b]For new units. Older units have some allowances made for regulations in effect at manufacture.
[c]Details for such units should be found in 21CFR1020 or in the particular state's regulations.
[d]SID is source to image intensifier distance.

Table 23. Characteristics of a Mammography System as Required by the Mammography Quality Standards Act

Item	Criterion
Type of equipment	Specially designed for mammography
Motion of tube-image receptor	Tube-image receptor may be fixed and remain so if power fails.
Image receptor size and grid	i. Screen-film units shall have a minimum of $18 \times 24\,cm^2$ and $24 \times 30\,cm^2$ and moving grids. ii. Magnification units can operate without the grid.
Light fields	Units with light fields shall have an average illumination of not less than 160 lux at 100 cm or the maximum source-image receptor distance, whichever is less.
Magnification	i. Units used for non-interventional problem solving shall have radiographic magnification capability ii. Units with magnification shall provide at least one magnification value between 1.4 and 2.0
Focal Spot selection	Unit indicates focal spot size and material selected prior to exposure, unless determined by algorithm during, where displayed after.
Compression	Unit shall have compression: i. power driven by hands-free controls on each side of the patient, including fine control;[a] ii. Compression paddle size shall match the full-field receptor size, and shall be level with the breast-support table to < 1 cm, except when designed otherwise; iii. The chest-wall edge shall be strain and parallel to the edge of the receptor, and may be curved for comfort if out of the field.
Technique factor selection and display	Has manual selection of mAs; ii. technique factors set display before exposure; iii. In automatic exposure control mode the technique used for the exposure displays afterwards.
Automatic exposure control (AEC)	i. Screen-film systems shall provide an AEC mode that is operable in all combinations of equipment; ii. positioning of detector shall permit flexibility in the placement under the target with the size and available positions of the detector marked on the input surface of the paddle, and the selected position of the detector indicated; iii. there shall be means to vary the selected optical density from the normal.
Film-Intensifying screens, if used	i. Film shall be designed for mammography; ii. screens shall be designed for mammography and the film used; iii. processing chemicals use as per manufacturer; iv. hot-lights and film masking devices shall be available;

[a]Applies to units built after 10/02.

NIR into three spectral regions: optical: ultraviolet (UV), visible, and infrared (IR); microwave RF; extremely low frequency (ELF) and static fields. The relationship between wavelength, frequency, and photon energy for these spectral regions is shown in Fig. 4.

Emission limits for specific optical sources combine with exposure limits to protect workers and the general public. These limits are further organized according to the type of source: lamps and other optical sources; and lasers. This separate consideration of non-laser sources and lasers necessarily reflects the different qualities of these sources. While lamps and other optical sources typically present a broad spectrum (i.e., the radiation is spread over many wavelengths) and widely divergent emission, lasers emit just one or at most a few discrete wavelengths in a very narrow, highly collimated beam.

The exposure limits for broad band non-laser optical sources are generally expressed in terms of some sort of spectral weighting scale to account for the fact that some wavelengths more efficiently cause injury than others. In the UV region, the International Commission on Non-ionizing Radiation Protection (ICNIRP) (29,30) and the National Institute for Occupational Safety and Health (NIOSH) (31) support the spectral weighting function and exposure limits set forth by the American Conference of Governmental Industrial Hygienists (ACGIH) (32) several decades ago. In this scheme, the weighting function is normalized to the peak of the spectral effectiveness curve at 270 nm, with an effective spectrally weighted limit of $30\,J \cdot m^{-2}$ over the region from 180 to 400 nm. Eye and skin exposure limits for monochromatic UV sources can be found in tables provided by ACGIH, ICNIRP, or NIOSH : The

Table 24. Quality Assurance Required for a Mammography System by the Mammography Quality Standards Act

Daily Quality Control Tests for Film Systems

Film processor control	i. Base plus fog density within 0.03 of the established level; ii. mid-density on test strip within 0.15 of the established level; iii. density difference within 0.15 of the established level.

Weekly Quality Control Tests for Film Systems

Phantom density and contrast	With approved phantoms, optical density$\geq$1.2 from typical exposure, varies <0.2 from normal, passes imaging for phantom, and contrast changes <0.05 with standard test.

Quarterly Quality Control Test for Film Systems

Film fixer clearance	Residual fixer in film <5 μgm·cm^{-2};
Reject analysis	Repeat or reject rate changes $<2\%$ of the total films in analysis (otherwise determine the reason for change, corrective actions recorded and the results assessed).

Semiannual Quality Control Tests for Film Systems

Darkroom fog	Darkroom fog shall not exceed 0.05 for 2 min exposure on the counter top
Screen-film contact	40 mesh screen on cassette shows no appreciable blurring.
Compression device performance	Device provides >111 Nt force (between 111 and 200 Nt[a])

Annual Quality Control Tests for Film Systems

Automatic exposure control performance	i. The AEC maintains optical density within 0.30 (or 0.15[a]) of mean (thickness varied from 2 to 6 cm and kVp varied appropriately for thicknesses); ii. optical density in center of phantom image >1.2.
kVp accuracy and reproducibility	i. Indicated kVp accurate within 5% at: the lowest clinical kVp that can be measured by a kVp test device, the most commonly used clinical kVp, and the highest available clinical kVp; ii., the coefficient of variation of reproducibility of the kVp ≤ 0.02 at the most commonly used clinical settings
System resolution	High contrast pattern resolves 11 line pair/mm with bars perpendicular to anode–cathode axis, and 13 when parallel; pattern 4.5 cm above breast support, centered, 1 cm of chest edge; test performed for each focal spot and target material.
Half-value layer (HVL)	HVL in mm$\geq$kVp/100
Breast entrance air kerma and AEC reproducibility	Coefficient of variation for both air kerma and mAs$\leq$0.05
Dosimetry	Average glandular dose (cranio-caudal view standard breast) ≤ 3 mGy/exposure.
X-ray field/light field/image receptor/compression paddle alignment	i. System has beam-limiting devices that allow the entire edge field to extend to the chest wall edge of the receptor and assure that the x-ray field does not extend beyond any edge of the receptor $>2\%$ of the SID; i. misalignment of the light and x-ray field $\leq 2\%$ of the SID; iii. chest wall edge of compression paddle $<1\%$ of the SID beyond the chest wall edge of the receptor.
Uniformity of screen speed	i. Difference between the maximum and minimum optical densities of all screens≤ 0.30; ii. screen artifacts shall also be evaluated during this test.
System artifacts	System artifacts shall be evaluated for all available focal spot sizes and target filter combinations with a sheet of homogeneous material to cover the, for all cassette sizes used.
Radiation output	System[a] can produce >7 mGy·s^{-1} air kerma at 28 kVp in Mo target/Mo filter mode at any SID with a detector 4.5 cm above the breast support surface with the compression paddle in place, over 3 s.
Automatic decompression, if included	System provides override capability to allow maintenance of compression, a continuous display of the override status, and a manual emergency compression release that can be activated in the event of power or automatic release failure.

[a]For units after built 10/02.

Table 25. Some Exemptions from Regulation Control for Naturally Occurring Radionuclides

Incandescent gas mantles	Marine compasses
Vacuum tubes	Timepieces, dials (various limits)
Welding rods	Lock illuminators
Electric lamps (<50 mg thorium)	Precision balances
Glassware ($<10\%$t source material)	Automobile shift quadrant
Outdoor or industrial germicidal lamps, sunlamps, lamps (<2 gm thorium)	Thermostat dials
Glazed ceramic tableware (glaze $<20\%$ source material)	Uranium as counterweights in aircraft, rockets, projectiles, missiles (labeled)
Piezoelectric ceramic($<2\%$ source material)	Electron tubes
Photographic film, negatives, and prints	Spark gap irradiators
Finished optical lenses ($<30\%$ thorium)	Gas and aerosol detectors
Aircraft engine parts (nickel-thoria alloy $<4\%$ thorium)	

limits are very similar for all three organizations. In addition to this general exposure limit, standard setting groups have established several hazard-specific limits spanning the entire optical region. The standard of ANSI and the Illumination Engineering Society of North America (IESNA) provides a typical treatment (33). This standard describes the application of exposure limits for the following hazards:

200–400 nm Skin and Eye Exposure Limit: very similar to the ACGIH, ICNIRP, and NIOSH ultraviolet limit.

320–400 nm Eye Exposure Limit: 1 mW·cm^{-2} for exposure durations $>1\,000$ s, and 1 J·cm^{-2} for shorter exposure durations.

400–1400 nm Retinal Thermal Hazard Exposure Limit: an exposure duration dependent limit based on an associated burn hazard spectral weighting function table.

400–700 nm Retinal Blue Light Hazard Exposure Limit: a limit that is time dependent for exposure durations $<10,000$ s, and exposure rate dependent for longer durations, with an associated blue light hazard spectral weighting function table.

Retinal Blue Light Hazard Exposure Limit—Small Source: basically the Retinal Blue Light Hazard limit modified to accommodate sources subtending an angle <11 mrad

Aphakic Eye Hazard Exposure Duration: extends the Retinal Blue Light Hazard down to 305 nm and

replaces the blue light hazard spectral weighting function with an aphakic hazard weighting function.

770–3,000 nm Infrared Hazard Exposure Limit: 0.1 W·cm^{-2} for exposure durations <1000 s and $1.8t^{-0.75}$ W·cm^{-2} for shorter exposure durations.

770–1400 nm Infrared Radiation Hazard Exposure Limit – Weak Visual Stimulus: where the visual stimulus may not activate the aversion response.

400–3,000 nm Skin – Thermal Hazard Exposure Limit: provides limits for thermal skin exposure.

The various references caution users that to recall that several compounds (e.g., tetracycline and its congeners, porphyrins, Retin-A.) can make individuals more susceptible to the photochemical damage associated with UV radiation and blue light. Some common drugs that promote photochemical reactions are listed in Table E1 of the ANSI Z136.1-2000 standard. (34)

Lasers present a unique set of optical radiation hazards and consequently detailed exposure limits have been developed. Laser exposure limits are based on wavelength and various beam characteristics (e.g., beam diameter and divergence, plus pulse characteristics for pulsed lasers). While the OSHA occupational safety and health standards (Part 1910 of Title 29 CFR) do not specifically address lasers, Safety and Health Regulations for Construction (29 CFR Part 1926.54) provides some very general requirements (e.g., documented training of laser operators, posting of laser use areas, laser protective eyewear for workers in areas where the laser radiation level could exceed 5 mW·cm^{-2}). The OSHAs construction standard also specifies non-wavelength specific (though presumably visible "light") laser radiation exposure limits of 0.001 mW·cm^{-2} for direct staring, 1 mW·cm^{-2} for incidental viewing, and 2.5 mW·cm^{-2} for diffuse reflections. Figure 5 illustrates the strong wavelength dependence of the ANSI Z136.1-2000^{k} eye maximum permissible exposure (MPE) irradiance for continuous wave lasers under default exposure durations. Pulsed laser MPEs are more complicated to calculate, especially for repetitively pulsed lasers. See Thomas et al. for guidance on performing these repetitively pulsed laser MPE calculations(35). The ANSI Z136.1-2000 standard provides a framework for calculating the MPE for lasers occupying the wavelength region between 0.18 μm and 1 mm. The time domain covered extends to pulse

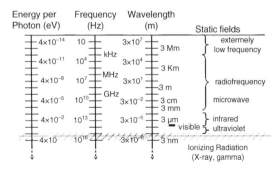

Figure 4. Nonionizing radiation: Wavelength, frequency, and photon energy for optical, microwave, radio frequency, and ELF spectral regions.

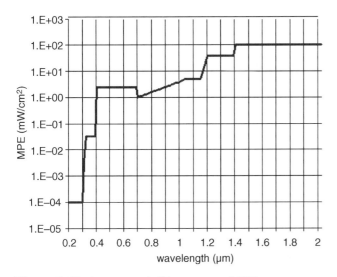

Figure 5. Maximum permissible exposure (MPE) versus wavelength for continuous wave lasers, assuming extended viewing duration for UV wavelengths ($<0.4\,\mu$m), only accidental viewing (<0.25 s) for visible wavelengths (0.4–$0.7\,\mu$m), and 10 s exposure for wavelengths ($>0.7\,\mu$m). For clarity this chart extends only to $2\,\mu$m, but the MPE remains $100\,$mW·cm^{-2} for wavelengths from 1.4 to 1 mm.

durations as short as 10^{-9} s for UV ($<0.4\,\mu$m) and far IR ($<1.4\,\mu$m), and down to 10^{-13} s for the visible and near-IR regions (i.e., 0.4–$1.4\,\mu$m). However, laser technology has pushed beyond these boundaries, with pulse durations in the 10^{-15} s range and shorter wavelength laser-like sources operating near the threshold of ionizing radiation now attainable in the laboratory. Subsequent editions of the Z136.1 standard are expected to address these gaps.

The FDA Laser Product Performance Standard (21CFR1040.10&11) requires laser manufacturers to classify laser products sold in the United State into one of four hazard classes according to the laser's ability to cause injury. The ANSI Z136 series of standards then specifies the required control measures appropriate for each class of laser. The IEC/EN 60825 series of standards establishes a similar laser hazard classification scheme for both manufacturers and application of control measures. Table 26 summarizes, in very general terms, the classes and subclasses for each of these organizations. In an apparent nod to pending harmonization, the recent Z136.3-2005 Standard for the Safe Use of Lasers in Health Care Facilities utilizes the IEC classification designations instead of the ANSI/FDA scheme (36). The laser hazard classes established by each of these three standard setting organizations also have associated accessible emission limits (AEL), but the AEL for each hazard class may not be identical for each organization.

The microwave region of the frequency spectrum is generally considered to extend from 300 MHz to 300 GHz, while the RF portion covers from 3 kHz to 300 MHz. (32,37) An opportunity for nomenclature confusion arises from the fact that some organizations, including ICNIRP (38), consider this microwave region to be a subset of an RF range extending from 300 Hz to 300 GHz. See Figure 4 for a sense of relative position of these regions in the electromagnetic spectrum. The FDA/CDRH Microwave Oven standard (21CFR1030.10) defines a microwave oven as a device designed to heat, dry, or cook food using electromagnetic radiation from 890 to 6000 MHz (most commercial microwave ovens operate at either 915 or 2450 MHz), and requires microwave oven manufacturers to limit radiation levels to $<5\,$mW·cm^{-2} at any point 5 cm or more from any external surface. OSHAs Construction standard (29 CFR 1926.54) limits microwave exposure to $10\,$mW·cm^{-2} (no averaging), but offers no other guidance in this region. The NIR section of OSHA's General Industry standard (29 CFR 1910.97) specifies a $10\,$mW·cm^{-2}, 6 min average for exposure limit for the 10 MHz–100 GHz range, with no spatial averaging, and prescribes RF warning sign appearance. This OSHA section applies to all radiations originating from radio stations, radar equipment, and other possible sources of electromagnetic radiation (e.g., as used for communication, radio navigation, and industrial and scientific purposes), but does not apply to the deliberate exposure of patients by health care providers. In addition, section 29 CFR 1910.268 [Telecommunications] provides guidance for "lock out tag out" type securing and grounding prior to working on 3–30 MHz radio station broadcast antennas, prohibits employers from allowing employees to look into energized microwave waveguides, requires compliance with the 29 CFR 1910.97 exposure limits, and requires posting warning signs with specific wording in accessible areas where those limits could be exceeded. Table 27 gives some sample exposure limits.

Several standards setting organizations have developed exposure limits for microwave and radiofrequency radiation. These standards show considerable consistency for the spectral region extending from 300 GHz (i.e., near the boundary between optical and microwave radiation) to ~ 15 GHz, where most standards recognize an exposure limit of $10\,$mW·cm^{-2}. At lower frequencies, additional considerations come into play, as indicated in Table 28. At frequencies below ~ 10 MHz, the longer wavelengths necessitate separate limits for magnetic and electric field exposures. This distinction arises because at longer wavelengths the emitted energy from an antenna passes first through a *near field* region, in which the electric and magnetic fields are not directly coupled as they are under the *far field* conditions normally associated with electromagnetic radiation. In the far field, the electric field strength is always directly proportional to the magnetic field strength, and the radiation levels generally decrease with the inverse square of the distance. In the near field, the electric and magnetic fields are not coupled, more complicated relationships with distance exist, and many far field assumptions no longer hold. The distance that the near field extends from the emitter is often defined as the wavelength divided by 2π, although ICNIRP defines the near field simply as the distance within one wavelength from a radiating antenna (38). By using the ICNIRP expression and consulting Fig. 3, we see, for example, that while at 10 MHz the near field extends only 30 m, at 100 kHz it is 3 km. The ICNIRP exposure limits (reference levels) appear in Table 34 (38,39), expressed as separate electric and magnetic field strengths (the ICNIRP guidance actually expresses exposure limits in terms of

Table 26. Summary of Laser Classification Schemes

FDA/CDRH 21CFR1040.10	ANSI Z136	IEC/EN 80625
Class I—levels of laser radiation are not considered hazardous	Class 1—no hazard; exempt from all control measures	Class 1—no risk, even with viewing instruments Class 1M[a]—no risk except perhaps to eye when viewed through viewing instruments (eye loupes or binoculars)
Class IIa—levels of laser (applies to visible only) radiation are not considered hazardous if viewed $\leq 1000\,s$ but are considered a chronic viewing hazard for any period of time $>1000\,s$ Class II—levels of (visible only) laser radiation considered a chronic viewing hazard	Class 2—visible $(0.4–0.7\,\mu m)$ lasers not considered hazardous for momentary viewing $(<0.25\,s)$, but for which the Class 1 accessible emission limit may be exceeded for longer exposure durations; avoid prolonged staring	Class 2—no eye risk for short term exposures, even with viewing instruments; no risk to skin (applies to visible lasers only) Class 2M[a]–no eye risk for short term exposures, except perhaps with viewing instruments; no risk to skin (visible only)
Class IIIa—levels of laser radiation are considered, depending upon the irradiance, either an acute intrabeam viewing hazard or chronic viewing hazard, and an acute viewing hazard if viewed directly with optical instruments	Class 3a—with "Caution" label: does not exceed the appropriate irradiance MPE, except perhaps when viewed through collecting optics (e.g., microscopes, telescopes)— with "Danger" label: may exceed the appropriate irradiance MPE	Class 3R[b]—low risk to eyes, no risk to skin
Class IIIb—levels of laser radiation are considered to be an acute hazard to the skin and eyes from direct radiation	Class 3b—emit greater than Class 3a limits and pose an acute eye hazard; more rigorous controls are required to prevent exposure of the unprotected eye	Class 3B—medium to high risk to eyes, low risk to skin
Class 4—levels of laser radiation are considered an acute hazard to the skin and eyes from direct and scattered radiation	Class 4—acute eye and skin hazard, plus ignition source (fire) and laser-generated airborne contaminants hazards; strict control measures required	Class 4—high risk to eyes and skin

[a]The "M" designation in the IEC classification scheme is derived from "magnifying" optical viewing instruments.
[b]The "R"designation in the IEC classification scheme is derived from reduced or relaxed requirements for manufacturers (no key switch or interlock connector required) and users (usually no eye protection required).

quantities that are not easily measured directly, such as currents induced in the body; the references levels then extrapolate those limits via models to the directly measurable field strengths given).

The microwave and RF exposure limits are generally based on protection against known adverse health effects of NIR exposure, but these limits may not be sufficiently protective for individuals with implanted medical devices (e.g., pacemakers, insulin pumps, cochlear implants, etc.) (38). Also, in the ELF region (defined here as by the ICNIRP as $<300\,Hz$ (40), although some organizations and regulations specify slightly different spectral boundaries), considerable ambiguity marks the basis underlying the generally accepted exposure limit values. As Petterson and Hitchcock note, exposure guideline derivation should

ideally stem from accepted mechanisms of interaction, dose response studies in animals, and epidemiological evidence of similar effects in humans; none of this has occurred for ELF fields (41). This view was reiterated in a 2002 update of the American Industrial Hygiene Association (AIHA) White Paper on ELF Fields (42). Finally, accurately assessing non-ionizing radiation hazards is complicated by the difficulty in properly measuring RF and ELF field strength. The NIOSH *Manual for Measuring Occupational Electric and Magnetic Field Exposures* provides some insight into this subject (43).

Because of its shallow tissue penetration and ease of control, ELF electrical fields are generally not of much concern below $\sim 15\,kV{\cdot}m^{-1}$. However, individuals with implanted medical devices (e.g., cardiac pacemakers

Table 27. Sample of Microwave and Radiofrequency Radiation Exposure Limits, 10 MHz–300 GHz[a]

Source	Reference	Frequency	Limit, mW·cm^{-2} Occupational	Public	Additional Criteria
OSHA	29 CFR 1926.54(l) [Construction]	0.3–300 GHz	10		(no averaging)
OSHA	29 CFR 1910.97(a)(2)(i) [General Industry]	0.01–100 GHz	10		Power (duration $\geq$ 0.1 h): 10 mW·cm^{-2} Energy (duration < 0.1 h): 10 mW·hr·cm^{-2} in any 0.1 h (no spatial averaging)
ICNIRP	Guidelines up to 300 GHz (1998)[n]	2–300 GHz	50	10	Averaged over any 20 cm^2 exposed area and, above 10 GHz, any $68/f^{1.05}$ (f = frequency in GHz) minute period; Limit peak exposures averaged over 1 cm^2 < 20 times these limits
		400–2,000 MHz	$f/40$	$f/200$	f = frequency in MHz
		10–400 MHz	10	2	Between 100 kHz and 10 GHz, averaged over any 6 minute period

[a]We thank the International Commission on Non-Ionizing Radiation Protection, ICNIRP, for the permission to reprint part of its guidelines in the present article. We also thank Health Physics, where the guidelines were first published.

Table 28. ICNIRP Electrical and Magnetic Field Exposure Limits, Static—10 MHz[a]

Reference	Frequency	Reference Level (Exposure Limit) Occupational Electrical V·m^{-1}	Magnetic A·m^{-1}	Public Electrical V·m^{-1}	Magnetic A·m^{-1}	Additional Criteria
ICNIRP Guidelines up to 300 GHz (1998)[b]	1–10 MHz	$610/f$	$1.6/f$	$87/f$	$0.73/f$	f = frequency in MHz
	0.065–1 MHz	610	$1.6/f$			f = frequency in MHz
	0.15–1 MHz			$87/f$	$0.73/f$	f = frequency in MHz
	0.82–65 kHz	610	24.4			
	3–150 kHz			87	5	
	0.025–0.82 kHz	$500/f$	$20/f$			f = frequency in kHz
	0.025–0.8 kHz			$250/f$	$4/f$	f = frequency in kHz
	8–25 Hz	20,000	$20,000/f$	10,000	$4,000/f$	f = frequency in Hz
	1–8 Hz	20,000	$163,000/f$	10,000	$32,000/f$	f = frequency in Hz
	0– Hz	Use electrical safety procedures to avoid electric shock	163,000	Avoid spark discharges	32,000	
ICNIRP Guidelines for Static Magnetic Fields (1994)[t]	Static	Average: 160,000 Maximum: 1,600,000 Limbs: 3,980,000		General: 33,000 Electronic medical implants: 400		

[a]We thank the International Commission on Non-Ionizing Radiation Protection, ICNIRP, for the permission to reprint part of its guidelines in the present article. We also thank Health Physics, where the guidelines were first published.
[b]1 tesla (T) = 796,000 amperes per meter (A/m) in air, vacuum and biological materials.

and insulin pumps) should avoid ELF electric fields above 1 kV·m^{-1}. The static electric and magnetic field designation is generally applied to any radiator with a frequency below 1 Hz (37). Static electrical fields are relatively simple to shield with grounded conducting enclosures. Static magnetic fields are more difficult to shield. The NIR environment surrounding nuclear magnetic resonance imaging facilities (primarily a high strength static magnetic field, but also RF radiation) poses a particular concern. The American College of Radiology's White Paper on MR Safety (44) offers some useful guidance on control of the hazards associated with MRI systems. While the ACR White Paper has served to increase awareness and foster discussion, the user community has not achieved consensus on these recommendations (45,46).

The FDAs 21CFR892.1000 identifies MRI systems as Class II medical devices, meaning among other things that institutions must implement safety programs rather than merely following the manufacturer's recommendations. In addition, MRI manufacturers generally must provide the customers with information, for example, indicating the extent of the static magnetic field (i.e., the location of the 5 G line). Attachment B of the FDAs *Guidance for the Submission of Premarket Notifications for Magnetic Resonance Diagnostic Devices* lists the elements of a MRI safety program, including patient screening, appropriate levels of patient monitoring and supervision, emergency procedures and shutdowns, noise control measures, access restrictions, control of cryogenic hazards, adherence to the IEC operating mode guidelines, use of MRI compatible equipment, fire precautions, and so on (47). The 5 G line around MRI facilities is generally posted with warning signs to avoid harmful effects on medical device wearers. There is no generally accepted format for MRI warning signs, though Shellock offers some helpful suggestions (48). The FDA guidance indicates that a MRI procedure performed under any of the following conditions constitutes a significant risk as defined in 21CFR812.3(m)(4), triggering the requirement for a FDA investigational device exemption, as well as the institutional review board (IRB) approval specified in 21CFR56 and the informed consent rules of 21CFR50 (49,50): procedures utilizing >8 tesla (T) for adults, children and infants older than 1 month, or >4 T for neonates (infants <1 month old); specific absorption rate (SAR) $\geq 4\,W\cdot kg^{-1}$ for 15 min or more for the whole body, $3\,W\cdot kg^{-1}$ for 10 min or more for the head, $8\,W\cdot kg^{-1}$ for 5 min or more per gram of tissue for the head or torso, or $12\,W\cdot kg^{-1}$ for 5 min or more per gram of tissue for the extremities; or any time rate of change of gradient magnetic fields sufficient to produce severe discomfort or painful nerve stimulation.

The IEC 60601-2-33 standard specifies three modes of operation relating to RF energy-induced heating and gradient magnetic field (51):

Normal operating mode: suitable for all patients and requires only routine monitoring;

First level controlled (FLC): may cause undue physiological stress; requires medical supervision and operator confirmation to enter this mode.

Second level controlled (SLC): may produce significant risk; requires IRB approval, and manufacturers are required to restrict operator access to this mode (e.g., password, key lock).

Although the FDA does not currently require MRI manufacturers to incorporate provisions for these three IEC modes, in the current global market all MRI manufacturers already include such provisions, and the FDA has indicated their intention to adopt these IEC 60601-2-33 guidelines (47,50).

CONCLUSIONS

Understanding codes, regulations, and license conditions can be the least exciting but most challenging part of a medical physicist's job! The federal codes may be the basis for state codes, but state codes are not necessarily identical to federal codes, even in NRC agreement states or OSHA-approved states. Compliance with myriad regulations and license conditions is a challenge. However, by knowing the codes and regulations one can write a better license or structure a radiation safety program with which it is easier for one to comply. To be forewarned is to be fore-

armed! We encourage readers to thoroughly study those codes and regulations applicable to their own facilities and to stay current with continuingly changing codes and regulations.

ACRONYMS AND DEFINITIONS

Most acronyms are discussed in the text.

AAPM.   The American Association of Physicists in Medicine.

ACGIH.   American Conference of Governmental Industrial Hygienists.

ACMP.   American College of Medical Physics.

ACR.   American College of Radiology.

Agreement States.   Those states that have entered into a formal agreement with the NRC to take over certain of its regulatory authority within that state.

AIHA.   American Industrial Hygiene Association.

AMP.   Authorized medical physicist.

ANSI.   American National Standards Institute, the voluntary standards-setting organization in the United States and its representative to the International Organization for Standardization.

AU.   Authorized user, a physician authorized to use radioactive materials in the health professions.

BEIR.   Biological Effects of Ionizing Radiation, often referring to reports bearing that acronym from the BRER.

BRER.   Board on Radiation Effects Research, a board of the National Academy to coordinate activities of the National Research Council involving the biological effects of radiation.

Byproduct material.   Any radioactive material (except special nuclear material) yielded in or made radioactive by exposure to the radiation incident to the process of producing or utilizing special nuclear material.

CDRH.   Center for Devices and Radiological Health, and agency of the Food and Drug Administration.

CGMP.   Current good manufacturing practices.

CRCPD.   Coference of Radiation Control Program Directors.

Department of Defense

DOE.   Department of Energy.

DOT.   Department of Transportation.

DS.   Dosimetry system.

EDE.   Effective dose equivalent.

ELF.   Extremely low frequency.

EPA.   Environmental Protection Agency.

FC.   Full calibration.

FDA.   Food and Drug Administration.

HDR.   High dose-rate brachytherapy.

HPS.   Health Physics Society.

IAEA.   International Atomic Energy Agency, an arm of the United Nations.

ICNIRP.   International Commission on Nonionizing Radiation Protection.

ICRP.   International Commission on Radiation Protection.

ICRU.   International Commission on Radiation Units and Measures.

IEC.   International Electrotechnical Commission, an organization that establish standards mostly pertaining to industry and manufacturers.

IR.   Ionizing radiation.

IRB.   Institutional Review Board, an institutions committee that evaluates new drugs, procedures or devices for proposed human use.

JCAHO.   Joint Commission on Accreditation of Healthcare Organizations.

LDR.   Low dose-rate brachytherapy.

MDR.   Medium dose-rate brachytherapy.

ME.   Medical event (see Table 14).

MQSA.   Mammography Quality Standards Act.

NARM.   Naturally occurring and accelerator-produced radioactive materials.

NCRP.   Nation Council on Radiation Protection and Measurement.

NIOSH.   National Institute for Occupational Safety and Health.

NIR.   Non-ionizing radiation.

NRC.   Nuclear Regulatory Commission.

NSSDR.   National Sealed Source and Device Registry, a registry maintained by the NRC of information on approved devices.

OAS.   Organization of Agreement States.

OSHA.   Occupational Health and Safety Administration.

PD.   Prescribed dose.

PDR.   Pulsed dose-rate brachytherapy.

PMA.   Premarket approval.

RAM.   Radioactive materials.

RAU.   Remote afterloading unit for brachytherapy.

RF.   Radiofrequency.

RSO.   Radiation safety officer.

SAU.   Substitute authorized user, a physician working under the supervision of an authorized user.

SC.   Spot check.

SDE.   Shallow dose equivalent.

Source Material.   Uranium, thorium or any combination thereof, in any physical or chemical form, or ores that contain by weight one-twentieth of 1% of them, excluding special nuclear material.

Special nuclear material (SNM).   Plutonium, uranium-233, uranium enriched in isotopes 233 or 235, or any material artificially enriched by any of these.

SSR.   Suggested State Regulations, guidelines for state radiation rules assembled by the CRCPD.

UNSCEAR.   United Nations Committee on the Effects of Atomic Radiation, a committee under the United Nations to study the biological effects of radiation.

UV.   Ultraviolet.

BIBLIOGRAPHY

1. Deye JA. Codes and Regulations, Radiation. In: Webster JG, editor Encyclopedia of Medical Devices and Instruments. New York: John Wiley & Sons, Inc; 1988.
2. U.S. Nuclear Regulatory Commission. 10 CFR 19 (Notices, Instructions, and Reports to Workers; Inspections), 1981 [Online]. Nuclear Regulatory Commission. Available at http://www.nrc.gov/reading-rm/doc-collections/cfr/part 0.19. Accessed 2004, December 4.
3. U.S. Nuclear Regulatory Commission. 10 CFR 20 (Standards for Protection Against Radiation; Final Rule,) 1991. [Online]. Nuclear Regulatory Commission. Available at http://www.nrc.gov/reading-rm/doc-collections/cfr/part020. Accessed 2004, December 4.
4. U.S. Nuclear Regulatory Commission. 10 CFR 20, 32, and 35 (Medical use of byproduct material: final rule.) Washington DC, Federal Register; Vol.67, No.79 (April 24): 20250-20397, 2002; [Online]. Nuclear Regulatory Commission. Available at http://www.nrc.gov/reading-rm/doc-collections/cfr/part020/part032/035. Accessed 2004, December 4.
5. U.S. Nuclear Regulatory Commission. 10 CFR 35 (*Medical use of byproduct material-Recognition of Specialty Boards; Final Rule*) Washington DC, Federal Register; Vol.70, No.60 (March 30): 16366-16367, 2005; [Online]. Nuclear Regulatory Commission. Available at http://www.nrc.gov/reading-rm/doc-collections/cfr/part35. Accessed 2005, April 19.
6. Institute of Medicine. Radiation Medicine: A Need for Regulatory Reform. In: Gottfried K-LD, Penn G. editors. Committee for Review and Evaluation of the Medical Use Program of the Nuclear Regulatory Commission, Institute of Medicine: 1996 ISBN-0-309-58875-8.
7. Memorandum of Understanding Between the OSHA and U.S. Nuclear Regulatory Commission. CPL 02-00-086–CPL 2.86 (1989, December 22) [Online]. Occupational Safety & Health Administration. Available at http://www.osha.gov/pls/osha-web/owadisp.show_document?p_table=DIRECTIVE&p_id=1658. Accessed 2005, April 29.
8. Federal Register [07/23/2004] 69: 44068-44069; Supporting Statement for the Information-Collection Requirement for the Ionizing-Radiation Standard (29CFR1910.1096) OMB Control No. 1218-0103(2004)(June2004).
9. Occupational Safety and Health Administration. 29 CFR 1910.1096 (Ionizing Radiations), [Online] Occupational Safety and Health Administration. Available at http://www.osha.gov/pls/oshaweb/owadisd.show. Accessed 2005, April 14.
10. Conference of Radiation Control Program Directors. Available at http://www.crcpd.org/free_docs.asp. Accessed 3 October 2005.
11. Curtis R. Non-ionizing radiation: standards and radiation (PowerPoint presentation). OSHA-Salt Lake Technical Center. [Online] Available at http://www.osha-slc.gov/SLTC/radiation_lectures/nir_stds_20021011.ppt. Accessed April 22, 2005.
12. Rockwell RJ, Parkinson J. State and local laser safety requirements. J Laser Appl 1999;11:225–231.
13. Henderson R, Schulmeister K. Laser Safety Philadelphia: Institute of Physics Publishing; 2004.

14. Center for Devices and Radiological Health. Laser Products – Conformance with IEC 60825-1, Am. 2 and IEC 60601-2-22; Final Guidance for Industry and FDA (Laser Notice 50), 2001.

15. News and Notices IAEA action plan to combat nuclear terrorism Health Phys 2002;82:908–909.

16. News and Notices "IAEA and UPU join forces to protect mail" Health Phys 2003;84:129–130.

17. Nuclear Energy Institute, Fact Sheet, August 2004.

18. U.S. Nuclear Regulatory Commission, U.S. Department of Transportation, U.S.-Specific Schedules of Requirements for Transport of Specified Types of Radioactive Material Consignments, RAMREG-002/U.S. Nuclear Regulatory Commission, NUREG-1600, January 1999.

19. Hazardous Materials Regulations; Compatibility With the Regulations of the International Atomic Energy Agency, RSPA-99-6283 (HM-230); Final rule; Published 01/26/2004; Effective Date: Oct 1, 2004; 69 FR 3631.

20. International Atomic Energy Agency, Regulations for the Safe Transport of Radioactive Material, 1996 Edition, Safety Standards Series/Requirements, ST-1, Vienna, Austria; International Atomic Energy Agency, 1996.

21. International Commission on Radiation Protection, 1990 Recommendations of the ICRP: ICRP Publication 60, Ann of the ICRP 21: 1-3 Oxford: Pergamon Press; 1991.

22. USNRC Potential for Erroneous Calibration, Bulletin 97-01. Dose Rate, or Radiation Exposure Measurements with Certain Victoreen Model 530 and 530SI Electrometer/Dosemeters (April 30, 1997).

23. Glasgow GP. Nuclear Regulatory Commission regulatory status of approved intravascular brachytherapy systems. Cardiovascular Rad Med 2002;3:1–11.

24. USNRC, 2004.

25. U.S. Nuclear Regulatory Commission . NRC Information Notice 2003-21: High-dose rate remote afterloader equipment failure. November 24, [Online] Available at http://www.nrc.gov/materials/miau/med-use-toolkit/info-notices.html. Accessed December 4, 2004.

26. U.S. Nuclear Regulatory Commission. Newsletter NUREG/BR-0117/04-2: Nuclear Material Safety and Safeguards [Online] Available at http://www.nrc.gov/reading-rm/doc-collections/nureg/brochures/br0117/04-2.pdf. Accessed December 4, 2004.

27. USNRC, Compatibility Categories and Health and Safety Identification for NRC Regulations and Other Program Elements—Procedure No. SA-200, 2004 [Online] Available at http://www.hsrd.ornl.gov/nrc/procintro.htm. Accessed 3 October 2005.

28. Conference of Directors of Radiation Control Programs, Suggested State Radiation Control Regulations. Avaliable at http://www.crcpd.org/SSRCRs. Accessed 3 October 2005.

29. International Commission on Non-Ionizing Radiation Protection. Guidelines on Limits of Exposure to Ultraviolet Radiation of Wavelengths Between 180 nm and 400 nm (incoherent optical radiation Health Phys 2004;87:171–186.

30. International Commission on Non-Ionizing Radiation Protection. Guidelines on UV Radiation Exposure Limits. Health Phys 1996;71:978.

31. National Institute for Occupational Safety and Health. Criteria for a Recommended Standard–Occupational Exposure to Ultraviolet Radiation; DHHS (NIOSH) Publication No. 73–11009, 1972.

32. American Conference of Governmental Industrial Hygienists. 2004 TLVs® and BEIs®. Cincinnati, OH: ACGIH, 2004.

33. American National Standards Institute/Illuminating Engineering Society of North America. RP-27.I-96: Recommended Practice for Photobiological Safety for Lamp & Lamp Systems - General Requirements. New York: IESNA; 1996.

34. American National Standards Institute, Inc., ANSI Z136.1-2000, American national standard for safe use of lasers. Orlando: (FL): Laser Institute of America; 2000.

35. Thomas RJ, et al. A procedure for multiple-pulse maximum permissible exposure determination under the Z136.1-2000 American National Standard for the Safe Use of Lasers. J Laser Appl 2001 13:134–139.

36. American National Standards Institute, Inc., ANSI Z136.3-2005, American national standard for safe use of lasers in health care facilities. Orlando, FL: Laser Institute of America; 2005.

37. Hitchcock RT. Radio-Frequency and Microwave Radiation. In: DiNardi SR. editor. The Occupational Environment—Its Evaluation and Control. Fairfax, VA: AIHA Press; 1998.

38. International Commission on Non-Ionizing Radiation Protection. Guidelines for Limiting Exposure to Time-Varying Electric, Magnetic, and Electromagnetic Fields (Up to 300 GHz). Health Phys 1994;74:494–522.

39. International Commission on Non-Ionizing Radiation Protection. Guidelines on Limits of Exposure to Static Magnetic Fields. Health Phys 1994;66:100–106.

40. International Non-Ionizing Committee of the International Radiation Protection Association. Review of Concepts, Quantities, Units and Terminology for Non-ionizing Radiation Protection. Health Phys 1985;49:1329–1362.

41. Patterson RM, Hitchcock RT. Extremely Low Frequency (ELF) Fields. In: DiNardi SR, editor. The Occupational Environment–Its Evaluation and Control. Fairfax, (VA): AIHA Press; 1998.

42. American Industrial Hygiene Association. AIHA White Paper on Extremely Low Frequency Fields, Fairfax, VA: AIHA; 2002.

43. Bowman JD, Kelsh MA, Kaune WT. Manual for Measuring Occupational Electrical and Magnetic Field Exposures [DHHS (NIOSH) Publication No. 98-154]. Cincinnati, (OH): National Institute for Occupational Safety and Health Publications Dissemination; 1998.

44. Kanal E, et al. American College of Radiology white paper on MR safety. Am J Radiol 2002;178:1335–1347.

45. Shellock FG, Crues JV. MR Safety and the American College of Radiology White Paper. Am J Radiology 2002;178:1349–1352.

46. Kanal E, et al American College of Radiology White Paper on MR Safety: 2004 Update and Revision. Am J Radiol 2004;182: 1111–1114.

47. Center for Devices and Radiological Health. Guidance for the Submission of Premarket Notifications for Magnetic Resonance Diagnostic Devices. CDRH, [Online]. Available at http://www.fda.gov/cdrh/ode/95.html. Accessed 5 October 2005.

48. Shellock FG. MR Safety Signs. RT Image 16(3): 2003. Available at http://www.rt-image.com/content=8702J84E489CAE9040A-240441. Accessed 30 Sept 2005.

49. Center for Devices & Radiological Health. Guidance for Industry and FDA Staff - Criteria for Significant Risk Investigations of Magnetic Resonance Diagnostic Devices. CDRH, 2003.

50. International Commission on Non-Ionizing Radiation Protection. Medical magnetic resonance (MR) procedures: protection of patients. Health Phys 2004;87:197–216.

51. Zaremba LA. FDA Guidance for magnetic resonance system safety and patient exposures: current status and future G considerations. In: Shellock FG, editor. Magnetic resonance procedures: health effects and safety, Boca Raton, (FL): CRC Press LLC; 2001.

See also CODES AND REGULATIONS: MEDICAL DEVICES; IONIZING RADIATION, BIOLOGICAL EFFECTS OF; NONINONIZING RADIATION, BIOLOGICAL EFFECTS OF; RADIATION PROTECTION INSTRUMENTATION.

COGNITIVE REHABILITATION. See Rehabilitation, computers in cognitive.

COLORIMETRY

Li-Jiuan Shen
National Taiwan University
Taipei, Taiwan

Richard Mandel
Boston University
Boston, Massachusetts

Wei-Chiang Shen
University of Southern
California,
Los Angeles, California

INTRODUCTION

Light can be characterized as a wave with frequency and wavelength λ. This wave has energy E, which is proportional to its frequency

$$E = hc/\lambda = h\nu$$

where h is Planck's constant and c is the velocity of light. When such a wave of light encounters a molecule, it will either be absorbed (i.e., its energy will be transferred to the molecule) or scattered (i.e., its direction of propagation will be changed). The probability of occurrence of each process will depend on the nature of the molecule encountered. If the electromagnetic energy is absorbed, the molecule is said to be excited. A molecule or part of a molecule that can be excited by absorption is called a chromophore. The absorption of light generally excites the electrons of the molecule or chromophore from its ground electronic state to one of its excited electronic states. Absorption of light is likely to occur if the energy of the light is equal to the difference in the energy between the ground (E°) and excited (E^*) electronic state.

$$\lambda = hc/(E^* - E^\circ)$$

These transitions have rather diffuse energies because excited states occur in both the vibrational and rotational as well as electronic states, giving rise to broadened energy ranges. A plot of the probability of absorption versus wavelength is called an absorption spectrum. The excitation energy is usually released as radiant energy in the form of kinetic energy and heat. Under certain conditions, some molecules rapidly reemit their energy as visible or ultraviolet (UV) light. This is known as fluorescence.

The wavelengths that give rise to electronic transitions in molecules are generally in the visible (700–400 nm) and UV (400–200 nm) region. These transitions are often characteristic of specific molecules and can be used to assay biological and biochemical samples. Transitions at longer wavelengths and lower energies in the infrared (IR) or near-infrared (NIR) generally correspond to excitation of vibrational states alone and are characteristic of specific functional groups such as carbonyl oxygen bonds or carbon–carbon bonds. Even though this region of the absorption spectrum does not find many biomedical

applications, IR and NIR spectroscopy have been used for qualitative and quantitative analysis of different forms (crystal and amorphous) of pharmaceutical solids (1,2). It is further developed into an on-line process monitoring and control entitled process analytical technology (PAT) by the collaboration of U. S. Food and Drug Administration (FDA) and the pharmaceutical industry (3). On the other hand, transitions at shorter wavelengths and higher energies, in the vacuum UV and X-ray region, cannot be carried out on biological samples in aqueous solution and have limited usefulness in biomedical assays.

The probability of light absorption at a single wavelength is described by the Beer–Lambert law. It has been observed that the passage of light through any given thickness of any substance results in the absorption of a constant fraction of that incident light. In differential form, this equation can be written

$$dI/I = -KCdl$$

where dI/I is the fraction of light absorbed by a layer of thickness dl, K is a constant that depends on the properties of the substance, and C is the concentration of the absorbing substance. Thus, for a given beam of light passing through a sample of finite thickness, the amount absorbed at each point in that sample is proportional to the incident intensity at that point. If, for example, in the first millimeter of passage through a sample 50% of the light is absorbed, then in the second millimeter of passage 50% of the remaining light or 25% of the initial intensity will be absorbed, and in the third millimeter 50% of the remainder or 12.5% of the initial intensity will be absorbed. In total, 87.6% of the incident light beam will be absorbed by the 3 mm path length. Mathematically, this is calculated by integrating the above equation, which yields

$$ln(I_0/I) = K'Cl \quad \text{or} \quad I/I_0 = 10^{-K'Cl}$$

where $K' = K/2.303$, l is the path length, I_0 is the initial intensity, and I the final intensity of the light beam after having passed the sample. The left-hand side of the equation defines the optical density and is a useful quantity, because it is directly proportional to the concentration of the absorbing substance and the path length. Absorbance, also called optical density (OD), is designated by A. The transmittance of a solution, designated T, is the fraction (I/I_0) and is related to the absorbance by

$$A = -\log T$$

In the above example, the absorbance through a single millimeter path length is $-\log 0.5 = 0.301$, while the total absorbance through 3 mm is $3 \times 0.301 = 0.903 = -\log 0.125$.

A plot of absorbance as a function of concentration (called a Beer's law plot) is ideally a straight line, while a plot of transmittance is a hyperbolic curve (Fig. 1). Deviations from linearity will be discussed below. Absorbance is a unitless quantity. Therefore, the constant (K') in the above equation must be in units of reciprocal concentration and reciprocal length (typically $cm^{-1} \cdot M^{-1}$). The absorbance is proportional to both concentration and

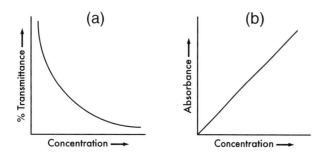

Figure 1. Plots of concentrations versus (a) transmittance (A) and (b) absorbance (B).

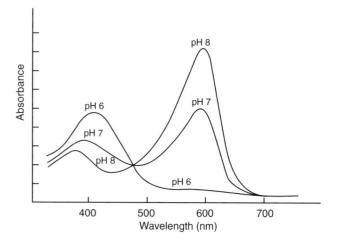

Figure 2. Absorption wavelength of phenol red under various pH conditions. A constant absorption wavelength is shown here at 479 nm, which is the isosbestic point.

length of light path, while the constant specifies a characteristic property of the absorbing chromophore. This constant, which is a function of wavelength, is called the absorption coefficient or extinction coefficient of the material. Absorption coefficients may be expressed on either a weight or a molar basis. The molar extinction coefficient, designated by e specifies the total absorption of a 1 M solution through a 1 cm path length. A molar extinction coefficient of 10,000 or greater is characteristic of a strongly absorbing substance.

Plots of absorbance as a function of concentration are linear if the absorbing chromophore remains the same over the concentration range studied, the chromophore is uniformly distributed, and the orientation of dichroic absorbers (i.e., substances having directional asymmetry) is random. Changes in the nature of the chromophore with concentration, such as ionization, hydration, aggregation, or disaggregation, will alter the absorption spectrum of the solution. There are no general rules as to whether the absorption will increase or decrease. An example of an absorption change resulting from molecular interactions is that of deoxyribonucleic acid (DNA). When the absorption of a solution of native DNA is compared with that of the same concentration of DNA bases, the DNA shows a 30–40% lower extinction coefficient at the 260 nm peak. Heating of the DNA to disrupt the orderly stacking of the bases in the double helix raises the extinction coefficient to that of its constituent bases. The DNA is said to be hypochromic with respect to its bases, while heating of the DNA is said to display a hyperchromic effect. This principle is also widely applied to determine the thermodynamic parameters of nucleic acids using colorimetric measurements. Often some parts of the spectrum will show increased absorbance and other parts will show decreased absorbance. In such a case and when there is an equilibrium between two different molecular forms, there will always exist a wavelength that shows invariant absorption. This point, known as an isobestic point, can be used to quantify concentration of compounds that can exist as different molecular forms. For example, phenol red changes from a protonated to an ionized form between pH 6 and 8, with a switch of the absorption wavelength from 432 (yellow) to 559 nm (red). The absorption at 479 nm, the isosbestic point, is constant regardless the molecular form of this compound and, therefore, this wavelength can be used to measure the concentration of phenol red without a concern to the pH of the solution (Fig. 2).

Non-uniform distributions of chromophore in solution will lead to decreased absorption. This is most readily understood by considering a situation in which one-half of the cross-section of a light beam goes through a transparent solvent, while the other one-half traverses an absorbing solution. Even if the solution is totally absorbing, at least one-half of the light will be transmitted through the sample leading to a limiting absorbance of $\log(I/I_0) = \log 2 = 0.301$. Thus, a plot of absorbance versus concentration will lead to a curve asymptotically reaching 0.301 as the concentration increases. An example of nonuniform distributions leading to changed absorbance is illustrated by the absorption of Acridine Orange in the presence of cellular organelles. Acridine Orange is a weak base that accumulates in acidic cellular compartments, such as endosomes, lysosomes, and Golgi. When it is concentrated in such organelles, the apparent absorbance of the solution decreases. Since the acid gradient is generated by a membrane-bound enzyme that requires adenosine triphosphate (ATP) as its energy source, addition of ATP to a subcellular suspension containing Acridine Orange will cause a decrease in absorption. In this way, changes in the absorbance of Acridine Orange can be used to monitor the activity of the ATPase (4).

The most common biochemical or clinical use of the absorption properties of chromophores is to measure concentration. In brief, the absorption spectrum, that is, a graph of absorbance as a function of wavelength, is recorded from which a suitable wavelength is chosen. Generally, the wavelength chosen is at or near the peak in absorption of the chromophore. This maximizes the sensitivity of the absorption measurement and decreases possible error resulting from incorrect wavelength calibration of the instrument. Other wavelengths may be chosen if another chromophore present in the solution interferes with the measurement. The extinction coefficient is determined, or a calibration curve of the absorbance is run at a number of different concentrations of chromophore. Concentration is determined by weighing a standard or from the extinction coefficient, if it is accurately known. The plot is examined for linearity. Subsequent concentrations can be determined either directly from the extinction coefficient or from the calibration curve of absorbance versus concentration. For example, the concentration of DNA can be readily determined by measuring the absorbance of a solution at 260 nm. At this wavelength, an absorbance of 1.0 corresponds to a concentration of 50 and 33 μg·mL^{-1} for double- and single-stranded DNA, respectively, at a 1 cm path length.

Different substances can often be distinguished by the use of spectrophotometric measurements if they exhibit different absorption spectra. In an ideal solution, we can assume that the total absorption at any wavelength, is equal to the sum of the individual absorptions. This principle has been used to determine the protein and nucleic acid content of cell extracts and cell fractions or to assess the purity of nucleic acids. At 260 nm, ribonucleic acid (RNA) has a peak absorption of 50.8 mL·mg^{-1}·cm^{-1}, while at 280 nm, its absorption is 24.8. Proteins containing an average proportion of aromatic amino acids have a small peak of 2.06 mL·mg^{-1}·cm^{-1} at 280 nm, which decreases to 1.18 mL·mg^{-1}·cm^{-1} at 260 nm (5). By comparing the ratio of absorbance at these two wavelengths, the relative RNA and protein concentration can be determined. The method assumes that the total absorbance at 260 and 280 nm is due to the sum of the absorbances of the constituent protein and nucleic acid. Solving the simultaneous equations yields the concentrations as follows:

Protein concentration (mg/mL)

$$= 0.674 \times A_{280} - 0.33 \times A_{260}$$

Nucleic acid concentration (mg/mL)

$$= -0.016 \times A_{280} + 0.027 \times A_{260}$$

This relationship varies for different proteins and nucleic acids. For accurate work, a calibration must be carried out to determine the actual coefficients for the samples studied. For an accurate determination, specific reagents can selectively react with DNA, RNA, and protein, respectively, for its own quantification to avoid the interference from each other (6,7). For example, the bicinchoninic acid (BCA) assay, Coomassie Blue-G 250 dye binding assay (the Bradford), and the Lowry method are commonly used colorimetic methods for protein quantification (7).

Sometimes samples consist of light-absorbing particles in suspension rather than molecules in solution. For example, while many bacteria do not have chromophores that absorb visible light, suspensions consisting of intact bacteria display apparent absorption due to the scattering of light (Rayleigh scattering). Rayleigh equation demonstrates the reciprocal forth-power relationship between light scattering intensity and wavelength. This apparent absorption has been used to quickly measure bacterial or viral concentrations. A calibration curve is determined by comparing absorbance with viable cell count. For example, the concentration of *Salmonellatyphimurium* in suspension can be determined from the absorption at 440 nm, since the absorption is linear with concentration such that 1.07×10^8 bacteria mL^{-1} yields an apparent absorbance of

1.0, for a 1 cm path length. Methods that utilize the apparent absorption of scattering solutions are often referred to as turbidimetry. Turbidimetric methods can also be used to monitor the kinetics of enzymatic processes that cause changes in the level of light scattering. For example, the enzyme rennin will coagulate milk, leading to an increase in its scattering (monitored at 600 nm) (8). The initial rate of change of apparent absorbance can therefore be measured and used to determine the concentration of active enzyme added. In some cases, the material being measured consists of light-absorbing particles in suspension with chromophores. For example, the absorbance of bacteriophages is primarily due to DNA. Since scattering always begins at wavelengths removed from the absorption, its contribution can be subtracted by measuring the absorbance at longer wavelengths than that of the chromophore and by linearly extrapolating the scattering from the absorption peak. For kinetic measurements on scattering solutions, difference spectra are often used, where the wavelengths are chosen such that one is at the chromophore maximum, while the other is at wavelengths where the only absorption is due to scattering. Specialized instruments that allow for absorbance measurement at two different wavelengths simultaneously are required for such measurements.

Most of the approaches to detect specific DNA sequence are expensive and time-consuming, such as fluorescent microarrays. An inexpensive and rapid assay for the identification of DNA sequence and single nucleotide polymorphisms has been recently described by using a colorimetric method with nanotechnology (9). In this assay, the color of negatively charged gold nanoparticles (Au-nps) is very sensitive to the degree of aggregation. Single-stranded DNA can stabilize Au-nps and prevent the salt-induced aggregation. This method has been applied into clinical samples and successfully identified the relationship between a specific gene and a fatal arrythmia (9).

Due to the contribution of nanotechnology, Au-nps can be utilized as ideal color reporting groups as the colorimetric biosensors. For example, it can provide the qualification and quantification of Pb^{2+} (10), and the detection of polynucleotides (11).

INSTRUMENTATION

Presently, most colorimetry is done using spectrophotometers rather than colorimeters. In principle, spectrophotometers and colorimeters are similar, with the latter being very simple and inexpensive versions of spectrophotometers. Both instruments consist of a light source, a means of selecting wavelength, a sample compartment, and a detector of transmitted light, as shown in a single-beam spectrophotometer (Fig. 3). The source generally consists of an incandescent tungsten lamp for measurements in the visible range between 350 and 700 nm and a hydrogen deuterium, or xenon arc lamp for measurements in the UV range down to 190 nm. Thermal lens spectrophotometry, a new high sensitivity method utilizes a laser as its source (12). A stabilized current is generally provided, especially for measurements in the UV range to

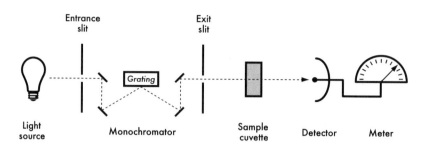

Figure 3. A single-beam spectrophotometer.

prevent fluctuations in the intensity of the lamp. Light from the lamp is collimated or focused by a lens onto a wavelength selector. In the case of the simplest colorimeters, the wavelength selector is a broadband filter that yields a distribution of wavelengths with a width of ~40 nm. In spectrophotometers and more sophisticated colorimeters, the wavelength is chosen with a monochromoter by either reflecting the beam off a grating or by passing it through a prism. In a spectrophotometer, the bandwidth of light can be set by adjusting the exit slit. As the slit opening decreases, the light incident on the sample decreases, but the spectral resolution improves. Excessively wide bandwidth can lead to decreased peak absorption and interference from other substances absorbing at nearby wavelengths. In a colorimeter, the bandwidth is fixed, depending on the properties of the filters. While it is possible to obtain narrow bandpass filters if necessary, they may cost as much as the colorimeter.

The monochromatic light is then focused on the sample, contained in a transparent rectangular cuvette, a test tube, or a microplate with 6–384 wells. In some spectrophotometers, the sample compartment has been designed to accommodate with 96–1536-well microplate for high throughput screening (13). Colorimeters generally use test tubes made of any clear materials, such as glass or plastics, as long as they allow any required chemical reactions to be carried out in the same disposable container in which the measurement is carried out. Spectrophotometers and automated colorimeters designed to determine large numbers of samples usually hold rectangular cuvettes and microplates. These are generally made of glass or plastics for measurements within visible light wavelengths, and of quartz for measurements < 350 nm in the UV range. The transmitted light is incident on a phototube that records the intensity of the light reaching it. Some spectrophotometers are double-beam instruments, which electro-

nically subtract an absorption blank. There are two types of the double-beam spectrophotometers. One is called the double -beam in-time spectrophotometer (Fig. 4). In this instrument, a single beam of light from the monochromator is alternately switched between the sample and the reference cuvettes. The two alternate beams then reach a single detector in separate time to provide an alternating signal with an amplitude that is proportional to the difference of the light intensities between the sample and the reference. The other type is called the double-beam in-space spectrophotometer (Fig. 5). In this instrument, two separate light pathways are created by a beam splitter and mirrors. One beam passes through the sample cuvette, and the other beam the reference cuvette. The light intensities of the sample and the reference cuvettes are measured by two separate detectors and the difference is recorded as the absolute photometric measurement.

Many spectrophotometers are still designed as single-beam instruments (Fig. 3); the solvent absorption is separately determined, stored, and subtracted by an interfaced microcomputer. In colorimeters and other single-beam instruments, the absorption is set to zero or the transmission is set to 100% when the reference blank is placed in the sample compartment. The detector is designed to give a direct reading of either transmission or absorbance. The newer spectrophotometer is a multifunctional instrument with a triple-mode cuvette port and microplate reading capability. The detection modalities include absorbance, fluorescence intensity, fluorescence polarization, time-resolved fluorescence, and luminiscence. With a dual-monochromator, the filters for specific wavelengths are not required in these spectrophotometers. (e.g., Spectra-Max M5 by Molecular Devices Corp.). The function of spectrophotometer also has improvement in the rate of data acquisition. It can be as fast as 50 μs, which is beneficial in rapid kinetics.

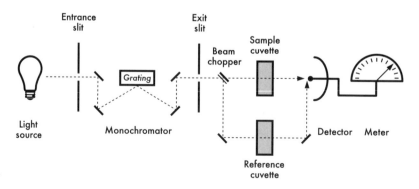

Figure 4. A double beam in-time spectrophotometer.

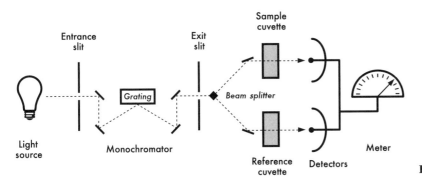

Figure 5. A double-beam in-space spectrophotometer.

CLINICAL APPLICATIONS OF COLORIMETRIC METHODS

Colorimetry has seen wide applications in clinical and medicinal chemistry and in biochemistry. Colorimetric methods are used to measure concentration or enzyme activity. Lately, colorimetric methods are also applied in genomics and proteomics research, such as the single nucleotide polymorphism analysis (14), the detection of protein microarrays (15), and so on. The concentration of a substance can be determined directly and immediately if the unknown has a distinct absorption band that does not overlap other substances in the assay mixture. If not, the unknown may be analyzed indirectly on completion of one or more chemical reactions, yielding a colored compound with a definite stoichiometric relationship to the unknown compound. The reaction must be rapid and specific and must react completely with the unknown to be of use. In contrast to quantitative measurements, kinetic measurements determine initial reaction kinetics by the rate of appearance of an absorbing product or by the rate of disappearance of an absorbing reactant. Kinetic measurements are carried out either directly by monitoring changes in the absorption of substrates or cofactors or indirectly by coupling a second, third, or additional enzyme reactions to the primary one. Each of these cases will be discussed and illustrated later.

Due to the development of new automated diagnostic instruments, the manual colorimetric assays have become less useful in most clinical laboratories. However, most tests performed in automated instruments are based on simple colorimetry. Therefore, the principles in colorimetry are essential for the improvement and development of new automated assays. Furthermore, manual colorimetric methods are still used in handling small numbers of samples, such as in small clinical laboratories or for special diagnostic purposes.

There are several general precautions that should be taken in colorimetric measurements of clinical samples. These are as follows:

1. Sample preparation. Most clinical samples (e.g., blood and urine) contain a large amount of components other than the analyte. The results of the test are only meaningful if the background reading of the sample has been subtracted. The background must be obtained by using proper controls. Usually, when a measurement is made by the addition of a reagent, an identical sample without the reagent can be used as a control or blank. When an enzyme activity is measured, a control or blank can be obtained by excluding the specific substrate or by including a specific inhibitor for the enzymatic reaction. Because of instability, many specimens must be handled immediately after they reach the laboratory and cannot be stored. The turbidity of a sample will cause false readings in colorimetry. This problem generally can be eliminated either by centrifugation, filtration, or comparing with identical controls.

2. Buffer solutions. The absorbances, especially in the visible wavelength range, are usually pH dependent. Furthermore, the type and concentration of ions in the buffer may also influence the absorbance of a molecule or the rate of reaction. Therefore, careful attention must be paid to the solvent and buffer system. In addition, the rates of enzymatic reactions are highly pH dependent. It is therefore important to choose a buffer with an ionization potential (pK) appropriate to the optimal pH for the assay. For example, imidazole or triethanolamine buffer is commonly used in the pH range of 6.5–7.0.

3. Temperature. Kinetic measurements use enzymes to catalyze the reactions of intensely colored substances. Enzymatic reactions are highly temperature dependent. It is therefore necessary to use isothermal control when carrying out such assays. If not, it is necessary to know precisely the reaction temperature. If the reaction is carried out in the colorimeter, care must be taken to ensure that the temperature of the reaction mixture is not affected by the heat produced in the sample chamber.

4. Inhibitors. Enzymatic reactions are subject to competitive inhibition or to the action of specific inhibitors leading to apparent decreases in the measured activity or concentration of the unknown. It is therefore extremely important to eliminate these interfering substances or to include the proper controls during the assay. In cases of product inhibition, the initial rate should be used for the determination of the enzymatic activity.

In the following sections, several colorimetric measurements of clinical importance will be discussed. They are used primarily as examples to indicate the standard procedures of colorimetry in clinical laboratories. The number of applications of colorimetry is too vast to present an extensive survey in this article.

Quantitative Measurement

Direct Measurement. Direct colorimetric measurement for quantitative assays are limited to very few cases in which the compound of interest itself has an intense absorption and in which there is no interference from other components in the samples. One example is the determination of hemoglobin in blood. Since oxygenated and deoxygenated hemoglobins, carboxyhemoglobin, and hemoglobin derivatives have different absorption spectra, the concentration of a specific type of hemoglobin, as well as the oxygen-binding capacity, can be determined by measuring the ratio of absorption (A) at two different wavelengths [16] (e.g., A_{562} : A_{540} for carboxyhemoglobin and A_{560} : A_{506} or A_{650} : A_{825} for oxygen-binding capacity), similar to that described above for the protein–nucleic acid measurement.

Bilirubin measurement in blood samples can also be done by direct measurement at 460 nm [17]. The absorbance is compared with that of a standard solution of potassium dichromate and is reported in term of units of icterus index. One unit of icterus is equivalent to a 1:10,000 solution of potassium dichromate. This determination is only approximate, because there are many other components in blood, mostly carotenoid pigments, which will interfere with the readings.

Indirect Measurements

Complex Formation. One of the most common applications of colorimetry is in the determination of protein concentration. Colorimetric determination of the total serum protein by the Biuret reaction is still used in many clinical laboratories. In this method, proteins react with copper sulfate in alkaline solution to form a Biuret complex that can be determined by the intensity of the violet color with an absorption at 555 nm [18]. The Biuret reaction is a relatively straightforward, precise, and accurate method. Usually, 0.1 mL serum is added to 5 mL of Biuret Reagent solution, which contains sodium, potassium tartrate, potassium iodide, and copper sulfate in $0.2\,M$ NaOH. After incubation at 30–32 °C for 10 min or at room temperature for 20 min, the reaction mixture is read at 555 nm. The sample absorbance is generally compared to human or bovine serum albumin (BSA) standards. The color of the Biuret complex is stable for several hours after it reaches the maximum intensity. The measurement of protein by Biuret measurement is subject to interference by other substances in serum. In serum separated from moderately hemolyzed blood or serum with high bilirubin content potassium cyanide is usually included in the reaction mixture to correct the difference. Dextrans may also interfere by causing turbidity. For these determinations, appropriate blanks should be used instead of just saline or water, as described in most general procedures.

Many other methods use nearly colorless chemical reagents that complex with the protein to form highly colored compounds. One of them, the Lowry method [19],

is the single most utilized assay in biochemistry research and is the most cited reference in the biochemical literature. However, it is not widely used in clinical applications. A variety of dyes, which form strong binding complexes with protein, is also used to quantitate protein concentration. For example, Coomassie Brilliant Blue R-250 is commonly used to visualize and quantify proteins in gels separated by electrophoresis and to quantify proteins eluted from gels [20]. To visualize proteins, gels are stained with dye in 50% methanol–10% acetic acid solvent, after which excess dye is removed by the solvent. To accurately quantitate protein concentration, dye bound to protein is removed by electroelution or chemical elution, and absorbance of the solutions at 595 nm is measured. Coomassie Brilliant Blue stained electrophoresis gels are quantitated by direct scanning colorimetry at the same wavelength. This dye is most useful to quantify proteins present in the amount range of 0.5–50 µg. The binding of silver to protein has been used to visualize and to quantitate proteins separated in gels [21]. While this method is ~ 100 times more sensitive than Coomassie Brilliant Blue staining, it is also much more expensive due to the cost of the silver reagent. Recently, several visible dyes have been developed for the staining of protein in electrophoresis [22]. Many of them can reach the sensitivity at the ng level, and some of them, such as 3, 3′-diethyl-9-methyl-4,5,4′,5′-dibenzothiacarbocyanine (Stain-All), can simultaneously stain DNA and RNA with the appearance of different colors.

A complex between the dye, Methylene Blue, and protein has been used to quantitate the growth of cells in culture and their inhibition by cytotoxic drugs. Cultures are stained, destained, then solubilized overnight in a 1% aqueous Sarkosyl solution, and quantified at 620 nm. This assay can be automated by using 96-well microtiter plates and commercially available spectrophotometers designed for enzyme-linked immunosorbent assays (ELISA) [23]. The dye complex can also show specificity and can be used to assay a single protein in a specimen. For example, serum albumin can be determined by its high affinity binding to Bromocresol Green [24].

Another important colorimetric method by complex formation in clinical laboratories is the measurement of metal ions by chelators. For example, serum iron levels are measured by the complex formation between ferrous and ferrozine. Serum iron can be determined either with or without the removal of proteins by trichloroacetic acid precipitation. Ferric is usually reduced to ferrous by ascorbic acid. The complex of Fe^{2+}-ferrozine has a dark violet color with an absorbance at 562 nm. Advantages of this method are its simplicity, sensitivity (molar absorbance $= 27,900$), and the constant absorbance in a wide range of pH (from pH 4–10) [25].

Colorimetry is commonly used for water testing, including specific impurities, such as ammonia, calcium, chlorine, copper, iron, nitrite, phenol, phosphate, sulfate, and sulfide, or total hardness. These tests are available in kit form and can be used with portable colorimeters for field testing. The methods generally utilize either complex formation or derivatization of the inorganic impurity to yield a highly colored compound.

Derivatization. Besides complex formation, substances can be detected colorimetrically after chemical modification to give colored products. One example is the diazotization of bilirubin. As mentioned previously, the direct measurement of bilirubin by its yellow color is inaccurate, because there are many pigments in serum that can interfere with the detection. For a more reliable assay, bilirubin is reacted with diazotized sulfanilic acid to give azobilirubin, which is red-violet in moderately acid solution and blue in strongly acid or alkaline solution. Most of the methods currently used in clinical laboratories (e.g., the Jendrassik–Grof method and the Malloy–Evelyn method) are based on this diazotization reaction.

Bilirubin exists in serum in two forms, the glucuronide-conjugated and the free form. The conjugated form (clinically known as "direct bilirubin") is more soluble in water and can be detected by a direct reaction with the diazo reagent. The free form (clinically known as "indirect bilirubin") is less soluble in water and can be detected by diazo reagent only if other reagents are also included. In the Malloy–Evelyn method, ethanol is added to increase the solubility of free bilirubin. In the Jendrassik–Grof method, caffeine–benzoate is added to displace serum protein-bound bilirubin. Therefore, both free and conjugated bilirubin can be detected by the diazo reagent. The Jendrassik–Grof method is generally considered the method of choice for the measurement of bilirubin in most clinical laboratories (26). The diazo reagent in this method is a mixture of sulfanilic acid and sodium nitrite in hydrogen chloride. This reagent should be prepared within 30 min of use. Serum is reacted with the diazo reagent for exactly 10 min, and the reaction is stopped by the addition of ascorbic acid solution. A strongly alkaline solution (tartrate in 1 N NaOH) is then added, and the color developed is compared with a standard curve of absorbance at 600 nm. Bilirubin is extremely sensitive to light. Sunlight can markedly decrease the bilirubin content in samples (as much as 50%/h). However, serum specimens can be stored for many weeks or months without appreciable change of bilirubin content if they are kept in the dark in a freezer. Other examples of derivatization include the determination of cholesterol by Liebermann–Burchard reaction. In this reaction, cholesterol is reacted with a mixture of acetic anhydride, acetic acid, and sulfuric acid to give a bluish green product that can be measured colorimetrically.

Enzymatic Conversion. Enzymatic conversion can be used to determine concentrations by the measurement of either the loss of substrate, the creation of product, or the change of the cofactor (coenzyme). Nicotinamide adenine dinucleotide (NAD) and nicotinamide adenine dinucleotide phosphate (NADP) and their respective reduced forms, NADH and NADPH are cofactors that have been extensively utilized for analytical purposes in colorimetric assays. These cofactors serve as the natural oxidizing and reducing agents in a wide variety of enzyme systems. With the appropriate enzyme, they can selectively oxidize or reduce a single substrate in the presence of innumerable other compounds, making possible the analysis of a single compound in a complex mixture. Both NADH and NADPH have identical absorption bands with peak absorption at

340 nm. The compounds NAD and NADP do not absorb at this wavelength. Therefore, changes in oxidation or reduction can be measured colorimetrically. In addition, the reduced forms can be completely destroyed at acidic pH without affecting the oxidized forms and the oxidized forms can be completely destroyed at basic pH without affect the reduced forms (27).

The measurement of glucose-6-phosphate in the range of 20–200 μM can be done in one step by monitoring the reduction of NADP to NADPH. This reaction is carried out by the oxidation of glucose-6-phosphate to 6-phosphate gluconolactone by the enzyme glucose-6-phosphate dehydrogenase, for which NADP is the coenzyme and is reduced to NADPH. The reaction is quantitated by measuring the 340 nm absorption of NADPH after completion of the reaction. Biological substances that do not react directly with nicotinamide nucleotides can be analyzed in this system by one or more additional enzymes. For example, glucose can be determined by the following two-step reaction that convert NADP to NADPH.

$$\text{Glucose} + \text{ATP} \xrightarrow{\text{hexokinase}} \text{Glucose-6-phosphate} + \text{ADP}$$

$$\text{Glucose-6-phosphate} + \text{NADP} \xrightarrow[\text{dehydrogenase}]{\text{glucose-6-phosphate}}$$

$$\text{6-Phosphate gluconolactone} + \text{NADPH} + \text{H}^+$$

An example of a three-step assay is the enzymatic analysis of inorganic phosphate (P_i), carried out with three simultaneous reactions, ends in the conversion of NADP to NADPH (28).

$$\text{Glycogen} + \text{P}_i \xrightarrow{\text{phosphorylase } \alpha} \text{Glucose-1-phosphate}$$

$$\text{Glucose-1-phosphate} \xrightarrow{\text{phosphate glucomutase}} \text{Glucose-6-phosphate}$$

$$\text{Glucose-6-phosphate} + \text{NADP} \xrightarrow{\text{glucose-6-phosphate dehydrogenase}}$$

$$\text{6-Phosphate gluconolactone} + \text{NADPH}$$

Hydrogen peroxide can be readily quantitated colorimetrically by its peroxidase-catalyzed reaction with a colorless chromogenic oxygen acceptor to form an intensely colored product. This reaction can therefore be used to measure the concentration of any organic compound that, on reaction, will produce hydrogen peroxide. For example, glucose levels in serum can be determined by indirect detection using the enzyme glucose oxidase. This enzyme catalyzes the oxidation of glucose to gluconic acid and hydrogen peroxide. The amount of hydrogen peroxide produced can be detected by reaction with peroxidase, and most commonly o-dianisidine. The oxidized product of o-dianisidine has a strong absorbance ~ 540 nm. The overall reactions are as follows:

$$\text{Glucose} + \text{O}_2 + \text{H}_2\text{O} \xrightarrow[\text{oxidase}]{\text{glucose}} \text{Gluconic acid} + \text{H}_2\text{O}_2$$

$$\text{H}_2\text{O}_2 + o\text{-Dianisidine} \xrightarrow{\text{peroxidase}} \text{H}_2\text{O} + \text{Oxidized } o\text{-dianisidine}$$

Since this measurement is dependent on the color formation of the oxidized chromogen, other substances in the sample can interfere with this reaction by competing with chromogens for hydrogen peroxide and by reducing the final color intensity. Some of the interfering substances in serum are creatine, uric acid, ascorbic acid, bilirubin, and glutathione. Therefore, results obtained directly from serum tend to be lower than the true values. Serum samples, especially with red cells, or extensive hemolysis require precipitation of the protein to remove interfering enzymes, with measurements carried out on the protein-free filtrates. Direct measurements of glucose from urine specimens cannot be done by the glucose oxidase method because of the presence of enzyme inhibitors.

Glucose oxidase is highly specific to β-d-glucose. In aqueous solutions, glucose exists 36% in the α form and 64% in the β form. The same ratio of the two forms is also found in serum. Crystalline glucose, however, can be either the α or β form, depending on the conditions of crystallization. In order to correct the difference in standard solutions, some commercial preparations of glucose oxidase contain another enzyme, mutarotase, which accelerates the conversion of α form to β form during the assay. Alternatively, standard solutions from crystalline glucose can be prepared 2 h before the determination to allow the mutarotation to reach equilibrium. The final assay solution contains the following components: glucose oxidase ($5\,U{\cdot}mL^{-1}$), peroxidase ($16\,U{\cdot}mL^{-1}$), and o-dianisidine ($0.6\,\mu mol{\cdot}mL^{-1}$) in a pH 7.0 phosphate buffer. Under these assay conditions, glucose can be measured at a range of up to 250 mg·dL^{-1}. Samples with higher glucose concentrations should be diluted before the determination. Note that most chromogens used in the peroxidase reaction (e.g., o-dianisidine and o-tolidine) are potential carcinogens, and precautions should be taken when handling these compounds.

Colorimetric assays have found substantial applications in determination of concentration by ELISA. For example, the previously described assay to measure glucose concentration has been adapted as follows (29). The production of hydrogen peroxide from glucose is catalyzed by glucose oxidase, an enzyme consisting of apoglucose oxidase and flavin adenine dinucleotide (FAD) cofactor. If a ligand is bound to FAD and antibodies to that ligand are added to the solution, the enzyme will be inactive. Competition by free ligand will make conjugated FAD available for apoglucose oxidase. Since this reaction is enzymatic, excess glucose can be added, which leads to amplification of the signal. Therefore, very low concentrations of ligand can be assayed colorimetrically by the oxidation of glucose.

Besides glucose, the production of hydrogen peroxide as an intermediate has been employed to measure a number of clinically important substances including lecithin, high density lipoprotein cholesterol, phospholipids, digoxin, and triglycerides in human serum (30). In the last example, triglycerides are hydrolyzed to glycerol and fatty acid with lipase. The resulting glycerol is phosphorylated by l-α-glycerophosphate oxidase to produce hydrogen peroxide, which reacts in the peroxidase-catalyzed coupling of 4-aminoantipyrine and sodium 2-hydroxy-3,5-dichlorobenzene sulfonate to form an intense red product.

Kinetic Measurement

Kinetic measurements are generally used to determine the activity of enzymes. The expression of enzyme activity units (U) has been extremely arbitrary and inconsistent over the years. Historically, activity has been measured colorimetrically, and often activity for many common enzymes is expressed as an initial rate of absorbance change of a particular chromogen at the appropriate wavelength per unit time. More recently, enzyme activities have been expressed on a molar basis (i.e., the molar rate of destruction, or creation of substrates or products per unit time). Kinetic measurement of enzyme activity requires more control than quantitative measurements, since activity is usually a function of pH, ionic strength, temperature, substrate concentration, and the presence of activators and inhibitors.

Direct Measurement. An enzymatic reaction can be monitored by the measurement of either the rate of loss of the substrate, the formation of the product, or the rate of change of the cofactor (coenzyme) concentration. All of these methods are commonly used to determine enzyme activity.

Substrate. In an enzymatic reaction, which converts a substrate to a colored product or a colored substrate to a colorless product, enzymatic activity can be detected colorimetrically. An example of this type of measurement is prostatic acid phosphatase determination (31). Acid phosphatase is present in many tissues, including bone, liver, kidney, erythrocytes, and platelets. Its level in serum is measured as a diagnostic for the detection of metastatic, prostatic carcinoma. Therefore, substrates specific to prostatic acid phosphatase isoenzyme are most desirable for this assay. One of the most commonly used substrates is thymolphthalein monophosphate, which, upon hydrolysis by the enzyme, produces thymolphthalein, a compound with intensive absorbance at 590 nm in alkaline solution.

$$\text{Thymolphthalein monophosphate} \xrightarrow[\text{phosphatase}]{\text{acid}}$$

$$\text{Thymolphthalein} + HPO_4^{2-}$$

The course of this reaction can be monitored by a direct colorimetric measurement of product formation. Note that thymolphthalein monophosphate is not completely specific for prostatic acid phosphatase. However, unlike acid phosphatase from other sources, prostatic acid phosphatase can be inhibited by tartrate. Thus, from the assays of the enzyme activity in the presence or absence of tartrate, one can determine the phosphatase activity specific to the prostatic secretion. Acid phosphatase is extremely labile at the neutral pH of normal serum. Therefore, specimens for the enzyme measurement should be handled on ice and delivered to the laboratory as rapidly as possible. Serum separated in the laboratory should be acidified by acetate buffer to a pH range (pH 5–6) to stabilize the enzyme. Samples should be stored in the freezer if they are not to be assayed on the same day.

Acid phosphatase activity is highly dependent on pH and the specific ion in the buffer. Small changes in these parameters may cause a large difference in the enzyme activity measurements. To avoid a discrepancy between assays, a standardized procedure has to be followed carefully. Additional factors, such as the ratio of serum sample to the volume of final mixture, surfactant (Brij-35) for activation of enzyme, and the source and concentration of the substrate, must be consistent. For example, thymolphthalein monophosphate obtained from different commercial sources has been found to give as much as a 40% deviation in the enzyme activity by the colorimetric measurement. A typical assay procedure uses a substrate solution consisting of 0.6 mL of 0.15 M acetate buffer (pH 5.4), 1 mM thymolphthalein monophosphate and 1.5 g·L^{-1} Brij-35, a sample volume of 50 μL serum, a 30 min incubation at 37 °C, and final color development in 1 mL of 0.1 M NaOH and 0.1 M Na$_2$CO$_3$. The absorbance at 590 nm is then measured. In controls, the substrate solution is incubated without serum sample. Acid phosphatase activity is determined by subtracting the reading of the control from that of the sample.

The activity of a large number of proteolytic enzymes can be measured directly using specific synthetic substrates designed for that purpose. For example, the activity of leucine aminopeptidase can be directly measured by the hydrolysis of l-leucyl-β-naphthylamide to β-naphthylamine, which can be read at 560 nm. The clinically important γ-glutamyl transpeptidase can be measured by its reaction with glutamyl nitroanilide to form nitroaniline, which can be read at 405 nm.

Cofactors. The cofactors NAD and NADP find wide use in the kinetic measurement of enzyme activity, as well as in the previously discussed quantitative measurement of substrate concentration. One of the most important enzymes to be measured in clinical laboratories is lactate dehydrogenase. Marked increase of the enzyme level has been found following myocardial infarction. Elevations have also been reported in leukemia, in anemias, and in some liver diseases. This enzyme catalyzes the interconversion of lactate and pyruvate in the presence of NAD or NADH.

$$\text{Lactate} + \text{NAD} \underset{\text{dehydrogenase}}{\overset{\text{lactate}}{\rightleftarrows}} \text{Pyruvate} + \text{NADH}$$

In clinical laboratories, the reaction can be detected in either direction (i.e., using either lactate or pyruvate as substrate). A simple colorimetric method to detect the change of absorbance at 340 nm is sufficient to measure the enzyme activity (32). Generally, a small aliquot of serum sample is added to a cuvette and is mixed with a substrate solution containing either lactate and NAD or pyruvate and NADH. The reaction mixture in the absence of enzyme is set up, and the absorbance at 340 nm is measured. A dilution of enzyme is added and the OD$_{340}$ is either monitored continuously or measured every minute or two for an appropriate time interval. The initial rate of increase in absorbance will be linear in time and proportional to the quantity of enzyme. The enzyme activity will therefore be proportional to the rate of change of the absorbance. A standard assay procedure is performed at 30 °C. When the absorption change is measured in a spectrophotometer without constant temperature equipment, the effect of temperature on the enzyme activity must be considered. For example, the enzyme activity measured at 30 °C is 1.44-fold higher than that at 25 °C. Any enzymes that require the NADs as cofactors can be assayed in this manner. In clinical laboratories, lactate dehydrogenase is usually determined with samples of serum. Normally, serum contains no inhibitors or interfering substances for this enzyme assay. However, blood samples with marked hemolysis should be avoided, because they may give false elevations of this enzyme in the serum. It has been reported that commercial NADH contains inhibitors for the dehydrogenase reactions. Therefore, when pyruvate is used as the substrate for lactate dehydrogenase determination, inhibitors in the NADH solution should be considered. Many other enzymes can be determined by direct colorimetric measurement of NAD–NADH or NADP–NADPH conversion. Some of these enzymes with clinical importance are isocitrate dehydrogenase, glutamate dehydrogenase, and glucose-6-phosphate dehydrogenase.

Indirect Measurements. The activity of an enzyme that is not NAD or NADP dependent can nonetheless be measured with this system if its products are substrates for other enzymatic reactions requiring pyridine nucleotides. For example, phosphate glucomutase can be determined as follows

$$\text{Glucose-1-phosphate} \xrightarrow{\text{phosphate glucomutase}} \text{Glucose-6-phosphate}$$

$$\text{Glucose-6-phosphate} + \text{NADP} \xrightarrow[\text{dehydrogenase}]{\text{glucose-6-phosphate}}$$

$$\text{6-Phosphate gluconolactone} + \text{NADPH}$$

To prevent interfering side reactions, this assay is often carried out in two steps. The phosphate glucomutase is allowed to react for a measured time period, without added NADP or glucose-6-phosphate dehydrogenase after which the reaction is stopped with heat. The quantity of glucose-6-phosphate produced during that time period is then determined by carrying out the second reaction. Even though NADPH can be directly monitored at 340 nm, it can also be measured indirectly by the interaction with a tetrazolium salt, 3-(4,5-dimethylthiazol-2-yl)-2,5-diphenyltetrazolium bromide (MTT), to reduce the latter to a more intensely colored form that absorbs in the visible rather than the UV range. The reducing reaction occurs only in the presence of an intermediate electron carrier, such as phenazine methosulfate. The final color of the reduced tetrazolium salt has an absorption ~520 nm. This same reagent is now used in place of [^{3}H]thymidine to measure cell proliferation or complement mediated cytotoxicity in lymphocytes. 3-(4,5-Dimethylthiazol-2-yl)-2,5-diphenyltetrazolium bromide is cleaved to a dark blue formazan product only if the cell has active mitochondria while even newly dead cells show no color change.

The enzyme creatine kinase is measured in clinical laboratories for the detection of diseases related to skeletal or heart muscles. In the presence of magnesium ions, this enzyme catalyzes the reaction of creatine phosphate with adenosine diphosphate (ADP):

$$\text{Creatine phosphate} + \text{ADP} \xrightarrow{\text{creatine kinase}} \text{Creatine} + \text{ATP}$$

This reaction can be detected by coupling with two other enzymes (33). The first enzyme, hexokinase, can use ATP, one of the products of the previous reaction, to convert d-glucose to d-glucose-6-phosphate, which is subsequently detected with the second enzyme, d-glucose-6-phosphate dehydrogenase, by measuring the conversion of NADP to NADPH as the increase of absorbance at 340 nm. The reactions catalyzed by the auxiliary enzymes are as follows:

$$\text{ATP} + \text{D-glucose} \xrightarrow{\text{hexokinase}} \text{ADP} + \text{glucose-6-phosphate}$$

$$\text{Glucose-6-phosphate} + \text{NADP} \xrightarrow{\text{glucose-6-phosphate}}$$

$$\text{D-Glucose-6-phosphate lactone} + \text{NADPH}$$

Since creatine kinase in serum is rapidly inactivated by oxidation, reducing agents are required to reactivate the enzyme in each assay. The most commonly used reactivating compound is *N*-acetylcysteine. Other sulfhydryl compounds can also be used if they do not interfere with the measurement (i.e., absorbance at 340 nm, solubility, and odor). Serum with extensive hemolysis should be avoided, because it may give falsely elevated values. One of the factors from hemolyzed red cells is the enzyme, adenylate kinase, which can generate additional amounts of ATP from ADP without creatine phosphate. To eliminate this artifact, 3 mM of adenosine monophosphate (AMP) should be included in the assay solution to inhibit adenylate kinase activity. Higher concentrations of AMP are required in markedly hemolyzed serum samples. However, at high concentrations, AMP can also inhibit creatine kinase activity. The false activity contributed from enzymes other than creatine kinase can be detected by running an appropriate control, without creatine phosphate in the final assay solution. The coupling of three reactions in this manner leads to a complicated assay that requires 10 different reagents added to the sample. These include imidazole buffer at pH 6.7 (100 mM), creatine phosphate (30 mM), ADP (2 mM), Mg^{2+} (10 mM), d-glucose (20 mM), NADP (2 mM), AMP (5 mM), *N*-acetylcysteine (20 mM), glucose-6-phosphate dehydrogenase (1.5 U·mL^{-1}), and hexokinase (2.5 U·mL^{-1}).

FUTURE DEVELOPMENTS

Colorimetric methods will continue to be important analytical tools in research and clinical laboratories. Applications that replace the need for radioactive compounds will continue to be developed because of the increasing costs and difficulties associated with the processing and disposal of radioactive materials. In the same way that the ELISA has replaced radioimmunoassay (RIA), colorimetric methods will continue to replace applications requiring radioisotopes. For example, in recent developments of microarray technology, colorimetry is still one of the major detection methods for the measurement of the signals (14,34). However, the sensitivity of colorimetric methods is limited by the absorption of the chromophores, new reagents with very high extinction coefficients will need to be developed for direct colorimetric measurements (35). This limitation is also overcome by the coupling of a second reaction to the primary measurement. Such amplifying systems will be increasingly used as a means to increase the colorimetric sensitivity. Modified colorimetric methods, such as the kinetic spectrophotometric method, can also

markedly improve the sensitivity of the determination. For example, trace concentrations of iron can be determined by the oxidation of *o*-tolidine (36). The rate of oxidation, which can be measured colorimetrically at 440 nm for the oxidized *o*-tolidine derivative, is proportional to the concentration of ferric ion. This method can detect as low as $4 \times 10^{-7}M$ iron in acidic solutions. The use of lasers as light sources in colorimetry has opened a new frontier of analytical chemistry. The method, called "thermal lens spectrophotometry" is based on the direct measurement of the absorbed radiant energy by the "thermal lens effect" and can detect a very small absorption by increasing the power of the heating laser (12). This method can determine accurately an absorbance that is lower than 0.001.

Since the sensitivity of fluorimetry is often greater than that of colorimetry, it will probably find more new applications than colorimetry especially in biomedical measurements. However, the instrumentation and the stability of fluorimetric measurements are generally more expensive and more complex than colorimetry. In addition, new colorimetric instruments have been developed that are capable of measuring several wavelengths simultaneously, of processing large numbers of samples automatically, and of reading the results of samples contained in multiwell or multicuvette containers. Furthermore, colorimeters with computer interfaces will automatically calculate results, will average multiple determinations, and will print and store the final data. This kind of instrumentation, already widely used in ELISA and microarrays, will be increasingly borrowed for use in other colorimetric assays. Therefore, colorimetry will continue to be used because of its speed, simplicity, versatility, and low cost. It can be anticipated that more sophisticated methods of colorimetric measurement, in combination with other advanced technologies such as nanotechnology (11), will continue to be developed into new analytical methods with various biomedical and clinical applications.

BIBLIOGRAPHY

1. Bugay DE. Characterization of the solid-state: spectroscopic techniques. Adv Drug Deliver Rev 2001;48:43–65.
2. Stephenson GA, Forbes RA, Reutzel-Edens SM. Characterization of the solid state: quantitative issues. Adv Drug Deliver Rev 2001;48:67–90.
3. Yu LX, Lionberger RA, Raw AS, D'Costa R, Wu H, Hussain AS. Applications of process analytical technology to crystallization processes. Adv Drug Deliver Rev 2004;56:349–369.
4. Stone DK, Xie X-S, Racker E. An ATP driven proton pump in clathrin-coated vesicles. J Biol Chem 1983;258:4059.
5. Warburg O, Christian W. Isolation and crystallization of enolase. Biochem Z 1941;310:384.
6. Morozkin ES, Laktionov PP, Rykova EY, Vlassov VV. Fluorometric quantification of RNA and DNA in solutions containing both nucleic acids. Anal Biochem 2003;322:48–50.
7. Sapan CV, Lundblad RL, Price NC. Colorimetric protein assay techniques. Biotechnol Appl Biochem 1999;29:99–108.
8. McMahon DJ, Brown RJ. Milk coagulation time—linear relationship with inverse of rennet activity. J Dairy Sci 1983;66: 341.
9. Li H, Rothberg LJ. Label-free colorimetric detection of specific sequences in genomic DNA amplified by the polymerase chain reaction. J Am Chem Soc 2004;126:10958–10961.

10. Liu J, Lu Y. Accelerated color change of gold nanoparticles assembled by DNAzymes for simple and fast colorimetric Pb^{2+} detection. J Am Chem Soc 2004;126:12298–12305.

11. Elghanian R, Storhoff JJ, Mucic RC, Letsinger RL, Mirkin CA. Selective colorimetric detection of polynucleotides based on the distance-dependent optical properties of gold nanoparticles. Science 1997;277:1078–1081.

12. Long ME, Swofford RL, Abrecht AC. Thermal lens technique: A new method of absorption spectroscopy. Science 1976;191: 183.

13. Sittampalam GS, Kahl SD, Janzen WP. High-throughput screening: advances in assay technologies. Cur Opin Chem Biol 1997;1:384.

14. Ihara T, Tanaka S, Chikaura Y, Jyo A. Preparation of DNA-modified nanoparticles and preliminary study for colorimetric SNP analysis using their selective aggregations. Nucleic Acids Res 2004;32:e105.

15. Liang RQ, Tan CY, Ruan KC. Colorimetric detection of protein microarrays based on nanogold probe coupled with silver enhancement. J Immunol Methods 2004;285:157–163.

16. Van Kampen EJ, Zijlstra WG. Spectrophotometry of hemoglobin and hemoglobin derivatives. Adv Clin Chem 1983;23:199.

17. Henry RJ, Golub OJ, Berkman S, Segalove M. Critique on the Icterus index determination. Am J Clin Pathol 1953;23:841.

18. Kingsley GR. Procedure for serum protein determinations. Stand Methods Clin Chem 1972;7:199.

19. Lowry OH, Rosebrough NS, Farr AL, Randall RI. Protein measurement with the Folin phenol reagent. J Biol Chem 1951;193:265.

20. Rylatt DB, Parish CR. Protein determination on an automatic spectrophotometer. Anal Biochem 1982;121:213.

21. Peats S. Quantitation of protein and DNA in silver stained gels. Anal Biochem 1984;140:178.

22. Jin L-T, Choi J-K. Usefulness of visible dyes for the staining of protein or DNA in electrophoresis. Electrophoresis 2004;25: 2429.

23. Finlay GJ, Baguley BC, Wilson WR. A semiautomated microculture method for investigating growth inhibiting effects of cytotoxic compounds on exponentially growing carcinoma cells. Anal Biochem 1984;139:272.

24. Webster D. The immediate reaction between bromcresol green and serum as a measure of albumin content. Clin Chem (Winston-Salem, NC) 1977;23:663.

25. Stookey LL. Ferrozine—a new spectrophotometric reagent for iron. Anal Chem 1970;42:779.

26. Jendrassik L, Grof P. Simplified photometric methods for the determination of the blood bilirubin. Biochem Z 1938; 297:81.

27. Lowry O, Passonneau J. A Flexible System of Enzymatic Analysis. New York: Academic Press; 1972. Chapt. 5.

28. Fawaz EN, Roth L, Fawaz G. The enzymatic estimation of inorganic phosphate. Biochem Z 1966;344:212.

29. Morris DL, Ellis PB, Carrico HJ, Yeager RM, Schroeder HR, Schroeder JP, Albarella JP, Boguslaski RC. Flavin adenine dinucleotide as a label in homogeneous colorimetric iminunoassays. Anal Chem 1981;53:658.

30. Fossati P, Prencipe L. Serum triglycerides determined colorimtrically with an enzyme that produces hydrogen peroxide. Clin Chem (Winston-Salem, NC) 1982;28:2077.

31. Ewen LM, Spitzer RW. Improved determination of prostatic acid phosphatase (sodium thymolphthalein monophosphate substrate). Clin Chem (Winston-Salem, NC) 1976;22:627.

32. Babson AL, Philips GE. A rapid colorimetric assay for serum lactic dehydrogenase. Clin Chim Acta 1965;12:210.

33. Szasz G, Gruber W, Bernt E. Creatine kinase in serum. 1. Determination of optimum reaction conditions. Clin Chem (Winston-Salem, NC) 1976;22:650.

34. Hoever M, Zbinden P. The evolution of microarrayed compound screening. Drug Discover Today 2004;9:358.

35. Hargis LG, Howell JA, Sutton RE. Ultraviolet and light absorption spectrometry. Anal Chem 1996;68:169R–183R.

36. Rooze H. Kinetic-spectrophotometric determination of trace levels of iron. Anal Chem 1984;56:601.

Further Reading

Evenson MA. Chapt. 3: Spectrophotomtric Techniques. In: Burtis CA, Ashwood ER, editors. Tietz Textbook of Clinical Chemistry. 3rd ed. Philadelphia (PA): W.B. Saunders; 1999. p 75–93. A widely used reference and textbook for clinical chemists. Chapter 3 covers both the concepts and the instrumentation of spectrophotometry.

Hargis LG, Howell JA, Sutton RE. Ultraviolet and light absorption spectrometry. *Anal Chem* 1996;68:169R–183R. A review includes a comprehensive list of reagents used for the spectrophotometric analysis of organic and inorganic compounds. It also includes a brief description of data processing and instrumentation. A total of 440 literatures are cited in this article.

Mikkelsen SR, Corton E. Bioanalytical Chemistry. Hoboken (NJ): John Wiley & Sons Inc., 2004. Chapt. 1. Spectroscopic Methods for Matrix Characterization. Chapt. 2. Enzyme. These two chapters focus on the biochemical and biomedical applications, especially to the spectroscopic measurement of protein concentrations and enzyme kinetics.

Pesce AJ, Frings CS, Gauldie J. Chapt. 4. Spectral Techniques. In: Kaplan LA, Pesce AJ, Kazmierczak SC, editors. Clinical Chemistry: Theory, Analysis, Correlation. 4th ed. St. Louis (MO): Mosby; 2003. p 83–106. This chapter includes a good description on the design of various types of spectrophotometers.

Skoog DA, West DM, Holler FJ, Crouch SR. Fundamentals of Analytical Chemistry. Brooks/Cole: 2004. A textbook in analytical chemistry with extensive coverage of both the principles and applications. Part V, Spectrochemical Analysis, includes: Chapt. 24. Introduction to Spectrochemical Analysis; Chapt. 25. Instruments for Optical Spectroscopy; and Chapt. 26: Molecular Absorption Spectroscopy.

See also ANALYTICAL METHODS, AUTOMATED; FLUORESCENCE MEASUREMENTS.

COMPUTERS IN CARDIOGRAPHY.   See ELECTROCARDIOGRAPHY, COMPUTERS IN.

COLPOSCOPY

MOSTAFA A. SELIM
ABDELWAHAB D. SHALODI
Cleveland Metropolitan
General Hospital
Palm Coast, Florida

Today, fewer women die annually from carcinoma of the cervix then in any period in modern history. Increased utilization of cervical cytology (Pap smears) and better assessment and evaluation of patients with abnormal Pap smears are important reasons for the progress in this field.

The Papanicolaou stained cytology smear is an invaluable tool for detecting cervical carcinoma. Such a test depends on collection of exfoliating cells from the cervix, spreading the material on a glass slide, and staining with special stains in order to identify the microscopic structures in nuclei and the cytoplasm of the cell. An abnormal smear is called positive and a normal smear is called negative for malignancy. However, in recent years the incidence of false-negative smears has been recognized to be significant, varying between 15 and 22% (1–4). This high incidence of false-negative smears emphasizes the need for histological study of the abnormal cervix before any therapy is started, regardless of the findings from the cytology smear.

In order to fill the serious gap in the screening of cervical cancer and its precursors created by the high incidence of false-negative smears, colposcopy is utilized (3–6).

THE COLPOSCOPE

The word colposcopy simply means viewing the vagina. The colposcope is a binocular, long-focal-length, wide-field microscope with which it is possible to examine the epithelium and subepithelial vascular pattern of the cervix at magnifications varying from 6 to 40×. Figure 1 shows a typical instrument. A beam of light is projected between the objectives so that the cervix is well illuminated. To help the visualization of blood vessels, a green filter may be fitted to the light source. This adds contrast in viewing microvessels against the tissue. Photographic apparatus consisting of a prefocused lens system, a 35 mm camera, and a flash tube are attached to the colposcope in order to document the findings. The photographs obtained using

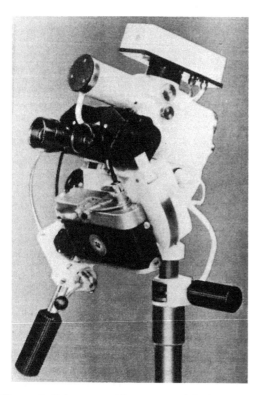

Figure 1. Colposcope with camera and flash attached.

35 mm film are two dimensional (2D). To obtain three-dimensional (3D) images, the instrument may be fitted with a stereocamera that produces image pairs. These pairs give a stereoscopic effect when viewed in a special viewer. The colposcope, the camera, and the flash tube are mounted on a fixed, sturdy base with rack-and-pinion drives for fine positioning and focusing. Modern colposcope is equipped with digital video camera interfaced microcomputer.

FINDINGS

A colposcopic examination is considered satisfactory whenever it is possible to view all of the critical portions of cervical anatomy, and if an abnormal lesion is seen, the upper margin of this lesion must be adequately visualized. Specifically, it is necessary to view the entire transformation zone. The transformation zone is the lower segment of the endocervical canal. This area is normally covered by columnar epithelium, although through the process of metaplasia some areas or the entire zone can be transformed into squamous epithelium. The normal transformation zone is recognized by the presence of small glandular openings, nabothian cysts, and a normal pattern.

The procedure can be performed quickly and easily on out-patients without anesthesia. However, the gynecologist needs intensive training in order to master the interpretation of what they sees.

The colposcope was first devised by Hinselmann in 1925 at Hamburg, Germany. The instrument became popular in the German speaking and Latin nations. It was accepted in the English speaking nations in the 1960s. One of the reasons for the delay in accepting colposcopy was that all of the original studies and the terminology were written in German (7,8).

INDICATIONS FOR COLPOSCOPY (3–6)

Ideally all patients, during their gynecological examination, at one time or another ought to be examined colposcopically. However, since colposcopy is time consuming and requires specialized training, this is usually not possible, and thus the following indications are recommended:

1. All patients with abnormal cervical cytology.
2. All patients with an abnormal lesion on the cervix, vulva, or vagina.
3. Cases of persistent cervicitis, vulvitis, or vaginitis.
4. Pregnant patients with unexplained vaginal bleeding.
5. All offspring of diethylstilbestrol-exposed pregnancies.

Colposcopy plays an essential role in evaluating these patients. The colposcope helps to pinpoint the abnormal area. However, the final diagnosis has to await histopathological diagnosis. This tissue diagnosis is obtained through endocervical curettings, punch biopsy, and/or cone biopsy according to the individual case, as illustrated in Fig. 2.

Recent studies confirmed that colposcopy could be utilized in follow-up in conservative management of low grade

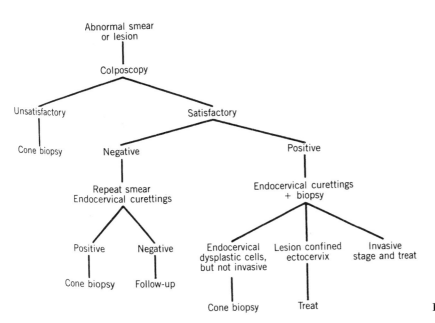

Figure 2. Protocol for management of patients.

precancerous lesions, as most of them regress on its own. In addition, the colposcope can help in detecting and guiding treatment of genital infections.

A movable, counterweighted arm is used for rough positioning of the colposcope. The arm is fixed to the examining table or to a heavy wheeled base (5,6).

There are several commercially available colposcopes. Manipulation, magnification, length, intensity, and type of green filter vary from one instrument to another. For detailed inspection of the vascular pattern, the magnification ought to be not <14× and preferably 16× (7,8).

TECHNIQUE OF COLPOSCOPY

1. The cervix and the upper vagina are examined, at a magnification of not < 14×, after excess mucus is removed by cotton swab moistened by physiological saline, which allows the subepithelial architecture to be seen in greater detail. To enhance visibility the green filter should be used.

2. Acetic acid (3%) is gently applied by a cotton swab. Abnormal epithelium will become whitish and sharply demarcated. The normal squamous epithelium appears pink, and columnar epithelium will have a grape-like appearance. The effect of the acetic acid will last ∼ 30–40 s.

3. Endocervical curetting and punch biopsies should be done for all unusual-appearing areas. Specimens should be put in separate formalin-containing bottles.

4. Bleeding is usually minimal and does not need any packing. However, during pregnancy, when vascularity is increased, Oxycel, Surgicel packing, or the use of Munsell's solution may be necessary.

Normal pattern of blood vessels. An example of a satisfactory colposcopic finding is shown in Fig. 3. The sketch identifies the various features of this view of the cervix. An example of an unsatisfactory view of the cervix is shown

in Fig. 4. In this case the transformation zone as indicated in the sketch is not fully visualized; the columnar epithelium is not seen and/or the upper margin of the lesion extends into the endocervical canal and thus cannot be fully visualized.

In evaluating satisfactory colposcopic findings, many factors must be considered. It is beyond the scope of this article to provide a detailed description of the different patterns seen in colposcopy, but a general idea of what is looked for can be presented. In order to reach an adequate diagnosis, the colposcopist must consider the following criteria in observing the uterine cervix: (1) vascular pattern, (2) intercapillary space, (3) color and texture, (4) surface pattern, and (5) sharpness of the line of demarcation between the lesion and the rest of the cervix.

The lower segment of the endocervical canal that extends to the visible part of the cervix is the transformation zone. This area is originally and normally covered by columnar epithelium. Through a process of metaplasia, some areas or the whole zone can be transferred into squamous epithelium. The normal transformation zone is recognized by the presence of small glandular openings, nabothian cysts, and a normal pattern of blood vessels (Fig. 3).

Vascular Pattern

Normal epithelium vessels appear as fine dots or a network of capillaries. However, in abnormal pathology, the individual vessel becomes prominent, leading to punctuation (Fig. 5). A mosaic pattern is due to increased communication between the individual vessels, arranged parallel to the surface epithelium (Fig. 6). Atypical vessels are capillaries that are irregular in shape, size, course, and arrangement (Fig. 7).

Intercapillary Space. Due to rapid proliferation of the abnormal epithelium, the intercapillary distance between the vessels is increased. The intercapillary distance of normal capillaries varies between 50 and 250 μm with an average of 100 μm. The intercapillary distance in early

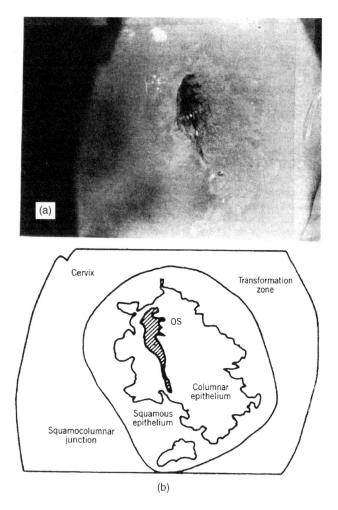

(b)

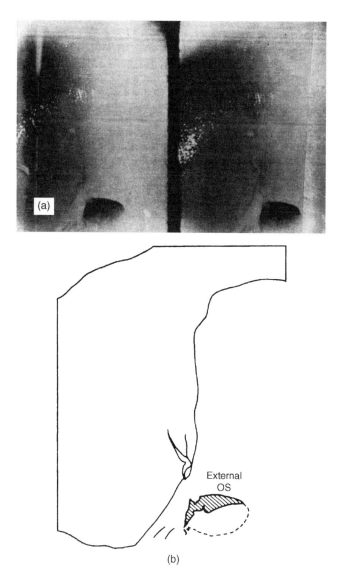

(b)

Figure 3. (a) Areas of glandular epithelium extending to ectocervix became whitish upon addition of acetic acid. Note the grape-like appearance of the glandular epithelium; no abnormal vessels or lesions are seen, and the lower end of the cervical canal is adequately visualized. (b) Sketch of the transformation zone, consisting of columnar and squamous epithelium. The squamous–columnar junction is part of the transformation zone. In the lower left corner of the diagram there is an island of glandular epithelium in the middle of the squamous epithelium. Note that the color of the squamous epithelium in the transformation zone is identical to that of the original squamous epithelium.

intraepithelial neoplasia (CIN 1) is 200 μm, while in severe intraepithelial neoplasia (CIN 3) it is 450–550 μm (6) (Figs. 5 and 6).

Color Tone. Abnormal epithelium is darker than normal epithelium, and upon addition of acetic acid it temporarily becomes whitish (Figs. 5 and 6).

Surface Epithelium. Abnormal lesions are uneven, granular, papillomatous, or nodular (Figs. 5–7). Normal squamous epithelium is smooth, whereas columnar epithelium has a grape-like appearance (Fig. 3).

Line of Demarcation. The abnormal epithelium is usually raised and well demarcated from the normal epithelium, especially after addition of acetic acid (Figs. 3–7).

Figure 4. (a) Unsatisfactory view of the cervix and (b) accompanying sketch. This cervix is covered completely by squamous epithelium; no glandular epithelium is seen, including the lower endocervical canal. The left side of the cervix has a large red region. This region is covered by newer squamous epithelium, which is more transparent so that the normally arranged vessels are seen though it. The entire field of the figure is covered by squamous epithelium. Neither glandular epithelium nor the squamocolumnar junction can be seen. This represents an unsatisfactory colposcopic examination.

ACCESSORY INSTRUMENTS TO COLPOSCOPY

The best examination table is one that can be adjusted for height and tilt, is in the lithotomy position, and is preferably electrically operated. Punch biopsy forceps ought to be sharp with square jaws (Fig. 8) in order to obtain well-oriented, adequate tissue for pathological examination. In addition, in order to prevent loss of fragments of tissue obtained by endocervical curettings, the curette must be a closed-sided instrument to create negative suction (Fig. 9).

To aid in visualization of the endocervical canal, which cannot be seen by a colposcope, a Hamou microcolpohysteroscope can be utilized (Fig. 10). The technique utilizes

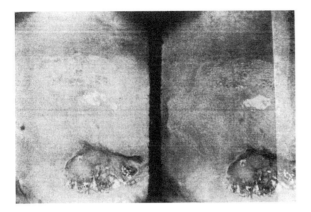

Figure 5. Punctation. Large, well-defined areas of the transformation zone became white after addition of acetic acid. The areas are covered by punctate of blood vessels with wide intercapillary distances. These abnormal lesions do not extend into the canal, and the lower canal columnar epithelium is seen and is normal. Some glandular openings are seen in the lower right part of the figure.

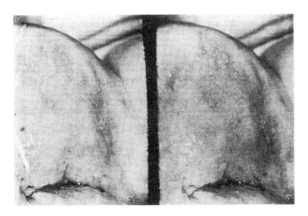

Figure 6. Large transformation zone covered by a mosaic pattern, inside well-defined borders. The areas became whitish after addition of acetic acid. The upper margin of the abnormal lesion can be adequately seen. The columnar epithelium is seen and appears to be normal. These findings are typical for carcinoma *in situ*, a precancerous lesion.

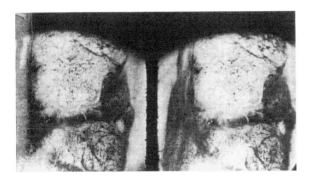

Figure 7. Highly abnormal cervix covered by atypical vessels with increased intercapillary distance. It is nodular and shows an exophytic growth pattern with a whitish and glossy appearance. These findings are indicative of invasive carcinoma.

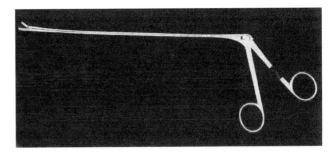

Figure 8. Punch biopsy forceps. Note the square jaw necessary to obtain an adequate and well-oriented biopsy for pathological examination.

Figure 9. Curette. Note that it is serrated and open only from one side in order to prevent loss of tissue. The stem is hollow to create a suction effect.

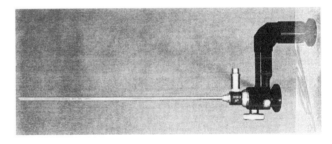

Figure 10. Hamou microcolpohysteroscope. Note that it has the same caliber as the curette and has an outlet for a fiberoptic light source.

the contact technique, thereby achieving $\sim 60\times$ magnification, and does not need any transmission media other than the normal mucus secretion of the endocervix. This method can aid in assessing the extent of the disease and may aid in defining the extent of cone biopsy, whenever it is needed to investigate or to treat the disease (9).

Digital Colposcopy (10–12)

Digital colposcopy is an improvement on regular colposcopy. Integrating video camera interfaced microcomputer and using real time image achieve this. Such arrangement allows computerized manipulation of the image signal. In order to manipulate the image by the computer it needs to be converted into matrix number. Each number is representing one point in image matrixes, called pixel. Each pixel has a value that corresponds to discrete gray level. The computer then converts the number back to analogue for display on the video.

Regular colposcope uses green filter to enhance visualization of blood vessels and discoloration. However, this has its limitation. Digital colposcopy has different ways to improve on this (e.g., spatial filtering the image). These are pixel-by-pixel process that creates a new image, where the pixels values are determined, taking into account the value of neighboring pixels.

A very significant advantage to digital colposcopy is that it allows storage of colposcopic images in digital format on different media. This gives an unlimited ways of storage and filing.

Digital colposcopy is useful in metric measurement of the lesions and accuracy of measurement improved by the new techniques and software. This measurements and the improvement in digital filing helps the clinician in accurate follow-up of the lesions for regression or progression. Recent clinical studies confirmed that it is possible to transmit the images for consultation: which proved of help to areas lacking specialists in the field of oncology. In addition digital colposcopy helps in quality control, follow-up and in teaching.

SUMMARY

The manuscript describes the history, the instrument of colposcope and the advances into the digital age. The significant clinical application of this instrument in diagnosis and treatment of precancerous,cancerous lesions of the cervix is described. The terminology used to explain the vascular changes in the cervix is defined.

BIBLIOGRAPHY

1. Selim MA, So-Bosita JL, Blair OM, Little BA. Cervical biopsy versus conization. Obstet Gynecol (NY) 1973;41:177–182.
2. Selim MA, So-Bosita JL, Neuman MR. Carcinoma *in situ* of the cervix uteri. Surg Obstet Gynecol 1974;139:697–700.
3. Selim MA, Vasquez HH, Masri R. Indications and experience in colposcopy in management of cervical neoplasia Surg Obstet Gynecol 1977;149:529–532.
4. Selim MA, Razi A. Cryosurgery for intraepithelial neoplasia of the cervix. Cancer (Philadelphia) 1980;46:2315–2318.
5. Sootra-Gartaux, Carter I, Jourdau-DeSilva N, Decremax P. Regression of low grade epithelial neoplasia. Obst Gynecol 2004;104:751–755.
6. Norman JE, et al. An evaluation of economic and suitability of screening for chlamydia trachomatis infection in women attending antenatal, abortion, colposcopy and family planning clinic in Scotland UK BJOG 2004;111:1261–1268.
7. Kolshad P, Stafl A, editors. Atlas of Colposcopy. Baltimore: University Park Press; 1972.
8. Jordan JA. Colposcopy in the diagnosis of cervical cancer and precursor. Clin Obstet Gynecol 1985;12:67–76.
9. Soutter WP, Fenton DW, Gudgeon P, Shoup P. Quantitative microcolpohysteroscope assessment of the extent of endocervical involvement by cervical intraepithelial neoplasia. Br J Obstet Gynecol 1984;91:712–715.
10. Craine BL, Craine ER. Digital imaging colposcopy. Basic concepts and application. Obs Gynecol 1993;82:869–873.
11. Craine BL, Craine ER, O'Toole CJ, Ji Q. Digital imaging colposcopy: corrected are measurements using shape- from -shading. IEEE Trans Med Imaging 1998;17:1003–1010.
12. Schadell D, et al. Suitability of digital colposcopy for telematic applications. Biomed Tech 2004;49:157–162.

See also Cryosurgery; cytology, automated; sexual instrumentation.

COMMUNICATION AIDS FOR THE BLIND. See Blind and visually impaired, assistive technology for.

COMMUNICATION DEVICES

Albert Cook
University of Alberta

INTRODUCTION

Augmentative and alternative communication (AAC) systems supplement, but do not replace other modes of communication such as speech, gestures, vocalizations, or facial expressions. The need for AAC, may be *congenital in utero* or *perinatal* (e.g., cerebral palsy or developmental disability) or *acquired* (neurological conditions). The severity of need varies from mild to moderate to severe based on physical, cognitive and linguistic involvement. The overall prevalence from mild to severe for AAC needs is 0.2–0.6% of the total population (1). The age range encompasses three distinct ranges: infant to preschool, school age to teenage, and teenage to adult. The selection of systems to meet the needs of individuals in these age ranges is influenced by the experience that the individual brings to the use of AAC and the degree to which language and speech have been developed prior to the need for AAC. If the person has developed speech and language, and then subsequently lost those abilities, it is very different from the child born without speech and language who has never had the opportunity to develop those skills. Alternative communication needs may also change over time. For example, the needs of children change as they develop cognitive and language skills. In contrast, some disorders are degenerative and result in loss of function and decrease in skills (e.g., amyotrophic lateral sclerosis, multiple sclerosis). Using current technologies, we are able to meet the AAC needs for children who have cerebral palsy or developmental disabilities, individuals with good cognitive skills, and adults with degenerative diseases. Our current approaches are less effective for individuals who have mental retardation, are ambulatory, have dual sensory impairment, traumatic brain injury or are elderly.

NEEDS SERVED BY AAC

There are two basic communication needs that lead to the use of augmentative and alternative communication systems: conversation and graphics (2). These two needs differ in many important aspects. Conversational needs are those that would typically be accomplished using speech if it were available. Examples are an informal conversation with a friend, a formal oral presentation to a group of people, a telephone conversation, or a small group discussion. Much of conversational use focuses on interaction between two or more people. Light (3) describes four types of communicative interaction: (1) expression of needs and wants, (2) information transfer, (3) social closeness, and (4) social etiquette. Expression of needs and wants is the most basic of AAC needs and allows requests for objects or people to be made. Information transfer allows expression of ideas, discussion and meaningful dialogue. Social closeness refers to the ways in which communication serves to connect individuals to each other, regardless of the content of the conversation. Social etiquette is used to describe those formalities that we adapt to our listener. For example, students will speak differently to their peers than to their teacher. Graphic communication describes all the things that we normally do using a pencil and paper, computer, calculator, and other similar tools, and it includes writing, drawing, mathematics, and Internet access.

Rates of communication using speech vary between 150 and 250 words/min (4). In contrast, many AAC devices use a keyboard to generate messages. This can result in significantly lower rates of communication than for speech. For example, a trained, nondisabled typist can generate typed text during transcription at a rate of nearly 100 words/min. However, this is still only about two-thirds of the rate of speech. If this same typist is asked to compose rather than transcribe, then their rate will drop by 50% to a maximum of 50 words/min (5). Many people who have disabilities must rely on single-finger typing, and they may only be able to type at a maximum rate of 10–12 words/min. Scanning (see below) reduces the maximum rates to as low as 3–5 words/min. The great disparity in rates of communication between a speaking person and an AAC system user often results in the speaking person's dominating a conversation with a nonspeaking person. This renders the individual using an AAC device to a passive role in the conversation.

There are three types of graphic communication: writing, mathematics, and drawing/plotting. Since each of these types of graphic communication serves a different need, AAC devices designed to meet each type also have different characteristics. Writing results in an electronic (soft copy) or paper (hard copy) output. However, writing does not need to be via letters combined into words on a page. Other symbols (e.g., line drawings, pictures) can also be printed on paper and used, in place of written output. Since some devices allow the selection of whole words, which are then output to a printer, spelling is not a prerequisite for the generation of written output. Alternatively, a nonspeaking person can point to letters on a board and have an attendant write down the letters to accomplish writing tasks.

There are three types of writing: note taking, messaging, and formal writing, the requirements of which all differ (6). Portability is important for note taking since the needs may be in many different locations (e.g., home, library, school, job, or meetings). Note taking may require writing at a high rate in order to keep up with the speaker. Just as nondisabled persons typically use abbreviations and other shorthand notations, so do users of AAC writing devices. Words and phrases stored in an AAC device may further increase the rate of text entry. The difference between note taking and messaging is the recipient of the written output. Messaging typically results in a note made for another person to read. This affects the types of abbreviations and shorthand notations that are used. The writing rate may also be slower than note taking since the person receiving the message is not present and waiting for

it. Individuals who have intellectual disabilities and use symbols for communication can also use messaging. For example, individuals living in group homes who wish to send messages to their families, but are unable to use voice communication over the telephone can use rubber stamps with the appropriate AAC symbols (7). These stamped messages are then FAXed to their significant others as a message. The family can respond by using a second set of rubberstamps with the symbols on them.

The most demanding type of writing is formal writing, including reports, school homework, writing for publication, and similar applications. Here, accuracy is of prime importance, with rate becoming secondary. Word processing allows accurate entry of written material, but some users may take up to several hours to create one page of written material. However, with an entry rate of 1.5 to 5 words/min, it can take 2.5 h or more, and the use of abbreviations and other input acceleration techniques is necessary to allow an individual to keep up with the demands of work or school.

The AAC systems for mathematics focus on the written manipulation of numbers required for arithmetic and higher mathematics, as opposed to calculator functions. The goal is to for the user to learn mathematics in a manner that is as close as possible to that used by nondisabled peers. Cursor movement is a key difference between written English (the cursor always moves left to right and moves down one line at the right margin) and mathematics, where the cursor moves left to right as we enter numbers to be added, but once we have a column of numbers, the cursor moves right to left as we enter the sum. Carry or borrow are concepts when learning to add or subtract, and children are taught to cross out the number at the top of the adjacent column and substitute the borrowed or carried value. These types of cursor movement are also required in a math worksheet for children who cannot use pencil and paper. Algebra requires special symbols (e.g., Greek letters) and the use of superscripts and subscripts. Statistics or calculus adds the need for special symbols such as summation signs and integral signs and the formatting of problems. Some commercial AAC devices include some or all of these mathematical functions.

The final type of graphic output is drawing used to convey information, to help us clarify our thinking, and for creative expression. Typically, we use pencils, paint and brush, or computer programs to draw. Many people who are unable to use these conventional means of drawing because of motor limitations utilize computer programs designed for drawing or plotting

CHARACTERISTICS OF AUGMENTATIVE COMMUNICATION SYSTEMS

The characteristics of AAC devices can be grouped into three major components shown in Fig. 1: (1) control interface, (2) processor, and (3) activity output (6). The control interface links to a selection method, selection set, and an optional user display. The processor is further broken down into components of (1) selection technique, (2) rate enhancement and vocabulary expansion, (3) vocabulary storage, (4) text

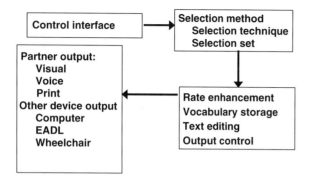

Figure 1. Characteristics of AAC systems.

editing, and (5) output control. The activity outputs to the communication partner include visual display, speech, and printer. A control port for external devices (e.g., computers or electronic aids to daily living) is sometimes included. Not every device includes all the individual functions shown in Fig. 1. In some cases, the functions shown in Fig. 1 are implemented in software using portable computers. Others are based on special purpose computers designed specifically for use as AAC devices.

Control Interface

The control interface is the way in which the user is able to make entries into the AAC device to generate a communication utterance or output. There are various types of interfaces based on the number of independent choices that can be made by an anatomical site or a combination of anatomical sites (e.g., hand, arm, head, chin, leg, foot).

Keyboards, single or dual switches, and joysticks or multiple switch arrays are the most commonly used control interfaces for augmentative communication devices. A variety of switches used with AAC systems are shown in Fig. 2 (8).

Various types of keyboards are typically used for AAC devices. When a standard keyboard is not accessible due to limited fine motor control, enlarged keyboards are used with gross muscle movements (e.g, hand, foot, elbow activation). When range of motion prevents reaching all the keys, a contracted keyboard is used with finger or mouth stick activation. Keyboards may also be modified by the use of key guards that prevent accidental activation of keys. When keyboards of any type cannot be used, single or multiple switches are used. Some examples of these are shown in Fig. 3.

Selection Set

The selection set of the augmentative communication device presents the symbol system and possible vocabulary selections to the user. One type of selection set is the label on the keys of a keyboard. The selection set may include individual letters, words, phrases, or other symbols. The mode of presentation to the user may be display-based, chart-based, or memory-based (9). Display- and chart-based approaches present a list of vocabulary choices (e.g., words, letters, symbols), and the user chooses the item of interest from that list. Display-based methods are electronically generated and built into the device. Chart-based approaches are typically on a separate sheet that is used for prompting and often not needed after training. Both of these types require only recognition memory, that is, the ability to recognize the correct item when a list is provided. Memory-based presentation of the selection set does not include a prompting list and relies on recall memory. This is a more difficult task than recognition

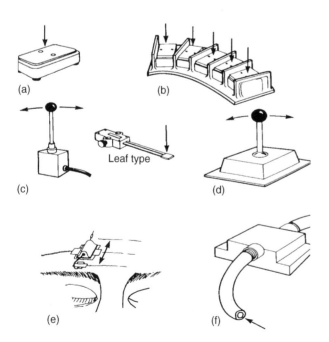

Figure 2. Typical interfaces for augmentative communication systems, (a) paddle (tread) switch; (b) array of paddle switches (slot switch); (c) wobble switches; (d) joystick; (e) brow wrinkle switch; (f) sip and puff switch (pneumatic). (From Electronic Devices for Rehabilitation, New York: John Wiley & Sons, 1984, with permission.)

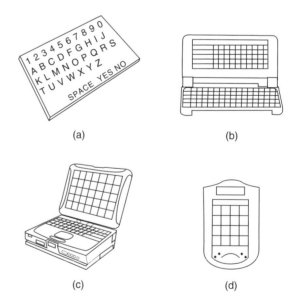

Figure 3. Direct selection communication devices. (a) Pointing boards with letters or other symbols, (b) hand-held keyboard, notebook-sized keyboard, and (d) palm-sized keyboards for high portability when good fine motor control exists.

for individuals with cognitive limitations due to developmental disabilities, stroke, brain injury, or similar conditions. Even in the absence of such disabilities, only ~ 200–300 vocabulary items and corresponding recall codes can be easily remembered and used. Thus, the use of display-based selection sets is desirable. However, display-based selection sets are often static, that is, there is one set of elements from which the user can choose. While this set of elements can be combined to access a larger set of vocabulary through codes or abbreviations, this converts a display-based selection set to a memory-based one.

One way of avoiding the problem of limited vocabulary choices associated with memory-based systems is the use of dynamic communication displays. An AAC device that has a dynamic display is shown in Fig. 2d. Dynamic displays change the displayed selection set when a new level is selected. Since the user's selection set is always updated on the display panel, it can be altered easily depending on previous choices. For example, a general selection set may consist of categories such as work, home, food, clothing, greetings, or similar classifications. If one of these is chosen, either by touching the display surface directly, using a mouse-driven cursor or by switch access, then a new selection set is displayed. This retains the display-based selection set while dramatically increasing the functional size of the selection set. For example, a variety of food-related items and activities (eat, drink, ice cream, pasta, etc.) would follow the choice of "foods" from the general selection set. The symbols on the display can be varied, and this changes the targets for the user. Since each new selection set is displayed, the user can depend on recall, not recognition memory.

A significantly different approach to the presentation of vocabulary choices to the Individual who uses AAC is implemented in Visual Scene Displays (VSDs) (10). Visual scene displays capture events in person's life on a screen. Hotspots are then liked to text messages that describe events, invite discussion or serve as prompts for conversational use. These VSDs offer a greater degree of contextual information to the Individual who uses AAC and communication partner information in order to support interaction. They enable communication partners to participate more actively in an interactive conversation and may represent either a generic context (e.g., a person's home) or a very specific personalized event (e.g., a birthday party). The screen contains pictures of the activity, place or event. A caregiver typically enters the vocabulary associated with the screen elements, although it could be the user who enters the text. A comparison of characteristics of a traditional grid AAC display containing vocabulary elements and a VSD is shown in Table 1. The biggest advantages of the VSD are the type of material that can be included in the display, the degree of personalization, the management of the display, and the methods available for concept retrieval. The functional uses of a traditional display focus on communication of needs, wants, and information exchange. The VSDs provide greater conversational support by allowing an interaction to be a shared activity and a potential learning environment for the user. While also including communication needs, wants, and information exchange, VSDs add a real element of social closeness. The VSDs can be applied to stimulate conversation between interactants, support play, shared experiences, and telling of stories. They also facilitate active participation of interactants during shared activities and can provide instruction to both the user and the communication partner through specific information or prompts. Specific populations that can be served by VSDs include those with cognitive limitations (e.g., Down's syndrome) and those with language limitations (e.g., aphasia, autism).

Another approach that allows dependence on recognition memory though a display-based approach is word prediction or word completion. These approaches can be used with any selection technique (11). In this case, there is a window on the screen that displays the most likely words based on the letters entered. To complete the word, the user selects the code (e.g., a number listed next to the word). If the desired word is not listed, the user continues to enter letters and the displayed words change as more letters are entered. For example, if the word "what" is entered, a word completion device may list "time", "is" "are" "can", etc., as choices to follow "What". The display is a display-based selection set that is dynamic based on user input. There are two approaches to the storage of items to be presented in a predictive system. Fixed dictionaries use a preselected stored word vocabulary that never changes. We can have different vocabularies for different contexts such as school, work or recreation. Other systems offer a menu of words using an adaptive vocabulary that is altered based on the user's own selections and is constantly updated. Character prediction phrase prediction are similar approaches using non-orthographic symbols or phrases instead of letters or words.

Selection Method

Two basic selection methods, direct selection or indirect selection are used in AAC systems (6). Direct selection is the fastest and easiest selection method to understand and use because each possible choice in the selection set is

Table 1. Comparison of Standard Grid AAC Display and Visual Scene Displays

Variable	Typical AAC Grid	VSD
Type of representation	Symbols, TO, line drawings	Digital photos, line drawings
Personalization	Limited	High
Amount of context	Low	High
Layout	Grid	Full or partial screen, grid
Display management	Menu, pages	Menu pages, navigation Bars
Concept retrieval	Select gird space, pop ups	Hotspots, speech key, select grid space

available at all times and the user merely chooses the one that he or she wants. Several examples of direct selection AAC devices are shown in Figure 2. Indirect selection is used to provide access for individuals who lack the motor skills necessary to use select directly and involves one or more intermediate steps between the user's action and the entry of the choice into the device. A variety of indirect selection AAC devices are shown in Figure 4. There are three types of indirect selection are scanning, directed scanning and coded access. All scanning-approaches rely on the basic principle of presenting the selection set choices to the user sequentially and having the user indicate when his or her choice is presented. Typically, the indication is by the activation of a switch by a single movement of any body part. A hybrid method called directed scanning allows the user to first activate the control interface (typically a joystick or other array of switches) to select a direction for cursor movement (vertically or horizontally) and then to stop the cursor movement with a switch at the desired element. Several types of scanning are used. In step scanning, the scan advances one step each time the switch is pressed. With autoscanning, the scan steps automatically until the switch is hit. An inverse scan reverses the process and advances continuously as long as the switch is depressed and stops when the switch is released. Each of these has advantages for different types of users.

Scanning is inherently slow, and there have been a number of approaches that increase the rate of selection (6). Just by placing the most frequently used characters near the beginning of the scan, the rate can be increased by as much as 30%. Many of the rate enhancement methods involve selecting groups of characters first to narrow the choices, then selecting the desired item from the selected group. Several types of adaptations are employed with

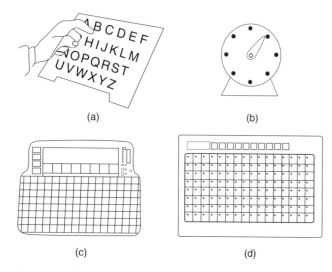

(a)   (b)

(c)   (d)

Figure 4. Scanning communication systems. (a) Simple pointing boards in which the listener points to each element and waits for a response, (b) circular scanning with up to 32 elements, (c) electronic row-column matrix scanning with picture display and voice output, (d) alternative electronic row-column matrix scanning with voice output. Parts (c) and (d) can have a variety of symbols (letters, words, pictures, line drawings) in the scanned elements.

scanning to increase the rate of selection. With group item scanning, items are clustered in groups. The first scan is through the groups and a switch press selects one. Then the items in the group are scanned sequentially. A row column scan is a group-item scan with the elements arranged in a matrix of rows and columns. The rows are scanned top to bottom, and then the columns are scanned left to right. A halving scan presents each half of the display until a switch is hit, then scans within that half.

In coded access, the individual uses a single switch or an array of switches to generate a unique sequence of movements to select a code that corresponds to each item in the selection set. The most common form of coded access used in AAC devices is the Morse code. Here the selection set is the alphabet, and an intermediate step [e.g., holding longer (dash) or shorter (dot)] is necessary in order to make a selection. Two-switch Morse code is also used in which one switch sends dots and the other sends dashes. Morse code was developed to be efficient and fast, and these features are exploited in its AAC applications. However, Morse code is a memory-based technique in general (although some display-based approaches have been used) and this can result in additional constraints on the user. The memory-based nature also makes it useful for individuals who have visual limitations that prevent them from using a display-based approach.

Output Formats

Most AAC devices and some assistive technology applications rely on voice output. There are two types of speech output, differing in the manner by which the speech is electronically produced. These are (1) digital recording, and (2) speech synthesis. Digitized Speech is similar to a tape recorder. The speech is electronically compressed with up to a few minutes of speech (2–3 s/utterance) stored at any one time. A care provider can record any voice (male, female, child) or sound (e.g., laughter) and the user can play it back using either direct or indirect selection. One drawback is that all vocabulary to be spoken must be recorded and the user cannot produce a new or novel utterence.

Voice synthesis is electronic generation of speech using a mathematical model of the vocal tract realized in software or hardware. The two types of sounds in speech are voiced and unvoiced (a hissing sound similar to unvoiced sounds such as s or f). Both of these types of speech signal are used as sound sources for the vocal tract model. The parameters of the model are varied to produce speech in a manner similar to the variation of the tongue, teeth, lips, and throat during human speech. The parameters are either derived from actual speech samples (e.g., linear predictive coding-base synthesis) or from a set of parameters representing each phoneme (~64 phonemes are required for intelligible speech synthesis) or morpheme (combinations of phonemes-over 1500 are required for intelligible synthesis). Text can also be converted directly to speech by using a set of rules. The conversion of text characters into the parameters required by the vocal tract model in the speech synthesizer is accomplished by text-to-speech software. The synthesizer combines the phonemes or morphemes into words. There are ~400 rules necessary for letter-to-phoneme conversion. Some systems also use morphonemic rules. About 8000 morphemes can generate 95% of all words. The major advantage of speech synthesis is that there is unlimited vocabulary as long as it can be represented in text strings.

Internet Access

Another common output for AAC devices is Internet access using the AAC device. While some AAC devices include basic computer functions (word processing, spreadsheet, presentation software, web browser), many do not. In the latter case, the AAC device is connected to a computer wither via hard wire or infrared link. In either case, any stored vocabulary or special access methods used with the AAC device are then available for on-line use. Further, many commercial AAC systems utilize portable computers with AAC software, and they can also function as Internet workstations. One of the most important communication functions accomplished via the internet is e-mail. Aside from the benefits that we all receive from e-mail (global access, low cost) people who have disabilities also have the benefit of composition independently and at a slower speed than face-to-face communication since the recipient reads it at a later time. The major advantage is that the person's disability is not immediately visible, and individuals who use AACs report that they enjoy establishing relationships with people who experience them first as a person and then learn of their disability (12). Information retrieval, socialization (e.g., chat rooms), development of literacy through large amounts of reading and writing, booking of airline or theater reservations, and general conduct of business from home without traveling to a place of business are other advantages of Internet access for individuals who use AACs.

AAC ASSESSMENT

Effective use of an augmentative communication device requires skills in several domains. These include gross and fine motor control required to make selections; visual, auditory, and tactile sensory capabilities; and cognitive and language abilities (e.g., the use of some symbolic representation).

Needs Assessment

The first step in an assessment aimed at the selection of an AAC system is the documentation of the individual's communication needs. The second step is to determine how many of these can be met through the individual's current communication methods. Finally, an AAC system is selected that will reduce as many of the unmet communication needs as possible through a systematic intervention.

The Participation Model (4) is a framework that focuses on identification of opportunity and access barriers to AAC use. Opportunity barriers are classified as policy, practice, attitude, knowledge, or skill (of support personnel). Access barriers include natural supports, environmental considerations, and user competencies. These are evaluated in terms of their potential to increase natural abilities (speech, gestures, vocalizations), improvement of communication through environmental adaptations or strategies, and profiles of the individual's abilities and skills. These are all measured and related to AAC use. Interventions are developed to address barriers resulting from opportunity, natural abilities, environment, or lack of an AAC system. Once identified, an intervention plan is derived for each identified barrier. For example, a policy may need to be changed, a teacher's attitude toward AAC altered, or the staff of a facility needs to be trained to develop their skills and knowledge of AAC. Natural ability interventions may involve speech-language pathology services to increase loudness or intelligibility of speech. Environmental adaptation interventions may result in a change in the layout of a classroom to place the individuals who use AAC in the center of the class rather than the periphery. The AAC system interventions determine the skills the user has and identify a system or systems that will be of use to that individual.

Physical–Motor Assessment

Physical assessment consists of identifying an anatomic site that can be used to indicate choices from the selection set. The primary sites are the upper limb (L/R), head and/or neck, eyes, leg, foot, or arm. These may be used directly or with a hand pointer to increase precision, or a mouth or head stick. Adaptations such as a keyguard or a hand splint may also be used. A variety of methods for using head control are shown in Fig. 5.

Sensory Assessment

Visual, auditory, and tactile abilities must also be assessed. Visual skills include acuity, tracking, and visual scanning. Auditory thresholds and speech perception should also be evaluated. Tactile response may be limiting if it is too sensitive or not sensitive enough. In the former case, the

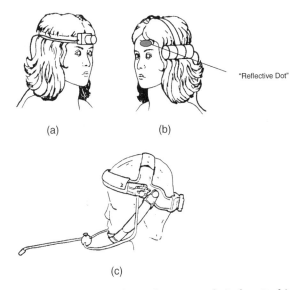

Figure 5. The head may be used as a control site by attaching a light beam (a): a reflective dot used with an infrared transmitter/receiver on the device (b), of a mechanical head pinter (c). part a (From Electronic Devices for Rehabilitation, New York: John Wiley & Sons, 1984, with permission, part c from Zygo Industries, Inc.)

person may be tactually defensive and not be able to use a switch or other interface that has a rough surface. If the person is insensitive, then they may not be able to sense when a switch or key is activated.

Cognitive and Language Assessment

A variety of symbols are used in AAC selection sets. These include real objects (including miniatures of object such as doll house furniture), colored drawings, line drawings, words, or letters. The symbol type for any given individual is selected through a formal assessment process (4). Language assessment related to AAC use is very difficult due to lack of expressive ability (i.e., the very reason for the need for AAC). Two approaches are employed. The first is single word vocabulary testing that measures vocabulary comprehension in relation to the individual's level of functioning. This assessment includes relationship concepts that have no real-world referent as well as traditional language sampling. The second type of language evaluation is a literacy assessment. This can include a reading evaluation in which both word recognition and reading comprehension are used (4). Several types of spelling tests are employed. Recognition spelling requires the user to pick the correct choice from a list of options. This requires only recognition memory and is easily accommodated into AAC devices. The second approach relies on first letter of word spelling and is related to AAC word completion techniques. Spontaneous spelling is the typical letter-by-letter text generation we all learn in school. This is the most flexible and powerful spelling skill since it results in the generation of vocabulary limited only by the persons knowledge, not by the features of the device.

A cognitive assessment may also be conducted to determine how the individual understands the world and how communication can be best facilitated within this understanding. No formal tests predict the ability of an individual to meet the cognitive requirements of various AAC techniques (4). Many cognitive tests require expressive language via AAC itself in order to accurately assess cognitive ability. For this reason, the individual's cognitive ability must be estimated to assess the probability of an AAC device being successful. There are several basic cognitive skills relevant to the use of an AAC system. These include: alertness, attention span, cause and effect, vigilance (the ability to visually and auditorially attend to a task and process information), expression of preferences, making choices, symbolic representation, and understanding of object and/or pictorial permanence.

Educational Assessment

The SETT model was developed to aid in the effective selection and use of assistive technologies, including AAC, in education (13). It consists of four elements: student, environment, task, tools. Each element Includes a set of questions that focus on the interrelationship between the elements to enhance classroom use of AAC. Student-related questions include What does the student need to do?, What are the student's special needs?, What are the student's current abilities? Environment questions comprise: What materials and devices are available?, What is the physical layout? Are there special issues?, What is the instructional arrangement?, Are changes planned?, What supports are available to the student?, What resources are available to those supporting the student? Task questions focus on: What activities take place?, What activities support the student's curriculum?, What are critical elements of activities?, How can activities be modified to meet student needs?, How can technology support the student's participation in the activities? Finally, tools questions include: What strategies might be used to increase student performance?, What no-tech, low tech, and high tech options should be considered? How might tools be tried with student in environments where they will be used? The SETT framework promotes team building. Consensus is built by using clearly understood language, requiring broad-based participation, and valuing input from all perspectives. Exploring environments and tasks strengthens the links between assessment and intervention. It is also necessary to develop a system of tools to enhance the student's abilities to address the tasks and build competency. The SETT framework can address and overcome many of the obstacles that lead to marginal student inclusion, general dissatisfaction, and device abandonment. It can also increase opportunities for success that come with assistive technology systems that are well matched to the student's needs and abilities.

TRAINING INDIVIDUALS WHO USE AAC FOR COMMUNICATIVE COMPETENCE

There are some disturbing statistics regarding AAC use. For example, Magilei and Sandoval (14) reported that 87% of parents reported that their child had access to assistive technology, but they had no training or technology assistance. Also, 33% of all assistive devices are abandoned, largely due to failure to meet the user's needs in community settings. Thus, it is essential that individuals who use AAC devices receive adequate training.

There are four basic types of AAC competence that must be developed in order for the user to be an effective communicator (15). Operational competence refers to understanding of the mechanics of making a selection, turning the device on or off, battery charging, and so on. This involves both the user and the caregiver. Linguistic competence is the understanding of the language elements (symbols, rules of language) that are included in the system. Social competence is the effective use of the AAC system in a functional manner to convey a message to a listener in a given context. Strategic competence deals with the strategies necessary to determine when to use one AAC mode rather than another and how to use it most effectively with a given partner. Communication partners must also be also trained. For children, the training of parents to recognize communication attempts and to understand the operational, linguistic, strategic, and social competencies is also important.

Both physical and communication skills are required for the use of an augmentative communication device. Communicative competence can be developed once sufficient skills are available in both of these domains to allow basic communication. Physical skills are required to make choices from the selection, and these skills may need to be developed in order for the individual to use the device effectively.

For operational competence development, training may be through tutorials built into a device, supplied separately by a manufacturer on a CD/DVD or made available through a company website. A provider (speech-language pathologist, teacher, rehabilitation engineer) may also provide this training face-to-face. In order for the user to develop linguistic competence the AAC device user must understand the symbol system and rules of organization. The individual who uses AAC often must be competent in two languages: the spoken language of the community and the language of their AAC device (15). Development of linguistic competence may require many hours of practice, often built around a functional task.

Many users of AAC devices have little or no experience in social discourse, and training in social competence is required. Rules of conversation are altered for AAC use, and the perception of the individual by their communication partners is also different than conversations between two speakers. In order to be socially competent, the user must have knowledge, judgment, and skills in both sociolinguistic (e.g., turn taking, initiating a conversation, conversational repair) and sociorelational areas (e.g., understanding of interaction between individuals) (15). These skills are best taught in the contexts in which they are to be used, that is, training should occur in the community at school, work, shopping mall, rather than in a clinic setting. Training should be motivational, educational, and functional through the use of age and environmentally appropriate activities, such as playing a board game, which allows participation and multiple communication turns around a topic. The incorporation of creativity and fun activities into therapy sessions can lead to carryover of desired skills–goals and limit the amount of drill-like exercises.

Every user of an AAC device develops strategies that make use more effective. Strategic competence describes the degree to which the user of the device is able to develop adaptive methods to make the most of the device. For example, a child's speech may be better understood at home than at school. He will rely on the electronic AAC device more in school, but they will also develop strategies to make maximum use of both systems. One approach to strategic competence training is to simulate a situation, model the types of interaction likely to occur, and have the user "practice" the strategies and skills necessary to make it a success, followed by an actual situation in which the user goes into the community. If the provider accompanies the user they can then prompt, encourage, and help to clarify when necessary. This combination of clinic-based practice and community-based skill development is often very effective.

A seven-step process for developing communication competence in individuals who use AACs has been developed (16). These seven steps are (1) specify the goal, do baseline observations, (2) select vocabulary, (3) teach the facilitators how to support the individual who uses AAC in developing the target skill, (4) teach the skill to the individual who uses AAC, (5) check for generalization, (6) evaluate outcomes, and (7) complete maintenance checks.

VOCABULARY SELECTION

There are a variety of types of vocabulary that serve various needs. For conversational messages, greetings, small talk, information sharing, wrap-ups, farewells, and conversational repair are important types of vocabulary to have available. Small talk is used for initiating and maintaining conversations and is transition between the greeting and information sharing. It builds social closeness where content is less important than connection to another person. Sometimes scripts (complete dialogues or stories that are stored and replayed bit by bit) are stored in the device so they can be used over and over. For an adult, a story might be about a film that the user saw and enjoyed and wants to talk about. For a child, the story could be about a trip to the circus. Generic small talk is more general and can be used in different conversations with different people. The needs of the AAC vocabulary vary by context, communication mode, and user characteristics.

For preliterate users, a coverage vocabulary is needed to communicate essential messages including greetings, requests for objects and information, comments (e.g., "that's cool", "wow", "that is terrible"), emotional states (e.g., happy, sad, angry), and needs (e.g., "I feel sick", help me"). In order to increase linguistic competence, developmental vocabulary is also required (4). This vocabulary includes words and concepts that are not yet understood. The vocabulary is not selected for functional purposes, but rather to encourage language and vocabulary growth. In some cases, the individual may learn or memorize the location of utterances and become more functionally communicative than their cognitive or linguistic skills would suggest.

For literate users, there are a variety of vocabulary resources (4). Core vocabulary is used by a variety of individuals and occurs frequently. There are word lists based on successful patterns as well as those based on a specific user. In one study, Individuals who use AACs who were operationally and socially competent used a list of 500 words that covered 80% of total utterances for all users (4). Fringe vocabulary refers to words and messages that are unique to the individual. These include names of people, places, activities, and preferred expressions. Fringe vocabulary personalizes the AAC system and compliments the core vocabulary list. Items for this list are identified by informants, usually the user, family, and friends. Fringe vocabulary is selected based on initial items of high interest to the user that have the potential for frequent use. This vocabulary provides ease of production by the user and interpretation by the partner. One method of selecting vocabulary is to use environmental inventories. These inventories attempt to document the individual's experiences by noting precipitating events and subsequent consequences in communicative interactions. A pool of vocabulary items is reduced to a list of the most critical words that the user can manage. Another approach is communication diaries and checklists in which informants record the words and phrases needed by an Individual who uses AAC.

Privacy Issues in AAC

When collecting data on vocabulary used by an individual there is a real risk of invading their privacy, since stored information may include intimate utterances or vocabulary, personal data, or confidential information (16). The use of automatic monitoring devices that capture all of what the user generates by collecting AAC device output in an electronic form during use to assess compliance with a training plan is also a potential violation of privacy. On the other hand, the collection of data reflecting the experience of the individual user of AAC can provide important information related to the individual and more broadly to AAC needs. The user should be able to make the choice as to whether their data is collected or not, and should be able to turn off a data logger when a private conversation is desired.

BIBLIOGRAPHY

1. Blackstone S. Populations and practices in AAC. Augment Commun News 1990;3(4): 1–3.
2. Cook AM. Communication devices. In: Webster JG, editor. Encyclopedia of Medical Devices and Instrumentation. New York: John Wiley & Sons; 1988.
3. Light J. Interaction involving individuals using augmentative and alternative communication systems: State of the art and future directions. Augment Altern. Commun 1988; 4(2): 66–82.
4. Beukelman DR, Mirenda P. Augmentative and alternative communication, management of severe communication disorders in children and adults. 2nd ed. Baltimore MD: Paul H. Brooks; 1998.
5. Foulds RA. Communication rates for non-speech expression as a function of manual tasks and linguistic constraints. Proc Int Conf Rehab Eng 1980;83–87.
6. Cook AM, Husswy SM. Assistive Tecnologies: Principles and Practice. St. Louis: Mosby; 2002.
7. Brodin J. Facsimile transmission for graphic symbol users. Euro Rehab 1992;2:87–92.
8. Electronic Devices for Rehabilitation. New York: John Wiley & Sons; 1984.
9. Vanderheiden GC, Lloyd LL. Communication systems and their components. In: Blackstone S, Bruskin D, editors. Augmentative communication: an introduction. Rockville, MD: American Speech-Language and Hearing Association; 1986.
10. Blackstone S. Visual Scene Displays. Augment Commun News 2004;16(2):1–5.
11. Swiffin AL, Arnott JL, Pickering AA, Newell AF. Adaptive and predictive techniques in a communication prosthesis. Augment Altern Commun 1987;3(4):181–191.
12. Blackstone S. The Internet: what's the big deal? Augment Commun News 1996; 9(4):1–5.
13. Zabala J. The SETT Framework: Critical Areas to Consider When Making Informed Assistive Technology Decisions. Proc Closing Gap Conf 1995.
14. Magilei A, Sandoval L. Creative Therapy Activities Using AAC for Adolescents and Adults. Proc CSUN Conf 2003.
15. Light J. Toward a definition of communicative competence for individuals using augmentative and alternative communication systems. Augment Altern Commun 1989;5(2):137–144.
16. Light JC, Binger C. Building communicative competence with individuals who use augmentative and alternative communication. Baltimore, MD: Paul H. Brookes Publishing; 1998.
17. Wiliams M. Privacy and AAC. Alt Speaking 2000;5(2):1–2.

See also COMMUNICATIVE DISORDERS, COMPUTER APPLICATIONS FOR; LARYNGEAL PROSTHETIC DEVICES.

COMMUNICATION DISORDERS, COMPUTER APPLICATIONS FOR

CYNTHIA J. CRESS
JORDAN R. GREEN
University of Nebraska-Lincoln
Lincoln, Nebraska

INTRODUCTION

A variety of relative inexpensive, portable, and easy-to-use computer devices are now available to educators, researchers, clinicians, clients, and other independent users. This article includes discussion of software and hardware applications for reprogrammable computer devices as well as discussion of devices with fixed programs that evolved from programmable computer technology. A communication disorder is defined as an impairment in the ability to (1) receive and/or process a message or symbol system; (2) represent concepts, messages or symbol systems; and/or (3) transmit and use messages or symbol systems (1,2). This impairment is observed in disorders of speech (use of the oromotor system to produce the movements and sounds of speech), language (comprehension and/or use of spoken, written, or other symbol systems), and hearing (detection, perception and processing of auditory system information, and associated language, speech, or interpersonal factors).

Types of communication disorders may affect specific areas of speech, language, or hearing, as well as broader aspects of communication and social interaction. Persons may have multiple disorders within these areas, as well as closely related disabilities such as swallowing impairments, physical impairments, cognitive impairments, and/or sensory impairments. Disorders may be congenital (present from birth) or acquired (through illness, injury, or late-onset disorders). Some disorders that are presumed to be associated with factors present at birth may not be demonstrated until later childhood. Children may experience speech and/or language delays in early childhood that are resolved with experience and not diagnosed as a disability. Some communication disorders are considered organic (associated with a physical or neurological cause), while other disorders may be functional (attributed to experiential or unknown causes). Communication disorders that are associated with systemic and/or neurological symptoms may continue to affect communication skills throughout the lifespan, although many persons develop compensatory skills to effectively produce, process, and comprehend spoken and written information.

Speech impairments tend to be grouped into one of three categories, depending on the aspects of speech most affected: disorders of articulation (difficulty producing speech sounds), fluency (unusual interruption in speaking rhythm, rate and/or repetition), or voice (difficulty producing appropriate loudness, pitch, quality or nasality) (3). Specific speech disorders that can affect multiple aspects of speech in children and/or adults include the following:

Apraxia: Neuromuscular deficits in planning and programming speech movements; if seen in children, usually called Childhood Apraxia of Speech.

Cleft Palate–Lip: Incomplete fusion of oral structures prenatally that may result in articulation and/or nasality problems.

Dysarthria: Speech problems that are caused by abnormal function of the muscles or nerves that control articulation, phonation, resonance, and/or respiration. Dysarthria is often associated with weakness or imprecision of movements; with severe dysarthria, as in some persons with cerebral palsy, the person may be considered "nonspeaking" and rely primarily on alternative forms of communication to speech *(Augmentative and Alternative Communication, or AAC).*

Laryngeal Pathology: Changes in voice quality due to injury, nodules, or tumors; a removal of the larynx (or laryngectomy) requires alternative forms of voicing and/or communication.

Language disorders are commonly distinguished as either developmental (congenital), or acquired. Three general categories of acquired language disorders in adults and older children include the following (3):

Aphasia: Impairment in understanding and/or producing language after language is developed, resulting from brain damage such as due to a stroke.

Dementia: General loss in mental functions, including language, due to pathological deterioration of the brain; this includes Alzheimer's disease.

Brain Injury: Acquired injury to the brain from trauma (Traumatic Brain Injury) or illness (e.g., encephalitis, meningitis) that results in a variety of communication, language, and/or cognitive processing disorders (e.g., memory, attention).

Developmental language disorders are presumed to result from conditions present at birth, and are usually manifested in early childhood. Common developmental disorders that affect language and may affect other cognitive or social processes include the following (3):

Specific Language Impairment: Significant deficits in language comprehension and use that are not attributable to hearing, cognitive, or motor disabilities.

Learning Disability: Difficulties in acquisition and use of listening, speaking, reading, writing, reasoning, and/or mathematical skills.

Dyslexia: Specific reading disorder with difficulties representing and analyzing sounds and print representations of words.

Central Auditory Processing Disorder: Difficulty identifying interpreting or organizing auditory information with normal hearing acuity skills.

Mental Retardation: Impaired cognitive function that affects a variety of skills including communication, self-help skills, and independence.

Autism Spectrum Disorders: Impairments in social interaction, communication, and restricted interests/routines that affect skills at relating appropriately to people, events and objects.

Categories of hearing impairment relate to the types and severity of hearing functions affected (4). Dysfunction of the outer or middle portions of the ear is considered conductive hearing loss, and dysfunction of the cochlea or auditory nerve is called sensorineural hearing loss. Hearing loss from either cause can be slight, mild, moderate, severe, or profound, depending on the sensitivity of the person's hearing in decibels. Persons with mild-to-severe hearing impairments in their better ear are considered hard of hearing, and often benefit from hearing aids to help amplify spoken language. Persons with profound hearing loss in both ears are usually called deaf, and may have speech or language delays secondary to the hearing impairment. Some deaf persons may communicate primarily through signed languages (such as American Sign Language) and learn English as a second language through visual and/or residual auditory input. Some persons with severe to profound hearing loss may elect to have a cochlear implant, a surgically implanted electromagnetic device to stimulate sensory input to the cochlea.

This article is organized according to the general type of computer application, then by specific applications to disorders of speech, language, and hearing. Types of computer applications in communication disorders include clinic administration and management, analysis of normal function, assessment, intervention, and assistive devices. Tables in each section present potential benefits and limitations of computer applications for each type of application discussed. These benefit and limitation tables are intended to briefly highlight some of the issues involved in computer applications and are not organized thematically or listed in any hierarchical order of significance. For consistency across different sections, the general term *client* has been used to refer to the user of the computer applications, since many programs developed for one population (such as children) can also be used with other populations in limited circumstances.

Computer applications listed are intended to overlap somewhat between different sections. For example, many of the techniques developed for analysis of normal function are also used to assess atypical function across all areas of communication disorders. Also, similar techniques have been developed in computer-aided assessment of some speech and hearing functions. As far as possible, programs and devices are described as types of applications rather than specific programs, which are updated and revised too frequently be considered in a reference chapter. Specific examples of software and hardware applications can be derived from the references and websites listed.

ADMINISTRATION–INFORMATION PROCESSING

Overview

Computer software designed for administration may apply to other functions within communication disorders; for example, word processing programs are potential tools for language intervention, and statistical programs may be linked to other analysis functions in research. This section will concentrate on devices and programs that facilitate, but do not directly accompany clinical, teaching, or research processes.

Clinical Management Applications

Programs are available to collect, organize, store, and report clinical information. Examples of functions addressed by available management software include the following (5–7): (1) maintaining mailing and billing logs, (2) collecting client case history, (3) maintaining correspondence and form letters, (4) graphing and organizing client performance data, (5) retrieving or compiling cumulative information, (6) analyzing cost effectiveness or distribution of clinic activities, (7) coordinating inventory control or budget management, (8) maintaining attendance or scheduling for activities, (9) creating electronic spreadsheets, (10) managing a calendar or clinical team interaction, (11) generating individualized evaluation plans (IEPs), and (12) writing and editing reports.

Computer-based applications are also used to create or adapt clinical materials for speech-language pathology (5). Many of these applications use standard multimedia tools such as digital resource libraries, authoring programs, and integrated drawing, video and/or web resources (8). For example, clinicians might create customized articulation or vocabulary exercises for their clients to practice. Clients might interact with commercially available story programs and interface to individualized prompts for additional narrative retelling or speech tasks. Adults and children might create self-prompting materials or applications to remind them of sequences or strategies for difficult tasks (9). Persons with literacy difficulties might create visual and/or auditory supplements for reading and writing materials to provide multisensory information.

Teaching Application

Programs assisting educators in the field of communication disorders supplement textbook materials, provide shared electronic information files, or simulate aspects of communication disorders for demonstration and practice. In a classroom, computers can be used to visually demonstrate lecture points by producing speech waveforms, demonstrating language analysis, or manipulating audio signals during a lecture. Many programs that simulate or demonstrate testing procedures or communication problems can be utilized in the classroom or for individual student practice. Students may individually respond to group tasks during lectures, allowing for personalized feedback on concept acquisition (10). University networking systems allow for web-based student input and discussion, creation and posting of assignments, feedback and dialog with instructors, and regular updating of course materials and progress (11–13). These networking systems have created opportunities for interactive learning through distance education courses and programs in communication disorders (14,15)

Research Applications

Research-based software can collect or manage information through electronic databases, statistical analysis, data manipulation or integration, and grant or report writing assistance. Large bibliographic databases such as Medline or ERIC search for references for a requested topic, title, or description. The Child Language Data Exchange System (CHILDES) collects and maintains electronic transcripts of typical and disordered language samples available for open access and analysis by language researchers (16). Community or professional listservs and websites can share software, information, messages, or reports, usually for access to all interested persons.

Also, computer networks can link researchers as well as clinicians or clients to other computer users, information centers, or databases. Improved computer conferencing technologies allow real-time shared work environments at a distance, including joint editing of documents, spreadsheets, or data analysis. Web-based video conferencing technologies allow real-time interactions among researchers at multiple locations.

Summary for Administration–Information Management Functions

Potential Benefits for Computer Applications

Allows easy storage, recall, and modification of information.

Provides standard format for reports.

Facilitates documentation of results for report or grant writing.

Sorts information across categories or levels.

Can coordinate or code information for multiple variables of interest.

Provides clean copy of output.

Links computer users in real time or through listservs.

Allows shared workspaces and interactions among persons at multiple locations.

Enables quick access to large bases of information.

Allows shared research databases and remote control of computer functions.

Allows students to practice analysis procedures independently.

Computer-based coding requires researchers to convert subjective decisions into rules.

Potential Limitations of Computer Applications

May require more total time on clerical tasks instead of passing routine functions to clerical staff.

Large data systems have more to lose if the system "crashes".

Potential for data corruption or transmission problems, particularly during periods of high web traffic or network overload.

Information may be entered into computer by users with little clinical knowledge.

Limited monitoring and/or maintenance of many web-based listservs or websites.

Large databases may be difficult to manage and access.

Standard format may discourage originality in report writing.

Computer organization may promote biased or poorly designed research.

Clinicians may design activities to utilize new technologies rather than primarily to meet a client's communication goals.

Students may need more direct interaction or coaching for some types of activities.

Potential for completion of distance education class assignments that do not reflect independent student work.

ANALYSIS OF NORMAL FUNCTION

Overview

Many advances in the assessment and intervention of communication disorders have resulted from extensions of the study of normal speech, hearing, and language functions. Computer technology has expanded the scope of acoustic, aerodynamic, and physiological studies of typical speech. Computer technology has also expanded the scope, accuracy, and speed of analysis of some aspects of language and hearing performance. Many of the applications discussed for analysis of normal function are also used to detect atypical functions, and may overlap with information in the Assessment section.

Speech Analysis Applications

Generate Stimuli. In psychoacoustic studies, digitized speech signals can be created to experimentally manipulate the acoustic features of a stimulus (17). Programs can also randomize the presentation order of acoustic stimuli manipulations, or synchronize stimuli with other experimental. Digitized signals are also converted to analog signals for synthesizing non-naturally occurring signals, or controlling peripheral devices such as pure tone or waveform generators. Adaptive techniques have been devised to randomly present ordered stimuli, organize intervals between events, and modify stimulus levels according to previous values and subject responses. Such programs can respond differentially to indications of subject fatigue or performance variation.

Record Data. Multichannel digital recorders are used to simultaneously capture multiple analog signals from a speaker including acoustic, airflow, air pressure, lung volume, muscle activity, and oral force and movement. The number of channels that can be recorded is only limited by the recorder and available storage memory. Specialized analog transducers and amplifiers convert each signal type into a voltage that can be digitized. These signals provide an objective means to study the physiologic processes that underlie speech production and to establish normative reference data, which can be used to gauge the degree of impairment in individuals with disordered speech. Most modern speech laboratories rely heavily on analog-to-digital conversion. Digitized signals can be easily accessed for editing, analysis, modification, and integration with other data.

Digital audio recordings of speech sounds are the most widely studied aspects of speech. Commercial sound cards provide a sufficient sampling rate and signal/noise ratio for most analysis applications. Typically, a preamplifier is used to increase signal gain prior to digitization.

Computer applications have also been essential for recording speech and swallowing movements. Most knowledge of speech movements has been derived from strain gauge devices, that were mounted directly to the upper lip, lower lip, and jaw [e.g., (18,19)]. Optically based motion capture (OBMC) systems register facial motion in three dimensions (3D) during speech and are rapidly replacing strain gauge techniques. These systems use computer-based visual pattern recognition to extract the motions of passively illuminated markers that are attached to the face. The OBMC systems offer a number of significant advantages over other methods for registering speech motion. Specifically, subjects are not required to maintain a specific posture while speaking nor are they encumbered by wires or metal beams extending from the mouth. At present, OBMC systems provide the only suitable method for studying orofacial movements in young children (20). One significant limitation of OBMC is, however, that they can only record the motions of superficial facial structures.

Most of the information regarding the performance of the articulators inside the mouth during speech and swallowing (i.e., tongue, velum, and the pharynx) has been obtained using four technologies: cineradiography, videofluoroscopy, X-ray microbeam, and Electromagnetic Midsagittal Articulography (EMA). Cineradiography is a filmed X-ray and is typically no longer performed because of concerns regarding radiation exposure. Videofluoroscopy is an X-ray technique that reduces radiation exposure and is usually only performed on individuals undergoing a clinical assessment of swallowing (21,22). X-ray microbeam minimizes X-ray exposure by using predictive algorithms to maintain a focused X-ray beam on pellets attached to moving structures such as the tongue (23). Electromagnetic Midsagittal Articulography tracks the motion of electromagnetic sensors through a calibrated magnetic field surrounding the midsagittal plane of the head (24).

Manipulate Data. Once digitized, the data representing the various aspects of speech performance are displayed and analyzed using custom or commercially available software. Computer programs can extract spatial, temporal, and spectral information from digitized signal input. Signal conditioning techniques are used to filter, rectify, and differentiate digital signals. These manipulations can be performed by dedicated hardware or computer software.

Several manufactures have developed commercial software for capturing speech recordings and for synthesizing speech (25–27). For example, Barlow and colleagues (28,29) have developed commercial software and hardware for recording and analyzing speech airflow, orofacial force, and orofacial reflexes. In contrast, commercial products for analyzing speech movements and muscle activity have not been developed. Presently, programming environments such as MATLAB (30) and Lab View (31) provide researchers who only have a moderate degree of programming experience a means to develop custom analyses to meet their specific needs.

Hearing Analysis Applications

Research in hearing science utilizes many of the same techniques of stimulus generation, recording, manipulation, storage, and control as speech science. For example, computer-generated tones or speech can be used in audiometry to maximize control over stimulus characteristics such as frequency or timing (32,33). Computer models of auditory processes can simulate relative effects or differences in intelligibility in response to systematic changes in the auditory system (34,35). Computer systems can produce randomly ordered sounds that are necessary for some types of auditory physiology and psychoacoustic research, as well as store and index massive databases of data.

Signal-averaging is essential for computer analysis techniques, such as acoustic emittance or auditory evoked response (AER), that require separation of a signal from background responses (36). The AER systems analyze brainstem responses to sound by averaging EEG measurements for repeated sound presentations. Modifications can include nonlinear gating of signal initiation that is not possible with analog signals.

Language Analysis Applications

Analysis of normal linguistic function with computers utilizes descriptive techniques to track developmental progress and maintain databases for language clients, or databasing systems to compare language development patterns observed to those of other populations (37). Descriptive analyses currently available summarize lexical and syntactic functions in normal or disordered language development, such as the Systematic Analysis of Language Transcripts (SALT) (38). Over the past 20 years, SALT analysis has provided standards for language transcript coding and analysis in comparison to normative expectations, including school-based language assessments (39).

The relative speed and standardization of these transcript and analysis systems enables the collection of large and accessible databases of language behavior and variation. One of the largest databases in the study of language development is the Child Language Data Exchange System (CHILDES) (16,40). Continuing development of databases provides accessible digitized audio and/or video transcripts for comparison of performance of children and adults, not only with typical and multicultural development, but Down syndrome, specific language impairment, aphasia, or focal brain injury (37).

Summary for Analysis Applications

Potential Benefits of Computer Analysis

Makes possible complex transformation of a speech signal.

Enables precise timing manipulations for stimulus intervals and randomization.

Allows precise editing, analysis, modification, and recombination of signal or subarrays of signals, without splicing of tapes.

Easier to incorporate probability theory and statistics in analysis process.

Enables selective analysis of separate muscle groups.

Can be used to record speech movements.

Can compare multiple sequences, continuous data points, or multiple plots of data directly with point-by-point analysis.

Can compute online analysis as study is conducted.

Digitally generated speech and/or auditory signals are less distorted and more easily controlled.

Digitally stored data does not degrade as fast as analog.

Signal averaging is necessary for evoked potential study because potentials of interest are small and contaminated by noise.

Possibility for richer and more flexible stimuli.

Analysis techniques can more easily borrow models from other disciplines.

Standardization of transcript collection and coding makes analysis across individuals easier, more reliable.

Computer databases can be collected and shared across research facilities.

Quicker analysis of routine elements of language sample frees more time for other types of analysis.

Language transcripts can be searched for items of interest quickly.

Potential Limitations of Computer Analysis

Limited and nonintelligent algorithms may reduce research questions to analyses of narrow aspects of behavior.

Assumptions for analyzing, storing, or manipulating signals or behaviors are implicit in programs and difficult to retrieve.

Time savings in analysis may be offset by increased transcription or data-entry time.

Created stimuli may distort relevant features of the natural speech event.

Analog-to-digital converters code information via interpretive process, producing an averaged rather than exact representation of continuous signal.

Limited normal or disordered databases restrict interpretive power.

Promotes tendency for researchers to limit field of study to only what the computer can analyze.

Faster analysis may only provide faster mistakes.

ASSESSMENT

Overview

Assessment for communication disorders involves (1) describing an individual at a given time with measures of communication functions, (2) comparing an individual to normative data, (3) extrapolating information to predict behavior over time, (4) profiling abilities or deficits to aid in planning intervention, and (5) providing an objective means of following and recording progress. The format for most computer-administered assessment programs follows the sequence of presenting the stimulus, accepting and evaluating the response, and providing a detailed report of results with optional storage of baseline scores. More complicated programs will offer variable depth testing in which the program evaluates client responses on line and adjusts test content to either extend testing in a problem area or skip questions in areas of high success. Some analysis programs do not test client performance directly but accept information on test performance and provide detailed analysis of that information, often with recommendations for further evaluation or intervention.

Speech

Dedicated speech devices (see also assessment procedures in the section Assistive Devices) have been used to collect acoustic, aerodynamic, and physiological data for dysarthric and apraxic speech (28,29). These types of data are used for both assessment of current function and tracking of intervention progress in children and adults (41,42). Recently, semiautomated algorithms (43) and automatic speech recognition technologies have been used to identify speaking characteristics of typical or disordered speech in children (44).

The vibratory characteristics of the vocal folds are assessed using acoustic voice analysis software (45) or direct imaging (46). Acoustic voice analysis systems have become commonplace in speech laboratories and clinics (42). Increased affordability combined with the continued need to objectify treatment outcomes ensures that these systems will be even more prevalent in the near future. Common acoustic voice measures include (1) fundamental frequency, (2) jitter (short-term variations in the period of the fundamental frequency), (3) shimmer (short-term variations in the amplitude of the fundamental frequency), and (4) a measure of glottal noise (47).

Many of the types of biofeedback used for articulation and voice assessment have also been applied to fluency disorders. For example, computer-based technology can track client speaking rate, pitch, intensity, stuttering frequency and duration, and percentage of utterances without stuttering (48). Other programs provide a user interface for clinicians to score perceptual judgments such as stuttering severity or number of syllables with stuttering [see (5)]. Available software to address fluency issues is listed at www.stutteringhomepage.com.

Both EMG and motion analysis technology can be used to quantify the movement deficits associated with impaired speech. Some abnormal features that have been identified using this technology include problems with force, endurance, movement displacement and velocity, interarticulator coordination, and movement pattern stability. The clinical implementation of motion analysis technology to assess speech motor problems has significantly lagged behind its application to assess gait and posture problems by physical therapists. Obtaining useful normative data for speech movements has been challenging because speech performance is highly variable across individuals. The cost and maintenance of this equipment are also problematic. These present challenges are not insurmountable, and it is

likely that, with more empirical research, motion analysis will become an integral part of speech assessment in the not-too-distant future.

Many commercially available speech assessment programs provide computer analysis of clinician-entered data rather than direct assessment of function. Computer-based measures such as the Sentence Intelligibility Test (49) present word or sentence stimuli and record client response, but the clinician judges the intelligibility of units within the test items relative to nondisabled individuals. From this input, the program analyzes and presents the percentage and severity of intelligibility impairment. Other programs, such as the Logical International Phonetic Programs (LIPP) to analyze prelinguistic vocalizations of infants, interface between listener judgments and presentations of digitized infant vocal samples (50). Automatic computer analysis of infant vocalization samples can analyze relative complexity and variety of infant sound production, but does not yet directly correspond to phonological categories of listener-perceived judgments (51).

Hearing

Computer advances in hearing assessment (see also hearing aid evaluation in the section Assistive Devices) include improvement of test signal generation, different emphasis in audiological tests, and improved automated measurement of hearing (52). Computerized assessments of hearing provide direct measurement of acoustic characteristics across separate frequency components of a complex signal, including hearing performance while a hearing aid is worn (53,54). Computer-based audiometers can measure tinnitus or pitch problems, loudness problems, reaction times, and masking properties. Speech intelligibility under various listener conditions can be estimated from digital speech samples and automatically scored (55,56). Information gathered from these various audiologic tests can be used to optimize a client's hearing aid performance.

Other audiologic tests that are slow or impractical with traditional audiometry can be facilitated with computers. Evoked potential audiometry to screen or test thresholds of hearing or other features of the auditory system is made possible with computer signal averaging of the brain wave signals recorded at sound presentations (52). Automatic screening of such otoacoustic emissions has become standard universal practice for infants, particularly newborns (57–59). Automated visual response audiometry provides information on user spontaneous response to sounds for young children and persons with cognitive disabilities (60–62). Some kinds of auditory response tests have been computerized for more consistent administration and scoring with older children and adults (63).

Language

Computerized language-assessment tools may probe client linguistic production, comprehension, or problem-solving skills, elicit or describe language production, administer or score language tests, or analyze syntactic, semantic, lexical, or pragmatic elements of language samples. Many language assessments are designed for particular aspects of specific language impairment or language–learning disabilities. However, language measures designed for one population can be applied to persons not in the target population who exhibit similar language characteristics if comparable normative data are available.

Several protocols are available for analyzing features of language transcripts; these protocols differ according to input format, number and type of linguistic analyses, speed, costs, and profile, training, editing, and search capabilities (37,64). Language sample tools can assist with two primary functions: data retrieval and tallying of codes entered by the clinician, and symbiotic programs that use algorithms to mark linguistic elements, with corrections by the clinician as needed (5). Most of these analysis programs are intended to profile language development and are standardized only for typically developing children, but some extensions of norms have been developed for children with language delays and disorders.

Each of the computerized language sample analyses has a standard input format and a range of options for language computations commonly used in the field, such as a Mean Length of Utterance (65). For example, various programs provide information on language characteristics such as conversational functions, grammar, semantic relations, vocabulary, narrative, and phonological patterns (66). Computer analysis can dramatically reduce assessment time while maintaining acceptable range of accuracy in clinical decisions to human-generated language analyses (67). Language sample analysis programs include the following: additional programs and reviews of features are available in Cochran (5):

1. Child Language Analysis (CLAN) (68).
2. Systematic Analysis of Language Transcripts (SALT) (38).
3. Computerized Profiling (69).

Computer-based phonological assessment programs provide analyses of standardized test results, systematic analysis of phonology, or phonological errors in a spoken transcript. Available phonology analysis programs fall into three types: analysis of phonemes and errors in different positions, distinctive feature analysis, or phonological process examination. These computer-based analyses are directly comparable to standard hand-scored measures, but can be substantially faster even for relatively novice users (64). Most programs require input of results of a particular language sample, and provide summary sheets of numbers or types of errors and characteristics of the systems impaired. Some also provide suggestions for target sounds in intervention. Features of available software addressing these goals, such as Programs to Examine Phonetic and Phonologic Evaluation Records (PEPPER) (71), Computerized Profiling (CP) (70), or the Interactive System for Phonological Analysis (ISPA) (71) are listed in Masterson, Long and Buder (72), and Cochran (5).

Other computer-based language tools have adapted strategies for collecting samples and/or scoring standardized paper assessments. Computers can facilitate assessment of language competence during conversational interaction by providing a dynamic context with shared reference to elicit spontaneous language (5). Programs have been developed to facilitate language sample analysis, test administration/scoring, and reading/writing skills in adults (41). Assessment techniques utilizing the computer for administration and scoring of paper language tests including vocabulary inventories, receptive vocabulary probes, and targeted probes of language fundamentals (5). Also, literacy assessments such as reading or spelling tests can be administered and/or scored by computer, and used to assist planning in language and literacy intervention (5).

Several standardized language assessments have been converted to computer-based administration and tested for equivalency to traditional administration. Results show variable influence of the computer interface, potentially related to the variables associated with the disability, task, access, and presentation. Some test administrations with simple recognition behaviors show equivalent results with preschool and school-aged children, using a variety of input devices (73). Other test adaptations indicate that computer administration may introduce variability and cognitive load from the task and input requirements, and that separate norms may be needed for computerized administrations of standardized tests (74,75). Selection of computerized language measurements should be consistent with the assessment principles of the clinician, and integrated with other direct measurements and observations of client language skills (76). Dynamic capabilities allow the potential for computer-based assessment of active language learning, rather than sampling of language skills already achieved. Jacobs (77,78) has developed a computer-based screening test to identify risk of language delay in children from multicultural backgrounds. Because the program uses video and audio stimuli to present and test language concepts in an invented language, it both samples children's dynamic potential to learn new language uses and avoids cultural bias of experience in their familiar languages. Continued progress in statistical analysis techniques for dynamic assessment and developmental data will facilitate the ability to base language assessment measures on predictions of language change over time (79).

Computer-assisted neuropsychological testing has been implemented with adults, for skills such as visual attention, memory, response speed, and motor tracking (41). Most language functions have not been sufficiently studied in typical communicators to identify language deficits associated with damage or dysfunction in specific neural regions (80,81). Additional integration of neural imaging and function measures may improve the diagnostic and descriptive potential of language-based assessments (82).

Summary of Assessment Issues

Potential Benefits of Computer Assessment
Computers prompt for missing information without making assumptions about nontested variables.

Standardizes presentation and analysis of assessment procedures.

Simplifies longitudinal assessment of function or learning.

Direct input of signals can improve accuracy and precision of measurement.

Computers are more reliable than humans at determining boundaries in acoustic or physiologic signals.

Possible to modify assessment techniques in response to change in client behavior response or test conditions.

Systems encourage accountability for decision making.

Analysis and assessment can be completed in real time, continuously as the program operates.

Larger databases can be accessible for test outcomes.

Easier to store data and track client progress.

Can temporally analyze and compare verbal and nonverbal aspects of transcript.

Signals can be generated, stored, and ordered for assessment presentation.

Evoked potential research is particularly useful for hard-to-test individuals, such as infants.

Simplifies routine assessment functions to encourage more clinician time spent in other types of assessment.

Dedicated specific language analysis tools can adapt rapidly to changing theories of language processes.

Enables assessment of comprehension with nonverbal stimuli and responses.

Comprehension of dynamic concepts can be assessed dynamically.

Computer can react to subject variables difficult to recognize, like fatigue, by pattern of response.

Artificial intelligence models can consolidate expertise from multiple clinicians and situations.

Computers can do multifactorial comparisons impossible by hand.

Enables noninvasive measures of internal functions.

Increase quantification of perceptual judgments.

Quantitative analysis is accessible to a wider range of clinicians with computer-based techniques.

Potential Limitations of Computer Assessment

Computer transfers experimental biases to implicit level, which is often inaccessible, and subject to misinterpretation.

Humans are better pattern recognizers than computers and are able to integrate a greater variety of factors.

Many assessment programs ignore or average out variability that may be one of the most interesting or useful aspects.

Need to determine if assessment tools themselves are valid before introducing computer for test administration and analysis.

Standardized tests are often designed to administer only one task.

Computerized assessments may overlook important factors or variations that are observed by clinicians.

Measurement techniques may be too time consuming to be clinically useful.

Techniques may require time, equipment, skills, and money not available to clinicians in general.

Clinicians may avoid responsibility for interpreting data and simply report computer results to clients.

Clinicians may be overwhelmed by the volume of data output from assessments, and potentially misinterpret results.

Comparative analysis is limited without normative database for the population assessed.

Danger for undetected errors in program's computations and output.

Clinicians may be tempted to entirely substitute computerized tests for informal tests.

If input to the assessment protocol is only estimated, computer will erroneously interpret that as fact.

Uniform assessment techniques may be uniformly mediocre without benefit of clinical intuition.

INTERVENTION

Overview

Note that relevant applications for clients with hearing impairments are distributed across three different sections according to intervention techniques addressing speech, language, or specific communication aspects of hearing impairment. Habilitation of hearing itself with hearing aids is discussed in the Assistive Devices section.

Applications of computers to intervention in communication disorders range from direct training and modification of measurable behaviors, to practice and feedback of speech/language functions, to presentation and teaching of facts or concepts, many of which utilize software and techniques originally targeted for regular education. Potential roles for the computer in fulfilling these tasks include the following [adapted from (83)]:

1. Tutor: Present information or tasks in sequences designed to achieve a desired behavior, provide drill, and practice.
2. Eyeglasses: Provide tools for amplifying and extending abilities (e.g., learning to influence the environment).
3. Mirror: Provide sensitive feedback on behavior.
4. Stimulus: Provide stimulating material for learning.
5. Access tool: Provide materials, real or simulated capabilities, or activities appropriate for tasks.
6. Communication: Input or output of spoken or written communication.

Tasks of the teacher or therapist in providing computer assisted instruction (CAI) include assessment of current and past performance and learning style, task analysis of what and how to teach, identification of instruction level, and evaluation of ongoing progress (84). Specific elements of tutorial programs can include (84,85): (1) pretest of requisite knowledge, (2) presentation of stimulus (text, graphics, videos, and/or speech), (3) acceptance and evaluation of response, (4) provision of prompts or cues for responses, (5) feedback or reinforcement of correct responses, (6) change in difficulty based on evaluation of response, and (7) record keeping and string of data on client's performance.

Knowledge-based or intelligent tutoring systems utilize artificial intelligence models for describing skills and rules for deriving given behaviors, which are applied to normal and deviant behavior and modified through cumulative knowledge and experience. Research and therapeutic applications of artificial intelligence include the following (86): (1) simulation of perceptual, communicative, and cognitive processes to examine the nature of normal and disordered processes and recommend and test assessment or intervention techniques; (2) development of interactive systems which can evaluate, creatively adapt to responses, and vary stimulus or measurement techniques; and (3) development and testing of models of behavior that attempt to objectify rules and systematic influences of subjective behaviors.

Speech Intervention

Except for a few specific techniques or tests targeted at a given disability, computer applications will be discussed by general qualities of the speech signal addressed in intervention. Many of the techniques discussed here have been applied to various disorders, including voice, articulation, fluency, dysarthria, apraxia, and speech consequences of hearing disorders or physical handicaps.

Visible Speech. Techniques derived for the analysis and display of speech characteristics are used not only for research but also for intervention for speech disorders. Clients receive visible feedback about the qualities of their speech production as a cue to improve specific qualities of the production. Types of information displayed include the following: (1) analysis and display of acoustic parameters; fundamental frequency, intonation, amplitude, spectrum, and nasality (42,87,88); (2) direct representation of speech waveform characteristics (89); (3) interpretation of perceptual analyses or other classifications of speech parameters (90); (4) information on the progress and efficacy of the swallowing response (89,92), and (5) information on closeness of fit of client production to a template along acoustic or perceptual parameters (88,90,93).

Speech Feedback. Delayed auditory feedback, in which a speaker's voice is replayed to headphones after a preset interval, has been implemented in both stand-alone and in-the-ear devices to improve fluency (93). Other modifications of feedback to improve fluency include altered frequency, amplitude, or speech masking of the client's speech (93). Some types of speech feedback may be supplemented by tactile feedback sensors, for clients with limited hearing. Feedback on a wide variety of speech qualities can also presented in game or simple visual presentations for young children and other clients with basic communication skills (90,95).

Speech Movement Indicators. Real-time displays of the speech acoustic signal are the most widely used method to deliver feedback to patients regarding their speech performance. These displays are used to training features of

speech such as loudness, rate, rhythm, intonation, or pitch (42). Acoustic voice analysis systems provide a convenient and noninvasive method to measure and track vocal characteristics throughout treatment (96,97).

A number of devices display aspects of speech movement in real-time, which can be used to train appropriate speech behaviors. Strain gauge and optically based motion transduction systems can be used to provide feedback to a patient of their lips and jaw movements (98). Coordination of speech breathing can be visualized using devices such as the Inductotrace (2) that transduce movements of the rib cage and abdomen. Dynamic palatography provides a means to display tongue to palate contact patterns in real-time during connected speech (99,100). Movement of the vocal folds can be obtained real-time through high speed stroboscopy (26,89). EMG is used to provide patients information about muscle activity patterns and may be used to increase or decrease muscle activity levels.

Speech Cueing. For clients with rate difficulties in speech, including fluency disorders, computer-based metronomes and cueing systems may facilitate fluent speech (93). These rate programs may be integrated with reading highlighting tasks, to reduce the difficulty of both the speech and language aspects of reading aloud. The feedback from a voice output assistive device may also provide cueing for clients with limited speech to elicit more frequent and/or complex speech signals (102,103).

Counseling and Support. Behavioral and emotional responses can directly affect speech disorders, and technology has been used to facilitate client counseling. For example, in cases of psychogenic aphonia (loss of voice for nonphysical reasons), direct feedback of the vocal function has been helpful in altering clients' perceptions of their own voice functions (89). Similarly, a biofeedback program was used in combination with behavioral treatment activities to increase client perception of control over stuttering episodes and reduce likelihood of relapse after treatment (102).

Hearing Intervention

While intervention with hearing-impaired persons utilizes techniques of both speech and language training (52,103,105), some communicative function training is unique to hearing-impaired persons. For example, several different programs, CDs, or websites use computer graphics and animation to teach sign language, finger spelling, speech reading, or simultaneous communication (8). Websites, such as www.deafed.net, provide coordinated resources for specialized interventions for hearing and associated speech/language skills.

Some intervention programs have utilized dynamic graphic displays to support practice or drill of specific speech and language skills associated with hearing impairments. Intervention systems have been established to reinforce therapy drills, provide remote-accessible tutorial programs, encourage speechreading practice and interaction with other deaf and hearing persons, and practice constructing language output with dynamic, visual input

(52,104). Aural rehabilitation for children with cochlear implants can be presented with specialized video games targeting specific listening and/or speech tasks. For persons with some hearing, computer speech output devices have been used to provide stimuli for auditory perceptual training. Direct intervention of hearing function through technology tends to be provided with assistive listening devices (see later section), although some systems have been constructed to target specific hearing features in listeners (107).

Balance and dizziness intervention can be assisted by computer-based presentation and evaluation of response to specific tasks. For example, technologies are becoming available to directly sample video images of a person's eye movements, standing balance, and walking (105). In therapy, computer technology can present virtual reality of environmental situations that gradually increase the balance challenge in response to the user's reactions in target behaviors.

Language Intervention

Language intervention utilizes many procedures and software programs in ways similar to computer-assisted instruction in general education. Programs designed to address general functions, such as problem solving, teaching school subjects, or playing video games, can be applied to language intervention as specific visual and auditory perception, reading, or sequencing practice tasks. The following are some of the ways computers can stimulate language development (5,83):

1. Language as subject matter: A wide variety of general and special education software is available which provides drills, tutorials, or examples of language skills such as spelling, grammar, vocabulary, and writing.

2. Language as currency: Language can be required as input or output in order to perform other activities such as games, problem-solving activities, or simulations; with some computer programs, the clinician can structure the program to accept either more or less complex language input, depending on the client's language skills.

3. Computer as a tool for linguistic communication: Clients with limited experience with either spoken or written communication can use computer systems such as word processors, computer networks, interactive terminals, speech output devices, or any of the augmentative communication devices to practice language and communication skills.

4. Computer as a tool for literacy: Computers can provide multisensory input and output to facilitate client reading and writing, as well as providing structure and expert feedback for problem-solving tasks such as editing grammar or creating a narrative.

5. Computer as a topic to generate conversation or interaction: Computers can serve not only as an instructional tool but as a source of problem solving and conversational topics embedded in the social context.

6. Computer as an instructor: Computer-assisted instruction is not limited to interactions between clients and computers alone in tutorial or drill-and-practice programs, but also as a facilitator of instruction as a triad with the clinician or educator.

7. Computer as an interactant: With some artificial intelligence programs, computers can react creatively to some aspects of client input and can function as an interactive learning tool; the programs vary in how much control the client has over learning and how the program adapts its linguistic responses or tasks.

8. Computer as a tool or reinforcer: Some types of video and accessory programs can present or manage tasks in ways that support the user's memory, cognition, language skills, and/or motivation.

Many general education software packages can be adapted for intervention with language disabilities. These educational software programs can be integrated with other technological resources such as websites and search engines to develop theme-based instruction with language support around an academic topic (11). Some modifications of standard educational software recommended for language and learning disabilities are (106) (1) expanding application of software beyond stated purpose, such as using alphabet programs for visual recognition or story programs for retelling a sequential narrative; (2) individualizing program characteristics for clients, such as level of difficulty and duration of stimulus, multimodal input or output, use of graphics or printed text, or tasks subdivision; (3) providing supplemental activities such as retelling a story or answering written or spoken questions to apply program information; and (4) individualizing content of entire programs for specific practice or focus by using authoring programs.

Some dedicated programs have been developed to practice specific domains of language skill in relatively focused decontextualized environments. Tasks addressed by software targeted for direct language practice include drills or tutorials for teaching vocabulary, figurative language, visual memory, phonological awareness, and grammar (90,109–111). Programs have been developed to modify characteristics of auditory speech input that are proposed to enhance skills in auditory and language comprehension, but data on the efficacy of these programs are mixed (112–114).

Most applications of technology with young children rely on computers or other technology as one element of an interactive or academic experience. For example, computers may assist in presenting or reinforcing concepts, supporting access to task materials, searching for information, or presenting stories or other written information to nonreaders (115). Children may develop cognitive concepts such as the connection between causative actions and effects through operation of a variety of switches that simplify the motor activities necessary to activate environmental devices or toys (116). Dedicated programs for early language intervention with preschool children address concepts similar to other language programs, usually embedded into teacher–child or child–child interactions: vocabulary, following directions, differentiating letters–shapes–colors, and numerical concepts (117,118). Early intervention software and devices for infants and toddlers have been designed to engage the child in meaningful and cognitively appropriate play, with maximum technical support to allow easy access to social interaction, play, and tasks such as speech or writing (116,119).

Computer technology can provide ongoing support for reading and writing as well as tool for learning a wide variety of literacy skills. Early literacy skills can be promoted with programs verbalizing letters, words, and phrases as they are typed, highlighting text as it is read, and animating stories and words (120,121). Programs can provide scaffolding of conventional writing tasks to simplify writing or typing tasks and compensate for poor grammar and spelling, thereby encouraging maximum language-output capabilities (6,122–124). Voice capabilities of programs can support reading of books and text out loud for multisensory output, as well as specific word prompts for literacy tasks (125,126). Voice recognition programs can provide limited spoken input to writing programs, although most still rely on user correction of errors, even when combined with standard or custom text analysis programs (121,127). Programs to support and teach spelling and/or phonological awareness skills tend to use a combination of prompts, voice feedback, and systematic cues to identify and associate letters and phonemes appropriately (88,90,128).

Many persons with language impairments have additional congenital or acquired disabilities in cognition, memory, social interaction, and/or organization. Programs designed or adapted for clients with other developmental disabilities can address skills such as perceptual-motor skills, cognition, vocational skills, creative arts, self-help, memory-cuing skills, or social skills (129–132). Recommended ways of structuring computer-based tasks for clients with developmental disabilities include (133): (1) providing proper incentives for the client, (2) structuring programs to adapt a task to the client's behavior, (3) adjusting speed of presentation and amount of repetition, and (4) evaluating progress longitudinally rather than by initial performance. Because persons with cognitive or other developmental disabilities have difficulty generalizing experiences from one context to another, it is important to embed technology experiences into real life functional and social interactions, particularly for persons with social impairments, such as Autism Spectrum Disorders (134,135).

Available rehabilitative software designed specifically for adults with language and cognitive impairments addresses impairments of functions similar to other language disabilities in different instructional formats and styles. Some of these programs utilize alternative input systems when visual/perceptual or motor skills interfere with performance, such as for supported reading and writing (136,137). Examples of specific areas addressed for cognitive rehabilitation include concept training, memory, personal organization, planning, and perceptual processing (138–140). Technology can also serve as a memory or organizational tool for persons with brain injury, using common off-the-shelf devices such as handheld personal organizers and integrated video, and/or cell phone technology.

Summary of Intervention Issues

Potential Benefits of Computer Intervention

Computers are objective and reliable.

Potential for undivided attention, client-paced instruction from program.

Reinforcement immediate, contingent on user response.

Possible to present information multimodally.

Context and tool for exploration, problem solving, networking concepts.

Motivates interaction with computer as dynamic context in group interaction.

Provides experiences not otherwise possible, such as voice output of original text.

Intervention elements can be regulated or altered by clinician.

Allows intervention with using nonlinguistic and visual stimuli.

Can present dynamic stimuli difficult to represent in other ways, such as verbs or sign language.

Programs can respond immediately to subtle variations in speakers that are difficult for listeners to detect.

Can incorporate more than one instructional strategy, simultaneously or in sequence.

Collects, analyzes performance data on line.

Allows additional practice time outside of therapy, even at a remote distance.

Intelligent programs can have accumulated experience across clients and experts.

For responses that are variable, the user's best production can be the target production rather than a set template.

Potential Limitations for Computer Intervention

Danger that computers oversimplify a task and are used as electronic flashcards.

May artificially separate memorization from integration of knowledge rather than memory through integration.

Automaticity of drill responses is not optimal for facilitating language development.

Computer interaction may detract from social interaction, particularly in persons with specific social impairments, such as in Autism Spectrum Disorders.

Computer may prompt clinicians or educators to teach things that children do not need to learn.

Programs can make mistakes more interesting.

Programs must introduce enough variability to build generalization skills.

Clients must understand program instructions, particularly if using self-paced exercises.

Computers and access strategies can introduce independent cognitive load to task.

Input and output modalities may divide perceptual resources.

More difficult problems in everyday life are solved by analysis rather than drill.

Software may legitimately do what is advertised, but concept taught may not be valid or valuable.

Predetermined programs may limit originality of clinician interpretation of behavior.

Computer tasks may introduce methods or cognitive loads that exceed the purpose of the original therapeutic goals.

Should teach rewards of communication and interaction for its own sake, not for computer reinforcement alone.

Users may expect technology to serve as a stand-alone system for client–computer interaction, rather than embedding technology use into interaction and functional use.

Users may expect that intervention technology can replace language, speech, or hearing functions that are inherently irreplaceable.

May not be able to simplify task enough to overcome difficulties of interacting with computer or symbolic mode.

May perceive that computer-based techniques are more effective than they really are Computer program may teach skills according to philosophies or techniques that are contrary to the clinician's or the client's values.

Programs may be designed to rehabilitate skills that are best served by support that bypasses impaired functions.

Users may adapt educational or intervention goals to suit the capabilities of the technology, rather than using technology to support intervention.

Computers cannot be equivalent in function to clinicians, and cannot be used to justified reduced clinical staff.

ASSISTIVE DEVICES

Overview

Computer technology has dramatically improved the functional capabilities of physically, sensory, and communicatively impaired individuals. Functions addressed by assistive computer technology include communicative output (speech or written), communicative input (visual or auditory), environmental control, and access to standard computer capabilities (both personal and public computer devices). Many aspects of adapted computer access and universal design have already been incorporated into standard computers, software, and consumer devices. Key resources for further information about assistive devices and augmentative and alternative communication include the following: Beukelman and Mirenda (141), Cook and Hussey (142), the AAC Web site (http://aac.unl.edu), and the Co-Net assistive device inventory (http://trace.wisc.edu/tcel/). Topics addressed within this section will be organized according to communicative input, output, and other applications for language, hearing, and speech–physical impairments.

Language

Input (Language Learning). Persons who rely on augmentative communication (AAC) may face additional difficulties over typical communicators in both developing and using language skills through these alternative means. Early application of an augmentative system can help structure linguistic intervention by providing a motivating means for language practice, alternative modes of communicative output, and a vocabulary access and expansion system which can translate individual word units into acceptable grammatical form (143–145). Persons with physical or visual impairments often have limited access to standard written materials, and computers can provide access to reading and text production to promote receptive and expressive language skills (146,147). Still, young children who are nonspeaking have multiple additional sources of cognitive and interactive difficulty that influence early language development, such as difficulty of parents and children recognizing atypical movements as communicative or linguistic (148). Current research has addressed ways to adapt AAC language input techniques to be more cognitively accessible (149–151).

Output (Language Use). Conversational language output using AAC is slower than spoken language and often relies on users switching back and forth between spelled and prestored vocabulary to complete messages (152). Older AAC users may alter linguistic features of their output such as grammar and word choice because of the limitations of their available system; access to fully generative language systems is a critical element of language development in AAC (153). Fully competent language users who rely on AAC may still modify their language output because of rate and partner limitations (154). An important part of being linguistically competent in AAC is being able to determine strategies for when and how to use different language content and modalities for different purposes (155). An essential part of intervention with AAC devices, particularly for young children, involves strategies for partner training and support of the augmented communicator (156).

Most effective assistive technology applications for persons with acquired language and/or cognitive impairments focus on control and support of existing skills, rather than direct rehabilitation of impaired skills (157,158). This support includes low tech output such as gestures or communication boards, high tech devices for message formulation and retrieval, and partner-supported techniques such as written choice or partner drawing (158–160). Computerized linguistic cuing and speech output devices have been applied to clients with acquired language disabilities with only limited success (137,161). For most users, continued technology use after an acquired injury depends on the environmental and partner support as much or more than the type of technology used (162).

Other. With the current proliferation of web access and simulation programs, persons with language and other disabilities can also have access to language and functional learning experiences for which they are physically not suited, like laboratory problems or driving simulation. Computers and other distance technology may also be used to supplement language therapy for homebound, rural, or physically disabled clients who cannot attend regular therapy sessions at a clinic (87,163). Most telecommunication systems require written language and computer access skills, which are problem areas for many language-impaired individuals with perceptual and/or symbolic deficits.

Hearing

Input. Digitized hearing aids have become the industry standard, with capacities to maximize high-frequency gain, filter out nonspeech distortion or noise, and program aids to mirror a patient's hearing loss (107,164,165). There have been significant advances in the use of computer technology for the direct evaluation and improvement of hearing aids and cochlear implants (53,54). New developments in hearing aid technology can allow users to alter the characteristics of the aid remotely to match various listening situations. Similarly, cochlear implants that provide electrocochlear stimulation have advanced signal-processing capabilities to improve speech intelligibility (33). Further research is needed to determine more accurate predictions of the types of benefits or limitations of cochlear implants expected for different populations (166).

Output. Other assistive devices for speech–language output for persons with hearing impairments provide dynamic translation of spoken information in face-to-face as well as remote interactions. For instance, interpreter services now offer real-time captioning, in which a transcriber records auditory information in a classroom or meeting to be immediately transmitted to the user's own display or a whole-room display. Standard text messaging and video transmission technologies allow quick access to direct interaction that can substitute for telephone access for deaf or hard of hearing users. If telephone services are used, TTY (teletypewriter) connections allow typed interactions between telecommunication devices using standard phone lines. If only one user has access to a TTY, telephone relay services are available to provide voice translations to and from TTY signals as needed. Similarly, fiberoptic Polycom video systems can be used either for direct signed conversations or remote ASL interpretation of spoken input when the interpreter is not physically present at the location. Text to speech and voice recognition systems are not yet sophisticated enough to translate spoken to written communication accurately in real time (167), but are used for limited purposes for persons with hearing impairments.

Other. Computer-based environmental control systems can be established in homes or workplaces to give a visual, automatic signal for telephones, doorbells, alarms, or emergency signals. See Cook (this volume, COMMUNICATION DEVICES) for more information on access to environmental control devices.

Speech and other Physical Impairments

Individuals with severely impaired control, coordination, or strength of oral-motor systems may require an assistive device to augment or replace functional speech. Types of disability or disease that may result in this condition include cerebral palsy, spinal cord injury, muscular dystrophy, multiple sclerosis, amyotrophic lateral sclerosis, stroke, traumatic brain injury, and spastic dysphonia. Individuals with motor impairments affecting their hands and arms may also require assistive devices to facilitate writing or access to a computer. Augmentative aids are designed to fit a client's motor, linguistic, cognitive, social, and educational skills and needs, and may address several different communication or access functions. Computer-based programs may be used to determine likely matches between technology and user needs, available either as stand-alone software or integrated software within dedicated communication devices (141,142). Many applications of technology for these users are covered under the chapter on computer access (Cook, this volume, COMMUNICATION DEVICES). This section will concentrate on technology to support speech input and output from persons who rely on AAC.

Input (Voice Recognition). Two types of uses are common for voice input systems with present technology. For persons who have good use of speech but severely limited physical movement, such as persons with high spinal cord injuries, voice input may be a primary strategy for controlling computers for a wide variety of writing, computing, and environmental control functions (167,168). Available voice input systems rely on voice recognition systems that are calibrated to the client's speaking style, and users typically must speak at a reduced rate and monitor the system for errors in recognizing words. Future development is necessary before voice recognition systems can be used seamlessly for writing or distance communication without error correction. For users with some speech with limited intelligibility, such as persons with cerebral palsy, voice recognition systems have been adapted to recognize and respond to atypical speech productions and provide support for writing and other communication tasks (169). Such systems need improvement in algorithms for recognizing subtle variations in vocalizations as well as triangulating intended utterances of users with high variability in speech production.

Output (Controlling Voice Output). Electronic voice output is an alternative or supplement to speech for many people with severe speech and/or physical impairments, who have difficulty being understood using their own voice. For individuals who cannot use standard keyboards or input devices, two general techniques are used to activate voice output devices: direct selection and scanning, both with options for encoding the input signal. The choice of input techniques depends on the range and type of behavior that the person can control, individual preferences, and communicative needs. Most of these input techniques could operate devices for speech, writing, and/or computer access functions. For direct selection, a variety of peripheral devices are available which utilize different motions to select communication units from an array of choices, including adapted keyboards, pointing devices, light pointers, switches, eyegaze, and speech input (141). For individuals who only have a small set of discrete motions, encoding techniques can be used to provide a larger selection set. For example, persons who can operate only a limited number of keys or a joystick can select a larger number of words or letters by using sequences of two or three movements or selections for each word.

If an individual has only a single controllable movement, fatigues easily, or does not use encoding techniques, a scanning input method may be appropriate. A series of items or groups of items are presented sequentially, and the individual makes a single signal at the proper time to select the desired item. More sophisticated scanning techniques systematically narrow down the field of choice to speed the process. Since the user of a scanning aid must wait until a desired item is highlighted by the device, scanning is often a slow method (170). For users with additional cognitive and/or language limitations, scanning can introduce cognitive load that may require specific training or interfere with skilled technology access (171,172).

Since most alternative input techniques are slower than speech or typing, special acceleration programs have been developed that can provide the client with the ability to input the same text with fewer necessary input selections or keystrokes. Whole words and phrases can be coded, by the user and/or manufacturer, which can be accessed by abbreviated letter and number codes, picture symbol codes, or word prediction. For conversation needs, computer devices can provide visible correctable displays and voice synthesizers. For writing, correctable displays can provide printed output, with the potential for portable printing and writing aids as well as handwriting facsimile output. See Beukelman and Mirenda (90) for more detailed information on these techniques.

Other. Functions other than speaking or writing that computer devices can provide for individuals with severe disabilities include environmental control and access to standard computers. A variety of computer-based devices can adjust controls of appliances or lights, often linked through the same device used for speech and writing. Devices such as robotic arms can potentially handle light books, turn pages, insert computer disks, and provide other manipulative functions. Current efforts to improve universal design for easy access of environments to all users will improve the usefulness of technology-based solutions for environmental control and access (173). See Cook (this volume, COMMUNICATION DEVICES) for more information on these resources.

Another rapidly growing need for persons with disabilities is the ability to access and use standard computers and computer programs in the daily living environment. Individualized adaptive devices for accessing a personal communication aid will not necessarily allow direct access to any standard computer, particularly if the client must use an alternative input device. Interfacing units (including keyboard emulators) can allow the user to operate virtually any computer software program with their specially chosen adaptive devices by making the output of the personal adaptive device look like the information that the computer program expects to receive from the standard keyboard. However, the ability to use most software with one's own adaptive computer system at home does not necessarily mean that the same adaptive equipment can be used to operate other computers, bank teller machines, or library terminals. Limited compatibility between different dedicated and nondedicated hardware and software limits the range of technology a person can access with any given tool. See Cook (this volume, COMMUNICATION DEVICES) for more information on these resources.

Summary of Assistive Device Issues

Potential Benefits of Assistive Devices

Provides voice output to supplement or substitute for impaired speech skills.

Provides language input as well as output to children learning language.

Provides support for the development of language, writing, and cognitive skills.

Voice output allows more complex interaction between speaking and nonspeaking children who are preliterate.

Enables user-programmable vocabularies.

Communication aids can be made portable with speech, writing, or computer access capabilities.

Allows access and control of a wide variety of activities via computer.

Allows privacy of writing without need for translator.

Enables user to alternate between communicative and computer modes.

Potentially greater vocabulary storage and access techniques than nonautomated aids.

Provides mechanism to reach larger audience or range of interactants.

Augmentative communication over a distance (e.g., listservs) can compensate for unequal spoken to nonspoken speed in face-to-face interaction.

Distance technologies allow for complete writing and vocational tasks to be completed at the user's pace as a standard feature of that interaction mode.

Less expensive for some control and communication functions than personal aide.

Provides means for organizing thought visually.

Provides speech systems with intelligible and variable output.

Can expand simple mechanical operation such as a switch activation to perform complex functions.

Potential Limitations of Assistive Devices

Augmented communication is not as fast, flexible, or as varied as standard spoken or written communication.

Technology cannot replace original function and is always less easily adaptable.

Difficulty with compatibility of various devices, particularly between standard technology and dedicated software and hardware for communication.

Computer technology may present additional cognitive load for some users.

Short product life; computer devices are quickly out of date.

Other nonautomated techniques may fulfill same function at less cost with greater control over modifications.

Nonautomated techniques may be more interactive with the listener.

Time, money, effort expenditures may outweigh benefits.

People who have difficulties processing symbolic information will also have difficulties using a symbolic communication device.

Objective measures of relative efficiency and effectiveness of communication aids for individuals are currently limited.

Nonautomated aids may be more portable.

Sensory aids, particularly cochlear implants, can affect physiological function.

Problems with client, family, clinician acceptance, and/or understanding of device.

Difficulties with listener acceptance of AAC within communicative interaction.

Success with technology is the degree to which technology adapts to client needs instead of client adapting to technology capability.

FUTURE DIRECTIONS

Research, development, and clinical application of computer technology will continue in the field of communication disorders. Some of the directions of future development are suggested by progress in current research or clinical applications. New developments that are being addressed for computer applications in speech, hearing, and language disorders, and assistive devices are listed below.

New Directions in Speech

Improved databases of quantitative indices of typical and disordered speech performance.

Clinical implementation of motion analysis technology to assess and/or predict speech motor problems in children and adults with disabilities.

Improved commercial products for analyzing speech movements and muscle activity.

Modeling of the relationship between oral movements and speech sounds in early development.

Use of imaging technology such as magnetic resonance imaging (MRI) to assess and monitor vocal tract anatomy and physiology.

Clinical implementation of automatic analysis of vocal samples for variety and complexity of sounds produced.

Computerized intelligibility assessments appropriate for children as well as adults.

Inexpensive and clinically feasible acoustic analysis for assessment of voice and speech production.

Instrumental feedback to reinforce vocal productions of young children.

Instrumental feedback for treating problems with vocal loudness and unintelligible speech.

Instrumental feedback of "easy onset" productions by clients to improve fluency.

Better integration of speech and other behavioral or language intervention technologies.

Genetic testing of congenital speech disorders.

Electrical stimulation of neuromotor centers and pathways using direct current or magnetic induction.

New Directions in Hearing and Hearing Aids

Modeling of highly complex auditory processes, such as stochastic processing of speech (different possible outputs from the same input) at the auditory nerve level.

Automated computer assessment of routine aspects of hearing tests, and better data on user interface effects for different populations (e.g., elderly users).

Computer enhancements of neurosurgery for hearing, including robotics, microsurgery, and interoperative monitoring of surgery with technology.

Customized presentation and adaptation of therapy activities for balance, such as matching user head or eye movement to computer-based targets.

User-adaptable hearing aids.

Improved text to speech and voice recognition systems for real-time translation of spoken to written communication for persons with hearing impairments.

Increased standardization and improvement of video capabilities of telecommunication devices for persons with hearing impairments.

Enhancements of unintelligible auditory signals in hearing aids using more complex signal detection theory.

Implantable hearing aids that are user adjustable to improve fidelity of sound in selective environments.

Brainstem implants for persons who have eighth nerve damage to their hearing (e.g., neurofibromatosis).

Implantable prosthetic devices for improving vestibular function, including mixed sensory input such as tactile feedback for the position of the user's head in space.

New Directions in Language

Improved databases of language development and patterns of children and adults with various language disabilities.

More extensive use of imaging techniques such as fMRI to determine brain systems involved in different language and communication tasks.

Simultaneous analysis of multiple aspects of language behaviors in samples.

Dependable fully automated language sample analysis programs, using artificial intelligence algorithms to avoid common errors in current programs.

Computerized techniques for describing and predicting relationships between linguistic variables observed.

Computer-prompted sampling of language behaviors in targeted contexts and topics, and integration with cognitive responses to those language concepts.

Voice recognition of input to assist with some aspects of broad transcription of language samples.

Clinical consultation and language sampling through distance technology, including personal handheld devices and video/audio conference interactions.

Universal design of word processing software with literacy support, analogous to the integration of computer access technology supplied with standard computers.

Improvement of voice input technologies for text creation and editing by persons with language and/or literacy impairments.

Customizable language and grammar correction systems that can recognize particular nonstandard error patterns for clients with language impairments.

Improvements in scanning and voice output technology for seamless multimedia access and control of written information.

Simple and small personal reminders and navigation tools for persons with head injuries or other developmental impairments, that minimize the need for complex user interface.

New Directions in Assistive Devices

Improvements in synthesized voices for voice output devices.

Improvements in speed and organization of vocabulary access for persons relying on AAC.

Fully portable and durable voice output technology for young children.

AAC devices for young children and persons with cognitive impairments that are organized by conceptual or visual association rather than linguistic categories.

Digital photo and virtual scene interfaces for children and persons with cognitive impairments.

Better understanding of the language development processes in children who rely on AAC throughout their lifespan.

Better understanding of the grammatical and conversational modifications to provide maximum clarity, speed, and naturalness in persons using AAC.

Voice recognition for input by severely dysarthric speakers.

Strategies for storing voice samples from persons with degenerative diseases to use for later personalized voices in their AAC systems.

Integrated small units of voice output that can be incorporated singly into activities and natural interaction, as well as gathered into a single communication device.

Use of data transfer and distance access technology to support telework for persons with disabilities, including the potential for shared positions through telework.

Intelligent agents within AAC devices that can prompt communication and social tasks, such as for persons with autism.

ACKNOWLEDGMENTS

Part of the data collection for this article and the previous edition was the interviewing of persons involved with the application of computer technology in communication disorders. These persons were asked to comment on functions within their areas of specialty that are addressed by computer technology. We greatly appreciate the cooperation and expertise of the following persons (in alphabetical order by chapter edition): *Second Edition:* David Beukelman, T. Newell Decker, Malinda Eccarius, Charles Healey, Karen Hux, and Neil Shepard. *First Edition:* Claudia Blair, Diane Bless, Robin Chapman, Michael Collins, Stanley Ewanowski, Robert Goldstein, Linda Hesketh, Carol Hustedde, Raymond Karlovich, Jesse Kennedy, Raymond Kent, Marilyn Kertoy, Mikael Kimelman, Pamela Mathy-Laikko, Vicky Lord-Larson, Richard Lehrer, Malcolm McNeil, Jon Miller, Linda Milosky, Lois Nelson, Katharine Odell, Mary Jo Osberger, Barry Prizant, Ann Ratcliff, John Peterson, Margaret Rosin, Lawrence Shriberg, Dolores Vetter, Francisco Villarruel, Gary Weismer, and Terry Wiley.

BIBLIOGRAPHY

1. Coleman JG. The Early Intervention Dictionary. Bethesda (MD): Woodbine House; 1993.

2. Nicolosi L, Harryman E, Kresheck J. Terminology of Communication Disorders: Speech-Language-Hearing. 4th ed. Philadelphia: Lippincott, Williams & Wilkins; 1996.

3. Gillam RB, Marquardt TP, Martin FN. Communication Sciences and Disorders: From Science to Clinical Practice. San Diego: Singular; 2000. p 85–98.

4. Herer GR, Knightly CA, Steinberg AG. Hearing: Sounds and silences. In: Batshaw ML, editor. Children with Disabilities. 5th ed. Baltimore: Paul H. Brookes; 2002.

5. Cochran PS. Clinical Computing Competency for Speech-Language Pathologists. Baltimore: Paul H. Brookes; 2004.

6. Cochran PS, Bull GL. Integrating word processing into language intervention. Top Lang Dis 1991;11(2):31–49.

7. Smith SW, Kortering LJ. Using computers to generate IEPs: Rethinking the process. J Special Educ Technol 1996;13(2): 81–90.

8. Lieberth AK, Martin DR. Authoring and hypermedia. Language Speech Hearing Ser Schools 1995;26(3):241–250.

9. Epstein JN, Willis MG, Conners CK, Johnson DE. Use of a technological prompting device to aid a student with attention deficit hyperactivity disorder to initiate and complete daily tasks: An exploratory study. J Special Educ Technol 2001;16(1): 19–28.

10. Foegen A, Hargrave CP. Group response technology in lecture-based instruction: Exploring student engagement and instructor perceptions. J Special Educ Technol 1999; 14(1):3–17.

11. Gardner JE, Wissick CA, Schweder W, Canter LS. Enhancing interdisciplinary instruction in general and special education: Thematic units and technology. Remedial Special Educ 2003;24(3):161–172.

12. Glaser CW, Rieth HJ, Kinzer CK, Colburn LK, Peter J. A description of the impact of multimedia anchored instruction on classroom interactions. J Special Educ Technol 1999;14(2): 27–43.

13. Roberson L. Integration of computers and related technologies into deaf education teacher preparation programs. Am Ann Deaf 2001;146(1):60–66.

14. Hasselbring TS. A possible future of special education technology. J Special Educ Technol 2001;16(4):15–21.

15. Masterson JJ. Future directions in computer use. Lang Speech Hearing Ser Schools 1995;26(3):260–262.

16. Sokolov JL, Snow CE. Transcript analysis using the Child Language Data Exchange System. In: Sokolov JL, Snow CE, editors. Handbook of Research in Language Development using CHILDES. Hillsdale (NJ): Lawrence Erlbaum Associates; 1994.

17. McGuire RA. Computer-based instrumentation: Issues in clinical applications. Lang Speech Hearing Ser Schools 1995;26(3):223–231.

18. Barlow SM, Cole KJ, Abbs JH. A new head-mounted lip-jaw movement transduction system for the study of motor speech disorders. J Speech Hearing Res 1983;26:283–288.

19. Müller EM, Abbs JH. Strain gauge transduction of lip and jaw motion in the midsagittal plane: Refinement of a prototype system. J Acoust Soc Am 1979;65:481–486.

20. Green JR, Moore CA, Higashikawa M, Steeve RW. The physiologic development of speech motor control: Lip and jaw coordination. J Speech Language Hearing Res 2000; 43:239–255.

21. American Speech-Language-Hearing Association, Guidelines for speech-language pathologists performing videofluoroscopic swallowing studies. ASHA Suppl 2004;24:

22. Logemann JA. Manual for the Videofluoroscopic Study of Swallowing. 2nd ed. Austin (TX): ProEd; 1993.

23. Hirose H, Kiritani S, Ushijima T, Yoshioka H, Sawashima M. Patterns of dysarthric mocments in patients with Parkinsonism. Folia Phoniatrica 1981;33:204–215.

24. Perkell J, Cohen M, Svirsky M, Matthies M, Garabieta I, Jackson M. Electro-magnetic midsagittal articulometer (EMMA) systems for transducing speech articulatory movements. J Acoust Soc Am 1992;92:3078–3096.

25. Boersma P, Weenink WD. Pratt, (Version 4.2) [Computer software]. Amsterdam: University of Amsterdam; 2004.

26. Kay Elemetrics Corp, 2 Bridgewater Lane, Lincoln Park, NJ, 07035. 2004. Tel: (973) 628-6200.

27. Milenkovic P. CSpeech [Computer software]. 118 Shiloh Drive, Madison, WI, 53705, 2004. Tel: (608) 833-7956. Available at http://userpages.net.chorus/cspeech.

28. Barlow SM, Suing G. Aerospeech: Automated digital signal analysis of speech aerodynamics. J Comput Users Speech Hearing 1991;7:211–227.

29. Barlow SM, Suing G, Andreatta RD. Speech aerodynamics using AEROWIN. In: Barlow SM, editor. Handbook of Clinical Speech Physiology. San Diego: Singular; 1999. p 165–189.

30. The Mathworks, 3 Apple Hill Drive, Natick, MA, 01760. 2004. Tel: 508-647-7000. Available at www.mathworks.com.

31. National Instruments, 11500 N. Mopac Expressway, Austin, TX, 78759. Tel: (888) 280-7645. Available at www.ni.com.

32. Chan JS, Spence C. Presenting multiple auditory signals using multiple sound cards in Visual Basic 6.0. Behav Res Methods Instrum Comput 2003;35(1):125–128.

33. Henry JA, Flick CL, Gilbert A, Ellingson RM, Fausti SA. Reliability of computer-automated hearing thresholds in cochlear-impaired listeners using ER-4B Canal Phone earphones. J Rehabil Res Dev 2003;40(3):253–264.

34. Walsh T, Demkowicz L, Charles R. Boundary element modeling of the external human auditory system. J Acoust Soc Am 2004;115(3):1033–1043.

35. Yao J, Zhang YT. The application of bionic wavelet transform to speech signal processing in cochlear implants using neural network simulations. IEEE Trans Biomed Eng 2002;49(11): 1299–1309.

36. Schneider U, Schleussner E, Haueisen J, Nowak H, Seewald HJ. Signal analysis of auditory evoked cortical fields in fetal magnetoencephalography. Brain Topog 2001;14(1):69–80.

37. Evans JL, Miller J. Language sample analysis in the 21st century. Sem Speech Lang 1999;20(2):101–116.

38. Miller J, Chapman R. 2000. SALT: Systematic Analysis of Language Transcripts (SALT). Version 6.1 [Computer software]. Madison (WI): Language Analysis Laboratory. Waisman Center. University of Wisconsin. Available at http://waisman.wisc.edu/salt/.

39. Miller J, Freiberg C, Rolland MB, Reeves M. Implementing computerized language sample analysis in the public school. Top Lang Dis 1992;12(2):69–82.

40. MacWhinney B. The CHILDES Project: Tools for Analyzing Talk. 3rd ed. Mahwah (NJ): Lawrence Erlbaum; 2000a.

41. Hallowell B, Katz R. Technological applications in the assessment of acquired neurogenic communication and swallowing disorders in adults. Sem Speech Lang 1999; 20(2):149–168.

42. Violin RA. Microcomputer-based systems providing biofeedback of voice and speech production. Top Lang Dis 1991; 11(2):65–79.

43. Green JR, Beukelman DR, Ball LJ. Algorithmic estimation of pauses in extended speech samples. J Med Speech Hearing Res 2004;12:149–154.

44. Hosom JP, Shriberg L, Green JR. Diagnostic Assessment of Childhood Apraxia of speech using automatic speech recognition (ASR) methods. J Med Speech Hearing Res 2004; 12.

45. Titze IT. Summary Statement for the Workshop on Acoustic Voice Analysis. National Center for Voice and Speech: Iowa City; 1995.

46. American Speech-Language-Hearing Association, Vocal tract visualization and imaging. ASHA 1992;34(March, 7 Suppl): 37–40.

47. Case JL. Technology in the assessment of voice disorders. Sem Speech Lang 1999a;20(2):169–184.

48. Bakker K. Technical solutions for quantitative and qualitative assessments of speech fluency. Sem Speech Lang 1999a;20(2): 185–196.

49. Yorkston K, Beukelman D, Tice R. Sentence Intelligibility Test (Version 1.0). [computer software]. Lincoln (NE): Communication Disorders Software, 1996.

50. Masterson JJ, Oller KD. Use of technology in phonological assessment: Evaluation of early meaningful speech and prelinguistic vocalizations. Sem Speech Lang 1999; 29(2):133–148.

51. Fell JH, MacAuslan J, Ferrier LJ, Chenausky K. Automatic babble recognition for early detection of speech related disorders. Behav Inf Technol 1999;18(1):56–63.

52. Mendel LL, Wynne MK, English K, Schmidt-Troike A. Computer applications in educational audiology. Lang Speech Hearing Schools 1995;26(3):232–240.

53. Johnson CE, Danhauer JL, Krishnamurti S. A holistic model for matching high-tech hearing aid features to elderly patients. Am J Audiol 2000;9(2):112–123.

54. Newman CW. Digital signal processing hearing aids: Determining need on an individual basis. Arch Otolaryngol Head Neck Surg 2000;126(11):1397–1398.

55. Bradley S. A spreadsheet for calculating the articulation index and displaying the unaided and aided speech spectrum. J Comput Speech Hearing 1991;7:357–361.

56. Leavitt R, Flexer C. Speech degradation as measured by the rapid speech transmission index (RASTI). Ear Hearing 1991; 12:115–117.

57. Iley KL, Addis RJ. Impact of technology choice on service provision for universal newborn hearing screening within a busy district hospital. J Perinatol 2000;20:S122–S127.

58. Johnson MJ, Maxon AB, White KR, Vohr BR. Operating a hospital-based universal newborn hearing screening program using transient evoked otoacoustic emissions. Sem Hearing 1993;14:46–55.

59. Stone KA, Smith BD, Lembke JM, Clark LA, McLellan MB. Universal newborn hearing screening. J Family Practice 2000;49:1012–1016.

60. Allen B, Lambert G. Computerization of V.R.O.A: A double blind hearing screening technique. Aust J Audiol 1990;1 2:11–15.

61. Eilers RE, Widen JE, Urbano R, Hudson TM, Gonzales L. Optimization of automated hearing test algorithms: A comparison of data from simulations and young children. Ear Hearing 1991;12:199–204.

62. Schmida MJ, Peterson JH, Tharpe AM. Visual reinforcement audiometry using digital video disc and conventional reinforcers. Am J Audiol 2003;12(1):35–40.

63. McCullough JA, Cunningham LA, Wilson RH. Auditory-visual word identification test materials: Computer application with children. J Am Acad Audiol 1992;3:208–214.

64. Long SH. About time: A comparison of computerized and manual procedures for grammatical and phonological analysis. Clin Linguistics Phonetics 2001;15(5):399–426.

65. Miller J, Chapman R. The relation between age and mean length of utterance in morphemes. J Speech Hearing Res 1981;24:154–161.

66. Long SH. Technology applications in the assessment of children's language. Sem Speech Lang 1999;20(2):117–132.

67. Long SH, Channell RW. Accuracy of four language analysis procedures performed automatically. Am J Speech-Langauge Pathol 2001;10:180–188.

68. MacWhinney B. Child Language Analysis (CLAN) Manual. Available at http://childes.psy.cmu.edu/pdf/clan.zip. Accessed 2000b.

69. Long SH, Fey ME, Channell RW. Computerized Profiling (CP). Version 9.2.7. [Computer program]. Cleveland, OH: Department of Communication Sciences, Case Western Reserve University. Also available at http://www.cwru.edu/artsci/cosi/cp.htm. Accessed 2000.

70. Shriberg L. Program to Examine Phonetic and Phonological Evaluation Records (PEPPER). Version 4.0. Hillsdale (N.J.): Lawrence Erlbaum Associates; 1986.

71. Masterson J, Bernhardt B. Computerized articulation and phonology evaluation system (CAPES) [computer program]. San Antonio: The Psychological Corporation, 2001.

72. Masterson JJ, Long SH, Buder EH. Instrumentation in clinical phonology. In: Bernthal JE, Bankson NW, editors. Articulation and Phonological Disorders. 4th ed. Boston: Allyn & Bacon; 1998.

73. Haaf R, Duncan B, Skarakis-Doyle E, Carew M, Kapitan P. Computer-based language assessment software: The effects of presentation and response format. Lang Speech Hearing Ser Schools 1999;30(1):68–74.

74. Shriberg LD, Kwiatkowski J, Snyder T. Articulation testing by microcomputer. J Speech Hearing Dis 1986;51(4):309–324.

75. Wiig EH, Jones SS, Wiig ED. Computer-based assessment of word knowledge in teens with learning disabilities. Lang Speech Hearing Ser Schools 27:21–28.

76. Cochran PS, Masterson JJ. NOT using a computer in language assessment/intervention: In defense of the reluctant clinician. Lang Speech Hearing Ser Schools 1995;26(3):213–222.

77. Jacobs EL. The effects of adding dynamic assessment components to a computerized preschool language screening test. Commun Dis Quart 2001;22(4):217–226.

78. Jacobs EL, Coufal KL. A computerized screening instrument of language learnability. Commun Dis Quart 2001;22(2):67–75.

79. van Geert P, van Dijk M. Focus on variability: New tools to study intra-individual variability in developmental data. Infant Behav Dev 2002;25:340–374.

80. Clancy B, Finlay B. Neural correlates of early language learning. In: Tomasello M, Bates E, editors. Language Development: The Essential Readings. Malden (MA): Blackwell; 2001.

81. Neville H, Mehler J, Newport E, Werker J, McClelland J. Special issue: The developing brain. Section 4: Language. Dev Sci 2001;4(3):293–312.

82. Bhatnagar SC, Andy OJ. Diagnostic techniques and neurological concepts. In: Bhatnagar SC, Andy OJ , editors. Neuroscience for the Study of Communicative Disorders. Baltimore: Williams & Wilkins; 1995. p 314–332.

83. Goldenberg EP. Computers in the special education classroom: What do we need, and why don't we have any?" In: Mulick J, Mallory B, editors. Transitions in Mental Retardation. Vol. 1. Norwood (NJ): Ablex Press; 1984.

84. Behrmann M. Handbook of Microcomputers in Special Education. San Diego: College Hill Press; 1984.

85. Rushakoff G. Clinical applications in communication disorders. In: Schwartz A, editor. Handbook of Microcomputer Applications in Communication Disorders. San Diego: College Hill Press; 1984.

86. Mahaffey R. An overview of computer applications. Top Lang Dis 1985;6:1–10.

87. Katz RC, Hallowell B. Technological applications in the treatment of acquired neurogenic communication and swallowing disorders in adults. Sem Speech Lang 1999; 20(3): 251–270.

88. Ruscello DM. Visual feedback in treatment of residual phonological disorders. J Commun Dis 1995;28:279–302.

89. Case JL. Technology in the treatment of voice disorders. Sem Speech Lang 1999b;20(3):281–295.

90. Masterson JJ, Rvachew S. Use of technology in phonological intervention. Sem Speech Lang 1999;20(3):233–250.

91. Crary MA, Groher ME. Basic concepts of surface electromyographic biofeedback in the treatment of dysphagia: A tutorial. Am J Speech-Language Pathol 2000;9(2):116–125.

92. Logemann JA, Kahrilas PJ. Relearning to swallow after stroke—application of non-invasive biofeedback. A case study. Neurology 1990;40:1136–1140.

93. Bakker K. Clinical technologies for the reduction of stuttering and enhancement of speech fluency. Sem Speech Lang 1999b;20(3):271–280.

94. Stuart A, Xia S, Jiang Y, Jiang T, Kalinowski J, Rastatter MP. Self-contained in-the-ear device to deliver altered auditory feedback: Applications for stuttering. Ann Biomed Eng 2003;31(2):233–237.

95. Fell H, Cress C, MacAuslan J, Ferrier L. VisiBabble for reinforcement of early vocalization. Presentation at the ASSETS '04 Conference, Atlanta, GA, 2004, October.

96. Hirano M, Kurita S, Sakaguchi S. Ageing of the vibratory tissue of human vocal folds. Acta Oto-Laryngolog 1989; 107:428–433.

97. Kasuya H, Ogawa S, Kikuchi Y. An acoustic analysis of pathological voice and its application to the evaluation of laryngeal pathology. Speech Commun 1986;5:171–181.

98. Fletcher SG, Hasegawa A. Speech modification by a deaf child through dynamic orometric modeling and feedback. J Speech Hearing Dis 1983;48:178–185.

99. Fletcher SG. Visual articulation training through dynamic orometry. Volta Rev 1989;91:47–64.

100. Ambulatory Monitoring, Inc., 731 Saw Mill River Road, Ardsley, N.Y. 10502. Tel: (800) 341-0066. Available at www.ambulatory-monitoring.com/index.html. Accessed 2004.

101. Hardcastle WJ, Gibbon FE, Jones W. Visual display of tongue-palate contact: Electropalatography in the assessment and remediation of speech disorder. Br J Dis Commun 1991;26:41–74.

102. Blischak DM. Increases in natural speech production following experience with synthetic speech. J Special Educ Technol 1999;15(2):44–53.

103. Hustad KC, Shapley KL. AAC and natural speech in individuals with developmental disabilities. In: Light JC, Beukelman DR, Reichle J, editors. Communicative Competence for Individuals who use AAC: From Research to Effective Practice. Baltimore: Brookes; 2003. p 41–62.

104. Blood GW. A behavioral-cognitive therapy program for adults who stutter: Computers and counseling. J Commun Dis 1995;28:165–180.

105. Pratt SR, Heintzelman AT, Deming SE. The efficacy of using the IBM speech viewer vowel accuracy module to treat young children with hearing impairment. J Speech Hearing Res 1993;36(5):1063–1074.

106. Tye-Murray N. Laser videodisc technology in the aural rehabilitation setting: Good news for people with severe and profound hearing impairments. Am J Audiol 1992; 1(2):33–36.

107. Uziel A, Mondain M, Hagen P, Dejean F, Doucet G. Rehabilitation for high-frequency sensorineural hearing impairment in adults with the symphonix vibrant soundbridge: A comparative study. Otol Neurotol 2003;24(5):775–783.

108. Shepard NT, Solomon D, Ruckenstein M, Staab J. Evaluation of the vestibular (balance) system. In: Ballenger JJ, Snow JB, editors. Otorhinolaryngology Head and Neck Surgery. 16th ed. San Diego: Singular; 2003.

109. Steiner S, Larson VL. Integrating microcomputers into language intervention with children. Top Lang Dis 1991; 11(2): 18–30.

110. Nippold MA, Schwarz LE, Lewis M. Analyzing the potential benefit of microcomputer use for teaching figurative language. Am J Speech Lang Pathol 1992;1:36–43.

111. Rose MO, Cochran PS. Teaching action verbs with computer-controlled videodisc vs. traditional picture stimuli. J Comput Users Speech Hearing 1992;8:15–32.

112. Diehl SF. Listen and Learn? A software review of Earobics. Lang Speech Hearing Ser Schools 1999;30(1):108–116.

113. Friel-Patti S, DesBarres K, Thibodeau L. Case studies of children using Fast Forword. Am J Speech-Language Pathol 2001;10:203–215.

114. Gillam RB, Crofford JA, Gale MA, Hoffman LM. Language change following computer-assisted language instruction with Fast Forword or Laureate Learning Systems software. Am J Speech-Language Pathol 2001;10:231–247.

115. Judge SL. Computer applications in programs for young children with disabilities: Current status and future directions. J Special Educ Technol 2001;16(1):29–40.

116. Kinsley TC, Langone J. Applications of technology for infants, toddlers, and preschoolers with disabilities. J Special Educ Technol 1995;12(4):312–324.

117. Cochran PS, Nelson LK. Technological applications in intervention for preschool-age children with language disorders. Sem Speech Lang 1999;20(3):203–218.

118. Howard J, Greyrose E, Kehr K, Espinosa M, Beckwith L. Teacher-facilitated microcomputer activities: Enhancing social play and affect in young children with disabilities. J Special Educ Technol 1996;13(1):36–47.

119. Cress CJ, Marvin CA. Common questions about AAC services in early intervention. Augment Alter Commun 2003;19 (4):254–272.

120. Howell RD, Erickson K, Stanger C, Wheaton JE. Evaluation of a computer-based program on the reading performance of first grade students with potential for reading failure. J Special Educ Technol 2000;15(4):5–14.

121. Wood LA, Masterson JJ. The use of technology to facilitate language skills in school-age children. Sem Speech Lang 1999;20(3):219–232.

122. Daiute C, Morse F. Access to knowledge and expression: Multimedia writing tools for students with diverse needs and strengths. J Special Educ Technol 1994;12(3):221–256.

123. Montgomery DJ, Karlan GR, Coutinho M. The effectiveness of word processor spell checker programs to produce target words for misspellings generated by students with learning disabilities. J Special Educ Technol 2001;16(2):27–40.

124. Sturm JM, Rankin JL, Beukelman DR, Schutz-Meuhling L. How to select appropriate software for computer-assisted writing. Intervention School Clinic 1997;32:148–162.

125. Wise BW. Computer speech and the remediation of reading and spelling problems. J Special Educ Technol 1994;12(3): 207–220.

126. Wood LA, Rankin JL, Beukelman DR. Word prompt programs: Current uses and future possibilities. Am J Speech-Language Pathol 1997;6(3):57–65.

127. Higgins EL, Raskind MH. Speaking to read: The effects of continuous vs. discrete speech recognition systems on the reading and spelling of children with learning disabilities. J Special Educ Technol 2000;15(1):19–30.

128. Edwards BJ, Blackhurst AE, Koorland MA. Computer-assisted constant time delay prompting to teach abbreviation spelling to adolescents with mild learning disabilities. J Spec Educ Technol 1995;12(4):301–311.

129. Hutinger P, Johanson J, Stoneburner R. Assistive technology applications in educational programs of children with multiple disabilities: A case study report on the state of practice. J Special Educ Technol 1996;13(1):16–35.

130. Morgan RL, Gerity BP, Ellerd DA. Using video and CD-ROM technology in a job preference inventory for youth with severe disabilities. J Special Educ Technol 2000;15(3):25–33.

131. Nelson KE, Heimann M, Tjus T. Theoretical and applied insights from multimedia facilitation of communication skills in children with autism, deaf children, and children with other disabilities. In: Adamson LB, Romski MA, editors. Communication and Language Acquisition: Discoveries from Atypical Development. Baltimore: Brookes; 1997. p 295–325.

132. Norman JM, Collins BC, Schuster JW. Using an instructional package including video technology to teach self-help skills to elementary students with mental disabilities. J Special Educ Technol 2001;16(3):5–18.

133. Rostron A, Sewell D. Microtechnology in Special Education. Baltimore: Johns Hopkins University Press; 1984.

134. Light JD, Roberts B, Dimarco R, Greiner N. Augmentative and alternative communication to support receptive and expressive communication for people with autism. J Commun Dis 1998;31:158–180.

135. Mirenda P, Wilk D, Carson P. A retrospective analysis of technology use patterns of students with autism over a five-year period. J Special Educ Technol 2000;15(3):5–16.

136. Katz RC. Computer applications in aphasia treatment. In: Chapey R, editor. Language Intervention Strategies in Aphasia and Related Neurogenic Communication Disorders. 4th ed. Philadelphia: Lippincott Williams & Wilkins; 2001. p 718–741.

137. Katz RC, Wertz RT. The efficacy of computer-provided reading treatment for chronic aphasic adults. J Speech Lang Hearing Res 1997;40(3):493–507.

138. Herrmann D, Yoder CY, Wells J, Raybeck D. Portable electronic scheduling/reminding devices. Cog Technol 1996;1: 19–24.

139. Kaasgaard K, Lauritsen P. The use of computers in cognitive rehabilitation in Denmark. Am J Speech-Language Pathol 1995;4:5–8.

140. Robinson I. Does computerized cognitive rehabilitation work? A review. Aphasiology 1990;4:381–405.

141. Beukelman DR, Mirenda P. Augmentative and Alternative Communication: Management of Severe Communication Disorders in Children and Adults. Baltimore: Brookes; 1998.

142. Cook AM, Hussey SM. Assistive Technologies: Principles and Practice. St. Louis: Mosby; 1995.

143. Goossens C, Kraat A. Technology as a tool for conversation and language learning for the physically disabled. Top Lang Dis 1985;6:56–70.

144. Reichle J, Hidecker MJC, Brady NC, Terry N. Intervention strategies for communication: Using aided augmentative communication systems. In: Light JC, Beukelman DR, Reichle J, editors. Communicative Competence for Individuals who use AAC: From Research to Effective Practice. Baltimore: Brookes; 2003. p 441–477.

145. Romski MA, Sevcik RA, Hyatt AM, Cheslock M. A continuum of AAC language intervention strategies for beginning communicators. In: Reichle J, Beukelman DR, Light JC, editors. Exemplary Practices for Beginning Communicators: Implications for AAC. Baltimore: Brookes; 2002. p 1–24.

146. Justice LM, Chow SM, Capellini C, Flanigan K, Colton S. Emergent literacy intervention for vulnerable preschoolers: Relative effects of two approaches. Am J Speech-Lang Pathol 2003;12:320–332.

147. Sandberg AD. Reading and spelling: Phonological awareness, and working memory in children with severe speech impairments: A longitudinal study. Augment Alter Commun 2001;17:11–26.

148. Cress CJ. Expanding children's early augmented behaviors to support symbolic development. In: Reichle J, Beukelman DR, Light JC, editors. Exemplary Practices for Beginning Communicators: Implications for AAC. Baltimore: Brookes; 2002. p 272–291.

149. Fallon KA, Light J, Achenbach A. The semantic organization patterns of young children: Implications for augmentative and alternative communication. Augment Alter Commun 2003;19(2):74–85.

150. Light JD, Drager KDR, Nemser JG. Enhancing the appeal of AAC technologies for young children: Lessons from the toy manufacturers. Augmentative and Alter Commun 2004;20 (3):137–149.

151. Wilkinson KM, Jagaroo V. Contributions of principles of visual cognitive science to AAC system display design. Augment Alter Commun 2004;20(3):123–136.

152. File P, Todman J. Evaluation of the coherence of computer-aided conversations. Augment Alter Commun 2002;18: 228–241.

153. Blockberger S, Sutton A. Toward linguistic competence: Language experiences and knowledge of children with extremely limited speech. In: Light JC, Beukelman DR, Reichle J, editors. Communicative Competence for Individuals who use AAC: From Research to Effective Practice. Baltimore: Brookes; 2003. p 63–106.

154. Smith MM, Grove NC. Asymmetry in input and output for individuals who use AAC. In: Light JC, Beukelman DR, Reichle J, editors. Communicative competence for individuals who use AAC: From research to effective practice. Baltimore: Brookes; 2003. p 163–195.

155. Mirenda P, Bopp KD. Playing the game: Strategic competence in AAC. In: Light JC, Beukelman DR, Reichle J, editors. Communicative Competence for Individuals who use AAC: From Research to Effective Practice. Baltimore: Brookes; 1993. p 401–437.

156. Cress CJ. AAC and language: Understanding and responding to parent perspectives. Top Lang Dis 2004;24(1): 28–38.

157. Garrett KL, Kimelman MDZ. AAC and aphasia: Cognitive-linguistic considerations. In: Beukelman DR, Yorkston KM, Reichle J, editors. Augmentative and Alternative Communication for Adults with Acquired Neurologic Disorders. Baltimore: Brookes; 2000. p 339–374.

158. Hux K, Manasse N, Weiss A, Beukelman DR. Augmentative and alternative communication for persons with aphasia. In: Chapey R, editor. Language Intervention Strategies in Aphasia and Related Neurogenic Communication Disorders. 4th ed. Philadelphia: Lippincott Williams & Wilkins; 2001. p 675–687.

159. Garrett KL, Beukelman DR. Augmentative communication approaches for persons with severe aphasia. In: Yorkston K, editor. Augmentative Communication in the Medical Setting. Tucson: Communication Skill Builders; 1992. p 245–321.

160. Lasker J, Hux K, Garrett K, Moncrief E, Eischeid T. Variations on the written choice communication strategy for individuals with severe aphasia. Augment Alter Commun 1997;13:108–116.

161. Colby K, Christinaz D, Parkinson S, Graham S, Karpf C. A word-finding computer program with a dynamic lexical-semantic memory for patients with anomia using an intelligent speech prosthesis. Brain Lang 1981;14:272–281.

162. Lasker JP, Bedrosian JL. Acceptance of AAC by adults with acquired disorders. In: Beukelman DR, Yorkston KM, Reichle J, editors. Augmentative and Alternative Communication for Adults with Acquired Neurologic Disorders. Baltimore: Brookes; 2000. p 107–136.

163. Vaughn GR, Kramer JO, Ozley CF. Tel-communicology for clinician and computer outreach. Commun Dis 1983;8: 75–88.

164. Taylor RS, Paisley S, Davis A. Systematic review of the clinical and cost effectiveness of digital hearing aids. Br J Audiol 2001;35(5):271–288.

165. Yueh B. Digital hearing aids. Arch Otolaryn Head Neck Surg 2000;126(11):1394–1397.

166. Wilson BS, Lawson DT, Muller JM, Tyler RS, Kiefer J. Cochlear implants: Some likely next steps. Ann Rev Biomed Eng 2003;5:207–49.

167. Koester HH. User performance with speech recognition: A literature review. Assis Technol 2001;13(2):116–130.

168. Noyes J, Frankish C. Speech recognition technology for individuals with disabilities. Augment Alter Commun 1992;8: 297–303.

169. Ferrier L, Shane H, Ballard H, Carpenter T, Benoit A. Dysarthric speakers' intelligibility and speech characteristics in relation to computer speech recognition. Augment Alter Commun 1995;11:165–174.

170. Ratcliff A. Comparison of relative demands implicated in direct selection and scanning: Considerations from normal children. Aug Alter Commun 1994;10:67–74.

171. (a) Light JC. Teaching automatic linear scanning for computer access: A case study of a preschooler with severe physical and communication disabilities. J Special Educ Technol 1993;12(2):125–134.

172. Wehmeyer ML. Assistive technology and students with mental retardation: Utilization and barriers. J Special Educ Technol 1999;14(1):48–58.

173. Rose D. Walking the walk: Universal design on the web. J Special Educ Technol 2000;15(3):45–49.

See also COMMUNICATION DEVICES; ENVIRONMENTAL CONTROL; REHABILITATION, COMPUTERS IN COGNITIVE.

COMPOSITES, RESIN-BASED.   See RESIN-BASED COMPOSITES.

COMPUTED RADIOGRAPHY.   See DIGITAL RADIOGRAPHY.

COMPUTED TOMOGRAPHY

MICHAEL J. DENNIS
Medical University of Ohio
Toledo, Ohio

INTRODUCTION

"Computed tomography...measures the attenuation of x-ray beams passing through sections of the body from hundreds of different angles, and then, from the evidence of these measurements, a computer is able to reconstruct pictures of the body's interior." That is the basic description of Computed Tomography (CT) as given by Sir Godfrey Hounsfield in his 1979 Nobel Lecture (1).

Computed tomography was a breakthrough in the implementation and acceptance of digital computers into clinical diagnostic imaging. It's commercial birth in the early 1970s was an amalgamation of X-ray imaging, detector development, mathematical methods, along with the developing computer capabilities of the time to produce a whole new way of peering into the human body.

The attenuation properties of X and γ rays are well known. The logarithmic scaled values of transmission measurements through a body yields a line integral or summation of the attenuation properties along the path of the beam. The linear attenuation coefficient, or the probability of interaction per microscopic distance traveled, is directly related to the density of the material and the effective atomic number of the material. A computer utilizes a mathematical algorithm to determine what the distribution of attenuation coefficients within the body must be to produce the measured set of transmission values. By unfolding the data in this way, tissues of interest within the patient are not obscured by anatomy above and below it. Consequently, structures may be accurately localized within the body and small, previously invisible differences in density or attenuation (< 1%) were seen.

Generally, the reconstructed data is calculated and presented as a series of cross-sectional slices. Each slice in the computer is represented by a two-dimensional (2D) matrix of numbers. The numbers in this array are scaled values of the linear attenuation coefficient, referred to as CT numbers or Hounsfield units. The individual data elements in a CT image are referred to as pixels or picture elements. The measurements through the body, however, are not along infinitely thin planes. The X-ray beam and resultant measurements have a finite width or thickness. A 2D picture element corresponds to a box shaped volume within the patient, referred to as a volume element or voxel (Fig. 1).

Advances in diagnostic medical imaging over the past half century have been phenomenal, in particular with the development and implementation of computed tomography

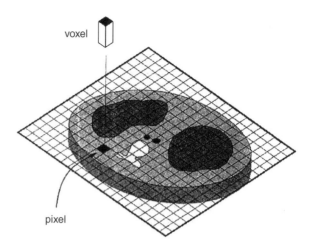

Figure 1. An axial CT image is composed of a 2D array of CT numbers with each picture element or pixel corresponding to a volume element or voxel within the patient.

and magnetic resonance imaging. The ability to virtually slice and dice a living human body to see the internal anatomical structures has eliminated the previously common practice of exploratory surgery, and has enabled more accurate diagnosis and improved effectiveness of medical treatment.

BASIC PRINCIPLES OF COMPUTED TOMOGRAPHY

The basic technique of computed tomography as illustrated in Fig. 2 is to probe a thin slice of the patient with a thin beam of radiation, which is attenuated as it passes through the patient. The fraction of the X-ray beam that is attenuated is directly related to the density, thickness, and composition of the material through which the beam has traveled and to the energy of the X-ray beam. Computed

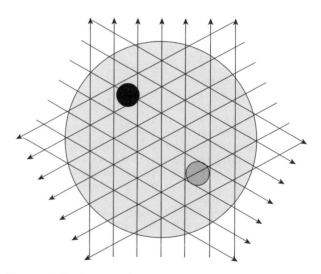

Figure 2. Basic principle of CT is that X-ray transmission measurements are taken along many rays through a thin slice of the patient from many different angles. The measured values are then used to map the distribution of attenuating material that produced the measured transmission values.

tomography utilizes this information from many different angles, to determine cross-sectional configuration with the aid of a computerized reconstruction algorithm. This reconstruction algorithm quantitatively determines the point-by-point mapping of the relative radiation attenuation coefficients for the set of transmission measurements.

The CT scanning system contains a radiation source and radiation detector along with precision mechanics to scan a cross-sectional slice through the patient. The X-ray detector is usually a linear array of detectors, that is, a series of individual X-ray sensors arranged in a line. Current multidetector CT (MDCT) systems utilize multiple rows of detectors in order to acquire the data in less time. The X-ray source is collimated to form a thin fan beam that is wide enough to expose the detector array. In a single-slice CT, the narrow beam thickness defines the thickness of the cross-sectional slice. The MDCT system slice thickness is determined by detector widths or the grouping of the linear arrays of detectors. The data acquisition system (DAS) reads the signal from the individual detectors, converts these measurements to numeric values, and transfers the data to a computer to be process. This processed is repeated as the X-ray source is rotated around the patient to acquire a full set of transmission measurements (2).

The CT image reconstruction algorithm generates 2D images from the set of measured transmission measurements. There are a number of reconstruction algorithms that can be used to generate the CT image. These mathematic algorithms can be divided into two general categories: analytical or transform techniques, and iterative reconstruction techniques. The transform techniques are generally based on the theorem of Radon (3), which states that any 2D distribution can be reconstructed from the infinite set of its line integrals through the distribution. The line integrals in CT are the sums of the linear attenuation coefficients along a line through the patient determined from the X-ray transmission measurements. The filtered-backprojection 2D reconstruction techniques used in most clinical CT scanners, as well as the cone-beam volumetric reconstruction algorithms based on Feldkamp's method (4) are analytical methods.

Iterative methods are rarely used in medical X-ray computed tomography, but are commonly used in nuclear medicine single-photon-emission computed tomography (SPECT) and positron emission tomography (PET) imaging. These methods are often more tolerant of limited or irregular data, and may use additional *a priori* information to improve the reconstructed results. Iterative techniques are generally algebraic methods that reconstruct the image by performing a series of iterative corrections on a guess of the image distribution (5–7).

EVOLUTION OF THE TECHNOLOGY

Although the mathematical principle of computed tomography was developed early in the twentieth century by Radon, application of the technology occurred much later. Techniques were independently developed in the 1950s for radioastronomy (8) and experimental work progressed through the 1960s, primarily in nuclear tracer imaging

and electron microscopy (9,10). Cormack addressed the problem relative to determine X-ray attenuation coefficient information, with the interest of using this information for improved radiation therapy calculations (11). In the late 1990s and early 1970s, Hounsfield at EMI, Ltd in England developed the first commercial X-ray CT system, also known as computer assisted tomography or CAT scanning (12). The initial prototype head scanner was installed in 1971 at Atkinson Morley's Hospital in Wimbleton, England, and commercial systems began delivery the following year.

Due to its unique capability of demonstrating anatomical information the medical interest and demand for CT grew rapidly in spite of the high costs and technical challenges. Numerous manufacturers entered the market with designs to decrease the scan time and to expand the use of CT to body imaging.

First Generation: Translate–Rotate

The initial clinical systems utilized an X-ray beam collimated to a small pencil beam mechanically linked to a detector on the opposite side of the patient. The mechanics translates the tube and detector across the full width of the patient, and then rotates one degree. This process is repeated until a full 180° of data is acquired (Fig. 3a). Two detectors were utilized in the initial EMI scanner in order to acquire two slices simultaneously, which was useful since each scan took over four minutes. One of the innovations utilized by Hounsfield to reduce the necessary dynamic range of the radiation detector, and also minimize certain artifacts, was to have the patient's head push into a elastic membrane into a water-filled box. The box was linked to the X-ray source and detector such that the X-ray beam always traversed through 24 cm of water and anatomy. This was quite effective, but impractical for expanding into body imaging.

Second Generation: Multidetector Translate–Rotate

To reduce the time to acquire the data, additional detectors lying within the scan plane were added and a narrow fan beam was used to cover this detector array. The system translates and rotates like the first generation systems, however, the rotation may be 20 or 30° between translations (Fig. 3b). In this way, the scan time could be as low as 20 s, and body size scan could be performed. While not used any more for medical CT scanners, translate–rotate data acquisition provides considerable flexibility regarding scan field of view and sample spacing, but at the cost of longer scan times. This approach is still used for some research and industrial testing systems which may be designed for samples of several millimeters, or of several meters (13).

Third Generation: Rotate–Rotate

A faster scan approach, which is still the basis for most current clinical scanners, is to utilize a linear array that fully encompasses the width of the patient. Mechanically the tube and detector rotates around the patient to acquire a series of fan beam views > 360° (Fig. 3c). An data set at a particular angle or view with this approach resulted in a fan shaped set of rays with the apex at the X-ray source.

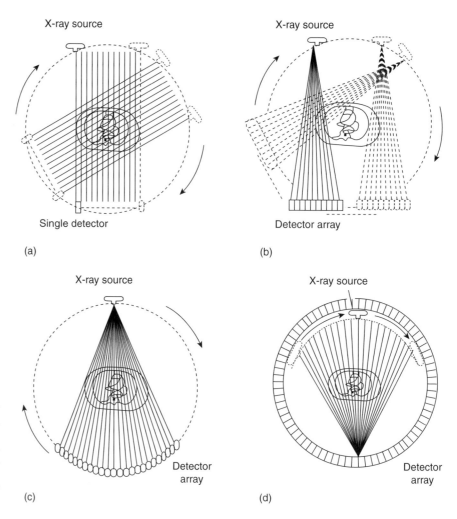

Figure 3. Data acquisition configurations or geometries used in CT. (a) First generation translate–rotate, (b) Second generation narrow fan beam translate–rotate, (c) Third generation rotate–rotate, and (d) Fourth generation fixed–rotate scanning systems. Most clinical systems utilize a rotate–rotate design.

The data sampling flexibility is restricted since the ray spacing is determined in large part by the size and spacing of the detector elements in the linear detector array. The number of views acquired, however, is determined by the number of samples taken over the 360° rotation. To distinguish these systems from the translate–rotate data acquisition systems, manufacturers labeled these rotate–rotate systems as third generation scanners.

Fourth Generation: Fixed–Rotate

Around the same time frame in the mid-to-late 1970s a data acquisition approach using a fixed ring of detectors was used. This requires the X-ray tube to rotate within the circle of detectors, or the use of a mechanism to tilt the detector out of the way of the X-ray beam (Fig. 3d). Usually the acquired data is rebinned or grouped such that a view data set or projection set consists of all transmission measurements made by a single detector as the X-ray tube rotates around the patient. This results in a fan shaped data set, but with the detector at the apex of the fan. Using this scheme the number of detectors determines the number of views acquired, but the ray spacing between views is determined by the data sampling rate. Predictably, the manufacturers of these fixed-rotate scanners labeled them as fourth generation. Further developments including

nutating or oscillating ring of detectors, steerable electron beams, 2D detector arrays and helical data acquisition are sometimes given generation numbers, but not in a consistent manner.

Electron Beam CT: Fixed–Fixed

These third and fourth generation rotate-only systems reduced the scan time initially to 10 s, with current scanners capable of rotating around the patient in < 0.5 s. In order to reduce scan times further, especially for rapid dynamic and cardiac imaging, electron beam cine CT system (EBCT) was developed by Imatron Corporation (Fig. 4) (14). This system uses a fixed detector system, but has the X-ray tube target encircling the patient. The unique X-ray tube uses an electron gun and deflection electronics to steer the electron beam within a large cone shaped vacuum enclosure to one of four target rings partially encircling the patient. The X-ray tube ring is opposed by a 240° double ring of fixed detectors. The system has no moving parts since the X-ray source location is changed by the steering of the electron beam. The X-ray source can rapidly move around the patient, and the data for an image acquired in 50 ms or less. The use of four separate target rings and two detector rings permitted the acquisition of eight separate axial planes without moving the patient.

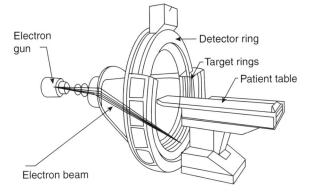

Figure 4. The electron beam CT (EBCT) scanner developed by Imatron (San Francisco) requires no moving parts, but rapidly moves the location of the X-ray source by steering the electron beam in the large cone-shaped X-ray tube to the desired source location.

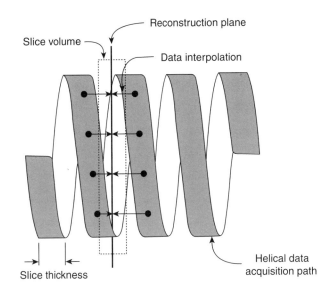

Figure 5. During helical CT data acquisition the patient is moved through the scanner while the X-ray source continuously rotates around the patient. To reconstruct a particular axial slice through the patient, the data at each angular position are interpolated to create a 360° data set corresponding to that slice.

It should be noted that an alternative research approach to cardiac imaging was developed at Mayo Clinic called the Dynamic Spatial Reconstructor. This system utilized a series of 14 X-ray tube-image intensifier pairs rotating around the patient to rapidly acquire the volume data. The system was designed to enable the use of 28 imaging system pairs (15).

Helical CT

The CT systems through the 1970s and 1980s were generally limited to a single rotation of the X-ray tube around the patient per data acquisition due to the need to have the high voltage cables connected to the tube. In the early 1990s, this changed with the advent of systems that utilized slip rings to transfer the power to the tube, permitting continuous rotation of the source. This continuous rotation allowed for more rapid dynamic scanning where a series of images of a single slice are sequentially acquired, allowing the characterization of motion or to evaluate the flow of a highly attenuating contrast material flowing into the tissue. More importantly, this continuous rotation permitted the ability to rapidly acquire a series of images covering a volume of the patient (16–18).

Normal axial scanning is performed in a step-and-shoot fashion, where the tube rotates around the patient within the plane to be imaged. The acquired data set is reconstructed to form the axial image at this location. The slice location and slice thickness are well defined by the X-ray beam. The patient table is incremented to the next location to be imaged and the process is repeated. The average time per image is the scan time plus the time to increment the table.

With the continuously rotating capability data can be acquired in a helical data acquisition mode. In this mode the table is moved continuously while the tube rotates around the patient. Since there is not a full set of X-ray views through a specific plane of the patient, the data for each angular position around the patient is interpolated from the nearby data acquired at that angle (Fig. 5) (19). Not only is the data acquisition faster, but one can also

arbitrarily select the locations of the planes to be reconstructed since the data is not fixed to a particular acquisition plane. For example, one could have a collimated slice thickness of 5 mm and generate contiguous or adjacent images every 5 mm, or one could reconstruct images from the same data set every 3 mm (or other arbitrary spacing), however the slice thickness would remain 5 mm.

Multidetector CT

In the latter part of the 1990s, systems were being marketed that contained more than one linear array of detectors. In these multidetector CT systems (MDCT), the slice width of the measured data does not correspond with the overall X-ray beam width, but on the width of the linear detector arrays used to acquire the data. The detector array generally consists of a number of narrow width or thin slice detectors that may be grouped together to generate a thicker effective slice. This detector array allows the acquisition of multiple slices in the axial mode. In the helical mode the overall X-ray beam width is larger than the image slice thickness defined by the detector rows. This permits the acquisition of more data in less time, allowing for faster scan times and the practical scanning with thin slice thicknesses (20,21).

As the number of rows of detectors increase from a few to 64 or 256 and beyond, the data acquired per scan rotation becomes a significantly sized volume. The diverging rays from the X-ray source form a cone of radiation striking the 2D area detector. Reconstruction algorithms developed to deal with these volume reconstructions, as opposed to the axial slice approach on earlier scanners, are sometimes referred to as cone beam scanning and reconstruction. Cone beam scanning can provide rapid information on a volume and is particularly useful for acquiring a rapid sequence of images of a volume to evaluate dynamic processes.

Table 1. Typical CT Number Values

Tissue	CT Number
Air	**−1000**
Fat	−60
Water	**0**
Cerebral spinal fluid	10
Brain edema	20
Brain white matter	30
Brain gray matter	38
Blood	42
Muscle	44
Hemorrhage	80
Dense bone	∼1000

CT SCANNER COMPONENTS

The CT scanners are a union of several component systems to provide clinical imaging capability. Outward mechanics include the table system and the gantry located in a radiation shielded scan room. The gantry contains the X-ray source and detector system. Computers are needed to control data acquisition, reconstruction and display of the images, and for the user interface or control console to allow operation of the system. This may be augmented with additional display and archival capabilities with a picture archiving and communication system (PACS), and workstations for additional display processing and print or filming capabilities Table 1.

Table 1 that the patient lays upon is a fairly basic component. It is typically a cantilevered design with the tabletop extending out from the pedestal, so that only the patient and the tabletop are in the X-ray beam. The tabletop must be strong enough to hold large patients, yet should not provide much attenuation of the X rays. Carbon composite materials are typical used for their strength and radiation transmission properties. Extensions to the table are used for a patient headholder, or for mounting of test and calibration phantoms.

Gantry

The gantry is the donut shaped main body of the computed tomography system and contains the x-ray source and detector system, as well as the mechanics for moving these devices as needed to perform the scan. The patient is extended on the table into the gantry aperture or hole in the gantry so that the X-ray source may rotate around the areas to be scanned (Fig. 6). The scannable region within the gantry is somewhat smaller than the gantry aperture or hole size. Typical is a 50 cm scan field of view within a 70 cm gantry aperture.

The entire gantry is usually pivoted to allow the top of the gantry to tilt toward or away from the table by 30° or more. This allows acquisition of images that are aligned or oriented to specific anatomy, such as the aligned with the disk in the lumbar spine. This feature is being used less, however, with the increasing use of thin slice data acquisition with MDCT systems permitting high quality computer generated images of alternate planes. The gantry also has localizer lights or lasers, and the table and gantry tilt controls to assist the technologist in posi-

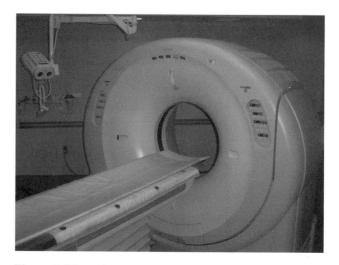

Figure 6. The major system components in the scan room are the patient table, and the scanner gantry which houses the X-ray source, detector, and mechanical drive components.

tioning the patient. The mechanics within the gantry include a large turret bearing, larger than the gantry aperture, to permit rotation of the rotating components of the system, and motor drives and controllers to actuate the scanning motions. Slip rings and data transponders are used to transmit power and data between the stationary and rotating system components. The gantry may also include active or passive cooling devices to prevent heat buildup.

X-Ray Source

An X-ray tube and generator are needed to produce the radiation for the scan. In the X-ray tube a negatively charged hot filament or cathode emits electrons that are accelerated by a high voltage. The high energy electron strike a target that is part of the positively charged anode and produce X rays along with a considerable amount of heat. The X-ray technique is characterized by the specifying the tube current or mA and the tube voltage or kilovolts, which determine the amount and energy of the X-ray photons emitted. The generation of X rays is the same as is found in other radiographic imaging systems. A notable difference is the workload these tubes endure in clinical imaging. Consequently, the X-ray tubes in CT scanners are often the big-brother to the tubes found in general radiography, with a super-sized anode capable of holding the considerable heat developed during the scans. X-ray tubes designed for CT systems may have other features to prevent anode wobbling, which can cause artifacts, or to be more effective at removing the heat generated. Important parameters for the X-ray tube include its focal spot size, the heat capacity and the cooling rate of the anode. A small focal spot or X-ray source size can provide better image resolution, but a small size may limit the X-ray output that can be obtained. Since the X-ray tubes utilize a rotating anode, it is important that the axis of this anode is parallel to the axis of rotation of tube around the patient, otherwise considerable gyroscopic torque would be placed on the tube.

The X-ray generator includes the high voltage transformer used to create the high voltages necessary for X-ray production. A key requirement for CT systems is to have a highly stable voltage with little ripple or variation. Older systems often used bulky three-phase transformers and voltage rectifiers in order to produce a constant high voltage. Current systems tend to use high frequency single-phase generators. These systems take the utility supplied power and process it to produce a high frequency electrical source with frequencies typically in the range from 1000 to 2000 Hz. The higher frequency has several advantages. High voltage transformer efficiencies are much better at high frequency, and since it is single phase, only one pair of coils is required, making for a much smaller transformer package. Single-phase power is normally associated with 100% ripple as the voltage varies from zero to its peak value. At high frequencies, however, a minimal amount of capacitance in the system smoothes this voltage ripple to produce a nearly uniform voltage. This transition to high frequency transformers has been an enabling technology for helical scanning. In order to continuously rotate the tube around the patient, the high voltage X-ray power cables had to be eliminated. With helical imaging systems, a low voltage of a couple of hundred volts is transferred to the rotating portion of the gantry through an electrical slip ring. The high voltage transformer is mounted on the rotating portion and circles the patient along with the X-ray tube, thereby eliminating the constraint of a single rotation on the older systems. Even with the smaller and lighter generator package, there is considerable mass rotating around the patient, and considerable G forces on these components, especially with the subsecond rotation times.

Collimation and Beam Filtration

Since high energy X rays cannot readily be focused like light, a collimator blocks the X rays coming from the X-ray tube that are not directed at the detector. The X-ray beam is shaped by tungsten or lead plates into its fan beam shape. The width of the fan beam may be varied allowing the technologist to select the slice thickness. On single slice CT scanners, with a single linear array of detectors, the tube side collimation determines the slice thickness. On MDCT systems, the width of the detector or the averaged grouping of detectors determines the slice width. The nominal slice width or thickness is the thickness of the reconstructed voxel at the center of the scanner.

Additional X-ray beam filtration is also in the X-ray beam. Beam filtration is material the X rays pass through before getting to the patient. Legally a certain amount of filtration is required in order to remove the soft or low energy X rays that contribute significantly to the patient dose with little chance of passing through the patient contribute to the transmitted signal. The CT scanner beams are generally heavily filtered, not only to reduce patient dose, but it also reduces beam-hardening artifacts. Most scanners also utilize a bowtie or compensating X-ray filter. This is a filter that has a variable thickness along the length of the fan beam, being thinner at the center of the field and thicker toward the edges of the scan field, thus

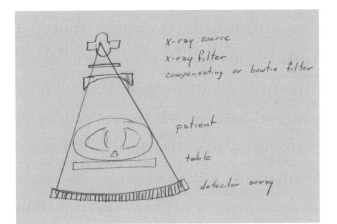

Figure 7. The X-ray beam is filtered to reduce the low energy X rays, passes through a bowtie shaped compensating filter that reduces peripheral dose to the patient, passes through the patient to the radiation detector array.

looking like a bowtie. This filter helps reduce the peripheral dose to the patient and also can help reduce beam hardening variations by adding attenuating material to the portions of the beam that are going through thinner portions of the patient (Fig. 7).

X-Ray Detector and Data Acquisition System

The X-ray detector is a critical component of the scanner system. It should be efficient at absorbing the X-ray beam energy, and converting the X rays into the detected signal, and should have a rapid response time to allow for rapid data acquisition. The detector size, along with the X-ray tube focal spot size, limits the potential image resolution (22).

Scintillation Detectors. The detector found in Hounsfield's original scanner was a sodium iodide (NaI) scintillation crystal linked to a photomultiplier tube (PMT). These types of devices are commonly used in nuclear medicine counting systems. The X rays are absorbed in the scintillation crystal where they are converted into a number of light photons. The PMT is a very sensitive detector of light and measures the light output. In nuclear medicine counting, the number of high energy photons entering the scintillation crystal is limited and each photon is analyzed and counted separately. With X-ray systems the rate at which photons are entering the detector is much faster than the ability of the system to detect separate distinguishable scintillations or flashes of light. The CT scintillation detectors are operated in a current mode rather than a pulse mode and measure the overall intensity of light produced instead of individual pulses of light.

A number of different scintillating or fluorescent materials have been used in CT scanners including cesium iodide (CsI), cadmium tungstate ($CdWO_4$), and fluorescing materials using rare earth elements, such as gadolinium and ytterbium. Important characteristics of the detector material include its X-ray absorption efficiency, the energy conversion efficiency, and its temporal response. The X-ray absorption efficiency depends on the density and atomic number of the material, as well as the thickness of the detector. Conversion efficiency is the ability of the fluorescing material to take the energy that is absorbed and convert it to light that can be measured by the light sensitive detectors. When the X ray is absorbed the light is emitted over a short period of time. If this time to emit the light is too long, then this afterglow may influence subsequent measurement. This is one of the reasons that NaI(Tl) is not used in current fast scanners.

Additional factors affecting the detector efficiency is the effectiveness of getting the produced light to the light detecting element and the efficiency of this light detector. The photomultiplier tubes of early scanners have been replaced by photodiode arrays. These components do not have the inherent amplification found in PMTs, but they enable the manufacture of small, closely spaced detectors and the implementation of 2D or multirow arrays.

Gas-Filled Detectors. Gas-filled ionization detectors are another type of detector system that was widely used in CT systems. These detectors operated on the principle that the X rays passing through matter, such as the gas in the detector, causes ionizations or free electrons. A voltage can be placed across the gas to collect the electrons and determine the number of ionizations and the amount of radiation. This type of detector is used in many X-ray survey meters. In order to increase the fraction of the radiation that interacted with the gas and increase the signal level, high pressure xenon gas is used. The electrodes are tungsten plates that are oriented toward the position of the X-ray source. This directional chamber limits the detector sensitivity to radiation coming at an angle from these tungsten plates, thereby providing a capability to reject some of the scatter radiation entering the detector. These systems have been supplanted by the solid-state, scintillation detector systems, especially with the advent of MDCT.

Multiple Row and Area Detectors. The scintillation material in the MDCT detectors is mounted onto a photodiode array chip. The scintillation crystal is diced or sawed to form a series of individual elements. The sawed surfaces, or with the assistance of a reflective coating, help direct the light produced to the light sensitive component directly beneath this element. The width of the detector, in the slice thickness direction, is typically ~0.5 mm. The number of rows of data that may be acquired is often limited by the data acquisition system (DAS). The signal from a series of rows may be combined to produce an effective slice thickness that is some multiple of this value. This may be done prior to the digitization, allowing for a thicker slab of tissue to be scanned per rotation, or may be done as a postprocessing technique to reduce image noise.

An example may be a four slice CT scanner with a series of 1.0 mm detector rows covering a total width of 20 mm. It is limited to acquiring four rows of data by its data acquisition system. A scan may be performed with 4×1 mm detectors for a total beam width of 4 mm, or 4×2 mm for a beam width of 8 mm, up to a 4×5 mm for a beam width of 20 mm (Fig. 8). The first approach would give the best interslice resolution, while the latter would allow one to scan a given volume in less time.

The DAS must provide a highly accurate digitization of the signal and is handling a tremendous amount of data. As an example a 64-slice scanner may have 64-active rows of detectors each containing 1000 elements. As the scanner rotates around the patient in 0.5 s, 1000 measurements are made from each of these elements. That results in 128 million precision measurements made each second. This value increases as more rows of detectors are added and area array detectors for cone beam scanning are used. Data acquired during the scan is transmitted by a telemetry system to the fixed portion of the gantry. The data is sent to a computer that utilizes array processors for rapid data

Single slice detector array

Multi-slice detector array

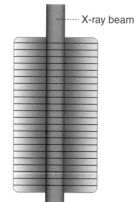

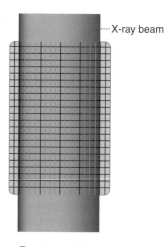

Figure 8. (a) In a single slice CT scanner the entire width of the detector is active and the slice width is determined by the collimated width of the X-ray beam. (b) Multidetector CT slice width is determined by the effective detector width. Individual detector elements may be grouped to yield a larger effective slice thickness. In this example a detector array consists of 20 rows each 1 mm wide. A four slice CT system may use groupings of four rows yielding 4 rows of 4 mm wide detectors as shown, or can use other groupings of the 20 rows.

One 3 mm slice
Determined by x-ray beam collination

Four 4 mm slices
16 mm collimated x-ray beam

reconstruction, and manages the storage and display of the resultant images.

Computer and Operator Console

The operator console utilizes an interactive computer display and dedicated function buttons to allow the procedure setup, scan initiation, image display, and data storage. The demographic information for the patient may be received from the facilities radiology or hospital information system (RIS or HIS) or entered by the technologist. Editable routine scan protocols facilitate scan setup and preview radiographic type image is used to identify the specific volume of the patient to be scanned. The images are displayed and some image processing and measurement features are available. A limited amount of the raw transmission data is stored on the system and may be used to for additional reconstructions from the data with alternate parameters. The reconstructed images may be filmed and the image data archived on the scan computer system, or transferred to a central PACS system for storage and for remote image display.

SCAN PITCH AND EFFECT ON PATIENT DOSE

One of the technique parameters set when performing a helical scan is the scan pitch, which is like the pitch on a screw. This refers to the ratio of the distance the table moves per $360°$ rotation of the X-ray source to the thickness of the X-ray beam. If the table moves the same distance as the beam width each rotation, then the scan pitch is one. The radiation dose with a pitch of one is similar to that obtained in a step-and-shoot axial mode where the table incrementation between scans equals the slice or beam thickness. With these contiguous axial images the entire surface of the patient within the area scan is struck once with the primary, unattenuated X-ray beam. Having a pitch < 1 indicates that overlapping data is acquired with a commensurate higher average dose, and a pitch greater than one results in gaps between primary exposed areas and a lower average radiation dose. A pitch < 1 requires less data interpolation and yields sharper slice profiles, while pitch > 1 reduces dose, but may blur the slice thickness profile and is more subject to certain image artifacts. With some CT systems this change in the effective technique and average dose as a result of the selected pitch is reported as the effective mAs. The effective mAs is the X-ray tube current (mA) times the time per rotation divided by the pitch.

The concept of pitch gets a little more complicated with MDCT systems. With single-slice CT systems the slice thickness corresponded to the detector width. In MDCT systems, there are multiple rows of detectors covering the width of the X-ray beam. This leads to two separate, but related pitch values. There is the collimator pitch that relates the table motion to the overall X-ray beam width, and the detector pitch that relates the table motion to the width of the individual detector rows (or their combined width when rows are combined prior to digitization) (23).

Consider an example with the four slice scanner with 20 rows of 1.0 mm wide detectors described above with a scan time per $360°$ tube rotation of 1 s. If the data acquisition mode is 4×2 mm, that is to simultaneously acquire four sets of data from detectors having a detector width of 2 mm, then four pairs of 1.0 mm physical detectors will be combined to produce four detector rows each with an effective 2 mm detector width, and the overall collimated beam width is 4×2 mm or 8 mm. If the table incrementation speed is $6 \text{ mm} \cdot \text{s}^{-1}$ or 6 mm/rotation, then the collimated pitch is

$$\text{Collimator pitch} = \frac{\text{Table travel per tube rotation}}{\text{Collimated beam width}} \quad (1)$$

$$\begin{aligned} &\text{Collimator pitch} \\ &= \frac{\text{Table travel per tube rotation}}{\text{Number of detector rows} \times \text{Detector width}} \\ &\text{Collimator pitch} = \frac{6 \text{ mm/rotation}}{4 \times 2 \text{ mm}} \\ &\text{Collimator pitch} = 0.75 \end{aligned} \quad (2)$$

A collimator pitch < 1 indicates that the radiation fields are overlapping, which will result in a patient radiation dose higher than an equivalent set of contiguous slices or a pitch of one. The detector pitch in this example is given by

$$\begin{aligned} &\text{Detector pitch} = \frac{\text{Table travel per tube rotation}}{\text{Detector width}} \\ &\text{Detector pitch} = \frac{6 \text{ mm/rotation}}{2 \text{ mm detector width}} \\ &\text{Detector pitch} = 3 \end{aligned} \quad (3)$$

CT SCAN TECHNIQUES

Preview Digital Radiograph

There are several scan modes or types of data acquisition that a CT scanner may be to acquire data. One commonly used technique is the acquisition of a scout or preview scan. These scans are basically a digital radiographs that are used to set up the tomographic imaging sequence, or may be used to visually locate the position of an axial slice on a radiographic reference image. To acquire the preview image the X-ray source and detector remain stationary. The detector sees a single line of an X-ray transmission image. As the table and patient are moved through the X-ray fan beam, the series of transmission lines acquired generate the radiographic image.

From the preview scan the technologist can define a range or volume within the patient to be scanned. Lateral preview scans can be used to determine the proper gantry angulation to orient the tomographic slices with desired anatomical structures, such as to the intervertebral disks in the spine.

Axial CT Scan

An axial scan is a basic CT scan, normally implying the data being acquired without the table moving during data acquisition (Although an axial image also refers to any image orientated transverse across the patient, as opposed to sagittal or coronal plane orientations.) Prior to the advent of the continuously rotating helical scanners, all CT scans were acquired with a stationary table. For a single-slice scanner the slice thickness or slice profile is defined by the collimation of the X-ray beam with the detector width being somewhat larger than this beam. For a multidetector CT the effective detector width of the rows of detectors tends to be the primary factor in determining the slice thickness. The effective detector width may be the summation of several physical rows of detectors. The grouping of detector rows may be to form thicker slices in order to reduce the image noise and number of images generated, or may be due to data acquisition system limitations.

The MDCT systems may be limited in their ability to acquire axial images due to the divergence of the fan beam. With a single detector row all of the transmission rays passed through a particular plane within the patient. The beam divergence along the slice thickness orientation caused some variation in the detected slice profile, but it was relatively minor. With the MDCT systems the data seen by the row of detectors on the ends is not consistently within a single plane due to the angulation of the diverging X-ray beam. This can cause inconsistencies in the data and may cause image artifacts or errors.

Helical CT Scan

The primary advantage to the continuously rotating source and detector is the ability to do helical or spiral CT scanning. Data is acquired as the patient is moved through the beam. There is no set of measurements where one has transmission data from all angles around the patient, but adjacent measurements are used to estimate the data corresponding to a particular plane. This mode allows for the rapid acquisition of data though a patient, and the ability to reconstruct images at any location within this volume. This rapid scanning allows procedures to be done quicker, often allows data to be acquired within a single breathhold, minimizing motion blurring, and facilitates the ability to perform contrast enhanced angiography studies to evaluate major blood vessels.

Dynamic Scan, Fluoro CT, and Triggered Scan Start

Another mode of data acquisition is dynamic scanning. In this mode a series of images are sequentially obtained at a single location. This can be used to analyze motion, or more commonly to evaluate the flow of contrast material into a tissue. This capability prior to continuously rotating systems was limited to one scan every few seconds since the tube had to stop and reverse motion between scans. Continuously rotating systems not only can acquire a sequence of images with no time gap between them, but also can obtain images overlapping in time where the time spacing between images is shorter than the data acquisition time for the image. Dynamic scanning can produce a series of images to assist in evaluating a tumor or mass by how it enhances or changes as iodine contrast material flows into

the tissue, or it may be used for quantitative analysis of the tissue perfusion.

A variation of this is fluoro or fluoroscopy mode CT. Here a series of images at a location are dynamically scanned and rapidly reconstructed to allow the technologist or physician to see the image in real time. This may be used to assist in a CT guided invasive procedure. Note that another approach is to have a conventional X-ray fluoroscopy system adjacent to the CT where the fluoroscopy is used for needle or catheter guidance and the CT is used to verify and evaluate results. Computed tomography fluoroscopy may also be used to visualize the arrival of injected contrast material into a vessel. This information may be used to initiate a scan sequence to catch the maximum concentration of the contrast media in the vessels of interest for CT angiography. Angiography scan starts may also be assisted using a feature where the computer evaluates the transmission data through a defined vessel and triggers scan start when a sufficient attenuation increase is detected.

Cardiac Gated CT

Physiologic motion can degrade the image quality. Fast helical and MDCT techniques allow for single breathhold studies. With the exception of the electron beam CT systems, a full set of data cannot be acquired of the heart without motion. In order to freeze the cardiac motion, the data is acquired and characterized relative to the cardiac cycle and is selectively grouped to obtain images without the typical motion blurring. This gated imaging requires an electrocardiogram (EKG) or similar input from the patient to define the cardiac cycle (24). Since the heart is relatively stationary during the longer diastolic rest phase than during systolic contraction, the gating may also be used to eliminate or minimize the systolic data to produce a diastolic only image. Alternatively, data can be acquired over many cardiac cycles and binned to produce images for various portions of the cardiac cycle. This multi-phase imaging process is similar to what is done in gated nuclear medicine and MRI studies. The series of images may be viewed in a movie mode to visualize the beating heart, and may be analyzed regarding wall motion and cardiac output.

CT NUMBERS

The linear attenuation coefficient is scaled into an integer pixel value. Medical systems utilize an offset scale that is normalized to water. This scale assigns air a CT value of -1000, water is at 0 and a material twice as attenuative as water has a CT value of $+1000$, and so on. The CT number is an integer relating to the attenuation properties of the tissue by the following formula.

$$\text{CT number} = \frac{(\mu_{\text{tissue}} - \mu_{\text{water}})}{\mu_{\text{water}}} \times 1000$$

Where μ_{tissue} and μ_{water} are the linear attenuation coefficients for the tissue in the particular voxel, and of water, respectively. Typical CT number values for some common tissues and test objects are listed in Table 1.

Display Window and Level

The CT numbers are commonly stored in the computer as 12 bit integers covering a CT number range from −1000 to +3000 (or −1023 to 3072). To display the full possible range of data one needs > 4000 shades of gray or displayed intensity. The human visual system, however, is limited, and we generally can discern something closer to 30 different shades. A common technique with all of the digital imaging methods is to use a viewer selectable mapping of the digital numbers representing the image to the various displayable intensities. There are a number of variations and processing methods that can be applied, but one of the most basic and most used methods is to define a display window level and window width.

The window level value defines the CT value that will be mapped as the middle gray intensity. The window width is the range of CT numbers that will have a range of gray values from black to white. Everything below the lower range value (the window level minus one-half of the window width) will be black and everything above the upper range level (the window level plus one-half of the window width) will be white. By adjusting these levels one can ignore the air-like CT densities, and display all the dense structures, such as bone as white, while obtaining a relatively high contrast view of the a narrow range of CT numbers corresponding to the soft tissue densities within the body. On the computer display one can easily vary these settings to look at low density structures in the lung or the high density detail of the bone if desired. Example of the effect of display window settings on the displayed image is seen in Fig. 9.

Other variations to this gray scale mapping function can also be performed, such as histogram equalization, where the resultant display will have an equal number of pixels for each gray level. The display may also be done in color where each CT number is mapped to a particular color. This is sometimes referred to as psuedo-color to emphasize that the displayed color is not that of the object, but some arbitrary assigned color. Clinical CT generally does not use color for basic cross-sectional image viewing. Color is commonly used, however, for processed data displays, such as 3D surface imaging where one views the surface of organ structures or of the vascular tree, or may be used as an overlay over the gray scale anatomical image with the color representing some functional feature, such as blood perfusion.

DISPLAY TECHNIQUES

Film and Soft-Read Workstations

Traditionally the cross-sectional images generated in a clinical procedure are windowed as appropriate for the tissues of interest, and then photographed or printed onto a large 14×17 in. $(356 \times 531\,mm)$ transparent film. If necessary, two sets of films may be made to have the window level and width adjusted for two different CT number ranges, such as for soft tissue and for the lower densities within the lung. The films provided a highly portable record of the study that can be illuminated with any X-ray film viewbox, and provides a medical record of the procedure. This was a manageable process producing a handful of films when used with single slice scanners acquiring relatively thick slices (3–10 mm) through a volume of interest.

With the fast MDCT systems one can rapidly scan through the same volume of the patient with thin slices. This results in hundreds to thousands of images for a single procedure. This would result in many dozens of films per study, which is not only expensive, but also unwieldy for physician review. This has been one of the drivers to implement a picture archiving and communication system (PACS) (see PACS topical entry), which enables the use of computerized soft-read workstations for the primary analysis of the image set.

Analyzing the images from the computer display provides a number of interactive tools for the reviewer. The ability to interactively change the window level and width is a powerful function for evaluating subtle features. One can measure the area and average CT number within a region-of-interest (ROI), measure distances and angles, and magnify regions of the image. One can rapidly page through a stack of images providing a better view of the continuity of structures from slice-to-slice.

Alternative Image Plane Display

A number of processing techniques are available for analyzing and presenting the volume information contained in a stack of axial slices (Fig. 10a). An alternative plane through the patient can be generated through this volume. This can be a coronal (frontal) plane, a sagittal (lateral) plane (Fig. 10b and c), or an arbitrary oblique plane. A series of parallel oblique planes may be reconstructed in a batch mode using the multi-planar reformation feature of the scanner or workstation, or the location may be interactively defined and displayed. The reformatted slice may be generated with a definable slice thickness down to the voxel size of the data set.

Maximum Intensity Projection

The displayed data in the oblique plane display is an average of the voxels contributing to each of the reformatted pixels. An alternative is to display an intensity value that corresponds with the largest voxel value in these contributing pixels. This type of display is referred to as maximum intensity projection (MIP). The slab thickness for the MIP image can include the entire volume scanned or a thinner slab (Fig. 10d). The MIP image gives a 3D type presentation for viewing dense structures such as bone or contrasted blood vessels, especially when rotating the viewing angle.

3D Surface Imaging

Another volume viewing technique is the 3D surface imaging. If structures can be characterized by their CT number range, the contours of the structure can be defined and surface view formed. These images have much in common with the visualization techniques used by computer-aided design or in computerized animation in the entertainment

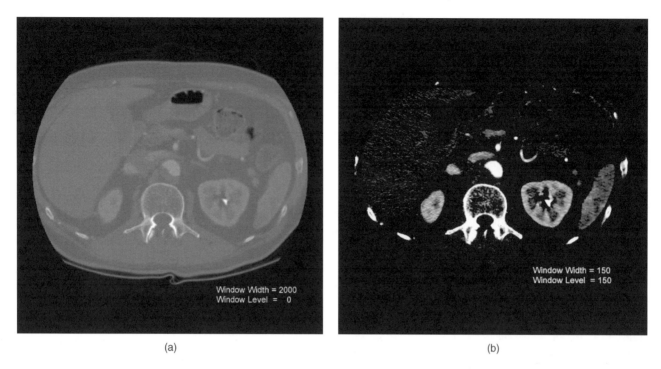

(a)   (b)

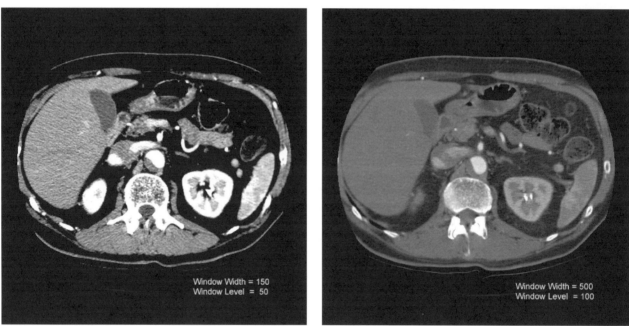

(c)   (d)

Figure 9. Various display window width and window level settings for the same abdominal CT image. (a) A wide window (2000 CT numbers) shows nearly the full range of CT numbers, but without discernable contrast between soft tissue structures. (b) A narrow window (WW = 150) yields high contrast between structures, but the window level centered at 150 results in most soft tissue being black because they are below the window range, with only the bone and structures containing iodine contrast media being seen. (c) A narrow window (WW = 150) centered at 50 results in a high contrast visualization of the soft tissue. (d) A typical display window (WW = 500, WL = 100) may compromise to provide good contrast while displaying a wider range of structures with lower displayed noise.

industry. The surface defining points are connected and plated with tiles or surface segments (25). The display software will project the nearest surfaces to the displayed image and use features such as distance from the viewer and angulation of the surface to define the brightness intensity. Light source position and coloration may also add to the display. Several different structures with differing CT number ranges can be simultaneously displayed with different color schemes for each structure. In this way, bone may be shades of white while a tissue or vascular structure may be red (Fig. 10e and f). The structure may also be given the property of transparency allowing visualization of deeper features, such as visualizing the ventricles of the brain through a visible but transparent skull. With specialized displays a stereoscopic pair of images may be viewed, enhancing the 3D effect, however, the ability to use motion and rotate the structures on the display is very effective at producing a 3D view.

Three-dimensional views may be enhanced with some computerized surgery. The user can select certain structures to be eliminated from the image. This selection can be made by defining cut planes or surfaces from various views and erasing structures outside of a volume of interest. Connectivity tools may also be used where the cursor is used to identify a structure, and then all surface points that are contiguously connected to the seed point are either selected or erased. In this way, one may select a vascular tree that is otherwise obscured by bony structures with

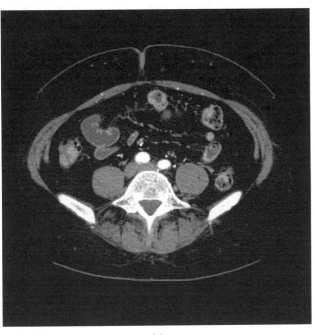

(a)

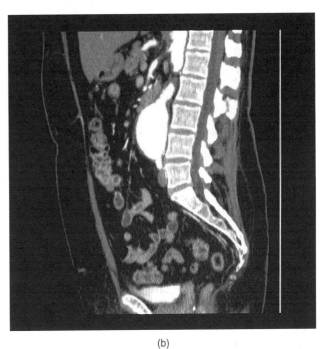

(b)

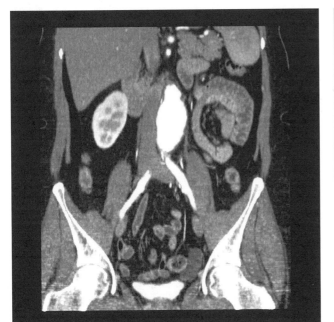

(c)

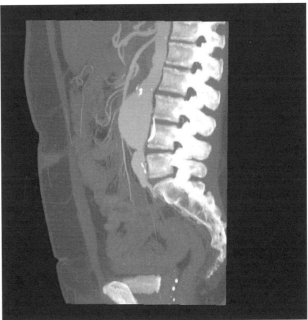

(d)

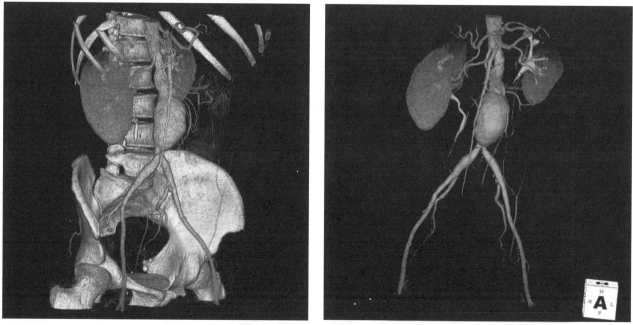

(e)   (f)

Figure 10. Alternative image sets can be generated from a series of closely spaced axial images. The image in (a) is one of 266 2.5 mm thick axial images. Iodine contrast media has been injected into the patient to make the major blood vessels visible, including the balloon shaped aortic aneurysm. From this data set the computer can generate (b) a sagittal image, or (c) a coronal image, or (d) a maximum intensity projection (MIP) image of a sagittal slab containing the spine and aneurysm. Structures may be identified by their CT number range to generate 3D surface images of bone and contrasted vessels (e), and structures may be removed to produce a vascular tree image.

similar CT numbers. Problems with this approach occur when the two structures touch making a connectivity bridge between them.

SPECIAL CLINICAL FUNCTIONS

Surgical Planning

The ability to produce a 3D visualization of structure surfaces can be used in several ways. It is useful for general viewing and obtaining an overview of certain structures. This may be useful in seeing areas that should be scrutinized more closely, and appears useful in communicating anatomical findings to surgeons and other physicians that are more familiar with the physical anatomy rather than a series of cross-sectional slices through it. In some cases data may be obtained from these images to assist in surgical planning, including the repositioning of bone fragments or the appropriate type and size prosthetic hardware to use.

CT Angiography, Virtual Colonoscopy, and CT Perfusion Imaging

Computed tomography angiography (CTA) is the procedure used to visualizing the blood vessels (26–28). Iodine contrast media is injected into the patient to increase the attenuation and increase the CT number of the blood within the vessels (Figs. 10e, f, and 11). Timing of the

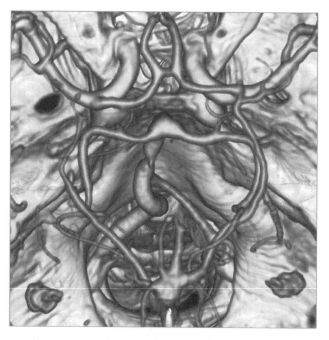

Figure 11. Three-dimensional view of a CT angiogram of the brain arteries including the Circle of Willis along with the skull structures as viewed from the top of the head.

CT scans is important in these procedures since the contrast media will return through the venous system and obscure the visualization of the arterial system, hence the use of some of the previously described scan start techniques. Besides general 3D viewing of a vascular tree to produce a CTA, other related techniques may provide additional information. Two points in a vascular tree may be identified and the computer can locate the line within the scan volume corresponding to the center of the vessels connecting them. The vessel along this line may be analyzed producing a plot of the vessel diameter or cross-sectional area. A stenosis appears as a reduce area, while an aneurysm may be seen as a greatly enlarged area. Since the vessel wall is defined as the surface between the high X-ray densities within the vessel to the water-like tissue densities outside the vessel, one can use 3D visualization techniques with the viewer located inside of the vessel. The viewer may travel or fly-through the vessel and visualize the structure of the lumen surface.

This fly-through technique is also the basis of virtual colonoscopy. In a clinical colonoscopy, the bowel is prepped to remove residual feces. An endoscope is inserted into through the anus into the colon and a camera and light source allows visualization of intestinal surface. If suspicious polyps or lesions are located, devices may be guided through the endoscope to remove or sample the tissue. In virtual colonoscopy bowel preparation is still needed and the colon inflated with air to produce a well-defined interface at the wall surface. A series of CT slices are acquired and 3D visualization and fly-through techniques are used to view the structure of the colon surface (Fig. 12).

Besides visualizing the vessel by CTA, the vascular condition of the tissue may be analyzed using a CT

perfusion procedure, especially in evaluating the brain. Acquiring the data in a perfusion study requires obtaining a series of images of selected slices over a short period of time as iodine contrast media or inhaled xenon gas in the blood flows into and washes out of the tissue. Various parameters may be measured, such as mean transit time (MTT) showing how fast the blood reaches the tissue. This may provide some indication of blood shunting or obstruction. The enhancement curve may be analyzed to obtain a relative cerebral blood volume (rCBV) and relative cerebral blood flow (rCBF) images. This information may be useful in evaluating strokes and obstructive disease.

Quantitative Analysis: Bone Density and Calcium Scoring

In general clinical CT scanners are not designed to produce highly accurate attenuation data, but rather high quality diagnostic images with minimal artifact. A number of factors can affect the calculated CT number value within a pixel, including its location and the size of the patient. With this caution, however, there are several applications where the analysis of the CT numbers is valuable. In general image interpretation, the CT value of a tissue lesion may be made to help determine if it is a mass, a cyst, edema, or a hemorrhage. Other scans may be performed specifically for the quantitative analysis. One screening procedure is calcium scoring. The calcium plaques in blood vessels will increase the CT value of the corresponding pixels. An evaluation of the amount of calcium in coronary arteries is an indicator of cardiac risk and may be measured by CT (29).

Osteoporosis is the loss of calcium bone mass, especially prevalent in postmenopausal women. The CT technique may be used to analyze the calcium content of the bone. Usually the trabecular bone in the middle of the spinal vertebra is analyzed due to their large surface area and sensitivity to bone loss. In order to produce reproducible data, the measurements made of the bone are compared to other reference densities within the image or in a comparable image. This may be done by having the patient lie on a phantom containing known reference materials, or comparing the measurements to other tissue in the image with known CT values (30–32). Most bone mineral densitometry, however, is performed with dedicated systems rather than using CT scan procedures.

Radiation Therapy Treatment Planning

Another group that would like quantitative information is the radiation oncologist for use in radiation therapy treatment planning. Some CT scanners are dedicated for radiation oncology use and these CT simulators may have special features and software to assist in radiation therapy simulation. In radiation therapy treatment planning it is important to be able to define the target tissue to be irradiated and the adjacent sensitive tissues, and to have them in the same position and orientation that they will be at the time of treatment. Since most linear accelerators used for treatment have flat tables, a hard flat table pad should be used for the corresponding CT scan. Likewise the body position, such as the position of the arms, should be as

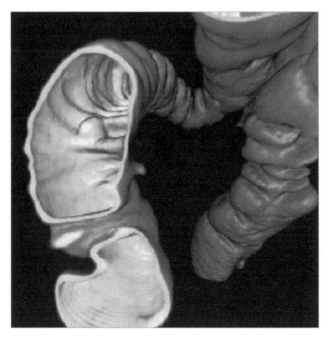

Figure 12. A 3D surface image of the interior wall of the colon allows a virtual colonoscopy fly-through to inspect the intestinal wall for polyps.

it will be during treatment. Skin markers and CT visible fiducial markers may be used to orient and register the images with the treatment plan.

Besides seeing the pathology and anatomy to identify the targets for the treatment planning, obtaining information on the attenuating properties of the various tissues to the high energy photon and electron beams is useful for accurate treatment planning. The problem is that diagnostic CT scans are acquired at relatively low photon energies as compared to that used in therapy. The diagnostic CT X-rays are much more sensitive to the atomic number of the materials within the voxels than are the high energy therapy beams. Characteristics of known tissues are used along with the measured CT numbers to estimate the physical or electron density of the tissue and its high energy attenuating properties.

Dual-Energy Scanning

One approach that can be used for quantitative imaging, in particular to determine effective atomic number and density of the tissue is dual energy scanning. Using two different X-ray beams will produce data corresponding to the attenuating properties at the two separate effective energies. At diagnostic X-ray energies the primary attenuation processes are photoelectric absorption and Compton scattering. Photoelectric absorption is highly dependent on the atomic number of the material and the probability of interaction falls off rapidly with increasing photon energy. Compton scattering is relatively independent of the atomic number and falls off at a much slower rate. That is, the probability of photoelectric absorption is proportional to Z^3/E^3, while Compton scattering falls off with $1/E$. This information may be used to take the CT measurements and calculate an alternate pair of basis images, such as effective atomic number and density (33). This is preferably done with the transmission data, but may be performed with the reconstructed images. Challenges exist in these calculations, however, due to other factors in the imaging process, and methods used to correct for other systemic errors.

Stereotactic Surgery Planning

Stereotactic surgery utilizes a hard fixed frame to the head to direct a needle to a very particular location in the brain. The base frame usually is attached to the skull with screws or pins. During CT scanning a localizing frame is attached to the base. When the CT scans are analyzed, the target location is identified in the images. The location of various frame components are also identified and recorded. This information is used to localize the target in the 3D frame space. During the surgical procedure the localizing frame is replaced with a needle guide that can be set for insertion to the target spot. The same approach is used for stereotactic radiosurgery where thin radiation therapy beams are used to irradiate specific targets in the brain.

PET–CT Image Fusion and PET-CT Scanners

Positron emission tomography (PET) scanning is a specialized nuclear medicine technique for generating cross-sectional images of the distribution of positron emitting radioactive tracers in the body. In this task, it is very sensitive at presenting this information. It is, however, relatively poor at presenting high resolution detailed anatomy. The PET images may be fused with corresponding CT images to delineate the structures containing the radioactive tracer. This is usually displayed as a color PET image overlaid onto a grayscale CT image. The alignment of the two data sets may be performed manually or automatically by various computer algorithms. A key aspect of this image fusion is the patient being in identical positions for both data sets. This can present problems including different table shapes, arm position, flexure of neck and back, or changes in the patient between the scans.

Much of the difficulties in image fusion are eliminated by the use of a specialized system that contains both the PET scan capability and CT scan capability (34,35). Typically these are two relatively independent scanners with a connected gantry and utilizing a single patient table system. One of the steps in PET scanning is to acquire transmission measurements through the patient in order to perform accurate attenuation correction of the data. On a stand-alone PET scanner, this is acquired by use of a radioactive source that emits similar photon energies. With PET–CT the CT image data may be used to determine the PET attenuation correction. Note that the CT scan represents X-ray attenuation properties at diagnostic X-ray energies, which are much lower than the 0.511 MeV photons from the PET radionuclides, and appropriate corrections must be applied.

RADIATION DOSE

Computed tomography is an X-ray procedure with an associated radiation dose. X rays are ionizing radiation, meaning that the X-ray photons have sufficient energy to rip orbital electrons from atoms. As a consequence, small amounts of absorbed energy can cause biochemical actions that may have biological consequences. Radiation dose is the amount of energy absorbed per mass of tissue at a defined location and is measured in rads or preferably in the SI unit of grays, where

$$1\,\text{Gy} = 1\,\text{J} \cdot \text{kg}^{-1} = 100\,\text{rads} \qquad (4)$$

Related are units of effective dose, the rem and sievert, which estimate the whole-body dose that has an equivalent long-term risk as an actual dose to just part of the body (36). Risk from radiation exposure can be divided into a couple of categories. Nonstochastic or deterministic effects are those that will happen if a certain radiation dose is received. Most relevant to diagnostic imaging are skin effects, such as erythema, the reddening of the skin. These effects require several gray of dose, which is significantly higher than doses normally encountered in CT. Stochastic or statistical biological effects are of some concern and should be part of the risk–reward evaluation for the procedure. The principal stochastic effect is the increase risk of getting cancer as a consequence of the radiation exposure. The risks are relatively small, but unwarranted radiation exposures should be avoided. Since developing embryos and fetuses are especially sensitive to radiation, special cautions are often taken to minimize *in utero* exposures.

Computed tomography is a bit different from standard radiographs relative to the total dose received. The maximum entrance dose to the skin may be quite similar between a CT scan and a radiograph, but in radiography the intensity of the radiation decreases due to attenuation as it passes through the patient. Consequently, the dose to deep structures is much less than the surface dose, and the dose at the exit surface can be orders of magnitude less than the entrance dose (37,38).

In CT, the X-ray source rotates around the patient, such that the entrance surface is not just on one side of the patient. This results in a more uniform dose and considerably more total energy deposited in the patient. In a typical head scan, the dose across the imaged slices is fairly uniform, and for body sections the midline dose is approximately half that of the surface dose. This results in a much higher effective dose to the patient. It is estimated that CT accounts for $\sim 10\%$ of the radiology imaging procedures, but amounts to around two-thirds of the total effective dose patients receive, and these values are likely to increase with the increasing utilization of CT.

Measuring radiation dose in CT presents some challenges. The X-ray beam is a narrow fan beam and may not even be constant across its width. Bow-tie compensating filters may further vary the beam intensity along the length of the fan beam. We see that slice width and spacing, and helical scan pitch, as well as the patient size, are also factors affecting the average dose.

Dose across the slice thickness, either in air or within a plastic phantom that simulates the patient, may be measured with a stack of thermoluminescent dosimeter (TLD) chips, with a radiation sensitive dosimetry film, or with a photoluminescent dosimeter strip. This data can be useful in characterizing the dose profile and the amount of scatter radiation present. Acquiring this data is cumbersome, and the use of this information to estimate a dose from a series of scans can be complex.

An alternative is to measure the CT dose index or CTDI. If one considers a series of contiguous slices, where the distance between the centers of adjacent slices is equal to the slice thickness, then the dose to a particular point in the patient is equal to the primary dose from the slice containing that point, plus the scatter radiation from the other slices. The CTDI is effectively this multiple slice average dose. It is measured with a long thin cylindrical chamber, typically 100 or 140 mm in length, about the size and shape of a pencil. It is exposed with a single axial scan. If the slice thickness is 5 mm, then the center 5 mm of the chamber receives the primary exposure. The adjacent 5 mm segments encounter the exposure for the adjacent slices, and the next 5 mm the scatter dose two slices away, and so on for the full length of the chamber. If one normalizes the measurement for to the 5 mm primary segment length, the overall measurement is the exposure from the primary beam plus scatter this location would receive from CT scans of the surrounding slices. In general,

the CTDI is given by

$$\text{CTDI} = (\text{measured exposure}) \times (f - \text{factor})$$
$$\times \frac{\text{chamber length}}{N \times \text{slice thickness}} \quad (5)$$

The *f*-factor is the Roentgen exposure to the rad or gray dose conversion factor for the material being exposed. Alternatively, the ionization chamber measurement may be calibrated in air kerma or air dose and the appropriate air kerma to dose conversion factor for the material is used. The value N is the number of detector rows being used, and N times the slice thickness is the width of the X-ray beam. The ratio of chamber length to the beam width is one over the fraction of the chamber that is exposed with the primary beam.

The measurement of CTDI is a straightforward measurement with the proper equipment. The phantoms used to simulate the attenuation by the patient are typically acrylic cylinders with a diameter of 16 cm for the head phantom and a diameter of 32 cm for the body phantom. The phantom has a hole in the center and at four locations near the periphery of the phantom for insertion of the pencil shaped CTDI ionization chamber.

A composite measurement of the center and peripheral value is the weighted CTDI or CTDI_w. It is given by

$$\text{CTDI}_\text{w} = \left(\frac{1}{3} \times \text{CTDI}_\text{center} \right) + \left(\frac{2}{3} \times \text{CTDI}_\text{peripheral} \right) \quad (6)$$

These CTDI values should correspond to the multiple slice average dose from a series of contiguous axial scans. This value should also correspond to the dose from an equivalent helical scan with a collimator pitch of one. If the axial slices overlap or the helical pitch is less than one, the dose will be higher. If the axial slices have gaps between them or the helical pitch is > 1, then the average dose will be lower. The volume CTDI or CTDI_vol dose estimate adjusts the CTDI_w for the slice spacing or pitch.

$$\text{CTDI}_\text{vol} = \text{CTDI}_w$$
$$\times \frac{\text{detector width} \times \text{number of detector rows}}{\text{table increment per } 360° \text{ tube rotation}}$$
$$\text{CTDI}_\text{vol} = \frac{\text{CTDI}_\text{vol} \text{CTDI}_w}{\text{pitch}} \quad (7)$$

A number of factors affect the patient dose. Some are defined by the design of the scanner, such as tube to patient distance, and the X-ray beam filtration. The patient size affects the attenuation and the subsequent dose for a given technique. Others are selectable by the technologist, (e.g., the kVp, mA, rotation speed, and pitch or slice increment). Since the X-ray output is directly proportional to the tube current or mA, then the total output per rotation, hence the dose, is directly proportional to the mA and the time per rotation. The dose is also related to the tube accelerating voltage or kV, but proportional to the kV to a power of ~ 3. Note that even though the dose goes up with kV, often some of the other dose factors can be reduced for comparable image quality, especially for large patients. The average patient dose is inversely proportional to the collimator pitch in helical (39). For comparable image quality, a thick-body section requires a higher X-ray technique than does a thin-body section since a larger percentage of the incident X-rays are absorbed. Techniques are typically reduced for pediatric cases due to the smaller body size and higher concern for radiation exposure. Instead of selecting one technique to be used for all slices, the scanner may have a type of dose modulation or automatic exposure control that reduces the mA for less attenuating body sections. This may be performed based on data from the preview scan, or may be determined by the attenuation found in the previous rotation (40,41).

CT RECONSTRUCTION METHODS

There are several different reconstruction methods or mathematical algorithms that can be used to estimate the cross-sectional distribution of attenuation coefficients that results in the measured set of X-ray transmission values (5,6,39). Knowledge of the basic elements of the reconstruction method can help determine elements relating to the image quality, and artifacts. Reconstruction methods can be categorized into two basic approaches: analytic and iterative reconstruction techniques. The primary reconstruction method in medical CT systems is the filtered backprojection method, an analytic reconstruction technique.

Projection Data

What is the relationship between the measured transmission data and the CT data values? It is necessary to normalize the measured transmission data and convert these values into projection values that correspond to the objects attenuation values. The projection data values for a narrow, monoenergetic beam of X radiation can be determined by considering Lambert's law of absorption

$$I = I_0 e^{-\mu s}, \quad \text{or} \quad (8)$$

$$I/I_0 = e^{-\mu s} \quad (9)$$

where I is the intensity of the transmitted beam, I_0 is the initial intensity or intensity of the beam with no attenuating material present, μ is the linear attenuation coefficient of the absorber material, and s is the thickness of the absorber. The linear attenuation coefficient corresponds to the fraction of the radiation beam that a thin absorber will absorb or scatter. This coefficient is dependent on the atomic number of materials present, the physical density, and the energy of the X-ray beam.

If instead of a single homogenous absorber there are a series of absorbers, each with thickness s, the overall transmitted intensity is

$$I/I_0 = e^{-\mu_1 s} \times e^{-\mu_2 s} \times e^{-\mu_3 s} \times e^{-\mu_4 s} \times \cdots \quad (10)$$

$$I/I_0 = e^{-(\mu_1 + \mu_2 + \mu_3 + \mu_4 + \cdots)s} \quad (11)$$

$$I/I_0 = e^{-\sum \mu_i s_i} \quad (12)$$

where μ_i is the linear attenuation coefficient of the ith absorber.

Considering a 2D section through an object of interest, the linear attenuation coefficients of the material distribu-

tion in this section can be represented by the function $\mu(x,y)$, where x and y are the Cartesian coordinates specifying the location within the section. The integral equivalent to the above equation is then

$$I/I_0 = e^{-\int \mu(x,y)ds} \quad (13)$$

integrated along the line, s, from the X-ray source to the detector.

The objective of the reconstruction program in a CT system is to determine the distribution of $\mu(x,y)$ from a series of intensity measurements through the section.

Inverting both sides of the equation to eliminate the negative sign, and taking the natural log of both sides to eliminate the exponential yields what is called the projection value, given by

$$p = \ln(I_0/I) = \int \mu(x,y)ds \quad (14)$$

This equation is the basis of the Radon transformation that is fundamental to the CT process. The inversion of this transform, going from the projection data to the 2D distribution was solved in 1917 by Radon (3). He showed that the distribution could be determined analytically from an infinite set of line integrals through the distribution.

The projection values are based on several assumptions that are not necessarily true in making practical measurements. This may require certain corrections to the data for these systemic errors, or may result in artifacts or degradations in the image. Some of these will be discussed relative to image quality and image artifacts.

Iterative Reconstruction Techniques

One of the broad categories of reconstruction is the iterative reconstruction techniques. With this approach an initial guess is made of the density distribution of the object. The computer then calculates the projection data values that would be measured for this assumed object in a process referred to as forward-projection. Each calculated value is compared to the corresponding measured projection data value, and the difference between these values is used to adjust the assumed density values along this ray path. This correction to the assumed distribution is applied successively for each measured ray. An iteration is completed when the image has been corrected along all measured rays, yielding an improved estimate of the object. The process is repeated and with each iteration the estimated object or reconstructed image improves its correspondence to the object distribution.

There are numerous variations of iterative processing that may be used. One of the most popular is the Algebraic Reconstruction Technique (7), or ART, which in itself is an offshoot of the Kaczmarz technique for inverting large ill-conditioned matrices (42). The variations include additive or multiplicative error correction, weighted or unweighted data, restricted or unrestricted values, the order in which one corrects the rays, and whether to apply the error corrections along a ray after each ray, or all at once for all rays.

Iterative techniques are rarely used in X-ray computed tomography. They require all data to be collected before completion of even the first iteration, and they are very process intensive. Iterative techniques may be useful for selected situations where the data is limited or distorted, working with incomplete data sets, or with irregular data collection configurations. Known information on the object or object values may be incorporated into these techniques and reconstruction dependent corrections, such as for beam hardening, may be incorporated into the process. While not normally used for medical X-ray computed tomography, iterative techniques are commonly used in nuclear medicine SPECT and PET imaging. Variations may also be used in X-ray tomosynthesis that is a partial angle data acquisition technique used to generate planar images through an object, but without complete elimination of overlying structures (43).

Analytical Reconstruction Techniques

If one can analyze and solve a series of equations directly, it is an analytical technique. Radon in 1917 mathematically determined that a solution existed for determining the distribution of an object from a series of line integrals through it. Interesting though, is that Radon's work was not utilized in the development of CT, but it was noted afterwards that it encompassed the analytic reconstruction methods. Applied developments of the principles used were often driven outside of X-ray imaging, including radio astronomy (8), electron microscopy (9), and nuclear medicine (10), and discovered methods were often not implemented due to computational requirements in the precomputer age.

Direct Fourier Reconstruction

The Fourier transform is a mathematical operation that converts the object distribution defined in spatial coordinates into an equivalent distribution of sinusoidal amplitude and phase values in spatial frequency. The one-dimensional (1D) Fourier transformation of a set of projection data at a particular angle θ is given as

$$P(\rho, \theta) = \int p(r, \theta) e^{-2\pi \rho r} dr \qquad (15)$$

where r is the spatial position along the set of projection data and ρ is the corresponding spatial frequency variable.

The direct Fourier reconstruction technique, as well as the filtered backprojection method, are based on a mathematical relationship know as the central projection theorem or central slice theorem. This theorem states that the Fourier transform of a 1D projection through a 2D distribution is mathematically equivalent to the values along a radial line through the 2D distribution of the original distribution.

Taking the Fourier transform of one set of projection data measurements through an object at a particular angle provides data values along one spoke in the object's 2D Fourier transform frequency space. Repeating this process for a number of angles defines the 2D Fourier transform of the object distribution in polar coordinates (Fig. 13). Taking the inverse Fourier transform of this data yields the reconstructed image of the object.

Direct Fourier technique is potentially the fastest method for image reconstruction, however, it generally does not achieve the image quality of the filtered backprojection method due to data interpolation difficulties. Typical computer methods and display systems are based on rectangular grids rather than polar distributions, and direct Fourier reconstructions generally require an interpolation of the data from polar coordinates to a Cartesian grid, usually performed in the frequency domain. Consequently, these methods are generally not used in commercial medical scanners.

Convolutions and Filters

The filtered-backprojection technique is the most commonly used CT reconstruction algorithm. Before discussing this method, a brief review of filtering and simple backprojection methods is in order.

A convolution is a mathematical operation in which one function is smeared by another function. A common example is a presentation of a blurry out-of-focus projection of a text slide. A small dot, such as a period, instead of being small, sharp, and dark gets blurry with smooth edges and less contrast or darkness. This blurry spot is effectively the point spread function of the image. All of the lines and characters in the original slide can be considered as being made up of many points. Replacing each of these points with the

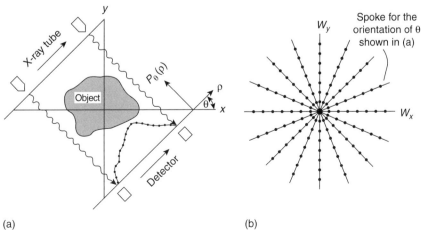

(a)   (b)

Figure 13. According to the central-slice theorem, the 1D Fourier transform of the projection values measured through an object in the spatial domain (a), correspond to the values of the 2D Fourier transform of the object along a diagonal line in the frequency domain. A series of transformed projection values yield the Fourier transform of the object, but in polar coordinates.

out-of-focus point results yields the overall blurry slide that is seen. The process of applying this blurry point to all points in the images is a convolution of the point spread function with the original object. Mathematically this is defined as

$$g(x) = f(x) \otimes h(x) \tag{16}$$

$$g(x) = \int_{-\infty}^{\infty} f(x)h(x - u)du \tag{17}$$

for 1D, or for 2D

$$g(x,y) = f(x,y) \otimes h(x,y) \tag{18}$$

$$g(x,y) = \int_{-\infty}^{\infty}\int_{-\infty}^{\infty} f(x,y)h(x - u, y - v)dudv \tag{19}$$

where the symbol $\otimes$ is the convolution operator between two functional distributions. If the system is a digital system with a discrete number of samples, the corresponding equation is

$$g_i = \sum_{k=-\infty}^{\infty} f_i h_{i+k} \tag{20}$$

or in 2D

$$g_{i,j} = \sum_{k=-\infty}^{\infty}\sum_{l=-\infty}^{\infty} f_{i,j} h_{i+k,j+l} \tag{21}$$

In the blurry slide example above, f may represent the original text (or image) distribution, h is the blur or point spread function, and g is the resultant blurred image.

Convolution operations can be used in image processing to smooth an image, as previously described, or to sharpen an image. Smoothing convolution filters are typically square (averaging) or bell shaped, while sharpening convolution filters often have a positive central value with adjacent negative tails.

Convolution Theorem

Reference was made to Fourier transforms in Eq. 15 and their ability to transform spatial data into corresponding spatial frequency data. Filtering operations, such as smoothing and sharpening, can readily be performed on the data in the spatial frequency domain. According to the convolution theorem, convolution operations in the spatial domain correspond to a simple functional multiplication in the spatial frequency domain (Fig. 14). This states that

$$g(x) = f(x) \otimes h(x) \tag{22}$$

is equivalent to

$$G(k_x) = F(k_x)H(k_x) \tag{23}$$

where $G(k_x)$, $F(k_x)$, and $H(k_x)$ are the Fourier transformed functions of $g(x)$, $f(x)$, and $h(x)$, where k_x is the spatial frequency conjugate of x. The functional multiplication in Eq. 23 is simply the multiplication of values of $F(k_x)$ and $G(k_x)$ at all values of k_x. The 2D convolution of Eq. 18 has its counterpart to Eq. 23 where $G(k_x,k_y)$, $F(k_x,k_y)$, and $H(k_x,k_y)$ are the 2D Fourier transforms of $g(x,y)$, $f(x,y)$ and $h(x,y)$. Note that the spatial frequency variables, k_x and k_y have units of 1/distance.

Since the convolution process is represented by a simple functional multiplication in the spatial frequency domain, the blurred image conceptually can be easily restored to the original object distribution. This can be accomplished by multiplying the Fourier transform of the blurred image, $G(k_x,k_y)$, by the inverse of the blurring function, that is $1/H(k_x,k_y)$. The result is the original object frequency

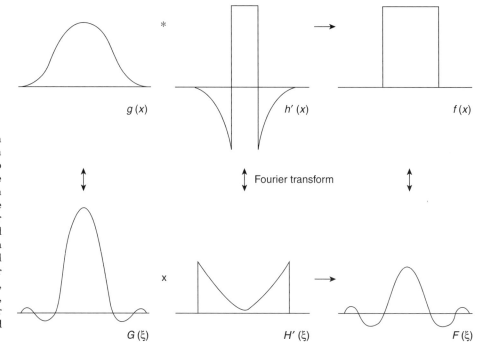

Figure 14. The convolution theorem states that a convolution operation in the spatial domain is equivalent to the functional multiplication of the Fourier transform of the functions in the spatial frequency domain. The filtering of the projection data for CT reconstruction may be performed by the convolving the projection data with a sharpening filter in the spatial domain, or by taking the Fourier transform of the projection data, multiplying by a ramp shaped filter, and taking the inverse Fourier transform to yield the filtered projection data.

distribution, $F(k_x,k_y)$. In practice this restoration is limited by the frequency limits of $H(k_x,k_y)$ leading to division by zero, and by the excessive enhancement of noise along with the signal at frequencies with small $H(k_x,k_y)$ values. This restoration process or deconvolution of the blurring function can likewise be performed as a convolution in the spatial domain.

Backprojection

Backprojection is the mathematical operation of mapping the 1D projection data back into a 2D grid. This is done intuitively by radiologists in interpreting X-ray films. If a high density object is visible in two or more radiographs taken at different angles, the radiologist mentally back-projects along the corresponding ray paths to determine the intersection of the rays within the patient and the location of the object.

Mathematically, this is done by taking each point on the 2D image grid and summing the corresponding projection value from each angular projection view. For that high density object the result is a line projected through the image from each view (Fig. 15a). This backprojection process yields a maximum density at the location of the object where the lines cross, but the lines form a star artifact emanating from the object. If an infinite number of views were used, the lines would merge and the density of the object would be smeared across the image with its amplitude decreasing with $1/r$ where r is the distance from the object. This simple backprojected image, f_b, can be represented by the convolution of the true image, f, with the blurring function $1/r$, or

$$f_b(r,\theta) = f(r,\theta)\otimes(1/r) \qquad (24)$$

With ideal data, this blurring function can be removed by a 2D deconvolution or filtering of the blurred image. The appropriate filter function can be determined by using the convolution theorem to transform Eq. 24 into it frequency domain equivalent, or

$$F_b(\rho,\theta) = F(\rho,\theta)(1/\rho) \qquad (25)$$

where the function $(1/\rho)$ is the Fourier transform of $(1/r)$ in polar coordinates. Dividing both sides by $(1/\rho)$ yields

$$F(\rho,\theta) = \rho F_b(\rho,\theta) \qquad (26)$$

The corrected image, f, can be obtained by determining a simple backprojection, f_b, taking its Fourier transform, filtering with the ρ function, and taking the inverse Fourier transform. Likewise this operation may be performed as a 2D convolution operation in the spatial domain. Equation 25 and 26 get more complicated when evaluated in rectangular coordinates rather than polar coordinates.

This approach of making a very blurred image through backprojection, and then attempting to sharpen the image tends to produce poor results with actual data. Filtering out this blurring function from the projection data prior to backprojecting, however, is quite effective and is the basis for the filtered backprojection reconstruction technique used in medical CT systems.

Filtered Backprojection Reconstruction Technique

According to the central slice theorem the Fourier transform of the 1D projection data is equivalent to the radial values of the 2D distribution Fourier transform of the distribution. Consequently, the filtering operation performed in Eq. 26 and illustrated in Fig. 14 can be performed on the projection data prior to backprojection. This is the conceptual basis for the filtered-backprojection reconstruction technique as illustrated in Fig. 15b (44,45).

As with other filtering operations, this correction can be implemented as a convolution in the spatial domain or as a functional multiplication in the frequency domain. Fourier filtered backprojection is performed by taking the measured projection data, Fourier transforming it into the frequency domain, multiplying by the ramp-shaped ρ filter, taking the inverse Fourier transform, and then backprojecting this filtered projection data onto the 2D grid.

If the filtering is performed in the spatial domain by convoluting the measured projection data with a spatial filter that is equivalent to the inverse Fourier transform of ρ, the process is often referred to as the convolution

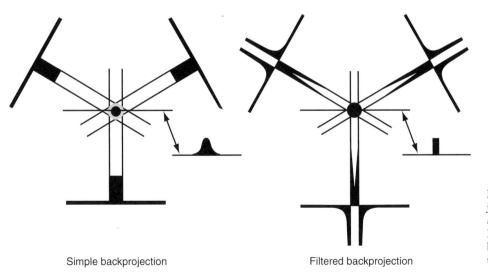

Simple backprojection   Filtered backprojection

Figure 15. (a) A simple backprojection from three views results in a highly blurred reconstruction. (b) Filter the projection data prior to backprojection corrects the data for the backprojection induced blurring.

filtered backprojection reconstruction method. The frequency filter, ρ, has the shape of a ramp and enhances high spatial frequencies of the projection data. The convolution function, or kernel, has the expected shape of a sharpening filter with a positive central value surrounded by negative tails that diminish in magnitude with distance from the center. Convolution and Fourier filtering techniques are mathematically equivalent and the general term filtered backprojection reconstruction technique may refer to either filtering approach.

Variations from this ideal ramp filter are normally used. Some of these may develop from applying the finite quantity of data and boundary assumptions to the mathematical derivation. Other variations, in particular frequency windowing, are applied on a more empirical basis. The high spatial frequency component of the projection data contains noise variations due to photon statistic along with diminishing amounts of signal data. The ramp filter greatly enhances these high frequency values, in particular the noise. Consequently, use of the ramp filter results in high resolution, but very noisy images. They are also more susceptible to image artifacts. If one wishes to see structures with only small differences in attenuation values, such as white versus gray brain matter, the noise must be reduced.

Medical CT scanners offer a variety of reconstruction algorithms or kernels from which to choose. In actuality they are not changing the reconstruction method, but the filter function used in the filtered backprojection technique. The ramp filter is modified by a windowing or apodizing filter that reduces the amplification of the higher frequency values. This has the same effect as smoothing the image. This smoothing is especially effective for CT imaging since the reconstruction process results in the noise frequency spectrum in the image following the reconstruction filter function, with most of the noise at the high spatial frequencies. In nuclear medicine they sometimes use the mathematical name for the windowing filter, such as cosine filter, Butterworth filter, or Hannings window. In X-ray CT the equipment manufacturers utilize different naming conventions for these filters kernels. Typically, they will have descriptive names, such as smooth, standard, sharp, bone, edge, or will have numerical values relating to its shape. The filter selection may also enable other features in the reconstruction process to minimize certain artifacts, such as motion in the body scans, or implement other needed data corrections.

The filtered backprojection reconstruction technique is the general method used in medical CT systems. This method is more tolerant of measured data imperfections than some of the other analytical techniques. This method provides relatively fast reconstructions and permits processing the data as projection views are obtained.

Cone Beam CT

Computed tomography scanning has progressed from a single row of detectors to the ability to use hundreds of rows of detectors for data acquisition. The CT reconstruction algorithms discussed above have generally considered all of the projection values being contained within a single plane through the patient. As one uses multiple rows of detectors, the projection data from the rows away from the

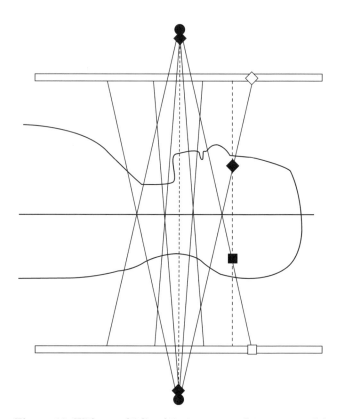

Figure 16. With a multislice detector or area detector, an axial (nonhelical) data acquisition results in rays outside of the center to pass through the patient at an angle rather than within a particular plane. This can cause artifacts in planar reconstructions. Specialized cone beam reconstruction algorithms may be used to help account for the angulated data.

center row pass through the patient at an angle. Objects within a view from one side may be outside to the view from the opposite view angle (Fig. 16). Generally, this angulation is ignored in the reconstruction, but data inconsistencies from the angulated data can cause artifacts in the image, especially around high density structures. Other algorithms are continuing to be developed that take into account the actual ray paths of the data (4). These algorithms are referred to as cone beam reconstruction methods.

With data acquired in an axial mode, with no table motion during data acquisition, large angulations cause significant degradation of the image quality, but may be useful in particular for high contrast structures such as bone or contrasted blood vessels. Considering the relationship between the object and the projection data, a simple rotation does not fully sample the Radon space needed for reconstruction. A helical data acquisition and certain other motion schemes do acquire a sufficient set of data for a potentially accurate cone beam reconstruction.

CT IMAGE QUALITY

A particular image will have limitations on the quality of the image and the types of structures that are visible. Two primary measures of image quality are the image resolution and noise (46). Resolution is the ability to see two separate small structures as two structures. It is somewhat different that detection. A very small dense structure considerably smaller than the voxel size may change the average attenuation within the voxel sufficiently to detect that a foreign object is present, however, the system would not have sufficient resolution to distinguish multiple objects within the voxel.

Resolution may be measured by using a test pattern of holes that are spaced with the center-to-center distance between holes being equal to twice the hole diameter. Alternatively, the pattern may be a series of lines or bars, or a pie shaped wedge of bars that get smaller toward the apex of the pattern. The resolution is usually stated as the hole or bar size of the smallest pattern clearly visible, or as a spatial frequency value in line pairs per millimeter ($lp \cdot mm^{-1}$). The resolution may be more precisely described by determining the modulation transfer function or MTF. This function indicates the level of signal loss for each of the spatial frequencies. The MTF is the normalized Fourier transform of the point spread function, and can be calculated by analyzing the image of a small dense object, or from measurements across a sharp boundary (47).

Geometric Blurring

A number of factors can affect the resolution in an image, some inherent in the design and construction of the system, and others selectable by the user. Limits on resolution for a system are generally determined by the detector width and the focal spot size of the X-ray source. Since the patient is approximately midway between the source and detector, there is necessarily some magnification of the object structures onto the detector. As a result of this magnification a sharp edge in the object will produce a blurry edge in the image due to the size of the X-ray source. This blurry area where only part of the X-ray source illuminates the detector is called the penumbra from the Latin for partial shadow, and the effect is referred to as geometric blurring. This sharpness can be improved by reducing the magnification, which is done by having the X-ray source farther from the patient and the detector as close to the patient as possible. It can also be improved by use of a smaller X-ray focal spot. Most X-ray tubes utilize a dual focal spot, which means they have two focal spot sizes from which to choose. Most scanning is done with the large focal spot. This allows operation at higher tube currents or mA, thereby allowing shorter scan times and less image noise. This is especially true for large field-of-view objects where this geometric blurring is not the limiting factor for resolution. Scanners will typically select the smallest focal spot that can be used with the mA and time parameters selected.

Ray Spacing

The detector size may also affect the resolution. For a rotate–rotate scan system, the in-plane detector-to-detector spacing defines the ray spacing within a view. Note that this ray spacing is variable due to the fan shape of the beam, but consider it at the center of rotation, which is normally the center of the patient. The detector size has decreased with the evolution of the scanners, resulting in the need for an increase in the number of detector channels and the amount of data gathered and processed. Since the physical detector spacing cannot readily be changed for a

given detector, a couple of alternative approaches have been used to change the effective ray spacing. An approach that has been used in the past maintains a fixed source–detector distance, but moves the X-ray tube closer to the patient, and the detector farther away when scanning smaller objects, such as the head. This increases the magnification and reduces the ray spacing, but puts more of a resolution burden on the size of the focal spot. No commercial systems still use this approach.

A common approach in rotate–rotate scanners is to improve sampling by using quarter–quarter offset detector shift. First generation translate–rotate scanners acquired data only $> 180°$ of rotation since the view at $0°$ is of the same data as the $180°$ view. If the ray spacing is not symmetric about the center of rotation, but the middle ray being displaced one-quarter of a detector width above the center of rotation, then the opposite view would have the ray one-quarter of a detector width below the center of rotation and the rays would be interleaved. It is not as obvious on a rotate–rotate scanner, but the same type of interleaving can be utilized. A rotate–rotate scanner only needs $180°$ plus a fan angle to sufficient set of data for a reconstruction. Certain fast scan modes will utilize this partial rotation data set, but typically the full $360°$ data set is used. In high resolution reconstructions, the full $360°$ data set should be used.

Pixel Size

The CT image is a finite array of values. Parameters selectable setting up the image reconstruction include the reconstructed field of view (FOV), and the image matrix size. Images are routinely reconstructed with a 512×512 matrix. The large FOV that would be used for a large body section is $\sim 50\,cm$. Therefore the generated pixels will be squares with sides of 50 cm/512 or $\sim 1\,mm$. In this case, the image would not be effective at resolving or differentiating objects smaller than this pixel size. Medical CT systems have the capability to provide better resolution than can be displayed with 1 mm pixels. To reduce the pixel size the image matrix size must be increased, or the reconstructed FOV reduced. The reduction in FOV is standard for imaging smaller body sections, such as the head where a 25 cm FOV may be used. This yields a 0.5 mm pixel size. If higher resolutions are desired, such as to evaluate the bones of the inner ear, then a sharp reconstruction filter is needed along with an even smaller pixel size in order to maximize the system resolution. Unfortunately these parameter changes also increase the image noise.

Image Magnification and Targeted Reconstructions

A number of display tools are available when an image is displayed. These include such useful features as the window level and width adjustments. One tool is image magnification where a portion of the original image can be magnified to fill the display or a digital magnifying glass can be moved around the screen. This function uses the data from the image and interpolates additional pixels to yield smaller pixel spacing. Since the source data is the original image, the magnified image does not contain any additional information, but the viewer may find it easier to see and interpret the image.

This image magnification is in contrast to a targeted reconstruction that goes back to the measured projection data and performs a reconstruction with new parameters such as reconstruction filter and pixel size. Consequently, a targeted reconstruction may yield information that was not present in the original image. In the case of helical scans and multislice detectors, the z-axis location of the reconstructed slices and the slice thickness, down to the acquisition detector row thickness, may also be specified.

z-Axis Resolution

The z-axis resolution is the resolution in the direction of the table motion or perpendicular to the axial image plane. The acquisition slice thickness and slice spacing dominate this resolution. The slice spacing limitation on z-axis resolution follows the same arguments as the pixel size does for the x–y resolution. The slice thickness is determined by the focal spot size and either the collimated beam thickness or the z-axis height of the detector. For multislice detectors, the detector rows may be utilized individually, or ganged together to yield a thicker slice, or both with the averaging of detector rows performed in the software. A major advantage to MDCT and cone beam CT is the ability to rapidly acquire many thin slices through the patient. This often results in isotropic resolution where the resolution in the z axis is equivalent to the in-plane resolution. These thinner slices and isotropic resolution improves the contrast of small high contrast structures, such as the blood vessels of the lung. The thin closely spaced slices greatly improve the quality of images that contain the z dimension, such as sagittal and coronal images, or 3D surface reformations.

Image Noise

Another significant limitation to image quality is image noise or graininess. Noise is caused by variations in the measured signal. These may be due to systemic causes, such as electronic noise, however, a well-designed system will not be limited by these sources. The primary source of noise in medical CT systems is due to quantum mottle or photon statistics. Due to the random nature of photon emission and absorption, repeated identical measurements will vary with a percent standard deviation proportional to one over the square root of the number of photons detected, that is

$$\sigma = \sqrt{N} \tag{27}$$

$$\%\sigma = 100\% \times \frac{\sqrt{N}}{N} \tag{28}$$

$$\%\sigma = 100\% \times \frac{1}{\sqrt{N}} \tag{29}$$

where σ is the standard deviation and N is the number of photons detected. The number of photons detected is determined by the output of the radiation source, the attenuation in the patient, and the efficiency of the detector in absorbing the radiation and converting it into a measurable signal. The source output is directly proportional to the selected tube current or mA, and the scan time per rotation and to the kVp to some power ~ 3. The beam is diminished by attenuation, hence noise is more prevalent scanning a large body section. Higher kVp settings are more penetrating and allow a larger fraction of the photons through the patient, but may also reduce the contrast of some structures.

Another factor is the size of the detector, in particular the detector height or the slice thickness. Thinner slices result in fewer X rays for a given technique. Helical scanning interpolates data from adjacent rotations, thereby increasing the effective slice thickness and reducing the noise to some degree. The image resolution and pixel size also can have an effect. One can consider the image noise to be related to the number of photons detected per voxel element. Reducing the voxel or resolution element size increases the noise.

Noise can be expressed as the standard deviation in the image of a uniform object. A more complete method of characterizing noise is to determine the noise power spectrum. The noise power spectrum defines the noise content in the image versus spatial frequency. The measure transmission data ideally contains white noise, that is, having a noise power spectrum at all frequencies. As with other imaging systems this ideal spectrum is modified by the modulation transfer function (MTF) or ability of the system to transfer signal (and noise) of various frequencies in the object to the image. The resolution limitations due to detector size and sampling reduces or eliminates some high frequency signal components. In CT, there is the additional step of the image reconstruction which modifies the noise power spectrum. The projection data is filtered with a ramp or a windowed ramp filter, reducing the low frequency content and enhancing the medium and high frequency content. For signal data where there is a correspondence in the data from various angles, the low frequency component is restored and a typical MTF response is generated. The noise content, however, is random and does not have a direct correspondence between angles. Consequently, the noise power spectrum will mimic the shape of the filtering function, enhancing noise at the higher spatial frequencies. Smoothing or the use of a windowed filter is especially effective in CT since there is a high level of noise relative to signal at these frequencies (48).

Object Contrast

Contrast is the difference in the measured value of a structure from its surroundings. The ability to distinguish structures that have attenuation differences of a fraction of a percent is one of the key imaging benefits of CT. The contrast between structures in CT is the fractional difference in attenuation coefficient, or more commonly the difference in attenuation coefficient relative to the attenuation coefficient of water. This is given by

$$\% \text{ Contrast} = \frac{|\mu - \mu_{\text{background}}|}{\mu_{\text{water}}} \times 100\% \tag{30}$$

When CT values have an offset of 1000 in order to normalize water to a value of zero, the comparable equation using the CT values is

$$\% \text{ Contrast} = \frac{|CT - CT_{background}|}{1000} \times 100\%$$

For example, if brain gray matter has a CT number of 40 and white matter 30, then the fractional contrast is 10/1000 or 1%.

One of the factors that often limits contrast is the partial volume effect. Voxels or resolution volumes that correspond to a particular pixel may contain a mixture of tissues or structures. The resultant CT number is generally a volume average of the contents of the voxel. Reducing the voxel size with thinner slice thickness or higher image resolution may reduce this volume averaging. For example, a 1 mm piece of bone that is twice as attenuating as water (CT number of + 1000) is surrounded by water (CT number of 0). If scanned with a 10 mm slice thickness, the bone would occupy 10% of the voxel volume and the resultant image would show a CT number for this volume of ∼ 100. Reducing the slice thickness to 2 mm would increase the CT number to ∼500. This is not an error or image artifact, but a limitation in the image quality and content. If structures are positioned in the voxel such that part of the ray goes through one material, and other portions of the ray go through other material, this may cause a measurement error and a partial volume artifact.

The linear attenuation coefficient values are a function of the density of the material and effective atomic number of the material, as well as the effective energy of the X-ray beam. Sometimes there is insufficient contrast between a structure of interest and its surroundings, or certain structures are to be highlighted. This may be done through the use of contrast agents or contrast media. This is a material that will change the X-ray absorption properties and the visibility of the structure. In X-ray studies the primary contrast agent used is iodine, which may be given orally to highlight the intestinal track, or injected to make blood vessels or highly vascular tissue visible. The iodine does not actually change the density of the blood significantly, but its higher atomic number of 56 versus 7 or 8 for tissue and water, and its k shell electron binding energy of 34 keV make it much more effective at stopping the X rays. Using contrast media the radiologist may determine the vascularity of a mass to help determine the type of tumor, or contrast agents can be used to enhance the blood vessels to evaluate blockages or aneurysms in CT angiography procedures.

Low Contrast Detectability

The ability to see small differences in contrast is a key feature of CT. What limits this ability, however, is noise. When the noise variations are of the same magnitude as the contrast between the structures of interest, the structure will not be visible. This is especially true for small objects, in that the human vision will effectively average the signal over an area and one may be able to distinguish larger low contrast objects. The limit to low contrast detectability is noise and resolution has relatively little impact

(Fig. 17c–e). Conversely, high contrast resolution is not greatly affected by noise since the structures contrast is much larger than the noise.

ARTIFACTS

The ideal imaging system reproduces a faithful image of the object. Limitations in resolution, noise, and low contrast detectability are not errors, but are definable limitations of particular imaging systems. Errors do occur, however, and structures or patterns may appear in the image that do not correspond to the patient or object being scanned. These false structures in an image are referred to as artifacts (49). Artifacts often are readily identifiable because of their characteristics, such as a bright line extending through and beyond the boundaries of the patient. These artifacts may cause problems by obscuring parts of the image. Less common, but of significant concern, are artifacts that can appear similar to pathologies.

Artifacts occur in all imaging systems, but the reconstruction process of CT enhances the opportunity for producing artifacts. Mathematically a perfect reconstruction of the object should be obtainable. This can be done with an infinite amount of perfect data, however this is not available in a practical system. In general, image artifacts are caused by an insufficient amount of data, or insufficient quality of data.

Insufficient Data Quantity

Modern CT scanners acquire millions of transmission measurements and insufficient data quantity is not a limiting factor for routine studies, but still may pose challenges in studies attempting to reduce the data acquisition time or those pushing the image resolution. Nyquist sampling theorem was mentioned as a requirement for sampling an object in order to capture the high spatial frequencies. When the object contains higher frequencies than the sampling rate can characterize, the high frequencies reappear in the data and take on the alias of a lower spatial frequency. The resulting error is referred to as aliasing. It can take on the appearance of ripples in the reconstructed object parallel to a sharp edge.

Aliasing may occur due to insufficient sampling of rays within a view, or may be due to an insufficient number of views. Ray aliasing may appear as oscillations or ripples parallel to high contrast. View aliasing typically appears a considerable distance from a high contrast object, and appear as a radial pattern of light and dark bands emanating from the high contrast object.

The occurrence of aliasing is greatly reduced by the data smoothing effect that occurs due to the finite size of the X-ray detector. The transmission data is not measuring a series of infinitely thin rays, but columns through the patient. The variations in intensity between the various paths within a single ray or detector measurement get averaged out and lost, and are not available to cause aliasing.

A related artifact is the Gibb's phenomena. The range of the frequency domain is limited by the sample spacing. Using the full ramp function results in an abrupt cutoff of the filter function or transfer function at this limit, or the

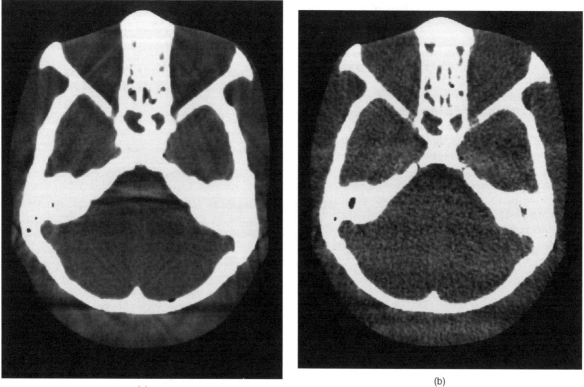

(a)   (b)

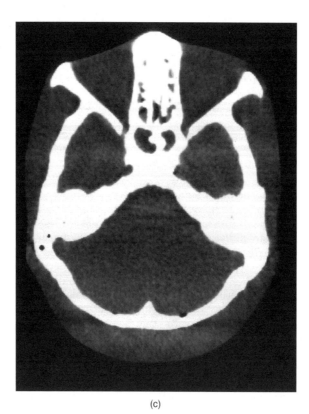

(c)

Figure 17. (a) A 5 mm thick CT slice through the posterior fossa can cause beam hardening and partial volume artifacts as seen in this acrylic phantom containing a skull. (b) Use of a thinner 1 mm slice thickness reduces the partial volume artifact from the angled bony structures, but increases the image noise or graininess. (c) Adding five 1 mm slices together (or the five sets of projection data after the logarithmic scaling) results in reduced artifact, but comparable image noise as the 5 mm slice.

filter may be designed with a sharp cutoff. This sharp edge in the frequency domain causes an overshoot and ringing artifact along edges in the image. This can be reduced by windowing the filter function with a function that smoothly goes to zero without an abrupt change.

Inconsistent Data

The CT reconstruction process is based on obtaining a consistent set of transmission views of the patient from various angles. The reconstruction algorithms effectively correct for errors caused in the backprojection from one view with corrections in other views. If the object or data is not consistent, then improper corrections are applied resulting in artifacts. There are several factors that can cause data inconsistencies and result in artifacts.

Motion

One of the most obvious inconsistencies is patient motion, whether squirming, coughing, fluid level movements in the stomach, breathing, or even the beating of the heart. Here the views from different angles are not of the same cross-sectional object. Motion artifacts are less of a problem in head imaging since with a cooperative patient the head can remain motionless for a considerable period of time. Body imaging can be more problematic causing image artifacts in addition to the motion blurring in the image.

The artifact is most pronounced when there is an abrupt change in the data from one view to the next usually resulting in streaks across the image. For a 360° data acquisition this abrupt change in the continuity of the data is likely to occur between the first and last view of the rotation. There are several ways to minimize this. One can overscan, collecting > 360° of data, and perform a weighted average of the data in the overlap region. Partial angle scanning, scanning 180° plus a fan angle, reduces this interface effect and reduces scan time, but also reduces image quality. The data may use a variable weighting of the data at the first–last view interface region reducing artifacts, but may increase noise some and result in the noise structure having a directional pattern toward this start–stop angle. This may be seen on some body scans but typically would not be used or present in head scans.

One of the obvious improvements in reducing motion artifacts is the use of shorter scan times. In early scanners it took tens of seconds to minutes to scan a single image, much less a volume through the patient. With MDCT and cone beam scanning, one can scan the entire chest within a breathhold. If one is looking at cardiac structures, however, the motion is rapid and less controllable. At times the motion may cause the image of a structure at one point in time overlaid on the image of the structure at a different point in time. This can cause artifacts that may mimic a pathology and special care is needed regarding this type of artifact. One example is imaging of the aorta as it changes diameter with the beating of the heart may create the image of a double vessel wall, which could look similar to a dissecting aneurysm (50). The electron beam CT systems were developed to provide very fast, freeze-action scans, and the use of cardiac gating to selectively collect data through the heart during defined portions of the cardiac cycle also reduce motion artifacts and blurring.

Partial Volume Artifact

Equation 12 in the discussion of the attenuation of X rays shows that a beam passing through a series of objects yields a relative transmission value that is the exponential of a sum of μs values. Taking the logarithm of this value yields the projection data that is the sum or integral of attenuation values along a line through the object. The X-ray beam that strikes a single detector is not a single infinitely slender beam, but a beam of some finite cross-sectional area. If part of the beam passes through one structure, and the other part of the beam passes through another structure, then the resulting transmission values corresponds to a sum of two or more exponentials, rather than an exponential of a sum. Taking the logarithm of this does not yield the same results. This can be seen by the example of a beam passing through a series of alternating dense and radiolucent layers will eventually get infinitesimally small. The same beam, however, having half of the beam transverse the dense material, and half the beam travel through the radiolucent material will always transmit at least 50% of the beam through the radiolucent path. This difference in attenuating material for different parts of a measured ray is commonly present along long linear boundaries. If the difference in density of the two materials is small, then there is little effect, but if there is a large difference in attenuation between the materials, such as for bone or a metal structure, then these inconsistencies result in streaks emanating as an extension of the edge (Fig. 17a).

This partial volume artifact is also produced by structures that penetrate only part way through the slice thickness, and may cause streaks between such partially penetrating structures. Partial volume streak artifacts may be reduced by reducing the slice thickness, which reduces the cross-sectional area of the measured ray. This can be especially useful in high resolution imaging of structures with bony prominences (Fig. 17b and c), or in the presence of dense metal structures (Fig. 18). The routine use of thinner detectors in MSCT systems, even when the detectors are averaged together as long as it is after the logarithmic scaling, reduces this artifact.

A related artifact is the windmill artifact due to insufficient sampling in the z axis in helical scanning. This presents as fan shaped lines emanating from edges. Increased sampling or lower pitch can reduce this artifact. Alternatively, view spacing in the axial direction may be reduced by using a flying focal spot approach that rapidly moves the position of the X-ray source in the X-ray tube. This uses an oscillating magnetic field to alter the path of the electrons and the location on the X-ray tube target where the electrons strike.

Beam Hardening Artifact

The attenuation parameters used in Eqs. 10–14 assumed a monoenergetic X-ray source yielding consistent values of μ for a given material. However, X-ray sources produce

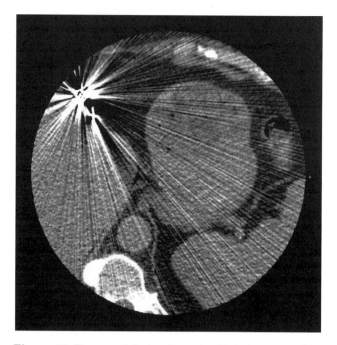

Figure 18. Dense metal structures in the body can produce pronounced streaking artifacts. Partial volume edge artifacts, beam hardening, scatter, and motion can all contribute to inconsistencies in the data causing the artifacts.

radiation with a range of photon energies up to the maximum energy of the electrons striking the X-ray tube target. This maximum photon energy corresponds to the kV or kVp voltage applied to the tube. The lower energy or soft photons are attenuated to a greater degree than the more penetrating high energy photons. Consequently, the effective energy of a beam passing through a thick object section is higher than that of a beam going through less material. This preferential transmission of the higher energy photons and the resulting increase in effective energy of the X-ray beam is referred to as beam hardening.

The linear attenuation coefficient is a function of the beam energy and changes in beam energy cause variations in the CT number for a material. Likewise, if rays passing through an object from different angles have different effective energies, then the inconsistencies can also cause artifacts. X-rays that pass through the center of a cylindrical object will have higher effective energy than those traversing through the edges. This additional attenuation and beam hardening of the central rays result in lower CT number values in the center, corresponding to the higher energy beam. This appearance of lower CT numbers in the center of an object is referred to as a cupping artifact.

Another type of beam hardening artifact occurs between two dense structures. The rays passing through two dense structures have increased beam hardening, and tissues between these structures will appear to have a lower CT value. This appears as a dark band between the structures, and may be present along with partial volume artifacts appearing as fine streaking from edges. This is often seen in head scans as dark bands between the petrous ridges (Fig. 17a).

The effect of beam hardening can be reduced by several techniques. The original EMI scanner used a constant-length water bath that resulted in a relatively uniform degree of attenuation from the center to the periphery. The addition of X-ray beam filtration reduces the soft X rays and reduces the degree of beam hardening (51). Bow-tie shaped compensating X-ray filters are often used and attenuate the beam more toward the periphery. Normalizing the data with a cylindrical object of similar size and material also reduces beam hardening as well as minimizing detector variation errors.

Beam hardening is also compensated for in the processing software. If the material being scanned is known, the measured transmission value can be empirically corrected. Difficulty occurs when the object consists of multiple materials with different effective atomic numbers, such as the presence of bone or metal in the tissue. Iterative beam hardening corrections can be used, where the initial reconstruction identifies the dense structures, and the rays through these structures are corrected for a second reconstruction. Dual energy scanning techniques can also eliminate beam-hardening effects.

Scatter Radiation

Detected scatter radiation produces a false detected signal that does not correspond to the transmitted intensity along the measured ray. Scatter is a factor in all radiographic measurements. The amount of scatter radiation detected in CT is much lower than encountered with large area radiographs due to the thin fan beam normally used in CT. Most of the scatter is directed outside of the fan beam and is not detected. The sensitivity of computed tomography, and the need for consistent data makes even the low level of scatter detected a potential problem. The amount of scatter and this problem becomes more challenging as the collimated beam width gets larger with MDCT and cone beam systems.

The scatter contribution across the detector array tends to be a slowly varying additive signal. The effect on the measured data is most significant for highly attenuated rays where the scatter signal is relatively large compared to the primary signal. The additional scattered photons detected make the materials along the measured ray appear less attenuating, in a manner similar to beam hardening. Because of the similarity of these effects, scattering artifacts are similar to and are often associated with beam hardening. Likewise, the basic beam hardening correction performed provides some degree of compensation for scatter. Using thin beams and large distances between the patient and detector can minimize scatter. The use of directional dependent detectors such as the xenon gas ionization detectors with their focused tungsten plates can also reduce the detection of scatter. As detectors get larger with cone beam CT applications, scatter is significant and antiscatter radiographic grids can be used to reduce the detected scatter. Since scatter varies slowly with distance, special reference detectors outside of the primary radiation beam can be used to measure the level of the scatter signal for more effective correction.

Cone Beam or Divergence Errors

The X-ray fan beam does not only diverge or fan out in the x–y plane, perpendicular to the rotational axis, but also a slight divergence in the z direction, across the detector row (Fig. 16). A high density structure toward the edges of the patient may be in the x-ray beam from one angle, but be missing the beam from the opposite angle. This inconsistency causes an artifact that may include diffuse streaking or smearing of the density from the edges of the object. This may also occur in a helical data acquisition that interpolates the data to produce a data set corresponding to a given slice. Alternative variations in the reconstruction process that utilize a cone beam reconstruction approach or backproject along the actual ray paths reduce this effect.

Note that a single axial rotation around the patient with a widely diverging beam with an area detector does not contain a complete set of projection data sufficient to perform a reconstruction. Mathematically, it does not fully sample the Radon space. Reconstructions can be performed with various means to estimate the missing data, with the difficulty increasing as the divergence perpendicular to the plane of rotation gets larger.

Other Systemic Errors

There are a lot of things that can happen to the signal between the X-ray source and the image. Computed tomography scanners are complex electromechanical devices that are sensitive to relatively small variations in the measurements. A number of factors can cause accuracy or inconsistency errors and result in artifacts, and considerable effort is taken by the manufacturers to minimize these problems and artifacts. These inconsistencies may be in the measured signal or may be the result of poor characterization of the spatial position or the measured ray path. Geometric errors can result in wobble of the center of rotation due to the mechanical limitations of the large turret bearing holding the rotating mechanism, or may be due to small changes in the focal spot position as the X-ray tube gets hot, or variations in the spacing or position of the individual detector elements.

Signal variations occur due to fluctuations in the X-ray tube output, and differences in response characteristics among the individual detector elements. Periodic calibration of the detectors is required and radiation sensors are used to monitor variations in the X-ray tube output. Typically the reference detectors are at the ends of the linear detector array. A large patient, or body part outside of the normal scan field of view can block the reference detector, causing an inaccurate normalization of the measurements. Alternatively, an X-ray sensor may be placed adjacent to the X-ray tube where the measurement cannot be blocked. Vibrations in components can also affect the readings, as well as a number of other factors.

Detector calibration and characterization is particularly important with rotate-rotate data acquisition systems. The ray paths for a single detector all are tangential to a circle around the center of rotation. A relatively small consistent error in a detector can accumulate to form a ring artifact in

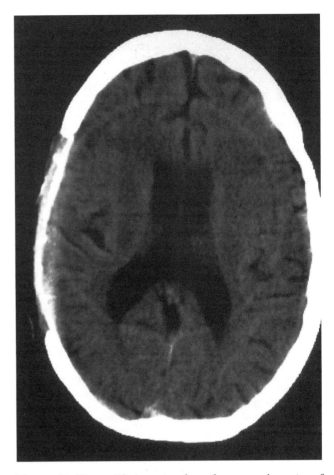

Figure 19. Ring artifacts centered on the scanner's center of rotation can be caused by the miscalibration or error in a single radiation detector on a rotate–rotate data acquisition CT scanner.

the image (Fig. 19). The commercial X-ray CT systems are designed to be as stable and consistent as practical, however characterization of the various components and software correction of the measured transmission data is an important step in clinical CT imaging.

BIBLIOGRAPHY

1. Hounsfield GN. Computed medical imaging. J Computer Assist Tomogr 1980;5:665–674.
2. Brooks R, DiChiro G. Principles of computer assisted tomography (CAT) in radiographic and radioisotopic imaging. Phys Med Biol 1976;21:689–732.
3. Radon J. Uber die bestimmung von funktionen durch ihre Integralwerte langs gewisser Mannigfaltigkeiten. Ber Verhandlung 1917;69:262–277.
4. Feldkamp LA, Davis LL, Kress JW. Practical cone-beam algorithm. J Opt Soc Am 1984;1:612–619.
5. Herman GT. Image Reconstruction from Projections: Implementation and Applications. New York: Springer-Verlag; 1979.
6. Parker JA. Image Reconstruction in Radiology. Boca Raton: CRC Press; 1990.
7. Gordon R, Bender R, Herman GT. Algebraic reconstructioni techniques (ART) for three-dimensional electron microscopy and x-ray photography. J Theor Biol 1970;29:471–481.

8. Bracewell RN. Strip integration in radio astronomy. Aust J Phys 1956;9:198–217.

9. DeRosier DJ, Klug A. Reconstruction of three-dimensional structures from electron micrographs. Nature London 1968; 217:130–134.

10. Kuhl DE, Edwards RQ. Image separation radioisotope scanning. Radiology 1963;80:653–662.

11. Cormack AM. Representation of a function by its line integrals with some radiological applications. J Appl Phys 1963;34:2722–2727.

12. Hounsfield GN. Computerized transverse axial scanning tomography: Part I: Description of system. Br J Radiol 1973;46:1016–1022.

13. Dennis MJ. Industrial Computed Tomography. Metals Handbook. Metals Park, (OH): ASM International; 1989. p 358–386.

14. Boyd DB, Lipton MJ. Cardiac computed tomography. Proc IEEE 1983;71:298–307.

15. Ritman EL, Robb RA, Harris LD. Imaging physiological functions: Experience with the dynamic spatial reconstructor Philadelphia: Praeger; 1985.

16. Mori I. Computerized tomographic apparatus utilizing a radiation source. US Patent 4,630,202. 1986.

17. Nishimura H, Miyazaki O. CT system for specially scanning subject on a moveable bed synchronized to x-ray tube revolution. US Patent 4,789,929. 1988.

18. Kalender WA, Klotz W, Vock E. Spiral volumetric CT with single breath-hold technique, continuous transport, and continuous scanner rotation. Radiology 1990;176:181–183.

19. Kalender WA. Computed Tomography: Fundamentals, System Technology, Image Quality, Applications. Munich: Publicis MCD Verlag; 2000.

20. Hu H. Multi-slice helical CT: Scan and reconstruction. Med Phys 1999;26(1):5–18.

21. Fishman EK, Jeffrey RB, editors. Multidetector CT: Principles, techniques, and clinical applications. Philadelphia: Lippincott, Williams & Wilkins; 2003. p 560.

22. Yester MW, Barnes GT. Geometrical limitations of computed tomography (CT) scanner resolution. Appl Opt Instr Med VI Proc SPIE 1977;127:296–303.

23. IEC. International Electrotechnical Commission: Medical electrical equipment–60601 Part 2-44: Particular requirements for the safety of X-ray equipment for computed tomography. Geneva, Switzerland: 1999.

24. Kachelriess M, Kalender WA. ECG-correlated image reconstruction from subsecond spiral CT scans of the heart. Med Phys 1998;25(12):2417–2431.

25. Cline HE, et al. Two algorithms for the three-dimensional reconstruction of tomograms. Med Phys 1988;15(3):320–327.

26. Schoepf UJ, et al. Multislice CT angiography. Eur Radiol 2003;13(8):1946–1961.

27. de Feyter PJ, Kresin GP, editors. Computed Tomography of the Coronary Arteries. New York: Taylor & Francis Group; 2004. p 208.

28. Vrtiska TJ, Fletcher JG, McCollough CH. State-of-the-art imaging with 64-channel multidetector CT angiography. Perspect Vasc Surg Endovasc Ther 2005;17(1):3–10.

29. Ulzheimer S, Kalender WA. Assessment of calcium scoring performance in cardiac computed tomography. Eur Radiol 2003;13(3):484–497.

30. Kalender WA, Klotz W, Suss C. Vertebral bone mineral analysis: an integrated approach with CT. Radiology 1987;164: 419–423.

31. Lang TF, et al. Assessment of vertebral bone mineral density using volumetric quantitative CT. J Computer Assisted Tomogr 1999;23(1):130–137.

32. Braillon PM. Quantitative computed tomography precision and accuracy for long-term follow-up of bone mineral density measurements: a five year in vitro assessment. J Clin Densitom 2002;5(3):259–266.

33. Alvarez RE, Macovski A. Energy selective reconstructions in x-ray computerized tomography. Phys Med Biol 1976;21: 733–744.

34. Vogel WV, et al. PET/CT: Panacea, redundancy, or something in between? J Nucl Med 2004;45(Suppl 1): 15S–24S.

35. Bockisch A, et al. Positron emission tomography/computed tomography—imaging protocols, artifacts and pitfalls. Mol Im Biol 2004;6(4):188–199.

36. Limitation of Exposure to Ionizing Radiation. NCRP Report No. 91. National Council on Radiation Protection; 1993.

37. Morin RL, Gerber TC, McCollough CH. Physics and dosimetry in computed tomography. Cardiol Clinics 2003;21(4):515–520.

38. McCollough CH, Schueler BA. Calculation of effective dose. Med Phys 2000;27(5):828–837.

39. Barrett HH, Swindell W. Radiological Imaging: The Theory of Image Formation, Detection, and Processing. New York: Academic Press; 1981.

40. Greess H, et al. Dose reduction in computed tomography by attenuation based on-line modulation of tube current: Evaluation of six anatomical regions. Eur Radiol 2000;10(2):391–394.

41. Kalender WA, et al. Dose reduction in CT by on-line tube current control: principles and validation on phantoms and cadavers. Eur Radiol 1999;9(2):323–328.

42. Kaczmarz S. Angenaherte Auflosung von Systemen linearer Gleichungen. Bull Acad Polonaise Sci Lett 1937;A35:355–357.

43. Grand DG. Tomosynthesis: A three-dimensional radiographic imaging technique. IEEE Trans Biomed Eng 1972;BME-19(1): 20–28.

44. Ramachandran GN, Lakshminarayanan. Three-dimensional reconstruction from radiographs and electron micrographs: III. Description and application of the convolution method. Indian J Pure Appl Phys 1971;9:997.

45. Shepp LA, Logan BF. The Fourier reconstruction of a head section. Trans IEEE 1974;NS-21:21–43.

46. McCollough CH, et al. The phantom portion of the American College of Radiology (ACR) Computed Tomography (CT) accreditation program: Practical tips, artifact examples, and pitfalls to avoid. Med Phys 2004;31(9):2423–2442.

47. Blumenfeld SM, Glover G. Spatial resolution in computed tomography. In: Newton TH, Potts DG, editors. Radiology of the Skull and Brain, Vol 5, Technical Aspects of Computed Tomography. New York: C.V. Mosby Company; 1981.

48. Joseph PM. Image noise and smoothing in computed tomography (CT) scanners. Opt Eng 1978;17:396–399.

49. Joseph PM. Artifacts in computed tomography. Phys Med Biol 1978;23:1176–1182.

50. Parry CK, Rajagopalan B. Characterization of artifact simulating aortic dissection in computed tomography imaging. J Digital Imaging 2001;14(2 Suppl. 1):220–221.

51. McDavid WD, et al. Spectral effects on three-dimensional reconstruction from x-rays. Med Phys 1975;2(6):321–324.

See also IONIZING RADIATION, BIOLOGICAL EFFECTS OF; MAGNETIC RESONANCE IMAGING; ULTRASONIC IMAGING.

COMPUTED TOMOGRAPHY SCREENING

DAVID J. BRENNER
Columbia University Medical Center
New York, New York

INTRODUCTION

Computed tomography (CT), developed by Hounsfield and colleagues in the 1970s (1,2) has revolutionized much of

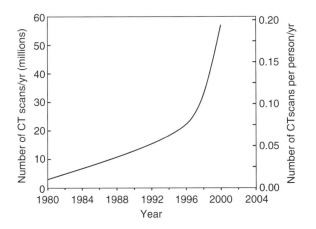

Figure 1. Graph shows the increase in the estimated number of CT scans performed in the United States between 1980 and 2000. (Based on data in Refs. 86–89.)

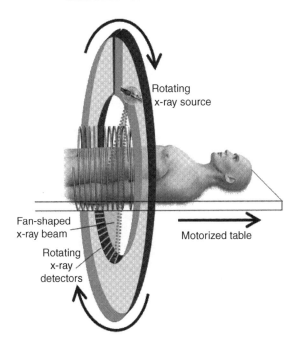

Figure 2. Schematic of helical (spiral) CT scanning. Both the X-ray source and, on the other side of the patient, the X-ray detectors, rotate around the patient. If the table were not moving, a single slice of the patient would be imaged (axial CT). Because the table is moving at the same time as the source–detector combination is rotating, the result is a helical or spiral CT scan of the patient, as depicted here. Shown in this schematic is a single row of detectors; modern multidetector scanners have several rows of detectors alongside each other, which allow both for thinner slice widths, and shorter scan times.

medical imaging by allowing a three dimensional (3D) view of the organ or part of the body of interest. Since its inception, the use of CT has increased very rapidly, both in the United States and other countries. At present, ~60 million CT scans are being performed each year in the United States. As seen in Fig. 1, this increase has occurred roughly over 20 years. It has largely been driven by the major technical advances in CT technology, in particular the development of helical multidetector CT scanners, as discussed below, which allow CT scans to be made in 1 s or less.

The basic principle of helical, or spiral, CT scanning is shown in Fig. 2. Essentially, the patient is moved through a continuously rotating X-ray source–detector combination. A more modern version is the multidetector CT, which gives the advantage of short scan times, coupled with potentially very thin slice widths. The result is a series of many images of "slices" of the organ or part of the body of interest, which can then be combined by computer-based mathematical techniques, to provide 3D views.

The use of CT for mass screening is a more recent innovation, driven in part by the increased availability and convenience of CT scans. Three applications, each of which will be discussed, have been suggested for CT-based screening: for early stage lung cancer in smokers and exsmokers; for lesions in the colon (virtual colonoscopy); for general screening for many diseases in the whole body (full-body screening).

All three modalities, as of 2005, are quite new, and a general consensus has not yet been reached about the efficacy of any of them. The general issues regarding efficacy of these new modalities are, in essence, the same as for all other potential mass screening modalities, for example, mammography, pap smear screening, and colonoscopy. However, as discussed, there is an added issue for CT-based screening modalities, namely, the significant X-ray radiation exposures involved.

The more general issues of screening relate to (a) whether the screening modality truly produces a stage shift (i.e., allows detection of more early-stage cancers and less late-stage cancers), (b) whether the screening

modality produces overdiagnosis (identifying lesions that the individual would die with, rather than die of); and (c) false positives, the possibility of mistakenly identifying a tumor, with the attendant possibility of subsequent unnecessary procedures. All these issues feed in to the general question of whether the overall mortality rate from the disease in question is significantly reduced by the screening test.

By contrast, the radiation exposure issues that relate to CT-based mass screening are unique. It is, of course, true that mammography also involves the use of X rays, but, as we will discuss, the radiation doses involved are generally much higher for CT-based screening compared to mammography. Thus the potential benefits of any CT-based screening procedure must, in addition to the more general efficacy issues discussed above, have to significantly outweigh any potential harm from repeated low dose X-ray exposures. In the next section, what is currently known about the hazards of low doses of X rays is reviewed.

CANCER RISKS ASSOCIATED WITH EXPOSURE TO LOW X-RAY DOSES

Some typical radiation doses associated with common radiological examinations are shown in Table 1. The biological effects of low dose X-ray exposures have been investigated and debated for more than a century (3). There is little

Table 1. Typical Organ Doses from Various Radiological Examinations[a]

Examination	Relevant Organ	Relevant Organ Dose, mSv
Dental X ray	Brain	0.005
PA Chest X ray	Lung	0.01
Lateral chest X ray	Lung	0.15
Screening mammogram	Breast	3
Adult abdominal CT	Stomach	10
Neonate abdominal CT	Stomach	25

[a]Radiation dose, a measure of ionizing energy absorbed per unit mass, has units of Gy (gray) or mGy ($1\,Gy = 1\,J\cdot kg^{-1}$); it is often quoted as an equivalent dose, in units of Sv (Sievert) or mSv. For X rays, which are the radiations produced in CT scanners, $1\,mSv = 1\,mGy$.

question that intermediate and high doses of ionizing radiation, say above 100 mSv, given acutely or over a prolonged period, produce deleterious effects in humans, the most significant being cancer induction (4). At lower doses, however, the situation is less clear. Compared to higher doses, the risks associated with low doses of radiation are lower, and progressively larger epidemiological studies are required to quantify the cancer risk to a useful level of precision. The reason is, as the dose goes down, the signal (radiation risk) to noise (natural background risk) ratio decreases.

Most of the quantitative information that we have regarding radiation-induced cancer risks comes from studies of A-bomb survivors. A-bomb survivor cohorts are generally used as the basis for predicting radiation-related risks to a general population because (a) they are the most thoroughly studied (over many decades) large exposed population; (b) the cohorts are not selected for disease; (c) all age groups are covered; and (d) a substantial subcohort of ~25,000 survivors, typically those who were ~2–3 km from the explosion hypocenters (5), received radiation doses comparable to those of concern here.

Key questions here are as follows: What is the lowest dose of X rays for which there is convincing evidence of significantly elevated cancer risks in humans? What is the most appropriate way to extrapolate these risks to still lower doses? What is the dependence of cancer risks on age at exposure? These issues have recently been extensively reviewed (3).

Effects of Radiation Dose on Cancer Risk

In summary, there is good epidemiological evidence of increased cancer risk for children exposed to an acute dose of 10 mSv (or higher), and for adults exposed to acute doses of 50 mSv (or higher) (3). As discussed below, relevant organ doses for CT exams are of the order of 15 mSv or less.

Extrapolation of Risks to Lower Radiation Doses

The issue here is how to estimate risks at doses somewhat (though not a great deal) lower than those for which there is statistically significant evidence of increased cancer risks. The current consensus (6) is that the measured risks can reasonably be linearly extrapolated to somewhat lower

doses, though as the dose of interest becomes progressively lower, the uncertainties inherent in this extrapolation become progressively greater. Relatively small extrapolations from epidemiological data are required (e.g., from 50 down to 15 mSv), however, to estimate cancer risks at the doses relevant to single CT examinations.

Effect of Dose Fractionation

If individuals receive multiple CT screenings over a period of years, the radiation dose will, of course, increase proportionately. The most likely case is that any radiation risks will also increase proportionately. Specifically, at high doses, theory (7), animal data (8), and epidemiological data (9), suggest that fractionating a radiation exposure decreases the overall risk at a given dose, but at the low doses of relevance here, both theory (7) and animal data (8) suggest that the risks are roughly independent of fractionation.

Effect of Age at Exposure

Regarding age at exposure, as can be seen in Fig. 3, radiation risks generally decrease markedly with age. The reason is because (a) sensitivity is related to the proportion of dividing cells in an organ, which decreases with increasing age; and (b) other competing risks play an increasing role with increasing age.

RADIATION DOSES FROM SCREENING CT

The radiation dose from CT scans depends on a number of factors: The most important are the tube current; the scan time; the pitch [For helical CT scans, the speed that the patient table moves relative to the rotation speed of the X-ray tubes–detectors will be an important determinant of the radiation dose; it is defined through the pitch, which is the linear table motion feed per 360° rotation, divided by the total beam width (the slice width × the number of detectors).]; the tube voltage; the number of detectors; and the particular scanner design (10). For a given CT scanner operating at a given voltage, the organ dose is

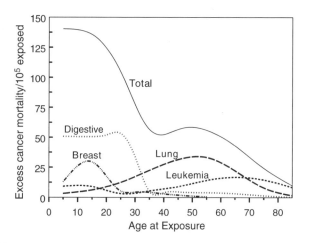

Figure 3. Estimated radiation-related absolute cancer mortality risk per 10^5 individuals in the United States exposed at different ages to a whole-body dose of 10 mSv (63).

proportional to the mAs [current (mA) × rotation time], and is inversely proportional to the pitch. It is always the case, however, that the relative noise in CT images will increase as the radiation dose decreases; thus there will always be a tradeoff between the need for low noise images and the desirability of using low radiation doses (11). As discussed later, the amount of noise that can be tolerated depends very much on the application. Thus, for example, more noise (i.e., lower doses) can probably be tolerated for virtual colonoscopy because of the radiological contrast of colonic polyps projecting into an air filled lumen), compared with whole-body screening (where most of the potential lesions show less radiologic contrast).

A relatively new and very promising radiation dose reduction technique for CT is automatic tube current modulation (Fig. 4) (12–15), now available from all the major scanner manufacturers: these systems continuously lower or raise the X-ray tube current to compensate for different instantaneous levels of attenuation of the X-ray beam by the patient. For example, when the beam is aimed in the posterior–anterior direction, fewer X rays are needed (for the same image quality) compared to the lateral–medial direction; or when the beam is passing through the region of the transverse colon, fewer X rays are needed compared to the pelvic bone region.

COMPUTED TOMOGRAPHY COLONOGRAPHY (VIRTUAL COLONOSCOPY)

There is no doubt (a) that colonoscopy-driven polypectomy can result in a significantly decreased incidence of colorectal cancer (16,17), and (b) that there is suboptimal compliance with current guidelines for colorectal cancer screening (18,19). Screening using CT colonography, often referred to as "virtual colonoscopy" (VC), was first suggested in 1983 (20), but has only recently become a potential option for mass screening (21–23).

In the most common current usage of VC, after bowel preparation, the colon is inflated with air or CO_2, and the colon is CT scanned. The resulting data can then be analyzed for polyps, based on two-dimensional (2D) images, or using a 3D endoluminal view. Virtual colonoscopy is an excellent application of CT because of the radiological contrast exhibited by colonic polyps projecting into a gas-filled lumen (20,21,24), and a National CT colonography trial is underway in the United States.

Virtual colonoscopy may well have the potential to increase colorectal cancer screening compliance, largely because of the possibility that it can be performed with noncathartic preexamination bowel preparation. Current compliance with screening guidelines is clearly poor: At most, about one-third of adults >50 in the United States have had an endoscopic examination within the past 10 years (18,19).

From a technological perspective, VC is not quite ready for use in mass-screening programs. The three main outstanding issues, all of which seem relatively close to solution, are as follows:

1. The sensitivity and specificity of VC for detecting lesions in the size range from 5 to 10 mm: VC sensitivity and specificity for lesions >10 mm in diameter are generally well over 90% (about as good as those for conventional optical colonoscopy) (25). There is evidence that a well-designed VC screening program can achieve at least 90% sensitivity and specificity in the size category from 7 to 10 mm (22,26), but not all studies have achieved this (27).

2. The use of noncathartic preexamination bowel preparation regimens: In general it may be less the invasive nature of conventional colonoscopy that results in poor compliance, but more the necessity for cathartic bowel preparation (28–32). Virtual colonoscopy offers the potential for noncathartic bowel preparation, through the use of barium or iodinated tagging agents, which impart a high density to both stool and residual fluid, allowing increased contrast with soft-tissue polyps. Recent results with noncathartic VC have been very encouraging (23,33–35).

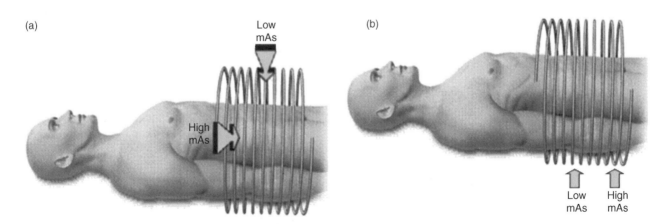

Figure 4. Principles of automatic tube current modulation: (a) Angular modulation, where the X-ray tube current is lowered as the X rays are aimed in the anterior–posterior directions, and increased when the X rays are aimed in the lateral–medial directions, when there will be more X-ray attenuation. (b) z-axis modulation where, for example, fewer X rays are required in the abdominal region superior to the pelvic bones, compared with the pelvic region.

3. Optimization and standardization of CT parameters: Just as mammographic examinations are now well standardized (36) and regulated (37), so VC should be optimized and standardized, if it is to be used for mass screening. Particularly until Points 1 and 2 are settled, it is probably premature to consider standardizing VC scanner parameters.

If VC were to become a standard screening tool for the >50s, the potential "market" in the United States will soon be >100 million people. Even if the recommended VC frequency were to be that currently recommended for optical colonoscopy (every decade), this would imply that several million VC scans might be performed each year. Should the relative simplicity of the VC tests result in the recommended examination frequency being increased, then several tens of millions of these VC scans might be expected to be performed in the United States each year. It is pertinent, therefore to consider the radiation exposure and any potential radiation risk to the population from such a mass screening program.

Because of the advantageous geometry of a VC scan, the dose–noise tradeoff can be very much weighted toward low dose, higher noise images (24,25,38–40). Several studies have come to the conclusion that more noise (and thus a lower dose) can be accepted in a VC scan compared to other CT scans, while still maintaining sensitivity and specificity, at least for polyps greater than ~7 mm in diameter (25,26,38,41,42).

It is important to note that, in general, paired VC exams are given, one in the supine and one in the prone position. Several studies have suggested that this technique improves colonic distention (43–45), decreasing the number of collapsed colonic segments.

Table 2 (46) shows estimated organ doses for one of the more common CT scanners (GE LightSpeed Ultra). The scanner parameters were taken from a recent Mayo Clinic study by Johnson and co-workers (26), and are toward the low dose end of published VC protocols (38). To provide an estimate of scanner-to-scanner dose variations, Table 3 (46) shows the radiation dose to the colon estimated for five of the more common CT scanners in use today, using identical scanner parameters in each case; the coefficient of variation of the dose to the colon is ~20%.

It can be seen from Table 2 that typical organ doses are <20 mSv, even for organs directly in the X-ray beam, for example, the colon, stomach, bladder, and kidneys. The subcohort of ~25,000 A-bomb survivors (5) that received comparable radiation doses (A-bomb dose range 5–50 mSv, mean 20 mSv) does show a slight increase in cancer mortality compared to the control population (4), but this increase is of marginal statistical significance ($p = 0.15$). It is also pertinent to point out that this A-bomb subcohort consists of individuals covering all age groups, and thus it is reasonable to assert that there is no direct statistically significant evidence from A-bomb survivor data that a single VC exam increases cancer risks in adults. It does not follow, of course, that the radiation risk is zero, rather that it is likely to be small. It also follows that there is persuasive evidence that a *series* of virtual colonoscopy examinations would result in increase in cancer risk due to the radiation exposure: The issue being how much of an

Table 2. Typical Organ Doses and Estimated Additional Absolute Lifetime Cancer Risks Associated with a Paired VC Screening Examination of a Healthy 50 Year Old[a,b]

	Organ Dose from Paired CTC Scans,[b] mSv	Additional Absolute Lifetime Cancer Risk because of Paired CTC Scans at Age 50, %
Colon (male)	13.2	0.044
Colon (female)	13.2	0.038
Bladder (male)	16	0.025
Bladder (female)	16	0.016
Stomach (male)	14.8	0.013
Stomach (female)	14.8	0.031
Kidney (male)	16.1	0.012
Kidney (female)	16.1	0.017
Liver (male)	13.8	0.016
Liver (female)	13.8	0.005
Leukemia (male)	6.6	0.032
Leukemia (female)	6.6	0.018
Lung (male)	2.2	0.006
Lung (female)	2.2	0.008
Total (male)		*0.15*
Total (female)		*0.13*

[a]See Ref. 46.
[b]Paired VC examinations at 65 mAs, 120 kVp, 10 mm collimation, pitch 1.35.

increase and how it compares with the potential benefit of the VC screening.

Table 2 also shows the estimated absolute lifetime cancer risks associated with the radiation exposure from paired VC scans in a 50 year old (46). As expected, the main organs at risk are the colon, stomach, and bladder, as well as the leukemic cancers. All the estimated absolute radiation risks are relatively small, the largest being <0.05% (1 in 2000). Summed over all the organs at risk, the estimated absolute lifetime risk of cancer induction from a pair of VC scans (with the scanner parameters from Table 2) in a 50 year old is ~0.14%, ~1 in 700. Estimated risks for cancer mortality would, of course, be considerably less.

Several points need to be considered regarding the estimated risks in Table 2:

1. The risks are highly dependent on the scanner settings used, particularly the mAs and the pitch. The settings used in Table 2 are on the low dose side of

Table 3. Estimated Colon Doses from Paired VC Scans Using the Same Machine Settings with Different CT Scanners[a]

Scanner	Colon Dose from Paired CTC Scans,[b] mSv
GE LightSpeed Ultra	13.2
GE QX/I, LightSpeed, LightSpeed Plus	11.6
Phillips Mx8000	9.0
Siemens Volume Zoom, Access	8.6
Siemens Sensation 16	7.6

[a]Ref. 46.
[b]Paired VC examinations at 65 mAs, 120 kVp, 10 mm collimation, pitch 1.35.

those used in current reported studies (38), but there is good evidence (23,26,40) suggesting that the mAs and thus the dose could be decreased further, by at least a factor of 5 (and perhaps as much as a factor of 10) from these settings, while still maintaining sensitivity and specificity for polyps larger than ~5 mm. Still further reductions of up to 50% in VC doses may be possible through the use of automatic tube current modulation (Fig. 4) (14,15), now available from all the major CT scanner manufacturers (14).

2. The estimated absolute cancer risks are highly age dependent. Thus, for example, the estimated radiation-associated absolute lifetime risk for colon cancer induction decreases from 0.044% for a VC scan at age 50 to 0.022% for a scan at age 70.

3. There are quantifiable uncertainties involved in the radiation risk estimates shown in Table 2. The largest is the uncertainties in "transferring" risk estimates from a Japanese population to a U.S. population, but there are also uncertainties associated with the extrapolation of risks from somewhat higher doses, where the risks are statistically significant, and uncertainties associated with the reconstructed dosimetry estimates at Hiroshima and Nagasaki (47). Based on Monte Carlo simulations of the various uncertainties (48), the upper and lower 90% confidence limits of the radiation risk estimates are about a factor of 3 higher and lower, respectively, than the point estimates.

In summary, because the geometry for VC is highly advantageous (soft-tissue polyps projecting into an air or CO_2 filled lumen), it can be performed using lower radiation doses than almost any other CT examination. The cancer risks associated with the radiation exposure from VC are unlikely to be zero, but they are small. A best estimate for the absolute lifetime cancer risk associated with the radiation exposure using typical current scanner techniques is ~0.14% for paired VC scans for a 50 year old, and about one-half of that for a 70 year old. These values could probably be reduced by factors of 5 or 10, with optimized protocols. Thus it seems clear that, in terms of the radiation exposure, the benefit/risk ratio is potentially large for VC.

LOW DOSE CT SCREENING FOR EARLY STAGE LUNG CANCER IN SMOKERS AND EXSMOKERS

Lung cancer is the number one cancer killer in the United States. Thus, there is increasing interest in the possibility of using low dose CT scans for annual screening of smokers and former smokers for early-stage lung cancer. In part, this is the result of the failure of earlier attempts to screen this population with conventional chest X rays (49). The logic is that these earlier screening modalities failed because of their inability to detect sufficiently small (typically <10 mm) lesions—and low dose lung CT has been demonstrated to have a greater sensitivity for detecting small pulmonary lesions (50). A National Lung Cancer Screening Trial is now underway (51).

As with virtual colonoscopy, the geometry for lung CT is quite advantageous, and this allows the use of a relatively low dose (i.e., noisier) image, while still maintaining good sensitivity for detecting small pulmonary lesions (52).

The potential mortality benefits of lung cancer screening have been much debated (53–56), and it is fair to say that, at the very earliest, the issue will not be resolved until the completion of the National Lung Cancer Screening Trial in 2009. Several relatively small pilot studies have already taken place (50,56–60), suggesting that low dose lung CT does have the potential for detecting more early stage tumors than other lung screening modalities. Whether this represents a meaningful shift toward earlier detection of potentially fatal tumors, or whether it is largely associated with an overdiagnosis of nonfatal lesions, is yet to be established—as have the potential risks of invasive procedures resulting from false positives (61).

Less attention has been paid to the potential radiation risks, specifically radiation-induced lung cancer, associated with radiation from these CT scans. In part, this is because the screening technique involves "low dose", rather than standard, CT lung scans, and in part this is because excess relative risks of radiation-induced cancer generally decrease markedly with increasing age (62).

There are, however, several indications that the radiation risk to the lung associated with this screening technique may not be insignificant:

1. Cancer risks from radiation are generally multiplicative of the background cancer risk (63), which is, by definition, high for lung cancer in the target population here; this general observation has been born out in terms of the interaction between radiation and smoking, which most authors have suggested is near-multiplicative (64–70) although an intermediate interaction between additive and multiplicative has also been suggested for radon exposure (71), and there is one report of an additive interaction (72).

2. As shown in Fig. 3, while radiation-related cancer risks generally decrease markedly with increasing age at exposure, radiation-induced lung cancer does not apparently show this decrease in risk with increasing age (62,63).

These considerations suggest that risk of radiation-induced lung cancer associated with the radiation from repeated low dose CT scans of the lung in smokers may not be negligible. A recent estimate (73) suggests that a 50 year old smoker planning an annual lung screening CT would incur an estimated radiation-related lifetime lung-cancer risk of 0.5%, in addition to their otherwise expected lung cancer risk of ~14% (the radiation-associated cancer risk to any other organ is far lower). If 50% of the ever-smoking current U.S. population aged between 50 and 75 received annual CT screens, the estimated number of lung cancers associated with the radiation from these scans would be ~14,000. These estimated risks set a baseline of benefit that annual CT screening must substantially exceed. This risk–benefit analysis suggests that mortality benefits from annual CT screening of considerably $> 3\%$ would be necessary to outweigh the potential radiation risks (73).

FULL-BODY CT SCREENING

There is increasing interest, particularly from independent radiology clinics, in the use of full-body CT screening of healthy adults (74–76). The technique is intended to be an early detection device for a variety of diseases including lung cancer, coronary artery disease, and colon cancer. At present, the evidence for the utility of this technique is anecdotal, and there is considerable controversy (77–79) regarding its efficacy: to date, no studies have yet been reported indicating a life-prolonging benefit (80). Because of the nature of the scan, the false positive rate is expected to be high (81), and a small study on full-body CT screening (80) found that 37% of those screened were recommended for further evaluation, whereas the overall evaluable disease prevalence is probably $\sim 2\%$ (79). Other estimates suggest that the false positive rate may be as high as 90% (79).

Another aspect that is important in assessing the technique is the potential risk from the radiation exposure associated with full-body CT scans. Typical doses from a single full-body scan are ~ 9 mGy to the lung, 8 mGy to the digestive organs, and 6 mGy to the bone marrow (82). The effective dose, which is a weighted average of doses to all organs (83), is ~ 7 mSv. If, for example, five such scans were undertaken in a lifetime, the effective dose would be ~ 35 mSv, that is, five times larger. Note that even with the same settings different scanners will produce somewhat different organ doses. In particular, the estimated dose to the lung is 8.9 mGy for the Siemens scanner, 9.2 mGy for the Philips, and 12.2 mGy for the GE scanner (82). To put these doses in perspective, a typical screening mammogram (see Table 1) produces a dose of ~ 2.6 mGy to the breast (36), with a corresponding effective dose of ~ 0.13 mSv (a factor of ~ 50 times less).

It is important to note that these CT doses and the corresponding risk estimates are based on a particular published protocol (84). Even for the same CT settings, different scanners will produce different doses and therefore risks (varying by up to 50%). Full-body scan protocols are by no means standardized at this time, and larger mAs settings will result in correspondingly larger doses and therefore larger risks.

The estimated lifetime cancer mortality risks from a single full-body scan are $\sim 4.5 \times 10^{-4}$ (~ 1 in 2200) for a 45 year old, and $\sim 3.3 \times 10^{-4}$ (~ 1 in 3000) for a 65 year old. To put these values in perspective, the odds of an individual dying in a traffic accident in the United States during the single year 1999 were ~ 1 in 5900 (85).

Of course, there is uncertainty in the radiation risk estimate: It is estimated (82) that the 95% credibility limits for the radiation risk estimate are about a factor of 3.2 in either direction: thus the lifetime risk from a full-body scan to a 45 year old could be as low as 1.4×10^{-4} or as high as 1.4×10^{-3}. The dominant potential radiation-induced cancer is of the lung; this is not unexpected because, as illustrated in Fig. 3, while radiation-related cancer risks generally decrease markedly with increasing age at exposure, radiation-induced lung cancer does not apparently show this decrease in risk until approximately age 55 (62,63,82).

The risk estimates for multiple scans, which would be necessary if full-body CT screening was to become a useful screening tool, are correspondingly larger. For example, a 45 year old who plans on undergoing 10 three-yearly full-body scans would potentially accrue an estimated lifetime cancer mortality risk of 0.33% (~ 1 in 300) (82). Again to give a comparison risk, this is comparable to the lifetime risk that a healthy 45 year old faces of dying of a brain tumor.

CONCLUSIONS

The increased availability and ease of use of CT scanners makes them attractive options for screening. The three CT screening modalities discussed here, virtual colonoscopy, lung screening, and full-body screening, are all comparatively new, and none have yet undergone definitive trials to assess their efficacy. However, both virtual colonoscopy, and low dose CT lung screening, but not full body screening, are currently undergoing national clinical trials in the United States and elsewhere. These trials should provide insight into the efficacy of virtual colonoscopy and low dose CT lung screening in terms of the potential mortality gain and the false positive rate. The trials will not, however, be able to assess the potential radiation risks, because of the long latency period between radiation exposure and development of a clinically recognizable malignancy. Nevertheless, because of the nontrivial doses associated with CT screening, the potential radiation risks will need to be factored into the overall risk–benefit analysis.

BIBLIOGRAPHY

1. Hounsfield GN. The E.M.I. scanner. Proc R Soc London B Biol Sci 1977;195:281–289.
2. Hounsfield GN. Computerized transverse axial scanning (tomography). 1. Description of system. Br J Radiol 1973;46: 1016–1022.
3. Brenner DJ, et al. Cancer risks attributable to low doses of ionizing radiation: assessing what we really know. Proc Natl Acad Sci USA 2003;100:13761–13766.
4. Preston DL, et al. Studies of mortality of atomic bomb survivors. Report 13: Solid cancer and noncancer disease mortality: 1950–1997. Radiat Res 2003;160:381–407.
5. Preston DL, et al. Effect of recent changes in atomic bomb survivor dosimetry on cancer mortality risk estimates. Radiat Res 2004;162:377–389.
6. NCRP, Evaluation of the linear-nonthreshold dose-response model for ionizing radiation, Report No. 136, NCRP; 2001.
7. NCRP. Influence of dose and its distribution in time on dose-response relationships for low-LET radiations, in: NCRP Report No. 64, National Council on Radiation Protection and Measurements, Washington (DC); 1980.
8. Ullrich RL, Jernigan MC, Satterfield LC, Bowles ND. Radiation carcinogenesis: time-dose relationships. Radiat Res 1987; 111: 179–184.
9. Howe GR. Lung cancer mortality between 1950 and 1987 after exposure to fractionated moderate-dose-rate ionizing

radiation in the Canadian fluoroscopy cohort study and a comparison with lung cancer mortality in the Atomic Bomb survivors study. Radiat Res 1995;142:295–304.

10. McNitt-Gray MF. AAPM/RSNA Physics Tutorial for Residents: Topics in CT. Radiation dose in CT. Radiographics 2002;22:1541–1553.

11. Martin CJ, Sutton DG, Sharp PF. Balancing patient dose and image quality. Appl Radiat Isot 1999;50:1–19.

12. Lehmann KJ, Wild J, Georgi M. Clinical use of software-controlled x-ray tube modulation with "Smart-Scan" in spiral CT. Aktuelle Radiol 1997;7:156–158.

13. Hundt W, et al. Dose reduction in multislice computed tomography. J Comput Assist Tomogr 2005;29:140–147.

14. Keat N. CT scanner automatic exposure control systems. Medicines and Healthcare Products Regulatory Agency, Report 05016, London; 2005.

15. Kalra MK, et al. Techniques and applications of automatic tube current modulation for CT. Radiology 2004;233:649–657.

16. Winawer SJ, et al. Prevention of colorectal cancer by colonoscopic polypectomy. The National Polyp Study Workgroup. N Engl J Med 1993;329:1977–1981.

17. Citarda F, et al. Efficacy in standard clinical practice of colonoscopic polypectomy in reducing colorectal cancer incidence. Gut 2001;48:812–815.

18. Seeff LC, et al. Patterns and predictors of colorectal cancer test use in the adult U.S. population. Cancer 2004;100:2093–2103.

19. Subramanian S, Amonkar MM, Hunt TL. Use of colonoscopy for colorectal cancer screening: evidence from the 2000 national health interview survey. Cancer Epidemiol Biomarkers Prev 2005;14:409–416.

20. Coin CG, et al. Computerized radiology of the colon: a potential screening technique. Comput Radiol 1983;7:215–221.

21. Hara AK, et al. Detection of colorectal polyps by computed tomographic colography: feasibility of a novel technique. Gastroenterology 1996;110:284–290.

22. Pickhardt PJ, et al. Computed tomographic virtual colonoscopy to screen for colorectal neoplasia in asymptomatic adults. N Engl J Med 2003;349:2191–2200.

23. Iannaccone R, et al. Computed tomographic colonography without cathartic preparation for the detection of colorectal polyps. Gastroenterology 2004;127:1300–1311.

24. Hara AK, et al. Reducing data size and radiation dose for CT colonography. AJR Am J Roentgenol 1997;168:1181–1184.

25. Macari M, et al. Colorectal neoplasms: prospective comparison of thin-section low-dose multi-detector row CT colonography and conventional colonoscopy for detection. Radiology 2002;224:383–392.

26. Johnson KT, et al. CT colonography: determination of optimal CT technique using a novel colon phantom. Abdom Imaging 2004;29:173–176.

27. Cotton PB, et al. Computed tomographic colonography (virtual colonoscopy): a multicenter comparison with standard colonoscopy for detection of colorectal neoplasia. JAMA 2004;291:1713–1719.

28. Weitzman ER, Zapka J, Estabrook B, Goins KV. Risk and reluctance: understanding impediments to colorectal cancer screening. Prev Med 2001;32:502–513.

29. Ristvedt SL, McFarland EG, Weinstock LB, Thyssen EP. Patient preferences for CT colonography, conventional colonoscopy, and bowel preparation. Am J Gastroenterol 2003;98:578–585.

30. Akerkar GA, Yee J, Hung R, McQuaid K. Patient experience and preferences toward colon cancer screening: a comparison of virtual colonoscopy and conventional colonoscopy. Gastrointest Endosc 2001;54:310–315.

31. Harewood GC, Wiersema MJ, Melton LJ, 3rd. A prospective, controlled assessment of factors influencing acceptance of screening colonoscopy. Am J Gastroenterol 2002;97:3186–3194.

32. Gluecker TM, et al. Colorectal cancer screening with CT colonography, colonoscopy, and double-contrast barium enema examination: prospective assessment of patient perceptions and preferences. Radiology 2003;227:378–384.

33. Callstrom MR, et al. CT colonography without cathartic preparation: feasibility study. Radiology 2001;219:693–698.

34. Lefere PA, et al. Dietary fecal tagging as a cleansing method before CT colonography: initial results polyp detection and patient acceptance. Radiology 2002;224:393–403.

35. Zalis ME, Perumpillichira J, Del Frate C, Hahn PF. CT colonography: digital subtraction bowel cleansing with mucosal reconstruction initial observations. Radiology 2003;226:911–917.

36. Kruger RL, Schueler BA. A survey of clinical factors and patient dose in mammography. Med Phys 2001;28:1449–1454.

37. Monsees BS. The Mammography Quality Standards Act. An overview of the regulations and guidance. Radiol Clin N Am 2000;38:759–772.

38. van Gelder RE, et al. CT colonography at different radiation dose levels: feasibility of dose reduction. Radiology 2002;224:25–33.

39. Hara AK, et al. CT colonography: single- versus multi-detector row imaging. Radiology 2001;219:461–465.

40. Iannaccone R, et al. Detection of colorectal lesions: lower-dose multi-detector row helical CT colonography compared with conventional colonoscopy. Radiology 2003;229:775–781.

41. Taylor SA, et al. Multi-detector row CT colonography: effect of collimation, pitch, and orientation on polyp detection in a human colectomy specimen. Radiology 2003;229:109–118.

42. Wessling J, et al. CT colonography: Protocol optimization with multi-detector row CT—study in an anthropomorphic colon phantom. Radiology 2003;228:753–759.

43. Chen SC, Lu DS, Hecht JR, Kadell BM. CT colonography: value of scanning in both the supine and prone positions. AJR Am J Roentgenol 1999;172:595–599.

44. Morrin MM, et al. CT colonography: colonic distention improved by dual positioning but not intravenous glucagon. Eur Radiol 2002;12:525–530.

45. Fletcher JG, et al. Optimization of CT colonography technique: prospective trial in 180 patients. Radiology 2000;216:704–711.

46. Brenner DJ, Georgsson MA. Mass screening with CT colonography: Should the radiation exposure be of concern? Gastroenterology 2005;129(1):328–337.

47. NCRP. Uncertainties in fatal cancer risk estimates used in radiation protection. Report 126, National Council on Radiation Protection and Measurements, Bethesda, MD; 1997.

48. Land CE, Gilbert E, Smith JM. Report of the NCI-CDC Working Group to Revise the 1985 NIH Radioepidemiological Tables. NIH Publication 03-5387. See also, available at www.irep.nci.nih.gov, NIH, Bethesda (MD); 2003.

49. Fontana RS. The Mayo Lung Project: a perspective. Cancer 2000;89:2352–2355.

50. Henschke CI, et al. Early lung cancer action project: Overall design and findings from baseline screening. Lancet 1999;354:99–105.

51. Vastag B. Lung screening study to test popular CT scans. JAMA 2002;288:1705–1706.

52. Rusinek H, et al. Pulmonary nodule detection: low-dose versus conventional CT. Radiology 1998;209:243–249.

53. Aberle DR, et al. A consensus statement of the Society of Thoracic Radiology: screening for lung cancer with helical computed tomography. J Thorac Imaging 2001;16:65–68.

54. Miettinen OS, Henschke CI. CT screening for lung cancer: coping with nihilistic recommendations. Radiology 2001;221: 592–596.
55. Patz EF, Jr. Black WC, Goodman PC. CT screening for lung cancer: not ready for routine practice. Radiology 2001;221: 587–591.
56. Swensen SJ, et al. Screening for lung cancer with low-dose spiral computed tomography. Am J Respir Crit Care Med 2002;165:508–513.
57. Sone S, et al. Results of three-year mass screening programme for lung cancer using mobile low-dose spiral computed tomography scanner. Br J Cancer 2001;84:25–32.
58. Nawa T, et al. Lung cancer screening using low-dose spiral CT: results of baseline and 1-year follow-up studies. Chest 2002; 122:15–20.
59. Garg K, et al. Randomized controlled trial with low-dose spiral CT for lung cancer screening: Feasibility study and preliminary results. Radiology 2002;225:506–510.
60. Sobue T, et al. Screening for lung cancer with low-dose helical computed tomography: anti-lung cancer association project. J Clin Oncol 2002;20:911–920.
61. Mahadevia PJ, et al. Lung cancer screening with helical computed tomography in older adult smokers: a decision and cost-effectiveness analysis. JAMA 2003;289:313–322.
62. Thompson DE, et al. Cancer incidence in atomic bomb survivors. Part II: Solid tumors, 1958–1987. Radiat Res 1994;137: S17–67.
63. NRC. Health effects of exposure to low levels of ionizing radiation: BEIR V. Washington (DC): National Academy Press; 1990.
64. Gilbert ES, et al. Lung cancer after treatment for Hodgkin's disease: focus on radiation effects. Radiat Res 2003;159: 161–173.
65. Tokarskaya ZB, et al. Interaction of radiation and smoking in lung cancer induction among workers at the Mayak nuclear enterprise. Health Phys 2002;83:833–846.
66. Melloni B, Vergnenegre A, Lagrange P, Bonnaud F. Radon and domestic exposure. Rev Mal Respir 2000;17:1061–1071.
67. Morrison HI, Villeneuve PJ, Lubin JH, Schaubel DE. Radon-progeny exposure and lung cancer risk in a cohort of Newfoundland fluorspar miners. Radiat Res 1998;150:58–65.
68. Neugut AI, et al. Increased risk of lung cancer after breast cancer radiation therapy in cigarette smokers. Cancer 1994; 73:1615–1620.
69. Pershagen G, et al. Residential radon exposure and lung cancer in Sweden. N Engl J Med 1994;330:159–164.
70. Samet JM, et al. Lung cancer mortality and exposure to radon progeny in a cohort of New Mexico underground uranium miners. Health Phys 1991;61:745–752.
71. Hornung RW, Deddens J, Roscoe R. Modifiers of exposure-response estimates for lung cancer among miners exposed to radon progeny. Environ Health Perspect 1995;103(Suppl 2): 49–53.
72. Pierce DA, Sharp GB, Mabuchi K. Joint effects of radiation and smoking on lung cancer risk among atomic bomb survivors. Radiat Res 2003;159:511–520.
73. Brenner DJ. Radiation risks potentially associated with low-dose CT screening of adult smokers for lung cancer. Radiology 2004;231:440–445.
74. FDA. Full-body CT scans: What you need to know. DHHS Publication FDA (03)-0001. Available at www.fda.gov/cdrh/ct/ctscansbro.html, U.S. Food and Drug Administration, Rockville (MD); 2003.
75. Brant-Zawadzki M. CT screening: why I do it. AJR Am J Roentgenol 2002;179:319–326.
76. Illes J, et al. Self-referred whole-body CT imaging: current implications for health care consumers. Radiology 2003;228: 346–351.
77. Holtz A. Whole-body CT screening: Scanning or scamming? Oncol Times 2003;25:5–7.
78. Berland LL, Berland NW. Whole-body computed tomography screening. Semin Roentgenol 2003;38:65–76.
79. Beinfeld MT, Wittenberg E, Gazelle GS. Cost-effectiveness of whole-body CT screening. Radiology 2005;234:415–422.
80. Casola G, et al. Whole body CT screening: Spectrum of findings and recommendations. Radiology 2002;225(Suppl.):317.
81. Casarella WJ. A patient's viewpoint on a current controversy. Radiology 2002;224:927.
82. Brenner DJ, Elliston CD. Estimated radiation risks potentially associated with full-body CT screening. Radiology 2004 232:735–738.
83. ICRP. 1990 Recommendations of the International Commission on Radiological Protection: Publication 60. Oxford: Pergamon; 1991.
84. Fishman EK, Horton KM. What application should you offer in a whole body CT screening center? Available at www.screeningctisus.com/articles/screeningctisus.html.
85. Hoyert DL, et al. Deaths: final data for 1999. Natl Vital Stat Rep 2001;49:1–113.
86. Evens RG, Mettler FA. National CT use and radiation exposure: United States 1983. AJR Am J Roentgenol 1985;144: 1077–1081.
87. Bahador B. Trends in diagnostic imaging to 2000. London: Financial Times Pharmaceuticals and Healthcare Publishing; 1996.
88. UNSCEAR. Sources and effects of ionizing radiation: United Nations Scientific Committee on the Effects of Atomic Radiation: UNSCEAR 2000 report to the General Assembly. United Nations, New York; 2000.
89. Linton OW, Mettler FA, Jr. National conference on dose reduction in CT, with an emphasis on pediatric patients. AJR Am J Roentgenol 2003;181:321–329.

See also BONE DENSITY MEASUREMENT; COMPUTER-ASSISTED DETECTION AND DIAGNOSIS.

COMPUTED TOMOGRAPHY SIMULATOR

XIANGYANG TANG
GE Healthcare Technologies
Waukesha, Wisconsin

GE WANG
University of Iowa
Iowa City, Iowa

INTRODUCTION

The development of conformal radiation therapy (RT) and computerized treatment planning dates back to the late 1950s (1,2). The first milestone of computerized treatment planning is the invention of beam's eye view (BEV) display in the late 1970s (3,4). Up to now, along with surgery and chemotherapy, conformal RT has become the most effective measure for curative or palliative management of cancers at various anatomic sites (5). The biological mechanism underlying RT is that radiation damages crucial structures, such as deoxyribonucleic acid (DNA), of a cell, resulting in cell death, whether the cell is cancerous or normal, when the biologically damaged cell cannot be repaired by itself (5). The radiation dose delivered by RT is toxic to normal biological tissues and organs while it kills abnormal cells to cure cancer or palliate the local symptom of a malignant tumor (5). If the radiation dose distribution is made sufficiently concentrated on a targeted cancerous volume and tolerable over nearby normal anatomic structures, the cells within the cancerous volume may be killed while those within the normal organs or tissues survive. Hence, the ultimate goal of radiation therapy is to cure or control the cancer by precisely delivering an adequate and a homogeneous radiation dose to the targeted cancerous volume while maintaining the unavoidable dose to surrounding biological structures, particularly those critical organs or tissues that are very sensitive to radiation dose, such as eye, testis, lung, spinal cord, brain, and so on, below the biological toxicity tolerance. To improve the therapeutic ratio while decreasing the occurrence of acute or chronic side effects caused by radiation toxicity as much as possible, an administration of fractionated radiation therapy process has been clinically proven to be more efficient than a "lump sum" radiation delivery (5).

A typical conformal RT process consists of diagnostic data acquisition, simulation, treatment planning, treatment verification, and treatment delivery, although its implementation can be customized based upon available personnel and resources. Intuitively, a malignant tumor cannot be cured or its local symptom cannot be controlled if it misses the adequate radiation dose prescribed by a radiation oncologist. Meanwhile, the normal cells of surrounding tissues or organs can be fatally damaged if it absorbs the radiation dose supposed to be delivered to a targeted cancerous tumor. Consequently, among all factors compromising the success of a radiation therapy process, the geometrical imprecision or inaccuracy caused by patient localization and immobilization, as well as inadequate dose during treatment delivery, play dominant roles. Moreover, since a radiation therapy process is usually administrated in a fractionated manner, the patient position reproducibility between treatment delivery fractions is also of crucial importance. The maintenance of geometrical precision, accuracy, and reproducibility, which can never be overemphasized, are the tasks of simulation, while the warranty of delivering an adequate radiation dose at a homogenous distribution are the tasks of radiation treatment planning.

Conventionally, being carried out by a physician on a simulator in which two-dimensional (2D) imaging techniques are usually utilized, the simulation in conformal RT is an interactive and iterative process. Instead of using photons at the MeV energy level that are utilized in radiation treatment delivery, photons at the keV energy level, that is, X-ray, are utilized in the simulation to provide fluoroscopy for beam designing, in which an image intensifier or flat panel imager is usually employed. Moreover, X-ray source and film-based radiography are usually utilized for portal verification. In the beginning of the simulation, a radiation oncologist identifies the clinical target volume (CTV) by initially specifying the gross target volume (GTV) with a clinical margin added by taking the possible surrounding metastases and spreads into account (5). A planning target volume (PTV) is eventually defined by taking an extra margin into account to compensate for systematic geometrical mismatch between the simulator and the treatment machine, as well as geometrical error caused by the beam setup uncertainty and internal organ motion (6). It is important to state that, such an extra margin is crucial to the success of a radiation therapy, because the internal organ of a patient is always moving, no matter how perfect the patient localization, immobilization, and the geometric match between simulator and treatment machines can be achieved. By mimicking the geometry of a radiation therapy machine, such as a linear accelerator (LINAC), and with the availability of those 2D imaging techniques mentioned above, the conventional simulation generates a delineation of PTV and layout of radiation fields, including the number and orientation of beams, aperture of collimator, shape and size of field, as well as the specification of beam blocker, wedge, or compensator and markers. Since the geometry of a conventional simulator is exactly the same as that of a treatment machine, a spatial and geometrical integrity between them can be achieved if radioopaque markers are placed on a patient's skin or at appropriate anatomic landmarks, for example, the sternum for a patient with breast cancer. The conventional simulation in conformal radiation therapy is an online iterative process. Such an on-line process is inefficient in terms of patient throughput, since it requires a patient to remain in the simulator couch during the entire simulation process. It is very tough, if not impossible, to obtain an optimized conventional simulation in conformal radiation therapy, because this tedious and time-consuming simulation process is greatly dependent on the skill or experience of the physician committed to the task.

Intrinsically, the anatomic structure of a human being is three-dimensional (3D), and can be mapped to 3D models representing its geometric, pathologic, and physiologic characteristics, respectively. The advent of clinical X-ray computed tomography (CT) in the early 1970s was a revolution over 2D imaging technology, making the 3D modeling of a patient a reality. However, the X-ray CT scanner in its early stage was not capable of providing high quality, especially high spatial resolution, tomographic images of a patient volume within an acceptable or tolerable time. Nevertheless, the potentiality offered by X-ray CT technology for RT was well recognized then. In the 1980s, CT technology evolved dramatically while great progress simultaneously had been accomplished in 3D computer visualization technologies. By combining the state-of-the-art CT technology with modern 3D computer visualization, the concept of CT simulator or virtual simulator was formed in the late 1980s (7–9). Since then, enormous effort and resources were invested in the research and development (R&D) of CT or virtual simulator, resulting in numerous CT or virtual simulators commercially available in the market. For convenience, a CT simulator hereafter refers to either a CT simulator or a virtual simulator unless otherwise specified.

As shown in Fig. 1, a CT simulator consists of a radiation therapy dedicated CT scanner (viz., RT-dedicated CT scanner) and a software workstation. The RT-dedicated CT scanner is to provide a 3D model of the patient to be treated by acquiring contiguous tomographic images over a volume of interest. The software workstation is to carry out simulation based on the 3D anatomic structures and clinical

Figure 1. A schematic diagram showing the process of 3D conformal RT using a CT simulator: (a) Three-dimensional patient data acquisition through a RT dedicated CT scanner; (b) CT simulation, treatment planning and verification; (c) Treatment delivery via a LINAC.

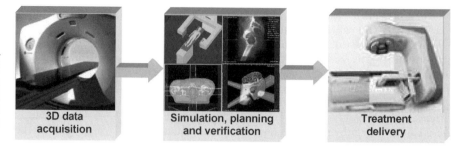

3D data acquisition

Simulation, planning and verification

Treatment delivery

information revealed by the 3D model of the patient. In general, the simulation software of a CT simulator can be either stand-alone or an embedded part of a treatment planning system. By integrating an RT-dedicated CT scanner and simulation software, a CT-simulator generally conducts the following tasks: patient positioning, patient immobilization, 3D patient data set acquisition by CT scanning, identification of a target volume (GTV, CTV, and PTV) and surrounding vital normal tissues and organs, placement of beams (number, orientation, isocenter, and collimator aperture), field design (beam shaper, blocker, wedge and compensator), as well as generation of portal images, such as digitally reconstructed radiography (DDR) and other instructions exported to a treatment planning and RT machine (10–17). In the following sections, these tasks are described in detail.

THREE-DIMENSIONAL COMPUTED TOMOGRAPHY PATIENT DATA ACQUISITION

A CT scanner is the cornerstone of a CT simulator for treatment planning and delivery using 3D conformal RT. With a focus on diagnostic imaging, CT technology has been substantially improved over the past three decades and became the most popular tomographic imaging modality in clinics (18,19). With the impressive progresses in CT technology, particularly the image reconstruction methods for multidetector row CT and volumetric CT (20–22), a state-of-the-art diagnostic CT scanner has well satisfied the requirement posed by RT. Unfortunately, however, a start-of-the-art diagnostic CT scanner usually cannot be readily used in a CT simulator for RT treatment planning, because the gantry aperture diameter of a diagnostic CT scanner is usually ~70.0 cm, which cannot guarantee a smooth accommodation of a patient and necessary immobilization devices, for example, the arm holder utilized in breast cancer radiation therapy (14). As a result, RT-dedicated CT scanners have been developed by major medical CT scanner manufacturers and are currently available on the market. The gantry aperture diameter of an RT-dedicated CT scanner ranges from 80 to 85 cm with a display field of view (DFOV) between 60 and 65 cm, which can handle virtually all clinical situations. Technical parameters of a typical RT-dedicated CT scanner are listed in Table 1. The performance of an RT-dedicated CT scanner should be periodically calibrated and verified (14,17). It is interesting to note that, although its overall performance is inferior to that of a state-of-the-art diagnostic CT scanner, a currently available RT-dedicated CT scanner can serve

radiation therapy very well (12). One of the major reasons for such a technical delay is a relatively small market volume of RT-dedicated CT scanners in comparison to that of diagnostics CT scanners. Considering the fast evolution of the diagnostic CT scanner and the technical catching up of the RT-dedicated CT scanner in X-ray detector z coverage and scanning speed, it is very hopeful for a future RT-dedicated CT scanner to have the overall performance of a current diagnostic CT scanner.

In general, the protocols used for diagnostic CT imaging can be used correspondingly by an RT-dedicated CT scanner to acquire a 3D patient data set for CT simulation, although a trade-off between the spatial resolution along the longitudinal direction of a patient (image slice thickness) and the total number of tomographic images has to be made in practice. As shown in Table 1, whereas the thinnest available slice thickness in a commercial RT-dedicated CT scanner is between 1.0 and 2.0 mm, an image slice thickness thinner than 3.0 mm is usually acceptable to render DRR for portal verification (23).

PATIENT POSITIONING AND IMMOBILIZATION

It has to be emphasized that, to guarantee a geometrical match between the CT scanning and the treatment delivery, the patient has to be in exactly the same position with all the immobilization devices, such as foam body casts, thermoplastic head masks (24), and stereotactic frames (25), in place. In addition to an enlarged gantry aperture to accommodate a patient and immobilization devices, an RT-dedicated CT scanner has a flat top patient couch that is exactly the same as the one used in a treatment machine so that an exact geometrical match and reliable patient position reproducibility can be achieved between the RT-dedicated CT scanner and the treatment system. As illustrated below, all the tasks accomplished by a CT simulator are solely dependent on the spatial integrity of the 3D patient data set acquired by an RT-dedicated CT scanner. Also, the importance of geometric accuracy and match between the CT scanner and RT machine, as well as the reproducibility of patient position, can never be overstated.

In principle, the orthogonal laser beams existing in an RT-dedicated CT scanner gantry can be employed to mark the isocenter of a target volume. However, to assure a convenient and efficient marking process on the patient's skin or other anatomic landmarks, it is preferable to have external laser devices installed on the side wall and ceiling of the room where the RT-dedicated CT scanner is installed (14). The laser beams on the side wall (transaxail and

Table 1. Primary Technical Parameters of a Typical Radiation Therapy Dedicated CT Scanner

Scan mode	Axial; Helical	
Scan speed (s/360° gantry rotation)	1.0, 2.0, 3.0, 4.0	
Number of detector row	4	
Diameter of gantry bore (mm)	800.00	
Display FOV (mm)	650.0	
X-ray tube	Current: 10–440 mA, 5mA increments	
	kVp: 80, 100, 120, 140	
	Power: 0.8–53.2 (kW)	
	Heat capacity: 6.3 MU	
Gantry tilt	±30°	
Image slice thickness (mm)	0.625, 1.25, 2.5, 3.75, 5.0, 7.5, 10.0	
Dose (mGy/100 mAs)	CTDI:	
		Head: 16.5 (center); 17.2
	(surface)	
		Body: 5.6 (center); 11.0
	(surface)	
	$CTDI_{100}$:	
		Head: 17.0 (center); 18.4
	(surface)	
		Body: 5.4 (center); 11.8
	(surface)	
	$CTDI_{vol}$:	
		Head: 17.9
		Body: 9.7
In-plane spatial resolution (lp/cm)	Standard	Hi-resolution
	4.2 at 50% MTF	10.5 at 50% MTF
	6.8 at 10% MTF	13.9 at 10% MTF
	8.5 at 0% MTF	15.4 at 0% MTF
Low contrast resolution [on 8 in. CATPHAN phantom]	5mm at 0.3% at 13.3 mGy	
	3mm at 0.3% at 37.6 mGy	
Noise(on AAPM water phantom)	0.32% +/−0.03% at 25.2 mGy (2.52 rad)	
Image generation speed (s/frame)	0.17	
Image matrix	512×512	
CT number scale	− 31743–31743 HU	

coronal) are fixed to identify the isocenter by moving the patient couch longitudinally and vertically. The laser beam on the ceiling (sagittal) has to be moveable laterally to reach the isocenter of a target volume since the patient couch of a CT scanner is generally not movable laterally. A schematic diagram showing the deployment of external laser beams on an RT-dedicated CT scanner is illustrated in Fig. 2.

There exist two ways to identify the isocenter of a target volume (14). The first way is to place marks on a patient's skin to obtain the absolute location of the isocenter in the coordinate system of the RT-dedicated CT scanner. Since the coordinate system of the RT-dedicated CT scanner is aligned with that of a treatment machine by geometrical calibration and verification, the isocenter coordinates obtained in such a way are ready for utilization by the treatment machine as long as the patient is appropriately marked. The primary advantage of this manner is the avoidance of any errors due to intermediate geometrical transforms. However, this method requires the involvement of radiation oncologists and physicists simultaneously on site while the patient remains in the couch of

the RT-dedicated CT scanner. Such a process is called a synchronized mode because the identification of a target volume has to be accomplished in real-time as described above. The second way is relatively flexible and can be carried out off-line. The patient with a few radioopaque marks placed on the skin or preferably solid anatomic landmarks goes through a regular CT scanning, and then can immediately leave the hospital. The radioopaque landmarks can be easily placed with the availability of external laser beams of an RT-dedicated CT scanner as introduced above. In the simulation, the isocenter of a target volume can be readily determined by a physicist under the guidance of a radiation oncologist. This way is in an asynchronous mode and provides substantial flexibility for physicians to improve patient throughput and patient comfort.

COMPUTED TOMOGRAPHY SIMULATION

With an RT-dedicated CT scanner, a set of transaxial images covering a volume of interest are obtained and

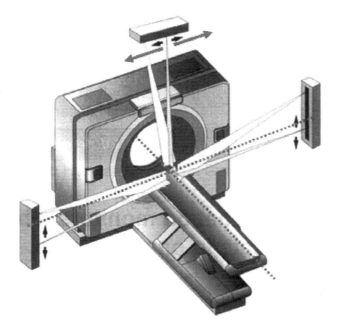

Figure 2. A schematic diagram showing an RT-dedicated CT scanner and its associated external laser beams for patient positioning in 3D conformal RT using a CT simulator for treatment planning (Courtesy Robert L. Steinnhauser, Gammex rmi.)

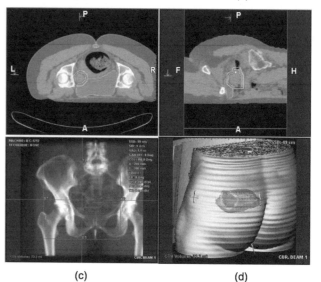

Figure 3. A target volume and its nearby normal critical organ are identified using slice-wise contouring in transaxial image plane (a), sagittal image plane (b), beam's eye view (c) and operator's eye view via segmented semitransparent surface rendering (d). In baseline image planes, the CTV is outlined by red, PTV by yellow and the nearby critical organ at risk by green contours, respectively, while colors fill each volume correspondingly in beam's eye view and operator's eye view. [Courtesy Dr. Georgios Sakas and Dr. Evelyn A. Firle, Fraunhofer Institute for Computer Graphics (IGD).] Please see online for color figure.

stored in the simulation workstation. Generally speaking, the 3D data set contains much more information about the patient than can be viewed in the transaxial plane only. For example, with the availability of multiplanar reformatting in other baseline planes (coronal, sagittal, and even oblique), physicians can attain more freedom in visualizing a targeted tumor volume and its geometrical relationship with adjacent normal anatomic structures. The 3D patient data set can be conceived as a virtual patient or a 3D model of the patient to be treated, and a 3D model fiducially represents both anatomic and physical properties of the patient. With modern 3D image processing and visualization techniques that will be described below, the anatomic and physical information of the patient, can be exploited in much more details (13).

Structure Identification

The purpose of a CT simulator is not only to identify the extent of a tumor, but also to provide an unambiguous delineation of the relationship between the targeted tumor volume and its neighboring normal tissues or organs. The basic, but most effective, tool in a CT simulator to identify a tumor is the use of contouring to define the GTV, CTV, and PTV of a targeted tumor volume, in contiguous transaxial tomographic images. The GTV of a cancerous tumor is initially specified by a radiation oncologist via outlining the boundaries of the GTV manually or semiautomatically. As mentioned above, during the target volume defining process, it is necessary to expand the GTV into a CTV by including clinical margins surrounding the GTV. The aim of adding clinical margins is to assure that possible spreading or metastases be included in the targeted volume. Furthermore, recognizing the imperfect geometrical reproducibility over treatment fractions, such as systematic

geometric errors caused by beam setup uncertainty and geometrical mismatch between an RT-dedicated CT scanner and a treatment machine, the random and/or periodic geometrical errors caused by patient and organ motions, such as breathing and swallowing, it is crucial to include an adequate margin into the planed target volume (5,6).

A manual contouring process is conducted by defining the boundary of the GTV, CTV, and PTV, respectively, in every image slice. The image slice can be in any baseline plane, though the transaxial slice is preferably utilized in practice. An example of such a process is illustrated in Fig. 3. The manual contouring operation can also be done in several other slices sequentially, and the boundaries in interim images can be automatically obtained by linear or nonlinear interpolation techniques, as long as the gap between adjacent contoured image slices are within certain threshold. An even more efficient way to carry out the manual contouring operation is to define the boundaries of the target volume on the central transaxial, coronal, and sagittal planes, respectively, and then the target volume can be obtained by automatic 3D linear or nonlinear interpolation techniques available in a CT simulator. On the other hand, a semiautomatic contouring can be accomplished by just placing a seed within the targeted volume, and the boundaries are then determined by thresholding over CT numbers in Housfield units or other characteristics of the voxels in an image. Subsequently, the targeted volume can be obtained through automatic image processing techniques, such as volume growing over contiguous image slices.

Beam Design

Once the targeted cancerous tumor is defined by the techniques presented above, the next step in the simulation is beam design: to determine the number and orientation of beams, aperture of collimators, as well as shape and size of the beam field. Prior to the beam design, a physician has to specify a coordinate system convention based on which simulation is to be conducted, the treatment machine type, and its associated characteristics, such as modality (photon or electron) and energy level of the beam intended for radiation treatment. Usually, a CT simulation workstation provides versatile and powerful 2 and 3D visualization tools and utilities for beam manipulation, and a typical beam designing process in central baseline planes is illustrated in Fig. 4. In the process, the orientation or deployment of beams is specified by machine angle, collimator angle, and patient couch angle, with the beam manipulation tools, such as creating, adding, deleting, mirroring, duplicating, or renaming, in the CT simulator. The isocenter of an identified target volume can be adjusted, and the resultant coordinate system shifts are recorded and exported to a treatment machine. With versatile 2 and 3D visualization tools and utilities, the collimator aperture can be adjusted interactively by clicking and dragging the X- and Y-jaw of a collimator on the display screen of the simulation workstation. The beam field can conform the boundaries of the identified target volume by manipulating various beam shapers (either aperture or shield). A shaper can be either available in the standard set offered by a CT simulator or defined by a physician according to the structure to be conformed. The definition of a new shaper

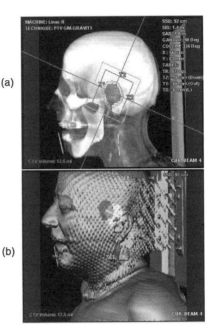

Figure 5. Modern 3D visualization techniques are extensively utilized in CT simulator to design beams via beam shaper for 3D conformal RT: (a) Beam's eye view, in which the CTV is filled in red, the PTV is outlined by green contour, the collimator aperture is defined by red rectangle, and the organ at risk (eye) is filled with green; (b) Operator's eye view using segmented semitransparent surface rendering, in which the PTV is in red, the beam field on the surface in yellow, and the organ at risk in green. [Courtesy Dr. Georgios Sakas and Dr. Evelyn A. Firle, Fraunhofer Institute for Computer Graphics (IGD).] Please see online for color figure.

can be either manual by clicking and dragging on the display screen or automatic by conforming the shaper to the boundaries of the targeted volume. Markers defining the corner of the field can be placed on a patient's skin or anatomic landmarks via clicking and dragging or typing into corresponding entries on the display screen. Furthermore, a typical CT simulator usually provides numerous utility functions for geometry measurement, such as distance, angle, area and volume, and the measurement results are displayed on the screen while textual annotations can be manipulated simultaneously.

Alternatively, the procedures introduced above can be carried out in the so-called BEV: the most fundamental and powerful 3D visualization technique employed in the simulation. Being illustrated in Fig. 5, the BEV is obtained via DRR by mimicking an X-ray source emanating from exactly the same location corresponding to the radiation focal spot of a treatment machine. With the availability of BEV, the following operations can be readily carried out: adjustment of the beam's isocenter, adding makers to the corners of a beam field, and manipulating the leaves of a multileaf collimator (MLC). Note that a number of modern image processing techniques can be combined with 3D visualization to highlight interested anatomic features while the interference from noninterested anatomic structures is being removed from visualization. For example, as shown in Fig. 6, by thresholding voxel gray values in an appropriate range, DRRs of the lungs, fat, muscle, other

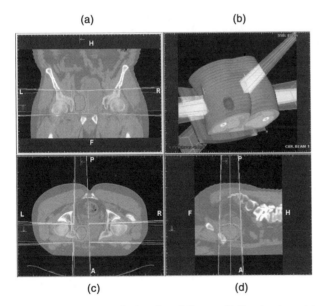

Figure 4. Beams are designed to deliver radiation treatment to the target volume identified through the process shown in Fig. 3, with the beam manipulating functions provided by a CT simulator in coronal plane (a), operator's eye view via 3D segmented semitransparent surface rendering (b), transaxial plane (c), and sagittal plane (d). [Courtesy Dr. Georgios Sakas and Dr. Evelyn A. Firle, Fraunhofer Institute for Computer Graphics (IGD).] Please see online for color figure.

(a)   (b)

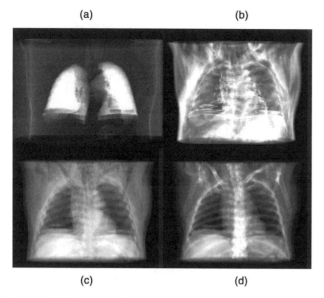

(c)   (d)

Figure 6. Other examples of modern 3D visualization techniques provided by a typical CT simulator for treatment planning in 3D conformal RT, in which only the interested organs or tissues are displayed: DRR of lung (a); DRR of fat (b); DRR of muscle (c); DRR of all anatomic structure and tissues (d). [Courtesy Dr. Georgios Sakas and Dr. Evelyn A. Firle, Fraunhofer Institute for Computer Graphics (IGD).]

tissues and organs can be exclusively displayed to facilitate target volume identification and contouring, beam deployment, and field designing. Another example of 3D visualization techniques for CT simulation is depth controlling, in which only a slab containing the target volumes and their surrounding structures are displayed. All these DRR-based 3D visualization techniques can be combined wherever they can make a simulation process more productive and reliable. Moreover, other modern 3D visualization techniques, such as surface rendering to generate operator's eye view or room's eye view, enable an even smoother and more efficient beam designing process. It is interesting to mention that the transparency in surface rendering can be adjusted to facilitate the simulation (e.g., Figs. 3 and 4).

In practice, various beam blockers to shape the beam field can be made of lead, depleted uranium, or low melting point lead alloy (5,26). Usually, these shapers are customized for individual patients. Such a customization process is time consuming, inefficient, and expensive. Instead of customizing beam shapers for each target volume of each patient, a beam field can be shaped by a multileaf collimator in which two banks of leaves are installed oppositely with each leaf being driven by a computer-controlled step motor (27–29) (see Fig. 7). Through a coordinated movement of these leaves, the beam field can conform various target volumes, provided that the thickness of a leaf is sufficiently small, such as 0.5 or 1.0 cm. Thus, the multileaf collimator is a much more general and efficient beam shaping technique, since it can be repeatedly and unlimitedly utilized in 3D conformal radiation therapy for any targeted volume at any location within a patient. Consequently, the facility for beam shaper fabrication, such as block cutting and compensator making, are no longer

Figure 7. A multileaf collimator used for beam shaping in 3D conformal radiation therapy consists of two banks of metal leaves that are installed oppositely, and each leaf is driven by a computer-controlled step motor or pneumatic device individually.

needed. To suppress the interleaf radiation leakage as much as possible, there usually exists interlocking between adjacent and opposite leaves of a multileaf collimator, which is implemented in a tongue–groove structure. All the beam design techniques introduced above can be readily employed with a multileaf collimator. An example is shown in Fig. 8, in which the deployment of all leaf is illustrated.

It is not an exaggeration to state that, without the aforementioned 3D visualization-based functionalities,

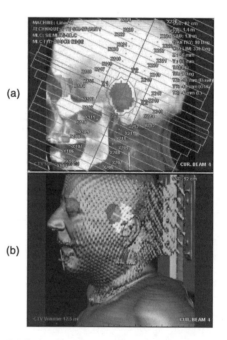

(a)

(b)

Figure 8. Modern 3D visualization techniques are extensively utilized in CT simulator to design beams via multileaf collimator for 3D conformal radiation therapy: (a) Beam's eye view, in which the CTV is filled in red, the PTV is outlined by green contour, the collimator aperture is defined by red rectangle, and the organ at risk (eye) is filled with green; (b) Operator's eye view using segmented semitransparent surface rendering, in which the PTV is in red, the beam field on the surface in yellow, and the organ at risk in green. [Courtesy Dr. Georgios Sakas and Dr. Evelyn A. Firle, Fraunhofer Institute for Computer Graphics (IGD).] Please see online for color figure.

the beam design using a multileaf collimator would be very difficult, if not impossible. Moreover, the optimization of the beam design, in which many plans have to be investigated and compared, via either beam shapers or a multileaf collimator can only be possible with the aid of these functionalities that are provided by a CT simulator.

Protocol Export

Once targeted volumes are identified, beams and their fields are designed, the last step is to export the simulation results to a treatment machine. The simulation results consist of both textual and pictorial information. The textual information usually includes a list of identified structures and their types (targeted cancerous volume, surrounding critical normal organ or tissue, and, etc.), a list of beam setting up parameters corresponding to each targeted volume (machine angles, patient couch angles, isocenter coordinates, field markers, and etc.), and a list of sequential actions to be executed in the treatment delivery process. The pictorial information refers to the pictures and images, such as DRRs, that can be used in the treatment process for planning verification by comparing them with the portal images acquired on the treatment machine.

COMPUTED TOMOGRAPHY SIMULATION FOR ADVANCED RADIATION THERAPY

The methodology for designing radiation beams in a CT simulator introduced so far is quite straightforward. For example, with respect to a PTV, all beams are coplanar, and the intensity distribution within the field of each beam is constant. Furthermore, usually only two to three beams are engaged in treatment delivery. As shown in the left column of Fig. 9, such a very limited number of coplanar beams with uniform intensity distribution are generally

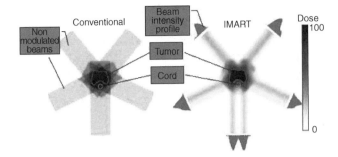

Figure 10. A CT image in transaxial plane shows a concave CTV outlined by the yellow boundary, the corresponding PTV contoured by the red boundary, and the nearby critical organ (spine cord) within the concavity. (Courtesy of Michelle Lee, Eleka AB.) Please see online version for color figure.

not able to conform a 3D target volume tightly, no matter how accurately the beam and its field are designed. Hence, the conventional 3D conformal radiation therapy is at most a coarse approximation to the ideal 3D conformity.

In reality, a targeted cancerous tumor can be much more complicated than just a regular and convex volume. For example, a target tumor can be irregular, concave with nearby normal critical organs or tissues within its cavity (see Fig. 10), or the targeted cancerous tumor even wraps itself around nearby normal organs or tissues. To make sure all cancerous tumors be included in a PTV, a physician has no choice but to take extra margins surrounding the identified CTV into account. Consequently, the PTV can be unfortunately quite large due to the conservative strategies exercised, resulting in an undesirable radiation dose to the surrounding normal organs or tissues, and significantly compromising the success of radiation therapy.

Given the complexity of the 3D shape of a targeted malignant tumor and its geometrical relationship with nearby normal structures, true 3D conformity is desired in RT to deliver the needed radiation dose accurately and avoid unnecessary damages to the cells of surrounding normal organs or tissues. Actually, the combination of the modern CT technology and 3D visualization methods has paved the way toward true 3D conformal radiation therapy. Aiming at improving the therapeutic index, tremendous efforts have been devoted in the radiation therapy community to exploring more effective RT solutions. What follows is a brief introduction to the exciting progresses made up to date.

Computed Tomography Simulation For Intensity Modulated Radiation Therapy

One of the innovations over conventional 3D conformal RT is the so-called intensity modulated radiation therapy (IMRT) (30–33). The two primary differences between the IMRT and conventional 3D conformal RT are (a) beams do not have to be coplanar; (b) the radiation intensity distribution within a beam is not uniform. In practice, noncoplanar beams can be readily realized by horizontally rotating the patient couch of a treatment machine. With respect to beam intensity, there exist two ways to modulate the distribution. One way is to partition the beam field into a few subfields, and various radiation blockers, such as

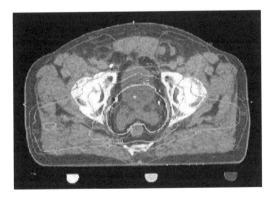

Figure 9. Schematic diagrams illustrates the mechanisms underlying conventional 3D conformal RT (a) and modern intensity modulated RT (b). The volume of the targeted tumor is concave, and the critical organ (spine cord) is within the concavity. The spine cord would undertake almost the same radiation dose as the target tumor volume if the conventional 3D conformal RT is administrated. However, with intensity modulation RT, the radiation dose to the spine cord can be substantially reduced to be below tolerable toxicity level. (Courtesy of Sabbe Mollio, Ph.D., University of California at Irvin.) Please see online for color figure.

wedge, partial blocker, or compensator, are employed to modulate the intensity of each subfield. Such an IMRT implementation is called the "step-and-shoot" mode. The other way is to move the leaves of a computer-controlled multileaf collimator dynamically and individually. The intensity distribution within each beam field is modulated by appropriately opening and shutting the leaves of a multileaf collimator. Such an implementation is called the "sliding window" mode, because each leaf of a multileaf collimator is continuously sliding during the treatment delivery. The radiation beams in these two modes can three-dimensionally encompass an identified target volume with improved conformity, and deliver a radiation dose distribution with a significantly improved accuracy than that allowed by the conventional 3D conformal radiation therapy, no matter how complicated the shape of the identified target volume is.

To illustrate the mechanism of 3D conformity in IMRT, the 2D in-plane conformity achieved by intensity-modulated coplanar beams is shown in the right column of Fig. 9, while the 3D conformity using noncoplanar intensity-modulated beams is not hard to deduce. As the conformity significantly is improved, either an escalated radiation dose can be delivered to the target volume at the same dose level absorbed by surrounding normal organs or tissues, or the same dose can be delivered to the target volume with less radiation toxicity to surrounding structures. It means that, IMRT not only conforms a high dose delivery to the targeted tumor volume, but also minimizes the radiation dose to the nearby structures.

To achieve an improved 3D conformity, the IMRT requires accurate 3D geometrical delineation of an identified target volume and its surrounding structures, beam orientation, field partition, intensity modulation via blockers, wedges, compensators, or a multileaf collimator. It is not difficult to imagine that, without the previously described functionalities offered by a CT-simulator, the implementation of IMRT would be out of the question. In fact, the dependence of the IMRT upon the CT simulator is so profound that the CT simulator is no longer separable from a radiation treatment planning system for which the IMRT is designed and verified. At present, the boundary between a CT simulator and a treatment planning system is fading. More and more radiation treatment planning systems are incorporating the functionalities that used to be provided by a stand-alone CT simulator as an integrated part in the IMRT-based 3D conformal radiation therapy.

With an improved 3D conformity, the success of the IMRT-based 3D conformal RT relies on the geometry accuracy and reproducibility to a significantly larger extent than that in conventional 3D conformal RT. Any geometry uncertainty (6) due to beam setup (mechanical and optical allowances), imperfect reproducibility, or organ motion induced errors can significantly compromise the outcome of the radiation therapy, as the intensity modulated beams conform the target volume very compactly. In addition to 3D conformity, the homogeneity of the dose distribution delivered to a target volume is also a requirement of a successful radiation therapy. However, with an emphasis on the 3D conformity, the dose homogeneity is usually compromised, posing more challenges to the RT process.

Meanwhile, the verification of the radiation dose distribution in IMRT becomes difficult, demanding much more complicated devices and phantoms to accomplish quality assurance (34,35). As a result, caution or discretion must be exercised in the clinic while a decision is being made on adopting IMRT for 3D conformal radiation therapy, although clinical reports showing favorable results of IMRT over conventional 3D conformal RT have been growing in the literature. In all fairness, not all cancer sites are suitable for the application of IMRT, and a general rule is that for a relatively regular target tumor, conventional 3D conformal RT is preferable, whereas IMRT is preferable in circumstances where the target tumor is irregular, concave with critical organs or tissues within the concavity, or even wraps itself around the nearby normal critical organs or tissues. Meanwhile, there may exist too much freedom in IMRT to conform an irregularly shaped target volume, leading to more difficulties in the optimization with respect to candidate treatment plans to achieve various objective functions while satisfying certain constraints. At present, numerous basic and clinical investigations are under way in the RT community to attain IMRT protocols for 3D conformal RT.

Computed Tomography Simulation For Tomotherapy

With the freedom in increasing the number of beams, specifying the beam shape and orientation, and modulating the beam intensity distribution, IMRT technology can be advanced even further (30,31). A more aggressive implementation of IMRT is the so-called tomotherapy (36–38). As implied by its name, tomotheraphy is inspired by the scanning mode of X-ray CT that has been extensively utilized in diagnostic imaging over the past three decades. Interestingly, the evolution of tomotherapy is similar to that of diagnostic CT. The geometry of an initial tomotherapy system is virtually the same as that of a single slice CT scanner prior to the introduction of helical–spiral CT. In such a tomotherapy system, the target volume is irradiated by a small fan beam moving along an arc trajectory (37). Consequently, the number of beams is increased dramatically while the field size of each beam is accordingly decreased. At each angular position, the radiation intensity of the small fan beam is dynamically modulated by a binary multileaf collimator that is driven pneumatically. In a "slice-by-slice" fashion, a target volume can be irradiated with a significantly improved 3D conformity, provided that the thickness of the small fan beam is sufficiently small. Similar to the CT evolution from the circular scanning mode to the helical–spiral scanning mode, tomotherapy has evolved into helical–spiral tomotherapy, in which the patient couch proceeds at a constant speed while the radiation source is rotated in the gantry around a patient (36,38). At each angular position, the intensity of the small fan beam is temporally modulated by a pneumatically driven multileaf collimator using an appropriate opening and shutting scheme. Through helical–spiral scanning, the radiation dose can conform a targeted volume very tightly, regardless of whether it is regular or irregular, concave or convex, delivering an adequate treatment dose to the targeted volume while depositing a minimum dose to the surrounding normal organs and tissues.

As in conventional IMRT, tomotherapy, either in the arc or the helical–spiral mode, requires an accurate geometrical delineation of a target volume and a reliable reproducibility of the patient positioning. The latest helical–spiral tomotherapy system has even evolved into an "all-in-one" system: a radiation therapy machine that includes all the needed modules: CT simulator, treatment planning, treatment delivery, as well as a real CT scanner (called tomoimage CT scanner) using therapeutic photons as the tomographic imaging means. Although its contrast resolution is significantly degraded in comparison to that of a diagnostic CT scanner or an RT-dedicated CT scanner, the tomoimage CT scanner can provide acceptable spatial resolution for accurate patient repositioning, playing a role of portal imaging, such as DRR, used in the conventional 3D conformal RT to assure geometry accuracy and reproducibility. With self-consistency among anatomic imaging, treatment planning, treatment delivery, and verification, such an "all-in-one" tomotherapy system can significantly improve the geometrical integrity, which is one of the most critical challenging tasks of IMRT (36,38).

Computed Tomography Simulation For Stereotactic Radiosurgery

In addition to 3D conformal RT, a state-of-the-art stereotactic radiosurgery (39–41) also relies entirely on the capabilities of a CT simulator in which either protons or high energy X-ray photons are employed. Rather than being administrated in a fractionated manner, stereotactic radiosurgery is a one-session or "lump sum" radiation treatment delivery procedure just like a surgery operation. Because of single treatment delivery, the radiation intensity must be elevated to the highest tolerable degree, and such a greatly elevated radiation intensity can damage cells in surrounding normal organs or tissues. Hence, stereotactic radiosurgery is usually employed in the management of small (2–3 cm) cancerous tumors. Moreover, with a radiation intensity that is greatly larger than that of 3D conformal RT, stereotactic radiosurgery demands a much more accurate geometric delineation of targeted tumors, and any geometrical uncertainty due to patient position, immobilization, and organ motion could result in treatment disaster. Consequently, stereotactic radiosurgery is only employed to cure small tumor in patient's head or neck, where patient immobilization devices can be applied reliably. In the early stage of stereotactic radiosurery, 2D imaging techniques, such as fluoroscopic angiography, were employed to identify target volumes and determine the number and orientation of radiation beams. Recently, similar to the evolution of 3D conformal RT, state-of-the-art stereotactic radiosurgery procedure increasingly rely on 3D patient data acquired by a CT scanner and the treatment planning in which the functionalities of a CT simulator are integrated.

THE FUTURE OF COMPUTED TOMOGRAPHY SIMULATOR

To achieve a full 3D conformity in radiation therapy, an ultimate solution would be the utilization of a very large number of pencil beams with their intensity being modulated instantly (42–44). However, a full 3D conformity demands perfect geometrical accuracy, since the dose gradient at the boundary of the 3D conformed target volume would be extremely sharp. Any geometrical uncertainties due to inaccurate patient position, unreliable immobilization, beam set up errors, misregistration between 3D patient data acquisition and treatment delivery, and particularly organ motion, could lead to treatment failure: underdosing-to-target volumes or overdosing to nearby normal critical organs or tissues, or both.

All the 3D conformal RT techniques introduced thus far are implemented in a "forward" way that is, once a set of 3D patient data is acquired, the simulation, treatment planning, planning verification, and delivery are implemented sequentially. In the forward mode, the simulation and treatment planning processes are conducted in a semiautomatic and iterative way with the involvement of radiation oncologists, physicists, therapists, and dosmetrists. To a great extent, the success of such a strategy is dependent on the experience or skill of the physicians engaged. Encouraged by the vigorous technological progresses made over the past three decades, it is believed that modern 3D conformal RT will continue progressing with the development of medical imaging and 3D visualization technologies: delivering treatment to target volumes with a full 3D conformity and adequate dose while surrounding normal structures are minimally damaged. It is not hard to imagine that future RT would not achieve an optimized solution if the treatment planning is only performed in the forward mode, because the parameter space to be exhaustively searched in the optimization process is really large.

Recognizing the complexity of modern IMRT, a more reasonable way to accomplish radiation treatment planning is to treat it as an inverse problem. Starting from specifications on a desirable radiation dose distribution and certain constraints, such as the avoidance of critical anatomic structures, inverse treatment planning can be performed based on a set of patient CT data with the help of a CT simulator (30,45). There exist two major tasks in the inverse treatment planning: (a) selection of an objective function and constraints; (b) development of an efficient algorithm to solve the optimization problem (46). A number of objective functions and constraints, such as dose (47), dose-volume (48,49), equivalent uniform dose (50,51), generalized equivalent dose (52), biological indices (53), and their combinations, have been proposed to date (54). Meanwhile, a few optimization approaches, such as simulated annealing (55), gradient (48), active set (56), genetic (57), maximum likelihood (58), dynamically penalized likelihood (59), and fuzzy logic (60) algorithms, have been investigated (54). It is underlined that inverse treatment planning is mathematically an ill-posed optimization problem, and that certain regularization techniques ought to be used in the optimization process (46). Considering the inhomogeneous dose distribution associated with a full 3D conformity, the simulation process for the inverse treatment planning is significantly different from that for the forward treatment planning, but the basic functionalities, such as target volume defining and 3D visualization of a CT simulator, are still the same. Currently, inverse radiation treatment planning is still an open research and development area that has attracted the major attention of the RT community.

The ultimate solution to guarantee geometry accuracy and reproducibility is to track targeted volumes and their motion dynamically for a full 3D conformity during the radiation treatment delivery. A tracking process can be realized through gating techniques employing various mechanical or optical sensors while a patient is scanned by an RT dedicated CT scanner (61–63). This is related to the concept of four-dimensional (4D) CT, by which variation of a 3D model of a patient is revealed instantly. Such a 4D planning strategy can be implemented either off- or on-line. In the off-line mode, a target volume and its surrounding normal organs or tissues are tracked by gating during the data acquisition by an RT-dedicated CT scanner, and the motion of the patient and organs are recorded and exported to a treatment machine for radiation delivery. In the on-line mode, a patient is scanned by an RT-dedicated CT scanner during the treatment delivery. Apparently, the on-line mode needs to integrate an RT-dedicated CT scanner into a treatment machine (64,65). It is believed that with the development of the 4D CT technology, the 3D conformal RT can be administrated in an adaptive and well-controlled manner, leading to a significantly improved therapeutic index.

The majority of simulation in treatment planning for RT is currently implemented based up on 3D patient datasets acquired by a CT scanner, because of its merits in data acquisition, image generation, superior spatial resolution, as well as the capability of estimating the electronic density for dose calculation and verification. In addition to CT scanners, other modern 3D imaging modalities, such as magnetic resonance imaging (MRI), positron emission tomography (PET), and ultrasonic imaging, can also be incorporated into a virtual simulator for 3D conformal RT. There is no doubt that, in the predictable future CT will remain the modality of chioce for 3D conformal RT. However, with the development of modern image registration and data fusion techniques, images acquired by MRI, PET, and ultrasound are becoming more and more relevant, suggesting chances for them to be utilized in 3D conformal RT. Finally, it should be pointed out that all the CT simulation and virtual simulation techniques we have covered in this article are also applicable for RT using other high energy particles, such as protons and neutrons (5).

BIBLIOGRAPHY

1. Wright K, et al. Field shaping selective protection in megavolt radiation therapy. Radiology 1959;72:101.
2. Tsien K. The application of automatic computing machines to radiation treatment planning. Br J Radiol 1955;28:432.
3. Reinstein Le, et al. A computer-assisted three-dimensional treatment planning system. Radiology 1978;127:259–264.
4. McShan DL, et al. A computerized three-dimensional treatment planning system utilizing interactive colour graphics. Br J Radiol 1979;52:478–481.
5. Smith Rp, Mckenna WG. The Basics of Radiation Therapy. In: Abeloff MD, et al. editors. Clinical Oncology 3rd ed. Elsevier; Churchill Livingstone: 2004.
6. Jones B, et al. United Kingdom radiation oncology 1 conference (UKRO 1): Accuracy and uncertainty in radiotherapy. Br J Radiol 2002;75:297–305.
7. Sherouse GW, et al. Virtual simulation: concept and implementation. The 9th International Conference of the use of computers in radiation therapy (ICCR). North Holland Publishing Co.; The Netherland: 1987.
8. Sherouse GW, Novins K, Chaney EL. Computation of digitally reconstructed radiographs for use in radiotherapy treatment design. Int J Radiat Oncol Biol Phys 1990;18:651–658.
9. Sherouse GW, Bourland JD, Reynolds K. Virtual simulation in the clinical setting: some practical considerations. Int J Radiat Oncol Biol Phys 1990;19:1059–1065.
10. Mutac S, et al. Quality assurance for computed-tomography simulators and the computed-tomography-simulation process: Report of the AAPM radiation therapy committee Task Group No. 66. Med Phys 2003;30:2762–2792.
11. Gerber RL, Purdy JA. Quality assurance procedures and performance testing for CT-simulators. In: Purdy JA, Starkschall G, editors. A Practical Guide to 3-D Planning and Conformal Radiation Therapy. Advanced Medical Publishing, Inc.; Middleton (WI): 1999.
12. Conway J, Robinson MH. CT virtual simulation. Br J Radiol 70 (Suppl.):1997;S106–S118.
13. Aird EGA, Conway J. CT simulation for radiotherapy treatment planning. Br J Radiol 2002;75:937–949.
14. Fraass B, et al. American Association of Physicists in Medicine Radiation Therapy Committee Task Group 53: Quality assurance for clinical radiotherapy treatment planning. Med Phys 1998;25:1773–1829.
15. Van Dyk J, Taylor JS. CT-simulators. In: Van Dyk J. editor. The Modern Technology for Radiation Oncology: A Compendium for Medical Physicists and Radiation Oncologists. Medical Physics Publishing; Madison (WI): 1999.
16. Kutcher GJ, et al. Comprehensive QA for radiation oncology: report of AAPM Radiation Therapy Committee Task Group 40. Med Phys 1994;21:581–618.
17. AAPM, Report No. 39. Specification and Acceptance Testing of Computed Tomography Scanners. American Institute of Physics; New York: 1993.
18. Kalender WA. Computed Tomography: Fundamentals, System Technology, Image Quality, Applications. 2nd ed. Wiley; New York: 2004.
19. Hsieh J. Computed Tomography: Principles, Design, Artifacts, and Recent Advances. SPIE Press; Bellingham (WA): 2003.
20. Wang G, Crawford CR, Kalender WA. Multi-row-detector and cone-beam spiral/helical CT. IEEE Trans Med Imag 2000; 19:922–929.
21. Tang X, Hsieh J. A filtered backprojection algorithm for cone beam reconstruction using rotational filtering under helical source trajectory. Med Phys 2004;31:2949–2960.
22. Tang X, Hsieh J, Nilsen RA, Dutta S. A helical cone beam filtered backprojection (CB-FBP) reconstruction algorithm using three-dimensional (3D) view weighting. SPIE Proc 2004;5535:577–587.
23. Langmack KA. Portal Imaging. Br J Radiol 2001;74:789–804.
24. Verrellen D, Linthout N, Berge DVD, Bel A, Storme G. Initial experience with intensity modulated therapy for treatment of the head and neck region. Int J Radio Oncol Bio Phys 1997;39:99–114.
25. Grant W, Woo SY. Clinical and financial issues for intensity-modulated radiation therapy delivery. Semin Radia Oncol 1999;9:99–107.
26. Korba A, et al. Pseudoblocks and portal localization. Radiology 1977;122:260–261.
27. Convery DJ, Rosenbloom ME. The generation of intensity modulated fields for conformal radiotherapy by dynamic collimation. Phys Med Biol 1992;37:48–59.

28. Stein J, Bortfeld T, Dorschel B, Schlegel W. Dynamic X-ray compensation for conformal radiotherapy by means of multi-leaf collimation. Radiother Oncol 1994;32:163–173.

29. Boyer AL, Yu CX. Intensity-modulated radiation therapy with dynamic multi-leaf collimators. Semin Radiat Oncol 1999;32:48–59.

30. Nutting C, Dearnaley DP, Webb S. Intensity modulated radiation therapy: a clinical review. Br J Radiol 2000;73:459–469.

31. Intensity Modulated Radiation Therapy Collaborative Working Group, Intensity-modulated radiotherapy: current status and issues of interest. Int J Radiat Oncol Biol Phys 2001; 51:880–914.

32. Purdy JA. Advances in three-dimensional treatment planning and conformal dose delivery. Semin Oncol 1997;24:655–672.

33. Boyer AL, Xiang L, Xia P. Beam shaping and intensity modulation in modern technology of radiation oncology. Van Dyk J, editors. Modern Technology of Radiation Oncology. Medical Physics Publishing; Madison (WI): 1999.

34. Wang X, et al. Dosimetric verification of intensity-modulated fields. Med Phys 1996;23:317–327.

35. Xing L, et al. Dosimetric verification of a commercial inverse treatment planning system. Phys Med Biol 1999;44:463–478.

36. Mackie TR, et al. Tomotherapy: a new concept for the delivery of conformal radiotherapy. Med Phys 1993;20:1709–1719.

37. Yu CX. Intensity-modulated arc therapy with Dynamic multi-leaf collimation: an alternative to tomotherapy. Phys Med Biol 1995;40:1435–1449.

38. Mackie TR, et al. Tomotherapy. Semin Radiat Oncol 1999; 9:108–117.

39. Svensson R, Lind BK, Brahme A. A new compact treatment unit design combining narrow pencil beam scanning and segmental multileaf collimation. Radiothr Oncol 1999;51:S21.

40. Schweikard A, Tombropoulos R, Adler JR. Robotic radiosurgery with beams of adaptable shape. In: Ayache N, editor. Computer Vision and Robotics in Medicine. Springer-Verlag; Berlin (Heidelberg): 1995.

41. Webb S. Conformal intensity-modulated radiotherapy (IMRT) delivered by robotic linacs—testing IMRT to the limit? Phys Med Biol 1999;44:1639–1654.

42. Woo SY, et al. A comparison of intensity modulated conformal therapy with a conventional external beam steoreotactic radiosurgery system for the treatment of single and multiple intracranial lesions. Int J Radia Oncol Biol Phys 1996; 35:593–597.

43. Kramer BA, et al. Dosmetric comparison of stereotactic radiosurgery to intensity modulated radiotherapy. Radia Oncol Invest 1998;6:18–25.

44. Cardinale RM, et al. A comparison of three stereotactic radiosurgery techniques: arcs vs non-coplanar fixed fields vs intensity modulation. Int J Radia Oncol Bio Phys 1998;42: 431–436.

45. Botfeld T. Optimized planning using physical objects and constraints. Semin Radiat Oncol 1999;9:20–34.

46. Chvetsov AV, Calvetti D, Sohn JW, Kinsella T. Regularization of inverse planning for intensity-modulated radiotherapy. Med Phys 2005;32:501–514.

47. Sauer OA, Shepard DM, Mackie TR. Application of constrained optimization to radiotherapy planning. Med Phys 1999;26:2359–2366.

48. Spirou SV, Chui CS. A gradient inverse planning algorithm with dose-volume constraints. Med Phys 1998;25:321–333.

49. Hristov DH, Stavrev P, Sham E, Fallone BG. On the implementation of dose-volume objectives in gradient algorithms for inverse treatment planning. Med Phys 2002;29:848–856.

50. Wu Q, Mohan R, Niemierko A, Schmidt-Ullrich R. Optimization of intensity-modulated radiotherapy plans based on the equivalent uniform dose. [comment]. Int J Radia Oncol Biol Phys 2002;52:224–235.

51. Das S, et al. Beam orientation selection for intensity-modulated radiation therapy based on target equivalent uniform dose maximization. Int J Radia Oncol Biol Phys 2003;55:215–224.

52. Choi B, Deasy JO. The generalized equivalent uniform dose function as a basis for intensity-modulated treatment planning. Phys Med Biol 2002;47:3579–35894.

53. Wang XH, et al. Optimization of intensity 3D conformal treatment plans based on biological indices. [comment]. Radiothre Oncol 1995;37:140–152.

54. Baydush AH, Marks LB, Das SK. Penalized likelihood fluence optimization with evolutionary components for intensity modulated radiation therapy treatment planning. Med Phys 2004;31:2335–2343.

55. Webb S. Optimization by simulated annealing of three-dimensional conformal treatment planning for radiation fields defined by a multileaf collimator. Phys Med Biol 1991;36:1201–1226.

56. Hristov DH, Fallone BG. An active set algorithm for treatment planning optimization. Med Phys 1997;24:1455–1464.

57. Ezzell GA. Genetic and geometric optimization of three-dimensional radiation therapy treatment planning. Med Phys 1996;23:293–305.

58. Olivera GH, et al. Maximum Likelihood as a common computational framework in tomotherapy. Phys Med Biol 1998;43: 3277–3294.

59. Llacer J, Solberg TD, Promberger C. Comparative behavior of the dynamically penalized likelihood algorithm in inverse radiation therapy planning. Phys Med Biol 2001;46:2637–2663.

60. Yan H, Yin F, Guan H, Kim JH. Fuzzy logic guided inverse treatment planning. Med Phys 2003;30:2675–2685.

61. McKezie AL. How should breathing motion be combined with other errors when drawing margins around clinical target volumes? Br J Radiol 2000;73:973–977.

62. Bergstrom P, Lofroth PO, Widmark A. High precision conformal radiotherapy (HPCRT) of prostate cancer—a new technique for exact positioning of the prostate at the time of treatment. Int J Radiat Oncol Biol Phys 1998;42:305–311.

63. The BS, et al. Intensity modulated radiation therapy (IMRT) following prostateectomy: more favorable acute genitourinary toxicity profile compared to primary IMRT for prostate cancer. Int J Radiat Oncol Biol Phys 2001;49:465–472.

64. Matsinos E. Current status of the CBCT project at Varian. Proc SPIE 2005.

65. Colbeth RE, Roos PG, Mollov IP. Flat panel CT detector for sub-second volumetric scanning. Proc SPIE 2005.

See also PHANTOM MATERIALS IN RADIOLOGY; RADIATION THERAPY SIMULATOR.

COMPUTED TOMOGRAPHY, SINGLE PHOTON EMISSION

FREDERIC H. FAHEY
Children's Hospital Boston

HARVEY A. ZIESSMAN
Johns Hopkins University

INTRODUCTION

Single photon emission computed tomography (or SPECT) provides a three-dimensional (3D) representation of the distribution within a patient's body of a radiopharmaceutical that was given as part of a diagnostic nuclear medicine study. Diagnostic nuclear medicine provides a unique way of imaging physiology and function. One can administer to a patient a pharmaceutical with a radioactive marker, and then use external detectors to determine where that "radiopharmaceutical" has distributed within the patient's body. The distribution of the radiopharmaceutical in the body depends on its specific biology. For example, suppose a patient is given a small amount of radioactive iodine (e.g., iodine-131 or ^{131}I). Since the thyroid naturally metabolizes iodine, some portion of the radioactive iodine will be preferentially incorporated into the thyroid with the rest going to other organs with lower concentrations. The amount of ^{131}I that will go to the thyroid will depend on whether the thyroid is functioning in a normal, hyperactive, or hypoactive fashion. By acquiring an image from this patient, one can determine whether certain regions of the thyroid are more active than others. For example, there may be hyperactive nodules within the thyroid. In this manner, nuclear medicine provides a unique opportunity to view the patient's physiology and not just the anatomy. Devices known as gamma cameras are used to provide images of the *in vivo* distribution (i.e., the distribution within the body) of the radiopharmaceutical. From these image data, the patient's specific physiology can be inferred.

The images produced by gamma cameras are two dimensional (2D) representations of a 3D object. In some cases, this is adequate to interpret the study. However, in many cases, the ambiguity introduced by activity in the overlying and underlying tissue can make it very difficult to infer the *in vivo* distribution of the radiopharmaceutical appropriately. In these cases, a 3D representation is necessary. Single photon emission computed tomography provides such a 3D representation. For example, SPECT can make it easier to determine whether the activity is reduced in the basal or apical aspect of the inferior wall of the myocardium or whether the activity in a tumor seen in the chest is in a rib or in the lung. As will be discussed, the 3D SPECT images are generated by a computer from a series of images taken about the patient at different angles. Since the photons used to generate the images are emitted from within the patient's body, SPECT is considered *emission* computed tomography, which is in contrast to *transmission* computed tomography (CT) where the X-ray photons emanate from an X-ray tube and are transmitted through the patient. In the early development of SPECT, the word "single" was used to distinguish SPECT from positron emission tomography (PET), which uses two photons to localize each event and requires different instrumentation. For these reasons, the generation of a 3D representation of the *in vivo* distribution of radiopharmaceutical that is not a positron emitter is referred to as single photon emission computed tomography or SPECT.

In 1963, a method of nuclear tomographic imaging was developed by Kuhl and Edwards (1). This method used a specially designed scanning device along with a processing method called simple backprojection to generate its 3D images. This early research in SPECT predates Hounsfield's work in CT by ∼10 years. Kuhl and Edwards (2) subsequently developed a dedicated SPECT device for imaging the brain known as the Mark IV scanner. During the 1970s, several devices were developed that were dedicated to SPECT, specifically for brain imaging. However, techniques were also being investigated to use the gamma camera that was used for all other nuclear imaging applications for SPECT as well. This required the gamma camera to rotate around the patient's body in order to acquire different angular views. Keyes et al. (3) developed the first prototype of the rotating gamma camera by mounting a standard gamma camera head to the gantry of a decommissioned cesium-137 radiation therapy unit. At about the same time, Jaszczak et al. (4) developed a commercial version of the rotating gamma camera. Although the development of dedicated SPECT devices continues, by far the majority of devices used clinically are rotating gamma cameras. For this reason, this chapter will devote most of its attention to the use of the rotating gamma camera.

SPECT DATA ACQUISITION AND PROCESSING

For SPECT data acquisition, a series of images are obtained at a number of different viewing angles. Typically, these images are acquired in a circular arc about the patient with the axis of rotation parallel to the long axis of the patient's body. This acquisition geometry is shown in Fig. 1. The images acquired at each viewing angle are referred to as "projection images" or "projections". These projections are acquired at a number of evenly spaced viewing angles over either a 360 or a 180° arc. There is a minimum number of projections that will assure adequate angular sampling at the periphery of the object being imaged which will lead to high quality, tomographic images. For SPECT, this number is between 50 and 150, depending on the size of the object and the spatial resolution of the system. The gamma camera is usually equipped with a multihole, parallel-hole collimator that assures that the photons interacting in the radiation detector of the gamma camera traveled from the patient to the detector on a ray that is perpendicular to the detector surface. Thus, if one knows where the photon interacted in the detector, one can assume that the photon was emitted from a point that was along this ray. In Fig. 1, we refer to this ray as the "line of origin".

Consider a small, high contrast feature in the center of a cylindrical object that contains some radioactivity as shown in Fig. 2. We can acquire some number of projections (3 in the example shown in Fig. 2). If we consider the tomographic plane passing through the center of the tumor, each of the projections will look reasonably similar with a single intensity corresponding to the tumor. The process by which we generate a tomographic image from this series of projections is referred to as "tomographic reconstruction". We can generate a tomographic image of this simple object by filtering and adding the projections, each oriented at the angle associated with that specific projection. This reconstruction technique is referred to as "filtered backprojection". In many cases, additional smoothing is applied using a windowing filter that can control the sharpness and noise associated with

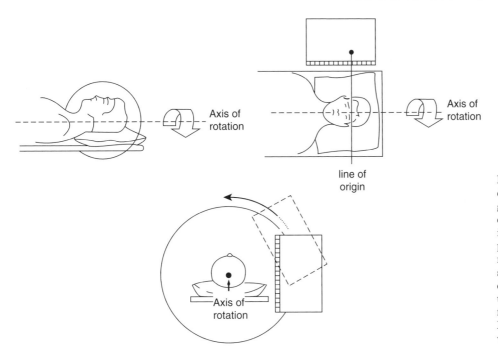

line of
origin

Figure 1. The geometry for SPECT data acquisition with a rotating gamma camera is shown. The camera rotates acquiring projection images at different angles about the patient. These data are subsequently reconstructed into cross-sectional slices that indicate the 3D *in vivo* distribution of the radiopharmaceutical within the patient. (Reprinted from Henkin R., editor, Nuclear Medicine. P 235, Copyright © 1996 with permission from Elsevier.)

the reconstructed image. The windowing functions that are typically used include the Hanning, Hamming, Shepp-Logan, and Butterworth filters. Depending on the signal and noise content of the underlying projection data, one can choose an appropriate windowing filter for the best image quality. Figures 3a–c shows three images that are filtered backprojection reconstructions of the same raw data (in this case, a brain study) using three different windowing filters (sharp filter, moderate filter and smooth filter, respectively). One can note the differences in image quality that one can attain by simply varying the filtering.

In addition, different clinical applications (e.g., brain vs. cardiac SPECT) may require different reconstruction filters. Therefore, it is very important to select the most

appropriate reconstruction filter for each clinical application of SPECT.

Although filtered backprojection has traditionally been the most common means of SPECT reconstruction, iterative reconstruction methods are currently available on newer SPECT systems. These methods utilize a "feedback" approach to generate the tomographic data. These methods start with an initial estimate of the object. This estimate may assume the object is totally uniform or it may use a filtered backprojection reconstruction of the object. From this estimate, a series of projections are calculated along the same viewing angles as the "real" projections, that is the acquired, raw data. If the estimate is close to the true object, then the calculated projections would be very similar to the real projections. If they are not similar, then the variations between the two are determined (either as a ratio or a difference) and used to alter the initial estimate. The process is then repeated. A new set of calculated projections are generated and again compared to the real projections. Presumably, each iteration provides a better estimate of the true object. In other words, the estimates should "converge" to a good representation of the true object. Typically, some statistic such as the "likelihood" or the "entropy" is used to determine how well the method is converging or when to stop the iterative process. In many cases, it can take tens or even hundreds of iterations before the reconstructed estimate converges. Thus, these methods have traditionally been quite slow, much slower than filtered backprojection. However, an advantage of these methods is the ability to take into account the physics of data collection and the statistics of the noise in the images to provide a more accurate reconstruction. Another advantage is that these methods are not as susceptible to some of the artifacts that one encounters in filtered backprojection. Much of the research in this area has centered on the development of methods that are more efficient or converge more

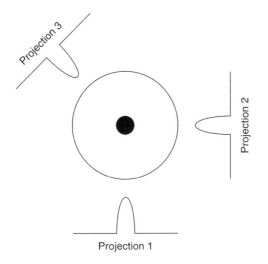

Figure 2. The object in the middle has a central region of high intensity. Three projections about the object are also shown. In a typical SPECT study, 50–150 projections are acquired about the object over 180 or 360°.

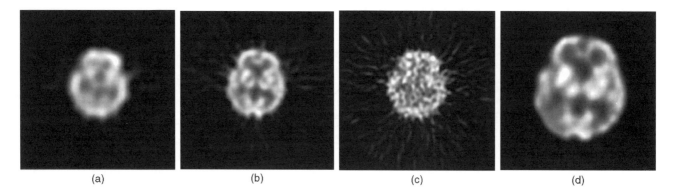

(a)   (b)   (c)   (d)

Figure 3. The SPECT brain study with four different reconstructions. Figures 3a–c are reconstructed with three different filters (smooth, moderate and sharp filter, respectively). Figure 3d is reconstructed with an iterative method known as OSEM.

quickly. With the increasing speed of computers and the development of more efficient iterative algorithms, the clinical application of these methods has become feasible and many newer SPECT systems provide iterative reconstruction methods as an option. Figure 3 shows a comparison with the same raw SPECT data reconstructed with both an iterative method (3d) and filtered backprojection (3a–c).

Consider a radiopharmaceutical that basically distributes uniformly within the body. Those photons that are emitted from deep within the body will have to travel through more tissue to reach the gamma camera than those emitted on the periphery of the body. In turn, those photons that must traverse more tissue are more likely to be absorbed within the body and, therefore, are less likely to be detected than those emitted at the periphery. In other words, the signal from deep-seated tissues is "attenuated" as compared to the surface tissues due to self-absorption of the photons within the object. To measure the amount of activity at different locations within the object accurately, one must apply a correction for photon attenuation. There are two basic approaches to attenuation correction in SPECT, one that assumes that the attenuation is uniform within the object, and one that considers the fact that different tissues may have varying absorption characteristics making the attenuation nonuniform.

In much of the body, the composition and density of the soft tissue is basically constant and thus one can assume a single attenuation property for all of the tissue. This assumption is apt, for example, in brain imaging. To apply uniform attenuation correction, one need only know where the body outline is located relative to where the radiopharmaceutical has distributed. For a particular point of interest within the object, the mean distance to the body outline is estimated and, using this estimate, the expected amount of photon attenuation from that point is determined. By multiplying the amount of radioactivity at that point by the reciprocal of the calculated photon attenuation at that point, one can estimate the amount of signal one would have received from that location if there were no photon attenuation. This correction can, in turn, be applied for every location within the object. If this correction is applied to the object with a uniform radioactivity distribution, a uniform signal throughout the object would be obtained. This method, referred to as the first-order Chang correction for photon attenuation, is the most common approach to applying uniform attenuation correction in clinical SPECT (5).

This uniform assumption, however, is not at all appropriate for the thorax where there is lung tissue, spine, and mediastinum in addition to soft tissue, all of which have very different attenuating properties. If a uniform attenuation assumption was used, one would overcorrect the regions of the lungs and undercorrect the regions near the spine and mediastinum. This is of particular concern for myocardial SPECT. For these reasons, no attenuation correction was applied traditionally for cardiac SPECT since no correction was considered better than a poor correction. However, over the past 10 years, a number of investigators have implemented various approaches to non-uniform attenuation correction. In these cases, one needs to not only know the body outline, but also must know the types of tissues within that outline and their attenuation characteristics. To determine this, one acquires a transmission image using an external, photon-emitting source. These data indicate which regions within the body are highly attenuating and which yield less attenuating. This knowledge is incorporated into the SPECT reconstruction process to correct for the non-uniform attenuation. A variety of ways have been developed for the acquisition of the transmission data, several of which will be discussed in the section on instrumentation. In addition to attenuation correction, several other corrections have been developed for SPECT in order to provide more quantitative reconstructed data including those for scatter and resolution recovery.

SPECT INSTRUMENTATION

SPECT requires the ability to acquire projection images from a number of viewing angles about the patient. Thus the gamma camera must be mounted onto a gantry that allows the camera to rotate about the patient in a circular or elliptical orbit. One of the limitations of the rotating gamma camera is the inherently low sensitivity of the system. Since the cameras must utilize absorptive collimation to determine the directionality of the interacting

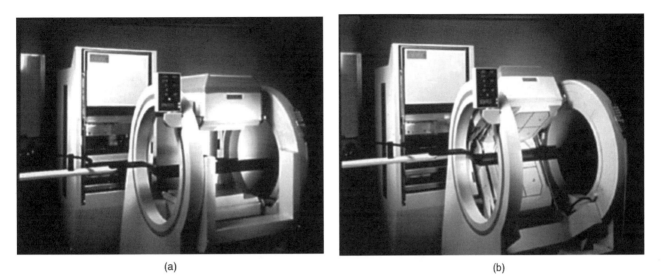

(a)   (b)

Figure 4. Dual-detector SPECT camera. With this camera, the two detectors can be oriented at 180° to each other for whole body imaging (a) or 90° for cardiac imaging (b).

photons, only a very small fraction of the photons emitted from within the patient will actually be detected by the gamma camera. More stringent collimation can be used to improve the spatial resolution but at a cost of lower sensitivity leading to a higher level of noise in the images. One straightforward approach to improving the sensitivity is to increase the number of detectors. Thus, both dual-detector and triple-detector SPECT systems have been developed. Figure 4a shows a modern dual-detector SPECT system.

Since the heart is located in the left anterior portion of the chest and low energy radiopharmaceuticals, such as thallium-201, are routinely used in this application, it is common to acquire cardiac SPECT data only over 180° (from right anterior oblique to left posterior oblique) rather than over 360°, since most of the data that is acquired in the right posterior projection only adds noise and poor resolution to the image. This being the case, the use of two opposing detectors does not reduce the total imaging time, since one will still need to rotate the gantry over 180°. For this reason, a number of manufacturers have designed their dual-detector cameras such that the data can be acquired with the detectors either opposing each other (180° orientation) and in a 90° orientation as shown in Fig. 4b. This allows for the same amount of SPECT data to be acquired in half the time. Increasing the number to three detectors improves the sensitivity of the SPECT device even further. Such triple-detector devices are excellent for acquiring SPECT but lack the flexibility for other nuclear imaging and thus tend to be less popular than the dual-detector systems.

As discussed previously, a transmission scan must be acquired in order to perform non-uniform attenuation correction. Several different approaches have been developed for the acquisition of the transmission image. Collimated, radioactive line sources can be scanned over an area of the patient during the acquisition of the emission data. Since the radionuclide in the sources (^{153}Gd) emits a gamma ray with a slightly different photon energy than the radiopharmaceutical administered to the patient, the two data sets (emission and transmission) can be acquired simultaneously. In an alternate method, a series of smaller line sources are used. These also contain ^{153}Gd and thus again the emission and transmission data can be acquired simultaneously.

In the past several years, hybrid SPECT–CT systems have been developed. In these devices, a helical CT study is acquired in conjunction with the SPECT study on the same device. The CT scan is used to characterize the material within the body such that a non-uniform attenuation correction can be applied. It typically requires a transformation to be performed between the CT values and the attenuating coefficients for SPECT because the energies of the photons in CT are different that those for SPECT. In some cases, the device provides a diagnostic quality CT that can be interpreted either in conjunction with the SPECT study or independently. In the case of one SPECT–CT device, the CT provided is not of diagnostic quality, and it is used only for attenuation correction and gross anatomical correlation.

With the newest developments in molecular medicine has come the desire to image small animals such as rodents. Thus a number of investigators have developed methods of performing SPECT imaging in rodents with very high spatial resolution. One approach that is straightforward is the use of very small pinhole collimators. The size of pinhole is only ~1 mm as compared to the 4–6-mm pinholes that are typically used for clinical imaging. This method can provide SPECT images that have spatial resolution that is better than what is typical in clinical SPECT by almost a factor of 10. These high resolution approaches cannot be applied in clinical SPECT because these very small pinholes are very inefficient and thus an inordinate image time on the order of several hours would be required in order to obtain human images of sufficient image quality.

CLINCIAL APPLICATIONS IN SPECT

In this section, we will review several of the most common clinical applications of SPECT.

Cardiac

The most common clinical indication for SPECT is myocardial (cardiac) perfusion imaging. Atherosclerotic heart disease is manifested by coronary artery narrowing, which limits blood flow to the region of the heart supplied by that artery, producing chest pain or myocardial infarction. SPECT can confirm or exclude significant coronary disease. If abnormal, invasive coronary angiography may be performed in anticipation of intervention, for example, coronary angioplasty or bypass surgery.

The radiopharmaceutical, thallium-201 or newer technetium-labeled cardiac radiotracers, is delivered by the individual coronary arteries to the myocardium where it is extracted. The SPECT images depict the 3D blood flow to each region of the myocardium supplied by its coronary artery.

The study is performed in two stages, at rest and with stress. Treadmill exercise is the usual stress, although pharmacologic methods are used in those unable to exercise. In a normal heart, SPECT will show good blood flow at rest and stress. In a patient with a prior myocardial infarction, no blood flow will be seen in the nonviable region at either stage. A patient with significant coronary artery stenosis, without infarction, will have a normal rest study, but an abnormal exercise study (Fig. 5).

Adequate blood flow and oxygen can be delivered to a resting heart even with a high grade coronary artery obstruction, however, the increased demand for oxygen and blood flow required at stress cannot be met and myocardial uptake will be decreased in the myocardium fed by that artery.

Tumors

SPECT with various radiopharmaceuticals provide valuable information regarding the extent of disease and distribution in the body. This information is used for initial diagnosis, staging, preoperative localization, and evaluating response to therapy.

Gallium-67 citrate (^{67}Ga) has been used for several decades for tumor imaging. It binds nonspecifically to a variety of tumors. Gallium-67 SPECT is used most commonly for imaging malignant lymphoma, a disease of lymphatic tissue seen in adults and children. Images depict the extent of disease and the effectiveness of therapy. This SPECT imaging is more accurate than CT or magnetic resonance imaging (MRI) for determining the effectiveness of therapy.

^{111}In OctreoScan is a somatostatin receptor peptide imaging agent, most useful for imaging tumors of neuroendocrine origin, (carcinoid, gastrinoma, neuroblastoma, pheochromocytoma, etc.) SPECT cross-sectional imaging makes it possible to see small tumors that are often not detected with CT or MRI.

^{111}In ProstaScint is radiolabeled monoclonal antibody directed against antigens on the surface of prostate cancer cells. It can detect the site of recurrence of prostate cancer,

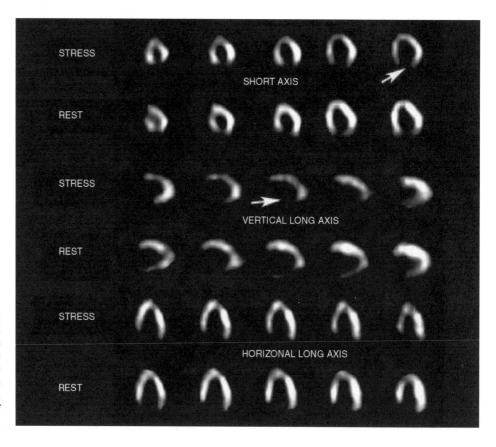

Figure 5. The SPECT exercise and rest cardiac study. Short axis, vertical long axis, and horizontal long axis cross-sectional sequential slices are shown. Arrows point to the inferior wall where there is no perfusion. Since this finding is unchanged between stress and rest, this is diagnostic of an inferior wall myocardial infarction.

suggested by a rising prostate serum antigen (PSA) level. Adequate imaging is not possible without SPECT because of the relatively high background activity.

Brain Imaging

SPECT is mandatory for brain imaging in order to visualize the various overlying and underlying convoluted regions of the brain.

Seizures are often caused by small regions of scar tissue in the brain. Patients not responding to drug therapy for their seizure disorder may be helped with resection of the seizure focus. Proper localization of the seizure site is critical for effective surgery. Brain wave studies (electroencephalogram or EEG) are only moderately successful in locating a seizure site. Even then, additional studies are needed for confirmation. The traditional method is to place electrodes on the surface of the brain and record electrical activity. However, this requires a neurosurgical operation and is associated with some morbidity.

SPECT brain blood flow radiopharmaceuticals, ^{99m}Tc hexamethyl propylene amine oxime (^{99m}Tc HMPAO) and ^{99m}Tc ethyl cysteinate dimer (^{99m}Tc ECD), show the distribution of blood flow in the brain. During seizure activity, the small area of the brain responsible has increased metabolism and increased blood flow. Between seizures, the abnormal site has decreased metabolism and blood flow. SPECT imaging can localize the seizure site by detecting these focal abnormal blood flow patterns. An example of a SPECT study in a seizure patient is shown in Fig. 6.

The second indication is to determine the cause for dementia. Characteristic patterns of abnormal perfusion are seen with certain types of dementia, for example, Alzheimer's disease, frontal lobe dementias, and multi-infarction dementia (strokes).

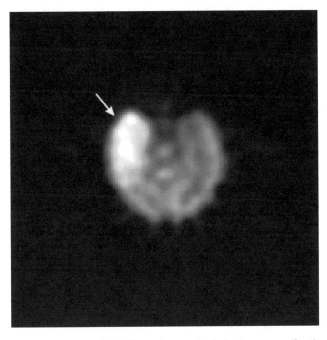

Figure 6. Brain SPECT study acquired during an epileptic seizure. The arrow indicates a region of high blood flow that corresponds to the part of the brain where the seizure originated.

Bone

SPECT is used for a variety of indications. It can help detect small sites of tumor, fracture, or infection not easily seen or localized with traditional two-dimensional bone scanning. Precise localization of a bone radiopharmaceutical uptake (e.g., ^{99m}Tc diphosphonate), can help differentiate benign from malignant processes, confirm small fractures as the cause of pain, and detect sites of infection.

SUMMARY

SPECT is a 3D imaging approach for evaluating the physiology and function of the patient. The patient is injected with a small amount of a radiopharmaceutical and a series of projection images are acquired at different angles about the patient. These data are reconstructed into a series of cross-section views of the *in vivo* distribution of the radiopharmaceutical. Depending on the radiopharmaceutical used, the nuclear medicine physician can infer essential information regarding the patient's physiologic condition. The rotating gamma camera is the most common device used to acquire SPECT data. Recent developments include the incorporation of CT into a hybrid SPECT–CT imaging device and the use of very small pinhole collimators for the imaging of small animals. With the advancements of molecular approaches to medicine, SPECT will continue to be a very important approach to medical imaging.

BIBLIOGRAPHY

1. Kuhl DE, Edwards RQ. Image separation radioisotope scanning. Radiology 1963;80: 653–661.
2. Kuhl DE, Edwards RQ. The Mark 3 Scanner: a compact device for multiple-view and section scanning of the brain. Radiology 1970; 96:563–70.
3. Keyes JW, Jr., Orlandea N, Heetderks WJ, Leonard PF, Rogers WL. The Humongotron—a scintillation-camera transaxial tomograph. J Nucl Med 1977; 18:381–387.
4. Jaszczak RJ, Murphy PH, Huard D, Burdine JA. Radionuclide emission computed tomography of the head with ^{99m}Tc and a scintillation camera. J Nucl Med 1977; 18:373–380.
5. Chang LT. A method for attenuation correction in radionuclide computed tomography. IEEE Trans Nucl Sci 1978; NS-25: 638–643.

Further Reading

Tsui BM. The AAPM/RSNA physics tutorial for residents. Physics of SPECT. *Radiographics* 1996; 16:173–183.
Miller TR. The AAPM/RSNA physics tutorial for residents. Clinical aspects of emission tomography. *Radiographics* 1996; 16:661–668.
Tsui BM, Frey EC, LaCroix KJ, Lalush DS, McCartney WH, King MA, Gullberg GT. Quantitative myocardial perfusion SPECT. *J Nucl Cardiol* 1998; 5:507–522.
Madsen MT. The AAPM/RSNA physics tutorial for residents. Introduction to emission CT. *Radiographics* 1995; 15:975–991.
King MA, Tsui BM, Pan TS. Attenuation compensation for cardiac single-photon emission computed tomographic imaging: Part 1. Impact of attenuation and methods of estimating attenuation maps. *J Nucl Cardiol* 1995; 2:513–524.

King MA, Tsui BM, Pan TS, Glick SJ, Soares EJ. Attenuation compensation for cardiac single-photon emission computed tomographic imaging: Part 1. Attenuation compensation algorithms. *J Nucl Cardiol* 1996; 3:55–64.

Groch MW, Erwin WD. SPECT in the year 2000: basic principles. *J Nucl Med Technol* 2000; 28:233–244.

Cherry SR, Sorenson JA, Phelps ME. Physics in Nuclear Medicine. 3rd ed. Philadelphia: Saunders: 2003. p 299–324.

See also ANGER CAMERA; NUCLEAR MEDICINE INSTRUMENTATION; POSITRON EMISSION TOMOGRAPHY; RADIOPHARMACEUTICAL DOSIMETRY.

COMPUTER-AIDED RADIATION DOSE PLANNING. See RADIATION DOSE PLANNING, COMPUTER-AIDED.

COMPUTER-ASSISTED DETECTION AND DIAGNOSIS

ROBERT M. NISHIKAWA
The University of Chicago
Chicago, Illinois

INTRODUCTION

Since their discovery, X rays have been used to make images (radiographs) that allow the internal condition of the human body to be examined. Reading or interpreting these images has been without exception performed solely by humans until very recently. As our society depends more and more on automation or assistance from automated systems, so too has the interpretation of radiographs, although still in a very limited way at the present time.

In the early 1960s, researchers attempted to automate the interpretation of radiographs. The first published study was by Winsberg et al. who developed an automated computerized scheme to diagnosis breast cancer from a radiograph of the breast (mammogram) (1). This attempt, like others from that time period, was largely unsuccessful. Compared to current technology, these studies suffered from poor quality film digitizers (all images were recorded on film), insufficiently powered computers that had severely limited memory and storage space, and only a rudimentary armament of image processing, pattern recognition and artificial intelligence techniques. Clearly, the goal of automating the interpretation of radiographs was beyond the technical capabilities of that era.

After those initial attempts, there was a period of inactivity. In the late 1980s, a new approach was developed called computer-aided diagnosis (CAD) (2–5). The goal here was not to automate the interpretation of radiographs, but to give assistance to radiologists when they read images. The seminal paper was published in 1990 by Chan et al. (6). They conducted an observer study, where radiologists read mammograms once without the computer aid and once with the computer aid. For this study, the computer aid was a computer-aided detection (CADe) scheme that detected microcalcifications on mammograms. Microcalcifications

are tiny deposits of calcium that can be an early indicator of breast cancer. Chan et al. (2) found a statistically significant improvement in the performance of radiologists in detecting microcalcifications when the radiologists used the computer aid. This study was the first CAD algorithm of any kind shown to be a benefit to radiologists and it validated the concept of the computer as an aid to the radiologist. It has spurred CAD research in mammography and in other organs, and in other imaging modalities.

Since that study, the field has grown rapidly. There are now at least four commercial systems available [two for detecting breast cancer from mammograms, and one each for detecting lung cancer from computed tomography (CT) scans and from chest radiographs]. Further, CAD is being developed for a wide variety of body parts (breast, thorax, colon, bone, liver, brain, heart, vasculature, and more) and several different imaging modalities [principally, radiography, magnetic resonance imaging (MRI), CT, and ultrasound]. Since mammography is the most mature area of CAD development, most of the examples used in this article to illustrate the principals of CAD will be drawn from mammographic CAD.

Mammography can detect breast cancer before it appears clinically (i.e., before it can be palpated or some other physical sign appears, such as a nipple discharge or breast pain). It has been shown to reduce breast cancer mortality. Typically, when a woman receives a mammogram, two X-ray images from two different angles are taken of each breast. One view is taken in the head to toe direction and the other at a 45° angle to the first. To detect breast cancer mammographically, radiologists look principally for two types of lesions. The first are masses, which are typically 0.5–5 cm in size. They appear slightly brighter than the surrounding tissue, but often have low contrast. Further, their appearance can be obscured by the normal structures of breast making them even more difficult to detect. Compounding their detection are normal tissues that appear to be a mass-like. This can occur if overlapping tissues in the breast are projected from a three-dimensional (3D) volume into the two-dimensional (2D) image. Taking two different views of each breast can help the radiologist to differentiate actual masses from a superposition of normal tissues, but actual masses are sometimes only seen in one view. The second type of lesion are microcalcifications, which are tiny deposits of calcium salts ranging in size from 10 μm up to 1 mm in size, however, mammographically microcalcifications < 200 μm are usually not detectable in a screening mammogram. Microcalcifications are difficult to detect because of their small size. Unlike masses that appear with low contrast, the principal limitation to detecting microcalcifications is the presence of image noise (i.e., a low signal/noise ratio).

Computer-aided diagnosis is well suited to mammography for several reasons. First, the only purpose for mammography is to detect breast cancer: Up to as many as 100 different abnormal conditions can be discovered from a chest X ray. Second, mammography is used as a screening modality, so the number of mammograms to be read is high. It is one of the most common radiological procedures performed. Third, breast cancer is difficult to

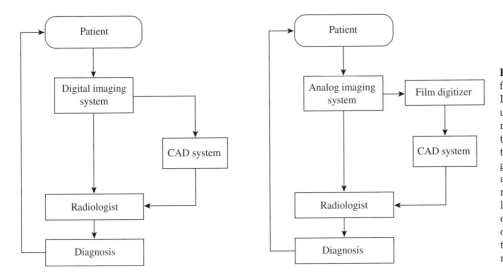

Figure 1. Schematic representation for the clinical implementation of CAD. If the image is acquired digitally (e.g., using CT, MRI, ultrasound, nuclear medicine, or a digital X-ray detector) the image can be used directly as input to a CAD scheme. If the patient is radiographed using an analog system (i.e., a screen-film system), then the image needs to digitized before it can be analyzed by the CAD system. In either case, the radiologist views the image or images and then views the output of the CAD systems, after which the radiologist makes their decision.

detect at an early stage mammographically. It requires that the radiologist careful check each image with a magnifying glass. Fourth, the prevalence is low: Only ~4 in every 1000 women screened has breast cancer. If a radiologist read 50 mammographic exams a day, they would see on average only 1 patient with breast cancer every week. Under such conditions, it is difficult to remain vigilant and alert. Fifth, mammography is one of the most common sources of malpractice lawsuits in the United States, so that missing a breast cancer can have severe consequences both for the patient and for the radiologist.

COMPUTER-AIDED DETECTION AND COMPUTER-AIDED DIAGNOSIS

What is Computer-Aided Diagnosis

The formal definition of CAD as first stated by Doi et al. is a diagnosis made by a radiologist who incorporates the output of a computerized analysis of the radiograph when making their decision (7). This definition emphasizes the distinction between CAD and automated diagnosis. The CAD is used as an aid to the radiologist, whereas the goal in automated diagnosis is to replace the radiologist.

There are in general two main types of CAD schemes. The first is CADe, where suspicious regions in the image are located. The second is CADx, where a suspicious region is classified (e.g., malignant vs. benign). Unfortunately, there is a possible confusion in nomenclature between the field as a whole and a specific type of CAD algorithm for distinguishing between different disease states (characterization or classification). To avoid this problem, the term CADe will be used for computer-aided detection, CADx for computer-aided diagnosis (classification), and CAD when referring to the whole field of study, which encompasses both CADe and CADx.

How Does Computer-Aided Diagnosis Work?

In CAD, the computer is used as an aid to the radiologist. It provides a second opinion to the radiologist. Double

reading of mammograms has shown to increase the detection of breast cancer by as much as 15% (8,9). However, double reading is expensive since two highly trained individuals must read the images instead of one, and it is logistically difficult to implement efficiently. As a result, double reading is not commonly practiced, especially in the United States. It is believed that CAD could be an effective and efficient method for implementing double reading (10).

Figure 1 shows schematically how CAD can be implemented clinically. A radiograph is made of the patient. If the radiograph was acquired digitally, it can be sent directly to a CAD system for analysis. If the radiograph was recorded on film, then the image needs to be digitized first. After the computer analyzes the image, the output of the CAD system is displayed to the radiologist. The radiologist then views this information and considers their personal opinion with the computer output before giving a diagnosis. In all cases, the radiologist has the final diagnostic decision; the computer acts only as an aid. The CADe schemes have been compared to a spell checker tool in word processing software. In situations where double reading is employed, that is, each case is reviewed by two radiologists in a sequential manner, each radiologist may use the computer aid independently. Even in situations where double reading by two radiologists is employed, CADe can still detect cancers that both radiologists missed (11). It is also possible to modify the double reading paradigm when CAD is used. The first reader after reading with CADe can assign only cases that are not either clearly normal or clearly abnormal for the second reader (12).

There are several different methods of displaying the CAD output to the radiologist. Figure 2 shows a CT scan that is annotated with a circle indicating an area that the computer deemed suspicious. This is typical for a CADe system. Different symbols are used by different commercial systems. Figure 3 shows a chest radiograph that is annotated with different symbols to indicate different types of disease states. The symbols also have different sizes indicating the severity of the disease. This type of display can be used for both CADe and CADx systems. Figure 4 shows a simple interface that can be used by a CADx system in

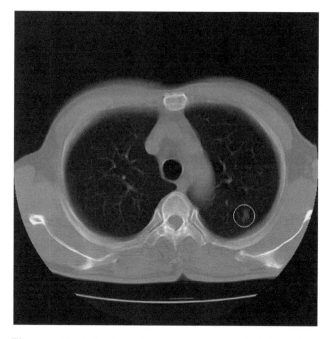

Figure 2. Typical output of a CADe system. The circle indicates a region that the computer deemed suspicious, in this case for the presence of a lung cancer in a slice from a CT scan of the thoracic.

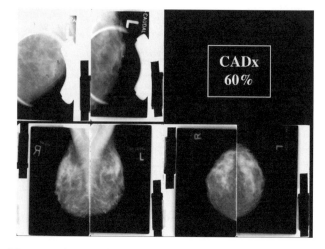

Figure 4. An example of a simple interface that could be used by a CADx scheme. Along with the images showing the lesion being analyzed, the display also shows the computer's estimated likelihood that the lesion is malignant. The images in the top left corner are spot magnification views and the images along the bottom are the standard screening views.

which the likelihood of malignancy is shown. Figure 5 shows a more advanced interface that can be used by CADx systems. Here the probability that a lesion is malignant is given numerically and in graphically form what this number means in terms of likelihood of being a malignant or benign lesion. In addition, this interface shows lesions similar to the unknown lesion that are selected from a library of lesions with known pathology. This provides a pictorial representation of the computer

result that to the radiologist may be more intuitive than just a number.

In CADe, lesions or disease states are identified in images. The output of CADe schemes is typically locations of areas that the computer considers suspicious. These areas can then be annotated on the digital image. For CADe, the computer can detect actual lesions (true-positive detection), miss an actual lesion (false-negative detection), or detect something that is not an actual lesion (false-positive detection). The objective of the computer algorithm is to find all the actual lesions while minimizing the

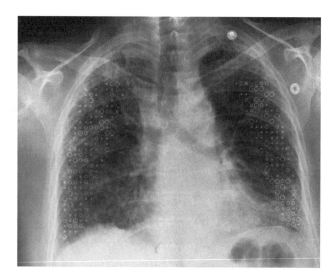

Figure 3. This chest radiograph is annotated with symbols indicating the presence of different appearances or patterns associated with interstitial chest disease. Crosses indicate a normal pattern; squares indicate a reticular pattern, circles represent a nodular pattern; and hexagons are honeycomb or reticulonodular patterns.

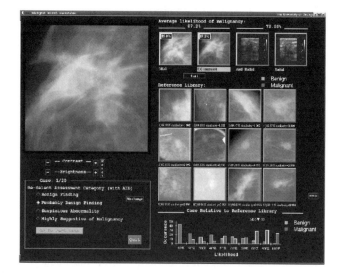

Figure 5. An example of an advanced CADx display that shows in addition to the likelihood of malignancy, a graphical representation and a selection of lesions similar to the lesion being examined recalled from a reference library containing lesions with known pathology. The red frame around a lesion indicates a malignant lesion and a green frame indicates a benign lesion.

number of false-positive detections. In practice, there exists a tradeoff between having high sensitivity (high true-positive detections) and the number of false detections per image. This tradeoff is important because the clinical utility of CADe depends on how well the algorithm works. In particular, if the false-positive rate is high (the exact definition of high is unknown) then the radiologist will spend unneeded time examining computer false-detections. Further, there is a possibility that a computer false detection could cause the radiologist to make an incorrect interpretation, if the radiologist should incorrectly agree with the computer. If the sensitivity of the CADe scheme is not high enough, then the radiologist could lose confidence in the CADe scheme's ability to detect actual lesions, reducing the potential effectiveness of the CADe scheme.

In CADx, the computer classifies different pathologies or different types of diseases. The most common application is to distinguish between benign and malignant lesions, to assist radiologists in deciding which patients need biopsies. The output of a CADx scheme is the probability that a lesion is malignant. If one sets a threshold probability value above which the lesion is considered malignant, the computer classification can be a true-positive (an actual malignant lesion is classified as malignant), a false-negative (an actual malignant lesion is classified as benign), a false positive (an actual benign lesion is classified as malignant), or a true negative (an actual benign lesion is classified as benign). The goal is to maximize the true-positive rate while minimizing the false-positive rate. Both false positives and false negatives could have serious consequences. A computer false positive could influence the radiologist to recommend a biopsy for a patient who does not need one. A computer false negative could influence the radiologist not to recommend a biopsy for a patient who has cancer.

There are important differences between CADx and CADe. In CADe, the results are presented as annotations on an image, so the radiologist examines image data, which they are trained and accustomed to do. In CADx, the output can be in numerical form, which radiologists are neither trained nor accustomed to using. Therefore, it may be more difficult for radiologists to use CADx effectively compared to using CADe. In addition, the consequences of inducing an error by the radiologist is much more severe in CADx than CADe. In most situations, the next step in a positive detection (CADe case) is more imaging, whereas in the CADx case, the next step is usually a biopsy or some other invasive procedure.

Why Is Computer-Aided Diagnosis Needed?

The goal of CAD is threefold: (1) to improve the accuracy of radiologists; (2) to make a radiologist more consistent (reduce intrareader variability) and to reduce discrepancies between radiologists (reduce interreader variability); and (3) to improve the efficiency of radiologists.

Since the interpretation of a radiograph is subjective, even highly trained radiologists make mistakes. The radiographic indication for the presence of a disease is often very subtle because variation in appearance of normal tissue often mimics subtle disease conditions. Another reason a radiologist may miss a lesion is that there is some other feature in the image that attracts their attention first. This is known as satisfaction of search and can occur when, for example, the radiologist's attention is focused on identifying pneumonia in a chest radiograph so that they do not notice a small lung cancer. Furthermore, in many instances the prevalence of disease in a population of patients is low. For example, in screening mammography only 0.4% of women who have a mammogram actually have cancer. It can be difficult to be ever vigilant to find the often-subtle indications of malignancy on the mammogram. Consequently, the missed cancer rate in screening mammography is between 5% and 35% (13–15) and, in chest radiography, the missed rate for lung cancer is ~30% (16).

Radiologists often have to decide whether an abnormal area is an indication of malignancy or of some benign process, or they may have to differentiate between different types of diseases that could give rise to the abnormality. This differentiation is often difficult because the radiographic indications between different disease types are not distinct. In diagnosing breast cancer, radiologists will recommend between 2 and 10 breast biopsies of benign lesions or normal tissue for every 1 cancer biopsied (17,18). The 50–90% of women who do not have breast cancer, but undergo a breast biopsy, suffer physical and mental trauma, and valuable medical resources were used unnecessarily. It is estimated that there is a 18.6% chance of receiving a biopsy for a benign condition after 10 years of screening for women in the United States (19). In general, the European false-positive rates are lower than in the United States (20). This is due in part to having a national programs and screening policies in Europe (21) and the higher likelihood of having malpractice lawsuit for a false-negative diagnosis in the United States.

Variability in performance of radiologists will reduce the effectiveness of the radiographic exam and, further, it will undermine the confidence that the public and medical community has in the technology. Again, because of differences in ability and differences in opinions, there exists variability between radiologists. In mammography, the variability between radiologists is well documented and can be very large (up to 40% variation in sensitivity in one national study) (22). Even a given radiologist is not always internally consistent. That is, the same radiologist may give a different interpretation when reading the same case a second time (23).

As the cost of the health care system increases, there is growing pressure to improve the efficiency of the system. In radiology, technology improvements have been introduced to improve workflow. Two examples are digital imaging systems, which can acquire images faster than conventional film systems and picture archive and retrieval systems (PACS), which can streamline the process of storing and recalling images. These and other technology improvements have increased the radiologists' workload. While not yet proven, it is hoped that CAD will allow the radiologist to read faster without reducing their accuracy. This will probably occur when the CAD schemes reach a certain level of accuracy (the exact level is unknown), so that the radiologist is not spending too much

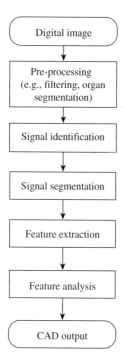

Figure 6. A flowchart of a generic CAD scheme. Most CADx and CADe schemes follow this template, although there are widely varying techniques for implementing each step.

time examining– considering computer false- positives. The CAD may also increase the radiologists' confidence in their interpretation, which should lead to faster reading times.

COMPUTER-AIDED DIAGNOSIS ALGORITHMS

There are probably thousands of publications on CAD, with hundreds of different techniques being developed. It is not practical to describe them all. There is, however, commonality between most approaches. Most techniques, whether they are for CADe or for CADx, can be described generically, as outlined in Fig. 6, as consisting of five steps: preprocessing, identification of candidate lesions (signals), segmentation of signals, feature extraction, and classification to distinguish actual lesions from false lesions or to differentiate between different types of pathologies or diseases (e.g., benign and malignant).

Radiographs can either be acquired using digital technology or screen-film systems. In digital systems, the image is acquired as a 2D array of numbers. Each element of the array is a pixel and corresponds to the amount of X-ray energy absorbed in the detector at that pixel location. This in turn is related to the number of X rays transmitted through the patient. Thus, a radiograph is a map of the X-ray attenuation properties of the patient. Different tissues in the body and different tissue pathologies have different attenuation properties, although the differences can be small. In a screen-film system, the image is recorded on X-ray film. The screen converts the X rays into visible light and the light is recorded by the film. For CAD purposes, this film must be digitized so that the image can be analyzed. A film digitizer basically shines light through the film and measures the amount of light transmitted. The resulting image is again a 2D array of numbers related to the X-ray attenuation properties of the patient.

One complicating property of screen-film systems is that they respond nonlinearly to X-ray exposure: digital systems respond linearly. Figure 7 shows the characteristic curve for screen-film system and a digital system. The contrast of objects in the image is proportional to the slope of the characteristic curve. For a screen-film system, at high and low exposures, slope approaches zero. Thus, for screen-film systems the contrast in the image is reduce in bright and dark areas of the image.

In this article, one approach to accomplishing these steps will be illustrated. In general, the techniques described are relatively simplistic, but are used to illustrate the concepts involved. More advanced and effective techniques are described in the literature; in particular,

Figure 7. Comparison of a linear and a nonlinear detector. In a nonlinear detector (e.g., a screen-film system, the response of the system is X-ray exposure dependent. That is, the amount of darkening on the film (called film-optical density) depends on the X-ray exposure to the detector. At low exposures and high exposures, the image contrast (difference in film optical density) is lower than at optimal exposures. For a linear detector (e.g., most digital X-ray detectors), the response of the system is linearly dependent on the X-ray exposure to the detector. As a result, the contrast is independent of X-ray exposure.

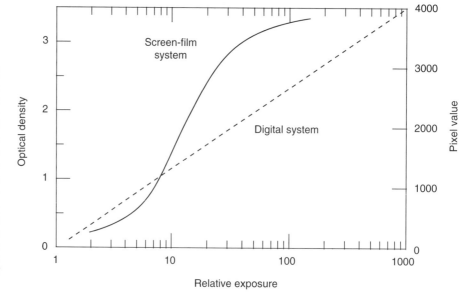

review articles (24–29), and conference proceedings, such as the Proceedings of the SPIE Medical Imaging Conference, International Workshop on Digital Mammography (30–35), and the Proceedings of Computer Applications in Radiology and Surgery (CARS).

Using a mammogram, the steps outlined in Fig. 6 are illustrated in Figs. 8–14 using a somewhat simple procedure. Figure 8a shows the mammogram with an arrow indicating the location of a cluster of calcifications. An enlargement of the cluster is shown in Fig. 8b. In a radiograph, dark regions correspond to areas where there are more X-rays incident on the image and these are assigned high pixel values; and bright areas correspond to areas with fewer X-rays incident and they are assigned low pixel values. Within the breast area of the image, dark areas correspond to predominately fatty areas of the breast and bright areas correspond to fibroglandular tissues (those involved in milk production and the physical support of the breast to the chest wall). In the following sections, this image will be used to illustrate some of the basic concepts in developing CAD schemes.

Most of the initial CAD research was applied to 2D images, in which the 3D body part is projected into a 2D plane to produce an image. While being relatively simple, fast, and inexpensive, 2-D imaging methods are limited by the superposition of tissue. Lesions can be obscured by overlapping tissues or the appearance of a lesion can be created where no lesion actually exists. Three-dimensional imaging [ultrasound, CT, magnetic resonance imaging (MRI), positron emission tomography (PET), and single-photon-emission competed tomography (SPECT)] produces a 3D image of the body, which is often viewed as a series of thin slices through the body. These images are more costly to produce and take significantly longer to complete the exam, however, they can produce vastly superior images. These image sets also take longer to read. In some cases, the radiologist needs to review up to 400 image slices for each exam. With such large datasets, it is believed that CAD may be beneficial for radiologists.

In developing these techniques, researchers use information that a radiologist uses, information that cannot be visualized by radiologists, and information based on the radiographic properties of the tissue and of the imaging system.

Preprocessing

The first step in a CAD scheme is usually preprocessing. Preprocessing is employed mainly for three reasons: (1) To reduce the effects of the normal anatomy, which acts as a camouflaging background to lesions of interest: (2) To increase the visibility of lesions or a particular feature of a lesion: (3) To isolate a specific region within the whole image to analyze (e.g., the lung field from a chest radiograph). This reduces the size of the image that needs to be analyzed reducing computation time, reducing the chances

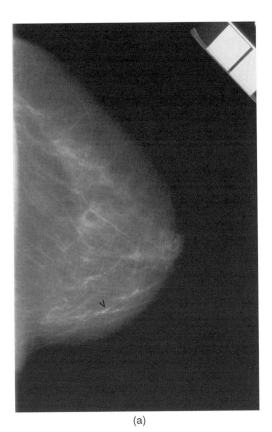

(a)

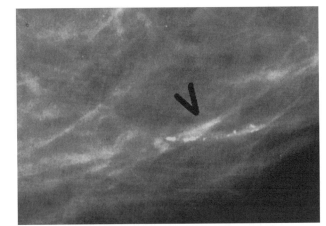

(b)

Figure 8. (a) A portion of a digitized mammogram. The arrow indicates the location of a cluster of calcifications. (b) An enlargement of the area containing calcifications.

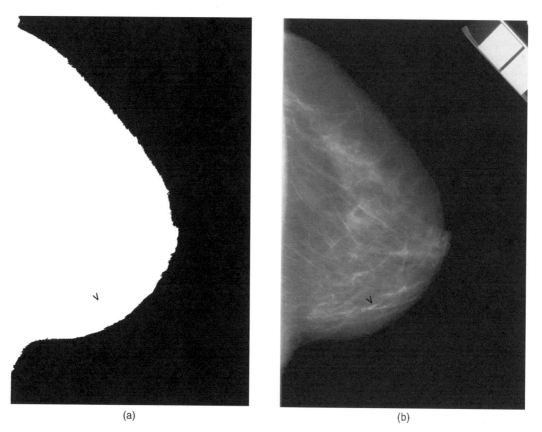

(a)   (b)

Figure 9. (a) A gray-level threshold was applied to the image shown in Fig. 8 in order to determine the breast boundary. Above the threshold value the image is made white and below the boundary the image is made black. The interface between the two areas is the skinline or breast boundary. (b) The determined breast boundary superimposed on the original image.

of a false detection, and reducing the complexity of the analysis, since only the area of interest is analyzed and extraneous image data are eliminated.

In our example CAD scheme, two procedures are performed. The first is to identify the border of the breast and the second is to process the image to reduce the background structure of the breast so as to highlight small bright areas that could be calcifications. The border of the breast was determined, so as to include only breast area in subsequent analysis. Outside of the breast, the image is black, corresponding to a large number of X rays incident on the film. Within the breast, X rays are absorbed or scattered so that the number of X rays incident on film is decreased and the image appears brighter. In this simple example, the breast outline is determined by thresholding the image so that below a threshold pixel value, all the pixel values are black and equal to above the threshold all pixels are white (Fig. 9a). The border between the black and white pixels is chosen as the border of the breast. The result is shown in Fig. 9b. This method produces a suboptimal result. At the top left corner of the image, there is an area included that does not contain breast tissue that was considered to be part of the breast. When the film was digitized, where there was a sharp transition from bright to dark, the response of the digitizer was slow, so that some of the bright area "bleeds" into the dark area.

In our example, the image is next processed by using a technique called the difference of Gaussians (DoG). In this technique, the image, which is typically 18×24 cm, is filtered twice by Gaussian filters, one with a small width (fwhm value, of 0.155 mm) and the other with a wider width (fwhm of 0.233 mm) (36). The first filter keeps small bright signals in the image (Fig. 10a) and the second eliminates the small bright signals (Fig. 10b). The two images are subtracted producing an image where large structures are eliminated and small signals are retained (Fig. 10c). Another useful outcome of this difference method is that the background of the subtracted image is more uniform compared to the original mammogram. This is important for the next step where potential calcifications are identified using a gray-level threshold method.

Note that for images that are recorded not on film but are acquired in digital format (i.e., a digital mammogram) now undergo preprocessing before the radiologist views the image (37). The goal is to improve the visibility of abnormalities for the human viewer. The techniques used in this type of preprocessing can be similar to those used in preprocessing for CAD purposes. The main difference is that a CAD preprocessed image would be considered overprocessed to the radiologist because they are designed to highlight only one type of lesion, whereas a radiologist needs to look for several to many different types of abnormalities in the image.

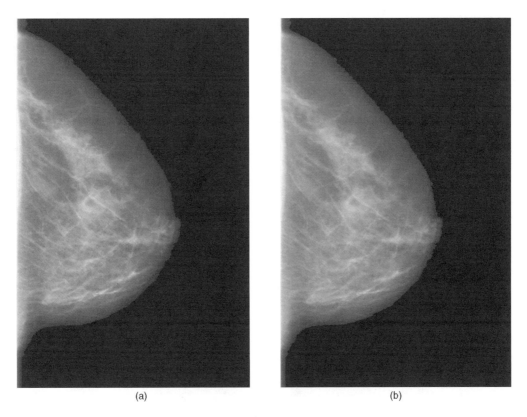

(a)   (b)

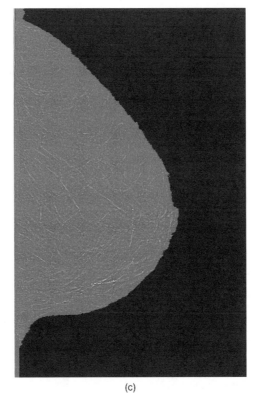

(c)

Figure 10. An example of a preprocessing method using the difference of Gaussian technique. (a) The image shown in Fig. 8 is filtered by a Gaussian filter that has a full width at half-maximum (fwhm) value of 0.155 mm. This filter enhances small objects on the order of 0.5 mm in diameter. (b) The image shown in Fig. 8 is filtered by a Gaussian filter that has a fwhm value of 0.233 mm. This filter degrades small objects on the order of 0.5 mm in diameter. (c) The image in Fig. 10b is subtracted from Fig. 10a producing an image with enhancement of small objects on the order of 0.5 mm in diameter and a reduction in the normal background structure of the breast.

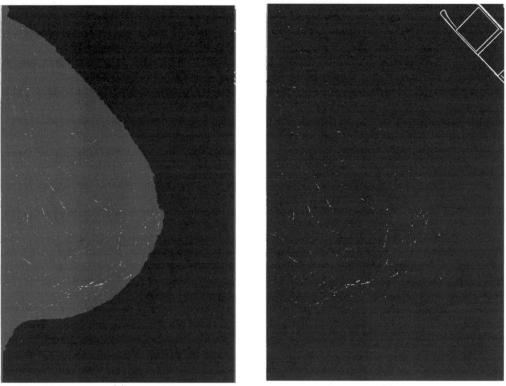

(a)   (b)

(c)

Figure 11. An illustration of signal identification using gray-level thresholding. Using the preprocessed image (Fig. 10c), a gray-level threshold is chosen so as to keep only a fraction of the brightest pixels. Three different threshold levels are shown in a–c. If the threshold value is too low as in part a, too many false signals (noncalcifications) will be kept. If the threshold is too high, as in part c, some of the actual calcifications are lost, even though most of the false signals are eliminated.

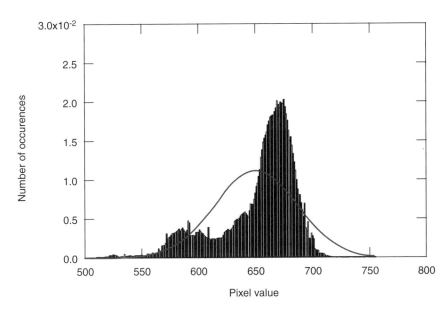

Figure 12. A histogram of pixel values from a hypothetical lesion. From a histogram like this one, features can be determined (e.g., mean pixel value and kurtosis). See text for details.

If the system does not respond linearly to X-ray exposure, then, the background level will affect the image contrast of lesions (see Fig. 7). In bright or dark areas of the image, the contrast is reduced compared to regions that are optimally exposed. This reduces the effectiveness of pixel-value-based and contrast-dependent techniques.

A limitation of these types of filtering is that lesions can often range in size. For example, lung nodules can as small as 0.5 cm or less and 5 cm or larger. A fixed-sized filter cannot be optimal for the full range of sizes possible. To accommodate a large size range, many researchers have developed multiscaled approaches. Wavelets are one class of multiscale filters. Multiple numbers of bandpass filters can be chosen based on which daughter wavelet decomposition are used in reconstructing the image. The principle of applying wavelets to medical images is discussed by Merkle et al. (38). In mammography, a weighted

sum of the different levels in the wavelet domain is performed to enhance calcifications (39). Different approaches differ in their choice of wavelets and the selection of which levels to use in the reconstruction. A list of different wavelets used for processing microcalcifications on mammograms is given in Table 1.

Identification of Lesions

Once the image has been preprocessed, candidate lesions or signals need to be identified. The goal in CADe is to maximize the number of actual lesions identified even if a large number of false signals are detected. The false detections are reduced in the feature analysis step.

Signal identification is sometimes accomplished simultaneously with lesion segmentation (see next section). However, there are circumstances in which it is not. The

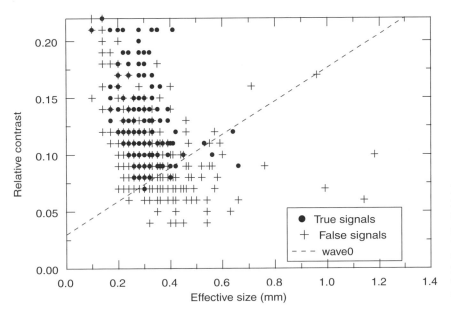

Figure 13. An example of feature analysis in which two features, contrast and size, are used to differentiate actual calcifications from computer-detected false detections. A threshold can applied to reduce the number of false detections, without eliminating many actual calcifications. In this example, the broken straight line shows the threshold values.

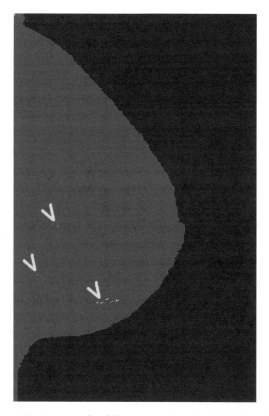

Figure 14. An example of the output of a simple CADe method illustrated in Figs. 8–13. This simple method detected the actual cluster of calcifications, but also two false clusters. More sophisticated methods can have much better performance over a wide variety of cases.

most obvious case is when a human indicates the location of signals. When functioning clinically a CADx scheme needs to know the location of the lesion that needs to be classified. This can be done using a CADe scheme. However, the CADe scheme may not detect a lesion that the radiologist is scrutinizing because either the CADe scheme may have failed to detect the lesion or the area that the radiologist is examining does not contain a real lesion. In either case, the radiologist would have to mark the location of the lesion in order for the CADx scheme to analyze it. When this occurs, the process is no longer automated.

Table 1. List of Different Mother Wavelets Used for Processing Mammograms Containing Microcalcifications

Lead Investigator	Reference	Wavelet
Brown	40	Undecimated spline
Chen	41	Morlet
Chitre	42	Daubechies' 6- and 20- coefficient
Kallergi	43	12-coefficient Symmlet
Laine	44	Dyadic
Lo	45	Daubechies 8-tap
Strickland	46	Biorthogonal B-spline
Wang	47	Daubechies' 4- and 20- coefficient
Yoshida	48	8-tap least asymmetric Daubechies

In X-ray imaging, the presence of a disease state is usually detected because the lesion is either more attenuating or less attenuating than the surrounding tissue. That is the lesion that will appear as a bright or a dark spot in the image. This makes gray-level thresholding a potentially useful method for segmentation. In Fig. 11, our example image has undergone gray-level thresholding using different thresholds. As the threshold is increased the total number of detections decrease. If the threshold is too high, some actual calcifications are lost. The signals that are detected that do not correspond to actual calcifications are caused by primarily image noise (due to the statistical fluctuations in the number of X rays incident on the patient per unit area) and normal breast tissue. Unfortunately, radiographs are a 2D projection of a 3D object, so that summation of overlapping tissues can produce areas in the image that mimic actual lesions. While gray-level thresholding can be effective on a given image, over a large number of images this simple thresholding method is not optimal. Calcifications can appear differently in different images, due in part to differences in the X-ray exposure used to make the image. Therefore, a single threshold applied to a cross-section of images will not be effective.

One method to improve the gray-level thresholding method is to make it adaptive. For example, one can use the statistics of a small region (e.g., $5 \, mm^2$) centered on the calcification to select the appropriate threshold (2). A threshold can be set to the mean pixel value within the region plus a multiple (typically 3.0–4.0) of the standard deviation within the region. In this way, differences in X-ray exposure and differences in image noise can be accounted for.

Lesion Segmentation

Once the location of a lesion has been determined, the exact border of the lesion needs to be determined. This is necessary to extract features of the lesion. Again, gray-level threshold can be employed. One simple method is to find the highest pixel value near the identified pixel (e.g., in a $0.3 \times 0.3 \, mm$ region) and then to find the mean pixel value in a local area surrounding the signal (e.g., in a $1 \times 1 \, cm$ region). A threshold can be taken to be the mean pixel value, plus 50% of the difference between the maximum pixel value and the mean pixel value. Finally, all pixels above the threshold that are connected to the pixel with the maximum value are considered to be part of the calcification. Starting from a seed point, in this case the pixel with the highest value, a region is grown that contains all connected pixels that are above threshold. Connected pixels can be defined as the pixel above, below, to the right, and to the left of a given pixel, so-called four-point or eight-point connectivity, the same four pixels plus the four corner pixels.

This region growing method can be improved by first correcting the appearance of the calcification in the image for the degradation caused by the imaging system. The imaging will blur the image and further, the imaging system can response nonlinearly to X-ray exposure. Jiang et al. (49), and Veldkamp and Karssemeijer (50) independently developed a segmentation technique based on background-trend correction and signal dependent thresholding. In these two approaches, corrections for the nonlinear response and the blurring of the calcification by the screen-film system and film digitizer are performed. At low and high X-ray exposures to the screen, the contrast, which is proportional to the slope of the characteristic curve, is reduced (see Fig. 7). That is, the inherent contrast of the calcifications is reduced when recorded by the screen-film system. Therefore, the image or radiographic contrast will depend on the background intensity. If a correction is not made for this nonlinearity, then it becomes extremely difficult to segment accurately calcifications in dense and fatty regions of the image simultaneously with calcifications in other regions of the breast. Similarly, the smaller the calcification, the more that its contrast is reduced due to blurring. This can be corrected based on the modulation transfer function of the screen-film system (51).

The above segmentation method will work well as long as there is sufficient separation between the calcifications. If two or more calcifications are too close, then they may be segmented as one large calcification. In such situations, more sophisticated methods are more effective. Besides the pixel value, thresholding can be applied based on other features of the image, such as the texture and gradients. These more advanced techniques are also important in applications where the border of the lesions is not well-defined visually in the image.

Feature Extraction

To reduce the number of false detections (i.e., to differentiate true lesions from false detections) or to classify a lesion (e.g., benign versus malignant) features are extracted and subsequently used by a classifier. The strategy in CADe is to segment as many actual lesions as possible. This will include a large number of false detections.

There are probably thousands of features that can be used and these are dependent on the imaging task. Further, the optimum set of features is not known for any imaging task. Therefore, a large number of different features are being used by different investigators. The features are based on those that a radiologist would use and those that a radiologist would not use. As an example, a radiologist uses the brightness of the lesion to determine whether a lesion is present in the image. This can be determined conceptually by plotting a histogram of pixel values (see Fig. 12). The brightness is related to the mean pixel value, M, and is given by

$$M = \frac{1}{N} \sum_{i=1}^{N} f(i) p(i) \tag{1}$$

where f is the frequency of occurrence of pixel value $p(i)$, N is the total number of pixels in the region being analyzed, and i is the index in the histogram, $f(i)$. An example of a feature not used by a radiologist is kurtosis, K, which is defined as

$$K = \frac{\frac{1}{N} \sum_{i=1}^{N} (f(i) - M)^4 p(i)}{\left[\frac{1}{N} \sum_{i=1}^{N} (f(i) - M)^2 p(i) \right]^2} \tag{2}$$

The kurtosis describes the flatness of the histogram. Histograms that are very peaked have high kurtosis. Kurtosis can also be thought of as comparing the histogram to a Gaussian distribution. A value of 3.0 indicates that the histogram has a Gaussian distribution.

A large, but incomplete, list of features used by different investigators for the detection of calcifications in mammograms is given in Table 2. The features used for other applications of CADe and CADx will differ than the ones in Table 2, but the categories will be the same: pixel intensity-based, morphology-based, texture-based, and others. For example, for 3D images, such a CT scan of the thorax, instead of a circularity feature, the sphericity of a lesion would be calculated. The drawback of having a large number of features to choose from is that the selection of the optimum set of features is difficult to do, unless a very large number of images are available for feature selection (69). These images are in addition to the images needed for training and the images needed for testing the technique.

Table 2. List of Different Features Used for Distinguishing Actual Calcifications from False Detections

Pixel-Value Based	Morphology Based	Derivative Based	Other
Contrast (52–57)	Area (52,54,56,58,59)	Mean edge gradient (45,54,55,57,60–64)	Number of signals per cluster (52,59,65)
Average pixel value (53,66)	Area/maximum linear dimension (61)	Standard deviation of gradient (62)	Density of signals in cluster (52,58,59)
Maximum value (45,67)	Average radius (67)	Gradient direction (64)	Distance to nearest neighbor (65)
Moments of gray-level histogram (68)	Maximum dimension (62)	Second derivative (55)	Distance to skin line (65)
Mean background value (45)	Aspect ratio (65)		Mean distance between signals (59)
Standard deviation in background (45,57,58)	Linearity (63)		First moment of power spectrum (6)
	Circularity (54,59,67)		Effective thickness (49)
	Compactness (61,53,55)		Peak contrast/area (67)
	Sphericity (contrast is the third dimension) (67)		Mean distance from center of mass (42)
	Convexity (68)		

Most of the features listed in Table 2 use standard techniques for determining their value. One feature, effective thickness of the calcification developed by Jiang et al., is calculated using a model of image formation (49). That is, what thickness of calcification will give rise to a given measured contrast in the digital image? To do this, corrections for the blurring of the digitizer and the screen-film system are performed, along with corrections for the characteristic curves of the digitizer and the screen-film system. The assumption is that, in general, calcifications are compact, so their diameter and thickness should be comparable. Film artifact (e.g., dust on the screen), will have a very high thickness value compared to its size, and therefore can be eliminated. Similarly, detections that are thin compared to their area are likely to be false positives due to image noise.

Highnam and Brady take this one step further. For every pixel in the image they estimate the corresponding thickness of nonfatty tissue (essentially fibroglandular tissue) by making the corrections described in the preceding paragraph and in addition corrections for X rays that are scattered within the breast, for the energy and intensity distribution of the X-ray beam, and for other sources (70). This in principal produces an effective image that is independent of how the image was acquired or digitized.

Most features are extracted from either the original image or a processed image that has sought to preserve the shape and contrast of the calcifications in the original image. Zheng et al. used a series of topographical layers ($n = 3$) as a basis for their feature extraction. The layers generated by applying a 1, 1.5, and 2% threshold using equation 2. This allows for features related to differences between layers (e.g., shape factor in layer 2 and shape factor in layer 3) and changes between layers (e.g., growth factor between layers 1 and 2) to be used.

Classification

Once a set of features has been identified, a classifier is used to reduce the number of false detections, while retaining the majority of actual calcifications that were detected. Several different classifiers are being used: simple thresholds (2,52–54,60–62,71), artificial neural networks (40,72–74), nearest-neighbor methods (54,75,76), fuzzy logic (45,68), linear discriminant analysis (77), quadratic classifier (40), Bayesian classifier (55), genetic algorithms (78), and multiobjective genetic algorithms (79).

The objective of the classifier is to find the optimal threshold that separates the two classes. Shown in Fig. 13 is an example problem where two features of candidate lesions have been extracted and plotted (2D problem). The classifier determines the boundary between the two classes (normal and abnormal). The optimal boundary is one that maximizes the area under the receiver operating characteristic (ROC) curve (see the section CADx Schemes). There are different types of classifiers: linear, quadratic, parametric, and nonparametric. A linear classifier will produce a boundary that is a straight line in a 2D problem, as shown in Fig. 13. Quadratic classifiers can produce a quadratic line. More complex classifiers (e.g., k nearest

neighbor, artificial neural networks, and support vector machines) can produce very complex boundaries.

Alternatives to Feature Extraction

In lieu of feature extraction, or in addition to feature extraction, several investigators have used the image data as input to a neural network (80–83). The difficulty with this approach is that the networks are usually, quite complex (several thousand connections). Therefore, to properly train the network and to determine the optimum architecture of the network requires a very large database of images.

Another limitation of these and similar approaches is that extremely high performance is needed to avoid having a high false-positive rate. The vast majority of mammograms are normal and the vast majority of the areas of mammograms that are abnormal do not contain calcifications. Mammographically, the average breast is ~100 cm^2 in area. For a 50 μm pixel, there will be 4 million pixels to be analyzed. If there are 13 calcifications of 500 μm in diameter, then only 0.01% of pixels will belong to a calcification. Therefore, a specificity of 99.9% will give rise to 10 false ROIs per image, which is more than 100 times higher than a radiologist.

Computation Times

The computation times are not often stated by most investigators, perhaps under the belief that this is not an important factor since computers will always get faster. In general, times range from 20 s (84) up to several tens of minutes (inferred from description of other published techniques), depending on the platform and pixel size of the image. For any of the technique to be used clinically, they must be able to process images at a rate that is useful clinically. For real-time analysis (e.g., for diagnostic mammography) there is approximately one patient approximately every 20 min per X-ray machine. This means there is 5 min available per film, including the time to digitize the film. However, many clinics have several mammography units, and some clinics have mobile vans that image up to 200 women off-site. In these situations, computation times of < 1 min per film may be necessary or multiple CADe systems would need to be employed. This also assumes that only one detection scheme is run. In mammography, at least one other algorithm, for the detection of masses, will be implemented, cutting the available time for computation in half.

In most centers, screening mammograms are not read to the next day. This allows for processing overnight and computation time becomes less critical, but still important. To analyze 20 cases (80 films) in 15 h (overnight) is 11 min per film for at least two different algorithms. For a higher volume center (40 cases), this gives < 3 min for each algorithm.

In other applications, the results of the computer analysis are needed immediately. For example, if a radiologist marks in the image an area that they would like analyzed by the computer (e.g., to determine the likelihood of malignancy of a lesion) then the radiologist, who is busy and under pressure to read the images quickly and accurately, does not want to wait to see the computer results. If the computer takes more than a few seconds to return a result to the radiologist, then the radiologist may choose not to use the computer because it is reducing down their productivity. Computation time is not a trivial matter.

EVALUATION OF CAD ALGORITHMS

CADx Schemes

The performance of a CADx scheme can be measured in terms of sensitivity (the fraction of actual abnormal cases that are called true) and specificity (the fraction of actual normal cases called normal). This pair of values is easy to compute. However, in general, as the sensitivity increases, the specificity decreases. This can give rise to the problem of comparing two CADx schemes, one with high sensitivity, but low specificity, and the other with lower sensitivity, but higher specificity. This problem is solved by using ROC analysis (85,86).

The CADx schemes that involve a differentiation between two categories or classes (e.g., benign and malignant) can be evaluated using ROC analysis. In a two-class classification, objects are separated into one of two classes. This is usually done based on feature vector that characterized the object. Conceptually, the problem reduces to sorting lesions into one of two classes based on the feature vector as illustrated in Fig. 15. In any problem of interest, the population of feature vectors for actual negatives (e.g., benign or computer-detected false-positives) overlaps with the population of feature vectors for actual positives (e.g., malignant or actual lesions). Depending on the selection of a threshold based on a given feature vector value, some actual negatives will be classified as positive and vice versa. Depending on the selected threshold, different pairs of true-positive fraction (the fraction of actual positive cases that are classified as positive) and false-positive fraction (the fraction of actual negative cases that are classified as positive) will be obtained. By selecting all possible thresholds, a set of TPF and FPF pairs will be obtained and they will form an ROC curve, as shown in Fig. 16. In practice, the populations shown in Fig. 15 are not smooth, because only a finite number of cases are used in evaluating a CAD scheme. Therefore, there is some uncertainty as to the true shape of the ROC curve. Fortunately, statistical methods exist to fit a curve to the TPF-FPF pairs (86). The most common summary metric to compare different ROC curves is the area under the curve (AUC). The AUC is the average TPF for all possible FPFs. Since there is uncertainty to the exact shape of the ROC curve, there is uncertainty in the true AUC value. Again, statistical methods have been developed to allow hypothesis testing between curves (86).

The most advanced of the ROC analysis methods can allow generalization of the results to any population of cases that are represented by the testing cases and to a pupulation of readers (87). This is, so-called multiple-reader-multiple-case (MRMC) model is important when analyzing observer studies (see the section Observer Studies).

If the CADx scheme is not a two-class problem, then currently there is no validated method for evaluating the CADx scheme in general (88). One can, however, reduce the problem to a series of two-class problems or make some assumptions to simplify the problem to allow two-class ROC analysis to be used (89–91).

While the AUC is a useful metric, it assumes that all TPF are equally desirable. Often, this is not the case. In diagnostic mammography, where the radiologist must

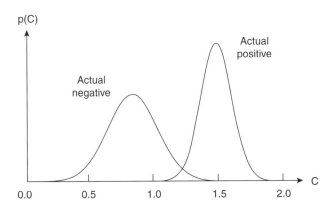

Figure 15. Conceptual representation of a two-class classification problem. The task is to classify lesions correctly into either actually negative lesions (e.g., benign or computer-detected false positive) or actually positive lesions (e.g., malignant or an actual lesion). The two probability distributions represents the probability of an positive or a negative lesion having a given value of a feature C (contrast). For a small range of contrasts (1.1–1.4), the two distributions overlap and misclassification can occur. To reduce the overlap regions, multiple features, instead of a single feature, are used.

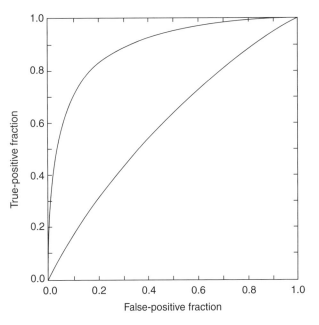

Figure 16. A typical ROC curve. The higher curve indicates higher performance than the lower curve. The upper curve represents the performance of radiologists in classifying masses as benign or malignant and it has an area under the curve of 0.91. The lower curve represents the performance of radiologists in classifying calcifications and it has an area under the curve of 0.62.

decide whether a known breast lesion is benign or malignant, the "penalty" for missing a cancer (classifying a malignant lesion as benign) is considered greater than classifying a benign lesion as malignant. In such circumstances, it may be more appropriate to consider only the region of the ROC curve in the high sensitivity (TPF) region (92,93). Partial AUC can be estimated, for example, considering the area under the curve for TPF > 0.80. This would imply that operating at a TPF < 0.80 is clinically unacceptable.

Finally, radiologists, when reading clinically, operate at one point on an ROC curve. In fact, two radiologists could have a difference of opinion on a number of cases, but operate on the same ROC curve. If one radiologist is more aggressive than the other, than the aggressive radiologist will have a higher TPF and a higher FPF than a more conservative radiologist and thus would be operating at a point higher up on the curve. Given that AUC is used as a figure of merit, then two radiologists having the same ROC curve have the same performance, although one may detect more disease than the other, because the other radiologist will have a lower false-positive rate (i.e., higher specificity).

CADe Schemes

The CADe schemes are not amenable to ROC analysis because CADe schemes address detection of lesions not the classification of a known lesion as with CADx schemes. That is, lesion classification is often a binary problem (e.g., benign or malignant), and therefore ROC analysis can be used. The CADe schemes involve the identification of the location of a lesion in the image and thus it is not a binary problem. The CADe schemes in general involve a tradeoff between TPF and false-positive rate (the number of false detections per image). Under such situations, free-response operating characteristic (FROC) curves are plotted (94). An example curve is shown in Fig. 17. For the task of detecting a lesion, it is important to identify the location of the lesion correctly. This raises the issue of how to score a CADe detection as correct or a miss (see the section Scoring Criteria).

In an ROC plot, the two axes range between 0 and 1.0, since they represent the true and false positive fractions

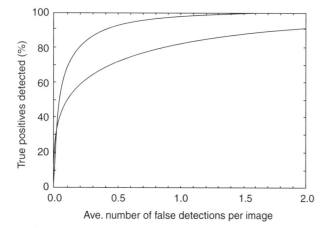

Figure 17. Typical FROC curves. The upper curve represents better performance.

(i.e., they are the ratio of two numbers). In an FROC plot, the y axis ranges from 0 and 1.0, but the x axis can range between 0 and 10 or higher, since there can be multiple detections on one image. That is, the upper limit for the average number of false positives per image is not fixed. Because of this difference between FROC and ROC curves, the statistical methods to analysis ROC data cannot be applied to FROC data. Recently, Chakraborty developed a statistical method for comparing two FROC curves (95), although the method still has some limitations.

In general, CADe schemes are not as accurate as radiologists, because the computer's false-positive rate is much higher than that of radiologists. Radiologists recall ~ 10% patients who receive a screening mammogram. Since only ~ 0.5% of the screened population will have breast cancer, the radiologist has a false-positive rate of ~ 0.0125–0.025 per image (the higher number is assuming that the lesion was seen in both views). The CADe schemes have ~ 10 times higher false-positive rates.

Technical Issues in Evaluation

The measured performance of a CAD scheme is important to compare different CAD schemes. The measured performance of a CAD scheme depends on several factors: (*1*) the inherent performance of the scheme; (*2*) the cases used to evaluate the scheme; (*3*) the scoring criteria used; (*4*) the quality of the truth data; (*5*) the method used to develop, train, and evaluate the scheme.

Ideally, one would like factors 2–5 above to be identical in all measurements of performance, but this does not exist at present, since there are no universal standards for these factors. Factors 2–5 are described below.

Database. The measured performance of a CAD scheme depends of the difficulty of cases used to evaluate performance. For a detection scheme, if the cases contain a high fraction of obvious abnormalities, then the measured performance will be high (96). In fact, differences in the cases used for evaluation can easily mask difference in true performance between two different CADe schemes.

The same problem exists for CADx schemes. Here an obvious case would be one that belongs clearly to one of the two classes. Often researchers will use a consecutive series of cases that went to biopsy as an evaluation dataset. These in general, represent difficult cases, especially for the actually benign cases, since at least one radiologist thought the cases was suspicious. However, since different radiologists have different skill levels, a set of consecutive biopsied cases from one institution may be different in difficulty from a set of consecutive biopsied cases from another institution. Currently, there is no method available to compare the difficulty of cases.

Public databases are becoming available (97). There are two websites where images can be downloaded: one for mammograms (http://marathon.csee.usf.edu/Mammography/Database.html) and the other for CT scans of the lungs (http://imaging.cancer.gov/reportsandpublications/reportsandpresentations/firstdataset). These in principal allow a common set of cases to be used in evaluating CAD schemes. This development is important in allowing comparison of different CAD schemes, but is insufficient in of itself.

Truth. Truth is elusive in CAD research, but vitally important. To optimize and to train a CAD scheme properly, accurate truth information is needed. Truth is needed at several levels. First, the pathology of actual lesions needs to be known (i.e., benign, malignant, or normal tissue). If a biopsy is performed, then it is possible to determine whether a lesion is malignant. However, there is some variability between pathologists, particularly on "borderline" lesions (those on the line between being malignant and premalignant). Further, studies have shown that pathologists, while being highly accurate, make mistakes ~5% of the times. Researchers usually accept this level of accuracy since it is too time consuming and costly to have the pathology of all biopsies redone.

There are lesions that are potentially malignant, but do not go to biopsy. In these patients, the lesion is followed and either a biopsy is performed in the future when the lesion becomes more suspicious, or the lesion is judged to stable after following the patient with more imaging in subsequent years. However, in a number of instances, the patient moves or changes hospitals and their follow-up is then incomplete. These patients are usually eliminated from CAD research use.

The next level of truth is where is the exact boundary of a lesion. It is not possible to know the answer to this question. Even though the boundary can be well defined by a pathologist, it is not possible to correlate the pathology boundary from the excised tissue to the boundary of the lesion in the image. To optimize (e.g., the segmentation of a lesion), the boundary of the lesion needs to be known exactly, that is, whether a given pixel in the image is part of the lesion or not. The current method to determine "truth" for the boundary of a lesion is to have a radiologist draw the outline of the lesion while viewing the image on a computer screen. Since this is subjective, the variability can be reduced by getting a number of radiologists to outline the lesion.

The last level of truth is whether a lesion is present or not. There are two notable examples of this problem: microcalcifications in a mammogram and lung nodules

in a CT scan. It is possible to know with certainty that a cluster of microcalcifications is present in a mammogram: The tissue containing the cluster that can be seen mammographically is excised and a pathologist can verify that calcifications are present in the tissue. It is not possible, however, to determine the location of every individual calcification in the cluster, even if the pathologist were able to count the number of calcifications in the cluster. When cancer metastasizes to the lungs, it is not possible to determine the exact number of nodules present, because some of the nodules are very subtle and difficult to detect in the images. With metastatic lung cancer, it is not necessary clinically to biopsy every nodule present.

To deal with this uncertainty, a number of radiologists, often called a panel of experts, can mark the location of individual microcalcifications or lung nodules. However, there are a number of problems with using a panel of experts. First, there is wide variability between radiologists, so that a large number of radiologists is needed to form the panel. The minimum number, if not known, is likely > 3. Panels can also be biased. Members of the panel can discuss their markings to reduce variability. However, one member of the panel could unduly influence other members of the panel, for example, if one member is a world-renowned expert (98).

Scoring Criteria. For CADx schemes, scoring the computer output is straightforward when ROC analysis is used. The truth for each case is known (either benign or malignant). Then it is a matter of selecting a threshold (applied to the CADx output) for which a lesion is considered malignant. For a given threshold, if the output of the CADx scheme is greater than the threshold value, then the lesion was classified as malignant.

For CADe output, scoring is much more problematic. There are many different criteria used for judging whether the computer found the lesion or whether the detection is a false positive (a computer detection not corresponding to an actual lesion). Most scoring criteria rely on some combination of the location of the center of the computer-detected signal and the actual lesion and the border of the computer-detected signal and the actual lesion. Figure 18 illustrates

Figure 18. An illustration of some different criteria used to score computer detections of actual lesions. The actual lesion is shaded light blue and the computer detection is given by the circle with the dashed line. In the right most figure, the computer is scored as a false detection, since the center of the computer-detected lesion is not within the boundary of the actual lesion. In the center part of the figure, the computer detection is scored as a true positive, since the center of the actual lesion is within the boundary of the computer detection. In this method, there is no penalty when the computer detects a very large lesion, even though the actual lesion may be small. Both of these methods are sensitive to the size of the actual lesion and the computer detected lesion. The method on the right corrects for these effects. In this method the area of the overlap of the computer-detected lesion and the actual lesion (light blue area) is divided by the area of the union of the two lesions (the gray and light blue areas). If this value is greater than a threshold (e.g., 40%), then the computer detection is scored as a true positive.

several different scoring methods. There is no consensus on which is the best method. In addition to the objective methods illustrate in the figure, many clinical studies rely on the subjective opinion of a radiologist to determine whether the computer correctly detected the lesion. Given the variability among radiologists, this method has severe limitations.

Methodology. As the field of CAD matures, evaluation methodology is becoming more refined, reducing possible biases that were common in earlier work. It is now known that different sets of images are needed to develop a technique, to train the technique, and to test the technique (69,99,100). However, unlike some other type of image analysis problems, medical images that have verified truth are not easily obtainable. Therefore, while it is desirable to have three independent datasets to develop, train and test, it is usually not possible.

When images are scarce, training and testing can be accomplished through cross-validation or bootstrapping (101–105). These methods divide a single dataset into two subsets: one for training and the other for testing. If the cases are divided such that all images from a given patient are all in one or the other subset, then bias due to commonality of images between testing and training can be eliminated. When dividing the dataset into subsets there are two contrasting goals: maximize the number of training images to properly train the algorithm versus maximizing the number of test images to minimize the variance in the measured performance.

There are three common cross-validation methods used in CAD research:

1. *K*-fold. The dataset is divided into *K* subsets of equal size. One of the *K* subsets is reserved for testing while the other *K* − 1 subsets are used for training. The process is repeated until each of the *K* subsets is used for testing. The mean performance and the variance can then be computed.
2. Leave-one-out. This is the extreme of the *K*-fold, where *K* equals the number of cases.
3. Jackknife. This is a generalization of the leave-one-out, where the number left out can be one to half the number of cases.

Many researchers believe that the 0.632+ bootstrap is the optimum method for evaluating CAD schemes (105). In this method, training samples are selected at random from the *N* cases in the dataset. After each sample is selected, a new sample is selected again from the full dataset; this is called random selection with replacement. After a total of *N* samples are selected, a fraction of the dataset will not have been selected. These are used to testing. Statistically speaking, 63.2% of the cases will be selected for training while 36.8% will have not been selected for training and form the testing set. This is repeated many times and the mean performance and variation can be computed. This technique is believed to minimize both the selection bias and the variance in the measured performance compared to other techniques.

While all these methods are effective in reducing some of the biases associated with measuring the performance of a

CAD scheme, they have one limitation. In all these methods, the CAD scheme is trained and tested multiple times using a different mixture of cases. This permits the mean performance (and the standard deviation in performance) to be calculated. However, this mean performance does not correspond to any specific CAD scheme, since it is an average of many trained schemes. Researchers must arbitrarily choose one trained scheme, often taking the best performing scheme, but that scheme may not have the best performance on different set on cases.

Evaluation of CAD Schemes as an Aid

The ultimate test of CAD efficacy will be whether patient outcomes are improved when CAD is used and whether it is cost-effective in doing so. Since CAD is still a new technology, it has not been widely use for a sufficient amount of time to address this question properly. Clinical studies are a prerequisite to patient outcome studies. Recently, a few clinical studies have been reported examining the effect of CADe on screening mammography with mixed results. However, intermediate measures of diagnostic efficacy can be measured. Studies have been conducted to show that CADe can find cancers that are missed clinically and observer studies, which simulate clinical reading conditions, have been performed. These are described below.

Missed Lesions. In theory, the actual performance of CAD on a set of cases is not important. Since CAD is to act as an aid to the radiologist, in principle, CAD only needs to be correct on cases the radiologist makes an error. For example, in CADe for mammography, the CADe scheme need only detect those cancers that the radiologist misses. Since radiologists miss between 5–35% of cancers, the sensitivity of the CADe scheme could be as low as 35% and still be an effective aid. Similarly, in theory, the CADe scheme could have 95% sensitivity and be of no use to the radiologist in detecting missed cancers. Therefore, it is necessary to measure the ability of a CADe scheme to detected clinically missed disease. In reality, if the CADe scheme does not have high sensitivity, the radiologist will lose confidence in the ability of the CADe to find cancer, thus reducing the CADe scheme's effectiveness as an aid.

There have been several studies in mammography and one in thoracic CT of the effectiveness of CADe to detected missed cancers. These studies show that between 50 and 80% of mammographically missed cancers can be detected by a CADe scheme (106–109) and 84% of missed lung nodules can be detected by a CADe scheme on lung CT scans (110).

These studies by themselves do not prove that CADe will be an effective clinical tool. It is still necessary to show that a radiologist will recognize an area detected by CADe as a diseased area. It is possible that the cancers detected by CADe will not be above the radiologist's threshold for what is a cancer. To test this, observer studies are conducted.

Observer Studies. Observer studies are conducted to measure the ability of a CAD scheme to improve radiologists' performance. They are designed to simulate clinical reading conditions and thus are considered an indication of how CAD may affect radiologists in their clinical work. In a typical CAD observer study, between 6 and 15 radiologists read a set of cases under two different reading conditions: without the computer aid and with the computer aid. The number of cases ranges from 60 to up to 1000. There are two different types of CAD observer studies: independent reading and sequential reading. In independent reading, each case is read twice, once under each of the two reading conditions, with each reading separate by time period, typically a few weeks, to reduce the chance that the radiologist will remember the case. With a large number of cases, multiple reading sessions are required and readers are required to read under both conditions in each session. Half the cases are read first with aid and then without and in the other half the order is reversed. In sequential reading, the without aid condition is always first and after the radiologist has given their opinion without, the computer result is shown and the case is reassessed by the radiologist. In this way, the radiologist views each image once. The sequential reading method more closely resembles how CAD is used clinically. In addition, the power of the sequential method is higher because the two reading (with and without aid) are correlated.

The first CAD observer study was conducted by Chan et al. in 1990 (6). They showed the potential clinical benefits of CAD for the first time. Using 15 radiologists and 60 mammograms, half with a cluster of calcifications and half without, they showed that a CADe scheme design to detected clustered microcalcifications improved the radiologists performance at a statistically significant level: the AUC increase from 0.94 without aid to 0.97 with aid ($p < 0.001$). Their CADe scheme had a sensitivity of 85% with four false-positive detections per image.

The first observer study involving CADx was conducted by Getty et al. (111). In their study, 6 radiologists read 150 cases containing a benign lesion and containing a malignant lesion. They found that when using CADx radiologist, performance in classifying breast lesions increased significantly. The radiologists' sensitivity increased from 0.51 to 0.69 at a false-positive fraction of 0.1. Further, they showed that general radiologists (those who did not read mammograms full-time) when using the computer aid had comparable performance to expert radiologists (those who read mammograms full-time). In their technique, the radiologists subjectively extracted information from the image and this information was used as input to the CADx scheme.

A similar technique was developed by Wu et al. (112). They should that the CADx scheme could outperform the radiologists in classifying breast lesions. That is, using the information extracted by the radiologists, the CADx scheme was more accurate than the radiologists were. This suggests that radiologists can extract useful information from the mammogram, but they cannot synthesize the information to produce the correct classification. This was the first study to show that a CAD scheme could outperform a radiologist.

The first observer study involving CAD where the computer extracted the features was performed by Jiang et al. (113). In their study, 10 radiologists read 104 cases containing clustered calcifications (54 malignant and 60 benign). They found that the radiologist increased their AUC from 0.61 to 0.75 when they used the computer aid ($p < 0.0001$). In practical terms, they found that on average each radiologist recommended biopsies on 6.0 more malignant clusters ($p = 0.0006$) and 6.4 fewer benign clusters ($p = 0.003$). They also found that their CADx scheme outperformed the unaided radiologists (AUC of 0.80 vs. 0.61, $p < 0.0001$).

Using this observer study, Jiang et al. showed that the variability between radiologists was reduced when they used the CADx scheme (114). They further showed, theoretically, that the radiologists using CADx outperformed independent double reading with two radiologists. In independent double reading, two radiologists read each case independently and if either considers the case abnormal, then it is called abnormal. Another form of double reading is either having the two radiologist discuss the case if they initially disagree and reach a mutual decision or have a third radiologist break the tie. Jiang et al. also simulated double reading with a tiebreaker where they assumed that when two radiologists disagreed, the third radiologist always made the correct decision. This represents a theoretical upper limit on the performance of double reading. When compared to this type of double reading, a single radiologist using CADx had nearly comparable performance. This provides evidence that CADx could be used effectively to implement double reading.

There have been several other observer studies, using ROC analysis, for mammography, breast ultrasound and chest radiography that show the potential for CAD to improve radiologists' performance (115–127).

Clinical Studies. While observer studies provide evidence for the benefits of CAD, actual clinical studies need to be performed to prove that CAD is useful clinically.

The first reported study was conducted by Freer and Ulissey. Using a CADe system for screening mammography, they read 12,860 cases first without aid and then immediately after with aid (sequential reading). They found an additional eight cancers after viewing the computer detections: a 19% increase. This increase was not statistically significant, however, in part because of the smaller number of patients with cancer in the screened population. The use of CADe also increased the callback rate (the number of women consider abnormal) from 6.5 to 7.7% (statistically significant). Since the prevalence of cancer is low, most of the recalls are false positives. Gur et al. conducted the largest clinical study to date (128). They reported on 115,571 screening exams, 56,432 read before the implementation of CADe and 59,139 read after the implementation of CADe. In their study, they compared two time periods before and after CADe was implemented. They found that the number of cancers detected per 1000 women screen when using CADe increased only slightly from 3.49 to 3.55 (not statistically significant) and the callback rate also increased only slightly from 11.39 to 11.40% (not statistically significant). In their study, 17 radiologists were reading clinically. When the data were analyzed by the number of cases read, those reading fewer cases, and by definition less experienced readers, had a increase in cancer detection from 3.05 to 3.65 (not statistically significant) while their callback rate changed from 10.5 to 12.0% (statistically significant, $p < 0.001$) (129).

Clinical evaluation of CAD is difficult, particularly when examining screening mammography. Since the prevalence of cancer in screening mammography is only 4/1000 women, large numbers of women need to be screened to collect enough cancer cases to have a statistically meaningful study. For example, if 25,000 women were screened, ~ 100 cancers would be present, of which ~ 20 maybe missed mammographically. If CADe can detect 50% of the missed cancers and the radiologist when using CADe recognize 100% of those as a miss, then 10 additional cancers would be detected. Measuring this small number of additional cancers is difficult statistically because there are several sources of variability. The number of cancers in screened population, the number of missed cancers, and the number of misses detected by the radiologist when using CADe can all fluctuate statistically speaking. To increase the statistical power of the study, a large number of cancers need to be detected. This can be accomplished by including more women in the study. As the number of women screened increases, it is likely that the results from more than one radiologist need to be included in the study. However, since there is a large variation between radiologists, having more than one observer will introduce another source of statistical uncertainty into the results. These issues need to be carefully addressed when planning a clinical study.

The ultimate endpoint for measuring the benefits of CAD are in a reduction in mortality and morbidity. Randomized controlled clinical trials are ideal for measuring this type of endpoint. For a CADe study, half the women would have their mammograms read without CADe and half would them read with CADe. The women would then be followed for a number of years and the number of deaths from breast cancer in the two groups can be measured and compared. Unfortunately, to have enough statistical power to measure a statistically significant decrease in mortality when using CADe is a very large number of women would need to participate in the trial (on the order of 50,000 or more) and these women would need to be followed for at least 5–10 years. Such a study is cost prohibited. Therefore, studies that do not require long-term follow up are the more likely ones to be preformed to measure the benefits of CAD.

ADVANCED APPLICATIONS

Radiologists rarely consider only a single image in making an interpretation. Most radiologic procedures involve multiple views of the body part of interest. Often there are images from more than one type of modality. Correlation of CT, chest X ray, and a nuclear medicine scan; and of multiple mammograms, ultrasound and breast MRI are just two examples of multimodality imaging. Further, images are often correlated with any clinical findings or patient history to improve the accuracy of the diagnosis. To be a more integrated tool, CAD schemes must be designed to help radiologist with multiimage, multimodality imaging.

Multiimage CAD

The classic example of multi-image CAD is in mammography, where there are two views taken of each breast. Radiologists compare images from the left and right breasts, since the breasts have a natural symmetry and normally they look similar. In addition, any finding in one image is correlated to possible findings in the corresponding second view. This helps the radiologist to determine if a finding is a superposition of normal tissue (a possibility if it is seen in only one view) or a real finding (if seen in both views).

Paquerault et al. developed a method to compare CADe detections between the craniocaudal and the mediolateral oblique views (130). They use a combination of geometric location and morphological and textural features to correlate computer detections between the two views. A major problem with this approach is that actual masses are not always detected in both views because they are not visible in both views or the computer misses the lesion in one of the two views.

For CADx combining results is somewhat simpler, since the output of a CADx scheme is a number that is related to the likelihood that the lesion is malignant. One could show the likelihood value for each image or combine them in some manner. There are different strategies for combining the two values, for example, averaging or choosing the maximum or minimum. The optimum strategy has been studied by Liu et al. (131). They found that the maximum performance in combining two estimates depends on the accuracy of the ROC curves that underlie the two estimates.

In CADe for chest imaging, Li et al. compare the right side to the left side of the chest radiograph, since, with some exceptions, there is symmetry between the two sides of the thorax (132). The comparison is done by subtracting the left and right sides of the chest radiograph, after correcting for the patient not being perfectly upright. The image is also warped to improve the match between structure in the left and right sides. This technique can be used to find any type of disease that appears as an asymmetry (e.g., lung cancer, pneumothorax, pneumonia, and emphysema).

Multimodality CAD

Currently, multimodality CAD only exists for breast imaging and only for classifying breast lesions, CADx. If a woman has an abnormal screening mammogram or some physical symptoms that indicate that she may have breast cancer, she receives a diagnostic work-up, which may include additional specialized mammograms, ultrasound, and a breast MRI. The radiologist interprets these images along with any clinical findings and patient data (e.g., family history) to decide whether or not to recommend the patient have a breast biopsy. The CADx, to be fully integrated into this process and provide the maximum amount of assistance, needs to be multimodality analyzing the lesion as it appears mammographically, sonographically, and from the magnetic resonance (MR) images that usually includes both spatial and temporal information.

CAD Server

The current paradigm for clinical implementation of CAD is for each site to have a CAD system or multiple systems. Since most radiographs are acquired in digital form, the possibility exists for CAD to be implemented over the internet. A central CAD server could serve a number of sites. This scenario offers a number of possibilities over the current paradigm:

1. The CAD schemes from multiple vendors or multiple versions from the same vendor could be made available and the user could specify which algorithm they wanted for a specific image. For example, a radiologist may prefer to have the mass detection scheme from vendor A and the microcalcifications scheme from vendor B run on the mammograms.

2. The CAD server could offer archiving services. This would be appealing to small clinical sites that do not want to or cannot afford to purchase and maintain their own image archive (known as a Picture Archiving and Communication System, PACS).

3. More powerful computer systems could be employed. This offers the possibility of using more power image processing and artificial intelligence techniques that would be time prohibited on a common desktop computer.

4. Patient confidentiality issues not withstanding, a centralized CAD server would analyze hundreds or even thousands of images daily. With some effort, it would be possible to use these images to provide additional training images for the CAD algorithms.

A project called the National Digital Mammography Archive (NDMA) developed such a concept and demonstrated an initial test bed (133). The NDMA was a scalable large-scale image storage and retrieval system providing clinical service for digital mammograms. It provided a number of other services, such as CAD functionality, image retrieval for research purposes (automatically de-identifying patient data), multilevel security, and a radiology teaching component for tele-education in mammography. This was done in real-time, using virtual private networks over the high bandwidth Next Generation Internet (Internet 2). Similar projects are underway in Europe: the eDiamond project (134) and the GPCALMA project (Grid Platform for Computer Assisted Library for MAmmography) (135).

FUTURE STUDIES

Computer-aided diagnosis is still an emerging field, far from being mature. However, much progress has been made in the past 20 years to point where commercial systems are now available for detecting breast cancer from mammograms and detecting lung nodules from CT scans of the thorax or a chest radiograph. The CADe systems for detecting colon polyps from colonography (CT of the colon,

as called virtual colonoscopy) are being developed. Not too surprisingly, all three applications are being used to screen for cancer. In mass screening, very few people screened have the diseased that is being screened for ($< 1\%$). Further, in mammography the radiologist views four views in total, while in CT of the chest and colon up to 400 slices of the body are obtained and viewed for each person. With this amount of information and with subtlety that a cancer can appear with, it is difficult for the radiologist to be ever vigilant. The CADe could play an important role in these situations.

While CADe schemes are available clinically, no clinical CADx system is available. In mammography, there is good evidence that CADx can be a useful aid to the radiologist, especially for differentiating between benign and malignant calcifications. As CT screening for lung cancer becomes more prevalent, there will be an increase in the number of lung biopsies that will need to be performed. Unlike a breast biopsy, however, a lung biopsy carries a bigger risk for complications. Therefore, a CADx scheme for distinguishing benign and malignant lung nodules would be very valuable. Two studies have shown that a CADx system can improve the performance of radiologists in deciding whether a lung nodule is malignant (136,137). It is anticipated that clinical CADx systems will be available in the future and these will need to undergo clinical evaluation.

BIBLIOGRAPHY

1. Winsberg F, et al. Detection of radiographic abnormalities in mammograms by means of optical scanning and computer analysis. Radiology 1967;89:211–215.
2. Chan H-P, et al. Image feature analysis and computer-aided diagnosis in digital radiography. 1. Automated detection of microcalcifications in mammography. Med Phys 1987;14: 538–548.
3. Fujita H, Doi K, Fencil LE, Chua KG. Image feature analysis and computer-aided diagnosis in digital radiography. 2. Computerized determination of vessel sizes in digital subtraction angiography. Med Phys 1987;14:549–556.
4. Giger ML, Doi K, MacMahon H. Image feature analysis and computer-aided diagnosis in digital radiography. 3. Automated detection of nodules in peripheral lung fields. Med Phys 1988;15:158–166.
5. Katsuragawa S, Doi K, MacMahon H. Image feature analysis and computer-aided diagnosis in digital radiography. Detection and characterization of interstitial lung disease in digital chest radiographs. Med Phys 1988;15:311–319.
6. Chan H-P, et al. Improvement in radiologists' detection of clustered microcalcifications on mammograms: The potential of computer-aided diagnosis. Invest Radiol 1990;25:1102–1110.
7. Doi K, et al. Computer-aided diagnosis (CAD): Development of automated schemes for quantitative analysis of radiographic images. Semin Ultrasound CT MR 1992;13:140–152.
8. Anderson ED, Muir BB, Walsh JS, Kirkpatrick AE. The efficacy of double reading mammograms in breast screening. Clin Radiol 1994;49:248–251.
9. Thurfjell EL, Lernevall KA, Taube AA. Benefit of independent double reading in a population-based mammography screening program. Radiology 1994;191:241–244.
10. Kopans DB. Double reading. Radiol Clin N Am 2000;38:719–724.
11. Destounis SV, et al. Can computer-aided detection with double reading of screening mammograms help decrease the false-negative rate? Initial experience. Radiology 2004;232: 578–584.

12. Astley S, et al. CADET: The computer-aided detection evaluation trial. In: Pisano E D, editor. Digital Mammography, 2004. Madison (WI): Medical Physics Publishing; 2005 (in press).

13. Lewin JM, et al. Comparison of full-field digital mammography to screen-film mammography for cancer detection: results of 4945 paired examinations. Radiology 2001;218:873–880.

14. Poplack SP,et al. Mammography in 53,803 women from the New Hampshire mammography network. Radiology 2000;217:832–840.

15. Smith-Bindman R, et al. Physician predictors of mammographic accuracy. J Natl Cancer Inst 2005;97:358–367.

16. Forrest JV, Friedman PJ. Radiologic errors in patients with lung cancer. West J Med 1981;134:485–490.

17. Barlow WE, et al. Performance of diagnostic mammography for women with signs or symptoms of breast cancer. JNCI 2002;94:1151–1159.

18. Kopans DB. The positive predictive value of mammography. AJR 1992;158:521–526.

19. Elmore JG, et al. Ten-year risk of false positive screening mammograms and clinical breast examinations. N Engl J Med 1998;338:1089–1096.

20. Elmore JG, et al. International variation in screening mammography interpretations in community-based programs. JNCI 2003;95:1384–1393.

21. de Wolf CJ, et al. European guidelines for quality assurance in mammography screening. 2nd ed Luxembourg: European Commission, Europe Against Cancer Programme; 1996.

22. Beam CA, Layde PM, Sullivan DC. Variability in the interpretation of screening mammograms by US radiologists. Findings from a national sample. Arch Intern Med 1996;156:209–213.

23. Ciccone G, Vineis P, Frigerio A, Segnan N. Inter-observer and intra-observer variability of mammogram interpretation: a field study. Eur J Cancer 1992;28A:1054–1058.

24. Giger ML, Huo Z, Kupinski MA, Vyborny CJ. Computer-aided diagnosis in mammography. In: Sonka M, Fitzpatrick JM, editor. Handbook of Medical Imaging. Bellingham (WA): The Society of Photo-Optical Instrumentation Engineers; 2000.

25. Karssemeijer N. Detection of masses in mammograms. In: Strickland RN, editor. Image-Processing Techniques in Tumor Detection. New York: Marcel Dekker; 2002.

26. Karssemeijer N, Hendriks JH. Computer-assisted reading of mammograms. Eur Radiol 1997;7:743–748.

27. Nishikawa RM. Detection of microcalcifications. In: Strickland RN, editor. Image-Processing Techniques in Tumor Detection. New York: Marcel Dekker; 2002.

28. Sampat MP, Markey MK, Bovik AC. Computer-aided detection and diagnosis in mammography. In: Bovik AC, editor. The Handbook of Image and Video Processing. New York: Elsevier; 2005.

29. Astley SM, Gilbert FJ. Computer-aided detection in mammography. Clin Radiol 2004;59:390–399.

30. Gale AG, Astley SM, Dance DR, Cairns AY. Digital Mammography Amsterdam: Elsevier; 1994.

31. Doi K, Giger ML, Nishikawa RM, Schmidt RA. Digital Mammography '96. Amsterdam: Elsevier Science; 1996.

32. Karssemeijer N, Thijssen M, Hendriks J, van Erning L. Digital Mammography Nijmegen 98. Amsterdam: Kluwer Academic Publishers; 1998.

33. Yaffe MJ. Digital Mammography 2000 Madison (WI): Medical Physics Publishing; 2000.

34. Peitgen HO. Digital Mammography IWDM 2002 Berlin: Springer-Verlag; 2003.

35. Pisano ED. Digital Mammography 2004 Madison (WI): Medical Physics Publishing; 2005.

36. Zheng B, et al. Computer-aided detection of clustered microcalcifications in digitized mammograms. Acad Radiol 1995;2:655–662.

37. Pisano ED, et al. Radiologists' preferences for digital mammographic display. Radiology 2000;216:820–830.

38. Merkle R, Laine AF, Smith SJ. Evaluation of a multiscale enhancement protocol for digital mammography. In: Strickland RN, editor. Image-Processing Techniques for Tumor Detection. New York: Macel Dekker; 2002.

39. Zhang W, Yoshida H, Nishikawa RM, Doi K. Optimally weighted wavelet transform based on supervised training for detection of microcalcifications in digital mammograms. Med Phys 1998;25:949–956. Yoshida H, et al. An improved computer-assisted diagnostic scheme using wavelet transform for detecting clustered microcalcifications in digital mammograms. Acad Radiol 1996;3:621–627.

40. Brown S, et al. Development of a multi-feature CAD system for mammography. In: Karssemeijer N, Thijssen M, Hendriks J, van Erning L, editors. Digital Mammography Nijmegen 98. Amsterdam: Kluwer Academic Publishers; 1998.

41. Chen CH, Lee GG. On digital mammogram segmentation and microcalcification detection using multiresolution wavelet analysis. Graphical Models Image Processing 1997;59:349–364.

42. Chitre Y, et al. Classification of mammographic microcalcifications using wavelets. Proc SPIE 1995;2434:48–55.

43. Kallergi M. Computer-aided diagnosis of mammographic microcalcification clusters. Med Phys 2004;31:314–326.

44. Laine A, Song S, Fan J. Adaptive multiscale processing for contrast enhancement. Proc SPIE 1993;1905:521–532.

45. Lo S-CB, et al. Detection of clustered microcalcifications using fuzzy modeling and convolution neural network. Proc SPIE 1996;2710:8–15.

46. Strickland RN, Hahn H. Wavelet transforms for detecting microcalcifications in mammograms. IEEE Trans Med Imaging 1996;15:218–229.

47. Wang TC, Karayiannis NB. Detection of microcalcifications in digital mammograms using wavelets. IEEE Trans Med Imaging 1998;17:498–509.

48. Zhang W, Yoshida H, Nishikawa RM, Doi K. Optimally weighted wavelet transform based on supervised training for detection of microcalcifications in digital mammograms. Med Phys 1998;25:949–956. Yoshida H, et al. An improved computer-assisted diagnostic scheme using wavelet transform for detecting clustered microcalcifications in digital mammograms. Acad Radiol 1996;3:621–627.

49. Jiang Y, et al. Method of extracting microcalcifications' signal area and signal thickness from digital mammograms. Proc SPIE 1992;1778:28–36.

50. Veldkamp WJ, Karssemeijer N. Accurate segmentation and contrast measurement of microcalcifications in mammograms: a phantom study. Med Phys 1998;25:1102–1110.

51. Barnes GT. Radiographic mottle: a comprehensive theory. Med Phys 1982;9:656–667.

52. Lefebvre F, et al. A fractal approach to the segmentation of microcalcifications in digital mammograms. Med Phys 1995;22:381–390.

53. Spiesberger W. Mammogram inspection by computer. IEEE Trans Biomed Eng 1979;26:213–219.

54. Carman CS, Eliot G. Detecting calcifications and calcification clusters in digitized mammograms. In: Doi K, Giger ML, Nishikawa RM, Schmidt RA, editors. Digital Mammography '96. Amsterdam: Elsevier Science; 1996.

55. Bankman IN, et al. Automated recognition of microcalcification clusters in mammograms. Proc SPIE 1993;1905:731–738.

56. Chan H-P, et al. Computer-aided detection of microcalcifications in mammograms: Methodology and preliminary clinical study. Invest Radiol 1988;23:664–671.

57. Kobatake H, Takeo H, Nawano S. Microcalcification detection system for full-digital mammography. In: Karssemeijer N, Thijssen M, Hendriks J, van Erning L, editors. Digital

Mammography Nijmegen 98. Amsterdam: Kluwer Academic Publishers; 1998.

58. Zhao D, Shridhar M, Daut DG. Morphology on detection of calcifications in mammograms. Proc SPIE 1993;1905:702–715.

59. Fukuoka D, et al. Automated detection of clustered microcalcifications on digitized mammograms. In: Karssemeijer N, Thijssen M, Hendriks J, van Erning L, editors. Digital Mammography Nijmegen 98. Amsterdam: Kluwer Academic; 1998.

60. Kobatake H, Jin H-R, Yoshinaga Y, Nawano S. Computer diagnosis of breast cancer by mammogram processing. In: Lemke H, Inamura K, Jaffe C, Felix R, editors. Computer Assisted Radiology. Berlin: Springer-Verlag; 1993.

61. Davies DH, Dance DR. Automatic computer detection of clustered calcifications in digital mammograms. Phys Med Biol 1990;35:1111–1118.

62. Fam BW, Olson SL, Winter PF, Scholz FJ. Algorithm for the detection of fine clustered calcifications on film mammograms. Radiology 1988;169:333–337.

63. Ema T, et al. Image feature analysis and computer-aided diagnosis in mammography: Reduction of false-positive clustered microcalcifications using local edge-gradient analysis. Med Phys 1995;22:161–169.

64. Fujita H, et al. Automated detection of masses and clustered microcalcifications on mammograms. Proc SPIE 1995;2434:682.

65. Mascio LN, Hernandez JM, Logan CM. Automated analysis for microcalcifications in high resolution digital mammograms. Proc SPIE 1993;1898:472–479.

66. Shen L, Rangayyan RM, Desautels JEL. Detection and classification of mammographic calcifications. Int J Pat Recog Artif Intell 1993;7:1403–1416.

67. Bottema MJ, Slavotinek JP. Detection of subtle microcalcifications in digital mammograms. In: Karssemeijer N, Thijssen M, Hendriks J, van Erning L, editors. Digital Mammography Nijmegen 98. Amsterdam: Kluwer Academic; 1998.

68. Magnin IE, El Alaoui M, Bremond A. Automatic microcalcifications pattern recognition from X-ray mammographies. Proc SPIE 1989;1137:170–175. Cheng H-D, Lui YM, Freimanis RI. A novel approach to microcalcification detection using fuzzy logic technology. IEEE Trans Med Imaging 1998;17:442–450.

69. Kupinski M, Giger ML. Feature selection with limited datasets. Med Phys 1999;26:2176–2182.

70. Highman R, Brady M. Mammographic Image Analysis. Dordrecht, The Netherlands: Kluwer Academic Publishers; 2000.

71. Zheng B, et al. Computer-aided detection of clustered microcalcifications in digitized mammograms. Acad Radiol 1995;2:655–662.

72. Diahi JG, et al. Evaluation of a neural network classifier for detection of microcalcifications and opacities in digital mammograms. In: Karssemeijer N, Thijssen M, Hendriks J, van Erning L, editors. Digital Mammography Nijmegen 98. Amsterdam: Kluwer Academic Publishers; 1998.

73. Nagel RH, Nishikawa RM, Doi K. Analysis of methods for reducing false positives in the automated detection of clustered microcalcifications in mammograms. Med Phys 1998;25:1502–1506.

74. Lure FYM, Gaborski RS, Pawlicki TF. Application of neural network-based multi-stage system for detection of microcalcification clusters in mammogram images. Proc SPIE 1996;2710:16–23.

75. Davies DH, Dance DR. The automatic computer detection of subtle calcifications in radiographically dense breasts. Phys Med Biol 1992;37:1385–1390.

76. Hojjatoleslami SA, Kittler J. Detection of clusters of microcalcifications using a K-nearest neighbour rule with locally optimum distance metric. In: Doi K, Giger ML, Nishikawa RM, Schmidt RA, editors. Digital Mammography '96. Amsterdam: Elsevier Science; 1996.

77. Cernadas E, et al. Detection of mammographic microcalcifications using a statistical model. In: Karssemeijer N, Thijssen M, Hendriks J, van Erning L, editors. Digital Mammography Nijmegen 98. Amsterdam: Kluwer Academic Publishers; 1998.

78. Anastasio MA, et al. A genetic algorithm-based method for optimizing the performance of a computer-aided diagnosis scheme for detection of clustered microcalcifications in mammograms. Med Phys 1998;25:1613–1620.

79. Anastasio MA, Kupinski MA, Nishikawa RM. Optimization and FROC analysis of rule-based detection schemes using a multiobjective approach. IEEE Trans Med Imaging 1998;17:1089–1093.

80. Wu YZ, Doi KN, Giger ML, Nishikawa RM. Computerized detection of clustered microcalcifications in digital mammograms - Applications of artificial neural networks. Med Phys 1992;19:555–560.

81. Sajda P, Spence C, Pearson J. Learning contextual relationships in mammograms using a hierarchical pyramid neural network. IEEE Trans Med Imaging 2002;21:239–250.

82. Stafford RG, Beutel J, Mickewich DJ. Application of neural networks to computer-aided pathology detection in mammography. Proc SPIE 1993;1898.

83. El-Naqa I, Yang Y, Nishikawa RM, Wernick MN. A support vector machine approach for detection of microcalcifications. IEEE Trans Med Imaging 2002;21:1552–1563.

84. Nishikawa RM, et al. Computer-aided detection of clustered microcalcifications on digital mammograms. Med Biol Engin Comput 1995;33:174–178.

85. Metz CE. Basic principles of ROC analysis. Semin Nucl Med 1978;8:283–298.

86. Metz CE. Fundamental ROC Analysis. In: Beutel H, Kundel J, Van Metter RL, editors. Handbook of Medical Imaging. Bellingham (WA): SPIE; 2000.

87. Dorfman DD, Berbaum KS, Metz CE. Receiver operating characteristic rating analysis. Generalization to the population of readers and patients with the jackknife method. Invest Radiol 1992;27:723–731.

88. Edwards DC, et al. Estimating three-class ideal observer decision variables for computerized detection and classification of mammographic mass lesions. Med Phys 2004;31:81–90.

89. Chan H-P, et al. Design of three-class classifiers in computer-aided diagnosis: Monte carlo simulation study. Proc SPIE 2003;5032:567–578.

90. Dreiseitl S, Ohno-Machado L, Binder M. Comparing three-class diagnostic tests by three-way ROC analysis. Med Decis Making 2000;20:323–331.

91. Mossman D. Three-way ROCs. Med Decis Making 1999;19:79–89.

92. Jiang Y, Metz CE, Nishikawa RM. An ROC partial area index for highly sensitive diagnostic tests. Radiology 1996;201:745–750.

93. McClish DK. Analyzing a portion of the ROC curve. Med Decis Making 1989;9:190–195.

94. Chakraborty DP. The FROC, AFROC and DROC Variants of the ROC analysis. In: Beutel J, Kundel H, Van Metter R, editors. Handbook of Medical Imaging. Bellingham (WA): SPIE; 2000.

95. Chakraborty DP, Berbaum KS. Observer studies involving detection and localization: modeling, analysis, and validation. Med Phys 2004;31:2313–2330.

96. Nishikawa RM, et al. Effect of case selection on the performance of computer-aided detection schemes. Med Phys 1994;21:265–269.

97. Nishikawa RM. Mammographic databases. Breast Dis 1998; 10:137–150.

98. Revesz G, Kundel HL, Bonitatibus M. The effect of verification on the assessment of imaging techniques. Invest Radiol 1983;18:194–198.

99. Fukunaga K, Hayes RR. Effects of sample size on classifier design. IEEE Trans Pattern Anal Machine Intell 1989;11: 873–885.

100. Sahiner B, et al. Feature selection and classifier performance in computer-aided diagnosis: the effect of finite sample size. Med Phys 2000;27:1509–1522.

101. Chan HP, Sahiner B, Wagner RF, Petrick N. Effects of sample size on classifier design for computer-aided diagnosis. Proc SPIE 1998;3338:846–858.

102. Chen DR, et al. Use of the bootstrap technique with small training sets for computer-aided diagnosis in breast ultrasound. Ultrasound Med Biol 2002;28:897–902.

103. Zheng B, Chang YH, Good WF, Gur D. Adequacy testing of training set sample sizes in the development of a computer-assisted diagnosis scheme. Acad Radiol 1997;4: 497–502.

104. Tourassi GD, Floyd CE. The effect of data sampling on the performance evaluation of artificial neural networks in medical diagnosis. Med Decis Making 1997;17:186–192.

105. Efron B, Tibshirani RJ. An Introduction to the Bootstrap New York: Chapman and Hall; 1993.

106. Brem RF, et al. Improvement in sensitivity of screening mammography with computer-aided detection: a multiinstitutional trial. AJR 2003;181:687–693.

107. Burhenne LJW, et al. Potential contribution of computer-aided detection to the sensitivity of screening mammography. Radiology 2000;215:554–562.

108. Moberg K, et al. Computed assisted detection of interval breast cancers. Eur J Radiol 2001;39:104–110.

109. te Brake GM, Karssemeijer N, Hendriks HCL. Automated detection of breast carcinomas not detected in a screening program. Radiology 1998;207:465–471.

110. Armato SG III, et al. Performance of automated CT nodule detection on missed cancers from a lung cancer screening program. Radiology 2002;225:685–692.

111. Getty DJ, Pickett RM, D'Orsi CJ, Swets JA. Enhanced interpretation of diagnostic images. Invest Radiol 1988;23:240–252.

112. Wu Y, et al. Artificial neural networks in mammography: Application to decision making in the diagnosis of breast cancer. Radiology 1993;187:81–87.

113. Jiang Y, et al. Improving breast cancer diagnosis with computer-aided diagnosis. Acad Radiol 1999;6:22–33.

114. Jiang Y, et al. Potential of computer-aided diagnosis to reduce variability in radiologists' interpretations of mammograms depicting microcalcifications. Radiology 2001;220: 787–794.

115. Awai K, et al. Pulmonary nodules at chest CT: effect of computer-aided diagnosis on radiologists' detection performance. Radiology 2004;230:347–352.

116. Brown MS, et al. Computer-aided lung nodule detection in CT: results of large-scale observer test. Acad Radiol 2005;12:681–686.

117. Chan HP, et al. Improvement of radiologists' characterization of mammographic masses by using computer-aided diagnosis: an ROC study. Radiology 1999;212:817–827.

118. Fraioli F, et al. Evaluation of effectiveness of a computer system (CAD) in the identification of lung nodules with low-dose MSCT: scanning technique and preliminary results. Radiol Med (Torino) 2005;109:40–48.

119. Hadjiiski L, et al. Improvement in radiologists' characterization of malignant and benign breast masses on serial mammograms with computer-aided diagnosis: an ROC study. Radiology 2004;233:255–265.

120. Huo Z, Giger ML, Vyborny CJ, Metz CE. Effectiveness of computer-aided diagnosis: Observer study with independent database of mammograms. Radiology 2002;224:560–568.

121. Kakeda S, et al. Improved detection of lung nodules on chest radiographs using a commercial computer-aided diagnosis system. AJR Am J Roentgenol 2004;182:505–510.

122. Kegelmeyer WP Jr, et al. Computer-aided mammographic screening for spiculated lesions. Radiology 1994;191:331–337.

123. Kobayashi T, et al. Effect of a computer-aided diagnosis scheme on radiologists' performance in detection of lung nodules on radiographs. Radiology 1996;199:843–848.

124. Li F, et al. Radiologists' performance for differentiating benign from malignant lung nodules on high-resolution CT using computer-estimated likelihood of malignancy. AJR Am J Roentgenol 2004;183:1209–1215.

125. MacMahon H, et al. Computer-aided diagnosis of pulmonary nodules: results of a large-scale observer test. Radiology 1999;213:723–726.

126. Shah SK, et al. Solitary pulmonary nodule diagnosis on CT: results of an observer study. Acad Radiol 2005;12:496–501.

127. Shiraishi J, Abe H, Engelmann R, Doi K. Effect of high sensitivity in a computerized scheme for detecting extremely subtle solitary pulmonary nodules in chest radiographs: observer performance study. Acad Radiol 2003;10:1302–1311.

128. Gur D, et al. Changes in breast cancer detection and mammography recall rates after the introduction of a computer-aided detection system. JNCI 2004;96:185–190.

129. Feig SA, Sickles EA, Evans WP, Linver MN. Re: Changes in breast cancer detection and mammography recall rates after the introduction of a computer-aided detection system. JNCI 2004;96:1260–1261.

130. Paquerault S, et al. Improvement of computerized mass detection on mammograms: fusion of two-view information. Med Phys 2002;29:238–247.

131. Liu B, Metz CE, Jiang Y. An ROC comparison of four methods of combining information from multiple images of the same patient. Med Phys 2004;31:2552–2563.

132. Li Q, et al. Contralateral subtraction: a novel technique for detection of asymmetric abnormalities on digital chest radiographs. Med Phys 2000;27:47–55.

133. Schnall MD, et al. National digital mammography archive. In: Karellas A, Giger ML, editors. RSNA 2004 Categorical Course: Advances in Breast Imaging: Physics, Technology, and Clinical Applications. Oak Brook (IL): Radiological Society of North America; 2004.

134. Lloyd S, et al. Digital mammography: a world without film? Methods Inf Med 2005;44:168–171.

135. Cerello P, et al. GPCALMA: a Grid-based tool for mammographic screening. Methods Inf Med 2005;44:244–248.

136. Shah S, et al. Computer-aided diagnosis of the solitary pulmonary nodule. Acad Radiol 2005;12:570–575.

137. Aoyama M, et al. Automated computerized scheme for distinction between benign and malignant solitary pulmonary nodules on chest images. Med Phys 2002;29:701–708.

See also BIOTELEMETRY; COMPUTERS IN THE BIOMEDICAL LABORATORY; MEDICAL EDUCATION, COMPUTERS IN; TELERADIOLOGY.

COMPUTERS IN CARDIOGRAPHY. See ELECTROCARDIOGRAPHY, COMPUTERS IN.

COMPUTERS IN THE BIOMEDICAL LABORATORY

GORDON SILVERMAN
Manhattan College

INTRODUCTION

Activities of daily living as well as within the industrial and scientific world have come to be domynated by machines that are designed to reduce the expenditure of human effort and increase the efficiency with which tasks are completed. The technology underlying these devices is heavily dependent on computers that are designed for maximum flexibility, modularity, and "intelligence". A familiar example of flexibility comes to us from the automobile industry where a car manufacturer may use a single engine design for a large variety of models. These automobiles may also contain microcomputers that can sense and "interpret" road surfaces and adjust the car's suspension system to maximize riding comfort. This type of design is characterized by the development of "functional" components that can be reused. The content of these elements, or "*objects*" as they are called, can be changed as technology improves, as long as access (or use) is not compromised by such changes. The characteristics of flexibility, modularity, and intelligence also exemplify instrumentation in the biomedical laboratory where, in recent years, laboratory sciences have become a matter of data and information processing. A user's understanding of the tools of data handling is as much a part of the scientist's skill set as performing a titration, conducting a clinical study, or simulating a biomedical model. Specific tasks in the biomedical (or other) laboratories have become highly automated through the use of intelligent instruments, robotics, and data processing systems. With the existence of >100 million internet-compatible computers throughout the world together with advanced software tools, the ability to conduct experiments "at-a-distance" opens such possibilities as remote control of such experiments and instruments, sharing of instruments among users, more efficient development of experiments, and report or publication of experimental data.

A new vocabulary has emerged to underscore the new computer-based biomedical laboratory environment. Terms such as *bioinformatics* (the scientific field that deals with the acquisition, storage, sharing and optimal use of information, data and knowledge) has come to characterize the biomedical laboratory culture. To understand contemporary biomedical laboratories and the role that the computer plays, it is necessary to consider a number of issues: the nature of data; how data is acquired; computer architectures; software formats; automation; and artificial intelligence.

Computers are instruments that process information, and within the biomedical laboratory they aid the analyst in recording, classifying, interpreting, and summarizing information. These outcomes encompass a number of specific instrument tasks:

- Data handling: acquisition, compression, reduction, interpretation, and record keeping.
- Control of laboratory instruments and utilities.
- Procedure and experiment development.
- Operator interface: control of the course of the experiment.
- Report production.

The biomedical laboratory environment is summarized in Fig. 1. Information produced by the experimental source

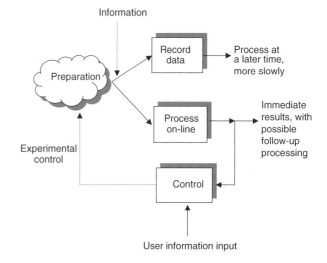

Figure 1. Information processing in laboratory environments.

may simply be recorded and the data analyzed at a later time. Alternatively, the data may be processed while the experiment proceeds and the results, possibly in reduced form, used to modify the experimental environment.

HISTORICAL ORIGINS

Scientific methodologies involving Biomedical laboratory research trace their origins to the experiments of Luigi Galvani in the 1789s with the study of "animal electricity" (1). This productive line of scientific investigation signals the start of the study of electrophysiology that continues to the present time. The "golden age" of electrophysiology began in the twentieth century (in particular starting ∼ 1920) and was led by such (Nobel recognized) scientists as Gasser, Adrian, Hodgkin, Huxley, Eccles, Erlanger, and Hartline, to name but a few (2). These scientists introduced the emerging vacuum tube technologies (e.g., triodes) to observe, record, and subsequently analyze the responses of individual nerve fibers in animal neural systems. The ingenious arrangements that the scientists used were based on an "analog computer" model. A sketch of such arrangements (3) for measuring action potentials is shown in Fig. 2. Of particular value was the cathode ray tube that could display the "rapid" electrochemical changes

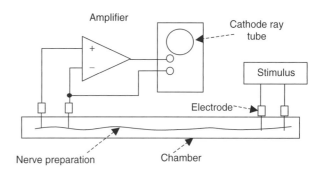

Figure 2. Sketch of an early laboratory arrangement for measuring response of nervous tissue.

from nervous tissue; this was a vast improvement over the string galvanometer that was previously used. A "highly automated" equivalent of this model with a digital computer architecture has been developed by Olansen et al. (4). Numerous electronic advances were made during the 1920s and 1930; Jan Toennies was one of the first bioengineers to design and build vacuum tube-based cathode followers and differential amplifiers. These advances found their way into the military technology of radar and other electronic devices of World War II. During the conflict, the electrophysiologists were pressed into service to develop the operational amplifier circuits that formed the basis of "computation" in the conduct of the war (e.g., fire- control systems). While numerous advances in speed, and instrumental characteristics (e.g., increased input impedance, noise reduction) characterize wartime developments, equipment available in 1950 to continue electrophysiological work precluded rapid analysis of results; it took many weeks to calculate experimental results, a task that was limited to "pencil and paper" computations aided by electromechanical calculators. All this changed with the introduction of digital technology and the digital computer. H.K. Hartline was one of the first of the Nobel Laureates to automate the electrophysiological laboratory with the use of the digital computer. A highly schematic representation of Hartline's experimental configuration is shown in Fig. 3 (5–7).

The architecture suggested in Fig. 3 became a fundamental model for information processing in the biomedical laboratory. However, within the digital computer, a number of (architectural) modifications have been introduced since the 1950s in order to improve informational *throughput*: the ability to complete data processing from acquisition to analysis to recording (as measured in "jobs/s").

DATA IN THE LABORATORY

An appreciation of the data underlying experiments in the biomedical laboratory is essential to successful implementation of a computer-based instrument system. The processing of experimental data is heavily dependent on the amount of information generated by the experimental preparation and the rate at which such data are to be processed by the computer. Both the quantity of data (e.g., the number of samples to be recorded) and the rate at which the data are to be processed can be estimated.

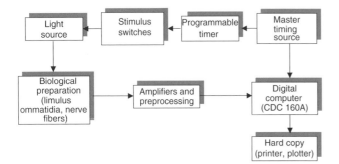

Figure 3. Early architecture of computer-automated biomedical laboratory environment. (After Hartline.)

These factors have important influence on the characteristics of the computer that is to be used in the design. A unit of information is the *bit* (as defined below), and the rate at which information is to be processed is determined by the *capacity* of the experimental environment (including any communications between the preparation and the rest of the system). The units of *system capacity* are bits/s. The following formulas are used to calculate these parameters:

$$\text{Information(bits)} = H = -\textstyle\sum_i p_i \log_2(p_i)$$
$$\text{Capacity(bits/s)} = C = H/T$$

In these formulas, p_i is the probability of experimental outcome i (the experiment may have a finite number of N outcomes) and $\log_2$ is the logarithm to the base 2. The "T" parameter is the time required to transmit the data from the preparation to the instrumental destination. (It must include any required processing time as well.) If all outcomes are equally likely (i.e., probable) then it can be readily shown that $2^H = N$, where N is the total number of possible outcomes (and H is the information content in bits). The informational bit (H) is not to be confused with the term bit associated with binary digits. [Correlation of information (H) and the number of required binary processing digits may be correlated after appropriate coding.]

As an example of an appropriate calculation, consider one scientific temperature-measuring system that can report temperatures from −50 to 150 °C in increments of 0.01 °C. The system can thus produce 20,000 distinct results. If each of these outcomes is equally likely or probable, then the amount of information that must be processed amounts to 14.29 bits. Further, if the data processing system requires 0.1 s to generate a reading, then the capacity of the system is 142.9 bits/s. (As one cannot realistically subdivide a binary digit, 15 binary bits would be needed within the processing system.)

The formulas noted above do not describe the format of the data. For example, how many decimal places should be included. Generally, experiments are designed either to confirm or refute a theory, or to obtain the characteristics of a biological element (for purposes of this discussion). Within the laboratory there may be many variables that affect the data. In experimental environments, the results may often depend on two variables: one is the independent variable and the second, which is functionally related to the independent variable, is specified as the dependent variable. Other variables may act as parameters that are held constant for any given experimental epoch. Examples of independent variables include time, voltage, magnetizing current, light intensity, temperature, frequency (of the stimulating energy source), and chemical concentration. Dependent variables may also come from this list in addition to others.

Two possibilities exist for the experimental variables: Their domains (values) may either be continuous or discrete (i.e., having a fixed number of decimal places) in nature. As a consequence, there are four possible combinations for the independent and dependent variables within the functional relationship; these are shown in Fig. 4, where the format "independent/dependent" applies to the axes. In Fig. 4, the solid lines represent actual values

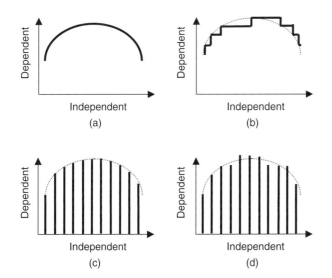

Figure 4. Underlying information formats for biomedical laboratory data: (a) continuous–continuous; (b) continuous–discrete; (c) discrete–continuous; (d) discrete–discrete.

Table 1. Binary Coded Outcomes of Experimental Data

Outcome	Binary Code
0	0000
1	0001
2	0010
3	0011
4	0100
5	0101
6	0110
7	0111
8	1000
9	1001
10	1010
11	1011
12	1100
13	1101
14	1110
15	1111

generated by the experimental preparation. The bold vertical lines emphasize the fact that measurements are taken only at discrete sampling times. "Staircase-like" responses indicate that only discrete values are possible for the dependent variable. Although any of the formats are theoretically possible, contaminants (noise for the dependent variable and bandwidth for the independent variable) usually limit the continuum of values. Thus, all instances of information in the biomedical laboratory are ultimately discrete–discrete (Fig. 4d) in nature. (In Figs. 4b–d, the dotted line reflects the original signal.)

The ability of a computer-based instrument system to discriminate between datum points is measured by its *resolution* and reflects the number of distinct values that a variable can assume. In the temperature-measuring example, there were 20,000 distinctly possible readings and consequently the resolution was 0.01 °C. When resolution is combined with the range of values that the variable can assume, the number of distinct experimental outcomes can be computed:

Number of unique experimental outcomes

= range/resolution

Summarizing the temperature-measuring system example in light of this relationship we conclude: resolution = 0.01 (°C); range = 200 (°C); number of outcomes = 20,000. Within a computer-based data acquisition (DAQ) system, the values may appear as coded representations of the underlying outcomes. The codes might be related to the equivalent decimal values of the original data. However, one could also assign an arbitrary code to each outcome. Because computer-based systems are designed to interpret codes that have two distinct states (or symbols), binary coding systems are normally used in laboratory applications. A variety of binary coding schemes are possible. A simple, but effective, code employs the binary number system to represent experimental outcomes. A number in this system is a weighted combination of the two symbols that are recognized in the binary number system, namely, 0 and 1. A complete representation of a binary number is given by

$$a_n 2^n + a_{n-1} 2^{n-1} + \cdots + a_1 2^1 + a_0 2^0 \cdot a_{-1} 2^{-1} + a_{-2} 2^{-2} + \cdots$$

where all coefficients (a values) are either 1 or 0. Starting with the least significant digit (2^0), the positional weights for the positive powers of 2 are 1, 2, 4, 8, 16, and so on. For the negative powers of 2, the weights in increasingly smaller values are 1/2, 1/4, 1/8, and so on. Table 1 contains a list of 16 possible outcomes from a (low resolution) laboratory experiment including both the binary and decimal equivalents. The outcomes might represent values of voltage, time, frequency, temperature, or other experimental variables.

The elements of a computer-based information processing system for biomedical laboratories are generally compatible with the binary system previously discussed. However, human users of such machines are accustomed to the decimal number system (as well as the alphanumeric characters of their native language). Within the system, internal operations are carried out using binary numbers and calculations. Binary results are often translated into decimal form before presentation to a user; numerical inputs, if in decimal format, are translated (by the computer) into binary format before use within the computer. Other number systems may be found in a computer application. These include octal systems (base 8) and the hexadecimal number system (base 16), where the base symbols include 0, 1,..., 9, A, B, C, D, E, F. A user may also be required to enter other forms of information such as characters that represent a series of instructions or a program. Several widely accepted codes exist for alphanumeric data, and some of these together with their characteristics are shown in Table 2.

Coded information such as that shown in Tables 1 and 2, may be passed (i.e., transmitted) between different elements of a computer-based laboratory information processing system. The communication literature provides a

Table 2. Partial List of Alphanumeric Codes

Name of Code	Number of Bits	Number of Available Code Combinations
Extended Binary code		
Decimal Interchange Code (EBCDIC)	8	256
American Standard Code		
For Information Interchange (ASCII)	7	128
ASCII-8		
8-bit extension of ASCII	8	256
Hollerith	12	4096

rather complete description of the technology (8,9), and while it is not immediately germane to many circumstances of this discussion, some elements need to be mentioned. For example, the "internet" should be noted as an emerging development in computer-based biomedical laboratories.

There are two general protocols for transmitting laboratory data from the information source to its destination. Each part of the coded information (i.e., the bit) may be passed via a single communication channel. The channel element is the media and it might be wire, fiber optic cable, or air (as in wireless). Since there is only one channel, the data is passed in serial fashion, one bit at a time. An alternative arrangement permits the bits to be passed all at once (in parallel), but this requires an independent path for each bit such as a multiwire architecture. Parallel transmissions have inherently greater capacity than serial schemes. For example, if it requires 1 μs (i.e., 10^{-6} s) to transmit a bit, then a parallel transmission, using an 8-bit code, can pass 8×10^6 bits/s (i.e., 8 Mbits/s). An equivalent serial system would only have a capacity of 1 Mbit/s as it would require 8 μs for complete transmission of the code representing one of the possible experimental outcomes. Note that serial systems have a decided economic advantage over parallel schemes.

There are circumstances when two-way communication between elements of a computer-based data processing system is necessary. One element (e.g., the computer) may initiate a measurement instruction to a remotely located laboratory instrument (e.g., a spectrophotometer); the remote unit, in turn, responds with a set of measurements. Instructions, data, and parameters may need to pass from the computer, and status information (e.g., a busy signal) or results must be able to pass from the instrument to the computer: all of this over a serial path. Serial architectures can occur over a one-way (i.e., single-lane highway) or a two-way (i.e., two-lane highway) link. Figure 5 summarizes these communication alternatives.

In the half-duplex case, a single path must suffice for two-way communication. For proper transmission, the path must be made ready for communication in an appropriate direction before communication starts.

OVERALL ARCHITECTURE OF A COMPUTER-BASED BIOMEDICAL LABORATORY SYSTEM

The hardware for contemporary computer-based laboratory instrument systems reflects an information-processing model as shown in Fig. 1. A broad representative computer-based DAQ, and processing system is shown in Fig. 6. The

elements depicted in the figure can be divided into several categories: the computer [PC, laptop, personal digital assistant (PDA)], sensors (or transducers), signal conditioning components, DAQ, software, and other elements for other aspects of computer-based environments (remote instrumentation, external processors, vision/imaging equipment, and motion control apparatus).

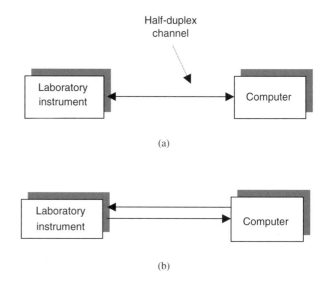

(a)

(b)

Figure 5. Serial communication alternatives: (a) half-duplex; (b) full-duplex.

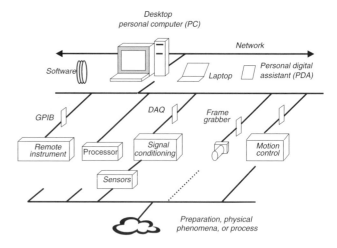

Figure 6. General arrangement of computer-based information processing system for biomedical laboratories.

Table 3. Sampling of Sensors for Biomedical Laboratory Applications

Application	Sensor Technology
Position	Resistive (potentiometric, goniometric), shaft Encoder, linear variable differential transformer (LVDT), capacitive, piezoelectric
Velocity	LVDT
Acceleration	LVDT, strain gauge, piezoelectric (attached to an elastic flexure), vibrometer
Force	Strain gauge, piezoelectric, LVDT, resistive, capacitative (all making use of a flexible attachment)
Pressure	Strain guage, piezoelectric, also LVDT and capacitative.
Flow	Measure pressure drop which is correlated to flow rate via a calibration function (Venturi tube, pitot tube)
Temperature	Thermoresistive (Seebeck), thermistor (semiconductor), resistive (platinum wire)
Light	Photocell, photoresistor, photodiode, phototransistor.

The computer in such systems has considerable impact on the maximum throughput and, in particular, often limits the rate at which one can continuously acquire data. New bus (communication) facilities in the modern computer have greatly increased speed capabilities. A limiting factor for acquiring large amounts of data is often the hard drive (secondary storage system in the computer). Applications requiring "real-time" processing of high frequency signals often make use of an external (micro)processor to provide for preprocessing of data. (The term *real-time* refers to a guaranteed time to complete a series of calculations.) With the rapid development of new technologies, and the reduction is size, laptops and PDAs have found their way into the laboratory, particularly when the data is to be accumulated in isolated sites as with many biological experiments. With appropriate application software, laptops can act as data loggers [simple recorders of source (raw) data are collected] and further processing subsequently completed on a PC (or other) computer. Even greater miniaturization now permits PDAs to collect and transmit data as well. With wireless technology, data can be e-mailed to a base station. Also noted in Fig. 6 is the potential to connect the computer to a network [local area network (LAN), wide area network (WAN), or Internet] with the possibility of conducting experiments under remote control.

Sensors (see sensors) or transducers (10,11) sense physical phenomena and produce electrical signals that DAQ components can ultimately accept after suitable signal conditioning. Transducers are grouped according to the physical phenomena being measured. These devices have an upper operating frequency above which they produce a signal that is no longer independent of the frequency of the source (phenomena) frequency. Position is the most common measurement and sensors normally translate such physical distortions into changes in the electrical characteristics of the sensor component. For example, capacitive transducers rely on the fact that the capacitance depends on the separation (or overlap) of its plates. Since the separation is a nonlinear function of the separation, capacitive sensors are usually combined with a conditioning circuit that produces a linear relationship between the underlying phenomena and the potential delivered to the DAQ. Changes in the dimensions of a resistor alter its resistance. Thus, when a thin wire is stretched, its resistance changes. This can be used to measure small displacements and generate a measure of strain. Inductive principles are employed to measure velocity. When a mova-

ble core passes through the center of a coil of wire, the (electromagnetic) coupling is altered in a way that can be used to determine its velocity. Table 3 summarizes several applications.

Electrical signals generated by the sensors often need to be modified so that they are suitable for the DAQ circuitry. A number of conditioning functions are carried out in the signal conditioning system: amplification (to increase measurement resolution and compatibility with the full-scale characteristic of the DAQ); linearization (to compensate for nonlinearities in the transducer such as those of thermocouples); isolation (of the transducer from the remainder of the system to minimize the possibility of electric shock); filtering (to eliminate noise or unwanted interference such as those frequencies that are erroneous (e.g., high frequency or those from the power lines); excitation provides external signal source requirements for the transducer (such as strain gages that require a resistive arrangement for proper operation).

The DAQs normally include an analogue-to-digital converter (ADC) for converting analog (voltage) signals into a binary (digital) quantity that can be processed by the computer. There are several well-developed techniques for performing the conversion (12). As a general principle, the concept shown in Fig. 7 can be used to explain the conversion process.

Resistors in a high-tolerance network are switched in a predetermined manner resulting in an output voltage that is a function of the switch settings. This voltage is compared to the unknown signal and when this reference equals the unknown voltage the switch sequence is halted.

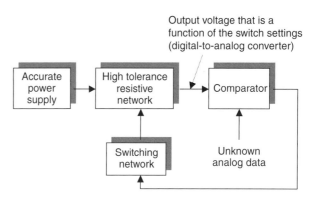

Figure 7. Principle of analog-to-digital conversion.

The pattern of switches then represents a digital quantity equivalent to the unknown analogue signal. (Other schemes are possible.)

Several characteristics of the DAQ need to be considered when specifying this part of a computer-based processing system.

Range: The spread in the value of the measurand (experimental input) over which the instrument is designed to operate.

Sensitivity: The change in the DAQs output for a unit change in the input.

Linearity: The maximum percentage error between an assumed linear response and the actual nonlinear behavior. (The user should be assured that the DAQ has been calibrated against some recognized standard.)

Hysteresis: Repeatability when the unknown is first increased from a given value to the limit of the DAQ (range) and then decreased to the same (given) value.

Repeatability: Max difference of the DAQ reading when the same input is repeatedly applied (often expressed as a percentage of the DAQs range).

Accuracy: Maximum degree to which an output differs from the actual (true) input. This summarizes all errors previously noted.

Resolution: Smallest change in the input that can be observed.

Time Constant: Time required for the DAQ to reach 63.2% of its final value from the sudden application of the input signal.

Rise Time: time required to go from 5 to 95% of its final output value.

Response Time: time that the DAQ requires reaching 95% of its final value.

Settling Time: Time that the DAQ requires to attain and/or remain within a given range of its final value (e.g., $\pm 2\%$ of its final value).

Delay Time: Time taken for the DAQ to reach 50% of its final value (not normally considered important).

Other sensor characteristics include: natural frequency, output impedance, mass, size, and cost.

Existing instruments may also be integrated into the computer-based environment if they include compatibility with a standard known variously as the General Purpose Interface Bus (GPIB) or IEEE 488: Originally developed by Hewlett-Packard (now Agilent) in 1965 to connect commercial instruments to computers (13). The high transfer rates (1MB/s) led to its popularity and it has evolved into an ANSI/IEEE Standard designated as 488.1, and subsequently as 488.2. The GPIB devices communicate with other such devices by sending device-dependent messages across the interface system (bus). These devices are classified as "Talkers", "Listeners", and/or "Controllers." A Talker sends data messages to one or more Listeners that receive the data. The Controller manages the flow of information on the bus by sending commands to all devices. For example, a digital voltmeter has the potential to be a Talker as well as Listener. The GPIB Controller is akin to the switching center of a telephone system. Such instruments may be connected to the computer system as seen in Fig. 6 as long as an appropriate component (card) is installed within the computer.

Images may be gathered from the biomedical laboratory using a (digital) camera and a suitable card within the computer. (See Olansen and Rosow in the *Reading List*.) Machine vision may be viewed as the acquisition and processing of images to identify or measure characteristics of objects. Successful implementation of a computer-based vision system requires a number of steps including:

Conditioning: Preparing the image environment including such parameters as light and motion.

Acquisition: Selecting image acquisition hardware (camera and lens) as well as software to be able to capture and display the image.

Analysis: Identification and interpretation of the image.

Computer-based (software) analysis of images takes into consideration the following:

Pattern Matching: Information about the presence or absence, number and location of objects (e.g., biological cells).

Positioning: Determining the position and orientation of a known object by locating features (e.g., a cell may have unique densities).

Inspection and *Examination*: Detecting flaws (e.g., cancer cells).

Gauging: Measuring lengths, diameters, angles and other critical dimensions. If the measurements fall outside a set of tolerance levels, then the object may be "discarded".

In addition to the acquisition of information from the laboratory, the computer system may be used to control a process such as an automated substance analysis using a robotic arm. (The motion control elements in Fig. 6 support such applications.) A sketch of such a system is shown in

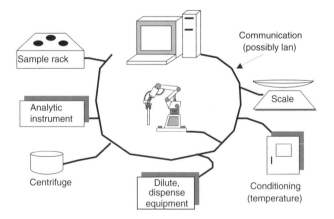

Figure 8. Computer-based architecture for an automated substance analysis system.

Fig. 8. The robotic arm would normally have several degrees of freedom (i.e., axes of motion) (14).

The system includes: a centrifuge (e.g., for analysis of blood samples); an analytic instrument (e.g., a spectrophotometer or chromatograph); a rack to hold the samples; a balance; a conditioning unit (possibly a stirrer or temperature oven); and instrumentation for dispensing, extracting and/or diluting chemicals. Various application programs within the computer could be used to precisely define the steps taken by the robotic arm to carry out a routine test. This program must also take into consideration the tasks to be carried out by each instrument (i.e., the *drivers*). When the computer does not obtain ongoing, continuous, status information from an instrument, the resultant arrangement is referred to as *open-loop* control of the particular instrument. In such cases, the program must provide for appropriate time delays (such as the time needed to position the robotic arm). Alternatively, the computer can receive signals from the various components that advise the program of their status; this is referred to as *closed-loop* control and is a generally more desirable mode of operation than the open-loop configuration.

COMPUTER BASICS

Personal computers are organized to carry out tedious, repetitive tasks in a rapid and error-free manner. The computer has four principal functional elements:

- Central processing unit (CPU) for arithmetic and logical operations, and instruction control.
- Memory for storage of data, results, and instructions (programs)
- Input/output components (I/O) for interaction between the computer and the external environment.
- Communication bus: or simply the bus, that allows the functional elements to communicate.

These elements are shown in Fig. 9 that comprises the functional architecture of the PC. Detailed descriptions, and operation of the PC and its components (e.g., secondary storage system—*hard drive*) are readily available (15). The architectures of computer-based instrument systems for laboratory environments generally fall into one of four categories; these are noted in Fig. 10.

A single-purpose (fully dedicated) instrument is shown in Fig. 10a. This arrangement is convenient because it is consistent with such things as existing building wiring, particularly the telephone system although emerging developments also lend this architecture to a wireless arrangement. Standard communication protocols (previously noted) allow manufacturers to develop instruments to accepted standards. This arrangement may be limited to a single PC and a single instrument, and the distance between the host (PC) and the instrument may also be constrained. By adding additional communication lines, other instruments can be added to the single PC.

Remote control of instruments is depicted in Fig. 10b and is accomplished by adding devices within the PC that support communication over a traditional (standard) telephone line (including use of the internet). Real-time operation in such circumstances may be limited because time is required to complete the communications between the PC and the remote instrument placing significant limits on the ability of the system to obtain complete results in a prescribed time interval. Delays produced by the PCs operating system (OS) must also be factored into information processing tasks.

With the development of, and need for, instruments with greater capacity, new architectures emerged. One configuration is shown in Fig. 10c and includes a single PC together with multiple instruments coupled via a standard (i.e., IEEE 488) or proprietary (communication) bus. While such architectures are flexible and new instruments can be readily added, the speed of operation can deteriorate to the point where the capabilities of the PC are exceeded. Speed is reduced because of competition for (access to) PC resources (e.g., hard disk space).

The arrangement shown in Fig. 10d is referred to as *"tightly coupled"*. In such cases, the instruments are integral to the PC itself. Communication between the PC and the instrument is rapid. Real-time (on-line) operation of the instrument is facilitated by a direct communication path (i.e., system bus) between the instrument and other critical parts of the PC such as its memory. No (external) PC controller is necessary and consequently the time delays associated with such functional elements do not exist. By varying the functional combinations, the system can be reconfigured for a new application. For example, functional components might include: data acquisition resources, specialized display facilities, and multiport memory for communication (message-passing) between the other elements.

Each of the arrangements in Fig. 10 includes a single PC. Additional operating speeds are possible (at relatively low cost) if more than one PC is included in the instrumental configuration. Such architectures are called multiprocessor-based instrument systems. Each processor carries out program instructions in its own right (16–18).

With the increasing complexity and capabilities of new software, a more efficient arrangement for computer-based laboratory systems has emerged. This is the *client–server* concept as shown in Fig. 11. The server provides services needed by several users (database storage, computation, administration, printing, etc.) while the client computer (the users) manage the individual laboratory applications (e.g., DAQ), or local needs (graphical interfaces, error checking, data formatting, queries, submissions, etc.).

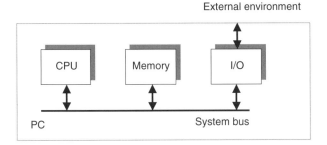

Figure 9. Basic architecture of the PC.

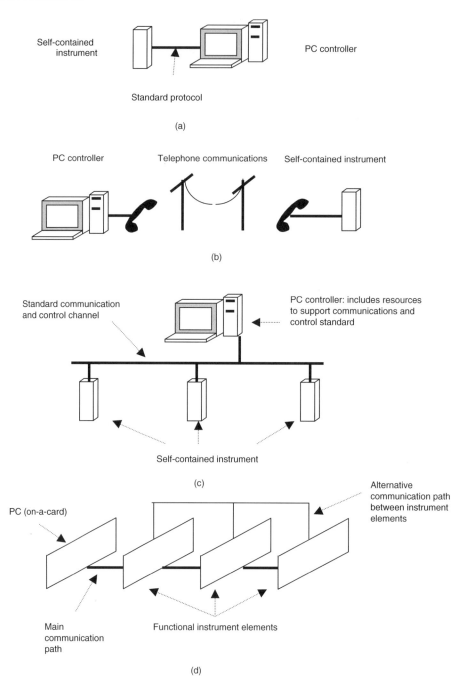

Figure 10. Single-PC instrument architecture (a) Dedicated system. (b) One form of remotely controlled arrangement. (c) Multiple instrument arrangement. (d) Tightly coupled architecture.

SOFTWARE IN COMPUTER-BASED INSTRUMENT SYSTEMS

Programming consists of a detailed and explicit set of directions for accomplishing some purpose, the set being expressed in some "language" suitable for input to a computer. Within the computer, the components respond to two signals: +5 V, or 0 V (ground). These potentials are interpreted as the equivalent of two logical conditions; normally the +5 V is viewed to mean logically true (or logical 1), and a 0 V is interpreted as logically false (or logical 0). (Note that this is not universally true, and in

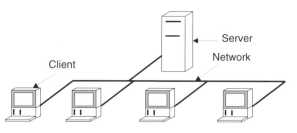

Figure 11. The client–server architecture for biomedical laboratory environments.

some situations the logical 0 signals "true" while the logical 1 signals "false", but this is normally an exceptional case.) During the early 1950s, laboratory computers were programmed by entering a series of logical 1s and 0s directly into the computer using a series of switches on the computer's front panel. A breakthrough occurred when English-like phrases could be used in place of these binary numbers. A series of programs within the machine called an *assembler* could be employed to interpret the instructions underlying the binary numbers. Within assembly language programs, English-like mnemonics are used in place of the numbers previously used to designate an instruction. The following example represents a series of assembly language instructions that might be used to add to quantities and store the results in one of the memory locations of the computer. (Text after the semicolon is considered to be a comment and not an instruction.)

```
MOV ACC, A      ;move the augend into the arithmetic
                 unit
ADD ACC, B      ;add the addend to the sum
MOV C, ACC      ;store the result in location "C"
```

During the 1950s, greater abstraction was introduced when text-based programming languages such as FORTRAN and COBOL were developed. Such languages are referred to as high level languages (HLLs). When individual versions are taken into account, there are literally hundreds of HLLs currently viable with languages such as C, C++, and JAVA being prominent. By using HLLS, the three lines of code shown above could effectively be replaced by a single instruction:

$$C = A + B$$

Statements such as these made problem solving and programming more abstract, readable, and reduced the time it took to develop software applications. The statements are entered into the computer using a program called an *editor*; the code is then *compiled* and *assembled* (translated using a *compiler* program and an *assembler* program) to reduce the original text to the binary numbers needed to control the computer: the only "instructions" that a computer really "understands".

Rather than having to "rewrite" a program each time it was required, HLLs provided a means to develop highly abstract "application programs". A key development of such powerful resources was the introduction of Visicalc, the first (primitive) spreadsheet program: It is progenitor of such widely used programs as Excel, LOTUS, and others. Development of automated spreadsheet programs was motivated by the need for them in business applications, but they have come to find considerable utility in biomedical laboratory environments, particularly for data and statistical analysis as well as for data acquisition.

Increasing levels of abstraction in which programming details are hidden have continued to drive developments in software. A most important transition was made when software entered the age of "visual" programming. Arrangements and interconnections of functional icons have come to replace text when developing software for the biomedical laboratory. A key example of this architecture is the "graphic programming language" (GPL) called LabVIEW, which stands for Laboratory Virtual Instrument Electronic Workbench. This programming scheme provides work areas (windows) that the programmer uses to develop the software. In particular, the windows include a "Front Panel" and a "Diagram". This software enables a user to convert the computer into a software instrument that carries out real tasks (when coupled to appropriate elements as shown in Fig. 6). The Panel displays the indicators and controls that a user would find if the investigator had obtained a separate instrument for the experimental setup. Figures 12 and 13 are representative of a LabVIEW (front) panel and diagram. They are suitably annotated to indicate, controls, indicators, and symbols to replace traditional programming constructs (19).

The HLL programming is characterized by a *control flow* model in which the program elements execute one at a time in an order that is coded explicitly within the program. The narrative-like statements of the program describe the sequential execution of "Procedure A" followed by "Procedure B", and so on. In contrast, a visual programming paradigm functions as a *dataflow* computer language. This depends on *data dependency*; that is, the object in the block (node) will begin execution at the moment when all of its inputs are available. (This reflects a "parallel" execution scheme and is consistent with a multitasking model.) After completing its internal operations, the block will present processed results at its output terminals. While one node waits for events, other processes can execute. This is in

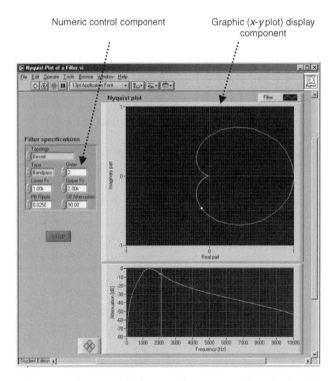

Figure 12. Front panel of a virtual instrument that obtains the attenuation and Nyquist characteristics (plot) of a filter that could be used in a Biomedical laboratory DAQ system. Control and display components are identified.

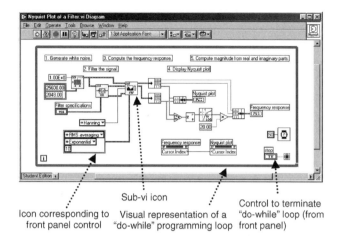

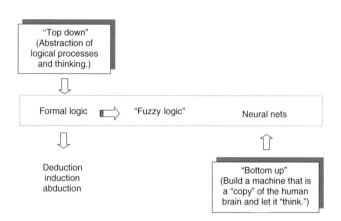

Figure 13. Diagram (Data Flow) of the virtual instrument for determining the frequency response and Nyquist plot of a filter.

Figure 14. Classification of artificial intelligence: Formal logic, Fuzzy Logic, Neural Nets.

some contrast to the control flow case where any waiting periods can create "dead time," and may reduce the system throughput (20).

The virtual instrument (vi) depicted in Figs. 12 and 13 computes the frequency response of a digital filter and displays the attenuation versus frequency (the independent variable), as well as the imaginary part of the response versus the real part (i.e., *Nyquist* plot). In this example, a white noise signal is used as the stimulus of the filter and the vi returns the frequency response of the filter (21).

EMERGING AND FUTURE DEVELOPMENTS FOR COMPUTER-BASED SYSTEMS IN BIOMEDICAL LABORATORIES

Data processing in the biomedical laboratory is coming to rely increasingly on artificial intelligence (AI) for analysis, pattern recognition, and scientific conclusions. The development of "artificially intelligent" systems has been one of the most ambitious and controversial uses of computers in the biomedical laboratory. Historically, developments in this area (biomedical laboratory) were largely based in the United States (22–24). Artificial intelligent can support both the creation and use of scientific knowledge within the biomedical laboratory. Human cognition is underscored by a complex and interrelated set of phenomena. From one perspective, AI can be implemented with computer systems whose performance is, at some level, indistinguishable from those of human beings. At the extreme of this approach, AI would reside in "computer minds" such as robots or virtual worlds like the information space found in the Internet. Alternatively, AI can be viewed as a way to support scientists to make decisions in complex or difficult situations. For example, anesthesiology requires the health provider to monitor and control a great many parameters at the same time. In such circumstances, dangerous trends may be difficult for the anesthesiologist to detect in "real time;" AI can provide "intelligent control." In science, AI systems have the capacity to learn, leading to the discovery of new phenomena and the creation of scientific knowledge. Modern computers and their associated appli-

cation software tools can be used to analyze large amounts of data, searching for complex patterns, and suggesting previously unexpected relationships (see Coiera in the Reading List). Simply stated, the goal of AI is to develop automata (machines) that function in the same way that a human would function in a given environment with a known complement of stimulants. In 1938, the British mathematician Alan Turing showed that a simple computational model (the Turing Machine) was capable of universal computation. This was one basis for the stored program model used extensively in modern computers.

An attempt to build an automaton that imitates human behavior falls into three broad categories: formal logic, "fuzzy" logic, and neural net technologies. These are depicted in Fig. 14.

While there is considerable overlap between human cognitive activities and machine technologies, in its most generic sense the relationships can be summarized in Table 4.

Historically, the first attempts at machine intelligence reflected formal logical thinking of which there are three kinds: deductive, inductive, and abductive. These are all built on a system of rules, some of which may be probabilistic in nature. Deductive reasoning is considered to be perfect logic : you cannot prove a false predicate to be true, or a true predicate to be true. The logic is built on the following sequence of predicates:

> If p then q
> p is true
> Therefore q is true

By using the classification of beats in the ECG signal based on QRS duration and RR interval, we can develop a simple

Table 4. Human Cognitive Activities and Corresponding Machine Technologies

Human Activity	Machine Technology
Pattern recognition	Neural Nets
Belief System and control	Fuzzy logic
Application of logic	Expert Systems: rules and generic algorithms

example of deductive reasoning:

All{beats (RR interval) falling between 1.0 and 1.5s having a QRS interval between 50 and 80 ms}are normal

Patient's {60th QRS complex occurs 1.25 s after the 59th complex with a duration of 60 ms}Patient's 60th QRS complex is normal

Inductive conclusions, which can be imperfect and produce errors, follow from a series of observations. This logic is summarized by the following series of predicate statements:

From : (P a), (P b), (P c), . . .

Infer : [forall(x)(P x)]

(P a), (P b), (P c), and so on, all signify that entities whose properties are a, b, c, and so on, belong to the category identified as P. We therefore conclude that any object whose properties are similar to those of a, b, c, and so on, belong to the category identified as P. For example, a physician may observe many patients who have had fevers and some of who have subsequently died. Postmortem examination may reveal that they all had a lung infection (labeled "pneumonia"). The physician may (erroneously) conclude that "all fevers must imply pneumonia", because he/she has a number of observations in which fever was associated with pneumonia.

Using cause–effect statements, abductive reasoning gather all possible observations (effects) and reaches conclusions regarding causes. For example, both pneumonia and septicaemia may both cause fever. The physician would then use additional observations (effects) to single out the "correct" cause. Abductive logic may also lead to false conclusions. Abductive logic follows from the argument that follows:

If p then q

q is true

Therefore p follows(i.e., is true)

Using the circumstances just cited, a physician may (erroneously) conclude that having observed a fever, the patient is suffering from septicaemia. (It may, of course, also be due to pneumonia.)

Machine-Based Expert Systems

These systems require machine-based reasoning methods just noted and are depicted in Fig. 15. In addition, they must include stylized or abstracted versions of the world. Each of the representations in the database must be able to act as a substitute or surrogate for the underlying object (or idea). In addition, these tokens may have metaphysical features that reflect how the system intends to "think about the world". For example, in one type of representation known as a script, a number of predicates may appear that describe what is to be expected for a particular medical test. An Expert System for "detecting" an asystole in an ECG (electrocardiogram) might invoke the following rules (see Coiera in Further Reading):

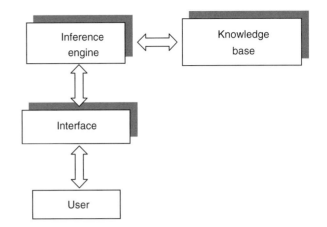

Figure 15. Block diagram of a typical Expert System.

Rule 1:

If heart rate = 0

Then conclude asystole

Rule 2:

If asystole and(blood pressure is pulsatile and in the normal range)

Then conclude retract asystole

Where knowledge is less certain, the rules might be modified. For example, for Rule 1, the conclusion might become, "conclude asystole with probability (0.8)".

A somewhat more informative example can be drawn from an interactive fragment from a contemporary medical Expert System (with similarities to the historical MYCIN software) (25):

The fragment does not represent a complete interactive session. The user would need to supply additional information to generate a potential diagnosis. An excellent demonstration of such systems can be found on the Internet:

http://dxplain.mgh.harvard.edu/dxp/dxp.sdemo.pl?/login=dems/cshome

While Expert System technologies have produced useful applications, they are confronted with a fundamental problem: How to determine what is "true" and what is "false". Contemporary systems address this in a variety of ways (e.g., providing a probabilistic result). This remains a problem for application software.

Fuzzy Logic Systems

These systems attempt to overcome the vagaries of truth and falsity and thus better reflect human thinking and may have some advantage over Expert Systems, where predicates are either true or false (or have some fixed probability of truth or falsity). Such systems were pioneered by Loti Zadeh in 1966 although exploitation began in earnest during the 1990s. [These are currently well over 2000 patents (many from Japan where this

[An asterisk (*) indicates physician responses. What follows ";" are explanatory comments]

Please enter findings
* sex male
* race white
* alcoholism chronic
* go
Disregarding:
    Exposure to rabbits
    Leg weakness
    Creatinine blood increased
Considering
    Age 26–55
Ruleout:
    Hepatitis chronic
    Alcoholic hepatitis
Abdomen pain generalized?
* no
Abdomen pain right quadrant?

;The program asks for facts about the
;patient
;There is a fixed vocabulary of symptoms
;that must be followed
;This starts processing in the Expert System
;The system finds a set of suspected diseases
;Symptoms not explained by these diseases
;are put aside.

;The system explains its reasoning

;and rules out certain disease

;It requests additional information to
;further refine its findings
;Fragment ends here.

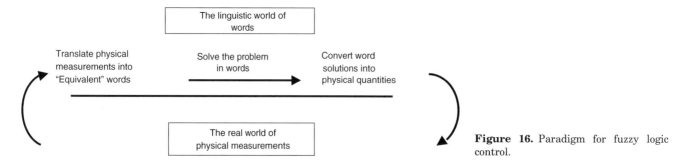

Figure 16. Paradigm for fuzzy logic control.

technology was first embraced) and billions of dollars of sales of fuzzy products.] The concept underlying fuzzy logic is shown in Fig. 16 (26). Measurements in the real world are translated into equivalent linguistic concepts; the resulting "word" problems are solved in the linguistic world and conclusions are reconverted into physical entities that control elements in the real world.

Translation from physical measurements is accomplished by using a "belief system" (so called "*membership functions*") that reflects the degree to which we accept the particular measurement. A given measurement will then determine the extent to which we interpret its meaning. A representative set of membership functions is shown in Fig. 17, shown together with outcomes for a particular physical measurement (input).

A similar set of functions is also constructed to represent outputs, or rules, as summarized below. The output functions determine the extent to which we should set a particular control parameter in the physical system. In Fig. 17, we interpret the measurement to mean that we

have a 50% belief that it has a low value; we also believe that this measurement could represent a "medium" quantity, but we only have a 25% confidence in this value. (One might say, "The measurement is somewhere between a low and a medium value. Membership functions provides a measure of such informality.)

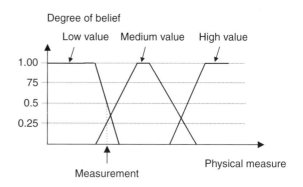

Figure 17. Fuzzy logic membership functions.

Table 5. Operations on Fuzzy Relations

Operation	Fuzzy Application[a]
Parameter 1 AND Parameter 2	Min (m1, m2)
Parameter 1 OR Parameter 2	Max (m1, m2)
NOT Parameter 1	1−m1

[a]The m stands for the membership or belief value as described in the narrative.

A system of rules also forms part of fuzzy control facilities. These rules have the following form:

> If (physical parameter 1 is low) AND (physical parameter 2 is high) THEN (apply Rule 1 with a low intensity).

> If (physical parameter 1 is medium) OR (physical parameter 2 is high) THEN (apply Rule 1 with a medium intensity)

The measurements and the logical operators (e.g., AND, OR, NOT) are employed according to the following set of (fuzzy) rules (as developed by Zadeh) (see Table 5).

For a given control problem, the measurement, application of the rules, and the invocation of actions leads to an overall profile of action; a typical result is shown in Fig. 18. From this a control value can be returned to the physical system: generating what is referred to as a *crisp result* (*Defuzzification*). There are several methods for obtaining this value from the curve noted in Fig. 18. Shown is the Center-of-Gravity method wherein the control value to be applied to the variable under control is the "balance point" of the curve.

Fuzzy control in a biomedical environment is exemplified by control of oxygen delivery to ventilated newborns in a neonatal intensive care environment (27). A sketch of the system is shown in Fig. 19.

For newborns requiring mechanical ventilation, oxygen toxicity is a potential danger that could result in chronic lung disease. Oxygen levels are also implicated in the development of retinopathy. Inspired oxygen concentration is commonly adjusted on an acute basis to control oxygen delivery and maintain patient saturation levels. The design of classic (engineering) control systems in such cases presents a significant challenge because of "transportation" delays. The alternative of manual control is also

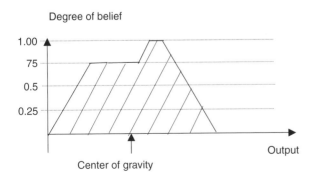

Figure 18. Defuzzification or generation of crisp results for a fuzzy logic control system.

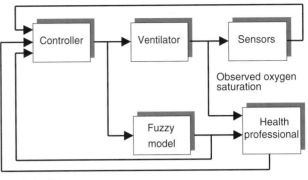

Figure 19. Fuzzy control of inspired O_2 concentration in ventilated infants.

unsatisfactory for two reasons: the patient might require increased O_2 requirement but the manual increase is delayed due to delay in human response (e.g., the clinician is not present); the patient has decreased O_2 requirement (clinical conditions improve) but the amount of O_2 is not immediately decreased (because there is a perception that the patient is "doing well" and does not require intervention).

The Fuzzy model included some 35 rules, of which the following is one example:

If {change in oxygen saturation is small-negative} AND {rate of change in oxygen saturation is medium-negative}
    THEN
{increase inspired oxygen concentration by a medium-positive amount}

The membership curves for the various parameters will impose specific amounts for each rule that is invoked. Each rule yields an "action" value according to a membership class or extent to which the rule should be applied. A weighted mean of all rule outputs produces a single value for inspired oxygen concentration. This system maintained a target oxygen saturation (the set point determined by the health professional) better than routine manual control. It reduced overall oxygen exposure. No complex mathematical models were required as might be the case for traditional control with predictive capabilities. The rules are easy to understand and modify; expert knowledge about the problem was utilized. The controller was easily designed for nonlinear system responses: a goal that is daunting for traditional control technologies.

Neural Net (NN)

These systems employ a combination of circuits that approximate the behavior of neurological cells. While not limited to such applications, NNs are particularly useful for pattern recognition (28,29). Figure 20 shows the model of a single neuronal element and as well as a network of neuronal elements (NN).

The design of NNs is definitely not a precise enterprise; it is decidedly an art. A NN is "trained" to recognize patterns and while there are a great variety of NN

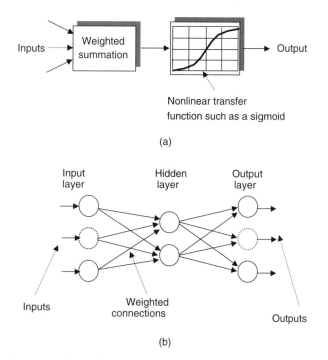

(a)

(b)

Figure 20. Neural Net architecture. (a) A single neuron. (b) A NN showing input, hidden, and output layers with weighted interconnections.

architectures and training paradigms, a general scheme works as follows:

- A NN is presented (at the input layer) with a set of "test" signals (e.g., samples of the input as a function of time). This is the so-called *training set*. If the NN produces an incorrect output (i.e., recognizing the wrong pattern), the weighted interconnections are automatically readjusted using an algorithm designated as *back propagation* (of error correction). The next time that the particular input sample is presented, the NN will *tend* to produce the correct answer.

- Training continues until the NN satisfactorily recognizes the members of the training set. This recognition does not have to be perfect: just as human experts might disagree on the interpretation of a pattern.

- The NN is then put into operation with inputs that it has not necessarily seen before; it will recognize whether or not the inputs have the same characteristics of the training set that it has been taught to recognize.

- This process may be iterative; if the operational results are unsatisfactory, additional samples may be added to the training set and retraining instituted.

The NNs have been used to advantage in an ever-growing set of circumstances (30–32). In one instance, the NN (*Hypernet*) used a multi-layer architecture in which subjects' anamnestic data, and 24-h diastolic and systolic blood pressure measurements were used as the parameters. System outputs included four 24-item arrays, whose values specify the hourly dosage to be administered to a patient for

each of the most common antihypertensive drugs. The presence and degree of hypertension (diagnosis) could be inferred from the drug dosage. No treatment was required for normal subjects whose records formed part of the training set. The system was evaluated on the basis of accuracy in both diagnosis and prescriptions. Hypernet correctly diagnosed 33 subjects out of 35, and 82% of the system's prescribed treatment was deemed correct or acceptable by a group of medical experts.

SUMMARY

Biomedical laboratory instrument technology has changed markedly over the last 50 years. During the 1970s, a few daring scientific investigators began using the computer for instrument control and data acquisition. Such features are now commonplace add-ons to most instrumentation. The modern desktop computer system is a window into computing resources available anywhere in the world. Increasing levels of abstraction within the computer and its application software signal the emergence of a data processing model for experimental design within the Biomedical laboratory. This will only accelerate in the future as new computational algorithms are developed. Artificial intelligence as a knowledge-based tool is certain to grow manifestly in this emerging culture.

BIBLIOGRAPHY

1. Nebeker F. Golden accomplishments in Biomedical Engineering. Charting the milestones of Biomedical. Engineering. Engineering in Medicine and Biology Society; 2003.
2. Schoenfeld RL. From Einthoven's Galvanometer to Single-Channel Recording. Charting the milestones of Biomedical Engineering, Engineering. Medicine and Biology Society; 2003.
3. Gasser HS, Erlanger J. A study of the active currents of nerve cells with the cathode ray oscilloscope. Am J Physiol 1922; 62: 406–524.
4. Olansen JB, Ghorbel F, Clark JW, Bidani A. Using virtual instrumentation to develop a modern biomedical engineering laboratory. Int J Eng Educ 2000; 16(3):244–254.
5. Schoenfeld RL. The role of a digital computer as a biological instrument. Ann New York Acad Sci 1964; 125(2):915–942.
6. Schoenfeld RL. How we know what the eye and the mind's eye sees. IEEE Eng Med Biol Mag 1992; 11:47–71.
7. Silverman G, Eisenberg L. Programmable parallel timing system. IEEE Trans Biomed Eng 1971; 18(3):201–205.
8. Stallings W. Data and computer communications. 5th ed. New York: Prentice Hall; 1997.
9. Rappaport TS. Wireless communications; principles and practice. New York: Prentice Hall; 1999.
10. Carr JJ. Sensors and circuits. New York: Prentice Hall; 1993.
11. Norton HN. Handbook of transducers. New York: Prentice Hall; 1989.
12. Tyler M. The ABCs of ADCs. Scientific Computing & Instrumentation, April 2001; p 32–34.
13. Institute of Electrical and Electronics Engineers. IEEE Standard Digital Interface for Programmable Instrumentation, Standard IEEE 488.1, IEEE 488.2.
14. Liscouski J. Laboratory and scientific computing: A strategic approach. New York: John Wiley & Sons; 1995.

15. Clements A. Principles of computer hardware. PWS-Kent; 1991.
16. Silverman G. Automation in the biomedical laboratory. IEEE Trans Biomed Eng 1884; BME-31:748–752.
17. Stromquist BR, Pavlides C, Zelano JA. On-line acquisition, analysis and presentation of neurophysiological data based on a personal microcomputer system. J Neurosci Methods 1990;35:215–222.
18. Beavis BC, Chait BT. Rapid, sensitive analysis of protein mixtures by mass spectrometry. Proc Natl Acad Sci USA 1990;87:6873–6877.
19. Wald M. Int J Eng Educ 2000;16:3.
20. Essick J. Advanced LabVIEW Labs. New York: Prentice Hall; 1999.
21. Ziemer RE, Traner WH, Fannin DR. Signals and Systems. 4th ed. New York: Prentice Hall; 1998.
22. Szolovits P. Artificial intelligence in medicine. AAAS Selected Symposia Series. Colorado: Westview Press; 1982.
23. Clancy WJ, Shortliffe EH editors. Readings in Medical Artificial Intelligence—The First Decade. Reading, MA: Addison-Wesley; 1984.
24. Miller PL. Selected Topics in Medical Artificial Intelligence. New York: Springer Verlag; 1988.
25. Buchanan BG, Shortliffe EH, editors. Rule-based expert systems: The MYCIN experiments of the Stanford heuristic programming project. Reading, Massachusetts: Addison-Wesley Publishing Company; 1984.
26. Ross TJ. Fuzzy logic with engineering applications. New York: McGraw-Hill; 1995.
27. Sun Y, Kohane I, Stark A. Fuzzy logic control of inspired oxygen concentration in ventilated newborn infants. www.chip.org/chip/projects/avent/sun-flpaper.html.
28. Fausett L. Fundamentals of neural networks. New York: Prentice-Hall; 1994.
29. Kohonen T. An introduction to neural computing. Neural Networks 1988;1:3–16.
30. Poli R, Cagnoni S, Livi R, Coppini G, Valli G. A nn expert system for diagnosing and treating hypertension. Computer March 1991.
31. Silverman G, Brudny J, Gage P. Artificial intelligence in rehabilitation medicine: an emerging technology. Proc 6th Annu Meet Am Telemedicine Assoc, Ft. Lauderdale, FL; June 3–6, 2001.
32. Hirsch S, Frank PI, Shapiro JL. Use of an artificial neural network in estimating prevalence and assessing underdiagnosis of asthma. Neural Computing and Appl 1997;5(2):134–128.

Further Reading

Liscouski J. Laboratory and scientific computing: A strategic approach. New York: John Wiley & Sons; 1995.
Webster JG, editor. Bioinstrumentation. New York: John Wiley & Sons; 2004.
Carr JC, Brown JM. Introduction to biomedical equipment technology. 4th ed. New York: Prentice Hall; 2001.
Olansen JB, Rosow E. Virtual bio-instrumentation. New York: Prentice Hall PTR; 2002.
Keener J, Sneyd J. Mathematical physiology. New York: Springer; 1998.
Paton BE. Sensors, transducers & LabVIEW. New York: Prentice Hall PTR; 1999.
Silverman G, Silver H. Modern Instrumentation: A Computer Approach. IOP Publishing; 1995.
Coiera E. Guide to Medical Informatics, the Internet and Telemedicine. Oxford University Press; 1997.

See also ANALYTICAL METHODS, AUTOMATED; COMPUTER-ASSISTED DETECTION AND DIAGNOSIS; CYTOLOGY, AUTOMATED; DIFFERENTIAL COUNTS, AUTOMATED.

COMPUTERS IN MEDICAL EDUCATION. See MEDICAL EDUCATION, COMPUTERS IN.

COMPUTERS IN MEDICAL RECORDS. See MEDICAL RECORDS, COMPUTERS IN.

COMPUTERS IN NUCLEAR MEDICINE. See NUCLEAR MEDICINE, COMPUTERS IN.

CONFOCAL MICROSCOPY. See MICROSCOPY, CONFOCAL.

CONFORMAL RADIOTHERAPY. See RADIOTHERAPY, THREE-DIMENSIONAL CONFORMAL.

CONTACT LENSES

ERIC R. RITCHEY
The Ohio State University
Columbus, Ohio

INTRODUCTION

Contact lenses are prescription medical devices applied to the anterior surface of the cornea for the temporary correction of refractive error. Contact lenses can be used successfully to correct a number of refractive error conditions, such as myopia, hyperopia, presbyopia, and aphakia. Currently, there are an estimated 38 million contact lens wearers in the United States and 125 million contact lens wearers worldwide, making contact lenses one of the most commonly prescribed medical devices available (1). The ability to wear contact lenses successfully is dependent on numerous factors, including but not limited to the contact lens design and materials, corneal health and physiology, the lens to epithelium interface, proper use of lens care solutions, patient compliance, professional fitting and follow-up care. Unacceptable contact lens fits can lead to deleterious results ranging from poor lens comfort to microbial keratitis and permanent loss of visual acuity.

HISTORY

The earliest origins of the theoretical application of a device to the anterior cornea for the correction of vision can be traced to Leonardo da Vinci in the early sixteenth century (2). Leonardo da Vinci described an experiment where the subject immerses his face in a transparent globe filled with water that effectively neutralizes the subject's refractive error. In 1636, Rene Descartes described the neutralization of refractive error through the use of a long water filled tube, called a hydrodiascope that was held against the anterior surface of the cornea (3). The ideas proposed by da Vinci and Descartes, however, could not be used in practical application for the correction of refractive error on a daily basis. The first true contact lens that could be worn on the eye was the scleral contact lens. The scleral contact lens is a large diameter lens (>13 mm) that features a scleral (also known as a haptic) segment of the lens that rests on the conjunctiva and a central corneal section that arched over the cornea. The first scleral contact lens was created Frederick Muller in 1887. Although the lens was a nonoptical lens, it showed that the placement of a lens on the eye was an achievable goal. In 1888, Adolph Fick of Germany and Euegene Kalt of France each developed optically corrective glass scleral contact lenses independently of one another (4). In 1936, William Feinbloom created a hybrid scleral lens with a poly(methylmethacrylate) (PMMA) scleral section and a glass corneal section (5). The change to PMMA as the lens material of choice was due to its superior durability compared to glass. In 1947, Kevin Touhy created the first PMMA corneal contact lens. The corneal contact lens featured an overall diameter smaller than the corneal diameter and lacked a scleral–haptic section that rests on the cornea. The Touhy design was the precursor to the modern rigid gas permeable contact lens used today (4,5).

In 1954, Czechoslovakias Otto Wichterle and Drahoslav Lim developed the first soft contact lens polymer called hydroxyethyl methacrylate (HEMA). The development of HEMA became the chemical backbone for soft hydrogel contact lenses and made the development of a soft contact lens possible (5,6). In 1971, Bausch and Lomb received the U.S. Food and Drug Administration (FDA) approval for the SofLens soft contact lens, the first soft contact lens released to the public for daily wear. The FDA granted the first approval for soft contact lenses overnight wear, also known as extended wear, in 1981 (6). Improvements in lens manufacturing technologies in the late 1980s allowed soft contact lens manufacturers to produce lenses, at a cost where frequent replacement, or disposable lenses were introduced to the marketplace. These disposable lens were developed and promoted for a number of replacement schedules including every 3 months, every month, every 2 weeks, and every day throughout the late 1980s and the 1990s (7).

In 1998, the first silicone hydrogel lens was released for public distribution. The development of a soft contact lens that successfully incorporates silicone into the lens matrix started with work with silicone elastomers in the 1950s (4). Silicone elastomers have the advantage of high oxygen permeability compared to HEMA-based contact lenses. The early silicone elastomer lenses had problems with wettability and comfort. Silicone hydrogel lenses provide the oxygen benefits of a silicone elastomer with the comfort of hydrogel materials. Currently, the FDA has approved some silicone hydrogel design contact lenses for continuous extended wear ranging from 1 week to 1 month dependent on the lens design (8,9).

CONTACT LENS OPTICS AND DESIGN

Contact lenses are designed to correct the refractive error of the patient by converging or diverging light entering the visual system. In myopia, or nearsightedness, the refractive power of the eye converges light excessively, causing incoming light to be focuses in front of the retina creating blur. To correct for myopia, a lens that diverges light,

indicated with a minus power, in combination with the optics of the eye will allow the incoming light to focus on the retina providing clear vision. In hyperopia, or farsightedness, the refractive power of the eye lacks sufficient convergence to focus incoming light on the retina, causing light to be focused behind the retina (10). To correct for hyperopia, a lens that causes light to converge, indicated with a plus power, in combination with the optics of the eye will allow the incoming light to focus on the retina providing clear vision. The contact lens must change the vergence of the incoming light to correct the patient's refractive error while maintaining positional stability, corneal health, and patient comfort.

POLY(METHYL METHACRYLATE)–RIGID GAS PERMEABLE CONTACT LENS DESIGN

Rigid contact lenses were initially manufactured from PMMA, a nonpermeable plastic material used for the first "hard" contact lens (4). Today rigid gas permeable (RGP) lenses, also referred to simply as gas permeable (GP) lenses, are manufactured from a number of gas permeable silicone acrylate or fluorosilicone acrylate materials. Silicone acrylate and fluorosilicone acrylate materials, although not as durable as PMMA lenses, have superior oxygen permeability and reduce corneal hypoxia associated with long term PMMA use (11,12). The manufacturing process for PMMA and rigid gas permeable lenses is essentially identical. The PMMA and gas permeable lenses are cut from a button of the lens material that is lathed with one of two basic curve designs. The traditional contact lens design is a tricurve or multicurve design with each peripheral curve flatter than the proceeding curve. The central curve is designated the base curve (BC), which is typically designed to match the central flat corneal curvature measured for the patient. The next two curvatures, referred to as the secondary curve and the peripheral curve, are each flatter than the preceding curve. The transitions between the curvatures, as well as the edge of the lens, are rounded and polished to prevent the development of sharp edges that will decrease patient comfort and cause physiological damage to the cornea (13). Rigid lenses may also be lathed in an aspheric curve design, which has a central base curve and a continuously flattening peripheral curvature without distinct secondary or peripheral curves. The purpose behind flattening the curvature of the lens as you move from the center of the lens for each design is to allow the gas permeable lens to match the progressively flattening aspheric corneal surface, promoting patient comfort and an improved physiological response to contact lens wear (12).

The rigid contact lens diameter can be subdivided into different zones that correspond to the curvature system described above. The central zone of the contact lens is referred to as the optic zone and the width of this zone is referred to as the optic zone diameter (OZD). The optic zone is the portion of the contact lens that contains the refractive power of the lens and is used for the correction of the patient's refractive error. The curvature of the lens at the optic zone is the base curve of the lens. The optic zone diameter constitutes the majority of the overall lens diameter (OAD). The width of the secondary and peripheral curves are designated the secondary curve width and the peripheral curve width (12,13). For aspheric designs, an optic zone diameter is typically not specified and measurement of the peripheral curve width is impractical due to the gradually changing curvature of this region. The typical gas permeable contact lens is 8–11 mm in diameter with an optic zone diameter between 6 and 9 mm (13).

SOFT CONTACT LENS DESIGN

Soft hydrogel contact lenses are produced by one of three techniques: lathe cut, spin casting, and cast molding. Lathe cut lenses are produced in a method similar to the production of rigid gas permeable lenses. A button of dehydrated plastic is cut to the desired shape using a computerized lathe. The lens is then hydrated in a saline solution bath, where it will expand into its final shape. Lathe cut soft contact lenses can be made in a wide variety of powers and curvatures. These lenses are typically replaced annually due to the cost of each lens produced. Because of the lathing and hydrations process, there can be some variability in the optics and curvatures observed between lenses (14,15).

Spin casting techniques utilize a spinning mold to distribute a liquid polymer in a thin consistent layer without the use of a lathe. The spinning cast mold features a concave surface that will hold the liquid polymer, and therefore determines the shape of the anterior surface of the lens. The posterior lens surface is determined by the centrifugal force generated by the spinning mold and the surface tension of the polymer. Once the liquid polymer has been distributed by the spinning mold, the lens is polymerized while the mold spins using ultraviolet (UV) light. Polymerization will solidify the liquid polymer and the lens is removed from the cast. Spin casting can be used to produce a number contact lens powers without the marks associated with lathing techniques (14,15).

A third method of soft contact lens production is the cast molding technique. Cast molding uses a two piece die or cast that when placed together will form the anterior and posterior surface of the lens. The lens polymer is injected between the two cast pieces and is then polymerized. The two cast pieces are then separated and the contact lens is finished and packaged for distribution (14,15). The majority of disposable contact lenses produced for the contact lens market in the United States are cast molded or spin casting designs.

FITTING PHILOSOPHY: RIGID GAS PERMEABLE CONTACT LENSES

Gas permeable rigid contact lenses are typically fit according to one of two fitting philosophies. One philosophy is the interpalpebral (IP) lens fit. Interpalpebral lens fit contact lenses feature small diameters lenses with the central base curve that parallels the flat central corneal curvature. This philosophy is designed to keep the contact lens centered on the corneal and promotes minimal interaction with the

upper lid when the patient blinks (16). The second philosophy is the lid attachment contact lens fit. The lid attachment fit philosophy utilizes large diameter contact lenses with a base curve that is equal or slightly flatter than the flattest corneal meridian measured by central keratometry. The lens edge will lie underneath the superior lid and will move in tandem with the lid while blinking (16). The peripheral curve is designed to maximize the contact of the eyelid with the lens. Lid attachment philosophy proponents argue that this design promotes superior tear exchange behind the contact lens and provides superior patient comfort as it keeps the lid from chronically moving over the edge of the contact lens (17). The mid-peripheral curve design for each technique aims to closely follow the curvature of the peripheral cornea to provide adequate lens movement and comfort. Adjustments are made to the final power of the contact lens to compensate for the tear lens created behind the lens with each fitting philosophy (16).

SOFT CONTACT LENS FITTING

The fitting of soft contact lenses is driven by the material properties of the lens. Due to the water content of soft contact lenses, a lens applied to the cornea will drape over the epithelial surface and assume the shape of the cornea. The goal of the practitioner fitting the lens is to avoid any adverse effects from lens wear while providing the patient with a comfortable lens fit and good visual acuity. The three primary considerations for the practitioner when evaluating the interaction of the contact lens and the cornea are the coverage of the cornea by the contact lens, the centration of the contact lens over the cornea, and the movement of the contact lens with the blink response. A successful soft contact lens fit consists of a lens that covers the entire corneal surface, referred to as paralimbal coverage, is centered over the corneal apex, and moves ∼ 0.25–1.00 mm with the blink (18). Soft contact lenses that consistently fail to cover and stay centered on the cornea or lack adequate lens movement may lead top potential adverse events with contact lens wear. The adverse events experienced from an inadequate soft contact lens fit can range in severity from poor lens comfort to microbial infection and a severe reduction in the patient's best corrected visual acuity. The success of a soft contact lens fit is also determined by the visual acuity obtained with the lens on eye and a patient history of lens comfort and adequate wear time throughout the day (7,18).

CONTACT LENS WEAR SCHEDULES

Contact lenses are prescribed by the practitioner with a specified wearing schedule. Wearing schedules can be divided into four categories: Daily wear, flexwear, overnight wear, and extended wear. Daily wear contact lenses are to be worn during waking hours and removed before sleep. The lenses are cleaned and stored overnight or discarded as determined by the recommended replacement schedule. Flexwear is a term used for contact lenses that are typically worn for daily wear but may be worn on a 24 h basis 1–2 days per week. Overnight wear contact lenses are to be worn while sleeping and removed upon awakening. The most common example of overnight wear are reverse geometry gas permeable contact lenses used for orthokeratology. Extended wear, also referred to as continuous wear, is where a lens is worn on a 24 h basis for a predetermined period of time such as 1 week or 1 month. Currently, the maximum allowable extended wear time permitted by the FDA is 30 continuous day dependent on the lens design and material. Soft hydrogel, silicone hydrogel, and rigid gas permeable contact lens materials have been approved by the FDA may be worn for varying lengths of extended wear. Contact lenses designed for extended wear are typically made from high oxygen permeability materials to compensate for 24 h lens wear.

CONTACT LENS CARE

Cleaning and disinfection of contact lenses is one of the most critical elements for successful contact lens wear. Failure to properly clean and disinfect the contact lens can lead to visually threatening adverse events. Lens care begins with proper hygiene. Hand washing is required before the handling of contact lens (19). Once the patient has washed their hands, lens cleaning and disinfection may begin. A number of different commercially available lens cleaning systems have been distributed for consumer use (20). The cleaning and disinfection of lens has traditionally included some form of digital cleaning (a.k.a lens rubbing), rinsing, and overnight storage of the lens in a preserved saline or disinfecting solution. Digital cleaning of the lens is important to free accumulated deposits such as proteins, lipids, and microbes from the lens surface. Rinsing the lens further removes the accumulated deposits and overnight storage in the contact lens solution disinfects and hydrates the contact lens (21). More recently, solutions have been developed to clean and disinfect contact lenses that do not require digital cleaning.

Cleaners are divided into surfactant cleaners and enzymatic cleaners. Surfactant cleaners are used to remove lipids, oils, and environmental pollutants. Enzymatic cleaners are used to remove proteins on the surface of the contact lens and may be derived from plants, animals, or bacteria. Currently, the most popular method for the cleaning and disinfection of soft contact lenses is the multipurpose contact lens solution (MPS). Multipurpose contact lens solutions are used as the cleaning solution during digital lens cleaning, the rinsing solution, and the overnight storage and disinfection solution. Several multipurpose contact lens solutions have been designated by the FDA as "no rub" cleaning solutions, where the patient simply rinses the lens for a designated period of time and then stores the lens in the solution overnight after removal from the eye. Regardless of no rub approval by the FDA, digital cleaning is recommended with the use of silicone hydrogel lens materials. Other cleaning and disinfection systems available for use with soft contact lenses include hydrogen peroxide based systems, thermal disinfection units, ultraviolet or microwave radiation disinfection units, or ultrasonic mechanical agitation units (19,20,22).

Cleaning and disinfection of rigid gas permeable contact lenses is similar to the care of soft contact lenses. The rigid gas permeable lens is removed from the eye, cleaned digitally, rinsed with tap water or saline solution, and stored in a preserved conditioning solution. Two bottle cleaning systems consist of a contact lens cleaning solution and a lens conditioning solution. The cleaning solution typically contains a surfactant cleaner and may also have an abrasive component, such as silica beads, to help remove deposits from the lens surface (23). After cleaning, the cleaning solution is rinsed from the lens surface using tap water or saline solution. The contact lens is then placed in conditioning or disinfecting solution for overnight storage. The use of tap water for rinsing rigid gas permeable contact lenses is a subject of much debate and should be avoided in situations where the quality of the water is suspect (24,25). The use of tap water should be avoided in patients with an overnight or extended wear schedule. Multipurpose one bottle cleaning systems are available for use with rigid gas permeable contact lenses. As with soft contact lens care systems, the multipurpose lens care solution is used as the cleaning, rinsing, and storage solution. Dependent on the formulation, the multipurpose solution may have to be rinsed from the lens surface with tap water or saline solution prior to lens insertion. Unlike multipurpose solutions used in soft contact lens care, multipurpose solutions used with rigid gas permeable lenses are not indicated for no rub usage (20).

CORNEAL PHYSIOLOGY AND RESPONSE TO CONTACT LENS WEAR

The cornea is a multilayered, avascular tissue that receives oxygen from the atmosphere through diffusion of oxygen in the preocular tear film. The nutrient supply for the cornea comes from the anterior chamber of the eye (26). When a contact lens is placed on eye, oxygen from the atmosphere must pass through the lens matrix or must be transported to the corneal epithelium by the tears pumped underneath the lens by the blink response. A chronic lack of oxygen, referred to as corneal hypoxia, will lead to swelling of the cornea called corneal edema (27). The permeability of a contact lens material, referred to as the Dk of the lens, is the ability of oxygen to permeate through a contact lens material. The Dk is determined by the chemical composition of a lens polymer. A more relevant measure the oxygen reaching the cornea through a contact lens is the oxygen transmissibility, referred to as Dk/t, where Dk is the oxygen permeability of the lens and t is equal to the average thickness of the contact lens (28). The first contact lens material, PMMA, had a Dk of 0 and relied solely on the pumping of tears (a.k.a the tear pump) beneath the contact lens with the blink to carry oxygen to the corneal surface (11). Insufficient oxygen to the cornea can lead to a number of adverse events, such as corneal edema or epithelial breakdown (27). Therefore much emphasis has been placed on the oxygen permeability of contact lenses used on a daily wear and an extended wear basis.

The ultimate goal traditionally has been the development of a contact lens material with oxygen permeability sufficient enough to mimic the conditions that occur when no lens is worn. In 1984, Holden and Mertz reported that a contact lens must have an oxygen transmissibility of $87.0 \pm 3.3 \times 10^{-9}$ (cm · mL O$_2$)/s · mL · mmHg) to limit the cornea to the 4% corneal edema noted after sleeping with no contact lens wear (29). A number of different hydrogel lens designs were utilized in the study with water content from 38.6 up to 75% as well as the Silsoft silicone elastomer lens. In 1999, Harvitt and Bonanno revisited oxygen permeability and corneal swelling to compensate for the effect of acidosis on oxygen consumption. Harvitt and Bonanno found that with decreasing contact lens oxygen transmissibility there was an increase in corneal stroma acidosis effectively reducing the amount of oxygen available to the cornea. The oxygen transmissibility required to prevent stromal anoxia after compensation for acidosis in closed eye conditions was found to be 125 barrer·mm^{-1} (1 barrer $= 10^{-10}$ cm^2·s^{-1}·cmHg^{-1} or 7.5005×10^{-18} m^2·s^{-1}·Pa^{-1}) (30). Oxygen transmissibility in traditional hydrogel lenses is limited the water content of the lens and the overall lens thickness. Oxygen transmissibility in hydrogel lenses is increased by increasing the water content while making a thinner contact lens. The theoretical best oxygen permeability for a hydrogel lens would be 80 barrer, the oxygen permeability of water (8). Thus, hydrogel lenses will never be able to obtain a level of oxygen transmissibility set forth by Holden and Mertz or Harvitt and Bonanno.

With the development of the Holden–Mertz criteria and subsequent adjustments by Harvitt and Bonanno, there was a renewed interest in silicone contact lens technology. The original patents for the siloxane hydrogel were filed in the late 1970s. The advantage of a marriage of silicone with hydrogel technology would be dramatically improved comfort and wettability compared to pure silicone elastomer lenses. The permeability of pure dimethylsiloxane is 600 barrer compared to 80 barrer for water (8). Thus, the incorporation of silicone into the hydrogel material greatly increases the oxygen permeability of the contact lens. The first silicone hydrogel lens approved for use in the United States was the Bausch and Lomb Purevision (Balafilcon A; 36% water) silicone hydrogel lens. The lens was approved for daily wear and later for up to 7 days of extended wear. The oxygen transmissibility of the lens was measured at 110. The CibaVision Focus Night and Day lens (Lotrafilcon A, 24% water) was approved for sale in the United States in October 2001. The CibaVision Focus Night and Day lens was the first contact lens approved for 30 days of continuous extended wear since 30 day extended wear approval was rescinded by the FDA in 1989. Purevision lenses received 30 day continuous wear approval one month later.

SAFETY

Contact lenses, when used properly under the care and supervision of a licensed practitioner, have been proven to be a safe and effective form of vision correction for millions of patients. However, as with any medical device, the risk of adverse events may occur with their use. With the approval of hydrogel lenses for extended wear in the 1980s, there were a number of severe adverse events reported. The most alarming were reports of increased incidence of corneal ulcers, also known as microbial keratitis or ulcerative keratitis, associated with extended wear schedules. Public debate on the safety of extended wear contact lenses grew as reports of serious complications rose in the media. The Contact Lens Institute sponsored two landmark studies on the incidence and relative risk of microbial keratitis with daily wear and extended wear contact lenses.

Poggio et al. reported in 1989 that the incidence of ulcerative keratitis was 4.1 cases per 10,000 people for daily wear hydrogel lenses. The incidence of ulcerative keratitis for extended wear hydrogel lenses was reported at 20.9 cases per 10,000 people per year (31). This compares to an incidence of 2.0 per 10,000 people per year for hard PMMA lenses and 4.0 per 10,000 people per year for RGP wearers. Poggio et al. reported a trend for decreasing incidence of ulcerative keratitis with extended wear lenses with increasing age that was not statistically significance ($p = 0.07$). Schein et al. (32) published a case-control study on the relative risk of ulcerative keratitis with the use of daily wear and extended wear of hydrogel contact lenses. They found that wearing extended wear lenses overnight produced 10–15 times the risk of ulcerative keratitis compared to wearing daily wear lenses. The race, sex, and age of the patient were not related to the relative risk of ulcerative keratitis. Smokers were found to have about three to four times the risk of developing ulcerative keratitis compared to nonsmokers. The study showed that the risk of ulcerative keratitis increased significantly with the number of consecutive days lenses were worn (32). Due to the results of these studies, the FDA rescinded the approval for 30 days of extended wear and applied a limit of 7 days maximum extended wear with hydrogel lens. The 7 day limit for extended wear was reexamined with the development of silicone hydrogel contact lens materials. Silicone hydrogel lenses were utilized in Europe prior to their approval for use in the United States. A number of articles were published on the clinical performance and safety with the use of these lenses. Iruzubieta et al. published a report in 2001 on the clinical performance of the CibaVision Night and Day lens. Seven patients discontinued lens wear due to lens discomfort and seven discontinued lens wear due to positive slit lamp findings out of 85 patients dispensed lenses. There were two cases of sterile peripheral ulcers and two cases of superior epithelial arcuate lesions. There were no cases of microbial keratitis in the study (33). Nilsson reported on a study of the Purevision contact lens used for 7 day verses 30 day extended wear. Nilsson randomized 504 patients into 7 day extended wear or thirty day extended wear. There was no statistically significant difference between the 7 day and 30 day group in the prevalence of objective findings and there were no incidents of microbial keratitis reported over a 12 month period (34).

Although microbial keratitis is the most severe potential adverse event possible with the use of contact lenses, a number of conditions may result from the use of contact lenses. Corneal changes from chronic oxygen deprivation, such as endothelial polymegethism, corneal striae, and epithelial microcysts have been associated with the development of microbial keratitis. Contact Lens-Induced papillary conjunctivitis (also known as Giant papillary conjunctivitis), sterile contact lens peripheral ulcers, contact lens acute red eyes (CLARE), and epithelial staining or abrasions are other significant adverse events associated with the use of contact lenses, particularly on an extended wear basis (27,35).

The use of contact lenses while swimming is contraindicated. If swimming in contact lenses is desired or unavoidable the patient should use of watertight goggles to prevent water from coming into contact with the lens. Rigid gas permeable contact lenses typically will displace from the eye during swimming as they lack adhesion to the corneal surface. Soft hydrogel and silicone hydrogels, while having the adhesion required to prevent lens loss while swimming, are prone to absorbing substances present in the water. Choo et al. published a study examining the establishment of bacterial colonies on soft hydrogel and silicone hydrogels after exposing lenses to chlorinated water while swimming. Of 28 lenses examined, 27 revealed bacterial colonization. The most prominent bacterial colony observed on the lenses was Staphylococcus epidermidis, which was found to be the most common bacteria in the water. Sixteen lenses examined that were not worn during swimming revealed only three lenses with bacterial colonies. No differences were observed in bacterial colonization for hydrogel lenses verses silicone hydrogel lenses (36). Given the potential for bacterial colonization and possible subsequent microbial infection, contact lenses should not be worn in conditions where the lens may be exposed to contaminated water.

ASTIGMATIC DESIGNS

Astigmatism is a condition where the cornea has a toric surface. A toric surface features two different curvatures located in meridians that are 90° apart. As a result, light is focused into two different line foci which causes blur (10). Astigmatism can be corrected with the use of rigid gas permeable or soft hydrophilic/silicone hydrogel lenses.

There are three rigid gas permeable lens designs utilized in the correction of astigmatism. Spherical gas permeable lens designs can correct astigmatism in patients where the astigmatism observed, as determined by refraction, matches the corneal astigmatism, determined by keratometry or corneal topography. The lens, when placed on the eye, corrects the patient's astigmatism through the use of a "tear lens". The tear lens is generated when the space beneath the contact lens created by the difference in curvature between the cornea and the contact lens is filled by the tear film. The tear lens refracts light to focus light from each meridian onto the retina providing the patient astigmatic correction. Spherical rigid gas permeable contact lenses can correct patients with two diopters or less corneal toricity (37). With corneal astigmatism of greater than two diopters, a spherical lens may flex, decenter, or fall out of the eye when the patient blinks. For patients with large amounts of corneal astigmatism, a spherical gas permeable contact lens design is not feasible. A rigid gas permeable lens can be designed with two different base

curvatures, anterior curvatures and refractive powers in each meridian (38). These lenses, referred to as back surface toric lenses or bitoric lenses are custom designed for each individual patient. Patients with minimal or no corneal toricity who require astigmatic refractive correction with rigid gas permeable contact lenses can be corrected with a lens that has a spherical base curve and toric anterior surface curvatures to correct the patient's refractive error (37). These lenses, referred to front surface toric prism ballasted gas permeable lenses, are used infrequently due to the popularity of soft toric contact lenses.

Soft hydrogel and silicone hydrogel contact lenses can be used to correct astigmatic refractive error in patients. Patients with low amounts of refractive astigmatism, typically under −0.75 diopters cylinder, can be corrected using the spherical equivalent refraction power (sphere power + (0.5 × cylinder power)) in a spherical contact lens. Patients with more than −0.75 diopters of refractive astigmatism should be corrected with a soft toric contact lens (37). The soft toric contact lens design may have toric front or back surfaces with the appropriate refractive power in each meridian. Soft toric contact lenses will assume the shape of the cornea; therefore there is no tear lens present to correct the astigmatism. Soft toric contact lenses must stay properly oriented on the cornea to correct the patient's astigmatism. The meridians of the soft toric contact lens should correspond with the meridians of the refractive error of the eye. Rotation of the contact lens will improperly align the contact lens meridians with the patient's refractive error and lead to a reduction of visual acuity. With increasing amounts of astigmatism, stability of lens rotation becomes more critical to ensure clear vision (39).

Lens stabilization can be achieved by a number of techniques. The most popular method of stabilization is the use of prism ballasting. A prism ballasted contact lens integrates prism into the shape of the lens. The contact lens is designed with a varying thickness profile where the top of the contact lens is thinner than the bottom of the contact lens. The interaction of the eyelids with the prism moves the lens into the proper orientation by a principle called the "watermelon seed" effect. Another method of soft toric lens stabilization is the dual thin zones, or double slab off, design. Lenses that use a dual thin zone design have thin superior and inferior portions of the lens with a thicker central area. The interaction of the eyelids with the dual thin zones stabilizes the lens and holds the lens in the proper orientation through the watermelon seed effect. Soft toric contact lens rotation stabilization methods, such as lens truncation, where a small portion of there inferior portion of the lens is removed so that the lens is stabilized by the lower lid margin and eccentric lenticularization may be used to ensure that the power meridians in the soft toric lens remain oriented in their proper position on the eye to ensure proper astigmatic correction (38,39).

PRESBYOPIA AND MONOVISION

Presbyopia is the reduction of accommodation that occurs with aging. Accommodation allows the intraocular lens to change shape, thus increasing the amount of plus power and allowing the patient to see near objects (10). Most patients will start to have a reduction in accommodation in their fourth decade. In spectacles, presbyopia is corrected through the use of bifocals. In contact lenses, presbyopia can be corrected with a number of multifocal contact lens designs or through monovision. Prior to the development of multifocal contact lenses the method of presbyopic vision correction with contact lenses was monovision. In monovision, one eye is corrected with the full refractive error needed to provide distance vision. The other eye is corrected with a lens that focuses the eye for a set near point. Monovision is most effective with patients that have low to moderate amounts of presbyopia (40). With increasing amounts of presbyopia, the disparity in refractive correction between the distance eye and the near eye becomes more significant with patients reporting more difficulty with binocular vision and stereopsis (41,42). Despite the development of multifocal contact lenses, monovision is still utilized with a single vision contact lens in one eye and a multifocal contact lens in the other eye in a technique referred to as modified monovision (41).

Translating bifocal, also known as alternating vision contact lenses correct presbyopia by having a distance and near corrective sections in the lens. The principle behind this correction is similar to bifocal spectacle lenses, where the upper portion of the lens provides distance correction and the lower bifocal portion provides the patient with near correction. The lower portion of the lens translates with eye movement. As the patient looks down, the bifocal section of the lens is pushed up over the pupil by the lower eyelid allowing the patient to see near objects (42). The orientation of the bifocal is maintained through the use of prism ballasting and the watermelon seed effect. As in spectacles, a translating trifocal design is available for patients who need more presbyopic correction. The majority of translating bifocal lens designs are rigid gas permeable lenses. Translating bifocal lenses work well with patients requiring high bifocal addition powers and have good eyelid apposition with the eye that will force the lens to translate with downgaze. The positioning of the lens optics is critical for translating bifocal designs. Translating bifocal lenses that do not translate in down gaze will provide no multifocal effect and poor near vision (41).

Another class of bifocal contact lenses is referred to as simultaneous vision bifocal lenses. Simultaneous vision lenses encompasses a number of lens designs, including aspheric lenses, concentric ring designs, and less commonly, diffractive optics (42). The unifying concept for each simultaneous vision bifocal design is the placement of images of the intended near target and the distance target on the retina at the same time. Simultaneous vision bifocal contact lens designs require the patient to ignore the image that is not relevant for the intended task. Simultaneous vision designs are best suited to patients who do not require high bifocal powers (41). Simultaneous vision contact lenses are particularly well suited for patients requiring intermediate vision correction for tasks such as computer use. Simultaneous vision lenses are the preferred bifocal contact lens design for patients who require near vision correction in straightforward gaze. Simultaneous

vision lenses work well for these tasks because translation of the lens is not required to provide the multifocal vision effect (42). The majority of soft hydrogel contact lenses are simultaneous vision designs.

ORTHOKERATOLOGY

With the advent of the corneal gas permeable contact lens, practitioners observed that patients would report spectacle blur at night upon removal of the contact lens. Use of the rigid gas permeable lens would change the shape of the cornea during daily wear and cause a subsequent change in the patient's refraction and spectacle blur after removing the contact lens (43). In 1962, Jessen described the Orthofocus technique, the first published report of a controlled attempt to change the refractive error of a patient through the use of rigid contact lens (44). The procedure of using a rigid contact lens to change the refractive error of a contact lens patient was given the name orthokeratology. Orthokeratology was defined by as "the reduction, modification, or elimination of refractive anomalies by the programmed application of contact lenses or other related procedures" (45). Early attempts to reshape the cornea were accomplished by fitting the contact lens much flatter than the corneal curvature of the patient. Orthokeratology failed to gain widespread acceptance over the next 20 years due to limitations in corneal lens lathe technology. In 1989, a major advancement in the field of orthokeratology took place. Wlodyga and Stoyan presented the concept of reverse geometry lenses for "accelerated" orthokeratology (43). Most previous fitting philosophies featured traditional lens designs with peripheral curves flatter than the base curve of the lens. Wlodyga and Stoyan proposed using a reverse geometry contact lens with a central base cure that was flatter than the midperipheral curve. The advantage of such a system is improved lens centration, a decrease in the time to reach maximum effect, a more consistent treatment effect and a possibility to correct patients with higher degrees of myopia. Despite the advancement in orthokeratology lens designs, orthokeratology was still limited by the lack of FDA approval for the overnight use of this technology. With the development of hyperpermeable contact lens materials, the use of reverse geometry orthokeratology contact lens for overnight wear became practical. In 2002, the FDA approved the Paragon Corneal Refractive Therapy (CRT) overnight orthokeratology lens for overnight wear to temporarily correct myopia up to −6.00 diopters (46). As of January 2005, overnight orthokeratology lenses approved by the FDA are produced by Paragon Vision Sciences (Mesa, AZ) and Bausch and Lomb/Polymer Technologies (Rochester, NY) along with their approved manufacturers.

SUMMARY

Contact lenses have been utilized by millions of people around the world to correct a variety of refractive error conditions including myopia, hyperopia, presbyopia, and astigmatism. The study of contact lenses is a dynamic pursuit as advances in manufacturing technologies and lens materials are incorporated into state of the art lens designs. Despite an excellent safety profile demonstrated through years of clinical practice, contact lenses remain medical devices that require expert fitting and monitoring by a licensed contact lens practitioner and patient compliance to ensure a successful outcome for the contact lens patient.

BIBLIOGRAPHY

1. Barr JT. 2004 Annual Report. Contact Lens Spectrum 2005; January.
2. Hofstetter H, Graham R. Leonardo and contact lenses. Am J Optom Arch Am Acad Optom 1953;41.
3. Enoch J. Descartes' contact lens. Am J Optom Arch Am Acad Optom 1956;33:77.
4. Barr JT. History and Development of contact lenses. In: Bennett ES, Weissman BA. editor Clinical Contact Lens Practice. Philadelphia: Lippincott Williams & Wilkins; 2005.
5. Mandell R. Historical development. In: Contact Lens Practice. 4th ed. Springfied: Charles C Thomas; 1988.
6. Barr JT, Bailey NJ. History of contact lenses. In: Bennett ES, Weissman BA, editors. Clinical Contact Lens Practice, revised ed. 1997 Philadelphia: Lippincott-Raven; 1997; chapt 11.
7. Chun M, Fox L, Zhou A. Disposable and frequent replacement hydrogel contact lenses. In: Bennett ES, Weissman BA, editors. Clinical Contact Lens Practice. Philadelphia: Lippincott Williams&Wilkins; 2005.
8. Nicolson P, Vogt J. Soft contact lens polymers: an evolution. Biomaterials 2001;22:3273–3283.
9. Landers R, Rixon A. Contact lens materials update: options for most prescriptions. Contact Lens Spectrum 2005(March).
10. Cline D, Hofstetter H, Griffin J, editors. Dictionary of Visual Science. 4th ed. Boston: Butterworth-Heinemann; 1997.
11. Cannella A, Bonafini J. Polymer chemistry. In: Bennett ES, Weissman BA, editors. Clinical Contact Lens Practice. Philadelphia: Lippincott Williams & Wilkins; 2005.
12. Phillips A. Rigid gas permeable corneal lens fitting. In: Phillips A, Speedwell L, editors. Contact Lenses. 4th ed. Oxford: Butterworth Heinemann; 1997.
13. Mandell R. Basic principles of rigid lenses. In: Contact Lens Practice. Springfield: Charles C Thomas; 1988.
14. Loran D. The verification of hydrogel contact lenses. In: Phillips A, Speedwell L, editors. Contact Lenses. 4th ed. Oxford: Butterworth Heinemann; 1997.
15. Yeung K, Weissman BA. Soft contact lens application. In: Bennett ES, Weissman BA, editors. Clinical Contact Lens Practice. Philadelphia: Lippincott Williams & Wilkins; 2005.
16. Mandell R. Fitting methods and philosophies. In: Mandell R, editor. Contact Lens Practice. Springfield: Charles C Thomas; 1988.
17. Bennett ES. Basic fitting. In: Bennett ES, Weissman BA, editors. Clinical Contact Lens Practice. Philadelphia: Lippincott Williams & Wilkins; 2005.
18. Uras R, Rah M. Sperical hydrophilic soft contact lenses. In: Mannis M, Zadnik K, Coral-Ghanem C, Kara-Jose N, editors. Contact Lenses in Ophthalmic Practice. New York: Springer; 2004.
19. Coral-Ghanem C, Bailey M. Maintenance and handling of contact lenses. In: Mannis M, Zadnik K, Coral-Ghanem C, Kara-Jose N, editors. Contact Lenses in Ophthalmic Practice. New York: Springer; 2004.
20. Tran L, Myung E. Contact lens care update. Contact Lens Spectrum 2005(April).
21. Atkinson K, Port M. Patient management and instruction. In: Phillips A, Speedwell L, editors. Contact Lenses. 4th ed. Oxford: Butterworth Heinemann; 1997.
22. Weisbarth R, Henderson B. Hydrogel lens care regiments and patient education. In: Bennett ES, Weissman BA, editors. Clinical Contact Lens Practice. Philadelphia: Lippincott Williams and Wilkins; 2005.
23. Bennett ES, Wagner H. Rigid lens care and patient education. In: Bennett ES Weissman BA, editors. Clinical Contact Lens Practice. Philadelphia: Lippincott Williams & Wilkins; 2005.
24. Koenig S, Solomon J, Hyndiuk R, Sucher R, Gradus M. Acanthamoeba keratitis associated with gas-permeable contact lens wear. Am J Ophthalmol 1987;103(6):832.
25. Shovlin J. Acanthamoeba keratitis in rigid lens wearers: the issue of tap water rinses. Int Contact Lens Clin 1990;17:47.
26. Mandell R. Anatomy and physiology of the cornea. In: Contact Lens Practice. 4th ed. Springfield: Charles C Thomas; 1988.
27. Kara-Jose N, Coral-Ghanem C, Joslin C. Complications associated with contact lens use. In: Mannis M, Zadnik K, Coral-Ghanem C, Kara-Jose N, editors. Contact Lenses in Ophthalmic Practice. New York: Springer; 2004.
28. Mandell R. Oxygen and the cornea. In: Contact Lens Practice. Springfield: Charles C Thomas; 1988.
29. Holden BA, Mertz G. Critical oxygen levels to avoid corneal edema for daily and extended wear contact lenses. Invest Ophthalmol Vis Sci 1984;25:1161–1167.
30. Harvitt D, Bonanno J. Re-evaluation of the oxygen diffusion model for predicting minimum contact lens Dk/t values needed to avoid corneal anoxia. Optom Vis Sci 1999;76(10):712–719.
31. Poggio E, et al. The incidence of ulcerative keratitis among users of daily-wear and extended-wear soft contact lenses. N Eng J Med 1989;321(12):779–783.
32. Schein O, Glynn R, Poggio E, Seddon J, Kenyon K. The relative risk of ulcerative keratitis among users of daily-wear and extended-wear soft contcat lenses. N Eng J Med 1989;321(12):773–778.
33. Iruzubieta JM, et al. Practical experience with a high Dk Lotrafilcon A fluorosilicone hydrogel extended wear contact lens in Spain. Clao J 2001;27(1):41–46.
34. Nilsson SE. Seven-day extended wear and 30-day continuous wear of high oxygen transmissibility soft silicone hydrogel contact lenses: a randomized 1-year study of 504 patients. Clao J 2001;27(3):125–136.
35. Binder PS. Complications associated with extended wear of soft contact lenses. Ophthalmology 1979;86(6):1093–1101.
36. Choo J, et al. Bacterial populations on silicone hydrogel and hydrogel contact lenses after swimming in a chlorinated pool. Optom Vis Sci 2005;82(2):134–137.
37. Twa M, Moreira S. Astigmatism and toric contact lenses. In: Mannis M, Zadnik K, Coral-Ghanem C, Kara-Jose N, editors. Contact Lenses in Ophthalmic Practice. New York: Springer; 2004.
38. Lindsay R, Westerhout D. Toric contact lens fitting. In: Phillips A, Speedwell L, editors. Contact Lenses. 4th ed. Oxford: Butterworth Heinemann; 1997.
39. Epstein A, Remba M. Hydrogel toric contact lens correction. In: Bennett ES, Weissman BA, editors. Clinical Contact Lens Practice. Philadelphia: Lippincott Williams & Wilkins; 2005.
40. Mandell R. Presbyopia. Contact Lens Practice. 4th ed. Springfield: Charles C Thomas; 1988.
41. Schornack M, Coral-Ghanem C, Pena AdS. Presbyopia and contact lenses. In: Mannis M, Zadnik K, Coral-Ghanem C, Kara-Jose N, editors. Contact Lenses in Ophthalmic Practice. New York: Springer; 2004.
42. Bennett ES, Jurkus J. Presbyopic correction. In: Bennett ES, Weissman BA, editors. Clinical Contact Lens Practice. Philadelphia: Lippincott Williams & Wilkins; 2005.
43. Winkler TD, Kame RT. Orthokeratology handbook. Boston: Butterworth-Heinemann; 1995; p 113.
44. Jessen G. Orthofocus techniques. Contacto 1962;6:200–204.

45. Kerns RL. Research in orthokeratology. Part I: Introduction and background. J Am Optom Assoc 1976;47(8):1047–1051.
46. Barr JT. Contact Lenses 2002: Annual Report. Contact Lens Spectrum. 2003(January).

See also BIOMATERIALS: POLYMERS; BLIND AND VISUALLY IMPAIRED, ASSISTIVE TECHNOLOGIES; LENSES, INTRAOCULAR; VISUAL PROSTHESES.

CONTINUOUS POSITIVE AIRWAY PRESSURE

DAVID M. RAPOPORT
NYU School of Medicine
New York, New York

RON S. LEDER
Universidad Nacional Autonoma de Mexico
Mexico, Distrito Federal

INTRODUCTION

Beginning in the 1970s, positive-end expiratory pressure (PEEP) began to be added to the pressure applied during inspiration in patients undergoing mechanical ventilation. The rationale was that when a patient had loss of alveolar surfactant, the alveoli tended to collapse during expiration. "Holding them open" by offsetting the increased elastic recoil with a pressure that did not return to atmospheric at the end of expiration was beneficial to gas exchange because it prevented complete collapse with resultant shunting of blood past airless lung. This process was applied to both infant lungs (neonatal respiratory distress syndrome, RDS) and to adult lungs (adult respiratory distress syndrome, ARDS) with improved oxygenation as the main endpoint.

As PEEP was more widely used, it was observed that at the time of removal of respiratory support, some patients (especially newborns) benefited from PEEP for oxygenation despite being able to ventilate. This suggested that the strategy of providing a constant distending pressure to the lung (CPAP) during BOTH inspiration and expiration without increasing the pressure during the inspiratory phase (i.e., ventilation) provided some transient benefit during the period before extubation. In addition, it proved possible to apply CPAP via a nose or face mask after extubation, with continued benefit to the lung (Table 1).

CPAP IN OBSTRUCTIVE SLEEP APNEA/HYPOPNEA SYNDROME (OSAHS)

Basic Circuit and Rationale for Use

CPAP was first introduced in 1981 as a treatment for obstructive sleep apnea/hypopnea syndrome (OSAHS). The concept was initially proposed by Collin Sullivan (Australian Patent AU-B83901/82) as a pneumatic upper airway splint and later shown to work even in the presence of chronic respiratory failure (chronic hypercapnia) by David Rapoport (U.S. Patent 4,655,213). In this application of CPAP, the effect of interest is that of continuous positive airway pressure and not its effect on the lung (as with PEEP), although this is necessarily always present. The critical rationale is the effect a positive pressure in the AIRWAY has on the collapsible upper airway (i.e., the posterior pharynx and hypopharynx), which is not relevant in the intubated patient. The pressure is applied via nose or mouth mask and distends the area that extensive physiologic work has shown to have a tendency to collapse during sleep (especially during the negative pressure of inspiration).

The original circuit proposed consisted of a nose mask to which was attached either a pressure-dissipating threshold valve or a restrictor that created a roughly constant backpressure due to a constant bias flow provided by a blower or other source of compressed air. Early in development, it became clear that fans and blowers had better characteristics than piston-type high-pressure compressors, due to their ability to deliver high flow rates to the mask with control via motor speed, little dependence of delivered pressure on the backpressure, low cost, and quieter operation.

The original concept described by Sullivan was that the CPAP (pressure) was needed continuously to "hold" the airway open against a natural tendency of the walls of the airway to collapse due to loss of active muscle tone during sleep and the suction caused by inspiration. There is a tendency for some degree of airway collapse during sleep in everyone. Patients with the obstructive sleep apnea/hypopnea syndrome tend to collapse their airway to excess. In all cases, the collapse and airway obstruction occurs because of loss of tone in the airway muscles, whose role is to stiffen the walls against the suction created by breathing during inspiration. Although there has been much debate, most models of this process of collapse and its treatment with CPAP have suggested that the treatment pressure needs to be relatively constant at the point of collapse unless the patient changes body position, head and neck position, sleep state, or wakes up. The point of collapse is usually found to be at the back of the throat or at the level of the soft palate.

Leak Circuit Modification

Until 1985, CPAP was delivered by means of a restrictor or mechanical valve that was placed on the patient's nose mask. This valve, through its design and its passive mechanical properties, held the pressure at a value that was as constant as the mechanics of the valve could achieve (a so-called "threshold" valve, which opens more to discharge air when pressure rises). It also provided a vent for exhaled CO_2 and excess humidity, as it was located near the patient; a side-effect of the constant dissipation of pressure was venting of the circuit, including the exhaled gas from the patient. In 1985, Rapoport proposed that the valve used to set the pressure in the circuit could be removed from the mask to increase patient comfort. However, this required that the "venting" function (removal of exhaled CO_2 from the circuit) be performed separately. The modification consisted of a small controlled leak deliberately introduced near the mask that did not significantly dissipate the pressure (previously this had been a large leak or a threshold valve). This modified circuit is the most widely used hose circuitry in both CPAP and noninvasive mask ventilation.

A further improvement was instituted in the mid-1980s, when it was observed that the use of a threshold valve was

Table 1. CPAP Definitions (From www.cpap.com)

IPAP	This stands for inhalation positive airway pressure. This is the pressure setting that is used when you inhale.
EPAP	This stands for exhalation positive airway pressure. This is the pressure setting that is used when you exhale. This setting is always lower than the IPAP, making exhalation easier or more comfortable.
Bi-Flex	This setting establishes a level of pressure relief that takes place at the end of inhalation and at the start of exhalation. Settings of "1", "2", or "3" will progressively provide increased pressure relief. You can adjust this setting to suit your comfort level.
Spontaneous	Spontaneous means the patient breathes without assistance from a respiratory rate set on the bilevel.
4–20 cm	4–20 cm is the pressure range that can be delivered to the patient. A CPAP (constant positive air pressure) uses one constant pressure from 4 cm to 20 cm. This pressure is measured in centimeters of water pressure (H_2O).
Optional DC Cable	A cable that plugs directly into a dc port on the CPAP machine. This allows the you to plug into a dc power source, such as a battery or car cigarette lighter.
Direct Battery Operation	This feature means the machine has a dc port on the back of the machine in which you can use the Puritan Bennett Battery Pack or a deep cycle marine battery.
Auto Altitude Adjustment	Auto altitude adjustment is the CPAP machine's ability to compensate for changes in altitude automatically.
Ramp	The ramp feature allows the user to start treatment at a lower pressure, and as they fall asleep, the pressure slowly rises. This is a comfort setting and can be from 0 to 45 minutes on most CPAP machines
Hour/Session	This feature records the hours of usage and the sessions the machines is used longer than 4 hours.
Optional Software	Software is an option on some CPAP machines. The software can give details, compliance, and performance. The patient or physician downloads the data from the CPAP machine and uses it to determine how long a patient has used their machine each night and how well the machine is working to stop apnea/hypopnea events.
Leak Compensation	The CPAP machine compensates for mask leak, to keep the CPAP pressure accurate.
Heated Humidifier	This is an optional feature that can be added to the machine. Some machines have heated humidifiers designed to integrate with the machine while all can be used with stand-alone heated humidifiers like the Fisher & Paykel HC150.
Passover Humidifier	This is an optional feature that can be added to the machine. The Passover humidifier is a chamber filled with cool water. The CPAP machine tubing is routed through this chamber, and cool humidity soothes your nasal passages.
Data Card	A data card is a small card the same size as a credit card, that stores information to be placed into a Data Card Reader, downloaded to a computer, and read with optional software. Depending on the model of machine, the data card will hold either compliance data, performance data, or both.
Auto ON/OFF	This feature turns the machines OFF and ON when putting on or taking off the mask. When you put your mask on, the machines senses you breathing and turns itself ON. Take off the mask, and the machine turns OFF.

optional. This was because the blower could be designed to have a sufficiently flat flow-to-pressure relationship at a given speed of rotation to maintain a near-constant pressure during the increased flow of inspiration and decreased flow of expiration and changing amount of mask leak. Since then, CPAP blowers have either been entirely passive (set at one blower speed for each prescribed CPAP) or had some type of speed control that adjusted the speed in response to sensed pressure feedback. A few devices still use a threshold valve, but these have tended to replace the passive mechanical valves with active electronically controlled stepper motor-driven valves.

Variations in Delivered Pressure

Because active control of pressure is necessitated by removal of the threshold valve from the mask, there has been gradually increasing attention to modifying the pressure contour provided to the patient interface. In particular, various techniques have been used to keep a particular pressure constant. Conceptually, two distinct targets for stabilization of the pressure are either pressure at the blower or pressure at the mask. Initially, CPAP systems were designed to have a constant pressure at the blower,

neglecting that this constant pressure at the blower would cause fluctuations at the patient mask (see above). More recently, attention has been directed to maintaining a constant pressure at other points in the circuit.

As air flows through a closed tube, it is driven by the pressure drop, which occurs progressively in the direction of flow. This implies that in any system with a nonzero resistance, there will always be a difference in pressure as one travels in the direction of flow along the tube. Specifically, as one travels from the blower along the tubing toward the patient's most collapsible airway point, the airway pressure will fall from that set at the blower and will differ depending on the rate of flow through the system and on where it is measured. Pressure differences between points along this route also depend on the direction of airflow (inspiration and expiration) and the size of the bias flow (e.g., through intentional or unintentional leaks at the mask). Thus, during inspiration, pressure is always higher by some small amount at the machine end of the tubing than it is at the patient's nose, and during expiration, it is often lower at the machine end of the tubing than at the nose if flow reverses. The amount of pressure difference between the machine end of the tubing and the patient depends on the resistance of the tubing connecting the two and on the flow through the system, which is the sum of the patient's breathing airflow and any leak that occurs at the mask.

As the purpose of the CPAP is to maintain a therapeutic pressure that prevents upper airway collapse, a strategy to control the variations in this pressure must be established. Different approaches have been taken by different devices intended to deliver what is called CPAP. In the earliest CPAP devices, the valve located at the mask controlled the pressure; this intrinsically adjusted for changes in leak and reversal of flow from inspiration to expiration, and the only requirement of the blower was to provide an excess (not necessarily constant) flow to the valve located near the patient. When the valve was removed from the mask, pressure control shifted away from the patient to a point in the circuit near the blower. At least under some conditions, pressure can differ considerably from the desired therapeutic pressure as felt by the patient. The following is a list of some strategies adopted by current CPAP devices to deal with this (in terms of the original therapeutic intent, constant pressure at the mask during inspiration is key):

1. The controller sets a constant pressure at the machine (constant blower speed). This pressure must be slightly in excess of the patient's need; i.e., it must be sufficiently high to allow some fall during inspiration under maximal leak conditions, or the patient will be under-treated at this critical time in inspiration. This strategy necessarily implies that pressure at the patient will be in excess of the required therapeutic pressure at all other times, and this may contribute to patient discomfort.

2. The controller is driven by active feedback from the pressure as measured in the mask. This feedback will cause the blower to continuously vary pressure (at the blower) so as to maintain it constant at the mask. Either blower speed or valve opening may be varied, but pressure as sensed at the mask is the controlled variable. Until leak at the mask becomes enormous, this will be the closest to the original concept of a mask CPAP proposed by Sullivan and implied by the ventilator uses of PEEP and CPAP.

3. Control of pressure as exerted at the machine is based on assumptions about how the pressure will change as it travels along the tubing that connects the blower to the patient. Some devices assume a known pressure drop across the tubing and just add this to the desired therapeutic pressure. Other devices use the flow (or some estimate of flow such as blower speed) to calculate a predicted drop in pressure between machine and patient, creating a deliberate but variably higher pressure than prescribed—in an attempt to deliver the constant therapeutic target.

The above strategies handle changes in flow through the system (e.g., changing leaks), but they may not adequately address changes in backpressure during each breath related to breathing. This is because the intrinsic properties of blowers (fans) are such that at a fixed rotational speed, these devices tend to produce a flow (not pressure) that is heavily influenced by backpressure (e.g., the tubing resistance and the difference in magnitude and direction of flow between inspiration and expiration). As a result, fans tend to produce a relatively constant pressure against a wide range of loads (because of the changes in delivered flow). Thus, blowers result in a pressure profile during breathing that is close in their behavior to that of a circuit with a threshold valve. The result of this pattern of response to varying backpressures is that setting a constant blower speed results in a system that, to a first approximation, maintains a pressure that is relatively constant at the blower, independent of the patient's breathing pattern. However, blowers are not perfect in this regard. Blowers (fans), when kept at a constant speed within each breath, necessarily produce slight changes in delivered pressure (higher during expiration and lower during inspiration). Greater variability in breath size, and large leaks through the mask will all result in progressively greater pressure swings at the blower. Because of tubing resistance, even greater pressure changes will occur at the patient if the system is entirely passive. Specifically, pressure in the system and at the patient will fall during inspiration and rise during expiration to a value different from the treatment pressure.

The latest CPAP machines (U.S. Patent Application 2005/0188989) have begun to address these intrabreath pressure variations by modifying the pressure they deliver within individual breaths as a function of the instantaneous flow. The assumption is that this will improve pressure (exhalation) induced discomfort, which is reported by many users of CPAP, by limiting unnecessary rises in pressure above therapeutic during expiration. The simplest way to accomplish this pressure stabilization is to vary the blower speed in response to fluctuations detected in measured pressure. This type of control is a classic feedback system and is used to keep pressure constant at the blower by responding to any deviations or perturbations that occur in the desired constant pressure. Detected changes in pressure result in the controller changing the speed of the blower. Typically, pressure in the circuit varies as a result of changes in the patient's breathing (inspiration vs. expiration) or changes in the leak from the system at the mask, both of which produce changes in the backpressure felt by the blower.

As pointed out, varying the blower speed within a breath in response to the sensed instantaneous flow can also be done to vary the blower pressure profile such that it is maintained constant (without measurement) at the mask. An alternative to this is to reinsert an active threshold valve at the blower that accomplishes a similar function based on sensing flow in the circuit or some other measured variable that allows prediction of pressure in the mask. Much like the original CPAP circuit, pressure control is provided by driving the blower to produce a pressure in excess of that needed, and diverting ("bleeding off") some pressure in the system by variably opening the valve at a "threshold" pressure. However, instead of targeting a constant blower pressure, the valve is instructed to produce a pressure profile predicted to cause a constant mask pressure, by adjusting the opening and closing of the valve.

"BiLevel" PAP. Introducing a valve under microprocessor control provided an interesting opportunity to create patterns other than a constant pressure in the system. As soon as the pressure delivered to the patient begins to be significantly higher during inspiration than during expiration, however, this nonconstant pressure is fundamentally different from CPAP. In fact, this is similar to the behavior of artificial ventilation devices (ventilators). If the control system is made aware of when inspiration and expiration begin, the valve used in venting pressure can be adjusted to rapidly achieve higher and lower pressures in synchrony with the patient; this can assist or even fully replace patient breathing efforts and is the essence of mechanical ventilation. Whereas CPAP is the imposition of a control algorithm targeting a near-constant pressure in the system or at least at the patient, ventilation (sometimes referred to as "bilevel ventilation") is the imposition of a nonconstant waveform of pressure on the output of the blower so as to raise inspiratory pressure above expiratory pressure at the patient level. However, as the pressure profile (constant or variable at the patient) depends only on the programming of the valve controller, much confusion exists in the literature about whether a "bilevel" device is being used for "CPAP" or assisted ventilation.

In concept, patients with obstructive sleep apnea have no ventilatory control abnormality once the airway is open. Thus, assistance with ventilation (once the airway is splinted open) is not indicated. The original proposal for bilevel "CPAP" was not targeted at ventilation, but to date, there has been little in the published literature to support its use for "comfort" in patients with OSAHS alone. However, as a noninvasive ventilator, bilevel devices are very effective and deliver what is essentially a combination of PEEP and pressure support ventilation. Their use is clearly indicated in chronically hypercapnic patients and in those with nocturnal hypoventilation. Not only is there little logic to the use of this type of device for intermittent obstructive apnea, but also recent publications have suggested that they can exaggerate central apnea—presumably because increasing breath size (pressure support) will increase plant gain in the patient's respiratory control loop and tend to produce increased overshoot of the size of compensatory ventilatory efforts whenever there is instability of breathing, creating a classic "ringing" system.

Although the above discussion shows that bilevel ventilation is completely different in purpose and application from CPAP, current devices are such that they can deliver both modes with little change in their circuitry if they contain the means to rapidly change the pressure according to a prescribed algorithm. As a result, there continues to be confusion about what is being done when a physician prescribes a treatment for a patient. Clarification as to the algorithm being used by a setting on the machine requires a decision as to whether the device targets

- A pressure that is constant at the blower (the controller removes fluctuations at the blower). When the pressure is measured at the patient, there will necessarily be small fluctuations throughout breathing.

Pressure will be lower at the patient during inspiration and higher during expiration than at the blower. This is passive CPAP.

- A pressure that is constant at the patient (the controller removes fluctuations at the patient). To accomplish this, the pressure when measured at the blower will be slightly higher during inspiration and lower during expiration. This is the purest form of classic CPAP.

- A pressure that is higher during inspiration than during expiration both at the blower and the patient. This type of pressure oscillation has as a purpose to actively assist the patient in magnifying his breathing efforts. The pressure changes assist the patient's spontaneous muscular breathing efforts when these are weak. This is active ventilatory support.

In the last two above cases, the expiratory pressure has been lowered from the value it would have achieved during expiration if the system was left to behave passively in response to the patient's breathing backpressure. The difference between the two algorithms above is not in the direction of change applied to the output expiratory pressure, but in the purpose for which it is lowered and the amount that pressure at the output of the blower is made to fall during expiration through active control. If the pressure at the blower is not lowered, i.e., forced to be constant, then the pressure as measured at the patient will rise during expiration. If the pressure at the blower is forced to fall slightly during expiration, the pressure may remain constant at the patient. Finally, if the pressure is forced to fall sufficiently at the blower during expiration, pressure will also fall at the patient during expiration. Unlike the first two algorithms, this pattern at the patient of a fall in pressure during expiration when compared with inspiration produces actual assistance to the patient's breathing efforts, and it defines assisted ventilation; this type of ventilatory assistance is fundamentally different from CPAP, whose purpose is only to hold the airway open.

MONITORING/TITRATION ISSUES

Recording the Pressure

The clinical prescription of CPAP is usually given as a single therapeutic pressure value. Typically, this is derived from some type of titration in a recorded sleep study. As should be evident from the earlier discussion of pressure gradients, this prescription pressure should to be related to how pressure was measured during the titration, as well as to how it will be implemented by the patient's CPAP machine, but this is often overlooked. If it assumed that the prescription is a generic one for a therapeutic pressure to be delivered in the mask, then the mask pressure should be the one measured during the titration study. However, most CPAP machines used in the laboratory do not provide this pressure as an easily available electronic output because they do not measure it. Instead they measure the pressure at the blower, which may differ by up to 1–2 cm H_2O from that at the patient and vary with respiration actively or passively. Furthermore, this gradient, as discussed above, varies with the uncontrolled leak conditions at the mask and the amount and type of tubing circuitry, including whether a humidifier is in line. Furthermore, because many CPAP machines output an internally measured pressure as an analog or digital signal to facilitate laboratory recording during the sleep study, it is critical to know whether the actual mask pressure is being output or whether the output is a calculated estimate of pressure of a CPAP machine to that assumed to be present at the patient interface. In our laboratory, we prefer the actual measurement of pressure in the mask of the patient and provide this to the patient as his "prescription pressure." This should be independent of the brand of CPAP chosen for chronic use by the patient in its relation to adequacy of pressure if measured in the mask.

Algorithm for Deciding on the "Therapeutic" Pressure

When a patient undergoes a "CPAP titration," the pressure in the system during the period of monitoring is gradually increased until all evidence of upper airway obstruction disappears. Different laboratories titrate to different indices, but in principle most include trying to abolish evidence of both severe and partial obstruction as below:

Apneas (complete cessation of airflow caused by obstruction for at least 10 s) usually disappear first, so that at pressures above 8 to 10 cm H_2O, it is rare to find obstructive apneas. Central apneas (failure of respiratory effort to occur, but usually without obstruction) may appear, especially at higher pressures. These are usually distinguished from obstructive apneas by the absence of persistent respiratory movements (rib and abdomen movements) during the apnea.

Hypopneas (significant reductions in airflow lasting at least 10 s) tend to predominate once CPAP has been applied at low levels. These are easily identified as obstructive by the presence of a flattened inspiratory flow/time contour, which differs from the sinusoidal shape of normal inspiration and breaths with unobstructed reductions in effort ("central" hypopneas). This "flow limited" behavior of obstructive hypopneas is explained by a Starling resistor model of the upper airway where dynamic collapse of the airway occurs due to the negative intraluminal pressure of inspiration. Transient appearance and disappearance of the flattened contour of groups of individual breaths indicates recurrent obstructive apnea and indicates the need for increased CPAP. Most laboratory titrations will strive to eliminate these events by raising CPAP.

Evidence of sustained elevated upper airway resistance (in contrast to discrete "events") may remain after all apneas and hypopneas disappear. This evidence can consist of stable snoring (upper airway vibration induced by unstable airway tissue), sustained runs of breaths with an inspiratory contour suggesting Starling behavior ("flow limitation"), or other direct measures of elevated airway resistance (e.g., direct measurement of intrathoracic pressure, from an esophageal catheter probe, divided by flow). It is currently often assumed that this evidence of high upper airway resistance must be completely relieved by elevating CPAP, but there is controversy as to the benefits of this form of titration. In some cases, raising CPAP to eliminate all such evidence of elevated upper airway resistance results in further improvement of sleep structure and decrease in daytime sleepiness. By contrast, in other subjects, few if any symptoms occur when the patient has sustained elevated resistance, provided this occurs without causing repetitive arousal. In this latter setting, raising the CPAP is difficult to justify, although often done. Very limited studies attempting to justify this titration approach have not to date supported any benefit of one approach over another.

During CPAP titration, in addition to defining the events that should prompt raising the pressure, it is important to consider when the pressure may be too high (and thus needs to be lowered). Although it is generally assumed that the lowest effective pressure is most comfortable and excess pressure will disrupt sleep, this has not been shown by controlled trials. However, most titration studies should include periodic reductions in CPAP once breathing and sleep have been stabilized to test for the lowest pressure at which evidence of airway instability (see above) recurs. This pressure may be different at different times in the night, and it is almost always different in the supine position and during REM sleep. These observations challenge the concept of a single prescription of CPAP.

Auto-Titration

The above discussion has assumed that a single therapeutic pressure at which the upper airway is effectively splinted exists for a given patient, and that this pressure remains relatively unchanged over time (within each night and across nights). There is ample evidence that this is NOT true. Where it has been studied, it is strongly suggested that for many patients in the supine position, upper airway obstruction is more severe and/or takes more CPAP to treat (although these are not synonymous). There may also be differences in the CPAP needed during REM and non-REM sleep. Thus, when a single pressure is prescribed, most practitioners use the highest pressure needed during a prolonged period of titration (e.g., at night), knowingly over-treating during the rest of the time.

Beginning in about 1990, several investigators began to automate the titration algorithm for choosing a pressure. The concept evolved of a feedback loop that constantly adjusted the CPAP based on sensing either frank apnea, hypopnea, or indices of upper airway abnormality like snoring and/or the contour of the inspiratory airflow. These devices were called auto-titrating CPAP or Auto-CPAP. Two conflicting goals were suggested for optimizing their function—maximizing the efficacy of CPAP and improving patient compliance by reducing pressure to the minimal need at all times. The first of these was to respond to unexpected increases in need in order to prevent undertreatment. The latter was to prevent unnecessarily high values of pressure at a time they were not needed. As both the signal driving feedback and the time constants of the systems developed varied greatly, it is difficult to address the whole group of Auto-CPAP devices in a single study. In particular, the effectiveness of the decision process for raising and lowering the CPAP will dictate whether the final pressure profile is high or low compared with CPAP.

This is not the logical target by itself, and only an outcome such as quality of sleep, improved hours of use by patients, and ultimately, improved daytime function and reduced sleepiness, can be used to evaluate the punitive value of Auto-CPAP over constant pressure. To date, however, limited data support this in large groups of patients. Some data suggest improved compliance with specific devices.

Having said that, no data suggest that the more reasonable of these devices is any LESS effective than CPAP, but some Auto-CPAP devices occasionally show changes in pressure that do not bear any logical relationship to the patient's breathing (runaways), and there is every reason to assume these will impair sleep.

A logical approach to evaluating such devices needs to address several questions before beginning to ask whether long-term use is effective or better than traditional CPAP:

1. Which signal is driving the response of Auto-CPAP? Possible signals include the flow signal amplitude (apnea and hypopnea), shape (detection of starling resistor behavior in the form of "flow limitation shape" as described above), vibrations (e.g., snoring and airway instability), breathing pattern on a longer timescale, direct sound measurements, and direct measures of airway abnormality (e.g., measurement of impedance via forced oscillation technique). The existing devices are driven by different signals, and new devices appear frequently. When compared head-to-head these devices have different responses to breathing test-waveforms, and both bench and patient testing is not yet standardized.

2. What makes the pressure rise? Is a response sought to each abnormal breath or detection of abnormal impedance? Is the pressure adjusted after a "testing" protocol—e.g. a periodic deliberate lowering of the effective pressure to induce some endpoint of abnormality?

3. When is pressure lowered, and after how long? Is continuous testing possible (as with forced oscillations to measure impedance) to which pressure can be lowered when the control variable is low, or is "normal" a condition that, once achieved, provokes a prolonged period of constant pressure (e.g., what is the response to "normal breathing" when detected)? When pressure is lowered, is this a provocative test, or an attempt to detect over-treatment? How frequent are pressure decreases? The implications of these decreases and their endpoint are physiologically significant—"testing" too frequently with a nonsubtle endpoint (e.g., an apnea or an arousal) will disrupt sleep. Testing too infrequently for decreasing pressure will produce ever increasing therapeutic pressure because there will be insufficient compensation for unavoidable errors in the algorithm's detection of a need to raise pressure.

Furthermore, a constant tradeoff exists between the need to optimally set CPAP for a stable physiologic state (e.g., in stable stage 2 sleep in the supine position, a pressure of x cm H_2O may be appropriate for long periods) and the need to respond with a rapid change in CPAP to state changes affecting the airway (e.g., awakening, entering REM, or rolling from the lateral to the supine position). Each machine currently on the market and in development has made different decisions about the way to balance these needs, and the resultant behavior, although it can be described, is not clearly better or worse by simple criteria. Large numbers of patients are needed to show benefit in terms of daytime outcome or compliance with therapy, and these trials are not widely available, nor are the results from one machine easy to apply to another machine or even to a slightly modified algorithm.

This field is still in evolution, but there has been disappointment in the advantage of the approach as reflected in better therapy. Despite this, automation of titration may still have large benefits for patients, even if it is not "better" titration, or even "more comfortable" CPAP. This arises from a trusted algorithm being able to replace the costly CPAP titration study, which is currently usually done in an attended fully monitored laboratory setting. To date, only a few machines on the market have sufficiently reliable "auto-titration" that they can be left unattended and monitored on a first-time user of CPAP, with the resulting pressure behavior assumed to represent an accurate reflection of the patient's need for CPAP. Even the best available machines still over-treat and under-treat some patients, and it seems advisable to recommend that evaluation of the results of a titration study be reviewed (at least off-line) by an expert with more than an assessment of the pressure profile the machine chose.

Our laboratory chooses to review all Auto-CPAP titrations by examining the flow profile and looking to see if overall we agree with the induced rises in pressure. We also review the pressure profile for rapid uncontrolled and unexpected rises in pressure that end with an arousal of the patient, and usually assume these are erroneous.

Finally, if Auto-CPAP is used for titrating a patient's need with the intent of using a single pressure as a prescription, yet another "algorithm" must be invoked to translate a constantly fluctuating pressure into a single prescription. Review of the pressure and or flow tracings rarely results in a single pressure that is constant for much of the night. One must, on subjective grounds, discard excesses and ignore periods of inadequate therapy during the fluctuations. One proposal is to discard the highest pressures achieved during 5% to 15% of the night. There has been no testing of this approach by objective criteria of long-term benefit.

INDICATIONS FOR USE OF CPAP IN OSAS

Stated simply, CPAP is currently the first line of treatment and is indicated for reversal of sleep-induced abnormal upper airway behavior, provided it is severe and results in disruption of sleep with negative daytime consequences. When obstructive apneas and hypopneas occur very frequently and result in severe blood oxygen desaturations, it seems obvious that CPAP is needed. Formal trials of the benefits of CPAP have relatively conclusively shown benefit when more than 30 apnea/hypopneas occur per hour of sleep. This benefit is mostly in the form of reduced daytime sleepiness, although small studies have suggested reductions in blood pressure, improvement in daytime cognitive performance, or reaction time after weeks to months of therapy. CPAP is now near universally accepted as the most effective therapy (better than surgery or oral appliances) but not always as the most acceptable therapy from the patient. This has resulted in compliance rates among moderate–severe apneics (see above definition), which range from 50% to 80%, leaving many patients suboptimally treated, or anxious to change to other treatments as they become available.

However, a more contentious issue is how mild can the physiological abnormality be before treatment with CPAP is either unnecessary or unacceptable to patients. Two relatively large clinical trials are currently underway to answer this question, but no definitive statement can be made at present. A therapeutic trial of CPAP may answer the question in individual patients who show abnormal respiratory events during sleep and have an overt complaint (such as excessive daytime sleepiness). The trial is considered successful provided that patients see a noticeable improvement in symptoms. Better documentation of the validity of this approach is urgently needed as recent studies have shown a very large number of subjects in the general population who have apnea–hypopnea indices ranging between 10 and 30 events per hour (up to 25% of the population), some of whom are asymptomatic, and others who have significant symptoms that might be due to this pathology. Anecdotally, many patients improve on CPAP, but many cannot adapt to chronic therapy. Some of these may benefit from alternative therapy, but CPAP may remain the most effective and definitive way to perform a therapeutic trial for all treatments for OSAHS.

ISSUES IN COMFORT/COMPLIANCE FOR OSAHS AND ANCILLARY TREATMENT ISSUES

Interfaces/Masks

As comfort is the most perceived issue affecting patient compliance, it is clear that the mask must be an important contributor to the patient's willingness to use the device. Although this is accepted dogma, compliance rates over the years in which CPAP has been available are not clearly changing, and much of the willingness to use CPAP may also be affected by subjective patient perceptions of improvement (cost/benefit) and the reinforcement they get from the care provider. It is rare that a patient will use CPAP if the prescribing physician does not believe it works. Many types of nasal, oral, and full-face interfaces have been developed to maximize comfort. Nasal masks are currently most used, and details of material, shape, supporting extensions to relieve pressure points, and so on are beyond the scope of this discussion. Non-mask nasal interfaces also exist ("pillows" or "prongs") and may help address issues of claustrophobia, variant facial anatomy preventing a good seal with a mask, and personal preference. Oral interfaces are less common, but they have a devout following by some patients. Finally, for those with large leakage out of the mouth when the nose is pressurized, full-face masks may provide an alternative. Chin straps are frequently used to reduce mouth leak. It is clear that the technologist who knows the available masks and spends time trying multiple ones with a new patient will have greater success than one using the "one size fits all" approach.

Headgear

Like masks, a variety of headgear exist. These affect fit of the mask, pressure on the nasal bridge, tension of the straps, and even appearance. There is little published on the relative effect these have on compliance or patient preference, but it seems this is an important area.

Oxygen

Some patients (a minority) who use CPAP have a concomitant or related need for supplemental oxygen. As the oxygen is being delivered into a larger air stream, the rate of infusion (typically 2 to 10 L/min) may need to be different from that prescribed for a patient just breathing supplemental O_2. Furthermore, simple examination of the circuit will show that the leak (intentional and unintentional at the mask) will have a large effect on the delivered concentration of the O_2 bled into the air stream. A larger leak will change by a factor of 2–4 the final concentration of O_2 at the patient's nose. As CPAP masks are intrinsically leaky and variable, so it is predictable that the need for O_2 will change. In patients without evidence of hypoventilation and central regulatory abnormalities (usually marked by daytime hypercapnea, or arterial PCO_2 >45 mm Hg), giving too much is not a problem other than expense, so titration to the highest level needed to keep the oxygen saturation during all of sleep (including REM) >90% is the usual goal. However, in patients who tend to hypoventilate, excessive O_2 will worsen CO_2 retention and may lead to accumulation of serum bicarbonate, further depressing ventilatory drive even in the daytime. Thus, it is desirable to try to minimize O_2 use.

Finally, it is not often appreciated that the location at which O_2 is inserted into the CPAP circuit has a large effect. If the bleed is into the hose near the blower, the tubing promotes mixing and acts as a reservoir of a relatively fixed but lower O_2-enriched gas. Pattern of breathing, i.e., time in inspiration and expiration and tidal volume, may have less effect, but the degree of leak will still play a large role. In contrast, if the O_2 is bled directly into the mask, especially if this is beyond the leak in the circuit, the leak may have less effect. However, small changes in timing of breathing and mask size will have enormous effects on the inspired O_2 concentration as buildup of a small volume of near pure O_2 can accumulate during pauses and part of inspiration, whereas there is little volume to act as a reservoir and mixing chamber. This issue should be addressed by providing the location of O_2 connection in any prescription so that it will at least match the titration technique.

Humidity

Although at first glance it is not clear why humidity should be needed if breathing occurs through the normal nasal mucosal humidifying mechanisms, drying of the nose and nasal reactive obstruction are common complaints in CPAP users. Recent literature suggests that humidifying the inspired air is helpful for these complaints, and anecdotally this may improve compliance with CPAP. Several mechanisms may be involved, but the most likely is that any leak out of the mouth will result in a constant desiccating flow through the nose with air below 100% relative humidity, rather than the usual bidirectional flow of normal breathing that replenishes the humidity in the nasal mucosa on expiration that was lost on inspiration. This situation is most prominent in mouth leak exacerbated by high pressure, after palatal surgery, and certain anatomical variants that promote mouth leak. Humidity does address this, but the degree of humidification of air is proportional to temperature as well as to the efficacy of the humidifier. Thus, simple cold pass over humidification is rarely sufficient, and heated humidification has been shown to have advantages in many studies. Because cooling of the air as it travels down the CPAP circuit may cause "rain-out" and a water hazard, some advocate the use of insulation or heating of the CPAP tubing as well as heated humidification of the air before delivery.

WRAP UP

Obstructive sleep apnea has probably been around for as long as there has been sleep, although it has been treated as a clinical disorder and syndrome only in the last 25 years. Epidemiological studies prove it is a major health hazard.

The National Institutes of Health recognize that most OSAHS patients remain undiagnosed and that the principal therapeutic apporach is CPAP. Even though the medical device industry has produced a variety of CPAP technologies and enough different makes and models to ravial the automobile industy, the therapy remains somewhat cumbersome, and so it is not associated with optimal compliance rates in the long term. It is, however, relatively noninvasive, efficient, and definitive at demonstrating therapeutic value for a particular patient.

The advantages of CPAP are that if tolerated it provides a relatively risk-free route to symptomatic relief of a serious disorder. An alternative in severe cases is a surgical procedure to place a tracheostomy. CPAP units require a prescription from a physician, however, several Internal vendors will accept a facsimile via fax or e-mail and will send CPAP machines and related equipment via mail order. Patients can select from model features on websites. Some vendors show over 250 styles and sizes of masks.

As a result of the need to improve patient compliance with CPAP therapy, manufacturers have constantly been improving the comfort and self-regulating capability of the machines to delivery therapy as needed, to match the changing conditions that occur during a night of sleep and over time. To accomplish this, devices include diagnostic capabilities to tailor the pressure of the therapy to the patient's needs. Just as pulse oximeters were the first wave of medical diagnostic devices to incorporate advanced data handling and storage capabilities, it seems that CPAP machines have incorporated technology as it has become available and serve as a test bed for applications such as improved control algorithms, performance, usability features, comfort, and compliance records. All of this information can be transmitted to the physician via the Internet or a small piece of flash memory. CPAP machines are available in the United States at prices ranging from $450 to $3000. The higher priced units can function as ventilation assist devices.

The result is that the same technology being developed for the growing CPAP industry can be used in other medical devices that play a role in self-regulating home therapy and objective tracking of patient compliance and/ or progress.

The search continues for a less cumbersome method to splint the airway open either with surgery, an implanted stent, or a drug that stiffens the upper airway during sleep. CPAP will probably remain the first choice for a therapeutic trial.

See also RESPIRATORY MECHANICS AND GAS EXCHANGE; VENTILATORY MONITORING.

CONTRACEPTIVE DEVICES

MOLLIE KANE
Madison, Wisconsin

INTRODUCTION

Contraceptives and contraceptive devices are possibly the most widely used medical devices today. In the United States, 42 million women, or 7 in 10 women ages 15–44, are currently sexually active and do not want to become pregnant (1). In 2002, 98% of women who had ever had intercourse had used at least one method of contraception. Ninety percent had used a condom at least once. Of reproductive age women (15–44 years), 62% were currently using contraception. Over 10 million women in the United States had undergone female sterilization (2).

In fact, from a population perspective, there may be no other medical devices that have such a profound impact on quality of life. The implications of a person's ability to control when, or whether, they have a child are profound. For the individual, it will forever influence their health, educational and work options, and income. The typical U.S. woman, who desires two children, will use contraception for ~ 30 years of her life in an attempt to control both the timing and number of her pregnancies (1).

What is especially unique about contraceptive devices, compared to most other medical devices, is that they are used by healthy people. They are neither for the detection nor treatment of disease. Therefore, patients and clinicians alike may have different thresholds for the acceptability of side effects and complications of contraceptives compared to other medical devices. The effects of contraceptives are often incorrectly compared to the effect of no treatment, rather than to the effect of pregnancy.

Despite the wide variety of contraceptive methods available in the United States and the high rate of women who use them, the number of unintended pregnancies remains alarming. The Pregnancy Risk Assessment Monitoring System (PRAMS) study from the U.S. Centers for Disease Control and Prevention (CDC) looked at live births in 17 States in 1999 and found that 34–52% of live births were the result of unintended pregnancy (3). One-half of unintended pregnancies in the United States occur among women who were not using contraception (4). The other one-half is the result of contraceptive failures because of incorrect or inconsistent use. Of all unintended pregnancies, one-half end in therapeutic abortion (5). Worldwide ~190 million pregnancies occurred in 1995 and 1 in 3 of these ended in therapeutic abortion (6).

The high rate of unintended pregnancy could potentially be curtailed via changes in the contraceptive industry. A 2004 report by the National Academy of Sciences states that "those millions of women who choose to terminate a pregnancy, many submitting to unsafe and illegal procedures that can be life threatening, attest to the need for improved access to and utilization of existing contraceptive methods and the need for new and improved contraceptive options" (6). Research is beginning to focus on how to make contraceptives easier to obtain and simpler to use. Methods that are available over the counter may be more accessible to some women. There are experts who feel that many contraceptive methods are appropriate for sale without prescription (7). Research could determine whether this type of accessibility could lead to increased contraceptive success and decreased rates of unintended pregnancy. Emergency contraception could be dispensed prophylactically. To prevent pregnancy between the time that a method is dispensed and utilized, emergency contraception should be used as soon as possible. Marketing can continue to focus on "positive side effects" of contraceptives, such as decreased acne or increased sexual function.

Other new and creative methods of dispensing contraceptives may increase use and continuation rates. Planned Parenthood of Columbia/Willamette, OR is offering prescription of hormonal contraceptives via the internet. Women fill out an online questionnaire and are then contacted by phone by a nurse practitioner. Appropriate candidates may then receive pills, patches, or rings via overnight mail (7). Pharmacy access is another new method for prescription of contraceptives. Several states now allow pharmacists to prescribe emergency contraception, which may be expanded to other contraceptive types in the future.

In addition, the intrinsic properties of a contraceptive device affect whether a woman will be able to use it with every episode of intercourse or whether she will choose to discontinue use. Safety profiles and side effects affect a woman's willingness to start or continue a given method. For example, breakthrough bleeding, weight gain, nausea, or mood changes are side effects that are not dangerous, but that are likely to result in method discontinuation. Ease of use and disruption of intercourse may also affect a woman's willingness to continue use of a given contraceptive.

There is no perfect, completely effective contraceptive method. The best contraceptive choice for an individual or a couple will depend on many factors including, among others, their sexual attitudes and behavior. A thorough sexual history will assist the clinician in guiding a patient in choosing an appropriate contraceptive. Considerations include a woman's level of comfort with her own body, her likelihood of exposure to sexually transmitted infections (STIs), the degree of cooperation from the patient's partner, and the patient's need to have contraception occur separate from intercourse (8). Frequency of intercourse may have an effect on contraceptive choice. Those who have intercourse rarely may not want a method that requires daily action. However, those who have intercourse rarely are most likely to have intercourse unexpectedly and find themselves unprepared.

Individuals who will have a new partner or multiple partners may desire a barrier method of contraception to decrease risk of STI transmission. However, those who would find an unintentional pregnancy completely unacceptable may prefer the higher effectiveness rates of hormonal contraceptives or sterilization. Using both types of contraception, barrier, and hormonal–sterilization, provides the highest level of effective pregnancy prevention and provides protection from STIs. However, an inverse relationship is seen between the use of effective noncoital contraceptives and condoms. In one study of 12,000 U.S. high school students, pill use was the strongest predictor of failure to use condoms. It even had a more profound effect than use of alcohol or drugs or having multiple partners (9).

This article reviews concepts involved in contraceptive efficacy, informed consent, and contraceptive research and development. A brief description is provided of those contraceptives that can be considered medical devices. Much of the information in this article comes from the book Contraceptive Technology, 18th Revised ed. (10), as well as from peer-reviewed literature, online information from the CDC (http://www.cdc.gov/), literature from the Alan Guttmacher Institute, a nonprofit organization focused on sexual and reproductive health research, policy analysis and public education (http://www.agi-usa.org/index.html), and documentation from medical device manufacturers via the internet. This article will not address nondevice methods of contraception such as oral contraceptives, injectables, and natural family planning.

CONTRACEPTIVE EFFICACY

Contraceptive efficacy is based on many factors. Inherent efficacy of the method is only one aspect of whether pregnancy will occur. Factors that facilitate or interfere with proper use of the method also contribute to efficacy. Such factors can be method related, such as breakthrough bleeding, or interruption of spontaneity, or user related. Study populations that have high rates of intercourse will have higher rates of pregnancy, and therefore lower contraceptive effectiveness rates, than study populations with low rates of intercourse. Age of the user or of the study population will also affect efficacy rates. Fertility declines with age, thus reducing unintended pregnancies. Therefore, study populations containing higher populations of older women will find higher contraceptive efficacy rates. In addition, frequency of intercourse declines with age. Women with regular menstrual cycles will have higher rates of pregnancy and thus, higher rates of contraceptive failure, than women with irregular cycles (11). For example, one study of women using the Reality female condom found that women with regular menstrual cycles were 7.2 times more likely to experience a pregnancy than women whose cycles were < 17 days or > 43 days (12).

Typical use contraceptive failure rates reflect the rate of pregnancy among women who report that a contraceptive is her method. It does not mean that she uses the method every time, uses it correctly, or even uses it at all. Perfect use reflects the rate of pregnancy that will occur among women who use a contraceptive correctly and with every episode of intercourse. These rates are usually estimated by researchers or are based on one or two studies.

Contraceptive failure rates are often calculated using the PEARL index, calculated as the number of pregnancies occurring divided by the number of woman years of exposure. The PEARL index is highly affected by the length of the study. Pregnancy rates are higher during the early stages of use of a method. Therefore, a study of 100 women using a given method for 1 year will have a higher PEARL index than a study of 10 women using the same method for 10 years.

Studies that administer pregnancy tests each month will find more pregnancies due to early detection of conceptions that end in spontaneous abortion prior to being recognized. Patients lost to follow up may be more likely to have experienced pregnancy than those who remain in the study.

Approximately 85 out of 100 sexually active U.S. couples that do not use contraception will become pregnant in 1 year based on estimates from populations who have low contraception use or who are actively trying to conceive (13). Conversely, the definition of infertility is the absence of conception after 1 year of unprotected intercourse. Contraceptive effectiveness rates should be regarded in comparison to what the pregnancy rate would have been if no contraceptive technique was used. For example, if use of a given contraceptive results in 15 pregnancies/100 women in the first year, then ~70 pregnancies were prevented (85 minus 15).

INFORMED CONSENT

Informed consent for contraceptive device use should always be obtained. Individuals may not be willing to tolerate the same level of risk with contraceptive use as they would for other medical devices. The Department of Health and Human Services provides regulations regarding what constitutes appropriate informed consent for sterilization. It contains seven basic principles. These seven principles can help to guide appropriate informed consent for all types of contraceptives (14,15). The seven principles follow: (1) The patient should understand benefits of the method. (2) The patient should understand risks of the method (including risks of method failure). (3) The patient should know alternatives to the method (including abstinence and no method). (4) The patient should know that they have the right and responsibility to ask questions about the method. (5) The patient should know that they

have the ability to withdraw from the method at any time. (6) All of the above issues must be explained in a way the patient understands. (7) Documentation that the caregiver has ensured understanding of each of the first six points should occur.

The term "informed choice" has been gaining popularity in the field of family planning. It implies the idea that all contraceptives have side effects and risks and that an individual must choose which side effects and risks are acceptable to them. An individual should not use a product if they find any one of the side effects or risks unacceptable.

MALE CONDOM

The male condom is a thin sheath of latex, lambskin, or polyurethane, placed over the shaft and glans of the penis. Condoms prevent pregnancy by blocking the passage of sperm into the vaginal canal. Condoms provide the only well-documented method for prevention of transmission of sexually transmitted infections and human immunodeficiency virus (HIV) by blocking the exchange of blood, semen, and vaginal secretions.

Latex condoms are by far the most popular and widely used type of male condoms. Latex condoms are available in the United States in a wide variety of styles and brands. They are available with or without reservoir tips or nipple ends, straight or tapered, smooth or ribbed, transparent or in a variety of colors, odorless, scented, or flavored. Average condom length is ~7.5 in. (19 cm), with ranges from ~6.5 (16.5 cm) to 9.5 in. (24 cm). Widths of <2 in. (5.1 cm) are considered snug and width >2.125 in. (5.4 cm) is "baggy". Thickness averages 0.0027 in. (0.0069 cm). Condoms with thickness <0.0019 in. (0.0048 cm) are considered extra sensitive. Those with thickness >0.0027 in. (0.0069 cm) are considered extra strong and may help with premature ejaculation. "Climax control" condoms are available with a small amount of benzocaine in them to aid with premature ejaculation. A Taiwanese company, SakuNet International provides condoms in 55 different sizes ranging from 3 (7.6 cm) to 9.4 in. (24 cm) in length and from 1.6 (4.1 cm) to 2.5 in. (6.4 cm) in diameter. Men log onto the SakuNet website and print out a measurement card that assists them in ordering the appropriate sized condom.

Lambskin condoms are manufactured from the intestinal lining of lambs. They contain pores that allow for the passage of small particles including HIV, hepatitis B virus, and herpes simplex virus (OO). Lambskin condoms are effective at pregnancy prevention because sperm are too large to penetrate the pores. Lambskin condoms should be used for contraception only.

There are four synthetic (polyurethane) condoms that are U.S. Food and Drug Administration (FDA) approved and available for purchase in the United States. These include two Avanti condoms (Durex Consumer Products), Trojan Supra (ARMKEL), and the eZ-on (Mayer Laboratories). Synthetic condoms have many advantages over latex condoms. Compared to latex condoms they are thinner and better at transmitting, and offer a less restrictive fit. They are stronger and more resistant to deterioration. They are compatible with water- or oil-based lubricants and are acceptable for individuals with latex allergy. However, polyurethane condoms have not been well studied for their effectiveness in the prevention of STIs and HIV (16).

The majority of male condom research has been done on latex condoms. Latex condoms that are used correctly and consistently are effective at the prevention of pregnancy. Method failure (condom breakage) is quite rare. With perfect use as few as 2% of couples using condoms for 1 year will experience pregnancy (17). As much as 24–65% of condom breakage occurs prior to intercourse, therefore not increasing the risk of pregnancy (18). Reported rates of condom breakage or slippage during intercourse or withdrawal are variable, but low. In a study of 353 latex condoms used by sex workers in Nevada brothels, none broke or fell off during intercourse. Two (0.6%) slipped off during withdrawal (18).

Actual pregnancy rates with condom use are much higher than perfect use rates. This is due to the high likelihood of incorrect or inconsistent condom use. For typical condom use, ~15% of couples will experience a pregnancy over 1 year of use (17). This includes couples who fail to use the condom for every episode of intercourse. Other causes of condom failure include failure to use the condom throughout intercourse, partial or complete condom slippage, poor withdrawal technique, incorrect placement of the condom, and use of oil-based lubricants that degrade the condom.

Research has shown latex condoms to be extremely effective at preventing transmission of HIV during vaginal and anal intercourse. In a 1994 study published in the *New England Journal of Medicine* of 124 sero-discordant heterosexual couples, no sero-conversion occurred in 20 months of consistent and correct condom use. With inconsistent condom use, 10% of HIV-negative partners seroconverted >20 months and with no condom use 15% of partners sero-converted (19).

Variable data is available regarding the effectiveness of condoms in reducing the transmission of each STI. After reviewing all available data, the CDC concluded that "the lack of data about the level of condom effectiveness indicates that more research is needed—not that latex condoms do not work" in the prevention of STI transmission (20). They found that latex condoms, when used consistently and correctly, can reduce the risk of transmission of gonorrhea, chlamydia, and trichomoniasis. In addition, they reduce the risk of transmission of genital herpes, syphilis, chancroid, and human papillomavirus when the infected areas are covered by the condom. They also found that use of latex condoms reduces the risk of HPV associated diseases such as cervical cancer (20).

Condoms are regulated as medical devices by the FDA. Every condom manufactured in the United States is electronically tested for holes or weak areas. In addition, manufacturers are required to test samples from each lot of finished, packaged condoms. Tests that must be performed on the sampled condoms include the water leak test, the air burst test, and the tensile property test (21). Should failure rates of the sample condoms be unsatisfactory, the entire lot will be destroyed.

Condoms pose very few risks or side effects for their users. Latex condoms may not be used by individuals with latex allergy. Individuals who do not have latex allergy, but experience an allergic reaction to a condom, may be reacting to condom specific components such as lubricant, perfume, or other agents used in the manufacturing process (22). Some studies have found an increased risk of urinary tract infections in women using condoms with spermicide.

Latex condoms are vulnerable to heat and sunlight. They must be stored in a cool, dark place. They may be stored in wallets for up to 1 month. They must be used within 5 years of their manufacture date, or within 2 years if they are lubricated with spermicide (23).

THE FEMALE CONDOM

The female condom (FC), Reality (Female Health Company, UK, http://www.femalehealth.com/index.htm), is a 17 cm long, 7.8 cm wide, 0.05 mm thick sheath made of polyurethane. It has a ring at each end. The inner ring is at the closed end of the sheath and is used to insert the FC and to hold it in place behind the public bone. The outer ring is at the open end of the sheath and remains outside the vagina. The sheath loosely lines the vagina and covers some of the vulva. The FC is prelubricated with a nonspermicidal silicone lubricant. Additional oil- or water-based lubricant may be added. The FC prevents pregnancy and also protects the vagina, cervix, and external genitalia against STI. The same device is available in other countries under different names.

The polyurethane that makes up the FC is soft and odorless. It is stronger than the latex used to make male condoms. It conducts heat well, making sexual intercourse feel more natural. It does not contain latex and does not deteriorate at high temperatures or require any special type of storage. The expiration date is 5 years from the day of manufacture.

The FC may be placed up to 8 h prior to intercourse and removed at any time following intercourse. It does not require an erect penis to hold it in place. The FC should not be used together with a male condom as friction between the products could result in product failure.

The FC has been found to be effective in preventing STI transmission both *In vitro* (24) and *In vivo* (25,26). Clinical studies in the United States, Latin America, and Japan revealed pregnancy rates similar to those found with the use of other barrier methods. In a study looking at the use of the FC as the sole means of contraception among 328 monogamous couples at six sites in the United States, Latin America, and Japan, cumulative failure rates were ~20%. However, with "perfect" correct and consistent use, <10% of the subjects experienced an accidental pregnancy (27). The FC has no serious side effects, and it does not alter the original flora or cause significant skin irritation or vaginal trauma.

The FC is approved for single use only. A single Reality condom costs ~$2.00. The World Health Organization states that single use is preferable. However, in situations where female condoms are not available or affordable, evidence suggests that the FC may be used safely up to five times.

VAGINAL SPERMICIDES

Vaginal spermicides contain benzalkonium chloride, octoxynol 9, or nonoxynol 9. They are available without prescription as films, suppositories, gels, creams, foam, and on condoms. Currently, only Nonoxynol-9 is available in the United States. It is a surfactant that works by destroying the sperm cell membrane. The spermicide is inserted into the vagina prior to any genital contact.

Vaginal spermicides used alone are fairly poor at preventing pregnancy. Pregnancy rates of 10–28% in 6 months of use have been found in recent studies. The higher pregnancy rates were found in a population of young women who had frequent coitus (28). Spermicide used together with condoms or other barrier methods can significantly increase the efficacy of these methods.

Spermicide may be placed up to 1 h prior to intercourse and must be placed again for each repeated episode of intercourse. Suppositories, foaming tablets, and films require time to dissolve prior to intercourse. Gel, tablets, suppositories, and film need to make contact with the cervix in order to be effective. Appropriate placement may be challenging for some women. Spermicides have not been associated with an increased risk of vaginitis. They may increase the risk of urinary tract infection in some cases. No adverse effects to spermicides have been reported, but toxicology data is limited. Spermicides should not be used together with any other vaginal medications or with vaginal cleansing products or douches. Spermicides should usually be avoided in the presence of sores on the genitals or when there is vaginal irritation and should not be used in the presence of cervical cancer.

Spermicides should not be used for the prevention of STI or HIV transmission. In 2000, a letter by Helen Gayle, the Director of the National Center for HIV, STD, and TB Prevention of the CDC, to clinicians reviewed several studies including that of the Joint UN Program on AIDS (UNAIDS). It concluded that Nonoxynol-9 does not prevent the transmission of HIV or other STIs. In addition, it may increase the risk of HIV transmission by causing irritation or ulceration of the female genital mucosa. Increased HIV transmission with use of Nonoxynol-9 was seen among women with very frequent high risk exposure. The effects of Nonoxynol-9 on HIV transmission among lower risk women are not known (29).

CERVICAL CAP

The cervical cap is a small, firm latex or silicon dome that creates a barrier over the cervix and holds spermicide in place. The Prentif Cavity Rim Cervical Cap (Lamberts Ltd, UK) was FDA approved in 1988, although it has been used in Europe for >60 years. However, Cervical Cap Ltd. (http://www.cervcap.com/), the sole U.S. distributor, dissolved itself in March of 2005, and as of June 2005, there are no other U.S. distrubutors. The Prentif Cap is made of latex in the shape of a thimble. A small groove on the inside creates suction to hold the cap in place over the cervix. The dome of the cap may be filled one-third full of spermicide prior to insertion, providing both a mechanical and a chemical barrier to the entry of sperm into the upper genital tract.

The Prentif Cap is available in four sizes, 22, 25, 28, and 31 mm, based on the diameter of the caps interior. It must be fit by a trained provider and is available only by prescription. It costs $49–68 when obtained directly from the distributor.

Approximately 20% of women are poor candidates for the cervical cap. A good fit is not possible in those with a very long or very short cervix, or when the cervix is asymmetrical. The Prentif Cap is also contraindicated in women with a history of cervical laceration, current cervicitis or infection, history of toxic shock syndrome, latex rubber allergy, a vaginal septum, cervical or uterine malignancy, or unresolved abnormal Pap smear.

Fitting should be performed near the time of ovulation, when the cervix is most small and firm. The bowel and bladder should be empty. Several sizes should be tested in ensure the best fit. With proper fitting, the cervix is completely covered. There should be 1–2 mm of space between the cap and the cervix. A firm tug on the dome should not displace the cap. When the cap is left in place for 10 min or more, a suction rim should be visible and palpable after removal.

Initial fitting should be performed at least 6-weeks postpartum and at least 2 weeks after therapeutic or spontaneous abortion. Refit should be performed annually, after childbirth, therapeutic or spontaneous abortion, after cessation of lactation, following a weight gain or loss of 10 lb (4.53 kg) or more, or after any reported accidental dislodgement.

The patient may leave the cap in place for up to 48 h. It should be in place for at least 8 h following the last episode of intercourse. Additional spermicide does not need to be placed for more than one episode of intercourse. Overfilling of the cap with spermicide may lead to slipping and dislodgement. If the cap is used with a condom, the condom must be lubricated to prevent "grabbing" between the two devices, which may result in dislodgement of the cap. The patient should use a back-up method of contraception for the first menstrual cycle of use and should check the cap for good placement after intercourse.

There may be degeneration of the latex over time. The FDA recommends annual replacement of the Prentif Cap. Exposure to heat, light, or petroleum-based products will accelerate deterioration of the cap. Early signs of deterioration include dimpling or thinning of the dome of the cap, a sticky texture, or a mosaic pattern forming on the dome. After use, the cap should be washed in soapy water, dried, dusted with cornstarch, and stored in its original container.

A second cervical cap, the FemCap, was FDA approved in 2003. It has been available in Europe since 1999. It is a silicon cap that covers the cervix and forms a seal to prevent the entrance of sperm into the upper genital tract. The FemCap is shaped like a sailor's cap with a brim, a dome, a groove between the brim and the dome, and a strap to facilitate removal. The dome covers the cervix, the rim fits against the vaginal fornices and forms a seal with the vaginal wall. The groove stores spermicide and traps sperm. Pushing on the dome breaks the suction, allowing removal of the cap by hooking a finger around the strap.

Prior to insertion one-quarter teaspoonful of spermicide should be spread into the dome of the FemCap. An additional one-half teaspoon is placed in the groove. Because a majority of the spermicide remains in the groove and because the groove faces the vaginal side, exposure of cervical epithelium to spermicide is minimized.

The FemCap is available in three sizes. Sizing is based on parity alone, which eliminates the need for fitting by a healthcare provider. However, the patient must be trained to use the cap and a prescription is required. Sizing is based on the inner diameter of the rim of the cap. The 22 mm cap is for women who have never been pregnant. The 26 mm cap is for women whose pregnancies have resulted in spontaneous abortion, therapeutic abortion, or Cesarean section. Any woman who has delivered one or more full-term infants vaginally should use a 30 mm cap.

A smooth, symmetrical cervix is required for FemCap use. Therefore, it is contraindicated in women with a history of cervical laceration. It should not be used in the presence of infection of the upper or lower genital tract. The FemCap may be used by couples with latex allergy. It should not be used during menstruation.

The FemCap should be placed prior to any sexual arousal to allow optimal formation of a seal. It must be in for at least 6 h after the last episode of intercourse and may be in place for up to 48 h at a time. Additional spermicide should be inserted for repeat episodes of intercourse.

The Vimule cap was previously available in the United States, but was recalled by the FDA in 1983 due to a high incidence of vaginal lesions. It remains available in Europe and elsewhere. It is a bell-shaped cap with a flanged rim. It is available in 3 sizes: 42, 48, and 52 mm by external diameter.

Two additional caps are available in Europe, but are not FDA approved. The Dumas cap, by Lamberts (Dalston) Ltd., is a shallow, bowl-shaped latex cap available in five sizes: 50, 55, 60, 65, and 75 mm by external diameter. The Oves cap, by Veos, UK Ltd., is a clear, disposable, silicone cap with a loop on the rim to aid in removal. It comes in three sizes: 26, 28, and 30 mm.

LEA'S SHIELD

In 2002, the FDA approved Lea's Shield (YAMA, Inc.; Millburn, NJ, http://www.leasshield.com/about.htm), a reusable vaginal barrier contraceptive made of medical grade silicone rubber. The Lea's Shield is shaped like an elliptical bowl that covers the cervix. The posterior surface is thickened to improve fit in the posterior fornix of the vagina. There is an anterior loop to assist in removal. The bowl contains a centrally located valve to allow the passage of cervical secretions and air. Lea's Shield comes in one size designed to fit most women. It is available by prescription from Planned Parenthood and from the manufacturer.

A phase II clinical trial published, in 1996 by Mauck et al. (30), found efficacy rates for the Lea's Shield to be similar to those found in other studies of the diaphragm and cervical cap. Efficacy data were available for 146 women who used the Lea's Shield as their only method of contraception for 6 months. The adjusted 6 month life table pregnancy rates were 5.6% for users of Lea's Shield with spermicide and 9.3% for users of Lea's Shield alone (not a statistically significant difference with $p = 0.086$). This corresponds to 1-year failure rates of 9 and 14%, respectively. However, the study included a high percentage (84%) of parous women. For unknown reasons, barrier contraceptives in general are less effective among parous women. There were no pregnancies among the small number of nulliparous women in this study.

Studies are not available directly comparing the Lea's Shield's safety or efficacy to the diaphragm or cervical cap. It is not known whether uterine position effects efficacy. No adverse events have been reported with the use of Lea's Shield. Data on the effect of the Lea's Shield on cervical or vaginal epithelium are not currently available. In the study by Mauck et al. (30), discontinuation rates for device related reasons were low, with 84% of women, but only 55% of their partners, reported liking the Lea's Shield. This device may be used by individuals with latex allergy.

A clinician should check the initial fit of Lea's Shield. The device covers the entire cervix and the strap should be behind the symphysis. If these two criteria are met and the device feels comfortable to the woman, then it is considered a proper fit.

Prior to each use the bowl of the device should be filled with spermicide. The device may be inserted at any time prior to intercourse and may be left in for up to 48 h at a time. Additional spermicidal jelly is not needed for additional episodes of intercourse. Lea's Shield should be left in place for at least 8 h after the last episode of intercourse. It should not be used during menses due to a theoretical increased risk of toxic shock syndrome. After removal, the Lea's Shield should be washed with soap and water, air dried, and stored in its original silk pouch. The manufacturer recommends annual replacement of the device.

THE CONTRACEPTIVE SPONGE

The Today sponge (Allendale Pharmaceuticals, Allendale, NJ, http://todaysponge.com/.), is a disk shaped device made of disposable polyurethane foam infused with 1 g of nonoxynol-9 spermicide. The sponge was initially marketed in 1983, and then removed from the market in 1995. The discontinuation of the sponge was not based on any problems with safety or efficacy. Rather, routine inspection of the manufacturing plant revealed non-TSS causing bacteria in the water used to make the Today Sponge. The company decided that the cost of the modifications needed to continue production were prohibitive. However, these modifications have now been undertaken and the Today Sponge became available again in Canada and via the Internet in early 2005. The FDA approval, leading to U.S. sales, is expected in 2005. Approval in the United Kingdom is also imminent.

The proximal side of the Today sponge is concave to allow tighter fit over the cervix that decreases the chance of dislodgement. The opposite side has a woven polyester loop to aid in removal. The polyurethane foam creates a texture that mimics vaginal tissue. The sponge's main mechanism of action is to release spermicide and 125–150 mg are released into the vaginal vault during 24 h of use. It also functions as a physical barrier to entry of sperm into the upper genital tract, and it absorbs sperm into its polyurethane matrix. The Today Sponge has a pH 6–8. It expires 18 months after the manufacturing date.

Studies in the United States reveal a 12 month failure rate of ∼10% with perfect use. Actual use 12 month failure rates are higher, and accedental pregnancy will occur in 15–20% of couples using the sponge (31).

The Today Sponge is one size that fits all common. It is an over-the-counter device and requires no fitting or other clinician involvement. Prior to insertion the sponge is activated by generously wetting it with water, then squeezing it until it produces suds. It is then inserted deeply into the vagina to cover the cervix. It is effective immediately upon insertion. It may be left in for up to 30 h and may be used for more than one episode of intercourse during this time with no additional spermicide needed. It should be left in place for at least 6 h after the last episode of intercourse. After removal, it is discarded. There is a theoretical risk of toxic shock syndrome if the sponge is left in place >24–30 h.

The Today Sponge is held in place by the vaginal muscles. It is possible to experience expulsion with straining. If the sponge is found at the vaginal entrance within 6 h after intercourse, it should be pushed back in. If it is expelled entirely, a new sponge should be wetted and inserted immediately.

The Today Sponge may tear upon removal. Any residual pieces within the vagina must be removed. The Today Sponge will not dissolve over time. Retention of a piece of the sponge may increase the risk of toxic shock syndrome or vaginal infections.

The Today Sponge is contraindicated in women with a current upper or lower genital tract infection, and in women with a history of toxic shock syndrome. It should not be used during menstruation due to a theoretical increase in the risk of toxic shock syndrome. It should not be used during the first 8 postpartum weeks. Following spontaneous or elective abortion, the sponge should not be used until an examination has been performed by a clinician to ensure that cervical and vaginal tissues appear normal. The Today Sponge should not be used underwater (in pools or hot tubs), because large quantities of water entering the vagina may dilute the spermicide. It should not be used together with any vaginal medications as interactions have not been studied.

The Today Sponge is equally efficacious in both parous and nulliparous women (31,32). The Today Sponge cannot be used if either partner has an allergy to one of its components. One study found that 4% of U.S. women discontinued use due to allergic type symptoms (33). The Today Sponge contains a small amount of metabisulfite. No allergic reactions to this sulfa component of the Today Sponge have been reported. However, couples where either partner has a sulfa allergy should not use the Today Sponge.

The Today Sponge has not been shown to increase the risk of vaginal infections with normal use. Increased risk of vaginal candidiasis has been found with prolonged use of an individual sponge. Some women experience a white vaginal discharge during use of the Today Sponge. This is normal, but may be confused with vaginal candidiasis. Vaginal dryness may occur due to the sponge soaking up vaginal secretions. Adequate wetting of the Today Sponge prior to insertion should minimize this.

CONTRACEPTIVE DIAPHRAGM

The contraceptive diaphragm is a silicone or latex cup shaped device with a firm, yet flexible, rim and a dome that covers the cervix. The diaphragm provides a mechanical barrier to the entrance of sperm into the upper genital tract. In addition, the cup of the diaphragm provides a chemical barrier by acting as a receptacle for spermicide.

Reported effectiveness rates vary widely, from 70 to 99 % (34). With perfect use, the 1 year pregnancy rate has been found to be as low as 6% per 100 women (34). Diaphragm failure may include failure due to improper fitting.

Prior to insertion of the diaphragm, spermicide should be distributed over the dome of the device. The device is effective immediately after insertion and may be placed up to 6 h prior to intercourse. If >6 h have elapsed, and for more than one episode of intercourse, additional spermicide must be inserted in the vagina (without removing the diaphragm). The diaphragm must remain in place at least 6 h after the last episode of intercourse and should not be left in place for >24 h. After removal the diaphragm should be washed in warm, soapy water, and stored in a clean, dry container.

Fitting rings or fitting diaphragms may be obtained from the manufacturers. In order to fit the diaphragm, the patient is placed in dorsal lithotomy position. The clinician inserts the gloved middle and index fingers into the vagina until the middle finger reaches the posterior fornix. The spot where the index finger touches the inferior pubic arch should be marked with the thumb or an instrument. The fingers are withdrawn in this position. The fitting ring or diaphragm is then placed with the rim over the end of the middle finger and the opposite rim touching the mark of the inferior pubic arch.

The appropriate fitting ring or diaphragm should then be inserted. Fitting is appropriate if the anterior rim lies just behind the symphysis pubis, the posterior rim lies at the vaginal fornix, the rim touches both lateral walls, and the cervix can be felt through the dome. At least one size smaller and one size larger should always be fitted to check for the optimal fit. The best size is the biggest one that meets all fitting criteria, but cannot be felt by the patient. With a proper fit, the diaphragm will not dislodge or fall out with straining and will not be felt by the patient, even with ambulation.

Contraindications to the diaphragm include a history of toxic shock syndrome, current upper or lower genital tract infection or allergy to any of the components or to spermicide. Adequate fit may not be possible with a markedly anteverted cervix. A shallow vaginal shelf or poor vaginal tone may lead to dislodgement. The diaphragm should be used with caution in the presence of a rectocele or cystocele.

Adverse effects from the diaphragm are minimal. There is an increased risk of recurrent urinary tract infection. The risk may be reduced with the softer rim or flat spring diaphragms. In addition, there is an increased risk of toxic shock syndrome. This can be minimized by avoiding use during menses or immediately postpartum and limiting use to 24 h at a time.

There are multiple types of diaphragms available in the United States. The ALL-FLEX Arcing Spring Diaphragm (Ortho-McNeil Pharmaceutical, Inc.) is a latex, buff-colored dome with a flexible rim. The rim contains a spring that forms an arc no matter where the rim is compressed. It is available in sizes 55–95 mm with 5 mm increments. It is the most widely used diaphragm and has been available in the United States since 1940. The firm rim of this diaphragm makes it the easiest type to insert. It may be the best choice for women with decreased pelvic tone, rectocele, or cystocele.

The Ortho Coil Spring Diaphragm, Ortho-McNeil Pharmaceutical, Inc, is a latex diaphragm with a flexible rim that contains a tension-adjusted, cadmium-plated coil spring. This allows for compression in one plane only. It is also available in sizes 55–95 mm with 5 mm increments.

The Milex Wide Seal is a silicone diaphragm with an arcing spring. A small skirt around the inner rim of the device is meant to hold spermicide in place and to improve the seal. It is available in eight sizes, with diameters of 60–95 mm, in 5 mm increments. In the United States, it is available only from the manufacturer.

The Reflexions Flat Spring diaphragm is made of latex with a rim similar to the coil spring. It is thinner and more delicate than the other diaphragms. It is available in seven sizes from 65- to 95 mm diameter with 5 mm increments.

The SILCS (pronounced "silks") intravaginal barrier is being developed by SILCS, Inc., Middlesex, N. J., in collaboration with the Contraceptive Research and Development Program (CONRAD) and The Program for Appropriate Technology in Health (PATH). One size will fit all women. It is a silicone device with a unique shape to allow for easier insertion and removal. The device is to be used with spermicide. It is currently recommended that it be left in for at least 8 h post-coitus. A maximum period of time per use has not yet been determined. The device is currently in phase II and III clinical trials. Once it is FDA approved it is intended to be made available over the counter.

TRANSDERMAL CONTRACEPTIVE PATCH

Ortho-Evra is a combination transdermal contraceptive patch containing 6.00 mg of norelgestromin and 0.75 mg of ethinyl estradiol (EE). Norelgestromin 150 μg and EE 20 μg are released into the bloodstream per 24 h of use. Norelgestromin is the primary active metabolite produced following oral administration of norgestimate. Ortho-Evra is available in cartons of one cycle (three patches) as well as in cartons containing a single patch to be used if a patch is lost or damaged.

The medication is administered via a triple layer patch with a contact surface area of 20 cm². The patch consists of the backing layer, an adhesive middle layer, and a release liner. The backing layer is made of flexible beige colored film with a polyethylene outer layer and a polyester inner layer. It has the appearance of a band-aid. It provides structural support and protects the adhesive middle layer. The adhesive layer contains polyisobutylene–polybutene adhesive, crospovidone, polyester, and lauryl lactate. The active ingredients, norelgestromin and EE, are in this layer. The release liner is a transparent polyethylene terephthalate film with a polydimethylsiloxane coating that protects the adhesive layer during storage and is removed prior to application of the device.

Following application of the device, both norelgestromin and EE rapidly appear in the blood stream. They reach a plateau level at ∼48 h and then remain at steady-state levels for the remaining 5 days. Half-lives of norelgestromin and EE are 28 and 17 h, respectively. The FSH, LH, and Estradiol levels, suppressed during treatment, return to normal by 6 weeks after discontinuation. Release of norelgestromin has been found to be unaffected by exposure to saunas, whirlpools, exercise, and cold bath water. Release of EE is slightly increased by the sauna, whirlpool, or exercise, but levels remain within the desired range. Ortho-Evra bypasses the need for GI absorption. Therefore, it may be a good choice for women with some types of GI absorption disease.

In an open label study of 1672 reproductive age women using the patch for 10,994 cycles there were five pregnancies resulting from method failure and one resulting from user failure. This gives a PEARL index of 0.59 for method failure and 0.71 for user failure (F). A second study comparing Ortho-Evra to oral contraceptives found a PEARL index of 1.24 for Ortho-Evra versus 2.18 for the oral contraceptives (this difference was not statistically significant) (35). Ortho-Evra users also completed more cycles with perfect use compared to oral contraceptive users (88 and 78%, respectively) (35).

Studies have found that discontinuation of Ortho-Evra due to side effects is relatively low. A study by Smallwood et al. (32) found discontinuation rates of 1–2% each for side effects including skin irritation, nausea, emotional liability, headache, dysmenorrhea, and breast discomfort (35,36). Break through bleeding may be more common with Ortho-Evra, compared to oral contraceptives, during the initial cycles of use.

Ortho-Evra uses a 28-day cycle. A new patch is applied once a week for 3 weeks (21 days), followed by 7 days with no patch in place. Withdrawal bleeding occurs during the patch-free week. "Patch change day" should always be the same day of the week for a given woman. To start the patch, a woman should wait for her menses to begin. She may either place the patch on the first day of her menses or do a Sunday start. The day she applies her first patch will be her "patch change day". The patch must be applied to clean, dry, intact skin on the buttock, abdomen, upper outer arm, or upper torso (but not breasts). No lotions or other topical products should be in place on the skin where the patch will be placed.

To place the patch, it is first removed from its foil pouch. One-half of the release liner is then removed. This half of the patch is applied to the skin and the second half of the release liner is removed. The woman should then press down on the entire patch with the palm of her hand for 10 s. She should then check to make sure that all of the edges are well adhered. The patch should then be checked every day to make sure it is sticking well.

Detachment of Ortho-Evra occurs rarely, with an occurrence of <3% at the time in one study (35). If a patch becomes partially or completely detached, it should be reapplied to the same spot immediately. If it will not reattach or it has become soiled, a new patch should be placed right away. No back-up contraception is needed and the "patch change day" remains the same. If more than 1 day has lapsed since detachment or if length of detachment is unknown, the current patch cycle should be discarded. The first patch from a new cycle should be placed. This will now be the new "patch change day" for this woman. In this case, a back-up method of contraception should be used for the first 7 days of the cycle. Other adhesives should not be used to hold a patch in place.

Contraindications and risks of Ortho-Evra are the same as those for the use of other combination hormonal methods. These are discussed in depth in multiple other references (see resources). In addition, Ortho-Evra is unique among contraceptives in causing significant rates of skin irritation, redness, or rash. Should skin problems arise, the patch may be removed and a new one applied in a different location until the next "change day". Patients with skin abnormalities, including eczema, psoriasis, or sunburn should not use Ortho-Evra. Ortho-Evra may have a lower efficacy in women weighing >198 lb (89.8 kg).

CONTRACEPTIVE VAGINAL RING

NuvaRing (Organon USA, Inc., Roseland, NJ, http://www.muvaring.com/Consummer/) is a nonbiodegradable, flexible vaginal ring containing 11.7 mg of etonogestrel and 2.7 mg of ethinyl estradiol (11). When placed in the vagina each ring releases ~0.015 mg·day^{-1} of ethinyl estradiol and 0.120 mg·day^{-1} of etonogestrel per day. Etonogestrel is the biologically active metabolite of desogestrel. The NuvaRing is colorless and is composed of ethylene vinylacetate copolymers and magnesium stearate. The outer diameter is 54 mm. Each NuvaRing is packaged in a reusable aluminum laminate sachet. They are available in boxes of either one or three sachets.

NuvaRing is left in place for 21 days, followed by a 7 day hormone free period to allow for menstruation. Serum hormone levels required to suppress ovulation are achieved within the first 24 h of NuvaRing use, so there is no delay in the onset of contraception. Serum hormone levels are lower than those achieved by oral contraceptives and the contraceptive patch. The ring has a steady release rate, so serum hormone concentrations do not vary throughout the day as with oral contraceptives. Maximum serum drug concentrations are reached at 59 h for EE and at 200 h for etonogestrel. Half-lives are 29.3 and 44.7 h for etonogestrel and EE, respectively. If NuvaRing is broken it does not release a higher concentration of hormones, making overdose unlikely.

Contraindications and adverse reactions for NuvaRing are the same as those for combination oral contraceptives. In addition, local reactions specific to the ring can occur. These include vaginitis (5.6%), leukorrhea (4.6%), and vaginal discomfort (2.4%). One major clinical trial revealed a withdrawal rate of 15.1% due to problems such as the sensation of a foreign body, coital problems, and expulsion (37).

NuvaRing should be inserted between day 1 and 5 of the menstrual cycle. It should not be started later than day 5 even if bleeding is still occurring. Once inserted, NuvaRing should remain in place continuously for 21 days. The NuvaRing is inserted by folding it in half and inserting it high into the vagina. Exact positioning of the NuvaRing is not important for effectiveness.

NuvaRing is removed 3 weeks after insertion on the same day of the week that it was inserted. After removal, it should be placed in the foil pouch and discarded. After a 1 week break, a new ring is inserted. This will be on the same day of the week as the ring was inserted during the previous cycle. A withdrawal bleed will usually occur on day 2 or 3 after removal of the ring. The new ring must be inserted 7 days after removal of the previous ring to maintain contraceptive effectiveness. It should be inserted even if menstrual bleeding has not ended. A back-up method of contraception should be used until day 7 of use of the first NuvaRing.

If the NuvaRing is expelled or inadvertently removed, it should be washed in cool or warm (not hot) water, and reinserted within 3 h. If > 3 h have lapsed between expulsion and reinsertion a back-up method of contraception should be used until the NuvaRing has been in place for at least 7 continuous days. If the ring-free period has been 7 or more days, the possibility of pregnancy should be considered.

NuvaRing is meant to be in place for no >21 continuous days. If it is inadvertently left in place for up to 7 additional days it should be removed and left out for 1 week. A new ring should then be inserted. If NuvaRing has been left in place for > 4 weeks, then the risk of pregnancy should be considered. A new ring should be inserted, but a back-up method of contraception should be used until the new NuvaRing has been in place for 7 continuous days.

NuvaRing can be stored at room temperature (77 °F) for up to 4 months with excursions permitted from 59–86 °F. For longer storage it must be kept refrigerated at 36–46 °F.

CONTRACEPTIVE IMPLANTS

As of June 2005, there are currently no contraceptive implants available in the United States. Norplant (Wyeth, Madison, NJ, http://www.norplantinfo.com/) has been removed from the market, but Implanon (Organon USA, Inc., Roseland, NJ, http://www.organon.com/authfiles/index.asp) may soon become available in the United States. Implants function via the continuous release of a very low dose of progesterone. As with all progesterone only contraceptives, the implants inhibit ovulation via inhibition of the midcycle peaks of LH and FSH. In addition, progesterone only methods lead to thickening of the cervical mucous to prevent penetration by sperm and to development of an atrophic endometrium to inhibit implantation.

Contraceptive implants are not subject to user error. Once in place, they are effective for their life expectancy or until removed. Use of Norplant in the United States has shown that it has a higher continuation rate than other contraceptive methods (I). Other benefits of contraceptive implants are similar to those of any progesterone only methods. Most notably, the implants do not contain estrogen, eliminating the risk of estrogen related complications and side effects. Progesterone only methods lead to amenorrhea or scanty periods for some women. The contraceptive effect of implants is rapidly reversed with removal.

Risks and side effects of implants are similar to those of all progesterone only methods of contraception. In addition, because implants release such a low dose of progesterone, drug interactions become a more significant concern. Also unique to implants is the risk for local skin problems. With Norplant one analysis found that 0.8% of users experienced local infection, 0.4% experienced expulsion of a capsule, and 4.7% had local skin irritation. The majority of these complications occur soon after insertion, but a significant number of women will also experience them after two or more months of use (38,39).

Norplant was an implant of six flexible capsules of silicon rubber tubing containing 26 mg of levonorgestrel. Each capsule was 2.4 mm in diameter and 34 mm in length. After insertion, Norplant initially, released $85 \mu g \cdot day^{-1}$ of levornorgestrel. This decreased to $30 \mu g \cdot day^{-1}$ over time. Norplant was very effective, however, removal was quite difficult in many cases.

In August of 2000, Wyeth recalled Norplant System Kits distributed beginning October, 1999 due to an atypically low level of levonorgestrel release in routine shelf-life stability tests. In July of 2002, Wyeth announced that it did not plan to reintroduce Norplant due to limitations in the availability of components of the product. Because the system is meant to be in place for a maximum of 5 years there should be very few women left with Norplant in place at this time. Ideally, all Norplant Systems will be removed by August of 2005.

Implanon is a single rod contraceptive implant containing 68 mg of etonogestrel, which is released gradually into the bloodstream for 3 years. The rod is nonbiodegradable and semirigid. It is 4 cm long and 2 mm in diameter and is made of ethylene vinylacetate copolymer. The etonogestrel is released at an average rate of $40 \mu g \cdot day^{-1}$. In clinical trials Implanon took just 2.2 min to insert and 5.4 min to remove. Implanon can migrate, complicating removal (40). In the clinical trials no pregnancies occurred (41–43).

INTRAUTERINE DEVICES

Today's intrauterine devices (IUD) are highly effective, very convenient, and safe. However, the intrauterine device is used in fairly low rates in the United States, likely because of persisting misunderstandings on the part of many women and clinicians (44). Public opinion still reflects concern about the safety of intrauterine devices left over from the Dalkon Shield. The Dalkon Shield (A. H. Robins) was released in 1971 but episodes of septic abortions and other infections were reported, and the FDA recommended removal from the market in 1974. Litigation against A. H. Robins caused the company to declare bankruptcy in 1985. The cause of the infections was likely a multifilament string used only by the Dalkon Shield that allowed the product to be removed more easily. Bacteria could migrate up the string and cause the serious infections (45). In fact, clinical studies on current IUD models show excellent safety and effectiveness profiles.

The mechanism of action of IUDs is not completely understood. The two broad possibilities are preventing fertilization and destroying the early embryo, with prefertilization effects providing the contraception nearly all of the time (46–48). There are a number of proposed mechanisms of the IUD to prevent fertrilization of the ovum. The IUD distorts the hormonal and enzymatic environment of the female reproductive tract. It causes a chronic, sterile inflammatory environment of the endometrium. The copper found in IUDs is particularly toxic to sperm. Sperm are damaged by the environment, and are less likely to have the potential to fertilize the egg. The IUD also protects against ectopic pregnancy. One office visit is required for many years of contraception. Fertility returns quickly upon removal of the device. Women who are poor surgical candidates for sterilization are often excellent candidates for the intrauterine device.

Increasing evidence indicates that intrauterine devices protect against endometrial cancer. Of six studies examining the relationship between intrauterine devices and endometrial cancer, five found a protective effect, although only in two of the studies was this statistically significant (49,50). The mechanism of protection is unknown, but is thought to be related to effects of the IUD on the endometrium.

Initial cost of the intrauterine devices is high. In addition, there is a charge for insertion. However, this is a one time fee. Amortized over the life of the IUD it becomes one of the least expensive forms of contraception.

Pain and cramping may occur at the time of insertion of the IUD, but usually resolve within 15 min. Up to 10% of IUD users will spontaneously expel the IUD, sometimes without realizing it. Once a woman has expelled an IUD she has a 30% chance of expelling future IUDs as well (51). There is approximately a 1 in 1000 risk of uterine perforation associated with insertion of the IUD.

Early IUD research indicated significant increases in the risk of upper genital tract infection and tubal infertility associated with IUD use. It is now recognized that this research was flawed in several respects. Newer data and reanalysis of old data both indicate that risks of upper genital tract infection and tubal infertility are very low. There is a transient increase in the risk of upper genital tract infection for ~20 days after the insertion procedure (it is the insertion, and not the IUD, that causes the increased risk). Afterward, the risk returns to low levels. It is not known whether upper genital tract infection or tubal infertility are more common in women with STI who receive an IUD versus women with STI and no IUD (52).

Antibiotic prophylaxis prior to intrauterine device insertion is not currently recommended. A large randomized controlled trial in Los Angeles County found no benefit to the use of prophylactic azithromycin for IUD placement. The same study found that only about 1 woman in 100 developed salpingitis during the first months of IUD use (53).

The IUDs may be used by women who are nulliparous, but have had a previous spontaneous or elective abortion. Women who are nulligravid may have increased difficulty with IUD insertion and retention. However, The World Health Organization considers nulligravidity to be eligibility criteria category 2, implying that the benefits of the method generally outweigh any theoretical or proven risk.

Intrauterine device users should use condoms with new partners or whenever there is a risk of acquiring an STI. Should chlamydia or gonorrhea infection occur, there is no evidence to suggest that the IUD needs to be removed. Standard treatment of the STI is indicated.

Two types of IUDs are currently available in the United States. The Copper T 380A (ParaGard, FEI Women's Health LLC, New York) has been available since 1988. It is made of polyethylene in a T shape with barium sulfate to give X-ray visibility. Copper is wound around the polyethylene resulting in a copper surface area of $380 \pm 23\,\mathrm{mm}^2$. The device is 36 mm tall and 32 mm wide with a 3 mm bulb at the bottom to which a monofilament polyethylene string is attached. The Copper T 380 is approved for use for up to 10 years. Studies have found that it remains highly effective for up to 12 years (54).

The copper on the CopperT 380 increases the presence of copper ions, enzymes, prostaglandins, and white blood cells in the uterine and tubal fluids. This impairs sperm function, which prevents fertilization. The device is extremely effective. In World Health Organization trials the cumulative 12-year pregnancy rate with the CopperT 380 was 2.2 pregnancies per 100 women. The highest risk time for pregnancy was during the initial year of use (54). A disadvantage of the CopperT 380 is an increase in menstrual blood loss.

The LNG-IUS (Mirena, Berlex Laboratories, Montville, NJ) was approved for use in the United States in 2000, although it has been available in Europe for > 10 years. It is a T-shaped device with a polyethylene frame and a cylinder composed of a polydimethylsiloxane-levonorgestrel mixture molded around the vertical arm. The cylinder is covered by a membrane to regulate release of the hormone. The device is 32 mm in height and 32 mm in width. There is a dark monofilament polyethylene thread at the base to assist with removal.

The LNG-IUS releases levonorgestrel directly into the endometrial cavity. The initial release rate is $20\,\mu\mathrm{g\cdot day}^{-1}$, which diminishes to $14\,\mu\mathrm{g\cdot day}^{-1}$ after 5 years. The system is approved for 5 years of use. Data shows that it actually remains effective for at least 7 years (55). Some levonorgestrel is absorbed, leading to a mean plasma concentration of 5%. This is lower than the plasma concentrations reached by other progesterone only methods.

Effectiveness rates for the LNG-IUS have been determined in several studies. These have found a first year cumulative failure rate of 0.14 per 100 women, a 5-year cumulative failure rate of 0.71 per 100 women and a 7-year cumulative failure rate of 1.1 per 100 women (55).

The LNG-IUS causes a significant decrease in menstrual blood loss. In addition, $\sim 20\%$ of users will experience complete cessation of menses. This effect is especially important in women for whom anemia is a concern. The LNG-IUS can be used to treat heavy menses, sometimes even in place of endometrial ablation or hysterectomy.

BILATERAL TUBAL STERILIZATION

Tubal sterilization may be performed postpartum, after spontaneous or elective abortion, or as an interval procedure. Interval sterilizations may be performed at any time during the menstrual cycle, however, current pregnancy must first be ruled out. Postpartum sterilization is performed by minilaparotomy prior to any involution of the uterus. Postabortion sterilization may be done via minilaparotomy or laparoscopically. Tubal sterilization performed as an interval procedure is done via laparoscope. Sterilization may also be done using a transcervical or transvaginal approach.

Laparoscopic sterilization can be performed on an outpatient basis. It leaves barely visible scars and return to normal activity is rapid. The cost and upkeep of laparoscopic equipment is expensive. Trocar insertion may result in injury to the bowel, bladder, or major blood vessels. General anesthesia is required.

Minilaparotomy requires a 2–3 cm incision placed in relation to the location of the uterine fundus. Only basic instruments and training are required. Minilaparotomy may be done under local anesthesia with sedation, with regional anesthesia, or with general anesthesia.

In November, 2002 the FDA approved the use of Essure (Conceptus Inc., San Carlos, CA, http://www.essure.com/consumer/c_homepage.aspes), a sterilization device that is placed using hysteroscopy through a transcervical approach. No entry into the peritoneal cavity is required. Back-up contraception is then used for 3 months followed by a hysterosalpingogram to confirm bilateral tubal occlusion. Data on this method is promising, but long-term efficacy rates are not yet available.

The transvaginal approach to sterilization is rarely used. It is contraindicated in the presence of major pelvic adhesions or enlarged uterus. There are many potential complications including cellulites, pelvic abscess, hemorrhage, proctotomy, or cystotomy.

Methods of tubal occlusion include electrocoagulation, mechanical occlusion, ligation, and chemical sclerosing. Bipolar electrocoagulation is performed via the laparoscopic approach only. At least 3 cm of the isthmic fallopian tube is coagulated using at least 25 W delivered in a cutting waveform. A current meter most accurately indicates complete coagulation.

Ligation is the most common method of occlusion used during laparotomy or minilaparotomy. There are multiple methods for ligation including the Pomeroy, modified Pomeroy, Parkland, Uchida, and Irving methods. Chemical sclerosing agents are under investigation but none are approved for use in the United States (56).

Several mechanical occlusion devices are available in the United States. Because less of the fallopian tube is destroyed, microsurgical reversal is more likely to succeed following these methods. Mechanical occlusion requires a normal fallopian tube. If there are adhesions, thickening, or dilation of the tube proper application is more difficult and failure is more likely. Mechanical occlusion devices include the Falope ring, a silicon rubber band, the Hulka-Clemens clip, a spring-loaded clip, and the Filshie clip, a titanium clip lined with silicone rubber. Each device has its own type of applicator and each requires special training to use.

Efficacy for tubal sterilization is high. The CREST study found a 5-year cumulative life-table probability of pregnancy to be 13 per 1000 procedures for all types of tubal ligation in aggregate. Rates for each method include 5-year cumulative pregnancy rates of 6.3 per 1000 procedures for postpartum partial salpingectomy, 16.5 per 1000 for bipolar coagulation, 10 per 1000 for silicone band methods, and 31.7 per 1000 for spring clips. The risk of sterilization failure persists for many years. Therefore, 10-year cumulative pregnancy risk is >5 year cumulative risk for all methods. Women who are younger at the time of sterilization are more likely to experience method failure because of their greater fecundity (56).

When tubal sterilization failure occurs, the risk of ectopic pregnancy is significant. For the women in the CREST study, one-third of poststerilization pregnancies were ectopic (57). Other risks of tubal sterilization are rare and are mainly related to the need for general anesthesia.

CONTRACEPTION RESEARCH AND DEVELOPMENT

In 2004, the National Academy of Sciences found that "while the existing array of contraceptive options represents a major contribution of science and industry to human well-being, it fails to meet needs in significant populations and the costs of that failure are high, for societies, for families, and for individuals" (6). Dozens of contraceptive devices utilizing a wide array of mechanisms of action are currently under development. At least 14 new contraceptive methods have been approved by the FDA since 1998. Nevertheless, research and development of new contraceptive methods is difficult. The Institute of Medicine Committee on New Frontiers in Contraception estimates that development of a new contraceptive takes 10–14 years and $400–800 million. Large companies are reluctant to assume the risk involved in long-term development of devices that potentially will not be approved or will result in minimal income. Small companies lack sufficient resources for such large undertakings. The success of development of any given contraceptive is unpredictable and economically risky. The risk of liability is also a strong disincentive to contraceptive research and development. Therefore, as other fields of medicine make research advances in genetics, molecular biology, and immunology, contraceptive technology has been unable to keep up. Given the overwhelming effect of unintended pregnancy on people's lives, and the very high rates of unintended pregnancy in the United States, such low resources for contraceptive development must be taken seriously (58).

Future success in the research and development of contraceptives may be improved by collaborations between small and large industry and not-for-profit organizations. To this end, CONRAD supported the creation of the Consortium for Industrial Collaboration in Contraceptive Research (CICCR). The CICCR identifies potential project leads under investigation by not for profit organizations and encourages industry to collaborate with these not for profits. In addition, they provide additional funds to not-for-profit investigators.

Research and development has traditionally focused on creating contraceptives with high rates of efficacy intrinsic to the device and better safety profiles. It is now recognized that it is also important to design contraceptives that are easier to use consistently and correctly. More consumer-based research will help to determine what types of contraceptive development would be most desirable to women of different ages, races, or life styles. In addition, the need for more dual protection methods is great. Other than condoms, there is no single method that provides high rates of efficacy for both pregnancy and STI prevention. Current understanding of how existing methods affect STI transmission, including effects on the immune system, needs to be improved (59). "Dual packaging" of contraceptives is being investigated. For example, oral contraceptives could be packaged together with emergency contraceptives or condoms.

Many microbicides are currently under varying stages of development. A perfect microbicide would destroy STI pathogens and prevent pregnancy, without causing vaginal tissue damage. Spermicides could inactivate HIV or other pathogens, interfere with cell attachment or entry, or prevent viral replication. Pregnancy prevention could be accomplished via effects on sperm motility.

There are several vaginal barrier contraceptives under development. The Today Sponge has been reintroduced in Canada and on the internet and is awaiting FDA approval. Protectaid, containing nonoxynol-9 and benzalkonium chloride, and Pharmatex, containing benzalkonium chloride alone, are both contraceptive sponges that are available in Canada and Europe and may become FDA approved. Oves is a disposable silicone cervical cap. Research is also underway on self-fitting diaphragms and new types of female condoms as well.

Jadelle is another hormonal implants that is FDA approved, but not available in the U.S. market. Jadelle is an improved two-rod version of Norplant. Hormone receptor blocking agents, such as mifepristone, are being investigated for emergency contraception, low daily dosing, or monthly dosing. Research is being done on non-hormonal systemic methods for women. For example, a contraceptive could specifically inhibit implantation by blocking leukemia inhibitory factor, perimplantation factor, or leptin.

Immunocontraceptives offer a new and innovative approach to contraception as well as STI prevention. Possible areas of immunologic research include pursuit of immunogens to sperm, reproductive hormones, hormone receptors, and sexually transmitted pathogens, as well as research on the local immune response of the female reproductive tract (6). Immunocontraceptives under research include a vaccine to HCG and to sperm antigens. At this time, the HCG vaccine requires frequent boosters and has a potential cross-reaction with pituitary hormones. Boosting mucosal immunity to sperm may be possible via oral or vaginal vaccine administration.

Research is underway on several systemic methods for males. Efforts have been targeting the production of LH and FSH by the pituitary. Lack of LH and FSH blocks hormonal support for testicular cell function. This results in decreased sperm production, but also decreases testosterone. Low testosterone levels lead to a lack of libido. Human studies are in progress to look at drug combinations to suppress LH while replacing testosterone losses. Several forms of testosterone are being examined, but many require daily or weekly injections. Because high levels of testosterone cause multiple side effects, it is challenging to find the perfect balance.

Other systemic male methods are also under investigation. Gossypol, a derivative of cottonseed oil, is effective in sperm suppression while not effecting testosterone levels. However, in current protocols it is up to 20% irreversible. Some chemotherapy drugs, including lonidamine, are also being considered for male contraceptives.

BIBLIOGRAPHY

1. Facts in Brief: Contraceptive Use. 2004. Available at http://www.agi-usa.org/pubs/fb_contr_use.html.
2. Mosher WD, et al. Use of contraception and use of family planning services in the United States: 1982–2002. Adv Data 2004;350:1–36.
3. Beck LF, et al. Prevalence of selected maternal behaviors and experiences, pregnancy risk assessment monitoring system (PRAMS), 1999. MMWR 2002;51(SS02):1–26.
4. Burnhill MS, Contraceptive use: the US perspective. Int J Gynaecol Obstet 1998;62(Suppl 1):S17–S23.
5. Sharing Responsibilities: Women, Society and Abortion Worldwide. 1991. Available at http://www.agi-usa.org/pubs/sharing. pdf.
6. Harrison PF, Rosenfield A, editors, Contraceptive Research and Development: Looking to the Future. Washington (DC): National Academy Press; 2004. p 1–28.
7. The Unfinished Revolution in Contraception: Convenience, Consumer Access and Choice. 2004. Available at http://www.guttmacher.org/pubs/2004/09/20/UnfinRevInContra.pdf.
8. Haffner DW, Styton WR. Sexuality and Reproductive Health. In: Hatcher RA et al. editors. Contraceptive Technology 18th Revised Edition. New York: Ardent Media, Inc.; 2004. p 20.
9. Guest F. HIV/AIDS and Reproductive Health, in Contraceptive Technology 18th Revised Edition. In: Hatcher RA, et al. editors. New York: Ardent Media, Inc.; 2004. p 160.
10. Hatcher RA, et al. Contraceptive Technology, 18th Revised Edition. New York: Ardent Media, Inc.; 2004.
11. Trussell J. The essentials of contraception: efficacy, safety, and personal considerations. In: Hatcher RA, et al. editors. Contraceptive Technology 18th Revised Edition. New York: Ardent Media, Inc.; 2004. p 230.
12. Steiner MJ, et al. Influence of cycle variability and coital frequency on the risk of pregnancy. Contraception 1999; 60(3):137–143.
13. Trussell J. Contraceptive efficacy. In: Hatcher RA, et al. editors. Contraceptive Technology 18th Revised Edition. New York: Ardent Media, Inc.; 2004. p 774.
14. Guest F. Education and counseling. In: Hatcher RA, et al. editors. Contraceptive Technology 18th Revised Edition. New York: Ardent Media, Inc.; 2004. p 264.
15. Sterilization of persons in federally assisted family planning projects. Fed Reg 1978;43:52146–52175.
16. Hatcher RA, et al. editors. Contraceptive Technology, 17th revised edition, New York: Ardent Media, Inc.; 1998. p 326.
17. Warner L, Hatcher RA, Steiner MJ. Male condoms. contraceptive technology 18th Revised ed. In: Hatcher RA, et al. editors. New York: Ardent Media, Inc.; 2004. p 334.
18. Hatcher RA, et al. editors. Contraceptive Technology, 17th Revised Edition. New York: Ardent Media, Inc.; 1998. p 329.
19. de Vincenzi I. A longitudinal study of human immunodeficiency virus transmission by heterosexual partners. European study group on heterosexual transmission of HIV. N Engl J Med 1994;331(6):341–346.
20. Latex Condoms and Sexually Transmitted Diseases- Prevention Messages. 2001. Available at http://www.metrokc.gov/health/apu/std/condomefficacy.htm.
21. Workshop Summary: Scientific Evidence on Condom Effectiveness for Sexually Transmitted Disease (STD) Prevention. 2001 July 20, 2001 Available at http://www.niaid.nih.gov/dmid/stds/condomreport.pdf.
22. Warner L, Hatcher RA, Steiner MJ. Male condoms. In: Hatcher RA, et al. editors. Contraceptive Technology 18th Revised Edition. New York: Ardent Media, Inc.; 2004. p 346.
23. Warner L, Hatcher RA, Steiner MJ. Male condoms. In: Hatcher RA, et al. editors. Contraceptive Technology 18th Revised Edition. New York: Ardent Media, Inc.; 2004. p 348.
24. Drew WL, Blair M, Miner RC, Conant M. Evaluation of the virus permeability of a new condom for women. Sex Transm Dis 1990;17(2):110–112.
25. The Female Condom: A Review. 1997, World Health Organization.
26. Macaluso M, et al. Efficacy of the female condom as a barrier to semen during intercourse. Am J Epidemiol 2003;157(4):289–297.
27. Farr G, Gabelnick H, Sturgen K, Dorflinger L. Contraceptive efficacy and acceptability of the female condom. Am J Public Health 1994;84(12):1960–1964.
28. Cates WJ, Raymond EG. Vaginal spermicides, In: Hatcher RA, et al. editors. Contraceptive Technology 18th Revised Edition. New York: Ardent Media, Inc.; 2004. p 356.
29. Available at http://www.cdc.gov/hiv/pubs/mmwr/mmwr11aug 00. htm.
30. Mauck C, et al. Lea's Shield: a study of the safety and efficacy of a new vaginal barrier contraceptive used with and without spermicide. Contraception 1996;53(6):329–335.
31. McClure DA, Edelman DA. Worldwide method effectiveness of the Today vaginal contraceptive sponge. Adv Contracept 1985;1(4):305–311.
32. Edelman DA, North BB. Updated pregnancy rates for the Today contraceptive sponge. Am J Obstet Gynecol 1987; 157(5):1164–1165.
33. Edelman DA, McIntyre SL, Harper J. A comparative trial of the Today contraceptive sponge and diaphragm. Am J Obstet Gynecol 1984;150(7):869–876.
34. Trussell J, Strickler J, Vaughan B. Contraceptive efficacy of the diaphragm, the sponge and the cervical cap. Fam Plann Perspect 1993;25(3):100–105,135.
35. Audet MC, et al. Evaluation of contraceptive efficacy and cycle control of a transdermal contraceptive patch vs an oral contraceptive: a randomized controlled trial. JAMA 2001;285(18):2347–2354.
36. Smallwood GH, et al. Efficacy and safety of a transdermal contraceptive system. Obstet Gynecol 2001;98(5 Pt. 1):799–805.
37. Hatcher RA, Nelson A. Combined hormonal contraceptive methods. In: Hatcher RA, et al. editors. Contraceptive Technology 18th Revised Edition. New York: Ardent Media, Inc.; 2004. p 449.
38. Klavon SL, Grubb GS. Insertion site complications during the first year of NORPLANT use. Contraception 1990;41(1):27–37.

39. Hatcher RA. Depo-Provera injections, implants, and progestin-only pills (minipills). In: Hatcher RA, et al. editors. Contraceptive Technology. New York: Ardent Media, Inc.; 2004. p 471.

40. Le J, Tsourounis C. Implanon: a critical review. Ann Pharmacother 2001;35(3):329–336.

41. Kiriwat O, et al. A 4-year pilot study on the efficacy and safety of Implanon, a single-rod hormonal contraceptive implant, in healthy women in Thailand. Eur J Contracept Reprod Health Care 1998;3(2):85–91.

42. Croxatto HB. Clinical profile of Implanon: a single-rod etonogestrel contraceptive implant. Eur J Contracept Reprod Health Care 2000;5(Suppl. 2):21–28.

43. Croxatto HB, Makarainen L. The pharmacodynamics and efficacy of Implanon. An overview of the data. Contraception 1998;58(6 Suppl.):91S–97S.

44. Espey E, Ogburn T. Perpetuating negative attitudes about the intrauterine device: textbooks lag behind the evidence. Contraception 2002;65(6):389–395.

45. Cheng D. The intrauterine device: still misunderstood after all these years. South Med J 2000;93(9) 859–864.

46. Ortiz ME, Croxatto HB, Bardin CW. Mechanisms of action of intrauterine devices. Obstet Gynecol Surv 1996;51(12 Suppl.): S42–S51.

47. Stanford JB, Mikolajczyk RT. Mechanisms of action of intrauterine devices: update and estimation of postfertilization effects. Am J Obstet Gynecol 2002;187(6):1699–1708.

48. Rivera R, Yacobson I, Grimes D. The mechanisms of action of hormonal contraceptives and intrauterine contraceptive devices. Am J Obstet Gynecol 1999;181(5 Pt. 1):1263–1269.

49. Grimes DA. Intrauterine devices (IUDs). In: Hatcher RA, et al. editors. Contraceptive Technology 18th Revised Edition. New York: Ardent Media, Inc.; 2004. p 499.

50. Hubacher D, Grimes DA. Noncontraceptive health benefits of intrauterine devices: a systematic review. Obstet Gynecol Surv 2002;57(2):120–128.

51. Bahamondes L, et al. Performance of copper intrauterine devices when inserted after an expulsion. Hum Reprod 1995;10(11):2917–2918.

52. Grimes DA. Intrauterine devices (IUDs), In: Hatcher RA, et al. editors. Contraceptive Technology 18th Revised Edition. New York: Ardent Media, Inc.; 2004. p 501.

53. Walsh T, et al. Randomised controlled trial of prophylactic antibiotics before insertion of intrauterine devices. IUD Study Group. Lancet 1998;351(9108):1005–1008.

54. Long-term reversible contraception. Twelve years of experience with the TCu380A and TCu220C. Contraception 1997; 56(6):341–352.

55. Sivin I, et al. Prolonged intrauterine contraception: a seven-year randomized study of the levonorgestrel 20 mcg/day (LNg 20) and the Copper T380 Ag IUDS. Contraception 1991; 44(5): 473–480.

56. Peterson HB, et al. The risk of pregnancy after tubal sterilization: findings from the U.S. Collaborative Review of Sterilization. Am J Obstet Gynecol 1996;174(4):1161–1168; discussion 1168–1170.

57. Peterson HB, et al. The risk of ectopic pregnancy after tubal sterilization. U.S. Collaborative Review of Sterilization Working Group. N Engl J Med 1997;336(11):762–767.

58. Stewart F, Gabelnick HL. Contraceptive research and development, In: Hatcher RA, et al. editors. Contraceptive Technology 18th Revised Edition. New York: Ardent Media, Inc.; 2004. p 606.

59. Stewart F, Gabelnick HL. Contraceptive research and development, In: Hatcher RA, et al. editors. Contraceptive Technology 18th Revised Edition. New York: Ardent Media, Inc.; 2004. p 603.

See also COLPOSCOPY; SEXUAL INSTRUMENTATION.

CORONARY ANGIOPLASTY AND GUIDEWIRE DIAGNOSTICS

RUPAK K. BANERJEE
ABHIJIT SINHA ROY
University of Cincinnati
Cincinnati, Ohio

LLOYD H. BACK
California Institute of Technology
Pasadena, California

INTRODUCTION

Percutaneous Transluminal Coronary Angioplasty (PTCA) is an invasive procedure, where a blocked coronary artery is opened by inserting a pressurized balloon. Since its inception in the year 1964 by Dotter and Judkins (1), coronary angioplasty has undergone much development and is commonly performed in Cardiac Catheterization today. A statistic provided by the Center for Disease Control (CDC) shows that nearly one-half of a million PTCA procedures were conducted in United States alone in the year 2002. A typical coronary angioplasty procedure includes these basic components:

Guiding Catheter

A guiding catheter serves three broad purposes: It provides support and passage to the introduction of smaller diameter guidewires. It provides a conduit to the administration of drugs and external agents, such as contrast agent for angiography. It provides damping, due to heart motion, to guidewires inserted through them.

For passage of guidewires, a catheter, having a diameter at least twice that of the guidewire, is recommended. Most guidewires are made up of soft material in the tip so that instance of vessel injury is as minimal as possible. In modern practice, guiding catheter are available in different shapes and sizes, such as Judkins, Amplatz curves, pig tail. Selection of a suitable guiding catheter depends on the application. Small sized guiding catheters of size 6F and 7F are most commonly used in PTCA as their size is well suited for guidewires of size 0.014 in. (0.355 mm) and for passage of balloon catheters, appropriate for coronary dimensions in humans. Some guiding catheters may be designed with side holes at the end, which enables them to engage the coronary ostium, while maintaining continuous administration of fluoroscopy agent. The length of guiding catheters is ~ 90–100 cm. Figure 1 shows some of the guiding catheters being used today in coronary angioplasty.

Guidewires

Modern day guidewires are designed for tip stiffness, easy maneuverability, location control and visibility to angiography. A typical guidewire has a solid core, usually made up of stainless steel or nitinol and has a gradual taper from the proximal to distal end. This core is encapsulated in a spring coil and platinum in the distal section for improved radiographic visibility. The spring coil, which is a teflon coated stainless steel, is usually welded to directly or through a band to the tapered end of a guidewire so that

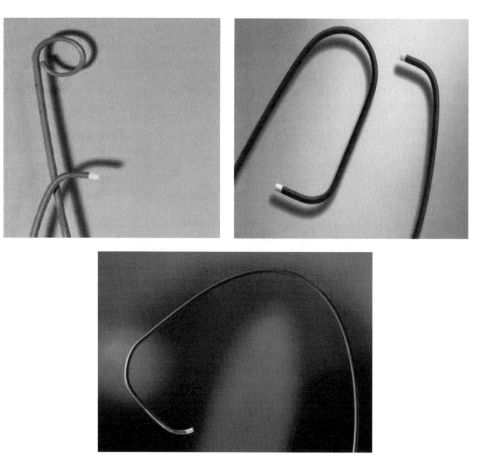

Figure 1. A few coronary angioplasty guiding catheters. (Courtesy of Boston Scien., MA.)

the user may bend the tip to access the desired artery. Guidewires are available in a wide range of sizes from 0.010 in. (0.254 to 0.457 mm) to 0.018 in. In modern day PTCA, 0.014 in. (0.355 mm) is the size most widely used. In some cases, double length (~300 cm) guidewires may also be used. These enable access to the diseased vessel while other devices, such as stent, dilation catheters, are being deployed, with minimal risk of vessel injury (2).

Several specialized guidewires are also available. These guidewires are designed for measurement of arterial pressure and flow to assess the ischemic severity of a stenosis. For pressure measurement, a piezoelectric pressure transducer is placed around the inner solid core of the 0.014 in. (0.355 mm) 3 cm from the tip (Fig. 2). This transducer facilitates measurement of transstenotic pressure drop and Myocardial Fractional Flow Reserve (FFR_{myo}) (3). These pressure sensors can measure pressure in the range of -30–300 mmHg (-3.9–39.9 kPa) with an accuracy of ± 1 mmHg (0.133 kPa). For phasic flow measurement, the technology most widely used is a Doppler-based flow sensor. The Doppler flow sensor is placed at the tip of the 0.014 in. (0.355 mm) guidewire (~175 cm long). For coronary flow measurements, usually 40 MHz piezoelectric transducer is used (Fig. 2). As a general rule, the smaller the vessel size, the larger is the frequency of the sensor. The ultrasound beam describes a conical beam in the distal vessel, thus obtaining a small sample volume. Doppler-based guidewire are capable of measuring translesional velocity, which is reported as Coronary Flow Reserve

(CFR = coronary flow at hyperemia / coronary flow at basal flow). Currently, however, Doppler guidewires are designed to measure average peak velocity, mean velocity for a cycle as well as diastolic/systolic flow ratio. A unit that measures both flow and pressure simultaneously in coronary vessels is shown in Fig. 3. Another technique used to measure CFR is based on coronary thermodilution. In this method, the wire has a microsensor at a location 3 cm from the floppy tip, which enables simultaneous recording of coronary pressure measurement and temperature, with an accuracy of 0.02 °C (4,5). The shaft of this wire can be used as a second thermistor, which provides the input signal at the coronary ostium of any fluid injection at a temperature different from blood. With this method, CFR is expressed as the ratio of mean transit time at basal flow to mean transit time at hyperemic flow. Experimentally, it has been shown that $CFR_{doppler}$ and CFR_{thermo} differ by $\sim 20\%$ (5). To facilitate simultaneous evaluation of epicardial and microvascular diseases, a single wire, having both pressure and flow sensor, is also available.

Balloon Catheter

Appropriate selection of a balloon catheter is a must for the success of coronary angioplasty. Beginning from the "over the wire" design, balloons have undergone many improvements. Present balloon catheters are both strong and flexible enough to handle tortuous vessel segment, with minimal intimal injury. Most balloon catheters have a silicone or hydrophilic coating, such as polyethylene, to

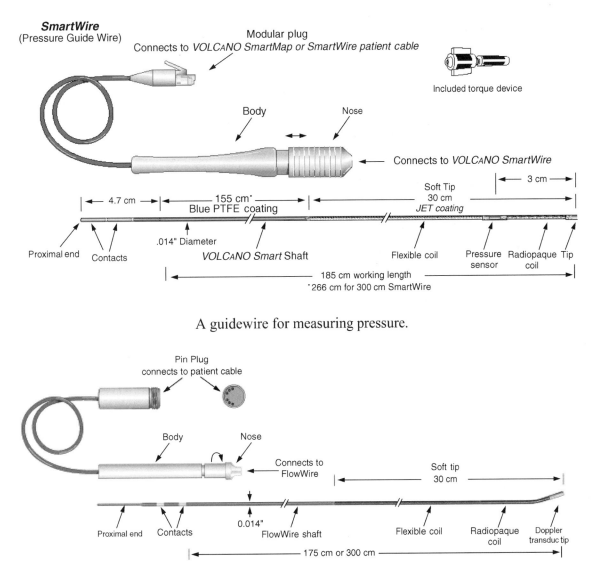

Figure 2. Guidewires for measuring pressure and flow in diseased coronary arteries. (Courtesy of Volcano Therapeutics Inc., CA.)

reduce friction. To dilate a balloon catheter, pressures upto 20 atm can be used. Figure 4a and b show a balloon catheter, having an *over the wire* design, with and without a stent. To generate the pressure, an indeflator is used. It consists of a cylinder, one end of which has a movable plunger and the other end is connected to the dilation catheter (Fig. 4c). The indeflator is partially filled with liquid and has an attached pressure gauge. Dilation catheters have a wide range of inflation diameters ranging from 1.5 to 4 mm depending on the artery dimension. Length of balloons varies from 10 to as much as 40 mm depending on the length of the atherosclerotic lesion. Materials used for manufacturing balloons can be polyethylene, polyolyfin, or nylon to name some, with the wall thickness varying from 0.0003 to 0.0005 in. (0.007 to 0.0127 mm). Additionally, all balloon catheters are sold with a *rated burst pressure*.

Besides the traditional design in which the catheter passes over the entire length of the guidewire, a design known as *monorail* is one in which the catheter passes just on its tip such that quick removal and insertion can be done. Some balloon catheters (*perfusion balloons*) are equipped with side holes at their tips in the shaft proximal and distal to the balloon to allow the blood to flow from the proximal to the distal vessel, when the balloon is inflated. This reduces the risk of ischemia in the heart. Nowadays, PTCA is usually combined with implantation of a stent in the clogged artery to keep the artery open, after balloon removal and reduce the risk of restenosis. The path of approach of catheters and guidewires into the coronaries is usually done via the femoral artery and vein with direction guidance being provided by fluoroscopy. Other paths of approach can be done via the axillary, brachial, or radial artery. For insertion of balloon as well as normal catheters and guidewires,

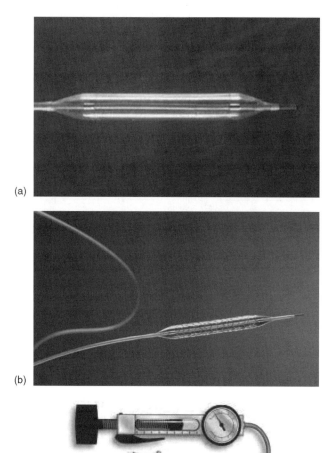

(a)

(b)

(c)

Figure 3. (a) An example of coronary dilation catheter; (b) with stent over it. This is an "over the wire design". (Courtesy of Boston Scientific, MA); (c) An indeflator, which is used to inflate the balloon. (Courtesy of Guidant Corp., IN.)

percutaneous needles, such as seldinger needle for femoral artery, Potts–Cournand needle (which is hollow from inside so that the user can know when the artery has been punctured) and, vascular sheaths are used. In case there is difficulty in detecting the arterial or venous pulse, a smart needle (PSG, Mountain View, CA) may be used, which has a Doppler crystal to direct the needle to the center of the vessel.

Vessel Closure Devices

On completion of PTCA, the punctured vessel needs to be closed to prevent any postprocedural bleeding followed by coagulation. The most commonly used device is a collagen

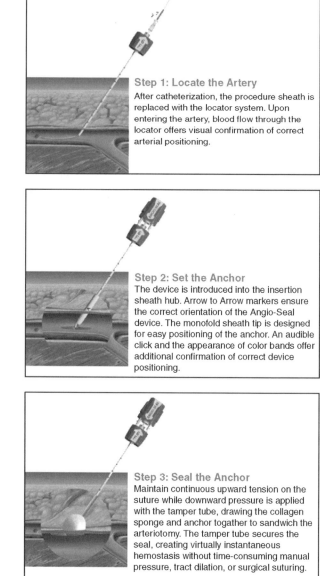

Step 1: Locate the Artery
After catheterization, the procedure sheath is replaced with the locator system. Upon entering the artery, blood flow through the locator offers visual confirmation of correct arterial positioning.

Step 2: Set the Anchor
The device is introduced into the insertion sheath hub. Arrow to Arrow markers ensure the correct orientation of the Angio-Seal device. The monofold sheath tip is designed for easy positioning of the anchor. An audible click and the appearance of color bands offer additional confirmation of correct device positioning.

Step 3: Seal the Anchor
Maintain continuous upward tension on the suture while downward pressure is applied with the tamper tube, drawing the collagen sponge and anchor togather to sandwich the arteriotomy. The tamper tube secures the seal, creating virtually instantaneous hemostasis without time-consuming manual pressure, tract dilation, or surgical suturing.

Figure 4. Vascular closure device, known as Angio-Seal. (Courtesy of Kensey Nash., PA.)

plug applied to skin outside the outer wall of the vessel (6). Another device, called the Hemostatic Puncture Closure Device, also known as Angio-Seal Vascular Closure device, (Kensey Nash, Exton, PA), uses an anchor on the inner wall of the vessel and uses an attached suture to raise a collagen plug to the outer wall of the vessel (7). Figure 5 shows the

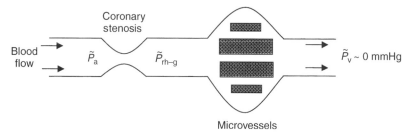

Figure 5. Measurement of myocordial fractional flow reserve is a coronary stenosis.

Table 1. Lesion Geometry and Hemodynamics Before, Intermediate, and After PTCA[a]

Lesion	A_m, mm^2	d_m, mm	% Area Stenosis	l_m, mm	CFR	$\tilde{p}_a$ mmHg	$\Delta\tilde{p}_h$ mmHg	$\tilde{p}_{rh}$ mmHg	FFR_{myo}
Before-PTCA	0.7 ± 0.1	0.95	90	0.75	2.3 ± 0.1	89 ± 3	34	55	0.62
Intermediate	1.43	1.35	80	0.75	3.3	86	14.3	70.4	0.82
After-PTCA	2.5 ± 0.1	1.80	64	3	3.6 ± 0.3	84 ± 3	7.4	75.2	0.89

[a]The parameters A_m, d_m, and l_m are the area, diameter, and length of the narrowest region of the stenoses; $\tilde{p}_a$ and $\tilde{p}_{rh}$ are the mean arterial pressure in the coronary ostium and distal to the stenosis under hyperemia; $\Delta\tilde{p}_h$ is the hyperemic pressure drop across the stenosis.

procedure followed to close the vessel using Angio-Seal. Another device, called the Prostar device (Perclose, Redwood City, CA) uses a sheath-like device to pass a suture around the puncture through the skin to close the puncture.

DIAGNOSTICS WITH GUIDEWIRES

With widespread use of PTCA, there has been a surge in the use of guidewires for evaluation of ischemic severity of focal and diffuse lesions before and after angioplasty. Since their inception, several authors (10–13) have validated the usefulness of guidewires in clinical procures, including PTCA. However, during coronary intervention, introduction of guidewire itself produces an additional resistance to blood flow that has not been well documented.

Issues

Current usage of guidewires for coronary stenoses diagnostics is for measurement of FFR_{myog} and CFR_g. By definition (4,12), FFR_{myog} is the ratio of mean distal pressure ($\tilde{p}_{rhg}$) to mean pressure proximal to the stenosis ($\tilde{p}_a$) at hyperemia, which is induced by administration of vasodilator agents (e.g., adenosine). Since pressure drop in normal epicardial vessels is very small, $\sim \tilde{p}_a$ mean aortic pressure. Likewise, CFR_g is the ratio of mean coronary flow at hyperemia ($\tilde{Q}_{hg}$) to mean coronary flow at basal (i.e., rest) ($\tilde{Q}_b$) (13).

A value of $FFR_{myog} = 0.75$ is assumed to accurately discriminate stenosis whether or not associated with inducible ischemia (4,12). Both FFR_{myog} and CFR_g increase after coronary angioplasty, thereby signifying an enhanced and normal blood supply to distal myocardium. However, in the presence of microvascular disease, FFR_{myog} and CFR_g measured alone cannot dissociate an epicardial coronary lesion from distal microvascular disease (12,14). To address the nonuniformity of microvascular circulation, application of Relative Coronary Flow Reserve, $rCFR_g$ (13,15) was proposed. However, in patients in whom a stenotic artery supplies an area of myocardial infarction, neither CFR_g nor $rCFR_g$ can differentiate flow impairment due solely to a stenosis.

Currently, guidewires of size 0.014 in. (0.355 mm) are capable of measuring both flow (CFR_g) and mean pressure drop, $\Delta\tilde{p}$ (and FFR_{myog}) across a stenosis. However, the introduction of a guidewire causes an obstructive effect, creating an "artifactual" stenosis (16–19). The threshold limit of $FFR_{myog} = 0.75$ is a measured value with guidewire. However, this measured value of 0.75 must be attributed to FFR_{myog} and not FFR_{myo}, the value for the lesion without guidewire obstruction. Limited information is

available on what degree of flow blockage exists with currently used guidewires, although clinical investigators have acknowledged the limitations of mean pressure drop and flow measurements because of flow obstruction produced by guidewires (20,21). In the following sections, the authors present a summary of their past studies on, (1) quantifying the flow obstruction effect of guidewires of diameter 0.014 in. (0.35 mm) and 0.018 in. (0.46 mm), which results in enhanced $\Delta\tilde{p}_h$ and reduced $\tilde{Q}_h$ in a significant, intermediate and moderate focal stenoses; and (2) corrections to be applied to FFR_{myog} and CFR_g to get true values of FFR_{myo} and CFR without guidewire, thus improving the diagnosis of focal coronary lesions.

To evaluate the flow obstruction effect, stenoses geometry of a focal pre-PTCA (22,23) and post-PTCA lesion (24,25) were obtained from the in vivo data set of Wilson et al. (21) in a 32 patient group. The patients had single-vessel, single-lesion coronary artery disease with unstable or stable angina pectoris. Dimensions and shape of the coronary stenosis before and after angioplasty were obtained from quantitative biplanar X-ray angiography. Biplane angiography of each lesion in orthogonal projections (60° left anterior oblique and 30° right anterior oblique) resolved vessel widths with cross-sectional area calculated from the equation for an ellipse, which were converted to mean diameters. The measured mean values $\pm$ SD of minimal area stenosis (A_m), mean pressure measured proximal to the stenoses at the ostium ($\tilde{p}_a$), CFR by Wilson et al. (21) and dimensions are summarized in Table 1 and Fig. 6. Patients with abnormalities that might affect the vasodilator capacity of the arteriolar vasculature were excluded from the study (21). Measured values of CFR with a 3F pulsed Doppler ultrasound catheter ($d_i = 1.0$ mm) with tip positioned proximal to the lesions (with minimal flow blockage) increased from $2.3 + 0.1$ to $3.6 + 0.3$ in the procedure; mean arterial pressure, measured in the coronary ostium, decreased from $89 + 3$ to $84 + 3$ mmHg (21). In the flow analysis, the residual composite lesion was assumed to have a smooth, rigid plaque wall, and round concentric shape. Additional dimensional data on the shape of similar size lesion are from Back and Denton (26).

Additionally, an intermediate stenosis (27) having maximal area blockage based on minimal diameter = 80% was used in this study [the dimensions of which are given in Table 1 and were obtained from Back and Denton (26)] to include a wide range of lesion sizes for obtaining the correlations. However, this intermediate lesion size was not measured by Wilson et al. (21). Further, for guidewire analyses, the guidewire was placed concentrically within the lesion. The concentric configuration of the guidewire within the lesion may give the largest pressure drop

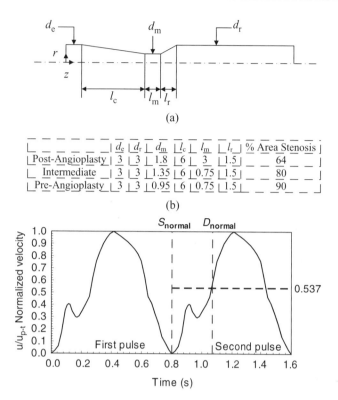

Figure 6. (a) Stenoses geometry showing the shape and dimensions in mm; (b): Normal coronary flow waveform $\tilde{u}/\bar{u}_{p-t}$ versus t, where S_{normal} indicates the beginning of systole and D_{normal} indicates the beginning of diastole.

(1). The geometry of the native, intermediate, and moderate lesion with and without guidewire is shown in detail in Fig. 7a–c .

The coronary velocity waveform $\bar{u}(t)$ (spatially averaged at each time across the cross-sectional area) used in the flow analyses was obtained in our laboratory from *in vitro* calibration (28), smoothing the fluctuating Doppler signal, and phase shifting the normal pattern for the proximal left anterior descending (LAD) artery. In Fig. 5b, the peak diastolic velocity $\bar{u}_{p-t}$ corresponds to a normalized velocity of 1.0, so that the mean peak velocity ratio $\tilde{\tilde{u}}/\bar{u}_{p-t}$ is 0.537, as shown by the dashed line.

With guidewire inserted concentrically, in the proximal vessel, the spatial velocity profile in the annular gap was taken to be the analogous Poiseuille flow relation for the axial velocity u (22,25):

$$u/2\bar{u} = [(1-(r/r_o)^2)\ln(r_o/r_i) + (1-(r_i/r_o)^2)\ln(r/r_o)]/ \tag{1}$$
$$[(1 + (r_i/r_o)^2)\ln(r_o/r_i)-(1-(r_i/r_o)^2)]$$

where u is a function of r and t. Without the guidewire in the proximal vessel, the spatial velocity profile was initially taken to be the Poiseuille flow relation for the axial velocity u:

$$u/2\bar{u} = (1-(r/r_o)^2) \tag{2}$$

The Carreau model, given by Eq. 3, was used for shear rate dependent non-Newtonian blood viscosity with the local shear rate (Eq. 4) calculated from the velocity gradient

through the second invariant of the rate of strain tensor (35).

$$\eta = \eta_\infty + (\eta_0 - \eta_\infty)\left[1 + (\lambda\dot{\gamma})^2\right]^{(n-1)/2} \tag{3}$$

$$\dot{\gamma} = \sqrt{\frac{1}{2}\left[\sum_i \sum_j \dot{\gamma}_{ij}\dot{\gamma}_{ji}\right]} \tag{4}$$

where $\eta_\infty = 0.00345$ Pa·s $\eta_0 = 0.056$ Pa·s, $\lambda = 3.313$ s and $n = 0.3568$.

Details of the numerical method used to calculate the pulsatile hemodynamics in coronary artery and lesions with and without guidewire were previously described by Banerjee et al. (22–24). A typical basal physiological value $\tilde{Q}_b = 50$ mL · min^{-1} for a coronary vessel of 3 mm size was used (30). The cycle time of 0.8 s and density of blood $\rho = 1.05$ g·cm^{-3} was used. In the Reynolds number (Re), a kinematic viscosity of $v = 0.035$ cm^2·s was used, a value near the asymptote in the Carreau model for blood ($\eta_\infty \rightarrow 0.00345$ Pa · s as $\dot{\gamma} \rightarrow \infty$), which gives $v_\infty \rightarrow 0.033$ cm^2·s. The Womersley number varied from 1.9 with guidewire size 0.35–2.25 mm in the pathophysiological scenario without guidewire.

CFR and $\tilde{p}_{rh}$ without guidewire (21–25) and computed values of CFR and $\tilde{p}_{rh-g}$ for the different lesions with guidewire (0.35 and 0.46 mm) were used to construct the maximal vasodilation-distal perfusion pressure curve, also known as CFR $- \tilde{p}_{rh}$ relationship. The maximal CFR $- \tilde{p}_{rh}$ curve was plotted by joining the measured (21), and computed values of CFR and $\tilde{p}_{rh}$ at hyperemia for the native, intermediate, and residual lesions after angioplasty since blood was supplied to the same distal vasculature, which was originally with marked arteriolar dilation. The CFR $- \tilde{p}_{rh}$ relationship was then used to construct the correlations between FFR$_{myo}$ and FFR$_{myog}$, and CFR and CFR$_g$ in native, intermediate, and residual lesions for the two guidewires. The linear CFR $- \tilde{p}_{rh}$ was also extrapolated toward its origin to estimate zero-coronary flow mean pressure ($\tilde{p}_{zf}$) for the Wilson et al. (21) patient group.

Diagnostics with Angioplasty Catheters

As an initial step, the authors calculated the $\Delta\tilde{p} - \tilde{Q}$ relationship post-PTCA (Curve J in Fig. 8), in conjunction with the pressure measurements, using the angioplasty catheter ($d_i = 1.4$ mm) before the development of small guidewire sensors (31). For resting conditions with the catheter present, flow was believed to be ~40% of normal basal flow in the absence of the catheter, and for hyperemia, ~20% of elevated flow in the patient group. Also, $\Delta\tilde{p}$ was significantly elevated in the tighter artifactual stenoses during the measurements. The above diagnostic measurements were compared with the pathophysiologic scenario, having no angioplasty catheter. The results of pathophysiologic conditions cannot be measured in lesions, and are descriptive of the unperturbed conditions that may have existed on average in the patient group after PTCA. In the absence of angioplasty catheter, the calculated $\Delta\tilde{p}$ was only ~1 mmHg (0.133 kPa) at basal flow, and increased moderately to ~7.4 mmHg for hyperemic flow measured proximally (CFR = 3.6) with minimal blockage. On the other hand, with the catheter, $\Delta\tilde{p}$ was ~28.7 mmHg for the basal flow showing an order of magnitude increase in $\Delta\tilde{p}$.

Increased Pressure Drop and Reduced Hyperemic Flow Due to the Presence of Guidewire

Tables 1 and 2 give the mean pressure drop $\Delta\tilde{p}_h$ and distal mean pressure $\tilde{p}_{rh}$ at hyperemic condition in native,

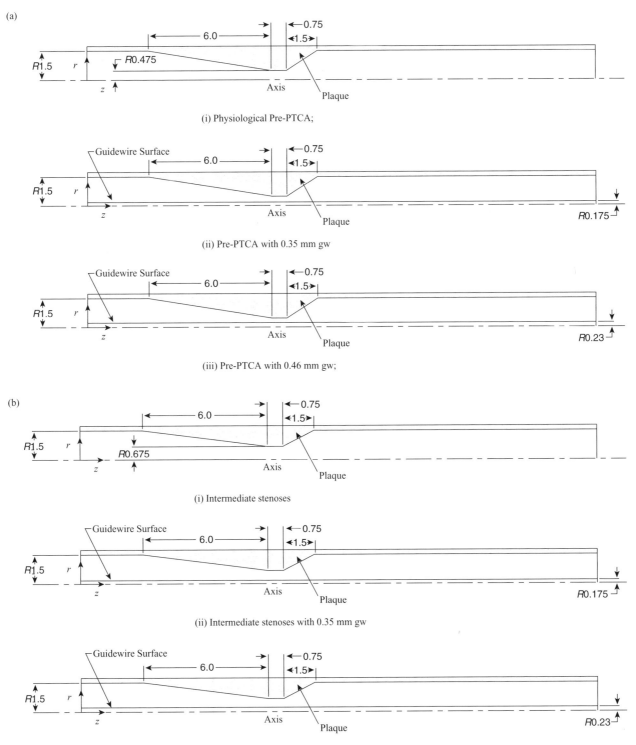

(a)

(i) Physiological Pre-PTCA;

(ii) Pre-PTCA with 0.35 mm gw

(iii) Pre-PTCA with 0.46 mm gw;

(b)

(i) Intermediate stenoses

(ii) Intermediate stenoses with 0.35 mm gw

(iii) Intermediate stenoses with 0.46 mm gw

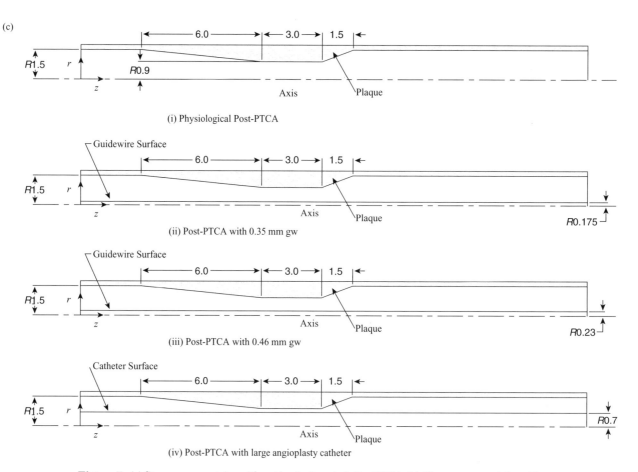

(i) Physiological Post-PTCA

(ii) Post-PTCA with 0.35 mm gw

(iii) Post-PTCA with 0.46 mm gw

(iv) Post-PTCA with large angioplasty catheter

Figure 7. (a) Stenoses geometries with guidewire inserted: Pre-PTCA; (b): Stenoses geometries with guidewire inserted: Intermediate; (c) Stenoses geometries with guidewire inserted: Post-PTCA.

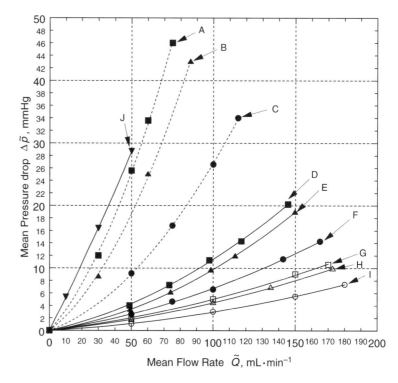

Figure 8. Mean pressure drop $\Delta \tilde{p}$ versus mean flow rate $\tilde{Q}$ in pre-PTCA, intermediate, and post-PTCA lesion with (0.35 and 0.46 mm) and without guidewire.

Table 2. Hyperemic Mean Pressure Gradients $\Delta \tilde{p}_{gh}$ and Distal Mean Pressure $\tilde{p}_{rh}$ Before, Intermediate, and After Intervention with Guidewire[a]

Lesion	$\tilde{p}_a$, mmHg	Guidewire size: $d_i = 0.35$ mm				Guidewire size: $d_i = 0.46$ mm			
		$\frac{d_i}{d_m}$	$\tilde{Q}_{gh}$ mL·min^{-1}	$\Delta \tilde{p}_{gh}$, mmHg	$\tilde{p}_{rh}$, mmHg	$\frac{d_i}{d_m}$	$\tilde{Q}_{gh}$, mL·min^{-1}	$\Delta \tilde{p}_{gh}$, mmHg	$\tilde{p}_{rh}$, mmHg
Before-PTCA	89	0.37	86.0	43.0	46.0	0.48	75.0	46.0	43.0
Intermediate	86	0.26	149.7	18.9	65.8	0.34	145.6	20.2	64.5
After-PTCA	84	0.19	172.5	9.9	72.8	0.26	170.0	10.6	72.1

[a]Values obtained from linear CFR-$\tilde{p}_{rh}$ curve.

intermediate and residual lesions with and without guidewire (22–26). The ratio of guidewire diameter d_i to throat diameter d_m for different stenoses are given in Table 2. From Tables 1 and 2, overall guidewires of size 0.35 and 0.46 mm caused 30–45% overestimation in $\Delta \tilde{p}_h$ in the three stenoses as hyperemic flow was reduced by 5–37% from the physiologic condition without guidewire obstruction. While diagnosis of severely blocked arteries is relatively easier and additional pressure drop due to the guidewire does not affect the diagnosis that much, it can be seen that additional pressure drop due to the guidewire could have an important role in clinical evaluation of moderate and intermediate stenoses. Further, it can be seen that with guidewire of size 0.35 and 0.46 mm, FFR$_{myog}$ < FFR$_{myo}$ and CFR$_g$ < CFR at hyperemia. The maximum decrease was observed in the native stenoses. The residual stenoses (i.e., post-PTCA) showed the least decrease. In summary, Fig. 8 shows the mean pressure drop from basal to hyperemic for the three stenoses with and without 0.35 and 0.46 mm guidewire.

Figure 9 shows the maximal CFR $-\tilde{p}_{rh}$ for the native, intermediate, and residual lesions after angioplasty with and without guidewire (27). Distal mean perfusion pressure for hyperemic flow $\tilde{p}_{rh}$ increased from 55 mmHg in pre-PTCA without guidewire to ~75 mmHg (9.99 kPa) in post-PTCA without guidewire. A $\tilde{p}_{rh}$ of 55 mmHg (7.33 kPa) is indicative of subendocardium ischemia.

Extrapolation of the nearly linear CFR $-\tilde{p}_{rh}$ relation toward its origin gave a zero-coronary flow mean pressure ($\tilde{p}_{zf}$) of ~20 mmHg (2.66 kPa). This value is near a measured value (32) of 18 mmHg (2.39 kPa), where myocardial blood flow ceased in the subendocardium layer of dog hearts, which were maximally dilated by infusion of adenosine.

FFR$_{myo}$-FFR$_{myo\text{-}g}$ and CFR–CFR$_g$ Correlations

Table 3 provides the values of CFR, CFR$_g$, FFR$_{myo}$, and FFR$_{myog}$ for the different stenosis configurations (27). Figures 10 and 11 show the correlation plots between

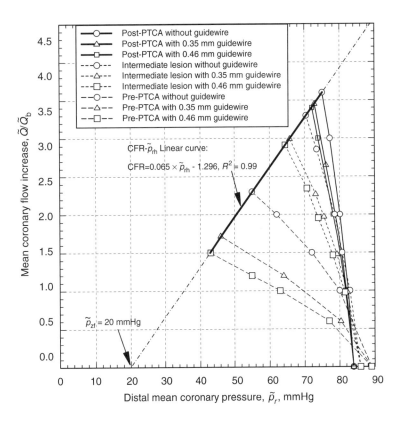

Figure 9. Relative mean coronary flow rate increase ($\tilde{Q}/\tilde{Q}_b$) versus distal mean coronary pressure $\tilde{p}_r$ before and after intervention. The maximum vasodilation-distal perfusion pressure relation (CFR- $\tilde{p}_{rh}$) is shown by the nearly linear solid line.

Table 3. Effect of the Presence of Guidewire on the Values of CFR and FFR$_{myo}$ at Hyperemic Conditions in Comparison to Values of CFR and FFR$_{myo}$ Under Pathophysiological Condition for Different Guidewire Sizes $d_i = 0.35$ mm (0.014 in.) and 0.46 mm (0.018 in.)^a

	Physiological		0.35 mm Guidewire		0.46 mm Guidewire	
Lesion	CFR	FFR$_{myo}$ with $\tilde{p}_v \sim 0$	CFR$_g$	FFR$_{myo\text{-}g}$ with $\tilde{p}_v \sim 0$	CFR$_g$	FFR$_{myo\text{-}g}$ with $\tilde{p}_v \sim 0$
Before-PTCA	2.3	0.62	1.72	0.52	1.50	0.48
Intermediate	3.3	0.82	2.99	0.76	2.91	0.75
After-PTCA	3.6	0.89	3.45	0.87	3.40	0.86

^{a}FFR$_{myo\text{-}g}$ is FFR$_{myo}$ measured with guidewire.

CFR and CFR$_g$, and between FFR and FFR$_g$ with their *linear* regression lines. In Fig. 10, CFR was related to CFR$_g$ by the equation: with guidewire size 0.46 mm, CFR = CFR$_g$ × 0.689 + 1.271 ($R^2 = 0.99$), and with guidewire size 0.35 mm, CFR = CFR$_g$ × 0.757 + 1.004 ($R^2 = 0.99$). With central venous pressure ($\tilde{p}_v$) ~ 0, as used in present clinical practice, FFR$_{myo}$ and FFR$_{myog}$ correlated equally well (Fig. 11): with guidewire size 0.46 mm, FFR$_{myo}$ = FFR$_{myog}$ × 0.737 + 0.263 ($R^2 = 0.99$), and with guidewire size 0.35 mm, FFR$_{myo}$ = FFR$_{myog}$ × 0.790 + 0.210 ($R^2 = 0.99$), which gave FFR$_{myo}$ = 0.8 for a measured FFR$_{myog}$ = 0.75. The study showed that the correlations for FFR and CFR for guidewire size 0.35 mm are closer to the ideal (expected) relationship, that is, CFR = CFR$_g$ and FFR = FFR$_g$ due to relatively lesser flow obstruction effect than guidewire size 0.46 mm.

Usefulness of FFR$_{myo\text{-}g}$-FFR$_{myo}$ and CFR-CFR$_g$ Correlations

Though there are uncertainties associated with measurements of CFR with guidewires in diagnostic procedures as these measurements are made distal to the stenosis, these corrections could be useful in measuring CFR and FFR simultaneously. This could provide useful information about the status of both the epicardial stenosis and distal microcirculation during the procedure. In light of this, CFR measurement proximal to the stenosis will be more accurate as this is the region of more stable flow. Further, the heterogeneity of stenoses could produce variations in the slope of the correlations. However, the correlations were obtained based on stenosis shape and size measured invasively by Wilson et al. (21) in a group of patients, and have shown a direct and stronger relation to the minimal stenosis area as compared to the overall length and shape of the stenosis. Further clinical evaluation with a larger patient group will be needed to confirm the dominant effect of minimal stenosis area on these correlations.

While several studies have focused on the relationship between CFR measured with guidewire and CFR measured with a noninvasive flow probe, for examples a Doppler cuff in animal studies, no similar study has been done to distinguish the effect of guidewire on FFR$_{myo}$ as pressure

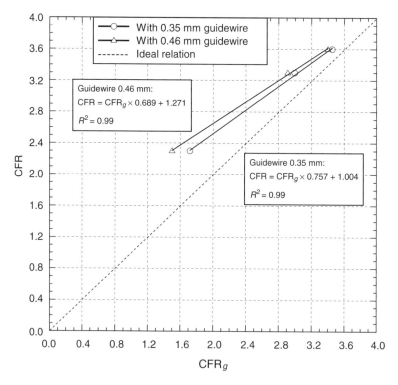

Figure 10. CFR versus CFR$_g$ correlation. Dotted line shows the ideal CFR versus CFR$_g$ relation.

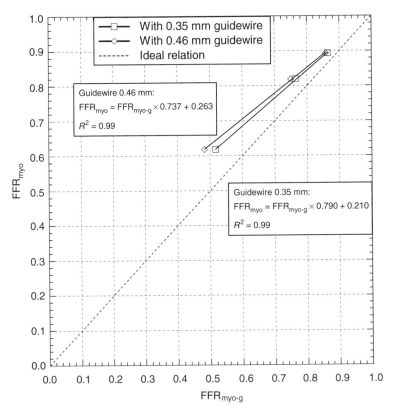

Figure 11. FFR_{myo} versus FFR_{myog} correlation (with $\bar{p}_v \sim 0$). Dotted line shows the ideal FFR_{myo} versus $FFR_{myo\text{-}g}$ relation.

can be measured invasively using guidewires only. Earlier *In vitro* measured Δp data reported by DeBruyne et al. (33) for steady flow of a saline solution through various blunt hollow plug stenosis models (50–90% area stenosis; $d_e = d_r = 4$ mm; $l_m = 10$ mm) with and without a small guidewire pressure sensor ($d_i = 0.38$ mm) also indicated appreciable overestimation of the true Δp as the severity of stenoses increased. In particular, for a 90% area stenosis ($[d_i/d_m] = 0.30$) at a flow rate $Q = 120$ mL·min^{-1}, the measured ratio ($\Delta p_g/\Delta p$) increased by the factor 23/16 mmHg (3.06/2.13 kPa) ($\sim 40\%$) due to the increased flow resistance with the guidewire spanning the model stenosis in the rigid plastic tube $d_e = 4$ mm. Our detailed hemodynamic analysis in a smaller coronary vessel ($d_e = 3$ mm; 90% area stenosis; larger ($[d_i/d_m] = 0.48$) also indicated significant increases in flow resistance with the guidewire present (Tables 1 and 2). Similarly, the stenosis resistance $\tilde{R}_h$, defined as ($\Delta p_h / \tilde{Q}_h$), decreased considerably from 0.29 to 0.04 mmHg·mL·min^{-1} from pre- to post-PTCA, respectively (27). With guidewire size 0.35 mm, $\tilde{R}_{hg}$ decreased from 0.5 for pre-PTCA to 0.06 for post-PTCA mmHg·mL·min^{-1} (27). While the 32 patient group in Wilson et al. (21) had normal microvascular function, the presence of microvascular impairment could produce different correlations between FFR_{myo} - FFR_{myog} and CFR–CFR_g with epicardial stenoses, and could be the focus of future studies on guidewire effect and microcirculation.

BIBLIOGRAPHY

1. Dotter CT, Judkins MP. Transluminal treatment of arteriosclerotic obstruction. Description of a new technic and a preliminary report of its application. 1964. Radiology 1989;172:904–920.
2. Baim DS. Percutaneous transluminal coronary angioplasty. Cardiac catheterization, angiography, and intervention. 5th ed. (MA): Williams & Wilkins; 2000. pp 537–580.
3. Dervan JP, McKay RG, Baim DS. The use of an exchange guide wire in coronary angioplasty. Cathet Cardiovasc Diagn 1985;11:207–212.
4. Pijls NH, et al. Measurement of fractional flow reserve to assess the functional severity of coronary-artery stenoses. N Engl J Med 1996;334:1703–1708.
5. Pijls NH, et al. Coronary thermodilution to assess flow reserve: validation in humans. Circulation 2002;105:2482–2486.
6. De Bruyne B, et al. Coronary thermodilution to assess flow reserve: experimental validation. Circulation 2001;104:2003–2006.
7. Siebes M, et al. Single-wire pressure and flow velocity measurement to quantify coronary stenosis hemodynamics and effects of percutaneous interventions. Circulation 2004;109:756–762.
8. Ernst SM, et al. Immediate sealing of arterial puncture sites after cardiac catheterization and coronary angioplasty using a biodegradable collagen plug: results of an international registry. J Am Coll Cardiol 1993;21:851–855.
9. Aker UT, et al. Immediate arterial hemostasis after cardiac catheterization: initial experience with a new puncture closure device. Cathet Cardiovasc Diagn 1994;31:228–232.
10. Gruentzig AR, Senning A, Siegenthaler WE. Nonoperative dilation of coronary artery stenosis: Percutaneous Transluminal Coronary Angioplasty. N Engl J Med 1979;301:61–68.
11. Ganz P, Harrington DP, Gaspar J, Barry WH. Phasic pressure gradients across coronary and renal artery stenoses in humans. Am Heart J 1983;106:1399–1406.
12. Ganz P, et al. Usefulness of transstenotic coronary pressure gradient measurements during diagnostic catheterization. Am J Cardiol 1985;55:910–914.

13. Anderson HV, et al. Measurement of transstenotic pressure gradient during percutaneous transluminal coronary angioplasty. Circulation 1986;73:1223–1230.

14. Gould KL, Kirkeeide R, Buchi M. Coronary flow reserve as a physiologic measure of stenosis severity, part 1: relative and absolute coronary flow reserve during changing aortic pressure and cardiac workload; part II: determination from arteriographic stenosis dimensions under standardized conditions. J Am Coll Cardiol 1990;15:459–474.

15. Gould KL, Lipscomb K, Hamilton GW. Physiologic basis for assessing critical coronary stenosis: instantaneous flow response and regional distribution during coronary hyperemia as measures of coronary flow reserve. Am J Cardiol 1974;33:87–94.

16. Gould KL. Coronary artery stenosis and reversing atherosclerosis. 2nd ed. London: Arnold Publishers; 1999.

17. Pijls NHJ, De Bruyne B. Coronary pressure. 2nd ed. The Netherlands: Kluwer Publishers; 1999.

18. Baumgart D, et al. Improved assessment of coronary stenosis severity using the relative flow velocity reserve. Circulation 1998;98:40–46.

19. Pijls NH, et al. Fractional flow reserve: a useful index to evaluate the influence of an epicardial coronary stenosis on myocardial blood flow. Circulation 1995;92:3183–3193.

20. De Bruyne B, et al. Fractional flow reserve in patients with prior myocardial infarction. Circulation 2001;104(2):157–162.

21. Meuwissen M, et al. Intracoronary pressure and flow velocity for hemodynamic evaluation of coronary stenoses. Expert Rev Cardiovasc Ther 2003;1:471–479.

22. Back LH. Estimated mean flow resistance increase during coronary artery catheterization. J Biomech 1994;27:169–175.

23. Kern MJ, et al. Translesional pressure—flow velocity assessment in patients: part I. Cathet Cardiovasc Diagn 1994;31:49–60.

24. Segal J, et al. Alterations of phasic coronary artery flow velocity in humans during percutaneous coronary angioplasty. J Am Coll Cardiol 1992;20:276–286.

25. Doriot P, Dorsaz P, Dorsaz I, Chatelain P. Accuracy of coronary flow measurements performed by means of doppler wires. Ultra Med Biol 2000;26:221–228.

26. Wilson RF, Laxson DD. Caveat emptor: a clinician's guide to assessing the physiologic significance of arterial stenoses. Cathet Cardiovasc Diagn 1993;29:93–98.

27. Banerjee RK, Back LH, Back MR. Effects of diagnostic guidewire catheter presence on translesional hemodynamic measurements across significant coronary artery stenoses. Biorheology 2003;40(6):613–635.

28. Banerjee RK, Back LH, Back MR, Cho YI. Physiological flow analysis in significant human coronary artery stenoses. Biorheology 2003;40(4):451–476.

29. Banerjee RK, Back LH, Back MR, Cho YI. Physiological flow simulation in residual human stenoses after coronary angioplasty. ASME J Biomech Eng 2000;122:310–320.

30. Sinha Roy A, Back LH, Banerjee RK. Guidewire flow obstruction effect on pressure drop-flow relationship in moderate coronary artery stenosis. Accepted for publication in J Biomech January, 2005.

31. Wilson RF, et al. The effect of coronary angioplasty on coronary flow reserve. Circulation 1988;77:873–885.

32. Back LH, Denton TA. Some arterial wall shear stress estimates in coronary angioplasty. Adv Bioeng ASME BED 1992;22:337–340.

33. Sinha Roy A, et al. Delineating the guidewire flow obstruction effect in assessment of fractional flow reserve and coronary flow reserve measurements. Accepted for publication in Am J Physiol: Heart Circ Physiol February, 2005.

34. Cho YI, Back LH, Crawford DW, Cuffel RF. Experimental study of pulsatile and steady flow through a smooth tube and an atherosclerotic coronary artery casting of man. J Biomech 1983;16:933–946.

35. Cho YI, Kensey KR. Effects of non-Newtonian viscosity of blood on flows in a diseased arterial vessel: Part 1. Steady flows. Biorheology 1991;28:241–262.

36. Back LH, Radbill JR, Crawford DW. Analysis of pulsatile viscous blood flow through diseased coronary arteries of man. J Biomech 1977;10:339–353.

37. Womersley JR. Method for the calculation of velocity, rate of flow and viscous drag in arteries when the pressure gradient is known. J Physiol 1955;127:553–563.

38. Banerjee RK, Back LH, Back MR, Cho YI. Catheter obstruction effect on pulsatile flow rate-pressure drop during coronary angioplasty. ASME J Biomech Eng 1999;121:281–289.

39. Brown BG, Bolson EL, Dodge HT. Dynamic mechanisms in human coronary stenosis. Circulation 1984;170:917–922.

40. Bache RJ, Schwartz JS. Effect of perfusion pressure distal to a coronary stenosis on transmural myocardial blood flow. Circulation 1982;65:928–935.

41. De Bruyne B, et al. Transstenotic coronary pressure gradient measurement in humans: in vitro and in vivo evaluation of a new pressure monitoring angioplasty guide wire. J Am Coll Cardiol 1993;22(1):119–126.

42. Fearon WF, Yeung AC. Evaluating intermediate coronary lesions in the cardiac catheterization laboratory. Rev Cardiovasc Med 2003;4:1–7.

See also ARTERIES, ELASTIC PROPERTIES OF; BRACHYTHERAPY, INTRAVASCULAR; HEMODYNAMICS; INTRAAORTIC BALLOON PUMP.

CPR.   See CARDIOPULMONARY RESUSCITATION.

CRYOSURGERY

ANDREW A. GAGE
State University of New York at Buffalo
Buffalo, New York

INTRODUCTION

Cryosurgery is a method of therapy that uses freezing temperatures to achieve effects on tissue. The term cryotherapy, often used interchangeably with cryosurgery, has broader connotations, including, for example, the application of cold packs to prevent tissue swelling after injury. Cryosurgery is one form of cryotherapy. The term cryoablation is commonly used also, especially in relation to the treatment of tumors.

Cryosurgery may be applied for several different purposes, some related to the adhesion of super-cold metal to tissue and some related to the response of tissue to freezing. For example, in the extraction of cataracts of the eye, a cold instrument or probe is used only to secure a hold on the lens of the eye and facilitate removal as the lens adheres to the probe, which functions as a handle. This technique can be used to facilitate extraction of tumors of diverse sites, including the brain, the eye, and the liver. Cryosurgery also can be used to produce an inflammatory response. For example, in the treatment of retinal detachment, fast freezing of the tissue for a few seconds will damage the tissue, rather than destroy it, and the resultant inflammatory response is expected to heal the detachment. Cryosurgery also can be used for the destruction of tissue, which might be selective, as in the treatment of non-neoplastic disease, or which can be complete, as is needed in the treatment of tumors. The destructive response is the major use of cryosurgery, and this type of response is emphasized in this article.

Cryosurgery is commonly used to destroy tissue by freezing *in situ*. The technique requires the use of cryosurgical apparatus to produce tissue temperatures in the freezing range. The freezing of the tissue is accomplished by the direct application of cryogenic agents or by the use of closed-probe systems in which the cryogen circulates and is not released on the tissue. The diverse techniques of cryosurgery range from the rather easy surface application of the cryogen for the treatment of skin disease to the more complex use by percutaneous application, as for prostatic disease. The diseases that can be treated by cryosurgery range from minor conditions, such as warts, to serious conditions such as advanced cancer. Wherever the disease, the goal of treatment by cryosurgery is controlled production of a predictable area of tissue necrosis.

HISTORICAL DEVELOPMENTS

Local tissue freezing for the treatment of cancer was first used by Dr. James Arnott of London, who described his technique in the year 1850. Using salt solutions containing ice (ca. $-12\,°C$), he produced local freezing by irrigation of advanced cancers in accessible sites, such as the breast and uterine cervix. He described diminution of the size of the tumor and amelioration of pain and drainage. Following his reports, some enthusiasm was generated for the use of cold as an anesthetic agent. Of course, the use of cold for relief of pain had been known since ancient times (1). In Arnott's time, general anesthesia had just been described in America, and its rapidly widening use sharply reduced the usefulness of cold as an anesthetic agent.

In the years from 1870 to 1900, the natural gases were liquefied and Dewar developed a vacuum flask to store cryogenic fluids. These advances permitted development of tissue-freezing techniques. In 1899, A. Campbell White of New York City described the use of liquid air to treat diverse types of skin lesions. Treatment was given by dipping a cotton swab into liquid air and applying the fluid quickly to the skin lesion, using repetitive application to freeze the entire lesion. White also suggested a wash bottle device that sprayed liquid air on skin lesions (Fig. 1).

In 1907, H. H. Whitehouse used freezing techniques for the treatment of a variety of disorders of the skin, including skin cancer. In the same year, Pusey reported similar varied uses of solidified carbon dioxide ($-78.5\,°C$), which was easier to handle and more easily obtained than the liquefied gases. For these reasons, carbonic snow remained in clinical use in succeeding years. In the 1930s, liquid oxygen ($-182.9\,°C$) had some use in the treatment of skin disease, but flammability and related safety considerations precluded general use.

In the 1930s and 1940s, the usefulness of solid carbon dioxide was increased by the development of new instru-

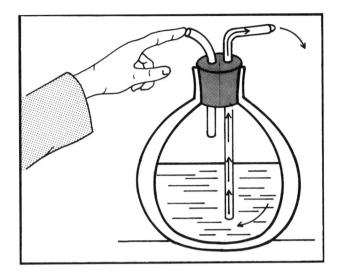

Figure 1. Dr. White's "wash bottle" spray method of ~1900, drawn as described in his article of that year. When the finger is placed over the air outlet, the pressure in the container produced by the boiling of liquid air in the container forces liquid air through the longer glass tube. The spray was then directed at the skin lesion.

ments, generally used to cool metal probes, sometimes with a mixture of solid carbon dioxide and acetone. These devices had little or no advantage over the use of the simple stick of solid carbon dioxide. Fluorinated hydrocarbons, commonly known as Freons, came into clinical use in the 1940s, but their freezing capacity was limited.

During these years, beginning in the 1940s with the extensive experiments of Temple Fay in Philadelphia with localized freezing of cancers by irrigation techniques, a number of reports dealt with the experimental production by local freezing of lesions in tissues, such as the liver or brain. In part, these experiments were made for purposes of physiological studies, but they required that an effort be made to develop instrumentation suitable for those purposes, and they showed the possibility of producing destruction of tissue by localized freezing, using diverse cryogenic agents.

In 1950, liquid nitrogen was introduced into dermatological practice by Allington, who described the use of cotton swabs dipped in this cryogen for the treatment of skin disease. As the availability of liquid nitrogen increased, cryosurgery gained in popularity. Nevertheless, cryosurgery remained a rather unimportant therapeutic modality because the freezing capability of cryogenic agents applied topically with cotton swabs was limited. Experimental studies showed that the depth of freezing with the swab techniques was ~2 mm when liquid nitrogen was used as the cryogen, and carbon dioxide was even less effective. This depth of freezing was about the same as the thickness of normal skin, so the technique was suitable only for superficial lesions of the skin.

Cryosurgery as a therapeutic technique received a major stimulus through the development of cryosurgical apparatus by Cooper and Lee in 1961 (2). The apparatus used liquid nitrogen ($-196\,°C$) in a closed system, which permitted continuous and rapid extraction of heat from tissues. The apparatus was originally designed to produce a cryogenic lesion in the brain for the treatment of Parkinsonism and other neuromuscular disorders. However, it was obvious that the apparatus had wider usefulness and, therefore, it was modified quickly and applied to the treatment of other types of diseases in diverse sites. In the following years, several types of cryosurgical apparatus using liquid nitrogen and other cryogenic agents were developed and found areas of usefulness in diverse benign and neoplastic conditions.

In the 1970s, some uses of cryosurgery were virtually abandoned, in part because of competition with other methods of local treatment, such as lasers and electrocoagulation. Other uses of cryosurgery became standard practice, especially in easily accessible areas, such as the skin and uterine cervix. In the 1980s, cryosurgical techniques for cardiac tachyarrhythmias were developed, but progress in other areas was slow.

In the 1990s, renewed interest in cryosurgery followed the development of intraoperative ultrasound, the improvement in cryosurgical apparatus, and the availability of percutaneous access techniques. The ultrasound image identified the site of the lesion, guided the placement of the probe into the lesion, and monitored the process of freezing. More types of cryosurgical apparatus were available, were well suited to percutaneous or endoscopic use, and permitted the simultaneous use of multiple probes. As a result, cryosurgical techniques have evolved into new applications, such as the treatment of visceral and other deep tumors. A recently written history of cryosurgery provides greater detail of its evolution, including the pertinent references (3).

EFFECT OF FREEZING ON TISSUE

All types of cells can be devitalized by freezing. The mechanisms of injury are related to crystallization of water, solute concentration in the cells, and irreversible changes in cell membranes. In the absence of cryoprotective agents, which are used when freezing is used to preserve cells, ice crystal formation produces damage that makes cell survival unlikely. Intracellular ice formation, which occurs when cells are frozen rapidly, is lethal. Extracellular ice formation, which occurs when cells are frozen slowly and water has sufficient time to leave the cell, is not considered as certainly destructive, but the cellular water loss does result in hypertonic damage to the cells. Under cryosurgical conditions, both mechanisms of injury are operative.

Damage from direct cellular injury is enhanced by the effect of freezing on the circulation. After freezing and thawing, the involved area becomes congested, and effective circulation through small vessels ceases within ~30 min. With stagnation of the microcirculation, the hypoxic cells die and necrosis follows. These effects are well known from studies of frostbite, in which the relative importance of direct cellular injury and vascular stasis have long been debated. The full scope of the effects of freezing on tissue is described in recent reviews (4–6). Though direct cell injury is important, in cryosurgery, microcirculatory failure

clearly is a major factor in cell death: The loss of blood supply deprives all cells in frozen tissues of any possibility of survival. This results in a uniform necrosis of the tissue, except at the periphery of the previously frozen area. At the periphery of the cryogenic lesion, where the freezing temperature is not sufficiently cold to kill all of the cells, some cells survive and some cells linger between life and death for days and may die showing signs of apoptosis, that is, gene-regulated cell death (7,8).

The cryogenic lesion is characterized by sharply circumscribed necrosis. As thawing takes place, the previously frozen area becomes edematous and discolored due to congestion and perivascular hemorrhage. At the periphery of the dark red area, which closely corresponds to the margins of the previously frozen tissue, a narrow, bright red zone due to hyperemia appears. Further evidence of injury develops slowly, and the later tissue response depends on the severity of freezing. If only the tissue is subjected to mild superficial freezing, as might be done for benign disease of the skin, then the response will range from inflammatory reaction to superficial necrosis. The more extensive freezing required by large tumors is followed by greater destruction of tissue. Sharply demarcated necrosis becomes apparent in ~2 days. The time required for slough of the necrotic tissue depends in part on its stroma. Cellular tissue sloughs quickly, but skin and other tissues with large quantities of fibrous stroma resist structural change and the necrotic tissue requires many days for separation. In the skin, the eschar requires two more weeks for separation, leaving a clean, granulating wound, which heals at a normal rate. The delay in healing that is characteristic of cryosurgical wounds is due to the time required for separation of the necrotic tissue. Healing is commonly favorable with rather little scarring. However, whenever full thickness of skin is lost, some scarring is inevitable. Hyperplastic scars are unusual. Depending on the severity of injury, the pigmentation of the treated area may be diminished or lost. Sometimes increased skin pigmentation at the periphery of the injured area is seen as a result of increased melanoblastic activity, but this is only temporary. In deep tissues, such as the viscera, the necrotic tissue is slowly absorbed over weeks or even longer, depending upon the volume of tissue frozen. Scarring is minimal.

Though tissues are devitalized by freezing, the matrix or structure of the tissue may be little changed, and this preservation of the framework is important in later repair. The resistance of the collagen fibers in the skin to damage by freezing is important to the reparative process (9). It is manifest in the favorable healing frequently seen in the treatment of skin disease and in the peripheral nerves after freezing. Though degeneration of axons and Schwann cells occurs, the perineurium is preserved, and this serves as a pathway for regrowth of axons, leading to eventual return of nerve function. Similar effects follow the freezing and repair of major blood vessels. Larger blood vessels, such as the aorta, femoral arteries, carotid artery, and portal vein, are devitalized by freezing *in situ*. With thawing, the previously frozen vessel is slightly dilated due to loss of tone, but the function as a blood conduit is unimpaired. The endothelium is lost, but the stroma of the vessel wall serves as the matrix for repair, commonly with some intimal thickening (10).

The effect of freezing on bone is of particular pertinence. Bone devitalized *in situ* by freezing is slowly resorbed and simultaneously replaced with new bone, a lengthy healing process that may take many months, depending on the volume of bone, similar to that which occurs with autogenous bone grafts. During repair, the devitalized bone maintains form and continues function, though bones subjected to considerable stress (as the femur) are susceptible to fracture in the first month or two when bone resorption is maximal (11). This favorable reparative response has permitted extensive freezing of bone tumors in order to avoid excision.

APPARATUS

A wide variety of cryosurgical apparatus, using diverse cryogens, such as liquid nitrogen, nitrous oxide, argon, and carbon dioxide, is available. Various Freons, which were used for cryosurgery in past years, are no longer used because of environmental concerns. The types vary from electronically controlled automated apparatus with probe heaters to inexpensive hand-held devices that are little more than thermos bottles with controls for cryogen flow. The cooling is produced by change in the phase of the cryogen, that is, evaporation of a liquid or solid, or by expansion of compressed gas through a small orifice [Joule–Thomson (J–T) effect]. Thermoelectric cooling (Peltier effect), produced by passing direct current through dissimilar metal junctions (thermocouples), has not been useful in cryosurgery.

Currently, most apparatus use liquid nitrogen ($-195.8\,°C$), argon ($-185.9\,°C$), or nitrous oxide ($-89.5\,°C$). Carbon dioxide, which sublimes at $-78.5\,°C$, though commonly used in past years, is little used in current times. Liquid nitrogen cools by change in phase, that is, changing from a liquid to a gas. Argon, nitrous oxide, and carbon dioxide are used as pressurized gases that cool by the J–T effect. The freezing capability of these cryogenic agents varies substantially, and this determines the choice of equipment for a particular disease. The cryogenic agents may be applied directly to the tissue, typically as a spray of liquid nitrogen, though nitrous oxide has been used in this way also. Freezing with liquid nitrogen applied via a cotton swab is another example of direct use. The cryogens may also be used in probes, which are a means of confining the cryogen in a closed system. At its tip, the metal probe has a heat-exchange surface that is applied to the tissue to be frozen. At this contact area, heat transfer to the probe results in tissue cooling. The freezing capacity varies with the size of the probe, the temperature of the probe, and the areas of the contact with the tissue. The heat removing capacity of probes varies from 10 to 100 W, depending on the features of probe construction and the cryogen that cools it.

The coldest cryogenic agent in clinical use is liquid nitrogen. It has the greatest freezing capability and is the best agent for destruction of large volumes of tissue, as is required in the treatment of cancer. Nevertheless, in current practice, pressurized argon is commonly used for the treatment of neoplastic disease. The use of multiple probes simultaneously compensates for the somewhat lesser freezing capability of argon. The other cryogens are useful for less serious lesions, such as nonneoplastic or benign neoplastic disease, for which lesser degrees of freezing will suffice (Table 1).

Liquid nitrogen is a clear fluid that is odorless and nonflammable. Its boiling point is $-195.8\,°C$ at atmospheric pressure. The liquid will expand to 750 times its volume under normal atmospheric pressure. It must be stored in a double-walled, vacuum-insulated container with provision for pressure relief and liquid nitrogen withdrawal. The most popular containers for office or clinic use have a capacity of 15–35 L, which provides a holding time of 60–90 days and, depending on the rate of use, will require refilling every 4–6 weeks. Liquid nitrogen will evaporate at a rate of a few percentage points per day, depending on the quality of the container. Withdrawal devices, basically spigots, are used to transfer the liquid nitrogen from the storage container to the cryosurgical instrument.

Hand-held cryosurgical apparatus, weighing ∼1 kg when empty, are commonly used, especially in dermatological practice (Fig. 2). They are basically small containers (thermos bottle construction) with storage capacities of 250–500 mL and with suitable on–off controls to initiate and control the spray of liquid nitrogen. Some devices allow pressure to build up in the container by means of a heat exchanger in the wall or top. Most have Luer lock fittings in the nozzle so that a variety of spray apertures, needles, or nozzles can be attached. These range in size from 24 to 15 gauge. The smaller aperture sizes are hindered by a tendency to become occluded by the development of frost in use, but this can be alleviated by bypass nozzle systems. Problems that must be solved in the construction of hand-held devices include the design of a delivery tube that permits adequate heat exchange and prevents drip of liquid nitrogen from the delivery system nozzle. Most hand-held devices can be fitted with probes of diverse shapes and sizes, including those suitable for treatment of oral or gynecological diseases. The hand-held devices, relatively heavy when the container is filled with liquid nitrogen, are somewhat more difficult to use with a probe than with a spray because it is cumbersome to hold the weight steady in the hand while the probe is adherent to the tissue. Motion may cause fracture of the bond between probe and tissue, and ineffective freezing may result. With the spray technique, the small movement caused by the weight in the hand is not important because there is no direct bond to tissue.

The automated apparatus, cooled by liquid nitrogen, available from several companies, is almost always used with probes. The feed lines leading to the probes are insulated, usually by vacuum (Fig. 3). The control of the flow of liquid nitrogen is achieved by regulators. Most systems require pressurization, which is facilitated and speeded with an internal heating device. A heater in the probe tip speeds release from the tissue at the conclusion of freezing.

The automated apparatus first available in the early 1960s provided for control of the probe temperature in the

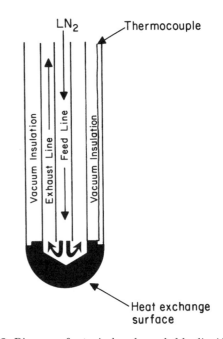

Figure 3. Diagram of a typical probe cooled by liquid nitrogen. The liquid nitrogen passes from a reservoir in the console, down the vacuum-insulated feed line, to the heat exchange surface at the probe tip. The probe tip is cooled by a change in phase (liquid to gas), then the gas is returned to the console. Probes of many sizes and shapes are available. A thermocouple may be placed in the probe tip.

Figure 2. Modern handheld liquid nitrogen cryosurgical unit of the type commonly used in the treatment of skin diseases. The taller unit is 11 in. (280 mm) in height, has a capacity of 16 oz (500 mL), and weighs 30 oz (846 g) when filled. The shorter unit is 8.5 in. (215 mm) in height, has a capacity of 10 oz (300 mL), and weighs 24 oz (618 g) when filled. The devices may be fitted with a large selection of cryogen spray tips with apertures ranging from 0.4 to 1 mm (shown in Fig. 2b) and probes ranging in size from 1 mm to 4 mm in diameter at the tip. The probes allow precise control of freezing and may be used for cutaneous or mucosal lesions. (Photographs by courtesy of Brymill Cryogenic Systems, Ellington, CT 06029.)

range of +36 to $-160\,°C$ (2). The apparatus was soon modified to enable treatment of diverse tumors (12). In the 1990s, an advance in cryosurgical equipment technology featured the development of a technique to use liquid nitrogen super cooled to ca. $-209\,°C$ in new equipment. The super cooling was achieved by passing pressurized liquid nitrogen through a heat exchanger immersed in a liquid nitrogen chamber ($-209\,°C$) held under vacuum. The supercold liquid nitrogen was circulated to the probe, producing probe surface temperatures in the range of -165 to $-185\,°C$, depending on the diameter of the probe (13). Used with multiple probes, this apparatus had substantially greater tissue-freezing capability than the earlier technology. Nevertheless, this new apparatus did not improve to a significant extent on the probe cooling rate which was only ~$20\,°C$/min at ~5 mm from the probe (14). The time required to reach a tissue temperature of $-50\,°C$ at a distance of 1 cm from the probe was 5–10 min (15). In comparison, the argon gas J–T apparatus cooled a probe much faster (16).

Recently, liquid nitrogen-cooled apparatus based on new technology equals the fast cooling rate of argon J-T apparatus and produces probe temperatures in the range of -170 to $-180\,°C$. This system uses a submersible nitrogen pump to generate the operating pressure to cool the probes with liquid nitrogen. Up to six vacuum-insulated probes may be cooled simultaneously (Cryo6. Erbe Co., Tubingen, Germany).

The cryosurgical apparatus that cool by the J–T effect are lightweight, portable, and quickly responsive in cooling or warming. Pressurized gas is passed through a small nozzle and expands, cooling the probe. This type of apparatus, using diverse kinds of gases, has been available for many years, but the cryogens commonly used in the modern apparatus are argon and nitrous oxide.

Argon, a colorless, odorless gas that boils at $-185\,°C$, has been used in J–T type apparatus since late in the 1960s. The gas in current devices is stored in steel cylinders that are pressurized at 3000 psi. In use, the probe temperature is ca. $-130\,°C$ at coldest. The cooling efficacy is pressure dependent. As the pressure in the cylinder falls, the cooling capacity diminishes. Argon permits the use of probes of very small diameter, as small as 17-gauge needles. In treatment, the use of such small probes requires that multiple probes be placed in the lesion. A larger probe, such as 3 mm in diameter, permits greater gas flow in the conduit, and will freeze a larger volume of tissue than the needle structures. Therefore, fewer probes may be used for the same volume of tissue. Since argon is a noble gas and is normally in the atmosphere, venting the gas from the apparatus into the operating area is not a safety concern.

Nitrous oxide ($-89.5\,°C$) is a colorless, nonflammable gas, commonly available in clinics and hospitals in the familiar "E" cylinders, which hold 2.72 kg (6 lb) of N_2O at a pressure of 5.1 MPa (740 psig). The withdrawal of the nitrous oxide gas depletes the gas pressure in the cylinder, which affects the rate of freezing. The gas cylinder may be enclosed in a warming jacket to provide some heat in order to maintain gas pressure at an appropriate level. There is a safety consideration. Cryosurgical units using nitrous oxide that do not provide for venting of the exhaust to the outside air may expose personnel to some ill effects, such as impaired performance and cognition. Such units should be used only in well-ventilated rooms. Older devices, which may exhaust into the room 20–90 L of N_2O/min, may be hazardous. Most new devices provide for gas scavenger systems to safely exhaust the nitrous oxide (Fig. 4).

In clinical use, the differences in cooling rate are important. Argon will cool a probe to $-100\,°C$ in ~1 min, and to $-130\,°C$ in ~2 min. Nitrous oxide will cool a probe to ca. $-80\,°C$ in the same time. In contrast, liquid nitrogen cools more slowly but will become colder, the probe temperature reaching -160 to $-180\,°C$ in ~5 min, the final temperature depending on the engineering features of the apparatus (14–16). However, the new type of liquid nitrogen apparatus with a design based on a reciprocating bellows pump will produce fast freezing of a probe, perhaps to $-180\,°C$ in < 1 min.

Sprays of cryogen can also be provided by nitrous oxide apparatus. If the fine droplets of liquid nitrous oxide are released on a surface, the droplets, instead of vaporizing to gas, recrystallize into solid nitrous oxide particles that lie on the surface or fly in all directions. To keep this partial-pressure effect from occurring, the droplets must be surrounded by pure nitrous oxide gas, necessitating an inverted-cup shield around the applicator tip.

Carbon dioxide is a colorless gas that is used as a cryogen in solid and gaseous form. Solid carbon dioxide ($-75\,°C$) has been used for direct application to tissue for ~100 years. Carbon dioxide is also available as a compressed gas contained in E cylinders. In J–T apparatus, it provides probe temperatures of ca. $-60\,°C$.

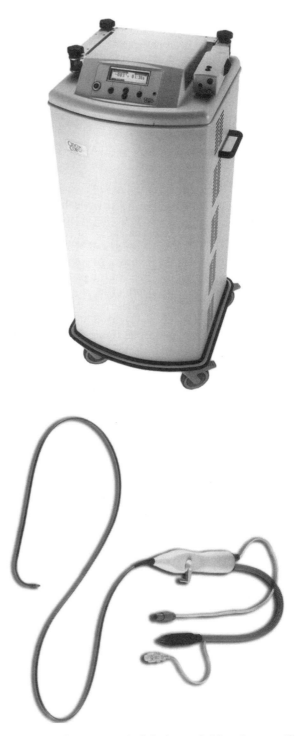

Figure 4. Modern cryosurgical device, cooled by nitrous oxide, used in cardiac cryosurgery for the treatment of atrial fibrillation and other arrhythmias. The console houses the primary gas supply in cylinders, the gas circuits, and the controlling electronics and software. The device cools the catheter probe by passing the pressurized gas through a restricting orifice at the probe tip. The probe is a steerable cryoablation catheter with leads to connect to the console for cryogenic gas flow, pressure monitoring at the catheter tip, and cardiac electrical signal recording (Fig. 3b). (Photographs by courtesy of CryoCor Inc., San Diego, CA 92121.)

Devices that cool by thermoelectric principles have limited freezing capability and can provide a probe tip temperature of ca. −20 to −30 °C at a low cooling rate. These devices have been satisfactory for ophthalmic use, but not for the treatment of tumors. The efficiency of these devices may be improved by combining with the technology of heat pipes, which then could provide probe temperatures of −50 to −60 °C (17). Nevertheless, the applicability of thermoelectric cooling to cryosurgery is limited.

TECHNIQUE

The two basic techniques for the use of cryosurgical apparatus are the direct application of the cryogen to the tissue, as in a spray of liquid nitrogen, or use of the cryogen in a closed system with probes. The choice of technique is in part a matter of personal preference and in part a matter of fitting the technique to the nature, size, and location of the disease. Probe techniques in general provide an easily controllable and a greater depth of freezing. The use of a probe is essential in freezing in less accessible areas of the body. Spray techniques, widely used in dermatological practice, permit easy application to accessible surfaces to achieve the superficial, and perhaps wide, freezing usually desired. Variants in these techniques blur the apparent sharp distinction between spray and probe techniques (Fig. 5).

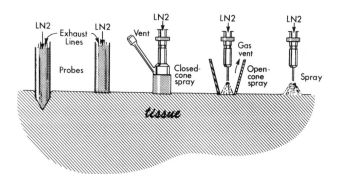

Figure 5. Diagrammatic representation of differences and similarities in probe, closed-cone, open-cone, and spray techniques, commonly used in the treatment of skin diseases. The two probes shown on the left are closed systems in which the cryogen does not come into direct contact with the tissues. In the closed systems, the liquid nitrogen, after change of phase, is vented as a gas somewhere along the return line, usually in the console of the apparatus. In the open systems, the three devices on the right, the cryogen is sprayed directly on the tissues. Surface freezing techniques by pressing a probe against the tissue or by spraying cryogen are commonly used. However, the pointed probe shown on the left may be inserted into the tissue to achieve deeper freezing. This penetration of the tissue causes a wound, which may cause some problems. In the cone techniques, the dispersion of the spray is limited and the effect of confining the spray is to produce an open probe. In the cones, the heat-exchange surface is on the skin and the venting of the gas is at the side or top of the device. The open-spray technique is effective in treating wide areas with irregular outline, but care has to be taken to avoid dispersion of the spray to unwanted areas.

The Spray Technique

Spraying of the cryogen, usually liquid nitrogen, from the nozzle of an appropriate apparatus is an efficient method of producing rapid freezing of tissue. When used as a spray, the cryogen is used at its coldest temperature, and it produces superficial freezing rather quickly and easily over wide areas if required. The more superficial the lesion, the more suitable is the use of a spray, especially if the surface is irregular, as over a bony surface, or if the area of disease is extensive. The spray often is used intermittently, especially for small lesions. Steady spraying, as in an effort to freeze deeply in one site, causes the liquid nitrogen droplets to strike the tissue without vaporizing and run off the frozen tissues to freeze in undesired areas (unless spray-limiting devices are used), especially if excessive pressure is being used in the apparatus. Other methods of control are to reduce the pressure in the spraying device, if possible, or to use a smaller nozzle size, but reduced cryogen flow results in lessened capability of freezing the tissues. Problems with the dispersion of the spray are best corrected by the use of spray-confining devices such as hollow cones placed over the lesion so that the liquid nitrogen may be sprayed into the hollow device (Fig. 6). The use of the cone devices also has the effect of creating an open probe and is a useful technique of improving depth penetration of spray techniques, as required for the treatment of invasive skin cancers. Similar devices, such as funnels and hollow cylinders, have been used to confine liquid nitrogen as it is poured over bone tumors after removal by curettement.

The Probe Technique

The cryogen is circulated in a closed system, using metal probes for contact with the tissue to provide a heat sink or heat-transfer surface. Various sizes and shapes of probes, ranging from rod-like probes 1 cm in diameter to 18 gauge needle size and catheter shapes, to fit diverse anatomical sites and disease dimensions are available. The probes may have flat surfaces for contact with the tissue or may be pointed to allow insertion into the tissue. Slip-on metal end pieces, which fit over the end of the probe to modify the freezing surface, increase the versatility but also slow the freezing capability. The lines that feed the liquid nitrogen to the probe tip are vacuum insulated or insulated with appropriate materials, which increases the efficiency of the cryogen and provides for the safety of the user. The J–T types of apparatus have thin cryogen feed conduits that require no insulation.

In use, the physician selects one or more probes as appropriate to the disease. The manner of use is to apply the freezing surface of the cryoprobe to the lesion and allow the cryogen to flow. The cold probe acts as a heat sink and produces tissue freezing by removing heat from the tissue faster than blood supply and conduction restores it. Surface contact freezing is performed by pressing the freezing surface of the probe firmly on the tissue in order to ensure a good contact for heat exchange. This contact is improved by the use of water-soluble hospital lubricating jelly between the probe and the tissue. Greater depth of freezing may be achieved by increased pressure on the probe or by penetration of a sharp, pointed probe into the tumor. The

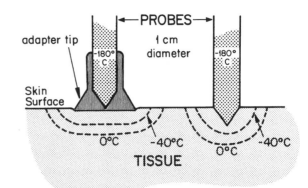

Figure 7. Diagram showing the effect of probe configuration on the shape of the frozen tissue. The probe on the right held with some pressure on the tissue causes indentation of the tissue, or minimal penetration, and produces a roughly hemispheric frozen zone. The same probe, fitted with a freezing tip adapter to form a wide, flat freezing surface, produces a wider but less deep frozen area. In each frozen zone, the −40 °C isotherm is shown to delineate its approximate location. The location of this isotherm varies slightly with the rate of cooling the tissue. Rapid cooling moves the isotherm slightly more toward the periphery.

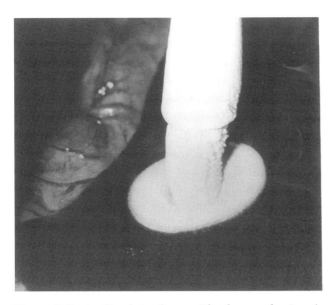

Figure 6. Canine liver being frozen with a large probe, 1 cm in diameter cooled with liquid nitrogen to −160 °C. The white frosted area is the tissue frozen after 3 min of contact. The depth of freezing is about the same as the lateral spread of frost from the side of the probe. With proper technique, the area of necrosis will closely approximate the visibly frozen area. (Reprinted with permission from J. Am. Med Assoc., **204**, 566, 1968.)

penetration technique has the disadvantage that a wound is produced, and this may later bleed. Care must be taken to avoid motion of the probe during freezing because this might fracture the bond with the tissue and interfere with heat exchange. Heat exchange also depends on the gradient of temperature between the tissue and the probe, so the probe is always used as cold as possible when tissue destruction is sought. The larger the gradient, the faster the rate of freezing.

Freeze–Thaw Cycles

As freezing progresses, the tissue turns white (frosted in appearance) and hard, a change that begins at the area of contact with the cryogen and extends to incorporate an increased volume of tissue as time passes (Fig. 6). As treatment continues, in the case of lesions in easily accessible areas, such as the skin, the extent of freezing is judged by inspection and palpation. Estimation of depth of freezing may be difficult, but physicians experienced in the techniques can make reasonably accurate estimates of depth. With surface contact freezing by probes, the shape of the frozen volume of tissue is roughly hemispheric, so the depth of freezing can be judged to be about the same as the lateral spread of freezing from the probe. Significant modification in the shape of the frozen area may be achieved by the selection of different probe shapes (Fig. 7). The shape of the frozen area is also modified by the amount of pressure placed on the probe because increased pressure compresses the tissue and increases the depth of freezing. Another factor influencing the shape of the frozen area is the presence of major blood vessels, which provide a source of heat. In freezing with the spray techniques, a more superficial freezing may be expected and often is desired.

However, if the spray is confined with a cone, the effect on the tissue is similar to probe freezing.

Freezing with the probe or spray continues until the desired volume of tissue, including a margin of apparently normal tissue, is treated or until the frozen area no longer enlarges. This is easy to observe in accessible tissues, such as the skin. In the use of a spray of liquid nitrogen, the nozzle of the device is moved about to distribute the cryogen evenly over the target, and this freedom of movement allows wide areas to be treated. On the other hand, probe freezing takes place from a selected site of application, and the rate of expansion of the frozen area slows as an equilibrium is established between the heat loss from the tissue via the probe, and heat brought to the area by the circulation of blood. For this reason, it is difficult to freeze tissue to a distance > 2–3 cm from the probe. This means that large lesions cannot be frozen completely in a single application of the probe. In these circumstances, the plan of treatment must include freezing from multiple sites with successive applications of the probe, which improves the width of the frozen area and also increases the depth of freezing in overlapped frozen areas to a slight extent. For large cancers, the simultaneous use of multiple probes is advantageous and time saving.

When used for the destruction of tissue, as for the treatment of tumors, cryosurgery must be performed in a manner that produces a predictable area of necrosis. The techniques to achieve this goal stress the rapid freezing of the tissue, slow thawing without assistance from warming devices, and immediate repetition of the freezing process in order to maximize destruction. Some modifications of these basic factors in technique are necessary for the diverse diseases in different parts of the body, especially if the intent is to destroy some cells while preserving others, but, in general, the cited basic technique provides the basis of effective therapy.

Rapid freezing promotes the formation of intracellular ice crystals, which are considered almost certainly lethal for cells. This occurs only in the tissue close to the freezing probe or in contact with the spray of cryogen. The rate of freezing of the tissue varies inversely with the distance from the site of application of the cryogen (Fig. 8). Close to the probe, temperature changes of the order of > 10 °C/min may be achieved, and this is considered rapid freezing. However, ~2 cm from the probe, the cooling rate is slow, perhaps of the order of 2–5 °C/min. The possibility of tissue survival is enhanced with slow freezing rates, but fortunately even slow freezing rates have lethal potential because of cellular dehydration and related deleterious effects.

The thawing rate should be slow and unassisted because this increases the time that the tissues spend in a phase when recrystallization phenomena can add to the cell injury. Tissue often thaws in about the same time as was required for freezing, but this depends on the volume of frozen tissue. Large volumes of frozen tissue warm slowly. In some therapeutic applications, a probe heater is used to speed release from the tissues after freezing. This practice also slightly speeds thawing of the tissues, but this probably does not increase the chance of cell survival if the

technique is otherwise correct. Certainly, no warming solutions should be used to thaw the tissue, because it is well established that quick thawing of tissues reduces injury from freezing, whether in cryopreservation, cryosurgery, or frostbite.

Repetition of freezing maximizes damage by subjecting the tissues again to the same mechanism of thermal injury. To take full advantage of the lethal effect of repetitive freezing, it is necessary that the frozen tissues be completely thawed (> 0 °C) before the next freezing cycle because then the entire volume of the tissue will pass through recrystallization phases with their attendant injurious effects. The second freezing cycle is usually faster and more effective in cooling the tissue than the first cycle and achieves a slightly greater depth of freezing because the latent heat in the previously frozen tissue is reduced and because the microcirculation has begun to fail in the interval between freezings, reducing the heat supply. The lethal isotherm is then deeper in the frozen tissue. For this reason, it is advisable to wait a few minutes between freezing–thawing cycles.

The freezing of tissue in cryosurgery is a heat-transfer process, and the mechanics of extraction of heat from the tissue directly influence cryosurgical technique. The duration of each freezing–thawing cycle of treatment is dependent on factors that affect the rapidity of tissue freezing, such as efficiency of heat exchange, gradient of temperature, blood supply to the area, attainment of a desired temperature goal, and on the size of the lesion. The temperature of the cryogen is also important: the larger the gradient between cryogen and tissue, the faster the cooling. It is also a function of the contact area between cryogen and tissue: a wide area for heat exchange cools tissues faster.

Thermal modeling techniques have been used to predict the size of the frozen area that may be produced by a probe. The blood flow, the probe temperature, the area of contact with the tissue, and the duration of application are known. This work has contributed to the understanding of the functions and cooling capacity of cryosurgical equipment, has provided a method of estimating the effect on the tissues, and has shown direction in equipment design and in the planning of treatment. The modeling techniques have been useful in the several aspects of cryobiology, including the mechanisms of injury to cells. Nevertheless, before clinical use, it is necessary for the physician to practice with the cryoprobes or with a spray of cryogen in test materials in order to develop the ability to predict the size and shape of the frozen field as a function of time, temperature gradient, and heat exchange surfaces. The heat diffusion equations that must consider the frozen area and a moving solid–liquid interface as freezing progresses are complex and are best reviewed in source material (5,18).

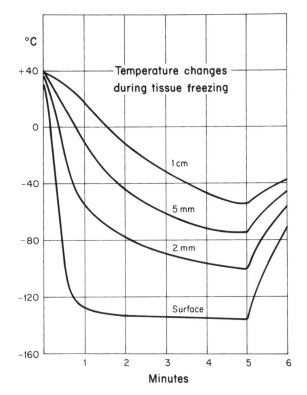

Figure 8. Graph showing temperature changes recorded from thermocouples inserted in the tissue at diverse distances from a probe cooled by liquid nitrogen. The tracing identified as "surface" is from a thermocouple placed in the contact area between the probe surface and the tissue. The other tracings show the temperature changes at 2, 5, and 7 mm from the edge of the probe. The most rapid and deepest cooling of the tissue is at the point of contact. The greater the distance from the probe, the slower the cooling, the less the depth of freezing, and the shorter the duration of freezing.

MONITORING TECHNIQUE

Many cryosurgical procedures, especially those for benign diseases and small cancers of the skin, are performed using only observation and palpation of the frozen tissue to determine the progress of treatment and judge its adequacy. In easily accessible sites, those physicians with considerable experience in cryosurgery can achieve satisfactory results without the use of monitoring techniques to guide therapy. However, the temperature of frozen tissue cannot be determined by its appearance: Frosted tissue looks the same at any freezing temperature. Equally important, the depth of freezing is difficult to judge in many situations, although the relationship between the depth of freezing and the lateral spread of freezing from a probe is an important clinical aid. The clinical evaluation, if not perfect, may produce an error in treatment that may be of critical importance in the treatment of malignant disease. Effective cryosurgery must yield predictable and certain necrosis.

To guide cryosurgical procedures and permit reasonable certainty of the death of tissues, methods of monitoring the freezing process have evolved. These methods include (1) the measurement of tissue temperature by thermocouples; (2) the measurement of electrical impedance on resistance in tissue; (3) the measurement of heat lost from the tissue by a heat flowmeter; (4) thermography and; (5) the imaging techniques, which are ultrasound, computerized tomography (CT); and magnetic resonance (MR).

Thermocouples

The most commonly used method of monitoring is by the insertion of needle-mounted thermocouples into the tissue at appropriate sites to measure tissue temperature (Fig. 9). Thermocouples are formed by the junction of two dissimilar conductors, commonly iron and constantan, or copper and constantan, in a closed electric circuit (Fig. 10). When the junction of the conductors is held at a temperature, an electromotive force (emf) proportional to the temperature difference will be generated. An instrument, such as a potentiometer or pyrometer, is used to measure this emf and provide a readout in terms of temperature.

Thermocouples perform several important functions in cryosurgical treatment. The insertion of thermocouples into the tissue in appropriate locations monitors the progress of the freezing treatment and ensures that temperatures destructive for tissues are attained. Thermocouples can also be used to ensure that tissues adjacent to a diseased area are not frozen and are hence safe from freezing injury. If used with a recorder, thermocouple measurements on the tracing provide written evidence of treatment. Thermocouples often are built into the probe tips, so that the temperature of the probe can be monitored. Though this provides useful information, confirming that the apparatus is working, probe temperature tells nothing about tissue temperature, except by estimations based on experience with the performance of the probe.

In cryosurgical treatment, especially for cancer, it is important to know that destructive temperatures are produced in the tissues. In the freezing–thawing cycles used in cryosurgery, tissue destruction is a multifaceted process involving the freezing rate, the tissue temperature, the duration of freezing, the thawing rate, and repetition of the freezing–thawing cycle. Separation of the relative destructive effects of these components is difficult, but the easiest to control and measure is the coldest temperature attained

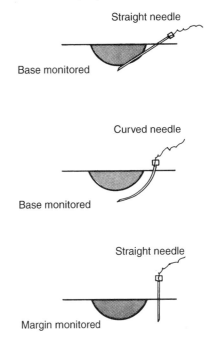

Figure 9. Diagram showing typical sites of thermocouple insertion. Three useful methods are shown. At the top is a thermocouple inserted at an angle from the side of the lesion through normal tissue. The thermocouple tip rests directly beneath the lesion and shows the temperature at the depth of the tissue. It probably is the most common method of thermocouple usage. In the center is shown a curved needle used for base monitoring. This method avoids passing the shaft of the thermocouple through the frozen tissue. At the bottom is shown an alternative method of thermocouple use. The thermocouple is inserted at the border of the lesion. With this technique, the temperature registered at this thermocouple is interpreted as being the same temperature that would be measured at the border beneath the tumor. This method assumes that the depth of freezing is approximately the same as the lateral spread of freezing from the probe. The advantage of this technique (and the use of a curved thermocouple) is that the thermocouple shaft remains outside of the frozen area until the advancing ice front incorporates the tip. More than one thermocouple may be used.

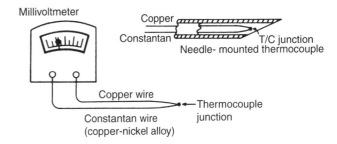

Figure 10. Diagram of a basic thermocouple-millivoltmeter circuit consisting of a copper and a constantan wire welded together to form a measuring junction, the thermocouple. The free ends are connected to the millivoltmeter, which measures the emf associated with the temperature at the thermocouple. In the upper right of the illustration is shown the mounting of the thermocouple in a hypodermic needle, which can be inserted in the tissue at appropriate sites to measure temperature changes during freezing–thawing cycles.

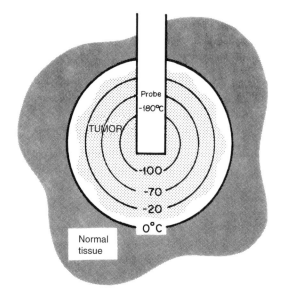

Figure 11. Drawing demonstrating the gradients in temperature that exist in tissue being frozen with a probe at −180 °C. The incorporation of a tumor in a frozen mass is depicted. The gradients are substantial, ranging from ∼0 °C at the edge of the frozen volume to nearly −180 °C adjacent to the probe. Cell Death would not occur in the entire frozen area. If treatment were to cease at this stage, the tissue in the 0 to −20 °C range would probably survive and the tumor would grown again.

in the cryosurgical treatment. Hence, a lethal temperature goal is used in cryosurgery.

Early in cryosurgical experience, it was thought that −15 or −20 °C was a proper lethal temperature goal. Substantial tissue damage results from a tissue temperature of −20 °C, but this temperature is not safe for the treatment of malignant disease. In nonneoplastic disease, usually conservative freezing is wise and, therefore, tissue temperatures of the order of −20 to −30 °C are satisfactory. Temperatures of −40 °C are satisfactory for superficial skin cancers. The treatment of more aggressive cancers, as in the oral cavity, requires repetitive freezing to tissue temperatures of −50 °C or colder if cure is to be achieved (Fig. 11).

The accuracy of thermocouples is a matter of interest since treatment depends in part on this measurement. Some minor error is inherent in the temperature recording system, which consists of the thermocouple, the electrical leads, and the readout device. The readout device is commonly a potentiometer with a digital readout, which is sufficiently accurate for the purpose of cryosurgery. An important source of error is from conductance of heat along the thermocouple needle shaft. If the needle shaft passes through a frozen area, the reading from the tip may be falsely low (Fig. 12). Under other conditions, heat may be added to the temperature-measuring system from extraneous sources anywhere between the thermocouple needle tip and the recorder. However, with proper use, the error due to heat conductance is minimal and does not interfere with the thermocouple function of supplementing clinical judgment. Based on experimental data, temperatures in the range of −20 to −50 °C, produced in short freezing cycles, may be associated with an error of ∼2 °C due to heat

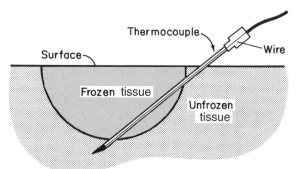

Figure 12. Diagram illustrating a mechanism for thermocouple error due to heat conduction. The thermocouple needle enters the tissue at an angle, just as commonly done in clinical practice. During freezing, the expanding ice front incorporates the needle shaft. The cooling of the shaft affects the thermocouple at the needle tip, and this may produce a falsely low reading. Also, the hub of the needle and the adjacent needle shaft in unfrozen tissue may be warmed by the tissue or the air, which may affect the reading in the opposite way. The magnitude of this possible error is a source of concern, but proper thermocouple use can avoid the error.

conductance (19). This differential is of little importance in cryosurgical techniques. The error due to heat conductance in thermocouples is less significant than the error produced by the positioning of the thermocouples in the tissue. A 1-mm variation in thermocouple placement in the tissue represents ∼10–15 °C difference in the temperature recorded in usual cryosurgical freeze–thaw cycles. Therefore, the important errors in thermocouple use are produced by the accuracy of placement rather than by heat conduction from extraneous sources.

Impedance–Resistance Measurements

Another method of quantification of freezing injury is a technique that measures the impedance or resistance changes to the passage of a small electric current through the tissues being frozen. Unfrozen tissue is a conductor of electricity because of the electrolyte content of the tissue fluid. During freezing, the formation of ice crystals in tissue and the removal of water from the tissue results in decreased electrical conductivity. When practically all of the extracellular water is crystallized, electrical impedance or resistance rises to the high levels. This change is interpreted as being associated with tissue death.

The techniques of impedance–resistance measurements require the insertion of needle electrodes in or about the tumor (Fig. 13). These conduct the small electric current from the line or battery-powered device to the tissue. The measurement is made between two electrodes. Both electrodes may be placed in the treated area, but it is preferable to measure between one electrode in the treated area and a distant reference electrode. The initial impedance–resistance in unfrozen tissue is of the order of 1–2 kΩ. When the entire measuring electrode is incorporated in frozen tissue, the impedance–resistance rises to megohm levels. This change seems to occur quickly, at least in comparison to the changes in temperature (Fig. 14).

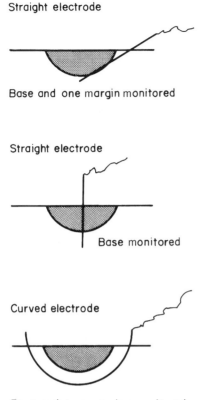

Figure 13. Diagram of electrical impedance–resistance monitoring system showing alternative methods of placement of the electrode. The entire electrode must be incorporated in the frozen tissue before the change in impedance–resistance rises to the megaohm level. Multiple electrodes may be used to increase the monitoring sites. The second electrode, which is placed in a more remote site to complete the electrical circuit, is not shown.

Impedance measurements were introduced into cryosurgical techniques by Le Pivert et al. (20), who measured electrical impedance in the tissue to the passage of a low frequency (1000 Hz) alternating current between electrodes. Impedance values of 0.5–1.0 MΩ were equated to −40 °C and were associated with tissue destruction. This destructive effect on the tissue was reported to be independent of repetition of the freezing–thawing cycles.

Other investigators, using Le Pivert's instrument in experiments on canine skin, found that an impedance of 1 MΩ corresponded to a tissue temperature of ca. −30 °C, but the range of temperatures about each impedance value was sufficiently great to cause concern about the possibility of tissue survival at the 1 MΩ value. Using a line powered 1000 Hz low current impedance meter, the relationship between tissue impedance and temperature and the subsequent development of tissue necrosis was investigated. An impedance of 10 MΩ was not always associated with tissue death, and the range of temperatures about any impedance value was considerable (Fig. 15). Comparison of the electrical characteristics of the tissue and the tissue temperature with the border of necrosis that was obvious in a few days showed that the tissue temperature was the more accurate predictor (21).

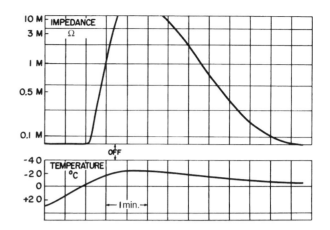

Figure 14. Graph illustrating simultaneous temperature and impedance changes. The temperature in the tissue was cooled to −22 °C at which time the cryosurgical apparatus was turned off. The tissue cooled further to −25 °C because the cold probe continued to function as a heat sink. Then the warming cycle began. The impedance rose steadily after the tissue cooled to −8 °C. When the impedance reached 10 MΩ ~40 s later, the cryosurgical apparatus was turned off. The impedance continued to rise and the trace passed off the chart, returning ~1 min later in the warming cycle. The temperature and impedance changes in thawing were more gradual in thawing than during freezing.

At this time, devices to measure tissue impedance–resistance during cryosurgical procedures are little used. The incorporation of an impedance electrode into the freezing surface of an endoscopic probe confirms that the probe is cooling properly, but yields no information about tissue temperature (22). Nevertheless, such a device would confirm fixation of the probe to the tissue. An impedance-measuring device could be useful in the determination of depth of freezing. The use of multiple electrodes about the circumference of a tumor to monitor cryosurgery is a use that requires further testing.

Whatever method of quantification of cryogenic injury is used to supplement clinical judgment in cryosurgery, whether thermocouples or electrodes are inserted in the

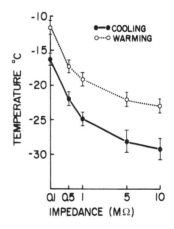

Figure 15. Graph showing the relationship between impedance and temperature. The data for the graph are taken from mean values during the cooling cycle. The standard error of the mean is shown.

tissues, it is important to recognize that their placement in itself may be a source of error. It is equally difficult to determine the relationship of the electrode or thermocouple to the margin of the disease. Both thermocouple and impedance devices should be viewed as adjunctive to clinical judgment in therapy.

Heat Flowmeter

The heat lost from the tissue can be measured by means of a heat flowmeter (23). This is done by attaching a differential thermocouple to the probe tip. As heat passes from the tissue into the probe, the temperature electromotive force, proportional to the temperature differential, develops across the disk. This may be recorded so that total heat exchange is used as the monitor of the cryosurgical procedure. Though the technique was used in the treatment of patients, it has not proven useful generally.

Thermography

The use of thermography to monitor the progress of freezing tissue is best considered as a research method. The thermograms, recorded in monochrome or in color, define isotherms down to $-40\,°C$. Bradley's experiments with thermography, comparing the freezing characteristics of different freezing techniques, have shown the faster freezing capability of the spray devices in comparison with the probe techniques and have confirmed the importance of pressure on the tissue from the probe in modifying the shape of the frozen area. Though *in vitro* tissue testing permits evaluation of depth of freezing, clinical use provides only surface freezing evaluation in most circumstances (24,25).

Ultrasound

Ultrasound used during cryosurgery provides a real-time image of the frozen volume of tissue. Since ultrasound provides a more global view of the frozen tissue than do thermosensors, this imaging technique has come into wide use for monitoring the freeze–thaw cycle, especially for visceral tumors. Ultrasonography offers the possibility of matching the extent of the neoplasm with the volume of tissue frozen in treatment. Frozen tissue is hypoechoic, so the ultrasonic image is black. The edge of the frozen tissue is hyperechoic and appears as a bright line. As freezing continues and the volume of frozen tissue increases, the hyperechoic rim moves away from the probe, leaving the hypoechoic zone behind it. Therefore the process of tissue freezing can be observed during the cryosurgical procedure. Experience with ultrasound monitoring in hepatic and prostatic cryosurgery is now substantial. The correlation between the ultrasound image and the actual diameter of the frozen tissue is excellent, though it has become evident that ultrasound overestimates the volume of tissue frozen. Much of the frozen volume is obscured by distortions and reflections of the image. Sonography does not provide an image beyond the near edge of the ice because of complete posterior acoustic shadowing (26,27). Some compensation for this limitation can be obtained by viewing the frozen volume from another angle. Three-dimensional (3D) ultrasound is in development (28).

The correlation between tissue temperature and the ultrasound image is of considerable importance because effective treatment depends on achieving an appropriate tissue temperature goal. The temperature of frozen tissue cannot be determined from its ultrasonic image. However, the hyperechoic rim of the image is $\sim0\,°C$ and the probe temperature is generally known, so inferences about the steep gradients of temperature in the tissue can be made. For example, using a probe at $-160\,°C$ on tissue for 5 min, the $-20\,°C$ isotherm is $\sim70\%$ of the distance from the probe to the periphery of the frozen tissue. The $-40\,°C$ isotherm, which is commonly used as a goal in tumor therapy, is $\sim60\%$ of the distance from the probe to the frozen boundary. A rapid cooling rate moves the isotherm toward the periphery.

Computerized Tomography and Magnetic Resonance

Other investigational methods of monitoring are radiological. The absorption of X rays in biological tissue is proportional to the tissue density. Water, which is the major component of most biological tissues, changes density upon freezing and can be visualized. Computerized tomography (CT) can show the entire cross-section of the lesion. The development of the frozen volume of tissue can be seen on a series of CT images taken at frequent intervals. Magnetic resonance (MR) will provide a 3D image of the frozen volume of tissue. When used with thermal models, the temperature changes in the tissue can be quantified (29,30). Magnetic resonance requires the use of MR-compatible probes, which have been developed and used to a limited extent. Electrical impedance tomography (EIT) has been proposed as a method to provide real time imaging and provides a global image by introducing low amplitude alternating current (ac) into the body and thereby measurs the electrical potentials on the surface of the body. These potentials are analyzed to create a tomographic image (31). At present, the high cost of apparatus and logistical considerations will limit the usefulness of these imaging techniques in cryosurgery. However, it seems likely that image guidance will achieve greater applicability as monitoring techniques during the freezing process.

To summarize, clinical judgment, the prime factor in the control of the freezing of tissue, requires assistance from the monitoring techniques. Depending on the clinical circumstances, including the nature and the site of the disease, either tissue temperature measurement by thermosensors or ultrasound imaging is commonly used. The limitations of these techniques in providing thermal information during cryosurgery are well defined. To compensate for the limitations, wherever practical, the use of both thermosensors and imaging techniques are advised.

CLINICAL USES

Cryosurgery has established a firm place in medical practice for the treatment of diverse diseases in different parts of the body. Lesions in accessible sites, such as the skin or oral cavity, may be treated with little need for anesthesia and without risk of hemorrhage. Treatment for many conditions can be given under local anesthesia in an office or outpatient clinic, avoiding the need for hospitalization, with its attendant inconvenience, cost, and risk. Cryosurgical treatment of viscera and other deep-seated disease is more difficult because of accessibility, but these techniques have become well developed in the past 15 years, benefiting from percutaneous technology and ultrasound imaging. Cryosurgical techniques deserve emphasis as an excellent choice of therapy on high surgical risk patients, especially those who have problems difficult to manage by other methods of treatment.

Skin Diseases

Cryosurgery is widely used in the treatment of skin diseases. Many non-neoplastic lesions, and a variety of benign tumors and cancers of the skin are successfully treated commonly with a hand-held device containing liquid nitrogen. Non-neoplastic skin lesions are easily treated by cryosurgery with little risk, but skin cancers require further comment.

Cryosurgery has become a standard technique of treatment for skin cancer in any location, and joins a variety of treatment methods, including surgical excision or radiotherapy as well as other special techniques. Most skin cancers are small, commonly not > 3 mm in depth and < 1 cm in surface diameter. These are easily treated by cryosurgery, and a cure rate in excess of 97% has been achieved. Certain types of skin cancer, called sclerosing or morphea type of basal cell carcinomas and tumors of the scalp, are difficult to cure by cryosurgery, perhaps because of difficulty in determining the extent of disease, or because of biologic behavior, or because of the richness of blood supply. These require aggressive freezing therapy in order to achieve a cure. In general, however, the histologic type of the cancer is not important, and either squamous cell or basal cell cancers can be treated. Melanomas are also easily destroyed by freezing, but most primary melanomas, with the possible exception of lentigo maligna, are better treated by excision in order to establish the diagnosis and to permit staging of the extent of the disease. Cryosurgery is well chosen for certain skin cancers in special situations. These include the following:

1. Multiple superficial small skin cancers, as commonly seen from excessive exposure to sunlight. A large number of cancers can be treated quickly and easily in comparison to multiple surgical excisions.
2. Cancers about the ears, nose, and eyes. The irregular contours of the head in these sites and the tightness of the skin over the underlying bone make treatment difficult at times. In these locations, cryosurgery offers ease of treatment and improved cosmetic results.
3. Cancers arising in irradiated skin. Such cancers require conservative management because new cancers will develop elsewhere in the damaged skin.
4. Cancers that persist after radiotherapy excision or other methods of treatment. These are often advanced cancers that present problems in management.

In these special situations, the results of cryosurgery compare favorably with surgical excision or radiation therapy.

Oral Diseases

Benign and malignant tumors and precancerous conditions of the oral cavity are well suited to treatment by cryosurgery (Fig. 16). An important use is for the treatment of leukoplakia, which is the general descriptive term for white patches in the mouth and includes a number of different pathological conditions. Some of these, such as epithelial dysplasia and papillary epithelial hyperplasia, are precancerous. A biopsy, preliminary to definitive treatment, is necessary to differentiate between those lesions that are relatively innocuous, those that are precancerous, and those that are carcinoma *in situ*. All may be treated by cryosurgery. Such disease is superficial, and either nitrous oxide or liquid nitrogen apparatus may be used because the tissue needs to be frozen only to a depth of ~3 mm. The results are excellent, and most patients will remain free of disease. Persistent disease or recurrent disease usually results from failure to correct the etiologic factors, that is, continuation of the use of alcohol and tobacco or continued irritation of the oral mucosa for other reasons. In dysplastic disease or carcinoma *in situ*, ~10% of patients will require additional treatment over a 2-year period.

Oral Cancer

Malignant tumors of the oral cavity are aggressive cancers that are difficult to cure by conventional treatment, that is,

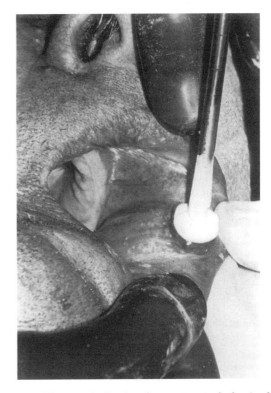

Figure 16. Photograph showing the treatment of a benign blood vessel tumor (hemangioma) by freezing *in situ*. A cryoprobe, 5 mm in diameter, is applied. The frozen zone is extending over the lesion. Freezing will continue until the entire tumor is encompassed. No thermocouple monitoring is necessary with this benign lesion.

surgical excision and/or irradiation. In general, the 5-year survival rate is only ~30% of all patients who acquire the disease. Cryosurgery can be used in special circumstances, especially for high surgical risk patients whose extensive associated disease would make general anesthesia and extensive operation prohibitive in risk. Cryosurgical techniques are also suitable when the cancer is on or adjacent to bone because the underlying bone limits the depth of penetration of a tumor and facilitates its destruction by freezing. In selected patients, the survival rate achieved by cryosurgical treatment appears comparable to that provided by surgical excision. However, the result can be achieved at a lessened cost to the patient in terms of operative mortality and postoperative functional disability. Nevertheless, cryosurgery is seldom used for oral cancer in current surgical practice.

The further away from direct vision, as is possible in the oral cavity, the more difficult is the application of cryosurgery. Cryosurgery has been used for cancers in the pharynx, larynx, bronchi, and esophagus with treatment given via endoscopy. Vision is limited, so accurate and extensive freezing is difficult. In the pharynx and larynx, occasional successes in attempts at cure have been achieved, but current experience provides little reason to choose cryosurgery in preference to radiotherapy or excision. Experience in the trachea and esophagus is limited to a few patients treated for palliation of symptoms. Obstruction can be relieved and tumor size kept under control by cryosurgery repeated every few months, but invasive growth continues. The presence of necrotic tissue in the larynx is a threat to the airway. In general, palliation of incurable cancer is provided better by radiotherapy or chemotherapy.

Bronchial Tumors

Considerable experience has matured in recent years with the cryosurgical treatment of bronchial tumors, including cancers, which produce symptoms by obstruction of the air passages. The treatment is via endoscopy, freezing the tumor with nitrous oxide-cooled probes, which then opens the airway and relieves the symptoms. With the removal of the obstructing tumor, irradiation of the remaining tumor may be used more safely (32).

Nose and Throat Diseases

Cryosurgery has been used for a wide variety of diseases of the nose and throat, including mucosal dysplasia, tonsillitis, nasal polyps, and rhinitis. However, generally other methods of treatment are chosen for these conditions. Special probes and techniques are necessary, and commonly the treatment can be done under local anesthesia. In chronic nasal airway obstruction due to hypertrophy of the nasal turbinates, freezing of the excess nasal tissue is followed by slough of the hypertrophic tissues, which improves the nasal airway and decreases the secretions. The treatment should be conservative so that normal tissue is not unnecessarily frozen. Healing occurs over a 2- or 3-week period and results in a reduced amount of turbinate tissue with a normal appearing mucosa and relief of symptoms.

Cryosurgery may be used for tonsillectomy. Cryosurgery is well chosen for adult patients who have blood dyscrasias or who are high surgical risks, because cryosurgery may be performed with little or no blood loss and low risk of postoperative bleeding. Nitrous oxide or liquid nitrogen apparatus may be used. The chance that tonsillar remnants may remain after freezing, which may lead to persistent symptoms, means that careful attention must be given to technique to ensure that all tonsillar tissue is frozen. Repetitive treatments may be used to eliminate tonsillar remnants.

Gynecological Diseases

Cryosurgical treatment has become common in gynecological practice in recent years. A wide variety of inflammatory and neoplastic diseases of the vulva, vagina, and cervix may be treated by freezing in situ. Since tissue diagnosis is important, it must be recognized that advances in colposcopy have improved the diagnostic ability of the physician to differentiate between inflammatory and neoplastic diseases.

Chronic cervicitis, which is inflammatory disease of the uterine cervix, is a principal use for cryosurgery. Careful evaluation of the extent of the disease must exclude the possibility of invasive carcinoma. Cryosurgery is effective treatment for dysplastic disease and carcinoma in situ. The differentiation between inflammatory disease and premalignant or malignant disease must be made by pap smear or biopsy before treatment is begun. The results of therapy are excellent, but if the disease was carcinoma in situ, persistent disease must be expected in ~9% of patients.

Cryosurgical techniques are seldom used in the treatment of invasive malignant disease of the vulva, vagina, and cervix. Though it has been used for cancer in some potentially curable patients who are not candidates for one reason or another for conventional therapy, sometimes with excellent results, too few patients have been treated to permit evaluation of its ability to cure cancer in these sites. More commonly, cryosurgery has been used to relieve pain due to cancer after other methods of treatment have failed. Under these conditions, cryosurgery will diminish the size of the cancer, reduce malodorous discharges, control bleeding, and ameliorate pain.

Proctological Diseases

Cryosurgery has been used to treat a variety of proctological disorders, ranging from relatively minor conditions such as anal fissures and hemorrhoids to large, incurable cancers. However, anal fissures and hemorrhoids may be treated successfully by a variety of different surgical methods. For example, injection of sclerosing agents into small to moderate sized hemorrhoids is a commonly used effective therapeutic method. These alternative techniques are good therapeutic methods, so cryosurgery has little place in the treatment of non-neoplastic proctological diseases.

Cancers of the anus and rectum may be treated by cryosurgery in selected patients who are at high surgical risk for conventional surgical excision. Best suited for treatment are carcinomas that are exophytic rather than ulcerated and infiltrating. In the rectum, the ideal cancer for cryosurgery is within reach of the examining finger and on the posterior and lateral walls where freezing is safer. Similar criteria have been used for selection of patients for therapy by electrocoagulation. It is best to reserve cryosurgery for patients who are difficult problems in management by excisional surgery.

Diseases of the Prostate Gland

Cryosurgical treatment of prostatic disease was introduced into clinical practice in the mid-1960s, had extensive trial in the 1970s, and then fell into disuse in the 1980s. However, after intraoperative ultrasound became available and when improved cryosurgical equipment was developed, then cryosurgery for prostatic cancer became a viable alternative to excisional surgery. The technique requires the passage of multiple thin probes, as few as six, as many as twenty, through the skin of the perineum into the prostate gland. The probes are placed in an appropriate spatial relationship to include the entire gland during the freezing. Then the probes are cooled, using pressurized argon in a J–T apparatus or using liquid nitrogen. The process of freezing is monitored by ultrasound imaging and commonly also by placement of thermosensors in the peripheral areas of the prostate (33). The cryosurgical treatment is suitable for most stages of prostatic cancer, including those patients who have persistent disease after irradiation. The long-term beneficial results are similar to those of excisional surgery.

The effect of prostatic cryosurgery on metastatic disease is controversial. Regression of metastatic deposits of prostatic cancer following repeated freezing of the prostate has been attributed to the release of tumor antigens from the frozen tissue and the subsequent development of prostatic tumor-specific antibodies. In patients with bone metastases, relief of bone pain often is achieved for several months, but radiological evidence of reduction of metastatic tumor is uncommon. In experimental animals bearing tumors, cryosurgery has been shown to produce an immune response, manifest in a lower incidence and a slower growth of recurrent tumor. Though objective evidence of benefit from a humoral or a cell-mediated immunological response in men following prostatic cryosurgery is lacking, considerable interest in the possibility of benefit by this mechanism is evident.

Visceral Tumors

In addition to the prostate gland, tumors in diverse organs have been treated by cryosurgery. In general, the tumor requires operative exposure, which can be acquired by minimally invasive percutaneous techniques in some cases. The freezing is done with ultrasound monitoring whenever feasible.

Tumors of the liver, either primary in the liver or metastatic from cancers in other organs, have received a wide clinical trial of cryosurgery. Conventional surgery of liver tumors is associated with two major problems, which are the technical difficulty of the operation, chiefly related to hemorrhage, and the fact that the majority of hepatic tumors are not suitable for surgical excision. With treatment by cryosurgery, prolonged survival may be achieved in ∼25% of patients, tumor recurs in the liver in ∼25%, and extra-hepatic disease becomes manifest in 50% of patients. Recurrence in the cryo-treated site occurs in 20% of patients (34). When one considers the fact that the patients selected for cryosurgery are those considered inoperable or unsuited for excisional surgery, the increased survival and chance of cure represents good results.

The application of cryosurgical techniques to kidney tumors is in clinical trial. The successful use of ultrasound to monitor the freezing process in tumors of the prostate and the liver has encouraged use in tumors of the kidney, especially when conservation of kidney function is critical to the patient. At the present time, in patients with marginal kidney functional reserve, partial nephrectomy is an option in therapy. In a similar manner, cryosurgical ablation offers the possibility of conserving renal tissue. The techniques and tissue effects of renal cryosurgery are similar to those in other viscera, such as the liver. Early clinical results indicate that cryosurgery is useful in the treatment of small tumors of the kidney in patients with associated disease (35).

Bone Tumors

Cryosurgical techniques have achieved a small place in the management of bone tumors. Cryosurgery is a conservative method of management and is better suited to the management of benign bone tumors rather than malignant tumors. The best technique requires that cryosurgery be used in combination with curettage, removing most of the soft tumor by scraping with a curette. This forms a cavity in the bone, and cryosurgery is used to freeze the cavity, destroying any residual tumor. In bone tumors, the freezing is performed with diverse techniques, including use of probes or by spraying or pouring liquid nitrogen into the bone cavity. When the cryosurgical treatment is finished, the bone cavity is filled with bone grafts or acrylic cement for added support. Cryosurgery is not yet used widely for bone tumors, but the results achieved by a few surgeons have been outstanding and have shown that amputation of selected long bones can be avoided (36).

Rhythm Disorders of the Heart

In recent years, cryosurgical techniques have been adapted to the treatment of arrhythmias of the heart, which is particularly useful for disabling abnormally fast heart rates that begin in the atrium. To prevent the ventricle from following the fast atrial rate, cryosurgical ablation of the atrioventricular node (AV node) is used to produce heart block. Then a pacemaker is implanted to maintain the cardiac rate. Similar techniques can be used to inactivate irritable foci in the ventricle or accessory pathways of conduction that produce dangerous fast heart rhythms. Cardiac cryosurgery is commonly done with catheter cryoprobes, passed into the heart from a peripheral blood vessel. The control of the freezing depends on electrophysiological monitoring that confirms that the desired freezing has been produced (37). Recent research related to cardiac disease has been in the direction of using warmer freezing temperatures to elicit a specific tissue response. The object is to stimulate angiogenesis as treatment of myocardial ischemia or to inhibit myogenesis as needed to control the smooth muscle response to the injury caused by balloon angioplasty (38). To achieve these therapeutic goals may require the use of adjunctive chemotherapy.

COMPARISON WITH OTHER METHODS OF TREATMENT

The advantages of cryosurgery are most evident when used to treat an accessible lesion without excising any tissue. When used in this way, the technique is quick, relatively painless, and associated with little or no blood loss. Since usually no tissue is excised, there is no opportunity for tumor cell implantation in the open wound. Conservation of tissue, especially bone, is possible. The lethal effect of freezing extends into bone and destroys any tumor cells that may be present, while the devitalized matrix remains as a structural basis for later repair. Most patients have little discomfort after surgery because of the desensitizing effect of cold on sensory nerves. Nerve function, lost as a consequence of freezing, commonly returns after several months. Wound healing is surprisingly good, but is delayed by the need to await necrosis and sloughing of devitalized tissue. Soft tissue wounds heal in a month. The healing of soft tissue is favorable, and extensive scarring is rare. Bone repair may require a year or longer for completion.

The principal disadvantages are that an entire specimen is not available for histologic examination and healing is slower than with excision and closure. The lack of a specimen is a circumstance shared by other methods of treatment, such as electrocoagulation and radiotherapy. In fact, only complete surgical excision yields the specimen for study as a whole. Slow healing is not entirely disadvantageous, since it provides time to study the open wound and is associated with favorable healing.

The chief limitation with cryosurgery, especially when dealing with cancer, is the difficulty in freezing sufficient tissue. The best presently available apparatus is barely adequate for large cancers because tissues are poor conductors of heat and are provided with a source of heat that limits the extension of freezing. The amount of tissue that can be frozen in a single application of a probe is small in comparison with the size of many cancers. Depth of freezing beyond 2 cm is difficult to achieve. Multiple applications of the probe, or the use of multiple probes and repetition of freezing are the methods of compensating for the difficulties of freezing sufficient tissue.

As a form of producing local necrosis of tissue, cryosurgery must be compared with the other methods of producing similar effects, including conservative local excision and electrocoagulation. Freezing *in situ* requires less anesthesia than electrocoagulation, and the freezing process is more easily controlled and yields a necrotic lesion of more predictable size. The scar after healing is less with cryosurgery than with electrosurgery. Conservative local excision is applicable to many cancers, but has greater risk of bleeding and infection than cryosurgery. Cryosurgery is better for lesions that rest on underlying bone because the extensive destruction can be achieved without excising bone and producing undesirable and unnecessary postoperative disability. In some advanced skin cancers that are difficult problems in management,

the results with Moh's technique of serial excision and immediate histologic monitoring are strongly competitive with cryosurgery.

Laser therapy is competitive with other methods of tissue destruction. The advantages of laser therapy closely resemble those of cryosurgery from the standpoint of healing, speed of treatment of multiple lesions, and usefulness in many diseases. The limitations of laser therapy are related to the expense of the equipment, considerations of safety in use, and the difficulty in determining the needed depth of treatment.

Radiation therapy may also be used for selected problems in management of tumors and has advantages in difficult areas, such as about the eyelids, the nose, and the anus. The results in these sites indicate that radiotherapy is a legitimate alternative choice in many patients; nevertheless, cryosurgery can achieve the same results, yet faster. Radiotherapy is not a good choice for lesions with bone invasion.

In comparison to many other tools used by the surgeon, cryosurgical apparatus is simple to use and the techniques of cryosurgery are easily learned. Nevertheless, as with any other technique that physicians use to cure disease, the key to success in therapy is in the selection of patients suitable to cryosurgical treatment and in the proper use of the technique. Considerable attention must be placed on technique that ensures good heat transfer from the tissues and an appropriate amount of destruction by freezing. Physicians who use cryosurgery for cancer should be familiar with alternate accepted methods of treatment so that an appropriate choice of therapy can be made and a change to a different method of therapy will be made when appropriate. Under these conditions, in the hands of an experienced physician, cryosurgery can be used to solve many difficult problems in cancer therapy and will yield surprisingly good results in carefully selected patients (39).

FUTURE DIRECTIONS

Cryosurgery has achieved a modest stature in medical, dental, and veterinarian practice. In the past 15 years, considerable progress has been made in cryosurgical techniques as a result of improved technology. With the availability of percutaneous and endoscopic probes, visceral and other deep lesions can be treated. Nevertheless, the available apparatus, even with the use of multiple probes, still has limited freezing capability. Improvements in equipment are needed. Current efforts are focused also on improvements in the imaging techniques, which will facilitate the control of the tissue freezing process.

The current direction of research explores cryosurgical techniques that are combined with other methods of treatment, such as radiotherapy and cancer chemotherapy, for the treatment of invasive cancers (40). Rather little experience has been reported concerning combinations of cryosurgery with radiotherapy, but the methods are complementary, with the effects of radiotherapy longer lasting in contrast to the quick destruction by freezing *in situ*. In advanced cancers, cryosurgery has also been used in combination with the systemic or local administration of cancer chemotherapeutic drugs. This use is based on the hypothesis that cells that are injured, but not necessarily killed by freezing *in situ*, might be more susceptible to complete destruction by an antineoplastic agent.

Current research shows continued interest in the specific immunological response against antigens of frozen tissue. The practical clinical benefit is in the possibility that freezing a cancer in its primary site would produce an immunological response that would destroy cancer in distant sites to which it had spread. Some reports suggest that such benefit has been achieved, especially in the treatment of advanced cancer of the prostate gland. Furthermore, a specific immunological benefit has been shown by several groups of investigators working with experimental tumor systems in animals. Unfortunately, evidence of clinical benefit remains unclear. Therefore, one can conclude only that the potential use of specific immunotherapy is an attractive feature of cryosurgery that requires further investigation.

Differences in the sensitivity to cold injury of the several types of cells are also of importance to the future development of cryosurgery (38). The cells of bone, osteocytes, are very susceptible to freezing and die at minimal subfreezing temperatures. In skin, the pigment-bearing cells, melanocytes, are highly sensitive to cold injury, and selective destruction of melanocytes can be achieved at temperatures of -4 to $-20\,°C$ in single short exposures. At this temperature range, the damage to other epidermal cells is minimal. Squamous cells of skin resist freezing injury at temperatures as cold as $-20\,°C$. The importance of these differences is in the possibility that selective destruction of cells may be possible and some therapeutic advantage may be gained. The recent demonstration of the occurrence of apoptosis in the periphery of the cryogenic lesion points the way to molecular-based optimization of therapeutic freezing techniques, whether in the direction of selective destruction, partial preservation, or more complete destruction of cells. This appears to be the most potentially rewarding direction of cryosurgical research.

BIBLIOGRAPHY

1. Henderson A. Cold: Man's assiduous remedy. Med Ann DC 1971;40:583–588.
2. Cooper I, Lee A. Cryostatic congelation: A system for producing a limited controlled region of cooling and freezing of biologic tissues. J Nerv Dis 1961;133:259–263.
3. Gage AA. History of cryosurgery. Sem Surg Oncol 1998;14:99–109.
4. Gage AA, Baust JG. Mechanisms of tissue injury in cryosurgery. Cryobiology 1998;37:171–186.
5. Rubinsky B. Cryosurgery. Annu Rev Biomed Eng 2000; 2:157–187.
6. Hoffmann NE, Bischof JC. The cryobiology of cryosurgical injury. Urology 2002;60:40–49.
7. Clarke DM, Hollister WB, Baust JG, VanBuskirk RG. Cryosurgical modeling: sequence of freezing and cytotoxic agent application affects cell death. Mol Urol 1999;3:25–31.
8. Yang WL, Addona T, Nair DG, Qi L, Ravikumar TS. Apoptosis induced by cryo-injury in human colorectal cancer cells is associated with mitochondrial dysfunction. Int J Cancer 2003;103:360–369.
9. Li AK, Ehrlich HP, Trelstad RL, Koroly MJ, Schattenkerk ME, Malt RA. Differences in healing of skin wounds caused by burn and freeze injuries. Ann Surg 1980;191:224–248.
10. Gage AA, Fazekas G, Riley EE. Freezing injury to large blood vessels in dogs. Surgery 1967;61:748–754.
11. Gage AA, Greene GW, Neiders ME, Emmings FG. Freezing bone without excision—an experimental study of bone cell destruction and manner of regrowth. JAMA 1966;196:770–774.
12. Gage AA. Cryosurgery in the treatment of cancer. Surg Gynecol Obstet 1992;174:73–92.
13. Baust JG, Gage AA, Ma H, Zhang CM. Minimally invasive cryosurgery—technological advances. Cryobiology 1997;34:373–384.
14. Saliken JC, Cohen J, Miller R, Rothert M. Laboratory evaluation of ice formation around a 3-mm Accuprobe. Cryobiology 1995;32:285–295.
15. Popken F, Seifert JK, Englemann R, Dutkowski P, Nassir F, Junginger T. Comparison of iceball diameter and temperature distribution achieved with 3-mm Accuprobe cryoprobes in porcine and human liver tissue and human colorectal liver metastases *in vitro*. Cryobiology 2000;40:302–310.
16. Hewitt PM, Zhao J, Akhter J, Morris DL. A comparative laboratory study of liquid nitrogen and argon gas cryosurgery systems. Cryobiology 1997;35:303–308.
17. Hamilton A, Hu J. An electronic cryoprobe for cryosurgery using heat pipes and thermoelectric coolers. J Med Eng Technol 1993;17:104–109.
18. Diller KR. Engineering-based contributions in cryobiology. Cryobiology 1997;34:304–314.
19. Gage A, Caruana J, Garamy G. A comparison of instrument methods of monitoring freezing in cryosurgery. J Dermatol Surg Oncol 1983;9:209–214.
20. Le Pivert P, Binder P, Ougier T. Measurement of intra tissue bio-electrical low frequency impedance: A new method to predict preoperatively the destructive effect of cryosurgery. Cryobiology, 1977;14:245–250.
21. Gage AA, Augustynowicz S, Montes M, Caruana J, Whalen D. Tissue impedance and temperature measurements in relation to necrosis in experimental cryosurgery. Cryobiology, 1985;22:282–288.
22. Homasson JP, Thiery JP, Angebault M, Outracht L, Maiwand O. The operation and efficacy of cryosurgical nitrous oxide-driven cryoprobe. Cryobiology 1994;31:290–304.
23. Harly S, Aastrup J, Elbrand O. Heat exchange in cryosurgery of Meniere's disease: Experimental and clinical studies. Cryobiology 1977;14:609–613.
24. Bradley P. Thermography as an aid to cryosurgery. Acta Thermographica 1977;2:83–90.
25. Pogrel MA, Yen CK, Taylor R. A study of infrared thermographic assessment of liquid nitrogen cryotherapy. Oral Surg Oral Med Oral Path 1996;81:396–401.
26. Brewer WH, Austin RS, Capps GW, Neifeld JP. Intraoperative monitoring and postoperative imaging of hepatic cryosurgery. Sem Surg Oncol 1998;14:129–155.
27. Saliken JC, Donnelly BJ, Rewcastle JC. The evolution and state of modern technology for prostate cryosurgery. Urology 2002;60:26–33.
28. Chin JL, Downey DB, Mulligan M, Fenster A. Three-dimensional transrectal ultrasound guided cryoablation for localized prostate cancer in nonsurgical candidates: A feasibility study and report of early results. J Urol 1998;159: 910–914.
29. Gilbert JC, Rubinsky B, Wong ST, Brennan KM, Pease GR, Leung PP. Temperature determination in the frozen region during cryosurgery of rabbit liver using MR image analysis. Magn Reson Imaging 1997;15:657–667.
30. Mala T, Edwin B, Tillung T, Kristian HP, Soreide O, Gladhaug I. Percutaneous cryoablation of colorectal liver metastases

potentiated by two consecutive freeze–thaw cycles. Cryobiology 2003;46:99–102.

31. Otten DM, Onik G, Rubinsky B. Distributed network imaging and electrical impedance tomography of minimally invasive surgery. Tech Cancer Res Treatment 2004;3:125–133.

32. Maiwand MO, Asimakopoulos G. Cryosurgery for lung cancer: Clinical results and technical aspects. Tech Cancer Res Treatment 2004;3:143–150.

33. Bahn DK, Lee F, Bodalament R, Kumar A, Greski J, Chernick M. Targeted cryoablation of the prostate: 7-year outcomes in the primary treatment of prostate cancer. Urology 2002;60:3–11.

34. Seifert JK, Junginger T. Cryotherapy for liver tumors: current status perspectives, clinical results, and review of the literature. Tech Cancer Res Treatment 2004; 3:151–163.

35. Spaliviero M, Moinzadeh A, Gill I. Laparoscopic cryotherapy for renal tumors. Tech Cancer Res Treatment 2004;3:177–180.

36. Bickels J, Meller I, Shmookler BM, Malawer MM. The role and biology of cryosurgery in the treatment of bone tumors. A review Acta Orthop Scand 1999;70:308–315.

37. Rodriguez LM, Timmermans C. Transvenous cryoablation of cardiac arrythmias. Tech Cancer Res Treatment 2004;3:515–524.

38. Gage AA. Selective cryotherapy. Cell Pres Tech 2004; 2:3–14.

39. Gage AA, Baust JG. Cryosurgery for tumors—a clinical overview. Tech Cancer Res Treatment 2004;3:187–199.

40. Baust JG, Gage AA. Progress toward optimization of cryosurgery. Tech Cancer Res Treatment 2004;3:95–101.

See also MINIMALLY INVASIVE SURGERY; TISSUE ABLATION.

CRYOTHERAPY.   See HEAT AND COLD, THERAPEUTIC.

CT SCAN.   See COMPUTED TOMOGRAPHY.

CUTANEOUS BLOOD FLOW, DOPPLER MEASUREMENT OF

LALITA KHAODHIAR
ARISTIDIS VEVES
Harvard Medical School
Boston, Massachusetts

INTRODUCTION

The necessity for measuring the skin blood flows occurs in many areas of physiology, pharmacology, and clinical medicine. Although the measurement of blood flow in the large blood vessels in the human body has been performed for centuries, the use of techniques to explore microcirculation have just evolved over the past 30 years. Available tests for assessing the skin microcirculation include tissue pH measurement, radioactive isotope clearance, capillary microscopy, plethysmography, transcutaneous oxygen tension, ultrasonic Doppler flowmetry, and laser Doppler flowmetry (1–5). Each method relies on different physiology principles and has its own advantages and disadvantages (Table 1). Currently, there is no gold standard test for the evaluation of skin blood flow and clinical observation remains the most acceptable method for assessing blood flow in the skin in clinical practice (6,7). The ideal blood flow measurement technique should be simple, noninvasive, reproducible, and able to provide a continuous measurement of skin blood flow.

LASER DOPPLER FLOWMETRY

Laser Doppler flowmetry is the most widely accepted technique currently used for evaluating blood flow in the skin microcirculation. The basic technology underlying laser Doppler was introduced in 1975 by Stern, who demonstrated that the use of laser Doppler shifted light to measure the moving blood cell in the skin microcirculation. This technique has been in clinical use since 1977 (8,9) and since then it has been extensively studied, particularly in the field of vascular surgery, rheumatology, and dermatology. Although the laser Doppler flowmetry technology and data processing have continued to evolved, it has yet to gain the widespread acceptance for clinical applications. In this article, the principle, instrumentation (laser probe and laser scanning), and the clinical applications are discussed.

PRINCIPLES

This technique depends on the Doppler principle, which is the alteration in the frequency of light that is emitted or reflected by a moving object. The Doppler frequency shift can be calculated using the following equation:

$$df = v/c \, f$$

where df is magnitude of the frequency shift; v is the velocity of the moving object with respect to the observer; c is the velocity of light, and f is the frequency of unshifted light. This means when light hits a moving object, it undergoes a frequency shift that is proportional to the velocity of the moving object (10).

Because of the movement of red blood cells in the skin microvascular network, low power light from a monochromatic stable laser is scattered and as a result is frequency shifted. Since the velocity of the red blood cells is ∼ 10 orders of magnitude smaller than the speed of light, it is impossible to measure this frequency-shifted light directly. The laser Doppler flowmetry, however, provides an indirect measurement of red cell velocity as follows: when the coherent laser light hits a surface, the light scattered from the red blood cells undergoes a frequency shift, but the light from the surrounding area remains at the same frequency as the transmitted light. The mixing of these two different light frequencies produces a beat frequency, that is, an oscillation of the measured light intensity. This beat frequency can be detected by the laser Doppler machine, and then analyzed to provide a skin blood flow measurement (4,11,12).

The term commonly used to describe blood flow measured by the Doppler techniques is flux, which is the amount relative to the product of the number of moving red cells in a given volume and their mean net velocity. The flux can be calculated using analogue circuitry or high speed digital processing. All laser measurements are usually expressed in volts and depend on the voltage difference created by the returned light to the computer. Higher blood flow at the skin

Table 1. Techniques for Assessing Skin Microcirculation

Technique	Advantages	Disadvantages
Photoplethysmography	Noninvasive, accurate, reproducible	Estimate based on changes in blood volume, not blood flow. Not useful in darkly pigmented skin
Transcutaneous oxygen tension	Noninvasive, provides physiologic and nutritional microcirculation assessment	Indirect estimate of blood flow, the measurement is affected by the affinity of blood for oxygen and the change in skin temperature
Thermometry	Noninvasive, correlated closely with capillary density	Flow estimated based on skin temperature, which can be influenced by ambient temperature, pain, anxiety
Capillary microscopy	Noninvasive, provides information on capillary size and number	Provides a relative estimate of blood flow based on visual characteristics
Ultrasonic doppler flowmetry	Noninvasive, measure nutritive perfusion	Probe is highly sensitive to motion,
Laser doppler flowmetry	Noninvasive, provides continuous real-time measurement of skin perfusion	Probe-poor reproducibility

level results in a higher amount of light picked by the laser Doppler and a higher voltage recorded by the computer.

Generally, the laser Doppler flowmetry measures blood flow in the very small blood vessels of the microvasculature, such as flow in the underlying arterioles and venules that regulate skin temperature and the low speed flows associated with nutritional blood flow in capillaries close to the skin surface. Thus, this technique does not differentiate between nutritional and non-nutritional skin perfusion (13).

INSTRUMENTATION

There are two types of laser Doppler flowmetry devices; a single-point laser probe, which evaluates the microvascular blood flow at one point of the skin, or a real-time laser scanner, which evaluates the blood flow in an area of skin.

Single-Point Laser Probe

In a single-point laser probe, laser light is transmitted to the tissue surface via optic fiber. The optic fiber terminates in an optic probe, which can be attached to the tissue surface. One or more light collecting fibers also terminate in the probe head and these fibers transmit a proportion of the scattered light to a photodetector and the signal processing electronics. Normal fiber separations in the probe head are a few tenths of a millimeter, so consequently blood flow is measured in a tissue volume of typically 1 mm³ or smaller. The measuring volume (depth) of laser penetration is generally ~ 1–1.5 mm, but it is dependent on many factors, such as probe configuration (14), laser light wavelength (3), and skin pigmentation. A light source with wavelength 543 nm has less penetration depth than 633 nm, which has less penetration depth than 780 nm (15). In clinical medicine, a wavelength of 633 nm, is generally used. The distance between the transmitting and receiving fibers (fiber separation) also influences the penetration depth, with the increasing depth with greater fiber separation.

The single-point laser probe is mainly used for evaluating the hypcremic response to a heat stimulus, or for evaluating the nerve-axon-related hyperemic response.

Heat-Related Hyperemic Response. To assess heat-related hyperemic response, the baseline blood flow is first measured. The skin is then heated to 44 °C for 20 min using a small brass heater. The measurement of the maximum blood flow is subsequently repeated to evaluate the magnitude of change from baseline.

Nerve-Axon-Related Hyperemic Response. The nerve-axon related-hyperemic response evaluates the integrity of the neurovascular function. In healthy subjects, the ability to increase blood flow depends on the existence of normal neurogenic vascular response, which is conducted through the C nociceptive nerve fibers. Stimulation of these nerve fibers leads to antidromic stimulation of adjacent C fibers, which secrete substance P, calcitonin gene-related peptide (CGRP) and histamine, causing vasodilatation and increased blood flow to the injured tissues, thereby promoting wound healing (Lewis' triple flare response) (Fig. 1). For this measurement, two single-point laser probes are applied (Fig. 2). One probe measures the blood flow to an area of skin, which is exposed directly to

Neurogenic vascular response

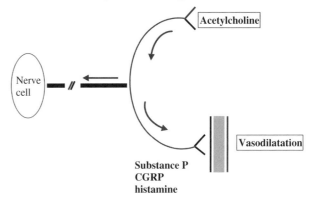

Figure 1. Stimulation of the C-nociceptive nerve fibers leads to antidromic stimulation of the adjacent C fibers, which secrete substance P, calcitonin gene related peptide (CGRP), and histamine that cause vasodilatation and increased blood flow.

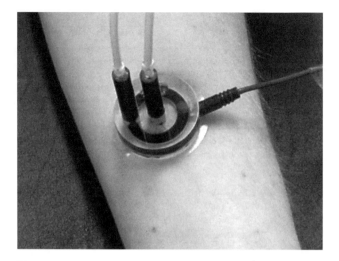

Figure 2. Measurements of direct and indirect effect of vasoactive substance using single-point laser probes: One probe is used in direct contact with the iontophoresis solution chamber (colored ring) and measures the direct response. The center probe measures the indirect response (nerve axon-related effect). A small quantity (<1 mL) of 1% acetylcholine chloride solution or 1% sodium nitroprusside solution is placed in the iontophoresis. A constant current of 200 mA is applied for 60 s achieving a dose of 6 mC cm^{-2} between the iontophoresis chamber and a second nonactive electrode placed 10–15 cm proximal to the chamber (black strap around the wrist). This current causes a movement of solution to be iontophorized toward the skin.

acetylcholine (Ach). The second probe, placed in close proximity (5 mm), measures the indirect effect of applied Ach. This indirect effect results from stimulation of C-nociceptive nerve fibers of the adjacent area and reflects the stability of the nerve-axon-related reactive hyperemia.

Our lab has examined the contribution of the nerve-axon reflex-related vasodilation to the total endothelium-dependent vasodilation at the forearm and the foot level in healthy adults and patients with diabetes mellitus (16). In healthy adults, the nerve-axon reflex-related response is approximately equal to one-third of the total response to Ach at both the forearm and the foot level. In diabetic patients with microvascular complications including diabetic neuropathy, Charcot arthropathy, and peripheral vascular disease, this contribution was significantly diminished. Another study demonstrated that the nerve-axon reflex-related vasodilatation is directly related to the function of the C-nociceptive fibers and is significantly associated with other nerve function measurements (17). As this method is an objective measurement, it is potentially useful as an alternative to currently employed techniques to evaluate small nerve fiber function.

Single-point measurements give a high temporal resolution (40 Hz data rates are typical) enabling rapid blood flow changes to be recorded. However, there are several limitations of the conventional laser Doppler probe. Because the probe is directly contacted to the skin, it can only measure the restricted area ~ 1 mm^2 at one time. Its pressure on the skin itself may also alter the skin blood flow (18). In addition, the probe is very sensitive to motion and vibration while it has poor reproducibility (3).

The Laser Scanning Method

The laser Doppler scanner–imager has been developed in response to the limitation of the laser Doppler probe. In the laser scanning method, a larger area of the skin can be studied while avoiding the contact between the scanner and the tissue being assessed. This technique is based on the same principle of measuring blood flow as the laser probe, but instead of the fiber optic probes, a system of mirrors and light-collecting lenses are used (14). This technique, the low intensity laser beam, is scanned across tissue surface in a raster fashion using a moving mirror. The scanner can scan the area from 5×5 cm to up to 50×50 cm. Light reflected back from the skin is then detected by a photodetector, which is connected to the computer enabling a mapping and a display of color-coded images of the blood flow. Regions of interest can then be defined and statistical data are calculated and recorded. This technique is also useful for the study of the skin microcirculation in response to various vasoactive substances.

To evaluate the endothelium-dependent and the endothelium-independent microvascular reactivity, the laser scanning method is used through the iontophoresis technique. The conditions associated with endothelial dysfunction are listed in Table 2.

The term iontophoresis denotes the introduction of soluble ions into the human skin by applying electric current. Using this technique, vasoactive substances can be applied to a localized area of the skin. The delivered dose depends on the current flowing and its duration. The test is noninvasive and avoids any systemic effects of the used drugs. By applying Ach chloride, the endothelium-dependent vasodilatation can be measured, while the use of sodium nitroprusside (SNP) measures the endothelium-independent vasodilatation.

In this technique, a delivery vehicle device is attached firmly to the skin with double-sided adhesive tape. The device contains two chambers that accommodate two single-point laser probes. A small quantity of (< 1 mL) of 1% Ach solution or 1% of SNP solution is placed in the iontophoresis chamber and a constant current of 200 mA is applied for 60 s, achieving a dose of 6 mC·cm^{-2} between the

Table 2. Conditions Associated with Impaired Endothelial Function

Atherosclerosis
Hypertension
Dyslipidemia; high LDL-C, low HDL-C, small dense LDL-C
Diabetes mellitus and impaired glucose tolerance
Metabolic syndrome
Obesity
Congestive heart failure
Preeclampsia
Vasculitis
Renal failure
Menopause
Family history of coronary heart disease
Family history of diabetes
Smoking
Inactivity

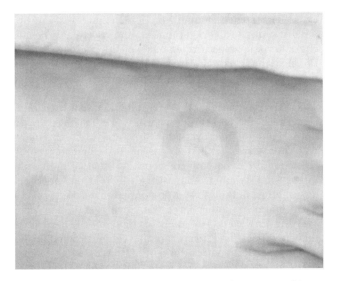

Figure 3. A normal response of blood flow in a skin to iontophoresis technique. Vasodilatation occurs in both the area that contact with the iontophoresis solution and area adjacent to, but not in direct contact with, the solution.

iontophoresis chamber and a second nonactive electrode placed 10–15 cm proximal to the chamber. This current causes a movement of solution to be iontophoresed toward the skin, resulting in vasodilatation (Fig. 3).

After the adhesive device has been removed, the localized area exposed to the vasoactive substances is scanned. The laser Doppler perfusion imager employs a 1 mW helium–neon laser beam of 633 nm wavelength, which sequentially scans an area of skin (Fig. 4). The maximum number of measured spots is 4096, and the apparatus produces a color-coded image of skin erythrocyte flux on a computer monitor. The scanner is set up to scan up to 32×32 measurement points over an area $\sim 4 \times 4$ cm.

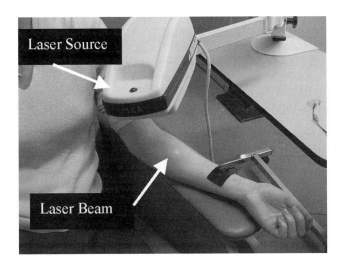

Figure 4. Laser doppler flowmetry: A helium–neon laser beam is emitted from the laser source to sequentially scan the circular hyperemic area (seen surrounding the laser beam) produced by the iontophorized vasoactive substance to a small area on the volar surface of the forearm.

Generally, the laser Doppler scanner is not a reliable method for a quantification of the nerve-axon reflex related vasodilatation. However, Krishnan and Rayman recently published a technique assessing the axon-reflex related vasodilatation using the laser Doppler imager (LDI) called LDIflare. This method involved skin heating to 44 °C to evoke flare followed by scanning the site using a laser Doppler imager. The LDIflare was markedly reduced in healthy control subjects after topical administration of anesthesia, confirming its neurogenic nature. Similarly, LDIflare was decreased in patients with diabetic neuropathy when compared to diabetic patients with no neuropathy or healthy controls. The authors suggested that the LDIflare may be a simple objective method to detect early neuropathy and may be useful in assessing therapeutic interventions aimed at preventing or reversing C-fiber function.

In summary, the laser Doppler scanning offers a more global assessment of skin perfusion than the laser Doppler probe, while avoiding a direct contact to the skin surface. The scanning technique is best suited for studying the relative changes in flow induced by variety of physiological maneuvers or pharmaceutical intervention procedures.

VALIDATION

Single-Point Laser Probe

The technique has been validated against direct measurements of the capillary flow velocity (19). The day-to-day reproducibility of the technique was evaluated in healthy subjects who were repeatedly tested at their foot and arm for 10 consecutive working days in our lab. The coefficient of variation (CV) for the baseline blood flow measurement obtained with the laser probe evaluating the response to heat was 44.0%, while that for the maximal response to heat was 27.9%. The indirect response to Ach, measured by a single-point laser probe, had a CV of 60.6% for the baseline measurements and 35.2% for the maximal hyperemic response after the iontophoresis. Note that the variability of the technique is mostly a spatial one, that is, the variability is mainly related to the high heterogeneity of the skin microcirculation and not to the technique itself. The reproducibility of the single-point laser Doppler can be improved when one pays attention to place the laser probe approximately in the same skin area for repeated measurement.

The Laser Scanner Method

The laser scanner method has a significantly better reproducibility than a single-point laser probe. The CV at the foot and forearm level is ranged between 14 and 19%. For the laser Doppler perfusion imaging measurements before and after the iontophoresis of Ach, the reproducibility of the technique was evaluated in our lab in five healthy subjects (four males and one female, age 23–39 years) who were repeatedly tested at the forearm for 10 consecutive working days. The coefficient of variation of the baseline measurement before the iontophoresis of Ach was 14.1% and during maximal hyperemic response after the iontophoresis it was 13.7% (16,20).

CLINICAL APPLICATIONS

Laser Doppler flowmetry have several clinical applications. It has been used to monitor to ischemic tissue (2), to follow the progress of atherosclerotic disease, to evaluate therapeutic effects of drugs or operations (21), and to study skin under both normal physiologic states and the pathogenetic mechanisms underlying pathologic skin disorders (22,23).

Atherosclerosis. Peripheral arterial disease (PAD) of the lower extremity is a common manifestation of the atherosclerotic process. As many as 10 million people in the United States have PAD with a prevalence > 10% in people aged > 60 years (24). Up to three-quarters of patients with PAD are asymptomatic (25,26). Those with symptoms usually present with intermittent claudication, resting ischemia or foot ulceration. Kvernebo et al. reported laser Doppler flowmetry values decreased progressively from healthy controls to patients with intermittent claudication to those with critical ischemia (21). The test was reproducible on a given population particularly with the local heating technique, but the reproducibility was poor for individual subjects. The possible explanations include changes in sympathetic vascular tone and different vascular architecture in the measuring volumes, which are only some few cubic millimeters (mm^3). These patients with PAD also demonstrate a reduced skin hyperemic response to iontophoresis of ACh and to SNP (27), suggesting both endothelium and smooth muscle dysfunction. In addition, when they exercise, a reduction of the leg skin hyperemic response to ACh delivery is more pronounced at peak of claudication than prior exercise. The authors suggested acute endothelial dysfunction occurs with exercise-induced leg muscle ischemia and threshold of claudication should not be exceeded during rehabilitation programs for PAD patients.

Laser Doppler flowmetry may also be useful for follow-up after therapeutic intervention in patients with PAD. Ray et al. (28) studied 41 patients who underwent technically successful revascularization for severe leg ischemia. Laser Doppler flowmetry performed before the procedure was better than toe and ankle systolic pressures in identifying patients who continued to have ischemic symptoms or required amputation after surgery. Another study reported the prevalence of high frequency flux waves, which increased in peripheral ischemia decreased after successful percutaneous angioplasty (29). One randomized, double-blind, placebo-controlled trial examining the effect of glutathione in patients with peripheral artery disease reported improvement in pain-free walking distance (PFWD) and microciculatory (laser Doppler flowmetry) test (30). Another study, the START-trial: STimulation of ARTeriogenesis using subcutaneous application of GM-CSF as a new treatment for peripheral vascular disease (31), is underway. In this study, the primary endpoint will be the change walking distance from day 0 to day 14 as assessed by an exercise treadmill test, while the cutaneous microcirculation alterations assessed by laser Doppler flowmetry will be one of the secondary endpoints.

Diabetes Mellitus. Over the last two decades, it has become apparent that metabolic alterations in diabetes cause both structural and functional changes in multiple areas within the arteriolar and capillary systems. The most characteristic structural changes of the capillary circulation in diabetic patients are a reduction in the capillary size and thickening of basement membranes, while the major functional change includes a marked limitation of microvascular vasodilation (32–34). Studies have shown that the structural changes are essentially responsible for the impairment of microvascular function. Microcirculatory test in patients with diabetes mellitus demonstrated a reduction in hyperemic response to heat stimulus and minor skin trauma, suggesting the role of endothelial dysfunction as the cause of the impaired vascular reactivity at microcirculatory level (35). Such dysfunction occurs early in the course of diabetes and may even predict diabetic micro- and macrovascular complications (36,37).

Several therapeutic interventions have been tested both *in vitro* and/or *in vivo* aiming to improve endothelial function in patients with diabetes mellitus. This includes insulin sensitizers (thiazolidenediones), angiotensin-converting enzyme (ACE) inhibitor, lipid lowering medications (statins), antioxidant, and exercise. To date, the studies have shown these interventions had no effects on the skin microcirculation in patients with diabetes, including both resting skin blood flow and blood flow after the iontophoresis of acetylcholine and sodium nitroprusside (38–41).

Skin Disease. Several skin diseases are associated with abnormal cutaneous blood flow, for example, psoriasis, skin inflammatory reaction, skin cancers, and scleroderma. Laser Doppler flowmetry has been used to determine differences in blood flow within psoriatic plaques, to identify the location of their leading edge, and to monitor the treatment (42,43). Laser Doppler perfusion imaging may also allow differentiation between different types of skin tumor. Stucker et al. found that malignant melanomas were significantly more perfused than basal cell carcinomas and tended also to be more so than melanocytic naevi (44,45).

Skin Burns. Laser Doppler flowmetry has been used to determine burn depth. Niazi et al. (46) demonstrated the Laser Doppler flowmetry flux values at 24 h after burn correlated with clinical assessment and histology reports, with low flux seen in deep dermal or full-thickness wound but high flux in superficial dermal wound (46). Park et al. (47) showed that laser Doppler flow measurements obtained within 72 h of burn injury correlated well with the depth of burn wound. The accuracy of laser Doppler in the assessment of burn depth is as high as 97%, compared with 60–80% for standard clinical methods (48). In addition, this technique has been used in concurrence with clinical judgment to objectively determine the need for excision of burns of indeterminate depth (49).

Skin Ulcer. Both ischemic and venous ulcers and their adjacent skin are associated with disorders in microcirculation. Together the deeper, subpapillary (thermoregulatory) network and the superficial (nutritive) network are affected. These network can be studied with laser Doppler flowmetry (subpapillary area) and capillary microscopy (superficial area). Gschwandtner et al. (50–53) examined the microcirculatory characteristics of ulcers. They described the remarkable local differences in subpapillary and nutritive perfusion in ischemic and venous ulcers and their surrounding skin (51,52). Ulcer areas without granulation tissue (sign of ulcer without healing) demonstrated low laser Doppler area flux and the very low capillary density. The lack of capillary in this area suggested the breakdown of nutritive microcirculation as an underlying cause of ulcers. Ulcer with granulation tissue (sign of wound healing) had high laser Doppler area flux and an intermediate capillary density, implying both thermoregulatory, and nutritive networks are essential for wound healing of ulcers. The skin area adjacent ulcers where healing process nearly completed had an intermediate laser Doppler area flux and highest capillary density, indicating an adequate thermoregulatory and nutritive perfusion (53). Bornmyr et al. (54) reported the use of laser Doppler perfusion imaging and digital photograph for postoperative evaluation of vascularized grafts and monitoring of treatment of chronic skin ulcers.

CONCLUSION

Laser Doppler flowmetry has already been an important tool for studying blood flow in the skin. The technique provides continuous, noninvasive real-time assessment of skin perfusion. It has been successfully used in many studies of the cutaneous blood flow in patients with atherosclerosis, diabetes, skin diseases, or skin ulcers.

BIBLIOGRAPHY

1. Chittenden SJ, Shami SK. Microvascular investigations in diabetes mellitus. Postgrad Med J 1993;69:419–428.
2. Furnas H, Rosen JM. Monitoring in microvascular surgery. Ann Plast Surg 1991;26:265–272.
3. Schabauer AM, Rooke TW. Cutaneous laser Doppler flowmetry: applications and findings. Mayo Clin Proc 1994;69:564–574.
4. Choi CM, Bennett RG. Laser Dopplers to determine cutaneous blood flow. Dermatol Surg 2003;29:272–280.
5. Fagrell B. Microcirculatory methods for the clinical assessment of hypertension, hypotension, and ischemia. Ann Biomed Eng 1986;14:163–173.
6. Hellner D, Schmelzle R. Laser Doppler monitoring of free microvascular flaps in maxillofacial surgery. J Craniomaxillofac Surg 1993;21:25–29.
7. Hirigoyen MB, Urken ML, Weinberg H. Free flap monitoring: a review of current practice. Microsurgery 1995;16:723–726; discussion 727.
8. Stern MD, Lappe DL, Bowen PD, Chimosky JE, Holloway GA Jr, Keiser HR, Bowman RL. Continuous measurement of tissue blood flow by laser-Doppler spectroscopy. Am J Physiol 1977;232:H441–H448.
9. Holloway GA Jr, Watkins DW. Laser Doppler measurement of cutaneous blood flow. J Invest Dermatol 1977;69:306–309.
10. Fischer JC, Parker PM, Shaw WW. Laser Doppler flowmeter measurements of skin perfusion changes associated with arterial and venous compromise in the cutaneous island flap. Microsurgery 1985;6:238–243.
11. Rendell M, Bergman T, O'Donnell G, Drobny E, Borgos J, Bonner RF. Microvascular blood flow, volume, and velocity measured by laser Doppler techniques in IDDM. Diabetes 1989;38:819–824.
12. Zinser G. Scanning laser Doppler flowmetry. In: Pillunat L, Harris A, Anderson D, Greve E, editors. Current concepts on ocularblood flow in glaucoma. The Hague: Kugler Publications; 1999: 197–204.
13. Rossi M, Carpi A. Skin microcirculation in peripheral arterial obliterative disease. Biomed Pharmacother 2004;58:427–431.
14. Essex TJ, Byrne PO. A laser Doppler scanner for imaging blood flow in skin. J Biomed Eng 1991;13:189–194.
15. Bonner RF, Nossal R. Principles of laser Doppler flowmetry. In: Shepherd AP, Oberg PA, editors. Laser Doppler Blood Flowmetry. Kluwer Academic Publishers; 1990.
16. Hamdy O, Abou-Elenin K, LoGerfo FW, Horton ES, Veves A. Contribution of nerve-axon reflex-related vasodilation to the total skin vasodilation in diabetic patients with and without neuropathy. Diabetes Care 2001;24:344–349.
17. Caselli A, Rich J, Hanane T, Uccioli L, Veves A. Role of C-nociceptive fibers in the nerve axon reflex-related vasodilation in diabetes. Neurology 2003;60:297–300.
18. Obeid AN, Barnett NJ, Dougherty G, Ward G. A critical review of laser Doppler flowmetry. J Med Eng Technol 1990;14:178–181.
19. Tooke JE, Ostergren J, Fagrell B. Synchronous assessment of human skin microcirculation by laser Doppler flowmetry and dynamic capillaroscopy. Int J Microcirc Clin Exp 1983;2:277–284.
20. Veves A, Akbari CM, Primavera J, Donaghue VM, Zacharoulis D, Chrzan JS, DeGirolami U, LoGerfo FW, Freeman R. Endothelial dysfunction and the expression of endothelial nitric oxide synthetase in diabetic neuropathy, vascular disease, and foot ulceration. Diabetes 1998;47:457–463.
21. Kvernebo K, Slagsvold CE, Stranden E, Kroese A, Larsen S. Laser Doppler flowmetry in evaluation of lower limb resting skin circulation. A study in healthy controls and atherosclerotic patients. Scand J Clin Lab Invest 1988;48:621–626.
22. Braverman IM. The cutaneous microcirculation. J Investig Dermatol Symp Proc 2000;5:3–9.
23. Braverman IM. The cutaneous microcirculation: ultrastructure and microanatomical organization. Microcirculation 1997;4: 329–340.
24. Criqui MH. Peripheral arterial disease—epidemiological aspects. Vasc Med 2001;6:3–7.
25. Hooi JD, Kester AD, Stoffers HE, Overdijk MM, van Ree JW, Knottnerus JA. Incidence of and risk factors for asymptomatic peripheral arterial occlusive disease: a longitudinal study. Am J Epidemiol 2001;153:666–672.
26. Stoffers HE, Rinkens PE, Kester AD, Kaiser V, Knottnerus JA. The prevalence of asymptomatic and unrecognized peripheral arterial occlusive disease. Int J Epidemiol 1996;25: 282–290.
27. Rossi M, Cupisti A, Perrone L, Mariani S, Santoro G. Acute effect of exercise-induced leg ischemia on cutaneous vasoreactivity in patients with stage II peripheral artery disease. Microvasc Res 2002;64:14–20.
28. Ray SA, Buckenham TM, Belli AM, Taylor RS, Dormandy JA. The predictive value of laser Doppler fluxmetry and transcutaneous oximetry for clinical outcome in patients undergoing revascularisation for severe leg ischaemia. Eur J Vasc Endovasc Surg 1997;13:54–59.
29. Bollinger A, Hoffmann U, Franzeck UK. Evaluation of flux motion in man by the laser Doppler technique. Blood Vessels 1991;28(1 Suppl): 21–26.
30. Arosio E, De Marchi S, Zannoni M, Prior M, Lechi A. Effect of glutathione infusion on leg arterial circulation, cutaneous microcirculation, and pain-free walking distance in patients with peripheral obstructive arterial disease: a randomized, double-blind, placebo-controlled trial. Mayo Clin Proc 2002; 77:754–759.

31. van Royen N, Piek JJ, Legemate DA, Schaper W, Oskam J, Atasever B, Voskuil M, Ubbink D, Schirmer SH, Buschmann I, Bode C, Buschmann EE. Design of the START-trial: STimulation of ARTeriogenesis using subcutaneous application of GM-CSF as a new treatment for peripheral vascular disease. A randomized, double-blind, placebo-controlled trial. Vasc Med 2003;8:191–196.

32. Jaap AJ, Shore AC, Stockman AJ, Tooke JE. Skin capillary density in subjects with impaired glucose tolerance and patients with type 2 diabetes. Diabetes Med 1996;13:160–164.

33. Jaap AJ, Pym CA, Seamark C, Shore AC, Tooke JE. Microvascular function in type 2 (non-insulin-dependent) diabetes: improved vasodilation after one year of good glycaemic control. Diabetes Med 1995;12:1086–1091.

34. Jaap AJ, Tooke JE. Pathophysiology of microvascular disease in non-insulin-dependent diabetes. Clin Sci (London) 1995; 89:3–12.

35. Sandeman DD, Shore AC, Tooke JE. Relation of skin capillary pressure in patients with insulin-dependent diabetes mellitus to complications and metabolic control. N Engl J Med 1992;327: 760–764.

36. Williams SB, Cusco JA, Roddy MA, Johnstone MT, Creager MA. Impaired nitric oxide-mediated vasodilation in patients with non-insulin-dependent diabetes mellitus. J Am Coll Cardiol 1996;27:567–574.

37. Stehouwer CD, Fischer HR, van Kuijk AW, Polak BC, Donker AJ. Endothelial dysfunction precedes development of microalbuminuria in IDDM. Diabetes 1995;44:561–564.

38. Economides PA, Caselli A, Tiani E, Khaodhiar L, Horton ES, Veves A. The effects of atorvastatin on endothelial function in diabetic patients and subjects at risk for type 2 diabetes. J Clin Endocrinol Metab 2004;89:740–747.

39. Hamdy O, Ledbury S, Mullooly C, Jarema C, Porter S, Ovalle K, Moussa A, Caselli A, Caballero AE, Economides PA, Veves A, Horton ES. Lifestyle modification improves endothelial function in obese subjects with the insulin resistance syndrome. Diabetes Care 2003;26:2119–2125.

40. Economides PA, Caselli A, Zuo CS, Sparks C, Khaodhiar L, Katsilambros N, Horton ES, Veves A. Kidney oxygenation during water diuresis and endothelial function in patients with type 2 diabetes and subjects at risk to develop diabetes. Metabolism 2004;53:222–227.

41. Caballero AE, Saouaf R, Lim SC, Hamdy O, Abou-Elenin K, O'Connor C, Logerfo FW, Horton ES, Veves A. The effects of troglitazone, an insulin-sensitizing agent, on the endothelial function in early and late type 2 diabetes: a placebo-controlled randomized clinical trial. Metabolism 2003;52: 173–180.

42. Speight EL, Essex TJ, Farr PM. The study of plaques of psoriasis using a scanning laser-Doppler velocimeter. Br J Dermatol 1993;128:519–524.

43. Speight EL, Farr PM. Calcipotriol improves the response of psoriasis to PUVA. Br J Dermatol 1994;130:79–82.

44. Stucker M, Hoffmann M, Memmel U, von Bormann C, Hoffmann K, Altmeyer P. [In vivo differentiation of pigmented skin tumors with laser Doppler perfusion imaging]. Hautarzt 2002;53:244–249.

45. Stucker M, Horstmann I, Nuchel C, Rochling A, Hoffmann K, Altmeyer P. Blood flow compared in benign melanocytic naevi, malignant melanomas and basal cell carcinomas. Clin Exp Dermatol 1999;24:107–111.

46. Niazi ZB, Essex TJ, Papini R, Scott D, McLean NR, Black MJ. New laser Doppler scanner, a valuable adjunct in burn depth assessment. Burns 1993;19:485–489.

47. Park DH, Hwang JW, Jang KS, Han DG, Ahn KY, Baik BS. Use of laser Doppler flowmetry for estimation of the depth of burns. Plast Reconstr Surg 1998;101:1516–1123.

48. Pape SA, Skouras CA, Byrne PO. An audit of the use of laser Doppler imaging (LDI) in the assessment of burns of intermediate depth. Burns 2001;27:233–239.

49. Jeng JC, Bridgeman A, Shivnan L, Thornton PM, Alam H, Clarke TJ, Jablonski KA, Jordan MH. Laser Doppler imaging determines need for excision and grafting in advance of clinical judgment: a prospective blinded trial. Burns 2003;29:665–670.

50. Gschwandtner ME, Koppensteiner R, Maca T, Minar E, Schneider B, Schnurer G, Ehringer H. Spontaneous laser doppler flux distribution in ischemic ulcers and the effect of prostanoids: a crossover study comparing the acute action of prostaglandin E1 and iloprost vs saline. Microvasc Res 1996;51: 29–38.

51. Gschwandtner ME, Ambrozy E, Fasching S, Willfort A, Schneider B, Bohler K, Gaggl U, Ehringer H. Microcirculation in venous ulcers and the surrounding skin: findings with capillary microscopy and a laser Doppler imager. Eur J Clin Invest 1999;29:708–716.

52. Gschwandtner ME, Ambrozy E, Schneider B, Fasching S, Willfort A, Ehringer H. Laser Doppler imaging and capillary microscopy in ischemic ulcers. Atherosclerosis 1999;142: 225–232.

53. Gschwandtner ME, Ambrozy E, Maric S, Willfort A, Schneider B, Bohler K, Gaggl U, Ehringer H. Microcirculation is similar in ischemic and venous ulcers. Microvasc Res 2001;62:226–235.

54. Bornmyr S, Martensson A, Svensson H, Nilsson KG, Wollmer P. A new device combining laser Doppler perfusion imaging and digital photography. Clin Physiol 1996;16:535–541.

See also BLOOD RHEOLOGY; HEMODYNAMICS.

CYSTIC FIBROSIS SWEAT TEST

WARREN J. WARWICK
University of Minnesota
Minneapolis, Minnesota

INTRODUCTION

Cystic fibrosis (CF, cystic fibrosis of the pancreas, mucovisidosis) is an autosomal recessive genetic clinical condition that occurs in ~ 1 in 2500 Caucasian newborn infants. The frequency ranges from ~1:200 in genetically isolated populations to 1:30,000 in certain ethnic and racial groups.

The CF mutations occur in a gene located in region q31.2 on the long (q) arm of human chromosome 7. The gene contains ~ 250,000 base (amino acids) pairs in 27 exons. These base pairs code for a 1480 amino acid molecule, which serves both as a transmembrane channel for the transfer of water and salt and other substances across the cell membrane and has been given the name: cystic fibrosis transmembrane regulator (CFTR) protein.

One mutation, the ΔF508 mutation, accounts for two-thirds of the mutations in Caucasian populations. Over 1338 mutations of the CFTR gene have been found in patients with one or more of the CF associated diseases although > 100 mutations have been found in patients with as yet none of the CF associated diseases. A few patients have been found with increased numbers of CF associated diseases, but with no mutations in the CFTR gene (1). These patients are presumed to have similar risk factors caused by other, as yet not identified, transmembrane channels. Such mutations or combinations of gene mutations, even in a large population, occur very rarely by chance (2). The most complete source of information about the CFTR gene is in the Cystic Fibrosis Mutation Database at http://www.genet.sickkids.on.ca/cftr/.

The CFTR protein controls sweat gland chloride ion transport as well as regulating other chloride secretory channels. This abnormality of chloride transport, which has been observed with almost all mutations, is an increased excretion of salt in the sweat. This elevation of chloride in sweat is seen in >98% of Caucasian CF patients. The clinical heterogeneity of CFTR expression is less obvious in the lung and other organs producing the diversity of clinical expression (phenotypes) of CF (3).

The reduced salt absorption in cystic fibrosis sweat glands is due primarily to poor chloride absorption with secondarily poor sodium absorption (4). A cyclic adenosine monophosphate (cAMP)-mediated sweating rate test has been developed that demonstrates a quantitative discrimination of CFTR function. This function may help distinguish between homozygous CF, CF carrier, and non-CF (5).

The measurement of the increased chloride in the sweat is the most reliable diagnostic test for the presence of two mutations of the CFTR mutation. This excessive sweat chloride is found in >98% of patients with the genetic potential to develop the clinical diseases associated with these CFTR mutations.

HISTORY

Although the folklore of many Caucasian peoples record that an infant who tastes salty will die young, the first modern confirmation of the excessive amount of salt in the sweat of children with CF was of seven patients with cystic fibrosis admitted to a New York City hospital with heat prostration (6). Although five other children also had heat prostration the unusual association with cystic fibrosis of the pancreas was investigated by di SantAgnese et al. (7) who found abnormal levels of sodium, chloride, and potassium in the sweat of these children. Their crude techniques for collection of sweat were inconsistently applied and produced nonstandard sweat samples for analysis, still all analyses consistently showed an increased amount of ions in the sweat.

Normal values were soon developed for children and adults without CF and parents of children with CF were shown to have slight, but significantly elevated, chloride and sodium in their sweat (8,9). The large amount of sweat required for these tests was obtained by sweating an arm or a leg of an adult or large child and required whole body sweating of small children. Such total body sweating led to some deaths due to heat exhaustion of some infants with genetic CF.

DEVELOPMENT OF A GOLD STANDARD

This uniform finding that children with CF of the pancreas have 10 times the amount of salt in their sweat created the need for a simple, cheap, rapid, and precise way to determine the salt content of sweat. Six years later L.C. Gibson, working in RE Cooke's laboratory, developed a technology

and technique (10) for sweat stimulation that (1) is simple and quick, (2) almost always produces a sufficient (>70 mg) amount of sweat, (4) has virtually zero risk of injury with equipment, and (5) could be built by any electrician associated with a hospital laboratory (11). Soon, laboratories worldwide adopted this technology and showed that >95% of normal patients had sweat chlorides <30 mmol/L, whereas almost all patients with CF had sweat chloride values >60 mmol/L. This method has been named the Gibson–Cooke Sweat Test (GCST) honoring the innovators and the Quantitative Pilocarpine Iontophoresis Test (QPIT) identifying and focusing on the key elements of the technology. Both names may be regarded as interchangeable and equally appropriate abbreviations of the excessively long "Quantitative Gibson–Cooke Pilocarpine Iontophoresis Sweat Test" (GCST/QPIT).

Over the subsequent 40 years the (GCST/QPIT) has been validated and confirmed as the only to be trusted sweat test technology for the laboratory diagnosis of CF. This GOLD Standard label has persisted despite the extreme care that must be taken to assure accurate results. The basis for that consensus is founded on three factors; (1) a known amount of sweat, (2) the chloride concentration provides the greatest discrimination, and (3) the arithmetic difference between unit measurements provides the same visual distance throughout the physiological clinical range of sweat ion concentrations.

Because of the many sources of technical error that exist with the GCST/QPIT, efforts have been made to develop alternative technology that are easier to do, have fewer risks of errors, and are simple enough to be used in general hospitals, clinics, and even physicians offices. So far, because of the potential for missed diagnoses, the best of these have only reached approval and are recommended for use only as screening tests (12).

The GCST/QPIT is the most valuable laboratory test for the diagnosis of CF. The primary indication for the GCST/QPIT is the presence of one or more of the common CF associated diseases, which include malabsorption, failure to thrive, recurrent pulmonary infection, chronic obstructive lung disease, nasal polyps, chronic sinusitis, male infertility, gallstones, unexplained cirrhosis, arthritis, diabetes, bleeding due to vitamin K deficiency, asthma, rectal prolapse, intussusception, meconium ileus, night blindness due to vitamin A deficiency, hyponatremia, bowel obstruction, volvulus, acute pancreatitis, and the child who tastes salty. Other required reasons for performing the sweat test include immediate family history (to include first cousins) of a patient having CF, a positive or suspicious newborn screening for cystic fibrosis, and the request of the parent for a sweat test.

The purity of the GCST/QPIT technology has been and continues to be maintained (13) by many CF Center Directors, national and international organizations, including Cystic Fibrosis Foundations, Directors of Clinical Laboratories, Cystic Fibrosis Center Directors, medical specialties including Pediatricians, Pulmonologists, Gastroenterologists, Clinical Biochemists and Pathologists. Their concerns to keep the GCST/QPIT technology pure and accurate have been the object of many papers and publications with the most complete being the 97 page *Guidelines for the Performance of the Sweat Test for the Investigation of Cystic Fibrosis in the UK*, Report from the Multi-Disciplinary Working Group with Representation from the Association of Clinical Biochemists, British Paediatric Respiratory Society, British Thoracic Society, Cystic Fibrosis Trust, Royal College of Paediatrics & Child Health, Royal College of Pathologists. UK National External Quality Assessment Schemes. This guideline has been formally appraised and endorsed by The Royal College of Paediatric and Child Health (http://www.acb.org.uk/Guidelines/sweat.htm) November 2003.

THE STATE OF THE ART OF SWEAT TESTS

While CF Center Directors are focusing on the precision of this test, as it is the most constant abnormality identifying CF, there have been many attempts to help the physician in practice to screen patients with some of the classic CF symptoms, and so to avoid the need, cost, and inconvenience of referring such patients to CF Centers for the approved GCST/QPIT best test.

Harry Shwachman, doyen of CF Center Directors, made the first screening test using agar plates filled with silver chromate (14). He used these agar plates on ward rounds and in clinics by placing the child's hand firmly on the silver chromate filled agar. Any salt on the hands would produce a strong white silver chloride image of the hand. This dramatic and immediately apparent test could give a false negative test if a CF patient had newly washed hands and could give a false positive test when a non-CF patient had been eating salty hand food such as potato chips. Because patients with positive screening tests and negative screening tests in patients who had classic CF related diseases still needed to be referred to the sweat test laboratory for the GCST/QPIT, this screening test is no longer used.

Over the 40 years the physiological basis for the GCST/QPIT has been known, efforts to develop a screening test that might be accepted as being as accurate as the GCST/QPIT, which might be done in a non-CF Center Laboratory, have received little support from CF physicians and Foundations.

Never-the-less three such tests have reached the attention of enough CF Center Directors to warrant a multi-CF Center study of their efficiency. These tests were (1) the CF Indicator System (15) a compact configuration of manufactured electrodes that dispense pilocarpine and a manufactured chloride sensor patch that collected a standard amount of sweat and was read as "normal", "CF" or "questionable" (16); (2) the chloride electrode *in situ* measurement of pCl after sweat stimulation by pilocarpine iontophoresis (17); and (3) the Macroduct system that used pilocarpine iontophoresis with visible collection of sweat in a plastic tube followed with conductivity or osmolarity measurements to match with a comparable concentration of sodium chloride (17).

The conclusions of the Cystic Foundation were that in the hands of community laboratories the potential for errors was so large that these tests should be considered only as screening sweat tests and that diagnostic GCST/QPIT tests for diagnosis should be done only at CF Center sweat testing laboratories.

Never-the-less the pressure to generalize sweat testing so CF diagnosis could be done before referral to CF Centers continued to be a powerful pressure for improving these three technologies. The CF Indicator was redesigned into a new integrated system that has been built, patented, and tested in one study (18). This study showed that compared to the GCST/QPIT the Quantum Patch (19) had equal sensitivity (94%) and specificity (99%), but differed in rate of failed tests, 1% for the Quantum Patch as compared to 15% for the GCST/QPIT. The Quantum test was faster, calculated the required amounts of sweat (3–10 mg) compared to the extra step of weighing and the larger amount of sweat (70 mg) required for the GCST/QPIT, was simpler to perform, required less equipment, less expensive equipment, and was less operator dependent. As of February 2005 the Quantum Patch technology, which includes a stimulator with disposable electrode, the Quantum Patch, and a scanner that simultaneous scans the patch and calculates the weight of the sweat and the chloride concentration, has not yet received FDA approval. The manufacturing standardization of the Quantum Patch test eliminates the substantial requirements for laboratory control and supervision and shortens the time for analysis to < 5 min compared to the GCST/QPIT required time of > 1 h. The simplification and the brief time needed may make this a diagnostic sweat test that could be done in a doctor's office or clinic. This rapid, simple, and quantitative pilocarpine iontophoresis sweat test has yet to be vetted by CF Scientists who have no financial interest in the product.

The chloride electrode (20–23) measures the pCl providing a true and immediate measurement of the chloride content of the sweat, but without a measure of the weight of the sweat. This technique might be resuscitated if an authoritative organization would mandate strict and inviolate guidelines for performance. If this could happen and if the CF Physicians and Organizations would accept such *in situ* measurements without knowing the weight of sweat or the rate of sweating, then diagnostic results could be known immediately. Given the strictness of the mandated strict and inviolate guidelines for performance this test might still be confined to CF Centers and other authorized Sweat Test Laboratories.

The Macroduct system is able to measure two abnormalities of the mixed-ion content of cystic fibrosis sweat. Both osmolarity and conductivity measure the ignored abnormality, the increased amount of all electrolytes in the sweat, and so have ignored what might be proven as an equal or better way to discriminate CF from non-CF subjects. Unfortunately, instead of adding new science to the study of this CF sweat abnormality, the manufacturers have reported the conductivity, or the osmolarity, of the sweat chloride concentration of a salt solution with that amount of conductivity or osmolarity. While such adjusted sweat chloride values can be used to discriminate between CF and non-CF subjects (24), the numbers are nonphysiological and so are offensive to and rejected by CF Center Directors and Sweat Test Laboratory Directors. In addition, the sweat chloride numbers confuse some practicing physicians who misdiagnosis some patients as carriers or even patients when the MacroDuct pseudosweat chloride numbers are in the GCST/QPIT intermediate or low CF ranges. Fortunately, two groups (25,26) have demonstrated that either conductivity or osmolarity can be used as new demonstrations of diagnostic alterations of sweat gland secretion. Such efforts should be encouraged.

THE POTENTIAL FOR A GCST/QPIT SUCCESSOR

The excellent work done so far has not closed the possibility that an improved sweat test cannot be developed (27). In a multistep procedure, nonstimulated sweat was collected for 10 min from the surfaces of either thumb for 10 min, while the rate of sweating was measured from the other thumb. The thumb collections are reversed and repeated a second time. Calculations "estimate the chloride concentration by dividing the amount of chloride per unit area of one finger by the amount of sweat per unit area of the other finger". The sweat chloride values were similar to GCST/QPIT sweat chloride values. Most laboratories will find this technique more tedious and filled with potential for errors than the GCST/QPIT method which, because of its reliability, accuracy, and dependability, remains the gold standard for the diagnosis of CF. The QPIT/GCST is and has been the worldwide standard because of its accuracy. However, because of the many sources of potential error and the detailed steps that are needed to ensure that accuracy the search will continue to find an equally accurate replacement.

At this time there are three candidates.

1. The Quantum Patch has been designed to have "all" sources of errors from the preparation of iontophoresis solutions, stimulation time, collection quantities, and computerized measurement of sweat weight and chloride concentration standardized by the manufacturer to give virtually 100% successful tests with sensitivity and specificity equal to the GCST/QPIT. Unfortunately, at this time the only published paper was published by the developers who have a financial interest in the product and the product is awaiting FDA 510K approval. If this product is well vetted by CF Centers and Sweat Test Laboratories it will be worthy of large scale tests by others. The Quantum Patch meets all the QPIT standards.

2. The chloride electrode test has had mixed reviews in the literature. It has the potential to deliver chloride concentrations within minutes after pilocarpine iontophoresis. However, because of inadequate attention to the details of performance the technology has done poorly in many hands. It has one major defect in that the weight of sweat or the rate of sweating are unknown. If standards for maintenance of equipment and of performing the test were to be simplified and standardized to CF Center and Sweat Test Laboratories requirements this technology could supersede the GCST/QPIT. Unfortunately, at this time there is little interest in such development. The chloride electrode test does not meet all of the QPIT standards.

3. The MacroDuct system is the only system currently manufactured, supported, and maintained. It is reliable and has acquired some supporters among CF Center Directors despite that it measures only conductivity or osmolarity which, with rare exceptions, is converted to the chloride content of NaCl solutions of similar conductivity or osmolarity. The novel collection system is excellent and provides the potential for quantifying both sweat rate and weight. The MacroDuct system is the only method that takes advantage of the initial description of the sweat abnormality, the increased sweat concentrations of Na, K, and Cl, but loses that advantage by reporting the mixed ion content of sweat by specimen, conductivity, or osmolarity, results equal to a solution of NaCl. The manufacturers and the CF Centers that favor this test, with two published exceptions, have failed to develop ion content as an alternative and as an equal or more specific abnormality characterizing CF. The pseudochloride concentration carries the risk of misdiagnosis, therefore this test will probably continue to carry the label of screening test in most CF physician's minds. New science, study of total ion content of sweat, might make this method a suitable replacement for the GCST/QPIT. There is no indication that the manufacturers are willing to make such an effort. As marketed and at this writing the MacroDuct system does not meet all of the QPIT standards.

BIBLIOGRAPHY

1. Groman JD, Meyer ME, Wilmott RW, Zeitlin PL, Cutting GR. Variant Cystic Fibrosis Phenotypes in the Absence of CFTR Mutations. N Eng J Med Aug 8, 2002;347:401–407.
2. Boyle MP. Nonclassic cystic fibrosis and CFTR related diseases. Curr Opin Pulm Med 2003;9(6):498–503.
3. Jiang Q, Engelhardt JF. Cellular heterogeneity of CFTR expression and function in the lung: implications for gene therapy of cystic fibrosis. Eur J Hum Genet 1998;6:12–31.
4. Reddy MM, Light MJ, Quinton PM. Activation of the epithelial Na(+) channel (ENaC) requires CFTR Cl(−) channel function. Nature (London) 1999;402:301–304.
5. Callen A, Diener-West M, Zeitlin PL, Rubenstein RC. A simplified cyclic adenosine monophosphate-mediated sweat rate test for quantitative measure of cystic fibrosis transmembrane regulator (CFTR) function. J Pediat 2000;137:849–855.
6. Kessler WR, Andersen DH. Heat prostration in fibrocystic disease of the pancreas and other conditions. Pediatrics 1951;8:648–56.
7. di Sant' Agnese PA, Darling RC, Perera GA, et al. Abnormal electrolyte composition of sweat in cystic fibrosis of the pancreas: clinical implications and relationship to the disease. Pediatrics 1953;12:549–563.
8. Darling RC, diSant'Agnese PA, Perera GA, Andersen DH. Electrolyte abnormalities of sweat in fibrocystic disease of pancreas. J Med Sci 1953;225:67–70.
9. Di Sant' Agnese PA, Darling RC, Perera GA, et al.Abnormal electrolyte composition of sweat in cystic fibrosis of the pancreas: clinical implications and relationship to the disease. Pediatrics 1953;12:549–563.
10. Gibson LE, Cooke RE. A test for concentration of electrolytes in sweat in cystic fibrosis of the pancreas utilizing pilocarpine electrophoresis. Pediatrics 1959;23:545–549.

11. Gibson LE, Cooke RE. A test for concentration of electrolytes in sweat in cystic fibrosis of the pancreas utilizing pilocarpine electrophoresis. Pediatrics 1959;23:545–549.

12. Smalley CA, Addy DP, Anderson CM. Does that child really have cystic fibrosis? Lancet 1978 Aug. 19; 2(8086):415–417.

13. Baumer JH. Evidence based guidelines for the performance of the sweat test for the investigation of cystic fibrosis in the UK. Arch Dis Childhood 2003;88:1126–1127.

14. Shwachman H, Mahmoodian A. Reappraisal of the chloride plate test as screening test for cystic fibrosis. Arch Dis Child 1981;56(2):137–139.

15. Warwick WJ, Huang NN, Waring WW, Cherian AG, Brown I, Stejskal-Lorenz E, Yeung WH, Duhon G, Hill JG, Strominger D. Evaluation of a cystic fibrosis screening system incorporating a miniature sweat stimulator and disposable chloride sensor. Clin Chem 1986;32(5):850–853.

16. Warwick WJ, Hansen LG, Werness ME. Quantification of chloride in sweat with the Cystic Fibrosis Indicator System. Clin Chem 1990;36(1):96–98.

17. Denning CR, Huang NN, Cuasay LR, Shwachman H, Tocci P, Warwick WJ, Gibson LE. Cooperative study comparing three methods of performing sweat tests to diagnose cystic fibrosis. Pediatrics 1980;66(5):752–757.

18. Warwick WJ, Hansen LG, Brown IV, Laine WC, Hansen KL. Sweat chloride: quantitative patch for collection and measurement. Clin Lab Sci 2001;14(3):155–159.

19. Warwick WJ, Hansen LG, Brown IV, Laine WC, Hansen KL. Sweat chloride: quantitative patch for collection and measurement. Clin Lab Sci 2001;14(3):155–159.

20. Hansen L, Buechele M, Koroshec J, Warwick WJ. Sweat chloride analysis by chloride ionspecific electrode method using heat stimulation. Am J Clin Pathol 1968;49(6):834–841.

21. Warwick WJ, Hansen LG. Measurement of chloride in sweat by use of a selective electrode and strip-chart recorder. Clin Chem 1978;24(2):381–382.

22. Warwick WJ, Hansen L. Measurement of chloride in sweat with the chloride-selective electrode. Clin Chem 1978;24(11):2050–2053.

23. Warwick WJ, Hansen L, Viela I, Matheson J. Comparison of the chloride electrode and gravimetric chloride titration sweat tests. Am J Clin Pathol 1979;72(2):142–145.

24. Hammond KB, Turcios NL, Gibson LE. Clinical evaluation of the macroduct sweat collection system and conductivity analyzer in the diagnosis of cystic fibrosis. J Pediatr 1994;124(2): 255–260.

25. Lezana JL, Vargas MH, Karam-Bechara J, Aldana RS, Furuya ME. Sweat conductivity and chloride titration for cystic fibrosis diagnosis in 3834 subjects. J Cyst Fibros 2003;2(1):1–7.

26. Barben J, Ammann RA, Metlagel A, Schoeni MH. Conductivity determined by a new sweat analyzer compared with chloride concentrations for the diagnosis of cystic fibrosis. J Pediatr 2005;146(2):183–188.

27. Naruse S, Ishiguro H, Suzuki Y, Fujiki K, Ko SB, Mizuno N, Takemura T, Yamamoto A, Yoshikawa T, Jin C, Suzuki R, Kitagawa M, Tsuda T, Kondo T, Hayakawa T. A finger sweat chloride test for the detection of a high-risk group of chronic pancreatitis. Pancreas 2004;28(3):e80–e85.

Further Reading

Warwick WJ. Cystic fibrosis sweat test for newborns. *JAMA* 1966 Oct 3; 198(1):177–180.

Warwick WJ, Hansen LG. Measurement of chloride in sweat by use of a selective electrode and strip-chart recorder. Clin Chem 1978;24(2):381–382.

Warwick WJ, Hansen L, Viela I, Matheson J. Comparison of the chloride electrode and gravimetric chloride titration sweat tests. Am J Clin Pathol 1979;72(2):142–145.

Warwick WJ, Viela I, Hansen LG. Comparison of the errors due to the use of gauze and the use of filter paper in the gravimetric chloride titration sweat test. Am J Clin Pathol 1979;72(2):211–215.

Denning CR, Huang NN, Cuasay LR, Shwachman H, Tocci P, Warwick WJ, Gibson LE. Cooperative study comparing three methods of performing sweat tests to diagnose cystic fibrosis. Pediatrics 1980;66(5):752–757.

Warwick WJ, Huang NN, Waring WW, Cherian AG, Brown I, Stejskal-Lorenz E, Yeung WH, Duhon G, Hill JG, Strominger D. Evaluation of a cystic fibrosis screening system incorporating a miniature sweat stimulator and disposable chloride sensor. Clin Chem 1986;32(5):850–853.

Warwick WJ, Hansen LG, Werness ME. Quantification of chloride in sweat with the Cystic Fibrosis Indicator System. Clin Chem 1990;36(1):96–98.

Warwick WJ, Hansen LG, Brown I. Improved correlation of sweat chloride quantification by the CF Indicator System and the Gibson-Cooke Sweat Test. Clin Chem 1993;39(8):1748.

Warwick WJ, Hansen LG, Brown IV, Laine WC, Hansen KL. Sweat chloride: quantitative patch for collection and measurement. Clin Lab Sci 2001;14(3):155–159.

See also BIOHEAT TRANSFER; FLAME ATOMIC EMISSION SPECTROMETRY AND ATOMIC ABSORPTION SPECTROMETRY.

CYTOLOGY, AUTOMATED

HARRY W. TYRER
University of Missouri-Columbia
Columbia, Missouri

INTRODUCTION

Cytology is the scientific study of cells, which includes their origin, structure, and function. Automated cytology arose from the important medical problem of classifying and enumerating cell types using instrumented means primarily for speed improvements. Automated cytology has also contributed to elucidating cell origin, structure, and function. It provides quantitative methodology directed to establishing relationships between variables to predict cellular behavior.

Understanding automated cytology requires knowledge of cells, the fundamentals of measurements on cells, and subsequent processing of these data. Present and future applications in automated cytology justify the value of automated cytology as well as define its usefulness and promise.

Cytology (Greek: *kytos*, hollow vessel; *logos*, word, reason) as a discipline primarily focuses on cell structure. Cell structure deals with those factors that define the shape and spatial distribution of the components within a cell. It is synonymous with cell morphology, which historically has been the primary method to describe cells. More generally, cytology must also encompass cellular function, which describes a cell's operational characteristics. For example, a red blood cell is a bag of hemoglobin that carries oxygen to all parts of the body, a lymphocyte produces antibodies as a result of stimulation, and squamous cells, which are on the surface of the skin, become leathery to protect the exposed skin from mechanical assault.

Automated cytology is an area of multidisciplinary specialization initially oriented toward the automated identification and classification of cells. The resulting measurement and computational methodologies provide a database from which relationships between the variables measured can be expressed. These relationships harbor the promise of establishing predictive relationships so that future structural and functional cellular characteristics can be determined. By extension then, the behavior of organs and organisms can then be determined.

CELLS

If we examine unstained cells with a microscope it is apparent that the cell is encased in a membrane. Within that membrane is a second structure encased within its own membrane. Thus the cell is subdivided into the nucleus and the cytoplasm. The nucleus is the control center and the cytoplasm carries out the function to which the cell differentiated. The work described here deals with nucleated or eukaryotic cells; incidentally cells without nuclei are called prokaryotic.

We now add a nucleophylic dye, such as hematoxylin, which binds preferentially to the nucleus. The nucleus is selectively stained and becomes much more apparent than the cytoplasm (Fig. 1). Note that both the nucleus and cytoplasm show spatial variation in light intensity. The cause of these variations can be analyzed by increasing resolution, which may include the use of electron microscopy. As we increase our ability to see into the cell either by optical or electron microscopy, we identify inclusions

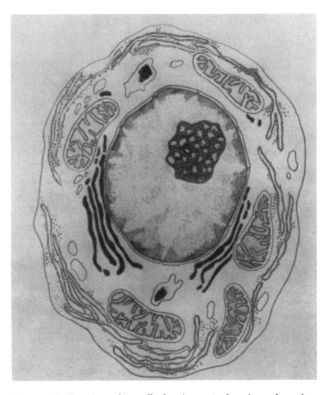

Figure 2. Drawing of a cell showing cytoplasmic and nuclear organelles. Cytoplasm: Drawing an imaginary line starting at 4 and ending at 10 o'clock, one traverses the following: cytoplasmic membrane, a granulated endoplasmic reticulum body, a single mitochondrion, a set of Golgi bodies, the nuclear membrane, and finally the nucleus. Continuing on simply reverses the order. Nucleus: On a line from 1 to 7 o'clock, one begins at the nuclear membrane to identify the nuclear elements: the nuclear membrane (note the infoldings at irregular intervals), the inner membrane, chromatin structure, nucleolus, the center of the nucleus, and finally the inner nuclear membrane.

within the cells, which are called organelles. From Fig. 2, we consider some well-defined organelles.

The nucleus, usually the largest organelle, consists almost entirely of nucleic acids. Both deoxyribonucleic (DNA) and ribonucleic acid (RNA) are present. By far, the major constituent is DNA. Within the nucleus and under appropriate conditions DNA molecules replicate giving rise to DNA synthesis and then to cell division. An organelle within the nucleus is the nucleolus. In Fig. 2, it is the darkened mottled body encased in the nucleus. It consists mostly of RNA. In the initial steps of transcription to make protein, DNA produces messenger RNA (mRNA) according to the genetic code. The mRNA then passes through the nuclear envelope into the cell cytoplasm.

The cytoplasm is the entity that performs the cell's function. These functions are carried out by several organelles. The same organelle of different cells may be substantially different so that the cell may focus its energy to carry out its usually single function.

The mytochondrion (in Fig. 2, the ovoid body with the complex curved inclusions) is an organelle that is responsible for producing the energy required by the cell. It has its own DNA and is able to replicate itself within the cell.

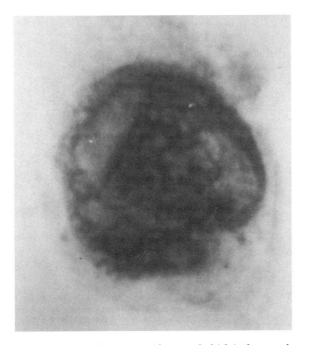

Figure 1. A cell, the lighter outside area of which is the cytoplasm and the darker of which is the nucleus.

The Golgi body complex (the long dark strands) provides temporary storage of secretory substances and connects to the endoplasmic reticulum.

The endoplasmic reticulum (light strand) is of two types in the same cell: with granules (the dots) and without granules. The granules are ribosomes. In the final stages of transcription (making protein from DNA), the mRNA travels from the nucleus to the ribosomes (probably in the endoplasmic reticulum) and attaches to the ribosome. The ribosome then binds the transfer RNA (tRNA) corresponding to the genetic code expressed in the mRNA. Each value of the genetic code has one tRNA, which in turn corresponds to one amino acid. One ribosome houses two tRNAs that connect the proper amino acids together to the protein chains. The resulting protein (called a polypeptide) passes through the Golgi complex to be secreted from the cell. Thus the function of a particular cell is to produce a particular protein. Additionally, binding a signal protein will alter the state of the cell, causing it to produce a different protein or to divide and function in a different manner.

As just shown, we can identify cell structure by binding chemicals to the structure to allow its observation. On the other hand, cellular products are determined by several methods. First, one can selectively poison a particular functional entity and compare the results that are produced from nondestroyed entities. Second, one can separate the entities and selectively return them until the tested functionality has been restored. These techniques are routinely performed on samples containing many cells. Unfortunately, the average population behavior masks the individual cell behavior, which may be of interest as a rare event.

Consequently, automated cytology has evolved to enumerate and provide quantitative information on the structural features as well as the functional capabilities of a single cell. Work is done on an intact single cell, which may be viable or nonviable. Viable means that the cell is able to perform its functional activities, whereas a nonviable cell cannot. A nonviable cell is preserved in an appropriate fixative. Such a fixed cell can withstand many of the rigors of experimental treatment compared to a viable cell. Unfortunately, fixation may alter properties that are to be measured. Thus, choice of fixation (including no fixation) requires careful consideration.

CYTOCHEMICAL PROBES

Cytochemistry is the chemical organization and activity of a cell. Numerous chemical probes have been reported for use in determining cell structure and function. The probe binds to the molecule to be detected. If a stochiometric relationship between the probe and detected molecule is maintained (i.e., conditions for binding are fixed so that the relationship between the amount of probe and detected molecule are fixed), then relative quantities of the detected molecule can be obtained. Furthermore, if the amount of detected molecule to quantity of probe molecule can be established, absolute standardization is achieved.

Biochemical probes are used in automated cytology to detect a wide range of structural and functional characteristics of cells. For example, let us consider the steps

involved in measuring DNA in single cells. Specifically we want to see the activation of DNA synthesis over several days. White blood cells withdrawn from fresh human blood, can be stimulated to replicate by placement in an appropriate tissue culture system. We can remove several cells from the tissue culture system every 24 h, and determine the relative amount of DNA in each cell. A cytochemical probe specific for DNA is propidium iodide (PI), which fluoresces in the red when excited by a 488-nm (blue) light. The fluorescence intensity of each cell is proportional to the amount of PI bound to the DNA, assuming care has been taken to minimize non-DNA binding of PI. From this, we form a histogram of the fluorescence intensity of the cells; that is, we plot for each value of fluorescence the number of cells (frequency) expressing that fluorescence value. The series of histograms in Fig. 3 shows that at first there is a

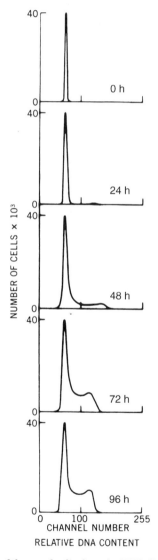

Figure 3. Each of the graphs displays the DNA distribution of cells in tissue culture. Cells were removed from culture and analyzed every 24 h. The top chart displays the DNA distribution of cells initially placed in culture. The succeeding charts show increasing numbers of cells with increasing amounts of DNA over the 4-day intervals sampled.

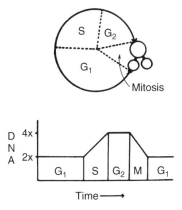

Figure 4. Cell cycle diagram and the corresponding graph of DNA quantity versus time. The Parameter G_1 is gap 1, a resting state; S is when the cell synthesizes DNA; G_2, gap 2, another resting state; and M is mitosis in which the cell divides.

single population of cells in the blood with a remarkably narrow range of DNA value. Within 24 h there is a barely visible set of cells with twice the normal amount of DNA; by 48 h and beyond, it is apparent that there are also cells with values intermediate to the two values. These distributions demonstrate DNA synthesis and cellular replication.

It is commonly known that cells replicate by dividing, creating daughter cells. In the process of division, from some resting state, the cell synthesizes DNA until two times the normal amount of DNA is present. At this point, the two nuclei are formed and the cell subsequently divides into two presumably identical parts. Figure 4 graphically illustrates the cycling of the cells. At the portion of the circle labeled G1 (resting state, Gap 1), a cell has two times the DNA shown in the lower part of Fig. 4. (Since cell DNA quantity is variable between species, we list the amount of DNA as X; furthermore, egg and sperm have $1X$ DNA, but most mature cells in the body normally have $2X$ DNA.) As time increases, the cell goes into the S phase of the cycle

where it synthesizes DNA. Finally, the cell has $4X$ DNA and waits some amount of time at G2 (Gap 2) until it divides. The cell divides (M, mitosis), generating two daughter cells in the G1 phase of the cycle.

These observations are based on total DNA of each cell. Had we also observed a cell in mitosis, under proper microscope and staining conditions, we would have observed the nucleolus condensing into well-defined bodies called chromosomes. Preparation of chromosomes is made from cells undergoing division. Such preparations are used to study genetic observations.

The same cytochemical principles are used to classify human chromosomes. DNA consists of chains of adenosine—thiamine (AT) or guanine—cytosine (GC) base pairs. A strategy to increase the information content in measuring the DNA of chromosomes is to use two dyes simultaneously; one with selective affinity to the AT bases the other with affinity to GC bases. Under proper excitation, each chromosome will produce a fluorescence intensity value from each of the two dyes. These two intensities are each proportional to the number of AT or GC bases of each chromosome. Histograms of these two parameters are two dimensional, and each point in the histogram is the number of chromosomes (or particles) with the same pair of fluorescence intensity values. For example, Chromomycin A3 (CA3) has an affinity for AT bases, whereas Hoechst bis-benzimidizole 33258 binds preferentially to GC bases. The increased information provided by the use of the second fluorescent stain increased the resolution for separating the 23 human chromosomes compared to each stain alone.

We have used DNA quantity as an example of the use of cytochemical probes. There are a large number of DNA probes in current use as shown in Table 1. The bibliography can be consulted for additional information on their properties and use.

Cytochemical probes are, of course, useful for detection and quantization of other cellular structures and functions. Table 2 displays a representative set of such probes.

Table 1. Biochemical Probes with DNA Specificity

| Probe[a] | Base-Pair Enhancement | Spectrum Maximum, nm | | Extinction Coefficient, wavelength, pH[b] |
		Excitation	Fluorescence	
Acridine Orange	None	490	530	50 (492, 7)
Ethidium bromide	None	546	610	517 (480, 7)
33258	AT	365	450	4 (338, 7)
Daunomycin	AT	500	600	14.8 (480, 7)
EK-4	AT	530	560	36 (570, 7)
LL-585	AT	500	600	32 (510, 7)
Proflavin	AT	460		34.9 (454, 7)
Quinacrine	AT	436	525	8.9 (424, 7)
VL-772	AT	500	600	32 (510, 7)
7-AAD	GC	546	610	21.5 (502, 7)
Chromomycin A3	GC	440	585	8 (490, 7)

[a]Abbreviations: 33258 = Hoechst bis(benzimidizole) No. 33258; 7-AAD = 7-aminoactinomycin D; A = adenine; C = cytosine; EK-4 = 26-diphenyl-4-(4-dimethylaminophenyl) pyriliumbisulfate; G = guanine; LL-585 = 6-benzothiazolyl-3-ethyl-2-(dimethylamino-stytyl)-benzothiazolium,-*p*-toluenesulfonate; T = thymine (T); VL-772 = 6-(dimethylamino)-2-[2,5-dimethyl-1-pheny-1*H*-pyrrol-3-yl ethcnyll]-1-methyl-quinolinium methosulfate.
[b]Extinction coefficient $\times 10^{-3}/(M\,cm)$; in parentheses are the values of measurement wavelength (in nanometers) and the pH, respectively.

Table 2. Fluorescent Probes to Assay Cell Structure and Function

Cell structure	Probe
RNA content	Acridine Orange, pyronin
Total protein	Fluorescein isothyocyanate
Cell cytoskeleton	NBD-phallacidin
Cell mitochondria	Fluorescent tetracyclenes
Cell function	
Membrane permeability (cell viability)	Fluorescein diacetate, propidium iodide, 33342[a]
Membrane potential (cytoplasm and mitochondria)	3,3'-Dihexyloxacarbocyanine
Calcium (Ca^{2+})	
Membrane bound	Chlortetracycline
Cytoplasmic	Quin-2
Enzyme activity	Fluorescein derivatives, methylumbelliferyll derivative, napthol derivatives
Intracellular pH	1,4-Diacetoxy-2,3-dicyanobsenzene, carboxyfluorescein
Phagocytosis	Fluorescent beads, fluorescent-stained bacteria and other particles

[a]Hoechst bis(benzimidizole) No. 33342.

Finally, the biochemical constituents of the cells have physical properties that are directly measurable. A well-known example is the pigment that makes blood cells red, hemoglobin. Furthermore, it is well known that the important metabolic constituent reduced nicotinemide adenine dinucleotide (NADH) fluoresces as does riboflavin, one of the 13 vitamins. These compounds have optical properties that are useful because of the convenience of use with existing equipment. This is not true of the majority of cell constituents whose physical properties may not be easily detected or measured. For example, proteins with aromatic amino acids, such as tryptophan, will fluoresce when excited by 257 nm light. Such a short wavelength is not conveniently handled because its propagation attenuation in standard optical materials is so severe.

MEASUREMENT OF CELLULAR PARAMETERS

Automated cytology implies that cellular parameters are to be measured. Since the visual quasiquantitative analyses described in the previous section cannot provide the accuracy and discrimination required, technologies have arisen that measure physical phenomena to describe cellular parameters. Commercially available instrumentation use optical or electrical phenomena whereas experimental devices have been used with acoustic, magnetic, and a variety of spectroscopic techniques.

Probes for nucleic acids abound. The probe binding mechanics to the nucleic greatly influences their use and application. Table 3 shows a list of such probes along with their excitation and emission wavelengths.

Protein probes tend to be nonspecific and usually stick to the protein. The stain can be removed by (sometimes) vigorous rinsing. Table 4 shows the excitation and emission

Table 3. Fluorescent Probes for Nucleic Acids with Excitation and Emission Wavelengths in nm

Probe[a]	Excitation	Emission
Hoechst 33342 (AT rich) (UV)	346	460
DAPI (UV)	359	461
POPO-1	434	456
YOYO-1	491	509
Acridine Orange (AO) (RNA)	460	650
Acridine Orange (DNA)	502	536
Thiazole Orange (vis)	509	525
TOTO-1	514	533
Ethidium Bromide	526	604
PI (UV/VIS)	536	620
7-Aminoactinomycin D (7AAD)	555	655

[a]Abbreviations: DAPI = 4',6-diamidino-2-phenylindole; POPO-1, YOYO-1,TOTO-1 are cyanine dimers available from Molecular Probes, Inc.; PI: Propidium Iodide; UV = ultraviolet, vis = visible.

wavelengths of various probes used for proteins. The probes are also bound to specific antibodies so that multi wavelength emission due to multiple antibody binding provides a multiparametric analysis. Common pairs are flouresceim isothyocianate (FITC) and Rhodamine (see Tables 4 and 13).

There are a variety of probes of importance to identifying cell parameters. These require care in their use. The following three tables (Tables 5–7) list such probes along with their excitation and fluorescent emission wavelengths. Ion probes are listed in Table 5, pH sensitive indicators appear in Table 6, and probes for oxidation states along with the oxidant appear in Table 7.

Finally, just as there is a dichromatic display for RNA and DNA with acridine orange, where AO can be seen to bind to nucleolus as a red fluorescence and to the nucleus as green fluorescence, so several probes have an affinity for specific organelles. Specifically, the Golgi bodies and mitochondria can be identified with the appropriate stain as Table 8 shows. Finally, the lipid stains help to identify the cell's membrane and other lipids.

Optical Measurements

By far, most measurements on cells are optical. Spectra obtained from a multiplicity of physical properties are

Table 4. Fluorescent Probes for Proteins with Excitation and Emission Wavelengths in nm

Probe[a]	Excitation	Emission
FITC	488	525
PE	488	575
APC	630	650
PerCP	488	680
Cascade Blue	360	450
Coumerin-phalloidin	350	450
Texas Red	610	630
Tetramethylrhodamine-amines	550	575
CY3 (indotrimethinecyanines)	540	575
CY5 (indopentamethinecyanines)	640	670

[a]Abbreviations: PE = Phycoerythrin; APC = allophycocyanin; PerCP = peridinin chlorophyll.

Table 5. Fluorescent Probes for Ions with Excitation and Emission Wavelengths in nm

Probe[a]	Excitation	Emission
INDO-1	350	405/480
QUIN-2	350	490
Fluo-3	488	525
Fura-2	330/360	510

[a]**INDO-1** = 1H-Indole-6-carboxylic acid, 2-[4-[bis[2-[(acetyloxy)methoxy]-2-oxoethyl]amino]-3-[2-[2-[bis[2- [(acetyloxy)methoxy]-2-oxoetyl]amino]-5-methylphenoxy]ethoxy]phenyl]-, (acetyloxy)methyl ester [$C_{47}H_{51}N_3O_{22}$], **FLUO-3** = Glycine, N-[4-[6-[(acetyloxy)methoxy]-2,7- dichloro-3-oxo-3H-xanthen-9-yl]-2-[2-[2- bis[2-[(acetyloxy)methoxy]-2- oxyethyl]amino]-5-methylphenoxy]ethoxy]phenyl]-N-[2- [(acetyloxy)methoxy]-2-oxyethyl]-, (acetyloxy)methyl ester.

Table 7. Fluorescent Probes for Oxidation States with Excitation and Emission Wavelengths in nm

Probe[a]	Oxidant	Excitation	Emission
DCFH-DA	(H_2O_2)	488	525
HE	(O_2^-)	488	590
DHR 123	(H_2O_2)	488	525

[a]DCFH-DA = dichlorofluorescin diacetate, HE = - hydroethidine 3,8-Phenanthridinediamine, 5-ethyl-5,6-dihydro-6-phenyl-, DHR-123 = dihydrorhodamine 123 Benzoic acid, 2-(3,6-diamino-9H-xanthen-9-yl)-, methyl ester.

the relative orientation of the cell to the source and detector can produce different scatter fields. Furthermore as the numerical aperture of detection increases the internal structure is increasingly detected. On some instrumentation, simultaneous detection of 0 and 90° scattering is used to discriminate cells based on internal structures. Light scattering under a highly coherent light source, such as a laser, is a very sensitive indicator of variations of index of refraction. Under normal conditions the cell is encapsulated in some medium and the index of refraction between air and the cellular medium contribute to the entire signal. Furthermore, under abnormal conditions, local differences in index of refraction, as a result of improper fluid mixing, contribute signals that approximate those produced by a cell.

Measurements of light scattering from particle suspensions provide a single datum from which estimates of total particles are obtained. Such measurements produce a signal that is dependent on the number, size, and shape of the suspended particles. Systems measuring either scattering or absorption of suspensions cannot provide discriminatory information for individual particles.

Optical Spectroscopy. When energy interacts with matter, a sequence of events occurs that is explained in terms of the atomic or molecular behavior. This interaction provides quantitative information as well as identification of the species involved. The well-known proportionality of energy E to electromagnetic frequency v is related by Planck's constant (h), which underscores the discrete nature of

useful in identifying and quantitating cellular biochemical and physical parameters. The most popular optical measurements, namely, light scatter, absorption spectroscopy, and fluorescence spectroscopy, and their use in automated cytology are discussed.

Light Scattering. Light scattering from a single cell has been used to characterize cell volume, shape, internal structures, and other cell properties that produce changes in index of refraction. Light scattering includes diffraction, refraction, and reflection of light from a single cell. From Maxwell's equations, a series expression was developed by Mie for plane wave propagation perturbed by a solid nonconductive homogeneous sphere with a diameter on the order of the wavelength of illuminating light. Although some cells are spherical and their diameter approximates the wavelength of visible light, cells are not homogeneous. In fact, it may be their optically heterogeneous structures that are important and, therefore, cannot be neglected. There is a large body of literature devoted to light scattering and some of it is directed to the problem of aerosol detection.

Measurements have established the relationship between the amount of light scattered and the size of the cell. For small-angle scattering ($< 2°$ numerical aperture), the signal intensity is proportional to the diameter cubed (in agreement with the Mie solutions). This proportionality is linear and monotonically increasing over a restricted size range. As the numerical aperture increases the relationship is no longer linear and is not monotonic over the entire possible range of sizes. Cell shape also influences scattering. Since most cells are not spherical,

Table 6. Fluorescent pH Sensitive Indicators with Excitation and Emission Wavelengths in nm

Probe[a]	Excitation	Emission
SNARF-1	488	575
BCECF	488	525/620
BCECF	440/488	525

[a]**SNARF-1** = Benzenedicarboxylic acid, 2(or 4)-[10-(dimethylamino)-3-oxo-3H- benzo[c]xanthene-7-yl]-, **BCECF** = Spiro(isobenzofuran-1(3H),9'-(9H)xanthene)-2',7'-dipropanoic acid, ar-carboxy-3',6'-dihydroxy-3-oxo-.

Table 8. Fluorescent Probes for Organelle with Excitation and Emission Wavelengths in nm

Probe[a]	Organelle	Excitation	Emission
BODIPY	Golgi	505	511
NBD	Golgi	488	525
DPH	Lipid	350	420
TMA-DPH	Lipid	350	420
Rhodamine 123	Mitochondria	488	525
DiO	Lipid	488	500
diI-Cn-(5)	Lipid	550	565
diO-Cn-(3)	Lipid	488	500

[a]BODIPY = borate-dipyrromethene complexes; NBD = nitrobenzoxadiazoxadiazole; DPH = diphenylhexatriene; TMA = trimethylammonium; DiO = diI-Cn-(5);diO-Cn-(3) = Carbocyanines (DiI, DiA, DiO), e.g., DiI 1,1'-dioctadecyl-3,3,3',3'-tetramethylindocarbocyanine perchlorate.

atomic behavior:

$$h = 6.6 \times 10^{-34} \text{ J} \cdot \text{s}$$
$$E = h\upsilon \qquad (1)$$

For example, gas discharge tubes supply energy by means of a high voltage between electrodes onto a sealed container with small amounts of gas. The resulting arc has an emission spectrum consisting of lines of light energy at discrete wavelengths that are characteristic of the electronic structure of that gas. The first correct explanation of this phenomenon arose from the Bohr model of the hydrogen atom. The calculated values of energy from the various discrete orbits that the electron could have about the nucleus correspond to the spectral lines obtained from the gas discharge series.

This characteristic spectrum can be used to identify and quantitate the amount of element or compound. The interaction of the constituent atoms of a molecule produces a spectrum that is substantially different from that of each atom. The spectrum is characteristic of the molecule.

The response of the material to light excitation may by considered to occur in two general steps (Fig. 5). In a molecule excited by light, an electron is elevated in energy to an excited state: This results in light absorption. The relaxation of this excited electron back to a lower energy state may result in (1) light being emitted (fluorescence and phosphorescence), (2) heating of the substance, or (3) some combination of the two. The light is not only absorbed but may be scattered (Rayleigh scattering) or be modulated to a different wavelength by the rotational and vibrational motion of the illuminated molecules (Raman scattering).

The emitted light frequency is different from the absorbed frequency υ_1, except for Rayleigh scattering. In Raman scattering, it is possible for the emitted frequency to exceed the exciting (or absorbed) frequency. In the case of fluorescence or phosphorescence, the loss of energy in the molecule (due to collisions or vibration) results in a lower frequency of emission compared to the excitation (absorbed) frequency. Similarly, because of energy decrease, phosphorescence frequency is less than fluorescence frequency. For the remainder of this article we will no longer deal with frequency of light, but with the more commonly used parameter, wavelength.

Absorption is obtained from the loss of light in a material at a particular wavelength. A homogenous material is irradiated by light of intensity I_r (r, reference). As a result of transmission through the material there is a loss of light resulting in light intensity I_s (s, sample) emanating from the material (see Fig. 6). The optical density or absorbance can be defined from these two quantities, which is

$$\log_{10}(I_r/I_s) = A \qquad (2)$$

The absorbance A (for this discussion, always an upper case A) is directly related to the light path thickness d, the concentration of the material C, and by the proportionality constant ε called the extinction coefficient, which is a property of the material. The resulting equality is the Beer–Lambert law, which may be expressed as

$$A = \varepsilon d C \qquad (3)$$

The self-absorption of the sample material reduces the sensitivity to values $0.1 < A < 1.2$. Thus the material is suspended in nonabsorbing solution to disperse it to satisfy the homogeneity assumption. In practice, the

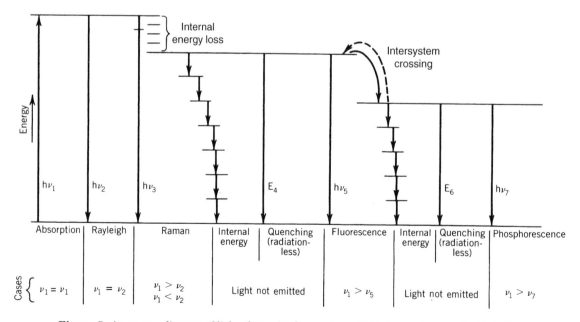

Figure 5. An energy diagram of light absorption by an atom. Light (energy) is provided into the system and is absorbed. It is then scattered (Rayleigh or Raman), dissipated in the system, or emitted by fluorescence or phosphorescence (h, Plank's constant; υ, frequency of light).

concentration of the material in solution can be obtained by using a known concentration of the sample (standard) and determining the ratio of absorbance of the unknown to the standard:

$$\frac{A_{\text{unknown}}}{A_{\text{standard}}} = \frac{C_{\text{unknown}}}{C_{\text{standard}}} \qquad (4)$$

For species differentiation, a spectrum is usually sufficient to uniquely identify that species.

It is useful to define the quantity "transmissivity" t, which is

$$t = I_{\text{s}}/I_{\text{r}} \qquad (5)$$

In practice, the input intensity I_{r} is fixed and the sample modulates I_{s}. So, if the sample does not absorb light, $t = 1$, and if the sample completely blocks out the light, $t = 0$. Thus, it is convenient to speak of the transmission T, which is a percentage, as

$$T = 100t \qquad (6)$$

The entire range of intensities is mapped between 0 and 100%.

A homogeneous mixture of particles and solutions can be expected to arise in conditions in which the measured geometries are very large compared to the absorbing particle size. In the microscopic environment, particle sizes are in the order of the measurement geometries, invalidating the homogeneous sample assumption. This is referred to as the distributional error.

The distributional error can be analyzed using Fig. 6 as follows: We wish to determine the true value of absorbance of the sample in solution. Assume that only the light passing through the area $[a_{\text{T}} = (l_1 + l_2)W]$ under consideration reaches a detector. Let us begin by assuming that the

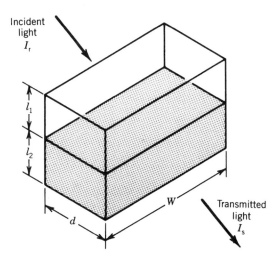

Figure 6. A uniform beam of light passing through a nonhomgeneous medium in a clear container. A material with extinction coefficient ε is suspended in a nonabsorbing medium. The material is uniformly distributed in the lower part of the container so it absorbs light by an amount $A_2 = dC_2$ over the surface l_2W. A different amount of the material is also uniformly distributed over the upper part of the container, absorbing light in the amount $A_1 = dC_1$ over l_1W.

entire mass m_{T} is distributed throughout the total volume (da_{T}). Express Eq. 3 as

$$A_T = \varepsilon d \left(\frac{m_{\text{T}}}{dWl} \right)$$

Now, if we assume, as shown in Fig. 6, an amount m_2 constitutes most of the absorbing material in the lower part of the volume and a different amount m_1 is in the upper part, we can substitute and cancel to get

$$A_T = \varepsilon \left(\frac{m_1 + m_2}{a_{\text{T}}} \right)$$

Now, we observe that the dry mass of the dissolved material is determined by $m = (A/)a$; that is, the viewing area a determines the measured dry mass, so that with further substitution and cancelation,

$$A_T = \frac{1}{a_T}(A_1 a_1 + A_2 a_2)$$

We can generalize: First, our derivation deals only with areas so that any arbitrary area could be used; second, we can expand the preceding equation to get

$$A_T = \frac{1}{a_T} \sum_{\text{all } i} A_i a_i \qquad (7)$$

Thus, the solution to the distributional error problem in quantitative microscopy is to measure the absorption over a small area and sum each measurement so that the entire object is covered. It is assumed that each area is so small that it is homogeneous. Thus, the absorption of a cell at a given wavelength is obtained by measuring the absorbance of contiguous areas over the boundaries of the cell.

Finally, we observe that had we summed the individual transmittances (see Eq. 2), a different and incorrect value would have been obtained. Summing the absorbencies is based on the conservation of mass: total absorbance depends on total mass regardless of its distribution.

Fluorescence properties, wavelength, and intensity also provide information on molecular species and concentration, respectively. The fluorescence intensity F is proportional to the absorbed light intensity, I_{a} (a, absorbed), which is the difference between incident I_{r} and exiting I_{s} light intensity. The constant of proportionality q is related to the efficiency of the conversion of absorbed energy to emitted energy and is called the quantum efficiency,

$$q = \frac{\text{number of quanta emitted}}{\text{number of quanta absorbed}}$$

So that we get

$$F = I_{\text{a}}q$$

Now inserting Eq. (3) into Eq. (2) and solving for I_{s}, we get

$$I_{\text{s}} = I_{\text{r}}\, 10^{-\varepsilon dC}$$

so that

$$F = qI_{\text{r}}(1 - 10^{-\varepsilon dC}) \qquad (8)$$

Again we have a parameter of material property and quantity relating to a measurable optical parameter. This relation using the McLaurin expansion, can be linearized by which assumes that $A < 0.05$ (e.g., small values of absorption). Now we can write

$$F = qI_\mathrm{r}(2.3\,\varepsilon\,dC) \qquad (9)$$

Again, if we have a standard and an unknown sample, the following relation holds, since we now assume that fluorescence is linearly related to concentration

$$\frac{F_{\text{unknown}}}{F_{\text{standard}}} = \frac{C_{\text{unknown}}}{C_{\text{standard}}}$$

where both values of F are measured, and C_{standard} is determined by the experimenter.

We have covered the fundamental issues of absorption and fluorescence spectroscopy directed to identifying and quantifying various molecular species. There are numerous other optical techniques that use biochemical probes to provide mechanistic information of cellular function. We will discuss only two: fluorescence polarization and resonance energy transfer.

Fluorescence polarization has been used to assess molecular motion with respect to the cell to which the molecule is bound. In general, the exciting light is polarized in a given direction and the absorbing probe is polarized in some other direction making an angle ϕ. The probability of absorption is proportional to the $\cos^2 \phi$, so that maximum absorption occurs in those probes parallel to the exciting light. The fluorescence emission is detected in two directions, parallel and perpendicular to the excitation source. These intensities can be used to form the quantity called emission anisotropy R,

$$R = \frac{I_\parallel - I_\perp}{I_\parallel + 2I_\perp}$$

We introduce the notion of an absorption and emission vector, each an independent directed quantity in space. In a rigid system, it can be surmised that the relative motion between the absorption and the emission vector is very small. However, in nonrigid systems there is motion between the absorption and emission vectors within the lifetime of the fluorescence emission. Thus, material probe motion occurring within the fluorescence lifetime is detected as an anisotropic increase in fluorescence.

Resonance energy transfer is used to assess the so-called nearest-neighbor distance. Energy transfer occurs between two resonating probes, that is, a donor transfers energy to an acceptor probe by nonradiative energy transfer. The prime condition is that there be a reasonable overlap between the donor emission spectrum and the acceptor absorption spectra. Obviously, quantum yields and donor–acceptor orientations must be satisfactory. This technique is used to determine the separation of the donor and acceptor probes. A critical parameter is R_0, which is defined as the distance for 50% energy transfer. Table 9 shows various acceptor and donor combinations and their respective values of R_0 that have been reported in the literature (4–7).

Table 9. Energy-Transfer Combinations[a]

Donor	Acceptor	Separation[b]
Fluorescein isothyocyanate	Rhodamine isothiocyanate	5.6
Quinacrine	Ethidium bromide	2.2
Quinacrine	7-Aminoactinomycin	3.0
33258[c]	Ethidium bromide	3.1
33258[c]	Daunomycin	4.0
33258[c]	Chromomycin A3	8.3
Chromomycin A3	Ethidium bromide	2.0

[a]For more information, see Refs. 4–7.
[b]Value of R_0 for 50% energy transfer in nanometers.
[c]Hoechst bis(benzimidizole) No. 33258.

Electrical Resistance

The Coulter effect is the name given to the phenomenon (Fig. 7) whereby current through a small aperture in an aqueous conducting medium is modulated by the passage of particles through it. This was first used to develop cell counters; later, it was observed that the change in resistance was related to cell volume. The large amount of data produced by the many blood cells is easily reduced if one uses a histogram of cell size, similar to those discussed previously. The substantially increased number of cells improved the counting statistics. This along with the improved speed measurements made such blood cell counting and sizing instrumentation the methodologies of choice.

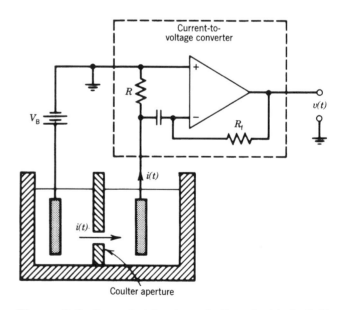

Figure 7. Coulter principle. A conducting physiologic fluid supports the current $i(t)$, resulting from the potential difference V_B in the immersed conducting plates. A particle passing through the Coulter aperture interrupts $i(t)$ by an amount depending on the relative size of the particle the feedback resistor R_f and the input resistor R convert the current $i(t)$ to the output voltage waveform $v(t)$. Processing of $i(t)$ provides data for counting and sizing particles.

Other Measurements

Various physical measurements have been (or should be) adapted for use in automated cytology. These include mechanical techniques such as acoustics, electromagnetic techniques (e.g., nuclear magnetic resonance, NMR), and other spectroscopic techniques. Acoustic microscopes have been constructed that are used for identifying cellular microstructure. On the other hand, NMR has been used to identify metabolic constituents in cells. Raman spectroscopy has been suggested to be useful for identifying chemical constituents in intact cells if the signal-to-noise ratio problems can be overcome. Finally, optical rotatory dispersion and circular dichroism have been used to study nucleic acids. Table 10 summarizes the measurement systems in common use and the cell features they measure.

Devices used in Automated Cytology for Cell Measurement and Isolation

Automated cytology makes use of instrumentation for cell analysis and for cell isolation. Cell analysis instrumentation is classified further into three types according to the information content obtained from the cell: (1) zero-resolution instrumentation, which generates a single datum for each parameter from a cell; (2) high resolution instrumentation, which generates a very large number of data from a single parameter measured from a cell; and (3) low resolution instrumentation, which obtains more information than zero-resolution devices from each cell as a result of a slight increase in resolution.

Instrumentation for isolation is of two general types. The first type is physical placement of the cell in a desired location in space. The second type identifies and localizes the cell with respect to some origin on a fixed medium such as a microscopic slide; cell identity and position data are stored in a computer.

Flow Cytometry and Sorting. Flow cytometry and cell sorting have been developed for the rapid identification and isolation of cells. By causing cells in a suspending medium to flow past detection devices, identification is accomplished. With appropriate electronics and data processing devices, cell analysis is effected. Based on the analysis, cells are isolated by physically moving the suspension medium.

Cell analysis includes detection of the parameter desired, appropriately converting the detected energy to some electrical signal, and processing that signal. A cell riding in the stream passes by the laser beam. The cell causes light to scatter and depending on its preparation may fluoresce. If the Coulter effect is implemented, the cell flows through the Coulter aperture and the modulating current flow is detected. Each cell causes a signal to appear at the output of the various detectors. In a sense, the cell has been identified according to the value of the detected parameter. This data is in essence real-time data; it can be used with further treatment for sorting (see below) or stored in the appropriate form for data reduction. In practice, all detection schemes have involved optical parameters or cellular resistance to measure cell size, cellular fluorescence, and other properties.

Cell sorting begins with identifying a detected cell as the one to sort. Essentially the suspending medium or stream is an electronically conducting fluid jet. Fortunately, the ions that imbue electronic conduction properties in the medium also provide the fluid with physiologically useful properties. The stream breaks up into droplets that contain the desired cells. The droplets are charged, and in turn are acted on by an electrostatic field that deflects the charged droplets containing the cells to appropriate containers. Sorting the droplets containing the desired cells and no other plays a major role in determining the purity of the sorted fraction. This is guaranteed by using acoustic energy vibrating the nozzle to create instability on the stream so that the stream breaks into droplets at a predictable point.

We demonstrate these principles with the aid of Fig. 8. The conducting physiologic fluid (sheath) flows into the nozzle and is ejected as a jet through a circular orifice of ~50 μm. A sample consisting of cells in suspension flows into the nozzle, is injected into the sheath, and is also ejected with the jet. The nozzle is designed to establish laminar flow conditions; this enables the sample to be accurately centered on the stream. If the radius of the sample in the sheath is on the order of the cell size, the cell is highly localized in the center of the stream. The acoustic drive assures the predictability of the location of droplet separation from the stream.

Table 11 lists lasers and their emission lines in common use. Argon ion lasers are capable of delivering several watts of laser energy in the blue-green region of the visible spectrum. Since they require high power input, and a very high discharge current, this laser is very bulky and emits a large amount of heat. Similarly, krypton ion lasers deliver several watts of laser energy in the visible spectrum, and

Table 10. Correlation of Selected Cell Measurements and Features

Measurement	Feature
Resistance (Coulter) orifice	Cell volume
Light scattering	
Low angle (2–20°)	Size and shape
Large angle (to 90°)	Size, shape, internal structure, and viability
Polarized	Macromolecular conformation
Acoustic energy	Cell compressibility and deformation
Pulse shape analysis	
Slit scan	Particle shape information; distribution of stain within the cell
Time of flight	Double-cell detection; particle shape information; particle diameter; resolution of cell structure
Time	Kinetics
Fluorescence intensity	Amount of fluorogen
Fluorescence depolarization	Macromolecular viscosity
Energy transfer	Nearest-neighbor detection (spatial separation); molecular mobility

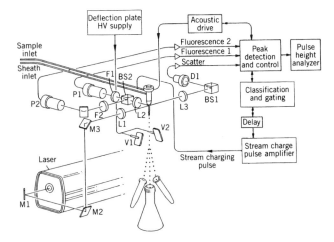

Figure 8. A simplified block diagram of a cell sorter. Mirrors directing the laser beam, M1, M2, and M3; lens to focus the laser beam, L1; lenses to image the laser illuminated cell on the detectors L2 and L3; beam splitters to deflect the light, BS1 and BS2; filters to separate the laser excitation from the emitted light of the particle; F1 and F2; photomultipliers to detect fluorescence, P1 and P2; and diode to detect forward scatter, D1. Courtesy FACS Division, Becton, Dickinson Electronics Laboratories, Mountain View, California.

require high power input and a very high discharge current. Mixtures of Argon and Krypton are sometimes referred to as "white light" lasers because of the coverage over the visible spectrum the combination of the two gases produces.

For a zero-resolution device implemented as shown in Fig. 8, each signal produces a pulse 10 μs wide, whose height is proportional to the total light energy detected. Thus, for each cell, the peak of three pulses is obtained: light scatter and two fluorescence wavelengths. Typically, 50,000 cells can be analyzed in 1 min. With 150,000 data points, powerful data reduction capability is required. For sorting, the cell requires ~250 μs to arrive at the point in the stream just prior to breaking up into droplets. This

Table 11. Lines of Laser Emission[a]

Helium–Cadmium	Argon Ion	Krypton Ion	Violet Diode	Krypton–Argon	Helium–Neon
325		350.7			
	351.1	356.4			
	364.8		405		
450					
	457.9				
	476.4				
	488.0			488	
	496.5				
	514.5				
		530.9			
		568.2		568.2	
					632.8
		648.1		648.1	
		752.5			

[a]Wavelengths of light in nanometers.

distance is programmed into the electronics causing the delay required for the identified cell to reach the end of the stream. When the cell reaches that point, the entire stream is charged by the conduction of the ions in the buffer (investigations indicate that <0.01 *M* salt solution can conduct satisfactorily). The drop containing the cell is separated from the stream and carries the charge imparted to it. The charged drop traverses between electrostatic plates; and due to the interaction of the electric field on the charge, the drop is deflected.

An interesting variation is to replace the electronics with a computer. The data placed into the computers will consist of scattered light intensity, two fluorescence intensities, and relative time of detection. The computer is used to analyze the data and within the 250 μs limit, outputs the sort word. If the sort word is variable so is the charge on the stream. Consequently, one can sort the droplets containing the cells at a defined location on a microscope slide. If the cell's data and position are stored in the computer, that cell can be retrieved by a computerized microscope (described later). The ability to correlate flow cytometric data with visual data is important in identifying the properties of rare cells, such as cancer cells. Analytical studies of the sort trajectory under conditions of drag show that a "knee" is produced when the horizontal velocity is zero (Fig. 9).

The cytometer just described illuminates and detects cells orthogonally to their traversing path. A different implementation is to place the illumination and detection optical path along the same axis as the path of cell traversal. In this way, the cell passes through the source and detector focal plane. After the cell has passed through the focal point, it is directed away from the optical axis.

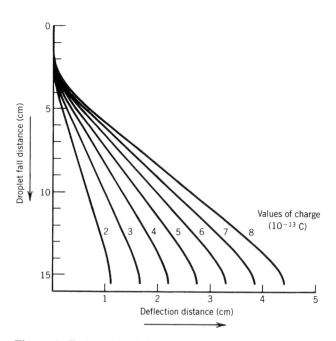

Figure 9. Trajectories of charged droplets for seven values of charge. The trajectories were obtained by solving the equations of motion, assuming that the drag on the droplet was a linear function of velocity.

Since the cell passes through the focal plane of the optical system, errors due to focusing and misalignment are reduced, particularly if the same objective is used for excitation and detection.

Most fluorescence spectra has uninteresting shape and the single value obtained by zero resolution systems is adequate. This limits the usefulness of spectral analysis of the fluorescent moiety. Nevertheless some effort has been carried out in this direction. The easiest is the use of a variable color filter and to select the resulting wave for shape (see slit scan systems below). Also some have used white light pulses and processing to identify the spectra. By far the solution has been to use multiple lasers for excitation and to detect the different emission on separate detectors. There are technical limitations as well, the need for UV light means expensive powerful lasers or substantial Xenon light sources, furthermore the speed of processing has now gone up an order of magnitude from the time between cells to the time that the cell is in the laser beam.

An interesting application is the use of fluorescence lifetimes or the so-called phase detection systems. One selects probes with overlapping spectra and different singlet excited states. Using a modulated laser beam and measuring the phase shift in the fluorescence signals gives a measure of differential lifetime.

Low Resolution Systems. In the flow system just described, the cell flowed past a laser beam and the detected energy resulted in a pulse that was peak detected. Clearly, any other parameters could be obtained from the pulse, including pulse area and pulse width, and each descriptor is a single number. Cells are analyzed and sorted based on a single value of a descriptor; thus, each parameter of a zero-resolution system takes a single value from each cell.

There is potential for greater information available from the pulse. For example, if the laser is flattened into a ribbon whose thickness is small compared to the length of the traversing cell, the resulting pulse will contain information concerning the shape of the cell. Proper analysis of the pulse can provide increased information about the cell because of the slight increase in resolution. This technology is referred to as "slit-scan". The very narrow laser beam is 4 μm in width and the flow system is slower than the normal 10 m/s.

Figure 10 shows the schematic diagram of a cell traversing through the laser beam. The figure on the left shows a zero-resolution wave form in which only the peak-detected value is used to represent the data from the cell. The low resolution system (right) shows some structure as the cell traverses through the laser beam: First the cytoplasm (C) is detected, then the nucleus (N), and then again the cytoplasm.

The development of automated cytology was largely motivated by the requirement to diagnose cancer by finding cancer cells in a sample. Cancer cells can be distinguished from normal cells based on their DNA content (as distinguished by, e.g., AO) and other parameters such as size and RNA. Patient samples for the purpose of detecting early cancers contain highly variable and usually very small numbers of diagnostic cells. Consequently, the problem is not in distinguishing cancer cells from normal cells, which is easily accomplished, but in distinguishing particles whose values fall in the space consistent with cancer cells, but are in fact not cancer cells. This requires additional information provided by low resolution systems, but not by zero-resolution systems.

High Resolution Cytometry. We now consider cell image analyzers usually implemented as a computer-controlled microscope. This high resolution instrumentation is characterized by the ability to obtain large amounts of data on a few cells and to perform complex analyses on those cells.

A microscope is connected to a computer so that data reduction and control of the object in the microscope can be performed. These data are used for automating cell recognition.

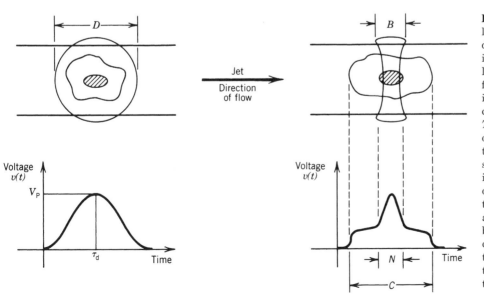

Figure 10. Axial view of a circular laser beam of diameter D and cylindrical laser beam of width B, both intersected by a cell. The circular laser beam produces a pulse signal from which a single parameter of information, such as peak height, can be obtained. For example, at time T_d, $v(t)$ has a peak value V_p corresponding to the cell in the center of the laser beam. The cylindrical-shaped laser beam is a ribbon of light inter-sected by a portion of the cell. The output signal, which is the convolution of cell structure with laser light, allows cell structure information to be retrieved easily. For example, the cytoplasm in the laser beam for time C is easily distinguished from the nucleus in the laser beam for time N.

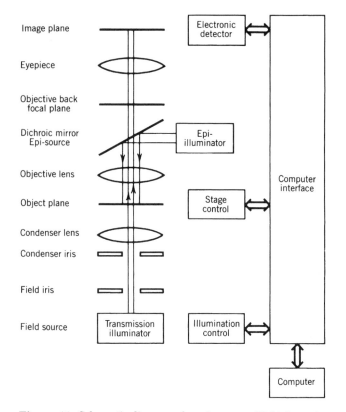

Figure 11. Schematic diagram of a microscope. Light from the transmission source traverses through to the image plane. Epi-illumination is reflected to the object that in turn emits fluorescence, which passes through the dichroic mirror onto the image plane. The computer provides scanning and illumination control and reads the electronically detected data.

The essential features of a computerized microscope are shown in Fig. 11. This is a schematic diagram of a modern-day microscope supporting transmission and reflection (fluorescence) microscopy. For transmission microscopy, the light path from the source is treated so that parallel light is transmitted through the object plane. The objective and eyepiece lenses function together to provide the visually detected image at the image plane. Fluorescence excitation (epi-illumination) is by short-wavelength light (violet), which is reflected by the dichroic mirror, then passes through the objective lens onto the object. The object's fluorescence emission passes through the dichroic mirror, which allows the long-wavelength light to pass; it is then detected in the image plane.

The computer interface provides data exchange between the computer and the controllers of stage motion, illumination, and detection. A photodetector at the image plane measures the intensity of light emanating from the object and a common implementation passes light through a pinhole on to the optical access. In this case, the moving stage controls the object so that its spatial variation of light intensity is acquired. The entire object is imaged one point at a time. Each point is a picture element (pixel) emitting light. Light from each object point produces a datum point, which is stored in a matrix such that the value of the ijth element is the intensity of the light at the ijth pixel.

We abstract the microscope to consist of a light source, an object plane, and detector. Any one moving with respect to the other two can produce the scanning required to image the object. Table 12 shows the characteristics of the scanning components of a microscope.

These systems have high degrees of flexibility since performance is controlled by means of software. In such a system, a cell is located for analysis, appropriately illuminated, and contiguous absorbance values of the image are detected. The image is made up of pixels (say 1×1 μm squares) and the pixel's intensities are digitized and stored. Software controls data acquisition, performs mathematical operations, and formats the data to a convenient form. For example, the absorbance values of a cell are converted to digital values, gray scale histograms are formed from which cluster analysis in a multiparameter space is performed. Consequently, statistical decision rules can be invoked to assess characteristics of a cell within some statistical confidence interval.

DATA ACQUISITION, PROCESSING, AND MODELING

In the previous section, we described cell measurement systems. Now, we concern ourselves with the acquisition and manipulation of the data produced by these systems. We will begin with data lists produced by zero-resolution devices and proceed to high resolution images produced by cell-scanning devices.

Zero-Resolution Systems

In zero-resolution systems, a cell produces a single value for each parameter measured. If only light scattering is measured, each cell produces only single values of scattering; if additionally two values of fluorescence are

Table 12. Scanning Components on Microscope

Component Performing Scan	Source	Characteristics of Components	
		Object	Detector
Source	Point scanner (laser illumination)	Stationary	Stationary
Object	Stationary (uniformly illuminated field)	Scanning stage	Stationary (pinhole)
Detector	Stationary (uniformly illuminated field)	Stationary	Point detector (television camera)

measured, each cell produces three data points. The acquired data are listed one point after the other as

Cell Number	Parameter 1	Parameter 2	Parameter 3
Cell 1	255	10	128
Cell 2	128	210	197
•	•	•	•
•	•	•	•
•	•	•	•
Cell n	37	196	212

Such list mode data can be stored and reduced into histograms of the data.

Histograms. A common form of data reduction is the production of histograms of the data. An n-parameter histogram is in $n+1$ dimensions, where the additional dimension is frequency of events.

A single-parameter histogram substantially reduces the amount of data compared to the list mode. For example, a 256-bin histogram (2^8) requires 256 words of storage compared to the 25,000 words required for storing the 50,000 bytes of data obtained from a single run. Examples of single-parameter histograms are shown in Fig. 3.

Dual-Parameter Histograms. Data from two parameters, P_1 and P_2, can be displayed so that each pair (p_1, p_2) contains the number of cells that express that pair of values. For data storage, the multiplicative effect of the number of dimensions reduces the histogram's value. A two-parameter histogram of 256 channels requires 256^2 (64 k) locations, which is approximately the number of data points acquired. Furthermore, the data will not cluster well when insufficient numbers of cells have been acquired. A common way around this problem is to reduce the resolution from 8 bits to 6 bits (256 to 64). Now, only 4096 channels are required for that histogram display. The increased ability to visually cluster the data is achieved with a loss in additional information.

Three-Dimensional Histograms. Although this is a four-dimensional system, a common method of displaying three-parameter data is to plot each triple for a given frequency. For example, Gaussian-distributed data in three parameters (with equal variances) is displayed as a spherical cloud in the three-parameter space at each frequency. Since the storage space for such a histogram is 256^3 (16 million), clearly storage of the list data is more efficient than storing the data in histogram form.

A useful method of analyzing data displays one or more parameters as a function of a specified set of values of another parameter. For example, it is convenient to analyze two fluorescence parameters resulting from the largest cells. The set of values representing the largest cells is called a window. For each value in the window, there corresponds one or more parameters that can be displayed.

The values within the window act as a gate to display the other parameters.

Finally, a physical interpretation of the histogram occurs when we consider its mean value. Assume we acquire data from a multiplicity of cells. Further, we wish to measure the presence of some molecule by causing a radioactive tag to bind to it. The resulting measurement of radioactivity is the sum of the contributions of the individual cells. If this is normalized to the number of cells, the result is a mean value. Now the histogram, of course, displays the number of cells that contain each value of data for all the values of data. So it provides substantially more information than the single mean-value datum. Furthermore, the mean of the histogram corresponds to the mean-value result. For example, the distribution of DNA in a cycling population is very uneven, a fact that would not be elucidated from mean values over the population.

Histogram Analysis. Parametric and Nonparametric Analysis of Histograms. Parametric analysis assumes a model or distribution is used to compare, or analyze, the data. Nonparametric analysis assumes no such model.

Nonparametric Analysis. It is useful to compare histograms so that the effect of different treatments can be statistically assessed; two techniques are generally used. The first one requires three or more identical histograms and sums up the individual channels to obtain a mean and variance of each channel. Based on this model a channel-by-channel t test is made of the sample histograms to decide if statistical significance is valid.

A second technique uses the Kolmogorov–Smirnoff technique (see Fig. 12). In this case the histogram [Fig. 12a] is normalized, then summed to a cumulative distribution [Fig. 12b]. The two cumulative distributions are compared. The comparison is in the form of the absolute value of the difference between the two histograms. If this value exceeds a certain critical value that determines the level of confidence, the two histograms are statistically significantly different.

Parametric Analysis. The majority of work in this area has been to develop models that will aid in finding the number of cells in the various compartments of the cell cycle from DNA histograms. It is assumed that the "true" DNA histogram consists of two impulses with DNA values at $2n$ (for G0/G1 cells), the other at $4n$ (for G2/M), and cells in S are in between. As a result of the imprecision of measurements, the impulses at $2n$ and $4n$ values of DNA have been broadened. More generally, we write the expression used to model single-parameter DNA distributions.

$$F(k) = F_1(2\pi\sigma_1)^{-1/2} \exp[-(k - \mu_1)^2/2\sigma_1^2]$$
$$+ F_2(2\pi\sigma_2)^{-1/2}\exp[-(k - \mu_2)^2/2\sigma_2^2] + s(k)$$

In parametric analyses of the DNA histograms, the broadened pulses are assumed to be Gaussian with means μ_1, μ_2 and variances σ_1, σ_2, respectively. The critical issue is the shape of the s-phase distribution, $s(k)$. In the earliest modeling systems a second-degree polynomial

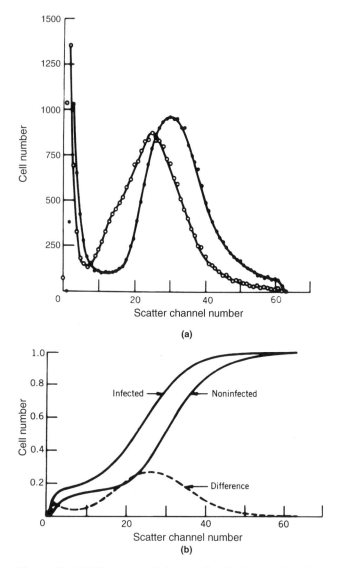

Figure 12. (a) Histograms of the size distribution of cells either infected (○) or not infected (●) with herpes virus. (b) The cumulative distribution of each of the above histograms.

Unfortunately, these models are good only for well-behaved populations whose value of DNA does not exceed $4n$, such as tissue culture cells. The DNA distribution of cells from cancer patients are complex due to cell clones arising with increased amounts of DNA, increased number of chromosomes (aneuploidy), and the presence of multi-nucleated cells.

Finally, these models seek only to fit curves to the values of DNA. They are not intended to elucidate the production of such distributions from fundamental principles, although some authors have tried such derivations, and achieved limited success.

Image Analysis

Cell data acquired by high-resolution systems consist of a list of photometric values of light intensity and the corresponding source point address. Scanning is sequential so the source addresses (or locations of the source points) need not be explicit. Each source point (pixel) on the cell is really an area of the cell that produces a uniform intensity. The pixel size determines the resolution of the image.

The light-intensity list can be manipulated, resulting in classification of the object by techniques referred to as pattern recognition. This is distinct from other operations on the image such as restoration. Image restoration uses the mathematical properties of the image generation system to obtain an inversion that will remove the errors introduced during image generation. Pattern recognition uses image features to distinguish between objects and to express that distinction in statistical terms. The general sequence of events for pattern recognition is shown in Fig. 13.

Pattern Recognition. The preprocessor operates on the digitized image, which is the list of pixel values. In preparation for feature extraction, sections of the image are defined (automatically or by human interaction) using edge detection filters to distinguish boundaries.

A second operation is the formation of a gray scale (and other) histograms. The gray scale histogram lists the number of pixels for each gray value (intensity value) for the entire dynamic range of gray values in the system. Image characteristics cluster about some value of gray in the histogram. For example, a cell typically displays a cluster value representing the nucleus and second cluster about the lesser dense values of the cytoplasm. Thus, a separation of the nucleus and cytoplasm can be effected depending on the degree of separation between the two clusters. The cell shown in Fig. 14 has had all its cytoplasmic values truncated to a particular gray value; this gives prominence to the highly variable but more dense values of the nucleus.

was assumed to be the fit. Other models have assumed Gaussian curves of various values to try and fit the data to the s values. The distinction between the cells that are in G1 and S is not sharp and these populations overlap each other. Consequently, methods that distinguish between G1 and S without models (e.g., graphical) usually underestimate G1 compared to S by 20%. The number of cells in S is small compared to the number of cells in G1; furthermore, the width of the G1 peak is also very small; consequently a small error in defining the G1 compartment can result in substantial errors in estimating S and G2.

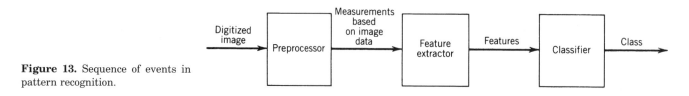

Figure 13. Sequence of events in pattern recognition.

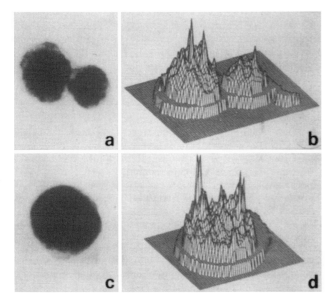

Figure 14. Display of a digitized cell showing the truncated values of the cytoplasm.

A second operation is equalization of the gray scale histogram. This is a nonlinear transformation resulting in a histogram with reduced number of gray levels at approximately equal frequency. Since the frequency of gray level occurrences contains no information about the texture of the image, this permits images to be placed into a consistent format prior to comparisons.

Feature Extraction. After preprocessing the image data to yield measures based on image data, it is possible to obtain cell characteristics, called features, that are based on geometric parameters, texture parameters, optical density, or gray level parameters.

For example, statistical texture analysis, such as is typically used in medical imaging, uses conditional probability matrices to obtain features. Picture analysis proceeds by forming a transition probability matrix from the image data. From this matrix the texture features used for classifying objects in the image are determined. Each element in the matrix, $P_L(I/J)$, is defined as the conditional probability of gray level occurring L picture points after gray level J occurs. Each of L matrices is square of dimension n, where n is also the number of gray values. Operations on these matrices can be performed to produce parameters, such as the second moment of the matrix or the moment of inertia about the diagonal of the matrix, for all values of I and J. Clearly, a very large number of parameters can be chosen in this manner. Texture analysis has been used successfully to classify cytological samples.

Classifier. The last step in the pattern recognition chain is classifying the cell based on the measured features. Each feature is a dimension in feature space. A range of values of each feature are obtained from a training set, that is, a set of objects whose classifications are known. This feature space is sectioned by boundaries defined by the class or members of the training set. These boundaries separate the objects into the different classification regions. The features of an unclassified or unknown object can then be compared to the classification obtained from the training set. The probability of the unknown object belonging to a particular class as then determined. An interesting problem is that the vectors produced by the features may not be orthogonal. If such vectors are correlated, the feature provides little new information in addition to its correlative feature. Consequently it is important to have features with different information content.

CURRENT USAGE

DNA Measurements

In previous examples, we discussed the measurement of DNA in intact cells, which are now providing new and useful information in both biology and medicine. We briefly mention three applications of interest: bacterial analysis, chromosome analysis, and sperm analysis.

Flow systems can be used to classify bacteria types by the difference in dye uptake by the DNA base pairs. Fluorescent CG- or AT-specific dyes are used to generate a two-dimensional histogram that can be used to differentiate bacterial species. This may be useful in a biology laboratory for rapid identification of bacterial types. In medicine, such a detection system would be useful in urine samples. However, with blood its usefulness is limited. First, the level of infection in blood is produced by very few bacterial cells. Their small size and the presence of overwhelming numbers of red blood cells makes detection extremely difficult. A second major difficulty is that the information of importance in medical practice is sensitivity: which and how much antibiotic kills the bacterium, not which bacterial cell is causing the infection.

Chromosome Analysis. As yet neither zero-resolution systems nor low resolution systems provide adequate separation between the human chromosomes to completely perform a human karyotype. It is interesting, though, that errors in karyotyping due to chromosomal particles produced by disease are constant and repeatable and thus can be expected to show exceptions to the normal.

Flow systems have been used to analyze sperm cells, primarily to measure the DNA of the cell types. Spermatids, which are spherical precursors to the mature sperm, are easily analyzed by flow cytometry and produce a separation between the X and Y chromosome. This separation implies that a distinction can be made between spermatids whose sperm will give rise to females (X chromosome) or males (Y chromosomes). These data were obtained with an axial flow analysis device and with a very small coefficient of variation (0.9%). Unfortunately, the mature sperm cells are not well-behaved spheres, but flattened bags of DNA with a comparatively long tail. Thus, an orientation artifact blurs the 2% difference in DNA resulting from the X–Y chromosome mismatch. However, flow analysis can be used for fertility determinations and where the number of sperm is an important parameter.

Hematology

An important area is hematology, the discipline concerned with the study of blood and its components.

Differential blood cell counters place white blood cells into five basic classes: lymphocyte, eosinophil, basophil, monocyte, and neutrophil. Since red blood cells outnumber the white cells by a thousandfold, they are usually excluded by preparation. These six-cell types have been classified by several commercial devices that use pattern recognition and high resolution image analysis for classification. Such classifications have occurred with varying levels of reliability. A major problem is a category required for "others", that is, for cells that cannot be classified into any of the six classes. This category reflects the fact that at any one time, blood cells are maturing into the different classes. Furthermore, the body's reaction to some diseases or insults is to produce increased numbers of cells from the bone marrow into the blood stream at various levels of maturation of the cells. Consequently, some workers have attempted to classify nucleated blood cells into 17 different types. Efforts at white blood cell classification using flow systems have not had the commercial success that scanning systems have had.

Immunology

An important application of flow systems has been to enumerate and classify the cells in the body that are part of the immune system. Briefly, the body produces antibodies to antigens. The antigens are usually foreign substances such as a bacterial cell surface; but in abnormal situations the body produces antibodies to its own antigens. The covalent binding of antibody to antigen initiates a set of reactions that results in the destruction of the foreign substance.

A specific kind of antibody, called a monoclonal antibody, reacts only with a single antigenic determinant. This is in contrast to the multiplicity of antigenic determinants that produce heteroclonal antibodies, which are usually produced by the body. Fluorescent stains such as FITC and PE are covalently bound to the monoclonal antibody of choice. Thus, simultaneous green and yellow fluorescence to distinguish the different immunological cell types can be produced (see Table 13).

One application of this technology is to differentiate the various classes of lymphocytes. Lymphocytes are divided into two general classes: B cells and T cells. In general, the B

Table 14. Ligand Binding

Immunologic
Cell surface receptor on lymphocytes
Cells identified by surface immunoglobulin
T cells identified by various T receptors including T4 for helper cell and T8 for killer cells
DNA synthesis: Antibody to bromodeoxyuridine
Indirect immunologic binding
Goat-anti-rabbit antibody with fluorescent tag binds to rabbit–anti-receptor molecule
Avidin tagged with fluorescent molecule binds to biotin attached to the antibody against the detected antigen
Hormonal
Estrogen receptor analog (17-fluorescein estradiol)

cells produce antibodies, whereas the T cells are direct protagonists in the immune response. By the use of monoclonal antibodies, T cells have been divided into many subsets. Some of the most important are T4 (helper cells) and T8 (suppressor cells). These cells are distinguished by T4 and T8 monoclonal antibodies. Table 14 shows the ligand-binding applications, including lymphocyte cell surface characteristics.

In medicine, determining the number of helper and suppressor cells is valuable in transplantation and cancer. The ratios of these cells and the changes in ratio as the patient undergoes treatment is an indicator of the patient's ability to respond to the treatment. Indeed the absence of helper cells due to viral destruction is the primary problem in acquired immune deficiency syndrome (AIDS).

Oncology

A major application in both flow and cell scanning systems has been in oncology. The primary thrust has been to classify cell types and to distinguish the malignant cells from nonmalignant cells. For example, nuclear texture has been used to classify uterine cervical cells into cancer cells and noncancer cells.

Studies involving the four major types of lung cancers have been performed to determine the ability of flow systems to first find and then sort the malignant cells. The criterion for malignancy is based on the morphological features used by pathologists to diagnose cancer. Cytological examination of lung cells begins with the sample (sputum). The cells in the sample are fixed to minimize disease contagion and for preservation. After fixation, the cells are stained with acridine orange, analyzed, and sorted for the highest value of green and red fluorescence. The Acridine Orange analysis of cancer cells provides information about the relative amount of RNA and DNA in cells. Cancer cells are found in regions showing the highest green fluorescence. By sorting, a substantial enrichment (65-fold) for cancer cells can be obtained. Concomitantly, there is a sharp reduction in normal cell types, such as lymphocytes and squamous cells.

A similar technique using acridine orange staining of cells has been applied to urinary cytology. This has resulted in a system that is demonstrably superior to human cytological analysis for earlier detection of recurrent cases of bladder cancer. Data on cells are placed in list

Table 13. Excitation and Emission of Selected Fluorescence Labels

	Excitation, nm	Emission, nm
Fluorescein isothyocyanate	488	530
(S)-Phycoerythrin	490	570
(R)-Phycoerythrin	498	575
Bodipy	503	511
Tetramethyl rhodamine	560	580
L-Rhodamine	572	590
B-Phycoerythrin	540	575
Texas Red	590	620
CY-5	649	666

mode for scatter, green fluorescence, and red fluorescence. The pulse width of scattering is used to find debris and eliminate it. The resulting "clean" data list then produces RNA and DNA histograms to find whether the number of cells in S>20%; if so, a recurrence is said to have occurred.

Many studies have been made that identify an aneuploid (aberrant number of chromosomes) population of cells. Such aneuploid cells are considered by some pathologists to be diagnostic of cancer.

In a clinical setting, a machine approach to cancer detection has been to analyze a very large number of cells without false positive indications. This is not possible with zero-resolution systems; however, low resolution systems have been developed with sufficient information content to identify the false alarms. These have been used in clinical trials and the results have been dramatic. High resolution instrumentation using cell-scanning techniques and texture analysis have been used to identify and classify cells according to cell type and denote the malignant cells. These systems have had success in identifying various normal as well as disease cell types.

The medical and biology literature abounds with examples of applications in flow cytometry and high resolution image analysis. The bibliography lists samples of this literature.

FUTURE PROSPECTS

Historically, the motivation for new instrumentation is the increased information. In automated cytology, instrumentation has increased information by improved optical resolution and the increased number of cells for improved reliability of statistics. Furthermore, the development of quantitative techniques to analyze cells makes possible the basis for a quantitative theory of biological phenomena.

In one sense these are learning tools; they provide basic measurements. In another sense, their increased speed is a basis for economic value since productivity is improved.

A further advantage is the use of computers. The measurement instrumentation can easily act as an input to a computer. The data handling power of computers makes possible analysis of enormous numbers of cells thereby establishing repeatable quantitative relationships.

The increase in speed and information and objectivity of this instrumentation makes it an economic force in the market. The present high expense of illumination equipment (primarily lasers) limits its economic practicality. Nevertheless, the present use of this equipment in diagnosis can be expected to increase. Furthermore, its use in therapy and prediction of disease are just now starting. With expected cost reduction, use of this instrumentation will increase.

BIBLIOGRAPHY

1. Mellors RC. Analytical Cytology. 2nd ed. New York: McGraw-Hill; 1959.
2. Melamed MR, Mullaney PF, Mendelsohn ML. Flow Cytometry and Sorting. New York: John Wiley & Sons; 1979.
3. Melamed MR, Lindmo T, Mendelsohn ML. Flow Cytometry and Sorting. 2nd ed. New York: Wiley-Liss; 1994.
4. Shapiro HM. Practical Flow Cytometry. New York: Alan R. Liss; 1985.

Further Reading

Cytometry, New York: original publisher Alan R. Liss. This is the journal of the International Society for Analytic Cytology. First issued in July 1980, it is a forum for automated cytology, with a thrust to basic measurements in the biology of single cells and particles. It is now published monthly by Wiley-Liss, Inc. Clinical Cytometry Started in 1994, and in 1997 produced the first issue of Current Protocols in Cytometry.

Analytical and Quantitative Cytology and Histology. St. Louis, MO: Science Printers and Publishers, Inc. This journal is sponsored by the International Academy of Cytology and the American Society of Cytology. It focuses primarily on automation and quantitative aspects of cytology and is also a forum for automated cytology.

See also ANALYTICAL METHODS, AUTOMATED; COMPUTERS IN THE BIOMEDICAL LABORATORY; DIFFERENTIAL COUNTS, AUTOMATED.

D

DECAY, RADIOACTIVE. See RADIONUCLIDE PRODUCTION AND RADIOACTIVE DECAY.

DECOMPRESSION SICKNESS, TREATMENT. See HYPERBARIC MEDICINE.

DEFIBRILLATORS

BRADLEY J. ROTH
Oakland University
Rochester, Michigan

INTRODUCTION

Ventricular fibrillation is a lethal malfunction of the heart. Normally the heart beats about once a second, and is controlled by electrical signals that occur in a predictable, periodic way. The heart's electrical activity, called the electrocardiogram (ECG), can be measured on the surface of the body. A normal ECG is shown in Fig. 1a. If the heart is in a state of ventricular fibrillation, the electrical control of the heart becomes disorganized and chaotic. Instead of producing a normal ECG, the fibrillating heart produces an ECG that looks more like random noise, as shown in Fig. 1b. Rather than contracting in unison, different regions of the heart contract independently, resulting in a quivering that is not effective in pumping blood.

Once the ventricles of the heart start to fibrillate, death follows in minutes. The American Heart Association estimates that in the United States 335,000 people die each year of sudden cardiac death, with most of the deaths attributed to ventricular fibrillation (1). The most effective way to prevent these deaths is to apply a strong electric shock to the heart within the first few minutes after the onset fibrillation (Fig. 2) (2,3). Devices that deliver such shocks are called defibrillators, and come in two types: external and internal. A physician, a paramedic, or even an untrained bystander can use an external defibrillator to apply a shock to an unconscious victim of ventricular fibrillation. The more sophisticated of these devices are automated so that the user need do little more than follow some simple instructions; such devices are called Automated External Defibrillators (AEDs). Internal defibrillators are similar to cardiac pacemakers, and are implanted

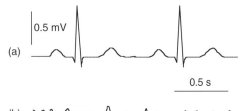

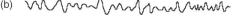

Figure 1. (a) A normal electrocardiogram (ECG). (b) The ECG during fibrillation.

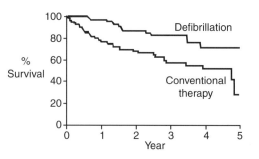

Figure 2. The percent survival for high risk patients when treated with conventional drug therapy or with an implanted defibrillator. [Modified and Reproduced with permission from Moss et al., Improved survival with an implanted defibrillator in patients with coronary disease at high risk for ventricular arrhythmia. *N. Engl. J. Med.*, 335: 1933–1940, 1996. see Ref. 2].

into patients who are at risk for ventricular fibrillation. They monitor the electrical activity of the heart and deliver a shock when necessary. Modern defibrillators can also function as pacemakers, and are called Implantable Cardioverter Defibrillators (ICDs).

EXTERNAL DEFIBRILLATORS

An external defibrillator works by applying a shock through electrodes on the surface of the body. Automated external defibrillators are becoming common in schools, on airplanes, and at other public places. A typical AED is shown in Fig. 3. Each electrode has an area of at least $50 \, \text{cm}^2$ and is attached to the skin by a self-adhesive pad. A conducting gel should always be placed between the skin and the electrode to reduce the skin resistance. The current passes from the electrodes through the entire torso, with only a fraction of it reaching the heart.

A defibrillator works by charging a capacitor to a high voltage and then discharging it through the patient's body (Fig. 4). When the switch S is to the left, a capacitance of about $200 \, \mu\text{F}$ is charged to $\sim 1500 \, \text{V}$, implying a stored charge of $0.3 \, \text{C}$ and a stored energy of $225 \, \text{J}$. Move the switch to the right, and the capacitor discharges through the resistance of the patient's body ($50 \, \Omega$ or more), generating a peak current of $30 \, \text{A}$ that decays exponentially with a time constant of $10 \, \text{ms}$.

An actual defibrillator circuit is more complicated than shown in Fig. 4. For example, the battery pack used to power an AED typically has a voltage of $\sim 12 \, \text{V}$. A high voltage power supply is needed to raise this voltage to the level necessary to charge the capacitor. Also, many defibrillators use a biphasic, truncated-exponential waveform, which is more effective for defibrillation than a monophasic wave form (Fig. 5). The biphasic waveform is produced by discharging the capacitor part way, then reversing the polarity of the leads, followed by further discharge. Switching circuitry that functions at high voltages is required.

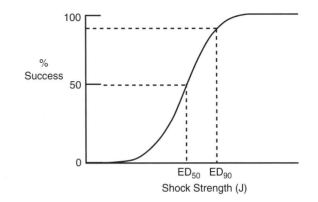

Figure 6. Defibrillation success follows a sigmoidal probability curve. The shock strength corresponding to a 50% success rate is called ED_{50} (for effective dose, 50%).

The word "automatic" in the term Automatic External Defibrillator means that the device can decide for itself if defibrillation is necessary. The AED monitors the electrocardiogram, and enough memory is included in the machine to store the ECG data. Also present are electronics that allow the device to analyze the ECG and decide if the ventricles are fibrillating. If they are, the AED will tell the caregiver to shock the patient. Most AEDs provide both written and oral instructions about how to attach the electrodes and operate the device. In theory, minimal training is required.

The success rate of defibrillation follows a probability curve like that shown in Fig. 6: the higher the shock energy, the higher the probability of defibrillation. A shock strength corresponding to a 50% success rate is known as "ED_{50}". To reduce the probability of a failed shock, physicians often use strengths of about ED_{90}. Unwanted side effects also increase with shock energy, so an AED usually shocks with a relatively low energy first, say 200 J. If that fails, it delivers shocks of increasing energy up to a maximum of ~ 360 J.

External defibrillators used in hospitals and ambulances are similar to AEDs, except that they are not automatic (the physician decides when to shock a patient rather than the device) and they can often be powered by plugging into the hospital's electric power grid instead of, or in addition to, relying on batteries.

IMPLANTABLE DEFIBRILLATORS

An implantable cardioverter defibrillator resembles a pacemaker, but its circuitry is similar to that in an AED. The battery, capacitor, and electronics are enclosed in a metal case (titanium or stainless steel), which is implanted under the skin in the chest (Fig. 7). The typical size of the case, or "can", is $\sim 50 \times 50 \times 15$ mm. The can often serves as one of the ICD electrodes.

The capacitors in an ICD are only slightly smaller ($\sim 125 \mu$F) than in an AED, but in an ICD the capacitor is charged to a voltage of only ~ 600 V, implying a charge of 0.075 C and an energy of 23 J. An ICD delivers about one-tenth the energy that an AED does, but in an ICD the shock is delivered through electrodes placed within the heart and is therefore just as effective for defibrillation. Tissue impedance for an ICD is at least 50 Ω, implying a discharge

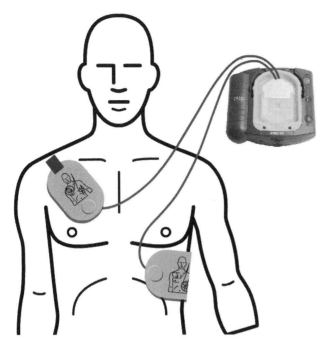

Figure 3. An automated external defibrillator (AED). This figure appears courtesy of Philips Medical Systems.

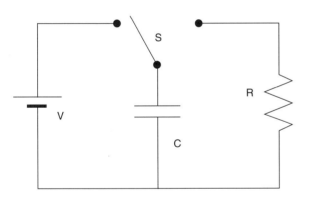

Figure 4. A simplified defibrillator circuit, where V is the voltage of the power supply, C is the capacitor, R is the resistance of the body, and S is a switch.

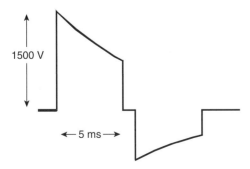

Figure 5. A typical biphasic, truncated exponential waveform used in many defibrillators.

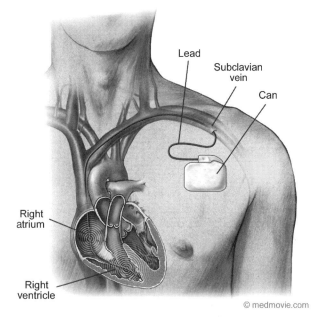

Figure 7. An implantable cardioverter defibrillator (ICD).

time constant of ~5 ms. Many ICDs contain two 250 μF capacitors charged in parallel, to give a total capacitance of 500 μF. When discharged, the connection of the capacitors is changed so they are in series, resulting in the capacitance of 125 μF mentioned earlier. One advantage of this technique is that when in parallel each capacitor needs to be charged to a voltage of only 300 V, which becomes a total voltage of 600 V when placed in series. Most ICDs have a maximum shock energy of ~30 J.

Lithium-type batteries, often lithium silver vanadium oxide, power ICDs. Two such batteries in series provide ~6 V. Since the capacitor voltage is ~600 V, the batteries are used to power a high voltage power supply. They are implanted in the patient's body, so changing them requires surgery, implying that battery lifetime is important. Lifetime is often measured in ampere-hours (A·h) (equivalent to a charge of 3600 C), and a typical battery is rated at ~3 A·h. If each time the capacitor is charged uses 0.075 C, the battery should be able to deliver thousands of shocks. However, the battery performance begins to decay before its total charge is exhausted, and also it must provide power for continuous monitoring of the ECG and other functions, so its observed lifetime is ~5 years. Another important property of a battery is the time required to charge the capacitor. Typically, the battery takes ~10–20 s to generate a full charge. If this time increased significantly, it would delay the delivery of the shock. The voltage decays gradually and predictably throughout the lifetime of a lithium silver vanadium oxide battery, so that the voltage can be used as an indicator of the battery's remaining useful life.

The electrodes and their leads are critical components of an ICD. Unlike the electrodes in an AED, ICD electrodes are implanted inside a beating heart and must continue to function there for years. Many ICD malfunctions arise because of problems with the leads. Like pacemaker leads, ICD leads are made from coils of wire to make them flexible and avoid breaks (Fig. 8). They are insulated, except at the

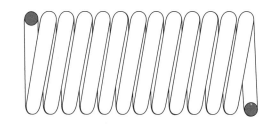

Figure 8. The conductor in the lead is often in the form of a coil to increase its flexibility and reduce mechanical stresses in the metal.

electrodes, by silicone rubber or polyurethane. A typical lead contains three electrodes: one for pacing and sensing, and two larger ones for defibrillation. An ICD lead is affixed to cardiac tissue on the inner (endocardial) surface of the heart. Often the tissue is damaged (inflammation, followed by fibrosis) in the area in contact with the lead tip. Steroid eluting leads minimize the tissue damage by slowly releasing the corticosteroid dexamethasone sodium phosphate. The ICD lead must be attached to the endocardial surface to prevent it from becoming dislodged. Some leads use a "passive" fixation technique consisting of plastic tines on the lead tip that become entangled in the trabeculae on the endocardial surface of the right ventricle (Fig. 9a). Other leads use an "active" fixation technique consisting of a metal helix, similar to a corkscrew, that is screwed into the endocardium (Fig. 9b). The defibrillation shock is delivered through a large electrode located many millimeters back from the lead tip. In some cases, current is passed through two electrodes (one in the right ventricle and one in the right atrium, as shown in Fig. 7), and in other cases the shock is delivered between one electrode and the defibrillator can.

The ICD recording lead senses the several-millivolt ECG signal within the heart. Two parameters that the ICD uses to detect abnormal arrhythmias are heart rate and arrhythmia duration. The ICDs use sophisticated algorithms to determine from the ECG if an arrhythmia is present, and these algorithms differ between manufacturers. Sufficient memory is included in the ICD to store ECGs before, during, and after a shock. Figure 1 shows that the ECG from ventricular fibrillation has a smaller amplitude than the normal ECG, making detection of fibrillation challenging. Information about the defibrillator,

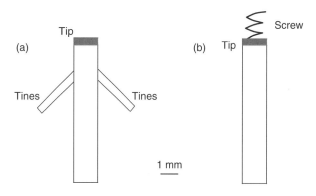

Figure 9. Distal end of a lead. (a) Passive fixation for attaching the lead to the endocardial surface of the heart using plastic tines. (b) Active fixation using a metal helix that is screwed into the heart wall.

e.g., the status of the battery and the lead impedance, as well as ECG traces, can be obtained through telemetry. Modern ICDs use the medical implant communications system radio frequency band (402–405 MHz). Most ICDs can be reprogrammed using telemetry.

When implanting an ICD, the physician must choose between a single-chamber and a dual-chamber defibrillator. Single-chamber devices have a single lead with the sensing electrode placed in the right ventricle. Their advantage is simplicity, longevity, and fewer complications. Dual-chamber devices have two leads: one sensing the right atrium and one sensing the right ventricle. Patients who rely on the ICD for pacing as well as defibrillation may benefit from the dual-chamber design. For example, a patient with a problem in the sinus node, which is located in the right atrium and serves as the heart's natural pacemaker, may respond best to atrial pacing. The atrial lead would then be for pacing, and the ventricular lead for defibrillation.

A cardiologist usually implants an ICD with the patient under local, not general, anesthesia. Typically, implantation requires an overnight hospital stay, although sometimes it is performed as an outpatient procedure. The ICD can is placed in a "pocket" under the skin in the upper chest, in the pectoral region. The lead is introduced into the subclavian vein, often by puncturing the vein with a needle, and then advanced into the right atrium under fluoroscopic view. A ventricular lead passes through the tricuspid valve between the right atrium and the right ventricle, and then is placed in contact with the endocardium near the apex of the heart (Fig. 7). After the implantation, the cardiologist tests the device by inducing fibrillation and then shocking the heart to check that defibrillation is successful.

The "C" in the term ICD stands for "cardioversion," which is a type of shock therapy for treating any rapid arrhythmia other than ventricular fibrillation. Typically, a physician uses cardioversion to treat atrial fibrillation or a ventricular tachycardia (a rapid but still organized beating of the ventricles), both serious abnormalities but neither immediately life threatening. The shock strength used during cardioversion may be weaker than or similar to that used for defibrillation, depending on the type of abnormality. Cardioversion often uses information from the ECG to time the shock optimally, whereas defibrillation shocks are delivered without any such timing.

MECHANISMS OF DEFIBRILLATION

Defibrillators have been developed empirically without a complete understanding of the mechanism of defibrillation. Many researchers are studying this important problem (4). The mechanism of defibrillation is closely tied to the mechanism of arrhythmia induction by an electrical shock. In order to initiate an arrhythmia, a shock must be given during the vulnerable period (during the time of repolarization of the ventricular action potential). A very weak shock during the vulnerable period has little effect. A stronger shock can induce an arrhythmia; the lowest strength that induces an arrhythmia is called the "lower limit of vulnerability". An even stronger shock will not cause

an arrhythmia; the lowest strength above the lower limit of vulnerability for which an arrhythmia is not induced is called the "upper limit of vulnerability". In order to defibrillate, a shock must be at least as strong as the upper limit of vulnerability. If not, a shock that would otherwise successfully defibrillate the heart could restart a new arrhythmia that might decay quickly into fibrillation.

One of the more difficult issues in defibrillation research is determining exactly where, when, and how a shock affects cardiac tissue (5). The crucial question is how the shock alters the transmembrane potential (the voltage across the cell membrane), because it is the transmembrane potential that opens and closes voltage gated ion channels, thereby causing an action potential and wave front propagation. A simple, one-dimensional (1D) model of current flow through the heart wall suggests that the transmembrane potential caused by a shock is large only within a few length constants (~1 mm) of the heart's surface. This model cannot be correct, because fibrillation occurs throughout the heart and the shock must affect a large fraction of the cardiac tissue—not just a thin surface layer—if it is to defibrillate. What then is the mechanism by which a shock alters transmembrane potential deep in the heart wall? This question has not yet been answered definitively, but recent evidence suggests that tissue anisotropy and fiber curvature plays a key role (6).

Scientists continue to study defibrillation even as engineers improve defibrillator designs empirically. How increased fundamental knowledge about defibrillation will improve defibrillators is an open question. Nevertheless, defibrillators contribute crucially to the treatment of cardiac arrhythmias. They represent the best, and often only, option for patients with ventricular fibrillation.

BIBLIOGRAPHY

1. American Heart Association. Heart Disease and Stroke Statistics—2005 Update. Dallas, TX.: American Heart Association; 2004.
2. Moss AL, et al. Improved survival with an implanted defibrillator in patients with coronary disease at high risk for ventricular arrhythmia. Multicenter Automatic Defibrillator Implantation Trial Investigators. N Engl J Med 1996;335 (26):1933–1940.
3. The Antiarrhythmias Versus Implantable Defibrillators (AVID) Investigators, A comparison of antiarrhythmic-drug therapy with implantable defibrillators in patients resuscitated from near-fatal ventricular arrhythmias. N Engl J Med 1997; 337:1576–1583.
4. Ideker RE, Chattipakorn TN, Gray RA. Defibrillation mechanisms: the parable of the blind men and the elephant. J Cardiovasc Electrophysiol 2000;11:1008–1013.
5. Roth BJ, Krrassowska W. The induction of reentry in cardiac tissue. The missing link: How electric fields alter transmembrane potential. Chaos 1998;8:204–220.
6. Trayanova NA. Concepts of ventricular defibrillation. Philos Trans R Soc London A 2001;359:1327–1337.

Further Reading

Jeffrey K. Machines in our Hearts: The Cardiac Pacemaker, The Implantable Defibrillator, and American Health Care. Baltimore: The Johns Hopkins University Press; 2001. An excellent history of the pacemaker/defibrillator industry.

Zipes DP, Jalife J. Cardiac Electrophysiology, From Cell to Bedside, 4th ed. Philadelphia: Saunders; 2004. A comprehensive, multi-author reference book that covers the entire field of cardiac electrophysiology, including chapters on defibrillators. New editions have been appearing about every 4 years.

Hayes DL, Lloyd MA, Friedman PA. Cardiac Pacing and Defibrillation: A Clinical Approach. Armonk, NY: Futura Publishing Co.; 2000. An excellent introduction to both pacing and defibrillation from the point of view of a medical doctor.

Bronzino JD. The Biomedical Engineering Handbook, 2nd ed. Boca Raton, FL: CRC Press; 2000. The definitive source for information about biomedical engineering, with chapters on pacemakers and defibrillators. New editions have been appearing about every 5 years.

See also ARRHYTHMIA ANALYSIS, AUTOMATED; CARDIOPULMONARY RESUSCITATION; PACEMAKERS.

DENTISTRY, BIOMATERIALS FOR.   See BIOMATERIALS FOR DENTISTRY.

DIATHERMY, SURGICAL.   See ELECTROSURGICAL UNIT (ESU).

DIFFERENTIAL COUNTS, AUTOMATED

DAVID ZELMANOVIC
JOLANTA KUNICKA
Bayer HealthCare LLC
Tarrytown, New York

INTRODUCTION

Blood is a tissue composed of a fluid medium called serum, which contains suspended formed elements called blood cells. These include red blood cells (RBCs), white blood cells (WBCs), and platelets. The white blood cells are subcategorized as neutrophils, lymphocytes, monocytes, eosinophils, and basophils. The relative concentrations of the white blood cell types, commonly referred to as white blood cell differential counts, or simply differentials, can provide important diagnostic information regarding the blood donor. In fact, the differential is one of the standard diagnostic tests, ordered by physicians frequently. Originally, differential counts were obtained by microscopic evaluation of 100 or 200 WBCs at 500- or 1000-fold magnification. In fact, the microscopic differential counting method remains the recognized reference method, as per NCCLS H20-A (1).

The first automated hematology analyzer was a cell counter based on the Coulter Principle, as described in Ref. 2. According to this principle, a suspension of particles diluted in an electrolyte-containing aqueous medium is drawn through minute apertures on either side of which charged electrodes are positioned. The electrolytic medium and electrodes are part of an electrical circuit, which can be direct current (dc) or radio frequency (RF). As a particle passes through the aperture it raises the impedance of the circuit because of its insulating properties. The momentary change in impedance is recorded as a signal pulse in the form of a voltage. The number of pulses is proportional to the particle concentration. Additional information about the particles can be obtained from pulse height analysis. White blood cell signals may be subcategorized based on pulse heights as lymphocytes + basphils, mid-range cells (monocytes + other large mononuclear forms), and granulocytes (neutrophils + eosinophils) (Fig. 1) in order to provide a so-called three-part differential counting method. This method is widely used in small laboratories around the world. Hematology analyzers providing three-part differentials are available from all major manufacturers of hematology instrumentation, and from many small manufacturers.

Since the advent of the Coulter principle, other techniques have been developed that permit the automated determination of the full five-part differential. Innovative technology of the current hematology analyzers permits assessment of the patient's clinical status through a combination of numeric results, morphology flags, cytograms and histograms. Significant technological improvements occurred to the automated analysis of WBCs since the first edition of the EMD (3). Additional automated information on abnormal blood cells combined with microscopic smear review provides an aid in disease diagnosis and patient monitoring. There is a continuing development of more sophisticated analysis to perform extended differential counts that would include differentiation of immature cells, but methods for standardization are not finalized yet.

This article includes a discussion of the cellular properties that are used to automatically analyze white blood cells and describes the techniques used to measure these properties. Examples are provided of the implementation of these techniques on commercial hematology analyzers. Finally, there is a brief discussion of the relative merits of the analysis methods in terms of result accuracy and laboratory efficiency.

MEASURABLE PROPERTIES OF WHITE BLOOD CELLS

Morphology

Cell Size.   White blood cells are distinguishable from red blood cells and platelets based on size. In normal individuals all WBCs are larger than red cells or platelets. The lymphocytes and basophils are usually smaller than neutrophils or eosinophils, which are in turn smaller than monocytes. Cell size is used by all automated hematology analyzers to distinguish WBCs from other cells in blood samples, and to further subcategorize WBC types.

Nuclear Size and Shape.   White blood cell nuclei may be classified as mononuclear (single lobed) or polymorphonuclear (multilobed). Lymphocytes and monocytes are mononuclear; neutrophils, eosinophils, and basophils are polymorphonuclear. In normal samples lymphocyte nuclei are round and monocyte nuclei are kidney shaped and larger than lymphocyte nuclei. The polymorphonuclear cells typically have two to five nuclear lobes. The nuclei of immature neutrophils may be band-shaped instead of having well-defined lobes. These properties are also used in all automated hematology analyzers.

3-Part differential histogram

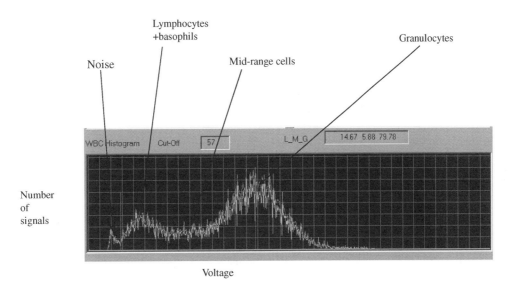

Figure 1. Coulter Principle—Three-part differential histogram.

Granule Number and Size. Cytoplasmic granularity is more pronounced for polymorphonuclear WBCs than mononuclear WBCs. Also, eosinophil granules are larger and more numerous than neutrophil or basophil granules. These properties are used by some automated analyzers to subcategorize WBC types.

Cytochemistry

RNA/DNA Staining. White blood cell nuclei and granules contain ribonucleic acid (RNA) and deoxyribonucleic acid (DNA), which are stained by nucleic acid staining dyes. These properties are used by some automated analyzers to subcategorize WBC types.

Enzyme Activity

Peroxidase. Granules in eosinophils, neutrophils, and to a lesser extent, monocytes, contain peroxidase enzyme, which catalyzes reactions between peroxide and various substrates that can result in deposition of dyes on the granules (4,5). Lymphocyte and basophil granules are peroxidase negative. This property is used in some automated hematology analyzers for subcategorization.

Esterase. Monocytes and neutrophils contain esterase enzymes, and these enzymes can catalyze reactions that result in deposition of dyes (6,7). This method can be used to distinguish monocytes and neutrophils from other cell types. This property is not currently used in automated hematology analysis.

Differential Lysis

Basophils. Basophils are more resistant to lysis than the other WBC types under certain pH conditions (8,9) and this property is used commercially to distinguish them from other WBC types.

Eosinophils. Eosinophils are more resistant to lysis than other WBC types under certain pH conditions in the presence of polyoxyethylene nonionic surfactants (10) and this property is used commercially to distinguish them from the other WBC types.

Immunocytochemistry

Multidimensional fluorescence flow cytometry uses the measurement of cell physical properties of cell size and cytoplasmic complexity, in addition to surface marker phenotyping using fluorescence-labeled monoclonal antibodies. Monoclonal antibodies have been produced that are specific to WBC generally, such as, CD45, and to specific WBC types, such as, CD3 for T-lymphocytes. These may be labeled with absorbent or fluorescent dyes for identification.

MEASUREMENT TECHNIQUES

Commercial analyzers use some combination of the following six measurement techniques to count WBCs and to distinguish among their subtypes:

Aperture Impedance

This technique uses the electrical insulation properties of WBCs in conjunction with the Coulter principle, described above. As stated, the amplitudes of WBC signals depend on size and to some extent, intracellular properties. Therefore, at least three distinct signal amplitude populations form on impedance signal frequency histograms.

Light Scattering

This technique uses the optical refraction properties of WBCs. White blood cells, suspended in a medium whose refractive index is significantly different (normally lower) than those of the cells, pass essentially in single file through a narrow aperture in a clear glass block. The suspension column is sheathed in a fluid whose refractive index matches that of the medium. A collimated beam of typically monochromatic visible light is incident on the glass cube in a direction that is perpendicular to the direction of the cell stream. As the cells interrupt the beam of light, they scatter it in a manner that is characteristic of their size and refractive index, and to a lesser extent of their internal properties, including granularity and nuclear configuration. Each signal pulse corresponds to an enumerated WBC. Light scattering intensities are a function of the scattering angle, and intensity measurements are usually made over two, three, or even four different scattering-angle intervals. The scattering intensity patterns are characteristic of the WBC subtypes. Two- or higher-dimensional scattering intensity plots form signal clusters that are associated with different WBC subtypes. These scatter–scatter plots are called scattergrams or cytograms, since they are multidimensional plots of the scatter signals generated by cells.

Optical Absorption

This technique uses the cytochemical properties of WBCs that permit the cells to be stained or to accept dye in a manner characteristic of the cell subtype, in conjunction with pretreatment of the cells. The measurement process is similar to that for light scattering, except that a characteristic decrease in light transmission is measured.

Fluorescence

Fluorescence is the reemission, at a lower frequency, of light absorbed at a given frequency. Fluorescence signals are generally larger than absorption signals. This technique uses cytochemical properties, as above, except that the stain is fluorescent rather than absorptive. This technique for obtaining WBC differentials should be distinguished from the immunofluorescence technique that relies on the lineage-specific expression of cell surface antigens to produce distinct so-called clusters of designation (CDs). Although immunofluorescence can be highly specific and therefore very accurate, it is not used in commercial hematology analyzers because of the high costs of the fluorescent-labeled monoclonal antibodies required to specifically tag the cell surface antigens.

Automatic Pattern Recognition

This method relies upon computerized pattern recognition algorithms to classify WBCs. Stained blood film slides are mounted on a microscope stage with motor-driven advancement of the slide in the plane perpendicular to the optical axis. Cellular images are captured on CCD arrays and the images are analyzed by pattern recognition techniques. A WBC differential can be reported based on the classifications. The basic technique was developed in the 1960s by a company called Geometric Data, in a product called the Hematrak (11). A current version of this technology, called Cellavision, is available from a company called CellaVision, Inc. (1555 Jupiter Park Dr., Suite 6, Jupiter, FL 33458, www.cellavision.com.

This technique is not widely used because the instrumentation, which provides a WBC differential, but not an absolute count, is expensive. The per-sample cost is also high. Further, until recently the throughput was low. Currently, Cellavision claims a throughput of 100 samples per hour, which is comparable to the throughput of major automated hematology analyzers.

One advantage of this technique is that the fixed, stained blood films, which are actual whole blood samples, can be stored for years. Also, since the positions of the imaged cells can be recorded, the cells can be recalled for future manual review. This is not possible with routine automated hematology analyzers, where the analyzed cell suspensions are disposed of immediately.

Image-in-Flow

This technology combines flow cytometry and image analysis. As suspended stained cell flow through a narrow aperture in an optical flow cell, their images are captured on CCD arrays. Pattern recognition algorithms analyze the images and classify the cells. In addition, these instruments act as flow cytometers and provide light scattering intensities and fluorescence intensities. The scattering and/or fluorescence data can be displayed on cytograms. As a result, it is possible to select a point on a cytogram and display the associated image. The Amnis Corporation ImageStream 100 is an example of such a device (12). Sysmex Corporation has reported on an experimental version of such a device, as well (13).

SAMPLE STABILITY

The physical and chemical properties of WBCs are subject to change *in vitro*. The extents of these changes depend on both storage time and temperature. First, WBC swell, then their membranes become leaky and they release their granules, and ultimately they autolyse. In addition, WBC nuclei undergo subtype-dependent configuration changes *in vitro*. Degradation is more rapid at ambient temperatures than under refrigeration. The rate of degradation differs according to the WBC subtype.

Swelling affects both cell size and refractive index. Also, since the granules are denser than their cytoplasmic medium, the cells become less dense upon granule release, even without consideration of swelling. Further, granule release affects the cytochemical properties of WBCs in a subtype-specific manner.

Generally, cell morphology is less stable over time *in vitro* than enzymatic properties, such as peroxidase and esterase activity. It is also less stable than nucleic acid staining capability. Therefore, automated analyzers that use only morphological properties to determine WBC differentials are more limited in terms of the *in vitro* age of samples that they can accept for analysis than are analyzers that use cytochemical properties. For example; automated analyzers that use light-scattering patterns to distinguish among WBC subtypes based on differences in morphology, use either fixed gates based on typical cell patterns to define cell populations, or combinations of gates and pattern recognition techniques, such as cluster analysis. Analyzers that use fixed gates provide inferior discrimination to those that include cluster analysis, because cell clusters shift as a result of morphological changes. Even analyzers using cluster analysis cannot distinguish well among cell populations once these begin to merge due to cell degradation. To the extent that cell cluster positions can be maintained, these limitations are overcome. Cytochemical staining based on enzymatic activity of cells, along with absorption or fluorescence measurements, provides added cluster position stability because of the relative stability of these cellular properties.

SAMPLE PREPARATION

White blood cell concentrations in peripheral blood samples normally range from $4 \times 10^3 - 11 \times 10^3$ per microliter (μL). Red blood cell concentrations normally range from $4 \times 10^6 - 5.5 \times 10^6 \cdot \mu L^{-1}$, and platelet concentrations from $150 \times 10^3 - 400 \times 10^3 \cdot \mu L^{-1}$. Individually, WBCs, RBCs, and platelets are mutually distinguishable by size alone. However, in undiluted whole blood samples the concentration of RBC is so large that electronic sensors can detect only a single prolonged signal due to the ever-present red cells. Given typical signal processing conditions for automated analyzers, this remains true even at 50-fold sample dilutions. It is not until approximately a 500-fold dilution that the signal interference from RBCs in a sample becomes manageable, but at this dilution only 100 or so WBCs are counted in a typical cycle in which 50,000 or so RBCs are counted. This is statistically inadequate for automated WBC differential determinations. Automated analyzers deal with this issue by selectively destroying (lysing) the RBCs in a whole blood sample, by adding surfactant and/or by reducing the osmolality of the suspension medium. This is usually done at a dilution ratio of 40–100:1 to maintain adequate WBC concentrations for counting purposes. In the absence of RBCs, at a dilution ratio of 50:1 as many as 100,000 events can be automatically analyzed within 10–20 s.

SIGNAL GENERATION

The number of WBC events analyzed by automated hematology systems during a measurement cycle must be controlled. Typically, 5000–10000 WBC events are counted during the cycle. In addition to controlling the number of events counted, the systems must control the quality of the observations made. This is necessary because in both optical and aperture impedance measurements the signal generated by a particle in the sensing zone depends on the position of the particle within the zone. To minimize particle position variability, the stream of cells in a WBC suspension is centered within the zone. The cell suspension is constricted to the center of the zone by enveloping it in a fluid cladding called a sheath. This constriction is often referred to as hydrodynamic focusing. Sheathing also serves to control event frequency.

The signals generated in aperture impedance systems and in light scatter systems are measures of cellular properties, such as size, density, granule content, nuclear size, and shape. The signals in systems using light absorption or fluorescence measurements are based on the labeling of cellular components by various dyes/stains.

Measurement artifacts that can interfere with signal generation include spurious signals resulting from high frequency electronic noise; often due to improper electrical grounding of electrical components. Spurious signals arising from light output instability, often due to uncontrolled switching from one laser light emission mode to another. Truncation of signals or short-term signal intensity variations associated with either poor hydrodynamic focusing of cells in the flow stream or misalignment of the stream in the signal generation path. Signals associated with particulate matter other than cells. These are often due to impurities in reagent containers or to precipitation of reagent components due to mishandling or improper storage.

SIGNAL DETECTION

Aperture Impedance

In both dc current and RF current versions, signals appear as voltages. Direct currents primarily probe cell size, whereas RF currents probe cell features, such as granularity, nuclear lobularity, and cell density. Impedance measurements are by themselves adequate for automated three-part differentials, but for automated five-part differential analysis they must be combined with at least one other measurement.

Light Scattering

Four scatter regions are usually associated with optical detectors for automated hematology analysis: axial or forward scatter (0–1°), low angle scatter (1–5°), high angle scatter (5–45°), and very high angle scatter (45–90°). The axial- or forward-scatter detectors are considered to be sensitive mainly to cell size. Low angle scatter is also associated mainly with cell size. The higher angle regions are associated with internal structure, mainly as a result of multiple scattering from numerous granules and/or from multilobed nuclei. Although there is an association of low angle scatter with cell size and high angle scatter with internal cell properties, both types of scatter depend on size, refractive index, and internal cell properties. Also, the angle cutoffs in parentheses are only by way of example and are not definitive, since smaller segments within these regions may be selected for optical detection, or the collection angle range may bridge the regions.

Light Absorption

Light absorption is detected as a loss of transmitted light. Detectors usually encompass up to ±20° of forward light scatter. The absorption is the difference between the light transmitted in the absence of and in the presence of a cell. This measurement technique is use in association with the selective uptake of an absorptive dye as a result of a distinguishing chemical feature of a WBC subtype.

Light loss is not identical to light absorption because it also involves light that is not absorbed, but that is instead scattered outside the collection cone of the detector. This is referred to as pseudoabsorption. In practice, pseudoabsorption contributes significantly to axial light loss for cells with large, numerous, and closely spaced granules that cause multiple light scattering to occur, with subsequent scatter at relatively large angles.

Light absorption signals are usually smaller than fluorescence signals and may also be smaller than low angle light scattering signals because the combination of dye extinction coefficient and the short path traversed through a cell usually results in only minor absorption. Certain combinations of concentration of granular material and absorptive dyes, such as eosinophil granules and 4-chloronaphthol, provide exceptions. Also, light absorption measurements have low signal/noise ratios because they are detected as fractional (1% range) reductions in light transmission values, and can have cell-specific pseudoabsorption contributions.

Fluorescence

Fluorescence detectors are normally placed at 90° to the incident beam to eliminate the contribution from the light source. This is not formally required, since the fluorescence signal is at a longer wavelength than that of the incident light. Therefore, a wavelength-selective beam splitter may be placed directly in the transmitted-light optical path, with the transmitted component following this path and the fluorescence component a diverted path.

Fluorescence signals are normally larger than absorption signals by their nature. Also, they have relatively high signal/noise because they are detected at different wavelengths than that of the incident radiation so that there is no interference from incident radiation. Also, they do not ride on much larger signals, as in the case of light absorption (Table 1).

EXAMPLES OF WBC DIFFERENTIAL ANALYSIS ON HEMATOLOGY SYSTEMS

The systems are listed alphabetically, by manufacturer name.

Abbott Cell-Dyn 4000 Hematology System and Bayer ADVIA 70 Hematology System

The Cell-Dyn 4000 and ADVIA 70 Systems use similar methodologies, and will therefore be described under the same heading. The ADVIA 70 will be described first, because it is of somewhat simpler design.

The ADVIA 70 Hematology Analyzer (Bayer Health-Care LLC, Diagnostics Division, Tarrytown, NY) combines the results of an optical channel and a dc electrical impedance channel to determine the WBC differential count.

In the impedance channel, a blood sample reacts with a basic reagent that contains a surfactant. The reagent lyses RBCs while maintaining the integrity of the WBCs. As stated previously, the WBC types that may be distinguished on this basis are lymphocytes + basophils, which generate the smallest WBC signals, so-called midsized cells, which usually contain monocytes, and granulocytes that include neutrophils and eosinophils. Although monocytes are expected to produce the largest impedance signals based on their size, they do not do so in this reaction system. This is because the reagent compromises the cellular integrity of the monocytes more than those of the neutrophils or eosinophils.

In the optical channel, a whole blood sample is diluted in a reagent that contains a surfactant to lyse RBCs and a fixative to maintain the integrity of WBCs. The reaction mixture is analyzed in the optical flow cell similar in design to the one described bwlow for the ADVIA 2120 Nuclear Density Channel. The light source is a laser diode emitting radiation at 633 nm. The analysis channel includes four silicon photodetectors, to determine the following: (1) Light extinction (Ex), which is the loss in transmission along the axis of incident radiation, (2) small-angle scattering intensity (Sa), (3) wide-angle scattering intensity (Wa), and (4) Super-wide-angle scattering intensity (Swa).

The analyzer displays two cytograms that are based on the signals from the four detectors: the Size cytogram is the graph of the small-angle versus wide-angle signal pairs and it is used to distinguish among neutrophils + eosinophils, monocytes, lymphocytes, and basophils. The distinctions are based on a combination of cell size, to which the Y-axis signals (small-angle) are most sensitive, and refractive index, to which the Y-axis signals (wide angle) are most sensitive. The other is called the Structure cytogram and is used to distinguish eosinophils from the other white cell types based on the eosinophils' internal structure. Eosinophils have larger, more numerous granules than the other white cell types. These granules scatter more light into larger angles than the other white cell types do because of multiple scattering of incident radiation among the granules. This diffuses the scattered radiation into larger angles than would result from single scattering.

Table 1. Measurement Techniques Used by the Major Manufacturers of Automated Hematology Analyzers for Routine Five-Part WBC Differential Determinations[a]

Hematology System/Manufacturer	dc Impedance	RF Impedance	Scatter	Absorption	Fluorescence
CD4000/Abbott	X		X	X	
Pentra 120/ABX	X		X	X	
ADVIA 2120/Bayer			X	X	
ADVIA 70/Bayer	(X)[a]		X		
LH750/Beckman Coulter	X	X	X		
XE-2100/Sysmex	X	X	X		X

[a]Techniques used for other than routine analyses are not included.

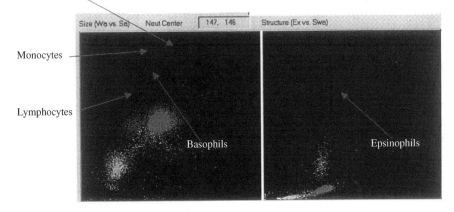

Neutrophils

Monocytes

Lymphocytes

Basophils

Epsinophils

Size (Wa vs. Sa)   Neut Center   147, 146   Structure (Ex vs. Swa)

Figure 2. Differentiation of WBC populations based on impedance measurements—ADVIA 70 cytograms.

The ADVIA 70 optical channel analyzes 10,000 cells for each sample and it provides the WBC percentages only. In combination with the absolute WBC count from the impedance channel, the system provides the absolute white cell type counts. The system also compares the lymphocyte and granulocyte percentages from the two channels as a crosscheck of results validity and to flag for the presence of abnormal WBC types.

Figure 2 shows associated Size and Structure cytograms. The populations in each figure are labeled and they correspond to lymphocyte signals, mid-range cell signals, and granulocyte signals, respectively. The lymphocyte signals occupy the region closest to the origin since they are the smallest cells and have relatively low refractive index values. The monocyte signals appear higher along the Y-axis, but only slightly to the right of lymphocytes because they are the largest cells, but also have the lowest refractive index values. The Neutrophil + eosinophil signals appear highest along the Y-axis and furthest along the X-axis because they are larger than lymphocytes and have higher refractive index values than either monocytes or lymphocytes. The basophil signals occupy the region below and to the right of monocyte signals, based on a combination of their sizes and refractive index properties. In the second panel of Fig. 2, the eosinophil signals are located higher along the Y-axis, which corresponds to super-wide angle signal intensity, than signals from the other cell types because of the multiple-scattering properties of the eosinophil granules.

The Cell-Dyn 4000 Hematology System (Abbott Laboratories, Abbott Park, Ill.), is similar in its routine WBC differential analysis technology to the ADVIA 70. It uses both impedance measurements and optical measurements for counting and differentiating WBC types (14–16).

The system differs from the ADVIA 70 system in two ways. First, it uses a 488 nm argon-ion laser instead of a 633 nm diode laser. Scattering patterns are sensitive to wavelength, so that scattering intensity at a given angle is different for 488 nm illumination than 633 nm illumination (16). Second, the CD4000 system distinguishes eosinophils from neutrophils based on 90° depolarized-light scatter "90D" versus 90° polarized-light scatter "90" (Figs. 3–4).

According to the manufacturer, 90°D measurement is especially sensitive to granularity, whereas 90° polarized light is sensitive to lobularity. Since eosinophil granules are generally larger and more numerous than those of neutrophils, they are expected to scatter more light into 90D than neutrophils.

The underlying concept in 90D measurement is the multiple scattering experienced within eosinophils. As the initially polarized light scattered from the first granule encountered is incident on subsequent granules, the polarizarion state of the light is progressively scrambled, so that the depolarized component of light increases at the expense of the polarized component.

ABX Pentra 120 Hematology System

The Pentra 120 Hematology System (ABX, Montpellier, France) uses a combination of dc impedance and light absorbance to determine the routine WBC differential count (17–21). Two reaction mixtures are prepared by

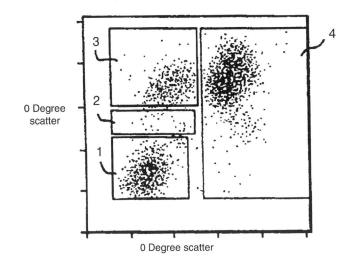

0 Degree scatter

0 Degree scatter

Figure 3. Differentiation of WBC populations based on light-scattering intensity Cell-Dyn 4000 0-degree vs. 10-degree scatter Cytogram.

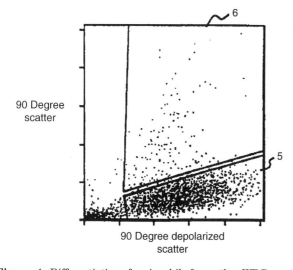

Figure 4. Differentiation of eosinophils from other WBC populations based on polarized vs. depolarized light scattering intensity-Cell-Dyn 4000 polarized light vs. depolarized light Cytogram.

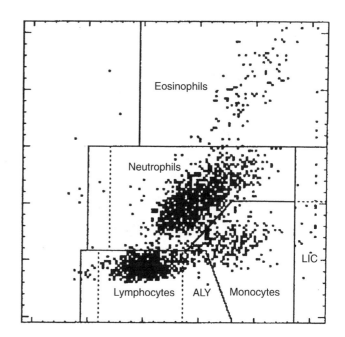

Figure 5. Differentiation of WBC populations based on DC impedance (to yield volume) and light absorption due to staining by chlorazol black-ABX Pentra 120 volume vs. light absorption cytogram.

the system. In one, the sample is first mixed with a reagent called Eosinofix, that lyses RBC and stains WBC differentially with chlorazol black, based on cell granularity. The mixture is subsequently diluted with a reagent that stops the reaction process in the first mixture. The suspension passes through a single flow cell that measures both the dc impedance of the WBC and their absorption of incident 488 nm radiation from the argon–ion laser. The resulting signals are displayed on a cytogram, where the X-axis corresponds to cell volume based on dc impedance, and Y-axis corresponds to light absorption (Fig. 5). On this cytogram, lymphocyte signals are clustered nearest to the bottom of the X-axis since they are typically the smallest WBC type. Neutrophil signals are higher along the X-axis since they are larger than lymphocytes. Monocyte and eosinophil signals are higher still, and at roughly the same height, based on their size. Lymphocytess signals are closest to the bottom of the Y-axis because they effectively do not stain with chlorazol black. Monocyte signals are slightly higher because monocytes stain weakly. Neutrophil signals are higher because they stain more heavily, and eosinophil signals are the highest because they stain the most.

This cytochemical staining method is similar to that used on the ADVIA 2120 system. Indeed, this cytogram is similar in appearance to the Peroxidase Channel cytogram of the ADVIA 2120 system when rotated through 90 and viewed in reflection.

In the second reaction mixture RBCs, as well as all WBCs except for basophils are lysed in a reagent called Basolyse (Roche Diagnostics). The resultant reaction mixture passes through the same flow cell as used for the first reaction mixtures and the dc impedance signals of the WBC are recorded (Fig. 6). The intact basophils produce distinctly larger signals than the other WBC types, and are enumerated on this basis. The basophil count is subtracted from the lymphocyte count obtained from the Eosinofix reaction mixture. This method of enumerating basophils is chemically similar to those used in both the Bayer ADVIA 2120 and Sysmex XE-2100 systems.

Bayer ADVIA 2120 Hematology System

In the ADVIA 2120 Hematology System (Bayer Health-Care LLC, Diagnostics Division, Tarrytown, NY), the results of two optical channels, the Peroxidase Channel and the Lobularity/Nuclear Density Channel, are combined to produce the white blood cell differential count (8,9,22,23).

The Peroxidase Channel measures the peroxidase activity inherent in the WBC types, along with differences in cell type size, to distinguish among the cell types. A whole blood samples is mixed first with a reagent that lyses the sample's RBCs and fixes the WBCs, and two additional reagents that contain hydrogen peroxide and the dye 4-chloronaphthol are added to the mixture. The cells' native peroxidase enzyme catalyzes the reaction of the peroxide with the naphthol, resulting in the precipitation of the dye on the cells' granules. After a few seconds of incubation, the cell suspension is then passed through an optical flow cell for analysis. Two silicon photodetectors

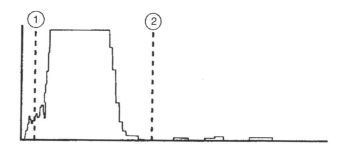

Figure 6. Differentiation of basophils from other WBC populations based on DC impedance (to yield volume)- ABX Pentra 120 volume frequency histogram.

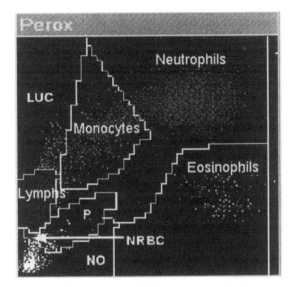

Figure 7. Differentiation of WBC populations based on cytochemistry – ADVIA 2120 Peroxidase Channel Cytogram.

are located such that one collects a component of the small-angle scattering intensity, and the other senses the drop in light transmission due to light absorption by the 4-chloronaphthol deposited on the cells' granules. A fixed volume at known dilution passes through the flow cell, so that an absolute concentration of cells can be determined. A peroxidase cytogram for a normal blood sample appears in Fig. 7.

The cell types are labeled, including "noise", which consists mainly of platelets. The positions of cells in certain regions of the peroxidase cytogram reflects cell morphology. The lymphocytes are typically the smallest cells and are also peroxidase-negative. Therefore, they appear along the lower part of the Y-axis, which represents the low angle scattering intensity and to the left along the X-axis, which represents light absorption and is sensitive to transmission loss. The LUCs (large unstained cells) are also peroxidase negative, but are larger than normal lymphocytes, so that their signals appear higher along the Y-axis, but at about the same position along the X-axis. Monocyte signals appear above and to the right of normal-sized lymphocytes. They are larger than lymphocytes and slightly peroxidase-positive. Neutrophil signals appear slightly above monocyte signals and to their right. They are typically smaller than monocytes but more peroxidase-positive. Finally, eosinophils, which are typically as large as or larger than neutrophils, have signals that appear to the right of and below the neutrophils.

The Lobularity/Nuclear Density Channel, also called Basophil channel uses the following two WBC features to distinguish among the cell types:

1. Basophils are significantly more resistant to lysis under acidic conditions than the other white cell types.
2. Mononuclear cell nuclei (MNs) scatter light in a different manner than polymorphonuclear nuclei (PMNs). The PMN scattering intensity depends on

the number of nuclear lobes; the more lobes per nuclear volume, the greater the low angle scattering intensity.

In this channel, a whole blood sample is mixed with a reagent that is acidic and contains a surfactant. The reagent lyses RBC and platelets, and strips all white cell types of their cytoplasm except for basophils. The reaction mixture is passed through an optical flow cell for analysis. The measurement includes light scattered at $\pm\,2$–$3°$ off the axis of the incident beam, and the other measuring light scattered at ± 5–$15°$.

The Lobularity/Nuclear Density Channel cytogram with cell types labeled is shown in Fig. 8. The basophils, which remain intact, scatter significantly more light than the much smaller nuclei of the other white cell types. Therefore they appear in the upper region of the Y-axis, which corresponds to 2–3° scattering. The nuclei appear near the bottom of the Y-axis due to their small size. However, since the nuclei have a higher refractive index than that of the intact basophils, at least some of them appear to the right of the basophils along the X-axis, which corresponds to 5–15° scatter. As noted above, scattering intensity is a nonlinear function of both size and refractive index. Further, it depends on the number of scattering particles encountered at one time in the sensing zone. The lobes of MNs are single scatterers, whereas the lobes of PMNs behave to a first approximation as multiple scatterers. The scattering pattern of the MNs is due to a combination of their single-lobed nature and their relatively low refractive index. The scattering pattern of the PMNs is due to a combination of their relatively high refractive index and the number of nuclear lobes.

The ADVIA 2120 system uses the Peroxidase Channel results to determine the percentages and absolute numbers of neutrophils, eosinophils, monocytes, lymphocytes + basophils, and LUCs, and to provide a WBC count for comparison with the Nuclear Density Channel WBC count.

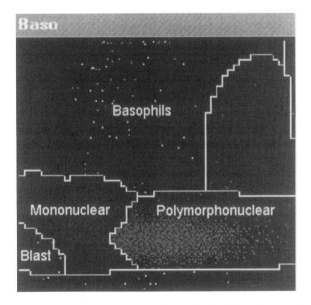

Figure 8. Differentiation of WBC populations based on scatter measurements – ADVIA 2120 Nuclear Density Channel Cytogram.

The primary WBC count determined and the number and percentage of basophils is determined in this channel. The absolute white cell subtype counts are computed as the product of the differential percentages and the primary WBC count. The system uses the additional information provided by the MN and PMN percentages and counts, and the Peroxidase Channel WBC count to cross-check the validity of the results of the two channels and to test for the presence of abnormal cells.

Beckman Coulter LH 750 Hematology System

The LH 750 Hematology System (Beckman Coulter, Hialeah, FL) uses VCS technology for routine WBC differential analysis (24–27). In the VCS technology (Volume/Conductivity/Scatter), the RBCs in an aliquot of blood are lysed in one reagent and then the WBCs are stabilized in a second reagent. The stabilized WBC suspension passes through a single quartz flow cell that is used for both electrical and optical measurements.

Cell volume (V) is determined based on dc impedance signals. The RF impedance signals provide information about internal structure, such as nuclear volume and internal chemistry. Light scattering signals from 10 to 70°, called median angle light scattering (MALS) provide information about cell granularity and lobularity. Since both RF signals and light scattering signals depend on cell size as well as on internal structure, whereas dc impedance is considered to be a function of size only, the RF signals and scattering signals are corrected for the contribution due to cell size, based on the dc signals. The resulting RF signals are called "opacity" because the signals are considered to be sensitive primarily to the density of internal components. For example, opacity is used to distinguish normal lymphocytes from variant lymphocytes, because of characteristic differences between the two cell subtypes in nuclear/cytoplasmic ratio. The volume-compensated MALS signals are called Rotated Light Scatter (RLS) signals. Volume compensation serves to better separate eosinophils from neutrophils on the one hand, and monocytes from lymphocytes on the other hand.

The WBC differential data are displayed as V (DC impedance) signal intensity versus RLS (compensated MALS) intensity signals (Fig. 9). In this cytogram, lymphocytes and monocytes appear to the left along the RLS axis, since they are mononuclear. Monocytes appear above lymphocytes along the V axis since they are larger. In fact, for normal samples they are highest along the V axis since they are the largest cells. Neutrophils and eosinophils appear to the right along the RLS axis, since their nuclei are polymorphonuclear and they have granularity. Eosinophils appear further to the right than neutrophils, because of their larger, more numerous granules. Basophils appear to the right of and somewhat above lymphocytes, based on size, lobularity, and granularity.

Sysmex XE-2100 Hematology System

The XE-2100 Hematology System (Sysmex Corporation, Kobe, Japan) combines fluorescence, forward and side scatter, and dc and RF impedance to determine the five-part WBC differential (Sysmex 28-36). It also provides

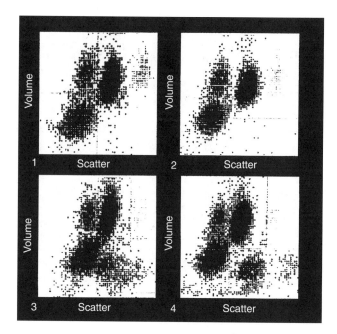

Figure 9. Differentiation of WBC populations based on DC impedance (to yield volume) and light scattering intensity-Beckman Coulter LH 750 volume vs. RLS cytograms.

absolute and differential counts of immature WBC, including bands, metamyelocytes, myelocytes, promyelocytes, and myeloblasts as part of the routine differential analysis (28). Three aliquots of whole blood are separately reacted and analyzed.

One aliquot is diluted with a reagent that lyses RBCs and compromises the integrity of WBC membranes, except for basophils. The suspension is passed through an optical flow cell, the suspended cells interrupt a beam of red light from a laser diode, and the forward scattering intensity and side scattering intensity are measured (Fig. 10). The basophils in the suspension produce larger forward scatter and side scatter signals than the other WBC types because they are larger, having retained their cellular integrity and their granules. This method for determining the basophil differential count is similar to that used by the ADVIA 2120 system.

A second aliquot is first diluted with a reagent that lyses the red blood cells and permeabilizes the membranes of the white blood cells to the passage of a red-fluorescent polymethine dye that stains RNA/DNA (29,30). A second reagent containing the dye is then added. As the suspended cells pass through the optical flow cell and interrupt the red laser-diode beam, the side fluorescence and side scatter signal intensities are measured (Fig. 11). Lymphocytes and monocytes produce larger fluorescence signals than neutrophils, basophils, or eosinophils in this reaction channel. This is presumably because the dye preferentially stains RNA, and lymphocytes and monocytes contain more cytoplasmic RNA than neutrophils, basophils, or eosinophils. On the other hand, lymphocyte and monocytes produce smaller side scatter signals than neutrophils and basophils, which in turn produce smaller side scatter signals than eosinophils. The mononuclear cells scatter least because they are less refractile than the polymorphonuclear

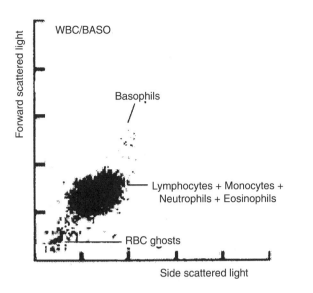

Figure 10. Differentiation of basophils from other WBC populations based on light scattering intensity. Sysmex XE-2100 forward scatter vs. side scatter cytogram.

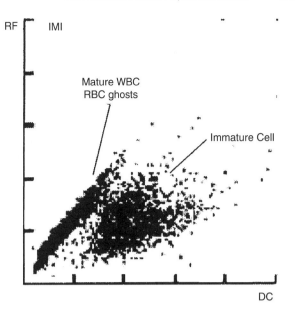

Figure 12. Differentiation of immature WBC populations from mature WBC populations based on RF and DC impedance Sysmex XE-2100 RF vs. DC cytogram.

cells. Neutrophils and basophils scatter less than eosinophils because they lack the side scatter component provided by eosinophils' large, numerous granules.

A third aliquot of blood is diluted with a reagent that lyses RBCs and maintains the cellular integrity of immature WBC types in preference to that of mature WBC. The cell suspension is passed through a narrow aperture on either side of which is an electrode. The electrical circuit, completed by the electrodes and the conductive reaction suspension medium, carries both a dc and RF current. The RF and dc impedance signals are measured for each cell as it passes through the aperture (Fig. 12). The mature WBCs and immature white cells produce RF signals of similar magnitude, but the immature cells produce larger dc signals. In this reaction mixture their cellular membrane

integrity is superior to that of the mature white cells, and the conductivity of the mature cells is reduced because of the compromised cell membrane integrity.

Abbott Cell-Dyn 1700, ABX Micros 60, Bayer ADVIA 60, Beckman Coulter Ac•T Diff, Sysmex KX-21N

The above are examples of aperture impedance devices for determining three-part differentials, as described above.

MEASUREMENT TECHNIQUE VERSUS ACCURACY OF RESULTS AND LABORATORY EFFICIENCY

All of the automated five-part differential analyzers described above provide an accurate five-part differential counts for fresh, normal samples as per the specifications listed in the respective operator's manuals. Differences in performance arise for samples stored *in vitro* longer than 8 h at room temperature or >24 h at 4° centigrade, as discussed above. Even larger differences may arise for samples with abnormal morphological and/or cytochemical properties. If the cells in the blood samples are morphologically abnormal, but retain their cytochemical properties, then analyzers using cytochemistry will provide accurate enumeration while analyzers using morphological properties will not. On the other hand, retention of cytochemical properties may mask an important underlying cause of abnormal morphology that morphological analysis may reveal through abnormal cytograms.

Laboratory efficiency can be evaluated in terms of throughput and cost. All of the major analyzers produce at least a hundred five-part differential results per hour. Also, the per-test cost, which includes reagents and instrumentation, as well as associated laboratory overhead, such

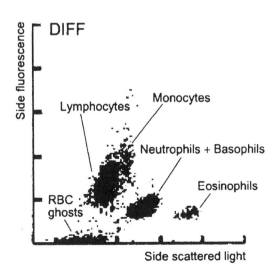

Figure 11. Differentiation of WBC populations based on fluorescence intensity and side scatter intensity. Sysmex XE-2100 side fluorescence vs., side scatter cytogram.

as space and lab personnel, does not vary significantly among manufacturers. However, throughput is affected by factors other than analyzer speed. Even though an analyzer can report 100 results per hour, the laboratory routinely does not immediately release the results without review. The reviews invariably trigger a re-analysis of some fraction of the results. Reanalysis rates, also called review rates, may vary from 10 to 60% depending on a number of factors, listed below. Effective management of review rates optimizes laboratory throughput by creating the proper balance between throughput and accuracy of results. This balance in turn bounds the cost of laboratory operations.

Since the review rate significantly affects laboratory efficiency, it is important to list the factors affecting the review rate, in order to compare the performance of different types of analyzers with respect to these factors. The factors include type of laboratory; donor population; sample age; review criteria.

If the laboratory is in a hospital, then the samples it receives will usually be fresh (<8 h postvenipuncture), so that differences in analyzer performance associated with sample stability will not be a factor. If it is a reference laboratory, which typically receives samples that are 4–48 h old, then these differences may significantly impact the review rate. Analyzers that use cytochemical properties of cells such as enzyme activity or nuclear staining to determine differentials may provide more reliable results and therefore reduce review rates than analyzers that use cell morphology alone to determine differentials.

If the donor population is comprised of mostly normal donors, such as occurs in labs that perform screens for insurance companies and the like, then the review rates can be expected to be low and any of the automated analyzers will provide good results. In this donor population, differences in the low review rate will depend more on differences in analyzer reliability than on differences in method accuracy. If the donors are from a general hospital population where a wide range of conditions apply, then technologies that are robust with respect to hematologic variations can be expected to produce a lower review rate than technologies that are suited to only narrow ranges of hematologic conditions. Hematologic variations may include wide swings in WBC concentration, wide swings in differential ratios, cell morphology abnormalities including size variations, changes to nuclear properties and changes to cytochemical properties. If the donor populations are well defined, such as oncology patients, newborns, end-stage-renal dialysis patients, thalassemics, sickle cell disease sufferers and so on, then analyzers whose measurement techniques are best suited to the given population should be selected, in order to optimize review rate. Although sample age correlates strongly to laboratory type, it also comes into play when samples that are normally run fresh must be stored for extended periods before being analyzed. In this case, analyzers that use more stable cellular properties for analysis and that are also accurate over a wide range of hematologic conditions are preferred.

Although review criteria are expected to vary widely based on differences in sample age and donor population, they can still vary widely even among laboratories of like type and among laboratories that handle the same types of patient populations. The reason is that there is wide latitude among laboratories in what is considered an accurate result and in what is considered an abnormal result. Therefore, variability in review rate is probably attributable more to the lab's choice of criteria than to any methodology-related factors.

BIBLIOGRAPHY

1. Reference Leukocyte Differential Count (Proportional) and Evaluation of Instrumental Methods: Approved NCCLS Document H20-A. Villanova (PA): National Committee for Clinical Laboratory Standards; 1992.
2. Wallace Coulter: Means for counting particles suspended in a fluid. US Patent 2656508. 10/20/1953.
3. Eggert AA. Differential counts, automated. In: Encyclopedia of Medical Devices, Webster, JG editor. John Wiley & Sons Inc.; 1988. pp 944–956.
4. Mansberg HP, Saunders AM, Groner W. The Hemalog –D white cell differential system. J Histochem Cytochem 1974;22: 711–724.
5. Saunders AM. Development of automation of differential leukocyte counts by use of cytochemistry. Clin Chem 1972;18: 783–788.
6. Gomori G. Histochemical differentiation between esterases. Proc Soc Exp Biol Med 1945;67:4.
7. Gomori G. Chloroacyl esters as histochemical substrates. J Histochem Cytochem 1953;1:469.
8. Cremins et al. Method for the determination of a differential white blood cell count. US patent 4,801,549. 1989 Jan 31.
9. Cremins et al. Leukocyte differentiation method. US Patent 5,518,928. 1996 May 21.
10. Hamaguchi et al. Reagent and Method for Measuring Leukocytes and Hemoglobin in Blood. US Patent 5116539. 1992 May 26.
11. Miller MN, et al. Pattern recognition system for generating hematology profile. US Patent 4307376. 1981 Dec 22.
12. George TC, et al. Distinguishing modes of cell death using the ImageStream multispectral imaging flow cytometer. Cytometry A 2004; Jun: 59(2):237–245.
13. Wang FS, Kubota F. A Novel Apoptosis Research Method with Image-Combined Flow Cytometry and HITC or IR-125 Staining. Cytometry (Clini Cytom) 2002;50:267–274.
14. Marshall PN. (to Abbott Laboratories); Flow Cytometric Lytic Agent and Method Enabling 5-Part Leukocyte Differential Count. US Patent 5,510,267. 1996 Apr 23.
15. Uptmore C, et al. Comparison of the Sysmex XE-2100 to the Abbott Cell-Dyn 4000, Automated Hematology Analyzer. Sysmex J Inter 2001;11(1):22–26.
16. CELL-DYN 4000 System Operation Manual, Revision 3-03.doc, Abbott Laboratories.
17. Pentra 120 SPS User Manual. Section 2: Description and Technology. P/N RAB 106 CA. ABX Horiba Diagnostics.
18. Lefevre et al. (to ABX); Apparatus For Counting And Determining At Least One Leucocytic Sub-Population. US Patent 5,196,346. 1992 Aug 11.
19. Lefevre et al. (to ABX); Reagent And Method Of Using Same For Automatically Counting Basphilic Leukocytes In The Blood In Resistivity Variation Measuring Apparatus. US Patent 5,196,346. 1993 Mar 23.
20. Lefevre et al. (to ABX); Reagent For Use In Automatic Analyzers For Distinguisher Leukocyte Sub-Populations In Blood Samples. US Patent 5,282,857. 1993 Aug 3.

21. Kass L. Staining of Granulocytic Cells by Chlorazol Black E. Am J Clin Pathol 1981;76:810–812.

22. ADVIA 2120 Operator's Guide V1.0.1.00, 2004. Bayer Health-Care LLC, Diagnostics Division, Tarrytown, NY.

23. Harris N, Kunicka J, Kratz A. The ADVIA 2120 Hematology System: Flow-cytometry-based analysis of blood and body fluids in the routine hematology laboratory. Lab Hematol 2005;11(1):47–61.

24. Fernandez T, et al. Performance Evaluation of the Coulter LH 750 Hematology Analyzer. Lab Hematol 2001;7:217–228.

25. Aulesa C, et al. Validation of the Coulter LH 750 in a Hospital Reference Laboratory. Lab Hematol 2003;9:15–28.

26. Coulter VCS Technology: Clinical Case Studies. Beckman Coulter Bulletin No. 3008.

27. Coulter Gen S System Enhanced VCS Technology: Clinical Case Studies. Bulletin 9165.

28. Fujimoto K. Principles of Measurement in Hematology Analyzers Manufactured by Sysmex Corporation. Sysmex J Inter 1999;9(1):31–44.

29. Sakata et al. (to Toa Medical Electronics Co, Ltd.); Reagent And Method For Classifying Leukocytes By Flow Cytometry. US Patent 5,928,949. 1999 July 27.

30. Uchihashi et al. (Sysmex Corporation); Reagent For Measurement of Leukocytes And Hemoglobin Concentration In Blood. US Patent 5,968,832. 1999 Oct 19.

Further Reading

Shibata et al. (to Toa Medical Electronics Co, Ltd.); Method And Apparatus For Determining A Particle Criterion And Particle Analyzer Using The Criterion. US Patent 5,690,105. 1997 Nov 25.

Sakata et al. (to Sysmex Corporation); Method For Classifying And Counting Immature Leukocytes. US Patent 5,958,776. 1999 Sept 28.

Shibata et al. (to Sysmex Corporation); Reagent and Method For Classification And Counting Of Leukocytes. US Patent 6,004,816. 1999 Dec 21.

Ruzicka K, et al. The New Hematology Analyzer Sysmex XE-2100: Performance Evaluation of a Novel White Blood Cell Differential Technology. Arch Pathol Lab Med 2001;125: 391–396.

Walters J, Garrity P. Performance Evaluation of the Sysmex XE-2100 Hematology Analyzer. Lab Hematol 2000;6:83–92.

Briggs C, et al. Performance Evaluation of the Sysmex XE-2100 Automated Haematology Analyzer. Sysmex J Inter 1999;9(2): 113–119.

See also BLOOD COLLECTION AND PROCESSING; CELL COUNTERS, BLOOD; CYTOLOGY, AUTOMATED.

DIFFERENTIAL TRANSFORMERS.   See LINEAR
VARIABLE DIFFERENTIAL TRANSFORMERS.

DIGITAL ANGIOGRAPHY

JAMES R. BENNETT
University of Iowa
Iowa City, Iowa

INTRODUCTION

The term "angiography" is derived from the Greek *angeio-*, meaning blood vessel, and *graphein*, meaning representation of a specified object (1). Thus, a very general definition of angiography would be "the representation of blood vessels". Angiography is a technique that allows visualization of any aspect of the human circulatory system. The principle of angiography is to increase the conspicuity of blood vessels during imaging by displacing the blood within the vessels of interest with a contrast medium, although this definition does not necessarily hold true for all imaging modalities. Whether it is the venous system, cardiac arteries, or the abdominal aorta, whenever diagnostic information is needed on a patient's vasculature, some form of angiography is likely to be employed. This article will be structured as follows: a brief history of angiography, the advent of digital subtraction angiography (DSA), non-catheter/noninvasive angiographic techniques, and finally a discussion on the future of vascular imaging.

A BRIEF HISTORY AND OVERVIEW

The first angiography was performed shortly after Wilhelm Roentgen's discovery of the X ray. In January of 1896, Mr. Hascheck and Dr. Lindenthal, of the Physicochemical Institute in Vienna, produced the first angiogram by injecting Teichmann's mixture into the arteries of a cadaver's hand and imaging the hand using X rays, shown in Fig. 1 (2). Born from this experiment was the field of vascular imaging, and for the next 90 years, angiography was the field's principal technique. In this section, the explanation of angiography are described in terms of X-ray imaging for the sake of simplicity. The overall principles of X-ray angiography generally hold true for all modalities. Yet, there are important deviations when utilizing other imaging technology, which are noted and explained in subsequent sections.

The principle behind all radiographic imaging is differential X-ray attenuation. As an X-ray beam passes through an object, the intensity of the beam is attenuated, or diminished, in proportion to the density and thickness of the object. This attenuation can be modeled by the following equation:

$$I = I_0 e^{-\mu t}$$

where I is the X-ray intensity after the original X-ray beam, with intensity I_0, passes through an object with a thickness of t and a linear attenuation coefficient of μ. For this article, assume that beam attenuation is logarithmically proportional to the attenuation coefficient, which in turn is linearly proportional to the material density.

Returning to radiographic imaging, materials with different density attenuate X-rays to differing degrees. Thus, a radiographic image is formed when an X-ray beam passes through an object, and is then captured on the other side by a radiosensitive fluorescent screen. The screen changes based on the X-ray intensity: The more intense the X-ray beam is, the greater the screen fluoresces and visa versa. However, there must be significant differences in material density to produce an image. For example, the femur (bone) is easily identified in an X-ray study of the leg, but differentiation between skin and muscle is nearly impossible. This is due to the fact that bone is significantly denser than the surrounding tissue: Skin and muscle have similar densities, and therefore attenuate X rays in a

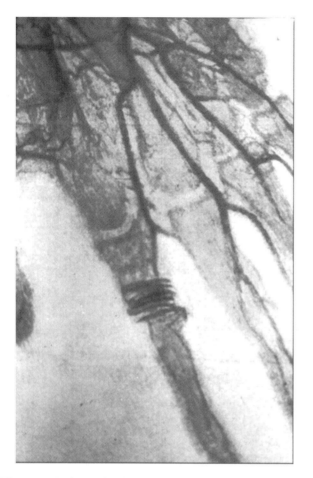

Figure 1. Radiograph of cadaver hand with contrast injection. (Copyrighted © by Radiology Centennial, Inc.)

nearly identical fashion. This attenuation homogeneity is true for many tissues in the body, including blood and blood vessels; hence the need for a dense contrast medium to discriminate the blood vessels from the adjacent tissues.

In Fig. 1, Teichmann's mixture was used as a contrast medium to image the blood vessels of a cadaver's hand. Teichmann's mixture is a dense liquid containing lime, mercury, and petroleum (2). Injection of the mixture displaced the blood within the vessels; as the X-ray beam passes through the hand, the material within the vessels significantly attenuates the beam, and results in the visualization of the vessels in the resultant image. For the next 90 years, the main changes to this technique would be the contrast material and its delivery technique. Iodinated agents have become the main contrast medium for X-ray angiography. Furthermore, catheter guided injections became the predominate means to deliver a contrast injection in a localized region of interest (ROI). This technique utilizes a small puncture to a main artery, where a guide wire is inserted and routed through the arterial system and parked at a location proximal (upstream) of the ROI, where the contrast agent is delivered. This technique will be explored further in the subsequent section.

The field of radiology has arguably incurred the greatest paradigm shift of any medical field resulting from the introduction of practical computing technology. Research in digital imaging began in the late 1960s and early 1970s and clinical applications were realized in the 1980s. Two-dimensional (2D) X-ray imaging was the first digital application on the clinical radiology scene. In a general sense, film was replaced with the digital capture of X-ray images. This change to digital imaging brought vast and immediately benefits, essentially falling into three categories: time, money, and quality. Digital imaging created near instantaneous X-ray images, eliminating the costly and time consuming process of developing film. Relative to angiography, this allowed radiologists to immediately ascertain whether or not the contrast injection was adequately captured during imaging, which was previously delayed by the lengthy film developing process. Postprocessing of the digital images had a revolutionary effect on image quality. Although the resolution of the digital images was about a fivefold decrease relative to film, the ability to digitally manipulate images far outweighed this loss in resolution. Radiologists could instantly adjust the contrast and brightness of an image, yet the most significant gain in the quality of angiographic imaging was the advent of digital subtraction angiography.

DIGITAL SUBTRACTION ANGIOGRAPHY

Digital subtraction angiography (DSA) undoubtedly revolutionized angiography and the field of vascular imaging as a whole. This is currently the gold standard against which all emerging vascular imaging techniques are compared, however, subtraction imaging is not a new technique to the field of radiology. It was first theorized in the 1930s that if a radiograph were taken before a contrast injection and another as the contrast passes through the imaging window, it would be possible to subtract out everything but the contrast within the vessels (3). The goal of subtraction is to remove the nondiagnostic artifacts in the X-ray image; organs, bones, surgical staples, metallic implants, and other fairly dense objects that can overshadow the contrast enhanced vessels as they intersect within the imaging plane. Essentially, when the enhanced vessel crosses a dense object, it becomes indistinguishable from the artifact on the radiographic image and can significantly impede interpretation. Manual film-subtraction angiography proved to be extremely valuable to the field of vascular imaging. Yet this process is expensive and time consuming: a single subtracted image involves complicated film development that usually requires multiple attempts before a usable image is produced.

The key principle behind image subtraction is the acquisition of a preinjection image, or mask image. Essentially, if two images are taken of the same motionless object, and the first image is subtracted from the second, the resultant image should be blank. If, however, there is any change to the object between the two image captures, only the difference will be visible when the first image is subtracted from the second. For example, in Fig. 2a, there is an image of an apple. Figure 2b represents the same apple, taken with the same camera in the same position, but with an interval of several days in between the two images. Within this time interval, the apple became spotted, while its shape did not change. Each image is a grayscale matrix of 200×200 pixels; each pixel has a grayscale value from 0 to 255. The first image matrix, or mask matrix, will be denoted by M and the second image will be denoted by I. To subtract the two images, it is necessary to subtract each matrix entry in the second image from its corresponding entry in the mask image, giving the subtracted image matrix, S, represented by this formula:

$$S = I - M$$

The resultant image, S, displays only the spots on the apple, while everything else in the image is black. Black represents zero on the grayscale image, thus all points in the two image matrices that did not become zero when the images are subtracted. The next step is to equalize the histogram of the resultant image.

A histogram is a plot representing the relative occurrence of pixel values in an image; the x axis contains the pixel values, 0–255, and the y axis contains the number of times that pixel value occurs in the image. An image that is very dark will have a histogram that is weighted toward the origin, as black would be the predominant pixel value and is represented by zero on grayscale images. Very light images are weighted toward the outer boundary because white is represented by 255. Upon close examination, the subtracted image in Fig. 2c does not exactly match the change in the two previous images. White spots were added to the second image and the subtracted image shows similar spots, but not perfectly white spots. It is necessary to adjust the histogram of the image after subtraction of the two images, due to the fact that subtraction does not result

in an image with an equalized histogram. An image with an equalized histogram has pixel values that are spread throughout the histogram. Figure 3a represents the histogram of the original apple image (Fig. 2a). The background of the image is white, and therefore the histogram is weighted toward the outer boundary. However, there is also a fairly equal spread of pixel values throughout the entire range. In comparison, Fig. 3b shows the histogram of the subtracted apple image (Fig. 2c). This histogram weighted heavily toward the origin and does not have pixel values throughout the entire range. The reason for this compressed histogram is beyond the scope of this article, but it is a general rule that the resultant image from subtraction will have a compressed histogram. The next step is to utilize an algorithm that equalizes the histogram by spreading the pixel values out over the entire range. Figure 3d shows the subtracted image after its histogram (Fig. 3c) was equalized. Now, the subtracted image is a true representation of the change in the original image.

Digital subtraction works well in optimal conditions where the camera and object do not move within the interval that the two images are taken. However, real-world implementation presents a host of issues, the most prevalent being patient motion. Imaging equipment is generally very precise in its positioning, that is, it does not deviate from its expected location. Patients tend to shift position in between the contrast and mask image capture, the side effect of which is the introduction of artificial artifacts in the subtracted image. An example of such artifact introduction is given in Fig. 4, where one of the two images has been shifted to the upper right quadrant. In some cases, the patient motion is due to discomfort, as the contrast medium tends to displace oxygenated blood within the vessels that may result in a burning sensation. Another source of patient motion can be organ movement, which can similarly introduce motion. In either case, motion results in image artifacts that can impede diagnostic interpretation. It is possible to compensate for these movements, given that the movement is a linear translation within the imaging plane. Simply shifting the contrast and mask images can realign the objects within the image, which will minimize the induced artifacts. In film-based image subtraction, this shifting can require multiple attempts before a usable image is produced because there is much time involved with developing the subtracted image. However, with digital imaging, it is possible to instantly manipulate the position of the two images. This technique is termed pixel shifting, as one generally shifts the images, pixel by pixel, until the objects are aligned within the images. Unfortunately, a significant amount of patient movement does not occur linearly within the imaging plane. Movement is nonlinear if the patient rolls their body, extends their limb, or any movement that is not translational within the imaging plane (the patient table can be representative of the imaging plane in most cases). A technique has recently been developed to adjust for such nontranslation movement. Although the details are beyond the scope of this article, essentially a map is created of objects both within the mask and contrast images. An algorithm compares the two maps and adjusts, or stretches, the images based on differences within the maps.

Figure 2. (a) Original picture of apple. (b). Apple after time elapsed and became spotted. (c) Digital subtraction of Fig. 2a from 2b.

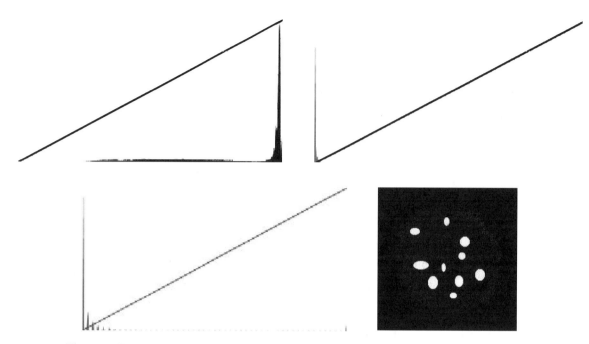

Figure 3. Histogram of Fig. 2a. (b) Histogram of Fig. 2c. (c) Histogram of Fig. 3d. (d) Figure 2c after histogram equalization algorithm applied.

Now that digital subtraction imaging has been covered, a brief explanation of DSA follows. A typical DSA procedure begins with the insertion of a catheter in the femoral artery, as shown in Fig. 5. A guidewire is inserted through the catheter and routed to a position proximal (upstream) to the vessels of interest, through which a contrast delivery sheath is inserted. The next step depends on the purpose of the study: if the clinician solely wishes to view one particular area, for example, the abdominal aorta, they will take a single mask image, inject and image the contrast, and then perform digital subtraction. If, however, the clinician wishes to perform a runoff sequence, where the peripheral vasculature is studied, they will program the imaging equipment to follow a series of imaging stations. These stations are required because the desired image is significantly larger than the actual imaging window. Thus, many frames, or stations, overlap each other and follow the vessels of interest until their termination at the extremities. Once

the stations are programmed, the imaging apparatus steps through the programmed sequence and captures a mask image at each station. Next, the machine returns to the first station, and when the clinician is ready to inject the contrast,

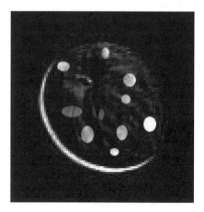

Figure 4. Example of digital subtraction when object shifts between original and mask frame.

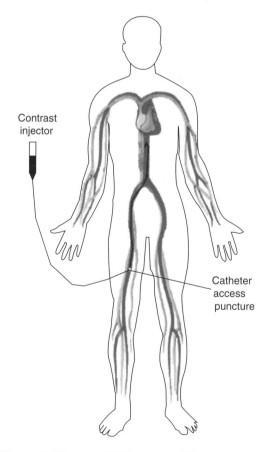

Figure 5. Diagram of DSA contrast delivery technique.

the machine begins imaging the first station at a set capture rate (typically 3–5 frames · s). The clinician injects the contrast and watches the viewing monitor for the contrast arrival at the first station. When the contrast reaches the lower portion of the first station, they triggers the machine to advance to the next station, which is always distal (downstream) to the previous station. Once the contrast passes through the second station, the clinician triggers the machine to the next station, and this process is repeated until the contrast reaches the final station. Digital subtraction is performed at each station, and the final step is merging the subtracted images from each station into a single image of the patient's peripheral vasculature. One of the most recent advances in DSA has come through flat panel technology. In traditional X-ray imaging systems, the X-ray beam is passed through the patient and captured with a scintillator, which is a screen that fluoresces in proportion to X-ray intensity. Typically, a camera captures the luminescence from the screen and transfers this image digitally to a computer. Flat panel technology replaces this system with complementary metal oxide semiconductor (CMOS) detector panels, which can be thought of as radiosensitive charge coupled device (CCD) chips. Essentially, the panel is a matrix of radiosensitive pixels, which monitor the X-ray beam intensity at each pixel. The intensity level from each pixel is converted to a digital signal, which is then reconstructed to form an X-ray image. This system bypasses the need for the scintillator screen and camera setup by capturing the X-ray beam directly. The benefits of flat panel technology are an increase in resolution and elimination of geometric distortion resulting from the nature of the optical camera system. Another advance in DSA technology is three-dimensional (3D) imaging, where the imaging system rotates around the patient table, capturing images at different angles, from which 3D images are reconstructed.

NON-CATHETER/NONINVASIVE ANGIOGRAPHY

Non-catheter/noninvasive imaging technologies, especially 3D technologies, have created the next paradigm shift in vascular imaging. The principles of such imaging technologies will not be discussed in this article, as such information can be found in this Encyclopedia within the respective modality articles. This section will begin with single detector-row computed tomography (CT) X-ray imaging, which became commercially available in the 1980s. This technology was able to reconstruct 2D axial slices (images) of the patient. In a similar technique to DSA, iodinated contrast agents were injected into the vasculature during CT imaging, which produced basic 2D contrast enhanced vascular images. This technique could be considered fairly rudimentary as compared to today's standards. Yet, it did allow clinicians to view patient's vasculature in relation to the rest of the body, instead of the enhanced vasculature being superimposed upon the rest of the body, as in DSA. As CT advanced, so did CT technology. Soon, multirow detector arrays were introduced to CT, which revolutionized vascular imaging. Multirow detectors increased longitudinal coverage and image resolution. It allowed for 3D images to

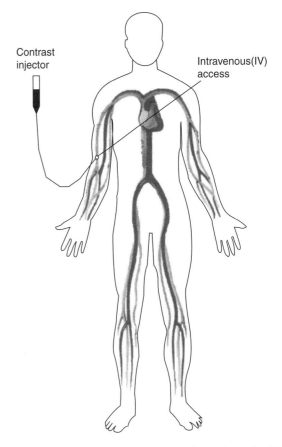

Figure 6. Diagram of computed tomography angiography (CTA) and gadolinium enhanced magnetic resonance angiography (MRA) contrast delivery technique.

be produced, along with increased visualization of small vessels. However, the greatest contribution of multirow detectors was the increased scanning speed. As scan time decreased, conspicuity increased because patient motion, and therefore the resulting artifacts, was less prevalent. Further benefits resulted from the noninvasive nature of this technique. Instead of catheter based contrast delivery in DSA, the iodinated contrast is injected intravenously, as shown in Fig. 6. The contrast travels from the venous system to the right side of the heart, through the pulmonary system, and then back to the left side of the heart where it is pumped throughout the arterial system. The contrast can be imaged wherever diagnostic information of vessels is needed. This technology greatly decreased the time and cost associated with an angiographic study, along with providing superior vascular images. There are several drawbacks associated with CT imaging. It requires an increased amount of radiation and iodinated contrast dose when compared with 2D angiography. The U.S. Food and Drug Administration (FDA) has recently classified ionizing radiation, used to create X-ray images, as a carcinogen. Computed tomography imaging delivers a substantial dose of ionizing radiation, especially in a full-body scan. The iodinated contrast material is nephrotoxic. In cases where a patient has borderline kidney function, the toxicity of a large dose of iodinated contrast may induce renal failure.

Another drawback to CT imaging is that metallic implants, such as hip prostheses, create artifacts in the images. Finally, calcium within the vasculature can limit evaluation of vessel patency (openness).

Magnetic resonance (MR) imaging can also be used to perform diagnostic imaging of blood vessels. To fully comprehend the nature of this modality, and therefore MR angiography, it would be recommended to review the article on MR imaging in this Encyclopedia. There are many methods for imaging the vasculature with this modality: time of flight (tof) angiography, phase-contrast (PC) angiography, and gadolinium enhanced angiography. In this section, tof and gadolinium enhanced MR angiography will be discussed, as they are the most prevalent in clinical use. Time of flight angiography utilizes constant radio frequency (RF) pulses within a presaturation slab. The RF pulses do not allow the atomic spins within that slab to realign with the magnetic field. Thus, these saturated atoms give off a very weak signal. However, when fresh blood from the arteries enters the imaging slice, the RF pulses have not saturated the spins, and therefore it gives off a much stronger signal when passing through the imaging slice. This technique works well for arterial visualization, the only drawback is that this technique does not work when the vessel, and therefore blood flow, is parallel with the imaging slice or if the blood flow is slow or turbulent. This is due to the fact that the blood becomes saturated as it travels within the saturation slab, and therefore does not give a strong signal, unlike the fresh blood (4). Gadolinium enhanced MRA does not have any of these drawbacks associated with tof angiography. Gadolinium enhanced MRA works in a similar fashion as DSA or CTA; however, the physics behind its principles are quite dissimilar. Gadolinium contrast agents work by shortening the $T1$ time of the blood and blood vessels (4). This enhances the blood and blood vessel visualization as the contrast enhanced blood passes through the imaging slice. Recently, there has been an emergence of advanced MRA techniques, such as fresh blood imaging and time-resolved MRA, which are well beyond the scope of this article. Magnetic resonance angiography overcomes many of the drawbacks associated with CTA; however, it is not without its own issues. Patients with pacemakers, surgical clips, metallic prostheses, or foreign bodies cannot undergo an MRA examination. Also, patients with claustrophobia cannot undergo MR imaging. It has a longer scan time as compared to CT imaging, and finally it does not produce images with adequate vessel wall definition.

Ultrasound (US) plays an important role in vascular imaging. Although most applications of this technique would not generally be classified as angiography, it is an extremely important tool nonetheless. The latest wide-scale medical application of ultrasound is termed triplex imaging. Triplex imaging combines grayscale, color, and Doppler information to form images of the vasculature with color coded Doppler blood flow data overlaid on the grayscale image. One of many recent developments in US imaging is intravascular ultrasound (IVUS). This technique works by inserting an extremely small US transducer into the blood vessel of interest via a catheter. The IVUS produces the most accurate vessel wall characterization of any currently available modality. Another recently developed technique involves injecting microbubbles into the bloodstream. This US technique would be the most likely to be classified as angiography. The principle behind microbubble US is that these tiny bubbles will create differences in density in the blood, which can be easily detected by a US transducer. Both IVUS and the microbubble technique are currently in their infantile stages, but have vast potential to improve vascular imaging. There are many advantages utilizing US as a vascular imaging modality, including the low cost and portability relative to MR and CT scanners. Also, vascular US imaging does not involve any harmful contrast medium or ionizing radiation.

FUTURE OF VASCULAR IMAGING

The need for vascular imaging will not diminish in the foreseeable future. The United States alone has seen obesity and diabetes climb to a near epidemic scale, factors that can significantly increase the incidence of circulatory disease. According to the World Health Organization, in 1997 there were 15.3 million deaths due to circulatory disease worldwide. It is unknown whether these deaths could have been prevented with medical intervention, but the main point is that circulatory disease will not be subsiding in the near future. Thus, it is probable that there will be an increasing need for diagnostic vascular imaging. There are many promising emergent modalities and techniques that will be included in the next generation of vascular imaging. Improvements such as 3D dimensional ultrasound and 256-row CT scanners are not far from implementation. One improvement that could have a vast impact on vascular imaging would be a nonnephrotoxic radioopaque contrast agent. The optimal solution would be a modality that combines the resolution and detail provided with CTA with the relatively risk-free imaging found in ultrasound and MRA.

BIBLIOGRAPHY

1. Merriam-Webster Online Dictionary. (2005). Merriam Webster Online. [Online]. Merriam-Webster. Available at http://www. merriam-webster.com Accessed 2005 April 18.
2. Sprawls P. 1996 Feb 1. The X-ray Century. [Online]. Emory University. Available at http://www.emory.edu/X-RAYS/century. htm. Accessed 2005; May 3.
3. Ziedses des Plantes B. Plantinigraphie en subtractie Roentgenographische differentiatiemethoden. Ph. D. dissertation, University of Utrecht; 1934.
4. Bakal CW, Silberzweig JE, Cynamon J, Sprayregen S. Vascular and Interventional Radiology: Principles and Practice. New York: Thieme Medical Publishers; 2002.

Reading List

Hagspiel KD, Matsumoto AH. The Radiological Clinics of North America. Philadelphia: W. B. Saunders, 2002.
Gonzalez RC, Woods RE. Digital Image Processing. Upper Saddle River (NJ): Prentice Hall; 2002.
World Health Organization. The World Health Report 1998. Geneva, Switzerland: World Health Organization; 1998.

DIVING PHYSIOLOGY.   See Hyperbaric medicine.

DNA SEQUENCING

Sotirios A. Tsaftaris
Aggelos K. Katsaggelos
Northwestern University
Evanston, Illinois

INTRODUCTION

The DNA molecule is one of the most important molecular structures in our planet. As an information carrier molecule it is used to encode the role and function of proteins that are later used to create complex organic structures (e.g., human cells).

Information is encoded using four nucleotides adenine, guanine, thymine, and cytosine, abbreviated, respectively, as A, G, T, and C. Nucleotides are joined together to form sequences, which encode certain functions. These sequences are translated into proteins, which dictate certain actions. It is therefore critical when examining those sequences to directly determine the exact sequence of nucleotides. Such an effort has been more publicly acknowledged with the Human Genome Project. The goal of this national effort was to extract deoxyribonucleic acid (DNA) sequences from human cells and decode the exact nucleotide sequence.

This process is called DNA sequencing and is commonly used in major research laboratories. There are many commercially available automated machines that can sequence DNA and output in a human readable format, usually through a computer, the exact nucleotide sequence of DNA.

Understanding how DNA sequencing works is the objective of this article, which is organized as follows. First, a very short introduction into the chemistry of the DNA molecule is provided. Subsequently, some of the basic principles used by common sequencing techniques are presented. This analysis is followed by a presentation of the most commonly available techniques and equipment. This article conclude with the presention of some of the most promising techniques for DNA sequencing in the future.

THE DNA MOLECULE

A double helix of DNA is made from two single strands of DNA, each of which is a chain of nucleotides (1). A nucleotide is an organic molecule made up of three basic parts: a phosphate group, a five-carbon sugar group, and a nitrogenous side group, which is more commonly called a base. Four different nucleotides occur in DNA: adenine, guanine, thymine, and cytosine. Nucleotides can be joined together in a linear chain to form a single strand of DNA.

A short single strand of DNA consisting of up to 100 or so nucleotides is called an oligonucleotide or oligo. It has a backbone of alternating sugar and phosphate groups with one of the four bases bound to each sugar group. The backbone gives an oligonucleotide a polarity, that is, it has two distinct ends, the 5′ and the 3′ end.

The chemical structure of the bases allows for the unique pairing between A-T (double hydrogen bond) and G-C (triple hydrogen bond). Each base in DNA has its unique Watson–Crick complement, which is formed by replacing every A with a T and vice versa, and every G with a C, and vice versa. Every oligonucleotide has a complementary sequence with opposite polarity (e.g., the complementary sequence of 5′-ATG-3′ is 3′-TAC-5′).

If two complementary sequences meet in a solution under appropriate conditions (temperature, pH, sequence length), they will attract each other and form a double-stranded structure. This process is called hybridization or annealing. Through hydrogen bonds and Van der Waals forces, these pairings are the basis for the exquisite molecular recognition, which allows DNA to act as an information-carrying molecule. There are two types of hybridization: (1) specific hybridization, which refers to cases where the two single strands are perfectly complementary at every position and the double-stranded molecule that is formed is perfect; and (2) nonspecific hybridization, for which the sequence may not be completely complementary, and thus it may contain mismatched base pairs. DNA melting or denaturation is the opposite of hybridization. When the temperature is raised, the chemical bonds break and the duplex breaks into the two single-stranded parts.

Ligation is the process of joining together double-stranded DNA with compatible sticky ends with the use of DNA ligase. A double-stranded DNA molecule can either have blunt ends or it can have single-stranded overhanging ends (called sticky ends) at one or both of its extremities. The enzyme DNA ligase, joins together, or ligates, the end of a DNA molecule to another molecule.

Restriction enzymes (endonucleases), recognize a specific short sequence of DNA, known as a restriction site and cut any double-stranded DNA at that location. Using enzymes called exonucleases, either double- or single-stranded DNA molecules may be selectively degraded from the ends in.

GEL ELECTROPHORESIS GENERICS

Gel electrophoresis is a technique used for the separation of nucleic acids and proteins (1). Separation of large (macro) molecules depends on two elements: charge and mass. When a biological sample (e.g., proteins or DNA) is mixed in a buffer solution and applied to a gel, these two factors act together. The electrical current from one electrode repels the molecules while the other electrode simultaneously attracts the molecules. The frictional force of the gel material acts as a molecular sieve, separating the molecules by size. During electrophoresis, macromolecules are forced to move through the pores when the electrical current is applied. Their rate of migration through the electric field depends on the strength of the field, size, and shape of the molecules, the relative hydrophobicity of the samples, and the ionic strength and temperature of the buffer in which the molecules are moving. After staining, the separated macromolecules in each lane can be seen in a series of bands spread from one end of the gel to the other, as seen, for example, in Fig. 1.

Some of the concepts of gel electrophoresis are used even in some of the most advanced commercially available

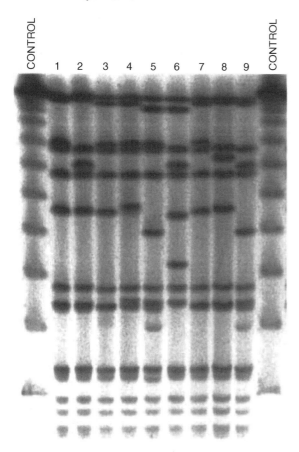

Figure 1. A photograph of a gel from an electrophoresis experiment. Spots higher on columns represent lighter molecules. The lanes at the left and right end are control lanes where the DNA used had known length.

techniques for DNA sequencing. The role of gel electrophoresis will be made clear in describing later Sanger's method.

TYPES OF DNA SEQUENCING

The inherent meaning of the word sequencing translates into finding the sequence of nucleotides of an unknown DNA strand. This type of DNA sequencing is usually referred to as de novo sequencing (de novo in Latin means from the beginning).

On the other hand DNA detection refers to the process of identifying known sequences of DNA within a sample. There are many laboratory techniques commonly used to perform such a task. Polymerase Chain Reaction (PCR) is used in amplifying (multiplying the concentration of) DNA sequences that contain certain primers (1). DNA microarrays is a technology used in gene expression profiling; it is a high throughput DNA detection mechanism where multiple DNA probes are simultaneously detected. These microarrays will be further analyzed in the following paragraphs. Biotin–streptavidin bead-based detection is a process that permits single-stranded DNA molecules containing a given subsequence to be filtered out from a heterogeneous pool of other DNA molecules (1). Strands

complementary to the subsequence are attached with biotin to streptavidin coated magnetic beads. The heterogeneous solution is passed over the beads and strands containing the subsequence anneal to the complementary sequence and are retained, while strands not containing it, pass through.

In many cases the DNA sequence under examination is largely known, but only small regions are of interest. This is the case in genotyping when it is desired to detect small variations in a whole genome (or gene) when compared with a known DNA sequence. Much of the variation in organisms originates from single-base changes in genes. These small changes can significantly affect the translation, and hence the role of the gene. This type of variation termed single nucleotide polymorphism (SNP, pronounced "snips") is of extreme interest in molecular biology. When a genome is examined for certain SNPs usually it is desired to detect subsequences of the form xxx...xYxxx...x, where xxx...x indicates known DNA bases and Y can be any base of A,T, G, or C. For a variation to be considered a SNP, it must occur in at least 1% of the population. The SNPs, which make up ~90% of all human genetic variation, occur every 100–300 bases along the 3-billion-base human genome. There exist >100 techniques for detecting known forms of SNPs. Many SNPs have no effect on cell function, but scientists believe others could predispose people to disease or influence their response to a drug. For more information on SNPs their significance and detection methods interested readers are directed to (3).

DNA SEQUENCING PRINCIPLES: THE SANGER METHOD

The foundations of DNA sequencing were laid in 1974. Two groups, a British headed by Sanger et al. (4) and an American lead by Maxam and Gilbert (5), independently discovered a technique that enables to break a fragment of DNA into smaller nested subfragments. Both groups shared the 1980 Nobel Prize in chemistry for their discovery. The method from the American team was based on a chemical cleavage protocol and used toxic chemicals and large amounts of radioactivity, whereas Sanger's method essentially mimics DNA replication as it takes place in cells. Sanger's method was eventually adopted by the industry and a form of it is still used today since it was simpler to implement in large-scale production sequencing.

For the Sanger method the following items are needed: (1) the unknown DNA; (2) a primer; (3) DNA polymerase; (4) a mixture of dNTPs (deoxynucleotide triphospates) and ddNTPs (di-deoxynucleotide triphosphates).

The unknown DNA, termed here template, is the fragment of DNA that needs to be sequenced. The fragment needs to be in a single-stranded form in the 3′-5′ direction. If in double-stranded form a single-stranded sequence can be obtained by melting (denaturing) the duplex. We also assume that the unknown template contains a known subsequence usually ~12–24 bases long. The complement of this subsequence in the 5′-3′ direction is called the primer. The primer is chemically synthesized. Once the primer is inserted in the solution containing the unknown DNA it will anneal (bind) to its complementary sequence

```
5'-GAATGTCCTTTCTCTAAG-3'
3'-GGAGACTTACAGGAAAGAGATTCAGGATTCAGGAGGCCTACCATGAAGATCAAG-5'
```

with hydrogen bonds (Fig. 2). The primer needs to be long enough to ensure that the annealing site is unique, but not very long such that the annealing is unstable.

Once the primer and the template have annealed the DNA polymerase starts reacting and catalyzes the DNA and extending from the 3′ end of the primer starts filling-in nucleotides (dNTPs) that are complementary to the template at each position. This serial addition of nucleotides is dependent on the bases of the template. The incoming nucleotide forms a covalent bond with the 3′ end of the previous sugar using its 5′ end. Under normal conditions, nucleotides are filled-in till the end of the template is reached, that is, the strand is fully extended. Sanger's idea was to modify this process such that it ceases before it reaches the end of the template.

By using a simple chemical modification, nucleotides can be transformed such that they prohibit the addition of another nucleotide in their 3′ end. The necessary chemical alteration is the substitution of the hydroxyl (OH) group on the 3′ end of the nucleotide with a hydrogen (H). Such modified nucleotides are called ddNTPs or dideoxynucleotide triphosphates and are usually termed as terminators. When such terminators are incorporated in the extension the replication stops, as shown in Fig. 3.

The DNA polymerase will stop extending when a ddNTP is incorporated. Now, if in the solution a certain mixture of dNTPs and ddNTPs is present at the end, DNA polymerase will create a mixture of strands of various lengths terminated by ddNTPs. To distinguish between the different ddNTPs (A,T,G, or C) a unique fluorescent label is attached to each one of them. In some implementations the label is attached to the primer, but assumes that the reaction is run in parallel in four tubes where each tube contains only one type of ddNTP. The relative concentrations of the dNTPs and ddNTPs are adjusted in such a way that we end up with about the same number of copies of fragments between 100 bp and 500 bp long, and a smaller number of shorter and longer fragments.

At the end the solution contains a random mixture of partially terminated double-stranded sequences. If the ddNTPs were labeled, and hence only one test tube was used, the sequences are denatured into single-stranded DNA molecules and are run on a polyacrylamide–urea gel in a single lane. The gel is dried onto chromatography paper (to reduce its thickness and keep it from cracking) and exposed to X-ray film. Since the template strand is not radioactively labeled, it does not generate a band on the X-ray film.

The fragments will be ordered on the gel lane according to length. A laser (for stimulating the emission of radiation) and a detector (for collecting the stimulated radiation) are placed at a certain distance away from the initial position. When a fragment is scanned by the laser, the fluorescent label attached to the terminator is excited, and a signal at a certain wavelength depending on the label will be emitted and sensed by the detector, as shown in Fig. 4. Multiple copies of each fragment will ensure high signal strength, which will hopefully be strong enough to be detected. By examining the peaks of the time sequence of the fluorescence intensity at different wavelengths, the bases of the unknown sequence can be determined.

The above procedure is similar if the label was attached on the primer or in the dNTPs and four tubes and separate reactions were run. In this case, the results of each tube correspond to a specific ddNTP. Each tube's contents are placed in a different lane (four in total) as seen in Fig. 5. With this setup only one fluorescent label is used, hence the excitation and detection mechanism is much more simplified. By examining the peaks of the intensity at each lane and working from bottom to top the base path or the "base ladder" can be determined.

It is evident that the resolution of the gel electrophoresis is rather critical and a single base resolution is usually a prerequisite. To improve the resolution of gel assays, the gels must be much large so that the molecules migrate further and are better resolved. They must contain a high concentration of urea (7–8 m) to prevent folding of the

```
5'-GAATGTCCTTTCTCTAAGTCCTAAG
3'-GGAGACTTACAGGAAAGAGATTCAGGATTCAGGAGGCCTACCATGAAGATCAAG-5'


5'-GAATGTCCTTTCTCTAAGTCCTAAGTCCTCCG
3'-GGAGACTTACAGGAAAGAGATTCAGGATTCAGGAGGCCTACCATGAAGATCAAG-5'


5'-GAATGTCCTTTCTCTAAGTCCTAAGTCCTCCGG
3'-GGAGACTTACAGGAAAGAGATTCAGGATTCAGGAGGCCTACCATGAAGATCAAG-5'


5'-GAATGTCCTTTCTCTAAGTCCTAAGTCCTCCGGATG
3'-GGAGACTTACAGGAAAGAGATTCAGGATTCAGGAGGCCTACCATGAAGATCAAG-5'


5'-GAATGTCCTTTCTCTAAGTCCTAAGTCCTCCGGATGG
3'-GGAGACTTACAGGAAAGAGATTCAGGATTCAGGAGGCCTACCATGAAGATCAAG-5'


5'-GAATGTCCTTTCTCTAAGTCCTAAGTCCTCCGGATGGTACTTCTAG
3'-GGAGACTTACAGGAAAGAGATTCAGGATTCAGGAGGCCTACCATGAAGATCAAG-5'
```

Figure 3. A mixture of the products of synthesis for the G ddNTP reaction.

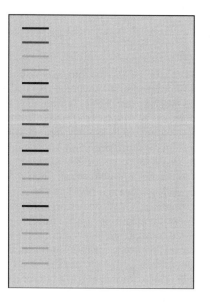

Figure 4. An example of a gel where labels are attached on the ddNTPs, and hence a single-lane gel is only used.

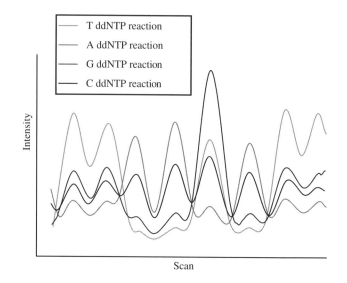

Figure 6. A chromatogram example from an experiment with four dyes. The curves correspond to intensity measurements of fluorescent emission at different wavelengths corresponding to the dyes used for each ddNTP reaction.

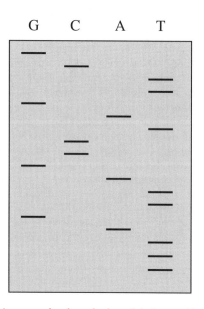

Figure 5. An example of a gel where labels are attached on the primer, and hence four lanes are used.

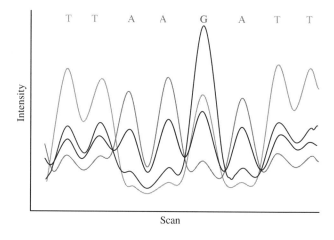

Figure 7. A decoded sequence from the chromatogram example of Fig. 6.

molecules and formation of DNA secondary structures by hydrogen bonding that would alter the mobility of the molecule. Similarly, the samples are denatured before they are loaded. The gels must run at higher temperature (~50 °C), to prevent hydrogen-bond formation.

From the above analysis, it is clear that in order to extract the bases from the gel reactions the peaks have to be identified. Traditionally the laser scanners output the fluorescence intensity into chromatograms (Fig. 6). Prior to the development of computers the chromatograms were interpreted by humans. The peaks were assigned to bases in a procedure known as base calling (Fig. 7). Ideally a periodic peak detection scheme would have been adequate if the signal was noiseless and perfect. There are certain

errors and limitations that make the base calling aspect of DNA sequencing a rather challenging task. Some of the sources of error are

1. Errors in fragment formation: (a) Abnormalities in primer extension (false stops, terminator is not incorporated, or conversely, several terminators accumulate at the same position). (b) Poor choice of relative concentrations of ddNTPs and dNTPs resulting in too many short or long fragments. (c) While DNA moves down the gel, secondary (e.g., hairpin) structures may form and change the mobility properties of the DNA fragments.

2. Convolution: Due to the stochastic nature of the DNA migration in the gel, the time scale of the chromatograms changes resulting in more elongated and less discreet peaks.

Table 1. Common Parameters Used when Evaluating DNA Sequencing Equipment

Parameter	Explanation
Technology used	Electrophoretic or nonelectrophoretic, capillary, and so on affects many of the other parameters in a DNA sequencing system
Length of the gel	Applies only to electrophoretic systems and refers to the length of the gel material. The longer the better since it increases resolution
Throughput	Measured in bases per cycle (or per day), illustrates the processing capability of the system
Cycle speed	The equipment may need replenishing of reagents after a run. The number of runs per day define the cycle speed
Read length	Maximum length of the DNA template that can be sequenced
Capacity	Number of different DNA templates that can be sequenced simultaneously
Sample volume	Defined as the volume of the template needed for a certain outcome quality (the less the better)
Error rates	Number of bases in error out of 1000 usually defines the error rate. Error rates are tightly bound to base calling quality assessment
Maintenance and operation cost	Number of dyes and labeling method used, have a direct impact on maintenance and operation cost

3. Intensity cross-talk: Due to the overlapping of the fluorescent response spectra of the fluorophores employed in the four-dye sequencing strategy there is a need for a transformation to recover the relative concentrations of the four dyes from the fluorescence intensities measured at four different wavelengths.

4. Measurement errors: White noise can originate from several sources, including background, detector, and other noise from the operating environment. Another type of noise encountered is low frequency variation due to slow changes in the background light level during collection. Such variations may be caused by deformation of the gel due to heating, the formation of bubbles in the path of the laser, variations in laser output power and other systematic changes in the environment.

Nowadays advances in statistics, signal processing, electronics, laser optics and software have lead to automated DNA sequencing and base calling capable of sequencing many different DNA templates.

EVALUATING DNA SEQUENCING TECHNIQUES

When deciding on DNA sequencing equipment, a prospective buyer has to evaluate certain aspects of the DNA sequencing scheme offered by the vendor. The buyer has to consider the traits shown in Table 1. All these parameters are critical, but their importance is weighted differently according to the application sought after by the buyer.

CURRENT COMMERCIAL STATE OF THE ART

Since the development of the early DNA sequencing methods the capabilities of the DNA sequencing equipment have improved dramatically. This change can be attributed to the radical advances in the fields of DNA chemistry, laser and optics, statistics, robotics, automation, and software. In many of the laboratories involved in the human genome

project, the high throughput DNA sequencing machines that were employed used robotic arms to move samples in and out of the machines and heavy automation to perform those tasks with minimal human intervention. Advances in laser optics led to even finer scanning and detection resolution with lower error rates. As seen in the previous section, one of the most critical aspect in DNA sequencing is the analysis of the chromatograms to determine the bases. Nowadays this task is performed by sophisticated software packages that employ statistics, digital signal processing, and adaptive algorithms that can identify the bases from the fluorescence graphs. A comparison of some of the most commonly used packages can be found in Table 2.

In the following section, sequencing devices are first presented that rely on electrophoretic principles followed by those that do not.

Electrophoretic-Based Methods

Slab-Gel. It was expected that the first automated DNA sequencers would be based on the Sanger method. Acrylamide slab gel electrophoresis until recently was the most widespread method of *de novo* sequencing (6). The Prism 373 by Applied Biosystems (ABI) Prism (Foster City, CA) was the first sequencer that could scan and detect such gels using a procedure very similar to the one described in the previous sections. Some of the drawbacks of slab gel instruments are gel casting (preparing the gel), gel loading (loading the gel into the device), and lane tracking (detecting lanes on the gel).

The Prism 373 underwent many changes in order to increase throughput and read length before it was replaced by ABI PRISM 377. The PRISM 377 is based on a four dye chemistry coupled with a CCD (charged couple device) imaging detector and can process up to 96 samples per cycle (9–11 h) with read lengths of 650–750 bases. Despite their drawbacks, slab gel systems are still preferred for applications with low throughput requirements but large read lengths. Reviews of experimental and commercial systems based on slab gels can be found in (7,8). Of such systems the following are worth noting since they are still used due to their unique properties.

Table 2. A Comparison of Base Calling and Sequence Analysis Software[a]

Name	Publisher	License	Short Description
Phred	University of Washington, Phil Green Laboratory	Free, Open Source	The phred software reads DNA sequencing trace files, calls bases, and assigns a quality value to each called base (1,3,4)
Autoseq	Reece Hart	Free, Open Source	Autoseq is a small package of base-calling software for ABI automated DNA sequencers (5)
Sequence Analyzer	GE Healthcare	Commercial	Usually bundled with MegaBASE sequencers (6)
Lasergene	DNAstar	Commercial	Comprehensive suite of easy-to-use sequence analysis software (7)
Sequencher	Gene Codes Corp	Commercial	Allows SNP detection (8)
Staden	R. Staden and other contributors	Free, Open Source	A suite of sequence assembly, editing, etc. (9)
Sequencing Analysis Software	Applied Biosystems	Commercial	Usually accompanies ABI Prism sequencers (10)
TraceTuner	Paracel	Discontinued	Another base calling application

The DNA 4300 System by LI-COR Inc (Lincoln, NE) uses two dyes at near-(IR) frequencies. Based on this innovation, the ability to operate and sequence from both ends of the template in parallel and coupled with an excellent software suite, the higher end version of the 4300 has read length of up to 1250 bases thus making it ideal for applications with large read length requirements.

The BaseStation from MJ Research Inc (Waltham, MA) uses a 75 μm thick polyacrylamide gel to improve heat dissipation in the gel thus reducing significantly the run time. Armed with robotic gel loading, a four-color photomultiplier with high sensitivity and a 100 sample capacity, the instrument offers a nice alternative to capillary systems when long read lengths are needed.

Capillary Systems. The persistent drawbacks of slab gel electrophoresis and the desire for faster sequencing runs and higher throughput led to the development of capillary array electrophoresis (CAE). Electrophoresis works in a way similar to slab gels, except that each capillary contains a single sample, and therefore tracking problems are eliminated. Furthermore, the high surface/volume ratio of a capillary allows for more rapid heat dissipation than is possible in slab gels, thus allowing higher operating voltages and faster run times.

Capillary electrophoresis uses capillaries usually 50 μm in diameter. Capillaries are very narrow tubes that based on the capillary action can draw liquid against gravity. Similarly to the technique used for manufacturing fiber optics, the capillaries are made from highly pure fused silica.

As seen in Fig. 8 the instrumentation is rather simple. The sample is injected into the capillary and high electrical field is applied to advance the sample into the capillary. Subsequently, the sample is replaced by a buffer solution and the field is reapplied to migrate the samples through the capillary. Since the capillary is filled with a sieving medium it allows for the separation of the DNA sequences according to length. The fragments pass through a laser-induced

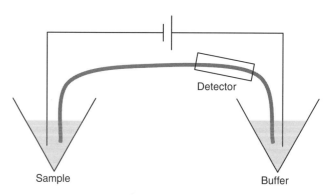

Figure 8. A single capillary electrophoresis device where a fused-silica capillary is used for the separation. The left end of the capillary is submerged into the sample solution while the other is dipped into a buffer-filled holding tank. At some point the capillary goes through a detection apparatus. High voltage is applied at each end using platinum electrodes.

fluorescence detector, which can record at different frequencies the fluorescence response of the four dyes similar to the slab gel techniques.

High voltage (double compared to slab gel techniques) allows for rapid separation, but certain phenomena limit the resolution at high read lengths (9). Although the ability to use high voltages is rather attractive, the most interesting aspect of capillaries is their flexibility, allowing them to be incorporated easily into automated systems.

Capillary array electrophoresis uses a collection of capillaries each of which is injected with a different sample. In some instruments the detector sequentially moves from capillary to capillary while in most advanced ones each capillary is scanned simultaneously using detectors attached to each one (10,20–22).

Some of the sequencers currently available in the market are Applied Biosystems PRISM 310 and 3730, Hitachi, BioRad/MJ Research (Hercules, CA) BaseStation, GE

Healthcare (Piscataway, NJ) MegaBASE 4000, Beckman Coulter (Fullerton, CA) CEQ 8000, SpectruMedix (State College, PA) Aurora, and RIKEN (Tsukuba and Wako, Japan) RISA.

Nonelectrophoretic-Based Methods

During the last 20 years, several techniques for sequencing have been discovered that do not rely on electrophoretic principles. Although these techniques have now reached the performance of CAEs for de novo sequencing they are mostly used for other types of sequencing described in a previous section.

Pyrosequencing. Pyrosequencing is a sequencing technique developed around real-time monitoring of the release of pyrophosphate (PPi) during polymerase assisted DNA synthesis (23). Similarly to electrophoretic methods a sequencing primer is hybridized to a single-stranded DNA template, and incubated with the enzymes, DNA polymerase, ATP sulfurylase, luciferase, and apyrase, and the substrates, adenosine 5'-phosphosulfate (APS) and luciferin. One of the four dNTPs is added to the solution. Assisted by polymerase the correct dNTP will be incorporated in the chain resulting in the release of PPi at a concentration analogous to the amount of incorporated nucleotides. The amount of PPi is constantly monitored by a coupled enzymatic reaction where PPi is converted to ATP by ATP sulfurylase. The ATP subsequently assists in the conversion of luciferin to oxyluciferin by firefly luciferase, which results in light emission. The process is repeated iteratively for the other dNTPs. A critical component for the success of the method is the removal of excess dNTP and ATP prior to a new dNTP addition. These can be achieved by attaching the template sequence on solid support that is washed prior to a new dNTP addition or by solution enzymatic reaction where apyrase is added to catalyze the remaining dNTPs.

The read length of pyrosequencing is smaller when compared to electrophoretic methods thus making pyrosequencing less advantageous for de novo sequencing. This is not true, however, for applications, such as genotyping of SNPs, resequencing, tag sequencing, microbial typing, and many others where pyrosequencing shines.

The technique, although in its infancy, can claim the first automated sequencer, the PSQ HS 96 from Pyrosequencing/Biotage (Uppsala, Sweden). It uses a disposable inkjet cartridge for precise delivery of small volume (200 nL) of six different reagents into a temperature-controlled microtiter plate and is widely used for SNP detection with a throughput of 96 samples per hour.

Sequencing by Hybridization: DNA Arrays, Microarrays.
Hybridization arrays or microarrays or DNA arrays were originally developed for *de novo* sequencing. Sequencing by Hybridization (SBH) requires annealing a labeled unknown DNA fragment to a complete array of short oligonucleotides (e.g., all 65,336 combinations of 8-mers) and decoding the unknown sequence from the annealing pattern (24). The array could be imaged using laser scanners and CCD devices or photomultiplier tubes. The computational complexity of decoding the annealed pattern limited the popularity of such systems for de novo sequencing. Nowadays, the key applications of DNA arrays are SNP and expression analysis (25).

The DNA microarrays are small, solid supports onto which the sequences from thousands of different genes are immobilized, or attached, at fixed locations. The supports themselves are usually glass microscope slides, of various sizes, but can also be silicon chips or nylon membranes. The DNA can be printed, spotted, or actually synthesized directly onto the support. It is important that the gene sequences in a microarray are attached to their support in an orderly or fixed way, because the location of each spot in the array identifies a particular gene sequence. The spots themselves can be DNA, cDNA, or oligonucleotides. In each microarray experiment two samples are tested simultaneously, each labeled with a different fluorescent dye. The control sample is labeled by the Cy5 dye and the test sample by the Cy3. Both samples are introduced in the microarray simultaneously and when excited by different laser frequencies each dye returns a distinct response, which can be recorded as intensity measurements. This process produces two images; the green image, which corresponds to intensity measurements of the Cy5 dye, and the red image, which corresponds to intensity measurements of the Cy3. These fluorescence intensities correspond to the levels of hybridization of the two samples to the DNA sequences spotted on the slide. An example of a microarray image is shown in Fig. 9.

The list of manufacturers of DNA arrays is rather exhaustive with >30 entries. Of those, the pioneer and first in market Affymetrix (Santa Clara, CA), Agilent (Palo Alto, CA), and Nimblegen (Madison, WI) should be mentioned. The war on density and throughput of microarrays is everlasting. Nimblegen, for example, can produce microarrays with >40,000 genes, with each spot being ∼80 μm in diameter.

GENOME SEQUENCING

As of now, the maximum read length permitted by today's commercial and experimental techniques does not exceed 2000 bases. The human genome is currently estimated to be >3 billion bases with 20,000–25,000 genes, while organisms such as the *Escheuchia coli* bacterium has ∼4.6 million bases. It is clear that in order to sequence whole genomes of organisms with the currently available techniques a method for combining sequencing results of smaller reads is needed (26).

Most of the techniques rely on shotgun sequencing, which is based on the idea of sequencing overlapping

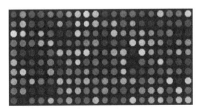

Figure 9. An example of a microarray image.

fragments of DNA (27). The genomic segment is sheared into overlapping fragments of DNA ~ 500 bases long and each fragment is then sequenced. The fragments are assembled into continuous sequences called contigs using complicated computer algorithms where they examine the overlapping sequences and try to order the fragments. There are two issues with this technique: gaps and errors. Since the shearing technique is random, and to avoid laboratory errors usually multiple shearing experiments are performed and the fragments are sequenced to assemble contigs. Another source of errors are repeats. If sequences appear in multiple positions throughout the genomic segment it will lead to errors when the fragments are overlapped. In some cases during contig assembly nonoverlapping sequences are formed that create gaps. Gaps are resolved with directed sequencing experiments with primers derived from the contigs that surround the gap.

The difference between the Human Genome Project (HGP) and Celera Corp was the origin of the target sequence. Human Genome Project used a directed sequencing method (also seen as hierarchical shotgun sequencing) for which the whole genome is first broken into long fragments. The fragments are then mapped into the genome, which is equivalent to finding their order (location) within the genome. Each fragment then is sequenced using shotgun sequencing. The advantage of this approach is the relatively easy assembly, while the disadvantages are the difficulty of building the library, mapping the long fragments, and the need for redundant sequencing.

Celera Corp relied on whole genome shotgun sequencing, which is essentially shotgun sequencing applied directly on the whole genome. The challenge is to assemble the whole genome from small 500 base fragments. While this technique overcomes the shortcomings of hierarchical sequencing and is faster and less expensive, the assembly is rather complicated and resolving the repeats requires sequencing of many clones.

THE FUTURE OF DNA SEQUENCING

Microscale Systems

Micro capillary Systems are microfabricated systems that in principle work similarly to CAE systems. Such systems have the potential of reducing cost while increasing speed and throughput. Due to their small size, lab-on-chip solutions are even considered where most of the sample preparation, amplification, and sequencing is all taking place on a single glass surface (chip). Their unique manufacturing methods allow for a large number of capillaries at low cost with arbitrary geometries, which are not possible with standard capillaries.

The Mathies group at University of California at Berkeley (28) published a method for fabricating capillary systems with complex architecture and layout on glass substrates using photolithography. One of their systems can achieve a read length of 500 bases with 99% accuracy and a cycle speed of 20 min (29).

A rather interesting technique is Massive Parallel Signature Sequencing (MPSS) (30). Using a microarray structure and beads, targets are sequenced iteratively at each cycle using a type IIs restriction enzyme that cleaves (cuts) within a target sequence, exposing a four-base-pair overhang. The overhang is identified using a sequence-specific ligation of a fluorescent linker. The method can read up to 20 bases (in 4–5 cycles) making it well suited for expression analysis.

Mass Spectrometry Based DNA Sequencing

Matrix Assisted Laser Desorption Ionization Time-of-Flight Mass Spectrometry (MALDITOF–MS) is the first MS based DNA sequencing technique (31). The method can replace the electrophoretic molecule separation step in DNA sequencing by a MS component. The mass of the molecule is estimated by measuring the time to travel of gas-phase DNA molecules within a flight tube that connects an excitation source (ultraviolet, UV laser) and an ion-to-electron conversion detector. The molecules collide at the detector thus registering the time to travel, which is analogous to molecular mass. The technique has the advantages of allowing the fast and parallel separation of a heterogeneous mixture of molecules without being affected by possible secondary structures that the DNA molecules have fallen into. Although the read lengths remain small, the availability of fast and autonomous MALDITOF–MS DNA sequencers (Sequenom Corporation in San Diego, CA) makes them good candidates for precise resequencing of small fragments useful in SNP detection.

DNA Sequencing at the Nanoscale

Most of the methods described below are based on manipulating properties of DNA at the nanoscale or utilizing properties of other materials at the nanoscale. Since such methods work with very low concentrations they can also be viewed as single-molecule sequencing methods and are suitable for applications when the template DNA is in very low concentration and amplification techniques could not be applied efficiently.

DNA Detection with Nanoparticles. Nanosphere Inc. (Northbrook, IL) has developed a method for rapid and low concentration detection of proteins and nucleic acids (32). Their technology is based on attaching oligonucleotide probes on nanoparticles. The probes attach to the target DNA and due to the unique properties of the used nanoparticles the event can be detected electrically, optically, or magnetically without amplification of the target sequence. The concentrations needed are below the operational threshold of PCR reactions. Although the techniques have not been extended to *de novo* sequencing the unique detection characteristics are proving very useful in detection scenarios.

In another effort from the founder of Nanosphere, Dr. Mirkin at Northwestern University (Evanston, IL), the electrical detection of DNA was first proposed (33). With this protocol the imaging aspect of microarray applications can be eliminated using gold nanoparticles that once hybridized onto the DNA probes and deposited with silver can close an electric circuit thus enabling detection of the hybridization event with electrical signals.

Sequencing with Atomic Force Microscopy. Atomic Force Microscopy (AFM) was invented at IBM Zurich Labs in 1986 and has completely revolutionized research at the nanoscale (34). A nanoscopic tip that is attached at the end of the cantilever interacts with the surface of the target material and records the tips deflections to create a topographic map of the surface. The AFM can be used to study the surface of duplex DNA and detect mismatches or it can be used as force measuring tool to study the mechanical properties of DNA. One application of particular interest is the AFM assisted unzipping of the DNA duplex, where a DNA duplex is suspended between a solid support and the AFM tip. Pulling the AFM tip further causes the duplex to unzip. The force needed to unzip depends on the percentage of the Gas Chromatography (GC) content, and hence can be used to estimate the GC content of an unknown target if needed in a more large-scale sequencing function (35). A very similar idea was proposed in Ref. 36 where optical traps are used to stretch DNA molecules and to measure force.

Nanopore Sequencing. Another interesting technique that uses features at the nanoscale is nanopore sequencing. As DNA passes through an 1.5 nm nanopore, different base pairs hinder the pore to different degrees, altering the electric conductivity of the pore (37). The pore conductance can be measured and monitored to identify the DNA sequence. The accuracy of base calling ranges from 60% for single events to 99.9% for 15 events. The technique has only been shown to work experimentally on certain sequences but exhibits a big potential for super fast sequencing without amplification of the target. It is evident that the evolution of this technique depends on nanopore engineering. To break apart from this restriction Visigen (Houston, TX) and Li-cor (Lincoln, NE: U.S. Patent 6,306,607) are in the process of engineering DNA polymerases or fluorescent labeled nucleotides that can provide real-time, base-dependent signals during the natural DNA synthesis process.

Sequencing by Fluorescence Microscopy. This is a new class of DNA sequencing methodologies, for which the fluorescence emitted during single molecule interactions is detected (38). The interactions most commonly referred to are single nucleotide incorporation during DNA polymerase replication or nucleotide digestion from an exonuclease. The change in fluorescence emission is detected using microscopes and CCDs. An enabling technology is fluorescence resonance electron transfer (FRET), where the fluorescence emission of two molecular dyes can be affected by their proximity. The research in the area is vast and already three companies, Nanofluidics (Menlo Park, California), Solexa (Essex, UK), and GenoVoxx (Lubeck, Germany), are developing products based on this technology for high throughput DNA detection and genotyping.

DNA Computing Based DNA Sequencing

Up to this point instruments and electronic computers were assigned the task of analyzing and processing DNA sequences. In 1994, the roles were reversed by the first proof of concept experiment by Adleman of using DNA to perform computations (39). This development led to the birth of the field of DNA computing [for a short introduction see (40)].

Landweber and Lipton were the first to suggest that DNA computing can be used to improve the performance of DNA sequencing (41). Their approach is based on DNA^2DNA computations, where nucleotides of an unknown sequence are translated into a new DNA sequence using a unique mapping transformation. A library of DNA oligonucleotides is synthesized and mixed in the solution containing the template DNA. The oligonucleotides then anneal to complementary parts. The partially double-stranded sequences are ligated and hybridized on a DNA chip. The reconstruction of the encoded sequence is achieved by analyzing the DNA array image. Although the technique was never implemented in large scale it points to potential future applications where instruments can be assisted by DNA computers.

The first proof of such development came a few years later in an announcement by Dr. Suyama from the University of Tokyo and Olympus Corp. (Japan), where they developed the first DNA-computer-assisted gene expression instrument (42). The instrument is a hybrid of a molecular computer and an electronic–digital computer. The molecular computer is in charge of DNA input–output, DNA reactions, capture of DNA results, and DNA detection while the electronic is responsible for information processing by means of DNA reaction calculations and result analysis.

In Ref. 43, a new laboratory protocol is proposed that assists in the faster sequencing of genomes. The distance between primers (probes) that have annealed on a target sequence can be estimated by measuring the intensity and color of light emission of specialized hybridization array. Although the method is not intended for de novo sequencing it is proposed as an alternative method of comparing genomes.

Bio-informatics, a subfield of computational biology, refers to processing, analyzing or storing DNA sequencing data with computers. The field of analyzing DNA sequences using digital signal processing theory has been known as genomic signal processing (44). The idea is to process the sequence of DNA as a digital signal and find certain characteristics. Recently, the application of DNA computing in digital signal processing, termed as DNA-based Digital Signal Processing, has been suggested (45). A future is envisioned where a DNA based digital signal processor can process DNA sequences and output certain characteristics in the form of DNA sequences that can be subsequently detected (or sequenced). This will allow researchers to process a vast amount of DNA sequences without prior sequencing.

BIBLIOGRAPHY

1. Watson JD, et al. Molecular biology of the gene. 5th ed. San Francisco: Pearson/Benjamin Cummings; 2004.
2. Shopsin B, Kreiswirth BN, Molecular Epidemiology of Methicillin-Resistant *Staphylococcus aureus*, [serial on the Internet], 2001; 7(2) Accessed 2005 July 14. Available at http://www.cdc.gov/ncidod/eid/vol7no2/shopsin.htm.

3. Weiner MP, Hudson TJ. Introduction to SNPs: discovery of markers for disease. Biotechniques 2002;32(Suppl.) S4–S13.

4. Sanger F, Nicklen S, Coulson AR. DNA sequencing with chain-terminating inhibitors. Proc Natl Acad Sci USA 1977;74:5463–5467.

5. Maxam AM, Gilbert W. A new method of sequencing DNA. Proc Natl Acad Sci USA 1977;74:560–564.

6. Studier FW. Slab-gel electrophoresis. Trends Biochem Sci 2000;25(12):588–590.

7. Meldrum D. Automation for genomics. Part Two: Sequencers, microarrays, and future trends. Genome Res 2000;10:1288–1303.

8. Huang GM. High-throughput DNA sequencing: a genomic data manufacturing process. DNA Seq 1999;10:149–153.

9. Viovy JL, Duke T. DNA electrophoresis in polymer solutions: Ogston sieving, reptation and constraint release. Electrophoresis 1993;14(4):322–329.

10. Zagursky RJ, McCormick RM. DNA sequencing separations in capillary gels on a modified commercial DNA sequencing instrument. Biotechniques 1990;9(1):74–79.

11. Green P (No date). Phrep, Phrap and Consed [Online]. University of Washington. Available at http://www.phrap.org/phredphrapconsed.html. Accessed 2005, June 29.

12. Ewing B, Green P. Basecalling of automated sequencer traces using phred. II. Error probabilities. Genome Res 1998;8:186–194.

13. Ewing B, Hillier L, Wendl M, Green P. Basecalling of automated sequencer traces using phred. I. Accuracy assessment. Genome Res 1998;8:175–185.

14. Hart C. (1997, August 1). Autoseq home page. [Online]. In-Machina. Availabel at http://www.in-machina.com/~reece/autoseq/ Accessed 2005, June 29.

15. Software Sequencing (No date). GE Healthcare - formerly Amersham Biosciences - Sequencing [Online]. GE Healthcare. Available at http://www5.amershambiosciences.com/aptrix/upp01077.nsf/Content/autodna_software_sequencing. Accessed 2005, June 29.

16. Lasergene (No date). DNASTAR. [Online]. DNASTAR, Inc. Available at http://www.dnastar.com/web/index.php. Accessed 2005, June 29.

17. Sequencher (No date). Gene Codes Corporation: Sequencher. [Online]. Gene Codes Corporation. Available at http://www.genecodes.com/sequencher/. Accessed 2005, June 29.

18. Staden Package (No date). Staden Package Home Page [Online]. SourceForge. Available at http://staden.sourceforge.net/ Accessed 2005, June 29.

19. Applied Biosystems Product Information Page (No date). Sequence Analysis Software [Online]. Applied Biosystems. Available at http://www.appliedbiosystems.com/. Accessed 2005, June 29.

20. Huang XC, Quesada MA, Mathies RA. DNA sequencing using capillary array electrophoresis. Anal Chem 1992;64(18):2149–2154.

21. Kambara H, Takahashi S. Multiple-sheathflow capillary array DNA analyzer. Nature(London) 1993;361(6412):565–566.

22. Crabtree HJ. Capillary array DNA sequencer based on a micromachined sheath-flow cuvette. Electrophoresis 2000;21:1329–1335.

23. Ronaghi M. Pyrosequencing Sheds Light on DNA Sequencing. Genome Res 2001;11:3–11.

24. Drmanac R, et al. DNA sequence determination by hybridization: a strategy for efficient large-scale sequencing. Science 1993;260:1649–1652; Erratum, Science 1994;163(5147):596.

25. Schena M, Shalon D, Davis RW, Brown PO. Quantitative monitoring of gene expression patterns with a complementary DNA microarray. Science 1995;270(5235):467–470.

26. Venter JC, et al. The Sequence of the Human Genome. Science 2001;291:1304–1351.

27. Sanger F, et al. Nucleotide sequence of bacteriophage lambda DNA. J Mol Biol 1982;162(4):729–773.

28. Woolley AT, Mathies RA. Ultra-High-Speed DNA Fragment Separations Using Microfabricated Capillary Array Electrophoresis Chips. Proc Natl Acad Sci USA 1994;91:11348–11352.

29. Simpson PC. High-throughput genetic analysis using microfabricated 96-sample capillary array electrophoresis microplates. Proc Natl Acad Sci USA 1998;95:2256–2261.

30. Brenner S, et al. *In vitro* cloning of complex mixtures of DNA on microbeads: physical separation of differentially expressed cDNAs. Proc Natl Acad Sci USA 2000;97:1665–1670.

31. Cantor CR, et al. DNA sequencing after the Human Genome Project. Nucleosides Nucleotides 1997;16:591–598.

32. Nam J-M, Park S-J, Mirkin CA. Bio-barcodes based on oligonucleotide-modified nanoparticles. J Am Chem Soc 2002;124:3820–3821.

33. Park SJ, Taton TA, Mirkin CA. Array-Based Electrical Detection of DNA Using Nanoparticle Probes. Science Feb. 2002;295(5559):1503–1506.

34. Binnig G, Quate CF, Gerber C. Atomic force microscope. Phys Rev Lett 1986;56:930–933.

35. Essevaz-Roulet B, Bockelmann U, Heslot F. Mechanical separation of the complementary strands of DNA. Proc Natl Acad Sci USA 1997;94:11935–11940.

36. Wang MD, et al. Stretching DNA with optical tweezers. Biophy J 1997;72:1335–1346.

37. Deamer DW, Branton D. Characterization of nucleic acids by nanopore analysis. Acc Chem Res 2002;35:817–825.

38. Braslavsky I, Hebert B, Kartalov E, Quake SR. Sequence information can be obtained from single DNA molecules. Proc Natl Acad Sci USA 2003;100:3960–3964.

39. Adleman L. Molecular computation of solutions to combinatorial problems. Science Nov. 1994;266:1021–1024.

40. Tsaftaris SA, Katsaggelos AK, Pappas TN, Papoutsakis ET. DNA computing from a signal processing viewpoint. IEEE Sig Proc Mag 2004;21(5):100–106.

41. Landweber LF, Lipton RJ. DNA2DNA Computations: A potential 'killer app'? Proceedings of the 24th International Colloquium on Automata, Languages and Programming (ICALP). Lecture Notes in Computer Science. New York: Springer-Verlag; 1997. 672–683.

42. Normile D. DNA-Based Computer Takes Aim at Genes. Science 2002;295(5557):951.

43. Mishra B. Comparing Genomes. Comp Sci Eng 2002;4(1):42–29.

44. Anastassiou D. Genomic Signal Processing. IEEE Sig Proc Mag 2001;18(4):8–20.

45. Tsaftaris SA, Katsaggelos AK, Pappas TN, Papoutsakis ET. How can DNA-Computing be applied in Digital Signal Processing?. IEEE Sig Proc Mag 2004;21(6):57–61.

Further Reading

The following two articles provide a well-rounded review of commercially available and experimental sequencing techniques.

Marziali A, Akeson M. New DNA sequencing methods. Annu Rev Biomed Eng 2001;3:195–223.
and

Shendure J, Mitra RD, Varma C, Church GM. Advanced sequencing technologies: methods and goals. Nature Rev Genet 2004;5(5):335–344.

The reader is suggested to study the projections into the future of sequencing technology of the first article and compare it to the presentation of the current status of the second article. The advancement in technology is rather interesting given that the papers are only three years apart.

A very informative presentation of the development of capillary array electrophoresis can be found in the following references.

Dovichi NJ, Zhang J. How capillary electrophoresis sequenced the human genome. Angew Chem Int Ed Engl 2000;39(24):4463–4468.
For a review of lab-on-chip methods for sequencing and genotyping please see the following reference.

Kan CW, Fredlake CP, Doherty EA, Barron AE. DNA sequencing and genotyping in miniaturized electrophoresis systems. Electrophoresis Nov. 2004;25(21–22):3564–3588.
An excellent review paper on DNA microarray technology is by the following references.

Venkatasubbarao S. Microarrays—status and prospects. Trends in Biotechnology 2004;22 (12):630–637.
Stears RL, Martinsky T, Schena M. Trends in microarray analysis. Nature Med 2003;9(1).
This article gives an interesting view of how microarrays can change the future of diagnostics.

Sauer S, et al. Miniaturization in functional genomics and proteomics. Nat Rev Genet 2005;6(6):465–76.
Finally, this article provides an overall picture on the effect of miniaturization of diagnostic and laboratory techniques in biology and medicine.

See also BIOINFORMATICS; MICROARRAYS; POLYMERASE CHAIN REACTION.

DOPPLER ECHOCARDIOGRAPHY. See

ECHOCARDIOGRAPHY AND DOPPLER ECHOCARDIOGRAPHY.

DOPPLER ULTRASOUND. See ULTRASONIC IMAGING.

DOPPLER VELOCIMETRY. See CUTANEOUS BLOOD

FLOW, DOPPLER MEASUREMENT OF.

DOSIMETRY, RADIOPHARMACEUTICAL. See

RADIOPHARMACEUTICAL DOSIMETRY.

DRUG DELIVERY SYSTEMS

DONATELLA PAOLINO
MASSIMO FRESTA
University of Catanzaro Magna Græcia
Germaneto (CZ), Italy

PIYUSH SINHA
MAURO FERRARI
The Ohio State University
Columbus, Ohio

PRINCIPLES OF CONTROLLED DRUG DELIVERY

A perspective drug delivery systems can be defined as mechanisms to introduce therapeutic agents into the body. Chewing leaves and roots of medical plants and inhalation of soot from the burning of medical substances are examples of drug delivery from the earliest times. However, these primitive approaches of delivering drugs lacked a very basic need in drug delivery; that is, consistency and uniformity (a required drug dose). This led to the development of different drug delivery methods in the later part of the eighteenth and early nineteenth century. Those methods included pills, syrups, capsules, tablets, elixirs, solutions, extracts, emulsions, suspension, cachets, troches, lozenges, nebulizers, and many other traditional delivery mechanisms. Many of these delivery mechanisms use the drugs derived from plant extracts.

The modern era of medicine development started with the discovery of vaccines in 1885 and techniques for purification of drugs from plant sources in the late nineteenth century, followed by the introduction of penicillin after its discovery in 1929, and a subsequent era of prolific drug discovery. The development and production of many pharmaceuticals involves the genetic modification of microorganisms to transform them into drug-producing factories. Examples are recombinant deoxyribonucleic acid (DNA), human insulin, interferon [for the treatment of acquired immunodeficiency syndrome (AIDS) related Kaposi's sarcoma, Hairy cell leukemia, Hepatitis B and C, etc.], interleukin-2 (Renal cell and other carcinomas), erythropoietin (for the treatment of anemia associated with chronic renal failure/AIDS/antiretroviral agents, chemotherapy-associated anemia in nomnyloid malignancy patient), and tissue plasminogen activator (1). It is now possible to produce oligonucleotide, peptide, and protein drugs in large quantities, while gene therapies also appear to be clinically feasible. Each of these therapeutic agents, by virtue of size, stability, or the need for targeting, requires a specialized drug delivery system (2). While the conventional drug delivery forms are simple oral, topical, inhaled, or injections, more sophisticated delivery systems need to take into account pharmacokinetic principles, specific drug characteristics, and variability of response from one person to another and within the same person under different conditions.

The efficacy of many therapeutic agents depends on their action on target macromolecules located either within or on the surface of particular cells types. Many drugs interact with enzymes or other macromolecules that are shared by a large number of cell types, while most often a drug exerts its action on one cell type for the desired therapeutic effect. Certain hormones, for example, interact with receptor mechanisms that are present in only one or a few cell types. An ideal gene delivery system should allow the gene to find its target cell, penetrate the cell membrane, and enter into the nucleus. Further, genes should not be released until they find their target and one has to decide whether to release the genes only once or repeatedly through a predetermined way (2). Thus, the therapeutic efficacy of a drug can be improved and toxic effects can be reduced by augmenting the amount and persistence of drugs in the vicinity of the target cells, while reducing the drug exposure to the nontarget cells.

This basic rationale is behind controlled drug delivery. A controlled drug delivery system requires simultaneous consideration of several factors, such as the drug property, route of administration, nature of delivery vehicle, mechanism of drug release, ability of targeting, and biocompatibility. These have been summarized in Fig. 1.

It is not easy to achieve all these in one system because of extensive independency of these factors. Further,

Figure 1. Design requirement for a drug delivery systems.

reliability and reproducibility of any drug delivery systems is the most important factor while designing such a system. The emphasis here is on the need for precision of control and to minimize any contribution to intraand intersubject variability associated with the drug delivery system. There are many different approaches for controlled drug delivery applications (3). They are summarized in the following section.

Overview of the Development of Drug Delivery Systems

To obtain a given therapeutic response, the suitable amount of the active drug must be absorbed and transported to the site of action at the right time and the rate of input can then be adjusted to produce the concentrations required to maintain the level of the effect for as long as necessary. The distribution of the drug-to-tissues other than the sites of action and organs of elimination is unnecessary, wasteful, and a potential cause of toxicity. The modification of the means of delivering the drug by projecting and preparing new advanced drug delivery devices can improve therapy. Since the 1960s, when silicone rubber was proposed as an implantable carrier for sustained delivery of low molecular weight drugs in animal tissues, various drug delivery systems have been developed.

At the beginning of the era of controlled drug delivery systems, a controlled release system utilizes a polymer matrix or pump as a rate-controlling device to deliver the drug in a fixed, predetermined pattern for a desired time period (4). These systems offered the following advantages compared to other methods of administration: (1) the possibility to maintain plasma drug levels a therapeutically desirable range, (2) the possibility to eliminate or reduce harmful side effects from systemic administration by local administration from a controlled release system, (3) drug administration may be improved and facilitated in underpriviledged areas where good medical supervision is not available, (4) the administration of drugs with a short *in vivo* half-life may be greatly facilitated, (5) continuous small amounts of drug may be less painful than several large doses, (6) improvement of patient compliance, and (7) the use of drug delivery systems may result in a relatively less expensive product and less waste of the drug. The first generation of controlled delivery systems presented some disadvantages, that is possible toxicity, need for surgery to

implant the system, possible pain, and difficulty in shutting off release if necessary. Two types of diffusion-controlled systems have been developed. The reservoir is a core of drug surrounded with a polymer film. The matrix system is a polymeric bulk in which the drug is more or less uniformly distributed.

Pharmaceutical applications have been made in ocular disease with the Ocusert, a reservoir system for glaucoma therapy that is not widely used, and in contraception with four systems: (1) subdermal implants of nonbiodegradable polymers, such as Norplant (6 capsules of 36 mg levonorgestrel); (2) subdermal implant of biodegradable polymers; (3) steroid releasing intrauterine device (IUD); and (4) vaginal rings, which are silicone coated. Other applications have been made in the areas of dentistry, immunization, anticoagulation, cancer, narcotic antogonists, and insulin delivery. Transdermal delivery involves placing a polymeric system containing a contact adhesive on the skin.

Since the pioneering work in controlled drug delivery, it was demonstrated that when a pharmaceutical agent is encapsulated within, or attached to, a polymer or lipid, drug safety and efficacy may be greatly improved and new therapies are possible (5). This concept prompted active and intensive investigations for the design of degradable materials, intelligent delivery systems, and approaches for delivery through different portals in the body. Recent efforts have led to development of a new approach in the field of controlled drug delivery with the creation of responsive polymeric drug delivery systems (6). Such systems are capable of adjusting drug release rates in response to a physiological need. The release rate of these systems can be modulated by external stimuli or self-regulation process.

Different Approach for Controlled Drug Delivery

Localized Drug Delivery. In many cases, it would be desired to deliver drugs at a specific site inside the body to a particular diseased tissue or organ. This kind of regional therapy mechanism would reduce systemic toxicity and achieve peak drug level directly at the target site. A few examples of drugs that require this kind of therapy are anticancer drugs, antifertility agents, and antiinflammatory steroids. These drugs have many severe unintended side effects in addition to their therapeutic effects.

Targeted Drug Delivery. The best controlled mechanism would be delivery of drug exclusively to the targeted cells or cellular components. That means the development of delivery mechanisms that would equal or surpass the selectivity of naturally occurring effectors (e.g., peptide hormones). As in the case of hormone action, drug targeting would probably involve a recognition event between the drug carrier mechanism and specific receptors at the cell surface. The most obvious candidates for the targetable drug carriers are cell-type specific immunoglobulins. The concept of targeted drug delivery is different than localized drug delivery. The latter simply implies localization of the therapeutic agent at an organ or

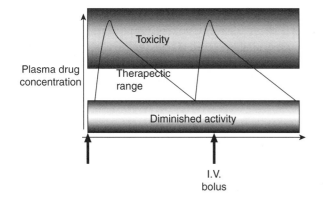

Figure 2. Plasma concentration versus time curve for intravenous (IV) drug administration showing first-order kinetic.

tissue site, while the former implies more subtle delivery to specific cell types.

Sustained Drug Delivery (Zero Order Release Profile). Injected or ingested drugs follow first-order kinetics, with initial high blood levels of the drug after initial administration, followed by an exponential fall in blood concentration. Toxicity often occurs when blood levels peak, while efficacy of the drug diminishes as the drug levels fall below the therapeutic range. This profile is shown in Fig. 2. and the drug kinetics is undesirable, especially in the case where the margin between toxicity and required therapeutic concentration levels is small. The importance of controlled-release drug delivery systems may be argued with reference to the goal of achieving a continuous drug release profile consistent with zero-order kinetics, wherein blood levels of drugs would remain constant throughout the delivery period. The therapeutic advantages of continuous-release drug delivery systems are thus significant, and encompass: *in vivo* predictability of release rates on the basis of *in vitro* data; minimized peak plasma levels, and thereby reduced risk of toxic effects; predictable and extended duration of action; reduced inconvenience of frequent dosing, thereby improving patient compliance (7,8).

Figure 3 illustrates the constant plasma concentration that is desired for many therapeutic agents.

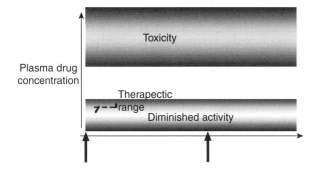

Figure 3. Plasma concentration versus time curve for sustained release profile of zero-order kinetics and pulsatile release profile.

The controlled release aspect of sustained drug delivery systems pertain to a reliable and reproducible system whose rate of drug delivery is independent of the environment in which it is placed. This requirement emphasizes the need for precision of control and elimination of undesired contribution associated with the drug delivery system.

Modulated Drug Delivery (Nonzero-Order Release Profile). A significant challenge in drug delivery is to create a delivery system that can achieve manipulable nonzero-order release profile. This could be pulsatile or ramp or some other pattern. In some cases it is also required that the release should be immediate. A pulsatile release profile within the therapeutic window is shown in Fig. 3.

Feedback Controlled Drug Delivery. The ideal drug delivery system is the feedback controlled drug delivery system that releases drug in response to a therapeutic marker. This can be classified into two classes: modulated and triggered device. A modulated device involves the ability to monitor the chemical environment and changes drug delivery rate continuously in response to the specific external marker, while in a triggered device no drug release takes place until it is triggered by a marker.

These different approaches of drug delivery can have different routes of administration. Some of the most preferred routes are oral, pulmonary inhalation, transdermal, transmucosal, and implantable systems.

Implantable Controlled Drug Delivery Devices. Although most controlled drug delivery systems are designed for transdermal, subcutaneous, or intramuscular uses, implantable devices are very attractive for a number of classes of drugs, particularly those that cannot be delivered via the oral route or are irregularly absorbed via the gastrointestinal (GI) tract (9). Implantable systems are designed to deliver therapeutic agents into the bloodstream. This replaces the repeated insertion of IV catheters. The basic idea behind this device is simple: The treatment of certain diseases that require the chronic administration of drug could benefit from the presence of implantable devices. These systems can also be used to deliver drug to the optimum physiological site. These systems are particularly suited for drug delivery requirements of insulin, steroids, chemotherapeutics, antibiotics, analgesics, contraceptives, and heparin. Implantable systems are placed completely under the skin (usually in a convenient, but inconspicuous location). Benefits include the reduction of side effect (drug delivery rate within the therapeutic window) caused by traditional administration techniques, and better control. Ideally an implantable system will have a feedback controlled release mechanism and will be controlled by electronics with a long-life power source to achieve zero-order or manipulable nonzero-order release profiles in a manner similar to a physiological release profile.

The focus of this research is on two major requirements of an implantable controlled drug delivery device:

1. One of the major requirements for implantable drug delivery devices is to allow controlled-release of therapeutic agents, especially biological molecules, continuously over an extended period of time. The goal here is to achieve a continuous drug release profile consistent with zero-order kinetics where the concentration of drug in the blood remains constant throughout the delivery period. As mentioned earlier, the therapeutic advantages of continuous release of drug by implantable delivery devices are significant: minimized adverse reactions by reducing the peak levels, predictable and extended duration of action, reduced inconvenience of frequent dosing and thereby improved patient compliance.

2. The second, and more important requirement, is to achieve a manipulable nonzero-order release profile, such as pulsatile or any other pattern required for applications in therapeutic medicine. Vaccines and hormones are examples that require pulsatile delivery (10,11). Gonadotropin releasing hormone, for example, is most effective when delivered in a pulsatile manner to female patients undergoing treatment for infertility.

A sequence of two implantable systems was developed to achieve the above mentioned goals. The first device that addresses the first goal is named nanochannel delivery system I (or nDSI), while the device that addresses the second goal is called nanochannel delivery system 2 (or nDS2).

The Economics of Drug Delivery Devices

The fact that drug delivery technology can bring both therapeutic and commercial value to healthcare products cannot be neglected. Big pharmaceutical companies have recently started losing their market share to generic competitors after their patents expired, and therefore they have started recognizing the importance of drug delivery companies. Pharmaceutical companies are looking to extend their patents lifetimes by making strategic alliances with drug delivery technology companies, by presenting old drugs in new forms. Most of the drug delivery products therefore reach the market as a result of strategic alliance between drug delivery companies and pharmaceutical companies. Pharmaceutical companies provide the drug that may not be delivered efficaciously with a conventional delivery mechanism, while the drug delivery companies provide the cutting edge technology to administer the drug more effectively. The joint venture not only offers considerable advantages over the R&D efforts to bring new drug into the market as drug delivery systems provide means to reformulate existing products, but it also protects the drugs from erosion by generics in the case of patented drugs. As a result, drug delivery technology companies seem to enjoy a good return on their investments in the form of increased revenues and market share (9,12).

The global drug delivery market grew between 1998 and 2002, with a compound annual growth rate (CAGR) of 13.7%, increasing from $39.6 billion to slightly > $66 billion. The market is expected to grow at a slightly lower CACR of 11.6% between 2002 and 2007 corresponding to a market value of $114.3 billion by 2007. One of the contributing factors in this growth is the use of drug delivery systems as strategy to expand the shelf-life of products (particularly blockbusters), enabling pharmaceutical companies to sustain the revenue streams from their best sellers.

The largest market for drug delivery systems in the world is in the United States, having captured 47.9% of the global market's revenue generation in 2002. This figure is forecast to fall to 41.9% by 2007 although the U.S. market will retain its position as the leading market. The U.S. market for drug delivery systems was worth $31.7 million in 2002, having experienced a CAGR of 12.6% during 1998–2002. Oral drug delivery systems had the largest market share, taking 47.7vo of the total market share. Transmucosal, injectable, and implantable systems together had 8.8% of the market share in 2002. The U.S. market value for drug delivery systems is expected to grow at a rate of 8.5% annually, reaching a value of $48 billion by 2007.

MICROELECTRO-MECHANICAL SYSTEMS

A number of devices have been developed to achieve controlled drug delivery. These devices utilize a different route of administration and different materials for device fabrication. Typically, each of these devices is targeted toward delivering one or a few of the therapeutics. The factors that need to be considered when designing a drug delivery device were previously discussed in great details (Fig. 1). This article begins with a brief history of implantable drug delivery devices. These include polymeric devices, osmotic pumps, micropumps, and microelectro-mechanical systems (MEMS) based devices. Since the drug delivery devices developed in this research are based upon MEMS technology, a good understanding of MEMS fabrication technology is needed, and therefore under the section MEMS for drug delivery devices, it is digressed from the topic implantable drug delivery devices and a more in-depth description on the use of MEMS for different drug delivery devices is presented. This includes MEMS for transdermal, oral, injectable, and *implantable* drug delivery. This article concludes with a critical analysis of implantable drug delivery devices.

A History of Implantable Drug Delivery Devices

The history of implantable devices goes back to May 1958 when the first implantable cardiac pacemaker was placed in an experimental animal (13). Later that year the first pacemaker was implanted in a human that operated for 3 h and then failed (14). The second unit operated for 8 h before failing, and the patient went unstimulated for 3 years before receiving a satisfactory implantable unit. The record shows that this patient was alive in 1991 and was using a pacemaker (15). The development of an implantable pacemaker revolutionized the field of biomedical science and engineering over the last 30 years providing many different implantable biomedical devices to the medical professionals for therapeutic and diagnostic use. Today, implantable cardioverter–defibrillators, drug delivery systems, neurological stimulators, bone growth stimulators, and other implantable devices make possible the treatment, of a variety of diseases.

Extensive research has been done on implantable drug delivery devices over the last 30 years. Different technologies have been developed with many breakthroughs in clinical medicine. The first such device that saw extensive clinical use was reported in the 1970s (15–18). This system used a bellows-type pump activated by partially liquefied Freon. The Freon was reliquefied with each transcutaneous refill of the implantable device, and the administration was constant. Later, extensive research started to develop more sophisticated devices that could offer better control and more clinical options. Another device was developed by Medtronic Company that has a peristaltic pump to deliver the drugs (19). The device was controlled by electronics. Another system developed by MimiMed Technologies employs a solenoid pump, a reservoir, and advanced electronic control (20). The Infusaid Company developed an advanced programmable implantable pump that employed a bellows-type pump and a solenoid valve set to control drug flow (21). Other technologies developed to achieve this goal are summarized in the following sections.

Polymeric Implants. Polymers have been used extensively in controlled drug delivery systems. These can be classified as (1) nondegradable polymeric reservoirs and matrices, and (2) biodegradable polymeric devices. The first kind of polymeric devices are basically silicone elastomers. This kind of drug delivery system is based upon the research conducted in the 1960s, when researchers recognized that certain dye molecules could penetrate through the walls of silicone tubing (22–24). This lead to the development of reservoir-based drug delivery system, which consisted of hollow polymer tubes filled with a drug suspension. The drug is released by dissolution into the polymer and then diffusion through the walls of the polymeric device. The two most commonly used nondegradable polymers are silicone and poly(ethylene-covinyl acetate) (EVAc). The Norplant 5 year contraceptive drug delivery system is based upon this technology. Some of the implantable reservoir systems are simple cylindrical reservoir surrounded by a polymeric membrane. The other variety in this first category is constructed of a solid matrix of nondegradable polymers. These systems are prepared by homogeneous dispersement of drug particles throughout the matrix (25). Drug release occurs by diffusion through the polymer matrix or by leaching or a combination of both (26). The matrix may be composed of either a lipophilic or hydrophilic polymer depending on the properties of the drug and the rate of release desired. However, it is difficult to achieve constant rates of drug release with nondegradable matrix systems, for example, the rate of release of carmustine from an EVAc matrix device drops continuously during incubation in buffered water (27). Constant release can sometimes be achieved by making the matrix as

a reservoir surrounded by a shell of rate-limiting polymeric membrane. In some cases, water soluble, cross-linked polymers can be used as matrices. Release is then activated by swelling of the polymer matrix after exposure to water (28). One other kind is a magnetically controlled system where magnetic beads are dispersed within the matrix (25). Drug is released by diffusion with a concentration gradient. The addition of an externally oscillating magnetic field causes the physical structure of the polymer to alter, creating new channels, and thus leading to further drug release.

Biodegradable polymeric devices are formed by physically entrapping drug molecules into matrices or microspheres. These polymers dissolve when implanted (injected) and release drugs. Examples of biodegradable polymers are poly(lactide-co-glycolide) (PLGA), and poly(p-carboxyphenoxypropane-co-sebacic acid) (PCPP-SA) (24). Some of the commercially available polymeric devices are Decapeptyl, Lupron Depot (microspheres), and Zoladex (cylindrical implants) for prostate cancer and Gliadel for recurrent malignant glioma. The half-life of therapeutics administered by microspheres is much longer than free drug injection. Polymers are also being investigated for treating brain tumors (29), and delivery of proteins and other macromolecules (30).

The above mentioned polymeric implants are utilized for sustained drug delivery. Methods have been developed to achieve controlled drug delivery profiles with implantable polymeric systems (31,32). These technologies include preprogrammed systems, as well as systems that are sensitive to (triggered or modulated by) modulated enzymatic or hydrolytic degradation, pH, magnetic fields, ultrasound, electric fields, temperature, light, and mechanical simulation. Researchers are also exploring the use of nontraditional MEMS fabrication techniques and materials that could be used to form microwell- or microreservoir-based drug delivery devices. For example, microwells of varying sizes (as small as 3fL/well) have been fabricated by micromolding of poly(dimethylsiloxane) (PDMS) on a photoresist-coated silicon wafer that is photolithographically patterned (33).

Osmotic Pumps. Osmotic pumps are energy modulated devices (9). These are usually capsular in shape. When the system is exposed to an aqueous environment, such as that after subcutaneous implantation, water is drawn to the osmotically active agent through a semipermeable membrane and pressure is supplied to the collapsible drug reservoir and drug is released through an orifice with precise dimension. The delivery mechanism is dependent on the pressure created and is independent of drug properties. The ALZET pumps (only for investigational purpose at this time, not for humans) have been used in thousands of studies on the effects of controlled delivery of a wide range of experimental agents, including peptides, growth factors, cytokines, chemotherapeutic drugs, addictive drugs, hormones, steroids, and antibodies (34). The ALZA Corporation built the DUROS implant based upon the foundation of the ALZET osmotic pump, the system of choice for implant drug delivery in research laboratories around the world for > 20 years. Viadur, a once-yearly implant for the palliative treatment of advanced prostate cancer, is the first approved product to incorporate ALZAs proprietary DUROS implant technology. A single Viadur implant continuously delivers precise levels of the peptide leuprolide for a period of 1 full year, providing an alternative to frequent leuprolide injections. Although most of the osmotic pumps are designed for sustained release profile, research is being conducted to modify this design for different patterns (9). Further, a catheter was attached to the exit port of an implantable osmotic pump to achieve site specific drug delivery at a location distant from site of implantation (35).

Micropumps. Micropumps have been actively investigated for drug delivery applications. Some micropumps are nonmechanical that utilizes electrohydrodynamic, electroosmotic, ultrasonic, or thermocapillary forces (36). However, most of the micropumps are mechanical, composed of mechanically moving membranes. A number of mechanical micropumps have been developed using various mechanisms, including piezoelectric (37), electrostatic (38), thermopneumatic (39), electromagnetic (40), bimettalic (41), shape memory alloy (SMA) (42), ionic conducting polymer films JCPF (43), and surface tension driven actuators (36). One example is the silicon piezoelectric micropump based on silicon bulk micromachining, silicon pyrex anodic bonding, and piezoelectric actuation (37). This can be used for application requiring low (typically $1\,\mu L \cdot min^{-1}$), precisely controlled flow rate. The whole system includes the refillable reservoir, control, and telemetry electronics and battery. This can be implanted in the abdomen and a catheter can be brought to the specific site. The Synchro-Med pump is an implantable, programable, battery-powered device commercially available by Medtronics (44). A large number of other implantable drug delivery devices have been developed in last decade utilizing the silicon microfabrication technology that was developed in integrated circuits (ICs) industries.

MEMS for Drug Delivery

Since the invention of silicon microfabrication technology in early 1960s, the IC has changed our world. During last 40 years, the semiconductor industry has come up with a fastest growing industry in our history. From a modest beginning, which allowed few transistors on a chip, we have reached an integration level of tens of millions of components in a square centimeter of silicon. The minimum feature size on silicon is reducing and thus the number of devices per square centimeter is increasing. Since the observation made in 1965 by Gordon Moore (45), co-founder of Intel, the number of transistors per square inch on integrated circuits had doubled every year since the integrated circuit was invented. Moore predicted that this trend would continue for the foreseeable future. In subsequent years, the pace slowed down a bit, but data density has doubled approximately every IS months, and this is the current definition of Moore's law.

This silicon fabrication technology was later extended to machining mechanical microdevices, which was later called MEMS. The pioneer work was done by Nathanson et al. in 1965 when they demonstrated the first micromachined structure to fabricate a free-standing gold beam electrode used in a resonant gate transistor (46). By late 1970s, there was an immense interest in silicon as a mechanical material (47,48). During 1980s and 1990s, many MEMS devices were fabricated, for example, micrometers (49–51), deformable mirrors (52,53), accelerometers (54–58), and comb-drive actuators (59).

In recent years, this fabrication technology has been extensively used for the development of microfluidic devices for biological and biochemical applications (these are called bio-MEMS) (60,61). Further, the integration of microfluidic devices and integrated circuits over the last decade has revolutionized the chemical and biological analysis systems, and has opened the possibility of fabricating devices with increased functionality and complexity for these applications (62–64). These tiny devices hold promise for precision surgery with micrometer control, rapid screening of common diseases and genetic predispositions, and autonomous therapeutic management of allergies, pain and neurodegenerative diseases (7). The development of retinal implants to treat blindness (65), neural implants for stimulation and recording from the central nervous system (CNS) (66), and microneedles for painless vaccination (67), are examples in which MEMS technology has been used. With microfabrication technology it is also possible to produce the novel drug delivery modalities with capabilities not present in the current systems. A variety of microfabricated devices, such as microparticles, microneedles, microchips, nanoporous membranes, and micropumps, have been developed in recent years for drug delivery applications (68–71). This section reviews various microfabricated devices. These have been categorized and described below as microfabricated devices for transdermal, oral, IV, and implantable drug delivery devices.

Microneedles for Transdermal Drug Deliver. Transdermal drug delivery is probably the most favored way of drug delivery since it avoids any degradation of molecules in the GI tract and first-pass effects of the liver, both of which are associated with the oral drug delivery, and eliminates the pain associated with IV injection (72–76). However, the major barrier for the transdermal delivery is the stratum corneum, the outermost dead layer of the skin. 1n human, it is 10–20 μm thick. A number of different approaches have been studied with two common goals: first is to disrupt stratum corneum structure in order to create "holes" big enough for molecules to pass through and the second goal is to develop microneedles that are long enough to provide transport pathways across the stratum corneum and short enough to reach nerves found in deeper tissues. These approaches include chemical–lipid enhancers (77,78), electric fields employing iontophoresis and electroporation (79), and pressure waves generated by ultrasound or photoacoustic effects (80,81).

MEMS technology has provided an alternative approach to transdermal drug delivery. The development of microneedles for transdermal drug delivery enhances the poor permeability of the skin by creating microscale conduits for transport across the stratum corneum (69,76). Needles of micron dimensions can pierce into the skin surface to create holes large enough for molecules to enter, but small enough to avoid pain or significant damage.

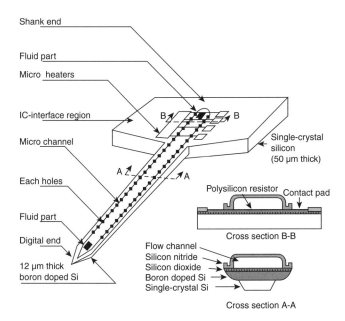

Figure 4. Schematic diagram of a silicon processed microneedles by Lin and Pisano (84).

Although the microneedles concept was proposed in the 1970s (82), it was not demonstrated experimentally until the 1990s (83). Since then, many different kinds of microneedles have been fabricated in several materials (e.g., silicon, glass, and metal). Further, these microneedles can be fabricated in-plane, where the needle lumen (flow channel) is parallel to the substrate surface, or out-of-plane, where the lumen is normal to the substrate. Some of these are summarized below.

Lin and Pisano (84) fabricated microneedles in silicon (Figs. 4 and 5). The primary structural material of these microneedles was silicon nitride, forming the top, and a bulk micromachined boron doped silicon base defined by etching the substrate in ethylenediamine pyrocatechol (EDP). This layer of silicon, which varied in thickness from $\sim 50\,\mu m$ at the shank to $12\,\mu m$ near the tip improved the structural strength. The lumen was defined by a sacrificial layer of phosphorous doped glass. These microneedles were 1–6 mm in length with lumens $9\,\mu m$ high and 30–50 μm wide.

The proximal ends of the microstructures had integrated polycrystalline silicon heater strips. The heater could generate bubbles, which were useful in pumping fluid down the lumen. Authors suggested that electrodes could also be patterned along the length of the needle by a slight process modification for the measurement of neural activity.

Other microneedles made out of polysilicon molding process were reported by Talbot and Pisano (85) (Fig. 6). The two halves of the mold are produced by bulk micromachining of silicon wafers followed by deposition of a $2\,\mu m$ phosphosilicate glass (PSG) release layer. The two halves are temporarily bonded together under nitrogen ambiance at $1000\,^\circ C$. After bonding, a $3\,\mu m$ layer of amorphous silicon is deposited by LPCVD through access holes in the top mold wafer. The mold along with the deposited film was then annealed at $1000\,^\circ C$. Deposition and annealing steps were repeated until the desired thickness of 12–18 μm was

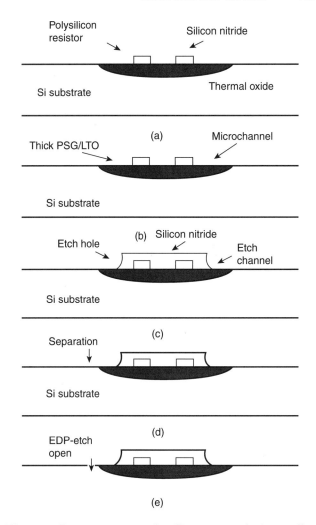

Figure 5. Process sequences of a silicon processed microneedles by Lin and Pissano (84).

obtained. Plasma etching was used to remove the polysilicon coating the funnel-shaped access holes in the top mold layer. The devices were released from the mold by etching in concentrated hydrofluoric acid, which selectively attacks the PSG. The mold could be used repeatedly by redepositing PSG, the release layer in order to minimize the cost. The resulting polysilicon microneedles are 1–7 μm long, 110–200 μm rectangular cross-section, and submicrometer tip radii.

Brazzle et al. (86–88) fabricated metal microneedles using a micromolding process. The fabrication process of the microneedles developed by Papautsky is shown in Fig. 7. A P+ etch stop layer was formed and backside anisotropic etching in KOH was performed to define a thin membrane. The lower wall of the microneedles consisted of deposited and patterned metal layers. A thick layer (5–50 μm) of positive photoresist was then spin coated and lithographically patterned on the top of the lower metal walls.

The dimensions of this sacrificial layer precisely defined the cross-section of the lumen. After sputter deposition of a Pd seed layer, the thick metal structure walls and top of the microneedles were formed by electrodeposition. The sacrificial photoresist was removed with acetone and the P+

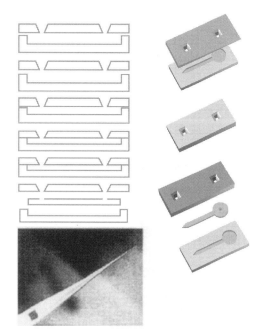

Figure 6. Microneedles fabricated from a polysilicon molding process using two silicon wafers (85).

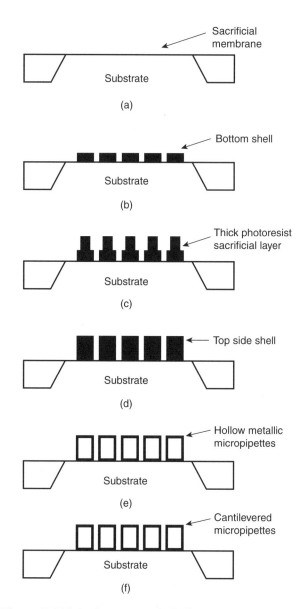

Figure 7. Fabrication process of a hollow in-plane microneedles (86).

membrane was etched away in an S176 plasma, resulting in a one-dimensional (1D) array of hollow microneedles released from the substrate.

Out-of-plane array of microneedles were fabricated by Stoeber and Liepmann (89,90). The fabrication process is summarized in Fig. 8. A double-sided polished wafer was oxidized. The lumen was etched through the wafer by plasma etching following a mask patterned at the backside. A silicon nitride film was then deposited across the backside and into the etched holes. Needle locations were photolithographically defined on the top surface on the wafer. The microneedle shaft was created by isotropic and etching on the silicon substrate. The isotropic etching forms a microneedle with a gradually increasing diameter along the shaft. By displacing the circular pattern for isotropic etching from the center of the lumen, a pointed needle shape was obtained. These microneedles were $200 \, \mu m$ tall, with a base diameter of $425 \, \mu m$ tapering to a $40 \, \mu m$ lumen. Individual needles were $750 \, \mu m$ apart. Fluid injection was demonstrated by delivering under the skin of a chicken thigh, a depth of $\sim 100 \, \mu m$.

Solid microneedles with no lumen were demonstrated by Henry et al. (76,91). The fabrication steps are shown in Fig. 9. A chrome mask was deposited on a silicon wafer and patterned into dots that have a diameter approximately equal to that of the base of the desired needles. A deep reactive ion etching was performed. Etching proceeded until the mask fell off from undercutting. The region protected by chromium remained and eventually became the microneedles. The tapering on the microneedles were controlled by adjusting the degree of anisotropy in the etch process. The resulting microneedles were $150 \, \mu m$ tall, and could be fabricated in dense arrays.

Gardeniers et al. (92) fabricated out-of-plane microneedles that employed reactive ion etching from both sides on

a (100) oriented silicon wafer (Fig. 10). A hole (feature a in Fig. 10), which becomes lumen and a slot (feature b) that defines the position of the needles tip and needle sidewalls, was etched at the top surface. These structures were aligned to the crystallographic planes of silicon so that anisotropic etching performed later produces the slanted structure. The connecting lumen (feature c) was etched from the back side. The substrate, including the sidewalls of the etched features were coated with the chemically vapor deposited silicon nitride. The nitride was removed form the top surface of the wafer and etched in KOH. The etch left a structure defined by (111) plane in the areas where the nitride slot walls were concave, but where the mask was convex, the etch found all of the fast etching planes. The nitride mask was stripped at the end of the process.

Microneedles have also been developed for gene delivery. One such structure was fabricated by Dizon et al. (93).

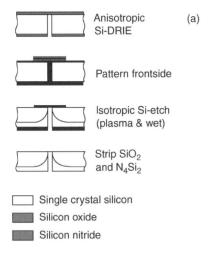

Anisotropic
Si-DRIE
(a)

Pattern frontside

Isotropic Si-etch
(plasma & wet)

Strip SiO$_2$
and N$_4$Si$_2$

☐ Single crystal silicon
▨ Silicon oxide
▨ Silicon nitride

(b)

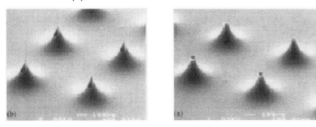

Figure 8. Out-of-plane array of microneedles. (a) Fabrication step, (b) Symmetric and asymmetric needles (90).

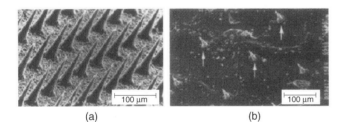

(a)   (b)

Figure 9. (a) Scanning electron micrograph (SEM) of microneedles made by reactive ion etching technique. (b) Micro-needle tips inserted across the epidermis. The underside of the epidermis is shown, indicating that the microneedles penetrated across the tissue and that the tips were not damaged. Arrows indicate some of the microneedle tips (91).

This structure was fabricated in dense array using a silicon bulk micromachining technique (Fig. 11), called Microprobes. The microprobes were ~80 μm high topped by a wedge-shaped tip with a radius of curvature <0.1 μm. The facets of the microstructure were fabricated utilizing fast etching (411) planes, produced by convex-corner undercutting in an anisotropic etching solution and a square mask. These microprobes can be coated with genes and pressed into cells or tissues. The sharp tips penetrate into cells and affect the transport of genetic material. Successful expression of foreign genes using this technique has been demonstrated in the nematode *Caenorhabditis elegans* (94), tobacco leafs (95), and mammalians cells (96).

Figure 10. Out-of-plane microneedles were fabricated that employed reactive ion etching from both sides on a (100) silicon wafer (92).

Figure 11. Solid silicon microprobe for gene delivery (93).

Mikszta et al. (67) used silicon micromachining technology for DNA and vaccine delivery to the epidermis. Figure 12 shows the microstructure, which they call micro-enhancer arrays (MEAs), that was fabricated by isotropic chemical etching of silicon wafers.

On the whole, existing microneedle-based drug delivery devices offer several advantages, such as the ability to inject drugs directly through the stratum corneurn at reproducible and accurate depth of penetration, minimal

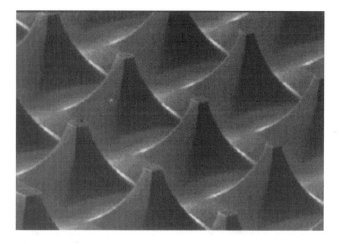

Figure 12. Silicon microenhancer arrays (MEAs) for DNA and vaccine delivery (67).

pain, and on-board ability to probe or sample the same device. Nevertheless, local irritation and low mechanical stability are some of the potential drawbacks that demand further investigation for alternate fabrication techniques and materials. Furthermore, improved fluid flow models that determine the most effective structural. fluidic, and biological design considerations for a given delivery application continue to be required.

Microparticles for Oral Drug Delivery. Oral route is a preferred method of drug delivery because of its ease of administration and better patient compliance. However, oral delivery of peptides and proteins has remained an illusive goal to date. The two main reasons why it is currently impossible are (1) destruction or inactivation due to enzymatic action, and the acidity of the upper GI tract; and (2) physiological permeation barrier, opposing penetration of large biological molecules through intestinal walls (71). These are mucosal layers and the tight junctions connecting intestinal epithelial cells, which restrict the possible passageways to be transcellular, and thus expose the diffusing biomolecule to enzymatic degradation. This method of drug delivery, therefore, leads to unacceptably low oral bioavailability. Consequently, various approaches based on the use of protective coatings (97), targeted delivery (98), permeation enhancers (99), protease inhibitors (100), and bioadhesive agents (101–103) have been explored in recent years. While all of these methods have been shown to increase the oral bioavailability of drug molecules, none of them offer a complete solution for adequate and safe oral delivery of peptides and proteins.

Microfabrication technology may address the shortcomings of the current oral drug delivery systems by combining the aforementioned approaches in a single drug delivery platform. Fabrication of microparticles of silicon and silicon dioxide has been conceptualized and demonstrated to achieve this (104–106). Unlike other spherical drug delivery particles, microfabricated devices may be designed to be flat, thin, and disk–shaped to maximize contact area with the intestinal lining and minimize the side areas exposed to the constant flow of liquids through the

intestines (107). The size of the particles (within thickness of 0.1–5 nm and diameters of 1–100 μm) can be selected to have good contact with the undulations of the intestinal wall and large enough to avoid endocytosis of the entire particle. Permeation enhancers, such as bile salts and metal chelating agents, can be added to loosen the tight junctions of the intestinal epithelium. Aprotinin, or other enzyme inhibitors, can also be added to protect the macromolecule from intestinal degradation. In addition, one can selectively attach bioadhesive agents onto the device surface using relatively simple surface chemical modification strategies. By replacing the specific markers attached to the microparticles, specific cell types and tissues can be targeted for therapy as well as imaging. This would allow for the high concentration of drug to be locally delivered while keeping the systemic concentration at a low level. Finally, these devices can have multiple reservoirs of desired size to contain not just one, but also many drugs–biomolecules of interest (108).

iMEDD Inc. in collaboration with Ferrari et al. (109) developed Oral MEDDS (Oral Micro-Engineered Delivery Devices), novel porous silicon particles that can be used as oral drug delivery vehicles. The microparticle dimensions ranged from 150 × 150 × 25–240 × 240 × 25 μm with a pore distribution of 20–100 nm (Fig. 13). Once prepared, the particles could be loaded with a liquid drug formulation through simple capillary action. Interstitial air is removed by vacuum aspiration, and the formulation is dried completely using vacuum or freeze-drying. OralMEDDS particles have been designed to target intestinal epithelial cells, adhere to the apical cell surface, and deliver a drug formulation containing a permeation enhancer that would open the local tight junctions of the paracellular transport pathway. The absorption of macromolecules and hydrophilic drugs, which are unable to undergo transcellular transport across lipid membranes, is largely restricted to this paracellular route. Therefore, the intestinal absorption of orally administered water-soluble drugs can be greatly enhanced through the utilization of OralMEDDs particles (110).

Micromachined silicon dioxide and PMMA microparticles designed by Desai and co-workers (70,111) can be best described as microparticles with reservoirs (Figs. 14 and 15). These microparticles are adaptable for use as a bioadhesive controlled release oral drug delivery system. Silicon dioxide microparticles were created by growing a thermal oxide under wet conditions followed by low pressure chemical vapor deposition to deposit a sacrificial layer of

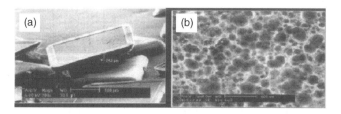

Figure 13. Scanning electron microscopy images of a porous silicon particle: (a) Demonstrating the thickness. (b) Particle demonstrating the pore size distribution of −20–100 nm (110).

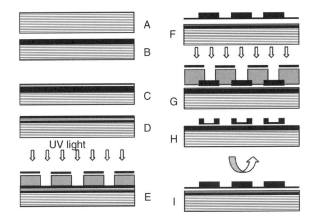

Figure 14. Process flow of the silicon dioxide microparticles (111).

(a)

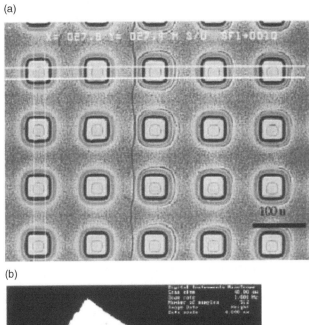

(b)

Figure 15. (a) 50 μ particles with 25 μ reservoirs. (b) AFM image of the particles (25 reservoir, 50 m particles) (111).

polycrystalline silicon (111). Next, a layer of low temperature silicon dioxide (LTO) was deposited to form the device layer. Positive lithography was carried out to define the shape of the device reservoir. A reactive ion etch (RIE) with S176 and 02 was used to fabricate the actual reservoir in the LTO device layer and any remaining photoresist was

then removed in negative photoresist remover. Negative lithography was carried out to define the device bodies. Reservoir features on the mask were aligned to the photomask features using front-side alignment. The unmasked area of the LTO layer was etched using RIE and the remaining photoresist was removed. These microdevices were then released into KOH solution by etching the sacrificial polysilicon layer. The particles were uniform and semitransparent due to their polycrystalline nature. Later, a lectin–biotin–avidin complex suited for binding these microparticles to the intestinal mucosa was developed. The Caco-2 cell line was used to examine the bioadhesive properties of microparticles *in vitro*. Bioadhesive silicon dioxide microparticles demonstrated greater adherence to Caco-2 cells as compared to unmodified particles.

Poly(methyl methacrylate) (PMMA) particles were fabricated by spinning PMMA (device layer) on to a clean silicon wafer (70). Positive lithography was carried out to define the device bodies followed by a reactive ion etching to carve the devices. Then a second mask positive photolithography was carried out to carve the device reservoir. The process flow of the device fabrication is shown in Fig. 16. The dimensions of the reservoir can be altered by changing the masked area and their depth can be modified by changing the time and/or flow rate of plasma in the RIE. By creating smaller reservoirs, a series of multiple reservoirs can be etched into the particles to create separate reservoirs for a combination of drugs or permeation enhancers. Since the PMMA is adherent to the surface of silicon by linkage to the native oxide layer, the wafer was soaked in basic solution to break this bonding and immediately release the particles. Bioadhesive properties were introduced to microfabricated PMMA microdevices by attachment of lectins, a group of proteins capable of specifically targeting cells in the GI tract. In this process, the PMMA microdevices were chemically modified by aminolysis to yield amine-terminated surfaces. Avidin molecules were covalently bound to the surface of the particles using a hydroxysuccinimide-catalyzed carbodiimide reagent and then incubated in an aqueous solution of biotinylated lectin. The bioadhesive characteristics of lectin-modified microdevices were successfully demonstrated *in vitro*.

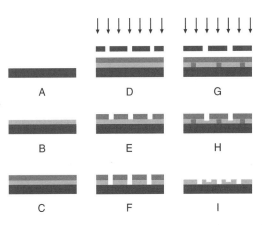

Figure 16. Process flow of PMMA microdevices with reservoir for oral drug delivery (70).

Microparticles for Intravenous Drug Delivery. The same microfabrication technology that has been used quite extensively for the fabrication of particles for oral drug delivery can be employed to develop precisely sized and shaped microparticles with high specific targeting abilities for IV delivery, especially for the treatment of diseases where oral and transdermal delivery are not effective. As an example, systemic chemotherapy using cytotoxic or biological treatment is the only treatment available for many patients with advanced metastatic cancer. While many tumors respond to initial courses of chemotherapy, after multiple courses and drugs, cancer cells become resistant to further therapy. In addition, growth of metastatic tumors is supported by factors, that are secreted by tumor cells themselves and cause angiogenic leaky vessels to grow. One strategy for preventing or treating metastatic tumors is to intervene in the process of angiogenesis by destroying the blood vessels that supply tumor cells rather than the tumor cells themselves (112). In such cases, precisely sized and shaped microparticles especially designed for IV delivery of cytotoxic biomolecules–drugs to the microvasculature of tumors with an improved safety profile could be employed. These have been described below.

Nonporous Microparticles. First generation of nonporous (solid) microparticles of silicon and silicon dioxide suitable for IV drug delivery (16,113), were rectangular shaped with thickness of 0.9 μm, and varied from 1 to 3 μm in length and width (Fig. 17). These microparticles were treated with amino- and mercaptosilanes, followed by coupling to human antibody (IgG) by using the heterobifunctional cross-linker succinimidyl 4-(N-maleimidolmethyl)-cyclohexane-l-carboxylate, to demonstrate their capability toward specific attachment of bioadhesive agents. These solid microparticles and their next generations are currently being explored for drug delivery and bioimaging applications (114).

Nanoporous Microparticles. Currently, porous silicon has begun to receive significant attention for biomedical usage. Nano- and microparticulates of this material have immense potential to be clinically and diagnostically significant both *in vivo* and *ex vivo* (115,116). Li et al. (113) demonstrated the incorporation, characterization, and release of cisplatin [*cis*-diammine dichloroplatinum(Il)],

carboplatin [cis-diammine (cyclobutane-1,1-dicarboxylato) platinum(Il)], and Pt(en)C12 [ethylenediamminedichloro platinum(Il)] within layers of calcium phosphate on porous Si–Si substrates for bone cancer treatment.

Superior control over particle dimensions, pore size, pore shape, and loading capacity is critical for microparticles for IR drug delivery (17,117). iMEDD Inc. has developed nanoporous microparticles (called IV-MEDDS or NK-MEDDS, where NK denotes the fact that the particles mimic Natural Killer cells) to treat systemically accessible solid tumors, specifically the multiple lesion sites associated with metastatic disease (71). The approach here is to kill the circulatory accessible endothelial cells that support the existing tumor capillaries using micromachined asymmetrical particles, that is, the top face of the particle contains a pore loaded with cytotoxic drugs, which is plugged with an erodible gelatinous material and layered with chemically grafted ligand (including growth factors, e.g., FGF, EGF and VEGF to bind endothelial or tumor cell receptors or folate and tumor-targeting RGD peptides to bind $\alpha_v \beta_3$ with high affinity) for targeting and protection. Designed to mimic the behavior of NK cells, a potent cytolytic agent, such as bee venom-derived melittin, can be plugged with a material designed to erode in 1–48 h. After injection, the particles circulate within the bloodstream for several minutes to several hours after that they are removed from the body's immune system. Bound particles should release their contents in the vicinity of the tumors and cause lysis and death of the target endothelial cells. Melittin peptides released by particles elsewhere in the body and not bound to endothelial target, are inactivated by binding to albumin and thus are not toxic to normal cells (71).

Based on the above-mentioned concept, Cohen et al. (118) prepared micron-sized particles with nanometer-sized pores out of porous silicon and porous silicon dioxide. The fabrication steps are shown in Figs. 18–20. The particles were fabricated with precise shapes and sizes. The size and thickness of these particles could be altered by changing the dimensions of the photolithography mask, the anodization time, and the electropolishing time. The

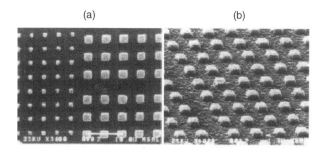

Figure 17. Scanning electron micrographs of microparticles. (a) Dimensions are $2.2 \times 2.1 \ (\pm 0.1)$ mm for the larger particles, and $1.2 \times 1.1 \ (\pm 0.05)$ mm for the smaller ones. (b) Shows tilted view of larger microparticles (104).

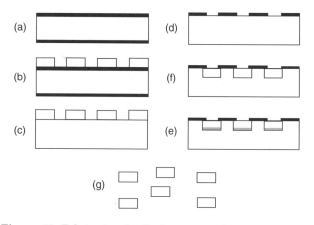

Figure 18. Fabrication details for porous silicon particles. (a) LPCVD silicon nitride deposition. (b) Photolithography. (c) Dry etch silicon nitride. (d) Piranha. (e) Anodization of silicon. (f) Electropolishing. (g) Particle release (118).

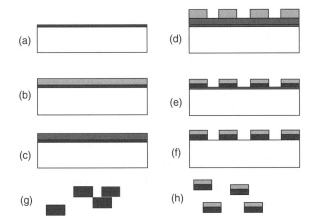

Figure 19. Process flow of porous SiO₂ particle fabrication. (a) Aluminum deposition. (b) Spun on mesoporous oxide film. (c) Baked mesoporous oxide film. (d) Photolithography. (f) Particle release in pirana. (g) Uncapped particles. (h) Particles capped with photoresist (118).

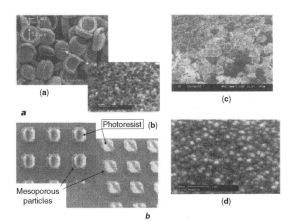

Figure 20. (a) SEM images of released porous silicon particles: Top image shows the shape and size of the particles. Bottom image demonstrates pores in the size range of 20–100 nm. (b) SEM images of mesoporous silicon oxide particles on wafer: (*a*) flat. (*b*) 45 tilt. (c) SEM images of released porous silicon dioxide particles. (d) SEM images of released porous silicon particles (118).

porous silicon dioxide particles were 4.7 μm squares with a thickness of 1.0 μm. The porosity of silicon dioxide particles was 52.5%. In order to determine safe particle size and concentration for IV drug delivery, a safety study was performed using solid silicon particles with various shapes, squares and circles, and varying sizes, 2, 5, and 10 μm. Results indicated that at concentrations of 1×10^7 particles per mouse, particles of size 2 and 5 μm safely circulate throughout the vasculature. No mice survived for any length of time when they were injected with 10 μm particles. Work is underway to demonstrate the coupling of EGF to porous dioxide particles that will allow for the particles to bind to the cells that express EGF receptors.

Smith et al. (114) prepared novel, controllably dual-sided, symmetric particulates of porous silicon from a polysilicon precursor. These particulates are precisely monodisperse on the scale of 1 μm (diameter and thickness) and may enable

unidirectional flow of transported drugs, proteins–peptides, nucleic acids, and so on. They may also facilitate controllably different intraparticle surface chemistries, and therefore potentially different types of antibodies, proteins, and so on, can be present on the same particle.

MEMS for Implantable Drug Delivery Devices. Implantable devices are preferred for the therapies that require many injections daily or weekly. The requirement and advantages of an implantable drug delivery device has been discussed above in greater detail. These devices can either be implanted into the human body or placed under the skin, consequently reducing the risk of infection by eliminating the need for frequent injections. Most of the implantable microsystems are expected not to cause pain or tissue trauma owing to their small size and are often virtually invisible. The advances in microfabricated implantable drug delivery device have been reviewed below.

Microreservoirs. Silicon microfabrication technology has been used to develop drug delivery device consisting of an array of microreservoirs (68,119,120) (Fig. 21). This device is currently being developed by MicroCHIPS, Inc., for use as external and implantable systems for the delivery of proteins, hormones, pain medications, and other pharmaceutical compounds (117). Each dosage is contained in a separate reservoir that is covered with a gold membrane. The membrane gets dissolved in the presence of chloride ions when anodic voltage is applied to the membrane of interest. This causes the membrane to weaken and rupture, allowing the drug within the reservoir to dissolve and diffuse into the surrounding tissues. This device allows the release of a potent substance in a pulsatile manner. Each microreservoir can be individually filled, so multiple substances can be delivered from a single MEMS device. Release of fluorescent dye and radiolabeled compounds has been demonstrated from these microreservoir devices *in vitro* in saline solution and serum (68).

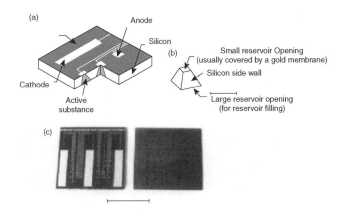

Figure 21. A schematic of a silicon microchip for controlled release. (a) Cut-away section showing anodes, cathodes, and reservoirs. (b) Shape of an individual reservoir. (c) Photograph of a prototype microchip: the electrode-containing frontside and the backside with openings for filling the reservoirs (15).

The release studies from this device demonstrated that the activation of each reservoir could be controlled individually, creating a possibility for achieving many complex release patterns. Varying amounts of chemical substances in solid, liquid, or gel form could be released into solution in either a pulsatile, a continuous, or a combination of both manners, either sequentially or simultaneously from a single device. Such a device has additional potential advantages including small size, quick response times, and low power consumption. In addition, all chemical substances to be released are stored in the reservoirs of the device itself, creating a possibility for the future development of autonomous devices. A microbattery, multiplexing circuitry, and memory could be integrated directly onto the device, allowing the entire device to be mounted onto the tip of a small probe, implanted, swallowed, integrated with microfluidic components to develop a laboratory-on-a-chip, or incorporated into a standard electronic package, depending on the particular application. Proper selection of biocompatible device materials may result in the development of an autonomous, controlled-release implant or a highly controllable tablet for drug delivery applications (68).

Nanoporous Silicon Membranes. Silicon nanopore membranes were developed by Ferrari and co-workers for application as immunoisolating biocapsules, and for molecular filtration (121–123). These membranes were shown to be sufficiently permeable to oxygen, insulin, and glucose, while at the same time impermeable to larger proteins, such as immunoglobulin G (IgG), which might lead to destruction of the transplanted cells (124). Since the diffusion through these membranes is linear, they can also be used for sustained drug delivery. This is currently being developed by iMEDD, Inc. (71,109). Over the years, nanopore technology has undergone continued improvements. Nevertheless, the basic structure and fabrication protocol for the nanopores has remained the same. The membrane area is made of thin layers of polysilicon, silicon dioxide, and/or single crystalline silicon depending on the design employed. The strategy used to make nano-size pores was based on the use of a sacrificial oxide layer sandwiched between two structural layers, for the definition of the pore pathways. The first design of nanoporous membranes consisted of a bilayer of polysilicon with L-shaped pore paths. The flow path of fluids and particles through the membrane is shown in (Fig. 22a) (125). As shown, fluid enters the pores through openings in the top polysilicon layer, travel laterally through the pores, make a 90° turn, and exit the pores through the bottom of the pore where both the top and bottom polysilicon layers lay on the etch stop layer. While this design performed well for preventing the diffusion of the larger, unwanted immune system molecules, its L-shaped path slowed down and, in some cases, prevented the diffusion of the smaller molecules of interest. The pores in this design were fairly long, which led to the slow diffusion of the desired molecules. Also, because of the large area per pore, it was difficult to increase the pore density and thus the diffusion rate. The next design had an improvement in the production of short, straight, vertical pores through a single-crystal base layer (Fig. 22b and c). This design had the advantage of direct flow paths. This

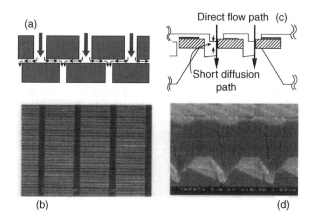

Figure 22. (a) Flow path through MI filters, with lateral diffusion through the nanopores defined by sacrificial oxide. (b) Cross-section of M2 design showing direct flow path. Scanning electron micrographs of microfabricated membrane: (c) top view detail; (d) side view detail (126).

direct path allows the smaller molecules of interest to diffuse much quicker through the membrane, while still size-separating the larger molecules. To further improve the reliability of the nanoporous membranes, several basic changes were made in the fabrication protocol from the previous membrane design to eliminate problems with the diffused etch stop layer (126). This design also incorporated a shorter diffusion path length, based on the thicknesses of the two structural layers. The design of a new membrane fabrication protocol incorporated several desired improvements: a well-defined etch stop layer, precise control of pore dimensions, and a lower stress state in the membrane. The new protocol also increased the exposed pore area of the membranes. The nanoporous membranes have been studied extensively for the use of drug delivery and the results are very encouraging.

Zero-Order Kinetics through Nanoporous Membrane. *In vitro* bovine serum albumin (BSA) release data through 13 nm pore is shown in Fig. 23. The experimental results show zero-order release profile (zero-order kinetics). Note that the zero-order kinetics does not follow Fick's law. Fick's laws are usually adequate to describe diffusion kinetics of solutes from a region of higher concentration to a region of lower concentration through a thin, semipermeable membrane. But, when the size of the membrane pores approaches that of the solute, an unexpected effect may occur, which deviate substantially from those predicted by Fick's laws. Diffusion of molecules in microporous media, such as zeolites, has led to experimental evidence of such unusual phenomena as molecular traffic control and single file diffusion (SFD) (127,128). Theoretical treatments and simulations suggest that in the case of SFD, solute molecules of equal size cannot pass each other in pores that approximate the dimensions of the molecule itself, regardless of the influence of concentration gradient, and thus their initial rate of movement (or flux) is underestimated by Fick's law (129–133).

The microfabricated nanopore channels are of molecular size in 1D, and therefore non-Fickian diffusion kinetics

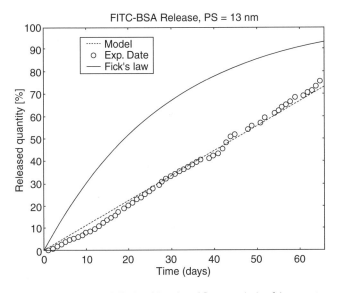

Figure 23. *In vitro* diffusion kinetics of fluorescein isothiocyanate (FITC) labeled BSA through 13 nm pore size: experimental data (o), Fick's law prediction (–), model-based simulation (...).

is observed. The observations are consistent with the diffusion reported for colloidal particles confined in closed 1D channels of micrometer scale where particle self-diffusion is non-Fickian for long time periods and the distribution of particle displacements is a Gaussian function (128). Zero-order flux is observed when a chamber filled with a solute is separated from a solute-free external medium by channels that are only several times wider than the hydrodynamic diameter of the individual molecules. The basic principle of diffusion as a mixing process with solutes free to undergo Brownian motion in three dimensions (3D) does not apply since in at least 1D solute movement within the nanopore is physically constrained by the channel walls. Experimental observations of colloidal particles in a density matched fluid confined between two flat plates reveal that particle diffusion becomes anisotropic near the interface; in this case leading to hindered diffusion as a consequence of constrained Brownian motion and hydrodynamic drag effects at distances close to the walls (134). In the case of nanoporous membranes, it is not entirely certain that the ordering of solutes imposed by the nanopore geometry will be as strict as true cylindrical pores, nor that the sequence of particles passing through the nanopores under the influence of the concentration gradient will remain unchanged over the time required to travel the 4 m length of the channel; particles could conceivably pass each other laterally. Whether a consequence of a SFD-like phenomenon or drag effects (or a combination of both), the nanopore membrane is rate limiting and, if properly tuned, restricts solute diffusion to a point that flux rates across the membrane are entirely independent of concentration gradient and follow zero-order kinetics.

In order to achieve further insight in the mechanisms involved in nanochannel diffusion, an experimental phenomenon in mathematical terms, thus yielding to the creation of a dynamical model, which makes it possible to simulate the diffusion experiments and fit the related data,

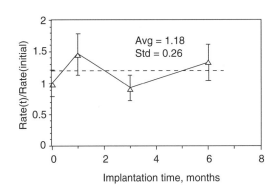

Figure 24. Ratio of post-preimplantation glucose diffusion rates.

is being investigated. A detailed description of such model is presented in Ref. 135.

Biocompatibility of Nanoporous Membranes. *In vivo* membrane biocompatibility was evaluated using glucose as a model molecule. Figure 24 shows the ratio of post-explantation glucose diffusion rate compared to its initial value. There was no noticeable change in glucose diffusion rates pre- and postimplantation illustrating that the silicon membranes did not foul over a 6-month implantation period. The membrane was placed on a titanium capsule and the entire assembly was placed subcutaneously in mice. The assembly was removed after 7 days and examined visually. There was no visible evidence of tissue binding to the surface. Figure 25 shows a photograph of the implant site after 30 days of implantation. As can be seen, only a thin vascular capsule forms around the implant as opposed to the avascular fibrous capsule. This minimal tissue response is supposed to be responsible for the comparable pre- and postimplantation glucose diffusion rates observed in this investigation.

Sandwich Design Filter. Nanochannels fabricated between two directly bonded silicon wafers were also developed for the applications as immunoisolating biocapsules, and molecular filtration (125,136–139). These devices possess high mechanical strength since the filtration occurs at the interface of two bonded silicon wafers instead of through a 1–10 μm thick membrane (in the case of silicon nanopores membrane). Well-developed bulk microfabrication

Figure 25. Photograph of implantation site after 30 days *in vivo*.

technology was used to fabricate these devices. With the use of a silicon dioxide sacrificial layer, pore sizes as small as 40 nm were fabricated with size variations < 4%. It was already established in the case of silicon nanopore membranes that the diffusion of molecules though nanopores is constant, and therefore the sandwich design filter can also be used for sustained drug delivery applications.

MOLECULAR DRUG DELIVERY SYSTEMS

This type of carrier, including cryptands, calixarenes (140), cyclophanes (141), spherands, cyclodextrins, and crown ethers, carry out chemical reactions that involve all intramolecular interactions where covalent bonds are not formed between interacting molecules, ions or radicals. Most of these reactions are of host–guest type.

Cyclodextrins: General Information

Between the several drug delivery systems, the molecular carrier have aroused great interest in the scientific world. Compared to all the molecular hosts mentioned above, cyclodextrins (CDs) are most important. As a result of molecular complexation phenomenon CDs are widely used in many industrial products, technologies, and analytical methods.

The CDs represent the more important molecular carrier today, in fact they are already strongly present in commerce for various types of drugs. Cyclodextrins have been discovered in the nineteenth century. They were produced for the first time by Villiers in 1881 by digesting the starch with Bacillus amylobacter, but only in 1903 was the cyclic structure of these compounds demonstrated by Schardinger.

Chemically, CDs are cyclic oligosaccharides, consisting of (α-1,4)-linked α-D-glucopyranose units. They are produced as a result of an intramolecular chain splitting reaction from degradation of starch by enzymes called cyclodextrins glucosyltransferases (CGTs) (142). In times past, only small amounts of CDs were generated and high production costs prevented their industrial application, but now most of the CGT genes have been cloned making the large scale production of this kind of carrier low cost.

The CDs are characterized by the presence of a lipophilic central cavity and a hydrophilic outer surface. The glucopyranose units are in the form of a chair and, for this reason, the CDs may be represented as a truncated cone. The OH groups are oriented with the primary hydroxyl groups of the various units of glucose on the narrow side of the cone and the secondary OH groups at the larger edge. The lipophilic character of the central cavity is determined by skeletal carbons and ethereal oxygens.

The CDs may contain even >15–16 glucopyranose units, but the must abundant natural CDs are α-cyclodextrin (α-CD), β-cyclodextrin (β-CD), and γ-cyclodextrin (γ-CD) containing six, seven, and eight glucopyranose units, respectively.

The CDs are chemically stable in alkaline solutions, but are susceptible to hydrolytic cleavage under strong acidic conditions, however, they are more resistant toward

Table 1. The Principal Physical Chemical Characteristics of Natural CDs

	α-CDs	β-CDs	γ-CDs
Molecular weight	972	1135	1297
Unites of glucopyranose	6	7	8
Internal diameter, Å	5	6	8
Solubility, mg · 100 mL^{-1}, 25 °C	14.2	1.85	23.2
Melting point, °C	250–255	250–265	240–245

acid-catalyzed hydrolysis than linear dextrins and the hydrolytic rate decreases with decreasing cavity size (143). The rate of both the nonenzymatic and enzymatic hydrolysis is decreased when the cavity is occupied by drug molecule.

Table 1 reports the principal physical–chemical characteristics of natural CDs. Natural CDs, in particular β-CD, have aqueous solubility much lower with respect to comparable linear or branched dextrins. This is probably due to the relatively strong binding of the CDs molecules in the crystal state (i.e., relatively high crystal lattice energy). Moreover, β-CD form intramolecular hydrogen bonding between the secondary hydroxyl groups that reduces the number of hydroxyl groups capable of forming hydrogen bonds with the surrounding water molecules (142). This low aqueous solubility may cause precipitation of solid CDs complexes.

The most important characteristics of CDs is their ability to form inclusion complexes both in solution and in the solid state, in which the guest molecule places its self in the hydrophobic cavity hiding from the aqueous environment. This leads to a modification of physical, chemical, and biological properties of the guest molecules, but principally of the aqueous solubility.

The β-CD is the most useful pharmaceutical complexing agent principally because of its low cost and easy production. It contains 21 hydroxyls groups, of which 7 are primary and 14 are secondary (Fig. 26). All the OH groups are reactive enough to be used as points of reaction for structural modifications, allowing the introduction of several functional groups in to the natural macrocyclic

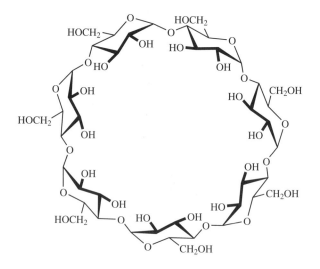

Figure 26. Structure of β-Cyclodextrin.

Table 2. Pharmaceutically Useful β-CD Derivatives[a]

Derivative	Position of Substituent	Substituent
Hydrophilic Derivatives		
Methylated β-CD	2,6-; 2,3,6-	$-O-CH_3$
Hydroxyalkylated β-CD	Random	$-O-CH_3 -CH(OH)-CH_3$
Branched β-CD	6-	$-Glucosyl, -maltosyl$
Hydrophobic Derivatives		
Ethylated β-CD	2,6-; 2,3,6-	$-O-C_2H_5$
Peracylated β-CD	2,3,6-	$-O-CO(CH_2)_n-CH_3$
Ionizable Derivatives		
Carboxyalkyl β-CD	Random	$-O-(CH_2)-COONa$
Carboxymethyl; ethyl	2,6-; 3-	$-O-CH_2-COONa;$ $-O-C_2H_5$
Sulfates	Random	$-O-SO_3Na$
Alkylsulfonates	Random	$-O-(CH_2)_n-SO_3Na$

[a]Obtained by substitution of the OH groups located on the edge of the CD ring. From Ref. 144.

ring. In the past few years, research has led to a great a number of modified CDs having better characteristics with respect to natural CDs (Table 2).

To date, ~100 different CD derivatives are commercially available, but only few of those derivatives have gone through toxicological evaluations and are available as bulk chemicals for pharmaceutical use. In particular, CDs currently used in drug formulation, derived from natural CD, include 2-hydroxypropyl-β-CD (HP-β-CD), randomly methylated βCD (RM-β-CD), sulfobutylether βCD (SBE-β-CD), maltosyl βCD (ML-β-CD), and (hydroxyethylβCD (HE-β-CD). The aqueous solubility of all these cited derivatives is $>50 \, g \cdot 100 \, mL^{-1}$.

Inclusion Complex Formation. The most important feature of CDs is their ability to form solid inclusion complexes (of the host–guest type) with a very wide range of solid, liquid, and gaseous lipophilic compounds by a phenomenon of molecular complexation (145). In these complexes, a guest molecule is kept within the cavity of the CD. Complex formation is a dimensional fit between the host cavity and the guest molecule (146). The lipophilic cavity of CDs molecules supplies a microenvironment where an appropriately sized lipophilic moietie can enter to form an inclusion complex (147). The formation of the complex never involves the formation or the breaking of covalent bonds (148). The main driving force for this kind of process is the replacement of enthalpy-rich water molecules contained in the cavity with more hydrophobic guest molecules present in the solution to attain an apolar–apolar association and decrease of CD ring strain resulting in a more stable lower energy state (142). The binding of guest molecules within the host CD is not permanent and is characterized by a dynamic equilibrium, whose strength depends on how well the host–guest complex fits together and on specific local interactions between surface atoms. More specifically, in aqueous solution the CD–drug complexes are constantly being formed and broken.

In particular, considering a 1:1 complexation, the association is usually described by the following equilibrium:

$$D_f + CD_f \overset{K}{\rightleftharpoons} DCD$$

The most important parameters that influence the inclusion process are the complexation strength or stability constant (K) defined by this equilibrium and equation 1, where (CD_f) and (D_f) are the concentrations of free CD and free drug, respectively; the other parameter is the lifetime (t) of the complex and equation 2, measured when the equilibrium is perturbed. The constants k_f and k_r are the forward and reverse rate constants, respectively, and k_{obs} is the observed rate constant for the reestablishment of the equilibrium after its perturbation.

$$K = \frac{k_f}{k_r} = \frac{[DCD]}{[D_f][CD_f]} \qquad (1)$$

$$K_{obs} = \frac{l}{\tau} = k_f([CD_f] + [D_f]) + k_r \qquad (2)$$

The CDs are able to complex the lipophilic substances both in solution, in this case water is the solvent of choice, or in the crystalline state. In some particular cases, the complexation may be performed also in the presence of any nonaqueous solvent, even if in this case a competition drug–solvent for the complexation may happen.

The inclusion of a drug in CDs lead to a profound change of its physicochemical properties as it is temporarily blocked within the host cavity giving rise to beneficial modifications of guest molecules, which are not achievable otherwise (149). In particular, the more influenced properties are enhanced solubility in water of highly insoluble guests, stabilization of labile guests against the degradative effects of oxidation, visible or ultraviolet (UV) light and heat, control of volatility and sublimation, physical isolation of incompatible compounds, chromatographic separations, taste modification by masking off flavors, unpleasant odors, and controlled release of drugs and flavors. Therefore, cyclodextrins may be used in several field: in food, pharmaceuticals, cosmetics, environment protection, bioconversion, packing, and textile industry.

The substances that may be complexed in CDs are quite varied and includes such compounds as straight- or branched-chain aliphatics, aldehydes, ketones, alcohols, organic acids, fatty acids, aromatics, gases, and polar compounds (e.g., halogens, oxyacids, and amines) (149).

Main Methods of Preparation of Drug–Cyclodextrins Complex. The CD complexes may be prepared with various methods. In solution, the complexes are prepared by addition of an excess amount of the drug to an aqueous CD solution. The suspension formed is equilibrated (for periods of up to 1 week at the desired temperature) and then filtered or centrifuged to form a clear CD–drug complex solution. Then, the water is removed from this solution by evaporation or sublimation, for examples, spray or freeze drying to obtain a solid complex.

Other methods applied to prepare solid CD–drug complex include kneading and slurry methods, coprecipitaion, neutralization, and grinding techniques (150). In some cases, the complexation efficiency is not very high, and therefore relatively large amounts of CDs are needed to complex small amounts of a given drug. Moreover, various vehicle constituents, such as surfactants, lipids, organic solvents, buffer salts, and preservatives, often reduce the efficiency. However, it is possible to enhance the efficiency through formation of multicomponent complex systems (151). For example, recent research demonstrated that water-soluble polymers are able to enhance the complexation efficacy of a wide variety of guest molecules, through stabilization of the CD–drug complex, and to increase the aqueous solubility of the natural cyclodextrins (152).

Analytical Methods Used to Detect the Complex Formation. Following the preparation of a drug–CD complex, a fundamental step is to verify this complexation, the stoichiometry of the complex, and its stability constant. All these parameters may be clarified by mean of several technique: thin-layer chromatography (TLC) (153), high performance liquid cromatography (HPLC) (154); gas chromatography–mass spectrometry (GC–MS) for the appraisal of the pattern of substitution (155); nuclear magnetic resonance (NMR) (156); circular dicroism (CD) (157) differential scanning calorimetry, X-ray diffraction, ultraviolet (UV) spectrometry (158), capillary zone electrophoresis (159); electrokinetic chromatography (160); GC stationary phase (161); light scattering and cryoelectronic microscopy (162).

Toxicological Profile. An important limitation for the pharmaceutical application of a substance (both drug or excipient) is the appearence of toxicity after its administration. For this reason, a fervent field of research has been the evaluation of a toxicological profile of CDs. Recently, a review has been published showing the adverse effects from CDs (163).

In general, oral administration of CDs does not show any toxic effect, due to lack of absorption from the GI tract.

Natural CDs, α- and β-CDs, as well as many of their alkylated derivatives, show significant renal toxicity, and for this reason are not used for parenteral use (164). A number of safety evaluations have shown that HP-β-CD, SB-E-βCD, ML-β-CD, γ-CD, and HP-β-CD appear to be suitable in parenteral, as well as oral formulations (164,165). However, the lack of available toxicological data will, more than anything else, hinder pharmaceutical applications of CDs.

Cyclodextrins Elimination. Both HP-β-CD and SBE -β-CD are quantitatively cleared unmodified by renal filtration. Following IV administration, these cyclodextrins have an half-life of 1 h (for humans, this value is specie dependent) with the major amount present in the urine between 1 and 4 h after administration.

The elimination of an unmodified CD–drug complex may lead to an increase in renal clearance of unchanged drug. The drug elimination may occur also following another mechanism. Water reabsorption physiologically occurs in the proximal and distal tubules leading to about a 100-fold increase in the concentration of filtered molecules. In this process, lipophilic drugs normally undergo passive reabsorption while polar molecules are only concentrated. This concentration, encourages the complex formation between the renal cleared cyclodextrins and any lipophilic molecule remaining in the kidney tubules. Since the complex is polar, the presence of the CDs is able to inhibit passive reabsorption of lipophilic drugs physiologically present resulting in greater renal clearance of lipophilic molecules.

Cyclodextrins as Drug Delivery Systems

The most classic application of CDs is in drug delivery. The CDs offer significant advantages over standard formulation. The CD–drug complexes can stabilize, enhance the solubility, bioavailability, and diminish the adverse effects of a drug. The bioadaptability and multifunctional characteristics of CDs make them able to minimize the undesirable properties of drugs in various routes of administration including oral, rectal, nasal, ocular, transdermal, and dermal. In Table 3, the role of CDs in drug formulation and delivery is reported in detail (166).

Cyclodextrins in Nasal Drug Delivery. Nasal administration of drugs is an important route of administration for several classes of drugs. Unfortunately, mucosa present both physical and metabolic barriers to drug permeation, restricting this therapeutically approach. The CDs may be used to overcome these obstacles. It has been demonstrated that CDs are able to reduce or to minimize the enzymatic activity of nasal mucosa (167).

Some morphological studies have shown that the methylated β-CDs are useful carriers for nasal drug delivery. Their effects on the mucosa are not significantly different to that of physiological saline and smaller than those of benzalkonium chloride, a worldwide used preservative for nasal drug formulations.

After nasal administration of a drug–CD formulation, only the drug permeates through the nasal epithelium, but not the highly water-soluble CD or its complex. In humans, DM-β-CD is hardly absorbed after nasal administration of a solution containing $\sim$ 5% of dimethyl-β-CD. Four percent of the nasally administered dose was recovered in the urine (168).The fraction of the CD dose that is not absorbed from the nasal cavity is removed by the nasal mucociliary clearance system.

Moreover, studies of permeation of various lipophilic drugs complexed in CDs demonstrate that they can largely improve the permeation through nasal mucosa of these substances both in the case of polypeptides and proteins (169).

Cyclodextrins in Ophthalmic Drug Delivery. Tear fluid contains a large variety and amount of enzymes that influence the permeation of topical applied drugs. Numerous studies have shown that CDs are useful additives in ophthalmic formulations because they are able to increase the aqueous solubility, stability of ophthalmic drugs, and to decrease drug irritation (170).

The CD complexation of water-soluble drugs (in order to modify an adverse property, e.g., increase their chemical stability or to decrease ophthalmic drug irritation) generally

Table 3. Role of Cyclodextrins in Drug Delivery[a]

Improved Drug Functions by CD Complexation	Example Drug	Type of CD
Increase in bioavailability (by increased solubility and stability)	Thalidomide	Natural CDs
As above	Nimuselide	β-CD, 2HP-β-CD
As above	Prednisolone	SBE-7-β-CD
As above	Oteprednol etabonate	γ-CD
As above	Sulfhamethazole	β-CD and HP-CD
As above	Tacrolimus	Natural and hydrophilic CDs
As above	Artemisin	β- and γ-CD
As above	Prostaglandin E1	Sulfobutly ether β-CD
Increase in solid-stability of amorphous drug	Quinapril	β-CDs
Increased absorption		
Oral delivery	Ketoconazole, testosterone	β-CD and HPβ-CD
Rectal delivery	Flurbiprofen, carmafur biphenyl acetic acid	2 HPβ-CD
Nasal delivery	Morphine, antiviral drug and insulin	2HPβ-CD
Transdermal delivery	Prostagalndin E1	6-*O*-(carboxymethyl) *O*-ethylβ-CD
Ocular delivery	Dexamethasone, Carbonicanhydrase inhibitors	2HPβ-CD β-CD
Protein and peptide delivery	Growth hormone, interleukin-2, aspartame, albumin and MABs	Different modified CDs
Reduction of local irritancy and toxicity	Pilocarpine, phenothiazine euroleptics, *all-trans*-retenoic acid	2 HPβ-CD (2,6-diOmethyl) β-CD and β-CD

[a]Modified from Ref. 162.

decrease the ophthalmic bioavailability (171); in the case of water-poor soluble drugs, the bioavailability strongly depends on the amount of CDs present: It is fundamental to use only the minimum quantity to form a complex (to solubilize) of the drug (172). In fact, the amount of CDs in excess reduces the free drug and, as a result the ocular permeation will decrease. An example is given in the research of Jarho et al. 1996 (173), which measured the permeability of arachidonyl-ethanolamide through isolated rabbit cornea. As reported in Fig. 27 the maximum value of permeability was found when the minimum amount of CD was used to maintain the drug in solution.

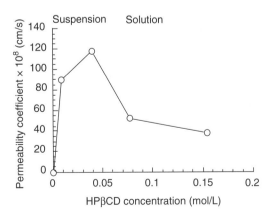

Figure 27. Permeability of arachidonylethanolamide through isolated rabbit cornea as a function of HP-β-CD concentration. The vehicle consisted of 0.5 mg · mL^{-1} suspension or solution of the drug in water containing 0–1.155 *M* HP-β-CD (173).

Moreover, this reduced bioavailability may be attributed to a too rapid ocular clearance (few minutes), and it has been demonstrated that by increasing the viscosity of the ophthalmic formulation, this obstacle may be reduced (174). In addition, a lot of substances used to increase the viscosity of aqueous solutions have been shown good ability to increase the complexation efficacy of CDs and, thus, the amount of CD needed to obtain adequate drug solubility can be decreased significantly when water-soluble polymer is present in the formulation (175).

Cyclodextrins in Dermal and Transdermal Drug Delivery. The CDs are relatively large molecules, and consequently both they and their complexes are not able to permeate through intact skin easily. Only 0.02% of the applied dose of radiolabeled HP-β-CD permeated through human skin. The principal barrier to the permeation of CDs is represented by the stratum corneum, since by stripping it, it is possible to enhance the percutaneous permeation by 24% (176). Lipophilic CDs (as DM-β-CD and RM-β-CD) are absorbed to a greater extent, but this absorption is still of little significance (0.3% of the applied dose).

The CDs are able to interact with some components of skin lipids. In particular, it has been demonstrated that pure aqueous solution of β-CDs and RM-β-CDs ad HPβCD are able to extract the lipids present in stratum corneum (177). Various studies (178,179) have shown that excess CDs, more than needed to complex the lipophilic drug, lead to a decrease of drug permeation through the skin (Fig. 28). When the drug (hydrocortisone) was in suspension, the increase of the cyclodextrin concentration lead to an increase of the flux through the skin. When all hydrocortisone was in solution,

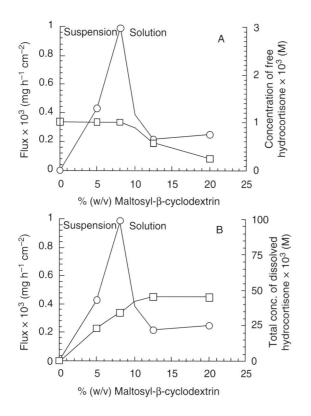

Figure 28. The effect of the maltosyl-β-cyclodextrin (ML-β-CD) concentration on the flux (○) of hydrocortisone through hairless mouse skin. The hydrocortisone concentration was kept constant at 0.045 *M*. (a) The flux in relation to the amount of free hydrocortisone (□) in the donor phase. (b) The flux in relation to the total amount of dissolved hydrocortisone (□) in the donor phase. The donor phase consisted of aqueous hydrocortisone suspension at ML-β-CD concentrations < 8% (w/v), but hydrocortisone solution at higher ML-β-CD concentrations (175).

the increase of the cyclodextrin content led to a decrease in flux. In all cases, maximum flux through the skin was obtained when just enough cyclodextrin was added to the vehicle to keep all hydrocortisone in solution. The mechanism has been already elucidated in a previous section.

A great number of topical drugs have been complexed with CDs (Tables 4 and 5). In every case, it has been demonstrated that CDs can markedly enhance the dermal delivery of lipophilic drugs (e.g., corticoids and NSAIDS). In particular, from a comparative evaluation (181) among the compounds DM-β-CD, β-CD, and HP-β-CD, resulted in HP-β-CD being more able to increase the percutaneous permeation (Fig. 29).

The effects of CDs on the permeation rates of drugs through the skin may be determined by both the increase of thermodynamic activity of drugs in a vehicle (in particular it is referred to the escaping tendency of drugs, and it is supposed that increasing this activity will lead to the augmentation of the permeation rate of drugs through the skin; moreover, the thermodynamic activity is proportional to the solubility of drugs in its vehicle and is maximal just in the saturated solution), the extraction of skin component, and the partition coefficient of the drug between skin and vehicle.

Recently, it has been evaluated for transdermal use in peracylated CDs with medium alkyl chain length (C_4–C_6) and in particular 2,3,6-tri-*O*-valeryl-β-CD (TV-β-CD). This particular type of CD shows the property of forming a film, and for this reason is very promising in transdermal preparations.

Cyclodextrins in Rectal Drug Delivery. The CDs have been studied to optimize the rectal delivery of drugs for systemic use. Table 6 reports the CDs and drugs investigated for rectal application. The effects of CDs on the rectal delivery of drugs depends on vehicle type (hydrophilic or oleaginous), physicochemical properties of the complexes, and an existence of excipients, such as viscous polymer. The enhancement of rectal permeation of lipophilic drugs made by CDs is generally attributed to the improvement of the release from vehicles and the dissolution rates in rectal fluids (Fig. 30). In the case of inabsorbable drugs, such as antibiotics, peptide, and proteins, the rectal delivery is based on the direct action of CDs on the rectal epithelial cells.

On the other hand, the prolonging effects of CDs on the drug levels in blood are caused by several factors: sustained release from the vehicles, slower dissolution rates in the rectal fluid, and retardation in the rectal absorption of drugs by an inabsorbable complex formation.

Another important aspect that has been evaluated is the stabilization of drugs in rectal delivery. The complexation of drugs with CDs has been used to improve the chemical stability in suppository bases according to a stabilizing effects (principally of β-CD and DM-β-CD) attributable to a poor solubilization of drugs in the oleaginous suppository base; this may lead to a difficult interaction of drugs with the base.

The CDs may inhibit the bioconversion of drugs in the rectum (182) leading to the alleviation of the rectal irritancy of the some drugs as NSAIDS (Fig. 31).

Cyclodextrins in Oral Drug Delivery. An important parameter to be considered for the oral administration of a drug is its water solubility. For poorly aqueous soluble drugs, CDs are able to increase the aqueous solubility and thus enhance its dissolution rate and the biopharmaceutical parameters. An example was observed with the celecoxib (a nonsteroidal antinflammatory drug). The complex celecoxib DM-β-CD showed an increased permeation respect to the free drug. This observed enhanced permeation was due to the fast dissolution rate of the included drug and to a destabilizing action exerted by the CD on the biomembrane (Fig. 32) (183).

The CD derivatives may be used in order to modify drug release of oral preparations. Table 6 reports some application of CDs.

The hydrophilic CDs are able to give an immediate release of the complexed drug, while hydrophobic CDs are useful for the prolonged release formulations. The use of *O*-carbossimethyl-*O*-ethyl-β-CD (CM-β-CD) gives a delayed release formulation.

The immediate release formulation is required in an emergency situations and in a particular way in the administration of analgesics, coronaric antipyretics, and

Table 4. The Use of Parent Cyclodextrins in Transdermal Route[a]

CDs	Abbreviation	Improvement	Drugs
α-Cyclodextrin	α-CD	Release and/or permeation Stability	Miconazole Tixoxortol 17-butyrate 21-propionate Betamethasone 4-biphenylacetic acid Chloramphenicol Ciprofloxacin Ethyl 4-biphenylyl acetate Flurbiprofen Hydrocortisone
β-Cyclodextrin	β-CD	Release and/or permeation	Indomethacin Nitroglycerin Norfioxacin Piroxicam Prednisolone Prostaglandin E_1 Sulfanilic acid
		Local irritation	Chlorpromazine hydrochloride Tretinoin
γ-Cyclodextrin	γ-CD	Release and/or permeation	Beclomethazone dipropionate Betamethasone Menadione Predonisolone

[a]Adapted from Ref. 180.

Table 5. The Use of Cyclodextrin Derivatives in Transdermal Route[a]

CD Derivatives	Abbreviation	Improvment	Drugs
Dimethyl-β-cyclodextrin	DM-β-CD	Release and/or permeation	4-Biphenylacetic acid Ethyl 4-biphenylyl acetate Indomethacin Predonisolene Sufanilic acid
		Local irritation	Chlorpromazine
Random methyl-β-cyclodextrin	RM-β-CD	Release and/or permeation	Acitretin Hydrocortisone Piribedil S-9977
Hydroxypropyl-β-cyclodextrin	HP-β-CD	Release and/or permeation	4-Biphenylacetic acid Dexamethasone 17β-estradiol Ethyl 4-biphenylyl acetate Hydrocortisone Liarozole Miconazole
Maltosyl-β-cyclodextrin	G_2-β-CD	Release and/or permeation	Hydrocortisone
β-Cyclodextrin polymer	β-CD polymer	Release and/or permeation	Tolnaftate Indomethacin
Diethyl-β-cyclodextrin	DE-β-CD	Release and/or permeation	Nitroglycerin
Carboxymethyl-β-cyclodextrin	CM-β-CD	Release and/or permeation	Hydrocortisone
Carboxymethyl-ethyl-β-cyclodextrin	CME-β-CD	Release and/or permeation	Prostaglandin

[a]Adapted from Ref. 180.

vasodilatators. Hydrophilic CDs have been used in order to improve the oral bioavailability of the previous mentioned drugs (184). The improvement is mainly dependent on the increase of solubility and wettability of drugs through the formation of inclusion complexes (185).

The oral bioavailability of a lipophilic drug from the CD complex may be optimized varying several factors, that influence the equilibrium of dissociation of the complex (166,186). The maximum improvement of the absorption is obtained when a sufficient amount of CD is used to complex all the molecules of the drug present in suspension, and the further addition of CDs lead to a reduction the free fraction of the drug and, therefore, reduces the bioavailability of the drug. Moreover, drug

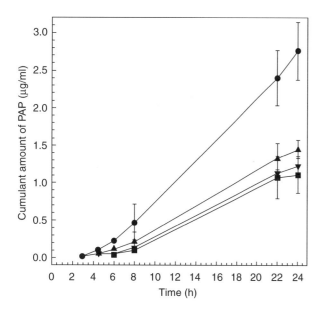

Figure 29. Total amount of free or complexed papaverine permeated through abdominal rat skin. Symbols: ν free PAP alone; ▲ PAP-HP-β-CyD; ● PAP-β-CyD; λ PAP-DM-β-CyD (179).

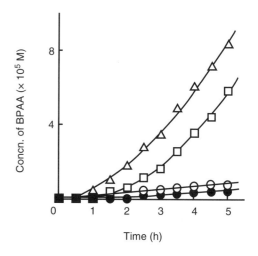

Figure 30. Profiles of BPAA permeation through isolated rat rectum after applications of suppositories containing EBA or its β-CD complexes in isotonic phosphate buffer (pH 7.4) at 37 °C. ● without β-CDs; ○ with β-CD; □ with DM-β-CD; Δ with HP-β- CD (182).

formulations contain a certain amount of excipients that may compete with the drug for the cavity of the CD. Competition may also happen with the endogenous substances present in the absorption site. The replacement

of the drug from the cavity of the CD from both endogenous and exogenous substances lead to an acceleration of the absorption of the drug (187,188). Early studies showed that in some cases the improvement of oral bioavailability is principally due to a stabilizing effect of CDs on labile drugs (189).

Table 6. The Use of Cyclodextrins in Rectal Delivery[a]

CDs	Improvement	Drugs
α-CD	Stability	Morphine hydrochloride
	Release and/or permeation	Cefmetazole
		G-CSF
		Morphine hydrochloride
	Stability	AD1590
		Carmoful
β-CD		Ethyl 4-biphenylyl acetate
	Release and/or permeation	4-Biphenylacetic acid
		Ethyl 4-biphenylyl acetate
		Naproxen
		Phenobarbital
		Piroxicum
γ-CD	Release and/or permeation	Diazepam
		Flurbiprofen
	Release and/or permeation	4-Biphenylacetic acid
		Carmoful
		Diazepam
DM-β-CD		Ethyl 4-biphenylyl acetate
		Flurbiprofen
		Insulin
	Local irritation	4-Biphenylacetic acid
		Ethyl 4-biphenylyl acetate
TM-β-CD	Release and/or permeation	Carmoful
		Diazepam
		Flurbiprofen
HP-β-CD	Release and/or permeation	4-Biphenylacetic acid
		Diazepam
		Ethyl 4-biphenylyl acetate
β-CD Polymer	Selective transfer into lymphatics	Carmofur

[a]Adapted from Ref. 180.

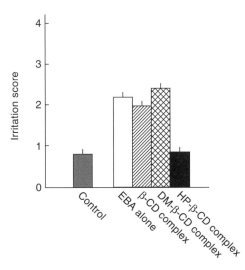

Figure 31. Irritation effect of suppositories containing EBA or its β-CD complexes (equivalent to BPAA 10 mg·kg⁻¹) on rectal mucosa in rats 12 h after multiple (four times at 12 h intervals) administration (182).

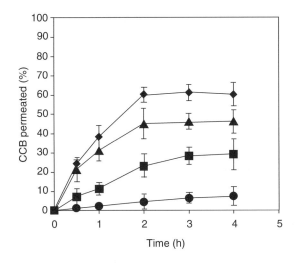

Figure 32. Permeation profiles of CCB alone or in the presence of the CCB-DM-β-CD complex in a different molar ratio. (●) CCB alone; (●) CCB-DM-β-CD 1:2 complex as suspension; (▲) CCB-DM-β-CD 1:5 complex as solution;(◆) CCB-DM-β-CD 1:10 complex as solution (181).

Others positive results may be obtained for the sublingual administration of drugs, complexed with CD (190,191). In this application, not only the drug permeates rapidly giving an immediate response, but it also avoid hepatic first pass metabolism.

The preparations characterized by a slow release are planned for having a zero-order release to guarantee a constant blood level for a long period of time. This type of formulation has many advantages, like the reduction of the administration frequency with an extension of the efficacy of the drug and the reduction of the toxicity associated with the administration of a simple dose. Hydrophobic CDs, as alkylated and acylated derivatives, are used as slow-release carriers for hydrophilic drugs. Between the

alkylated CDs, 2,6-di-O-ethyl-β-CD (DE-β-CD) and 2,3,6-tri-O-ethyl-β-CD (TE-β-CD) were the first used as slow-release carriers (192).

Another type of CD useful in oral formulation is 2,3,6-tri-O-butyryle-β-CD (TB-β-CD), whose bioadhesive property make it very advantageous in oral and transmucosal formulations.

Cyclodextrins in Parenteral Administration. Modern technology is trying to obtain semisynthetic CDs that have the following characteristics to use as parenteral drug delivery systems. For this application, the drug toxicity at high doses will need to be improved for chronic treatment; its inability to react with cholesterol, phospholipids, or others members of the cellular membrane, and its biodegradability in circulation and elimination of small molecular metabolites.

In general, for this kind of application only hydrophilic CDs, in particular HP-βCD are used. This has been, carefully studied by means of innumerable toxicological experiments and has been the object of numerous clinical tests on human. One formulation, based on the carrier Sporanox by Jassen (193) has been approved by the U.S. Food and Drug Administration (FDA). Another hydrophilic derivative is used for parenteral use is β-CD sulfobutylether. It is used under the name of Captisol.

The sulfate CD represent another class of soluble CD in water with a characteristic biological activity. It shows an antiangiogenic power that may be useful in new therapies against cancer. A few studies have demonstrated that the sulfate CD does not have any hemolytic properties at all, are not toxic, and protect against the nephrotoxicity induced by gentamicin without even reducing renal accumulation of this active principle (194).

Cyclodextrins in Anticancer Therapy. Cyclodextrins also play a vital role in the drug formulation design for cancer therapy. Bekers et al. (195) in 1991 studied the effect of cyclodextrins on the chemical stability of mitomycin C, a clinically useful anticancer drug able to generate severe dermatological problems after administration. The complexation of this drug with CD reduced the skin necrosis observed after the treatment with the free drug.

Real advantages were demonstrated in the delivery of paclitaxel, an anticancer agent used in breast, ovarian, lung, head and neck cancers, characterized by very low water solubility. For this reason, it must be formulated as a micellar solution made up of polyoxyethylated castor oil and 50% absolute ethanol. This formulation triggers severe acute adverse effects in both animals and humans. The complexation of paclitaxel in CDs, β-CD, DM-β-CD, and TM-β-CD, showed a modulation of the maintainance of the anticancer activity (196).

Cyclodextrins as Carrier for Biological Drugs. Besides drugs, different peptides and proteins (197), oligosaccharides, and oligonucleotides (198) are also delivered by the formation of inclusion complexes with cyclodextrins because of CDs ability of interacting with cellular membranes and giving rise to improved cellular uptake. The most recent usage of cyclodextrins lies in the ability of these agents to deliver agents, such as plasmids, viral vectors, and anti-sense constructs. The *in vitro* stability of antisense molecules is increased by binding to CDs, such as hydroxypropyl b-CD. A two- to threefold increase in the cellular uptake of antisense constructs by hydroxyalkylated b-CD has been noted in human T-cell leukemia H9 cells (199). Certain CDs modulate the intracellular distribution or activity of antisense molecules and they may be used for reversal of atherosclerosis (200). Cyclodextrins are also used to formulate the enhancement of the physical stability of viral vectors for gene therapy by suspending the adenovirus and adeno-associated virus in blends of CD, complex carbohydrates, and various surfactants (201). Three native CDs (α, β, and γ) were observed to improve the antiviral effect of ganciclovir on two human cytomegalovirus strains (202). Use of CDs as carriers of antiviral drugs appears to be a good alternative to traditional treatments as it allows the administration of lower doses and reduces the toxic effect of drug molecules.

Cyclodextrins in Colon Targeting.   Colon targeting may be classified as a delayed release with a fairly long time because the time required to reach the colon is ~8h in humans (203). When a formulation is administered orally, it will dissociate in the GI fluid and for this reason CD complexes are not suitable for colon delivery. For this reason, it was proposed to use CD–drug conjugates (a prodrug) that were able to survive the passage through the stomach and small intestine. In particular, the linkage of CD to biphelylyacetic acid (BPAA) has been investigated. It is interesting to note that the solubility of this type of prodrug is strictly related to the cavity size of the CD. Moreover, in the case of ester-type conjugates, drug release is the case of ester-type conjugates, drug release is triggered by the ring opening of CDs, which consequently provides site-specific drug delivery to the colon. On the other hand, the amide conjugates do not release the drug even in the cecum and colon, despite the ring opening of CDs. The amide linkage of the small saccharide–drug conjugates may be resistant to bacterial enzymes and poorly absorable from the intestinal tract due to high hydrophilicity. Therefore, the ester-type conjugate is preferable as a delayed release-type prodrug that can release a parent drug selectively in the cecum and colon (204).

SUPRAMOLECULAR AGGREGATES FOR DRUG DELIVERY

General Characteristics of Surfactants

Surfactants are molecules characterized by a polar head and an apolar tail region, the latter occupies the larger molecular volume, in a particular way for ionic surfactants. When dispersed in water, surfactants self-associate into a variety of equilibrium phases, the nature of which stems directly from the interplay of the various (forces inter- and intermolecular), as well as entropy evaluations. Surfactants also self-associate in nonaqueous solvents, particularly apolar liquids, such as alkanes. In this case, the orientation of the surfactant molecules are reversed compared to those observed in aqueous solution. This reorientation lead to a lowering of the free energy of the system

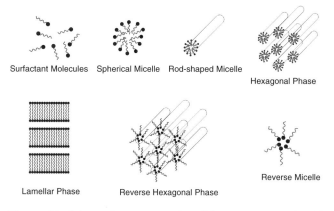

Figure 33. Schematic representation of the most commonly self association structures in water, oil or a combination thereof (205).

overall. When surfactants are incorporated into two immiscible solvents as oil and water, the surfactant molecules locate themselves at the oil–water interface. This arrangement is thermodynamically favorable.

Figure 33 reports a number of possible self-association structures that surfactant may form when placed in a oil and water (205).

Microemulsions

The microemulsion concept was introduced as early as the 1940s by Hoar and Schulman, who generated a clear single-phase solution by titrating a milky emulsion with hexanol (206). Later, in 1959, Schulman coined the term microemulsion (207). Today microemulsions are defined as A mixture of water, oil, and amphiphile substances forming a single optically, isotropic and thermodynamically stable liquid solution. The stability is the most important difference between emulsions and microemulsions. In fact, emulsions are fundamentally thermodynamically unstable and, even if they show an excellent kinetic stability, may undergo phase separation (208). Another important difference is related to their appearance. Emulsions are milky while microemulsions are clear or translucent. In addition, there is a noticeable difference in their method of preparation, since emulsions require a large input of energy while microemulsions do not. Microemulsions are dynamic systems in which the interface is continuously and spontaneously fluctuating (209).

Schematic representations of the three types of microemulsions are most likely formed are reported in Fig. 34. The structures shown are very different, but in each there is an interfacial surfactant monolayer separating the oil and water domains.

Three approaches have been proposed to explain the spontaneous microemulsion formation and their consequent stability: interfacial or mixed-film theories (210); solubilization theories (211); and thermodynamic treatments (212). In particular, the free energy of microemulsion formation reported in equation 3 is dependent on the extent to which surfactant is able to lower the surface tension of the oil–water interface and the change in

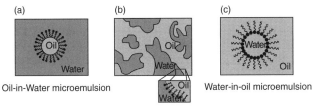

Figure 34. Schematic representation of three type of microemulsion microstructures: (a) oil-in-water, (b) bicontinuous, and (c) water-in-oil microemulsion (205).

entropy of the system such that,

$$\Delta G_f = \gamma \Delta A - T \Delta S \qquad (3)$$

where ΔG is the free energy of formation, γ is the surface tension of the oil–water interface, ΔA is the change in interfacial area on microemulsification, ΔS is the change in entropy of the system, and T is the temperature.

When a microemulsion is formed, the change in ΔA is very large due to the formation of a great number of very small droplets generated. Originally, it was proposed that to form a microemulsion a negative value of γ was required. It is now accepted that this value of g is always positive, but it is very small (of the order of fractions of $mN \cdot m^{-1}$), and is offset by the entropic component. The dominant favorable entropic contribution is the very large dispersion entropy arising from the mixing of one phase in the other in the form of large numbers of small droplets. Thus a negative free energy of formation is achieved when large reductions in surface tension are accompanied by significant favorable entropic change. In such cases, microemulsification is spontaneous and the resulting dispersion is thermodynamically stable.

The phase behavior of simple microemulsion systems comprising oil, water, and surfactant can be studied with the aid of a ternary phase diagram in which each corner of the diagram represents 100% of that particular component. More commonly, however, and in a special way in the case of microemulsions for pharmaceutical applications, the microemulsion contains additional components, such as a cosurfactant and/or drug. The cosurfactant is also amphiphilic with an affinity for both the oil and aqueous phases and partitions to an appreciable extent into the surfactant interfacial monolayer present at the oil–water interface. It has three functions: to provide very low interfacial tensions required for the formation of microemulsions and their thermodynamic stability; to modify the curvature of the interface based on the relative importance of their apolar groups; and to act on the fluidity of the interfacial film. If the film is too rigid, it prevents the formation of microemulsion and results in a more viscous phase. The existence of unsaturated bounds on the hydrocarbon chain of the surfactants equally increases the fluidity of the film. The cosurfactants used are small molecules, generally alcohols with the length of the carbon chain ranging from C2 and C10, or amines with short chains can also be used as cosurfactants. Moreover, a large number of drug molecules are themselves surface active and influence phase behavior.

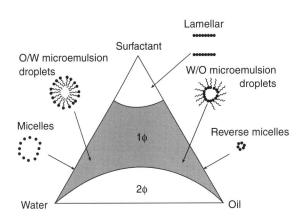

Figure 35. A hypothetical pseudo-ternary phase diagram of an oil–surfactant system with emphasis on microemulsion and emulsion phases. Within the phase diagram, existence fields are shown where conventional micelles or water–oil microemulsion, and oil–water microemulsions are formed along with the bicontinuous microemulsions. At very high surfactant concentrations two-phase systems are observed (205).

In the case where four or more components are present, pseudoternary phase diagrams are used where a corner will typically represent a binary mixture of two components, such as surfactant–cosurfactant, water–drug, or oil–drug. A highly schematic (pseudo) ternary-phase diagram illustrating various phases is presented in Fig. 35. Not every combination of various components produce microemulsions over the whole range of possible compositions, in some instances the extent of microemulsion formation may be very limited. The procedure most often employed to construct the phase diagrams is to prepare a series of (pseudo) binary compositions and titrate with the third component, evaluating the mixture after each addition. The temperature must be accurately controlled and the observations must not be made on metastable systems (213). Transitions between the various phases pictured in these phase diagrams can be driven by the further addition of one of the components, addition of a new component (drug or electrolyte), or by changing the temperature. Transitions from water/oil (w/o) to oil/water (o/w) microemulsions may occur via a number of different structural states including bicontinuous, lamellar, and also multiphase systems. In particular, microemulsions stabilized by nonionic surfactants are very susceptible to an increased temperature, leading to the phase inversion temperatures (PIT). The presence of PIT may cause problems especially when formulations are for parenteral application and must be sterilized by means of an autoclave. On the other hand, the presence of PIT may be use for the drug delivery directed to a specific site.

Advantages of Microemulsions as Drug Delivery Systems. Microemulsions present some important characteristics that make themselves very versatile carriers. In particular, they present a thermodynamic stability, optical clarity, and ease of preparation. The existence of microdomains of different polarity within the same single-phase solution allow the solubilization both water soluble and at the same time if this is so desired. Furthermore it is also possible to incorporate amphiphilic drugs into the microemulsion. It must be emphasized that the use of o/w microemulsions in drug delivery is more straightforward than it is with w/o microemulsions. The reason is because the droplet structure of o/w microemulsions is not broken following the dilution by a biological aqueous phase; this aspect make possible the oral as well as parenteral administration. The process of dilution will result in the gradual desorption of surfactant present at the droplet interface. This process is thermodynamically driven by the requirement of surfactant to maintain an aqueous phase concentration equivalent to its critical micelle concentration while maintaining temperature, pH, and ionic strength. The use of w/o microemulsions for oral or parenteral drug delivery is complicated by the fact that they are destabilized when diluted by biological aqueous fluids.

Applicative Potentialities of Microemulsions

Transdermal Application. Microemulsions represent an ideal vehicle for the topical administration of drugs because they combine the emulsion properties with those of solution. It is well known that surfactants produce stratum corneum dehydration and barrier compromise (214,215), and consequently the high levels of surfactant–cosurfactant present in the microemulsions may cause a disruption of the stratum corneum. Consequently, there is an enhancement in the permeation of drugs. However, the choice of the component is very important to minimize the alteration of the stratum corneum and the appearance of toxic effects. The choice of biocompatible components can guarantee an increased skin tolerability. For this reason, the potential application of highly biocompatible o/w microemulsions as topical drug carrier systems for the percutaneous delivery of antiinflammatory drugs (i.e., ketoprofen) was investigated (216). The components were triglycerides as the oil phase, a mixture of lecithin, and *n*-butanol as a surfactant–cosurfactant system, and an aqueous solution as the external phase. The topical carrier potentialities of lecithin-based o/w microemulsions were compared with respect to conventional formulationsn (i.e., a w/o emulsion, a o/w emulsion, and a gel).

The percutaneous adsorption of ketoprofen, evaluated through healthy adult human skin, delivered with microemulsions, showed an enhancement with respect to conventional formulations. No significant percutaneous enhancer

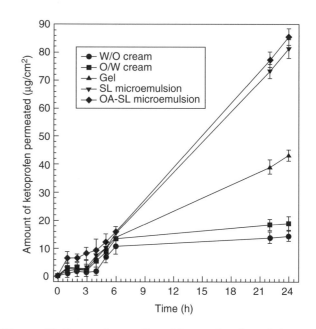

Figure 36. Permeation profiles of ketoprofen through human skin from various topical formulations. Each value is the mean value of three different experiments ± S.D (216).

effect was observed for ketoprofen-loaded oleic acid–lecithin microemulsions (Fig. 36). Moreover microemulsions showed a good human skin tolerability (Table 7).

Several reports have demonstrated that microemulsive vehicles may increase transdermal delivery of both lipophilic and hydrophilic drugs, compared to conventional formulations, depending on the constituents used for the microemulsive vehicle (217–219). These research papers suggested that microemulsion formulations may increase cutaneous drug delivery by means of the high solubility potential for both lipophilic and hydrophilic drugs, which creates an increased concentration gradient toward the skin and/or by using constituents with penetration enhancer activity (211). The incorporated ratio of the respective constituents influence in a significant way the percutaneous and transdermal drug delivery potential of the microemulsions. In every case, the enhancement of the drug delivery mechanism seems to be related to the drug mobility in the vehicle, and that measurement of self-diffusion

Table 7. Human Skin Irritancy Test of Various Topical Formulations After 24 h of Treatment

Sample	Irritation Evidence at 24 h								Score[a]
	Number of Cases[b]								
	Vesicles	Edema	Erythema	Flakiness	Dryness	Wrinkling	Glazing	No Visible Reaction	
OA 1%			3	7		2		18/30	10.17 ± 2.08
w/o						2	3	25/30	6.20 ± 2.77
O/W						2	1	27/30	4.67 ± 2.52
Gel						1	2	27/30	4.33 ± 1.15
SL-ME						1	1	28/30	3.50 ± 1.39
OA-SL-ME						2	1	27/30	4.67 ± 2.08

[a]Nonparametric variable Kruskal–Wallis test provided: $P < 0.001$ for OA (1% w/w) aqueous dispersion vs. all other samples; $P < 0.05$ for w/o cream versus all other samples; $P < 0.05$ for SL microemulsions versus o/w cream, gel, and OA-SL microemulsion.
[b]The value reported in each column represent the number of subjects who showed the skin reaction symptom.

coefficients is valuable to optimize the formulation of a given microemulsion vehicle, in order to maximize drug delivery.

Ophthalmic Application. The drug delivery system used in the ophthalmic field must overcome the disadvantages present in traditional formulations (e.g., a very low bioavailability, 1–10% of the drugs, and consequently frequent administrations are required during the day). Microemulsions represent an interesting alternative because their industrial production and sterilization are relatively simple and inexpensive; they are thermodynamically stable and permit us to solubilize both lipophilic and hydrophilic drugs.

With ophthalmic use, the choice of the various components is fundamental more than with any other topical application. The ionic surfactants are generally too toxic to be used for this application, therefore, nonionic surfactants are preferred (220). These surfactants are easily soluble in water due to the presence of either functional groups. The most used surfactants in the preparation of microemulsions are the poloxamers and polysorbates.

The choice of the oily phase is important because it conditions both the existance of the microemulsion and the solubilization of the drug. Polar oils, such as triglycerides with medium or long chains, are preferred instead of nonpolar oils, based on their solubility. The most often used consist of vegetable oils, such as soja oil, castor oil, or triglycerides, for which 95% of the fatty acids are made up of 8–10 carbon atoms, Myglyol 812s (triesters of glycerol, capric, and caprylic acids), isopropyl myristate, fatty acids, such as oleic acid, and esters of saccharose, such as mono-, di-, or tripalmitates of saccharose. As these excipients are well tolerated by the eye, their degree of purity must be high in order to prevent any contamination with potentially irritating substances.

Several additives, such as buffers, antibacterial, and isotonic agents, contained in the aqueous phase may affect the area of existence of the microemulsions, and therefore they must be studied in the presence of other constituents of the microemulsions. For example salinity influences the phase diagrams when ionic surfactants are added and decreases the phase inversion temperature (PIT) of the nonionic surfactants. Thiomersal and chlorobutanol are preservatives that are usually used in eye drop formulations, with concentrations of 0.01–0.2%, can be used without altering microemulsions structure (221).

The main advantage of the microemulsions is the increase in the solubilization of poorly soluble drugs. In a recent work, different o/w microemulsions containing indomethacin (an antiinflammatory drug). were evaluated *in vivo* by determining both the tolerability (Draize test) and the ocular drug bioavailability. This investigation showed that the colloidal carrier has a certain tolerability, eliciting only a slight irritation at the level of the conjunctiva. A positive effect regarding tolerability was exerted by jaluronic acid. In fact, by increasing the concentration of jaluronic acid present in the formulation up to 1% (w/v), an improved microemulsion ocular tolerability was observed with a substantial reduction of conjunctiva irritation (Table 8). *In vivo* ocular bioavailability of the microemulsion formulation containing indomethacin was evaluated

Table 8. Effect of Microemulsions on Ocular Structures[a]

Ocular Structure	Without Hyaluronic Acid	With Hyaluronic Acid
Conjuctiva Irritation	1.8	0.4
Conjuctiva Edema	0.4	0.2
Fluorescein Adsorption	0.8	0.2

[a]The scores were calculated awarding a value on scales from 0 to 3 at each observed reaction. All the assigned values were added and divided for the number of subjects.

by means of the Draize test. At various time intervals, the rabbits were killed, aqueous humor samples were collected and indomethacin content was determined by high performance liquid chromatography (HPLC). Indomethacin-loaded microemulsion was compared with an aqueous dispersion of the drug, containing the same drug concentration. The microemulsion-encapsulated indomethacin formulation showed a significant ($P > 0.005$) increase of drug levels compared with the free drug (Fig. 37). High colloidal properties of microemulsions may achieve a better interaction with the corneal epithelium in terms of paracellular transport or passage, thus leading to a greater drug transport into the ocular tissues. The microemulsion controlled drug release showed by ocular pharmacokinetics was probably elicited by the colloidal carrier mucoadhesion on the cell surface, thus allowing a prolonged ocular permanence and a release of the content directly into the cell (222).

Lecithin Organogel

A particular type of self-aggregate is represented from lecithin organogel. They were seen for the first time in 1988 by Scartazzini and Luisi (223), who noted that an addition of trace amounts of water into nonaqueous solutions of lecithin caused a sudden increase in the viscosity (~ 100 times) producing a transition of the initial nonviscous solution into a jelly-like state. In succeeding years, it was demonstrated that lecithin, when dissolved in at nonpolar solvent, forms spherical reversed micelles. The addition of water induces an uniaxial growth of the micelles. As a result, at the end of the preparation one will find cylindrical aggregates instead of the initial spherical ones. After reaching threshold length, the extended micelles begin overlapping, forming a temporal 3D network. This

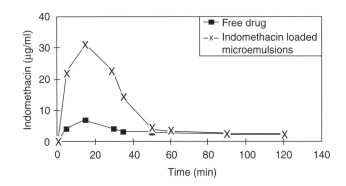

Figure 37. Bioavailability of free indomethacin or loaded microemulsions.

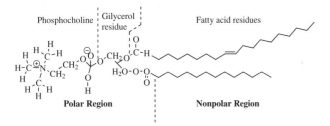

Figure 38. Structural formula of lecithin.

supramolecular structure from entangled micellar aggregates bears resemblance with that of uncrossed polymers in semidilute or concentrated solutions. For this reason, they are often called polymer-like micelles, wormlike, or threadlike micelles, or spaghetti-like structures.

The transition to polymer-like micelles is accompanied with a formation of hydrogen bonds between the phosphate group of a lecithin molecule (Fig. 38) and water.

The lecithin organogel is an optically transparent isotropic phase, appearing as the initial solution before the addition of water. The only difference between them is in the increased viscosity. This aspect is strictly dependent on oil, water, and lecithin concentrations, as well as on temperature (224). The amount of water needed to obtain the gel-like structure is a peculiar properties of any organic solvent (225). An important parameter for the organogel structure formation is the purity of the lecithin solution, in fact, commercial low purity lecithin is not able to form gels (226). The last component for the formation of lecithin organogel is water. This solvent can be substituted by polar organic solvents, such as glycerol, ethylene glycol, and formammide, or by a mix of ethanol–water in different ratios (227). The physical–chemical characteristics of the incorporated drug noticeably influence its release from organogel (144).

An important characteristic of this aggregate is its thermoreversibility, in fact, at 40 °C they become fluid, but by reducing the temperature they again reassemble a gel-like structure.

In this kind of carrier, it is possible to load hydrophilic drugs (localize themselves in aqueous, internal compartment), lipophilic drugs (in the hydrophobic environment), and amphiphilic substances (at the interface w/o).

The principal application of this carrier is its transdermal delivery, as first proposed in the early 1990s by Luisi's research group (225,226). Scopolamine, broxaterol, and propranolol were incorporated into lecithin organogels (containing cyclohexane, isooctane, or IPM as the oily phase). The permeation rates increased 10 fold compared to a solvent drug solution used as a control (180). The utility of lecithin organogels has been supported by *in vivo* human skin tolerability studies by means of a noninvasive technique as spectrophotometry of reflectance (228). *In vivo* percutaneous tolerability results showed no appearance of erythema even after 48 h of application. Certain amphiphilic lipids are characterized by lyotropic and thermotropic aggregation-phase transition. These supramolecular aggregates are under investigation to evaluate their potentialities as drug delivery systems (229).

COLLOIDAL DRUG DELIVERY DEVICES

The main scope of colloidal drug delivery systems is the modulation of the phamacokinetics and/or the tissue distribution of a drug in a beneficial way. The properties of colloidal drug delivery systems to target specific sites of action (organs or tissues) are related to the physicochemical and morphological properties of the carriers, namely, these parameters determine the destination and the fate of the drug entrapped within the carrier system, provided that a drug is released from the system at a suitably controlled rate (230,231). By using colloidal carriers, drugs can be selectively directed to specific sites by applying passive or active strategies of delivery, rather than allowing a free drug diffusion throughout the body by using conventional dosage forms. The carrier physicochemical properties (i.e., size and surface properties) are the main determining factors in passive targeting of colloidal drug carriers. On the other hand, the possibility to achieve a colloidal carrier with active targeting capacity is related to the possibility of inserting specific ligands on the carrier surface so as to achieve a specific receptor-mediated interactions with target cells (232,233).

The potential use of colloidal drug carriers in clinical therapy is strongly related to their *in vivo* fate. In particular, the rapid uptake (following a phagocytosis mechanism) of these carriers by the reticulum endothelial systems (RES), that is abundantly present at the level of the liver, spleen bone marrow, and lungs, is the only fate after their IV administration, thus leading to rapid removal from blood circulation. The phenomenon of opsonization, that is based on binding of some plasma proteins (opsonines) onto the surface of colloidal carriers, is the first step allowing the carrier recognition and binding promotion by phagocytes (234). Therefore, the opportunity to avoid the carrier opsonization is often translated into a deep change of the carrier biodistribution patterns.

In this attempt, colloidal carriers with the ability to avoid RES uptake have been developed, thus achieving long circulating properties (235). The so-called Stealth colloidal carriers are obtained by grafting their surface with hydrophilic macromolecules, mainly poly(ethylene glycols), that hamper the opsonization.

Colloidal drug delivery systems are not able to extravasate, except in tissues and/or organs in which the endothelium is discontinuous (i.e., liver, spleen, and bone marrow) or defective, such as in the case of tumors or in the sites of infection and/or inflammation. Therefore, the therapeutic uses of IV administered colloidal drug delivery devices can be grouped into three cases (235,236): (*1*) drug accumulation in macrophages; (*2*) *in vivo* drug distribution away from the sites of toxicity; (*3*) circulating reservoirs of labile or short blood half-life drugs.

The use of colloidal drug delivery systems has the following advantages: protection, duration, direction, internalization, and amplification.

Protection. Drugs entrapped within colloidal carriers can be protected against both environmental factors (i.e., temperature, UV radiation, moisture) and the action of detrimental factors of the host (i.e., degradative enzymes). Also, the patient can be protected against toxic effects of administered drugs.

Duration. These carriers can be suitably projected and prepared to achieve a perfectly controlled drug release to fulfill the therapeutic requirements, thus allowing the maintenances of therapeutic (but nontoxic) drug levels in the bloodstream or at the level of local administration sites for a prolonged time. This situation leads to a reduction of administration frequency, and hence to enhanced clinical safety and increased patient compliance.

Direction. As mentioned above, drugs may be passively or actively targeted to specific sites of action by colloidal delivery devices, thus providing an improvement of the drug therapeutic efficacy. These carriers can also provide a site-avoidance delivery, namely, the drug delivery away from sites of their toxicity.

Internalization. Colloidal carriers may be able to promote the intracellular delivery of drugs by ensuring different interaction pathways with target cells in comparison with the free drug that may not be able to reach the inner-cell due to unfavorable physicochemical parameters.

Amplification. In the case of antigen delivery, colloidal drug delivery systems can act as immunological adjuvant in vaccine formulations.

General Colloidal Carrier Classification

By considering the carrier features, colloidal drug delivery devices can be classified into conventional, long circulating and actively targeted systems (Fig. 39).

Conventional colloidal carriers (liposomes and nanoparticles) can be characterized by a wide differences both in terms of composition and physicochemical properties (i.e., size, size distribution, surface charge, number, and fluidity of phospholipid bilayers), in the case of liposomes, matrix compactness, in the case of nanoparticles. The modulation of these properties can influence technological properties, such as colloidal stability, drug loading, drug release rate, and to a certain extent the *in vivo* behavior of conventional colloidal carriers (i.e., blood stability, clearance, and distribution). However, some *in vivo* features are very consistent among different types of conventional colloidal carriers, by presenting a short blood circulation time when parenteraly administered due to a rapid RES uptake. A consequential successful therapeutic use of conventional colloidal carriers characterized by the accumulation at the level of the mononuclear phagocyte system is the delivery of antimicrobial agents to infected macrophages (237,238). Conventional colloidal carriers are also very effective as vaccine adjuvants against viral, bacterial, and parasitic infections (239).

Long-circulating colloidal delivery systems allow the therapeutic treatment of a wide range of diseases involving tissues other than liver and spleen (240). A common characteristic of all long-circulating systems is the presence along the surface of the colloidal carrier of hydrophilic macromolecular moieties, such as polyethylene glycol

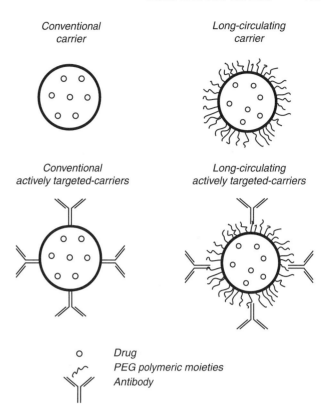

Figure 39. Schematic representation of the various kind of colloidal drug delivery devices. Conventional carriers are made up of a body matrix (phospholipid bilayers in vesicles or polymeric network in nanoparticles) with a hydrophilic colloidal surface (neutral, negatively, or positively charged). Long-circulating systems (the so-called Stealth devices) are coated by hydrophilic polymeric moieties (i.e., PEG) that provide a certain steric stability, and hence reduction of opsonization. Actively targeted carriers (i.e., antibody targeted) can be of conventional (targeting-agent conjugated directly to colloidal carrier surface) or sterically stabilized type (targeting-agent conjugated with a PEG moiety anchored to the surface of the colloidal carrier).

(PEG). Highly hydrated macromolecular moieties determine a steric barrier against interactions with molecular and cellular components in the biological environment, thus avoiding the opsonization phenomenon, and hence the RES organ uptake (235,240).

To obtain a certain specificity, actively targeted carriers can be obtained by conjugation of a colloidal drug delivery systems to specific antibodies, antibody fragments (e.g., Fab or single-chain antibodies), or small targeting agents (peptides, hormons, specific ligands), thus increasing target site binding and the delivery of the encapsulated drug. In the first generation of these kind of colloidal carriers, the active targeting agent was conjugated directly to their surface. This strategy led to a successful *in vitro* recognition and activity, but to a failure in *in vivo* applications due to the RES uptake. The last generation of actively targeted carriers is represented by long-circulating colloidal carriers with the PEG moieties conjugated with the targeting agent, thus presenting suitable *in vivo* features.

Colloidal Carrier Characterization

For routine measurements of particle sizes, two techniques are commonly used. Photon correlation spectroscopy (PCS) (also know as dynamic light scattering), which measures the fluctuation of the intensity of light when it is scattered by particles movements. The particle diameter range goes from a few nanometers to ~3 μm, so PCS is not useful in for lipid particles >3 μm. In these cases, a laser diffraction (LD) technique is used. This method is based on the relation between the diffraction angle and the particle radius, so that smaller particles cause more intense scattering at high angles compared to the larger ones.

In general, it is recommended to use both techniques simultaneously in order to obtain precise data. However, it should be kept in mind that both PCS and LD do not measure particle sizes directly, they only correlate light scattering to particle size.

To obtain direct information on particle sizes and shapes, electron microscopy (EM) is used. Electron microscopy extracts structural information carried by the scattered electrons; the most commonly used EM techniques are transmission electron microscopy (TEM) and SEM. Atomic force microscopy (AFM) is another microscopic technique that is getting increasing attention. This method is based on the interactive forces between a surface and a probing tip that leads to the imaging of particles. This technique has the clear advantage of simplicity of sample preparation, so that it is possible to conduct analysis directly on the hydrated, solvent containing samples (241).

The field-flow fractionation (FFF) is a technique recently used for measurements of solid lipid nanoparticle sizes. It is based on the different effect of a perpendicular applied field on particles in a laminar flow (242); the characterization of particles is based on the different nature of perpendicular fields, for example, sedimentation size (cross-flow FFF) or charge (electric field FFF). All these principles can be used combined together in order to obtain unique resolution.

The determination of a zeta potential is predictive of the storage stability of colloidal dispersions (243). In general, the greater the zeta potential value of a nanoparticulate system, the better the colloidal suspension stability due to a repulsion effect between charged nanoparticles. Nanoparticle stability can also be obtained by the addition of some polymers, such as PEG, which adhere to the particle surface stabilizing it. Surface characteristics are also important for the *in vivo* fate and the interaction with biological systems of colloidal carriers.

The characterization of the physical state of colloidal carriers (particularly vesicles and lipid-based particles) can be efficiently carried out by two techniques, DSC and X ray. The DSC method is based on the fact that different material polymorphic form possess different melting points and melting enthalpies (244) and that changes in thermotropic parameters of a systems are usually evidence of different spontaneous and/or induced arrangements. X-ray techniques allow the characterization of polymorphic forms and the determination of large and small spaces in an ordered matrix, such as the lipid grid of a solid lipid nanoparticle (245). The advantages of these two techniques

are the possibility of particle suspension analysis without drying the solvent, thus avoiding possible modifications of the carrier structure.

Also, NMR and electron spin resonance (ESR), are used for the investigation of dynamic phenomena in colloidal lipid dispersions. Nuclear magnetic resonance is based on the different proton relaxation times in the liquid and semisolid–solid state (246). The NMR technique can also be used to determine lamellarity in vesicular carriers (247). The ESR technique uses a paramagnetic spin probe to give a noninvasive characterization of the distribution of the spin probe between hydrophilic and hydrophobic phases. Both NMR and ESR are noninvasive methods and allow repeated measurements of the same sample.

Vesicular Drug Carriers

Drug delivery systems composed of lipidic compounds have gained great importance in medical, pharmaceutical, cosmetic, and alimentary fields. Formulations based on phospholipids and other excipients represent an interesting field of application in the novel research for delivery models.

Lipidic materials are characterized by their possibility to self-organize in different supramolecular arrangements as a function of some environmental factors (i.e., temperature, lipid concentration, type of medium, ionic strength, pH value, and presence of other compounds). Among the various supramolecular forms of aggregation, the bilayer structure, and hence the formation of vesicles (defined as a lipid bilayer surrounding an aqueous space) represents the most suitable device in terms of drug delivery. In fact, vesicles are boundary structures (Fig. 40), in which it is possible to have at the same time various microenvironments characterized by different physicochemical properties, namely, a highly hydrophilic region made up of the intravesicular aqueous compartment, a highly hydrophobic region of the bilayer core made up of the alkyl chains of the lipid constituent, and an amphipatic region at the level of the vesicular surface made up of the polar lipid headgroups. These peculiarities make vesicular systems a very versatile drug carrier being able to entrap and delivery hydrophilic (in the intravesicular aqueous compartment), hydrophobic (in the core of vesicular bilayers), and amphipatic (at the level of vesicular boundary zone) drugs.

An important feature that make vesicles a unique drug delivery system is the biomimetism of having the same supramolecular lipid organization of natural membrane living cells.

Therefore, the possibility to create a structure similar to the biological membrane for carrying out the delivery of drugs has represented an interesting challenge for a number of researchers. In particular, liposomes, ethosomes, transfersomes and niosomes have been extensively investigated and are up to now the main vesicular systems used in drug delivery.

Liposomal Carrier. The appearance of Banghman's vesicle in the mid-1960s, the so-called Liposome, represented a milestone in the field of innovative drug delivery. Liposomes are mostly made up of phospholipids, and for

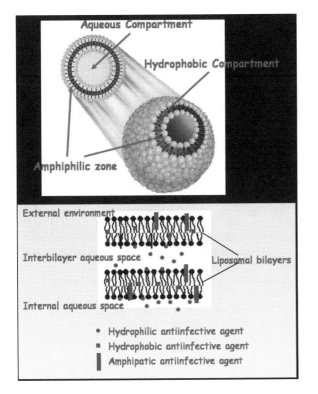

Figure 40. Schematic representation of a liposomal structure with the characteristic microenvironments.

this reason they are highly biocompatible and biodegradable. The liposomal carrier has the advantage of also being able to deliver macromolecules, such as enzymes, proteins, and genetic material (248).

From the morphological point of view (Table 9), liposome systems can be classified as a function of the number of bilayers and the mean size of the carrier in unilamellar, oligolamellar or multilamellar vesicles, and in small (<100 nm), medium (100–500 nm) and large (>1 μm) vesicles, respectively.

Lipid Component Used in Liposomal Formulations.
Lecithins and cholesterol (Chol) are the lipids most commonly used in the preparation of liposomes. Other components can be used in the liposome preparation, that is, steroid molecules, charged phospholipids, ganglioside,

and polymeric material to modulate the carrier properties as a function of the therapeutic requirements to be achieved (249). In fact, different components can modify the biodistribution, the surface charge, the release, and the clearance rate of the liposomal drug delivery system (249,250). The circulation lifetime of a liposome is also altered by the charge of the liposome surface that can influence the pharmacokinetic of the system (251).

It was demonstrated that the use of negatively charged lipids [i.e., phosphatic acid (PA), phosphatidylserine (PS), phosphatydilglycerol (PG)] are able to elicit a rapid clearance of the liposomal system from the blood stream mediated by the RES uptake (249,252).

Cholesterol plays a fundamental role in liposome formulations being, able to act as a vesicle membrane modulator as concern membrane fluidity. It has a stabilizing function on the liposome bilayers both *in vitro* and *in vivo*, allowing the protection of the vesicular structure by the action of blood high density lipoproteins (HDL) and hence the possibility of having a prolonged circulation of intact liposomes (253).

Similarly to cholesterol, some phospholipid components are also able to influence the physicochemical behaviors of liposomes to obtain a more rigid vesicular structure that is much more resistant to the phospholipid extraction effect mediated by blood HDL. In this attempt, both 1,2-distearoyl-3-*sn*-phasphatidylcholine (DSPC) and sphingomyelin (SM) have been used to maintain a certain vesicular carrier integrity following IV administration. A rigid vesicular structure of liposomes hampers an effective adsorption of opsonine and prolongs the plasmatic level of the drug carrier, that is, liposomes made-up of 1,2-distearoyl-3-*sn*-phasphatidylcholine–cholesterol (DSPE–Chol) showed higher half-time than liposomes prepared with phosphatidylcholine (PC) or 1,2-dipalmitoyl-3-*sn*-phasphatidylcholine (DPPC) (253). In particular, SM has an additional stabilizing effect on the liposome formulations when used together with Chol (254). In this case, SM can interact with cholesterol by forming intermolecular hydrogen bonds and eliciting an increased compactness of the liposomal bilayers that leads to an improved serum stability (249).

Since the appearance of liposomes, positively charged lipids were introduced in liposome composition to obtain a vesicular system characterized by a net positive charge

Table 9. Main Characteristics of the Various Liposome System

Liposome Type	Abbreviation	Properties
Multi-lamellar vesicles	MLVs	Vesicles constituted by 7–15 bilayers with a mean size > 1.5 μm
Multi-vesicular vesicles	MVVs	Vesicles constituted by 3–5 vesicles contained within a bigger one. The mean size is > 1.5 μm
Oligo lamellar vesicles	OLVs	Vesicles constituted by 2–5 bilayers with a mean size ~ 1 μm
Giant unilamellar vesicles	GUVs	Vesicles constituted by only one bilayers with mean size ≥ 1 μm
Large unilamellar vesicles	LUVs	Vesicles constituted by only one bilayers with mean size ranging from 400 to 800 nm
Medium unilamellar vesicles	MUVs	Vesicles constituted by only one bilayers with mean size ranging from 200 to 400 nm
Small unilamellar vesicles	SUVs	Vesicles constituted by only one bilayers with mean size ranging from 30 to 100 nm

along the liposomal surface. In the last decade, positively charged liposomes have gained much more interest than in the past due to their potential application as carriers for genetic material delivery (255). In this attempt, the most frequently used cationic lipids are DMRIE, N-(2-hydroxyethyl)-N,N-dimethyl-2,3-bis(tetradecyloxy)-1-propanaminium bromide; dioctadecyl amino glycyl spermine (DOGS); dioleoylphosphatidyletethanolamine (DOPE); 2,3-dioleyloxy-N-[2(spermine carboxaminino)-ethyl]-N,N-dimethyl-1-propanaminium trifluoroacetate (DOSPA); 1,2-dioleoyl-3-trimethylammonium propane (DOTAP); 2,3-bis(oleyl)oxipropyl-trimethylammonium chloride (DOTMA). Cationic liposomes composed of DOTMA and DOPE became commercially available as a transfection reagent designated Lipofectin.

The above mentioned cationic lipid components of this particular kind of liposomes are able to interact with, and neutralize, the negatively charged DNA or ribonucleic acid (RNA). This interaction leads to a genetic material condensation into a more compact structure. The resulting lipid–genetic material complexes (lipoplexes), rather than DNA or RNA encapsulation within liposomes, provide protection and promote cellular internalization and expression of the condensed plasmid (255).

Most recently, amphiphilic polymeric materials have been introduced in the composition of vesicles to cover their surface by inserting their hydrophobic domain in the liposomal bilayers (anchor moiety) and facing the hydrophilic domain toward the aqueous environment (shield moiety). This advance allowed a further modularity of the liposomal carrier by conjugating together the advances of colloidal drug delivery devices (carrier capacity) with those of macromolecules (fine chemical approach and infinite modulation potentiality) (249). The principal polymer used to cover the surface of liposome formulation was polyethylene glycol. This is a flexible-chained hydrophilic polymer of different molecular weight (i.e., PEG-750, PEG-2000, PEG-5000) conjugated to phosphatidyletethanolamine (PE) or distearoylphosphatidyletethanolamine (DSPE) (256). Liposomes containing PEG in their structure (the so-called pegylated liposomes) represented an important class of vesicular delivery systems that started the new generation of liposome carriers. The presence of this hydrophilic polymer on the surface of liposomes not only is able to reduce the RES uptake and to increase the blood circulation time (249), but it can also modulate some pattern of interaction with cultured cells, such as the intracellular drug delivery (257).

Another important aspect for drug delivery by liposomes is the possibility to achieve a triggered release of the encapsulated agent from the carrier following certain stimuli. Targeted drug delivery is based on the fact that upon attachment to the target site, or delivery into the target cell, the therapeutic agent must be released from the carrier to exert its action. When liposomes are taken up by the target cell through endocytosis, they come into contact with acidic conditions. For some drugs and biothecnological products (e.g., peptides and genetic material) it could be essential to escape from liposomes and endosomes, thus entering the cytosol before reaching the lysosomal structures with their highly efficient degradation machinery.

Liposome destabilization under acidic conditions and bilayer fusogenic properties are required to achieve lysosome escape. Besides the pH-dependent liposome release, other triggered releases may be accomplished for certain drug selectivity, namely, bilayer composition controlled release, destabilization by removal of bilayer components, complement-induced leakage, and temperature-induced destabilization of the liposomal bilayer structure. Therefore, to have a triggered liposomal carrier release, some compounds that are stimuli responsive must be introduced in the liposomal bilayer composition (e.g., DOPE, cholesteryl hemisuccinate, oleic acid, fusogenic peptides) (258,259).

Main Methods to Prepare Liposomes. As reported in Table 9, various types of liposomes exist, each of those with specific peculiarities that make them suitable for certain therapeutic applications. Although aqueous dispersions of phospholipids spontaneously lead to a self-aggregation into closed bilayers, vesicles, particular procedures must be carried out if a certain type of liposome has to be obtained. In fact, this type of liposome is mainly determined by the preparation procedure, and for these reason the main preparation methods are reported below.

Thin-Layer Evaporation (TLE). This method allows the formation of multilamellar vesicles. Basically, a mixture of lipid compounds is dissolved by an organic solvent (chloroform) or a mixture of two organic solvents (chloroform–methanol) in a round-bottomed flask. Other hydrophobic components (e.g., drugs) can be cosolubilized with the liposome-forming materials. The complete evaporation of the organic solvent by a rotavapor lead to the formation of a thin lipid film along the surface of the glass wall. This lipid film is then hydrated with an aqueous solution buffered to the desired pH value and solubilizing any hydrophilic component that should be entrapped within liposomes (e.g., water-soluble drugs). The hydration temperature is normally higher than the highest transition temperature (T_m) of lipids used in the film preparation. In some cases, to increase the surface of film deposition, and hence the surface undergoing buffer hydration, glass beads can be added during the TLE preparation procedures (260).

Reverse-Phase Evaporation Vesicles (REVs). This method allows us to obtain large unilamellar, oligolamellar, and multilamellar vesicles. A lipid film, formed as reported in the TLE method, is dissolved in an organic solvent (diethyl ether) and an aqueous solution is added. This two-phase mixture is energetically sonicated, thus obtaining an w/o emulsion. The organic solvent constituting the external phase of the w/o emulsion is gradually removed by a rotavapor up to the reversion of the phases with the appearance of an external hydrophilic phase. The total removal of the organic solvent leads to the formation of a gel-like highly concentrated liposome suspension that can be suitably diluted with a suitable aqueous buffer solution. This method represents the first approach used in the attempt to increase the amount of drug entrapped within vesicles (261,262).

Freeze and Thawed Multilamellar Vesicles (FAT–MLVs). A multilamellar liposome formulation obtained with the TLE method is subjected to a serious of cycles of freezing in liquid nitrogen and thawing in warm water (~40 °C). At the end of the procedure, liposomes are kept at room temperature to stabilize the bilayer. This procedure is carried out to obtain a multilamellar liposomes with a homogeneous distribution of solutes throughout the various multilamellar aqueous compartment (263,264).

Dehydration Rehydration Vesicles (DRVs). Multilamellar liposomes obtained with one of the previous methods are submitted to a freezing-drying process. The product of liophylization is resuspended in an aqueous solution (265). This method leads to the formation of oligolamellar or multilamellar liposomes with an high drug entrapment efficiency.

Vesicles by Extrusion Technique (VET). The reduction of the mean size of a colloidal liposomal suspension characterized also by a narrow size distribution can be achieved with the extrusion of multilamellar liposomes through polycarbonate membranes of different sizes (from 400 to 50 nm). Usually, 10 cycles of extrusion are carried out to obtain an homogeneous formulation. Both LUV and SUV are obtained following the VET method (266).

pH Gradient Loading Method. This method is used to increase the loading capacity of liposomes in regard to ionizable drugs. This method is based on the formation of a pH gradient between the innerliposomal aqueous phase and the external environment. This situation promotes the protonation or deprotonation of an entrapped drug thus favoring its accumulation within the vesicular carrier due to the incapability of a ionized molecule to freely diffuse through a lipid bilayer (Fig. 41) (267). Ammonium sulfate or ammonium citrate are used to obtain an acid pH environment while calcium acetate to have basic conditions (250,268). The efficiency of liposome

drug loading using the method of pH gradient is influenced by the drug partition coefficient between the aqueous phase and the lipid bilayer (269).

One of the most important parameters for an ideal drug delivery system is the drug loading capacity. The amount of drug encapsulated in liposome formulation is influenced by a serious of parameters, such as the preparation method, the size of the liposome, and the type of lipid used to form the lipid film (263). Therefore, to have a colloidal liposome system with particular carrier properties, it is often necessary to carry out two or more preparation procedures. Namely, the DRV or FAT procedure can be carried out to improve the liposome encapsulation capacity, and then the VET method to obtain a small mean size with a narrow size distribution. These two aspects (carrier capacity and mean size) are very important for liposomes to be proposed for certain therapeutic application (i.e., antitumoral chemotherapy).

The removal of untrapped drug is the last step in the preparation of a drug-loaded liposome colloidal suspension. Many lipophilic drugs exhibit a high affinity to the bilayer and are completely liposome associated. For compounds with an encapsulation < 100%, the nonencapsulated fraction of the drug may determine unacceptable side effects. The removal of the untrapped drug can be carried out by the following techniques: dialysis, ultracentrifugation, ultrafiltration, gel permeation chromatography, and ion exchange reactions.

Liposome Stability. An ideal drug delivery system should maintain its physicochemical characteristics during storage, that is, mean size, size distribution, thermotropic parameters, no lipid degradation (hydrolysis and/or peroxidation), no appearance of microbial flora, to be considered for practical applications. Liposomes are self-assembled colloidal carriers, and hence their stability can be strongly influenced by the component used for their preparation, considering that the presence of foreign molecules in the liposomal bilayers deeply influence their mode and strength of aggregation in a concentration-dependent manner. For this reason, in the case of drugs to be delivered by liposomes and characterized by liposomal bilayer localization, particular attention should be paid to the drug/lipid ratio. This is a very important parameter because the payload of the drug can be increased with a consequential reduction of the system stability. In some cases, the segregation of the lipid bilayer components in various microdomains can be observed (270).

The osmolarity of liposomes seems to be a very important factor to achieve a stable liposomal system. Some studies (271) showed that hypertonic conditions triggered a rapid drug release from Ara-C-loaded liposomes and that the release kinetic is characterized by a biphasic profile with a first step of very rapid and massive Ara-C release followed by a second phase of slow drug release (249).

The chemical stability of liposome formulations mainly depends on the chemical characteristics of both drugs and lipid component used for the carrier preparation (272). The

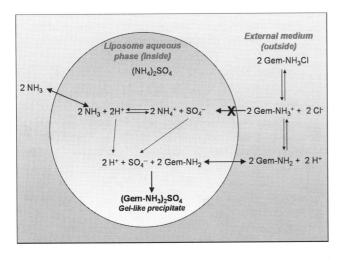

Figure 41. A schematic representation of the encapsulation of Gemcitabine in multilamellar liposomes by using a pH gradient method in the presence of ammonium sulfate 250 m*M* (267).

presence in the phospholipid bilayers of polyunsatured fatty acid moieties, such as arachidonic, linoleic, or linolenic acid, can favor the occurrence of peroxidation processes at the level of single or conjugated double bonds. The membrane lipid peroxidation can destabilize liposomal bilayers due to the formation of secondary oxidation products that can change the integrity of the liposome structure (249).

Main Therapeutic Applications of Liposome. The following criteria should be taken into account to evaluate the possibility of delivering a drug by using the liposomal carrier: (*1*) the chosen drug should be sufficiently active; (*2*) the drug should be efficiently entrapped within liposomes; (*3*) the drug must be compatible with the liposomal carrier.

A basic concept for the success of liposome drug delivery is the fact that the encapsulated agent may be released at a suitable rate to become bioavailable upon arrival at the action site. Liposomes protect drugs from metabolism and inactivation in plasma and also allow a reduction of the drug accumulation in healthy tissues and/or organs, due to size restrictions in the transport of large macromolecules and carriers across healthy endothelium (271). A number of pathologies (i.e., cancer, stroke, infections, and some methabolic diseases) are characterized by direct or mediated inflammation, which elicits discontinuities in the endothelium vasculature of the diseased zone. This thus increases the extravasation of colloidal carriers and, in combination with an impaired lymphatics and a high value of interstitial pressure, the accumulation of the therapeutic agent-loaded liposomes at the level of the diseased site (passive targeting). This situation, referred to as enhanced permeation and retention (EPR) phenomenon, consequently elicits an increase of the drug therapeutic index (249).

A successful therapeutic approach of the liposomal passive targeting is the efficacious delivery both *in vitro* and *in vivo* of various anticancer drugs (249). As shown in Fig. 42, the use of pegylated liposomes (Stealth liposomes) with a mean size of ~100 nm allows the passage of the carrier in the tumor tissue and a local accumulation of the encapsulated drug. Furthermore, liposomal chemotherapeutic agents display distinctive pharmacokinetic charac-

teristics, because they possess longer elimination half-lives, reduced clearance, and smaller volume of distribution with respect to corresponding free drugs. Taken together, these features lead to the highest levels of cytotoxic agents in tumors, as demonstrated in preclinical models and clinical trials, whereas healthy tissues are spared from toxicity. Liposomal anticancer drugs lead to improved clinical effectiveness and better toxicity profile with respect to corresponding free drugs when they are used for the treatment of metastatic tumors (e.g., breast and ovarian cancers). A successful example of antitumoral agent-loaded long-circulating liposomes is Doxil, a doxorubicin-loaded pegylated liposomes with a 100 nm mean size.

This innovative liposomal formulation is currently approved for use in AIDS-related Kaposi's sarcoma and refractory ovarian cancer. It has also shown activity in other tumors, including metastatic breast cancer. A preclinical toxicology study of IV administered doxorubicin-loaded stealth liposomes compared to the free drug showed that the drug liposomal formulation was less toxic (LD50 32 mg·kg^{-1}) than the free doxorubicin (LD50 17 mg·kg^{-1}). The organ specific toxicities seen with Doxil were qualitatively similar to those of free doxorubicin, but less severe (273). In addition, Doxil accumulates in tumor tissues to a large extent with respect to the free drug due to its capacity to escape macrophagic uptake (274). Reduced toxicity and selectivity are the reasons of the improvement of doxorubicin therapeutic index.

A recent and very active field of research in the liposomal anticancer chemotherapy is the active targeting of long-circulating liposomes (249,275,276). A high density of the targeting moiety on the surface of liposomes is very important to have an efficient binding to the target site, a specific antigen or receptor expressed on the surface of target cell. This interaction increases the amount of drug in the target site and it decreases the systemic side effects (275).

Antibodies, particularly monoclonal, are the more versatile ligands that can be conjugated on the liposome surface (the so-called immunoliposomes). In the past, an obstacle in the use of immunoliposomes was the antigenicity of murine antibodies that were easily available, however, the more recent availability of humanized forms should contribute to overcome this problem. Important

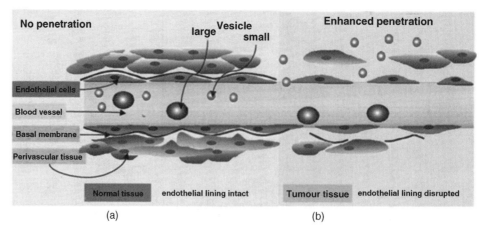

Figure 42. Schematization of the accumulation mechanism of long-circulating small unilamellar liposomes in solid tumor. Extravasation of liposomes through vascular endothelium of the tumor site (a); behaviour of liposomes in a normal tissue (b).

parameters for immunoliposomes are the ability to become selectively cell associated and the ability to deliver the loaded drug within target cells. In the case of immunoliposomes endocytosis seems to be the predominant mode of delivery to the cells, and hence has an efficient intracellular delivery. The mean size of immunoliposomes should be $\leq 100\,nm$.

Given a suitable antibody with high specificity and affinity for the target antigen, the critical factor is the *in vivo* accessibility of target cells to the immunoliposomes. To have an efficient target binding of the injected immunoliposomes, target cells should be located in the intravascular compartment and/or in accessible tissues and organs characterized by leaky vascular structures. Thus, in terms of targeting drug delivery by immunoliposomes, two anatomical compartments can be considered. One is a readily accessible intravascular site, such as the vascular endothelial surface, T cells, B cells, or a thrombus. The other is a much less accessible extravascular site, such as a solid tumor, an infection site, or an inflammation site, where the vascular structure is leaky (277).

Antibiotics encapsulation in liposomes is of great utility in the case of very potent drugs that can be administered intravenously and present a certain toxicity (i.e., nephro- and neurotoxicity). The toxicity of antibiotics limits their dosing, and hence the drug efficacy. Antimicrobial agent-loaded liposomes were used for the treatment of various obligate and facultative bacterial infections (i.e., *Salmonella*, *Listeria*, *Brucella*, *Mycobacterium*, *Staphylococcus* and *Escherichia coli*) (278). Obligate microbes are more difficult to eradicate due to the fact that they can multiply only within host cells, while facultative bacteria can be reached by the drug in the extracellular compartment. The conventional liposome biodistribution properties represent a noticeable advantage for treatment of infections in which bacteria are taken up and/or reside in the cells of the phagocytic systems. Another advantage of the liposome carrier is the capability to facilitate the entrance within infected cells of antimicrobial agents that are not able to cross cell membranes with a consequential intracellular drug accumulation (279) (Fig. 43). An intrabacterial antibiotic drug accumulation was also observed (280), thus showing that liposome formulations can contribute to overcome bacterial resistance phenomena due to drug impermeability (Fig. 44). In particular, in the case of intracellularly infected phagocytic cells (e.g., *Legionella pneumophila*, *Mycobacterium tuberculosis*, *Listeria monocytogenes*, and *Staphylococcus aureus*) a 10–100 times increased efficacy has been reported for the antibacterial agent-loaded liposome formulation compared to the free drug (278) both *in vitro* and *in vivo*. The specific macrophage targeting of liposomes can be further improved by grafting the surface of liposomes with carbohydrate moieties whose receptors are expressed along the surface of macrophages. This possibility may lead to an additional efficacy of the liposome delivery device in the *in vivo* treatment of intramacrophagic infections.

Long-circulating liposomes with a reduced size have the opportunity to accumulate in the infection site according to the mechanism reported in Fig. 42. In an experimental *in vivo* model, a large accumulation of long-circulating liposomes in the infected lung was observed; while no presence

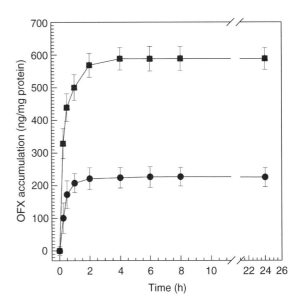

Figure 43. Accumulation profiles of ofloxacin into McCoy cells as a function of time. The biological assay was carried out at room temperature (20 °C) by adding $5.7\,\mu g \cdot mL^{-1}$ of free ($\bullet$) or liposome entrapped ($\blacksquare$) ofloxacin into confluent McCoy cells. Each point represents the average of nine different experiments $\pm$ standard deviation. Data from Ref. 279.

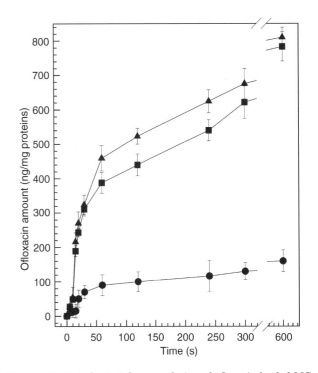

Figure 44. Intrabacterial accumulation of ofloxacin-loaded MC-Chol-DP ($4:3:4\,M$ ratio) unilamellar liposomes within *E.coli* ATCC 25922 ($\blacksquare$) and *E.coli* ATCC 35218 ($\blacktriangle$) versus the free drug (*E.coli* ATCC 35218 accumulation) ($\bullet$) as a function of time. Free drug accumulation within both *E.coli* strains is very similar (data not reported). The experiments were carried out at 37 °C. Each point represents the mean value of five different experiments $\pm$ S.D. Data from Ref. 280.

of long-circulating liposomes was noted in the noninfected lung. Interestingly, the accumulation extent seemed to be a function of the severity of the infection (281). Therefore, Stealth liposomes offer targeting to the deep tissues, which can harbor *Mycobacterium avium intercellulare*. The chance of using Stealth liposomes containing some new and potent antibacterial agents, (e.g., fluoroquinolones) can represent a real improvement in the therapy for the eradication of infections situated in organs and tissues other than the RES.

Liposomes can be suitable delivery devices in antiviral chemotherapy (282) due to their capability of delivering entrapped drugs across cell membranes (257,279). This aspect is of fundamental importance in antiviral chemotherapy, because the nature of virus action and proliferation is intracellular. In particular, the liposomal therapy of viral infections can be accomplished by two different approaches: (*1*) the encapsulation of the antiviral drug having a liposome-mediated antiviral activity; (*2*) the encapsulation of immunomodulators, such as lymphokines (macrophage activation factor, MAF), thus achieving an activation of the macrophages.

Liposomal antiviral chemotherapy can offer special targeting possibilities due to the natural ability of viruses to fuse with cellular membranes. In this case, various antiviral therapeutic approaches can be achieved by the following strategies: (*1*) the administration of drug-loaded long-circulating liposomes bearing cellular antigens that attract and destroy viruses; (*2*) the saturation of the cell receptor by binding other antigens delivered with liposomes; (*3*) the reconstitution of viral glycoproteins onto liposomes (the so-called virosomes), which are characterized by a very strong fusogenic activity depending also on the vesicle lipid composition. Such virosomes can bind to and fuse selectively with the infected cells. Therefore, this particular carrier can ensure a very effective and specific intracellular antiviral therapy.

Liposomal antiviral chemotherapy, for example, can be efficaciously used for the treatment of HIV infection. The encapsulation of gelonin (a plant toxin) allowed a selective killing of human immunodeficiency virus (HIV) infected cells (283). Another success with respect to HIV therapy was observed in the case of treatment with liposomes containing fragment A of diphtheria toxin, which was toxic to HIV infected cells, but not to uninfected cells (284).

Another application of liposomes in antiinfective chemotherapy can be the treatment of fungal infection. Invasive fungal infections are among the most important causes of morbidity and mortality in immunocompromised patients. Amphotericin B and nystatin are the most widely used drugs in the treatment of systemic fungal infections (285). These two drugs show some drawbacks when used *in vivo* in the treatment of mycosis (i.e., nephrotoxicity and side effects at the level of the CNS). In this case, liposomes are a suitable colloidal carriers for amphotericin B, not being able to accumulate in the kidneys (e.g., of the site avoidance mode of liposome action) and the nervous system and providing a smart system to efficaciously solubilize amphotericin B. As for other pathological situations, the most important advantage of the liposomal carrier is its ability to accumulate at the level of the same cells where

fungi are localized (mainly the RES). The improved selectivity and the reduced toxicity determined the noticeable increase of the amphotericin B therapeutic index. Considering the consistent therapeutic advantages of amphotericin B-loaded liposomes (286), a new liposomal formulation was produced and commercialized by Vestar, Inc., with the name of AmBisome. This pharmaceutical formulation is made up of phosphatidylcholine, cholesterol, distearoylglycerol, and amphotericin B (2:1:0.8:0.4 molar ration) with a 9.5 lipid/drug ratio. The mean size of these small unilamellar liposomes ranges from 45 to 80 nm.

Infective diseases caused by parasites are a great problem for developing countries. In these particular infections, especially for those pathologies where the infection agent is closely associated to the RES, the possibility of delivering already existing drugs by liposomes can represent a very attractive strategy. In fact, due to poor drug membrane penetration, *in vivo* treatments of these pathologies are often poorly effective, despite the *in vitro* effective activity of the drug. An interesting example of effective liposome treatment of protozoal diseases is leishmaniasis.

The parasites of leishmaniasis live almost exclusively in fixed macrophages at the level of the RES (liver, spleen, and the rest of the visceral). Antimonial derivatives (therapeutic index approaching 1) are the most effective drugs for this pathology. Liposomal formulations of these drugs can improve the therapeutic effectiveness up to a thousand times with respect to the free drug. Experiments showed that doses close to the lethal level of free potassium antimony tartrate were ineffective, but a single dose (40% of the previous dose; $20 \, \text{mg} \cdot \text{kg}^{-1}$) of drug-loaded liposomes completely eliminated the parasites (287).

Liposomes can also be used as immunoadjuvants for vaccines (288) and as macrophage activators against tumoral, viral, and microbial cells. For both applications, a substance is delivered to macrophages thus triggering immunization, immunomodulation, or activation by means of antigens. The presence in the liposome structure of a nonliposomal adjuvant, that is, muramyl tripeptide covalently coupled to phosphatidylethanolamine, can enhance the antibody response induced by liposome-associated antigens.

Liposomes can be efficaciously used to deliver to the CNS. Under some pathological conditions (i.e., tumors, ischemia, and traumatic shocks) a hypermeabilization of the blood–brain barrier can occur, thus allowing the passage of very small aggregates (< 100 nm). The CDP–choline loaded very small (50 nm) long circulating liposomes were used to treat successfully the cerebral ischemia (289,290). The drug-loaded liposome was able to increase the amount of drug that reached the brain and the survival rate of rats submitted to ischemia and reperfusion (Fig. 45). The liposomal formulation is also able to efficaciously antagonize the phenomenon of postischemic damage maturation that is the main reason of a poor neuronal recovery and hence of an enlargement of the damaged (291).

Liposome formulations resulted effective not only in systemic administration, but also in topical administration (e.g., dermal, mucosal, ocular, pulmonary).

The potential application of liposomes as dermal delivery systems has been extensively investigated, with regard

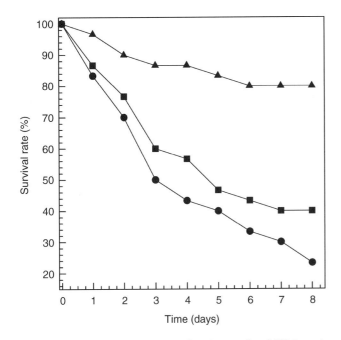

Figure 45. Survival rate of postischemic reperfused Wistar rats (320–350 g). The duration of the ischemic event was 30 min. The rats were treated with saline ●, with the free drug ■, or with CDP–choline loaded liposomes ▲. Unloaded liposomes showed no effect on rat survival (data not reported). The results are expressed as the percentage of the total number of animals in each group which survived ischemia as a function of time. Data from Ref. 290.

to vesicle composition and size (292–294). It was proposed (295) that the main advantages of using liposomes as topical drug formulations were due to their demonstrated ability: (1) to reduce serious drawbacks and incompatibilities that may arise from undesirably high systemic absorption of drugs; (2) to enhance accumulation of drugs at administration sites due to the high substantivity of liposomes with biological membranes; and (3) to the possibility to incorporate both hydrophilc and hydrophobic drugs. In addition, liposomes can be readily prepared on a large scale. The requirement of smart drug delivery systems for skin application comes from the necessity to have a modulation of both the administration rate and the skin permeation properties, namely, a sustained drug release strictly confined at the level of the skin with no systemic absorption or an enhanced transdermal effect to deliver the drug to some inner structures (e.g., joints) or to achieve a systemic effect are required as a function of the disease to be treated (296). By the use of quantitative skin autoradiography, it was demonstrated that small liposomes allowed the localization of a greatest amount of caffeine (hydrophilic drug) in the epidermis and a lowest amount in the dermis and appendages (297). In this case, liposomes ensured a drug skin accumulation three times greater that that observed for an aqueous drug solution prepared in the presence of penetration enhancers.

The liposome lipid composition and the thermodynamic state of the liposomal bilayers play a crucial role in the effect of this vesicular carrier on drug transport rate across the skin. In particular, incorporation of drugs in the liquid-state liposomes provides a higher skin permeation rate than that observed for drug-loaded gel-state (the so-called solid) liposomes (298). Liposomes made up of the same lipids usually present in the skin were prepared and referred to as skin-lipid liposomes (299). These kind of liposomes are able to provide a drug dermal delivery of the highest drug disposition within the deeper skin layers, that is, in the epidermis and dermis, while avoiding systemic drug adsorption (299). For example, skin-lipid liposomes can be a suitable topical carrier for chronic topical applications of corticosteroids by optimizing drug concentration at the site of action while minimizing systemic absorption and, as a consequence, possible side effects (300). In the case of transdermal drug delivery requirements, the high deformability of vesicular carriers seems to be a fundamental feature to achieve the intact vesicles penetration, thus also favoring the delivery of encapsulated drugs across the skin. Special liposomes characterized by an high bilayer elasticity have been developed, namely, ethosomes and transfersomes. Ethosomal systems are different from transfersomes by their structure and mechanism of action. As an example of different behavior, occlusion has no effect on skin permeation of molecules from ethosomes, while transfersomes are unable to enhance drug delivery under the occluded conditions. Ethosomal systems contain vesicles with fluid bilayers (soft vesicles) in a hydroethanolic milieu. Both components have a crucial role in the delivery of the active agent (301,302).

Liposomal colloidal carriers also can be applied as ophthalmic drug delivery devices to increase the bioavailability and the efficacy of drugs (303). Liposomes can enhance the ocular drug absorption and prolong the precorneal retention time (303), thus increasing drug effectiveness. In particular, the ocular application of positively charged small oligolamellar liposomes seems to be promising, considering that positively charged delivery devices may ensure a suitable bioadhesivity with the negatively charged corneal epithelium. As shown in Fig. 46, the acyclovir-loaded liposome showed a significant ($P < 0.005$) and noticeable increase of drug levels in the aqueous humor compared to the liposome–acyclovir physical mixture and the free drug (260). Several mechanisms can be proposed to elucidate the ocular effects of liposomes, but adsorption and/or lipid exchange seem to be most probably involved (303). Cornea permeability alteration due to liposomes may be discarded as a plausible explanation for enhanced drug penetration, since the presence of empty lipid vesicles added to drug solutions does not enhance the availability of the drug. Last, but not least, liposomes present a very good ocular tolerability showing no evidence of ocular inflammation or discomfort (260).

Niosomal Carrier. Niosomes are nonionic surfactant self-assembled vesicles that presents a structure similar to liposome (Fig. 47) and hence they can represent alternative vesicular systems with respect to liposomes, due to the niosome ability to encapsulate different type of drugs within their multienvironmental structure (304). The first application of niosomes was the cosmetic field followed by their use as drug delivery systems (305).

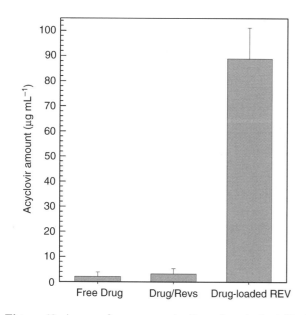

Figure 46. Aqueous humor concentrations of acyclovir at 30 min following topical instillation (50 μL) of acyclovir-loaded positively charged REVs (oligolamellar) liposomes (DPPC-Chol-DDAB 7:4:1 molar ratio), acyclovir–liposomes physical mixture and aqueous solution. Each bar represents mean values ±S.D. of four experiments. Data from Ref. 260.

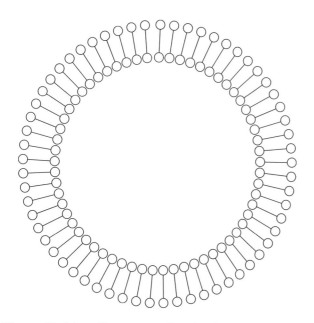

Figure 47. Schematic representation of a niosome structure. ◯, hydrophilic head group; —, hydrophobic tail (305).

Components Used in Niosome Preparation. The main components of niosomes are nonionic surfactants. Different types of self-assembling nonionic surfactant were proposed as starting material to prepare niosomes (i.e., the SPAN and the Brij series). The type of surfactant can influence the stability of the vesicular system being able to influence the fluidity of bilayer structures. In particular, the nonionic surfactant can influence the leakiness of the entrapped drug from niosomes with the

following increasing order, SPAN80 < SPAN20 < SPAN40 < SPAN60.

High niosomal concentration of soluble surfactant agents can influence the solubility of this vesicular colloidal carrier and determine the formation of micelles or complex aggregates. This phenomenon is observed with the presence of actylglucoside in the niosome formulation. This compound can destabilize the niosome bilayer and start a micellization phenomenon (305).

Another fundamental component for the preparation of niosomes is cholesterol. This molecule is used as an additive compound both to reduce the temperature of the vesicular gel to the liquid-crystal phase transition (305) and to decrease the overall HLB value of the surfactant mixture used for the preparation (306,307). Thus, cholesterol allows a more efficient aggregation of the nonionic surfactant component into a closed bilayer structure, and then a higher stability of the niosomal vesicles. The inclusion of cholesterol into niosomal formulation can reduce the leakiness of the membrane. A 1:1 molar ratio of cholesterol and nonionic surfactant is generally used for niosome preparation.

A parameter that should be taken into account in the choice of the niosome component is the physicochemical property of the encapsulated drug, due to a series of possible interactions occurring with the nonionic surfactant component leading to the formation of homogeneous dispersion or aggregate structure (305).

Methods of Niosome Preparation. Niosomes are prepared through the hydration of a mixture of nonionic surfactant–helper lipid (cholesterol) (1:1 molar ratio) at a temperature ranging from 40 to 70 °C followed by a suitable sizing process to obtain the required colloidal dispersion characteristics. The methods used to reduce the niosome mean size and to achieve an homogenous size distribution are similar to those used for liposomes, that is, extrusion through decreasing pore size polycarbonate filters, cycles of sonication, and high pressure homogenization (305,308). Similarly to liposomes, the mean size of niosome formulations is very important to reduce the RES uptake (305).

As concern the hydration of the nonionic surfactant–helper lipid mixture, some procedures reported for liposomes also can be used (e.g., the TLE method). In addition, other specific preparation methods have been developed for niosomes (305):

1. Injection of an organic solution (ether) of surfactant and cholesterol in a drug aqueous solution and heating of this mixture above the boiling point of the organic solvent;

2. Formation of an o/w emulsion between a drug aqueous solution and an organic solution of surfactant–cholesterol. Then, the organic phase is evaporated off and an aqueous niosomal colloidal dispersion is obtained;

3. Injection of the melted surfactant–cholesterol mixture in an aqueous heated solution of the drug under continuous stirring or vice versa injection of a

warmed aqueous drug solution into the niosomal component mixture.

The niosomal formulations obtained with the previous mentioned methods are generally micro size.

Considering the importance of the drug loading parameter, some procedures can be carried out to increase the amount of the ecapsulated drug within niosomes. There is evidence (305) that the DRV method, originally developed for the preparation of multilamellar liposomes with a high entrapment efficiency of water-soluble drugs (309), can also be used for niosomes with an increaes of their loading capacity from 3.3 to 64.4%. Another method successfully used to increase the amount of drug entrapped in niosomes is based on the formation of a pH gradient (305).

At the end of the preparation procedures, the excess of nonencapsulated drug is removed by dialysis, centrifugation, or filtration.

Toxicological Aspects of Niosomes. Considering that niosomes are made up of at least 50% synthetic nonionic surfactant, the toxicological profile of this carrier is very important for its application as a drug delivery system. Unfortunately, there are not many studies on niosome toxicity. An *in vitro* investigation, made on a model of ciliotoxicity to evaluate the influence of alkyl polyoxyethylene moiety of niosomes on the nasal mucosa, showed that increasing of the alkyl chain length of the nonionic surfactant determined a reduction of toxicity while the increase of the polyoxyethylene chain length pronounced the carrier ciliotoxicity. These findings seems to be correlated with the thermotropic state of niosomes, considering that the longer the alkyl chain the higher the transition temperature from gel-to-liquid phase, while the longer the polyoxyethylene chains the lower the transition temperature. This findings concluded that gel-state niosomes are less ciliotoxic than the liquid-state vesicles. On the contrary experiments on human keratinocytes showed on toxic activity related to both the alkyl chain length and the length of polyoxyethylene chain (310).

For the parenteral administration of niosomes, usually through the IV route, the evaluation of the vesicular system hemocompatibility is very important. The incubation of $C_{16}G_2$ and Span 60 niosomes with rat erythrocytes showed <5% hemolysis after 5 h. This level of hemolysis is not significant, considering that <2% of an injected dose of C16G2 niosomes is still present in the blood stream 5 h after dosing (305).

In the case of niosomal soluble surfactant components, a dose-dependent effect was observed. When low concentrations are used, the soluble surfactant is totally incorporated in the niosome structure and a drastic reduction of its intrinsic toxicity is achieved. The situation changes when the amount of soluble surfactants (e.g., Solulan C_{24}) is increased, because the formation of micelles occurs, and then the free monomers and/or micelles may exert their toxic action on cultured cells (311). Therefore, the whole niosomal carrier should be investigated for potential toxicity rather than the single components.

The issue of niosome toxicity is quite complex due to the fact that the presence of a drug can change the toxicological profiles of the unloaded carrier. For example, the inclusion of doxorubicine in $C_{16}G_2$ niosomes produce a severe dose-dependent inflammatory effect at the level of the lung within 24 h following intraperitoneal administration (305). After intraperitoneal administration of empty $C_{16}G_2$ niosomes or the free drug, such an effect on lungs is not observed. A possible explanation is the fact that doxorubicin-loaded niosomes are transported away from the peritoneum by the lymphatics via the thoracic duct allowing a higher dose in the main veins emptying into the heart. This hypothesis can be supported by the fact that 56% of a methotrexate-loaded niosome formulation is found in the thoracic lymph following intraperitoneal administration with respect to 12% observed for a free drug solution (312).

The modulation of drug toxicological effect is an important aim of the niosomal carrier. The encapsulation of vincristine in noisomes can reduce the free drug toxicological profile and improve the drug antitumoral activity in S-180 sarcome and Erlich ascites mouse models (313).

Niosomes in Complex Systems. The need for a more precise controlled drug release prompted the research of new and more sophisticated delivery systems. For this reason, niosomes based on Span surfactants were used to prepare a v/w/o (vesicle in water in oil system) niosomal formulation (314). The release rate of carboxyfluorescein, a hydrophilic fluorescet probe, showed the following increasing trend: v/w/o < w/o emulsions < niosome dispersion. Also, the nature of surfactant can influence the release of the fluorescent probe according to the following decreasin order: Span 20 > Span 40 > Span 60. The presence of Span 80 in the v/w/o system can drastically increase the probe release from the system due to its unsaturation in the alkyl chain, which generate a more leaky bilayer structure. While, the crystallization of Span 60 in the oil phase elicit the formation of an oil gel phase that can noticeable reduce the release rate from this vesicular system (314). A temperature-dependent release can be obtained in Span 60 v/w/o by adding Span 20 as a stabilizer, thus providing a faster probe release at 37 °C (305).

Niosome colloidal dispersion can be easily viscosized by the addition of hydrocolloids.

The addition of Solulan C24 in C16G2 niosomes determined the formation of the discome phase, that is a large vesicle (~60 μm) able to encapsulate hydrophilic compounds. These giant vesicles were found to be of two types: large vesicles that appear ellipsoid in shape and large vesicles that are truly discoid (305). The features of the discome structure prompt the used of this particular niosomal system as an ophthalmic drug delivery.

Therapeutic Applications of Niosomes. Niosomes can be used as a fine drug delivery systems being able to confer a certain selectivity to the entrapped drug as a function of their composition and physicochemical properties. After IV administration, niosomes show a high liver tropism (304,305). However, a niosomal formulation containing doxorubicin, composed of palmitoyl muramic acid, cholesterol, Solulan C_{24}, can escape from the liver uptake (305).

At the same time, a iopromide-loaded niosomal formulation extruded through a 220 nm filter and with the presence of stearylamine in its composition is able to accumulate in the kidneys (315). These findings showed that the presence of a positive charge on the surface of the niosomes can improve the targeting to the kidneys. The intraperitonel administration of niosomes with Span 80 in their formulation (312) can produce a lymphatics targeting, while $C_{16}G_2$ niosomes (305) after intraperitoneal administration can act as a depot system.

The presence in niosomal formulations of surfactant characterized by ester bonds can support the enzymathic degradation by esterases present in plasma, thus influencing the biodegradability, the residence time, and the stability of the system in the plasma. Moreover, the nature of the entrapped drug can influence the structure of the niosomal surface and the biodistribution of the system.

The first application of niosomes was as antiparasitic vesicular system for the treatment of leishmaniasis. The administration of a niosamal formulation containing stibogluconate was very useful to reduce the parasite disease because niosomes acted as a drug depot in the liver. In this case, the antiparasitic activity of niosomes regarding to the liver leishmania donovani can be correlated to the rapid uptake of the formulation in the liver after IV administration. However, this formulation cannot eradicate the parasite in the spleen and bone marrow. For this reason, different types of polyoxyethylene niosomes ($C_{16}EO_2$, $C_{16}EO_4$, $C_{16}EO_6$) are used to suppress the parasite in the spleen and bone marrow (305).

The IV administration of 100 nm of $C_{16}G_3$ niosomes containing methotrexate can improve the hepatic levels of the drug with serum levels of the drug higher than when it is administered in solution (316). In particular, a 23-fold increase in the area under the curve of metotrexate plasma level as a function of time is observed after IV administration of niosomes (4.5 μm mean size) containing Span 60 to tumor bearing mice (317), this finding is probably due to the great size of this vesicular system. Span 60 niosomes can further increase the plasma level of methotrexate if they are administered following the macrophages activation with mramyl dipeptide-gelatin derivatives (317). The oral and IV administration of $C_{16}G_3$ niosomal formulation encapsulating methotrexate can cross the blood–brain barrier and provide a sustained release of this drug at the level of the CNS (316). However, the delivery of drug to the brain with niosomes has not been successful.

The administration of doxorubicin-loaded $C_{16}G_3$ niosomes (850 nm mean size) in tumor bearing mice determined a high drug level in the tumor site, serum, and lung, but not in the liver (305,318). While, doxorubicin-loaded 240 nm niosomes made up of Span 60 increased plasma, liver, and tumor levels. The reduction of proliferation of the S-180 sarcoma in NMRI mice after IV administration of niosomal formulation containing doxorubicin demonstrated an increased drug anticancer activity after encapsulation in niosomes (Fig. 48). At the same time the side effects, in particular cardiotoxic activity, are reduced following entrapment in niosomal formulations (314). Niosomes can improve the antitumoral effect of vincristine in S-180 sarcoma well as other anticancer drugs (313).

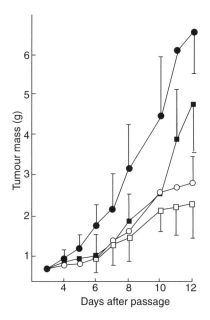

Figure 48. The growth in the mass of implanted tumor as a function of time after IV injection of (●) phosphate buffered saline pH 7.4, (■) doxorubicin solution (5 mg·kg^{-1}), (○) doxorubicin (5 mg·kg^{-1}) $C_{16}G_3$ niosomes, (□) doxorubicin (5 mg·kg^{-1}) $C_{16}G_3$:cholesterol (50:50) (305).

A diclofenac-loaded niosomal formulation composed by Span 60, cholesterol and DCP (22:73:5) produces a noticeable reduction of inflammatory processes in rat more efficaciously than the free drug. The improved activity of the drug can be determined by an increase in the area under the plasma time curve. Similar findings were obtained for niosomes-containing flurbiprofen, which showed an improved drug effect and bioavailability with a reduction of side effects produced by the free drug (305).

Niosomes can be used as agents for diagnostic imaging. Iopromide radioparque agent encapsulated in niosomes made up of $C_{16}G_3$, cholesterol, and stearylamine, can be concentrated in the kidney after IV administration (315). As mentioned above, the kidney targeting action is mediated by the positive charge on niosome surface.

Niosomes also can be effectively used for the oral delivery of drugs. The first application in this field was carried out with methrotrexate-loaded $C_{16}G_3$ niosomes characterized by a mean size of 100 nm (316). This investigation showed higher levels of methrotrexate in serum, liver, and brain after oral delivery using the niosomal formulation with respect to the free drug. A certain interest is focused on the possibility of using niosomes as carrier for the oral delivery of peptides and proteins. For example, ovoalbumine-loaded niosomes are able to increase the production of specific antibodies after oral administration (305).

Other successful applications of niosomes as delivery systems concern the topical administration of drugs and particularly the transdermal and ophthalmic delivery of drugs.

Niosomal formulations can increase the amount of drug permeated through the stratum corneum (319), even if the exact mechanism involved in the drug and/or carrier passage has to be investigated and elucidated in a more detailed way. A hypothetical mechanism of skin penetration is related to a possible reorganization of the niosomal membrane at the level of the stratum corneum (320). *In vitro* data showed an efficacious transdermal delivery of oestradiol when it is entrapped in $C_{18}EO_7$ and $C_{12}EO_7$ niosomes. The improved drug passage through the outer skin layer seems to be mediated by the high flexibility of the bilayer structure of some niosomal formulatons (319). Similarly, a niosomal formulation made-up of glyceryl dilaurates ($C_{16}EO_7$) and cholesterol can increase the passage through the stratum corneum and the penetration of cyclosporine A into the inner layer of the skin (305). Then, niosome can be used as a transdermal drug delivery system for both hydrophobic and hydrophilic drugs.

Niosomes were proposed as a potential ophthalmic drug delivery system. Cyclopentolate-loaded niosomes made-up of Span 20 and cholesterol can pass through the cornea in a pH dependant manner, that is, pH value 5.5 is optimal for the cyclopentolate penetration, while at pH 7.4 a decreased permeation was observed. However, the *in vivo* mydriatic response is irrespective of the pH of the niosomal formulation. The explanation of the increased corneal adsorption of cyclopentolate may be due to a niosome-induced modification of the permeability characteristics of the conjunctival and scleral membranes (321).

Similar to liposomes, niosomes can be used as a vaccine adjuvant. A niosomal formulation composed by 1-monopalmitoyl glycerol, cholesterol, diacetyl phosphate can be used to encapsulate antigenic compounds and this result is fundamental for the adjuvanticity (305). A v/w/o niosomal system containing Span 80 and cotton seed oil was evaluated as an immunological adjuvant using the antigen tetanus toxoid (314). An increased secondary response (level of IgG1) was observed when the v/w/o formulation was administered by the intramuscular route in comparison with the vesicle formulation and the free antigen.

Ethosomal Carrier. Ethosomes have been invented by Touitou (322–324). The low toxicity and the property of ethanol as a permeation enhancer (325) as well as the possibility to include ethanol in the liposomal formulation, has brought to the realization of a new vesicular system for transdermal delivery: ethosome (301).

Ethosomes presents interesting features correlated with its ability to permeate intact through the human skin due to its high deformability. In fact, ethosomes are soft, malleable vesicles tailored for enhanced delivery of active agents. It has been shown that the physicochemical characteristics of ethosomes allow this vesicular carrier to transport active substances more efficaciously through the stratum corneum into the deeper layers of the skin than conventional liposomes (326). This aspect is of great importance for the design of carriers to be applied topically both for topical and systemic drug administration. Furthermore, the ethosomal carrier is also able to provide an effective intracellular delivery of both hydrophilic and lipophilic molecules (327) and also the penetration of an antibiotic peptide (i.e., bacitracin) within fibroblast cells was facilitated (328).

Formulative Aspects of Ethosomes. Ethosomes are a vesicular system made up of a phospholpid component, ethanol, and water. Phospholpid is the lipid component that confers the shape of vesicle to the delivery system. Ethanol is an important component in ethosome due to its destabilizing action regarding the packed-ordered structure of conventional liposomes (326), thus conferring the characteristic elasticity and deformability to this vesicular carrier. There are a number of methods that can be used to prepare stable ethosomal formulations depending on drug and the target of drug delivery (322–324). Among these, a frequently used method to prepare ethosomes is based on the dissolution of phospholipids in ethanol (20–50% w/v). Then, an aqueous solution is added to the lipidic solution under stirring thus allowing the formation of ethosomes (327–329).

The ethanol/phospholipid ratio used for the preparation of ethosomes is a crucial factor influencing the mean size and size distribution of ethosomes (Fig. 49). Usually, ethosomes prepared with a great amount of ethanol ($\geq 40\%$ v/v) show a narrow vesicle size distribution. The size of ethosomes decreases with increasing ethanol concentration, while the concentration of phospholpid influenced the vesicle mean size in a different way, namely, the higher the phospholpid concentration the larger the ethosome mean size (301,330). The amount of ethanol used in the formulation can modify the superficial charge of ethosomes and the skin interaction (301). Normally, the presence of drugs have no significant influence on both mean size and size distribution. Ethosome composition can also influence the lamellarity as shown by electron transmission microscopy (Fig. 50), since the formation of either unilamellar or multilamellar ethosomes is a multifactor process.

Ethosomes can entrap hydrophobic and hydrophilic molecules in their structure. With respect to liposomes, where hydrophilic drugs are entrapped in the aqueous

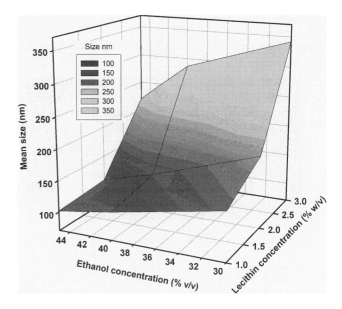

Figure 49. Influence of the amount of ethanol and lecithin used for the preparation of ethosomes on vesicle suspension mean size and colloidal polydispersity index. Data from Ref. 330.

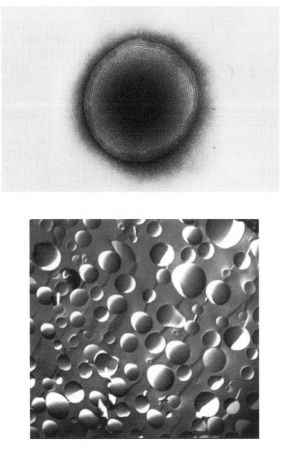

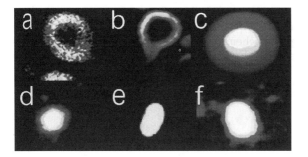

Figure 50. Transmission electron microscopy of ethosomal vesicles composed of 2% lecithin and 30% ethanol (a) (301). Freeze-fracture electron micrographs of ethosomes composed of 45% ethanol and 2% lecithin (b) (330).

compartment and hydrophobic drugs are in the lipid bilayer core, in ethosomal formulations drugs are homogenously present in ethosome structures in spite of drug physicochemical properties (301,327) (Fig. 51). This finding can be explained by the multilamellarity of the ethosomal

Figure 51. Entrapment of fluorescent probes by phospholipid vesicles determined by confocal scanning laser microscopy. Liposomes (a–c) or ethosomes (prepared with 2% lecithin and 30% ethanol) which (d–f) were prepared with one of three following fluorescent probes: rhodamine red, a highly lipophilic molecule (a,d); D-289, an amphiphilic molecule, (b,e); calceine, a hydrophilic molecule (c,f). White represents the highest concentration of a probe, followed by yellow, with red being the lowest probe concentration (301).

vesicles as well as by the presence of ethanol in the ethosome, which allows for better solubility of the lipophilic and amphiphilic probes (301). The ethosome composition can also influence the drug entrapment efficiency, that is, the amounts of ethanol and phospholpid used for ethosome preparation positively influence the loading capacity of the colloidal carrier. Namely, the higher the amount of ethanol and phospholpid the greater the drug entrapment within ethosomes (301,326,330), the values of drug entrapment efficiency are often higher than those expected for a conventional vesicle formulations (330). This fact can be explained by the presence of ethanol, which increases the drug solubility in the polar phase of ethosomes (301).

Therapeutic Potentialities of Ethosomes. The enhanced percutaneous permeation capability of ethosomes is due to the unique feature of this carrier that is able to interact with the stratum corneum and to elicit a reversible disorganization of the stratum corneum lipid packing order, thus increasing the skin permeability to drugs and vesicles (329).

An important characteristic to be evaluated before the proposal of a drug carrier as a potential topical drug delivery system is its *in vivo* skin tolerability on human subjects. *In vivo* reflectance spectrophotometry data (330) on volunteers showed that ethosomes elicit no induction of skin erythema, while a hydroethanolic solution with an equal water/ethanol ratio of ethosomes induces a remarkable skin erythema (Fig. 52). These results demonstrate that ethanol present in the ethosomal formulation is not

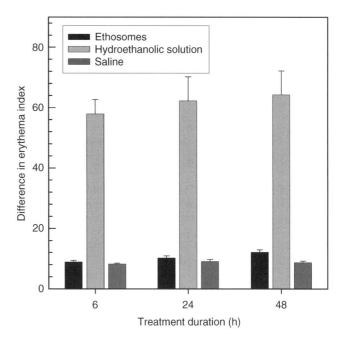

Figure 52. *In vivo* human skin tolerability of various topical formulations after 6, 24, or 48 h of treatment. Results are expressed as a mean value of the variation of the erythema index ($n = 6$) ± standard deviation. Legend keys: ethosomes, formulation containing 2% (w/v) Phospholipon 90 and 45% (v/v) ethanol; hydroethanolic solution, solution of water, and ethanol at a volume ratio of 55:45; saline, control saline (0.9% w/v NaCl in water) solution (330).

able to act as a skin erythema-inducing agent, even though it is present at a high concentration.

A wide range of drugs have been formulated in ethosomal carriers and tested *in vitro*, *in vivo*, and in clinical studies. These molecules comprise steroid hormones, antivirals, antibiotics, vitamins, peptides, and cosmeceutical agents. Moreover, ethosomes are very efficient carriers for targeting molecules to the pilosebaceous units and could be used for acne and alopecia treatment. Carrier consists of materials approved for pharmaceutical and cosmetic use (316,327–332).

An interesting example of the ethosome potential application as innovative topical carriers is represented by the transdermal delivery of cannabidiol (333), a new drug candidate for treatment of rheumatic diseases, that presents a number of drawbacks when administered orally. The ethosomal formulation is able to prevent the inflammation and edema induced by subplantar injection of carrageenan in ICR mice.

Often, the skin permeation enhancement observed for all ethosome-based formulations is much greater than that can be expected from ethanol alone. This behavior can be due to a synergistic mechanism between ethanol, phospholipid vesicles, and skin lipids (301).

Two different research groups reported an *in vivo* sustained release effect of ethosomes with a prolongation of the drug therapeutic activity, which can be related to an accumulation in the skin (330,333).

Then, recent findings on ethosomes are very encouraging and confirm that this carrier is very promising for the topical administration due to the enhanced delivery of drugs through the skin, thus prompting various opportunities for the development of suitable therapeutic strategies through the topical route.

Ultradeformable Vesicular Carrier. It is believed that liposomes, when administered on the skin, first disintegrate their structure, and then diffuse through the barrier in the form of small fragments or lipid monomers (334). Conventional rigid liposomes were shown to be unsuitable vesicular carriers to cross the skin barrier (335). Highly deformable vesicles were developed, the so-called transfersomes or ultradeformable liposomes were invented by Cevc (336). It was shown that the high deformability of vesicular carriers could allow them to penetrate intact skin if applied nonocclusively *in vivo* (336,337), thus favoring the delivery of encapsulated drugs across the skin (338,339). Ultradeformable vesicles seem to cross the skin without irreversible disruption, probably because they are much more elastic, and hence more deformable respect to classic liposomes. For the preparation of ultradeformable vesicles, the so-called edge activators were incorporated into the phospholipid bilayers at suitable amounts, namely, bile salts were often used for this purpose (340).

Formulative Aspects of Ultradeformable Liposomes.
Lecithins and a bile salt at different molar ratios are the main components of ultradeformable liposomes that can be prepared with the TLE method or any other used for liposome preparation (the section main methods to prepare liposomes) (336–340). Small amounts of ethanol ($\leq 7\%$ v/v)

are normally used for the preparation of ultradeformable liposomes.

Similarly to ethosomes, the ratio between the various components is a crucial factor for the determination of the physicochemical and drug-loading capacity properties of ultradeformable liposomes.

The DSC studies have demonstrated that the amount of the edge activator is related to the increased fluidity of the vesicular bilayers up to a certain values, beyond this value the formation of mixed micelles and other kinds of colloidal aggregates were observed (341). For this reason, large amounts of edge activator beyond a certain value hinder the transdermal drug delivery; in fact, mixed micelles and aggregates are much less effective transdermal carriers than ultradeformable liposomes. The formation of a coexistence region characterized by various phospholipid/bile salt aggregates (i.e., mixed vesicles, opened vesicles, mixed micelles, and rod-like mixed micelles) is evidenced by a reduction of the mean size and a concomitant increase of the polydispersity index values, thus showing the presence of a wide size distribution (341). The presence of bile salts in the composition of ultradeformable liposomes leads to a negative zeta-potential due to the increase of negative charge (carboxylate group of bile salts) along the surface of vesicle bilayers.

The ratios between the components and the type of edge activator used to prepare ultradeformable liposomes can influence the amount of drug entrapped within this carrier. When nonionic surfactants (i.e. Span and Tween) are used instead of bile salts, a decrease of the drug entrapment efficiency of the carrier is observed. In any case, a high concentration of edge activators cause a drastic reduction of the drug loading capacity due to the presence of other forms of aggregation than vesicles. The above mentioned aggregates have a poor drug loading capacity. The higher hydrophilic form of ultradeformable liposomes than conventional liposomes and their high flexibility avoids the aggregation and fusion of the transfersomal system providing a stable vesicular structure (342).

Similarly to other vesicular carriers, ultradeformable liposomes can entrap different types of molecules (i.e., lipophilic, hydrophilic, and amphipatic drugs). The release rate of entrapped drug from ultradeformable liposomes is mainly influenced by the carrier composition and the drug physicochemical properties. Generally, the release of water soluble drugs from ultradeformable liposomes is modulated by the concentration gradient between the inner and the outer compartment. The water gradient through the skin can trigger the release of the entrapped drug. The release of hydrophobic drug is slower than that of the hydrophilic drug, and it is confined to the contact and lipid exchange between ultradeformable liposomes and biological membranes (342). The slower release of the hydrophobic drugs is due to a strong interaction between drugs and lipid bilayers (343). Amphipatic drugs have intermediate release characteristic between hydrophobic and hydrophilic drugs.

Therapeutic Potentialities of Ultradeformable Liposomes.
Ultradeformable liposomes are characterized by a deformable structure that can pass intact through the skin using a

water active gradient, thus favoring the drug delivery through the skin without modifying the integrity of the cutaneous barrier (342,343).

In vitro and *in vivo* tests with ultradeformable liposomes showed that this vesicular system does not produce any toxic effects after topical application and it is well tolerated by the skin tissue (342).

Skin is a nanoporosus barrier that permits only the passage of small (nanometer size molecule) compounds (334,342). The nonocclusive topical application of ultradeformable liposomes undergoes water evaporation from the formulation and the consequent dying out of the vesicles (343). The elastic and hydrophilic properties of ultradeformable liposomes determine the movement of the vesicle through the skin pores by following the transdermal water gradient. The topical application of ultradeformable liposomes can increase the size of skin nanopore (78,344). After ultradeformable liposome percutaneous permeation, these vesicles can distribute in cells and after the bypassing of cutaneous capillary they can reach the subcutaneous tissue.

The most important example of the application of ultradeformable liposomes as transdermal drug delivery is represented by Transfenac, formulation of diclofenac in ultradeformable liposomes (338). Transfenac mediates the agent transport through intact skin and into the target tissues. Therapeutically meaningful drug concentrations in the target tissue are reached even when the administered drug dose in Transfenac is $< 0.5\,\mathrm{mg \cdot kg^{-1}}$ body weight. Diclofenac association with ultradeformable carriers permits it to have a longer effect and to reach 10 times higher concentrations in the tissues under the skin in comparison with the drug from a commercial hydrogel. The relative advantage of diclofenac delivery by means of ultradeformable liposomes increases with the treated muscle thickness and with decreasing drug dose, as seen in mice, rats, and pigs (338); this can be explained by assuming that the drug associated with carriers is cleared less efficiently by the dermal capillary plexus.

Transfenac, hence promises to be a useful formulation for the treatment of diseases of superficial tissues, such as muscles or joints, having the potentiality to replace combined oral–topical diclofenac administration in humans.

Particle Drug Carriers

Micro- and nanoparticles are solid colloidal suspensions in which the mean particle size is > 1 or $< 1\,\mu m$, respectively. Under the morphological point of view, two different types of particles can be distinguished: capsules and spheres (Fig. 53). Sphere systems are usually characterized by a porous matrix in which drugs are contained, while capsule systems are formed by a core containing drugs surrounded by a shell.

Polymeric Particles. These colloidal carriers are prepared from natural or synthetic polymers and, in dependence of the preparation method and of the polymer used, micro- or nanocapsules and micro- or nanospheres can be distinguished. Polymeric particles have become very important because of their ability to deliver a variety of

Figure 53. Schematic representation of spheres and capsules as potential drug delivery devices.

drugs to different areas of the body for sustained periods of time (345–347). Up to the end of 1980s, microparticles were extensively investigated. Now great interests are focused on the much smaller carriers (e.g., nanoparticle) that are able to ensure fine drug delivery opportunities both in terms of efficacy and selectivity. For this reason, this section will be mainly focused on polymeric nanosystems.

Concerning the materials used for the preparation of these colloidal carriers, natural polymers (i.e., proteins, polysaccharides, waxes) are not widely used because they present a huge variability in their purity and defined physicochemical properties. Furthermore, they often require a cross-linking procedure that may cause an alteration of the encapsulated drug. For these reason, mainly synthetic polymers have received attention and have been largely investigated for potential use in drug delivery devices.

The use of a large series of polymers is restricted and limited by their bioacceptability, which is also influenced by colloidal particle mean size. In fact, the diameter of the smallest blood capillaries is $\sim 4\,\mu m$, thus nanoparticles should have a smaller diameter than this to traverse all capillaries. As a general consideration, for the suitable choice of the appropriate macromolecular polymer to be used as a nanoparticle matrix, the colloidal particle size and the preparation method will first depend on the biocompatibility of the polymer, second on the physicochemical properties of the drug, as well as on the therapeutic goal to be reached. In this colloidal carrier, drugs can be adsorbed, attached, dissolved, entrapped, and/or encapsulated (348). Micro- and nanoparticles can be used to deliver both hydrophobic and hydrophilic molecules, proteins, vaccines, biological macromolecules. They can also be formulated for targeted delivery to all organs or made for long-term systemic circulation (235). Thus, a lot of synthesis procedures exists.

Preparation Methods and Formulative Aspects. Drugs can be incorporated into nanoparticles in a number of ways: (*1*) drug can be entrapped in the polymeric matrix; (*2*) it can be encapsulated in a nanoparticle core; (*3*) it can be chemically conjugated to the polymer; (*4*) it can be surrounded by a shell-like polymer membrane; (*5*) it can be adsorbed on particle surface.

Polymeric nanoparticles can be prepared using a lot of different techniques. One of the most used preparation methods is the emulsification–solvent evaporation technique. The polymer and the drug are solubilized in an organic solvent and an emulsion is prepared by adding water and a surfactant. Liquid nanodroplets are produced by sonication or homogenization, and then the organic solvent is evaporated in order to achieve the nanoprecipitation of the polymeric material in solid nanoparticles (345,349). Obviously, this procedures can be used only for hydrophobic drugs. To allow the encapsulation of hydrophilic molecules, a modification of this procedure led to the multiple emulsion technique (350).

Another method is the phase-inversion nanoencapsulation (PIN), which has been used to encapsulate insulin for oral administration (351). A limitation of these two techniques is the use of toxic and fluorinated solvents, which may cause drug degradation. For these reasons, other techniques, that do not compromise drug stability have been developed.

One of these is the emulsification–diffusion method. In this case, the polymer and the active compound are dissolved in a partially water-soluble solvent. This organic solution is added, and then emulsified in an aqueous phase containing a surfactant. To favor the precipitation of nanoparticles, additional water is added to the emulsion under stirring. At the end of the process, the solvent can be removed by centrifugation or dialysis.

The nanoprecipitation method involves the dissolution of the polymer and the drug in a freely water-miscible organic solvent (e.g., acetone) and then the addition of this organic solution into a water phase containing a nonionic surfactant (e.g., Pluronic F68). The organic solvent is then removed under reduced pressure by a rotavapor (352). This procedure leads to the formation of nanospheres, but if a biocompatible oily component is added in the organic solution nanocapsules are formed (353).

The formation of nanospheres or nanocapsules also can be achieved by an in situ polymerization process. The emulsion or micellar polymerization is the most used approach to achieve nanocapsules and nanospheres by starting from the polymeric monomer, respectively. In the case of micellar polymerization, reactions take place in the solvent phase. The following polymers can be prepared as nanosphere colloidal suspensions following this preparation procedure: PMMA, poly(ackyl cyanoacrylate), and acrylic copolymer. When the polymerization is carried out at the interface (interfacial polymerization) between an oil phase and an aqueous solution, nanocapsules are formed (354).

The determination of the loading capacity of nanoparticle colloidal suspensions can be carried out by separation of the untrapped material with ultracentrifugation followed by the drug analysis after dissolution of the pelleted polymeric matrix. Other reliable separation methods are ultrafiltration and gel permeation chromatography (345,346). The drug loading capacity also can be calculated by determining the drug content in the supernatant or in the filtrate. In fact, the amount of drug entrapped in nanoparticle colloidal systems can be obtained by subtraction of the untrapped drug amount from the total amount of drug present in the suspension.

The mechanisms of drug release from nanoparticle colloidal suspensions depends on the characteristics of the colloidal suspension, as well as on physicochemical properties of the drug. In particular, the release of a drug may occur by one of the following mechanisms or a cooperation of more than one of them: (1) drug desorption from the colloidal surface (both for nanospheres and nanocapsules); (2) drug diffusion through the polymeric network of the nanospheres; (3) drug diffusion through the polymeric shell of nanocapsules; (4) polymeric matrix erosion of nanoparticles. The drug release rate is dependent on the release mechanism, the diffusion coefficient, and polymer biodegradation rate. The nanoparticle drug release is also greatly influenced by the type of interaction with the biological substrate (345).

Besides the previous mentioned drug release mechanisms, it should be considered that the drug delivery function of nanoparticles also can be accomplished by a direct contact with the biological membranes, thus leading to an enhanced drug delivery through membranes with respect to a simple drug solution (355). As a consequence of this behavior, it may happen that the in vitro drug release profiles are poorly related to the in vivo drug delivery and release situation (356).

Size and zeta potential are important physicochemical parameters to be determined to achieve a suitable colloidal carrier. Nanoparticle size is influenced by the preparation technique and by the polymer used, that is, low molecular weight polymers form small-sized nanoparticles, but this fact reduces the amount of encapsulated drug. An increase of polymer concentration usually elicits an increase of both nanoparticle size and encapsulation efficiency (357,358).

The zeta potential is a measure of the surface electrical charge of the particles. As the zeta potential increases, the repulsion phenomenon between particles will be greater, thus leading to a more stable colloidal dispersion. The minimum zeta potential value to prevent particle aggregation and to have a stable nanosuspension was defined to be $\pm 30\,\mathrm{mV}$ (359).

Therapeutic Applications of Nanoparticles. The oral administration is one of the promising application of nanoparticles that have been administered either for achieving a systemic uptake or for having a local residence within the GI tract. The polymers used for peroral application are nondegradable polymers (cellulose, acrylate derivatives, etc.) and are designed not to be adsorbed (359).

Polymer nanoparticles for oral treatment may be formulated as an aqueous suspension or incorporated into traditional dosage forms. A lot of different nanoparticle formulations have been incorporated into tablets or capsules, and then compared with traditional dosage forms. In all cases, nanoparticles maintained the advantages of a colloidal carrier, such as an enhanced dissolution of lipophilic drugs and a prolonged and sustained release (360).

An innovative nanoparticle application in this field is oral chemotherapy, which can be a valid alternative, because it allows a continuous exposure of the cancer cells to anticancer drugs at a lower concentration, thus reducing or avoiding side effects. In addition, it is more convenient and better tolerated by the patients, especially for those with advanced metastatic cancers. Unfortunately, most anticancer drugs cannot be administered orally because of their poor solubility, stability, and permeability. It has been found that anticancer drug encapsulation into an oral formulation of nanoparticles has been able to play a key role in drug adhesion and interaction with cancer cells. For example, PEG-coated nanoparticles are able to adhere to intestinal cells and subsequently to escape from the multidrug resistance pump proteins (361).

Chitosan-coated nanoparticles are used for colon targeted drug delivery of diclofenac. Chitosan nanoparticles are microencapsulated in Eudragit L-100 or S-100 to form a gastroresitent reservoir system in which the drug release is triggered only in the basic (pH 8) environment of colon (362).

For parenteral delivery, nanoparticles can be formulated as aqueous dispersions or they are converted in lyophilized powders to be resuspended just before their administrations (345,359).

Cancer therapy is one of the most important applications of polymeric nanoparticles. Nowadays, the aim of any anticancer research is to improve patient survival after chemo- or radiotherapy. Unfortunately, traditional anticancer therapy is affected by a lot of side effects that involve healthy cells leading to an unsuitable quality of life for cancer patients. So the effectiveness of a treatment is related to the ability to target the cancer cells while affecting as few healthy cells as possible. Nanoparticles can provide an alternative solution for the site-specific delivery of anticancer drugs due to their small size and the possibility to escape RES recognition and uptake, thus leading to a prolonged blood circulation time.

Biodegradable nanoparticles made of PLGA have been used to incorporate paclitaxel, a microtubule-stabilizing agent that causes cell death by promoting the polymerization of tubuline during cell division. Paclitaxel was encapsulated to a very large extent ($\sim 100\%$ encapsulation efficiency) and this paclitaxel-loaded colloidal system showed a 70% loss of viability of human small-cell lung cancer cells at a drug concentration as low as $0.025 \mu g \cdot mL$. Paclitaxel also has been incorporated in poly(ethylene oxide) modified poly(β-amino ester) nanoparticles to obtain a sustained release into most solid tumors (363). Also, Tamoxifen (364) and verteporfin (365) have been encapsulated into PLGA or poly (ϵ-caprolactone) particles for *in vivo* studies against breast cancer. Doxorubicin, a widely used anticancer drug, has been encapsulated into PLGA nanoparticles (366) that presented the ability to release the drug up to 1 month. In addition, this carrier system avoids a lot of the undesiderable effects of doxorubicin (e.g., cardiotoxicity).

If a sustained release of the drug in the tumor site is required, then the nanoparticle surface must be modified in order to avoid RES macrophages, which recognize hydrophobic particles as foreign. To escape RES, the surface of nanoparticles is modified with hydrophilic molecules that form a steric barrier on the particle surface. Polyoxypropylene–polyoxyethylene (POP/POE) surfactants are suitable macromolecules to prevent nanoparticles from sticking to the blood vessel endothelium and to inhibit RES recognition. Indeed, among the various copolymer members, poloxamine and poloxamer have the best prolonged circulation time of nanoparticles (367). Unfortunately, poloxamer and poloxamine do not exhibit prolonged circulation times

when nanoparticles are made up of PLGA (368). Recently, the most common moiety used for nanoparticle surface coating to obtain the so-called Stealth nanoparticles is PEG and its derivatives (369,370). Attachment of PEG on the nanoparticle surface can be performed in different ways: (1) by adsorption, (2) by incorporation during the preparation process, (3) by covalent linkage with the nanoparticle polymeric matrix.

In the passive targeting, as with stealth liposomes, long circulating nanoparticles escape from the blood circulation through the fenestrations of capillaries perturbed by inflammatory processes or by tumors (235). Inflamed vessels present fenestration sized up to 700 nm, so improved colloidal nanoparticles (< 200 nm) are able to pass across, thus accumulating and releasing the drug just in the site of action. These particular characteristic of nanoparticles leads to an increased therapeutic index of the incorporated drug. Nowadays, chitosan-based particulate systems are attracting the most attention as potential long-term drug delivery systems because of mucoadhesive and long circulating properties of chitosan. Doxorubicin–dextran conjugates were encapsulated into chitosan nanoparticles to minimize the cardiotoxicity of the drug. This system reduced not only the drug side effects, but also improved the therapeutic efficacy of doxorubicin in the case of solid tumors (371).

The strategy of nanoparticle passive targeting is widely used for cancer therapy, but it presents a limitation due to a high resistance factor of some solid tumors that cannot be circumvented by PEG-coated nanoparticles. Therefore, an alternative approach is the use of temperature-sensitive nanoparticles, which are able to release its drug content only in hypertermic zones (235). Other approaches may involve the use of biochemical triggers, such as the pH-sensitive lipid-anchored copolymers, to generate fusogenic particles (249).

The strategy of active targeting increases the probability of having a selective direction of nanoparticles to a designed site. In this case, ligands that specifically bind to surface receptors of target sites are coupled to long-circulating particulates. Among polymers suitable for coupling to specific ligands, poloxamers and PEG have received the most attention. A derivatization is achieved between the end group of the poloxamer chains with pyridyl disulfide. After a disulfide exchange with a thiol-containing moiety on the peptide or antibody to be attached on the surface of the nanoparticle, long-circulating actively targeted particles are obtained (372).

The use of nanoparticles as drug delivery systems can overcome the barriers to the penetration of antiinfective drugs into cells, that is, strong protein binding, an unfavorable lipid–water distribution coefficient, an unfavorable pH gradient between different cellular compartments, and the existence of active transport pump mechanisms that prevent the accumulation of sufficient antibiotic concentrations in the interior of the infected cells (278). In particular, a suitable nanoparticle coating can promote the permeation of hydrophilic drugs through membranes (Fig. 54). In fact, the free drug, able to freely diffuse in the aqueous medium and to interact with the outer-hydrophilic zone of the membrane model, has to pass

Phase I   Phase II

Biological Membrane Model

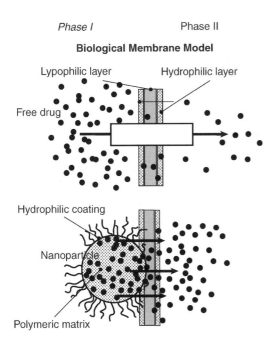

Figure 54. A model of the interaction between the aqueous phase containing a free hydrophilic antibiotic, or the drug-loaded PECA nanospheres and the biological membrane model. The permeation driving force is the different drug concentration between phases I and II (355).

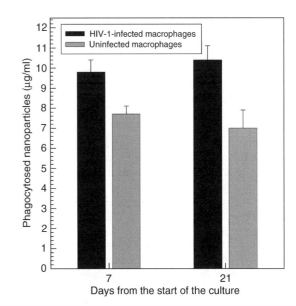

Figure 55. Influence of HIV infection on phagocytosis of poly(butylcyanoacrylate) nanoparticles by human macrophages. Cell cultures were infected with HIV-1 at day 1 after start of the culture. At day 7 or 21, the nanoparticles (200 nm diameter) were added at a final concentration of $0.5 \, \mathrm{mg \cdot ml^{-1}}$ to the infected cultures and incubated for 6 h. Data from Ref. 374.

through the lypophilic layer of the same membrane. This process could represent the limiting step in the diffusion through the membrane model of a hydrophilic molecule. In the case of nanoparticles, the outer hydrophilic shell of the particles (coated with nonionic surfactant) could ensure an interaction with the hydrophilic layer of the membrane, while the internal lypophilic core of the particle can ensure a close interaction with the hydrophobic layer of the membrane, providing a high permeation of the drug (354). The nanosphere-mediated increase of drug membrane permeation leads to an improvement of the antibacterial activity (373) and to an intrabacterial drug accumulation. The entrapment of antibiotics in nanospheres may prevent the bacterial resistance to drugs due to pleiotropic drug resistance and changes in the bacterial outer membrane leading to a decrease in OmpF porin, which probably causes decreased drug permeation.

Nanoparticles can be used for the treatment of various viral infection diseases localized at the level of the RES. For example, RES cells can be infected by both strains of the HIV, namely, HIV-1 and HIV-2. Monocytes and macrophages seem to have a fundamental role in the immunopathogenesis of the HIV infection, by behaving as virus reservoirs from which HIV can disseminate throughout the body and brain. Because nanoparticles can be easily phagocytosed by macrophages, they may represent a suitable and promising drug delivery system for the treatment of HIV infection persisting in these cells.

In vitro studies showed that human macrophages are able to phagocyte different types of polyacrylic and albumin nanoparticles (374). Interestingly, HIV-infected macrophages seem to have a higher phagocytotic activity concerning nanoparticle uptake than noninfected macrophages

(Fig. 55). This phenomenon can be due to an activated state of these infected cells, leading potentially to a preferential phagocytosis of drug-loaded nanoparticles, and hence to a targeted delivery of antiviral drugs to these cells.

The RES tropism of conventional nanoparticles also can be efficiently used for the treatment of protozoa infection (i.e., leishmaniasis) (348). Contrary to liposomes, empty poly(isobutylcyanoacrylate) nanospheres exhibit a certain *in vitro* and *in vivo* antiparasitic activity against (375). Probably, this action could be attributed to peroxide production following nanosphere phagocytosis, which led to a respiratory burst, more pronounced in infected than uninfected macrophages.

One of the major problems connected with ophthalmic therapy is drug loss after instillation of eyedrops. To improve ocular bioavailability, mainly nanoparticles have been used. Colloidal systems have the convenience of a drop and are able to maintain drug concentration and activity at its site of action, probably due to an improved ocular mucoadhesion. In fact, poly(butylcyanoacrylate) nanospheres were able to improve the amikacin ocular delivery (376) by increasing the corneal and aqueous humor concentration of amikacin with respect to the free drug and to other formulations. The surface coating of nanospheres can also be used for ophthalmic application and can be a crucial factor to achieve a well-tailored and efficient drug delivery. In spite of the nanoparticle polymer and the method of coating, the presence of PEG on the surface of nanoparticles improves the ocular drug permanence time and increases the drug level in various ocular structures compared with both conventional ocular formulations and uncoated nanoparticles (352,377). Ocular gene therapy is also possible with polymeric nanoparticles (378).

In recent years, mucosal surfaces (nasal, pulmonary, buccal, and ocular) have received much attention as alternative routes of systemic administration (379). Chitosancoated nanoparticles present mucoadhesive properties that can be useful to enhance mucosal drug adsorption. An example is the enhanced nasal absorption of insulineloaded chitosan nanoparticles that do not damage the biological system.

The buccal adsorption of vaccine encapsulated into nanoparticles is an alternative to the parenteral route of administration of vaccines. The oral or nasal delivery of ovalbumine from chitosan microparticles enhances the systemic and local immune response against diphteria toxoid vaccine (380).

Lipid-Based Nanoparticles.

Lipids instead of polymers can be used to obtain colloidal drug carriers. Solid lipid nanoparticles (SLN), nanostructured lipid carriers (NLC), and lipid drug conjugates (LDC) are nanoparticles with a solid lipid matrix and present an average diameter in the nanometer range. These innovative colloidal carriers have attracted increasing attention in recent years. They are regarded as an alternative carrier system to traditional colloidal systems (e.g., polymeric micro- and nanoparticles). General ingredients include solid lipids, emulsifiers, and water. Lipids include triglycerides, partial glycerides, fatty acids, steroids, and waxes. All excipients are generally recognized as safe (GRAS) substances, so a wide variety of compounds can be used for formulation purposes.

Solid Lipid Nanoparticles.

Thse nanoparticles are colloidal systems made up of solid lipids and are stabilized by surfactants. During the 1950s, lipid nanoemulsions were introduced for the parenteral nutrition and later they were used as carriers for lipophilic drugs. The major problem with nanoemulsions was the loss of drugs related to their liquid form. The SLN were developed to overcome this problem. In fact, the use of solid lipids instead of liquid oils is a very attractive idea to achieve controlled drug release, because drug mobility is much slower into a solid lipid than in a liquid oil.

The SLN can be prepared using different procedures. The main preparation process is the high pressure homogenization (HPH) method, in which a dispersion of the drug in the melted lipid is constricted through a narrow gap (in the range of few microns) under very high pressure, thus disrupting lipid particles down to the submicron range. Other methods are the solvent emulsification–evaporation method and the microemulsion-based preparation (381). The first is a method to prepare nanoparticles by precipitation in o/w emulsions. The second production method is based on the preparation and the subsequent dilution in cold water (2–3 °C) of a microemulsion made up of a low melting lipid, an emulsifier, a coemulsifiers, and water. Following preparation, as for other colloidal carriers, the determination of the mean size, size distribution, and zeta potential of SLN is necessary to define their physicochemical properties.

Formulative Aspects of SLN.

Evaluations on the degree of crystallinity and lipid modifications are very important for SLN. In fact, the organization of lipids into a crystalline reticulate is fundamental for the determination of the drug encapsulation and release rates. In general, if the lipid matrix is made-up of pure molecules (i.e., tristearin or tripalmitin), a perfect crystal with few imperfections, is formed. As incorporated drugs are located between fatty acid chains, in crystal imperfections, and between the lipid layers, a highly ordered crystal form cannot accommodate large amounts of drug (382). Also, the lipid modification influences the encapsulation and release degree of a drug. In fact, glycerides exist in three different polymorphic forms: α, β', and β. The degree of reticulate imperfection of the lipid decreases in this sequence: $\alpha < \beta' < \beta$. Lipid nanoparticles recrystallize at least partially in the α-form, but with increasing formation of the more stable β'/β modifications, the lattice is becoming perfected, the number of imperfections decreases, and drug is expelled from nanoparticles.

The coexistence of additional colloidal structures, such as micelles, liposomes, and supercooled melts, has to be taken into account after the preparation of SLN; the quantification of these additional structures is a serious challenge because of their size similarities with those of the SLN. Therefore, it would be desirable to use methods that are sensitive to the simultaneous presence of different colloidal species. Both NMR and ESR techniques meet these requirements.

Physical stability of SLN dispersions is generally >1 year (383), up to 3 years in the case of SLN made of glyceryl palmitate. The storage stability of SLN depends on two factors: (1) the physical modification of lipid structure ($\alpha \rightarrow \beta'/\beta$); (2) the presence of additional colloidal structures (liposomes, micelles, drug nanoparticles).

Gelation phenomena, that is, increase of particle sizes and drug expulsion from the lipid carrier, are the major problems of storage stability. Gelation is the transformation of a low viscosity SLN colloidal dispersion into a viscous gel. It occurs when SLN is put in contact with other surfaces and shear forces, and it is connected with crystallization processes. This destabilizing phenomenon can be retarded or prevented by the addition of coemulsifying surfactants with high dynamic mobility, such as glycocholate (381).

The increase of particle sizes is a consequence of particle aggregation and it is less significant when SLN have a zeta potential value of $-25\,mV$, while it becomes an important phenomenon when the zeta potential is $-15\,mV$ or less (381).

Drug expulsion is related to crystallization of lipids and their modification to the β'/β form, in which the lipid lattice is packed in a more ordered way with a reduction of imperfections.

To have an optimal storage conditions, SLN can be lyophilized or spray-dried. During lyophilization, a colloidal lipid dispersion is deprived of its solvent to guarantee a better chemical and physical stability. However, two transformations in the formulation occur during and after this process, which might be the source of additional stability problems: (1) passage of SLN from an aqueous dispersion to a powder with possible changes of osmolarity and pH; (2) resolubilization that favors particle aggregation. To overcome these problems, a cryoprotector (e.g., trehalose, sorbitol, mannose, and glucose) is added to the SLN dispersion before lyophilizing. These protective agents are used in a 10–15% (w/v) concentration and act to decrease the osmotic activity of water and favour a glassy state of the frozen sample (384).

Spray drying is an alternative method to transform an aqueous SLN dispersion into a dry product. It is cheaper and simpler than lyophilization, but has the disadvantage of needing high temperatures, which can cause particle aggregation. Therefore, it is recommended to use lipids with melting points >70 °C for spray drying. Also, in this case the addition of cryoprotective agents may be useful to prevent particle aggregation.

Drug incorporation into SLN is related to crystalline modification of the lipids and is inversely proportional to the β′/β modification of lipids. Depending on the drug/lipid ratio and solubility, the drug is located in the core of the particles, in the shell, or dispersed throughout the matrix, so that drug loading capacity of conventional SLNs is generally from 25 up to 50% (381). The drug-loading capacity is higher for lipid mixtures with different acyl chain lengths than for lipids that form a perfect crystal with few imperfections and cannot accommodate large amount of drug.

The release profiles could be modulated showing a burst release followed by a prolonged release, or generating systems without any burst release at all. The release kinetic can be controlled by modification of the preparation procedures and the type of surfactant and lipid material.

Therapeutic Applications of SLN. Due to their small sizes, SLN may be administered through every route: oral, transdermal, and IV administration can be possible.

The SLN for oral administration may include aqueous dispersions or conventional dosage forms (e.g., tablets, capsules, and pellets). Camptothecin-loaded particles are an example of orally administered SLN. This is a stearic acid/Poloxamer 188 formulation of SLN that present a zeta potential value of −69 mV and an encapsulation efficiency of 99.6%. The incorporation of camptothecin into SLN provided drug protection from hydrolysis (381). A better bioavailability, prolonged plasma levels, and lack of nephrotoxicity are observed for orally administered SLN encapsulating drugs, thus leading to the conclusion that SLN are a promising sustained release system for the oral administration of lipophilic drugs (381,385).

The SLN are formulated into creams, hydrogels, or ointments before their application onto the skin and form an adhesive film upon the skin, which is able to restore the protective action of the naturally occurring hydrolipidic skin film when it is damaged. Many different cosmetic ingredients have been encapsulated into SLN (i.e., coenzyme Q_{10}, vitamin E, and retinal) (386). A modern approach to an intelligent release of the drug from SLN to the skin is related to lipid modification from the α to the β form. An intelligent drug-loaded SLN is a colloidal system that maintains itself into the more energetic α-form during storage, while transforming into the β-form after application onto the skin, thus releasing its incorporated drug by expulsion from the lipid crystalline reticule. The SLN per se also have a sun protective effect (387) that is due to their particulate nature and their ability to scatter UV light. In this case, SLN show a synergistic effect if formulated with a molecular sunscreen, also showing better skin protection and a reduction in side effects.

It is possible to use SLN for pulmonary drug delivery since they maintain their particle size and polydispersion index after nebulization. Only very little aggregation could be detected, which is of no significance for pulmonary administration (388). In addition, SLN may be used as a powder for inhalation. The use of SLN instead of polymeric nanoparticles has many advantages, such as high tolerability, faster degradation, and passive targeting toward lung macrophages.

The SLN can be injected intravenously and used to target drugs to particular organs. They also can be administered intramuscularly or subcutaneously. When administered subcutaneously, SLN act as a depot of the drug, while they are cleared from the circulation by RES (liver and spleen) when administered intravenously. The incorporated drug is released upon erosion by diffusion from the particles or by enzyme degradation. Obviously, in the case of IV administration, the SLN size must be < 5 μm to avoid the possibility of embolism into the fine capillaries. Similar to any colloidal drug with a small mean size (≤ 200 nm), SLN have been coated by polyoxyethylene in order to achieve long circulating colloidal lipid particles (389). Stealth SLN can be prepared by using Pluronic F188, a block poly(oxyethylene/polyoxypropylene) copolymer that anchors its hydrophobic portion in the SLN lipid matrix, while the hydrophilic portion forms the hydrophilic coating of SLN. Stealth SLN can be used in tumor and antibacterial therapy. Paclitaxel-loaded SLN provided a higher and prolonged plasma level of the drug with respect to the cremophor EL-based commercial formulation with a consistent reduction of side effects (381). In all the investigated cases, SLN containing an anticancer drug showed higher blood levels with respect to the relative commercial drug formulations after IV injection.

The therapy at the level of the brain is always difficult due to the presence of the blood–brain barrier, and hence the possibility of having a suitable drug delivery system that is able to reach the brain can be extremely useful. Stealth SLN allow brain delivery of drugs that are not capable of passing through the blood–brain barrier (390).

The potential SLN toxicity has to be considered as a function of the administration route. Topical and oral administration are absolutely nonproblematic because all excipients used in SLN formulations are those currently employed for the formulation of traditional dosage forms or cosmetic products. The situation is slightly different for parenteral administration. In this case, only surfactants accepted for parenteral administration can be used (e.g., lecithin, Tween 80, PVP, Poloxamer 188). Up to now, there is no SLN product for parenteral use on the market. However, SLN show a very good tolerability both *in vitro* and *in vivo* (381).

Nanostructured Lipid Carrier. The NLC were introduced at the end of the 1990s in order to overcome some limitations of SLN: (*1*) too low payload for a number of drugs, (*2*) drug expulsion during storage, (*3*) high water content of SLN dispersions.

"Imperfect" type   "Amorphous" type   "Multiple" type

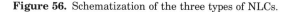

Figure 56. Schematization of the three types of NLCs.

The NLC are made up of very different lipid molecules mixed together, that is, a blend of solid with liquid lipids (oils). The resulting matrix of the lipid particles shows a melting point depression compared to the original solid lipid, but the matrix is still solid at body temperature. There are three different models of NLC, depending on the way of production and the composition of the lipid blend (Fig. 56). In the first model, called the imperfect type, spatially different lipids lead to larger distances between the fatty acid chains of the glycerides and to general imperfections in the crystal structure, thus allowing a greater extent of drug entrapment. The second type of NLC, the so-called amorphous type, contains liquid oil nanocompartments within the lipid particle matrix. In this way, crystallization can be avoided and the solid character of NLC can be maintained as shown by NMR and DSC measurements (391). The third type of NLC, the so-called multiple type, are produced by mixing a solid lipid with a high amount of oil. In this way, a phase separation occurs between solid lipids and oil molecules thus forming nanocompartments. This phenomenon occurs during the cooling process after the hot homogenisation method.

The NLC are produced successfully by the high pressure homogenization method and it is possible to obtain particle dispersions with a solid content of 50 or 60% (392). The particle dispersions thus produced have a high consistency with a cream-like or almost solid appearance.

Because of their high particle concentrations, NCL can be used for granulation or as wetting agents in pellet production. In addition, NLC can easier be processed to traditional oral dosage forms, such as tablets or capsules (393).

These carriers have been used for dermal delivery. Similarly to SLN, they can be incorporated in to existing products or formulated in a final product containing only NLC. When incorporated into an o/w emulsion saturated with the drug, the NLC disordered structure is preserved and the drug remains inside the particles. Following skin application, and then increased temperature, water loss and NLC lipid transition to a more stable polymorphic form triggers drug expulsion. The drug expelled from NLC is supersaturated with the drug already present in the emulsion. The supersaturation phenomenon can be used to increase the skin drug permeation as observed for cyclosporine (394).

In the future, an area of particular interest can be the prolonged release of drugs after subcutaneous or intramuscular injection (e.g. erythropoietin). Also, NLC dispersions for IV injection appear feasible (395).

To overcome the important limitation of SLN and NLC to incorporate only lipophilic drugs or very low concentrations of highly potent hydrophilic drugs, lipid drug conjugates were recently developed with drug loading capacities of up to 33% (396). In this type of lipid nanoparticle, an insoluble drug–lipid conjugate bulk is prepared using two different methods: (1) salt formation with a hydrophobic moiety (e.g., with a fatty acid) and (2) covalent linking (e.g., to ethers or ester).

FUTURE PRESPECTIVES

Drug delivery technology has had considerable advances, bringing many clinical products to the market. However, the major needs for drug delivery devices are still unmet and important classes of drugs have yet to benefit from these technological successes. The central focus of any controlled delivery devices is "control." This can b achieved if: (1) the size of the device can be modulated as accurately as possible; (2) the device can be produced with a certain reproducibility; (3) the device is stable enough for administration purpose; (4) the device is biocompatible; (5) the rate of drug delivery should be independent of the surrounding environment. To date, it is possible to have recourse to a lot of innovative approachs characterized by different features as a fucntion of the physicochemical properties of the delivered drug, the administration route, and therapeutic aims.

Many aspects of these drug delivery devices require an improvement to be applied to clinical use, but considering the *in vitro* and *in vivo* results, they seem to be very interesting.

From the appearance of the first drug delivery device to date, a constant improvement has been made, that is, from microtechnology we have passed to nanotechnology, and from an aspecific drug delivery, we have passed to a selective drug delivery. New challenges for the future are the feasibility of scaling-up processes to bring to the market quickly innovative therapeutic entities and the possibility of obtaining multifunctional devices that will be able to fulfill the different biological and therapeutic requirements.

BIBLIOGRAPHY

1. Park K. Controlled Drug Delivery: Challenges and Strategies. Washington (DC): American Chemical Society; 1997.
2. Park K, Mersny RJ. Controlled drug delivery: present and future. Controlled Drug Delivery: Designing Technology for the Future, 2–13. Washington (DC): American Chemical Society; 2000.
3. Juliano RL. Controlled delivery of drugs: an overview and prospectus. Drug Delivery Systems: Characteristics and Biomedical Application 310. Oxford University Press; 1980.
4. Langer R. Implantable controlled release systems. Pharmacol Ther 1983;21:35–51.
5. Langer R. Drug delivery and targeting. Nature (London) 1998;392:5–10.
6. Kost J, Langer R. Responsive polymeric delivery systems. Adv Drug Deliv Rev 2001;46:125–148.
7. Breimer DD. Future challenges for drug delivery. J Control Rel 1999;62:3–6.
8. Klausner EA, et al. Novel levodopa gastroretentive dosage form: in-vivo evaluation in dogs. J Control Rel 2003;88:117–126.

9. Ranade VV, Hollinger MA, editors. Drug Delivery Systems. CRC Press; 1996.

10. Powell MF. Drug delivery issues in vaccine development. Pharm Res 1996;13(12):1777–1785.

11. Grayson ACR, Shawgo RS, Li Y, Cima MJ. Electronics mems for triggered delivery. Adv Drug Deliv Rev 2004; 56:173–184.

12. Global drug delivery: Industry profile. Datamonitor; Aug 2003.

13. Schwan HP, Webb GN, editors. IRE Transactions on Biomedical Engineering.

14. Schwan HP, Webb GN. Biophys J 1967;7:978.

15. Greatbatch W, Holmes CF. History of implantable devices. IEEE Eng Med Biol Mag 1991;10(3):38–49.

16. Blackshear PJ, et al. A permanently implantable self-recycling low flow constant rate multipurpose infusion pump of simple design. Surg Forum 1970;21:136–137.

17. Blackshear PJ, Rohde TD, Prosl F, Buchwald H. The implantable infusion pump: a new concept in drug delivery. Med Prog Technol 1979;6:149–161.

18. Rupp WM, et al. The use of an implantable insulin pump in the treatment of type II diabetes. NEJM 1982;307:265–270.

19. Salkind AJ, et al. Electrically driven implantable prostheses. New York: Plenum Press; 1986.

20. Saudek CD, et al. A preliminary trial of the programmable implantable mediation system for insulin delivery. NEJM 1989;321:574–579.

21. Fogel H, et al. Treatment of iddm with a totally implantable programmable insulin infusion devise. Rev Euro Technol Biomed 1990;12:196.

22. Folkman J, Long DM, Rosenbaum R. Silicone rubber: a new diffusion property useful for general anesthesia. Science 1966;154:148–149.

23. Folkman J, Long D. The use of silicone rubber as a carrier for prolonged drug therapy. J Surg Res 1964;4:139–142.

24. Fung LK, Saltzman WM. Polymeric implants for cancer chemotherapy. Adv Drug Deliv Rev 1997;26:209–230.

25. Danckwerts M, Fassihili A. Optimization and development of a core-in-cup tablet for modulated release of theophylline in simulated gastrointestinal fluids. Drug Dev Ind Pharm 1991;17:1465–1502.

26. Hall EAH. The developing biosensor arena. Enz Microb Technol 1986;8:651–658.

27. Yang MB, Tamargo RJ, Brem H. Controlled delivery of 1,3-bis(2chloroethyl)-l-nitrosourea from ethylene-vinyl acetate copolymer. Cancer Res 1989;49:5103–5107.

28. Kim SW, Bae YH, Okano T. Hydrogels: swelling, drug loading, and release. Pharm Res 1992;9:283–290.

29. Gutman RL, Peacock G, Lu DR. Targeted drug delivery for brain cancer treatment. J Control Rel 2000;65:31–41.

30. Raiche AT, Puleo DA. Association polymers for modulated release of bioactive proteins. IEEE Eng Med Biol Mag 2003;22(5):35–41.

31. Emanuele AD, Staniforth JN. An electrically modulated drug delivery device. J Pharm Res 1991;8:913–918.

32. Sershen S, West J. Implantable, polymeric systems for modulated drug delivery. Adv Drug Deliv Rev 2002;54: 1225–1235.

33. Jackman RJ, et al. Fabricating large arrays of microwells with arbitrary dimensions and filling then using discontinuous dewetting. Anal Chem 1998;70:2280–2287.

34. Available at http://www.alzet.com.

35. Urquhart J, Fara J, Willis KL. Rate controlled delivery systems in drug and hormone research. Annu Rev Pharmacol Toxicol 1984;24:199–236.

36. Yun KS, et al. A surface-tension driven micropump for low-voltage and low-power operations. J Microelectromec Syst 2002;11:454–461.

37. Maillefer D, Van Lintel H, Rey-Mermet G, Hirschi R. A high-performance silicon micropump for an implantable drug delivery system. Proc 12th IEEE MEMS 1999; 541–546.

38. Zengerle R, Kluge S, Richter M, Richter A. A bidirectional silicon micropump. Proc IEEE MEMS 1995; 19–24.

39. Jeong OC, Yang SS. Fabrication and test of a thermopneumatic micropump with a corrugated p+ diaphragm. Sens Actuators A Phys 2000;83:249–255.

40. Zhang W, Ahn CH. A bidirectional magnetic micropump on a silicon wafer. Proc IEEE Solid-State Sensor Actuator Workshop 1996; 94–97.

41. Yang Y, Zhou Z, Ye X, Jiang X. A bimetallic thermally actuated micropump. J Microelectromec Syst 1996;59:351–354.

42. Benard WL, Kahn H, Heuer AH, Huff MA. Thin-film shape memory alloy actuated micropumps. J Microelectromec Syst 1998;7:245–251.

43. Guo S, Nakamura T, Fukuda T, Oguro K. A new type of micropump using icpf actuator. IEEEIASME Inter Conf Adv Intelligent Mechatronics 1997; 16–20.

44. Available at http://www.medtronic.com.

45. Moore G. VLSI, what does the future hold. Electron Aust 1980;42:14.

46. Nathanson HC. A resonant-gate silicon surface transistor with high-q band-pass properties. Appl Phys Lett 1965;7:84.

47. Petersen KE. Silicon as mechanical materials. Proc IEEE 1982;70(5):420.

48. Howe RT. Polycrystalline silicon micromechanical beams. I Electrochem Soc 1983;130(6):1420.

49. Mehregany M, Gabriel KJ, Trimmer WSN. IEEE Trans El Dev 1988;35:719.

50. Bart SF, et al. Design considerations for micromachined electric actuators sensors. Actuators 1988;14:269–292.

51. Fan LS, Tai YC, Muller RS. Ic-processed electrostatic micromotors. Sensors Actuators 1989;20:41–48.

52. Jaecklin VP, et al. Line-addressable torsional micromirrors for light modulator arrays. Sensors Actuators A 1994;41–42:324.

53. Fischer M, Graef H, von Munch W. Electrostatically deflectable polysilicon torsional mirrors. Sensors Actuators A 1994;44:83–89.

54. Tsang WK, Core TA, Sherman SJ. Fabrication technology for an integrated surface-micromachined sensor. Solid State Technol 1993; 36–39.

55. Chau KHL, et al. An integrated force-balanced capacitive accelerometer for low-g applications. Int Conf Solid-State Sensors Actuators 1995; 593.

56. Hierold C, et al. A pure cmos surface micromachined integrated accelerometer. IEEE Workshop Micro Electro Mech Syst 1996; 174.

57. Offenberg M, et al. Novel process for a monolithic integrated accelerometer. Int Conf Solid-State Sensors Actuators 1995; 589.

58. Burns DW, et al. Resonant microbeam accelerometers. Int Conf Solid-State Sensors Actuators 1995;2:659.

59. Hirano T, Furuhata T, Gabriel KJ, Fujita H. Design, fabrication, and operation of submicron gap comb-drive microactuators. J Microelectromec Syst 1995;1:52–59.

60. Suzuki H. Advances in the microfabrication of electrochemical sensors and systems. Electroanalysis 2000;12(9):703–715.

61. Wang J. Electrochemical detection for microscale analytical systems: a review. Talanta 2002;56:223–231.

62. Simpson PC, Woolley AT, Mathies A. Microfabrication technology for the production of capillary array electrophoresis chip. Biomed Microdev 1998;11:7–26.

63. Liu S, Guttman A. Electrophoresis microchips for dna analysis. Trends Anal Chem 2004;23(6):422–431.

64. Woolley AT, Lao K, Glazer AN, Mathies RA. Capillary electrophoresis chips with integrated electrochemical detection. Anal Chem 1998;70:684–688.

65. Liu W. Retinal implant: bridging engineering and medicine. Electron Devices Meeting, IEDM '02 Digest Inter 2002; 492–495.

66. Kipke DR, Vetter RJ, Williams JC, Hetke JY JF. Silicon-substrate intracortical microelectrode arrays for long-term recording of neuronal spike activity in cerebral cortex. IEEE Trans Neural Sys Rehab Eng 2003;11(2):151–155.

67. Mikszta JA, et al. Improved genetic immunization via micromechanical disruption of skin-barrier function and targeted epidermal delivery. Nature Med 2002;8(4):425419.

68. Santini Jr. JT, et al. Microchips as controlled drug delivery devices. Angew Chem Int Ed 2000;39:2396–2407.

69. McAllister DX, Allen MG, Prausnitz MR. Annu Rev Biomed 2000;2:289.

70. Tao S, Lubeley M, Desai TA. Bioadhesive poly(methyl methacrylate) microdevices for controlled drug delivery. J Control Rel 2003;88:215–228.

71. Lewis JR, Ferrari M. BioMEMS for drug delivery application, Lab-on-achip: Chemistry in miniaturized synthesis and analysis system. New York: Elsevier Science; 2003. pp 373–389.

72. Polla DL, et al. Microdevices in medicine. Annu Rev Biomed 2000;2:551–576.

73. Hadgraft J, Guy RH, editors. Transdermal Drug Delivery: Developmental Issues and Research Initiatives. New York: Marcel Dekker; 1989.

74. Smith EW, Maibach HI, editors. Percutaneous Penetration Enhancers. CRC Press; 1995.

75. Amsden BG, Goosen MFA. Transdermal delivery of peptide and protein drugs: an overview. AIChE J 1995;41:1972–1997.

76. Henry S, McAllister DX, Allen MG, Prausnitz MR. Microfabrication microneedles: a novel approach to transdermal drug delivery J Pharm Sci 1998;87:922–925.

77. Barry B, Williams A. Penetration enhancers. Adv Drug Deliv Rev 2003;56:603–618.

78. Cevc G. Lipid vesicles and other colloids as drug carriers on the skin. Adv Drug Deliv Rev 2004;56:675–711.

79. Preat V, Vanbever R. Skin electroporation for transdermal and topical delivery. Adv Drug Deliv Rev 2004;56:659–674.

80. Doukas A. Transdermal delivery with a pressure wave. Adv Drug Deliv Rev 2004;56:559–579.

81. Mitragotri S, Kost J. Low-frequency sonophoresis: a review. Adv Drug Deliv Rev 2004;56:589–601.

82. Gerstel MS, Place VA. Drug delivery device, US Patent 3,964,482, 1976.

83. Reed ML. Microsystems for drug and gene delivery. Proc IEEE 2004;92(1):56–75.

84. Lin L, Pisano AP. Silicon-processed microneedles. J Microelectromec Syst 1999;8(l).

85. Talbot NH, Pisano AP. Polymolding: Two wafer polysilicon micromolding of closed-flow passages for microneedles and microfluidic devices. Tech Dig Solid-State Sensor and Actuator Workshop 1998; 265–268.

86. Brazzle JD, Papautsky L, Frazier AB. Micromachined needle arrays for drug delivery or fluid extraction. IEEE Eng Med Biol Mag 1999;18:53–58.

87. Brazzle JD, Papautsky L, Frazier AB. Fluid-coupled hollow metallic micromachined needle arrays. Proc SPIE, Microfluidic Devices Systems 1998;3515:116–124.

88. Brazzle JD, Mohanty S, Frazier AB. Hollow metallic micromachined needles with multiple output ports. Proc SPIE, Microfluidic Devices Systems 1999;11, 3877:257–266.

89. Stoeber B, Liepmann D. Two-dimensional arrays of outofplane needles. Proc ASME Int Mechanical Engineering Congr Exposition 2000; 355–359.

90. Stoeber B, Liepmann D. Fluid injection through out-of-plane microneedles. Proc Ist Annu Int IEEE-EMBS Special Topic Conf Microtechnologies Medicine Biology 2000; 224–228.

91. Henry S, McAllister DV, Allen MG, Prausnitz MR. Micromachined needles for the transdermal delivery of drugs. Proc IEEE 11th Annu Int Workshop Micro Electro Mechanical Systems 1998; 494–498.

92. Gardeniers JGE, et al. Silicon micromachined hollowmicroneedles for transdermal liquid transfer. Proc IEEE Conf MEMS 2002; 141.

93. Dizon R, Han H, Russell AG, Reed ML. An ion milling pattern transfer technique for fabrication of three-dimensional micromechanical structures. J Microelectromech Syst 1993;2(4): 151–159.

94. Ling P, et al. Genetic transformation of nematodes using arrays of micromechanical piercing structures. J Microelectromec Syst 1995;19(5):766–770.

95. Trimmer W, et al. Injection of dna into plant and animal tissues with micromechanical piercing structures. Proc 8th Int Workshop Micro Electra Mechanical Systems 1995; 111–115.

96. Feldman MD, et al. Stent-based gene therapy. J Long-Term Effects Med Implants 2000;10:47–68.

97. Saffran M, et al. Biodegradable azopolymer coating for oral delivery of peptide drugs. Biochem Soc Trans 1990;18:752–754.

98. Fara JW, Myrback RE, Swanson DR. Evaluation of oxprenolol and metoprolol Oros systems in the dog: comparison of in vivo and in vitro drug release, and of drug absorption from duodenal and colonic infusion sites. Br J Clin Pharmacol 1985;19:91–95.

99. Fasano A, Uzzau S. Modulation of intestinal tight junctions by Zonula occludens toxin permits enteral administration of insulin and other macromolecules in an animal model. J Clin Invest 1997;99:1158–1164.

100. Schwarz UI, et al. P-glycoprotein inhibitor erythromycin increases oral bioavailability of talinolol in humans. Int J Clin Pharmacol Ther 2000;38:161–167.

101. Arangoa MA, et al. Bioadhesive potential of gliadin nanoparticulate systems. Eur J Pharm Sci 2000;11:333–341.

102. Lehr CM. Lectin-mediated drug delivery: the second generation of bioadhesives. J Control Rel 2000;65:19–29.

103. Bies C, Lehr CM, Woodley JF. Lectin-mediated drug targeting: history and applications. Adv Drug Deliv Rev 2004;56: 425–435.

104. Nashat AH, Moronne M, Ferrari M, Detection of functional groups and antibodies on microfabricated surfaces by confocal microscopy. Biotechnol Bioeng 1998;60:137–146.

105. Ferrari M. Therapeutic microdevices and methods of making and using same, US Patent 6,107,102, 2000.

106. Ferrari M, et al. Particles for oral delivery of peptides and proteins, US Patant 6,355,270 131, 2000.

107. Tao SL, Desai TA. Microfabricated drug delivery systems: from particles to pores. Adv Drug Deliv Rev 2003;55:315–328.

108. Martin FJ, Grove C. Xlicrofabricated drug delivery systems: concepts to improve clinical benefit. Biomed Microdev 2001;3:97–108.

109. Smith BR, et al. Nanodevices in Biomedical Applications. Lee A, Lee J, Ferrari M, editors. BioMEMS and Biomedical Nanotechnology, Vol. I: Biological and Biomedical Nanotechnology. New York: Springer; 2005.

110. Foraker AB, et al. Microfabricated porous silicon particles enhance paracellular delivery of insulin across intestinal Caco-2 cell monolayers. Pharm Res 2003;20:110–116.

111. Ahmed A, Bonner C, Desai TA. Bioadhesive microdevices with multiple reservoirs: a new platform for oral drug delivery. J Control Rel 2002;81:291–306.

112. Ishida O, Maruyama K, Sasaki K, Iwatsuru M. Size-dependent extravasation and interstitial localization of polyethyleneglycol liposomes in solid tumor-bearing mice. Int J Pharm 1999;190:49–56.

113. Li X, et al. Porosified silicon wafer structures impregnated with platinum anti-tumor compounds: Fabrication, characterization, and diffusion studies. Biomed Microdev 2000; 2:265–272.

114. Smith BR, et al. A biological perspective of particulate nanoporous silicon. Mat Technol 2004;19:16.

115. Canham LT, et al. Derivatized mesoporous silicon with dramatically improved stability in simulated human blood plasma. Adv Mater 1999;11:1505–1507.

116. Stewart MP, Buriak JM. Chemical and biological applications of porous silicon technology. Adv Mater 2000;12:859–869.

117. Microchips Inc. Available at www.mchips.com.

118. Cohen MH, et al. Microfabrication of silicon-based nanoporous particulates for medical applications. Biomed Microdev 2003;5:253–259.

119. Santini Jr. JT, Cima MJ, Langer R. A controlled-release microchip. Nature (London) 1999;397:335–338.

120. Richards AC, et al. A biomems review: Mems technology for physiologically integrated devices. Proc IEEE 2004; 82:6–21.

121. Ferrari M, et al. Proc Mat Res Soc. 1995;414:101–106.

122. Chu WH, Ferrari M. Micromachined filter and capsule having porous membranes and bulk support, US Patent 5,570,076, 1996.

123. Kellar CG, Ferrari M. Microfabricated particle filter, US Patent 5,651,900, 1997.

124. Desai TA, Hansford D, Ferrari M. Characterization of micromachined silicon membranes for immunoisolation and bioseparation applications. J Membr Sci 1999;159:221–231.

125. Desai TA, et al. Nanoparous anti-fouling silicon membranes for implantable biosensor applications. Biosensors Bioelectronics 2000;15:453–462.

126. Leoni L, Boiarski A, Desai TA. Characterization of nanoporous membranes for immunoisolation: diffusion properties and tissue effects. Biomed Microdev 2002;4(2):131–139.

127. Clark LA, Ye GT, Snurr RQ. Molecular traffic control in a nanoscale system. Phys Rev Let 2000;84:2893–2896.

128. Wei Q, Bechinger C, Leiderer P. Single-file diffusion of colloids in onedimensional channels. Science 2000;287:625–627.

129. Auerbach SM. Theory and simulation of jump dynamics, diffusion and phase equilibrium in nanopores. Int Rev Phys Chem 2000;19:155–198.

130. Mao Z, Sinnott SB. A computational study of molecular diffusion and dynamic flow through carbon nanotubes. J Phys Chem B 2000;104:4618–4624.

131. Nelson P, Auerbach S. Self-diffusion in single-file zeolite membranes is fickian at long times. J Chem Phys 1999; 110:9235–9243.

132. MacElroy JMD, Suh SH. Self-diffusion in single-file pores of finite length. J Chem Phys 1997;106:85–95.

133. Levitt DG. Dynamics of a single-file pore: Non-fickian behavior. Phys Rev A Gen Phy 1973;8:30–50.

134. Lin B, Yu J, Rice S. Direct measurements of constrained brownian motion of an isolated sphere between two walls. Phys Rev E 2000;62:3909–3919.

135. Cosentino C, et al. A dynamic model of biomolecules diffusion through two-dimensional nanochannels. J Phys Chem B 2005;109: to be published.

136. Tu JK, Huen T, Szema R, Ferrari M. Filtration of sub-100 run particles using a bulk micromachined, direct-bonded silicon filter. Biomed Microdev 1999;1:113–119.

137. Desai TA, et al. Nanopore technology for biomedical applications. Biomed Microdev 1999;2:11–40.

138. Leoni L, Desai TA. Micromachined biocapsules for cell-based sensing and delivery. Adv Drug Deliv Rev 2004;56:211–229.

139. Tu JK, Ferrari M, Microfabricated particle filter, US Patent 5,938,923, 1999.

140. Kim SJ, Kim BH. Syntheses and structural studies of calix[4]arene-nucleoside and calix[4]arene-oligonucleotide hybrids. Nucleic Acids Res 2003;31:272–274.

141. Murakami Y, Hayashida O. Supramolecular effects and molecular discrimination by macrocyclic hosts embedded in synthetic bilayer membranes. Proc Nat Acad Sci 1993;90: 1140–1145.

142. Szetjli J. Introduction and general overview of cyclodextrin chemistry. Chem Rev 1998;98:1743–1753.

143. Hirayama F, et al. Utilization of diethyl–ß cyclodextrin as a sustained-release carrier for isosorbide dinitrate. J Pharm Sci 1989;77:233–236.

144. Paolino D, Puglisi G, Ventura CA, Fresta M. Lecithin Organogels: Effect of Drug Physico-Chemical Characteristics on Matrix Release. European Conference on Drug Delivery and Pharmaceutical Technology, 97; 2004.

145. Eastburn SD, Tao BY. Applications of modified cyclodextrins. Biotech Adv 1994;12:325–339.

146. Muñoz-Botella S, del Castillo B, Martin MA. Cyclodextrin properties and applications of inclusion complex formation. Ars Pharm 1995;36:187–198.

147. Loftsson T, Brewster ME. Pharmaceutical applications of cyclodextrins: Drug solubilisation and stabilization. J Pharm Sci 1996;85:1017–1025.

148. Schneiderman E, Stalcup AM. Cyclodextrins: a versatile tool in separation science. J Chromatogr B 2000;745:83–102.

149. Schmid G. Cyclodextrin glucanotransferse production: yield enhancement by overexpression of cloned genes. Trends Biotechnol 1989;7:244–248.

150. Loftsson T, Fridriksdottir H, Sigurdardottir AM, Ueda H. The effect of water-soluble polymers on drug-cyclodextrin complexation. Int J Pharm 1994;110:169–177.

151. Redenti E, Szente L, Szejtli J. Drug/cyclodextrin/hydroxyacid multicomponent systems. Properties and pharmaceutical applications. J Pharm Sci 2000;89:1–8.

152. Loftsson T, Másson M, Sigurjónsdóttir JF, Methods to enhance the complexation efficiency of cyclodextrins. STP Pharma Sci 1999;9:237–242.

153. Jindrich J, et al. Regioselectivity of alkylation of cyclomaltoheptaose and synthesis of its mono-2-O-methyl, -ethyl, -allyl, and propyl, derivatives. Carbohydr Res 1995;266:75–80.

154. Gazpio C, et al. HPLC and solubility study of the interaction between pindolol and cyclodextrins. J Pharm Biomed Anal 2005;37:487–492.

155. Koizumi K, Kubota Y, Utamura T, Horiyama S. Analysis of heptakis 2,6,-di-O-methyl-ß-cyclodextrin by thin-layer chromatography, HPLC and gas chromatography/mass spectrometry. J Chromatogr 1986;368:329–337.

156. Uccello-Barretta G, et al. Combining NMR and molecular modelling in a drug delivery context: investigation of the multi-mode inclusion of trans-n-{4-[n'-(4-chlorobenzoyl)-hydrazinocarbonyl] cyclohexylmethyl}-4-romobenzenesulfonamide, a new chemotype of npy-5 antagonist, into β-cyclodextrin. Bioorg Med Chem 2004;12:447–458.

157. Bakirci H, Zhang X, Nau WM. Induced circular dichroism and structural assignment of the cyclodextrin inclusion complexes of bicyclic azoalkanes. J Org Chem 2005;70: 39–46.

158. Puglisi G, et al. Preparation and Physico-Chemical Study of Inclusion Complexes between Idebenone and Modified β-Cyclodextrins. J Inclus Phenom 1996;24:193–210.

159. Beaufour M, Morin P, Ribet JP. Chiral separation of the four stereoisomers of a novel antianginal agent using a dual cyclodextrin system in capillary electrophoresis. J Sep Sci 2005;28:529–533.

160. Wang Z, et al. Enantioseparation of chiral allenic acids by micellar electrokinetic chromatography with cyclodextrins as chiral selector. Electrophoresis 2005;26:1001–1006.

161. Liang M, Qi M, Zhang C, Fu R. Peralkylated-beta-cyclodextrin used as gas chromatographic stationary phase prepared by sol–gel technology for capillary column. J Chromatogr A 2004;1059:111–119.

162. Geze A, et al. Long-term shelf stability of amphiphilic beta-cyclodextrin nanosphere suspensions monitored by dynamic light scattering and cryo-transmission electron microscopy. J Microencapsul 2004;21:607–613.

163. Irie T, Uekama K. Pharmaceutical applications of cyclodextrins. III. Toxicological issues and safety evaluation. J Pharm Sci 1997;86:147–162.

164. Thompson DO. Cyclodextrins-enabling excipients: their present and future use in pharmaceuticals. Crit Rev Ther Drug Carrier Syst 1997;14:1–104.

165. Rajewski RA, et al. Preliminary safety evaluation of parenterally administered sulfoalkyl ether β-cyclodextrin derivatives. J Pharm Sci 1995;84:927–932.

166. Stella VJ, Rajewski RA. Cyclodextrins: their future in drug formulation and delivery. Pharm Res 1997;14:556–567.

167. Lopez RFL, Collett JH, Bentley MVLB. Influence of cyclodextrin complexation on the in vitro permeation and skin metabolism of dexamethasone. Int J Pharm 2000;200:127–132.

168. Reeuwijk HJEM, et al. Liquid chromatographic determination of b-cyclodextrin derivatives based on fluorescence enhancement after inclusion complexation. J Chromatogr 1993;614:95–101.

169. Kublik H, Bock TK, Schreier H, Muller BW. Nasal absorption of 17b-estradiol from different cyclodextrin inclusion formulations in sheep. Eur J Pharm Biopharm 1996;42:320–324.

170. Loftsson T, Masson M. Cyclodextrins in topical formulations: Theory and practice. Int J Pharm 2001;225:15–30.

171. Keipert S, Fedder J, Bohm A, Hanke B. Interactions between Cyclodextrins and pilocarpine-as an example of a hydrophilic drug. Int J Pharm 1996;142:153–162.

172. Davies NM, Wang G, Tucker IG. Evaluation of a hydrocortisone/hydroxypropyl-b-cyclodextrin solution for ocular drug delivery. Int J Pharm 1997;156:201–209.

173. Jarho P, et al. Increase in aqueous solubility, stability and in vitro corneal permeability of anandamide by hydroxypropyl-β-cyclodextrin. Int J Pharm 1996;137:209–217.

174. Loftsson T, Fridriksdottir H. The effect of water-soluble polymers on the aqueous solubility and complexing abilities of β-cyclodextrin. Int J Pharm 1998;163:115–121.

175. Sigurdardottir AM, Loftsson T. The effect of polyvinylpyrrolidone on cyclodextrin complexation of hydrocortisone and its diffusion through hairless mouse skin. Int J Pharm 1995;126:73–78.

176. Tanaka M, et al. Effect of 2-hydroxypropyl-β-cyclodextrin on percutaneous absorption of methyl paraben. J Pharm Pharmacol 1995;47:897–900.

177. Vitória M, et al. Characterization of the influence of some cyclodextrins on the stratum corneum from the hairless mouse. J Pharm Pharmacol 1997;49:397–402.

178. Loftsson T, Sigurjónsdóttir AM. The effect of polyvinylpyrrolidone and hydroxypropyl methylcellulose on HP-β-CD complexation of hydrocortisone and its permeability through hairless mouse skin. Eur J Pharm Sci 1994;2:297–301.

179. Masson M, Loftsson T, Masson V, Stefansson E. Cyclodextrins as permeation enhancers: some theoretical evaluations and in vitro testing. J Control Rel 1999;59:107–118.

180. Bhatnagar S, Vyas SP. Organogel-based system for transdermal delivery of propanolol. J Microencapsul 1994;11:431–438.

181. Ventura CA, et al. Biomembrane model interaction and percutaneous absorption of papaverine through rat skin: effects of cyclodextrins as penetration enhancers. J Drug Target 2001;19:379–393.

182. Arima H, Kondo T, Irie T, Uekama K. Enhanced rectal absorption and reduced local irritation of the anti-inflammatory drug ethyl 4-biphenylylacetate in rats by complexation with water-soluble b-cyclodextrin derivatives and formulation as oleaginous suppository. J Pharm Sci 1992;81:1119–1125.

183. Ventura CA, et al. Celecoxib-Dimethyl-β-Cyclodextrin inclusion complex. Characterization and in vitro permeation study. Eur J Med Chem 2005;40:624–631.

184. Brewster ME, Anderson WR, Estes KS, Bodor N. Development of aqueous parenteral formulations for carbamazepine through the use of modified cyclodextrins. J Pharm Sci 1991;80:380–383.

185. Sridevi S, et al. Enhancement of dissolution and oral bioavailability of gliquidone with hydroxy propyl-β-cyclodextrin. Pharmazie 2003;58:807–810.

186. Uekama K, Otagiri M. Cyclodextrins in drug carrier systems. CRC Crit Rev Ther Drug Carrier Syst 1987;3:1–40.

187. Tokumura T, et al. Enhancement of bioavaiability of cinnarizine from its ß-cyclodextrin complex on oral administration with D,L-phenylalanine as a competing agent. J Pharm Sci 1986;75:391–394.

188. Nakanishi K, Masada M, Nadai T, Miyajima K. Effect of the interaction of drug–ß-cyclodextrin complex with bile salts on the drug absorption from rat small intestinal lumen. Chem Pharm Bull 1989;37:211–214.

189. Uekama K, et al. Improvement of the oral bioavailability of digitalis glycosides by cyclodextrin complexation. J Pharm Sci 1983;72:1338–1341.

190. Stuenkel CA, Dudley RE, Yen SS. Sublingual administration of testosterone–hydroxypropyl–ß-cyclodextrin inclusion complex simulates episodic androgen release in hypogonadal men. J Clin Endocrinol Metab 1991;72:1054–1059.

191. Fridriksdottir H, et al. Design and in vivo testing of 17b-estradiol–hydroxypropyl–ß-cyclodextrin sublingual tablets. Pharmazie 1996;51:39–42.

192. Hirayama F, Uekama K. Cyclodextrin-based controlled drug release system. Adv Drug Deliv Rev 1999;36:125–141.

193. Szejtli J. 1997; Cyclodextrin News 11, Budapest Cyclolab.

194. Shiotani K, Irie T, Uekama K, Ishimaru Y. Cyclodextrin sulfates in parenteral use: Protection against gentamicin nephrotoxicity in the rat. Eur J Pharm Sci 1995;3:139–151.

195. Bekers O, et al. Effect of cyclodextrins on the chemical stability of mitomycins in alkaline solution. J Pharm Biomed Anal 1991;9:1055–1060.

196. Alcaro S, et al. Preparation, Characterization, Molecular Modeling and In Vitro Activity of Paclitaxel-Cyclodextrin Complexes. Bioorg Med Chem Lett 2002;12:1673–1641.

197. Irie T, Uekama K. Cyclodextrins in peptide and protein delivery. Adv Drug Deliv Rev 1999;36:101–123.

198. Redenti E, Pietra C, Gerloczy A, Szente L. Cyclodextrin in oligonucleotide delivery. Adv Drug Deliv Rev 2001;53:235–244.

199. Zhao T, Temsamani J, Agarwal S. Use of cyclodextrin and its derivatives as carriers for oligonucleotide delivery. Antisense Res 1995;5:185–192.

200. Dass CR, Jessup W, Apolipoprotiens AI. Cyclodextrins and liposomes as potential drugs for the reversal of atherosclerosis. J Pharm Pharmacol 2000;52:731–761.

201. Croyle MA, Cheng X, Wilson JM. Development of formulations that enhance physical stability of viral vectors for gene therapy. Gene Ther 2001;8:1281–1290.

202. Nicolazzi C, et al. Effect of the complexation with cyclodextrins on the in vitro antiviral activity of ganciclovir against human cytomegalovirus. Bioorg Med Chem 2001;9:275–282.

203. Hovgaad L, Broendsted H. Current applications of polysaccharides in colon targeting. CRC Crit Rev Ther Drug Carrier Syst 1996;13:185–223.

204. Minami K, Hirayama F, Uekama K. Colon-specific drug delivery based on a cyclodextrin prodrug: release behavior of biphenylylacetic acid from its cyclodextrin conjugates in rat intestinal racts after oral administration. J Pharm Sci 1998;87:715–720.

205. Lawrence MJ, Rees GD. Microemulsion-based media as novel drug delivery systems. Adv Drug Deliv Rev 2000;45: 89–121.

206. Hoar TP, Schulman JH. Transparent water-in-oil dispersions: the oleopathic hydro-micelle. Nature (London) 1943;152:102–103.

207. Schulman JH, Stoeckenius W. Prince LM. Mechanism of formation and structure of microemulsions by electron microscopy. J Phys Chem 1959;63:1677–1680.

208. Shinoda K, Lindman B. Organised surfactant systems: microemulsions. Langmuir 1987;3:135–149.

209. Lam AC, Schechter RS. The theory of diffusion in microemulsions. J Colloid Interface Sci 1987;120:56–63.

210. Kriwet K, Muller-Goymann CC. Diclofenac release from phospholipid drug systems and permeation through excised human stratum corneum. Int J Pharm 1995;125:231–242.

211. Schmalfuss U, Neubert R, Wohlrab W, Modification of drug penetration into human skin using microemulsions. J Control Rel 1997;46:279–285.

212. Osborne DW, Ward AJ, O'Neill KJ. Microemulsions topical drug delivery vehicles: in-vitro transdermal studies model hydrophilic drug. J Pharm Pharmacol 1991;43:450–454.

213. Rosano HL, Cavello JL, Chang DL, Whittham JH. Microemulsions: a commentary on their preparation. J Soc Cosmet Chem 1988;39:201–209.

214. Grunewald AM, Gloor M, Gehring W, Kleesz P. Damage to the skin by repetitive washing. Contact Dermatitis 1995;32:225–232.

215. Effendy I, Maibach HI. Surfactants and experimental dermatitis. Contact Dermatitis 1995;33:217–225.

216. Paolino D, et al. Lecithin microemulsions for the topical administration of ketoprofen: percutaneous adsorption through human skin and in vivo human skin tolerability. Int J Pharm 2002;244:21–31.

217. Trotta M, Morel S, Gasco MR. Effect of oil phase composition on the skin permeation of felodipine from o/w microemulsions. Pharmazie 1997;52:50–53.

218. Delgado-Charro MB, et al. Delivery of a hydrophilic solute through the skin from novel microemulsion systems. Eur J Pharm Biopharm 1997;43:37–42.

219. Kriwet K, Muller-Goymann CC. Diclofenac release from phospholipid drug systems and permeation through excised human stratum corneum. Int J Pharm 1995;125:231–242.

220. Attwood D. Colloydal Drug Delivery Systems. New York: Marcel Dekker; 1994.

221. Benita S, Muchtar S. Ophthalmic Compositions. Eur. Patent 0521,799,A1, 1992.

222. Vandamme TF. Microemulsions as ocular drug delivery systems: recent Developments and future challenges. Prog Retinal Eye Res 2002;21:15–34.

223. Scartazzini R, Luisi PL. Organogels from Lecithins. J Phys Chem 1988;92:595–596.

224. Yurtov EV, Murashova NM. Lecithin Organogels in Hydrocarbon Oil. Colloid J 2003;65:114–118.

225. Willimann HL, Luisi PL. Lecithin organogels as matrix for transdermal transport of drugs. Biochem Biophys Commun 1991;177:897–900.

226. Willimann HL, et al. Lecithin Organogels as Matrix for Transdermal Transport of Drugs. J Pharm Sci 1992;81: 871–874.

227. Shchipunov YA, Shumilina EV. Lecithin bridging by hydrogen bonds in the organogel. Mat Sci Eng 1995;3:43–50.

228. Paolino D, et al. *In Vivo* Evaluation of Lecithin Organogels for Transdermal Application, 30th Annual Meeting & Exposition of the Controlled Release Society. Glasgow (UK); 2003.

229. Shah JC, Sadhale Y, Chilukuri DM. Cubic phase gels as drug delivery systems. Adv Drug Deliv Rev 2001;47:229–250.

230. Chasin M, Langer R. Biodegradable Polymers as Drug Delivery Systems. New York: Marcel Dekker; 1990.

231. Rolland A. Pharmaceutical Particulate Carriers: Therapeutic Applications. New York: Marcel Dekker; 1993.

232. Schwab G, et al. Antisense oligonucleotides adsorbed to poly-alkylcyanoacrylate nanoparticles specifically inhibit mutated Ha-ras-mediated cell proliferation and tumorigenicity in nude mice. Proc Natl Acad Sci USA 1994;91:10460–10464.

233. Benns JM, Kim SW. Tailoring new gene delivery designs for specific targets. J Drug Target 2000;8:1–12.

234. Pastan I, Chaudhary V, Fitzgerald DJ. Recombinant toxins as novel therapeutic agents. Annu Rev Biochem 1992;61:331–354.

235. Moghimi SM, Christy Hunter A, Murray JC. Long-Circulating and Target-Specific Nanoparticles: Theory to Practice. Pharmacol Rev 2001;53:283–318.

236. Allen TM, Moase EH. Therapeutic opportunities for targeted liposomal drug delivery. Adv Drug Deliv Rev 1996;21:117–133.

237. Bakker-Woudenberg IAJM. Delivery of antimicrobials to infected tissue macrophages. Adv Drug Deliv Rev 1995; 17:5–20.

238. Ten Hagen TL, Van Vianen W, Bakker-Woudenberg IAJM. Modulation of nonspecific antimicrobial resistance of mice to Klebsiella pneumoniae septicemia by liposome-encapsulated muramyl tripeptide phosphatidylethanolamine and interferon-gamma alone or combined. J Infect Dis 1995;171: 385–392.

239. Russell-Jones GJ. Oral vaccine delivery. J Control Rel 2000;65:49–54.

240. Woodle MC, Storm G. Long Circulating Liposomes: Old Drugs, New Therapeutics. Berlin: Springer-Verlag; 1998.

241. B Ruozi, et al. Atomic force microscopy and photon correlation spectroscopy: Two techniques for rapid characterization of liposomes. E J Pharm Sci 2005;25:81–89.

242. Gimbert LJ, Haygarth PM, Beckett R, Worsfold PJ. Comparison of centrifugation and filtration techniques for the size fractionation of colloidal material in soil suspensions using sedimentation field-flow fractionation. Environ Sci Technol 2005;39:1731–1735.

243. K Thode, Muller RH, Kresse M. Two-time window and multi-angle photon correlation spectroscopy size and zeta potential analysis-highly sensitive rapid assay for dispersion stability. J Pharm Sci 2000;89:1317–1324.

244. Ford JL, Timmins P. Pharmaceutical thermal analysis techniques and applications. Southampton: John Wiley & Sons, Inc.

245. Venkateswarlu V, Manjunath K. Preparation, characterization and in vitro release kinetics of clozapine solid lipid nanoparticles. J Control Rel 2005;95:627–638.

246. Bower PV, et al. Solid-state NMR structural studies of peptides immobilized on gold nanoparticles. Langmuir 2005;21: 3002–3007.

247. Frohlich M, Brecht V, Peschka-Suss R. Parameters influencing the determination of liposome lamellarity by ^{31}P-NMR. Chem Phys Lipids 2001;109:103–112.

248. Lasic DD. Novel applications of liposomes. TIBTECH 1998; 16:307–321.

249. Drummond DC, et al. Optimizing liposomes for delivery of chemotherapeutic agents to solid tumors. Pharmacol Rev 1999;51:691–743.

250. Waterhouse DN, et al. Preparation, characterization and biological analysis of liposomal formulations of vincristine. Methods Enzymol 2005;391:140–157.

251. Park JW. Liposome-based drug delivery in breast cancer treatment. Breast Cancer Res 2002;4:95–99.

252. Senior JH. Fate and behaviour of liposomes in vivo: A review of controlling factors. CRS Crit Rev Ther Drug Carrier Syst 1987;3:123–193.

253. Senior J, Gregoriadis G. Stability of small unilamellar liposomes in serum and clearance from the circulation: the effect of the phospholipid and cholesterol components. Life Sci 1982;30:2123–2136.

254. Allen TM, et al. Liposomes containing synthetic lipid derivates of poly (ethylene glycol) show prolonged circulation half-live in vivo. Biochim Biophys Acta 1991;1066:29–36.

255. Pedroso de Lima MC, et al. Cationic lipid–DNA complexes in gene delivery: from biophysics to biological applications. Adv Drug Deliv Rev 2001;47:277–294.

256. Crosasso P, et al. Preparation, characterization and properties of sterically stabilized paclitaxel-containing liposomes. J Control Rel 2000;63:19–30.

257. Paolino D, et al. Tolerability and improved protective action of idebenone-loaded pegylated liposomes on ethanol-induced injury in primary cortical astrocytes. J Pharm Sci 2004;93:1815–1827.

258. Sudimack JJ, Guo W, Tjarks W, Lee RJ. A novel pH-sensitive liposome formulation containing oleyl alcohol. Biochim Biophys Acta 2002;1564:31–37.

259. Torchilin VP, Rammohan R, Weissig V, Levchenko TS. TAT peptide on the surface of liposomes affords their efficient intracellular delivery even at low temperature and in the presence of metabolic inhibitors. Proc Natl Acad Sci USA 2001;98:8786–8791.

260. Fresta M, et al. Characterization and in vivo ocular absorption of liposome-encapsulated acyclovir. J Pharm Pharmacol 1999;51:565–576.

261. Szoka Jr F, Papahadjopoulos D. Procedure for preparation of liposomes with large internal aqueous space and high capture by reverse-phase evaporation. Proc Natl Acad Sci USA 1978;75:4194–4198.

262. Duzgunes N. Preparation and quantitation of small unilamellar liposomes and large unilamellar reverse-phase evaporation liposomes. Methods Enzymol 2003;367:23–27.

263. Fresta M, Wehrli E, Puglisi G. Neutrase entrapment in stable multilamellar and large unilamellar vesicles for the acceleration of cheese ripening. J Microencapsul 1995;12:307–325.

264. Mayer LD, Hope MJ, Cullis PR, Janoff AS. Solute distributions and trapping efficiencies observed in freeze-thawed multilamellar vesicles. Biochim Biophys Acta 1985;817: 193–196.

265. Alino SF, et al. High encapsulation efficiencies in sized liposomes produced by extrusion of dehydration-rehydration vesicles. J Microencapsul 1990;7:497–503.

266. Hunter DG, Frisken BJ. Effect of extrusion pressure and lipid properties on the size and polydispersity of lipid vesicles. Biophys J 1998;74:2996–3002.

267. Celano M, et al. Cytotoxic effects of gemcitabine-loaded liposomes in human anaplastic thyroid carcinoma cells. BMC Canc 2004;4:63.

268. Clerc S, Barenholz Y. Loading of amphipatic weak acids into liposomes in response to transmembrane calcium acetate gradients. Biochim Biophys Acta 1995;1240:257–265.

269. Cullis PR, et al. Influence of pH gradient on the transbilayer transport of drugs, lipids, peptides and metal ions into large unilamellar vesicles. Biochim Biophys Acta 1997;1331:187–211.

270. Fresta M, Ventura CA, Mezzasalma E, Puglisi G. A calorimetric study on the idebenone-phospholipid membrane interaction. Int J Pharm 1998;163:133–143.

271. CB Hansen TM, Lopes de Menezes DE. Pharmacokinetics of long-circulating liposomes. Adv Drug Deliv Rev 1995;16:267–284.

272. Barenholz Y, et al. Stability of liposomal doxorubicin formulations: problems and prospects. Med Res Rev 1993;13:449–491.

273. Kanter PM, et al. Preclinical toxicology study of liposome encapsulated doxorubicin (TLC D-99): comparison with doxorubicin and empty liposomes in mice and dogs. In Vivo 1993;7:85–95.

274. Charrois GJ, Allen TM. Multiple injections of pegylated liposomal Doxorubicin: pharmacokinetics and therapeutic activity. J Pharmacol Exp Ther 2003;306:1058–1067.

275. Maruyama K, Ishida O, Takizawa T, Moribe K. Possibility of active targeting to tumor tissue with liposomes. Adv Drug Deliv Rev 1999;40:89–102.

276. Iden DL, Allen TM. In vitro and in vivo comparison of immunoliposomes made by conventional coupling techniques with those made by a new post-insertion approach. Biochim Biophys Acta 2001;1531:207–216.

277. Forssen E, Willis M. Ligand-targeted liposomes. Adv Drug Deliv Rev 1998;29:249–271.

278. Fresta M, Puglisi G. Colloidal drug delivery systems in anti-infective chemotherapy. Pandalai SG, editor. Recent Research Developments in Antimicrobial Agents and Chemotherapy Trivandrum-8. India: Research Signpost; 2000.

279. Fresta M, et al. Intracellular accumulation of ofloxacin-loaded liposomes in human synovial fibroblasts. Antimicrob Agents Chemother 1995;39:1372–1375.

280. Furneri PM, Fresta M, Puglisi G, Tempera G. Ofloxacin-loaded liposomes: their in vitro activity and effect on Bacterial drug accumulation. Antimicrob Agents Chemother 2000; 44:2458–2464.

281. Bakker-Woundenberg IAJM, Lokerse AF, Ten Kate MT, Storm G. Enhanced localization of liposomes with prolonged blood circulation time in infected lung tissue. Biochim Biophys Acta 1992;1138:318–326.

282. Duzgunes N, et al. Delivery of antiviral agents in liposomes. Methods Enzymol 2005;391:351–373.

283. Cudd A, et al. Specific interaction of CD4-bearing liposomes with HIV-infected cells. J Acquir Immune Defic Syndr 1990;3:109–114.

284. Lee JT, et al. Evaluation of cationic liposomes for delivery of diphtheria toxin A-chain gene to cells infected with bovine leukemia virus. J Vet Med Sci 1997;59:169–174.

285. Fielding RM, et al. Comparative pharmacokinetics of amphotericin B after administration of a novel colloidal delivery system, ABCD, and a conventional formulation to rats. Antimicrob Agents Chemother 1991;35:1208–1213.

286. Lopez-Berestein G, et al. Liposomal amphotericin B for the treatment of systemic fungal infections in patients with cancer: a preliminary study. J Infect Dis 1985;151: 704–710.

287. Proulx ME, et al. Treatment of visceral leishmaniasis with sterically stabilized liposomes containing camptothecin. Antimicrob Agents Chemother 2001;45:2623–2627.

288. Gregoriadis G. DNA vaccines: a role for liposomes. Curr Opin Mol Ther 1999;1:39–42.

289. Fresta M, Wehrli E, Puglisi G. Enhanced therapeutic effect of cytidine-5^I-diphosphate choline when associated with G_{M1} containing small liposomes as demonstrated in a rat ischemia model. Pharm Res 1995;12:1769–1774.

290. Fresta M, Puglisi G. Survival rate improvement in a rat ischemia model by long circulating liposomes containing CDP-choline. Life Sci 1997;61:1227–1235.

291. Fresta M, Puglisi G. Reduction of maturation phenomenon in cerebral ischemia with CDP-choline-loaded liposomes. Pharm Res 1999;16:1843–1849.

292. Mezei M, Gulusekharam V. Liposomes, a selective drug delivery system for the topical route of administration: Gel dosage form. J Pharm Pharmacol 1982;34:473–474.

293. Touitou E, et al. Liposomes as carriers for topical and transdermal delivery. J Pharm Sci 1994;83:1189–1203.

294. Mezei M, Gulusekharam V. Liposomes, a selective drug delivery system for the topical route of administration. Life Sci 1980;26:1473–1477.

295. Egbaria K, Weiner N. Liposomes as a topical drug delivery system. Adv Drug Deliv Rev 1990;5:287–300.

296. Foong WC, Harsanyi BB, Mezei M. Biodisposition and histological evaluation of topically applied retinoic acid in liposomal cream and gel dosage forms. Haning I, Pepeu G, editors. Phospholipids. New York: Plenum Press; 1990.

297. Touitou E, et al. Modulation of caffeine skin delivery by carrier design: liposomes versus permeation enhancers. Int J Pharm 1994;103:131–136.

298. Bouwstra JA, Honeywell-Nguyen PL. Skin structure and mode of action of vesicles. Adv Drug Deliv Rev 2002;54:41–55.

299. Fresta M, Puglisi G. Application of liposomes as potential cutaneous drug delivery. In vitro and in vivo investigation with radioactively labelled vesicles. J Drug Target 1996; 4:95–101.

300. Fresta M, Puglisi G. Corticosteroid dermal delivery with skin-lipid liposomes. J Control Rel 1997;44:141–151.

301. Touitou E, et al. Ethosomes–novel vesicular carriers for enhanced delivery: characterization and skin penetration properties. J Control Rel 2000;65:403–418.

302. Cevc G, Blume G. New, highly effcient formulation of diclofenac for the topical, transdermal administration in ultradeformable drug carriers, Transfersomes. Biochim Biophys Acta 2001;1514:191–205.

303. Gregoriadis G, Florence AT. Liposomes in drug delivery. Clinical, diagnostic and ophthalmic potential. Drugs 1993; 45:15–28.

304. Bouwstra JA, van Hal DA, Hofland HEJ, Junginger HE. Preparation and characterization of nonionic surfactant vesicles. Colloids Surface A 1997;123:71–80.

305. Uchegbu IF, Vyas PS. Non-ionic surfactant based vesicles (niosomes) in drug delivery. Int J Pharm 1998;172:33–70.

306. Carafa M, et al. Preparation and properties of new unilamellar non-ionic: ionic surfactant vesicles. Int J Pharm 1998;160:51–59.

307. Santucci E, et al. Vesicles from polysorbate-20 and cholesterol-α simple preparation and a characterisation. STP Pharm Sci 1996;6:29–32.

308. Gupta PN, et al. Non-invasive vaccine delivery in trasferosomes, niosomes and liposomes: a comparative study. Int J Pharm 2005;293:73–82.

309. Kirby C, Gregoriadis G. Dehydration-rehydration vesicles: a simple method for high yield drug entrapment in liposomes. Biotechnology 1984; 979–984.

310. Hofland HEJ, et al. Safety aspects of non-ionic surfactant vesicles-a toxicity study related to the physicochemical characteristics of non-ionic surfactants. J Pharm Pharmacol 1992;44:287–294.

311. Dimitrijevic D, et al. The effect of monomers and of micellar and vesicular forms of non-ionic surfactants (Solulan C24 and Solulan 16) on Caco-2 cell monolayers. J Pharm Pharmacol 1997;49:611–616.

312. Jain CP, Vyas SP. Lymphatic delivery of niosome encapsulated methotrexate. Pharmazie 1995;50:367–368.

313. Parthasarathi G, Udupa N, Umadevi P, Pillai GK. Niosome encapsulated of vincristine sulfate-improved anticancer activity with reduced toxicity in mice. J Drug Target 1994; 2:173–182.

314. Yoshioka T, Florence AT. Vesicle (niosome)-in-water-in-oil (v:w:o) emulsions-an in-vitro study. Int J Pharm 1994;108:117–123.

315. Erdogan S, et al. In-vivo studies on iopromide radiopaque niosomes. STP Pharma Sci 1996;6:87–93.

316. Azmin MN, et al. The effect of non-ionic surfactant vesicle (niosome) entrapment on the absorption and distribution of methotrexate in mice. J Pharm Pharmacol 1985;37:237–242.

317. Chandraprakash KS, Udupa N, Umadevi P, Pillai GK. Effect of macrophage activation on plasma disposition of niosomal ^{3}H-Methotrexate in sarcoma-180 bearing mice. J Drug Target 1993;1:143–145.

318. Rogerson A, Cummings J, Willmott N, Florence AT. The distribution of doxorubicin in mice following administration in niosomes. J Pharm Pharmacol 1988;40:337–342.

319. Vanhal D, et al. Diffusion of estradiol from non-ionic surfactant vesicles through human stratum-corneum in vitro. STP Pharm Sci 1996;6:72–78.

320. Junginger HE, Hofland HEJ, Bouwstra JA. Liposomes and niosomes: interactions with human skin. Cosmet Toilet 1991; 106:45–50.

321. Saettone MF, et al. Non-ionic surfactant vesicles as ophthalmic carriers for cyclopentolate a preliminary evaluation. STP Pharm Sci 1996;6:94–98.

322. Touitou E. Compositions for Applying Active Substances to or through the skin, US Patent 5,540,934, 1996.

323. Touitou E. Compositions for Applying Active Substances to or through the skin, US Patent 5,716,638, 1997.

324. Touitou E. Compositions and Methods for Intracellular Delivery. PCT/IL02/00516, 2002.

325. Berner B. editors. Percutaneous Penetration Enhancers. Boca Raton (Fl): CRC Press; 1995.

326. Dayan N, Touitou E. Carriers for skin delivery of trihexyphenidyl HCl: ethosomes vs. liposomes. Biomaterials 2000; 21:1879–1885.

327. Touitou E, et al. Intracellular delivery mediated by an ethosomal carrier. Biomaterials 2001;22:3053–3059.

328. Godin B, Touitou E. Mechanism of bacitracin permeation enhancement through the skin and cellular membranes from an ethosomal carrier. J Control Rel 2004;94:365–379.

329. Godin B, Touitou E. Ethosomes: new prospects in transdermal delivery. Crit Rev Ther Drug Carrier Syst 2003;20:63–102.

330. Paolino D, et al. Ethosomes for skin delivery of ammonium glycyrrhizinate: in vitro percutaneous permeation through human skin and in vivo anti-inflammatory activity on human volunteers. J Control Rel 2005;106:99–110.

331. Horwitz E, et al. A clinical evaluation of a novel liposomal carrier for aciclovir in the topical treatment of recurrent herpes labialis. Oral Surg Oral Med Oral Pathol Oral Radiol. Endod 1999;87:700–705.

332. Touitou E, Godin B, Weiss C. Enhanced delivery of drugs into and across the skin by ethosomal carriers. Drug Dev Res 2000;50:406–415.

333. Lodzki M, et al. Cannabidiol—transdermal delivery and anti-inflammatory effect in a murine model. J Control Rel 2003; 93:377–387.

334. Touitou E, Drug delivery across the skin. Expert Opin Biol Ther 2002;2:723–733.

335. Schreier H, Bouwstra J. Liposomes and niosomes as topical drug carriers: dermal and transdermal drug delivery. J Control Rel 1994;30:1–15.

336. Cevc G, Blume G, Schatzlein A. Transfersomes-mediated transepidermal delivery improves the regio-specificity and biological activity of corticosteroids in vivo. J Control Rel 1997;45:211–226.

337. Cevc G, et al. The skin: a pathway for systemic treatment with patches and lipid-based agent carriers. Adv Drug Deliv Rev 1996;18:349–378..

338. Cevc G, Blume G. New, highly effcient formulation of diclofenac for the topical, transdermal administration in ultradeformable drug carriers, Transfersomes. Biochim Biophys Acta 2001;1514:191–205.

339. Kim A, Lee EH, Choi SH, Kim CK. *In vitro* and *in vivo* transfection effciency of a novel ultradeformable cationic liposome. Biomaterials 2004;25:305–313.

340. Cevc G, et al. Ultraflexible vesicles, Transferosomes, have an extremely low pore penetration resistance and transport therapeutic amounts of insulin acrossthe intact mammalian skin. Biochim Biophys Acta 1998;1368:201–215.

341. Hildebrand A, et al. Solubilization of negatively charged DPPC/DPPG liposomes by bile salts. J Coll Interf Sci 2004; 279:559–571.

342. Cevc G, Transfersomes®-Innovative transdermal drug carriers. In: modified release drug delivery technology. New York: Marcel Dekker; 2002.

343. Cevc G, Blume G. Hydrocortisone and dexamethasone in ultradeformable drug carriers, Transfersomes®, have an increased in biological potency and reduced therapeutic dosages. Biochim Biophys Acta 2004;1663:61–73.

344. Cevc G. Transfersomes, liposomes and other lipid suspensions on the skin: permeation enhancement, vesicle penetration, and transdermal drug delivery. Crit Rev Ther Drug Carrier Syst 1996;13:257–388.

345. Vauthier C, et al. Poly(alkylcyanoacrylates) as biodegradable materials for biomedical applications. Adv Drug Deliv Rev 2003;55:519–548.

346. Couvreur P, et al. Nanocapsule technology: a review. Crit Rev Ther Drug Carrier Syst 2002;19:99–134.

347. Cui Z, Mumper RJ. Microparticles and nanoparticles as delivery systems for DNA vaccines. Crit Rev Ther Drug Carrier Syst 2003;20:103–137.

348. Kreuter J, editor. Nanoparticles, Colloidal Drug Delivery Systems. New York: Marcel Dekker; 1994.

349. Feng SS, Huang G. Effects of emulsifiers on the controlled release of paclitaxel (Taxol) from nanospheres of biodegradable polymers. J Control Rel 2001;71:53–69.

350. Li YP, et al. PEGylated polycyanoacrylate nanoparticles as tumor necrosis factor-alpha carriers. J Control Rel 2001;71: 287–296.

351. Carino GP, Jacob JS, Mathiowitz E. Nanosphere based oral insulin delivery. J Control Rel 2000;65:261–269.

352. Giannavola C, et al. Influence of preparation conditions on acyclovir-loaded poly-*d*,l-lactic acid nanospheres and effect of PEG-coating on ocular drug bioavailability. Pharm Res 2003;20:584–590.

353. Teixeira M, Alonso MJ, Pinto MM, Barbosa CM. Development and characterization of PLGA nanospheres and nanocapsules containing xanthone and 3-methoxyxanthone. Eur J Pharm Biopharm 2005;59:491–500.

354. Fresta M, et al. Preparation and characterization of polyethyl-2-cyanoacrylate nanocapsules containing antiepiletic drugs. Biomaterials 1996;17:751–758.

355. Cavallaro G, et al. Entrapment of β-lactames antibiotics on polyethylcyanoacrylate nanoparticles. Studies on the possible *in vivo* application of this colloidal delivery system. Int J Pharm 1994;111:31–41.

356. Harmia T, et al. Enhancement of the myotic response of rabbits with pilocarpine-loaded polybutylcyanoacrylate nanoparticles. Int J Pharm 1986;33:187–193.

357. Blanco MD, Alonso MJ. Development and characterization of protein-loaded poly(lactide-*co*-glycolide) nanospheres. Eur J Pharm Biopharm 1997;43:287–294.

358. Bala I, Hariharan S, Kumar MN. PLGA nanoparticles in drug delivery: the state of the art. Crit Rev Ther Drug Carrier Syst 2004;21:387–422.

359. Muller RH, Jacobs C, Kayser O. Nanosuspensions as particulate drug formulations in therapy. Rationale for development and what we can expect for the future. Adv Drug Deliv Rev 2001;47:3–19.

360. Ubrich N, et al. Oral evaluation in rabbits of cyclosporin-loaded Eudragit RS or RL nanoparticles. Int J Pharm 2005; 288:169–175.

361. Xu Z, et al. *In vitro* and *in vivo* evaluation of actively targetable nanoparticles for paclitaxel delivery. Int J Pharm 2005;288:361–368.

362. Lorenzo-Lamosa ML, et al. Design of microencapsulated chitosan microspheres for colonic drug delivery. J Control Rel 1998;52:109–118.

363. Potineni A, Lynn DM, Langer R, Amiji MM. Poly(ethylene oxide)-modified poly(beta-amino ester) nanoparticles as a pH-sensitive biodegradable system for paclitaxel delivery. J Control Rel 2003;86:223–234.

364. Shenoy DB, Amiji MM. Poly(ethylene oxide)-modified poly-(epsilon-caprolactone) nanoparticles for targeted delivery of tamoxifen in breast cancer. Int J Pharm 2005;293:261–270.

365. Konan-Kouakou YN, Boch R, Gurny R, Allemann E. *In vitro* and *in vivo* activities of verteporfin-loaded nanoparticles. J Control Rel 2005;103:83–91.

366. Yoo HS, Lee KH, Oh JE, Park TG. *In vitro* and *in vivo* antitumor activities of nanoparticles based on doxorubicin-PLGA conjugates. J Control Rel 2000;68:419–431.

367. Redhead HM, Davis SS, Illum L. Drug delivery in poly(lactide-co-glycolide) nanoparticles surface modified with poloxamer 407 and poloxamine 908: *in vitro* characterisation and *in vivo* evaluation. J Control Rel 2001;70:353–363.

368. Stolnik S, et al. Surface modification of poly(lactide-*co*-glycolide) nanospheres by biodegradable poly(lactide)-poly-(ethyleneglycol) copolymers. Pharm Res 1994;11:1800–1808.

369. Peracchia MT, et al. Stealth PEGylated polycyanoacrylate nanoparticles for intravenous administration and splenic targeting. J Control Rel 1999;60:121–128.

370. De Jaeghere F, et al. Cellular uptake of PEO surface-modified nanoparticles: evaluation of nanoparticles made of PLA:PEO diblock and triblock copolymers. J Drug Target 2000;8:143–153.

371. Mitra S, Gaur U, Ghosh PC, Maitra AN. Tumor targeted delivery of encapsulated dextran-doxorubicin coniugate using chitosan nanoparticles as carrier. J Control Rel 2001;74:317–323.

372. Nobs L, Buchegger F, Gurny R, Allemann E. Surface modification of poly(lactic acid) nanoparticles by covalent attachment of thiol groups by means of three methods. Int J Pharm 2003;250:327–337.

373. Fresta M, et al. Pefloxacine mesilate- and Ofloxacin-loaded polyethylcyanoacrylate nanoparticles. Characterization of the colloidal drug carrier formulation. J Pharm Sci 1995;84:895–902.

374. Schäfer V, et al. Phagocytosis of nanoparticles by human immunodeficiency virus (HIV)-infected macrophages: a possibility for antiviral drug targeting. Pharm Res 1992;9:541–546.

375. Lherm C, et al. Unloaded polyisobutylcyanoacrylate nanoparticles: efficiency against bloodstream trypanosomes. J Pharm Pharmacol 1987;39:650–652.

376. Losa C, et al. Improvement of ocular penetration of amikacin sulphate by association to poly(butylcyanoacrylate) nanoparticles. J Pharm Pharmacol 1991;43:548–552.

377. Fresta M, et al. Ocular tolerability and in vivo bioavailability of PEG-coated polyethyl-2-cyanoacrylate nanosphere-encapsulated acyclovir. J Pharm Sci 2001;90:288–297.

378. Gomes Dos Santos AL, Bochot A, Fattal E. Intraocular delivery of oligonucleotides. Curr Pharm Biotechnol 2005; 6:7–15.

379. Agnihotri SA, Mallikarjuna NN, Aminabhavi TM. Recent advances on chitosan-based micro- and nanoparticles in drug delivery. J Control Rel 2004;100:5–28.

380. van der Lubben IM, et al. Chitosan microparticles for mucosal vaccination against diphtheria: oral and nasal efficacy studies in mice. Vaccine 2003;21:1400–1408.

381. Muller RH, Mader K, Gohla S. Solid lipid nanoparticles (SLN) for controlled drug delivery—a review of the state of the art. Eur J Pharm Biopharm 2000;50:161–177.

382. Jenning V, Gohla SH. Encapsulation of retinoids in solid lipid nanoparticles (SLN). J Microencapsul 2001;18:149–158.

383. Westesen K. Novel lipid-based colloidal dispersions as potential drug administration systems-expectations and reality. Coll Polym Sci 2000;278:609–618.

384. Cavalli R, Gasco MR, Barresi AA, Rovero G. Evaporative drying of aqueous dispersions of solid lipid nanoparticles. Drug Dev Ind Pharm 2001;27:919–924.

385. Hu L, Tang X, Cui F. Solid lipid nanoparticles (SLNs) to improve oral bioavailability of poorly soluble drugs. J Pharm Pharmacol 2004;56:1527–1535.

386. Wissing SA, Muller RH. Cosmetic applications for solid lipid nanoparticles (SLN). Int J Pharm 2003;254:65–68.

387. Wissing SA, Muller RH. Solid lipid nanoparticles as carrier for sunscreens: *in vitro* release and *in vivo* skin penetration. J Control Rel 2002;81:225–233.

388. Videira MA, et al. Lymphatic uptake of pulmonary delivered radiolabelled solid lipid nanoparticles. J Drug Target 2002;10:607–613.

389. Zara GP, et al. Intravenous administration to rabbits of non-stealth and stealth doxorubicin-loaded solid lipid nanoparticles at increasing concentration of stealth agent: pharmacokinetics and distribution of doxorubicin in brain and other tissues. J Drug Target 2002;10:327–335.

390. Muller RH, Keck CM. Drug delivery to the brain-realization by novel drug carriers. J Nanosci Nanotechnol 2004;4:471–483.

391. Jenning V, Thünemann AF, Gohla SH. Characterization of a novel solid lipid nanoparticle carrier system based on binary mixtures of liquid and solid lipids. Int J Pharm 2000;199:166–177.

392. Muller RH, Radtke M, Wissing SA. Solid lipid nanoparticles (SLN) and nanostructured lipid carriers (NLC) in cosmetic and dermatological preparations. Adv Drug Deliv Rev 2002;54:131–155.

393. Bummer PM. Physical chemical considerations of lipid-based oral drug delivery-solid lipid nanoparticles. Crit Rev Ther Drug Carrier Syst 2004;21:1–20.

394. Zhang Q, et al. Studies on the cyclosporin A loaded stearic acid nanoparticles. Int J Pharm 2000;200:153–159.

395. Cavalli R, Caputo O, Gasco MR. Preparation and characterization of solid lipid nanospheres containing paclitaxel. Eur J Pharm Sci 2000;10:305–309.

396. Olbrich C, Geßner A, Kayser O, Müller RH. Lipid-drug conjugate (LDC) nanoparticles as novel carrier system for the hydrophilic antitrypanosomal drug diminazenediaceturate. J Drug Target 2002;10:387–396.

See also DRUG INFUSION SYSTEMS; PHARMACOKINETICS AND PHARMACODYNAMICS.

DRUG INFUSION SYSTEMS

SELAHATTIN OZCELIK
Texas A&M University
Kingsville, Texas

INTRODUCTION

An infusion system can be described as the process of delivering fluids and medications in solution to patients by way of an infusion device. Intravenous route is generally used for drug delivery; however, subcutaneous, epidural, and enteral routes are also used for special drug administration (1). Administration of medication into the patient by way of some kind of drug infusion device provides the desired level of medication in the patient and allows direct control over pharmacological variables, such as onset of drug effects and peak serum drug concentrations (1). This type of drug administration has been the choice especially for specific conditions, including the use of antibiotics for severe infection, chemotherapy for malignant conditions, cardiac medication in critical cases, and analgesics for relief of severe pain.

When curing all common medical disorders that justify therapeutic intervention, pharmacological therapy is the preferred and effective method of treatment. Device-based drug delivery systems for the administration of effective pharmacological therapy can be grouped as injection–infusion, transdermal patch-based, and inhalation systems (2). Among the established methods of injection and infusion systems are the needle and jet injection, intravascular, intraspinal, intraoperative site, and intraperitoneal–transperitoneal infusion systems. Major applications of drug infusion are anesthesia delivery, antibiotic–antiviral therapy, nutritional support, pain management, cardiovascular disease therapy, chemotherapy, diabetes management, hydration therapy, bone marrow and organ transplant support therapy, and transfusion therapy (2).

The use of powered infusion devices has grown enormously in the last two decades. Infusion pumps together with an appropriate administration set provide an accurate flow of fluids over a prescribed time period. The simplest devices are the gravity controllers, in which the flow of liquid under the force of gravity is regulated by clamping action. More complex infusion systems include a positive pumping action for infusion (3). Volumetric pumps posses a linear peristaltic pumping mechanism. Syringe pumps work by pushing the plunger of a disposable syringe along at a predetermined rate. The type of pump used depends on the required volume and speed of infusion (3).

Medication errors are a major concern of healthcare professionals and medical institutions and have been reported to contribute to between 7000 and 140,000 deaths only in the United States each year (4–9). The impact of medication errors was found to be more severe in pediatric patients (4). There are a wide variety of reasons for medication errors. Reports suggest that the most significant factor is the user errors; however, the contribution of device-related problems to medical errors cannot be underestimated. Many reports of incidents have also been received involving infusion pumps. These incidents are primarily due to over infusions and may result in patient harm or death (3). In practice, some of the common problems in drug infusion systems originate from syringe pumps. Most of the patient morbidity and mortality, being the most significant among all the problems, has happened when using syringe pumps. Another common problem with drug delivery is venous air embolisms, which can be caused by air ingress due to improper drug delivery technique, damaged equipment and tubing, leaking or loose tubing connectors, or failure to stop delivery prior to complete evacuation of IV bags. Venous air embolisms have been observed in central venous cannulation and pressurized intravenous infusion systems (3).

Today it has been proved that technology is essential in reducing risk in medication delivery. Computerized physician order entry (CPOE) and bar code applications for drug administration are such technologies that are capable of reducing medication errors. Unfortunately, most hospitals have not yet implemented these systems; therefore, many errors that otherwise might be eliminated continue to put patients at risk (10). Computerized intravenous (IV) infusion devices, so-called smart pumps include software that incorporates dosage limits established by the medical institution, warnings when dosage limits are exceeded, configurable settings by patient type, and access to transaction data for quality improvement efforts. Such systems make it possible to provide an additional verification at the point of care to help prevent IV medication errors (11). The Institute for Safe Medication Practices and the Emergency Care Research Institute recognize safety systems for IV medication as vital to reducing medication-related errors (10,12). A couple of examples utilizing a new technology is MEDLEY from Alaris Medical Systems and COLLEAGUE CX by Baxter Healthcare Corporation. These infusion systems allow hospitals to enter various drug infusion protocols into a drug library with predefined dose limits. For example, if a dose is outside the programmed range or clinical parameters, the pump halts or informs the physician by providing an alarm. Some pumps are even capable of integrating patient monitoring and other parameters, such as patient's age or clinical condition. More and more manufacturers are bringing similar devices to market (10).

The aim of this section is to provide a review on drug infusion systems, basic operational principles of pumps used in such systems, new infusion devices that are being developed, and recent developments for the control of infusion devices. With this goal in mind, this section is organized as follows: The section Common Infusion Systems presents most commonly used infusion systems and their operational principles. In the section, New Developments in Drug Infusion Systems, smarter, smaller, reliable, and cost effective new infusion devices, which are currently under development, are reviewed. Since the performance of any automated system highly depends on its controller structure, most recent research work on the control of drug infusion devices is reviewed in the section

Recent Advancements in Controller Design for Drug Infusion Systems.

COMMON INFUSION SYSTEMS

Any drug infusion system requires some kind of control unit in which necessary parameters are monitored continuously so that the drug is delivered to the patient in a desired manner. Infusion devices range from very simple mechanical devices based on elastic containers, springs, and flow restrictors to sophisticated microprocessor controlled pumps. The choice of an infusion device depends on both the type of therapy to be applied and patient characteristics. Some IV infusions can safely and effectively be delivered via gravity drip systems, while others require sophisticated microprocessor controlled pumps for more precise control, positive pressure, and greater flow rate range. Traditionally, these devices are used in hospitals for the controlled delivery of drugs and fluids. However, as these devices become smarter with the use of new technology, more and more patients are using them for home therapy.

Therefore, the use of infusion pumps is increasing in the community. To ensure that a patient receives the correct dose, the appropriate infusion device should be chosen for the drug. Syringe pumps are commonly used at low rates of infusion, but may not be suitable for drugs that require constant blood levels (13). It is required that any infusion system be able to reliably deliver the prescribed drug dose–volume to the patient, at pressures that overcome all baseline and intermittent resistance, while causing no harm to the patient. Additive resistances, such as the small bore and kinking potential of connecting tubing, cannula, needles, and patient vessels, make infusion flow difficult. Filters, viscous solutions, and syringe stiction can also adversely affect the infusion flow. Therefore, infusion pumps are required to overcome these resistances and accurately deliver prescribed drugs to patients. These pumps must be capable of delivering infusions at pressures of between 100 and 500 mmHg (2–10 psi or 13.79–68.95 kPa) (13). Ideally, pumps should also reliably detect the infusion pressure and the presence of air in the line close to the patient vessel being infused. Table 1 provides pressure ranges for IV pump pressure settings. Infusion devices can be classified according to their power source as gravity controllers and infusion pumps.

Table 1. Pressure Ranges for IV Pump Pressure Setting[a,b]

Pressure, mmHg	Example	Pressure, psi
2–20	Central venous pressure range	0–0.4
10–30	Peripheral venous pressure range	0.2–0.6
100	Extravasation risk	2
100–150	Systolic arterial pressure range	2–3
75	Gravity pressure of fluid 100 cm above cannulation site	1.5
500	Highest probable pressure required by an infusion pump	10
1000	Maximum modern–Ambulatory pump occlusion pressure setting	20
3000	Common max. pressures form older peristaltic pumps	60

[a]See Ref. 14.
[b]1 mmHg = 1 psi = 6.89 kPa.

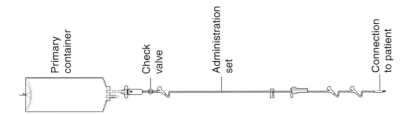

Figure 1. Common gravity drip infusion device.

Gravity Drip Systems

The simplest infusion device is the gravity drip system in which a bag or bottle is hung on a hook of a pole sufficiently high from the level of the patient. Figure 1 shows a typical gravity drip infusion system. The fluid flows by gravitational force down the line and into the catheter. They are quite suitable for lower risk applications, including fluid replacement therapy, provided that the required flow rate is achieved by the delivery pressure of the device (4). Gravity controllers are based on gravity to provide the infusion pressure. Therefore, in order to achieve the desired flow rate, the fluid container is placed sufficiently high above the patient's heart. A drop sensor monitoring the drip rate is attached to the drip chamber of the administration set. The rate of flow in a simple gravity drip system is controlled by a special clamp or valve on the line that can be manually adjusted to permit the prescribed amount of fluid to flow through (usually described in drops per minute).

These devices range in complexity and ease of operation from roller and slide clamps to more sophisticated rotating valves. Compared with slide and roller clamps, rotating valves are less awkward to manipulate and provide a more consistent flow rate. However, even the most sophisticated manual drip valve cannot offer precise flow control, due the viscosity of the solution being infused. Another factor to be taken into account is the second flow control caused by the size of the needle at the end of the line through which the fluid flows into the catheter. The smaller the needle is the slower the maximum rate of flow into the body (15). These types of devices are quite effective in controlling overinfusion; however, control of underinfusion would not be satisfactory due to increased resistance to flow. One way to avoid this problem is to use a drip rate controller with a visible flow status system (14). The pressure available from a bag of saline is equal to the height that the bag is above the patient's heart. Drip rate controller is a type of gravity controller, in which the desired flow rate is set in drops per minute and controlled by occlusion valves powered by electricity. All models of drip controllers have a drop sensor. More advanced models incorporate a flow status system, which gives a visual indication of resistance to flow (4). The required number of drops delivered by gravity controllers is controlled by the drop counting mechanism that is quite accurate. However, the actual amount of volume delivered to patients may vary because of error involved when converting the number of drops to milliliters (mL). Conversion chart values for drops · mL^{-1} are approximate and a small error made for each drop may result in a large difference in the entire volume of drug delivered.

The volume of fluid in a drop depends on several factors, some of which are the fluid's composition, temperature and surface tension, the drip rate set, the size, shape, and condition of the drop-forming orifice (3). Expected nominal volume for a drop is 20 drops · mL^{-1}. This nominal rate can easily be achieved for most simple aqueous solutions of electrolytes, lactates, or dilute sugars. However, due to the viscosity characteristics of parenteral nutrition mixtures, fat soluble vitamins and solutions containing alcohol, the drop volume will be lower than nominal resulting in longer infusion time (3). Naturally, with all fluids, the drop volume decreases as the delivery rate increases. These variations are acceptable for the majority of infusions. However, if the volumetric accuracy is critical, then an infusion pump must be used (3).

There is a standard formula for calculating the flow rate on any type of IV tubing as follows:

$$(V \times df)/t = \text{drops min}^{-1}$$

t = time to be infused (in min); V = volume of solution to be infused; df = drop factor of solution set (drops mL^{-1}); (mL × drop factor)(min^{-1}) = drops min^{-1}.

If the result of the calculation includes a decimal point, round-off to the nearest.

Electronic controllers provide better accuracy for the regulation of flow by controlling uneven or runaway flow of fluid in a gravity drip system. These electronic devices are equipped with a drop sensor to monitor flow rate and can detect infiltrations and mal-positioning of the catheter or IV tubing by measuring backflow. An alarm sounds when flow rate is altered or when backflow is detected.

The gravity drip is conceptually simple, inexpensive, and requires less equipment than most other infusion systems. In the home setting, however, it has some limitations. First, it is difficult to maintain a constant infusion rate in a gravity drip system due to factors, such as the decreasing volume of fluid in the bag (i.e., the infusion rate will decrease as the bag empties) and changes in the shape of the tubing around the clamp. Consequently, a gravity system may provide insufficient flow control for drugs that require a very slow, very precise, or very long infusion time. Second, errors in using the gravity drip that remain unnoticed can result in serious complications (15). In addition, a gravity drip system may be an inappropriate choice for certain patients due to functional limitations of the patients or their caregivers. Because the IV bag is suspended well above the catheter site in this system, patients with decreased mobility may have difficulty changing the bag. Ambulatory patients on continuous infusion may also find gravity drip frustrating because the system is not easily portable. Despite the drawbacks of this traditional method of IV administration, it does maintain some important functional advantages over more expensive electronic infusion devices discussed below. Because the drugs are forced into the vein under the pressure of gravity alone, there may be less irritation at the catheter site, especially peripheral catheter sites. Gravity drip systems may also be preferred for patients who are confused by and resistant to learning how to use more complex, computerized drug delivery systems (15).

Infusion Pumps

An electronically controlled device that could deliver constant and precise amounts of fluid over a specified time period was a major technological advance in infusion therapy. Although many therapies can be delivered safely and effectively via gravity drip systems, others require the highly precise and constant flow rate offered by electronic infusion devices (15). For example, intraarterial infusions usually require positive pressure pumps because the back pressure is higher in arteries than in veins (15). Volumetric or syringe pumps are the most common. Other methods include elastomeric, pneumatic, implantable, clockwork, or spring (3). They are used to accurately administer intravascular drugs, fluids, whole blood, and blood products. These pumps can administer up to 2000 mL of fluid (normally from a bag or bottle) at flow rates of $0.1–2000\,\mathrm{mL\cdot h^{-1}}$.

Volumetric Pumps.

Most volumetric pumps will perform satisfactorily at rates as low as $5\,\mathrm{mL\cdot h^{-1}}$. However, these pumps are generally not used for delivering drugs at rates $<1\,\mathrm{mL\cdot h^{-1}}$, even though the device can be set to such low rates. The rate is in milliliters per hour $(\mathrm{mL\cdot h^{-1}})$ or micrograms per kilogram per hour $(\mu g\cdot kg^{-1}\cdot h^{-1})$ (3).

Most volumetric pumps have the feature of automatic alarm and shut down in case air enters the system, an occlusion is detected, or the reservoir is empty. The device controls the total volume to be infused and provides digital read-out of volume infused. Some of the other features include automatic switching to keep the vein open (KVO) rate at the end of infusion; switch to internal battery operation automatically if the mains supply fails; micro and macro delivery modes; computer interface; operator call alarm; a drop sensor—used for monitoring and alarm purposes (e.g., as an empty container) rather than as a control of the delivery rate; primary and secondary infusion capability; technical memory log for incident analysis. Features, such as air-in-line detection or a mechanism that cannot pump air and comprehensive alarm systems, make IV infusion much safer (3).

Most infusion pumps work by peristaltic action, which is achieved by alternately squeezing and releasing the tube containing the fluid to force the fluid through at a predetermined rate. There are two types of volumetric pumps: peristaltic and dedicated cassette. Peristaltic mechanisms can be further classified as linear peristaltic and rotary peristaltic. Both mechanisms consist of fingers, cams, or rollers that pinch off a section of the set. In linear peristaltic mechanisms as seen in Fig. 2, cams are located on a camshaft. Required volume is delivered to the patient by pinching off each section as the shaft rotates. These mechanisms are commonly used. In rotary peristaltic mechanisms, as seen in Fig. 3, rollers are placed on a

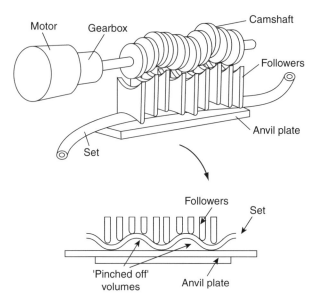

Figure 2. Schematic of a linear peristaltic pump (3) © CROWN COPYRIGHT.

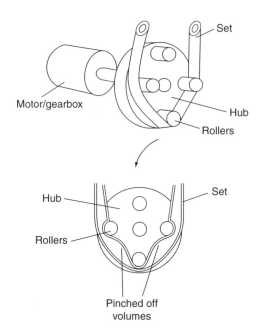

Figure 3. Schematic of a rotary peristaltic pump (3) © CROWN COPYRIGHT.

hub and as it rotates the volume in each pinched off section is delivered to the patient. The volume delivered varies according to the size of the cams, rollers, the tube, and the speed at which they rotate. These mechanisms are usually designed for a particular administration set (3).

Another mechanism used in infusion pumps is the dedicated cassette mechanism. Commonly, these types of pumps consist of a cassette body in which a valve and cylinder are placed, a piston, valve actuator, and a crank mechanism. This type of pump is depicted in Fig. 4. Drug is sucked to the cylinder from a bag or a container as the piston moves down and it is pumped to the patient through

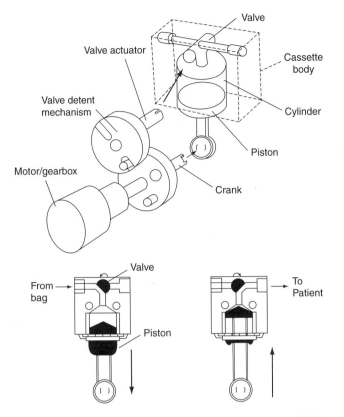

Figure 4. Dedicated cassette set (3) © CROWN COPYRIGHT.

a valve as the piston moves up (3). Although volumetric pumps can develop high pressures, they generally have a preset default value. In determining what pressure level is to be set, one needs to determine the factors of pressure raisers and calculate the needed pressure. However, it is important that the occlusion pressure should be set to the lowest possible value in order to observe early warning of occlusions.

A specific type of infusion set is required when using volumetric infusion pumps in order to achieve satisfactory drug delivery and to detect occlusion pressure. If an infusion set other than the required one is considered to be used, then extra care must be given when configuring the pump for that infusion set. Although using incorrect sets might seem to operate satisfactorily, this may be misleading and the actual performance and accuracy of drug infusion would be far from the desired level. This would lead to severe consequences. Air-in-line detectors use ultrasonic or optics for detecting air bubbles in the line. Air-in-line and occlusion detectors are designed for use with a particular infusion set. Therefore, these detectors may not be working properly if an incorrect set is used. Some other unwanted results are underinfusion due to very small inner-diameter of tube; overinfusion due to tubing material that is not flexible enough; and wear or rupture of tube from pumping action due to tubing material that is not strong enough. It is therefore important to use recommended sets for infusion. Specifications for testing of pumps at maximum flow rates are currently not given by the international standard for infusion devices. Therefore

some fall-off in performance at high flow rates should be expected (3).

Most infusion pumps used today are modern, sophisticated versions of one of these two types of pumps. With the development of small, portable pumps with specialized uses for particular types of therapies and adaptations, these pumps are being used commonly by nonprofessionals as part of home therapy. Because computerized pumps can deliver medication at a wide range of dose frequencies and intensities, they broaden the scope of therapies that can be safely and effectively administered at home.

Pumps specifically for the infusion of narcotics to treat cancer-related pain, for example, may have adaptations that provide a low level of ongoing infusion, but also permit patients to dose themselves with bursts of medication when pain becomes intense, up to a preprogrammed number of such extra doses per day. Other pumps, designed for the volume of fluid typical of most antibiotic therapy, can be preprogrammed to deliver infusions at standard intervals (e.g., four times per day), thus enabling patients to sleep undisturbed while receiving therapy. Pumps used for long-term IV nutrition administration, on the other hand, may be designed to administer the large volume of fluid required for the overnight infusions typical of patients receiving this therapy. Infusion pumps currently available range in complexity and sophistication. These pumps can range from very simple, single-medication stationary infusion pumps to fully programmable, ambulatory pumps.

Sophisticated pumps can deliver multiple medications and are equipped with a variety of alarms, bells, and other warning mechanisms. While stationary pumps may be appropriate for patients who are bedridden or whose medications are delivered over shorter periods of time, ambulatory pumps provide greater independence for patients on continuous, frequent, or long-term therapy regimens. Many pumps also have automatic piggyback mechanisms that control secondary infusions at an independent rate, decreasing the nursing time required for multiple infusions (15). Besides these benefits and advancements, infusion pumps do have certain disadvantages. If patients, caregivers, or even health professionals find the level of sophistication of these pumps confusing, the patients' safety could be jeopardized through misuse of equipment. Many patients, and the nurses who instruct and care for them, might prefer simpler models that are easier to operate. Even many hospital nurses are unfamiliar with or unaware of sophisticated features of pumps they use on a regular basis. Highly sophisticated pumps cost more and often require considerably more training for both the health professional and the patient than simpler models. New types of electronic infusion pumps are constantly evolving, widening the menu from which providers must choose and from patients and health professionals must learn to operate.

Syringe Pumps. In this type of pump, drug is pumped forward in the tubing by a syringe-type pushing action. Schematic drawing of a syringe pump is depicted in Fig. 5. The syringe is placed in a housing of the pump, while syringe plunger is attached to a moving carriage. The carriage is attached to the lead screw through a nut,

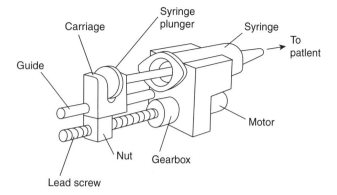

Figure 5. Typical syringe pump (3) © CROWN COPYRIGHT.

and the lead screw is attached to the motor through a gearbox. As the motor rotates, lead screw forces the carriage to slide on the guide, resulting syringe plunger to move forward in the syringe. This forward action delivers the drug to patient and empties the syringe. Controlled rate for the delivery of a drug is ensured by controlling the motor speed and rotation. This controlled rate may be in steps or continuous.

Advanced syringe pumps permit the simultaneous administration of several different therapies at different intervals, with dosages and administration regimens pre-programmed on a microchip that fits in the back of the pump. Syringe pumps can deliver small volumes of drugs at low flow rates. Single-use syringes are inexpensive and mass manufactured items, and are not meant to be highly accurate. When used in a syringe pump at low plunger speeds, the friction between the syringe plunger and the barrel causes a jerking effect and the fluid is delivered as a series of small boluses. The fit between the plunger and the barrel may vary from batch to batch and, consequently, the jerking effect may also vary. This problem is commonly known as stiction. In general, the bigger the syringe and the lower the flow rate, the more pronounced the stiction. Stiction may not be a problem with drugs having a long half-life or that do not require steady blood levels in the short term, such as heparin or insulin. In contrast, the delivery of powerful drugs with short half-life, like cate-cholamines, at rates under $5\,mL \cdot h^{-1}$ from large syringes (>30 mL) is not recommended.

Some currently available peristaltic pumps provide reasonably smooth flows at low delivery rates and should be considered as an alternative. Occasionally, the dimensions of a particular model of disposable syringe may be changed by the manufacturer. As a safe practice, only the syringe recommended by the manufacturer should be used with a syringe pump. These pumps are suitable for lower volume and low flow rate infusions. It is important to note that the actual drug delivered at the beginning of infusion process may be considerably less than the preset value. Due to the backlash, especially at low flow rates, it takes some time for the flow rate to reach steady-state regime.

Syringe pumps vary according to their functionality and so do their features. More advanced and expensive models have many features, including delivery pressure displays and in-line pressure monitoring. In most recent advanced

pumps, one can set occlusion alarm pressures to very low values. This feature helps patients to prevent hazards due to occlusions by the alarm signal from the system in shorter times. These advanced pumps may also prevent the delivery pressure rising to unwanted high values. Since these devices are powered externally, placing them approximately at the patient's level will suffice for the pump to work satisfactorily. In fact, if the pump is placed well above patient's level, some draining could result.

Implantable Pumps. Some therapies that require very small drug dosages can be administered by way of totally implantable pumps. Insulin delivery, continuous epidural morphine administration for chronic pain management, and continuous venous antineoplastic therapy infusion for liver cancer patients are some examples where implantable pumps are used. The only service directly related to infusion therapy for these devices is refilling of the pump's reservoir, which may be done weekly or even less frequently in a medical outpatient or home setting (15).

Patient Controlled Analgesia (PCA) Pumps. These pumps are designed specifically for use in PCA. Unlike a general-purpose infusion pump, these pumps allow the patients to deliver the drug on their own by operating a switch or pressure pad connected by a cord to the pump. It is important that free-flow is prevented. These pumps can be connected to a computer or printer and have a memory, where data in terms of usage is stored. This feature allows the clinician to review when, how often, and how much of drug infused by the patient. The PCA pumps are typically syringe pumps, since the required drug to be infused can usually be supplied in a single-use syringe. Some PCA pumps are based on volumetric designs, in which a battery powered volumetric pump has a disposable internal fluid reservoir. The PCA pumps can be disposable (pneumatic and elastomeric) or nondisposable (3). The PCA pumps can be programmed by clinical staff in different ways. Options include loading dose, continuous infusion (basal rate), continuous infusion with bolus on demand, bolus on demand only, with choice of units (mL or $\mu g \cdot mL^{-1}$, etc.) and variable lockout time, drug concentration. Once programmed, a key or software code is needed to access control of the pump. In some cases, patients are given limited access in order to change some parameters.

Elastomeric Infusers. Elastomeric infusers are devices that can be used as substitutes for infusion pumps. These infusers consist of disposable containers with inner-elastic bladders that can be filled with the medication. The devices are sold empty and are filled by the pharmacist through a port at the top of the bladder. The drug flows through an opening at the base of the bladder membrane and into the tube leading to the patient. The force of the flow, and thus the rate of infusion, is determined by the elasticity of the bladder and the concentration of the drug, regardless of whether the bladder is above, below, or on level with the IV site. Different drugs and dosages require devices of differing size and bladder membrane composition. Most devices currently on the market are designed for either antibiotic or antineoplastic therapy administration. They can be used for IV, intraarterial, and subcutaneous administration of drugs.

A patient on a twice-a-day regimen of home IV antibiotics would use two infusers per day, while a patient on continuous antineoplastic therapy might use a single device for several days at a time. Some devices allow patient-controlled administration of bolus doses above and beyond the continuous infusion rate. A disadvantage to the use of these devices for patient-controlled analgesia is the lack of a memory function that can record the frequency of patient-requested bolus doses, like that found in some electronic infusion pumps. Bladder devices are also not appropriate for multiple drug regimens. According to one home infusion provider, the availability of disposable elastomeric infusion devices has increased the feasibility of home-based care for disabled elderly patients. Like sophisticated electronic infusion pumps, these devices can deliver a precise dose over a specific period of time. However, because they are self-contained and much simpler to operate, they may be less confusing for patients who are uncomfortable with high tech equipment. The patient or caregiver need only hook the device to the catheter at dosing time and disconnect and dispose of it when the dose has been completed.

Anesthesia Pumps. These are also syringe-type pumps designed particularly for anesthesia infusion. Operating of these pumps is limited to theaters and high dependency areas. It is possible that the rate and other functions can be adjusted during infusion. Their flow rates are normally much higher than the typical syringe pumps rate. It therefore allows quick delivery in a single operation. These pumps can be interfaced with a computer and have built-in drug libraries. They are embedded with a drug-specific smart card and can be programmed for drug concentration and patient's body weight. The pump is automatically configured for the drug being infused. If the pump is to be used for other applications, automatically built-in features for specific application must be disabled.

Ambulatory Pumps. These pumps are designed so that patients can continue their drug therapy away from the hospital. These pumps allow patients to continue their normal life while treatment is being given. For ease of use and carry they are light and small in size, and are powered by battery. Alarming features of these pumps are not fully provided due to limitation in their size; therefore, their use in therapies in which precise flow is required for critical drugs is not recommended. The main mechanism used in these pumps is the syringe or cassette type.

Therapies that can be administered by ambulatory pumps include analgesia, continuous and PCA, antibiotic or antiviral infusions, chemotherapy, and hormone delivery. Back pressure, temperature of the flow-limiting element, temperature, and viscosity of the fluid are such factors that determine the accuracy of ambulatory pumps. One type of ambulatory pump manufactured by Baxter is given in Fig. 6. Depending on the pumping mechanism used, flow rates can range between 0.01 and 1000 mL·h^{-1}. Different models have different features. Flow rates can be set in millimeters per hour or day, milliliters per hour or day. They can also be programmed for different delivery modes (3). Ambulatory pumps are generally powered by electricity. Accuracy and alarming features will be limited if the pump is not powered by electricity. The pumping mechanism is generally the same mechanism used in volumetric and syringe pumps.

Some ambulatory pumps are reusable. They consist of a syringe that is operated by pressurized gas, usually carbon dioxide or a precompressed spring. In the case of a pressurized-gas-operated system, the force generated by the pressurized gas pushes the syringe plunger. As the syringe plunger moves forward, it infuses the drug to the patient. The infusion rate is determined by the pressure of the gas as well as the rate at which pressurized gas is released. When the infusion is completed, syringe and gas cartridges are thrown away and the rest of the device is kept for future use. Infusers and bolus-only analgesia devices controlled by the patient are of disposable devices. They

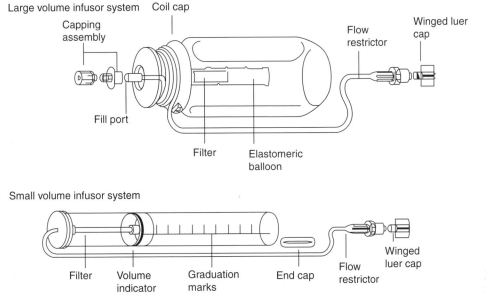

Figure 6. Ambulatory pumps manufactured by Baxter.

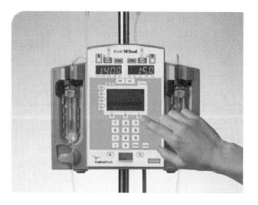

Figure 7. Signature Edition Gold infusion system by ALARIS. A range of infusion programs can be selected to save nursing time and meet sophisticated administration requirements including: Loading Dose, Multi-Dose and Multi-Step.

consist of a calibrated bolus chamber that is filled from an elastomeric reservoir or syringe by a capillary tube (3).

There are a number of companies that manufacture a variety of drug infusion systems. Some of the state-of-the-art products are given in Fig. 7–11. Abbott Laboratories' hospital products business (now Hospira), introduced the Plum A+ IV drug delivery medication management

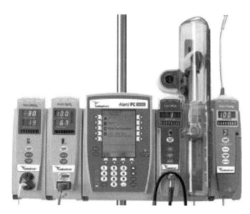

Figure 8. The Medley medication Safety System by ALARIS is a modular point-of-care computer that integrates infusion, patient monitoring and clinical best practice guidelines in a single platform for optimal outcomes.

system. This system is an innovative infusion system for electronic control of intravenous medication administration. It is used for standard, piggyback, or concurrent delivery and are suitable for a wide range of medical–surgical and critical care applications

NEW DEVELOPMENTS IN DRUG INFUSION SYSTEMS

Advances in science and technology result in new materials and devices. These materials and in particular electronic devices allow engineers to develop smarter, better performed, smaller, reliable, and cost-effective products. Therefore, new drug infusion systems are being developed and increasingly used in hospitals as well as in home therapies. Some of these developments are summarized in the following list.

Automated Syringe-Filling System: Stanford Research Institute's (SRI) drug delivery system expertise is to develop a compact, home device for diabetics that would help them fill their insulin syringes accurately. The system needed to handle both long- and fast-acting insulin formulations and needed to be easy to use and reliable for elderly and vision-impaired patients. The SRI developed an automated system that stored both types of insulin, automatically resuspended the long-acting insulin, checked for adequate drug supply, dispensed the proper amount of medication, and kept a dosage record. The system is under test and evaluation (16).

Disposable Drug Infusion Pump: Medical devices and pharmaceutical companies working with SRI to reengineer a disposable drug infusion pump design that reduced the number of parts by 30%, and reduced cost (16).

Tiny Drug Infusion System: A tiny meter in a belt will someday monitor dosages of up to 12 drugs needed around the clock by patients with diabetes, cancer, or acquired immune deficiency syndrome (AIDS). The dime-sized device is being developed by Integrated Sensing Systems Inc. The device will make sure patients are receiving the correct drug in the right volume at the right flow rate. It will hook into a drug controller that attaches to a patient's belt. The controller, ~2.5 × 1 in., (6.35 × 2.54 cm) will deliver drugs from an attached reservoir to the patient. The system, as envisioned, will

Figure 9. Outlook Safety Infusion System with DoseScan and DoseGuard by B|BRAUN technologies helping to ensure that the Right patient receives the Right medication in the Right dose from an authorized clinician at point of care.

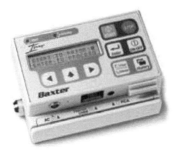

Figure 10. The Ipump Pain Management System by BAXTER can be programmed for epidural, IV, or subcutaneous delivery. The PCA doses can be set per hour. Control flow rates can be set in 0.1 mL·h^{-1} increments for maximum flexibility with continuous flow rates up to 90 mL·h^{-1}.

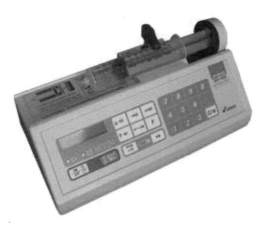

Figure 11. This AITECS by EO Systems is a multipurpose syringe pump with flow rates from 1 to 1500 mL·h^{-1}, can be used for any nuclear cardiology and nuclear medicine infusion.

simultaneously deliver up to 12 drugs in units as small as nanoliters, or billionths of a liter (17).

Bar-Coded Infusion System: B. Braun Medical has introduced the Horizon Outlook IV Safety Infusion System. Braun notes that the most common source of human error is inaccurate manual programming of intravenous pumps. The Braun infusion system uses bar code technology to ensure the right patient is receiving the right drug in the right dose from an "authorized" clinician. Its patented DoseScan bar code technology creates a primary level of safety, with automated checks and balances that augment the manual procedures in use today. Secondary protection is provided by its Dose-Guard software, which notifies clinicians if institution defined dose limits are exceeded (18).

Coronory Micro-Syringe: EndoBionics has created the first micromedical device to inject safely through vessel walls. Using standard interventional procedures, physicians will position the EndoBionics μSyringe (Micro-Syringe) in coronary or peripheral vessels. While the μSyringe is closed, the microneedle is hidden and does not injure vessel walls as it is maneuvered into place. When the μSyringe is opened, the microneedle slides through the vessel wall to inject drugs directly to the surrounding tissue. The drugs are deposited around the outside of the vessel and diffuse inward through the vessel layers. The microscopic puncture is so small that it heals almost immediately, limiting trauma and bleeding (19).

Needleless Injection: PowderJect Technologies has developed a technology that could be considered a hybrid of transdermal and parenteral (injection): a needleless injection. The company's device propels powder drugs with a supersonic jet of helium gas. A high pressure ampule of helium within the device is broken open, the gas flows through a cassette that is holding the powder between two membranes. The membranes rupture and the gas stream picks up the particles. The particles are propelled fast enough to penetrate the stratum corneum, the outer layer of the skin. The drug is targeted to the boundary between the epidermis and the dermis. Drugs then dissolve and either reach systemic circulation or exert a local effect. Vaccines can be picked up by antigen-presenting cells in the epidermis or by the lymph system (20).

Alza is another company that is developing technologies to deliver drugs through the skin. One of these technologies, called E-Trans, uses electrical current to deliver drugs across the skin, a process known as iontophoresis. The lead product is for the on-demand delivery of fentanyl, an opioid analgesic used for the treatment of acute pain. When a patient pushes a button on the device, current flows between two electrodes. As current flows, we get a predetermined amount of drug injected into the body. That gives a very reliable way of delivering a particular amount of drug into the body (20).

Alza is also developing what it calls Macroflux technology, which incorporates a thin titanium screen with microprojections to create mechanical pathways for drug transport. It expands the range of drugs amenable to transdermal delivery to include small hydrophilic molecules and macromolecules. It can be incorporated with the E-Trans technology or more traditional transdermal patches. One simple prototype in early exploration involves a Macroflux system where the projections have been coated with the therapeutic agent, such as a macromolecule. After application, the agent is rapidly absorbed into the skin (20).

Elan Pharmaceutical Technologies have a technology known as Medipad worn by the patient on the chest, back, or abdomen. The device is a small, plastic gas-driven pump with an adhesive backing. The adhesive is used to attach the device to the patient's body, and a button is pressed. A needle is deployed, which enters the subcutaneous space and then delivers the drug at a constant rate until the entire content of the reservoir is expended. The first applications for this device will be in chronic pain management and in the delivery of macromolecules that have inherently short biological half-lives (20).

A Novel Device for Flow Monitoring: A novel device for blockage detection in catheters during drug delivery is designed. This device consists of a low cost disposable microfluidic chip and a nondisposable detection unit. The microfluidic chip consists of a microstructured silicon layer bonded between two glass covers using anodic bonding technology. The flow monitoring is performed by a robust light transmission method. The main component of the microfluidic chip is a movable element coupled with a spring to a base. Depending on the drug flow state the element is blocking or vacating an optical path through the chip (21).

A High Performance Silicon Micropump: A new, low cost, high performance silicon micropump has been developed for a disposable drug delivery system (22). It is reported that the pump demonstrated linear and accurate ($\pm5\%$) pumping characteristics for flow rates up to $2 \, mL \cdot h^{-1}$ with intrinsic insensitivity to external conditions. The stroke volume of 160 nL was maintained constant by the implementation of a double limiter acting on the pumping membrane. The chip is a stack of three layers, two Pyrex wafers anodically bonded to the central silicon wafer. The technology is based on the use of Silicon On Insulator (SOI) technology, silicon Deep Reactive Ion Etching (DRIE), and the sacrificial etch of the buried oxide in order to release the structures (22).

An Implantable Microfabricated Drug Delivery System: A fully implantable drug delivery system capable of delivering hundreds of individual doses has been developed by MicroCHIPS (23). This product is intended for the controlled release of potent therapeutic compounds that might otherwise require frequent injections. The device is capable of storing therapeutic drugs in solid, liquid, or gel form. It allows individual storage of discrete doses for multiple-drug regimens. Device monitoring and therapy modification can be achieved via wireless communication with an external controller. Currently, a fully implantable device contains 100 individual doses. A future device intended for human clinical trials will contain 400 doses, enough for a daily release of drug for >1 year (23).

A Water-Powered Microdrug Delivery System: A plastic microdrug delivery system has been designed by utilizing the principle of osmosis without any electrical power consumption. The system has an osmotic microactuator and a polydimethylsiloxane (PDMS) microfluidic cover compartment consisting of a reservoir, a microfluidic channel, and a delivery port. The typical dimension of the microfluidic channel is 1 cm in length with a cross-sectional area of 30–100 μm^2 to minimize the diffusive drug flow while pressure drop remains moderate. Employing the net water flow induced by osmosis, the prototype drug delivery system has a measured constant delivery rate of 0.2 $\mu L \cdot h^{-1}$ for 10 h, with an accumulated delivery volume of 2 μL. Both the delivery rate and volume could be altered by changing the design and process parameters for specific drug delivery applications up to a few years (24).

Microflow Regulator for Drug Delivery Systems: A micromachined flow regulator has been designed to provide a constant liquid flow rate of $1 \, mL \cdot h^{-1}$ within a pressure difference of 100–600 mbar (0.01–0.06 kPa). At pressures >600 mbar (0.06 kPa) the device is designed to block the flow, preventing an overdelivery of medicine. One application of this device is the replacement of the flow restrictor in an elastomeric infusion system, which will increase the accuracy and safety of the drug delivery system. This pressure compensating flow regulator is passive; hence it needs no external energy source. The device is small, lightweight, and relatively inexpensive; therefore, it could be used as a disposable unit in a microfluidic system (25).

Nanoengineered Device for Drug Delivery: A high precision device has been developed to yield long-term zero-order release of drugs for therapeutic applications. The device contains nanochannels that were fabricated in between two directly bonded silicon wafers, and therefore poses high mechanical strength. Diffusion through the nanochannels is the rate-limiting step for the release of drugs (26).

Smartdose by PRO-MED AG: This device is a safe, accurate, and simple infusion system. It is a disposable prefilled drug delivery system for enteral or parenteral controlled infusion. SmartDose is equipped with its own source of energy (chemical reaction) to dispense liquid over a specific time with a predetermined administration rate. The administration accuracy and safety is comparable with those of electronic pumps, yet the ease of use is similar to a simple infusion bag. The system is especially convenient for emergency, ambulatory, and homecare therapy, as well as for hospitalized patients (27).

Biodegradable Polymeric Drug Delivery Systems: These systems are increasingly being used for the design of temporary drug delivery systems. As these polymers hydrolyze in the body into low molecular degradation products, which are either metabolized or excreted, biodegradable delivery systems do not have to be removed after completion of release. Poly(DL-lactide-co-glycolide) (PLGA) is the most widely investigated biodegradable polyester and is widely used as a carrier polymer in parenteral sustained release formulations, either as microspheres, microparticulates or injectable gels (28).

RECENT ADVANCEMENTS IN CONTROLLER DESIGN FOR DRUG INFUSION SYSTEMS

Closed-loop system control is a technological concept that may be applicable to several aspects of critical care practice. This is a technology in the early stages of evolution and much more research and data are needed before its introduction into usual clinical practice. Furthermore, each specific application and each device for each application are sufficiently different in terms of hardware and computer algorithms (29). Studies have shown that closed-loop infusion systems may have a role in critical care practice, improve clinical outcomes, eliminate errors due to poor performance of automated infusion devices, and provide precise, error-free drug administration. Some of the most recent works in advanced controller designs are reviewed below:

Huzmezan et al. (30) states that feedback control of drug administration is well suited to anesthetized surgical patients as well as the critically ill patients because of drugs with rapid onset times, short duration of action and small margins of safety are frequently used. The application of an adaptive predictive process control technology to drug administration will assist physicians in avoiding both overdosages and underdosages in their patients. An adaptive controller would avoid overdosing and underdosing by compensating for nonlinear drug responses as well as inter- and intrapatient variation (30).

Linkens has proposed the design of a fuzzy control for patient muscle relaxation (31). With advancements in sensor and instrumentation technology, automated drug infusion systems are also evolving into hierarchical systems. Research in this area has led to a variety of control strategies ranging from simple linear controllers to complex adaptive and rule-based schemes to handle inter- and intrapatient variability in drug responses (31).

For the assessment of depth of anesthesia, an intelligent system has been developed, which utilizes auditory evoked brain potentials, heart rate, and blood pressure measurements (32). Using wavelet analysis, the features within the auditory evoked signals are extracted and then fed to a learning neurofuzzy system, which in turn classifies the depth of anesthesia. In addition, the heart rate and blood pressure signals are used as a second measure based on a rule-based fuzzy logic system. The two measures are then fused to give a final indication of anesthetic depth. This is then fed back to a target controlled infusion (TCI) system for regulating the infusion of the drug Propofol for the maintenance phase of anaesthetic state (32).

A control strategy is developed by Bequette to regulate blood pressure and cardiac output during surgery (33). Adaptation is incorporated through a multiple model predictive control (MMPC) approach. A Bayesian-based estimator recursively updates weighting functions to find the best combination of models that describes the current input–output behavior; this weighted model is used for the output prediction (33).

A robust direct model reference adaptive controller (DMRAC) is developed by Palerm et al. (34) for plants with uncertainty in both the time delay elements and in the transfer function coefficients. The control of hemodynamic variables, particularly mean arterial pressure (MAP) and cardiac output (CO), is a challenging problem. A good controller is difficult to design, due to the complex, nonlinear behavior of the system. Adding to this are the significant changes in dynamics from one patient to another, and even variations in the patient's response to the drugs as his condition evolves (34).

A model predictive control strategy is developed and tested on a nonlinear canine circulatory model for the regulation of hemodynamic variables under critical care conditions (35). Different patient conditions, such as congestive heart failure, postoperative hypertension, and sepsis shock are studied in closed loop simulations. The model predictive controller, which uses a different linear model depending on the patient condition allows constraints to be explicitly enforced. The controller is initially tuned based on a linear plant model, then tested on the nonlinear physiological model; the simulations demonstrate the ability to handle constraints, such as drug dosage specifications, commonly desired by critical care physicians (35).

To evaluate the use of intelligent systems in the improvement of patient care, an agent was developed to regulate ICU patient sedation by Moore et al. in (36). A temporal differencing form of reinforcement learning was used to train the agent in the administration of intravenous propofol in simulated ICU patients. The agent utilized the well-studied Marsh–Schnider pharmacokinetic model to estimate the distribution of drug within the patient. A pharmacodynamic model then estimated drug effect. The agent demonstrated satisfactory control of the simulated patient's consciousness level in static and dynamic set-point conditions. It also satisfactorily demonstrated superior stability and responsiveness when compared to a well-tuned PID controller, which is a method of choice in closed-loop sedation control literature (36).

Advanced model-based controllers that can take into account the model of the patient and constraints on the state of the patient and the drug infusion rates have been developed (32). Delivery of insulin to type 1 diabetics, control of anesthesia, and chemotherapy for cancer patients are typical examples of drug delivery systems. The main objective of a drug delivery system is to provide effective therapy while minimizing the side effects. These controllers are based upon the theory of multiparametric programming. This theory allows an optimal division of the multidimensional space of the state of the patient into a set of regions and each region is characterized by a unique drug infusion law that is an explicit function of the state in the corresponding region. These developments simplify controller implementation and result in tighter control of drug infusion rates and better lifestyle for patients (37).

Parker et al. in (38) discusses closed-loop blood glucose regulation algorithms that use the intravenous route for insulin delivery to insulin-dependent diabetic patients. Classical control methods and advanced algorithms using implicit knowledge or explicit models (empirical, fundamental, or gray-box) of the diabetic patient are examined in (38). Current research on characterizing patient variability is presented, in the context of a model predictive controller able to adjust to changes in patient glucose and insulin sensitivity (38).

Linkens in (39) presents the control of on-line drug infusions to patients in an operating theater for regulating their muscle relaxation according to necessary surgical procedures. It is stated that fuzzy logic control (FLC) offers the advantages of model-free controller design for systems that are dynamically nonlinear, uncertain, and possibly time varying (39).

Rao et al. discusses the design of two different control methodologies for automated regulation of hemodynamic variables in (40). These controllers are designed to regulate

MAP and CO in critical care subjects using inotropic and vasoactive drugs. Both controllers account for inter – and intrapatient variability and handle drug infusion constraints. The first approach is a multiple model predictive controller (MMPC). The algorithm uses a multiple model adaptive approach in a model predictive control framework to account for variability and explicitly handle drug rate constraints. The second approach, a robust direct model reference adaptive controller (DMRAC) is developed for plants with uncertainty in both the time delay elements and in the transfer function coefficients, such as the drug infusion process. The controllers are experimentally evaluated on canines that are pharmacologically altered to exhibit symptoms of hypertension and depressed cardiac output (40).

Bequette (41) discusses the development of an artificial pancreas and current efforts in the control of complex systems. It is stated that advances in continuous glucose sensing, fast-acting insulin analogues, and a mature insulin pump market allow commercial realization of a closed-loop artificial pancreas. Model predictive control is discussed in-depth as an approach that is well suited for a closed-loop artificial pancreas (41).

Target controlled infusion (TCI) systems are discussed by Van Poucke et al. (42). In their work, a novel mathematical algorithm is proposed for controlling the effect site concentration using a TCI device. The algorithm limits the peak plasma concentration, thereby slowing the onset of anesthetic drug effect, but potentially ameliorating side effects. Simulations are used to examine the delay in time to peak effect for fentanyl, alfentanil, sufentanil, remifentanil, and propofol when the peak plasma concentration is limited by the algorithm. Results showed that the plasma overshoot can be reduced by 60% with only ~ 20% delay in the onset of drug effect (42).

McKinley et al. (43) compares the effectiveness of a new method of closed-loop control of blood pressure with usual manual control. In their work, closed-loop and manual drug administrations were studied. The target and observed MAP and drug infusion rate were recorded electronically. Time taken to achieve initial control; fidelity of control, and average drug dose administered were all measured. Results showed that closed-loop achieved faster initial control and greater fidelity as compared to manual control. There was no difference in average drug dose administered. It was concluded that the new closed-loop system is more effective than the usual manual control in managing acute blood pressure disturbances in the seriously ill patients (43).

The bispectral index (BIS) was used for automatic control of propofol anesthesia, using a proportional-integral-differential control algorithm (44). The performance of the controlled system was measured in patients undergoing minor surgery under propofol and remifentanil anesthesia. Anesthesia was manually induced with target-controlled infusions (TCI) of propofol and remifentanil. After the start of surgery, when anesthesia was clinically adequate, automatic control of the propofol TCI was commenced using the closed-loop system. The system provided adequate operating conditions and stable cardiovascular values in all patients during closed-loop control. The system was able to provide clinically adequate anesthesia in all patients (44).

Brock et al. in (45) presents a study to determine the relative advantage of computer-controlled couch movement versus manual repositioning to correct patient setup error measured using an electronic portal imaging device (EPID). The speed of setup adjustment and accuracy of corrected setup were determined. Computer-controlled setup adjustment was determined to be faster and slightly more accurate than manual correction (45).

Another comparison study between computer and manual control is presented in (46) by Hoeksel et al. They investigated the effects of computer-controlled blood pressures on hemodynamic stability when compared to conventional manual control. Systemic artery blood pressures were managed either by computer or by a well-trained anesthesiologist. Hemodynamic stability was determined from the standard deviation of the MAP samples and from the percentages of time that arterial pressure was hypertensive or hypotensive. The average standard deviation of the MAP samples was smaller for the computer-controlled than for the manually controlled group. The systemic artery pressure was less hypertensive and less hypotensive in the computer-controlled than in the manually controlled group. It was concluded that, compared with manual control, computer control of systemic hypertension significantly improved hemodynamic stability during cardiac surgery (46).

The clinical applicability of administering sodium nitroprusside by a closed-loop titration system compared with a manually adjusted system was evaluated. The MAP was registered and the results were then analyzed. It was reported that the computer-assisted therapy provided better control of MAP, was safe to use, and helped to reduce nursing demands (47).

Chitwood et al. in (48) states that hypertension is common after a cardiac operation and has been treated using manually controlled doses of intravenous sodium nitroprusside. To evaluate the clinical impact of an automated closed-loop administration system on patients after cardiotomy, a prospective trial was conducted. Patients with hypertension were managed by either manual nitroprusside titration or a closed-loop automated titration system. The automated group showed a significant reduction in the number of hypertensive episodes per patient. At the same time, the number of hypotensive episodes per patient was reduced with automated closed-loop titration. Chest tube drainage, percentage of patients receiving transfusion, and total amount transfused were all reduced significantly by the use of an automated titration system (48).

A nonprogrammable and programmable insulin external pump using regular insulin on glycemic stability, the risk of severe hypoglycemia, and metabolic control in type 1 diabetic patients was compared (49). The results of the study suggest that programmable external insulin pumps, although more complex and more expensive than nonprogrammable insulin pumps, significantly reduce fasting glycemia during the day without increasing the risk of severe hypoglycemia and are safer during the night (49).

In Ref. 50, it was argued that continuous improvements in microelectronics, as well as in the development of biomaterials and stable insulin solutions, led to the availability of implantable pumps that are able to infuse insulin by the peritoneal route, in a continuous and programmable way, for several years. These systems represent the most efficient and physiological mode of insulin therapy at the present time. It was demonstrated during clinical trials that intravascular, implantable, glucose sensors using glucose oxidase were able to measure with good accuracy real-time blood glucose for several months. In their study, they performed the first trials of closed-loop insulin delivery according to sensor signal for periods of 48 h in type 1 diabetic patients. This mode of functioning appeared to be feasible and able to establish glucose control closer to physiology than the use of implantable pumps in open loop (50).

Tamborlane et al. (51) states that while treatment of Type 1 diabetes mellitus (T1DM) in children and adolescents is especially difficult, recent technological advances have provided new therapeutic options to clinicians and patients. The urgency to achieve strict diabetes control and the introduction of new and improved insulin pumps have been accompanied by a marked increase in use of continuous subcutaneous insulin infusion (CSII) therapy in youth with diabetes. Results of clinical outcome studies indicate that CSII provides a safe and effective alternative to MDI therapy, even when employed in a regular clinic setting in a large number of children (51).

BIBLIOGRAPHY

1. Draft for Public Comment, Australian/New Zealand Standard. 2005. DR04547.
2. Medtech, Inc., U.S. Markets for Drug and Fluid Delivery Devices. 2001. Report No. RP-485004.
3. Infusion Systems, Medical Devices Agency. 2003. MDA DB 2003(2).
4. Sparks DR, et al. Preventing medication infusion errors and venous air embolisms using a micro-machined specific gravity sensor. Adv Deliv Devices 2004.
5. Kohn L, et al., editors. Report of Institute of Medical Committee on Quality Health Care in America, To err is human: building a safer health system. Washington (D.C.): National Academy Press; 2000.
6. Phillips J, et al. Retrospective analysis of mortalities associated with medication errors. Am J Health-Syst Pharm 2001;58:1831–1841.
7. Gorman R, et al. Prevention of medication errors in the pediatric inpatient setting. Pediatrics 2003;112(2):431–436.
8. Taxis K, Barber N. Ethnographic study of incidence and severity of intravenous drug errors. BMJ 2003;326:684–696.
9. Parshuram C, et al. Discrepancies between ordered and delivered concentrations of opiate infusions in critical care. Cri Care Med 2003;31(10):2483–2487.
10. Smart infusion pumps join CPOE and bar coding as important ways to prevent medication errors. ISMP Me Safety Alert 2002.
11. Wilson K, Sullivan M. Preventing medication errors with smart infusion technology. Am J Health-Syst Pharm 2004; 61(2):177–183.
12. Steingass SK. Beyond pumps: Smart infusion systems. Nursing Management 2004;35:10–10.
13. Ferrari R, Beech DR. Infusion pumps: guidelines and pitfalls. Aust Prescr 1995;18:49–51.
14. Davis WOM. Types of pumps in use, Infusion devices training tutorial (Online). Available at http://www.ebme.co.uk/arts/art11.htm. 2004.
15. Home drug infusion therapy under medicare. Congress of the United States, Office of Technology Assessment. 1992.
16. Intravenous therapy monitoring and discontinuing, (online) http://www.laras-lair.com/nursing/IVtherapy.pdf. RPN Self Learning Package. SRI International-Medical Development (Online). Available at http://www.sri.com/esd/med_devel/altdrugdel.html. 2005.
17. Infusion device could improve drug delivery (Online). Available at http://www.detnews.com/2004/technology/0402/16/b04-64095.htm. 2005.
18. Braun Medical introduces bar-coded infusion system (Online). Available at http://www.uspharmacist.com. 2002.
19. Endobionics, Inc. (online) http://www.endobionics.com/technology.html. 2005.
20. Henry C. Special delivery. Sci/Technol 2000;78(38):49–65.
21. Richter M, et al. A novel device for low flow monitoring in drug delivery systems. 8th Int Conf New Actuators 2002; 223–226.
22. Maillefer D, et al. A high performance silicon micropump for disposable drug delivery systems. Micro Electro Mechanical Systems (MEMS). The 14th IEEE International Conference on MEMS, 2001; p 413–417.
23. Maloney JM. An implantable microfabricated drug delivery system. Proc IMECE 2003;1–2.
24. Su Y, Lin L. A water-powered micro drug delivery system. IEEE J Microelectromechan Systems 2004;13(1):75–82.
25. Cousseau P, et al. Improved mico-flow regulator for drug delivery systems. IEEE Int Conf MicroElectroMechanical Syst 2001;527–530.
26. Sinha PM, et al. Nanoengineered device for drug delivery application. Inst Phys Publ Nanotechnol 2004;15:S585–S589.
27. Pro-Med AG. SmartDose controlled infusion systems. Business Briefing Pharmatech 2004;1–2.
28. Steendam R. SynBiosys: Biodegradable polymeric drug delivery system. Bus Briefing: Pharma Outsourcing 2005; 59–61.
29. Jastremski M, et al. A model for technology assessment as applied to closed loop infusion systems. Technol Ass Task Force Soc Crit Care Med 1995;23(10):1745–1755.
30. Huzmezan M, Dumont GA, Zikov T, Bibian S. Advances in automatic drug delivery for general anesthesia. Peter Wall Institute Exploratory Workshop, Automation and Robotics: The Key For Computer Integrated Health Care Delivery. University of British Columbia. 2002.
31. Linkens DA. Fuzzy control for patient muscle relaxation. Erudit News Letter. Vol. 2, No. 1 (online). Available at http://www.erudit.de/erudit/newsletters/news_21/page_5.htm, 2003.
32. Linkens DA. An intelligent system for drug delivery in control of anesthesia. Peter Wall Institute Exploratory Workshop, Automation and Robotics: The Key For Computer Integrated Health Care Delivery. University of British Columbia. 2002.
33. Bequette BW. A multiple model approach for adaptation during drug delivery. Peter Wall Institute Exploratory Workshop, Automation and Robotics: The Key For Computer Integrated Health Care Delivery. University of British Columbia. 2002.
34. Palerm C, Bequette BW, Ozcelik S. Robust control of drug infusion with time delays using direct adaptive control: experimental results. Am Control Conf 2000.
35. Rao R, Huang J, Bequette BW, Kaufman H. Control of a nonsquare drug infusion system-a simulation study. Biotechnol Prog ACS Pub 1999;15(3):556–564.
36. Moore B, Sinzinger E, Quasny T, Pyeatt L. Intelligent control of closed-loop sedation in simulated ICI patients. Am Assoc Artificial Intelligence 2004.

37. Dua V. Explicit model based control for drug delivery systems, CoMPLEX seminar (Online). Available at http://www.ucl. ac.uk/CoMPLEX/dua_abstract.html. 2004.

38. Parker R, Doyle F, Peppas N. The intravenous route to blood glucose control. IEEE Eng Med Bio Mag 2001;20(1):65–71.

39. Linkens DA. Fuzzy control for patient muscle relaxation. Erudit News Letter Vol. 2, No. 1, (Online). Available at http://www.erudit.de/erudit/newsletters/news_21/page_5.htm. 2003.

40. Rao RR, Palerm CC, Aufderheide B, Bequette BW. Automated regulation of hemodynamic variables. IEEE Eng Med Bio Mag 2001;20(1):2438.

41. Bequette BW. A Critical Assessment of Algorithms and Challenges in the Development of a Closed-Loop Artificial Pancreas. Diab Technol Therapeut 2005;7:28–47.

42. Van Poucke GE, Bravo LJB, Shafer SL. Target controlled infusions: targeting the effect site while limiting peak plasma concentration. IEEE Trans Biomed Eng 2004;51:1869–1875.

43. McKinley S, et al. Clinical evaluation of closed loop control of blood pressure in seriously ill patients. Crit Care Med 1991;19(2):166–170.

44. Absalom AR, Kenny GNC. Closed-loop control of propofol anesthesia using bispectral index: performance assessment in patients receiving computer-controlled propofol and manually controlled remifentanil infusions for minor surgery. Br J Anesthesia 2003;90:737–741.

45. Brock KK, McShan DL, Balter JM. A comparison of computer-controlled versus manual on-line patient setup adjustment. J App Clin Med Phys 2002;3(3):241.

46. Hoeksel SAAP, Blom JA. Compure control versus manual control of systemic hypertension during cardiac surgery. Acta Anesthesiol Scand 2001;45:553–557.

47. Bednarski P, et al. Use of a computerized closedloop sodium nitroprusside titration system for antihypertensive treatment after open heart surgery. Crit Care Med 1990;18(10):1061–1065.

48. Chitwood WR, III DMC, Lust RM. Multicenter trial of automated nitroprusside infusion for postoperative hypertension. Ann Thorac Surg 1992;54:517–522.

49. Catargi B et al. A randomized study comparing blood glucose control and risk of severe hypoglycemia achieved by nonprogrammable versus programmable external insulin pumps. Diab Metab 2001;27:323–328.

50. Renard E, Costalat G, Bringer J. From external to implantable insulin pump, can we close the loop? Diab Metab 2002;28(4 Pt. 2):19–25.

51. Tamborlane W, Bonfig W, Boland E. Recent advances in treatment of youth with type 1 diabetes: Better care through technology. Diab Metab 2001;18:864–870.

See also DRUG DELIVERY SYSTEMS; NUTRITION, PARENTERAL.